MW00837610

HANDBOOK OF ENERGY ENGINEERING CALCULATIONS

HANDBOOK OF ENERGY ENGINEERING CALCULATIONS

Tyler G. Hicks, P.E. Editor

International Engineering Associates
Member: American Society of Mechanical Engineers
United States Naval Institute

New York Chicago San Francisco Lisbon London Madrid
Mexico City Milan New Delhi San Juan Seoul
Singapore Sydney Toronto

The McGraw-Hill Companies

Cataloging-in-Publication Data is on file with the Library of Congress.

McGraw-Hill books are available at special quantity discounts to use as premiums and sales promotions, or for use in corporate training programs. To contact a representative please e-mail us at bulksales@mcgraw-hill.com.

Handbook of Energy Engineering Calculations

1 2 3 4 5 6 7 8 9 0 DOC/DOC 1 9 8 7 6 5 4 3 2 1

ISBN 978-0-07-174552-9
MHID 0-07-174552-1

This book is printed on acid-free paper.

Sponsoring Editor
Larry S. Hager

Editing Supervisor
David E. Fogarty

Project Manager
Tania Andrabi, Cenveo Publisher Services

Copy Editor
Sharon Green

Proofreader
Manish Tiwari, Cenveo Publisher Services

Production Supervisor
Richard C. Ruzycka

Composition
Cenveo Publisher Services

Art Director, Cover
Jeff Weeks

To energy-conscious engineers everywhere, and that includes all engineers: Your energy conservation concerns and efforts were often laughed at by people who thought our energy supply was limitless. Today, with energy prices soaring, and concerns over global warming, the energy-conscious engineer has the respect of almost everyone, everywhere. May this handbook help your energy design and energy conservation work to be more widely used, respected, and appreciated—worldwide.

ABOUT THE EDITOR

TYLER G. HICKS, P.E., is a consulting engineer and a successful engineering book author. He has worked in plant design and operation in a variety of industries, taught at several engineering schools, and lectured both in the United States and abroad. Mr. Hicks is the author of more than 20 books in engineering and related fields, including *Standard Handbook of Engineering Calculations*, *Handbook of Civil Engineering Calculations*, *Handbook of Mechanical Engineering Calculations*, and *Mechanical Engineering Formulas Pocket Guide.*

CONTENTS

Contributors and Advisers xi
Preface xiii
Acknowledgments xvii
How to Use This Handbook xix

Section 1. Energy Conversion Engineering 1.1

Section 2. Steam Power Generation 2.1

Section 3. Gas-Turbine Energy Power Generation 3.1

Section 4. Internal-Combustion Engine Energy Analyses 4.1

Section 5. Nuclear Energy Engineering 5.1

Section 6. Hydroelectric Energy Power Plants 6.1

Section 7. Wind-Power Energy Design and Application 7.1

Section 8. Solar-Power Energy Application and Usage 8.1

Section 9. Geothermal Energy Engineering 9.1

Section 10. Ocean Energy Engineering 10.1

Section 11. Heat Transfer and Energy Conservation 11.1

Section 12. Fluid Transfer Engineering 12.1

Section 13. Interior Climate Control Energy Economics 13.1

Section 14. Energy Conservation and Environmental Pollution Control 14.1

Index I.1

CONTRIBUTORS AND ADVISERS

In preparing the various sections of this handbook, the following individuals either contributed sections, or portions of sections, or advised the editor or contributors, or both, on the optimum content of specific sections. The affiliations shown are those prevailing at the time of the preparation of the contributed material or the recommendations as to section content.

In choosing the procedures and worked-out problems, these specialists used a number of guidelines, including: (1) What are the most common applied problems that must be solved in this discipline? (2) What are the most accurate methods for solving these problems? (3) What other problems might be met in this discipline? When the answers to these and other related questions were obtained, the procedures and worked-out problems were chosen. Thus, the handbook represents a cross section of the thinking of a large number of experienced practicing engineers, project directors, and educators.

To those who might claim that the use of step-by-step solution procedures and worked-out examples makes engineering "too easy," the editor points out that for many years engineering educators have recognized the importance and value of problem solving in the development of engineering judgment and experience. Problems courses have been popular in numerous engineering schools for many years and are still given in many schools. However, with the greater emphasis on engineering science in most engineering schools, there is less time for the problems courses. The result is that many of today's graduates can benefit from a more extensive study of specific problem-solving procedures.

Edmund B. Besselievre, P.E., *Consultant*, Forrest & Cotton, Inc.

Robert L. Davidson, *Consulting Engineer*

Stephen M. Eber, P.E., Ebasco Services, Inc.

Gerald M. Eisenberg, *Project Engineering Administrator*, American Society of Mechanical Engineers

V. Ganapathy, *Heat Transfer Specialist*, ABCO Industries, Inc.

Gregory T. Hicks, R.A., Gregory T. Hicks and Associates, Architects

Tyler G. Hicks, P.E., International Engineering Associates

Edgar J. Kates, P.E., *Consulting Engineer*

Max Kurtz, P.E., *Consulting Engineer*

Joseph Leto, P.E., *Consulting Engineer*

Jerome F. Mueller, P.E., Mueller Engineering Corp.

George M. Muschamp, *Consulting Engineer*, Honeywell, Inc.

Rufus Oldenberger, *Professor*, Purdue University

John S. Rearick, P.E., *Consulting Engineer*

Raymond J. Roark, *Professor*, University of Wisconsin

Lyman F. Scheel, *Consulting Engineer*

B. G. A. Skrotzki, P.E., *Power* magazine

S. W. Spielvogel, *Piping Engineering Consultant*

Kevin D. Wills, M.S.E., P.E., *Consulting Engineer*, Stanley Consultants, Inc.

As the handbook user will see, the editor also drew on the published works of numerous engineers and scientists appearing in many technical magazines and journals. Each of these individuals is cited in the calculation procedure in which their work appears. Their position title and affiliation are given as of the time of original publication of the cited procedure or data.

PREFACE

This handbook presents some 2500 calculation procedures in the field of energy engineering. Each calculation procedure is presented in both the USCS (United States Customary System) and the SI (System International). Thus, engineers and designers worldwide will find that each calculation uses familiar units. Therefore, this handbook is usable by engineers and designers in every country in the world. Further, many of the calculation procedures are unique—the engineer and designer will rarely find these procedures in other books, on the Internet, or in reports or technical papers. The author/editor is currently, and has been, preparing, using, studying, and collecting, unique and routine calculation procedures during his long active and varied engineering career.

Topics covered in this handbook include combustion of fossil fuels (the conversion of energy from one form to another usable form), steam-power plants, gas-turbine power generation, internal-combustion engine power plants, nuclear power generation, hydro power plants, and alternative energy generation via wind turbines, ocean tidal power, underwater tidal current power, solar power, geothermal energy power, and several other schemes for using nature's forces to generate power. This handbook also includes calculations of heat transfer and energy conservation, fluids transfer energy engineering, interior climate control energy economics, and energy conservation and environmental pollution control.

In preparing this handbook the author has been constantly aware of the effects on global warming produced by power generators of various kinds, especially those using fossil fuels. Some experts report that more than half of atmospheric greenhouse gases emitted today are produced by power generation using fossil fuels.

Because of the emphasis on global warming throughout the world, each calculation procedure in this handbook recognizes the impact the particular form of power generation has on greenhouse gases. Some methods of power generation directly produce undesirable atmospheric effluents. Other "clean" or "green" power generation methods are responsible for atmospheric pollution during their manufacturing process. This fact is pointed to by environmentalists as a negative fact to be held against the "clean" or "green" method of power generation.

The intense worldwide interest in greenhouse gases, atmospheric pollution, and global warming started a new vocabulary in the energy engineering field. Today we encounter terms such as "green" power, "clean" power, carbon footprint, carbon taxes, energy taxes, biofuel usage, caps on power-plant carbon dioxide emissions, energy audits, etc. Most these terms imply greater costs in generating power because of emissions reduction requirements. This handbook recognizes the new demands on energy-system engineers and designers and includes calculations that help the resulting design comply with modern global-warming-reduction requirements.

To reflect modem design conditions, this handbook starts with calculations of external combustion of fossil fuels used in steam power plants because this is the widest used (some 70 percent of the total in many areas) form of electric energy generation today. Next, the handbook presents gas-turbine calculations. Today, gas turbines are the favored motive power choice for steam-station expansion and topping services. With its small footprint, freedom from complex regulations, and use of low-emissions natural gas, the gas turbine is becoming more popular in energy engineering every year. Also, using heat-recovery steam generators (HRSG) in the exhaust of a gas turbine allows significant heat recovery while improving the overall cycle efficiency.

Internal-combustion engines—the diesel, natural-gas reciprocating engine, and other such power generators—comprise the Section 4 of the handbook. The diesel is a popular form of electric-power generator in rural areas around the world. And its carbon footprint is not as damaging as that of the traditional steam plant.

Natural-gas-powered reciprocating engines are popular in oil refineries and in areas having abundant supplies of natural gas. In some areas these engines can be run on either natural gas or diesel fuel, depending on which is cheaper.

Cogeneration, wherein waste heat is put to work to improve the overall cycle efficiency, and to wrest more energy from a given process, is popular with steam turbines, gas turbines, and internal-combustion engines. Since greenhouse gases can be reduced when fuel consumption is lowered, cogeneration is an important step in conserving, and reducing, fossil-fuel usage. This handbook presents a number of cogeneration calculations that can help engineers and designers reduce fossil-fuel consumption while increasing cycle efficiency.

Nuclear-power generation calculations are presented in Section 5 of this handbook. Nations around the world are rapidly adopting and building nuclear power plants of large capacity. In the United States, after more than 30 years of being ignored—nuclear power is now suddenly back in favor as a non-atmospheric-polluting source of electric power. Some large power companies are planning on shutting down older coal-fired plants in anticipation of new nuclear generating stations. But even though the steam is generated by a nuclear reactor, the steam- and power-generating cycles are much the same as in a conventional oil-, gas-, or coal-fired plant. So the thermodynamic analyses of the steam cycle follow traditional methods. A number of calculations presented in this handbook bring the engineer and designer "up to speed" in the nuclear field, which is growing in importance every day throughout the world.

The nuclear crisis that occurred in Japan in March 2011, though produced by natural causes, has the world reviewing its nuclear power plants. While many countries say they will not stop building new nuclear power plants, these countries add that they are fully reviewing all safety features of such plants. The likely outcome is that new nuclear power plants will have a variety of advanced safety features not seen in earlier plants, Some engineers are considering triple redundancy for cooling systems and other critical emergency controls.

Nuclear power-plant accidents are feared by almost everyone because of the radiation danger to people in nearby cities and towns. Since radiation cannot usually be seen, it is feared more than a danger that's visible. For this reason the public is more likely to reject a proposed nearby nuclear power plant with the famous retort "not on my backyard." Hence, engineers expect greater resistance to locating new nuclear power plants near populated areas.

The eventual outcome to such resistance is difficult to predict. But engineers familiar with the world's huge energy demand, and the projected future population growth, make a unanimous prediction: nuclear power plants are the only viable choice for supplying the electrical energy demanded by the world's growing population and expanding industries. Like it or not, nuclear power is the world's most likely future large-scale electrical energy power source.

Section 6 of this handbook covers hydroelectric generation of electrical power. Long a "clean" energy source, hydro power is limited by the number of suitable sites that can be found for new plants. Today, "small-scale" sites are popular. Calculation procedures given in this handbook show how such small-scale sites can be used and the types of turbines most suitable for such sites. Using the data provided, an engineer can safely choose a site, and its equipments, to generate "clean" energy.

Wind power is another "clean" energy source and it is covered in Section 7. Calculation procedures in this section cover the range of wind turbines available today. With wind farms on both land and sea throughout the world, a steady growth in the generation of power from this source is forecast. The capacity of wind turbines increases almost every year, providing larger output from each unit and each farm. With some states ordering electric utilities to generate 20 percent of their output from renewable energy sources by 2020, or sooner, the future of wind power is most promising.

Section 8 of this handbook covers solar-power energy applications and usage. A number of calculation procedures show how to select solar panels and other devices for efficient use of sun-generated electric power. Many new ways to use solar panels are being found throughout the world. Thus, automobile parking fields are being covered with solar panels to generate electric power for local structures—schools, shopping malls, office buildings, etc. Another advantage of such panels is that they can generate power to recharge electric-powered cars when both solar panels and electric-powered vehicles become more common. So, as with wind power, the future of solar power is also most promising.

Geothermal energy engineering is covered in Section 9 of this handbook. As with hydroelectric power, geothermal power is very dependent on available sites. There are very few suitable geothermal sites worldwide. But for those that are available, the calculation procedures in this section show how to make maximum use of the available energy. And the various types of steam-generating cycles that can be used are fully explored in the procedures presented. Any engineer can become reasonably conversant with geothermal energy engineering in a few hours using the calculation procedures in this section of the handbook.

Ocean-energy and tidal-stream energy calculation procedures are presented in Section 10. Again, as with hydro and geothermal energy engineering, ocean energy and tidal power depend on suitable sites, and the number of suitatable sites is relatively few, yet engineers continue to try to develop machinery that can be used in normal ocean waters to generate electric power. Calculation procedures given in this section show how to utilize suitable ocean sites to generate electric power reliably and economically. The search for more ways to harness the waves and tidal changes to generate power will continue for as long as civilization needs electricity.

Heat transfer and energy conservation calculation procedures are presented in Section 11 of this handbook. Since energy can be used and reused with proper heat transfer, as well as conserved, the calculation procedures in this section are very important to energy engineers worldwide. Energy conservation by reducing heat losses and conserving available heat are important topics covered in this section of the handbook.

Section 12 deals with the important topic of fluid transfer engineering. Thus, piping, pumps (of many different types), fluid viscosity, energy conservation in choice of pump, materials selection for pump parts, and economic competitive analyses are some of the topics covered in Part 1 of this section of the handbook. Part 2 of Section 12 focuses on choosing heat insulation for piping; orifice-meter selection; relative carrying capacity of piping; water-meter sizing, liquid siphons, water hammer, compressed-air and gas piping, design of steam transmission piping; desuperheater analysis; steam accumulator selection and sizing; choosing plastic piping; estimating costs of steam leaks; and line sizing for flashing steam condensate, plus many other important energy-related fluid transfer calculations.

Interior climate control—heating, ventilation, and air conditioning—is covered in Section 13 of the handbook. Since heating and air conditioning are amongst the largest energy consumers in today's commercial and residential buildings, a large number of calculation procedures are presented in this section. With the emphasis on LEED (Leadership in Energy and Environmental Design) Green Building Rating System™, there is intense interest in energy conservation in every type of new building being constructed today. Likewise, there is great interest in rehabbing existing buildings so they, too, become LEED rated.

An interesting study of the outcome of the LEED program was recently published. The study showed that LEED-rated buildings saved on energy costs during their entire life cycle. Further, the energy and cost savings were such that the owners earned more from each building even though the LEED program might have been more expensive during the construction of the building.

With the higher costs of energy today, engineers and designers are conscious of producing more efficient designs for heating, ventilating, and air conditioning. Calculation procedures in Section 13 of this handbook cover many topics, including energy equations of a variety of types, annual heating and cooling costs, energy savings using ice-storage systems, run-around heat recovery, heating capacity requirements for buildings with steam- and hot-water heating, carbon dioxide buildup in occupied spaces, centrifugal-compressor power and energy input, fan and pump energy performance, air-bubble enclosure energy analysis, expansion-tank sizing for hydronic systems, heat and energy-loss determination for buildings of all sizes and dimensions, heating apparatus steam and energy consumption, air-heating coil selection and energy analysis, energy and heat-load computation for air-conditioning systems, plus numerous other interior-climate and energy control calculations. Using the data given, any engineer or designer should be able to prepare a preliminary design and analysis quickly and easily for any structure anywhere in the world.

Section 14, the final one in this comprehensive handbook, covers energy conservation and environmental pollution control. Both of these topics are of key importance in today's competitive energy world. Calculation procedures covered include atmospheric control system energy investment analysis, energy-from-waste economic analysis, emissions reduction using flue-gas heat recovery, cogeneration heat savings, cost of cogeneration heat-recovery boilers, sizing an electronic precipitator for air-pollution control, explosive-vent sizing for industrial buildings, thermal pollution estimates for power plants, flash steam heat-energy recovery for cogeneration, environmental and safety-regulation rating for equipment, high-temperature hot-water heating to save energy, repowering options for power plants, energy aspects of cooling-tower choice, plus numerous other key calculation procedures for energy conservation and environmental pollution control.

Each section is introduced with technical parameters important to the topic of the section. These parameters also provide last-minute information on energy facts important to engineers designing, building, and operating energy-using facilities. Having these parameters on hand will help energy engineers stay familiar with current important developments.

Great care has been used throughout the long preparation of this handbook of energy engineering calculations to be exactingly accurate. But errors can, and do, occur. If a user finds an error the editor would appreciate being informed about it immediately. You can reach him personally on the Internet at: tyghicks@aol.com 24/7. Or you can write him in care of the publisher. In closing, it is the sincere hope of the editor that you find this handbook helpful in the daily work of your energy engineering activities.

TYLER G. HICKS, P.E.

ACKNOWLEDGMENTS

The contributors and advisers consulted hundreds of sources when preparing the material for inclusion in this handbook. Besides using the books and other publications listed as references in the "Related Calculations" for most procedures, the editor, contributors and advisers consulted and drew material from technical magazines and journals, trade-association standards, engineering and scientific papers, industrial and engineering catalogs, and a variety of similar publications. Most of these are noted in appropriate places throughout the handbook. Additional acknowledgments, listed in the order received, are given below.

Data and charts credited to the Hydraulic Institute are reprinted from the *Hydraulic Institute Standards*, copyright by the Hydraulic Institute, and from the *Pipe Friction Manual*, copyright by the Hydraulic Institute. Data on diesel engine cooling systems are reprinted from *Standard Practices for Stationary Diesel Engines*, copyright by the Diesel Engine Manufacturers Association. Data on minimum requirements for plumbing are drawn from the *American Standard National Plumbing Code* with permission of the publisher, The American Society of Mechanical Engineers.

Specific firms, trade associations, and publications that were extremely helpful in supplying data for various sections of the handbook include Martin Marietta Corporation; *Electronic Design* magazine; Dresser Industries Inc.—Dresser Industrial Valve and Instrument Division; Ingersoll-Rand Company; Anaconda American Brass Company; Waterloo Register Division—Dynamics Corporation of America; ITT Hammel-Dahl; *Mechanical Engineering*, a monthly publication of The American Society of Mechanical Engineers; McQuay, Inc.; The G. C. Breidert Co.; Modine Manufacturing Company; Rubber Manufacturers Association; Condenser Service & Engineering Co., Inc.; Armstrong Machine Works; American Air Filter Company; Crane Company; *Machine Design* magazine; The RAND Corporation; Texas Instruments Incorporated; McGraw-Hill Publications Company, McGraw-Hill, Inc.; Morse Chain Company; Grinnell Corporation; General Electric Company; The B. F. Goodrich Company; American Standard Inc.; the American Society of Heating, Refrigerating and Air-Conditioning Engineers; International Engineering Associates; Taylor Instrument Process Control Division of Sybron Corporation; Clark-Reliance Corporation; American Society for Testing and Materials; Acoustical and Insulating Materials Association; W. S. Dickey Clay Manufacturing Co.; Flexonics Division, Universal Oil Products Co.; Dunham-Bush, Inc.; Carrier Air Conditioning Company; National Industrial Leather Association; Worthington Corporation; Goulds Pumps, Inc. Illustrations and problems credited to Carrier Air Conditioning Company are copyrighted by Carrier Air Conditioning Company.

Individuals who were helpful to the editor of this handbook at one or more times before, and during its preparation, include Lyman F. Scheel, Consulting Engineer; Jack Jaklitsch, Editor, *Mechanical Engineering*; Spencer A. Tucker, Martin & Tucker; Paul V. DeLuca, Porta Systems Corp.; Professor Steven Edelglass, Cooper Union; Professor William Vopat, Cooper Union; Professor Theodore Baumeister, Columbia University; Frederick S. Merritt, Consulting Engineer; James J. O'Connor, Editor, *Power* magazine; Nathan R. Grossner, Consulting Engineer; Nicholas P. Chironis, *Product Engineering*; Franklin D. Yeaple, *Product Engineering* and *Design Engineering*; John D. Constance, Consulting Engineer; John R. Miller, Texas Instruments Incorporated; Rupert Le Grand, *American Machinist*; Ronald G. Kogan, United Computing Systems, Inc.; Al Brons, Flexonics Div.,

Universal Oil Products Company; Carl W. MacPhee, ASHRAE *Guide and Data Book*; Frank P. Anderson, Secretary, Hydraulic Institute; Joseph Mittleman, *Electronics*; Cheryl A. Shaver, E.E., who was a major help in metricating several sections of the handbook; Thomas F. Epley, Editorial Director, U.S. Naval Institute Press; Janet Eyler, *Electronics*; Charles R. Hafer, P.E.; Calvin S. Cronan, *Chemical Engineering* magazine; Nicholas Chopey, Executive Editor, *Chemical Engineering* magazine; Joseph C. McCabe, Editor-Publisher, *Combustion* magazine; Francis J. Lavoie, Managing Editor, *Machine Design* magazine; Donald E. Fink, Managing Editor—Technical, *Aviation Week & Space Technology* magazine; Richard J. Zanetti, Editor-in-Chief, *Chemical Engineering* magazine; Robert G. Schwieger, Editorial Director, *Power* magazine; Barbara LoSchiavo, Editorial Support, *Machine Design* magazine; Michael G. Ivanovich, Editor, *Heating/Piping/Air Conditioning* magazine; Calmac Manufacturing Corp. whose *Ice Bank* is a registered trademark of that corporation; Joseph Leto, P.E., Consulting Engineer; Gerald M. Eisenberg, Project Engineering Administrator, American Society of Mechanical Engineers, who contributed a number of new procedures and ideas; Stephen M. Eber, P.E., Ebasco Services, Inc., who also contributed a number of new procedures and ideas; Jerome Mueller, P.E., Mueller Engineering Corporation, who was most helpful with thoughts on applied calculation procedures; V. Ganapathy, Heat Transfer Specialist, ABCO Industries, Inc.; Kevin D. Wills, M.S.E., P.E., Consulting Engineer, Stanley Consultants, Inc.; Joseph B. Shanley, Mechanical Engineer, who metricated many illustrations and procedures; and numerous working engineers and scientists in firms and universities in the United States and abroad.

I owe an enormous thanks to my long-time Sponsoring Editor, Larry Hager, of McGraw-Hill Professional. Larry is always available, helpful, and understanding with his authors. He has put up with this author for 25+ years and has been a steady hand on the helm for me all these years. I will always be thankful to Larry for helping me deliver every book on schedule and within the size requirements of the publisher.

Completion of this handbook would not have been possible without the enormous help of my wife, Mary Shanley Hicks, a publishing professional, who worked thousands of hours on the computer and at many other tasks to put the gigantic manuscript into publishable condition. She—more than the editor—deserves whatever accolades this handbook earns. And she has the editor's grateful thanks for keeping him on the job until the manuscript was finished.

TYLER G. HICKS, P.E.

HOW TO USE THIS HANDBOOK

There are two ways to enter this handbook to obtain the maximum benefit from the time invested. The first entry is through the index; the second is through the table of contents of the section covering the discipline, or related discipline, concerned. Each method is discussed in detail below.

Index. Great care and considerable time were expended on preparation of the index of this handbook so that it would be of maximum use to every reader. As a general guide, enter the index using the generic term for the type of calculation procedure being considered. Thus, for the selection of a centrifugal pump to reduce energy consumption, enter at *pumps*. From here, proceed to the specific type of pump being considered—such as *centrifugal*. Once the page number of the specific calculation procedure is determined, turn to it to find the step-by-step instructions and worked-out example that can be followed to solve the problem quickly and accurately.

Contents. The contents of each section lists the titles of the calculation procedures contained in that section. Where extensive use of any section is contemplated, the editor suggests that the reader might benefit from an occasional glance at the table of contents of that section. Such a glance will give the user of this handbook an understanding of the breadth and coverage of a given section, or a series of sections. Then, when he or she turns to this handbook for assistance, the reader will be able more rapidly to find the calculation procedure he or she seeks.

Calculation Procedures. Each calculation procedure is a unit in itself. However, any given calculation procedure will contain subprocedures that might also be useful to the energy-concerned user. Thus, a calculation procedure on pump selection will contain subprocedures on pipe-friction loss, pump static and dynamic heads, etc. Should the user of this handbook wish to make a computation using any of such subprocedures, he or she will find the worked-out steps that are presented both useful and precise. Hence, the handbook contains numerous valuable procedures that are useful in solving a variety of applied energy engineering problems.

Another important point that should be noted about the calculation procedures presented in this handbook is that many of the calculation procedures are equally applicable in a variety of disciplines. Thus, a pump selection procedure can be used for energy-, civil-, mechanical-, chemical-, electrical-, and nuclear-engineering activities, as well as some others. Hence, the user might consider a temporary neutrality for his or her particular energy engineering specialty when using the handbook because the calculation procedures are designed for universal use.

Any of the energy engineering procedures herein can be programmed on a laptop or desktop computer. Such programming permits rapid solution of a variety of energy engineering design and operating problems. The fast, modern computing equipment available today provides greater speed and accuracy for nearly all complex energy engineering problems in today's complex world of energy control and use.

SI Usage. The technical and scientific community throughout the world accepts the SI (System International) for use in both applied and theoretical calculations. With such widespread acceptance of SI, every engineer must become proficient in the use of this system of units if he or she is to remain up-to-date. For this reason, every calculation procedure in this handbook is given in both the United States Customary System (USCS) and SI. This will help all engineers become proficient in using both systems of units. In this handbook the USCS unit is generally given first, followed by the SI value in parentheses or brackets. Thus, if the USCS unit is 10 ft, it will be expressed as 10 ft (3 m).

Engineers accustomed to working in USCS are often timid about using SI. There really aren't any sound reasons for these fears. SI is a logical, easily understood, and readily manipulated group of units. Most engineers grow to prefer SI, once they become familiar with it and overcome their fears. This handbook should do much to "convert" USCS-user engineers to SI because it presents all calculation procedures in both the known and unknown units.

Overseas engineers who must work in USCS because they have a job requiring its usage will find the dual-unit presentation of calculation procedures most helpful. Knowing SI, they can easily convert to USCS because all procedures, tables, and illustrations are presented in dual units.

Learning SI. An efficient way for the USCS-conversant engineer to learn SI follows these steps:

1. List the units of measurement commonly used in your daily work.
2. Insert, opposite each USCS unit, the usual SI unit used; Table 1 shows a variety of commonly used quantities and the corresponding SI units.
3. Find, from a table of conversion factors, such as Table 2, the value to use to convert the USCS unit to SI, and insert it in your list. (Most engineers prefer a conversion factor that can be used as a multiplier of the USCS unit to give the SI unit.)

TABLE 1 Commonly Used USCS and SI Units*

USCS unit	SI unit	SI symbol	Conversion factor—multiply USCS unit by this factor to obtain the SI unit
square feet	square meters	m^2	0.0929
cubic feet	cubic meters	m^3	0.2831
pounds per square inch	kilopascal	kPa	6.894
pound force	newton	N	4.448
foot pound torque	newton-meter	N·m	1.356
kip-feet	kilo-newton	kNm	1.355
gallons per minute	liters per second	L/s	0.06309
kips per square inch	megaPascal	MPa	6.89

*Because of space limitations this table is abbreviated. For a typical engineering practice an actual table would be many times this length.

TABLE 2 Typical Conversion Table*

To convert from	To	Multiply by	
square feet	square meters	9.290304	E − 02
foot per second squared	meter per second squared	3.048	E − 01
cubic feet	cubic meters	2.831685	E − 02
pound per cubic inch	kilogram per cubic meter	2.767990	E + 04
gallon per minute	liters per second	6.309	E − 02
pound per square inch	kilopascal	6.894757	
pound force	newton	4.448222	
kip per square foot	Pascal	4.788026	E + 04
acre-foot per day	cubic meter per second	1.427641	E − 02
acre	square meter	4.046873	E + 03
cubic foot per second	cubic meter per second	2.831685	E − 02

Note: The E indicates an exponent, as in scientific notation, followed by a positive or negative number, representing the power of 10 by which the given conversion factor is to be multiplied before use. Thus, for the square feet conversion factor, 9.290304 × 1/100 = 0.09290304, the factor to be used to convert square feet to square meters. For a positive exponent, as in converting acres to square meters, multiply by 4.046873 × 1000 = 4046.8.

Where a conversion factor cannot be found, simply use the dimensional substitution. Thus, to convert pounds per cubic inch to kilograms per cubic meter, find 1 lb = 0.4535924 kg, and 1 in^3 = 0.00001638706 m^3. Then, 1 lb/in^3 = 0.4535924 kg/0.00001638706 m^3 = 27,680.01, or 2.768 E + 04.

*This table contains only selected values. See the U.S. Department of the Interior *Metric Manual*, or National Bureau of Standards, *The International System of Units* (SI), both available from the U.S. Government Printing Office (GPO), for far more comprehensive listings of conversion factors.

4. Apply the conversion factors whenever you have an opportunity. Think in terms of SI when you encounter a USCS unit.
5. Recognize—here and now—that the most difficult aspect of SI is becoming comfortable with the names and magnitude of the units. Numerical conversion is simple, once you've set up *your own* conversion table. So think pascal whenever you encounter pounds per square inch pressure, newton whenever you deal with a force in pounds, etc.

SI Table for an Energy Engineer. Let's say you're an energy engineer and you wish to construct a conversion table for yourself. List the units you commonly meet in your daily work; Table 1 is the list compiled by one energy engineer. Next, list the SI unit equivalent for the USCS unit. Obtain the equivalent from Table 2. Then, using Table 2 again, insert the conversion multiplier in Table 1.

If you ever have any questions about more efficient use of this handbook, e-mail the author at: tyghicks@aol.com. Or write him in care of the publisher. You will receive a prompt, courteous, and helpful response.

HANDBOOK OF ENERGY ENGINEERING CALCULATIONS

SECTION 1
ENERGY CONVERSION ENGINEERING

Combustion Calculation Parameters 1.1
Coal Fuel Combustion in a Furnace 1.1
Percent Excess Air while Burning Coal 1.4
Fuel Oil Combustion in a Furnace 1.5
Natural Gas Fuel Combustion in a Furnace 1.7
Wood Fuel Combustion in a Furnace 1.11
Combustion Analysis by the Molal Method 1.13
Estimating the Temperature of the Final Products of Combustion 1.15
Determination of the Savings Produced by Preheating Combustion Air 1.16
Combustion Calculations Using the Million Btu (1.055 MJ) Method 1.17

COMBUSTION CALCULATION PARAMETERS

The focus in this section of the handbook is on external combustion of fossil fuels—such as in a boiler furnace. A large portion of fossil-fuel energy use takes place in the form of external combustion to generate steam for power generation and space heating. Internal combustion of fossil fuels—such as in gas turbines, and diesel and gas engines—is discussed within sections 3 and 4 of this handbook.

In today's world the greatest emphasis is on complete combustion to reduce greenhouse gases and the ensuing pollution when such gases are emitted into the atmosphere. Thus, in the midwestern United States, combustion of coal in steam power plants produces some 210 lb (95.3 kg) of CO_2 per million Btu (1.055 MJ) of coal used in a steam boiler. Contrast this with 120 lb (54.4 kg) of CO_2 produced per million Btu (1.055 MJ) of natural gas used in a steam boiler.

On the basis of power generated, 1.9 lb (0.86 kg) of CO_2 is produced per kWh generated by a coal-fired boiler and turbine. With natural-gas firing of a steam boiler, 1.25 lb (0.57 kg) of CO_2 is produced per kWh generated. These statistics indicate that more efficient combustion of fossil fuels reduces greenhouse gases and operating costs. Hence design and operating engineers are making strenuous efforts to improve combustion efficiency to comply with ever stricter regulations while reducing plant operating costs.

As indicated later in this section of the handbook, some states and countries are considering banning coal-fired power plants unless they meet much more stringent emission requirements. At this writing, the Environmental Protection Agency seeks to establish ozone levels of 0.060 to 0.070 ppm.

COAL FUEL COMBUSTION IN A FURNACE

A coal fuel has the following ultimate analysis (or percent by weight): C = 0.8339; H_2 = 0.0456; O_2 = 0.0505; N_2 = 0.0103; S = 0.0064; ash = 0.0533; total = 1.000 lb (0.45 kg). This coal is burned in a steam-boiler furnace. Determine the weight of air required for theoretically perfect combustion, the weight of gas formed per pound (kilogram) of coal burned, and the volume of flue gas, at the boiler exit temperature of 600°F (316°C) per pound (kilogram) of coal burned; air required with 20 percent excess air, and the volume of gas formed with this excess; the CO_2 percentage in the flue gas on a dry and wet basis.

Calculation Procedure:

1. ***Compute the weight of oxygen required per pound (kilogram) of coal***

To find the weight of oxygen required for theoretically perfect combustion of coal, set up the following tabulation, based on the ultimate analysis of the coal:

Element	×	Molecular-weight ratio	=	lb (kg) O_2 required
C; 0.8339	×	32/12	=	2.2237 (1.009)
H_2; 0.0456	×	16/2	=	0.3648 (0.165)
O_2; 0.0505; decreases external O_2 required			=	−0.0505 (0.023)
N_2; 0.0103 is inert in combustion and is ignored				
S; 0.0064	×	32/32	=	0.0064 (0.003)
Ash 0.0533 is inert in combustion and is ignored				
Total 1.0000				
lb (kg) external O_2 per lb (kg) fuel			=	2.5444 (1.154)

Note that of the total oxygen needed for combustion, 0.0505 lb (0.023 kg), is furnished by the fuel itself and is assumed to reduce the total external oxygen required by the amount of oxygen present in the fuel. The molecular-weight ratio is obtained from the equation for the chemical reaction of the element with oxygen in combustion. Thus, for carbon $C + O_2 \rightarrow CO_2$, or $12 + 32 = 44$, where 12 and 32 are the molecular weights of C and O_2, respectively.

2. ***Compute the weight of air required for perfect combustion***

Air at sea level is a mechanical mixture of various gases, principally 23.2 percent oxygen and 76.8 percent nitrogen by weight. The nitrogen associated with the 2.5444 lb (1.154 kg) of oxygen required per pound (kilogram) of coal burned in this furnace is the product of the ratio of the nitrogen and oxygen weights in the air and 2.5444, or (2.5444)(0.768/0.232) = 8.4228 lb (3.820 kg). Then the weight of air required for perfect combustion of 1 lb (0.45 kg) of coal = sum of nitrogen and oxygen required = 8.4228 + 2.5444 = 10.9672 lb (4.975 kg) of air per pound (kilogram) of coal burned.

3. ***Compute the weight of the products of combustion***

Find the products of combustion by addition:

Fuel constituents	+ Oxygen	→	Products of combustion		lb	kg
C; 0.8339	+ 2.2237	→	CO_2	=	3.0576	1.387
H; 0.0465	+ 0.3648	→	H_2O	=	0.4104	0.186
O_2; 0.0505; this is *not* a product combustion						
N_2; 0.0103; inert but passes through furnace				=	0.0103	0.005
S; 0.0064	+ 0.0064	→	SO_2	=	0.0128	0.006
Outside nitrogen from step 2			$= N_2$	=	8.4228	3.820
lb (kg) of flue gas lb (kg) of coal burned				=	11.9139	5.404

4. ***Convert the flue-gas weight to volume***

Use Avogadro's law, which states that under the same conditions of pressure and temperature, 1 mol (the molecular weight of a gas expressed in lb) of any gas will occupy the same volume.

At 14.7 lb/in^2 (abs) (101.3 kPa) and 32°F (0°C), 1 mol of any gas occupies 359 ft^3 (10.2 m^3). The volume per pound (kilogram) of any gas at these conditions can be found by dividing 359 by the molecular weight of the gas and correcting for the gas temperature by multiplying the volume by the ratio of the absolute flue-gas temperature and the atmospheric temperature. To change the weight analysis (step 3) of the products of combustion to volumetric analysis, set up the calculation thus:

Products	Weight, lb	Weight, kg	Molecular weight	Temperature correction		Volume at 600°F, ft^3	Volume at 316°C, m^3
CO_2	3.0576	1.3869	44	(359/44)(3.0576)(2.15)	=	53.6	1.518
H_2O	0.4104	0.1862	18	(359/18)(0.4104)(2.15)	=	17.6	0.498
Total N_2	8.4331	3.8252	28	(359/28)(8.4331)(2.15)	=	232.5	6.584
SO_4	0.0128	0.0058	64	(359/64)(0.0128)(2.15)	=	0.15	0.004
ft^3 (m^3) of flue gas per lb (kg) of coal burned					=	303.85	8.604

In this calculation, the temperature correction factor 2.15 = absolute flue-gas temperature, °R/absolute atmospheric temperature, °R = (600 + 460)/(32 + 460). The total weight of N_2 in the flue gas is the sum of the N_2 in the combustion air and the fuel, or 8.4228 + 0.0103 = 8.4331 lb (3.8252 kg). The value is used in computing the flue-gas volume.

5. *Compute the CO_2 content of the flue gas*
The volume of CO_2 in the products of combustion at 600°F (316°C) is 53.6 ft^3 (1.158 m^3), as computed in step 4, and the total volume of the combustion products is 303.85 ft^3 (8.604 m^3). Therefore, the percent CO_2 on a wet basis (i.e., including the moisture in the combustion products) = ft^3 CO_2/ total ft^3 = 53.6/303.85 = 0.1764, or 17.64 percent.

The percent CO_2 on a dry, or Orsat, basis is found in the same manner, except that the weight of H_2O in the products of combustion, 17.6 lb (7.83 kg) from step 4, is subtracted from the total gas weight. Or, percent CO_2, dry, or Orsat basis = (53.6)/(303.85 – 17.6) = 0.1872, or 18.72 percent.

6. *Compute the air required with the stated excess flow*
With 20 percent excess air, the air flow required = (0.20 + 1.00)(air flow with no excess) = 1.20 (10.9672) = 13.1606 lb (5.970 kg) of air per pound (kilogram) of coal burned. The air flow with no excess is obtained from step 2.

7. *Compute the weight of the products of combustion*
The excess air passes through the furnace without taking part in the combustion and increases the weight of the products of combustion per pound (kilogram) of coal burned. Therefore, the weight of the products of combustion is the sum of the weight of the combustion products without the excess air and the product of (percent excess air)(air for perfect combustion, lb); or, given the weights from steps 3 and 2, respectively, = 11.9139 + (0.20)(10.9672) = 14.1073 lb (6.399 kg) of gas per pound (kilogram) of coal burned with 20 percent excess air.

8. *Compute the volume of the combustion products and the percent CO_2*
The volume of the excess air in the products of combustion is obtained by converting from the weight analysis to the volumetric analysis and correcting for temperature as in step 4, using the air weight from step 2 for perfect combustion and the excess-air percentage, or (10.9672)(0.20)(359/28.95) (2.15) = 58.5 ft^3 (1.656 m^3). In this calculation the value 28.95 is the molecular weight of air. The total volume of the products of combustion is the sum of the column for perfect combustion, step 4, and the excess-air volume, above, or 303.85 + 58.5 = 362.35 ft^3 (10.261 m^3).

By using the procedure in step 5, the percent CO_2, wet basis = 53.6/362.35 = 14.8 percent. The percent CO_2, dry basis = 53.8/(362.35 – 17.6) = 15.6 percent.

Related Calculations. Use the method given here when making combustion calculations for any type of coal—bituminous, semibituminous, lignite, anthracite, cannel, or cooking—from any coal field in the world used in any type of furnace—boiler, heater, process, or waste-heat. When the air used for combustion contains moisture, as is usually true, this moisture is added to the combustion-formed moisture appearing in the products of combustion. Thus, for 80°F (26.7°C) air of 60 percent relative humidity, the moisture content is 0.013 lb/lb (0.006 kg/kg) of dry air. This amount appears in the products of combustion for each pound of air used and is a commonly assumed standard in combustion calculations.

Fossil-fuel-fired power plants release sulfur emissions to the atmosphere. In turn, this produces sulfates, which are the key ingredient in acid rain. The federal Clean Air Act regulates sulfur dioxide emissions from power plants. Electric utilities which burn high-sulfur coal are thought to produce some 35 percent of atmospheric emissions of sulfur dioxide in the United States.

Sulfur dioxide emissions by power plants have declined some 30 percent since passage of the Clean Air Act in 1970, and a notable decline in acid rain has been noted at a number of test sites. In 1990 the Acid Rain Control Program was created by amendments to the Clean Air Act. This program further reduces the allowable sulfur dioxide emissions from power plants, steel mills, and other industrial facilities.

The same act requires reduction in nitrogen oxide emissions from power plants and industrial facilities, so designers must keep this requirement in mind when designing new and replacement facilities of all types which use fossil fuels.

At the time of this writing, there are a number of states and countries considering phasing out coal-burning power plants at the end of their useful commercial lives if they do not install means to capture carbon dioxide and other greenhouse gases. These entities are also urging a shift to natural-gas fuel. Further, combined-cycle natural-gas-fueled power plants are being urged as replacements of coal-fired power plants.

PERCENT EXCESS AIR WHILE BURNING COAL

A certain coal has the following composition by weight percentages: carbon 75.09, nitrogen 1.56, ash 3.38, hydrogen 5.72, oxygen 13.82, sulfur 0.43. When burned in an actual furnace, measurements showed that there was 8.93 percent combustible in the ash pit refuse and the following Orsat analysis in percentages was obtained: carbon dioxide 14.2, oxygen 4.7, carbon monoxide 0.3. If it can be assumed that there was no combustible in the flue gas other that the carbon monoxide reported, calculate the percentage of excess air used.

Calculation Procedure:

1. ***Compute the amount of theoretical air required per lb_m (kg) of coal***

Theoretical air required per pound (kilogram) of coal, $w_{ta} = 11.5C' + 34.5[H'_2 - O'_2/8)] + 4.32S'$, where C', H'_2, O'_2, and S' represent the percentages by weight, expressed as decimal fractions, of carbon, hydrogen, oxygen, and sulfur, respectively. Thus, $w_{ta} = 11.5(0.7509) + 34.5[0.0572 - (0.1382/8)] + 4.32(0.0043) = 10.03$ lb (4.55 kg) of air per lb (kg) of coal. The ash and nitrogen are inert and do not burn.

2. ***Compute the correction factor for combustible in the ash***

The correction factor for combustible in the ash, $C_1 = (w_t C_f - w_r C_r)/(w_f \times 100)$, where the amount of fuel, $w_f = 1$ lb (0.45 kg) of coal; percent by weight, expressed as a decimal fraction, of carbon in the coal, $C_f = 75.09$; percent by weight of the ash and refuse in the coal, $w_r = 0.0338$; percent by weight of combustible in the ash, $C_r = 8.93$. Hence, $C_1 = [(1 \times 75.09) - (0.0338 \times 8.93)]/(1 \times 100) = 0.748$.

3. ***Compute the amount of dry flue gas produced per lb (kg) of coal***

The lb (kg) of dry flue gas per lb (kg) of coal, $w_{dg} = C_1(4CO_2 + O_2 + 704)/[3(CO_2 + CO)]$, where the Orsat analysis percentages are for carbon dioxide, $CO_2 = 14.2$; oxygen, $O_2 = 4.7$; carbon monoxide, CO = 0.3. Hence, $w_{dg} = 0.748 \times [(4 \times 14.2) + 4.7 + 704)]/[3(14.2 + 0.3)] = 13.16$ lb/lb (5.97 kg/kg).

4. ***Compute the amount of dry air supplied per lb (kg) of coal***
The lb (kg) of dry air supplied per lb (kg) of coal, $w_{da} = w_{dg} - C_1 + 8[H'_2 - (O'_2/8)] - (N'_2/N)$, where the percentage by weight of nitrogen in the fuel, $N'_2 = 1.56$, and "atmospheric nitrogen" in the supply air, $N_2 = 0.768$; other values are as given or calculated. Then, $w_{da} = 13.16 - 0.748 + 8[0.0572 - (0.1382/8)] - (0.0156/0.768) = 12.65$ lb/lb (5.74 kg/kg).

5. ***Compute the percent of excess air used***
Percent excess air = $(w_{da} - w_{ta})/w_{ta} = (12.65 - 10.03)/10.03 = 0.261$, or 26.1 percent.

Related Calculations. The percentage by weight of nitrogen in "atmospheric air" in step 4 appears in *Principles of Engineering Thermodynamics,* 2nd edition, by Kiefer et al., John Wiley & Sons, Inc.

FUEL OIL COMBUSTION IN A FURNACE

A fuel oil has the following ultimate analysis: C = 0.8543; $H_2 = 0.1131$; $O_2 = 0.0270$; $N_2 = 0.0022$; S = 0.0034; total = 1.0000. This fuel oil is burned in a steam-boiler furnace. Determine the weight of air required for theoretically perfect combustion, the weight of gas formed per pound (kilogram) of oil burned, and the volume of flue gas, at the boiler exit temperature of 600°F (316°C), per pound (kilogram) of oil burned; the air required with 20 percent excess air, and the volume of gas formed with this excess; the CO_2 percentage in the flue gas on a dry and wet basis.

Calculation Procedure:

1. ***Compute the weight of oxygen required per pound (kilogram) of oil***
The same general steps as given in the previous calculation procedure will be followed. Consult that procedure for a complete explanation of each step.

Using the molecular weight of each element, we find

Element	×	Molecular-weight ratio	=	lb (kg) O_2 required	
C; 0.8543	×	32/12	=	2.2781	(1.025)
H_2; 0.1131	×	16/2	=	0.9048	(0.407)
O_2; 0.0270; decreases external O_2 required			=	−0.0270	(−0.012)
N_2; 0.0022 is inert in combustion and is ignored					
S; 0.0034	×	32/32	=	0.0034	(0.002)
Total 1.0000					
lb (kg) external O_2 per lb (kg) fuel			=	3.1593	(1.422)

2. ***Compute the weight of air required for perfect combustion***
The weight of nitrogen associated with the required oxygen = (3.1593)(0.768/0.232) = 10.458 lb (4.706 kg). The weight of air required = 10.4583 + 3.1593 = 13.6176 lb/lb (6.128 kg/kg) of oil burned.

3. *Compute the weight of the products of combustion*
As before,

Fuel constituents	+	Oxygen	=	Products of combustion
C; 0.8543 + 2.2781	=	3.1324	=	CO_2
H_2; 0.1131 + 0.9148	=	1.0179	=	H_2O
O_2; 0.270; *not* a product of combustion				
N_2; 0.0022; inert but passes through furnace	=	0.0022	=	N_2
S; 0.0034 + 0.0034	=	0.0068	=	SO_2
Outside N_2 from step 2	=	10.458	=	N_2
lb (kg) of flue gas per lb (kg) of oil burned	=	14.6173 (6.578)		

4. *Convert the flue-gas weight to volume*
As before,

	Weight					Volume at	
Products	lb	kg	Molecular weight	Temperature correction		600°F, ft³	316°C, m³
CO_2	3.1324	1.4238	44	(359/44)(3.1324)(2.15)	=	55.0	1.557
H_2O	1.0179	0.4626	18	(359/18)(1.0179)(2.15)	=	43.5	1.231
N_2 (total)	10.460	4.7545	28	(359/28)(10.460)(2.15)	=	288.5	8.167
SO_2	0.0068	0.0031	64	(359/64)(0.0068)(2.15)	=	0.82	0.023
ft³ (m³) of flue gas per lb (kg) of oil burned					=	387.82	10.978

In this calculation, the temperature correction factor 2.15 = absolute flue-gas temperature, °R/absolute atmospheric temperature, °R = (600 + 460)/(32 + 460). The total weight of N_2 in the flue gas is the sum of the N_2 in the combustion air and the fuel, or 10.4580 + 0.0022 = 10.4602 lb (4.707 kg).

5. *Compute the CO_2 content of the flue gas*
CO_2, wet basis = 55.0/387.82 = 0.142, or 14.2 percent. CO_2, dry basis = 55.0/(387.2 – 43.5) = 0.160, or 16.0 percent.

6. *Compute the air required with stated excess flow*
The pounds (kilograms) of air per pound (kilogram) of oil with 20 percent excess air = (1.20)(13.6176) = 16.3411 lb (7.353 kg) of air per pound (kilogram) of oil burned.

7. *Compute the weight of the products of combustion*
The weight of the products of combustion = product weight for perfect combustion, lb + (percent excess air)(air for perfect combustion, lb) = 14.6173 + (0.20)(13.6176) = 17.3408 lb (7.803 kilogram) of flue gas per pound (kilogram) of oil burned with 20 percent excess air.

8. *Compute the volume of the combustion products and the percent CO_2*
The volume of excess air in the products of combustion is found by converting from the weight to the volumetric analysis and correcting for temperature as in step 4, using the air weight from step 2 for perfect combustion and the excess-air percentage, or (13.6176)(0.20)(359/28.95)(2.15) = 72.7 ft³ (2.058 m³). Add this to the volume of the products of combustion found in step 4, or 387.82 + 72.70 = 460.52 ft³ (13.037 m³).

By using the procedure in step 5, the percent CO_2, wet basis = 55.0/460.52 = 0.1192, or 11.92 percent. The percent CO_2, dry basis = 55.0/(460.52 – 43.5) = 0.1318, or 13.18 percent.

Related Calculations. Use the method given here when making combustion calculations for any type of fuel oil—paraffin-base, asphalt-base, Bunker C, no. 2, 3, 4, or 5—from any source, domestic or foreign, in any type of furnace—boiler, heater, process, or waste-heat. When the air used for combustion contains moisture, as is usually true, this moisture is added to the combustion-formed moisture appearing in the products of combustion. Thus, for 80°F (26.7°C) air of 60 percent relative humidity, the moisture content is 0.013 lb/lb (0.006 kg/kg) of dry air. This amount appears in the products of combustion for each pound (kilogram) of air used and is a commonly assumed standard in combustion calculations.

NATURAL GAS FUEL COMBUSTION IN A FURNACE

A natural gas has the following volumetric analysis at 60°F (15.5°C): CO_2 = 0.004; CH_4 = 0.921; C_2H_6 = 0.041; N_2 = 0.034; total = 1.000. This natural gas is burned in a steam-boiler furnace. Determine the weight of air required for theoretically perfect combustion, the weight of gas formed per pound (kilogram) of natural gas burned, and the volume of the flue gas, at the boiler exit temperature of 650°F (343°C), per pound (kilogram) of natural gas burned; air required with 20 percent excess air, and the volume of gas formed with this excess: CO_2 percentage in the flue gas on a dry and wet basis.

Calculation Procedure:

1. ***Compute the weight of oxygen required per pound of gas***

The same general steps as given in the previous calculation procedures will be followed, except that they will be altered to make allowances for the differences between natural gas and coal.

The composition of the gas is given on a volumetric basis, which is the usual way of expressing a fuel-gas analysis. To use the volumetric-analysis data in combustion calculations, they must be converted to a weight basis. This is done by dividing the weight of each component by the total weight of the gas. A volume of 1 ft^3 (1 m^3) of the gas is used for this computation. Find the weight of each component and the total weight of 1 ft^3 (1 m^3) as follows, using the properties of the combustion elements and compounds given in Table 1:

		Density		Component weight = column 1 × column 2	
Component	Percent by volume	lb/ft^3	kg/m^3	lb/ft^3	kg/m^3
CO_2	0.004	0.1161	1.859	0.0004644	0.007
CH_4	0.921	0.0423	0.677	0.0389583	0.624
C_2H_6	0.041	0.0792	1.268	0.0032472	0.052
N_2	0.034	0.0739	0.094	0.0025026	0.040
Total	1.000			0.0451725	0.723

$$\text{Percent } CO_2 = 0.0004644/0.0451725 = 0.01026, \text{ or } 1.03 \text{ percent}$$

$$\text{Percent } CH_4 \text{ by weight} = 0.0389583/0.0451725 = 0.8625, \text{ or } 86.25 \text{ percent}$$

$$\text{Percent } C_2H_6 \text{ by weight} = 0.0032472/0.0451725 = 0.0718, \text{ or } 7.18 \text{ percent}$$

$$\text{Percent } N_2 \text{ by weight} = 0.0025026/0.0451725 = 0.0554, \text{ or } 5.54 \text{ percent}$$

The sum of the weight percentages = 1.03 + 86.25 + 7.18 + 5.54 = 100.00. This sum checks the accuracy of the weight calculation, because the sum of the weights of the component parts should equal 100 percent.

TABLE 1 Properties of Combustion Elements*

Element or compound	Formula	Molecular weight	At 14.7 lb/in^2 (abs) (101.3 kPa), 60°F (15.6°C)		Nature			Heat value, Btu (kJ)	
			Weight, lb/ft^3 (kg/m^3)	Volume, ft^3/lb (m^3/kg)	Gas or solid	Combustible	Per lb (kg)	Per ft^3 (m^3) at 14.7 lb/in^2 (abs) (101.3 kPa), 60°F (15.6°C)	Per mole
Carbon	C	12	—	—	S	Yes	14,540 (33,820)	—	174,500
Hydrogen	H_2	2.02†	0.0053 (0.0849)	188 (11.74)	G	Yes	61,000 (141,886)	325 (12,109)	123,100
Sulfur	S	32	—	—	S	Yes	4,050 (9,420)	—	129,600
Carbon monoxide	CO	28	0.0739 (1.183)	13.54 (0.85)	G	Yes	4,380 (10,187)	323 (12,035)	122,400
Methane	CH_4	16	0.0423 (0.677)	23.69 (1.48)	G	Yes	24,000 (55,824)	1,012 (37,706)	384,000
Acetylene	C_2H_2	26	0.0686 (1.098)	14.58 (0.91)	G	Yes	21,500 (50,009)	1,483 (55,255)	562,000
Ethylene	C_2H_4	28	0.0739 (1.183)	13.54 (0.85)	G	Yes	22,200 (51,637)	1,641 (61,141)	622,400
Ethane	C_2H_6	30	0.0792 (1.268)	12.63 (0.79)	G	Yes	22,300 (51,870)	1,762 (65,650)	668,300
Oxygen	O_2	32	0.0844 (1.351)	11.84 (0.74)	G				
Nitrogen	N_2	28	0.0739 (1.183)	13.52 (0.84)	G				
Air‡	—	29	0.0765 (1.225)	13.07 (0.82)	G				
Carbon dioxide	CO_2	44	0.1161 (1.859)	8.61 (0.54)	G				
Water	H_2O	18	0.0475 (0.760)	21.06 (1.31)	G				

*P. W. Swain and L. N. Rowley, "Library of Practical Power Engineering" (collection of articles published in *Power*).
†For most practical purposes, the value of 2 is sufficient.
‡The molecular weight of 29 is merely the weighted average of the molecular weight of the constituents.

Next, find the oxygen required for combustion. Since both the CO_2 and N_2 are inert, they do not take part in the combustion; they pass through the furnace unchanged. Using the molecular weights of the remaining components in the gas and the weight percentages, we have

Compound	×	Molecular-weight ratio	=	lb (kg) O_2 required
CH_4; 0.0718	×	64/16	=	3.4500 (1.553)
C_2H_6; 0.0718	×	112/30	=	0.2920 (0.131)
lb (kg) external O_2 required per lb (kg) fuel			=	3.7420 (1.684)

In this calculation, the molecular-weight ratio is obtained from the equation for the combustion chemical reaction, or $CH_4 + 2O_2 = CO_2 + 2H_2O$, that is, 16 + 64 = 44 + 36, and $C_2H_6 + 7/2O_2 = 2CO_2 + 3H_2O_2$, that is 30 + 112 = 88 + 54. See Table 2 from these and other useful chemical reactions in combustion.

TABLE 2 Chemical Reactions

Combustible substance	Reaction	Mols	lb (kg)*
Carbon to carbon monoxide	$C + ½O_2 = CO$	1 + ½ = 1	12 + 16 = 28
Carbon to carbon dioxide	$C + O_2 = CO_2$	1 + 1 = 1	12 + 16 = 28
Carbon monoxide to carbon dioxide	$CO + ½O_2 = CO_2$	1 + ½ = 1	28 + 16 = 44
Hydrogen	$H_2 + ½O_2 = H_2O$	1 + ½ = 1	2 + 16 = 18
Sulfur to sulfur dioxide	$S + O_2 = SO_2$	1 + 1 = 1	32 + 32 = 64
Sulfur to sulfur trioxide	$S + 3/2O_2 = SO_3$	1 + 2/2 = 1	32 + 48 = 80
Methane	$CH_4 + 2O_2 = CO_2 + 2H_2O$	1 + 2 = 1 + 2	16 + 64 = 44 + 36
Ethane	$C_2H_6 + 7/2O_2 = 2CO_2 + 3H_2O$	1 + 7/2 = 2 + 3	30 + 112 = 88 + 54
Propane	$C_3H_8 + 5O_2 = 3CO_2 + 4H_2O$	1 + 5 = 3 + 4	44 + 160 = 132 + 72
Butane	$C_4H_{10} + 13/2O_2 = 4CO_2 + 5H_2O$	1 + 12/2 = 4 + 5	58 + 208 = 176 + 90
Acetylene	$C_2H_2 + 5/2O_2 = 2CO_2 + H_2O$	1 + 5/2 = 2 + 2	26 + 80 = 88 + 18
Ethylene	$C_2H_4 + 3O_2 = 2CO_2 + 2H_2O$	1 + 3 = 2 + 2	28 + 96 = 88 + 36

*Substitute the molecular weights in the reaction equation to secure lb (kg). The lb (kg) on each side of the equation must balance.

2. *Compute the weight of air required for perfect combustion*
The weight of nitrogen associated with the required oxygen = (3.742)(0.768/0.232) = 12.39 lb (5.576 kg). The weight of air required = 12.39 + 3.742 = 16.132 lb/lb (7.259 kg/kg) of gas burned.

3. *Compute the weight or the products of combustion*

				Products of combustion	
Fuel constituents	+	Oxygen	=	lb	kg
CO_2; 0.0103: inert but passes through the furnace			=	0.010300	0.005
CH_4; 0.8625	+	3.45	=	4.312500	1.941
C_2H_6; 0.003247	+	0.2920	=	0.032447	0.015
N_2; 0.0554; inert but passes through the furnace			=	0.055400	0.025
Outside N_2 from step 2			=	12.390000	5.576
lb (kg) of flue gas per lb (kg) of natural gas burned			=	16.800347	7.562

4. *Convert the flue-gas weight to volume*

The products of complete combustion of any fuel that does not contain sulfur are CO_2, H_2O, and N_2. Using the combustion equation in step 1, compute the products of combustion thus: $CH_4 + 2O_2 = CO_2 + H_2O$; 16 + 64 = 44 + 36; or the CH_4 burns to CO_2 in the ratio of 1 part CH_4 to 44/16 parts CO_2. Since, from step 1, there is 0.03896 lb CH_4 per ft^3 (0.624 kg/m^3) of natural gas, this forms (0.03896)(44/16) = 0.1069 lb (0.048 kg) of CO_2. Likewise, for C_2H_6, (0.003247)(88/30) = 0.00952 lb (0.004 kg). The total CO_2 in the combustion products = 0.00464 + 0.1069 + 0.00952 = 0.11688 lb (0.053 kg), where the first quantity is the CO_2 in the fuel.

Using a similar procedure for the H_2O formed in the products of combustion by CH_4, we find (0.03896)(36/16) = 0.0875 lb (0.039 kg). For C_2H_6, (0.003247)(54/30) = 0.005816 lb (0.003 kg). The total H_2O in the combustion products = 0.0875 + 0.005816 = 0.093316 lb (0.042 kg).

Step 2 shows that 12.39 lb (5.58 kg) of N_2 is required per lb (kg) of fuel. Since 1 ft^3 (0.028 m^3) of the fuel weights 0.04517 lb (0.02 kg), the volume of gas which weighs 1 lb (2.2 kg) is 1/0.04517 = 22.1 ft^3 (0.626 m^3). Therefore, the weight of N_2 per ft^3 of fuel burned = 12.39/22.1 = 0.560 lb (0.252 kg). This, plus the weight of N_2 in the fuel, step 1, is 0.560 + 0.0025 = 0.5625 lb (0.253 kg) of N_2 in the products of combustion.

Next, find the total weight of the products of combustion by taking the sum of the CO_2, H_2O, and N_2 weights, or 0.11688 + 0.09332 + 0.5625 = 0.7727 lb (0.35 kg). Now convert each weight to ft^3 at 650°F (343°C), the temperature of the combustion products, or:

Products	Weight, lb	Weight, kg	Molecular weight	Temperature correction		Volume at 650°F, ft^3	Volume at 343°C, m^3
CO_2	0.11688	0.05302	44	(379/44)(0.11688)(2.255)	=	2.265	0.0641
H_2O	0.09332	0.04233	18	(379/18)(0.09332)(2.255)	=	4.425	0.1252
N_2 (total)	0.5625	0.25515	28	(379/28)(0.5625)(2.255)	=	17.190	0.4866
ft^3 (m^3) of flue gas per ft^3 (m^3) of natural-gas fuel					=	23.880	0.6759

In this calculation, the value of 379 is used in the molecular-weight ratio because at 60°F (15.6°C) and 14.7 lb/in^2 (abs) (101.3 kPa), the volume of 1 lb (0.45 kg) of any gas = 379/gas molecular weight. The fuel gas used is initially at 60°F (15.6°C) and 14.7 lb/in^2 (abs) (101.3 kPa). The ratio 2.255 = (650 + 460)/(32 + 460).

5. *Compute the CO_2 content of the flue gas*

CO_2, wet basis = 2.265/23.88 = 0.947, or 9.47 percent. CO_2, dry basis = 2.265/(23.88 – 4.425) = 0.1164, or 11.64 percent.

6. *Compute the air required with the stated excess flow*

With 20 percent excess air, (1.20)(16.132) = 19.3584 lb of air per lb (8.71 kg/kg) of natural gas, or 19.3584/22.1 = 0.875 lb of air per ft^3 (13.9 kg/m^3) of natural gas. See step 4 for an explanation of the value 22.1.

7. *Compute the weight of the products of combustion*

Weight of the products of combustion = product weight for perfect combustion, lb + (percent excess air) (air for perfect combustion, lb) = 16.80 + (0.20)(16.132) = 20.03 lb (9.01 kg).

8. *Compute the volume of the combustion products and the percent CO_2*

The volume of excess air in the products of combustion is found by converting from the weight to the volumetric analysis and correcting for temperature as in step 4, using the air weight from step 2 for perfect combustion and the excess-air percentage, or (16.132/22.1)(0.20)(379/28.95)(2.255) = 4.31 ft^3 (0.122 m^3). Add this to the volume of the products of combustion found in step 4, or 23.88 + 4.31 = 28.19 ft^3 (0.798 m^3).

By the procedure in step 5, the percent CO_2, wet basis = 2.265/28.19 = 0.0804, or 8.04 percent. The percent CO_2, dry basis = 2.265/(28.19 – 4.425) = 0.0953, or 9.53 percent.

Related Calculations. Use the method given here when making combustion calculations for any type of gas used as a fuel—natural gas, blast-furnace gas, coke-oven gas, producer gas, water gas, sewer gas—from any source, domestic or foreign, in any type of furnace—boiler, heater, process, or waste-heat. When the air used for combustion contains moisture, as is usually true, this moisture is added to the combustion-formed moisture appearing in the products of combustion. Thus, for 80°F (26.7°C) air of 60 percent relative humidity, the moisture content is 0.013 lb/lb (0.006 kg/kg) of dry air. This amount appears in the products of combustion for each pound of air used and is a commonly assumed standard in combustion calculations.

WOOD FUEL COMBUSTION IN A FURNACE

The weight analysis of a yellow-pine wood fuel is: C = 0.490; H_2 = 0.074; O_2 = 0.406; N_2 = 0.030. Determine the weight of oxygen and air required with perfect combustion and with 20 percent excess air. Find the weight and volume of the products of combustion under the same conditions, and the wet and dry CO_2. The flue-gas temperature is 600°F (316°C). The air supplied for combustion has a moisture content of 0.013 lb/lb (0.006 kg/kg) of dry air.

Calculation Procedure:

1. *Compute the weight of oxygen required per pound (kilogram) of wood*

The same general steps as given in earlier calculation procedures will be followed; consult them for a complete explanation of each step. Using the molecular weight of each element, we have

Element	×	Molecular-weight ratio	=	lb (kg) O_2 required
C; 0.490	×	32/12	=	1.307 (0.588)
H_2; 0.074	×	16/2	=	0.592 (0.266)
O_2; 0.406; decreases external O_2 required			=	−0.406 (−0.183)
N_2; 0.030 inert in combustion				
Total 1.000				
lb (kg) external O_2 per lb (kg) fuel			=	1.493 (0.671)

2. *Compute the weight of air required for complete combustion*

The weight of nitrogen associated with the required oxygen = (1.493)(0.768/0.232) = 4.95 lb (2.228 kg). The weight of air required = 4.95 + 1.493 = 6.443 lb/lb (2.899 kg/kg) of wood burned, if the air is dry. But the air contains 0.013 lb of moisture per lb (0.006 kg/kg) of air. Hence, the total weight of the air = 6.443 + (0.013)(6.443) = 6.527 lb (2.937 kg).

3. *Compute the weight of the products of combustion*

Use the following relation:

Fuel constituents	+	Oxygen	=	Products of combustion, lb (kg)
C; 0.490	+	1.307	=	1.797 (0.809) = CO_2
H_2; 0.074	+	0.592	=	0.666 (0.300) = H_2O
O_2; not a product of combustion				
N_2; inert but passes through the furnace			=	0.030 (0.014) = N_2
Outside N_2 from step 2			=	4.950 (2.228) = N_2
Outside moisture from step 2			=	0.237 (0.107)
lb (kg) of flue gas per lb (kg) of wood burned			=	7.680 (3.458)

4. *Convert the flue-gas weight to volume*
Use, as before, the following tabulation:

	Weight					Volume at	
Products	lb	kg	Molecular weight	Temperature correction		600°F, ft^3	316°C, m^3
CO_2	1.797	0.809	44	(359/44)(1.797)(2.15)	=	31.5	0.892
H_2O (fuel)	0.666	0.300	18	(359/18)(0.666)(2.15)	=	28.6	0.810
N_2 (total)	4.980	2.241	28	(359/28)(4.980)(2.15)	=	137.2	3.884
H_2O (outside air)	0.837	0.377	18	(359/18)(0.837)(2.15)	=	35.9	10.16
Cu ft (m^3) of flue gas per lb (kg) of oil						233.2	6.602

In this calculation the temperature correction factor 2.15 = (absolute flue-gas temperature, °R)/(absolute atmospheric temperature, °R) = (600 + 460)/(32 + 460). The total weight of N_2 is the sum of the N_2 in the combustion air and the fuel.

5. *Compute the CO_2 content of the flue gas*
The CO_2, wet basis = 31.5/233.2 = 0.135, or 13.5 percent. The CO_2, dry basis = 31.5/(233.2 – 28.6 – 35.9) = 0.187, or 18.7 percent.

6. *Compute the air required with the stated excess flow*
With 20 percent excess air, (1.20)(6.527) = 7.832 lb (3.524 kg) of air per lb (kg) of wood burned.

7. *Compute the weight of the products of combustion*
The weight of the products of combustion = product weight for perfect combustion, lb + (percent excess air)(air for perfect combustion, lb) = 8.280 + (0.20)(6.527) = 9.585 lb (4.313 kg) of flue gas per lb (kg) of wood burned with 20 percent excess air.

8. *Compute the volume of the combustion products and the percent CO_2*
The volume of the excess air in the products of combustion is found by converting from the weight to the volumetric analysis and correcting for temperature as in step 4, using the air weight from step 2 for perfect combustion and the excess-air percentage, or (6.527)(0.20)(359/28.95)(2.15) = 34.8 ft^3 (0.985 m^3). Add this to the volume of the products of combustion found in step 4, or 233.2 + 34.8 = 268.0 ft^3 (7.587 m^3).

By using the procedure in step 5, the percent CO_2, wet basis = 31.5/268 = 0.1174, or 11.74 percent. The percent CO_2, dry basis = 31.5/(268 – 28.6 – 35.9 – 0.20 × 0.837) = 0.155, or 15.5 percent. In the dry-basis calculation, the factor (0.20)(0.837) is the outside moisture in the excess air.

Related Calculations. Use the method given here when making combustion calculations for any type of wood or woodlike fuel—spruce, cypress, maple, oak, sawdust, wood shavings, tanbark, bagesse, peat, charcoal, redwood, hemlock, fir, ash, birch, cottonwood, elm, hickory, walnut, chopped trimmings, hogged fuel, straw, corn, cottonseed hulls, city refuse—in any type of furnace—boiler, heating, process, or waste-heat. Most of these fuels contain a small amount of ash—usually less than 1 percent. This was ignored in this calculation procedure because it does not take part in the combustion.

Industry is making greater use of discarded process waste to generate electricity and steam by burning the waste in a steam boiler. An excellent example is that of Agrilectric Power Partners Ltd., Lake Charles, LA. This plant burns rice hulls from its own process and buys other producers' surplus rice hulls for continuous operation. Their plant is reported as the first small-power-production facility to operate on rice hulls.

By burning the waste rice hulls, Agrilectric is confronting, and solving, an environmental nuisance often associated with rice processing. When rice hulls are disposed of by being spread on land adjacent to the mill, they often smolder, creating continuous, uncontrolled burning. Installation of its rice-hull burning, electric-generating plant has helped Agrilectric avoid the costs associated with landfilling and disposal, as well as potential environmental problems.

The boiler supplies steam for a turbine-generator with an output ranging from 11.2 to 11.8 MW. Excess power that cannot be used in the plant is sold to the local utility at a negotiated price. Thus, the combustion of an industrial waste is producing useful power while eliminating the environmental impact of the waste. The advent of PURPA (Public Utility Regulatory Policies Act) requiring local utilities to purchase power from such plants has been a major factor in the design, development, and construction of many plants by food processors to utilize waste materials for combustion and power production.

COMBUSTION ANALYSIS BY THE MOLAL METHOD

A coal fuel has this ultimate analysis: C = 0.8339; H_2 = 0.0456; O_2 = 0.0505; N_2 = 0.0103; S = 0.0064; ash = 0.0533; total = 1.000. This coal is completely burned in a boiler furnace. Using the molal method, determine the weight of air required per lb (kg) of coal with complete combustion. How much air is needed with 25 percent excess air? What is the weight of the combustion products with 25 percent excess air? The combustion air contains 0.013 lb of moisture per lb (0.006 kg/kg) of air.

Calculation Procedure:

1. *Convert the ultimate analysis to moles*

A mole of any substance is an amount of the substance having a weight equal to the molecular weight of the substance. Thus, 1 mol of carbon is 12 lb (5.4 kg) of carbon, because the molecular weight of carbon is 12. To convert an ultimate analysis of a fuel to moles, assume that 100 lb (45 kg) of the fuel is being considered. Set up a tabulation thus:

	Weight			
Ultimate analysis, %	lb	kg	Molecular weight	Moles per 100-lb (45-kg) fuel
C = 0.8339	83.39	37.526	12	6.940
H_2 = 0.0456	4.56	2.052	2	2.280
O_2 = 0.0505	5.05	2.678	32	0.158
N_2 = 0.0103	1.03	0.464	28	0.037
S = 0.0064	0.64	0.288	32	
Ash = 0.0533	5.33	2.399	Inert	
Total	100.00	45.407		9.435

2. *Compute the mols of oxygen for complete combustion*

From Table 2, the burning of carbon to carbon dioxide requires 1 mol of carbon and 1 mol of oxygen, yielding 1 mol of CO_2. Using the molal equations in Table 2 for the other elements in the fuel, set up a tabulation thus, entering the product of columns 2 and 3 in column 4:

(1) Element	(2) Moles per 100-lb (45-kg) fuel	(3) Moles O_2 per 100-lb (45-kg) fuel	(4) Total moles O_2
C	6.940	1.00	6.940
H_2	2.280	0.5	1.140
O_2	0.158	Reduces O_2 required	−0.158
N_2	0.037	Inert in combustion	
S	0.020	1.00	0.020
Total moles of O_2 required	—	—	7.942

3. *Compute the mols of air for complete combustion*
Set up a similar tabulation for air, thus:

(1) Element	(2) Moles per 100-lb (45-kg) fuel	(3) Moles air per 100-lb (45-kg) fuel	(4) Total moles air
C	6.940	4.76	33.050
H_2	2.280	2.38	5.430
O_2	0.158	Reduces O_2 required	−0.752
N_2	0.037	Inert in combustion	
S	0.020	4.76	0.095
Total moles of air required		—	37.823

In this tabulation, the factors in column 3 are constants used for computing the total moles of air required for complete combustion of each of the fuel elements listed. These factors are given in the Babcock & Wilcox Company—*Steam: Its Generation and Use* and similar treatises on fuels and their combustion. A tabulation of these factors is given in Table 3.

TABLE 3 Molal Conversion Factors

	Mol/mol of combustible for complete combustion; no excess air					
	For combustion			Combustion products		
Element or compound	O_2	N_2	Air	CO_2	H_2O	N_2
Carbon,* C	1.0	3.76	4.76	1.0	—	3.76
Hydrogen, H_2	0.5	0.188	2.38	—	1.0	1.88
Oxygen, O_2						
Nitrogen, N_2						
Carbon monoxide, CO	0.5	1.88	2.38	1.0	—	1.88
Carbon dioxide, CO_2						
Sulfur,* S	1.0	3.76	4.76	1.0	—	3.76
Methane, CH_4	2.0	7.53	0.53	1.0	2.0	7.53
Ethane, C_2H_6	3.5	13.18	16.68	2.0	3.0	13.18

*In molal calculations, carbon and sulfur are considered as gases.

An alternative, and simpler, way of computing the moles of air required is to convert the required O_2 to the corresponding N_2 and find the sum of the O_2 and N_2. Or, $376O_2 = N_2$; $N_2 + O_2 =$ moles of air required. The factor 3.76 converts the required O_2 to the corresponding N_2. These two relations were used to convert the 0.158 mol of O_2 in the above tabulation to moles of air.

Using the same relations and the moles of O_2 required from step 2, we get $(3.76)(7.942) = 29.861$ mol of N_2. Then $29.861 + 7.942 = 37.803$ mol of air, which agrees closely with the 37.823 mol computed in the tabulation. The difference of 0.02 mol is traceable to roundings.

4. *Compute the air required with the stated excess air*
With 25 percent excess, the air required for combustion $= (125/100)(37.823) = 47.24$ mol.

5. *Compute the mols of combustion products*
Using data from Table 3, and recalling that the products of combustion of a sulfur-containing fuel are CO_2, H_2O, and SO_2, and that N_2 and excess O_2 pass through the furnace, set up a tabulation thus:

(1) Moles per 100-lb (45-kg) fuel	(2) Mol/mol of combustible	(3) Moles of combustion product per 100-lb (45-kg) fuel
CO_2; 6.940	1	6.940
H_2O; 2.280 + (47.24)(0.021 + 0.158)	—	3.430
SO_2; 0.020	1	0.202
N_2; (47.24)(0.79)	—	37.320
Excess O_2; (1.25)(7.942) − 7.942	—	1.986
	Total moles, wet combustion products = 49.878	
	Total moles, dry combustion products = 49.878 − 3.232 = 46.646	

In this calculation, the total moles of CO_2 is obtained from step 2. The moles of H_2 in 100 lb (45 kg) of the fuel, 2.280, is assumed to form H_2O. In addition, the air from step 4, 47.24 mol, contains 0.013 lb of moisture per lb (0.006 kg/kg) of air. This moisture is converted to moles by dividing the molecular weight of air, 28.95, by the molecular weight of water, 18, and multiplying the result by the moisture content of the air, or (28.95/18)(0.013) = 0.0209, say 0.021 mol of water per mol of air. The product of this and the moles of air gives the total moles of moisture (water) in the combustion products per 100 lb (45 kg) of fuel fired. To this is added the moles of O_2, 0.158, per 100 lb (45 kg) of fuel, because this oxygen is assumed to unite with hydrogen in the air to form water. The nitrogen in the products of combustion is that portion of the moles of air required, 47.24 mol from step 4, times the proportion of N_2 in the air, or 0.79. The excess O_2 passes through the furnace and adds to the combustion products and is computed as shown in the tabulation. Subtracting the total moisture, 3430 mol, from the total (or wet) combustion products gives the moles of dry combustion products.

Related Calculations. Use this method for molal combustion calculations for all types of fuels—solid, liquid, and gaseous—burned in any type of furnace—boiler, heater, process, or waste-heat. Select the correct factors from Table 3.

ESTIMATING THE TEMPERATURE OF THE FINAL PRODUCTS OF COMBUSTION

Pure carbon is burned to carbon dioxide at constant pressure in an insulated chamber. An excess air quantity of 20 percent is used and the carbon and the air are both initially at 77°F (25°C). Assume that the reaction goes to completion and that there is no dissociation. Calculate the final product's temperature using the following constants: Heating value of carbon, 14,087 Btu/lb (32.74×10^3 kJ/kg); constant-pressure specific heat of oxygen, nitrogen, and carbon dioxide are 0.240 Btu /lb_m (0.558 kJ/kg), 0.285 Btu/lb_m (0.662 kJ/kg), and 0.300 Btu/lb (0.697 kJ/kg), respectively.

Calculation Procedure:

1. *Establish the chemical equation for complete combustion with 100 percent air*
With 100 percent air: $C + O_2 + 3.78N_2 \rightarrow CO_2 + 3.78N_2$, where approximate molecular weights are: for carbon, $MC = 12$; oxygen, $MO_2 = 32$; nitrogen, $MN_2 = 28$; carbon dioxide, $MCO_2 = 44$. See the Related Calculations of this procedure for a general description of the 3.78 coefficient for N_2.

2. *Establish the chemical equation for complete combustion with 20 percent excess air*
With 20 percent excess air: $C + 1.2O_2 + (1.2 \times 3.78)N_2 \rightarrow CO_2 + 0.2O_2 + (1.2 \times 3.78)N_2$.

3. *Compute the relative weights of the reactants and products of the combustion process*
Relative weight = moles × molecular weight. Coefficients of the chemical equation in step 2 represent the number of moles of each component. Hence, for the reactants, the relative weights are: for C = $1 \times MC = 1 \times 12 = 12$; $O_2 = 1.2 \times MO_2 = 1.2 \times 32 = 38.4$; $N_2 = (1.2 \times 3.78)MN_2 = (1.2 \times 3.78 \times 28) = 127$. For the products, relative weights are: for $CO_2 = 1 \times MCO_2 = 1 \times 44 = 44$; O_2, $= 0.2 \times MO_2 = 0.2 \times 32 = 6.4$; $N_2 = 127$, unchanged. It should be noted that the total relative weight of the reactants equal that of the products at 177.4.

4. *Compute the relative weights of the products of combustion on the basis of a per unit relative weight of carbon*
Since the relative weight of carbon, C = 12 in step 3; hence, on the basis of a per unit relative weight of carbon, the corresponding relative weights of the products are: for carbon dioxide, $wCO_2 = MCO_2/12 = 44/12 = 3.667$; oxygen, $wO_2 = MO_2/12 = 6.4/12 = 0.533$; nitrogen, $wN_2 = MN_2/12 = 127/12 = 10.58$.

5. *Compute the final product's temperature*
Since the combustion chamber is insulated, the combustion process is considered adiabatic. Hence, on the basis of a per unit mass of carbon, the heating value (HV) of the carbon = the corresponding heat content of the products. Thus, relative to a temperature base of 77°F (25°C), $1 \times HVC = [(wCO_2 \times c_pCO_2) + (wO_2 \times c_pO_2) + (wH_2 + c_pN_2)](t_2 - 77)$, where the heating value of carbon, HVC = 14,087 Btu/lb_m (32.74×10^3 kJ/kg); the constant-pressure specific heat of carbon dioxide, oxygen, and nitrogen are $c_pCO_2 = 0.300$ Btu/lb (0.697 kJ/kg), $cpO_2 = 0.240$ Btu/lb (0.558 kJ/kg), and $c_pN_2 = 0.285$ Btu/lb (0.662 kJ/kg), respectively; final product temperature is t_2; other values as before. Then, $1 \times 14{,}087 = [(3.667 \times 0.30) + (0.533 \times 0.24) + (10.58 \times 0.285)(t_2 - 77)]$. Solving, $t_2 = 3320 + 77 = 3397$°F (1869°C).

Related Calculations. In the above procedure it is assumed that the carbon is burned in dry air. Also, the nitrogen coefficient of 3.78 used in the chemical equation in step 1 is based on a theoretical composition of dry air as 79.1 percent nitrogen and 20.9 percent oxygen by volume, so that 79.1/20.9 = 3.78. For a more detailed description of this coefficient see Section 3 of this handbook.

DETERMINATION OF THE SAVINGS PRODUCED BY PREHEATING COMBUSTION AIR

A 20,000 ft^2 (1858 m^2) building has a calculated total seasonal heating load of 2,534,440 MBH (thousand Btu) (2674 MJ). The stack temperature is 600°F (316°C) and the boiler efficiency is calculated to be 75 percent. Fuel oil burned has a higher heating value of 140,000 Btu/gal (39,018 MJ/L). A preheater can be purchased and installed to reduce the breeching discharge combustion air temperature by 250°F (139°C) to 350°F (177°C) and provide the burner with preheated air. How much fuel oil will be saved? What will be the monetary saving if fuel oil is priced at $1.10 per gallon?

Calculation Procedure:

1. *Compute the total combustion air required by this boiler*
A general rule used by design engineers is that 1 ft^3 (0.0283 m^3) of combustion air is required for each 100 Btu (105.5 J) released during combustion. To compute the combustion air required, use the relation CA = H/100 × Boiler efficiency, expressed as a decimal, where CA = annual volume of combustion air, ft^3 (m^3); H = total seasonal heating load, Btu/yr (kJ/yr). Substituting for this boiler, CA = (2,534,400)(1000)/100 × 0.75 = 33,792,533 ft^3/yr (956.329 m^3/yr).

2. *Calculate the annual energy savings*
The energy savings, ES = (stack temperature reduction, deg F)(ft^3 air per year)(0.018), where the constant 0.018 is the specific heat of air. Substituting, ES = (250)(33,792,533)(0.018) = 152,066,399 Btu/yr (160,430 kJ/yr).

With a boiler efficiency of 75 percent, each gallon of oil releases 0.75 × 140,000 Btu/gal = 105,000 Btu (110.8 jk). Hence, the fuel saved, FS = ES/usable heat in fuel, Btu/gal. Or, FS = 152,066,399/105,000 = 1448.3 gal/yr (5.48 m^3/yr).

With fuel oil at $1.10 per gallon, the monetary savings will be $1.10 (1448.3) = $1593.13. If the preheater cost $6000, the simple payoff time would be $6000/1593.13 = 3.77 years.

Related Calculations. Use this procedure to determine the potential savings for burning any type of fuel—coal, oil, natural gas, landfill gas, catalytic cracker offgas, hydrogen purge gas, bagesse, sugar cane, etc. Other rules of thumb used by designers to estimate the amount of combustion air required for various fuels are: 10 ft^3 of air (0.283 m^3) per 1 ft^3 (0.0283 m^3) of natural gas; 1300 ft^3 of air (36.8 m^3) per gal (0.003785 m^3) of No. 2 fuel oil; 1450 ft^3 of air (41 m^3) per gal of No. 5 fuel oil; 1500 ft^3 of air (42.5 m^3) per gal of No. 6 fuel oil. These values agree with that used in the above computation—i.e. 100 ft^3 per 100 Btu of 140,000 Btu per gal oil = 140,000/100 = 1400 ft^3 per gal (39.6 m^3/0.003785 m^3).

This procedure is the work of Jerome F. Mueller, P.E. of Mueller Engineering Corp.

COMBUSTION CALCULATIONS USING THE MILLION BTU (1.055 MJ) METHOD

The energy absorbed by a steam boiler fired by natural gas is 100-million Btu/h (29.3 MW). Boiler efficiency on a higher heating value (HHV) basis is 83 percent. If 15 percent excess air is used, determine the total air and flue-gas quantities produced. The approximate HHV of the natural gas is 23,000 Btu/lb (53,590 kJ/kg). Ambient air temperature is 80°F (26.7°C) and relative humidity is 65 percent. How can quick estimates be made of air and flue-gas quantities in boiler operations when the fuel analysis is not known?

Calculation Procedure:

1. ***Determine the energy input to the boiler***

The million Btu (1.055 MJ) method of combustion calculations is a quick way of estimating air and flue-gas quantities generated in boiler and heater operations when the ultimate fuel analysis is not available and all the engineer is interested in is good estimates. Air and flue-gas quantities determined may be used to calculate the size of fans, ducts, stacks, etc.

It can be shown through comprehensive calculations that each fuel such as coal, oil, natural gas, bagasse, blast-furnace gas, etc. requires a certain amount of dry stoichiometric air per million Btu (1.055 MJ) fired on an HHV basis and that this quantity does not vary much with the fuel analysis. The listing below gives the dry air required per million Btu (1.055 MJ) of fuel fired on an HHV basis for various fuels.

Combustion Constants for Fuels

Fuel	Constant, lb dry air per million Btu (kg/MW)
Blast furnace gas	575 (890.95)
Bagasse	650 (1007.2)
Carbon monoxide gas	670 (1038.2)
Refinery and oil gas	720 (1115.6)
Natural gas	730 (1131.1)
Furnace oil and lignite	745 – 750 (1154.4 – 1162.1)
Bituminous coals	760 (1177.6)
Anthracite coal	780 (1208.6)
Coke	800 (1239.5)

To determine the energy input to the boiler, use the relation $Q_f = (Q_s)/E_h$, where Q_f = energy input by the fuel, Btu/h (W); Q_s = energy absorbed by the steam in the boiler, Btu/h (W); E_h = efficiency of the boiler on an HHV basis. Substituting for this boiler, Q_f = 100/0.83 = 120.48 million Btu/h on an HHV basis (35.16 MW).

2. *Estimate the quantity of dry air required by this boiler*
The total air required $T_a = (Q_f)$(Fuel constant from list above). For natural gas, T_a = (120.48)(730) = 87,950 lb/h (39,929 kg/h). With 15 percent excess air, total air required = (1.15)(87,950) = 101,142.5 lb/h (45,918.7 kg/h).

3. *Compute the quantity of wet air required*
Air has some moisture because of its relative humidity. Estimate the amount of moisture in dry air in M lb/lb (kg/kg) from M = 0.622 $(p_w)/(14.7 - p_w)$, where 0.622 is the ratio of the molecular weights of water vapor and dry air; p_w = partial pressure of water vapor in the air, psia (kPa) = saturated vapor pressure (SVP) × relative humidity expressed as a decimal; 14.7 = atmospheric pressure of air at sea level (101.3 kPa). From the steam tables, at 80°F (26.7°C), SVP = 0.5069 psia (3.49 kPa). Substituting, M = 0.622 (0.5069 × 0.65)/(14.7 – [0.5069 × 0.65]) = 0.01425 lb of moisture/lb of dry air (0.01425 kg/kg).

The total flow rate of the wet air then = 1.0142 (101,142.5) = 102,578.7 lb/h (46,570.7 kg/h). To convert to a volumetric-flow basis, recall that the density of air at 80°F (26.7°C) and 14.7 psia (101.3 kPa) = 39/(480 + 80) = 0.0722 lb/ft^3 (1.155 kg/m^3). In this relation, 39 = a constant and the temperature of the air is converted to degrees Rankine. Hence, the volumetric flow = 102,578.7/ (60 min/h)(0.0722) = 23,679.3 actual cfm (670.1 m^3/min).

4. *Estimate the rate of fuel firing and flue-gas produced*
The rate of fuel firing = Q_f/HHV = (120.48 × 10^6)/23,000 = 5238 lb/h (2378 kg/h). Hence, the total flue gas produced = 5238 + 102,578 = 107,816 lb/h (48,948 kg/h).

If the temperature of the flue gas is 400°F (204.4°C) (a typical value for a natural-gas fired boiler), then the density, as in step 3, is: 39/(400 + 460) = 0.04535 lb/ft^3 (0.7256 kg/m^3). Hence, the volumetric flow = (107,816)/(60 min/h × 0.04535) = 39,623.7 actual cfm (1121.3 m^3/min).

Related Calculations. Detailed combustion calculations based on actual fuel gas analysis can be performed to verify the constants given in the list above. For example, let us say that the natural-gas analysis was: methane = 83.4 percent; ethane = 15.8 percent; nitrogen = 0.8 percent by volume. First convert the analysis to a percent weight basis.

Fuel	Percent volume	MW	Col. 2 × Col. 3	Percent weight
Methane	83.4	16	1334.4	72.89
Ethane	15.8	30	474	25.89
Nitrogen	0.8	28	22.4	1.22

Note that the percent weight in the above list is calculated after obtaining the sum under Column 2 × Column 3. Thus, the percent methane = (1334.4)/(1334.4 + 474 + 22.4) = 72.89 percent.

From a standard reference, such as Ganapathy, *Steam Plant Calculations Manual*, Marcel Dekker, Inc., find the combustion constants, K, for various fuels and use them thus: For the air required for combustion, A_c = (K for methane)(percent by weight methane from above list) + (K for ethane)(percent by weight ethane); or A_c = (17.265)(0.7289) + (16.119)(0.2589) = 16.76 lb/lb (16.76 kg/kg).

Next, compute the higher heating value of the fuel (HHV) using the air constants from the same reference mentioned above. Or HHV = (heat of combustion for methane)(percent by weight methane) + (heat of combustion of ethane)(percent by weight ethane) = (23,879)(0.7289) + (22,320)(0.2589) = 23,184 Btu/lb (54,018.7 kJ/kg). Then, the amount of fuel equivalent to 1,000,000 Btu (1,055,000 kJ) = (1,000,000)/23,184 = 43.1 lb (19.56 kg), which requires, as computed above, (43.1)(16.76) = 722.3-lb dry air (327.9 kg), which agrees closely with the value given in step 1, above.

Similarly, if the fuel were 100 percent methane, using the steps given above, and suitable constants from the same reference work, the air required for combustion is 17.265 lb/lb (7.838 kg/kg) of fuel. HHV = 23,879 Btu/lb (55,638 kJ/kg). Hence, the fuel in 1,000,000 Btu (1,055,000 kJ) = (1,000,000)/(23,879) = 41.88 lb (19.01 kg). Then, the dry air per million Btu (1.055 kg) fired = (17.265) (41.88) = 723 lb (328.3 kg).

Likewise, for propane, using the same procedure, 1 lb (0.454 kg) requires 15.703-lb (7.129-kg) air and 1 million Btu (1,055,000 kJ) has (1,000,000)/21,661 = 46.17-lb (20.95-kg) fuel. Then, 1 million Btu (1,055,000 kJ) requires (15.703)(46.17) = 725-lb (329.2-kg) air. This general approach can be used for various fuel oils and solid fuels—coal, coke, etc.

Good estimates of excess air used in combustion processes may be obtained if the oxygen and nitrogen in dry flue gases are measured. Knowledge of excess-air amounts helps in performing detailed combustion and boiler efficiency calculations. Percent excess air, EA = $100(O_2 - CO_2)/[0.264 \times N_2 - (O_2 - CO/2)]$, where O_2 = oxygen in the dry flue gas, percent volume; CO = percent volume carbon monoxide; N_2 = percent volume nitrogen.

You can also estimate excess air from oxygen readings. Use the relation, EA = (constant from list below)$(O_2)/(21 - O_2)$.

Constants for Excess-Air Calculations

Fuel	Constant
Carbon	100
Hydrogen	80
Carbon monoxide	121
Sulfur	100
Methane	90
Oil	94.5
Coal	97
Blast furnace gas	223
Coke oven gas	89.3

If the percent volume of oxygen measured is 3 on a dry basis in a natural-gas (methane) fired boiler, the excess air, EA = (90)[3/(21 – 3)] = 15 percent.

This procedure is the work of V. Ganapathy, Heat Transfer Specialist, ABCO Industries.

SECTION 2
STEAM POWER GENERATION

Steam Power Calculation Parameters 2.1
Energy Efficiency and Heat Rate of Steam Turbogenerator 2.2
Reheat-Regenerative Turbogenerator Energy Analysis 2.3
Moisture Content and Steam Enthalpy of Steam Turbine Exhaust 2.7
Steam Flow for Steam-Turbine No-Load and Partial-Load Operations 2.8
Energy Test Data for Steam Power Plant Performance 2.10
Steam Rate for Turbogenerator at Various Loads 2.11
Reheating-Regenerative Steam Turbine Cycle Energy Analysis 2.12
Reheat-Regenerative Cycle Steam Rate 2.13
Energy Efficiency Analysis for Binary Cycle Steam Plant 2.15
Cogeneration System Energy Efficiency for a Traditional Steam Power Plant Cycle 2.17
Regenerative Bleed-Steam Cycle General Energy Analysis 2.18
Bleed Regenerative Steam Cycle Analysis 2.22
Energy Performance of Reheat-Steam Cycle 2.25
Power Energy Output Analysis of Mechanical-Drive Steam Turbine 2.29
Energy Output Analysis of Condensing Steam Turbine 2.31
Steam-Turbine Regenerative-Cycle Performance 2.33
Reheat-Regenerative Steam-Turbine Heat Rates 2.36
kW Output of Extraction Steam Turbine 2.37

STEAM POWER CALCULATION PARAMETERS

Fossil-fired steam power-generating plants account for the majority of power generated today throughout the world. Such plants have been dominant in electric generation worldwide for more than a century. While nuclear generation is catching up with fossil-fuel generation in terms of generating capacity, it will be some years before the two are equal in output.

Meanwhile, fossil-fueled coal-fired plants have dual environmental challenges:

- The carbon dioxide and particulates produced during the combustion of coal are blamed for atmospheric pollution. Coal-fired plants have operated in an environmentally unregulated manner for a century because greenhouse gases (GHG) were accepted as part of the plant output by designers, operators, and most regulators. Today GHG occupy front and center in the design of every coal-fired power plant, as well as liquid-fuel- and gaseous-fuel-fired generating plants.
- The coal ash produced by combustion of coal is categorized as a toxic pollutant in landfills to which it is sent for disposal. Even though some of the coal ash produced by the combustion of coal is recycled in the manufacture of building construction materials and concrete, much of the coal ash is disposed of in landfills. Toxic lead and arsenic in the coal ash can pollute ground water in the landfills in which the ash is dumped. Published statistics show that some 1300 million tons of coal ash are produced annually by the combustion of coal in power plants in the United States alone.

Since the amount of fuel burned in a steam plant is a function of the steam cycle efficiency, this section provides a number of pertinent cycle and equipment efficiency calculations. With a higher cycle efficiency, less fuel is burned, and the plant's operating cost is lower. When a smaller amount of coal fuel is burned, less atmospheric and landfill pollution will be caused by the fossil-fueled generating plant. Oil and gas fuels contribute to GHG (at a lesser amount than coal) but not to landfill pollution.

Power magazine recently studied coal-fired steam plant thermal efficiency and found that ultra super critical (USC) plants in the United States are projecting a net thermal efficiency of 39 percent, based on the higher heating value (HHV) of the coal. *Power* also reports that seawater- and ocean-water-cooled coal-fired plants have up to 1.5 percent higher thermal efficiency than cooling-tower-cooled steam plants.

European coal-fired steam plants often report higher thermal efficiencies than those in the United States. Thus, one European plant reports 43.2 percent, based on the lower heating value (LHV) of the coal burned. Likewise, steam plants in other parts of the world report thermal efficiencies in the 40 percent plus range. Such efficiencies are usually based on the LHV of the coal fuel.

Because central-station power plants, and many industrial power plants, are built to function for 50 or more years, fuel-selection choices are extremely important. Thus, a fuel chosen today—coal, oil, gas, or any of the waste-product types—may become much costlier in future years if carbon dioxide or other GHG caps are lowered for the chosen fuel. So, engineers are becoming much more cautious about fuels they choose for all new power plants. With recent development of large shale-gas sources, this fuel is getting more attention than ever before for steam power plants.

Another area where steam plants are finding greater examination by regulatory bodies is in cooling water use. With the once-through cooling systems used for steam condensers, the major impact is the temperature rise of the water. There is little evaporation in once-through cooling systems because the temperature rise of the cooling water is modest. In recirculating cooling systems, such as with cooling towers, there are evaporative water losses. Also, water must be blown down or discharged from the system to reduce chemical buildup. The blow-down water is lost.

To reduce carbon dioxide emissions to the atmosphere, water is used to capture CO_2, either before (pre-) or after (post-) combustion. Department of Energy (DOE) studies show that pulverized-coal subcritical pressure steam boilers consume some 900 gal of water per MWh (gal/MWh) (3407 L/MWh) for carbon capture. Supercritical pressure boilers consume some 800 gal/MWh (3028 L/MWh) for carbon capture.

New coal-fired power-generating plants will be planned and designed in engineering design offices for many years to come, all over the world. Why? Because coal deposits are enormous in many parts of the world. Further, ignoring its GHG effects, coal is an abundant, low-cost fuel that plant operators are familiar with and know how to handle. The big unknown for everyone—from engineering designers to plant owners—is the fiscal impact of controls required to reduce GHG emissions from coal-fired electric-generating plants. The regulators could win, and end the century-long dominance of coal-fired electric-generating plants. Or engineers could preserve the coal-fired plants' dominance by finding low-cost ways to reduce substantially the GHG emitted by such plants. Either way, engineers will have a prominent role in the outcome.

ENERGY EFFICIENCY AND HEAT RATE OF STEAM TURBOGENERATOR

A 20,000-kW turbogenerator is supplied with steam at 300 lb/in^2 (abs) (2067.0 kPa) and a temperature of 650°F (343.3°C). The backpressure is 1 in (2.54 cm) Hg absolute. At best efficiency, the steam rate is 10 lb (25.4 kg) per kWh. (*a*) What is the combined thermal efficiency (CTE) of this unit? (*b*) What is the combined engine efficiency (CEE)? (*c*) What is the ideal steam rate?

Calculation Procedure:

1. *Determine the combined thermal efficiency*

(*a*) Combined thermal efficiency, CTE = $(3413/w_r)(1/[h_1 - h_2])$, where w_r = combined steam rate, lb/kWh (kg/kWh); h_1 = enthalpy of steam at throttle pressure and temperature, Btu/lb (kJ/kg); h_2 = enthalpy of steam at the turbine backpressure, Btu/lb (kJ/kg). Using the steam tables and

Mollier chart and substituting in this equation, CTE = (3413/10)(1/[1340.6 – 47.06]) = 0.2638, or 26.38 percent.

2. *Find the combined engine efficiency*

(*b*) Combined engine efficiency, CEE = $(w_i)/(w_e)$ = (weight of steam used by ideal engine, lb/kWh)/(weight of steam used by actual engine, lb/kWh). The weights of steam used may also be expressed as Btu/lb (kJ/kg). Thus, for the ideal engine, the value is 3413 Btu/lb (7952.3 kJ/kg). For the actual turbine, $h_1 - h_{2x}$ is used, where h_{2x} is the enthalpy of the wet steam at exhaust conditions; h_1, is as before.

Since the steam expands isentropically into the wet region below the dome of the *T-S* diagram, Fig. 1, we must first determine the quality of the steam at point 2 either from a *T-S* diagram or Mollier chart or by calculation. By calculation using the method of mixtures and the entropy at each point: $S_1 = S_2 = 0.0914 + (x_2)(1.9451)$. Then x_2 = (1.6508 – 0.0914)/1.9451 = 0.80, or 80 percent quality. Substituting and summing, using steam-table values, h_{2x} = 47.06 + 0.8(1047.8) = 885.3 Btu/lb (2062.7 kJ/kg).

(*c*) To find the CEE we first must obtain the ideal steam rate, w_i = $3413/(h_1 - h_{2x})$ = 3413/(1340.6 – 885.3) = 7.496 lb/kWh (3.4 kg/kWh).

Now, CEE = (7.496/10)(100) = 74.96 percent. This value is excellent for such a plant and is in a range being achieved today.

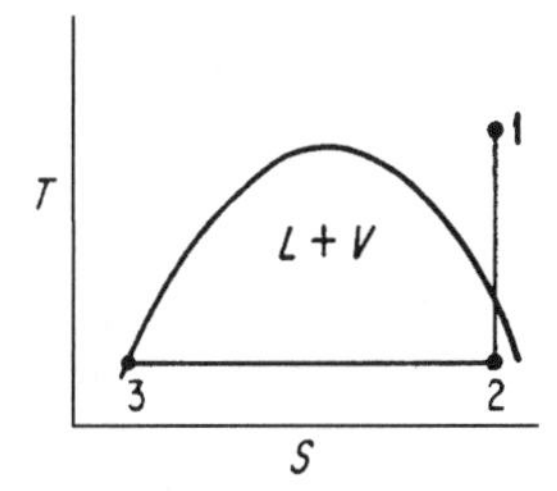

FIGURE 1 *T-S* diagrams for steam turbine.

Related Calculations. Use this approach to analyze the efficiency of any turbogenerator used in central-station, industrial, marine, and other plants.

REHEAT-REGENERATIVE TURBOGENERATOR ENERGY ANALYSIS

A turbogenerator operates on the reheating-regenerative cycle with one stage of reheat and one regenerative feedwater heater. Throttle steam at 400 lb/in^2 (abs) (2756.0 kPa) and 700°F (371.1°C) is used. Exhaust at 2-in (5.08-cm) Hg is taken from the turbine at a pressure of 63 lb/in^2 (abs) (434 kPa) for both reheating and feedwater heating with reheat to 700°F (371.1°C). For an ideal turbine working under these conditions, find: (*a*) Percentage of throttle steam bled for feedwater heating; (*b*) Heat converted to work per pound (kg) of throttle steam; (*c*) Heat supplied per pound (kg) of throttle steam; (*d*) Ideal thermal efficiency; (*e*) Other ways to heat feedwater and increase the turbogenerator output. Figure 2 shows the layout of the cycle being considered, along with a Mollier chart of the steam conditions.

Calculation Procedure:

1. *Using the steam tables and Mollier chart, list the pertinent steam conditions*

Using the subscript 1 for throttle conditions, list the key values for the cycle thus:

$$P_1 = 400 \text{ lb/in}^2 \text{ (abs) (2756.0 kPa)}$$
$$t_1 = 700°\text{F (371.1°C)}$$
$$H_1 = 1362.2 \text{ Btu/lb (3173.9 kJ/kg)}$$
$$S_1 = 1.6396$$
$$H_2 = 1178 \text{ Btu/lb (2744.7 kJ/kg)}$$
$$H_3 = 1380.1 \text{ Btu/lb (3215.6 kJ/kg)}$$
$$H_4 = 1035.8 \text{ (2413.4 kJ/kg)}$$
$$H_5 = 69.1 \text{ Btu/lb (161.0 kJ/kg)}$$
$$H_6 = 265.27 \text{ Btu/lb (618.07 kJ/kg)}$$

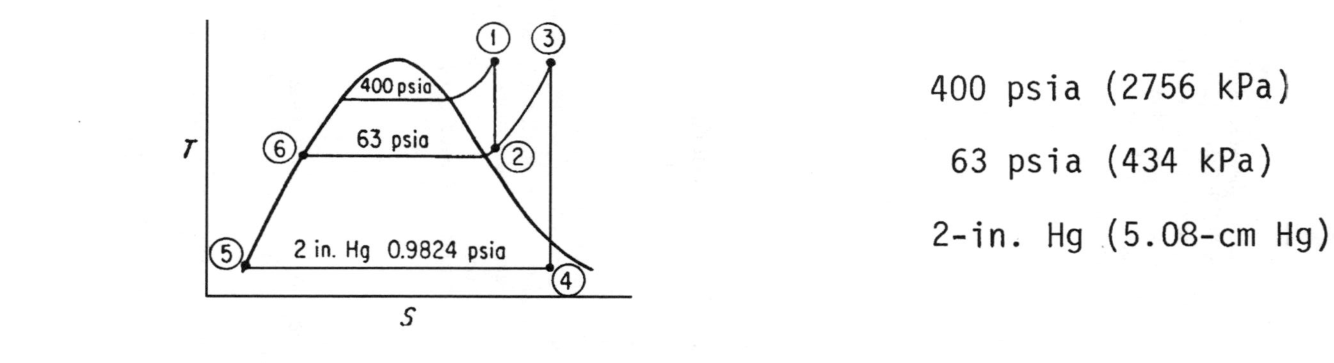

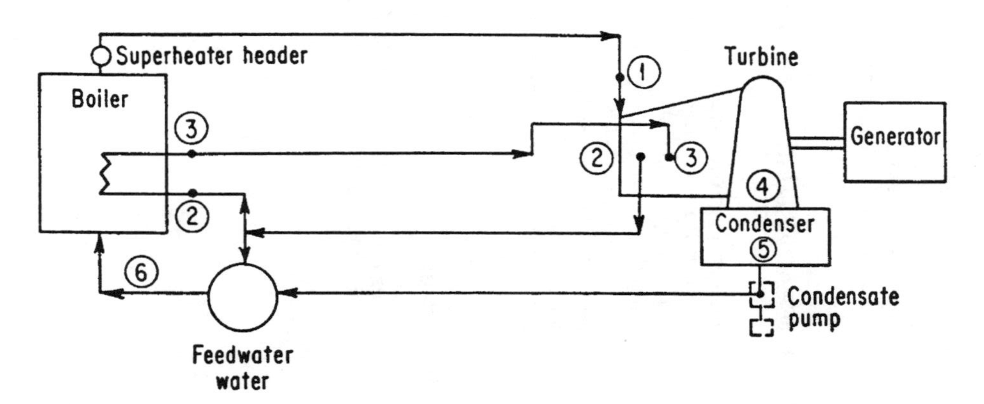

FIGURE 2 Cycle layout and *T-S* chart of steam conditions.

2. *Determine the percentage of throttle steam bled for feedwater heating*

(*a*) Set up the ratio for the feedwater heater of (heat added in the feedwater heater)/(heat supplied to the heater)(100). Or, using the enthalpy data from step 1 above, $(H_6 - H_5)/(H_2 - H_5)(100) = (265.26 - 69.1)/(1178 - 69.1)(100) = 17.69$ percent of the throttle steam is bled for feedwater heating.

3. *Find the heat converted to work per pound (kg) of throttle steam*

(*b*) The heat converted to work is the enthalpy difference between the throttle steam and the bleed steam at point 2 plus the enthalpy difference between points 3 and 4 times the percentage of throttle flow between these points. In equation form, heat converted to work $= H_1 - H_2 + (1.00 - 0.1769)(H_3 - H_4) = (1362.2 - 1178) + (0.0823)(1380.1 - 1035.8) = 467.55$ Btu/lb (1089.39 kJ/kg).

4. *Calculate the heat supplied per pound (kg) of throttle steam*

(*c*) The heat supplied per pound (kg) of throttle steam $= (H_1 - H_6) + (H_3 - H_2) = (1362.3 - 265.27) + (1380.1 - 1178) = 1299.13$ Btu/lb (3026.97 kJ/kg).

5. *Compute the ideal thermal efficiency*

(*d*) Use the relation, ideal thermal efficiency = (heat converted to work)/(heat supplied) = 467.55/1299.13 = 0.3598, or 35.98 percent.

6. *Show other ways to heat feedwater while increasing the turbogenerator output*

For years, central stations and large industrial steam-turbine power plants shut off feedwater heaters to get additional kilowatts out of a turbogenerator during periods of overloaded electricity demand. When more steam flows through the turbine, the electrical power output increases. While there was a concurrent loss in efficiency, this was ignored because the greater output was desperately needed.

Today steam turbines are built with more heavily loaded exhaust ends so that the additional capacity is not available. Further, turbine manufacturers place restrictions on the removal of feedwater heaters from service. However, if the steam output of the boiler is less than the design capacity of the steam turbine, because of a conversion to coal firing, additional turbogenerator capacity is available and can be regained at a far lower cost than by adding new generator capacity.

Compensation for the colder feedwater can be made, and the lost efficiency regained, by using a supplementary fuel source to heat feedwater. This can be done in one of two ways: (1) increase heat input to the existing boiler economizer, or (2) add a separately fired external economizer.

Additional heat input to a boiler's existing economizer can be supplied by induct burners, Fig. 3, from slagging coal combustors, Fig. 4, or from the furnace itself. Since the economizer in a coal-fired

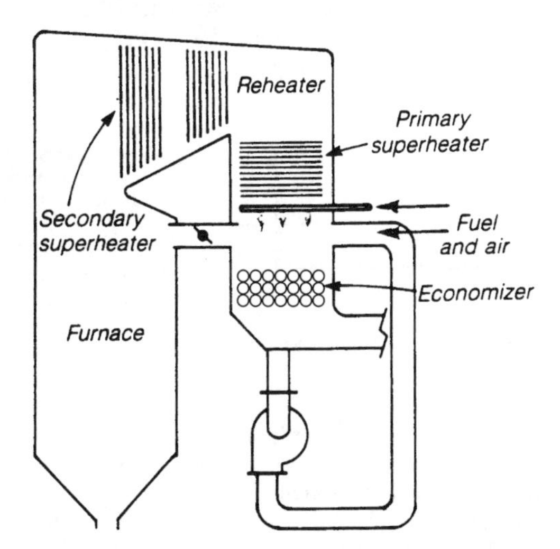

FIGURE 3 Heat input to the economizer may be increased by the addition of induct burners, by bypassing hot furnace gases into the gas path ahead of the economizer, or by recirculation. (*Power.*)

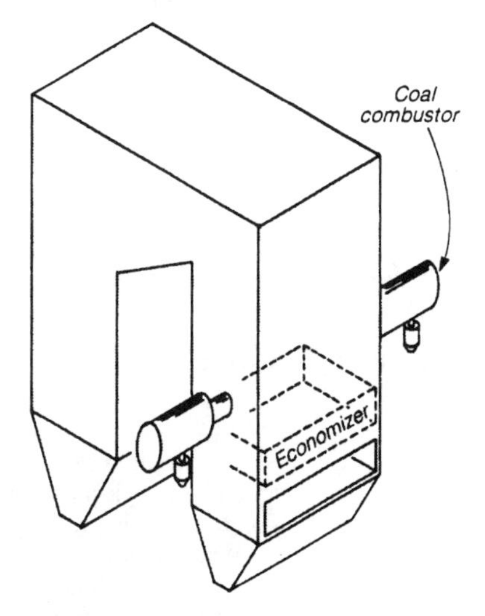

FIGURE 4 Slagging combustors can be arranged to inject hot combustion gases into gas passages ahead of economizer. (*Power.*)

boiler is of sturdier construction than a heat-recovery steam generator (HRSG) with finned tubing, in-duct burners can be placed closer to the economizer, Fig. 3. Burner firing may be by coal or oil.

Slagging coal combustors are under intense development. A low-NO_x, low-ash combustor, Fig. 4, supplying combustion gases at 3000°F (1648.9°C) may soon be commercially available.

To accommodate any of the changes shown in Fig. 3, a space from 12 (3.66 m) to 15 ft (4.57 m) is needed between the bottom of the primary superheater and the top of the economizer. This space is required for the installation of the induct burners or for the adequate mixing of gas streams if the furnace or an external combustor is used to supply the additional heat.

Another approach is to install a separately fired external economizer in series or parallel with the existing economizer, which could be fired by a variety of fuels. The most attractive possibility is to use waste heat from a gas-turbine exhaust, Fig. 5. The output of this simple combined-cycle arrangement would actually be higher than the combined capabilities of the derated plant and the gas turbine.

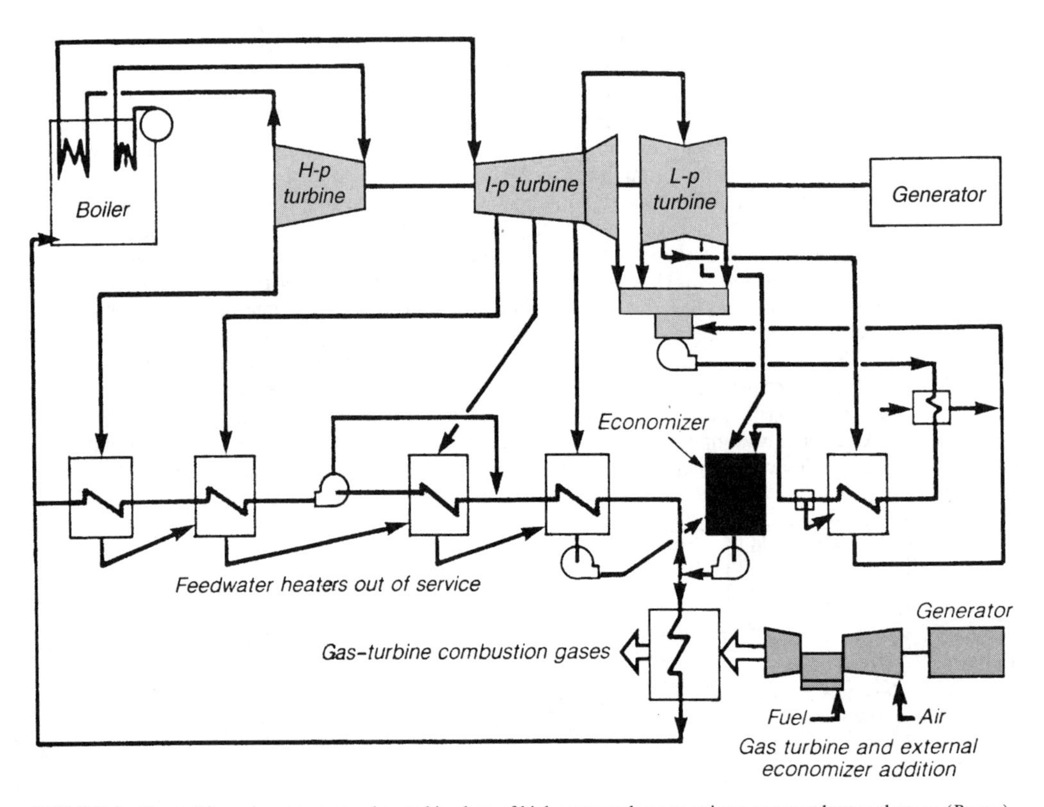

FIGURE 5 Gas-turbine exhaust gases can be used in place of high-pressure heaters, using a compact heat exchanger. (*Power.*)

The steam-cycle arrangement for the combined plant is shown in Fig. 5. Feedwater is bypassed around the high-pressure regenerative heaters to an external low-cost, finned-tube heat exchanger, Fig. 6, where waste heat from the gas turbine is recovered.

When high-pressure feedwater heaters are shut off, steam flow through the intermediate- and low-pressure turbine sections increases and becomes closer to the full-load design flow. The reheat expansion line moves left on the Mollier chart from its derated position, Fig. 7. With steam flow closer to the design value, the exhaust losses per pound (kg) of steam, Fig. 8, are lower than at the derated load.

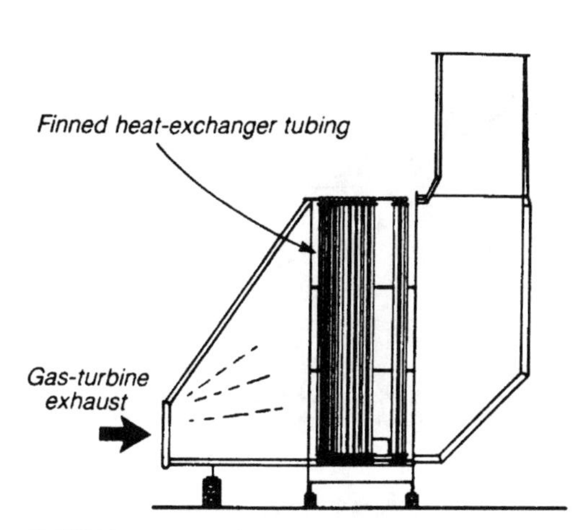

FIGURE 6 Gas-turbine heat-recovery finned-tube heat exchanger is simple and needs no elaborate controls. (*Power.*)

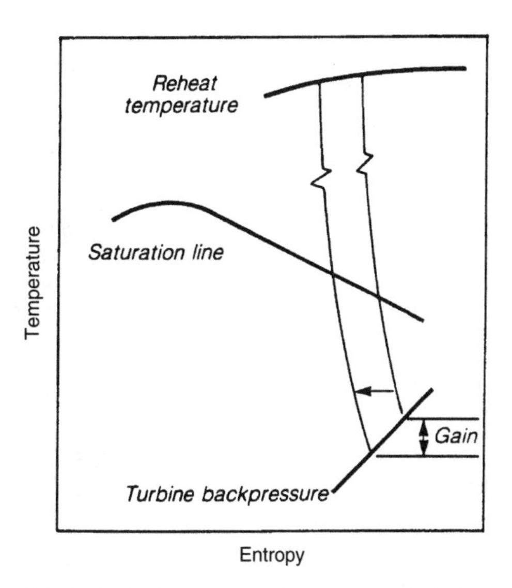

FIGURE 7 Reheat expansion line is moved to the left on the *T-S* chart, increasing power output. (*Power.*)

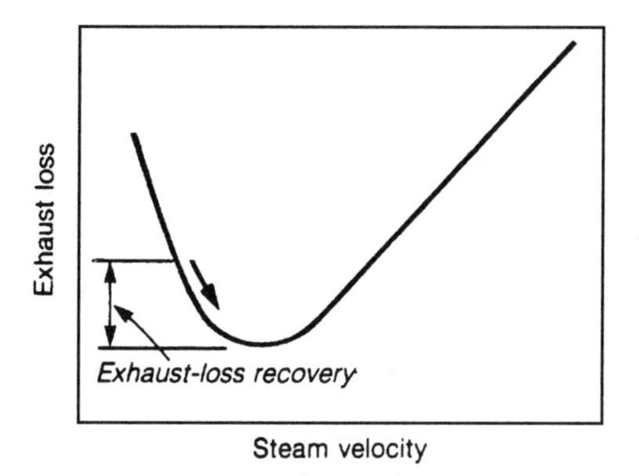

FIGURE 8 Heat loss in steam-turbine exhaust is reduced when operating at rated flow. (*Power.*)

The data and illustrations in this step 6 are based on the work of E. S. Miliares and P. J. Kelleher, Energotechnology Corp., as reported in *Power* magazine.

MOISTURE CONTENT AND STEAM ENTHALPY OF STEAM TURBINE EXHAUST

What is the enthalpy and percent moisture of the steam entering a surface condenser from the steam turbine whose Mollier chart is shown in Fig. 9? The turbine is delivering 20,000 kW and is supplied steam at 850 lb/in^2 (abs) (5856.5 kPa) and 900°F (482.2°C); the exhaust pressure is 1.5 in (3.81 cm) Hg absolute. The steam rate, when operating straight condensing, is 7.70 lb/delivered kWh (3.495 kg/kWh) and the generator efficiency is 98 percent.

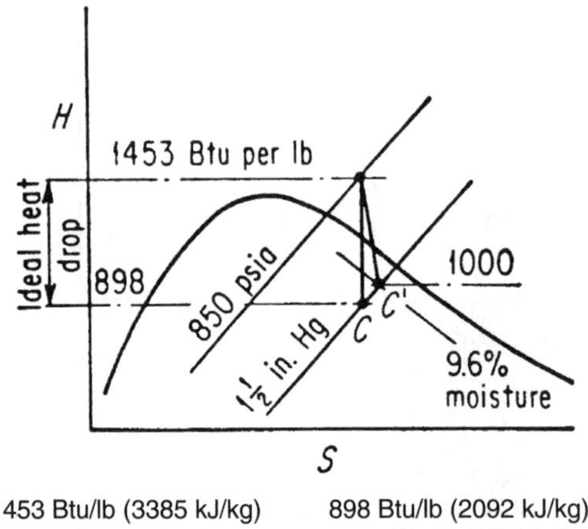

1453 Btu/lb (3385 kJ/kg)
850 psia (5856 kPa)
1.5 in. Hg (3.8 cm Hg)
898 Btu/lb (2092 kJ/kg)
1000 Btu/lb (2330 kJ/kg)

FIGURE 9 Mollier chart for turbine exhaust conditions.

Calculation Procedure:

1. ***Compute the engine efficiency of the turbine***
Use the relation $E_e = 3413/(w_s)(H_1 - H_c)$, where E_e = engine efficiency; w_s = steam rate of the turbine when operating straight condensing in the units given above; enthalpies H_1 and H_c are as shown in the Mollier chart. Substituting, $E_e = 3413/(7.7)(1453 - 898) = 0.7986$ for ideal conditions.

2. ***Find the Rankine engine efficiency for the actual turbine***
The Rankine engine efficiency for this turbine is: $(0.7986/0.98) = 0.814 = (H_1 - H_{c'})/(H_1 - H_c)$. Solving, $(H_1 - H_{c'}) = 0.814(555) = 452.3$ Btu/lb (1053.8 kJ/kg).

At the end of the actual expansion of the steam in the turbine, $H_{c'} = 1453 - 452.3 = 1000.7$ Btu/lb (2331.6 kJ/kg) enthalpy.

3. ***Determine the moisture of the steam***
Referring to the Mollier chart where $H_{c'}$ crosses the pressure line of 1.5 in (3.81 cm) Hg, the moisture percent is found to be 9.6 percent.

Related Calculations. The Mollier chart can be a powerful and quick reference for solving steam expansion problems in plants of all types—utility, industrial, commercial, and marine.

STEAM FLOW FOR STEAM-TURBINE NO-LOAD AND PARTIAL-LOAD OPERATIONS

A 40,000-kW straight-flow condensing industrial steam turbogenerator unit is supplied steam at 800 lb/in^2 (abs) (5512 kPa) and 800°F (426.7°C) and is to exhaust at 3 in (76 cm) Hg absolute. The half-load and full-load throttle steam flows are estimated to be 194,000 lb/h (88,076 kg/h) and 356,000 lb/h (161,624 kg/h), respectively. The mechanical efficiency of the turbine is 99 percent and the generator efficiency is 98 percent. Find (*a*) the no-load throttle steam flow; (*b*) the heat rate of the unit expressed as a function of the kW output; (*c*) the internal steam rate of the turbine at 30 percent of full load.

Calculation Procedure:

1. ***Find the difference between full-load and half-load steam rates and the no-load rate***
(*a*) Assume a straight-line rating characteristic and plot Fig. 10*a*. This assumption is a safe one for steam turbines in this capacity range. Then, the difference between full-load and half-load steam rates is 356,000 – 194,000 = 162,000 lb/h (73,548 kg/h). The no-load steam rate will then be = (half-load rate) – (difference between full-load and half-load rates) = 194,000 – 162,000 = 32,000 lb/h (14,528 kg/h).

2. ***Determine the steam rate and heat rate at quarter-load points***
(*b*) Using Fig. 10*b*, we see that the actual turbine efficiency, $E_t = 3413/(w_s)(H_1 - H_f)$, w_s = steam flow, lb/kWh (kg/kWh); H_1 = enthalpy of entering steam, Btu/lb (kJ/kg): H_f = enthalpy of condensate at the exhaust pressure, Btu/lb (kJ/kg).

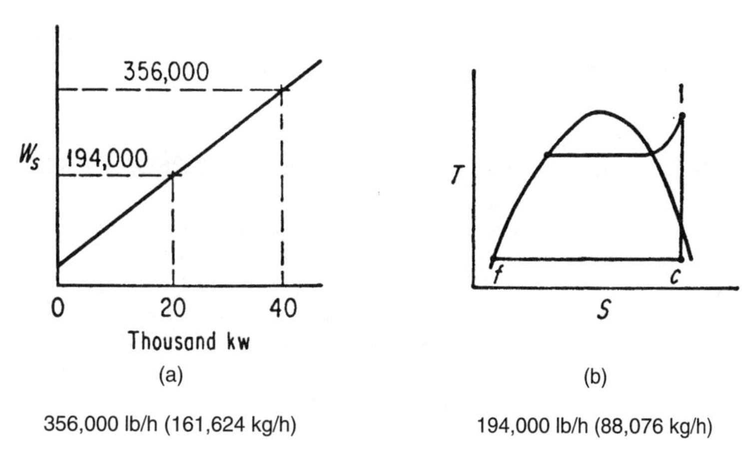

FIGURE 10 (*a*) Straight-line rating characteristic. (*b*) *T-S* diagram.

Further, turbine heat rate = $3413/E_t$ Btu/kWh (kJ/kWh) = $w_k(H_1 - H_f)$, where w_k = the actual steam rate, lb/kWh (kg/kWh) = $w = w_s$/kW output, where the symbols are as defined earlier.

Substituting w_s = 32,000 no-load throttle flow + (difference between full-load and half-load throttle flow rate/kW output at half load)(kW output) = 32,000 + (162,000/20,000)(kW) = 32,000 + 8.1 kW for this turbine-generator set. Also w_k = (32,000/kW) + 8.1.

Using the steam tables, we find H_1 = 1398 Btu/lb (3257.3 kJ/kg); H_f = 83 Btu/lb (193.4 kJ/kg). Then, $H_1 - H_f$ = 1315 Btu/lb (3063.9 kJ/kg). Substituting, heat rate = [(1315)(32,000)/(kW)] + (1316)(8.1) = 10.651.5 + (42,080,000/kW).

Computing the steam rate and heat rate for the quarter-load points for this turbine-generator we find:

At full load, w_s = 8.9 lb/kWh (7.04 kg/kWh)
At ¾ load, w_s = 9.17 lb/kWh (4.16 kg/kWh)
At ½ load, w_s = 9.7 lb/kWh (4.4 kg/kWh)
At ¼ load, w_s = 11.3 lb/kWh (5.13 kg/kWh)

At full load, heat rate = 11,700 Btu/kWh (27,261 kJ/kWh)
At ¾ load, heat rate = 12,080 Btu/kWh (28,146 kJ/kWh)
At ½ load, heat rate = 12,770 Btu/kWh (29,754 kJ/kWh)
At ¼ load, heat rate = 14,870 Btu/kWh (34,647 kJ/kWh)

3. *Determine the internal steam rate of the turbine*
(*c*) For the turbine and generator combined, $E_e = 3413/(w_k)(H_1 - H_c)$, where E_e = turbine engine efficiency; H_c = enthalpy of the steam at the condenser; other symbols as given earlier. Since, from the steam tables H_1 = 1398 Btu/lb (3257.3 kJ/kg) and H_c = 912 Btu/lb (2124.9 kJ/kg), then $(H_1 - H_c)$ = 486 Btu/lb (1132.4 kJ/kg).

From earlier steps, w_s = 356,000 lb/h (161,624 kg/h) at full-load; w_s = 32,000 lb/h (14,528 kg/h) at no-load. For the full-load range the total change is 356,000 – 32,000 = 324,000 lb/h (147,096 kg/h). Then, w_s at 30 percent load = [(32,000) + 0.30(324,000)]/0.30(40,000) = 10.77 lb/kWh (4.88 kg/kWh). Then, E_e = 3413/(10.77)(486) = 0.652 for combined turbine and generator.

If the internal efficiency of the turbine (not including the friction loss) E_i, then $E_i = 2545/(w_a)(H_1 - H_c)$. Thus $E_i = E_e$/(turbine mechanical efficiency)(generator efficiency). Or E_i = 0.652/(0.99)(0.98) = 0.672. Then, the actual steam rate per horsepower (kW) is $w_a = 2545/(E_i)(H_1 - H_c)$ = 2545/(0.672)(486) = 7.79 lb/hp (4.74 kg/kW).

Related Calculations. Use this approach to analyze any steam turbine—utility, industrial, commercial, marine, etc.—to determine the throttle steam flow and heat rate.

ENERGY TEST DATA FOR STEAM POWER PLANT PERFORMANCE

A test on an industrial turbogenerator gave these data: 29,760 kW delivered with a throttle flow of 307,590 lb/h (139,646 kg/h) of steam at 245 lb/in^2 (abs) (1688 kPa) with superheat at the throttle of 252°F (454°C); exhaust pressure 0.964 in (2.45 cm) Hg (abs); pressure at the one bleed point, Fig. 11*a*, 28.73 in (72.97 cm) Hg (abs); temperature of feedwater leaving bleed heater 163°F (72.8°C). For the corresponding ideal unit, find: (*a*) percent throttle steam bled, (*b*) net work for each pound of throttle steam, (*c*) ideal steam rate, and (*d*) cycle efficiency. For the actual unit, find: (*e*) the combined steam rate, (*f*) combined thermal efficiency, and (*g*) combined engine efficiency.

Calculation Procedure:

1. *Determine the steam properties at key points in the cycle*
Using a Mollier chart and the steam tables, plot the cycle as in Fig. 11*b*. Then, S_1 = 1.676; H_1 = 1366 Btu/lb (3183 kJ/kg); H_2 = 1160 Btu/lb (2577 kJ/kg); P_2 = 14.11 lb/in^2 (abs) (97.2 kPa); H_3 = 130.85 Btu/lb (304.9 kJ/kg); P_3 = 5.089 lb/in^2 (abs) (35.1 kPa); H_4 = 46.92 Btu/lb (109.3 kJ/kg); P_4 = 0.4735 lb/in^2 (abs) (3.3 kPa); H_5 = 177.9 Btu/lb (414.5 kJ/kg).

(*a*) The percent throttle steam bled is found from: $100 \times (H_5 - H_4)/(H_2 - H_4) = 100 \times (177.9 - 46.92)/(1106 - 46.92)$ = 12.41 percent.

2. *Find the amount of heat converted to work*
(*b*) Use the relation, heat converted to work, $h_w = H_1 - H_2 + (1 - m_2)(H_2 - H_7)$, where m_2 = percent throttle steam bled, H_7 = enthalpy of exhaust steam in the condenser. Substituting, heat converted to work, h_w = (1366 – 1106) + (1 – 0.1241)(1106 – 924.36) = 419.1 Btu/lb (976.5 kJ/kg).

3. *Compute the ideal steam rate*
(*c*) Use the relation, ideal steam rate, l_r = 3413 Btu/kWh/h_w. Or, l_r = 3413/419.1 = 8.14 lb/kWh (3.69 kg/kWh).

4. *Find the cycle efficiency of the ideal cycle*
(*d*) Cycle efficiency, C_e = (heat converted into work/heat supplied). Or $h_w/(H_1 - H_3)$; substituting, C_e = 419.1/(1366 – 130.85) = 0.3393, or 33.9 percent.

5. *Determine the combined steam rate*
(*e*) The combined steam rate for the actual unit is R_c = lb steam consumed/kWh generated. Or R_c = 307,590/29,760 = 10.34 lb/kWh (4.69 kg/kWh).

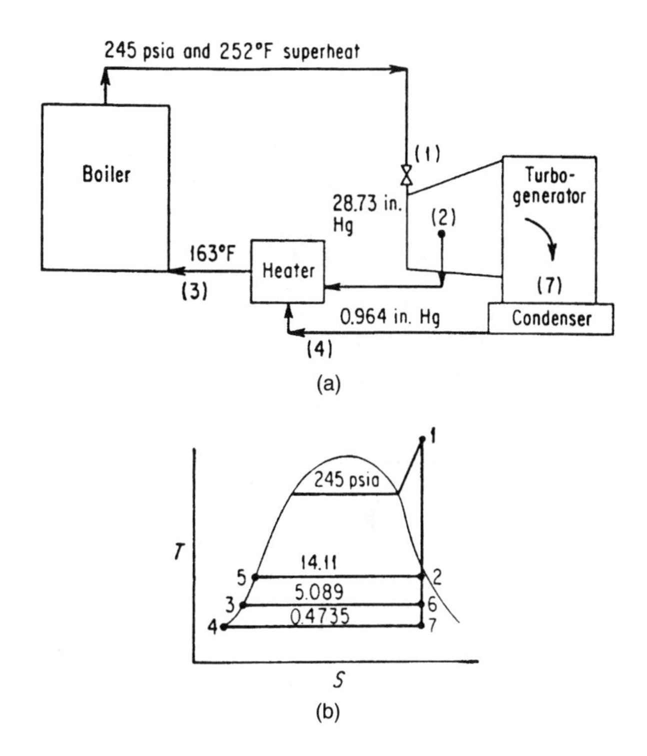

245 psia (1688 kPa)
252°F (454°C)
0.963 in. Hg (2.45 cm Hg)
28.73 in. Hg (72.97 cm Hg)
163°F (72.8°C)

FIGURE 11 (*a*) Cycle diagram with test conditions. (*b*) *T-S* diagram for cycle.

6. ***Find the combined thermal efficiency of the actual unit***

(*f*) The combined thermal efficiency, TE_c = 3413/heat supplied. Or $TE_c = 3413/10.34(H_1 - H_3) = 35413/10.34(1366 - 130.85) = 0.267$, or 26.7 percent.

7. ***Compute the combined engine efficiency***

(*g*) The combined engine efficiency TE_c/C_c, or 26.7/33.9 = 0.7876, or 78.76 percent.

Related Calculations. Use this general procedure to determine the percent bleed steam, net work of each pound of throttle steam, ideal steam rate, cycle efficiency, combined thermal efficiency, and combined engine efficiency for steam-turbine installations in central stations, industrial, municipal, and marine installations. Any standard set of steam tables and a Mollier chart are sufficiently accurate for usual design purposes.

STEAM RATE FOR TURBOGENERATOR AT VARIOUS LOADS

A 100-MW turbogenerator is supplied steam at 1250 lb/in^2 (abs) (8612.5 kPa) and 1000°F (537.8°C) with a condenser pressure of 2 in (5.08 cm) Hg (abs). At rated load, the turbine uses 1,000,000 lb (454,000 kg) of steam per hour; at zero load, steam flow is 50,000 lb/h (22,700 kg/h). What is the steam rate in pounds (kg) per kWh at 4/4, ¾, 2/4, and ¼ load?

Calculation Procedure:

1. *Write the steam-flow equation for this turbogenerator*
The curve of steam consumption, called the Willian's line, is practically a straight line for steam turbines operating without overloads. Hence, we can assume a straight line for this turbogenerator. If the Willian's line is extended to intercept the Y (vertical) axis for total steam flow per hour, this intercept represents the steam required to operate the turbine when delivering no power. This no-load steam flow—50,000 lb/h (22,700 kg/h) for this turbine—is the flow rate required to overcome the friction of the turbine and the windage, governor and oil-pump drive power, etc., and for meeting the losses caused by turbulence, leakage, and radiation under no-load conditions.

Using the data provided, the steam rate equation can be written as $[(50/L) + 9.5] = (F/L) = [50 + (1000 - 50)/100(L)]/(L)$, where F = full-load steam flow, lb/h (kg/h); L = load percent.

2. *Compute the steam flow at various loads*
Use the equation above thus:

Load fraction	Load, MW	Steam rate lb/kWh (kg/kWh)
¼	100 × ¼ = 25	50/25 + 9.5 = 11.5 (5.22)
2/4	100 × 2/4 = 50	50/50 + 9.5 = 10.5 (4.77)
¾	100 × ¾ = 75	50/75 + 9.5 = 10.17 (4.62)
4/4	100 × 4/4 = 100	50/100 + 9.5 = 10.00 (4.54)

Related Calculations. The Willian's line is a useful tool for analyzing steam-turbine steam requirements. As a check on its validity, compare actual turbine performance steam conditions with those computed using this procedure. The agreement is startlingly accurate.

REHEATING-REGENERATIVE STEAM TURBINE CYCLE ENERGY ANALYSIS

An industrial turbogenerator operates on the reheating-regenerative cycle with one reheat and one regenerative feedwater heater. Throttle steam at 400 lb/in^2 (abs) (2756 kPa) and 700°F (371°C) is used. Exhaust is at 2 in (5.1 cm) Hg (abs). Steam is taken from the turbine at a pressure of 63 lb/in^2 (abs) (434 kPa) for both reheating and feedwater heating. Reheat is to 700°F (371°C). For the ideal turbine working under these conditions, find: (*a*) percentage of throttle steam bled for feedwater heating, (*b*) heat converted to work per pound (kg) of throttle steam, (*c*) heat supplied per pound (kg) of throttle steam, (*d*) ideal thermal efficiency, (*e*) *T-S*, temperature-entropy, diagram and layout of cycle.

Calculation Procedure:

1. *Determine the cycle enthalpies, pressures, and entropies*
Using standard steam tables and a Mollier chart, draw the cycle and *T-S* plot, Fig. 12*a* and *b*. Then, $P_1 = 400$ lb/in^2 (abs) (2756 kPa); $t_1 - 700$°F (371°C); $H_1 = 1362.3$ Btu/lb (3174 kJ/kg); $S_1 = 1.6396$; $H_2 = 1178$ Btu/lb (2745 kJ/kg); $H_g = 1380.1$ Btu/lb (3216 kJ/kg); $S_g = 1.8543$.

(*a*) Percent throttle steam bled = $(H_6 - H_5)/(H_2 - H_5) = (196.15/1107.9) = 0.1771$, or 17.71 percent.

2. *Find the amount of heat converted to work per pound (kg) of throttle steam*
(*b*) The amount of heat converted to work per pound (kg) of throttle steam = $(H_1 - H_2) + (1 - 0.1771)(H_g - H_4) = 467.3$ Btu/lb (1088.8 kJ/kg).

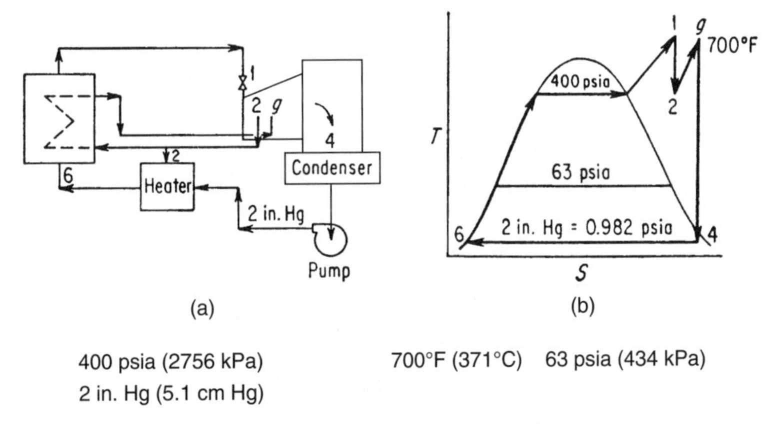

400 psia (2756 kPa) 700°F (371°C) 63 psia (434 kPa)
2 in. Hg (5.1 cm Hg)

FIGURE 12 (*a*) Cycle diagram. (*b*) *T-S* diagram for cycle in (*a*).

3. ***Compute the heat supplied per pound (kg) of throttle steam***
(*c*) The heat supplied per pound (kg) of throttle steam = $(H_1 - H_6) + (H_g - H_2)$ = 1299.1 Btu/lb (3026.9 kJ/kg).

4. ***Determine the ideal thermal efficiency***
(*d*) The ideal thermal efficiency = [heat recovered per pound (kg) of throttle steam]/[heat supplied per pound (kg) of throttle steam] = 467.3/1299.13 = 0.3597, or 35.97 percent. The *T-S* diagram and cycle layout can be drawn as shown in Fig. 12*a* and *b*.

Related Calculations. This general procedure can be used for any turbine cycle where reheating and feedwater heating are part of the design. Note that the enthalpy and entropy values read from the Mollier chart, or interpolated from the steam tables, may differ slightly from those given here. This is to be expected where judgment comes into play. The slight differences are unimportant in the analysis of the cycle.

The procedure outlined here is valid for industrial, utility, commercial, and marine turbines used to produce power.

REHEAT-REGENERATIVE CYCLE STEAM RATE

Steam is supplied at 600 lb/in^2 (abs) (4134 kPa) and 740°F (393°C) to a steam turbine operating on the reheat-regenerative cycle. After expanding to 100 lb/in^2 (abs) (689 kPa), the steam is reheated to 700°F (371°C). Expansion then continues to 10 lb/in^2 (abs) (68.9 kPa) but at 30 lb/in^2 (abs) (207 kPa) some steam is extracted for feedwater heating in a direct-contact heater. Assuming ideal operation with no losses, find: (*a*) steam extracted as a percentage of steam supplied to the throttle. (*b*) steam rate in pounds (kg) per kWh; (*c*) thermal efficiency of the turbine; (*d*) quality or superheat of the exhaust if in the actual turbine combined efficiency is 72 percent, generator efficiency is 94 percent, and actual extraction is the same as the ideal.

Calculation Procedure:

1. ***Assemble the key enthalpies, entropies, and pressures for the cycle***
Using the steam tables and a Mollier chart, list the following pressures, temperatures, enthalpies, and entropies for the cycle, Fig. 13*a*: P_t = 600 lb/in^2 (abs) (4134 kPa); t_1 = 740°F (393°C); P_2 = 100 lb/in^2

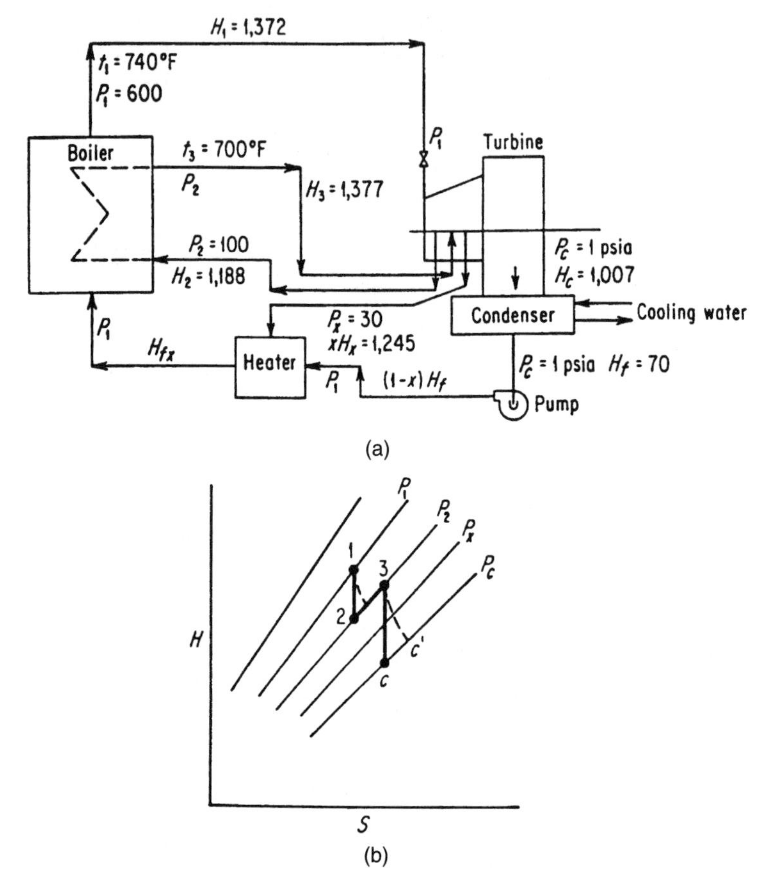

600 psia (4134 kPa) 740°F (393°C) 700°F (371°C) 100 psia (68.9 kPa)
1188 Btu/lb (2768 kJ/kg) 1372 Btu/lb (3197 kJ/kg) 1245 Btu/lb (2901 kJ/kg)
1 psia (6.89 kPa) 30 psia (207 kPa) 1007 Btu/lb (2346 kJ/kg) 70 Btu/lb (163 kJ/kg)

FIGURE 13 (*a*) Cycle diagram. (*b*) *H-S* chart for cycle in (*a*).

(abs) (689 kPa); t_3 = 700°F (371°C); P_x = 30 lb/in^2 (abs) (207 kPa); p_c = 1 lb/in^2 (abs) (6.89 kPa); H_1 = 1372 Btu/lb (3197 kJ/kg); S_1 = entropy = 1.605; H_2 = 1188 Btu/lb (2768 kJ/kg); H_3 = 1377 Btu/lb (3208 kJ/kg); S_3 = 1.802; H_x = 1245 Btu/lb (2901 kJ/kg); H_c = 1007 Btu/lb (2346 kJ/kg); H_f = 70 Btu/lb (163 kJ/kg); H_{fx} = 219 Btu/lb (510 kJ/kg). Plot Fig. 13*b* as a skeleton Mollier chart to show the cycle processes.

2. *Compute the percent steam extracted for the feedwater heater*

(*a*) The steam extracted for the feedwater heater, x, = $(H_{fx} - H_f)(H_x - H_f)$ = (219 – 70)/(1245 – 70) = 0.1268, or 12.68 percent.

3. *Find the turbine steam rate*

(*b*) For the Rankine-cycle steam rate, $w_s = 3413/(H_1 - H_c)$. For this cycle, $w_s = 3413/[(H_1 - H_2) + x(H_3 - H_x) + (1 - x)(H_3 - H_c)]$. Or, w_s = 3413/[1372 – 1188) + 0.1268(1377 – 1245) + (1 – 0.1268)(1377 – 1007)] = 6.52 lb/kWh (2.96 kg/kWh).

4. *Calculate the turbine thermal efficiency*

(*c*) The thermal efficiency, $E_t = [(H_1 - H_2) + x(H_3 - H_x) + (1 - x)(H_3 - H_c)]/[(H_3 - H_2) + (H_1 - H_{fx})]$. Or, $E_t = [(1372 - 1188) + (0.1268)(1377 - 1245) + (1 - 0.1268)(1377 - 1007)]/[(1377 - 1188) + (1372 - 219)] = 0.3903$, or 39 percent. It is interesting to note that in an ideal cycle the thermal efficiency of the turbine is the same as that of the cycle.

5. *Determine the condition of the exhaust*

(*d*) The engine efficiency of the turbine alone = (actual turbine combined efficiency)/actual generator efficiency). Or, using the given data, engine efficiency of the turbine alone = 0.72/0.94 = 0.765.

Using the computed engine efficiency of the turbine alone and the Mollier chart, $(H_3 - H_{c'}) = 0.765(H_3 - H_c) = 283$. Solving, $H_{c'} = H_3 - 283 = 1094$ Btu/lb (2549 kJ/kg). From the Mollier chart, the condition at $H_{c'}$ is 1.1 percent moisture. The exhaust steam quality is therefore 100 – 1.1 = 98.9 percent.

Related Calculations. This procedure is valid for a variety of cycle arrangements for industrial, central-station, commercial, and marine plants. By using a combination of the steam tables, Mollier chart, and cycle diagram, a full analysis of the plant can be quickly made.

ENERGY EFFICIENCY ANALYSIS FOR BINARY CYCLE STEAM PLANT

A binary cycle steam and mercury plant is being considered by a public utility. Steam and mercury temperature will be 1000°F (538°C). The mercury is condensed in the steam boiler, Fig. 14*a* at 10 lb/in^2 (abs) (68.9 kPa) and the steam pressure is 1200 lb/in^2 (abs) (8268 kPa). Condenser pressure is 1 lb/in^2 (abs) (6.89 kPa). Expansions in both turbines are assumed to be at constant entropy. The steam cycle has superheat but no reheat. Find the efficiency of the proposed binary cycle. Find the cycle efficiency without mercury.

Calculation Procedure:

1. *Tabulate the key enthalpies and entropies for the cycle*

Set up two columns, thus:

Mercury cycle	Steam cycle
$H_{m1} = 151.1$ Btu/lb (352 kJ/kg)	$H_{s1} = 1499.2$ Btu/lb (3493 kJ/kg)
$S_{m1} = 0.1194$	$S_{s1} = 1.6293$
$S_{me} = 0.0094$	$H_{sf} = 69.7$ Btu/lb (162.4 kJ/kg)
$H_{mf} = 22.6$ Btu/lb (52.7 kJ/kg)	

2. *Compute the quality of the exhaust for each vapor*

Since expansion in each turbine is at constant entropy, Fig. 14*b*, the quality for the mercury exhaust, x_m is: $0.1194 = 0.0299 + x_m(0.1121)$; $x_m = 0.798$.

For the steam cycle, the quality, x_s is: $1.6293 = 0.1326 + x_s(1.8456)$; $x_s = 0.81$.

3. *Find the exhaust enthalpy for each vapor*

Using the properties of mercury from a set of tables, the enthalpy of the mercury exhaust, $H_{me} = 22.6 + 0.798(123) = 120.7$ Btu/lb (281.2 kJ/kg). The enthalpy of the condensed mercury, $H_{mf} = 22.6$ Btu/lb (52.7 kJ/kg).

For the exhaust steam, the enthalpy $H_{se} = 69.7 + 0.81(1036.3) = 909.1$ Btu/lb (2118 kJ/kg), using steam-table data. The enthalpy of the condensed steam, $H_{sf} = 69.7$ Btu/lb (162.4 kJ/kg).

Assuming 98 percent quality steam leaving the mercury condenser, then the enthalpy of the wet steam leaving the mercury condenser, $H_{sw} = 571.7 + 0.98(611.7) = 1171.2$ Btu/lb (2728 kJ/kg).

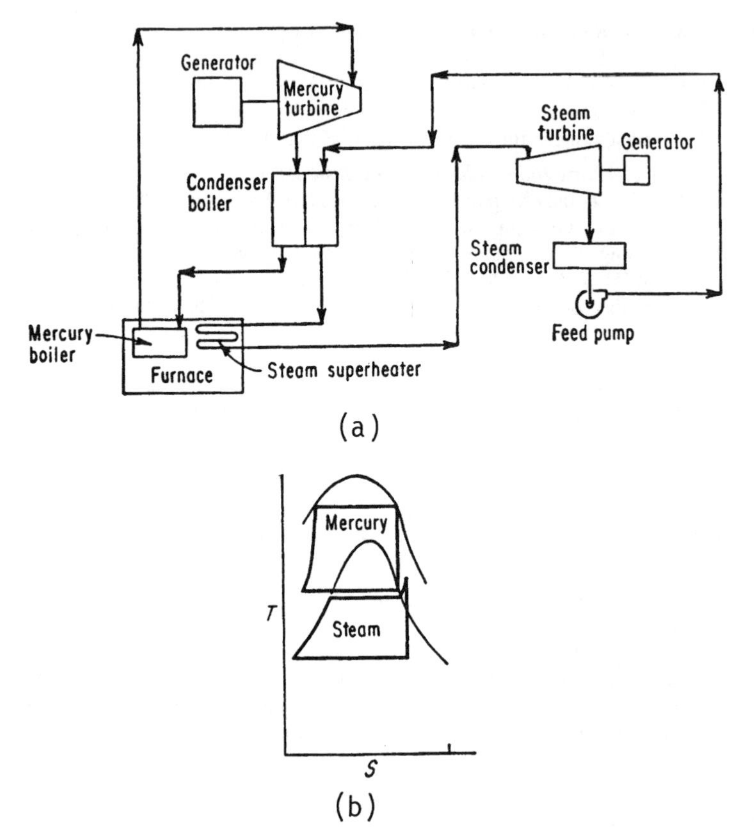

FIGURE 14 (*a*) Binary cycle. (*b*) *T-S* diagram for binary cycle.

4. *Write the heat balance around the mercury condenser*

The steam heat gain = $H_{sw} - H_{sf}$ = 1171.2 – 69.7 = 1101.5 Btu/lb (2566.5 kJ/kg). Now, the mercury heat loss = $H_{me} - H_{mf}$ = 120.7 – 98.1 = 98.1 Btu/lb (228.6 kJ/kg). The weight of mercury per pound (kg) of steam = steam heat gain/mercury heat loss = 1101.5/98.1 = 11.23.

5. *Determine the heat input and work done per pound (kg) of steam*

The heat input per pound of steam is: For mercury = (lb Hg/lb steam)($H_{m1} - H_{mf}$) = 11.23(151.1 – 22.6) = 1443.05 Btu (1522.5 J). For steam = ($H_{s1} - H_{sw}$) = 1499.2 – 1171.7 = 327.5 Btu (345.5 J). Summing these two results gives 1443.05 + 327.5 = 1770.55 Btu (1867.9 J) as the heat input per pound (kg) of steam.

The work done per pound (kg) of steam is: For mercury = (lb Hg/lb steam)($h_{m1} - H_{me}$) = 11.23(151.1 – 120.7) = 341.4 Btu (360.2 J). For steam = $H_{c1} - H_{se}$ = 1499.2 – 909.1 = 590.1 Btu (622.6 J). Summing, as before, the total work done per pound (kg) of steam = 931.5 Btu (982.7 J).

6. *Compute the binary cycle efficiency*

The binary cycle efficiency = [work done per pound (kg) of steam]/[heat input per pound (kg) of steam]. Or binary cycle efficiency = 931.5/1770.55 = 0.526, or 52.6 percent.

7. *Calculate the steam cycle efficiency without the mercury topping turbine*

The steam cycle efficiency without the mercury topping turbine = [work done per pound (kg) of steam]/($H_{s1} - H_{sf}$) = 590.1/(1499.2 – 69.7) = 0.4128, or 41.3 percent.

Related Calculations. Any binary cycle being considered for an installation depends on the effects of the difference in thermodynamic properties of the two pure fluids involved. For example, steam works under relatively high pressures with an attendant relatively low temperature. Mercury, by comparison, has the vapor characteristic of operating under low pressures with attendant high temperature.

In a mercury-vapor binary cycle, the pressures are selected so the mercury vapor condenses at a temperature higher than that at which steam evaporates. The processes of mercury vapor condensation and steam evaporation take place in a common vessel called the condenser-boiler, which is the heart of the cycle.

In the steam portion of this cycle, condenser water carries away the heat of steam condensation; in the mercury portion of the cycle it is the steam that picks up the heat of condensation of the mercury vapor. Hence, there is a great saving in heat and the economies effected reflect the consequent improvement in cycle efficiency.

The same furnace serves the mercury boiler and the steam superheater. Mercury vapor is only condensed, not superheated. And if the condenser-boiler is physically high enough above the mercury boiler, the head of mercury is great enough to return the liquid mercury to the boiler by gravity, making the use of a mercury feed pump unnecessary.

To avoid the high cost entailed with using mercury, a number of man-made solutions have been developed for binary vapor cycles. Their use, however, has been limited because the conventional steam cycle is usually lower in cost. And with the advent of the aero-derivative gas turbine, which is relatively low cost and can be installed quickly in conjunction with heat-recovery steam generators, binary cycles have lost popularity. But it is useful for engineers to have a comprehension of such cycles. Why? Because they may return to favor in the future.

COGENERATION SYSTEM ENERGY EFFICIENCY FOR A TRADITIONAL STEAM POWER PLANT CYCLE

An industrial plant has 60,000 lb/h (27,240 kg/h) of superheated steam at 1000 lb/in^2 (abs) (6890 kPa) and 900°F (482.2°C) available. Two options are being considered for use of this steam: (1) expanding the steam in a steam turbine having a 70 percent efficiency to 1 lb/in^2 (abs) (6.89 kPa), and (2) expand the steam in a turbine to 200 lb/in^2 (abs) (1378 kPa) generating electricity and utilizing the low-pressure exhaust steam for process heating. Evaluate the two schemes for energy efficiency when the boiler has an 82 percent efficiency on a HHV basis.

Calculation Procedure:

1. *Determine the enthalpies of the steam at the turbine inlet and after isentropic expansion*
Cogeneration systems generate power and process steam from the same fuel source. Process plants generating electricity from steam produced in a boiler and using the same steam after expansion in a steam turbine for process heating of some kind are examples of cogeneration systems.

Conventional steam-turbine power plants have a maximum efficiency of about 40 percent as most of the energy is wasted in the condensing-system cooling water. In a typical cogeneration system the exhaust steam from the turbine is used for process purposes after expansion through the steam turbine; hence, its enthalpy is fully utilized. Thus, cogeneration schemes are more efficient.

At 1000 lb/in^2 (abs) (6890 kPa) and 900°F (482.2°C), the enthalpy, h_1 = 1448 Btu/lb (3368 kJ/kg) from the steam tables. The entropy of steam at this condition, from the steam tables, is 1.6121 Btu/lb °F (6.748 kJ/kg K). At 1 lb/in^2 (abs) (6.89 kPa), the entropy of the saturated liquid, s_f, is 0.1326 Btu/lb °F (0.555 kJ/kg K), and the entropy of the saturated vapor, s_g, is 1.9782 Btu/lb °F (8.28 kJ/kg K), again from the steam tables.

Now we must determine the quality of the steam, X, at the exhaust of the steam turbine at 1 lb/in^2 (abs) (6.89 kPa) from (entropy at turbine inlet condition) = (entropy at outlet condition)(X) + (1 – X)(entropy of the saturated fluid at the outlet condition); or X = 0.80. The enthalpy of steam corresponding to this quality condition is h_{2s} = [enthalpy of the saturated steam at 1 lb/in^2 (abs)] (X) + [enthalpy of the saturated liquid at 1 lb/in^2 (abs)](1 – X) = (1106)(0.80) + (1 – 0.80)(70) = 900 Btu/lb (2124 kJ/kg).

2. *Compute the power output of the turbine*
Use the equation $P = (W_s)(e_t)(h_1 - h_{2s})/3413$, where P = electrical power generated, kW; W_s = steam flow through the turbine, lb/h (kg/h); e_t = turbine efficiency expressed as a decimal; h_1 = enthalpy of the steam at the turbine inlet, Btu/lb (kJ/kg); h_{2s} = enthalpy of the steam after isentropic expansion through the turbine, Btu/lb (kJ/kg). Substituting, P = (60,000)(0.70)(1448 – 900)/3413 = 6743 kW = (6743)(3413) = 23 MM Btu/h (24.26 MM kJ).

3. *Find the steam enthalpy after expansion in the cogeneration scheme*
The steam is utilized for process heating after expansion to 200 lb/in^2 (abs) (1378 kPa) in the back-pressure turbine. We must compute the enthalpy of the steam after expansion in order to find the energy available.

At 200 lb/in^2 (abs) (1378 kPa), using the same procedure as in step 1 above, h_{2s} = 1257.7 Btu/lb (2925.4 kJ/kg). Since we know the turbine efficiency we can use the equation, $(e_t)(h_1 - h_{2s}) = (h_1 - h_2)$; or (0.70)(1448 – 1257.7) = (1448 – h_2); h_2 = 1315 Btu/lb (3058.7 kJ/kg), h_2 = actual enthalpy after expansion, Btu/lb (kJ/kg).

4. *Determine the electrical output of the cogeneration plant*
Since the efficiency of the turbine is already factored into the exhaust enthalpy of the cogeneration turbine, use the relation, $P = W_s(h_1 - h_2)/3413$, where the symbols are as defined earlier. Or, P = 60,000(1448 – 1315)/3413 = 2338 kW.

5. *Compute the total energy output of the cogeneration plant*
Assuming that the latent heat of the steam at 200 lb/in^2 (abs) (1378 kPa) is available for industrial process heating, the total energy output of the cogeneration scheme = electrical output + (steam flow, lb/h)(latent heat of the exhaust steam, Btu/lb). Since, from the steam tables, the latent heat of steam at 200 lb/in^2 (abs) (1378 kPa) = 834 Btu/lb (1939.9 kJ/kg), total energy output of the cogeneration cycle = (2338 kW)(3413) + (60,000)(834) = 58 MM Btu/h (61.2 MM kJ/h).

Since the total energy output of the conventional cycle was 23 MM Btu/h (24.3 MM kJ/h), the ratio of the cogeneration output vs. the conventional output = 58/23 = 2.52. Thus, about 2.5 times as much energy is derived from the cogeneration cycle as from the conventional cycle.

6. *Find the comparative efficiencies of the two cycles*
The boiler input = (weight of steam generated, lb/h)(enthalpy of superheated steam at boiler outlet, Btu/lb – enthalpy of feedwater entering the boiler, Btu/lb)/(boiler efficiency, expressed as a decimal). Or, boiler input = (60,000)(1448 – 200)/0.82 = 91.3 MM Btu/h (96.3 MM kJ/h). The efficiency of the conventional cycle is therefore (23/91.3)(100) = 25 percent. For the cogeneration cycle, the efficiency (58/91.3)(10) = 63.5 percent.

Related Calculations. This real-life example shows why cogeneration is such a popular alternative in today's world of power generation. In this study the cogeneration scheme is more than twice as efficient as the conventional cycle—63.5 percent vs. 25 percent. Higher efficiencies could be obtained if the boiler outlet steam pressure were higher than 1000 lb/in^2 (abs) (6890 kPa). However, the pressure used here is typical of today's industrial installations using cogeneration to save energy and conserve the environment.

This procedure is the work of V. Ganapathy, Heat Transfer Specialist. ABCO Industries, Inc.

REGENERATIVE BLEED-STEAM CYCLE GENERAL ENERGY ANALYSIS

Sketch the cycle layout, *T-S* diagram, and energy-flow chart for a regenerative bleed-steam turbine plant having three feedwater heaters and four feed pumps. Write the equations for the work-output available energy and the energy rejected to the condenser.

Calculation Procedure:

1. ***Sketch the cycle layout***

Figure 15 shows a typical practical regenerative cycle having three feedwater heaters and four feedwater pumps. Number each point where steam enters and leaves the turbine and where steam enters or leaves the condenser and boiler. Also number the points in the feedwater cycle where feedwater enters and leaves a heater. Indicate the heater steam flow by m with a subscript corresponding to the heater number. Use W_p and a suitable subscript to indicate the pump work for each feed pump, except the last, which is labeled W_{pF}. The heat input to the steam generator is Q_a; the work output of the steam turbine is W_e; the heat rejected by the condenser is Q_r.

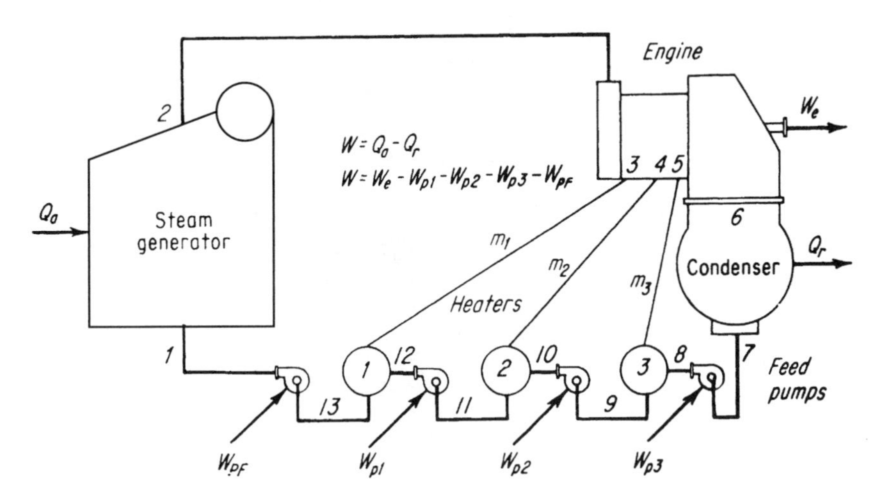

FIGURE 15 Regenerative steam cycle uses bleed steam.

2. ***Sketch the T-S diagram for the cycle***

To analyze any steam cycle, trace the flow of 1 lb (0.5 kg) of steam through the system. Thus, in this cycle, 1 lb (0.5 kg) of steam leaves the steam generator at point 2 and flows to the turbine. From state 2 to 3, 1 lb (0.5 kg) of steam expands at constant entropy (assumed) through the turbine, producing work output $W_1 = H_2 - H_3$, represented by area 1-*a*-2-3 on the *T-S* diagram, Fig. 16*a*. At point 3, some steam is bled from the turbine to heat the feedwater passing through heater 1. The quantity of steam bled, m_1, lb is less than the 1 lb (0.5 kg) flowing between points 2 and 3. Plot stages 2 and 3 on the *T-S* diagram, Fig. 16*a*.

From point 3 to 4, the quantity of steam flowing through the turbine is $1 - m_1$ lb. This steam produces work output $W_2 = H_3 - H_4$. Plot point 4 on the *T-S* diagram. Then, area 1-3-4-12 represents the work output W_2, Fig. 16*a*.

At point 4, steam is bled to heater 2. The weight of this steam is m_2 lb. From point 4, the steam continues to flow through the turbine to point 5, Fig. 16*a*. The weight of the steam flowing between points 4 and 5 is $1 - m_1 - m_2$ lb. Plot point 5 on the *T-S* diagram, Fig. 16*a*. The work output between points 4 and 5, $W_3 = H_4 - H_5$, is represented by area 4-5-10-11 on the *T-S* diagram.

At point 5, steam is bled to heater 3. The weight of this bleed steam is m_3 lb. From point 5, steam continues to flow through the turbine to exhaust at point 6, Fig. 16*a*. The weight of steam flowing between points 5 and 6 is $1 - m_1 - m_2 - m_3$ lb. Plot point 6 on the *T-S* diagram, Fig. 16*a*.

The work output between points 5 and 6 is $W_4 = H_5 - H_6$, represented by area 5-6-7-9 on the *T-S* diagram, Fig. 16*a*. Area Q_r represents the heat given up by 1 lb (0.5 kg) of exhaust steam. Similarly, the area marked Q_a represents the heat absorbed by 1 lb (0.5 kg) of water in the steam generator.

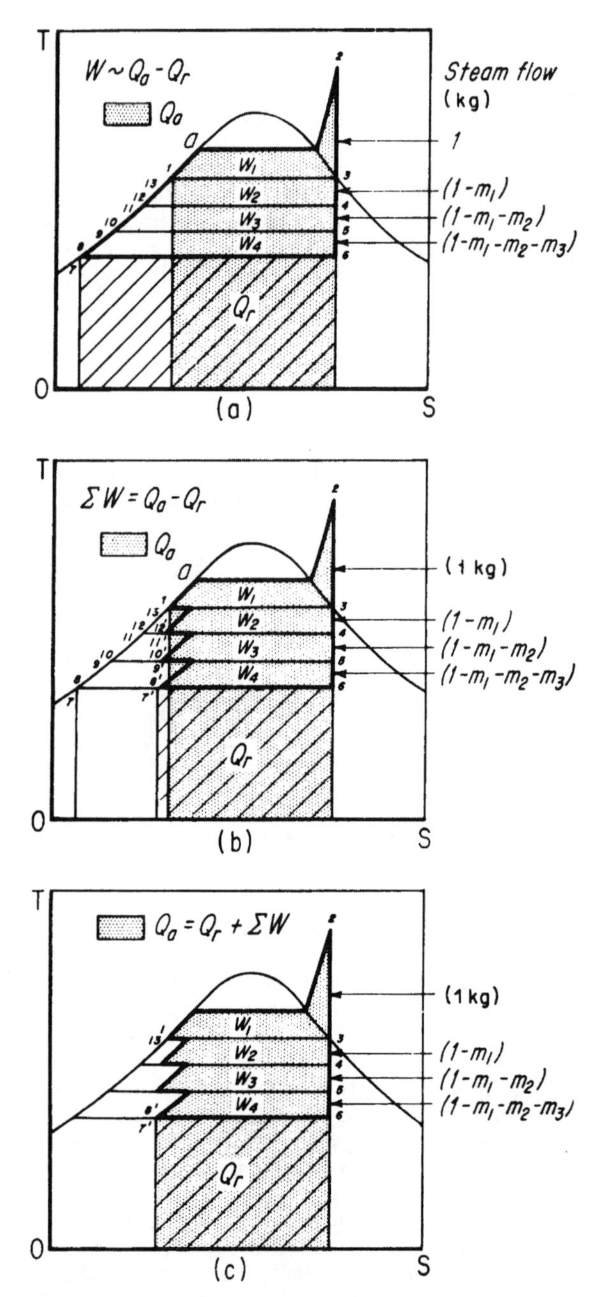

FIGURE 16 (*a*) *T-S* chart for the bleed-steam regenerative cycle in Fig. 15; (*b*) actual fluid flow in the cycle; (*c*) alternative plot of (*b*).

3. *Alter the T-S diagram to show actual cycle conditions*

As plotted in Fig. 16*a*, Q_a is true for this cycle since 1 lb (0.5 kg) of water flows through the stearn-generator and the first section of the turbine. But Q_r is much too large; only $1 - m_1 - m_2 - m_3$ lb of steam flows through the condenser. Likewise, the net areas for W_2, W_3, and W_4, Fig. 16*a*, are all too large, because less than 1 lb (0.5 kg) of steam flows through the respective turbine sections. The area for W_1, however, is true.

A true *proportionate-area* diagram can be plotted by applying the factors for actual flow as in Fig. 16*b*. Here W_2, outlined by the heavy lines, equals the similarly labeled area in Fig. 16*a*, multiplied by $1 - m_1$. The states marked 11′ and 12′, Fig. 16*b*, are not true state points because of the ratioing factor applied to the area for W_2. The true state points 11 and 12 of the liquid before and after heater pump 3 stay as shown in Fig. 16*a*.

Apply $1 - m_1 - m_2$ to W_3 of Fig. 16*a*. to obtain the proportionate area of Fig. 16*b*; to obtain W_4, multiply by $1 - m_1 - m_2 - m_3$. Multiplying by this factor also gives Q_r. Then all the areas in Fig. 16*b* will be in proper proportion for 1 lb (0.5 kg) of steam entering the turbine throttle but less in other parts of the cycle.

In Fig. 16*b*, the work can be measured by the difference of the area Q_a and the area Q_r. There is no simple net area left, because the areas coincide on only two sides. But area enclosed by the heavy lines *is* the total net work *W* for the cycle, equal to the sum of the work produced in the various sections of the turbine, Fig. 16*b*. Then Q_a is the alternate area $Q_r + W_1 + W_2 + W_3 + W_4$, as shaded in Fig. 16*c*.

The sawtooth approach of the liquid-heating line shows that as the number of heaters in the cycle increases, the heating line approaches a line of constant entropy. The best number of heaters for a given cycle depends on the steam state of the turbine inlet. Many medium-pressure and medium-temperature cycles use five to six heaters. High-pressure and high-temperature cycles use as many as nine heaters.

4. *Draw the energy-flow chart*

Choose a suitable scale for the heat content of 1 lb (0.5 kg) of steam leaving the steam generator. A typical scale is 0.375 in per 1000 Btu/lb (0.41 cm per 1000 kJ/kg). Plot the heat content of 1 lb (0.5 kg) of steam vertically on line 2-2, Fig. 17. Using the same scale, plot the heat content in energy streams m_1, m_2, m_3, W_e, W, W_p, W_{pF}, and so forth. In some cases, as W_{p1}, W_{p2}, and so forth, the energy stream may be so small that it is impossible to plot it to scale. In these instances, a single thin line is used. The completed diagram, Fig. 17, provides a useful concept of the distribution of the energy in the cycle.

Related Calculations. The procedure given here can be used for all regenerative cycles, provided that the equations are altered to allow for more, or fewer, heaters and pumps. The following calculation procedure shows the application of this method to an actual regenerative cycle.

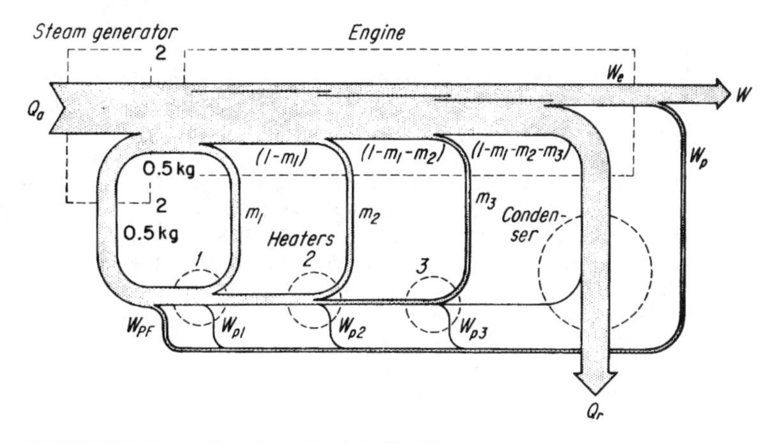

FIGURE 17 Energy-flow chart of cycle in Fig. 15.

BLEED REGENERATIVE STEAM CYCLE ANALYSIS

Analyze the bleed regenerative cycle shown in Fig. 18, determining the heat balance for each heater, plant thermal efficiency, turbine or engine thermal efficiency, plant heat rate, turbine or engine heat rate, and turbine or engine steam rate. Throttle steam pressure is 2000 lb/in^2 (abs) (13,790.0 kPa) at 1000°F (537.8°C); steam-generator efficiency = 0.88; station auxiliary steam consumption (excluding pump work) = 6 percent of the turbine or engine output; engine efficiency of each turbine or engine section = 0.80; turbine or engine cycle has three feedwater heaters and bleed-steam pressures as shown in Fig. 18; exhaust pressure to condenser is 1 inHg (3.4 kPa) absolute.

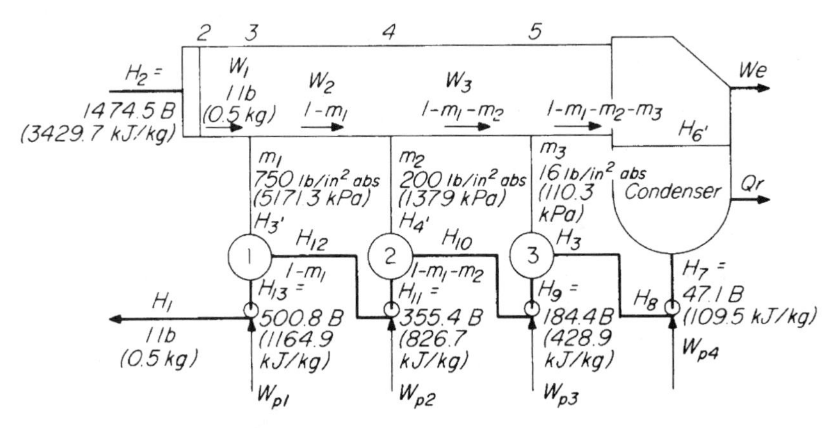

FIGURE 18 Bleed regenerative steam cycle.

Calculation Procedure:

1. ***Determine the enthalpy of the steam at the inlet of each heater and the condenser***

From a superheated-steam table, find the throttle enthalpy H_2 = 1474.5 Btu/lb (3429.7 kJ/kg) at 2000 lb/in^2 (abs) (13,790.0 kPa) and 1000°F (537.8°C). Next, find the throttle entropy S_2 = 1.5603 Btu/(lb · °F) [6.5 kJ/(kg · °C)], at the same conditions in the superheated-steam table.

Plot the throttle steam conditions on a Mollier chart, Fig. 19. Assume that the steam expands from the throttle conditions at constant entropy = constant S to the inlet of the first feedwater heater 1, Fig. 18. Plot this constant S expansion by drawing the straight vertical line 2-3 on the Mollier chart, Fig. 19, between the throttle condition and the heater inlet pressure of 750 lb/in^2 (abs) (5171.3 kPa).

Read on the Mollier chart H_3 = 1346.7 Btu/lb (3132.4 kJ/kg). Since the engine or turbine efficiency $e_e = H_2 - H_3/(H_2 - H_3)$ = 0.8 = 1474.5 – H_3/(1474.5 – 1346.7); H_3 = actual enthalpy of the steam at the inlet to heater 1 = 1474.5 – 0.8(1474.5 – 1346.7) = 1372.2 Btu/lb (3191.7 kJ/kg). Plot this enthalpy point on the 750-lb/in^2 (abs) (5171.3-kPa) pressure line of the Mollier chart, Fig. 19. Read the entropy at the heater inlet from the Mollier chart as $S_{3'}$ = 1.5819 Btu/(lb · °F) [6.6 kJ/(Kg · °C)] at 750 lb/in^2 (abs) (5171.3 kPa) and 1372.2 Btu/lb (3191.7 kJ/kg).

Assume constant-S expansion from $H_{3'}$ to H_4 at 200 lb/in^2 (abs) (1379.0 kPa), the inlet pressure for feedwater heater 2. Draw the vertical straight line 3′–4 on the Mollier chart, Fig. 19. By using a procedure similar to that for heater 1, $H_{4'} = H_{3'} - e_e(H_{3'} - H_4)$ = 1372.2 – 0.8(1372.2 – 1230.0) = 1258.4 Btu/lb (2927.0 kJ/kg). This is the actual enthalpy of the steam at the inlet to heater 2. Plot this enthalpy on the 200-lb/in^2 (abs) (1379.0-kPa) pressure line of the Mollier chart, and find $S_{4'}$ = 1.613 Btu/(lb · °F) [6.8 kJ/(kg · °C], Fig. 19.

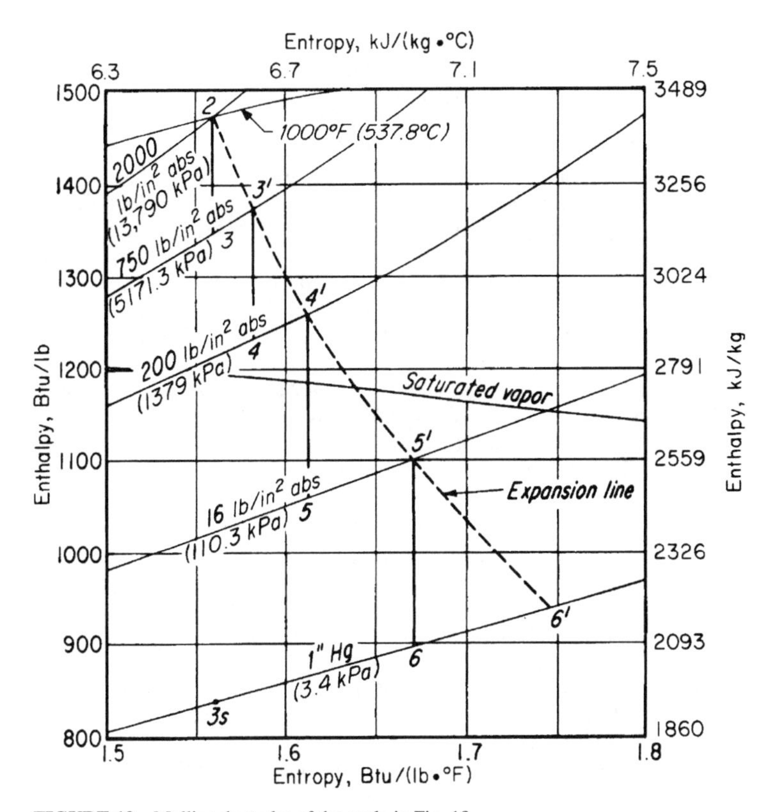

FIGURE 19 Mollier chart plot of the cycle in Fig. 18.

Using the same procedure with constant-S expansion from $H_{4'}$, we find H_5 = 1059.5 Btu/lb (2464.4 kJ/kg) at 16 lb/in^2 (abs) (110.3 kPa), the inlet pressure to heater 3. Next, find $H_{5'} = H_{4'} - e_e (H_{4'} - H_5)$ = 1258.4 – 0.8(1258.4 – 1059.5) = 1099.2 Btu/lb (2556.7 kJ/kg). From the Mollier chart find $S_{5'}$ = 1.671 Btu/(lb · °F) [7.0 kJ/(kg · °C)], Fig. 19.

Using the same procedure with constant-S expansion from $H_{5'}$ to $H_{6'}$ find H_6 = 898.2 Btu/lb (2089.2 kJ/kg) at 1 inHg absolute (3.4 kPa), the condenser inlet pressure. Then $H_{6'} = H_{5'} - e_e (H_{5'} - H_6)$ = 1099.2 – 0.8(1099.2 – 898.2) – 938.4 Btu/lb (2182.7 kJ/kg), the actual enthalpy of the steam at the condenser inlet. Find, on the Mollier chart, the moisture in the turbine exhaust = 15.1 percent.

2. *Determine the overall engine efficiency*

Overall engine efficiency e_e is higher than the engine-section efficiency because there is partial available-energy recovery between sections. Constant-S expansion from the throttle to the 1-inHg absolute (3.4-kPa) exhaust gives $H_{3s'}$, Fig. 19, as 838.3 Btu/lb (1949.4 kJ/kg), assuming that all the steam flows to the condenser. Then, overall $e_e = H_2 - H_{6'}/(H_2 - H_{3S})$ = 1474.5 – 938.4/1474.5 – 838.3 = 0.8425, or 84.25 percent, compared with 0.8 or 80 percent, for individual engine sections.

3. *Compute the bleed-steam flow to each feedwater heater*

For each heater, energy in = energy out. Also, the heated condensate leaving each heater is a saturated liquid at the heater bleed-steam pressure. To simplify this calculation, assume negligible steam pressure drop between the turbine bleed point and the heater inlet. This assumption is permissible when the distance between the heater and bleed point is small. Determine the pump work by using the chart accompanying the compressed-liquid table in Keenan and Keyes—*Thermodynamic Properties of Steam,* or the ASME—*Steam Tables.*

For heater 1, energy in = energy out, or $H_{3'}m_1 + H_{12}(1 - m_1) = H_{13}$, where m = bleed-steam flow to the feedwater heater, lb/lb of throttle steam flow. (The subscript refers to the heater under consideration.) Then $H_{3'}m_1 + (H_{11} + W_{p2})(1 - m_1) = H_{13}$, where W_{p2} = work donc by pump 2, Fig. 18, in Btu/lb per pound of throttle flow. Then $1372.2m_1 + (355.4 + 1.7)(1 - m_1) = 500.8$; $m_1 = 0.1416$ lb/lb (0.064 kg/kg) throttle flow; $H_1 = H_{13} + W_{p1} = 500.8 + 4.7 = 505.5$ Btu/lb (1175.8 kJ/kg), where W_{p1} = work done by pump 1, Fig. 18. For each pump, find the work from the chart accompanying the compressed-liquid table in Keenan and Keyes—*Steam Tables* by entering the chart at the heater inlet pressure and projecting vertically at constant entropy to the heater outlet pressure, which equals the next heater inlet pressure. Read the enthalpy values at the respective pressures, and subtract the smaller from the larger to obtain the pump work during passage of the feedwater through the pump from the lower to the higher pressure. Thus, $W_{p2} = 1.7 - 0.0 = 1.7$ Btu/lb (4.0 kJ/kg), from enthalpy values for 200 lb/in^2 (abs) (1379.0 kPa) and 750 lb/in^2 (abs) (5171.3 kPa), the heater inlet and discharge pressures, respectively.

For heater 2, energy in = energy out, or $H_{4'}m_2 + H_{10}(1 - m_1 - m_2) = H_{11}(1 - m_1)H_{4'}m_2 + (H_9 + W_{p3})(1 - m_1 - m_2) = H_{11}(1 - m_1)1258.4m_2 + (184.4 + 0.5)(0.8584 - m_2) = 355.4(0.8584)m_2 = 0.1365$ lb/lb (0.0619 kg/kg) throttle flow.

For heater 3, energy in = energy out, or $H_{5'}m_3 + H_8(1 - m_1 - m_2 - m_3) = H_9(1 - m_1 - m_2)$ $H_{5'}m_3 + (H_7 + W_{p4})(1 - m_1 - m_2 - m_3) = H_9(1 - m_1 - m_2)1099.2m_3 + (47.1 + 0.1)(0.7210 - m_3) = 184.4(0.7219)m_3 = 0.0942$ lb/lb (0.0427 kg/kg) throttle flow.

4. *Compute the turbine work output*

The work output per section W Btu is $W_1 = H_2 - H_{3'} = 1474.5 - 1372.1 = 102.3$ Btu (107.9 kJ), from the previously computed enthalpy values. Also $W_2 = (H_{3'} - H_{4'})(1 - m_1) = (1372.2 - 1258.4)(1 - 0.1416) = 97.7$ Btu (103.1 kJ); $W_3 = (H_{4'} - H_{5'})(1 - m_1 - m_2) = (1258.4 - 1099.2)(1 - 0.1416 - 0.1365) = 115.0$ Btu (121.3 kJ); $W_4 = (H_{5'} - H_{6'})(1 - m_1 - m_2 - m_3) = (1099.2 - 938.4)(1 - 0.1416 - 0.1365 - 0.0942) = 100.9$ Btu (106.5 kJ). The total work output of the turbine $= W_e = \Sigma W = 102.3 + 97.7 + 115.0 + 100.9 = 415.9$ Btu (438.8 kJ). The total $W_p = \Sigma W_p = W_{p1} + W_{p2} + W_{p3} + W_{p4} = 4.7 + 1.7 + 0.5 + 0.1 = 7.0$ Btu (7.4 kJ).

Since the station auxiliaries consume 6 percent of W_e, the auxiliary consumption = 0.6(415.9) = 25.0 Btu (26.4 kJ). Then, net station work $w = 415.9 - 7.0 - 25.0 = 383.9$ Btu (405.0 kJ).

5. *Check the turbine work output*

The heat added to the cycle Q_a Btu/lb $= H_2 - H_1 = 1474.5 - 505.5 = 969.0$ Btu (1022.3 kJ). The heat rejected from the cycle Q_r Btu/lb $= (H_{6'} - H_7)(1 - m_1 - m_2 - m_3) = (938.4 - 47.1)(0.6277) = 559.5$ Btu (590.3 kJ). Then $W_e - W_p = Q_a - Q_r = 969.0 - 559.5 - 409.5$ Btu (432.0 kJ).

Compare this with $W_e - W_p$ computed earlier, or $415.9 - 7.0 = 408.9$ Btu (431.4 kJ), or a difference of $409.5 - 408.9 = 0.6$ Btu (0.63 kJ). This is an accurate check; the difference of 0.6 Btu (0.63 kJ) comes from errors in Mollier chart and calculator readings. Assume 408.9 Btu (431.4 kJ) is correct because it is the lower of the two values.

6. *Compute the plant and turbine efficiencies*

Plant energy input $= Q_a/e_b$, where e_b = boiler efficiency. Then plant energy input = 969.0/0.88 = 1101.0 Btu (1161.6 kJ). Plant thermal efficiency $= W/(Q_a/e_b) = 383.9 = 1101.0 = 0.3486$. Turbine thermal efficiency $= W_e/Q_a = 415.9/969.0 = 0.4292$. Plant heat rate = 3413/0.3486 = 9970 Btu/kWh (10,329.0 kJ/kWh), where 3413 = Btu/kWh. Turbine heat rate = 3413/0.4292 = 7950 Btu/kWh (8387.7 kJ/kWh). Turbine throttle steam rate = (turbine heat rate)/$(H_2 - H_1)$ = 7950/(1474.5 − 505.5) = 8.21 lb/kWh (3.7 kg/kWh).

Related Calculations. By using the procedures given, the following values can be computed for any actual steam cycle: engine or turbine efficiency e_e; steam enthalpy at the main-condenser inlet; bleed-steam flow to a feedwater heater; turbine or engine work output per section; total turbine or engine work output; station auxiliary power consumption; net station work output; plant energy input; plant thermal efficiency; turbine or engine thermal efficiency; plant heat rate; turbine or engine heat rate; turbine throttle heat rate. To compute any of these values, use the equations given and insert the applicable variables.

ENERGY PERFORMANCE OF REHEAT-STEAM CYCLE

A reheat-steam cycle has a 2000 lb/in^2 (abs) (13,790-kPa) throttle pressure at the turbine inlet and a 400-lb/in^2 (abs) (2758-kPa) reheat pressure. The throttle and reheat temperature of the steam is 1000°F (537.8°C); condenser pressure is 1 inHg absolute (3.4 kPa); engine efficiency of the high-pressure and low-pressure turbines is 80 percent. Find the cycle thermal efficiency.

Calculation Procedure:

1. *Sketch the cycle layout and cycle T-S diagram*

Figures 20 and 21 show the cycle layout and *T-S* diagram with each important point numbered. Use a cycle layout and *T-S* diagram for every calculation of this type because it reduces the possibility of errors.

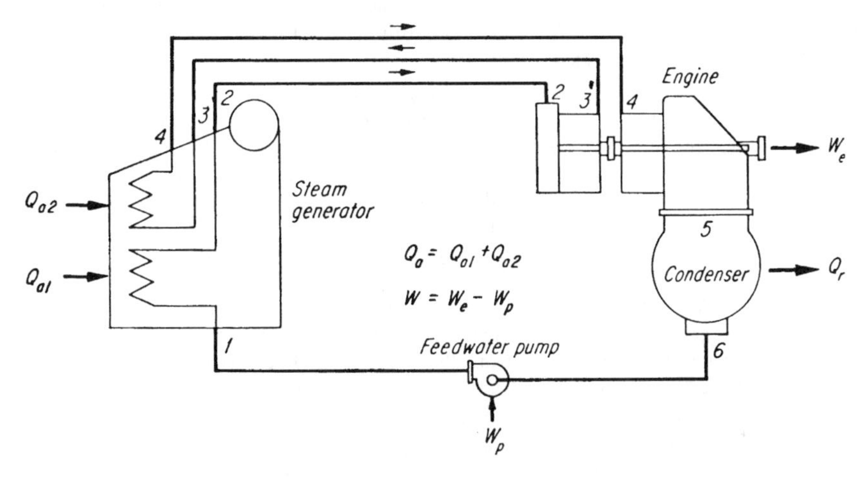

FIGURE 20 Typical steam reheat cycle.

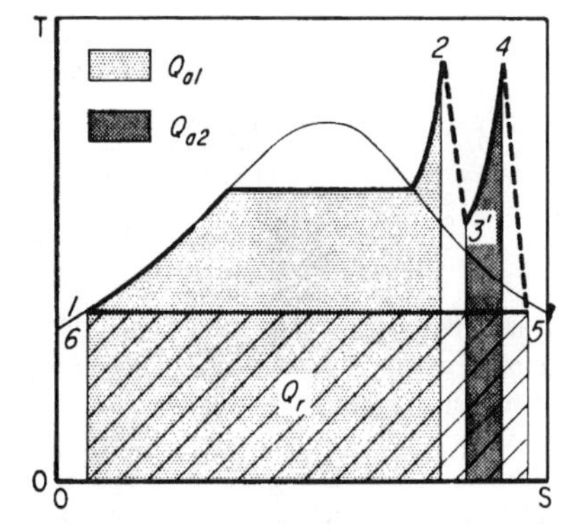

FIGURE 21 Irreversible expansion in reheat cycle.

2. *Determine the throttle-steam properties from the steam tables*
Use the superheated steam tables, entering at 2000 lb/in^2 (abs) (13,790 kPa) and 1000°F (537.8°C) to find throttle-steam properties. Applying the symbols of the *T-S* diagram in Fig. 21, we get H_2 = 1474.5 Btu/lb (3429.7 kJ/kg); S_2 = 1.5603 Btu/(lb · °F) [6.5 kJ/(kg · °C)].

3. *Find the reheat-steam enthalpy*
Assume a constant-entropy expansion of the steam from 2000 to 400 lb/in^2 (13,790 to 2758 kPa). Trace this expansion on a Mollier (*H-S*) chart, where a constant-entropy process is a vertical line between the initial [2000 lb/in^2 (abs) or 13,790 kPa] and reheat [400 lb/in^2 (abs) or 2758 kPa] pressures. Read on the Mollier chart H_3 = 1276.8 Btu/lb (2969.8 kJ/kg) at 400 lb/in^2 (abs) (2758 kPa).

4. *Compute the actual reheat properties*
The ideal enthalpy drop, throttle to reheat = $H_2 - H_3$ = 1474.5 – 1276.8 = 197.7 Btu/lb (459.9 kJ/kg). The actual enthalpy drop = (ideal drop) (turbine efficiency) = $H_2 - H_{3'}$ = 197.5(0.8) = 158.2 Btu/lb (368.0 kJ/kg) = W_{e1} = work output in the high-pressure section of the turbine.

Once W_{e1} is known, $H_{3'}$ can be computed from $H_{3'} = H_2 - W_{e1}$ = 1474.5 – 158.2 = 1316.3 Btu/lb (3061.7 kJ/kg).

The steam now returns to the boiler and leaves at condition 4, where P_4 = 400 lb/in^2 (abs) (2758 kPa); T_4 = 1000°F (537.8°C); S_4 = 1.7623 Btu/(lb · °F) [7.4 kJ/(kg · °C)]; H_4 = 1522.4 Btu/lb (3541.1 kJ/kg) from the superheated-steam table.

5. *Compute the exhaust-steam properties*
Use the Mollier chart and an assumed constant-entropy expansion to 1 inHg (3.4 kPa) absolute to determine the ideal exhaust enthalpy, or H_5 = 947.4 Btu/lb (2203.7 kJ/kg). The ideal work of the low-pressure section of the turbine is then $H_4 - H_{5'}$ = 1522.4 – 947.4 = 575.0 Btu/lb (1338 kJ/kg). The actual work output of the low-pressure section of the turbine is $W_{e2} = H_4 - H_{5'}$ = 575.0(0.8) = 460.8 Btu/lb (1071.1 kJ/kg).

Once W_{e2} is known, $H_{5'}$ can be computed from $H_{5'} = H_4 - W_{e2}$ = 1522.4 – 460.0 = 1062.4 Btu/lb (2471.1 kJ/kg).

The enthalpy of the saturated liquid at the condenser pressure is found in the saturation-pressure steam table at 1 inHg absolute (3.4 kPa) = H_6 = 47.1 Btu/lb (109.5 kJ/kg).

The pump work W_p from the compressed-liquid table diagram in the steam tables is W_p = 5.5 Btu/lb (12.8 kJ/kg). Then the enthalpy of the water entering the boiler $H_1 = H_6 + W_p$ = 47.1 + 5.5 = 52.6 Btu/lb (122.3 kJ/kg).

6. *Compute the cycle thermal efficiency*
For any reheat cycle,

$$e = \text{cycle thermal efficiency}$$
$$= \frac{(H_2 - H_{3'}) + (H_4 - H_{5'}) - W_p}{(H_2 - H_1) + (H_4 - H_{3'})}$$
$$= \frac{(1474.5 - 1316.3) + (1522.4 - 1062.4) - 5.5}{(1474.5 - 52.6) + (1522.4 - 1316.3)}$$
$$= 0.3766, \text{ or } 37.66 \text{ percent}$$

Figure 22 is an energy-flow diagram for the reheat cycle analyzed here. This diagram shows that the fuel burned in the steam generator to produce energy flow Q_{a1} is the largest part of the total energy input. The cold-reheat line carries the major share of energy leaving the high-pressure turbine.

Related Calculations. Reheat-regenerative cycles are used in some large power plants. Figure 23 shows a typical layout for such a cycle having three stages of feedwater heating and one stage of reheating. The heat balance for this cycle is computed as shown above, with the bleed-flow terms *m* computed by setting up an energy balance around each heater, as in earlier calculation procedures.

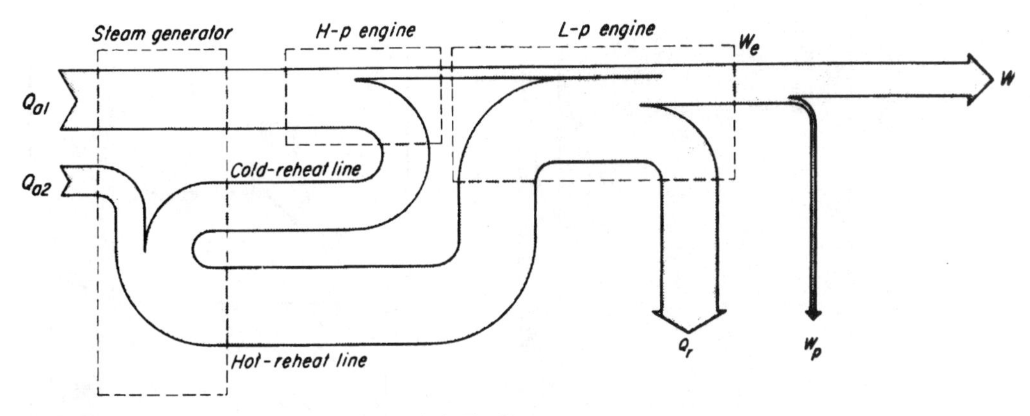

FIGURE 22 Energy-flow diagram for reheat cycle in Fig. 20.

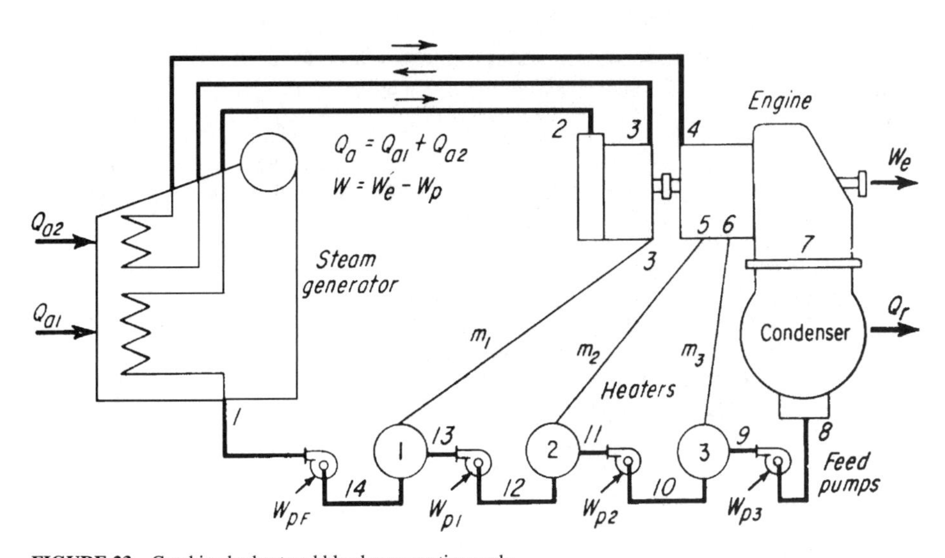

FIGURE 23 Combined reheat and bleed regenerative cycle.

By using a *T-S* diagram, Fig. 24, the cycle thermal efficiency is

$$e = \frac{W}{Q_a} = \frac{Q_a - Q_r}{Q_a} = 1 - \frac{Q_r}{Q_{a1} + Q_{a2}}$$

Based on 1 lb (0.5 kg) of working fluid entering the steam generator and turbine throttle,

$$Q_r = (1 - m_1 - m_2 - m_3)(H_7 - H_8)$$

$$Q_{a1} = (H_2 - H_1)$$

$$Q_{a2} = (1 - m_1)(H_4 - H_3)$$

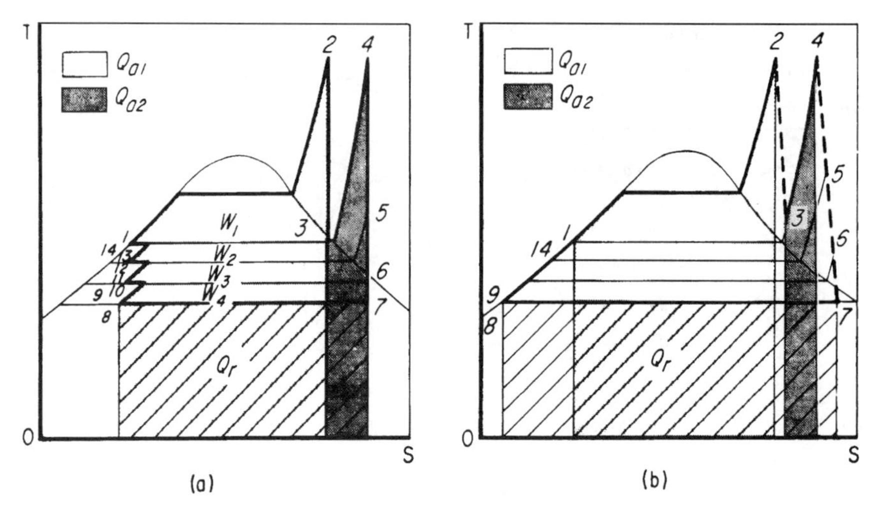

FIGURE 24 (*a*) *T-S* diagram for ideal reheat-regenerative-bleed cycle; (*b*) *T-S* diagram for actual cycle.

Figure 25 shows the energy-flow chart for this cycle.

Some high-pressure plants use two stages of reheating, Fig. 26, to raise the cycle efficiency. With two stages of reheating, the maximum number generally used, and values from Fig. 26.

$$e = \frac{(H_2 - H_3) + (H_4 - H_5) + (H_6 - H_7) - W_p}{(H_2 - H_1) + (H_4 - H_3) + (H_6 - H_5)}$$

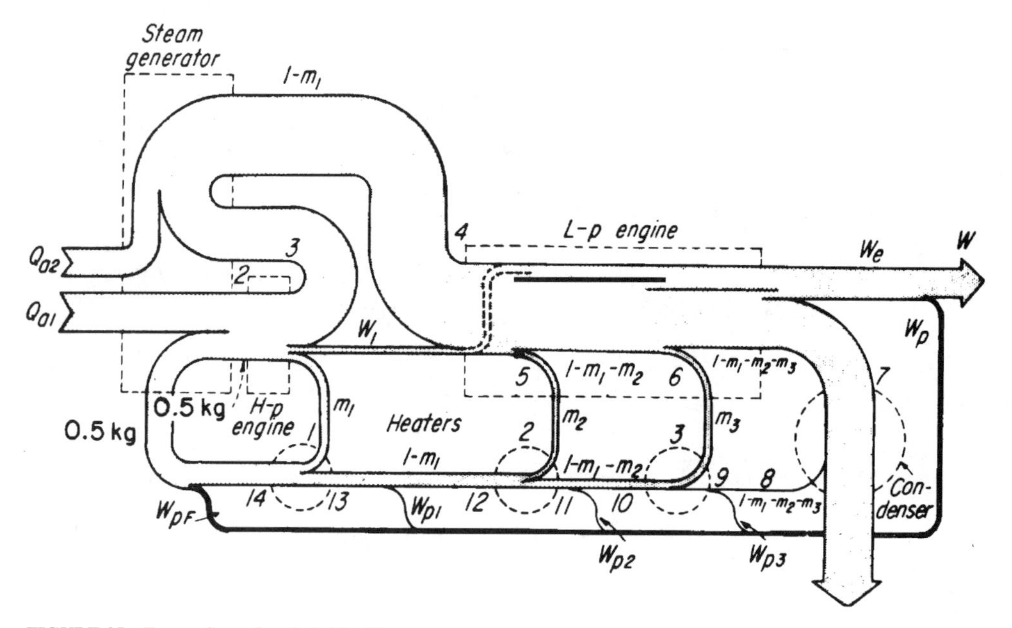

FIGURE 25 Energy flow of cycle in Fig. 23.

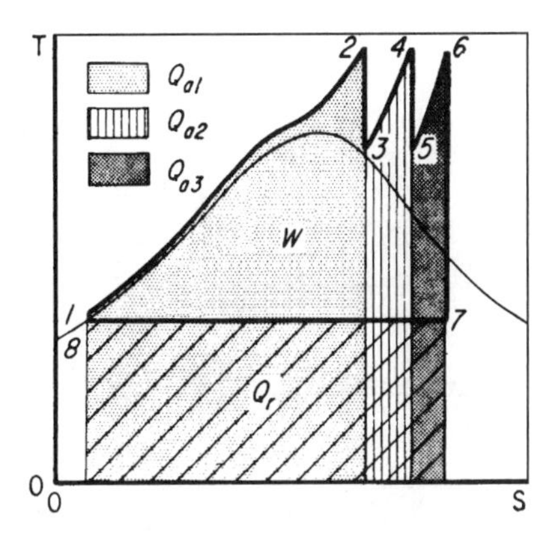

FIGURE 26 *T-S* diagram for multiple reheat stages.

POWER ENERGY OUTPUT ANALYSIS OF MECHANICAL-DRIVE STEAM TURBINE

Show the effect of turbine engine efficiency on the condition lines of a turbine having engine efficiencies of 100 (isentropic expansion), 75, 50, 25, and 0 percent. How much of the available energy is converted to useful work for each engine efficiency? Sketch the effect of different steam inlet pressures on the condition line of a single-nozzle turbine at various loads. What is the available energy, Btu/lb of steam, in a noncondensing steam turbine having an inlet pressure of 1000 lb/in^2 (abs) (6895 kPa) and an exhaust pressure of 100 lb/in^2 (gage) (689.5 kPa)? How much work will this turbine perform if the steam flow rate to it is 1000 lb/s (453.6 kg/s) and the engine efficiency is 40 percent?

Calculation Procedure:

1. ***Sketch the condition lines on the Mollier chart***

Draw on the Mollier chart for steam initial- and exhaust-pressure lines, Fig. 27, and the initial-temperature line. For an isentropic expansion, the entropy is constant during the expansion, and the engine efficiency = 100 percent. The expansion or condition line is a vertical trace from h_1 on the initial-pressure line to h_2, on the exhaust-pressure line. Draw this line as shown in Fig. 27.

For 0 percent engine efficiency, the other extreme in the efficiency range, $h_1 = h_2$ and the condition line is a horizontal line, Draw this line as shown in Fig. 27.

Between 0 and 100 percent efficiency, the condition lines become more nearly vertical as the engine efficiency approaches 100 percent, or an isentropic expansion. Draw the condition lines for 25, 50, and 75 percent efficiency, as shown in Fig. 27.

For the isentropic expansion, the available energy = $h_1 - h_{2x}$, Btu/lb of steam. This is the energy that an ideal turbine would make available.

For actual turbines, the enthalpy at the exhaust pressure $h_2 = h_1$ – (available energy) (engine efficiency)/100, where available energy = $h_1 - h_{2x}$ for an ideal turbine working between the same initial and exhaust pressures. Thus, the available energy converted to useful work for any engine efficiency = (ideal available energy, Btu/lb)(engine efficiency, percent)/100. Using this relation, the available energy at each of the given engine efficiencies is found by substituting the ideal available energy and the actual engine efficiency.

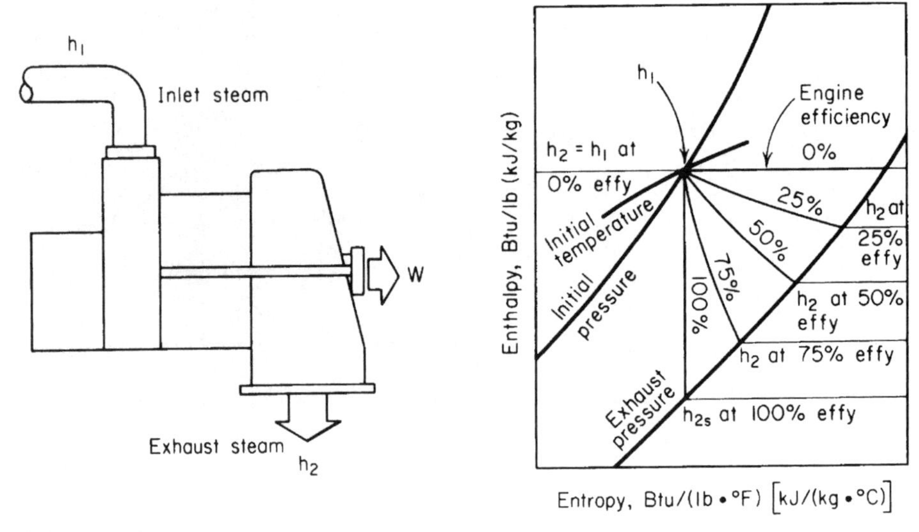

FIGURE 27 Mollier chart of turbine condition lines.

2. ***Sketch the condition lines for various throttle pressures***

Draw the throttle- and exhaust-pressure lines on the Mollier chart, Fig. 28. Since the inlet control valve throttles the steam flow as the load on the turbine decreases, the pressure of the steam entering the turbine nozzle is lower at reduced loads. Show this throttling effect by indicating the lower inlet pressure lines, Fig. 28, for the reduced loads. Note that the lowest inlet pressure occurs at the minimum plotted load—25 percent of full load—and the maximum inlet pressure at 125 percent of full load. As the turbine inlet steam pressure decreases, so does the available energy, because the exhaust enthalpy rises with decreasing load.

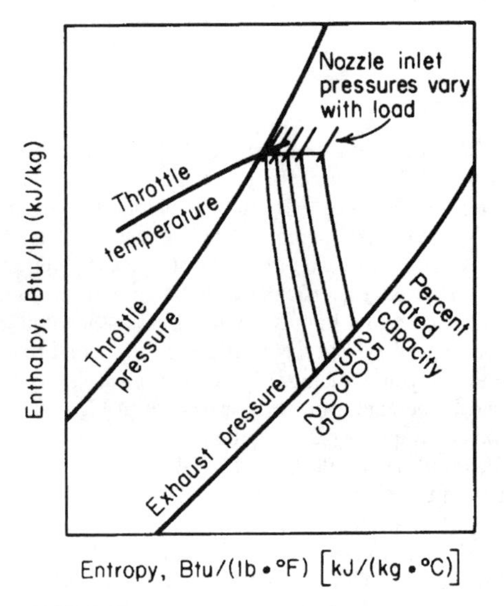

FIGURE 28 Turbine condition line shifts as the inlet steam pressure varies.

3. *Compute the turbine available energy and power output*
Use a noncondensing-turbine performance chart, Fig. 29, to determine the available energy. Enter the bottom of the chart at 1000 lb/in^2 (abs) (6895 kPa) and project vertically upward until the 100-lb/in^2 (gage) (689.5-kPa) exhaust-pressure curve is intersected. At the left, read the available energy as 205 Btu/lb (476.8 kJ/kg) of steam.

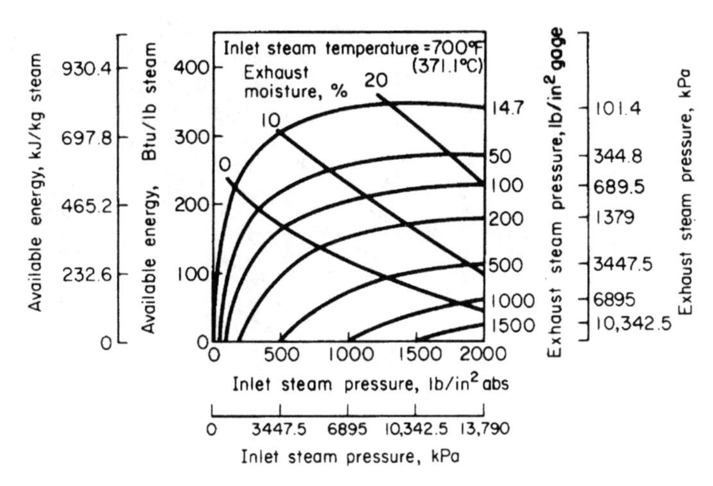

FIGURE 29 Available energy in turbine depends on the initial steam state and the exhaust pressure.

With the available energy, flow rate, and engine efficiency known, the work output = (available energy, Btu/lb)(flow rate, lb/s)(engine efficiency/100)/[550 ft · lb/(s · hp)]. [*Note:* 550 ft · lb/(s · hp) = 1 N · m/(W · s).] For this turbine, work output = (205 Btu/lb) (1000 lb/s) (40/100)/550 = 149 hp (111.1 kW).

Related Calculations. Use the steps given here to analyze single-stage noncondensing mechanical-drive turbines for stationary, portable, or marine applications. Performance curves such as Fig. 29 are available from turbine manufacturers. Single-stage noncondensing turbines are for feed-pump, draft-fan, and auxiliary-generator drive.

ENERGY OUTPUT ANALYSIS OF CONDENSING STEAM TURBINE

What is the available energy in steam supplied to a 5000-kW turbine if the inlet steam conditions are 1000 lb/in^2 (abs) (6895 kPa) and 800°F (426.7°C) and the turbine exhausts at 1 inHg absolute (3.4 kPa)? Determine the theoretical and actual heat rate of this turbine if its engine efficiency is 74 percent. What are the full-load output and steam rate of the turbine?

Calculated Procedure:

1. *Determine the available energy in the steam*
Enter Fig. 30 at the bottom at 1000-lb/in^2 (abs) (6895.0-kPa) inlet pressure, and project vertically upward to the 800°F (426.7°C) 1-in (3.4-kPa) exhaust-pressure curve. At the left, read the available energy as 545 Btu/lb (1267.7 kJ/kg) of steam.

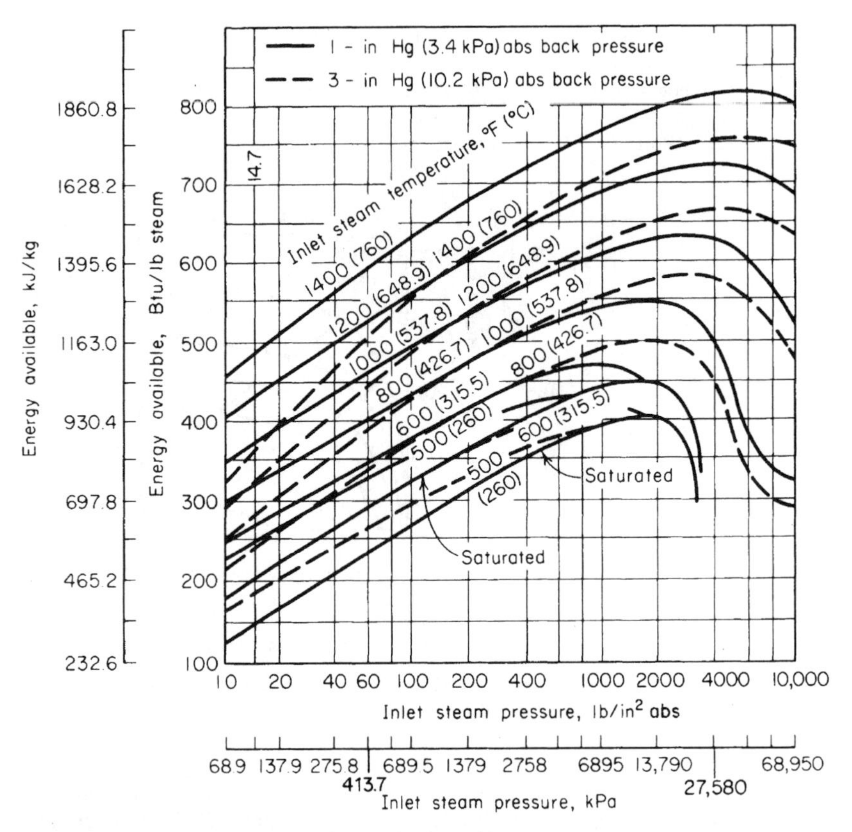

FIGURE 30 Available energy for typical condensing turbines.

2. ***Determine the heat rate of the turbine***

Enter Fig. 31 at an initial steam temperature of 800°F (426.7°C), and project vertically upward to the 1000-lb/in^2 (abs) (6895.0-kPa) 1-in (3.4-kPa) curve. At the left, read the theoretical heat rate as 8400 Btu/kWh (8862.5 kJ/kWh).

When the theoretical heat rate is known, the actual heat rate is found from: actual heat rate HR, Btu/kWh = (theoretical heat rate, Btu/kWh)/(engine efficiency). Or, actual HR = 8400/0.74 = 11,350 Btu/kWh (11,974.9 kJ/kWh).

3. ***Compute the full-load and steam rate***

The energy converted to work, Btu/lb of steam = (available energy, Btu/lb of steam) (engine efficiency) = (545) (0.74) = 403 Btu/lb of steam (937.4 kJ/kg).

For any prime mover driving a generator, the full-load output, Btu = (generator kW rating) (3413 Btu/kWh) = (5000)(3413) = 17,060,000 Btu/h (4999.8 kJ/s).

The steam flow = (full-load output, Btu/h)/(work output, Btu/lb) = 17,060,000/403 = 42,300 lb/h (19,035 kg/h) of steam. Then the full-load steam rate of the turbine, lb/kWh = (steam flow, lb/h)/(kW output at full load) = 42,300/5000 = 8.46 lb/kWh (3.8 kg/kWh).

Related Calculations. Use this general procedure to determine the available energy, theoretical and actual heat rates, and full-load output and steam rate for any stationary, marine, or portable condensing steam turbine operating within the ranges of Figs. 30 and 31. If the actual performance curves are available, use them instead of Figs. 30 and 31. The curves given here are suitable for all preliminary estimates for condensing turbines operating with exhaust pressures of 1 or 3 inHg absolute (3.4 or 10.2 kPa). Many modern turbines operate under these conditions.

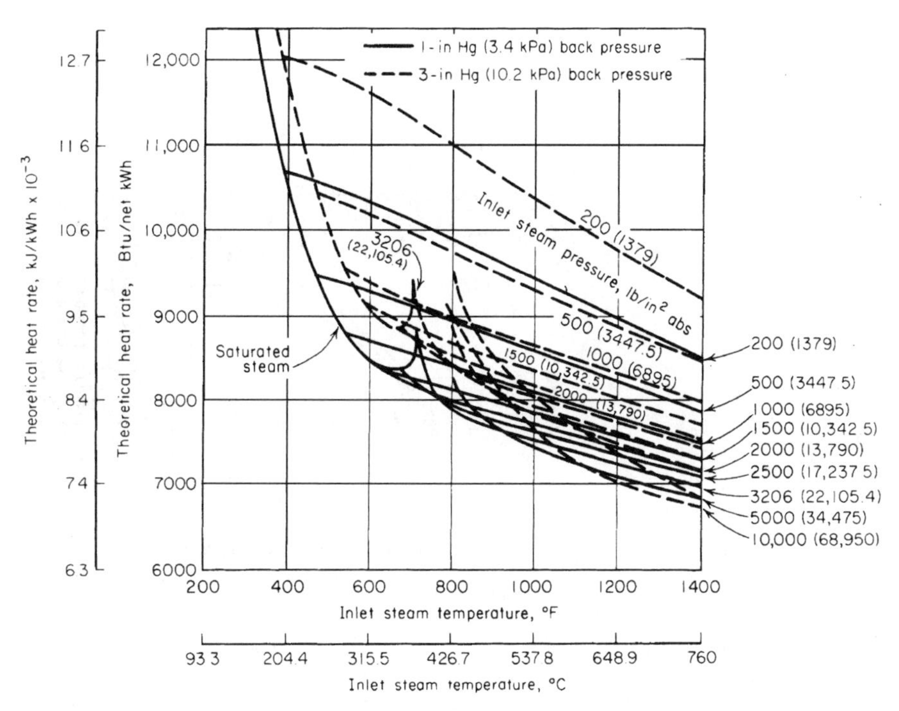

FIGURE 31 Theoretical heat rate for condensing turbines.

STEAM-TURBINE REGENERATIVE-CYCLE PERFORMANCE

When throttle steam is at 1000 lb/in^2 (abs) (6895 kPa) and 800°F (426.7°C) and the exhaust pressure is 1 inHg (3.4 kPa) absolute, a 5000-kW condensing turbine has an actual heat rate of 11,350 Btu/kWh (11,974.9 kJ/kWh). Three feedwater heaters are added to the cycle, Fig. 32 to heat the feedwater to 70 percent of the maximum possible enthalpy rise. What is the actual heat rate of the turbine? If 10 heaters instead of 3 were used and the water enthalpy were raised to 90 percent of the maximum possible rise in these 10 heaters, would the reduction in the actual heat rate be appreciable?

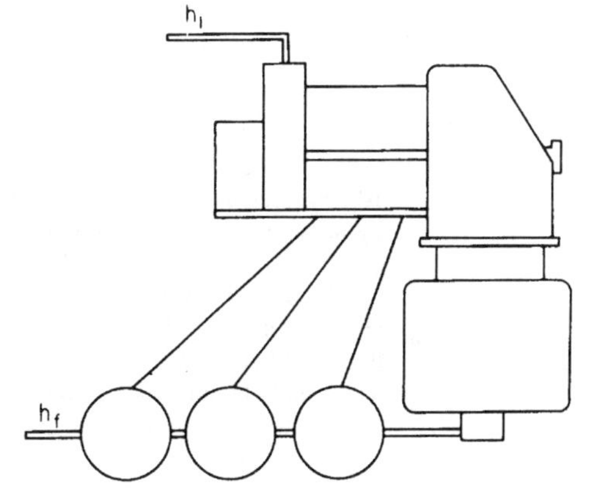

FIGURE 32 Regenerative feedwater heating.

Calculation Procedure:

1. *Determine the actual enthalpy rise of the feedwater*

Enter Fig. 33 at the throttle pressure of 1000 lb/in^2 (abs) (6895 kPa), and project vertically upward to the 1-inHg (3.4-kPa) absolute back-pressure curve. At the left, read the maximum possible feedwater enthalpy rise as 495 Btu/lb (1151.4 kJ/kg). Since the actual rise is limited to 70 percent of the maximum possible rise by the conditions of the design, the actual enthalpy rise = (495) (0.70) = 346.5 Btu/lb (805.9 kJ/kg).

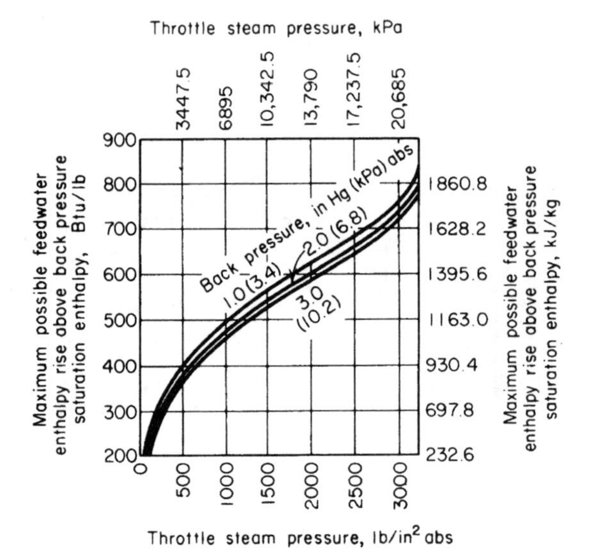

FIGURE 33 Feedwater enthalpy rise.

2. ***Determine the heat-rate and heater-number correction factors***
Find the theoretical reduction in straight-condensing (no regenerative heaters) heat rates from Fig. 34. Enter the bottom of Fig. 34 at the inlet steam temperature, 800°F (426.7°C), and project vertically upward to the 1000-lb/in^2 (abs) (6895-kPa) 1-inHg (3.4-kPa) backpressure curve. At the left, read the reduction in straight-condensing heat rate as 14.8 percent.

Next, enter Fig. 34 at the bottom of 70 percent of maximum possible rise in feedwater enthalpy, and project vertically to the three-heater curve. At the left, read the reduction in straight-condensing heat rate for the number of heaters and actual enthalpy rise as 0.71.

3. ***Apply the heat-rate and heater-number correction factors***
Full-load regenerative-cycle heat rate, Btu/kWh = (straight-condensing heat rate, Btu/kWh) [1 − (heat-rate correction factor) (heater-number correction factor)] = (13,350)[1 − (0.148)(0.71)] = 10,160 Btu/kWh (10,719.4 kJ/kWh).

4. ***Find and apply the correction factors for the larger number of heaters***
Enter Fig. 35 at 90 percent of the maximum possible enthalpy rise, and project vertically to the 10-heater curve. At the left, read the heat-rate reduction for the number of heaters and actual enthalpy rise as 0.89.

Using the heat-rate correction factor from step 2 and 0.89, found above, we see that the full-load 10-heater regenerative-cycle heat rate = (11.350) [1 − (0.148) (0.89)] = 9850 Btu/kWh (10,392.3 kJ/kWh), by using the same procedure as in step 3. Thus, adding 10 − 3 = 7 heaters reduces the heat rate by 10,160 − 9850 = 310 Btu/kWh (327.1 kJ/kWh). This is a reduction of 3.05 percent.

To determine whether this reduction in heat rate is appreciable, the carrying charges on the extra heaters, piping, and pumps must be compared with the reduction in annual fuel costs resulting from the lower heat rate. If the fuel saving is greater than the carrying charges, the larger number of heaters can usually be justified. In this case, tripling the number of heaters would probably increase the carrying charges to a level exceeding the fuel savings. Therefore, the reduction in heat rate is probably not appreciable.

Related Calculations. Use the procedure given here to compute the actual heat rate of steam-turbine regenerative cycles for stationary, marine, and portable installations. Where necessary, use the steps of the previous procedure to compute the actual heat rate of a straight-condensing cycle before applying the present procedure. The performance curves given here are suitable for first approximations in situations where actual performance curves are unavailable.

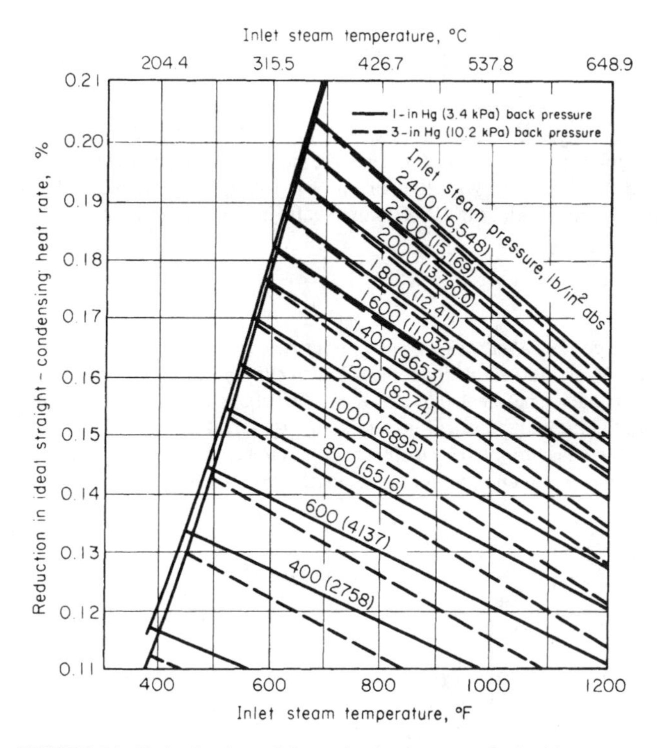

FIGURE 34 Reduction in straight-condensing heat rate obtained by regenerative heating.

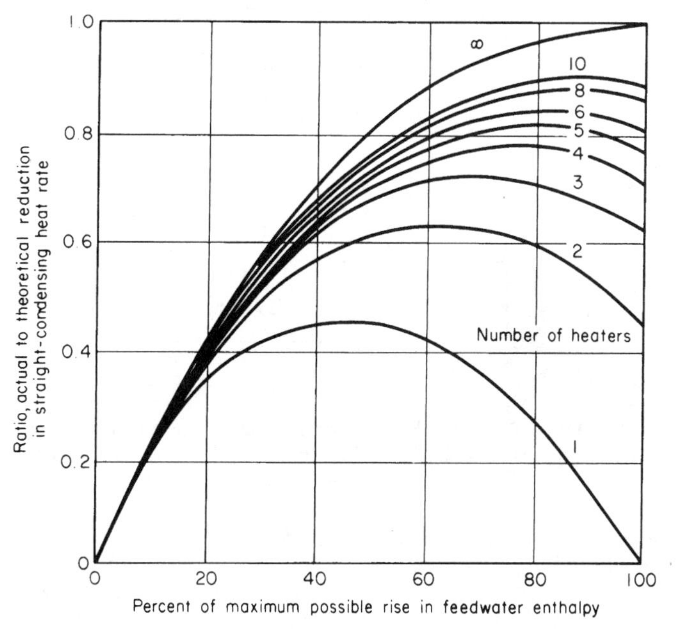

FIGURE 35 Maximum possible rise in feedwater enthalpy varies with the number of heaters used.

REHEAT-REGENERATIVE STEAM-TURBINE HEAT RATES

What are the net and gross heat rates of a 300-kW reheat turbine having an initial steam pressure of 3500 lb/in^2 (gage) (24,132.5 kPa) with initial and reheat steam temperatures of 1000°F (537.8°C) with 1.5 inHg (5.1 kPa) absolute backpressure and six stages of regenerative feedwater heating? Compare this heat rate with that of 3500 lb/in^2 (gage) (24,132.5 kPa) 600-mW cross-compound four-flow turbine with 3600/1800 r/min shafts at a 300-mW load.

Calculation Procedure:

1. *Determine the reheat-regenerative heat rate*

Enter Fig. 36 at 3500-lb/in^2 (gage) (24,132.5-kPa) initial steam pressure, and project vertically to the 300-mW capacity net-heat-rate curve. At the left, read the net heat rate as 7680 Btu/kWh (8102.6 kJ/kWh). On the same vertical line, read the gross heat rate as 7350 Btu/kWh (7754.7 kJ/kWh). The gross heat rate is computed by using the generator-terminal output; the net heat rate is computed after the feedwater-pump energy input is deducted from the generator output.

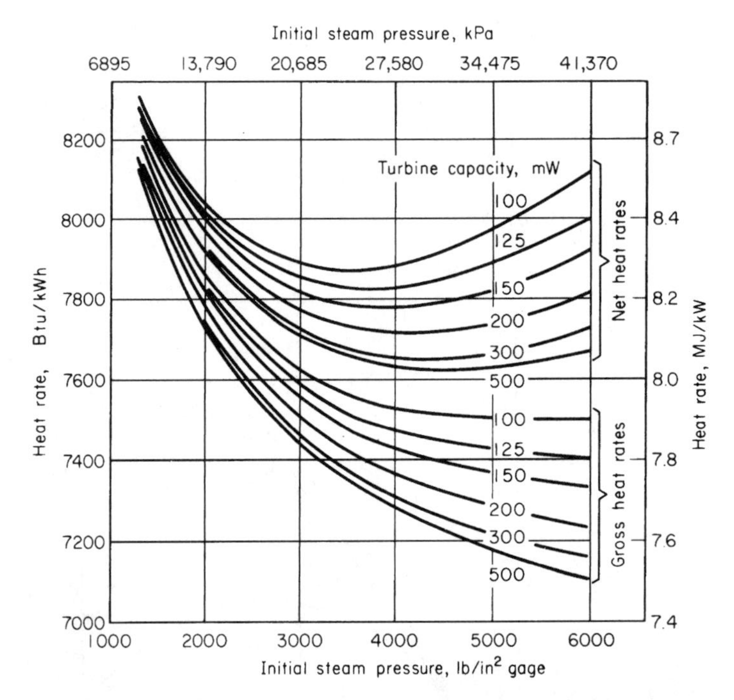

FIGURE 36 Full-load heat rates for steam turbines with six feedwater heaters, 1000°F/1000°F (538°C/538°C) steam, 1.5-in (38.1-mm) Hg (abs) exhaust pressure.

2. *Determine the cross-compound turbine heat rate*

"Enter Fig. 37 at 350 mW at the bottom, and project vertically upward to 1.5-inHg (5.1-kPa) exhaust pressure midway between the 1- and 2-inHg (3.4- and 6.8-kPa) curves. At the left, read the net heat rate as 7880 Btu/kWh (8313.8 kJ/kWh). Thus, the reheat-regenerative unit has a lower net heat rate. Even at full rated load of the cross-compound turbine, its heat rate is higher than the reheat unit.

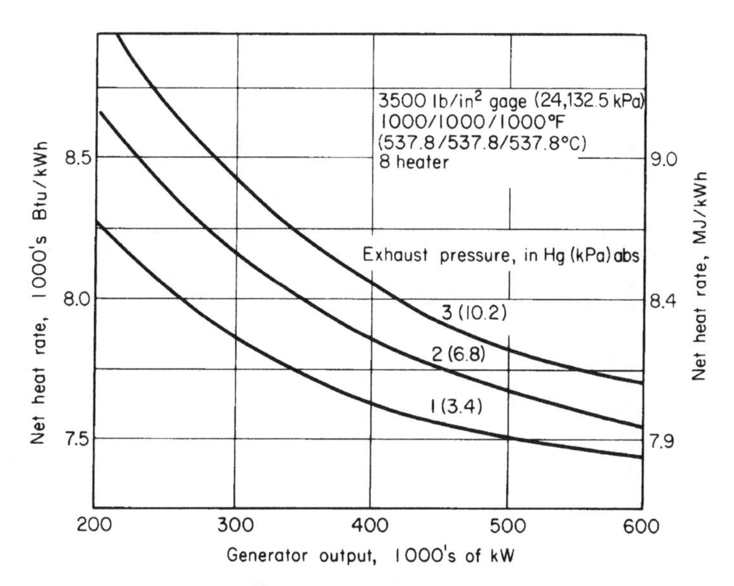

FIGURE 37 Heat rate of a cross-compound four-flow steam turbine with 3600/1800-r/min shafts.

Related Calculations. Use this general procedure for comparing stationary and marine high-pressure steam turbines. The curves given here are typical of those supplied by turbine manufacturers for their turbines.

With the price of all commonly used fuels on the rise, nuclear power is increasingly being looked at by engineers worldwide. A number of nations—the United States, Japan, France, England, and Germany—are currently generating portions of their electricity needs by using nuclear energy. Today's nuclear plants use fission to generate the steam for their turbines. Fusion is now being seriously investigated as the next source of nuclear power. In June 2005, France was selected as the country in which an experimental fusion reactor would be built. Test results will be provided to participating countries. Fusion reactors produce much less nuclear waste than fission-based reactors.

According to an article in *The New York Times,** using data from Princeton Plasma Physics Laboratory and The Energy Information Administration, the daily waste from a 1000-MW electric-generating plant using 9000 tons of coal per day is 30,000 tons of CO_2, 600 tons of SO_2, and 80 tons of NO. A nuclear fission plant using 14.7 lb (6.67 kg) of uranium produces 6.6 lb (2.99 kg) of highly radioactive material. The projected nuclear fusion plant using 1 lb (0.454 kg) of deuterium and 1.5 lb (0.68 kg) of tritium produces 4.0 lb (1.8 kg) of helium.

A large demonstration project is expected to begin operating in approximately 2030. The commercial fusion reactor is expected in approximately 2050. Thus, while the fusion reactor has great promise, its commercial utilization is expected to take many years to develop. At this writing, the consensus is that both fission and fusion nuclear power will find use during the 21st century.

kW OUTPUT OF EXTRACTION STEAM TURBINE

An automatic extraction turbine operates with steam at 400 lb/in^2 absolute (2760 kPa), 700°F (371°C) at the throttle; its extraction pressure is 200 lb/in^2 (1380 kPa) and it exhausts at 110 lb/in^2 absolute (760 kPa). At full load 80,000 lb/h (600 kg/s) is supplied to the throttle and 20,000 lb/h (150 kg/s) is extracted at the bleed point. What is the kW output?

**The New York Times,* June 29, 2005.

Calculation Procedure:

1. *Determine steam conditions at the throttle, bleed point, and exhaust*
Steam flow through the turbine is indicated by "enter" at the throttle, "extract" at the bleed point, and "exit" at the exhaust, as shown in Fig. 38*a*. The steam process is considered to be at constant entropy, as shown by the vertical isentropic line in Fig. 38*b*. At the throttle, where the steam enters at the given pressure, $p_1 = 400$ lb/in^2 absolute (2760 kPa) and temperature, $t_1 = 700°F$ (371°C), steam enthalpy, $h_1 = 1362.7$ Btu/lb (3167.6 kJ/kg) and its entropy, $s_1 = 1.6398$, as indicated by Table 3, Vapor of the Steam Tables mentioned under Related Calculations of this procedure. From the Mollier chart, a supplement to the Steam Tables, the following conditions are found along the vertical isentropic line where $s_1 = s_x = s_2 = 1.6398$ Btu/(lb · °F) (6.8655 kJ/kg · °C):

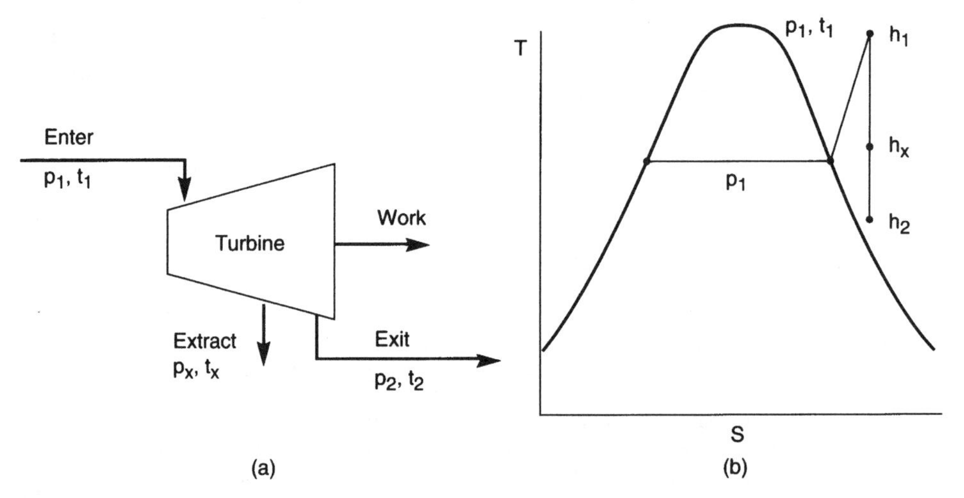

FIGURE 38 (*a*) Turbine cream flow diagram. (*b*) Temperature-entropy schematic for steam flow.

At the bleed point, where the given extraction pressure, $p_x = 200$ lb/in^2 (1380 kPa) and the entropy, s_x, is as mentioned above, the enthalpy, $h_x = 1284$ Btu/lb (2986 kJ/kg) and the temperature, $t_x = 528°F$ (276°C). At the exit, where the given exhaust pressure, $p_2 = 110$ lb/in^2 (760 kPa) and the entropy, s_2, is as mentioned above, the enthalpy, $h_2 = 1225$ Btu/lb (2849 kJ/kg) and the temperature, $t_2 = 400°F$ (204°C).

2. *Compute the total available energy to the turbine*
Between the throttle and the bleed point the available energy to the turbine, $AE_1 = Q_1(h_1 - h_x)$, where the full-load rate of steam flow, $Q_1 = 80{,}000$ lb/h (600 kg/s); other values are as before. Hence, $AE_1 = 80{,}000 \times (1362.7 - 1284) = 6.296 \times 10^6$ Btu/h (1845 kJ/s). Between the bleed point and the exhaust the available energy to the turbine, $AE_2 = (Q_1 - Q_2)(h_x - h_2)$, where the extraction flow rate, $Q_x = 20{,}000$ lb_m/h (150 kg/s); other values as before. Then, $AE_2 = (80{,}000 - 20{,}000)(1284 - 1225) = 3.54 \times 10^6$ Btu/h (1037 kJ/s). Total available energy to the turbine, $AE = AE_1 + AE_2 = 6.296 \times 10^6 + (3.54 \times 10^6) = 9.836 \times 10^6$ Btu/h (172.8×10^3 kJ/s).

3. *Compute the turbine's kW output*
The power available to the turbine to develop power at the shaft, in kilowatts, $kW = AE/(\text{Btu/kW} \cdot \text{h}) = 9.836 \times 10^6/3412.7 = 2880$ kW. However, the actual power developed at the shaft, $kW_a = kW \times e$, where e is the mechanical efficiency of the turbine. Thus, for an efficiency, $e = 0.90$, then $kW_a = 2880 \times 0.90 = 2590$ kW (2590 kJ/s).

Related Calculations. The Steam Tables appear in *Thermodynamic Properties of Water Including Vapor, Liquid, and Solid Phases,* 1969, Keenan, et al., John Wiley & Sons, Inc. Use later versions of such tables whenever available, as necessary.

SECTION 3

GAS-TURBINE ENERGY POWER GENERATION

Gas-Turbine Energy Parameters 3.1
Gas Turbine–Steam Turbine Cycle Analysis 3.2
Gas Turbine Combustion Chamber Inlet Air Temperature 3.5
Analysis of Regenerative-Gas-Turbine Cycle 3.7
Combined-Cycle Plant Heat-Rate Comparison 3.10
Selecting Most Efficient Method for Increasing Combined-Cycle Plant Output 3.11
Gas-Turbine Heat-Recovery Steam-Generator (HRSG) Choice 3.17
Analyzing Gas-Turbine Cycle Efficiency and Output 3.19
Industrial Gas Turbine Life-Cycle Cost-Model Best Relative-Value Determination 3.23
HRSG Tube-Bundle Vibration and Noise Analysis 26
Gas-Turbine-Plant Oxygen and Fuel Input Determination 3.30
Simulation of Heat-Recovery Steam Generators (HRSG) 3.32
Calculating Heat-Recovery Steam Generator (HRSG) Temperature Profiles 3.36

GAS-TURBINE ENERGY PARAMETERS

The gas turbine "came on the scene" at one of the most opportune times in engineering history. Just as the world began to worry about greenhouse gases (GHGs), global warming, and the air pollution coal-fired plants caused, the aero-derivative gas turbine showed up. The advantages of this new prime mover were many:

- It had already been proven in aircraft service to be reliable, relatively low cost, efficient, lightweight, and compact. Further, gas turbines can be started rapidly (as compared to hours for a steam plant), can be operated remotely, can be started and stopped unattended, and easily handle wide-load swings. Maintenance is simple; most gas turbines operate for long periods between overhauls. Coolants are usually not required for a gas turbine.
- The gas turbine's footprint for land service is small—few utilities wanting to install gas turbines for topping, standby, or combined cycles even have to expand existing structures, let alone put up a new building.
- Permit requirements are simple compared to getting a new coal-fired station approved.
- Heat-recovery steam generators (HRSG) hooked to the exhaust of a gas turbine provide additional steam in a cogeneration mode while recovering heat that would otherwise be wasted. The additional steam can be used to generate electric power in a steam turbine, or it can be used for space heating, industrial processes, district heating, and so on.
- Natural gas, a popular fuel for gas turbines, has concurrently become more readily available, and at lower prices. Further, additional sources of natural gas are being found using a new technique called "fracking." Light distillate oils are also used for gas turbines. Heavy residual oil can be used as well, but its corrosive nature must be allowed for in the manufacture of the combustor and turbine.

- When introduced into steam plants in a combined cycle, the gas turbine raises the overall thermal efficiency of the plant to a new level, without increasing the plant's GHG output by much. At their introduction, gas turbines were planned to handle 1000 to 2000 peaking hours per year. Since then, their successful performance has far exceeded these plans. Today, gas turbines are widely used in base-load service in combined-cycle plants. And the capacity of such combined-cycle plants can exceed 500 MW, a far cry from the 75 MW capacity when first introduced.
- Personnel requirements for operating the new gas turbines are minimal; most plants did not need to add operators to their existing staff.
- Manufacturers, worldwide, have added gas turbines to their list of products. And aero-gas-turbine manufacturers quickly increased the size and capacity of their product line to meet the needs of electric utilities for bigger gas turbines.
- Heat rates of 9600 to 10,400 kJ/kWh are reported for free-turbine-drive gas turbines operating at variable speed. Overall combined-cycle plant efficiency exceeding 60 percent has been reported by some plants.
- When gas turbines were first introduced in combined-cycle electric-generating plants, the ratio of the output of the steam turbine to the gas turbine was 8:1. Today, gas turbines are filling a much larger output role in combined-cycle plants worldwide.

The result of all this activity is that high-efficiency combined-cycle plants are now the norm for most electric utilities, and the gas turbine has become as popular as the steam turbine once was. Today's gas turbines show overall thermal efficiencies exceeding 30 percent. Such performance is highly desirable in times of increasing fuel costs and uncertainties in the dependability of certain fuel supplies.

GAS TURBINE–STEAM TURBINE CYCLE ANALYSIS

Sketch the cycle layout, *T-S* diagram, and energy-flow chart for a combined steam turbine–gas turbine cycle having one stage of regenerative feedwater heating and one stage of economizer feedwater heating. Compute the thermal efficiency and heat rate of the combined cycle.

Calculation Procedure:

1. *Sketch the cycle layout*
Figure 1 shows the cycle. Since the gas-turbine exhaust-gas temperature is usually higher than the bleed-steam temperature, the economizer is placed after the regenerative feedwater heater. The feedwater will be progressively heated to a higher temperature during passage through the regenerative heater and the gas-turbine economizer. The cycle shown here is only one of many possible combinations of a steam plant and a gas turbine.

2. *Sketch the T-S diagram*
Figure 2 shows the *T-S* diagram for the combined gas turbine–steam turbine cycle. There is irreversible heat transfer Q_T from the gas-turbine exhaust to the feedwater in the economizer, which helps reduce the required energy input Q_{a2}.

3. *Sketch the energy-flow chart*
Choose a suitable scale for the energy input, and proportion the energy flow to each of the other portions of the cycle. Use a single line when the flow is too small to plot to scale. Figure 3 shows the energy-flow chart.

4. *Determine the thermal efficiency of the cycle*
Since $e = W/Q_a$, $e = Q_a - Q_r/Q_a = 1 - [Q_{r1} + Q_{r2}/(Q_{a1} + Q_{a2})]$, given the notation in Figs. 1 through 3.

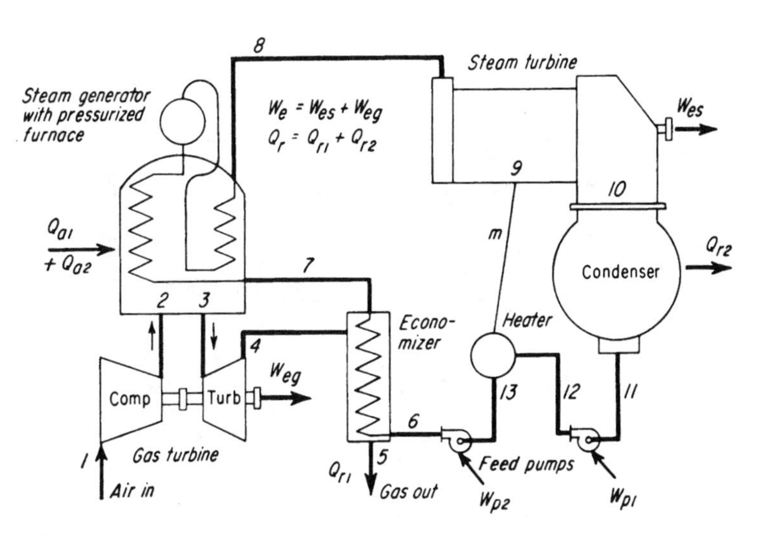

FIGURE 1 Combined gas turbine–steam turbine cycle.

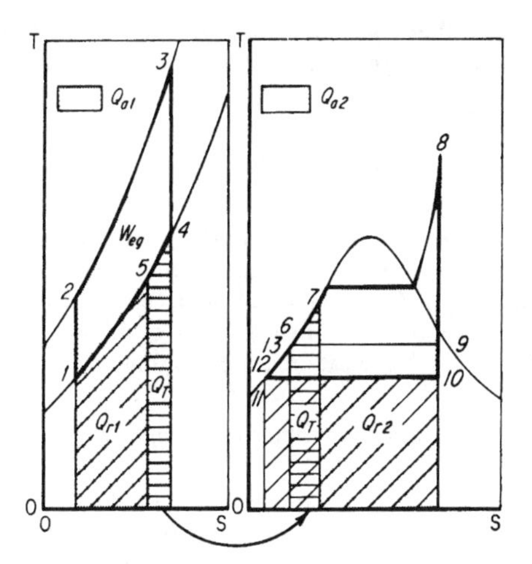

FIGURE 2 *T-S* charts for combined gas turbine–steam turbine cycle have irreversible heat transfer Q_7 from gas-turbine exhaust to the feedwater.

The relative weight of the gas w_g to 1 lb (0.5 kg) of water must be computed by taking an energy balance about the economizer. Or, $H_7 - H_6 = w_g(H_4 - H_5)$. Using the actual values for the enthalpies, solve this equation for w_g.

With w_g known, the other factors in the efficiency computation are

$$Q_{r1} = w_g(H_5 - H_1)$$

$$Q_{r2} = (1 - m)(H_{10} - H_{11})$$

$$Q_{a1} = w_g(H_3 - H_2)$$

$$Q_{a2} = H_8 - H_7$$

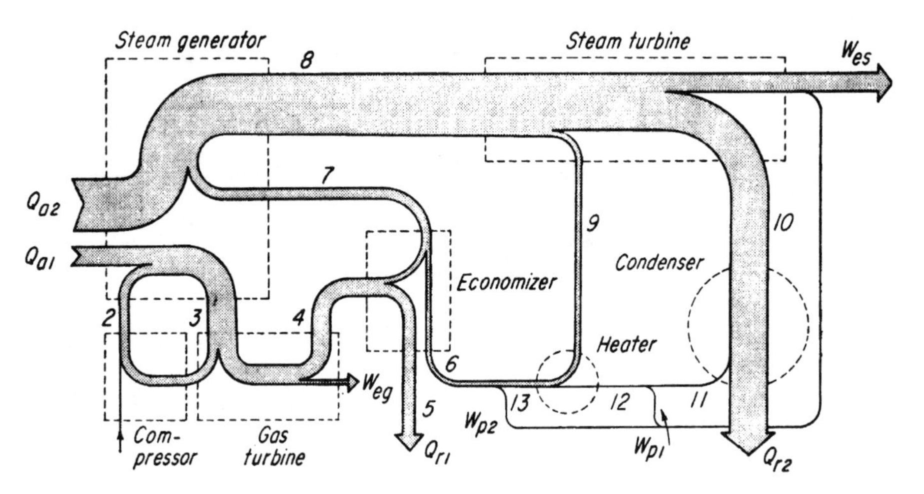

FIGURE 3 Energy-flow chart of the gas turbine–steam turbine cycle in Fig. 1.

The bleed-steam flow m is calculated from an energy balance about the feedwater heater. Note that the units for the above equations can be any of those normally used in steam- and gas-turbine analyses.

Calculation Procedure:

1. ***Find the amount of oxygen required for complete combustion of the fuel***

Eight atoms of carbon in C_8 combine with 8 molecules of oxygen, O_2, and produce 8 molecules of carbon dioxide, $8CO_2$. Similarly, 9 molecules of hydrogen, H_2, in H_{18} combine with 9 atoms of oxygen, O, or 4.5 molecules of oxygen, to form 9 molecules of water, $9H_2O$. Thus, 100 percent, or the stoichiometric, air quantity required for complete combustion of a mole of fuel, C_8H_{18}, is proportional to 8 + 12.5 moles of oxygen, O_2.

2. ***Establish the chemical equation for complete combustion with 100 percent air***

With 100 percent air: $C_8H_{18} + 12.5O_2 + (3.784 \times 12.5)N_2 \rightarrow 8CO_2 + 9H_2O + 47.3N_2$, where 3.784 is a derived volumetric ratio of atmospheric nitrogen, (N_2), to oxygen, O_2, in dry air. The (N_2) includes small amounts of inert and inactive gases. See Related Calculations of this procedure.

3. ***Establish the chemical equation for complete combustion with 400 percent of the stoichiometric air quantity, or 300 percent excess air***

With 400 percent air: $C_8H_{18} + 50O_2 + (4 \times 47.3)N_2 \rightarrow 8CO_2 + 9H_2O + 189.2N_2 + (3 \times 12.5)O_2$.

4. ***Compute the molecular weights of the components in the combustion process***

Molecular weight of $C_8H_{18} = [(12 \times 8) + (1 \times 18)] = 114$; $O_2 = 16 \times 2 = 32$; $N_2 = 14 \times 2 = 28$; $CO_2 = [(12 \times 1) + (16 \times 2)] = 44$; $H_2O = [(1 \times 2) + (16 \times 1)] = 18$.

5. ***Compute the relative weights of the reactants and products of the combustion process***

Relative weight = moles × molecular weight. Coefficients of the chemical equation in step 3 represent the number of moles of each component. Hence, for the reactants, the relative weights are: $C_8H_{18} = 1 \times 114 = 114$; $O_2 = 50 \times 32 = 1600$; $N_2 = 189.2 \times 28 = 5298$. Total relative weight of the reactants is 7012. For the products, the relative weights are: $CO_2 = 8 \times 44 = 352$; $H_2O = 9 \times 18 = 162$; $N_2 = 189.2 \times 28 = 5298$; $O_2 = 37.5 \times 32 = 1200$. Total relative weight of the products is 7012 also.

6. ***Compute the enthalpy of the products of the combustion process***

Enthalpy of the products of combustion, $h_p = m_p(h_{1600} - h_{77})$, where m = number of moles of the products; h_{1600} = enthalpy of the products at 1600°F (871°C); h_{77} = enthalpy of the products at 77°F (25°C). Thus, $h_p = (8 + 9 + 189.2 + 37.5)(15{,}400 - 3750) = 2{,}839{,}100$ Btu [6,259,100 Btu (SI)].

7. *Compute the air supply temperature at the combustion chamber inlet*
Since the combustion process is adiabatic, the enthalpy of the reactants $h_r = h_p$, where h_r = (relative weight of the fuel × its heating value) + [relative weight of the air × its specific heat × (air supply temperature – air source temperature)]. Therefore, h_r = (114 × 19,100) + [(1600 + 5298) × 0.24 × (t_a – 77)] + 2,839,100 Btu [6,259,100 Btu (SI)]. Solving for the air supply temperature, t_a = [(2.839,100 – 2,177,400)/1655.5] + 77 = 477°F (247°C).

Related Calculations. This procedure, appropriately modified, may be used to deal with similar questions involving such things as other fuels, different amounts of excess air, and variations in the condition(s) being sought under certain given circumstances.

The coefficient, (?) = 3.784 in step 2, is used to indicate that for each unit of volume of oxygen, O_2, 12.5 in this case, there will be 3.784 units of nitrogen, N_2. This equates to an approximate composition of air as 20.9 percent oxygen and 79.1 percent "atmospheric nitrogen," (N_2). In turn, this creates a paradox, because page 200 of *Principles of Engineering Thermodynamics,* by Kiefer, et al., John Wiley & Sons, Inc., 1930, states air to be 20.99 percent oxygen and 79.01 percent atmospheric nitrogen, where the ratio $(N_2)/O_2$ = (?) = 79.01/20.99 = 3.764.

Also, page 35 of *Applied Energy Conversion,* by Skrotzki and Vopat, McGraw-Hill, Inc., 1966, indicates an assumed air analysis of 79 percent nitrogen and 21 percent oxygen, where (?) = 3.762. On that basis, a formula is presented for the amount of dry air chemically necessary for complete combustion of a fuel consisting of atoms of carbon, hydrogen, and sulfur, or C, H, and S, respectively. That formula is: W_a = 11.5C + 34.5[H – (0/8)] + 4.32S, lb air/lb fuel (kg air/kg fuel).

The following derivation for the value of (?) should clear up the paradox and show that either 3.784 or 3.78 is a sound assumption which seems to be wrong, but in reality is not. In the above equation for W_a, the carbon hydrogen, or sulfur coefficient, $C_x = (MO_2/DO_2)M_x$, where MO_2 is the molecular weight of oxygen, O_2; DO_2 is the decimal fraction for the percent, by weight, of oxygen, O_2, in dry air containing "atmospheric nitrogen," (N_2), and small amounts of inert and inactive gases: M_x is the formula weight of the combustible element in the fuel, as indicated by its relative amount as a reactant in the combustion equation. The alternate evaluation of C_x is obtained from stoichiometric chemical equations for burning the combustible elements of the fuel, i.e., $C + O_2 + (?)N_2 \rightarrow CO_2 + (?)N_2$; $2H_2 + O_2 + (?)N_2 \rightarrow 2H_2O + (?)N_2$; $S + O_2 + (?)N_2 \rightarrow SO_2 + (?)N_2$. Evidently, $C_x = [MO_2 + (?) \times MN_2)]/M_x$, where MN_2 is the molecular weight of nitrogen, N_2, and the other items are as before.

Equating the two expressions, $C_x = [MO_2 + (? \times MN_2)]/M_x = (MO_2/DO_2)M_x$, reveals that the M_x terms cancel out, indicating that the formula weight(s) of combustible components are irrelevant in solving for (?). Then, $(?) = (1 - DO_2)[MO_2/(MN_2 \times DO_2)]$. From the above-mentioned book by Kiefer, et al., DO_2 = 0.23188. From *Marks' Standard Handbook for Mechanical Engineers,* McGraw-Hill, Inc., 1996, MO_2 = 31.9988 and MN_2 = 28.0134. Thus, (?) = (1 – 0.23188)[31.9988/(28.0134 × 0.23188)] = 3.7838. This demonstrates that the use of (?) = 3.784, or 3.78, is justified for combustion equations.

By using either of the two evaluation equations for C_x, and with accurate values for M_x, i.e., M_c = 12.0111; M_H = 2 × 2 × 1.00797 = 4.0319; M_s = 32.064, from *Marks' M.E. Handbook,* the more precise values for C_C, C_H, and C_S are found to be 11.489, 34.227, and 4.304, respectively. However, the actual C_x values, 11.5, 34.5, and 4.32, used in the formula for W_a are both brief for simplicity and rounded up to be on the safe side.

GAS TURBINE COMBUSTION CHAMBER INLET AIR TEMPERATURE

A gas turbine combustion chamber is well insulated so that heat losses to the atmosphere are negligible. Octane, C_8H_{18}, is to be used as the fuel and 400 percent of the stoichiometric air quantity is to be supplied. The air first passes through a regenerative heater and the air supply temperature at the combustion chamber inlet is to be set so that the exit temperature of the combustion gases is 1600°F (871°C). (See Fig. 4.) Fuel supply temperature is 77°F (25°C) and its heating value is to be taken as 19,000 Btu/lb_m (44,190 kJ/kg) relative to a base of 77°F (15°C).

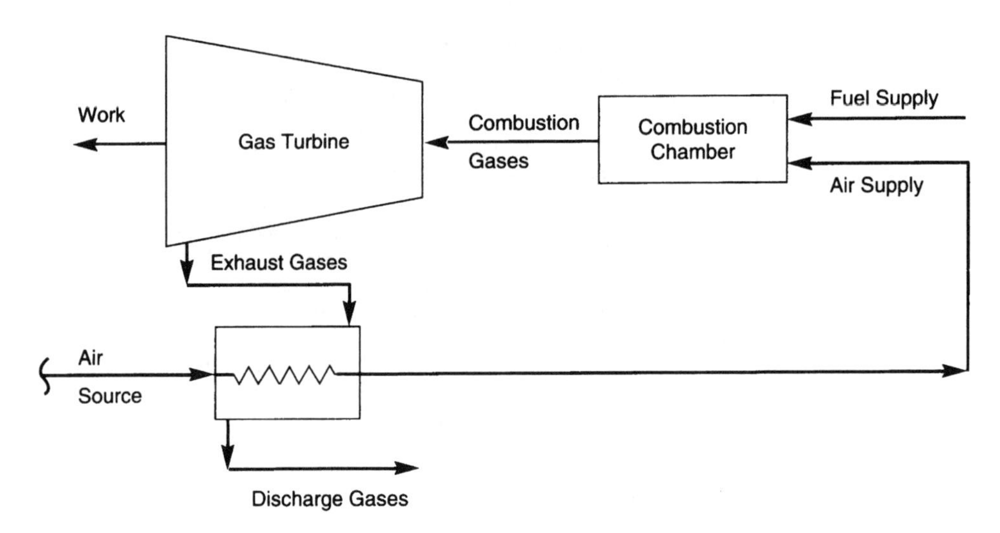

FIGURE 4 Gas-turbine flow diagram.

The air may be treated in calculations as a perfect gas with a constant-pressure-specific heat of 0.24 Btu/(lb · °F) [1.005 kJ/(kg · °C)]. The products of combustion have an enthalpy of 15,400 Btu/lb · mol) [33,950 Btu/(kg·mol)] at 1600°F (871°C) and an enthalpy of 3750 Btu/(lb · mol) [8270 Btu/(kg · mol)] at 77°F (24°C). Determine, assuming complete combustion and neglecting dissociation, the required air temperature at the inlet of the combustion chamber.

Calculation Procedure:

1. *Find the amount of oxygen required for complete combustion of the fuel*

Eight atoms of carbon in C_8 combine with 8 molecules of oxygen, O_2, and produce 8 molecules of carbon dioxide, $8CO_2$. Similarly, 9 molecules of hydrogen, H_2, in H_{18} combine with 9 atoms of oxygen, O, or 4.5 molecules of oxygen, to form 9 molecules of water, $9H_2O$. Thus, 100 percent, or the stoichiometric, air quantity required for complete combustion of a mole of fuel, C_8H_{18}, is proportional to 8 + 12.5 mol of oxygen, O_2.

2. *Establish the chemical equation for complete combustion with 100 percent air*

With 100 percent air: $C_8H_{18} + 12.5O_2 + (3.784 \times 12.5)N_2 \rightarrow 8CO_2 + 9H_2O + 47.3N_2$, where 3.784 is a derived volumetric ratio of atmospheric nitrogen (N_2) to oxygen O_2 in dry air. The N_2 includes small amounts of inert and inactive gases. See Related Calculations of this procedure.

3. *Establish the chemical equation for complete combustion with 400 percent of the stoichiometric air quantity, or 300 percent excess air*

With 400 percent air: $C_8H_{18} + 50O_2 + (4 \times 47.3)N_2 \rightarrow 8CO_2 + 9H_2O + 189.2N_2 + (3 \times 12.5)O_2$.

4. *Compute the molecular weights of the components in the combustion process*

Molecular weight of $C_8H_{18} = [(12 \times 8) + (1 \times 18)] = 114$; $O_2 = 16 \times 2 = 32$; $N_2 = 14 \times 2 = 28$; $CO_2 = [(12 \times 1) + (16 \times 2)] = 44$; $H_2O = [(1 \times 2) + (16 \times 1)] = 18$.

5. *Compute the relative weights of the reactants and products of the combustion process*

Relative weight = moles × molecular weight. Coefficients of the chemical equation in step 3 represent the number of moles of each component. Hence, for the reactants, the relative weights are $C_8H_{18} = 1 \times 114 = 114$; $O_2 = 50 \times 32 = 1600$; $N_2 = 189.2 \times 28 = 5298$. Total relative weight of the reactants is 7012. For the products, the relative weights are $CO_2 = 8 \times 44 = 352$; $H_2O = 9 \times 18 = 162$; $N_2 = 189.2 \times 28 = 5298$; $O_2 = 37.5 \times 32 = 1200$. Total relative weight of the products is 7012 also.

6. *Compute the enthalpy of the products of the combustion process*
Enthalpy of the products of combustion, $h_p = m_p(h_{1600} - h_{77})$, where m_p = number of moles of the products; h_{1600} = enthalpy of the products at 1600°F (871°C); h_{77} = enthalpy of the products at 77°F (25°C). Thus, $h_p = (8 + 9 + 189.2 + 37.5)(15{,}400 - 3750) = 2{,}839{,}100$ Btu [6,259,100 Btu (SI)].

7. *Compute the air supply temperature at the combustion chamber inlet*
Since the combustion process is adiabatic, the enthalpy of the reactants $h_r = h_p$, where h_r = (relative weight of the fuel × its heating value) + [relative weight of the air × its specific heat × (air supply temperature – air source temperature)]. Therefore, $h_r = (114 \times 19{,}100) + [(1600 + 5298) \times 0.24 \times (t_a - 77)] + 2{,}839{,}100$ Btu [6,259,100 Btu (SI)]. Solving for the air supply temperature, $t_a = [(2{,}839{,}100 - 2{,}177{,}400)/1655.5] + 77 = 477$°F (247°C).

Related Calculations. This procedure, appropriately modified, may be used to deal with similar questions involving such things as other fuels, different amounts of excess air, and variations in the condition(s) being sought under certain given circumstances.

The coefficient, (?) = 3.784 in step 2, is used to indicate that for each unit of volume of oxygen, O_2, 12.5 in this case, there will be 3.784 units of nitrogen, N_2. This equates to an approximate composition of air as 20.9 percent oxygen and 79.1 percent "atmospheric nitrogen" (N_2). In turn, this creates a paradox, because page 200 of *Principles of Engineering Thermodynamics,* by Kiefer et al., John Wiley & Sons, Inc., 1930, states air to be 20.99 percent oxygen and 79.01 percent atmospheric nitrogen, where the ratio N_2/O_2 = (?) = 79.01/20.99 = 3.764.

Also, page 35 of *Applied Energy Conversion,* by Skrotzki and Vopat, McGraw Hill, Inc., 1945, indicates an assumed air analysis of 79 percent nitrogen and 21 percent oxygen, where (?) = 3.762. On that basis, a formula is presented for the amount of dry air chemically necessary for complete combustion of a fuel consisting of atoms of carbon, hydrogen, and sulfur, or C, H, and S, respectively. That formula is $W_a = 11.5C + 34.5[H - (0/8)] + 4.32S$, lb air/lb fuel (kg air/kg fuel).

The following derivation for the value of (?) should clear up the paradox and show that either 3.784 or 3.78 is a sound assumption which seems to be wrong, but in reality is not. In the above equation for W_a, the carbon, hydrogen, or sulfur coefficient, $C_x = (MO_2/DO_2)M_x$, where MO_2 is the molecular weight of oxygen, O_2; DO_2 is the decimal fraction for the percent, by weight, of oxygen, O_2, in dry air containing "atmospheric nitrogen," (N_2), and small amounts of inert and inactive gases: M_x is the formula weight of the combustible element in the fuel, as indicated by its relative amount as a reactant in the combustion equation. The alternate evaluation of C_x is obtained from stoichiometric chemical equations for burning the combustible elements of the fuel, i.e., $C + O_2 + (?)N_2 \rightarrow CO_2 + (?)N_2$; $2H_2 + O_2 + (?)N_2 \rightarrow 2H_2O + (?)N_2$; $S + O_2 + (?)N_2 \rightarrow SO_2 + (?)N_2$. Evidently, $C_x = [MO_2 + (? \times MN_2)]/M_x$, where MN_2 is the molecular weight of nitrogen, N_2, and the other items are as before.

Equating the two expressions, $C_x = [MO_2 + (? \times MN_2)]/M_x = (MO_2/DO_2)M_x$, reveals that the M_x terms cancel out, indicating that the formula weight(s) of combustible components are irrelevant in solving for (?). Then, $(?) = (1 - DO_2)[MO_2/[MN_2 \times DO_2)]$. From the above-mentioned book by Kiefer, et al., $DO_2 = 0.23188$. From *Marks' Standard Handbook for Mechanical Engineers,* McGraw-Hill, Inc., 1996, $MO_2 = 31.9988$ and $MN_2 = 28.0134$. Thus, $(?) = (1 - 0.23188)[31.9988/(28.0134 \times 0.23188)] = 3.7838$. This demonstrates that the use of (?) = 3.784, or 3.78, is justified for combustion equations.

By using either of the two evaluation equations for C_x, and with accurate values of M_x, i.e., $M_C = 12.0111$; $M_H = 2 \times 2 \times 1.00797 = 4.0319$; $M_s = 32.064$, from *Marks' M.E. Handbook,* the more precise values for C_C, C_H, and C_S are found out to be 11.489, 34.227, and 4.304, respectively. However, the actual C_x values, 11.5, 34.5, and 4.32, used in the formula for W_a are both brief for simplicity and rounded up to be on the safe side.

ANALYSIS OF REGENERATIVE-GAS-TURBINE CYCLE

What is the cycle air rate, lb/kWh, for a regenerative gas turbine having a pressure ratio of 5, an air inlet temperature of 60°F (15.6°C), a compressor discharge temperature of 1500°F (815.6°C), and performance in accordance with Fig. 5? Determine the cycle thermal efficiency and work ratio.

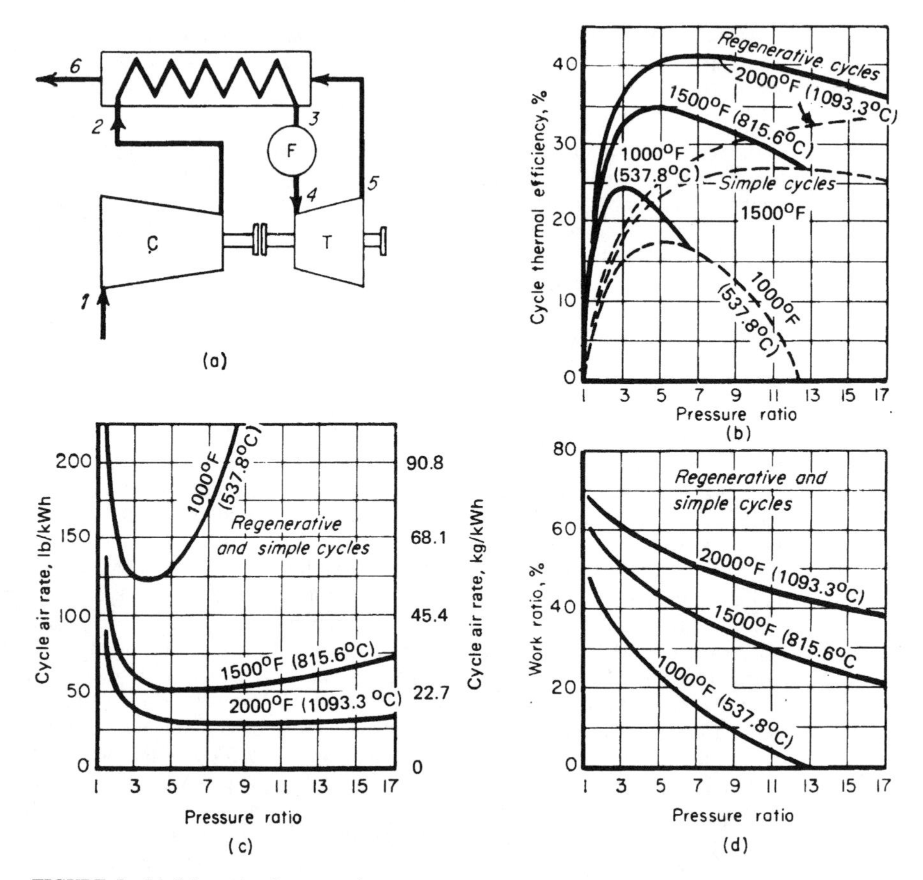

FIGURE 5 (*a*) Schematic of regenerative gas turbine; (*b*), (*c*), and (*d*) gas-turbine performance based on a regenerator effectiveness of 70 percent, compressor and turbine efficiency of 85 percent; air inlet = 60°F (15.6°C); no pressure losses.

What is the power output of a regenerative gas turbine if the work input to the compressor is 4400 hp (3281.1 kW)?

Calculation Procedure:

1. *Determine the cycle rate*

Use Fig. 5, entering at the pressure ratio of 5 in Fig. 5*c* and projecting to the 1500°F (815.6°C) curve. At the left, read the cycle air rate as 52 lb/kWh (23.6 kg/kWh).

2. *Find the cycle thermal efficiency*

Enter Fig. 5*b* at the pressure ratio of 5 and project vertically to the 1500°F (815.6°C) curve. At left, read the cycle thermal efficiency as 35 percent. Note that this point corresponds to the maximum efficiency obtainable from this cycle.

3. *Find the cycle work ratio*

Enter Fig. 5*d* at the pressure ratio of 5 and project vertically to the 1500°F (815.6°C) curve. At the left, read the work ratio as 44 percent.

4. *Compute the turbine power output*

For any gas turbine, the work ratio, percent = $100w_c/w_c$, where w_c = work input to the turbine, hp; w_t = work output of the turbine, hp. Substituting gives $44 = 100(4400)/w_t$; $w_t = 100(4400)/44 = 10{,}000$ hp (7457.0 kW).

Related Calculations. Use this general procedure to analyze gas turbines for power-plant, marine, and portable applications. Where the operating conditions are different from those given here, use the manufacturer's engineering data for the turbine under consideration.

Figure 6 shows the effect of turbine-inlet temperature, regenerator effectiveness, and compressor inlet-air temperature on the performance of a modern gas turbine. Use these curves to analyze the

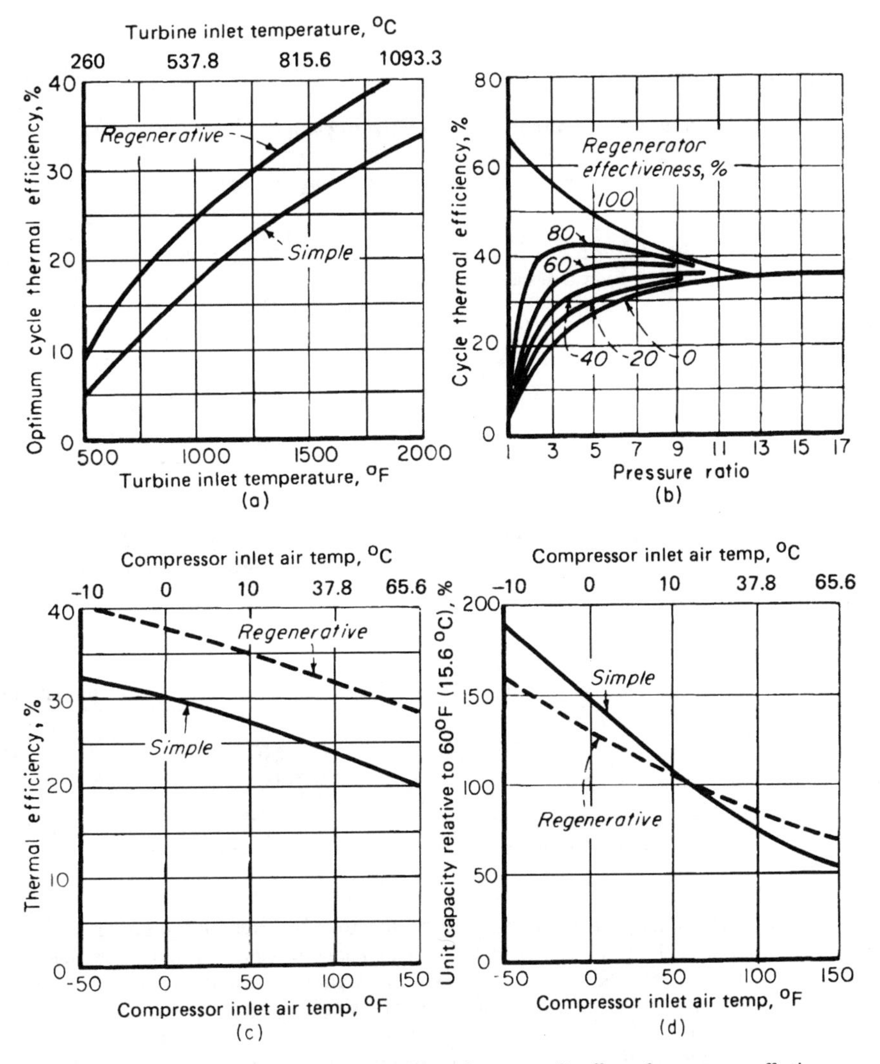

FIGURE 6 (*a*) Effect of turbine inlet on cycle performance; (*b*) effect of regenerator effectiveness; (*c*) effect of compressor inlet-air temperature; (*d*) effect of inlet-air temperature on turbine-cycle capacity. These curves are based on a turbine and compressor efficiency of 85 percent, a regenerator effectiveness of 70 percent, and a 1500°F (815.6°C) inlet-gas temperature.

cycles of gas turbines being considered for a particular application if the operating conditions are close to those plotted.

COMBINED-CYCLE PLANT HEAT-RATE COMPARISON

Compare the heat rate of a proposed conventional steam-power plant generating electricity with that of a combined-cycle gas-turbine plant of the same electrical capacity. Use the net plant heat rate at various generating capacity loads as the basis for the comparison.

Calculation Procedure:

1. ***Obtain the heat rate for the proposed steam plant at various percent rated capacities***
Use data from the proposed plant design calculations, or from an existing steam plant of the same output. Thus, for the proposed steam plant being considered here, the heat rate at various loads is:

Percent capacity, steam plant	Heat Rate, Btu/kWh	Heat rate, kJ/kWh
20	12,800	13,504
40	11,300	11,922
60	10,900	11,499
80	10,600	11,183
100	10,400	10,972

2. ***Determine the heat rate for the proposed combined-cycle plant***
Use data from the proposed combined-cycle plant or from an existing similar plant. Thus:

Percent capacity, combined cycle	Heat rate, Btu/kWh	Heat rate, kJ/kWh
20	13,950	14,717
40	11,000	12,132
60	10,200	10,761
80	9,800	10,339
100	9,700	10,234

3. ***Plot the heat rates to show the comparison between the two plants***
Figure 7 shows the heat rates for the two plants. As can be seen, the combined-cycle plant has a heat rate of 700 Btu/kWh (738.5 kJ/kWh) less than the steam plant. This is a saving of 6.7 percent. Such a saving can be significant in a time of rising fuel prices and increased concern over atmospheric pollution.

At about 30 percent of capacity the heat rates for the two types of plants are equal. But once this capacity is exceeded, the combined-cycle plant is more efficient. This is why the gas turbine is proving so popular in today's central stations for both topping and combined-cycle duty. Figure 7 is from Skrotzki and Vopat—*Power Station Engineering and Economy*, McGraw-Hill 1960.

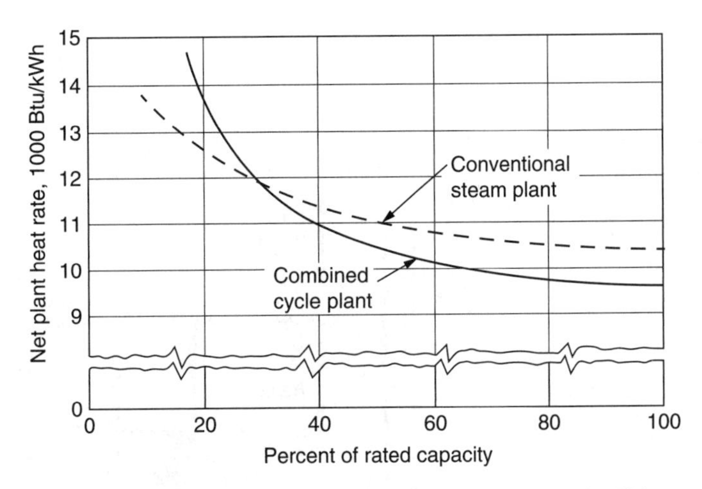

FIGURE 7 Combined gas-turbine–steam-turbine plant works with superior efficiency compared with conventional steam-turbine plant.

SELECTING MOST EFFICIENT METHOD FOR INCREASING COMBINED-CYCLE PLANT OUTPUT

Select the best option to boost the output of a 230-MW facility based on a 155-MW natural-gas-fired gas turbine (GT) featuring a dry low NO, combustor (Fig. 8). The plant has a heat-recovery steam generator (HRSG) which is a triple-pressure design with an integral deaerator. A reheat condensing steam turbine (ST) is used and it is coupled to a cooling-tower/surface-condenser heat sink turbine inlet. Steam conditions are 1450 lb/in^2 (gage)/1000°F (9991 kPa/538°C). Unit ratings are for operation at International Standard Organization (ISO) conditions. Evaluate the various technologies considered for summer peaking conditions with a dry bulb (DB) temperature of 95°F and 60 percent RH (relative humidity) (35°C and 60 percent RH). The plant heat sink is a four-cell, counterflow, mechanical-draft cooling tower optimized to achieve a steam-turbine exhaust pressure of 3.75 inHg absolute (9.5 cmHg) for all alternatives considered in this evaluation. Base circulating-water system includes a surface condenser and two 50 percent-capacity pumps. Water-treatment, consumption, and disposal-related O&M (operating & maintenance) costs for the zero-discharge facility are assumed to be \$3/1000 gal (\$3/3.8 m^3) of raw water, \$6/1000 gal (\$6/3.8 m^3) of treated demineralized water, and \$5/1000 gal (\$5/3.8 m^3) of water disposal. The plant is configured to burn liquid distillate as a backup fuel.

Calculation Procedure:

1. ***List the options available for boosting output***

Seven options can be developed for boosting the output of this theoretical reference plant. Although plant-specific issues will have a significant effect on selecting an option, comparing performance based on a reference plant, Fig. 8, can be helpful. Table 1 shows the various options available in this study for boosting output. The comparisons shown in this procedure illustrate the characteristics, advantages, and disadvantages of the major power augmentation technologies now in use.

Amidst the many advantages of gas turbine (GT) combined cycles (CC) popular today from various standpoints (lower investment than for new greenfield plants, reduced environmental impact, and faster installation and startup), one drawback is that the achievable output decreases significantly as the ambient inlet-air temperature increases. The lower density of warm air reduces mass flow

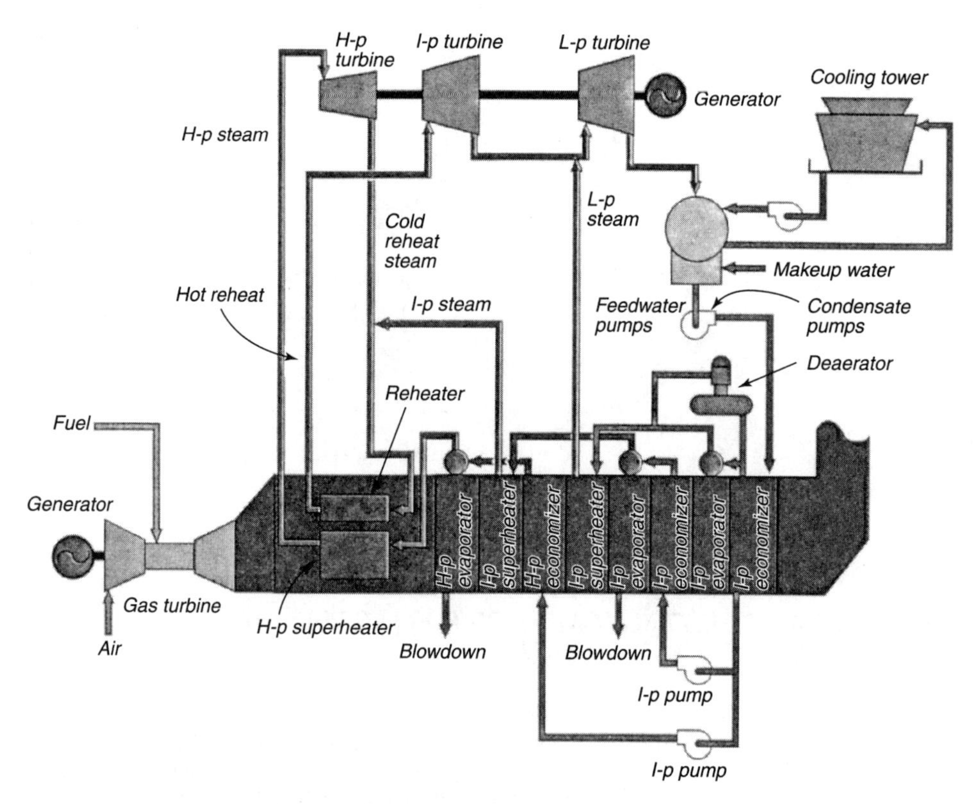

FIGURE 8 155-MW natural-gas-fired gas turbine featuring a dry low NO_4 combustor. (*Power*.)

TABLE 1 Performance Summary for Enhanced-Output Options

Measured change from base case	Case 1 Evap. cooler	Case 2 Mech. chiller	Case 3 Absorp. chiller	Case 4 Steam injection	Case 5 Water injection	Case 6* Supp.-fired HRSG	Case 7† Supp.-fired HRSG
GT output, MW	5.8	20.2	20.2	21.8	15.5	0	0
ST output, MW	0.9	2.4	−2.1	−13	3.7	8	35
Plant aux. load, MW	0.05	4.5	0.7	400	0.2	0.4	1
Net plant output, MW	6.65	18.1	17.4	8.4	19	7.6	34
Net heat rate, Btu/kWh‡	15	55	70	270	435	90	320
Incremental costs							
Change in total water cost, $/h	15	35	35	115	85	35	155
Change in wastewater cost, $/h	1	17	17	2	1	1	30
Change in capital cost/net output, $/kW	180	165	230	75	15	70	450

*Partial supplementary firing.
†Full supplementary firing.
‡Based on lower heating value of fuel.

through the GT. And, unfortunately, hot weather typically corresponds to peak power loads in many areas. So the need to meet peak-load and power-sales contract requirements causes many power engineers and developers to compensate for ambient-temperature-output loss.

The three most common methods of increasing output include: (1) injecting water or steam into the GT, (2) precooling GT inlet air, and/or (3) supplementary firing of the heat-recovery steam generator (HRSG). All three options require significant capital outlays and affect other performance parameters. Further, the options may uniquely impact the operation and/or selection of other components, including boiler feedwater and condensate pumps, valves, steam turbine/generators, condensers, cooling towers, and emissions-control systems.

2. *Evaluate and analyze inlet-air precooling*

Evaporative cooling, Case 1, Table 1, boosts GT output by increasing the density and mass flow of the air entering the unit. Water sprayed into the inlet-air stream cools the air to a point near the ambient wet-bulb temperature. At reference conditions of 95°F (35°C) DB and 60 percent RH, an 85 percent effective evaporative cooler can alter the inlet-air temperature and moisture content to 85°F (29°C) and 92 percent RH, respectively, using conventional humidity chart calculations, page 13.104. This boosts the output of both the GT and—because of energy added to the GT exhaust—the steam turbine/generator. Overall, plant output for Case 1 is increased by 5.8 MW GT output + 0.9 MW ST output—plant auxiliary load of 0.9 MW = 6.65 MW, or 3.3 percent. The CC heat rate is improved 0.2 percent, or 15 Btu/kWh (14.2 kJ/kWh). The total installed cost for the evaporative cooling system, based on estimates provided by contractors and staff, is $1.2 million. The incremental cost is $1,200,000/6650 kW = $180.45/kW for this ambient condition.

The effectiveness of the same system operating in less-humid conditions—say 95°F DB (35°C) and 40 percent RH—is much greater. In this case, the same evaporative cooler can reduce inlet-air temperature to 75°F DB (23.9°C) by increasing RH to 88 percent. Here, CC output is increased by 7 percent, heat rate is improved (reduced) by 1.9 percent, and the incremental installed cost is $85/kW, computed as above. As you can clearly see, the effectiveness of evaporative cooling is directly related to reduced RH.

Water-treatment requirements must also be recognized for this Case, No. 1. Because demineralized water degrades the integrity of evaporative-cooler film media, manufacturers may suggest that only raw or filtered water be used for cooling purposes. However, both GT and evaporative-cooler suppliers specify limits for turbidity, pH, hardness, and sodium (Na) and potassium (K) concentrations in the injected water. Thus, a nominal increase in water-treatment costs can be expected. In particular, the cooling water requires periodic blowdown to limit solids buildup and system scaling. Overall, the evaporation process can significantly increase a plant's makeup-water feed rate, treatment, and blowdown requirements. Compared to the base case, water supply costs increase by $15/h of operation for the first approach, and $20/h for the second, lower RH mode. Disposal of evaporative-cooler blowdown costs $l/h in the first mode, $2/h in the second. Evaporative cooling has little or no effect on the design of the steam turbine.

3. *Evaluate the economics of inlet-air chilling*

The effectiveness of evaporative cooling is limited by the RH of the ambient air. Further, the inlet air cannot be cooled below the wet-bulb (WB) temperature of the inlet air. Thus, chillers may be used for further cooling of the inlet air below the wet-bulb temperature. To achieve this goal, industrial-grade mechanical or absorption air-conditioning systems are used, Fig. 9. Both consist of a cooling medium (water or a refrigerant), an energy source to drive the chiller, a heat exchanger for extracting heat from the inlet air, and a heat-rejection system.

A mechanical chilling system. Case 2, Table 1, is based on a compressor-driven unit. The compressor is the most expensive part of the system and consumes a significant amount of energy. In general, chillers rated above 12-million Btu/h (3.5 MW) (1000 tons of refrigeration) (3500 kW) employ centrifugal compressors. Units smaller than this may use either screw-type or reciprocating compressors. Overall, compressor-based chillers are highly reliable and can handle rapid load changes without difficulty.

A centrifugal-compressor-based chiller can easily reduce the temperature of the GT inlet air from 95°F (35°C) to 60°F (15.6°C) DB—a level that is generally accepted as a safe lower limit for preventing

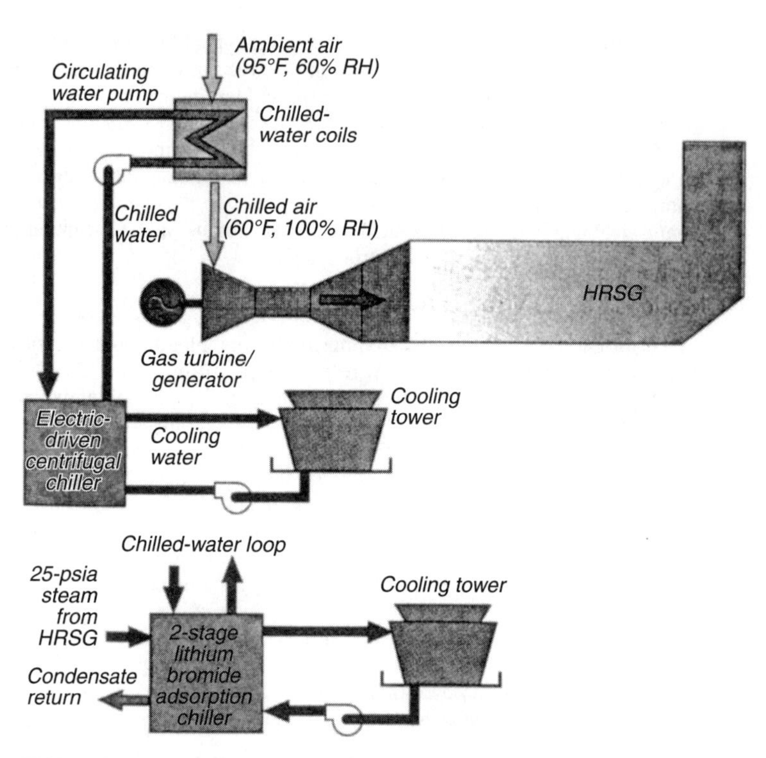

FIGURE 9 Inlet-air chilling using either centrifugal or absorption-type chillers boosts the achievable mass flow and power output during warm weather. (*Power.*)

icing on compressor inlet blades—and achieve 100 percent RH. This increases plant output by 20.2 MW for GT + 2.4 MW for ST – 4.5 MW plant auxiliary load = 18.1 MW, or 8.9 percent. But it degrades the net CC heat rate by 0.8 percent and results in a 1.5-in-(3.8-cm)-H_2O inlet-air pressure drop because of heat-exchanger equipment located in the inlet-air stream.

Cooling requirements of the chilling system increase the plant's required circulating water flow by 12,500 gal/min (47.3 m^3/min). Combined with the need for increased steam condensing capacity, use of a chiller may necessitate a heat sink 25 percent larger than the base case. The total installed cost for the mechanical chilling system for Case 2 is $3-million, or about $3,000,000/18,100 kW = $165.75/kW of added output. Again, costs come from contractor and staff studies.

Raw-water consumption increase the plant's overall O&M costs by $35/h when the chiller is operating. Disposal of additional cooling-tower blow-down costs $17/h. The compressor used in Case 2 consumes about 4 MW of auxiliary power to handle the plant's 68-million Btu/h (19.9 MW) cooling load.

4. ***Analyze an absorption chilling system***

Absorption chilling systems are somewhat more complex than mechanical chillers. They use steam or hot water as the cooling motive force. To achieve the same inlet-air conditions as the mechanical chiller (60°F DB, 100 percent RH) (15.6°C, 100 percent RH), an absorption chiller requires about 111,400 lb/h (50,576 kg/h) of 10.3-lb/in^2 (gage) (70.9-kPa) saturated steam, or 6830 gal/min (25.9 m^3/min) of 370°F (188°C) hot water.

Cost-effective supply of this steam or hot water requires a redesign of the reference plant. Steam is extracted from the low-pressure (l-p) steam turbine at 20.3 lb /in^2 (gage) (139.9 kPa) and attemperated until it is saturated. In this case, the absorption chiller increases plant output by 8.7 percent or 17.4 MW but degrades the plant's heat rate by 1 percent.

Although the capacity of the absorption cooling system's cooling-water loop must be twice that of the mechanical chiller's, the size of the plant's overall heat sink is identical—25 percent larger than the base case—because the steam extracted from the l-p turbine reduces the required cooling capacity. Note that this also reduces steam-turbine output by 2 MW compared to the mechanical chiller, but has less effect on overall plant output.

Cost estimates summarized in Table 1 show that the absorption chilling system required here costs about $4-million, or about $230/kW of added output. Compared to the base case, raw-water consumption increases O&M costs by $35 /h when the chiller is operating. Disposal of additional cooling-water blow-down adds $17/h.

Compared to mechanical chillers, absorption units may not handle load changes as well; therefore they may not be acceptable for cycling or load-following operation. When forced to operate below their rated capacity, absorption chillers suffer a loss in efficiency and reportedly require more operator attention than mechanical systems.

Refrigerant issues affect the comparison between mechanical and absorption chilling. Mechanical chillers use either halogenated or nonhalogenated fluorocarbons at this time. Halogenated fluorocarbons, preferred by industry because they reduce the compressor load compared to nonhalogenated materials, will be phased out by the end of the decade because of environmental considerations (destruction of the ozone layer). Use of nonhalogenated refrigerants is expected to increase both the cost and parasitic power consumption for mechanical systems, at least in the near term. However, absorption chillers using either ammonia or lithium bromide will be unaffected by the new environmental regulations.

Off-peak thermal storage is one way to mitigate the impact of inlet-air chilling's major drawback: high parasitic power consumption. A portion of the plant's electrical or thermal output is used to make ice or cool water during off-peak hours. During peak hours, the chilling system is turned off and the stored ice and/or cold water is used to chill the turbine inlet air. A major advantage is that plants can maximize their output during periods of peak demand when capacity payments are at the highest level. Thermal storage and its equipment requirements are analyzed in Section 13 of this handbook.

5. *Compare steam and water injection alternatives*

Injecting steam or water into a GT's combustor can significantly increase power output, but either approach also degrades overall CC efficiency. With steam injection, steam extracted from the bottoming cycle is typically injected directly into the GT's combustor, Fig.10. For advanced GTs, the steam source may be extracted from either the high-pressure (h-p) turbine exhaust, an h-p extraction, or the heat-recovery steam generator's (HRSG) h-p section.

Cycle economics and plant-specific considerations determine the steam extraction point. For example, advanced, large-frame GTs require steam pressures of 410 to 435 lb/in^2 (gage) (2825 to 2997 kPa). This is typically higher than the economically optimal range of h-p steam turbine exhaust pressures of 285 to 395 lb/in^2 (gage) (1964 to 2722 kPa). Thus, steam must be supplied from either the HRSG or an h-p turbine extraction ahead of the reheat section.

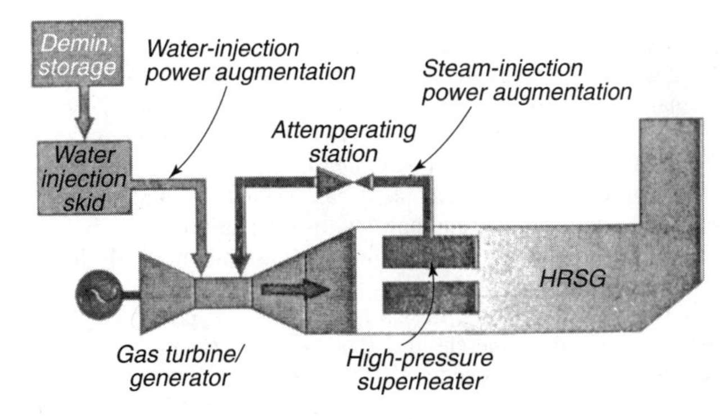

FIGURE 10 Water or steam injection can be used for both power augmentation and NO_x, control. (*Power.*)

Based on installed-cost considerations alone, extracting steam from the HRSG is favored for peaking service and may be accomplished without altering the reheat steam turbine. But if a plant operates in the steam-injection mode for extended periods, extracting steam from the turbine or increasing the h-p turbine exhaust pressure becomes more cost-effective.

Injecting steam from the HRSG superheat section into the GT increases unit output by 21.8 MS, Case 4 Table 1, but decreases the steam turbine/generator's output by about 12.8 MW. Net gain to the CC is 8.4 MW. But CC plant heat rate also suffers by 4 percent, or 270 Btu/kWh (256.5 kJ/kWh).

Because the steam-injection system requires makeup water as pure as boiler feedwater, some means to treat up to 350 gal/min (22.1 L/s) of additional water is necessary. A dual-train demineralizer this size could cost up to $1.5-million. However, treated water could also be bought from a third party and stored. Or portable treatment equipment could be rented during peak periods to reduce capital costs. For the latter case, the average expected cost for raw and treated water is about $130/h of operation.

This analysis assumes that steam- or water-injection equipment is already in place for NO_x, control during distillate-fuel firing. Thus, no additional capital cost is incurred.

When water injection is used for power augmentation or NO_x control, the recommended water quality may be no more than filtered raw water in some cases, provided the source meets pH, turbidity, and hardness requirements. Thus, water-treatment costs may be negligible. Water injection. Case 5 Table 1, can increase the GT output by 15.5 MW.

In Case 5, the bottoming cycle benefits from increased GT-exhaust mass flow, increasing steam turbine/generator output by about 3.7 MW. Overall, the CC output increases by 9.4 percent or 19 MW, but the net plant heat rate suffers by 6.4 percent, or 435 Btu/kWh (413.3 kJ/kWh). Given the higher increase in the net plant heat rate and lower operating expenses, water injection is preferred over steam injection in this case.

6. *Evaluate supplementary-fired HRSG for this plant*

The amount of excess O_2 in a GT exhaust gas generally permits the efficient firing of gaseous and liquid fuels upstream of the HRSG, thereby increasing the output from the steam bottoming cycle. For this study, two types of supplementary firing are considered—(1) partial supplementary firing, Case 6 Table 1, and (2) full supplementary firing. Case 7 Table 1.

There are three main drawbacks to supplementary firing for peak power enhancement, including 910 lower cycle efficiency, (2) higher NO_x and CO emissions, (3) higher costs for the larger plant equipment required.

For this plant, each 100-million Btu/h (29.3 MW) of added supplementary firing capacity increases the net plant output by 5.5 percent, but increases the heat rate by 2 percent. The installed cost for supplementary firing can be significant because all the following equipment is affected: (1) boiler feed pumps, (2) condensate pumps, (3) steam turbine/generator, (4) steam and water piping and valves, and (5) selective-catalytic reduction (SCR) system. Thus, a plant designed for supplementary firing to meet peak-load requirements will operate in an inefficient, off-design condition for most of the year.

7. *Compare the options studied and evaluate results*

Comparing the results in Table 1 shows that mechanical chilling, Case 2, gives the largest increase in plant output for the least penalty on plant heat rate—i.e., 18.1 MW output for a net heat rate increase of 55 Btu/kWh (52.3 kJ/kWh). However, this option has the highest estimated installed cost ($3-million), and has a relatively high incremental installed cost.

Water injection, Case 5 Table 1, has the dual advantage of high added net output and low installed cost for plants already equipped with water-injection skids for NO_x control during distillate-fuel firing. Steam injection. Case 4 Table 1, has a significantly higher installed cost because of water-treatment requirements.

Supplementary firing. Cases 6 and 7 Table 1, proves to be more acceptable for plants requiring extended periods of increased output, not just seasonal peaking.

This calculation procedure is the work of M. Boswell, R. Tawney, and R. Narula, all of Bechtel Corporation, as reported in *Power* magazine, where it was edited by Steven Collins. SI values were added by the editor of this handbook.

Related Calculations. Use of gas turbines for expanding plant capacity or for repowering older stations is a popular option today. GT capacity can be installed quickly and economically, compared to conventional steam turbines and boilers. Further, the GT is environmentally acceptable in most areas. So long as there is a supply of combustible gas, the GT is a viable alternative that should be considered in all plant expansion and repowering today, and especially where environmental conditions are critical.

GAS-TURBINE HEAT-RECOVERY STEAM-GENERATOR (HRSG) CHOICE

Choose a suitable heat-recovery boiler equipped with an evaporator and economizer to serve a gas turbine in a manufacturing plant where the gas flow rate is 150,000 lb/h (68,040 kg/h) at 950°F (510°C) and which will generate steam at 205 lb/in^2 (gage) (1413.5 kPa). Feedwater enters the boiler at 227°F (108.3°C). Determine if supplementary firing of the exhaust is required to generate the needed steam. Use an approach temperature of 20°F (36°C) between the feedwater and the water leaving the economizer.

Calculation Procedure:

1. ***Determine the critical gas inlet-temperature***
Turbine exhaust gas (TEG) typically leaves a gas turbine at 900–1000°F (482–538°C) and has about 13 to 16 percent free oxygen. If steam is injected into the gas turbine for NO_x control, the oxygen content will decrease by 2 to 5 percent by volume. To evaluate whether supplementary firing of the exhaust is required to generate needed steam, a knowledge of the temperature profiles in the boiler is needed.

Prepare a gas/steam profile for this heat-recovery boiler as shown in Fig. 11 TEG enters on the left at 950°F (510°C). Steam generated in the boiler at 205 lb/in^2 (gage) (1413.5 kPa) has a temperature

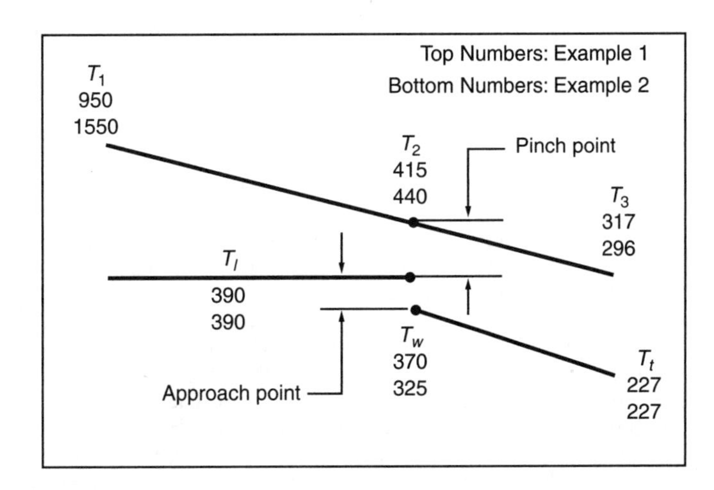

950°F (510°C)	1550°F (843°C)	390°F (199°C)	390°F (199°C)
415°F (213°C)	440°F (227°C)	370°F (188°C)	325°F (163°C)
317°F (158°C)	296°F (147°C)	227°F (108°C)	227°F (108°C)

FIGURE 11 Gas/steam profile and data. (*Chemical Engineering.*)

of 390°F (198.9°C), from steam tables. For steam to be generated in the boiler, two conditions must be met: (1) The "pinch point" temperature must be greater than the saturated steam temperature of 390°F (198.9°C), and (2) the temperature of the saturated steam leaving the boiler economizer must be greater than that of the feedwater. The pinch point occurs somewhere along the TEG temperature line, Fig. 11, which starts at the inlet temperature of 950°F (510°C) and ends at the boiler gas outlet temperature, which is to be determined by calculation. A pinch-point temperature will be assumed during the calculation and its suitability determined.

To determine the critical gas inlet-temperature, T_1, get from the steam tables the properties of the steam generated by this boiler: t_s = 390°F (198.9°C); h_1, heat of saturated liquid = 364 Btu/lb (846.7 kJ/kg); h_s, total heat of saturated vapor = 1199.6 Btu/lb (2790.3 kJ/kg); h_w, heat of saturated liquid of feedwater leaving the economizer at 370°F (187.8°C) = 342 Btu/lb (795.5 kJ/kg); and h_f, heat of saturated liquid of the feedwater at 227°F (108.3°C) = 196.3 Btu/lb (456.6 kJ/kg).

Writing an energy balance across the evaporator neglecting heat and blow-down losses, we get: $(T_1 - T_2)/(T_1 - T_3) = (h_s - h_w)/(h_s - h_f) = X$, where T_1 = gas temperature in boiler, °F (°C); T_2 = pinch-point gas temperature, °F (°C); T_3 = outlet gas temperature for TEG, °F (°C); enthalpy, h, values as listed above; X = ratio of temperature or enthalpy differences. Substituting, X = (1199.6 – 342)/(1199.9 – 196.3) = 0.855, using enthalpy values as given above.

The critical gas inlet-temperature, $T_{lc} = (t_s - Xt_f)/(1 - X)$, where t_s = temperature of saturated steam, °F (°C); t_f = temperature of feedwater, °F (°C); other symbols as before. Using the values determined above, T_{lc} = [390 – (0.855)(227)]/(1 – 0.855) = 1351°F (732.8°C).

2. *Determine the system pinch/point and gas/steam profile*

Up to a gas inlet-temperature of approximately 1351°F (732.8°C), the pinch point can be arbitrarily selected. Beyond this, the feedwater inlet temperature limits the temperature profile. Let's then select a pinch point of 25°F (13.9°C), Fig. 11. Then, T_2 the gas-turbine gas temperature at the pinch point, °F (°C) = t_f + pinch-point temperature difference, or 390°F + 25°F = 415°F (212.8°C).

Setting up an energy balance across the evaporator, assuming a heat loss of 2 percent and a blow-down of 3 percent, leads to: $Q_{evap} = W_e$ (1 – heat loss)(TEG heat capacity, Btu/°F) $(T_1 - T_2)$, where W_e = TEG flow, lb/h; heat capacity of TEG = 0.27 Btu/°F; T_1 = TEG inlet temperature, °F (°C). Substituting, Q_{evap} = 150,000(0.98)(0.27)(950 – 415) = 21.23 × 10^6 Btu/h (6.22 MW).

The rate of steam generation, $W_s = Q_{evap}/[(h_s - h_w)$ + blowdown percent × $(h_t - h_w)]$, where the symbols are as given earlier. Substituting, W_s = 21.23 × 10^6/[(1199.6 – 342) + 0.03 × (364 – 342)] = 24,736 lb/h (11,230 kg/h).

Determine the boiler economizer duty from Q_{econ} = (1 + blowdown)$(W_s)(h_w - h_f)$, where symbols are as before. Substituting, Q_{econ} = 1.03(24.736)(342 – 196.3) = 3.71 × 10^6 Btu/h (1.09 MW).

The gas exit-temperature, $T_3 = T_2 - Q_{econ}$/TEG gas flow, lb/h)(l – heat loss)(heat capacity, Btu/lb °F). Since all values are known, T_3 = 415 – 3.71 × 10^6/(150,000 × 0.98 × 0.27) = 317°F (158°C). Figure 11 shows the temperature profile for this installation.

Related Calculations. Use this procedure for heat-recovery boilers fired by gas-turbine exhaust in any industry or utility application. Such boilers may be unfired, supplementary fired, or exhaust fired, depending on steam requirements.

Typically, the gas-pressure drop across the boiler system ranges from 6 to 12 in (15.2 to 30.5 cm) or water. There is an important tradeoff: a lower pressure drop means the gas-turbine power output will be higher, while the boiler surface and the capital cost will be higher, and vice versa. Generally, a lower gas-pressure drop offers a quick payback time.

If ΔP_e is the additional gas pressure in the system, the power, kW, consumed in overcoming this loss can be shown approximately from $P = 5 \times 10^{-8}$ $(W_e \Delta P_e T/E$, where E = efficiency of compression).

To show the application of this equation and the related payback period, assume W_e = 150,000 lb/g (68,100 kg/h), T = 1000°R (average gas temperature in the boiler), ΔP_e = 4 in water (10.2 cm), and E = 0.7. Then $P = 5 \times 10^{-8}$ (150,000 × 4 × 1000/0.7) = 42 kW.

If the gas-turbine output is 4000 kW, nearly 1 percent of the power is lost due to the 4-in (10.2-cm) pressure drop. If electricity costs 7 cent/kWh, and the gas turbine runs 8000 h/yr, the annual loss will be 8000 × 0.07 × 42 = $23,520. If the incremental cost of a boiler having a 4-in (10.2-cm) lower pressure drop is, say $22,000, the payback period is about 1 year.

If steam requirements are not stated for a particular gas inlet condition, and maximum steaming rate is desired, a boiler can be designed with a low pinch point, a large evaporator, and an economizer. Check the economizer for steaming. Such a choice results in a low gas exit temperature and a high steam flow.

Then, the incremental boiler cost must be evaluated against the additional steam flow and gas-pressure drop. For example, Boiler A generates 24,000 lb/h (10,896 kg/h), while Boiler B provides 25,000 lb/h (11,350 kg/h) for the same gas pressure drop but costs $30,000 more. Is Boiler B worth the extra expense?

To answer this question, look at the annual differential gain in steam flow. Assuming steam costs $3.50/1000 lb (3.50/454 kg), the annual differential gain steam flow = 1000 × 3.5 × 8000/1000 = $28,000. Thus, the simple payback is about a year ($30,000 vs $28,000), which is attractive. You must, however, be certain you assess payback time against the actual amount of time the boiler will operate. If the boiler is likely to be used for only half this period, then the payback time is actually 2 years.

The general procedure presented here can be used for any type of industry using gas-turbine heat-recovery boilers—chemical, petroleum, power, textile, food, etc. This procedure is the work of V. Ganapathy, Heat-Transfer Specialist, ABCO Industries, Inc., and was presented in *Chemical Engineering* magazine.

When supplementary fuel is added to the turbine exhaust gas before it enters the boiler, or between boiler surfaces, to increase steam production, one has to perform an energy balance around the burner, Fig. 12, to evaluate accurately the gas temperature increase that can be obtained.

V. Ganapathy, cited above, has a computer program he developed to speed this calculation.

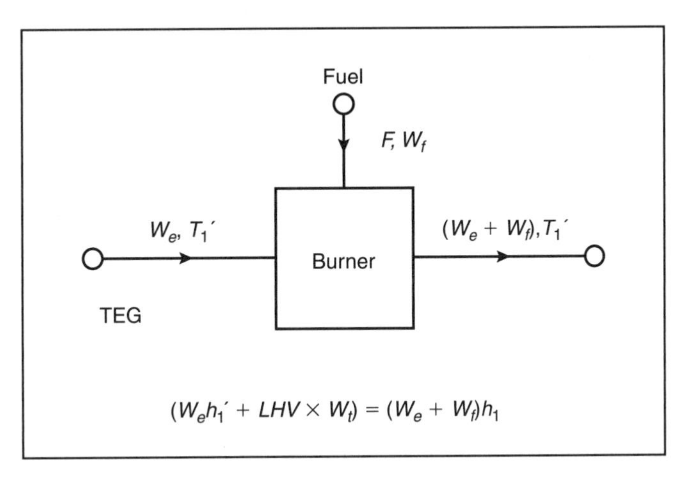

FIGURE 12 Gas/steam profile for fired mode. (*Chemical Engineering*.)

ANALYZING GAS-TURBINE CYCLE EFFICIENCY AND OUTPUT

A gas turbine consisting of a compressor, combustor, and an expander has air entering at 60°F (15.6°C) and 14.0 lb/in^2 (abs) (96.5 kPa). Inlet air is compressed to 56 lb/in^2 (abs) (385.8 kPa); the isentropic efficiency of the compressor is 82 percent. Sufficient fuel is injected to give the mixture of fuel vapor and air a heating value of 200 Btu/lb (466 kJ/kg). Assume complete combustion of the fuel. The expander reduces the flow pressure to 14.9 lb/in^2 (abs), with an engine efficiency of 85 percent. Assuming that the combustion products have the same thermodynamic properties as air, $c_p = 0.24$, and is constant. The isentropic exponent may be taken as 1.4. (*a*) Find the temperature after compression, after combustion, and at the exhaust; (*b*) Determine the Btu/lb (kJ/kg) of air supplied, the work delivered by the expander, the net work produced by the gas turbine, and its thermal efficiency.

Calculation Procedure:

1. ***Plot the ideal and actual cycles***

Draw the ideal cycle as 1-2-3-4-1, Figs. 13 and 14. Actual compression takes place along 1-2′. Actual heat added lies along 2′-3′. The ideal expansion process path is 3′-4′. Ideal work = c_p (ideal temperature difference). Actual work = c_p (actual temperature difference).

2. ***Find the temperature after compression***

Use the relation $(T_2/T_1) = (P_2/P_1)^{(k-1)/k}$, where T_1 = entering air temperature, °R; T_2 = temperature after adiabatic compression, °R; P_1 = entering air pressure, in units given above; P_2 = pressure after

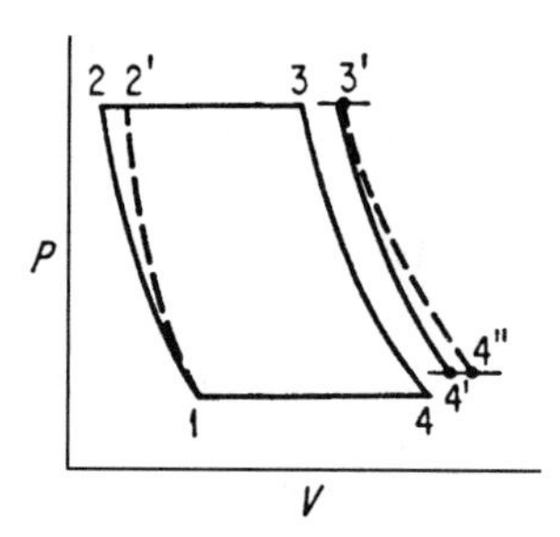

FIGURE 13 Ideal gas-turbine cycle, 1-2-3-4-1. Actual compression takes place along 1-2′; actual heat added 2′-3′; ideal expansion 3′-4′.

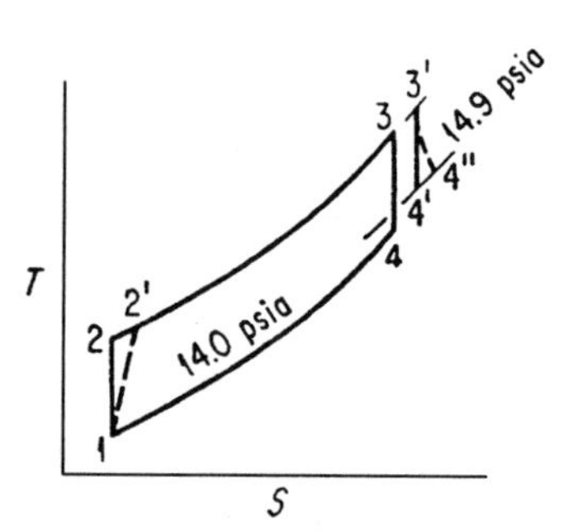

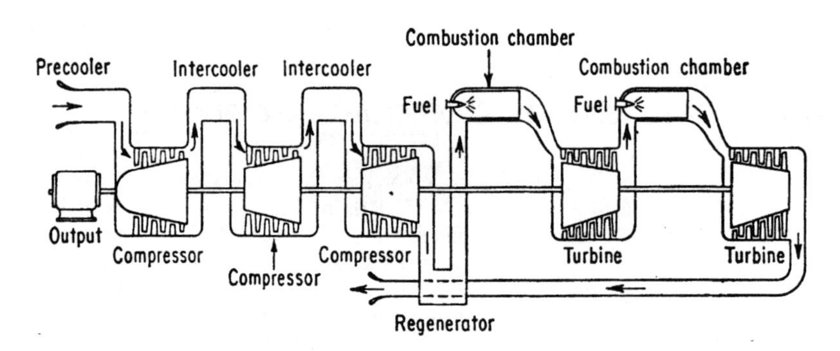

FIGURE 14 Ideal gas-turbine cycle *T-S* diagram with the same processes as in Fig. 13; complete-cycle gas turbine shown below the *T-S* diagram.

compression, in units given above; k = isentropic exponent = 1.4. With an entering air temperature, T_1 of 60°F (15.6°C), or 60 + 460 = 520°R, and using the data given, $T_2 = 520[(56/14)]^{(1.4-1)/1.4} = 772.7$°R, or 772.7 − 520 = 252.7°F (122.6°C).

(*a*) Here we have isentropic compression in the compressor with an efficiency of 85 percent. Using the equation, efficiency, isentropic = $(c_p)(T_2 - T_1)/(c_p)\,(T_{2'} - T_1)$, and solve for $T_{2'}$ the temperature after isentropic compression. Solving, $T_{2'}$ = 0.82 = 0.24(772.7 − 520)/0.24($T_{2'}$ − 520) = 828.4°R, or 368°F. This is the temperature after compression.

3. *Determine the temperature after combustion*
To find the temperature after combustion, use the relation heating value of fuel = $Q = c_p(T_{3'} - T_{2'})$, where $T_{3'}$ = temperature after combustion, °R. Substituting, 200 = 0.24($T_{3'}$ − 828). Solving, $T_{3'}$ = 1661.3°R; 1201.3°F (649.6°C).

4. *Find the temperature at the exhaust of the gas turbine*
Using an approach similar to that above, determine T_4 from $(T_{4'}/T_{3'}) = [(P_{4'}/P_{3'})]^{k-1/k}$. Substituting and solving for $T_{4'} = 1661[(14.9/56)]^{(1.4-1)/1.4}$ = 1137.9°R, or 677.8°F (358.8°C).

Now use the equation for gas-turbine efficiency, namely, turbine efficiency = $c_p\,(T_{3'} - T_{4''})/c_p\,(T_{3'} - T_{4'})$ = 0.85, and solve for $T_{4''}$, the temperature after expansion, at the exhaust. Substituting as earlier, $T_{4''}$ = 1218.2°R, 758.2°F (403.4°C). This is the temperature after expansion, i.e., at the exhaust of the gas turbine.

5. *Determine the work of compression, expander work, and thermal efficiency*
(*b*) The work of compression = $c_p\,(T_{2'} - \mathrm{T}_1)$ = 0.24(828 − 520) = 74.16 Btu (78.23 J).

The work delivered by the expander = $c_p(T_{2'} - T_1)$ = 0.24 (1661 − 1218) = 106.32 Btu (112.16 J).

The net work = 106.3 − 74.2 = 32.1 Btu (33.86 J). Then, the thermal efficiency = net work/heat supplied = 32.1/200 = 0.1605, 16.6 percent thermal efficiency.

Related Calculations. With the widespread use today of gas turbines in a variety of cycles in industrial and central-station plants, it is important that an engineer be able to analyze this important prime mover. Because gas turbines can be quickly installed and easily hooked to heat-recovery steam generators (HRSG), they are more popular than ever before in history.

Further, as aircraft engines become larger—such as those for the Boeing 787 and the Airbus 380—the power output of aeroderivative machines increases at little cost to the power industry. The result is further application of gas turbines for topping, expansion, cogeneration, and a variety of other key services throughout the world of power generation and energy conservation.

With further refinement in gas-turbine cycles, specific fuel consumption, Fig. 15, declines. Thus, the complete cycle gas turbine has the lowest specific fuel consumption, with the regenerative cycle a close second in the 6-to-1 compression-ratio range.

Two recent developments in gas-turbine plants promise much for the future. The first of these developments is the single-shaft combined-cycle gas and steam turbine, Fig. 16. In this cycle, the gas turbine exhausts into a heat-recovery steam generator (HRSG) that supplies steam to the turbine. This cycle is the most significant electric-generating system available today. Further, its capital costs are significantly lower than competing nuclear, fossil-fired steam, and renewable-energy stations. Other advantages include low air emissions, low water consumption, smaller space requirements, and a reduced physical profile, Fig. 17. All these advantages are important in today's strict permitting and siting processes.

Having the gas turbine, steam turbine, and generator all on one shaft simplifies plant design and operation, and may lower first costs. When used for large reheat cycles, as shown here, separate high-pressure (h-p), intermediate-pressure (i-p), and low-pressure (l-p) turbine elements are all on the same shaft as the gas turbine and generator. Modern high-technology combined-cycle single-shaft units deliver a simple-cycle net efficiency of 38.5 percent for a combine-cycle net efficiency of 58 percent on a lower heating value (LHV) basis.

The second important gas-turbine development worth noting is the dual-fueled turbine located at the intersection of both gas and oil pipelines. Being able to use either fuel gives the gas turbine greater opportunity to increase its economy by switching to the lowest-cost fuel whenever necessary. Further developments along these lines is expected in the future.

The data in the last three paragraphs and the two illustrations are from *Power* magazine.

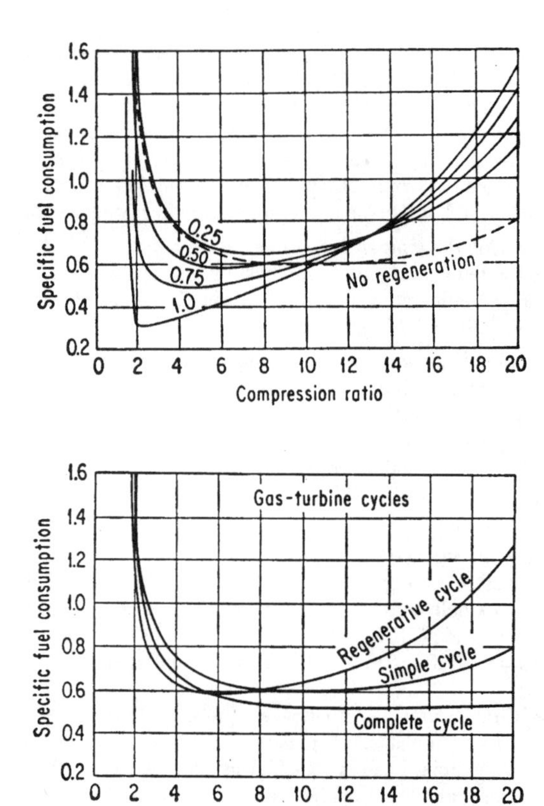

FIGURE 15 With further gas-turbine cycle refinement, the specific fuel consumption declines. These curves are based on assumed efficiencies with $T_3 = 1400°F$ (760°C).

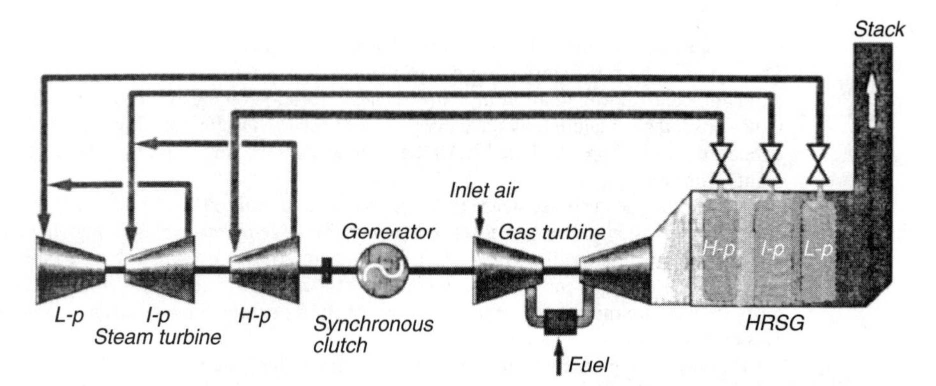

FIGURE 16 Single-shaft combined-cycle technology can reduce costs and increase thermal efficiency over multi-shaft arrangements. This concept is popular in Europe. (*Power.*)

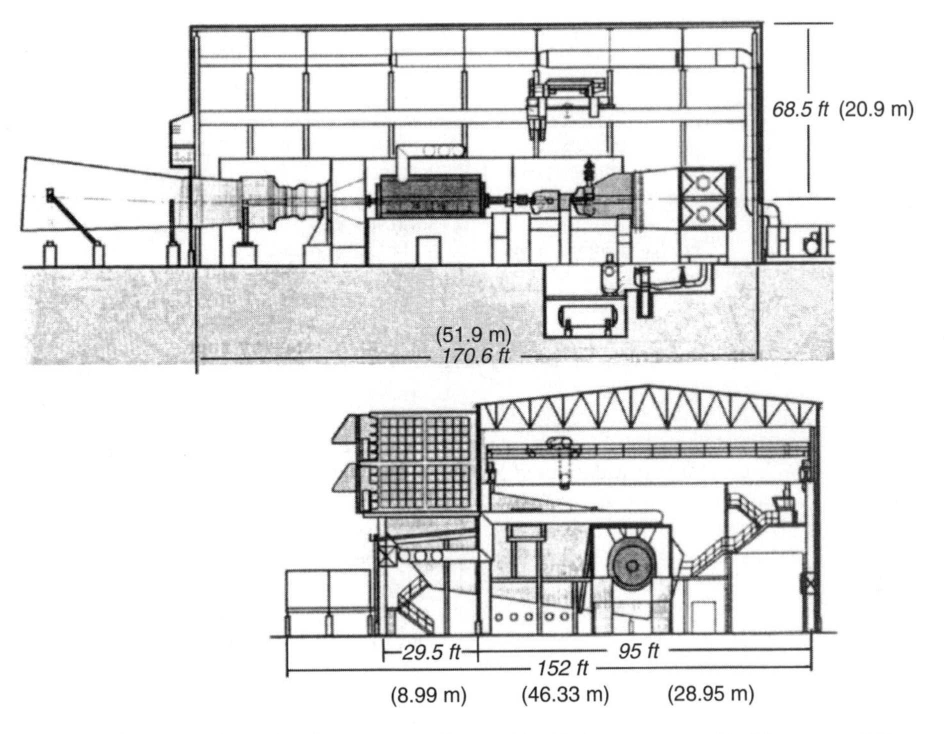

FIGURE 17 Steam turbine, electric generator, and gas turbine fit into one compact building when all three machines are arranged on a single shaft. Net result: Reduced site footprint and civil-engineering work. (*Power.*)

INDUSTRIAL GAS TURBINE LIFE-CYCLE COST-MODEL BEST RELATIVE-VALUE DETERMINATION

An industrial application requires a 21-MW continuous electrical output year-round. Five different gas turbines are under consideration. Determine which of these five turbines is the best choice, using a suitable life-cycle cost analysis.

Calculation Procedure:

1. ***Assemble the cost data for each gas turbine being considered***
Assemble the cost data as shown below for each of the five gas turbines identified by the letters A through E. Contact the gas-turbine manufacturers for the initial cost, \$/kW, thermal efficiency, availability, fuel consumption, generator efficiency, and maintenance cost, \$/kWh. List these data as shown below.

The loan period, years, will be the same for all the gas turbines being considered, and is based on an equipment life expectancy of 20 years. Interest rate on the capital investment for each turbine will vary, depending on the amount invested and the way in which the loan must be repaid and will be provided by the accounting department of the firm considering gas-turbine purchase.

Equipment Attributes for Typical Candidates*

Parameter	Gas-turbine candidates				
	A	B	C	D	E
Initial cost, \$/kW	205	320	275	320	200
Thermal efficiency, %	32.5	35.5	34.0	36.5	30.0
Loan period, yr	20	20	20	20	20
Availability	0.96	0.94	0.95	0.94	0.96
Fuel cost, \$/million Btu	4	4	4	4	4
Interest, %	6.5	8.0	7.0	8.5	7.5
Generator efficiency, %	98.0	98.0	98.0	98.0	98.0
Maintenance cost, \$/kWh	0.004	0.005	0.005	0.005	0.005

*Assuming an equipment life of 20 years, an output of 21 MW.

2. *Select a life-cycle cost model for the gas turbines being considered*
A popular and widely used life-cycle cost model for gas turbines has three parts: (1) the annual investment cost, C_p; (2) annual fuel cost, C_f; (3) annual maintenance cost, C_m. Summing these three annual costs, all of which are expressed in mils/kWh, gives C_T, the life-cycle cost model. The equations for each of the three components are given below, along with the life-cycle working model, C_T:

The life-cycle cost model (C_T) consists of annual investment cost (C_p) + annual fuel cost (C_f) + annual maintenance cost (C_m). Equations for these values are:

$$C_p = \frac{I\{i/[1-(1-i)^{-n}]\}}{(A)(\text{kW})(8760)(G)}$$

where

I = initial capital cost of equipment, dollars
i = interest rate
n = number of payment periods
A = availability (expressed as decimal)
kW = kilowatts of electricity produced
8760 = total hours in year
G = efficiency of electric generator

$$C_f = E(293)$$

where

E = thermal efficiency of gas turbine
293 = conversion of Btu to kWh

$$C_m = M/\text{kW}$$

where M = maintenance cost, dollars per operating (fired) hour.

Thus, the life-cycle working model can be expressed as

$$C_r = \frac{I\{i/[1-(1-i)^{-n}]\}}{(A)(\text{kW})(8760)(G)} + F/E(293) + M/\text{kW}$$

Where F = fuel cost, dollars per million Btu (higher heating value)

To evaluate the comparative capital cost of a gas-turbine electrical-generating package, the above model uses the capital-recovery factor technique. This approach spreads the initial investment and interest costs for the repayment period into an equal annual expense using the time value of money. The approach also allows for the comparison of other periodic expenses, like fuel and maintenance costs.

3. ***Perform the computation for each of the gas turbines being considered***
Using the compiled data shown above, compute the values for C_p, C_f, and C_m, and sum the results. List for each of the units as shown below.

Results from Cost Model

Unit	Mils/kWh produced
A	48.3
B	47.5
C	48.3
D	46.6
E	51.9

4. ***Analyze the findings of the life-cycle model***
Note that the initial investment cost for the turbines being considered ranges between \$200 and \$320/kW. On a \$/kW basis, only unit E at the \$200 level, would be considered. However, the life-cycle cost model, above, shows the cost per kWh produced for each of the gas-turbine units being considered. This gives a much different perspective of the units.

From a life-cycle standpoint, the choice of unit E over unit D would result in an added expenditure of about \$975,000 annually during the life span of the equipment, found from [(51.9 – 46.6)/1000] (8760 h/yr)(21,000 kW) = \$974,988; this was rounded to \$975,000. Since the difference in the initial cost between units D and E is \$6,720,000 – \$4,200,000 = \$2,520,000, this cost difference will be recovered in \$2,520,000/974,988 – 2.58 years, or about one-eighth of the 20-year life span of the equipment.

Also, note that the 20-year differential in cost/kWh produced between units D and E is equivalent to over 4.6 times the initial equipment cost of unit E. When considering the values output of a life-cycle model, remember that such values are only as valid as the data input. So take precautions to input both reasonable and accurate data to the life-cycle cost model. Be careful in attempting to distinguish model outputs that vary less than 0.5 mil from one another.

Since the predictions of this life-cycle cost model cannot be compared to actual measurements at this time, a potential shortcoming of the model lies with the validity of the data and assumptions used for input. For this reason, the model is best applied to establish comparisons to differentiate between several pieces of competing equipment.

Related Calculations. The first gas turbines to enter industrial service in the early 1950s represented a blend of steam-turbine and aerothermodynamic design. In the late-1950s/early-1960s, lightweight industrial gas turbines derived directly from aircraft engines were introduced into electric power generation, pipeline compression, industrial power generation, and a variety of other applications. These machines had performance characteristics similar to their steam-turbine counterparts, namely pressure ratios of about 12:1, firing temperatures of 1200–1500°F (649–816°C), and thermal efficiencies in the 23–27 percent range.

In the 1970s, a new breed of aeroderivative gas turbines entered industrial service. These units, with simple-cycle thermal efficiencies in the 32–37 percent bracket, represented a new technological approach to aerothermodynamic design.

Today, these second-generation units are joined by hybrid designs that incorporate some of the aeroderivative design advances but still maintain the basic structural concepts of the heavy-frame machines. These hybrid units are not approaching the simple-cycle thermal-efficiency levels reached by some of the early second-generation aeroderivative units first earmarked for industrial use.

Traditionally, the major focus has been on first cost of industrial gas-turbine units, not on operating cost. Experience with higher-technology equipment, however, reveals that a low first cost does not mean a lower total cost during the expected life of the equipment. Conversely, reliable, high-quality equipment with demonstrated availability will be remembered long after the emotional distress associated with high initial cost is forgotten.

The life-cycle cost model presented here uses 10 independent variables. A single-point solution can easily be obtained, but multiple solutions require repeated calculations. Although curves depicting simultaneous variations in all variables would be difficult to interpret, simplified diagrams can be constructed to illustrate the relative importance of different variables.

Thus, the simplified diagrams shown in Fig. 18, all plot production cost, mils/kWh, versus investment cost. All the plots are based on continuous operation of 8760 h/yr at 21-MW capacity with an equipment life expectancy of 20 years.

The curves shown depict the variation in production cost of electricity as a function of initial investment cost for various levels of thermal efficiency, loan repayment period, gas-turbine availability, and fuel cost. Each of these factors is an element in the life-cycle cost model presented here.

This procedure is the work of R. B. Spector, General Electric Co., as reported in *Power* magazine.

HRSG TUBE-BUNDLE VIBRATION AND NOISE ANALYSIS

A tubular air heater 11.7 ft (3.57 m) wide, 12.5 ft (3.81 m) deep, and 13.5 ft (4.11 m) high is used in a boiler plant. Carbon steel tubes 2 in (5.08 cm) in outer diameter and 0.08 in (0.20 cm) thick are used in inline fashion with a traverse pitch of 3.5 in (8.89 cm) and a longitudinal pitch of 3 in (7.62 m). There are 40 tubes wide and 60 tubes deep in the heater; 300,000 lb (136,200 kg) of air flows across the tubes at an average temperature of 219°F (103.9°C). The tubes are fixed at both ends. Tube mass per unit length = 1.67 lb/ft (2.49 kg/m). Check this air heater for possible tube vibration problems.

Calculation Procedure:

1. *Determine the mode of vibration for the tube bundle*

Whenever a fluid flows across a tube bundle such as boiler tubes in an evaporator, economizer, HRSG, superheater, or air heater, vortices are formed and shed in the wake beyond the tubes. This shedding on alternate sides of the tubes causes a harmonically varying force on the tubes perpendicular to the normal flow of the fluid. It is a self-excited vibration. If the frequency of the Von Karman vortices, as they are termed, coincides with the natural frequency of vibration of the tubes, then resonance occurs and the tubes vibrate, leading to possible damage of the tubes.

Vortex shedding is most prevalent in the range of Reynolds numbers from 300 to 200,000, the range in which most boilers operate. Another problem encountered with vortex shedding is acoustic vibration, which is normal to both the fluid flow and tube length observed in only gases and vapors. This occurs when the vortex shedding frequency is close to the acoustic frequency. Excessive noise is generated, leading to large gas pressure drops and bundle and casing damage. The starting point in the evaluation for noise and vibration is the estimation of various frequencies.

Use the listing of C values shown below to determine the mode of vibration. Note that C is a factor determined by the end conditions of the tube bundle.

	Mode of vibration		
End conditions	1	2	3
Both ends clamped	22.37	61.67	120.9
One end clamped, one end hinged	15.42	49.97	104.2
Both hinged	9.87	39.48	88.8

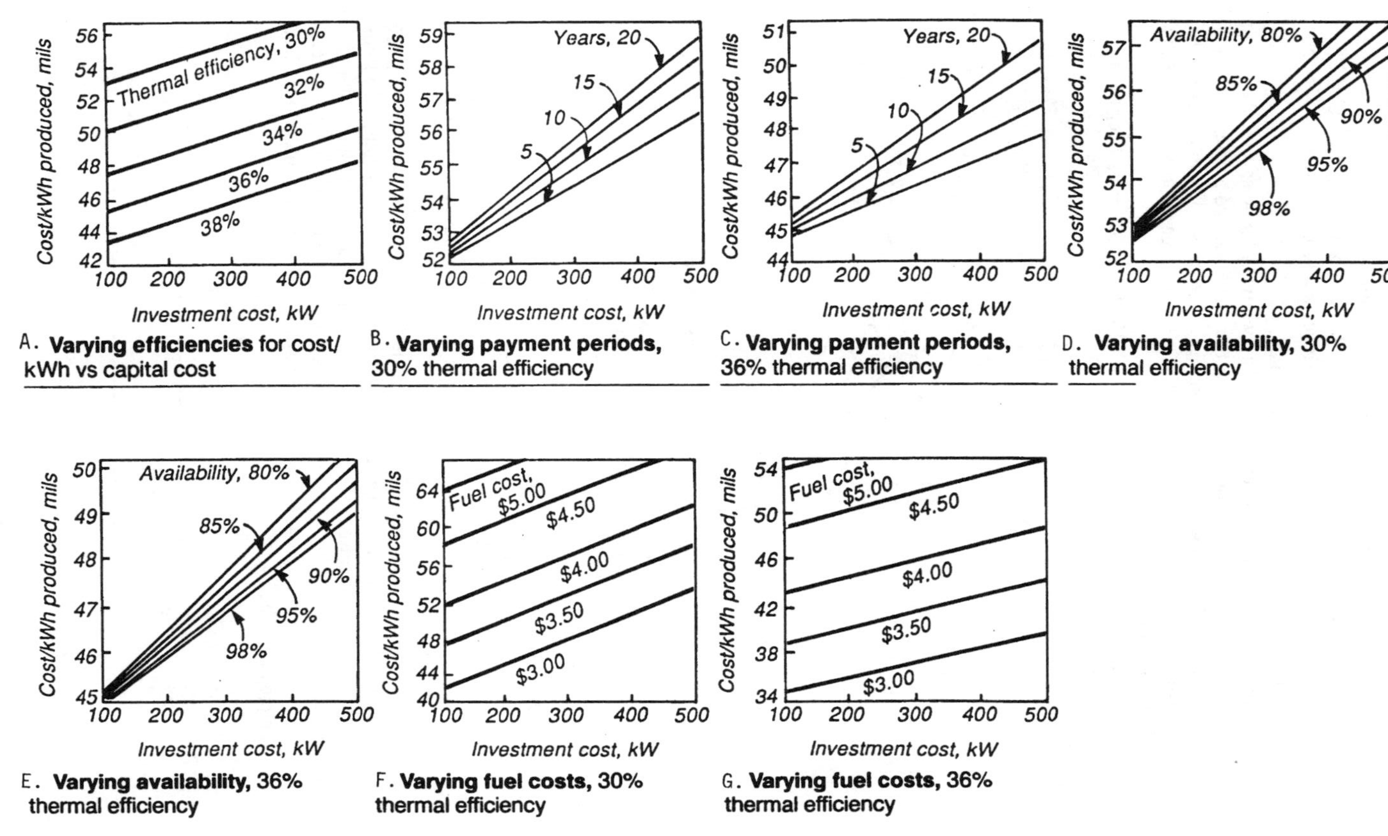

FIGURE 18 Economic study plots for life-cycle costs. (*Power*.)

Since the tubes are fixed at both ends, i.e., clamped, select the mode of vibration as 1, with $C = 22.37$. For most situations, Mode 1 is the most important case.

2. *Find the natural frequency of the tube bundle*
Use the relation, $f_n = 90C[d_o^4 - d_i^4]/(L^2 - M^{0.5})$. Substituting, with $C = 22.37$, $f_n = (90)(22.37)[2^4 - 1.84^4]^{0.5}/(13.5^2 - 1.67^{0.5}) = 18.2$ cycles per second (cps). In Mode 2, $f_n = 50.2$, as $C = 61.67$.

3. *Compute the vortex shedding frequency*
To compute the vortex shedding frequency we must know several factors, the first of which is the Strouhl number, *S*. Using Fig. 19 with a transverse pitch/diameter of 1.75 and a longitudinal pitch diameter of 1.5, we find $S = 0.33$. Then, the air density = 40/(460 – 219) = 0.059 lb/ft^3 (0.95 kg/m^3); free gas area = 40(3.5 – 2)(13.5/12) = 67.5 ft^2 (6.3 m^2); gas velocity, V = 300,000/(67.5)(0.059)(3600) = 21 ft/s (6.4 m/s).

Use the relation, $f_c = 12(S)(V)/d_o = 12(0.33)(21)/2 = 41.6$ cps, where f_c = vortex shedding frequency, cps.

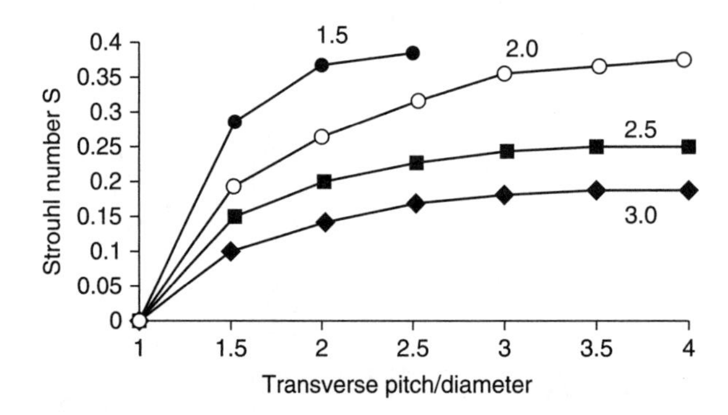

FIGURE 19 Strouhl number, *S*, for inline tube banks. Each curve represents a different longitudinal pitch/diameter ratio (Chen).

4. *Determine the acoustic frequency*
As with vortex frequency, we must first determine several variables, namely: absolute temperature = °R = 219 + 460 = 679°R; sonic velocity, $V_s = 49(679)^{0.5}$ = 1277 ft/s (389.2 m/s); wave length, $\lambda = 2(w)/n$, where w = width of tube bank, ft (*m*); n = mode of vibration = 1 for this tube bank; then $\lambda = 2(11.7)/1 = 23.4$ ft (7.13 m).

The acoustic frequency, $f_a = (V_s)/\lambda$, where V_s = velocity of sound at the gas temperature in the duct or shell, ft/s (m/s); $V_s = [(g)(\rho)(RT)]^{0.5}$, where R = gas constant = 1546/molecular weight of the gas; T = gas temperature, °R; ρ = ratio of gas-specific heats, typically 1.4 for common flue gases; the molecular weight = 29. Simplifying, we get $V_s = 49(T)^{0.5}$, as shown above. Substituting, f_a = 1277/23.4 = 54.5 cps. For $n = 2$; $f_a = 54.4(2) = 109$ cps. The results for Modes 1 and 2 are summarized in the tabulation below.

Mode of vibration n	1	2
f_n, cps	18.2	50.2
f_c, cps	41.6	41.6
f_a (without baffles)	54.5	109
f_a (with baffles)	109	218

The tube natural frequency and the vortex shedding frequency are far apart. Hence, the tube-bundle vibration problem is unlikely to occur. However, the vortex shedding and acoustic frequencies are close. If the air flow increases slightly, the two frequencies will be close. By inserting a baffle in the tube bundle (dividing the ductwork into two along the gas flow direction) we can double the acoustic frequency as the width of the gas path is now halved. This increases the difference between vortex shedding and acoustic frequencies and prevents noise problems.

Noise problems arise when the acoustic and vortex shedding frequencies are close—usually within 20 percent. Tube-bundle vibration problems arise when the vortex shedding frequency and natural frequency of the bundle are close—within 20 percent. Potential noise problems must also be considered at various turndown conditions of the equipment.

Related Calculations. For a thorough analysis of a plant or its components, evaluate the performance of heat-transfer equipment as a function of load. Analyze at various loads the possible vibration problems that might occur. At low loads in the above case, tube-bundle vibration is likely, while at high loads acoustic vibration is likely without baffles. Hence, a wide range of performance must be reviewed before finalizing any tube-bundle design, Fig. 20.

This procedure is the work of V. Ganapathy, Heat Transfer Specialist, ABCO Industries, Inc.

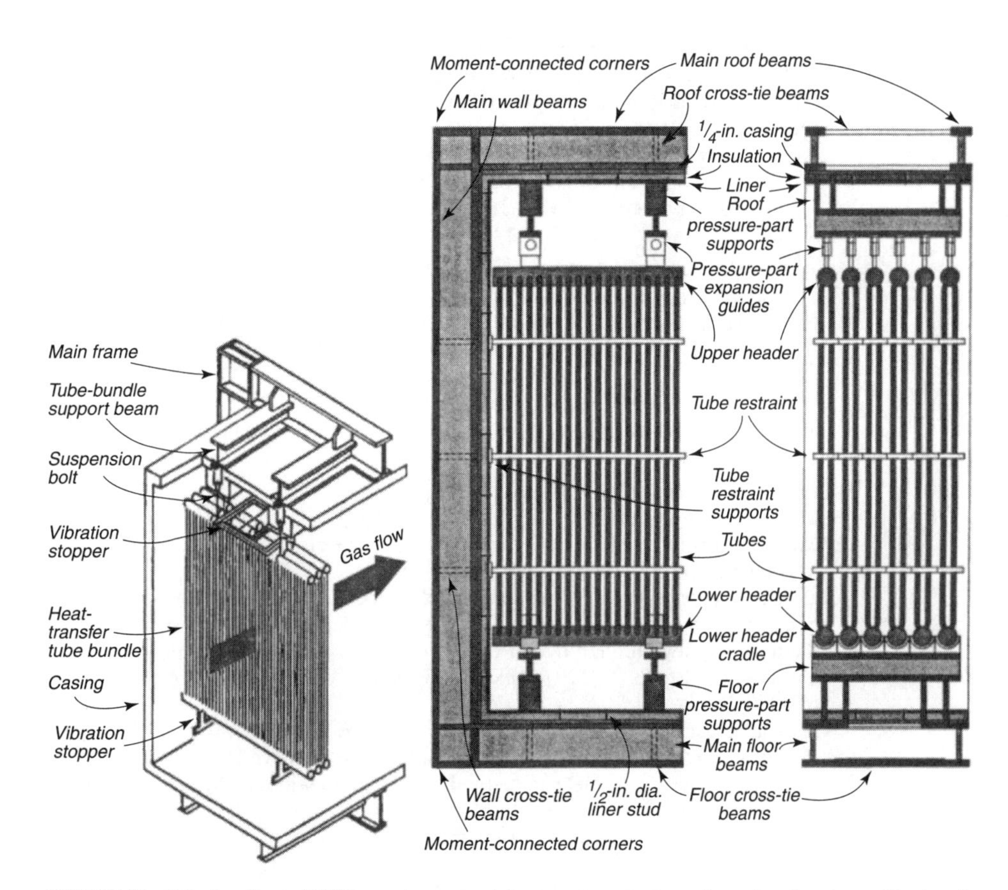

FIGURE 20 Tube bundles in HRSGs require appropriate support mechanisms; thermal cycling in combined-cycle units makes this consideration even more important. (*Power*.)

GAS-TURBINE-PLANT OXYGEN AND FUEL INPUT DETERMINATION

In a gas-turbine HRSG (heat-recovery steam generator) it is desired to raise the temperature of 150,000 lb/h (68,100 kg/h) of exhaust gases from 950°F (510°C) to 1575°F (857.2°C) in order to nearly double the output of the HRSG. If the exhaust gases contain 15 percent oxygen by volume, determine the fuel input and oxygen consumed, using the gas specific-heat method.

Calculation Procedure:

1. *Determine the air equivalent in the exhaust gases*

In gas-turbine–based cogeneration/combined-cycle projects, the HRSG may be fired to generate more steam than that produced by the gas-turbine exhaust gases. Typically, the gas-turbine exhaust gas contains 14 to 15 percent oxygen by volume. So the question arises: How much fuel can be fired to generate more steam? Would the oxygen in the exhaust gases run out if we fired to a desired temperature? These questions are addressed in this procedure.

If 0 percent oxygen is available in W_g lb/h (kg/h) of exhaust gases, the air-equivalent W_a in lb/h (kg/h) is given by: $W_a = 100(W_g)(32O_x)/[23(100)(29.5)] = 0.0417\ W_g(O)$. In this relation, we are converting the oxygen from a volume basis to a weight basis by multiplying by its molecular weight of 32 and dividing by the molecular weight of the exhaust gases, namely 29.5. Then multiplying by (100/23) gives the air equivalent as air contains 23 percent by weight of oxygen.

2. *Relate the air required with the fuel fired using the MM Btu (kJ) method*

Each MM Btu (kJ) of fuel fired (HHV basis) requires a certain amount of air, A. If Q = amount of fuel fired in the turbine exhaust gases on a LHV basis (calculations for turbine exhaust gases fuel input are done on a low-heating-value basis), then the fuel fired in lb/h (kJ/h) = $W_f = Q/\text{LHV}$.

The heat input on an HHV basis = $W_f(\text{HHV})/(10^6) = (Q/\text{LHV})(\text{HHV})/10^6$ Btu/h (kJ/h). Air required lb/h (kg/h) = $(Q\ /\text{LHV})(\text{HHV})(A)$, using the MM Btu, where A = amount of air required, lb (kg) per MM Btu (kJ) fired. The above quantity = air available in the exhaust gases, W_a = $0.0417\ W_g(O)$.

3. *Simplify the gas relations further*

From the data in step 2, $(Q/\text{LHV})(\text{HHV})(A)/10^6 = 0.0417\ W_g(O)$. For natural gas and fuel oils it can be shown that $(\text{LHV}/A_x\ \text{HHV}) = 0.00124$. For example, LHV of methane = 21,520 Btu/lb (50,055.5 kJ/kg); HHV = 23,879 Btu/lb (55,542.6 kJ/kg), and A = 730 lb (331.4 kg). Hence, $(\text{LHV}/A_x\text{HHV}) = 21{,}520/(730 \times 23{,}879) = 0.00124$. By substituting in the equation in step 1, we have $Q = 58.4\ (W_g)(O)$. This is an important equation because it relates the oxygen consumption from the exhaust gases to the burner fuel consumption.

4. *Find the fuel input to the HRSG*

The fuel input is given by $W_g + h_{g1} + Q = (W_g + W_f)(h_{g2})$, where h_{g1} and h_{g2} are the enthalpies of the exhaust gas before and after the fuel burner, respectively; W_f = fuel input, lb/h (kg/h); Q = fuel input in Btu/h (kJ/h).

The relation above requires enthalpies of the gases before and after the burner, which entails detailed combustion calculations. However, considering that the mass of fuel is a small fraction of the total gas flow through the HRSG, the fuel flow can be neglected. Using a specific heat for the gases of 0.31 Btu/lb °F (1297.9 J/kg K), we have, $Q = 150{,}000(0.31)(1575 - 950) = 29 \times 10^6$ Btu/h (8.49 kW).

The percent of oxygen by volume, $O = (29 \times 10^6)/(58.4 \times 150{,}000) = 3.32$ percent. That is, only 3.32 percent oxygen by volume is consumed and we still have 15.00 – 3.32 = 11.68 percent left in the flue gases. Thus, more fuel can be fired and the gases will not run out of oxygen for combustion.

Typically, the final oxygen content of the gases can go as low as 2 to 3 percent using 3 percent final oxygen, the amount of fuel that can be fired = (150,000)(58.4)(15 – 3) = 105 MM Btu/h (110.8 MM J/h). It can be shown through an HRSG simulation program (contact the author for more information) that all of the fuel energy goes into steam. Thus, if the unfired HRSG were generating 23,000 lb/h (10,442 kg/h) of steam with an energy absorption of 23 MM Btu/h (24.3 MM J/h), approximately, the amount of steam that can be generated by firing fuel in the HRSG = 23 + 105 = 128 MM Btu/h (135 MM J/h), or 128,000 lb/h (58,112 kg/h) of steam. This is close to a firing temperature of 3000 to 3100°F (1648 to 1704°C).

Related Calculations. Using the methods given in Section 1 of this handbook, one may make detailed combustion calculations and obtain a flue-gas analysis after combustion. Then compute the enthalpies of the exhaust gas before and after the burner. Using this approach, you can check the burner duty more accurately than using the gas specific-heat method presented above. This procedure is the work of V. Ganapathy, Heat Transfer Specialist, ABCO Industries, Inc.

Power magazine recently commented on the place of gas turbines in today's modern power cycles thus: Using an HRSG with a gas turbine enhances the overall efficiency of the cycle by recovering heat in the gas-turbine's hot exhaust gases. The recovered heat can be used to generate steam in the HRSG for either (1) injection back into the gas turbine, Fig. 21, (2) use in district heating or an industrial process, (3) driving a steam turbine-generator in a combined-cycle arrangement, or (4) any combination of the first three.

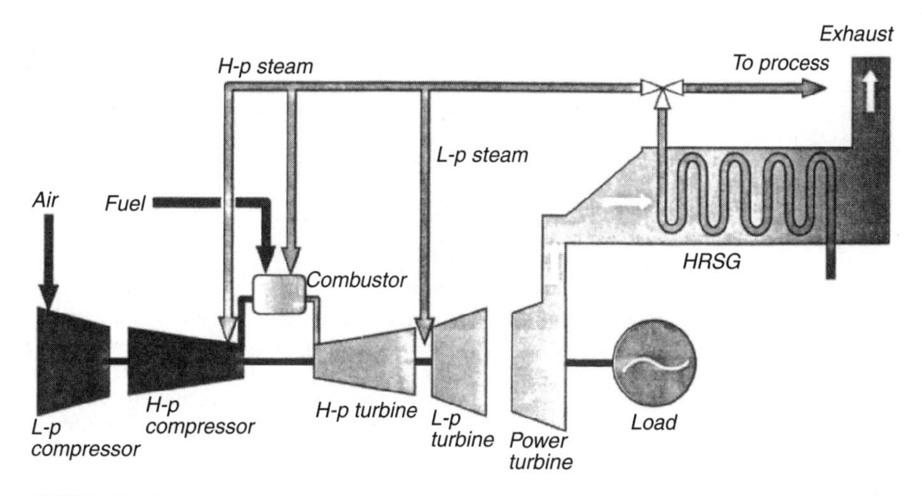

FIGURE 21 Steam injection systems offer substantial improvement in both capacity and efficient. (*Power*.)

Steam injection into the gas turbine has many benefits, including: (1) achievable output is increased by 25 percent or more, depending on the gas-turbine design, (2) part-load gas-turbine efficiency can be significantly improved, (3) gas-fired NO_x, emissions can be markedly reduced—up to the 15–45 ppm range in many cases, (4) operating flexibility is improved for cogeneration plants because electrical and thermal outputs can be balanced to optimize overall plant efficiency and profitability.

Combined-cycle gas-turbine plants are inherently more efficient than simple-cycle plants employing steam injection. Further, combined-cycle plants may also be considered more adaptable to cogeneration compared to steam-injected gas turbines. The reason for this is that the maximum achievable electrical output decreases significantly for steam-injected units in the cogeneration mode because less steam is available for use in the gas turbine. In contrast, the impact of cogeneration on electrical output is much less for combined-cycle plants.

Repowering in the utility industry can use any of several plant-revitalization schemes. One of the most common repowering options employed or considered today by utilities consists of replacing an aging steam generator with a gas-turbine/generator and HRSG, Fig. 22. It is estimated that within the next few years, more than 3500 utility power plants will have reached their 30th birthdays. A significant number of these facilities—more than 20 GW of capacity by some estimates—are candidates

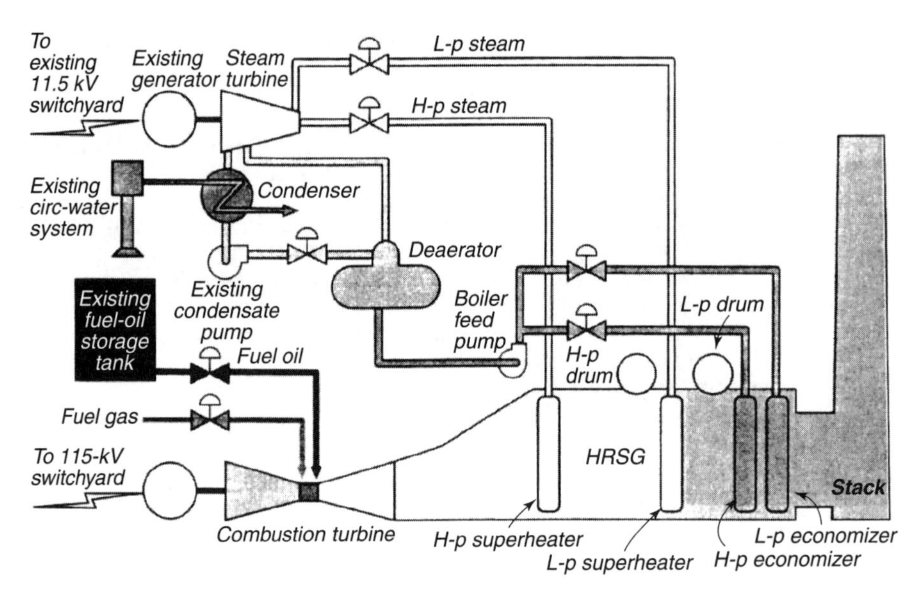

FIGURE 22 HRSG and gas turbine used in repowering. (*Power.*)

for repowering, an option that can cut emissions and boost plant efficiency, reliability, output, and service life.

And repowering often proves to be more economical, per cost of kilowatt generated, compared to other options for adding capacity. Further, compared to building a new power plant, the permitting process for repowering it is typically much shorter and less complex. The HRSG will often have a separate firing capability such as that discussed in this calculation procedure.

These comments from *Power* magazine were prepared by Steven Collins, Assistant Editor of the publication.

SIMULATION OF HEAT-RECOVERY STEAM GENERATORS (HRSG)

A gas turbine exhausts 140,000 lb/h (63,560 kg/h) of gas at 980°F (526.7°C) to an HRSG generating saturated steam at 200 lb/in^2 (gage) (1378 kPa). Determine the steam-generation and design-temperature profiles if the feedwater temperature is 230°F (110°C) and blowdown = 5 percent. The average gas-turbine exhaust gas specific heat is 0.27 Btu/lb °F (1.13 kJ/kg °C) at the evaporator and 0.253 Btu/lb °F (1.06 kJ/kg °C) at the economizer. Use a 20°F (11.1°C) pinch point, 15°F (8.3°C) approach point, and 1 percent heat loss. Evaluate the evaporator duty, steam flow, economizer duty, and exit-gas temperature for normal load conditions. Then determine how the HRSG off-design temperature profile changes when the gas-turbine exhaust-gas flow becomes 165,000 lb/h (74,910 kg/h) at 880°F (471°C) with the HRSG generating 150-lb/in^2 (gage) (1033.5 kPa) steam with the feedwater temperature remaining the same.

Calculation Procedure:

1. ***Compute the evaporator duty and steam flow***

Engineers should be able to predict both the design and off-design performance of an HRSG, such as that in Fig. 23, under different conditions of exhaust flow, temperature, and auxiliary firing without

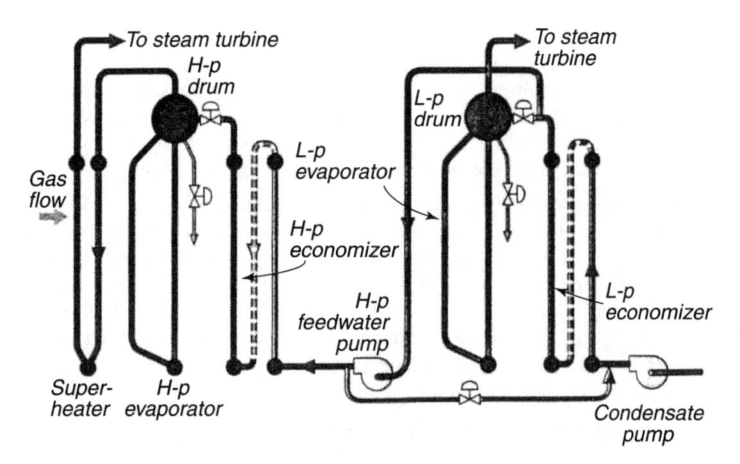

FIGURE 23 HRSG circuit shown is used by at least one manufacturer to prevent steaming in the economizer during startup and low-load operation. (*Power*.)

delving into the mechanical design aspects of tube size, length, or fin configuration. This procedure shows how to make such predictions for HRSGs of various sizes by using simulation techniques.

HRSGs operate at different exhaust-gas conditions. For example, variations in ambient temperature or gas-turbine load affect exhaust-gas flow and temperature. This, in turn, affects HRSG performance, temperature profiles, efficiency, and steam generation. The tool consultants use for evaluating HRSG performance under different operating conditions is simulation. With this tool, you can: (1) predict off-design performance of an HRSG; (2) predict auxiliary fuel consumption for periods when the gas-turbine exhaust-gas flow is insufficient to generate the required steam flow; (3) evaluate options for improving an HRSG system; (4) evaluate field data for validating an HRSG design; (5) evaluate different HRSG configurations for maximizing efficiency.

In this HRSG, using steam-table data, the saturation temperature of 200-lb/in^2 (gage) (1378-kPa) steam = 388°F (197.8°C). The gas temperature leaving the evaporator with the 20°F (11.1°C) pinch point = 388 + 20 = 408°F (208.9°C). Water temperature entering the evaporator = saturated-steam temperature – the approach point temperature difference, or 388 – 15 = 373°F (189.4°C).

Then, the energy absorbed by the evaporator, Q_1 = (gas flow, lb/h)(1.0 – heat loss)(gas specific-heat, Btu/lb °F)(gas-turbine exhaust gas HRSG entering temperature, °F – gas temperature leaving evaporator, °F). Or, Q_1 = (140,000)(0.99)(0.27)(980 – 408) = 21.4 MM Btu/h (6.26 MW). The enthalpy absorbed by the steam in the evaporator, Btu/lb (kJ/kg) = (enthalpy of the saturated steam in the HRSG outlet – enthalpy of the feedwater entering the evaporator at 373°F) + (blowdown percentage)(enthalpy of the saturated liquid of the outlet steam – enthalpy of the water entering the evaporator, all in Btu/lb). Or, enthalpy absorbed in the evaporator = (1199.3 – 345) + (0.05)(362.2 – 345) = 855.2 Btu/lb (1992.6 kJ/kg). The quantity of steam generated = (Q_1, energy absorbed by the evaporator, Btu/h)/(enthalpy absorbed by the steam in the evaporator, Btu /lb) = (21.4 × 10^6)/855.2 = 25,023 lb/h (11,360 kg/h).

2. *Determine the economizer duty and exit-gas temperature*

The economizer duty = (steam generated, lb/h)(enthalpy of water entering the economizer, Btu/lb – enthalpy – enthalpy of the feedwater at 230°F, Btu/lb)(l + blowdown percentage) = (25,023)(345 – 198.5)(1.05) = 3.849 MM Btu/h (1.12 MW).

The gas temperature drop through the economizer = (economizer duty)/(gas flow rate, lb/h) (1 – heat loss percentage)(specific heat of gas, Btu/lb °F) = (3.849 × 10^6)/(140,000)(0.99)(0.253) = 109.8°F (60.9°C). Hence, the exit-gas temperature from the economizer = (steam saturation temperature, °F – exit-gas temperature from the economizer, °F) = (408 – 109) = 299°F (148.3°C).

3. *Calculate the constant K for evaporator performance*
In simulating evaporator performance the constant K_1 is used to compute revised performance under differing flow conditions. In equation form, K_1, = ln[(temperature of gas-turbine exhaust gas entering the HRSG, °F – HRSG saturated steam temperature, °F)/(gas temperature leaving the evaporator, °F – HRSG saturate steam temperature, °F)]/(gas flow, lb/h). Substituting, K_1, = ln[(980 – 388)/(408 – 388)]/140,000 = 387.6, where the temperatures used reflect design condition.

4. *Compute the revised evaporator performance*
Under the revised performance conditions, using the given data and the above value of K_1 and solving for T_{g2}, the evaporator exit-gas temperature, ln[(880 – 366)/(T_{g2} – 366)] = 387.6(165,000)$^{-0.4}$; T_{g2} = 388 °F (197.8°C). Then, the evaporator duty, using the same equation as in step 1 above = (165,000)(0.99)(0.27)(880 – 388) = 21.7 MM Btu/h (6.36 MW).

In this calculation, we assumed that the exhaust-gas analysis had not changed. If there are changes in the exhaust-gas analysis, then the gas properties must be evaluated and corrections made for variations in the exhaust-gas temperature. See *Waste Heat Boiler Deskbook* by V. Ganapathy for ways to do this.

5. *Find the assumed duty, Q_a, for the economizer*
Let the economizer leaving-water temperature = 360°F (182.2°C). The enthalpy of the feedwater = 332 Btu/lb (773.6 kJ/kg); saturated-steam enthalpy = 1195.7 Btu/ lb (2785.9 kJ/kg); saturated-liquid enthalpy = 338.5 Btu/lb (788.7 kJ/kg). Then, the steam flow, as before, = (21.5 × 10^6)/[(1195.7 – 332) + 0.05 (338.5 – 332)] = 25,115.7 lb/h (11,043 kg/h). Then, the assumed duty for the economizer, Q_a = (25,115.7)(1.05)(332 – 198.5) = 3.52 MM Btu/h (1.03 MW).

6. *Determine the UA value for the economizer in both design and off-design conditions*
For the design conditions, *UA* = *Q*/(Δ*T*), where *Q* = economizer duty from step 2, above; Δ*T* = design temperature conditions from the earlier data in this procedure. Solving, *UA* = (3.84 × 10^6)/{[(299 – 230) – (408 – 373)]/ln(69/35)} = 76,800 Btu/h °F (40.5 kW). For off-design conditions, *UA* = (*UA* at design conditions) (gas flow at off-design/gas flow at design conditions)$^{0.65}$ = (76,800) (165,000/140,000)$^{0.65}$ = 85,456 Btu/h °F (45.1 kW).

7. *Calculate the economizer duty*
The energy transferred = Q_t = (*UA*)(Δ*T*). Based on 360°F (182.2°C) water leaving the economizer, Q_a = 3.52 MM Btu/h (1.03 MW). Solving for t_{g2} as before = 382 – [(3.52 × 10^6)/(165,000)(0.9)(0.253)] = 388 – 85 = 303°F (150.6°C). Then, Δ*T* = [(303 – 230) – (388 – 360)]/ln(73/28) = 47°F (26.1°C). The energy transferred = Q_t = (*UA*)(Δ*T*) = (85,456)(47) = 4.01 MM Btu/h (1.18 MW).

Since the assumed and transferred duty do not match, i.e., 3.52 MM Btu/h versus 4.01 MM Btu/h, another iteration is required. Continued iteration will show that when Q_a = Q_t = 3.55 MM Btu/h (1.04 MW), and the temperature of the water leaving the economizer, = 366°F (185.6°C) (saturation) and exit-gas temperature = 301°F (149.4°C), the amount of steam generated = 25,310 lb/h (11,491 kg/h).

Related Calculations. Studying the effect of gas inlet-temperature and gas flows on HRSG performance will show that at lower steam generation rates or at lower pressures that the economizer water temperature approaches saturation temperature, a situation called "steaming" in the economizer. This steaming condition should be avoided by generating more steam by increasing the inlet gas temperature or through supplementary firing, or by reducing exhaust-gas flow.

Supplementary firing in an HRSG also improves the efficiency of the HRSG in two ways: (1) The economizer acts as a bigger heat sink as more steam and hence more feedwater flows through the economizer. This reduces the exit-gas temperature. So with a higher gas inlet-temperature to the HRSG, we have a lower exit-gas temperature, thanks to the economizer. (2) Additional fuel burned in the HRSG reduces the excess air as more air is not added; instead, the excess oxygen is used. In conventional boilers we know that the higher the excess air, the lower the boiler efficiency. Similarly, in the HRSG, the efficiency increases with more supplementary firing. HRSGs used in combined-cycle steam cycles, Fig. 24, may use multiple pressure levels, gas-turbine steam injection, reheat, selective-catalytic-reduction (SCR) elements for NO_x control, and feedwater heating. Such HRSGs require extensive analysis to determine the best arrangement of the various heat-absorbing surfaces.

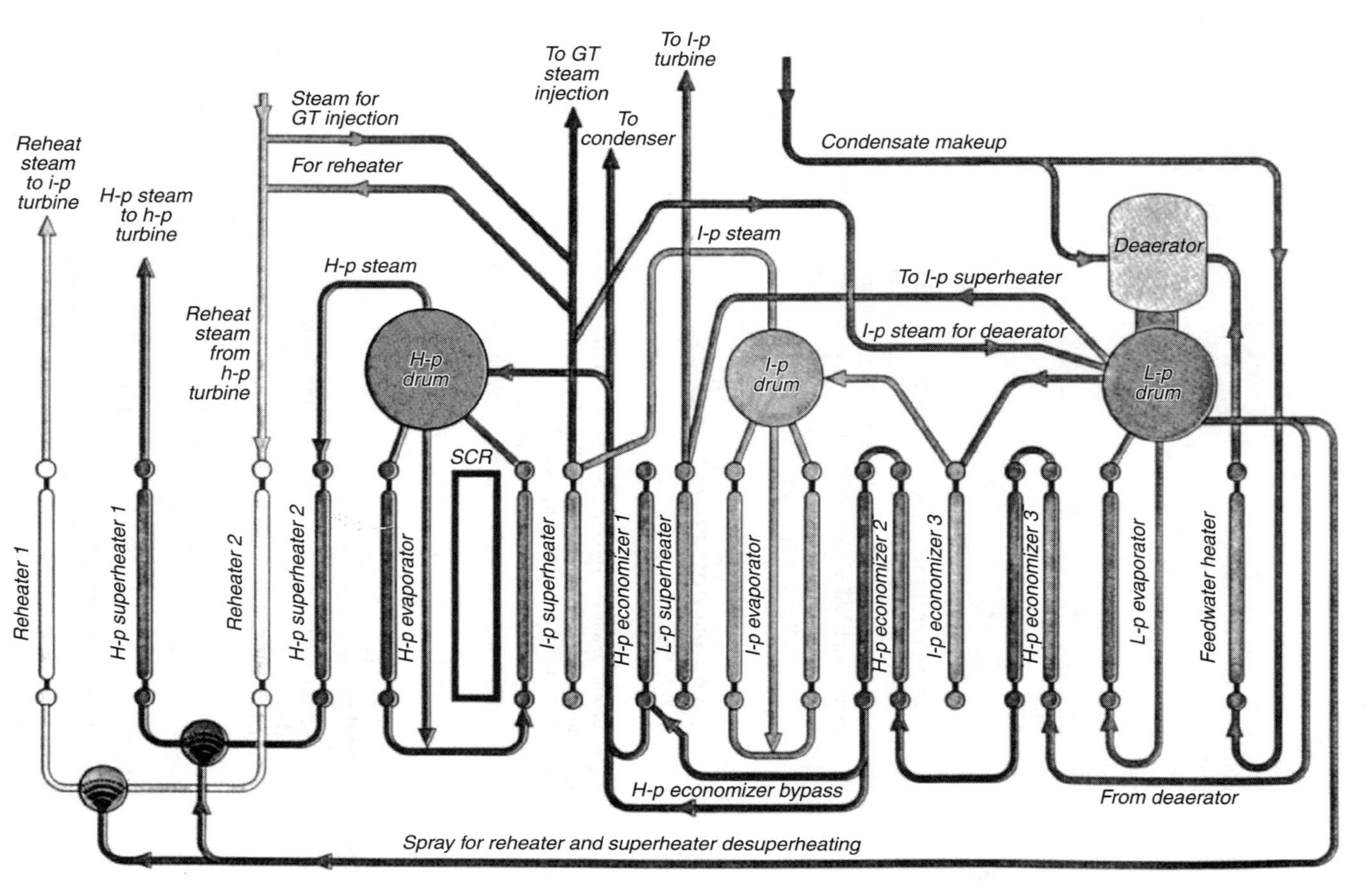

FIGURE 24 HRSGs in combined-cycle steam cycles are somewhat more involved when multiple pressure level, gas-turbine steam injection, reheat, SCR, and feedwater heating are used. (*Power*.)

TABLE 2 HRSG Performance in Fired Mode

Item	Case 1	Case 2	Case 3
Gas flow, lb/h (kg/h)	150,000 (68,100)	150,000 (68,100)	150,000 (68,100)
Inlet gas temp, °F (°C)	900 (482.2)	900 (482.2)	900 (482.2)
Firing temperature, °F (°C)	900 (482.2)	1290 (698.9)	1715 (935.0)
Burner duty, MM Btu/h (LHV)*	0 (0)	17.3 (5.06)	37.6 (11.01)
Steam flow, lb/h (kg/h)	22,780 (10,342)	40,000 (18,160)	60,000 (27,240)
Steam pressure, lb/in^2(gage) (kPa)	200 (1378.0)	200 (1378.0)	200 (1378.0)
Feedwater temperature, °F (°C)	240 (115.6)	240 (115.6)	240 (115.6)
Exit-gas temperature, °F (°C)	327 (163.9)	315 (157.2)	310 (154.4)
System efficiency, %	68.7	79.2	84.90
Steam duty, MM Btu/g (MW)	22.67 (6.64)	39.90 (11.69)	59.90 (17.55)

*(MW).

For example, an HRSG generates 22,780 lb/h (10.342 kg/h) of steam in the unfired mode. The various parameters are shown in Table 2. Studying this table shows that as the steam generation rate increases, more and more of the fuel energy goes into making steam. Fuel utilization is typically 100 percent in an HRSG. The ASME efficiency is also shown in the table.

This simulation was done using the HRSG simulation software developed by the author, V. Ganapathy, Heat Transfer Specialist, ABCO Industries, Inc.

CALCULATING HEAT-RECOVERY STEAM GENERATOR (HRSG) TEMPERATURE PROFILES

A gas turbine exhausts 150,000 lb/h (68,100 kg/h) of gas at 900°F (482.2°C) to an HRSG generating steam at 450 lb/in^2 (gage) (3100.5 kPa) and 650°F (343.3°C). Feedwater temperature to the HRSG is 240°F (115.6°C) and blowdown is 2 percent. Exhaust-gas analysis by percent volume is: $CO_2 = 3$; $H_2O = 7$; $O_2 = 15$. Determine the steam generation and temperature profiles with a 7-lb/in^2 (48.2-kPa) pressure drop in the superheater, giving an evaporator pressure of 450 + 7 = 457 lb/in^2 (gage) (3148.7 kPa) for a saturation temperature of the steam of 460°F (237.8°C). There is a heat loss of 1 percent in the HRSG. Find the ASME efficiency for this HRSG unit.

Calculation Procedure:

1. *Select the pinch and approach points for the HRSG*

Gas turbine heat-recovery steam generators (HRSGs) are widely used in cogeneration and combined-cycle plants. Unlike conventionally fired steam generators where the rate of steam generation is predetermined and can be achieved, steam-flow determination in an HRSG requires an analysis of the gas/steam temperature profiles. This requirement is mainly because we are starting at a much lower gas temperature—900 to 1100°F—(482.2 to 593.3°C) at the HRSG inlet, compared to 3000 to 3400°F (1648.9 to 1871.1°C) in a conventionally fired boiler. As a result, the exit-gas temperature from an HRSG cannot be assumed. It is a function of the operating steam pressure, steam temperature, and pinch and approach points used, Fig. 25.

Typically, the pinch and approach points range from 10 to 30°F (5.56 to 16.6°C). Higher values may be used if less steam generation is required. In this case, we will use 20°F (11.1°C) pinch point ($= T_{g3} - t_s$) and 10°F (5.56°C) approach ($= t_s - t_{w2}$). Hence, the gas temperature leaving the evaporator = 460 + 20 = 480°F (248.9°C), and the water temperature leaving the economizer = 460 – 10 = 450°F (232.2°C).

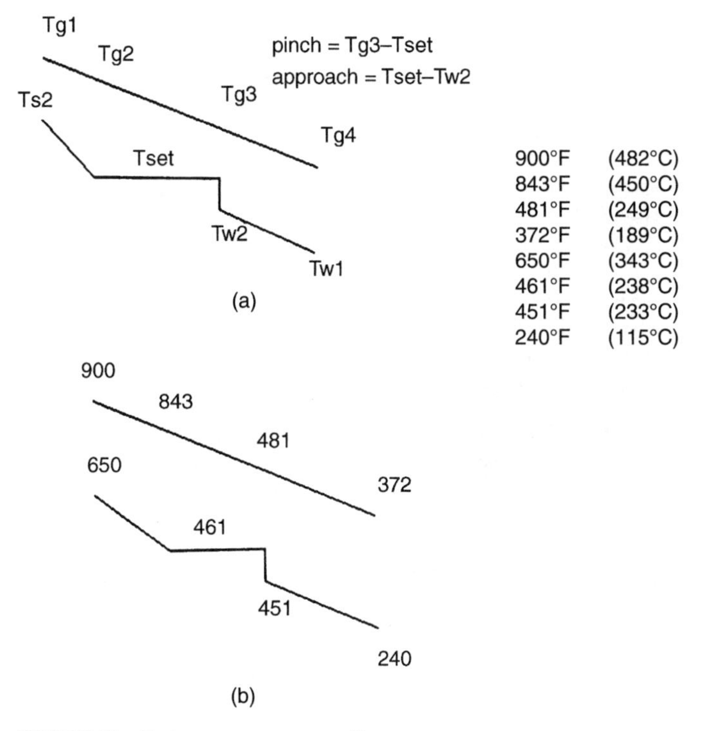

FIGURE 25 Gas/steam temperature profiles.

2. ***Compute the steam generation rate***
The energy transferred to the superheater and evaporator = $Q_1 + Q_2$ = (rate of gas flow, lb/h)(gas specific heat, Btu/lb °F)(entering gas temperature, °F – temperature of gas leaving evaporator, °F) (1.0 percent heat loss) = (150,000)(0.267)(900 – 480)(0.99) = 16.65 MM Btu/h (4.879 MW).

The enthalpy absorbed by the steam in the evaporator and superheater = (enthalpy of the superheated steam at 450 lb/in^2 (gage) and 650°F – enthalpy of the water entering the evaporator at 450°F) + (blowdown percentage)(enthalpy of the saturated liquid at the superheated condition – enthalpy of the water entering the evaporator, all expressed in Btu/lb). Or, enthalpy absorbed in the evaporator and superheater = (1330.8 – 431.2) + (0.02)(442.3 – 431.2) = 899.8 Btu/lb (2096.5 kJ/kg).

To compute the steam generation rate, set up the energy balance, 899.8(W_s) = 16.65 MM Btu/h, where W_s = steam generation rate.

3. ***Calculate the energy absorbed by the superheater and the exit-gas temperature***
Q_1, the energy absorbed by the superheater = (steam generation rate, lb/h)(enthalpy of superheated steam, Btu/lb – enthalpy of saturated steam at the superheater pressure, Btu/lb) = (18,502)(1330.8 – 1204.4) = 2.338 MM Btu/h (0.685 MW).

The superheater gas-temperature drop = (Q_1)/(rate of gas-turbine exhaust-gas flow, lb/h)(1.0 – heat loss)(gas specific heat) = (2,338,000)/(150,000)(0.99)(0.273) = 57.67°F, say 58°F (32.0°C). Hence, the superheater exit-gas temperature = 900 – 58 = 842°F (450°C). In this calculation the exhaust-gas specific heat is taken as 0.273 because the gas temperature in the superheater is different from the inlet gas temperature.

4. ***Compute the energy absorbed by the evaporator***
The total energy absorbed by the superheater and evaporator, from the above, is 16.65 MM Btu/h (4.878 MW). Hence, the evaporator duty = Q_2 = 16.65 – 2.34 = 14.31 MM Btu/h (4.19 MW).

5. *Determine the economizer duty and exit-gas temperature*

The economizer duty, Q_3 = (rate of steam generation, lb/h)(l + blowdown expressed as a decimal) (enthalpy of water leaving the economizer – enthalpy of feedwater at 240°F) = (18,502)(1.02) (431.2 – 209.6) = 4.182 MM Btu/h (1.225 MW).

The HRSG exit-gas temperature = (480, the exit-gas temperature at the evaporator computed in step 1, above) – (economizer duty)/(gas-turbine exhaust-gas flow, lb/h)(1.0 – heat loss)(exhaust gas specific heat) = 371.73°F (188.9°C); round to 372°F (188.9°C). Note that you must compute the gas specific heat at the average gas temperature of each of the heat-transmission surfaces.

6. *Compute the ASME HRSG efficiency*

The ASME Power Test Code PTC 4.4 defines the efficiency of an HRSG as: E = efficiency = (energy absorbed by the steam and fluids)/(gas flow × inlet enthalpy + fuel input to HRSG on LHV basis). In the above case, $E = (16.65 + 4.182)(10^6)/(150{,}000 \times 220) = 0.63$, or 63 percent. In this computation, 220 Btu/lb (512.6 kJ/kg) is the enthalpy of the exhaust gas at 900°F (482.2°C) and (16.65 + 4.182) is the total energy absorbed by the steam in MM Btu/h (MW).

Related Calculations. Note that the exit-gas temperature is high. Further, without having done this analytical mathematical analysis, the results could not have been guessed correctly. Minor variations in the efficiency will result if one assumes different pinch and approach points. Hence, it is obvious that one cannot assume a value for the exit-gas temperature—say 300°F (148.9°C)—and compute the steam generation.

The gas/steam temperature profile is also dependent on the steam pressure and steam temperature. The higher the steam temperature, the lower the steam generation rate and the higher the exit-gas temperature. Arbitrary assumption of the exit-gas temperature or pinch point can lead to temperature cross situations. Table 3 shows the exit-gas temperatures for several different steam parameters. From the table, it can be seen that the higher the steam pressure, the higher the saturation temperature, and hence, the higher the exit-gas temperature. Also, the higher the steam temperature, the higher the exit-gas temperature. This results from the reduced steam generation, resulting in a smaller heat sink at the economizer.

This procedure is the work of V. Ganapathy, Heat Transfer Specialist, ABCO Industries, who is the author of several works listed in the references for this section.

TABLE 3 HRSG Exit-Gas Temperatures versus Steam Parameters*

Pressure lb/in² (gage) (kPa)	Steam temp °F (°C)	Saturation temp °F (°C)	Exit gas °F (°C)
100 (689)	sat (170)	338 (170)	300 (149)
150 (1034)	sat (186)	366 (186)	313 (156)
250 (1723)	sat (208)	406 (208)	332 (167)
400 (2756)	sat (231)	448 (231)	353 (178)
400 (2756)	600 (316)	450 (232)	367 (186)
600 (4134)	sat (254)	490 (254)	373 (189)
600 (4134)	750 (399)	492 (256)	398 (203)

*Pinch point = 20°F (11.1°C); approach = 15°F (8.3°C); gas inlet-temperature = 900°F (482.2°C); blowdown = 0; feedwater temperature = 230°F (110°C).

SECTION 4
INTERNAL-COMBUSTION ENGINE ENERGY ANALYSES

Internal-Combustion Engine Parameters 4.1
Cogeneration Energy Economics Using I-C Engines 4.2
Energy Efficiency of Diesel Generating Unit 4.7
Diesel Engine Energy Efficiency and Characteristics 4.8
I-C Engine Horsepower and Mean Effective Pressure 4.8
Energy Analysis I-C Industrial Engine Choice 4.9
High-Temperature and High-Altitude I-C Engine Performance 4.10
Energy Analysis of I-C Engines 4.12
I-C Engine Characteristics Analyses 4.12
Cooling-Water Energy Needs of I-C Engines 4.13
I-C Engine Room Vent System Design 4.17
Bypass Cooling-System Design for I-C Engines 4.19
Energy Recovery via Hot Water from I-C Engines 4.24
Fuel Storage Capacity and Cost for I-C Engines 4.25
Energy Requirements for I-C Engine Cooling-Water and Lube-Oil Pumps 4.26
Choice of Lube-Oil Cooler for I-C Engines 4.27
Determining Solids Intake of I-C Engines 4.28
Energy Performance Factors for I-C Engines 4.29
Diesel-Engine Volumetric Efficiency 4.30
Air-Cooled I-C Engine Choice for Industrial Uses 4.33

INTERNAL-COMBUSTION ENGINE PARAMETERS

Internal-combustion (I-C) engines are widely used for power generation in smaller cities, towns, and isolated sites. Most I-C engines generate less greenhouse gas than equivalent-capacity coal-fired steam plants. Dual-fuel I-C engines can use either liquid or gaseous fuels. Such engines are popular in areas where more than one fuel is available. With dual-fuel capability, the engine operator can switch from a higher-priced fuel to a lower-cost fuel to reduce operating costs.

Diesel engines are the most popular I-C type in use today for stationary installations generating electricity. Spark-ignition gasoline and gas engines are popular in areas where such fuels are plentiful at low cost, but they are a minority compared to diesel engines, which are rated as high as 50,000 hp (37,300 kW).

Stationary diesel engines generating electricity have a typical fuel consumption of 0.08 gal/kWh (0.30 L/kWh) and a lube-oil consumption of 0.0005 gal/kWh (0.002 L/kWh). Such engines are often fitted with a waste-heat boiler to capture heat in the engine exhaust. Thus, the diesel engine was an early entrant in the cogeneration field, and remains prominent today.

A useful rule of thumb for diesel engines is that a diesel-engine waste-heat boiler can produce 1.9 lb/h (0.86 kg/h) of 100-psi (gage) (689-kPa) saturated steam at full load per rated horsepower. Steam pressures as high as 150 psi (gage) (1,034 kPa), and beyond, can be obtained from waste-heat boilers using diesel-engine exhaust gases. This heat represents about 70 percent of the available heat in the engine exhaust. About 30 percent of the heat content of the fuel at the average engine load can be recovered from the diesel-engine liquid (usually water) cooling system.

Large stationary diesel engines in electric generating service are normally operated at 80 percent of their maximum rated rpm. This allows a 20 to 25 percent overload capacity in the event it is needed in the generating station.

Diesel engines are often chosen for sites where there is a scarcity of cooling water for steam condensers used in steam-power plants. A diesel engine requires minimum amounts of cooling water for makeup purposes. And diesel engines are often used in existing steam plants to handle peak loads and provide a better overall heat balance for the plant. There are no standby losses with a diesel engine because when it is shut down its fuel use is zero. By comparison, a steam plant uses fuel when it is on standby.

Studies by the ASME Oil Engine Power Subcommittee show that diesel engines produce about 9 kWh per gallon of fuel (2.3 kWh per liter of fuel) at 30 to 40 percent load factor, and 12 to 13 kWh per gallon of fuel (3.2 to 3.4 kWh per liter of fuel) at 80 to 90 percent load factor. Further, diesel engines have a relatively small footprint and can often be installed in existing structures with no need for any major alteration of the building. Fuel consumption of diesels is relatively flat over the range of loads they handle. The fuel consumption of industrial-use diesel engines is 5 to 6 gal/h (18.9 to 22.7 L/h) per 100 rated horsepower (74.6 kW) of the engine.*

COGENERATION ENERGY ECONOMICS USING I-C ENGINES

Determine if an internal-combustion (I-C) engine cogeneration facility will be economically attractive if the required electrical power and steam services can be served by a cycle such as that in Fig. 1 and the specific load requirements are those shown in Fig. 2. Frequent startups and shutdowns are anticipated for this system.

Calculation Procedure:

1. ***Determine the sources of waste heat available in the typical I-C engine***
There are three primary sources of waste heat available in the usual I-C engine. These are: (1) the exhaust gases from the engine cylinders; (2) the jacket cooling water; (3) the lubricating oil. Of these

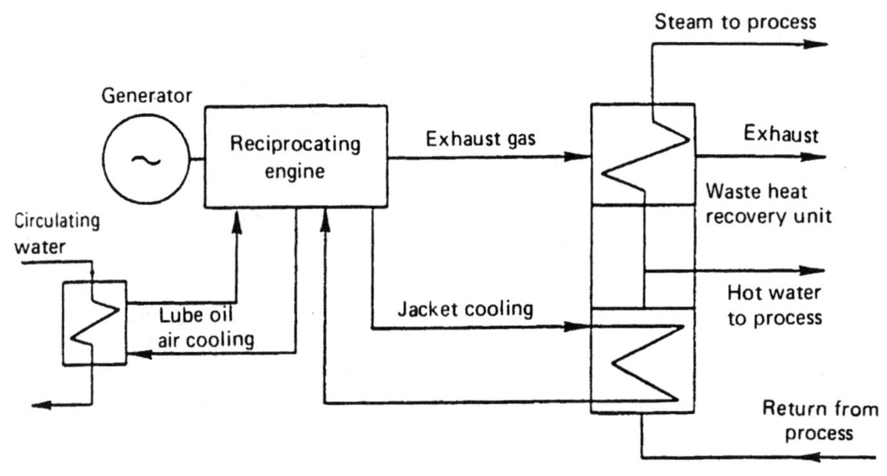

FIGURE 1 Reciprocating-engine cogeneration system waste heat from the exhaust, and jacket and oil cooling, are recovered. (*Indeck Energy Services, Inc.*)

*Anderson—*Diesel Engines,* McGraw-Hill, 1949.

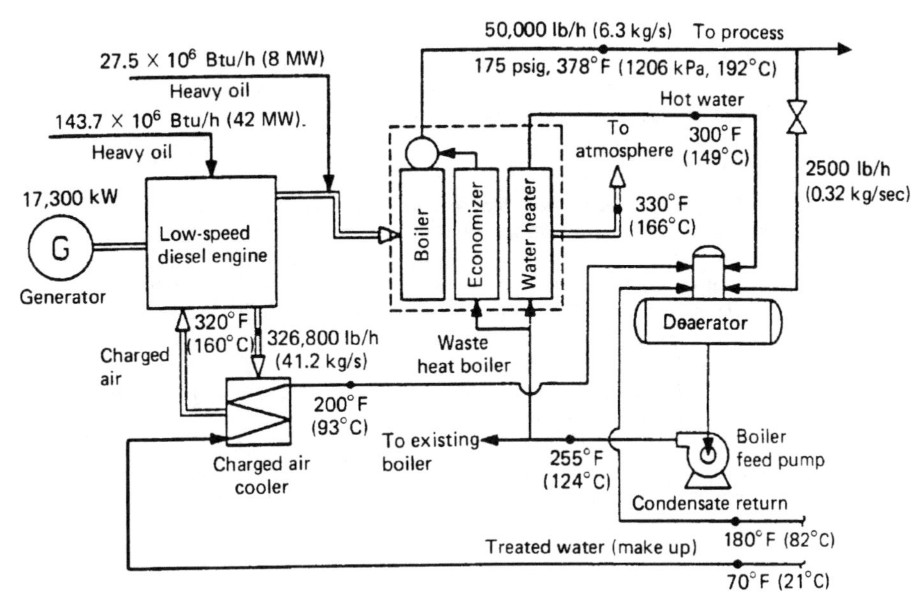

FIGURE 2 Low-speed diesel-engine cogeneration. (*Indeck Energy Services, Inc.*)

three sources, the quantity of heat available is, in descending order: exhaust gases; jacket cooling water; lube oil.

2. *Show how to compute the heat recoverable from each source*

For the exhaust gases, use the relation, $H_A = W(\Delta t)(c_g)$, where W_A = rate of gas flow from the engine, lb/h (kg/h); Δt = temperature drop of the gas between the heat exchanger inlet and outlet, °F (°C); c_g = specific heat of the gas, Btu/lb °F (J/kg °C). For example, if an I-C engine exhausts 100,000 lb/h (45,400 kg/h) at 700°F (371°C) to a HRSG (heat-recovery steam generator), leaving the HRSG at 330°F (166°C), and the specific heat of the gas is 0.24 Btu/lb °F (1.0 kJ/kg °C), the heat recoverable, neglecting losses in the HRSG and connecting piping, is H_A = 100,000(700 – 330)(0.24) = 8,880,000 Btu/h (2602 MW).

With an average heat of vaporization of 1000 Btu/lb (2330 kJ/kg) of steam, this exhaust gas flow could generate 8,880,000/1000 = 8880 lb/h (4032 kg/h) of steam. If oil with a heating value of 145,000 Btu/gal (40,455 kJ/L) were used to generate this steam, the quantity required would be 8,880,000/145,000 = 61.2 gal/h (232 L/h). At a cost of 90 cents per gallon, the saving would be $0.90(61.2) = $55.08/h. Assuming 5000 hours of operation per year, or 57 percent load, the saving in fuel cost would be 5000($55.08) = $275,400. This is a significant saving in any plant. And even if heat losses in the ductwork and heat-recovery boiler cut the savings in half, the savings would still exceed $100,000 a year. And as the operating time increases, so too do the savings.

3. *Compute the savings potential in jacket-water and lube-oil heat recovery*

A similar relation can be used to compute jacket-water and lube-oil heat recovery. The flow rate can be expressed in either pounds (kg) per hour or gallons (L) per minute, depending on the designer's choice.

Since water has a specific heat of unity, the heat-recovery potential of the jacket water *is* $H_w = w(\Delta t_w)$, where w = weight of water flow, lb/h (kg/h); Δt_w = change in temperature of the jacket water when flowing through the heat exchanger, °F (°C). Thus, if the jacket-water flow is 25,000 lb/h (11,350 kg/h) and the temperature change during flow of the jacket water through and external heat exchanger is 190 to 70°F (88 to 21°C), the heat given up by the jacket water, neglecting losses, is H_w = 25,000(190 – 70) = 3,000,000 Btu/h (879 MW). During 25 hours the heat recovery will be 24(3,000,000) = 72,000,000 Btu (75,960 MJ). This is a significant amount of heat which can be used in process or space heating, or to drive an air-conditioning unit.

If the jacket-water flow rate is expressed in gallons per minute instead of pounds per hour (L/min instead of kg/h), the heat-recovery potential, H_{wg} = gpm(Δt)(8.33) where 8.33 = lb/gal of water. With a water flow rate of 50 gpm and the same temperature range as above, H_{wg} = 50(120)(8.33) = 49,980 Btu/min (52,279 kJ/min).

4. *Find the amount of heat recoverable from the lube oil*

During I-C engine operation, lube-oil temperature can reach high levels—in the 300 to 400°F (149 to 201°C) range. And with oil having a typical specific heat of 0.5 Btu/lb °F (2.1 kJ/kg °C), the heat-recovery potential for the lube oil is $H_{wo} = w_o(\Delta t)(c_o)$, where w_o = oil flow in lb/h (kg/h); Δt = temperature change of the oil during flow through the heat-recovery heat exchanger = oil inlet temperature − oil outlet temperature, °F or °C; c_o = specific heat of oil = 0.5 Btu/lb °F (kJ/kg °C). With an oil flow of 2000 lb/h (908 kg/h), a temperature change of 140°F (77.7°C), H_o = 2000(140)(0.50) = 140,000 Btu/h (41 kW). Thus, as mentioned earlier, the heat recoverable from the lube oil is usually the lowest of the three sources.

With the heat flow rates computed here, an I-C engine cogeneration facility can be easily justified, especially where frequent startups and shutdowns are anticipated. Reciprocating diesel engines are preferred over gas and steam turbines where frequent startups and shutdowns are required. Just the fuel savings anticipated for recovery of heat in the exhaust gases of this engine could pay for it in a relatively short time.

Related Calculations. Cogeneration, in which I-C engines are finding greater use throughout the world every year, is defined by Michael P. Polsky, President, Indeck Energy Services, Inc., as "the simultaneous production of useful thermal energy and electric power from a fuel source or some variant thereof. It is more efficient to produce electric power and steam or hot water together than electric power alone, as utilities do, or thermal energy alone, which is common in industrial, commercial, and institutional plants." Figures 1 and 2 in this procedure are from the firm of which Mr. Polsky is president.

With the increased emphasis on reducing environmental pollution, conserving fuel use, and operating at lower overall cost, cogeneration—especially with diesel engines—is finding wider acceptance throughout the world. Design engineers should consider cogeneration whenever there is a concurrent demand for electricity and heat. Such demand is probably most common in industry but is also met in commercial (hotels, apartment houses, stores) and institutional (hospital, prison, nursing-home) installations. Often, the economic decision is not over whether cogeneration should be used, but what type of prime mover should be chosen.

Three types of prime movers are usually considered for cogeneration—steam turbines, gas turbines, or internal-combustion engines. Steam and/or gas turbines are usually chosen for large-scale utility and industrial plants. For smaller plants the diesel engine is probably the most popular choice today. Where natural gas is available, reciprocating internal-combustion engines are a favorite choice, especially with frequent startups and shutdowns.

Recently, vertical modular steam engines have been introduced for use in cogeneration. Modules can be grouped to increase the desired power output. These high-efficiency units promise to compete with I-C engines in the growing cogeneration market.

Guidelines used in estimating heat recovery from I-C engines, after all heat loses, include these: (1) Exhaust-gas heat recovery = 28 percent of heat in fuel; (2) Jacket-water heat recovery = 27 percent of heat in fuel; (3) Lube-oil heat recovery = 9 percent of heat in fuel. The Diesel Engine Manufacturers Association (DEMA) gives these values for heat disposition in a diesel engine at three-quarters to full load: (1) Fuel consumption = 7366 Btu/bhp · h (2.89 kW/kW); (2) Useful work = 2544 Btu/bhp · h (0.999 kW/kW); (3) Loss in radiation, etc. = 370 Btu/bhp · h (0.145 kW/kW); (4) To cooling water = 2195 Btu/bhp · h (0.862 kW/kW); (5) To exhaust = 2258 Btu/bhp · h (0.887 kW/kW). The sum of the losses is 1 Btu/bhp · h greater than the fuel consumption because of rounding of the values.

Figure 3 shows a proposed cogeneration, desiccant-cooling, and thermal-storage integrated system for office buildings in the southern California area. While directed at the microclimates in that area, similar advantages for other microclimates and building types should be apparent. The data presented

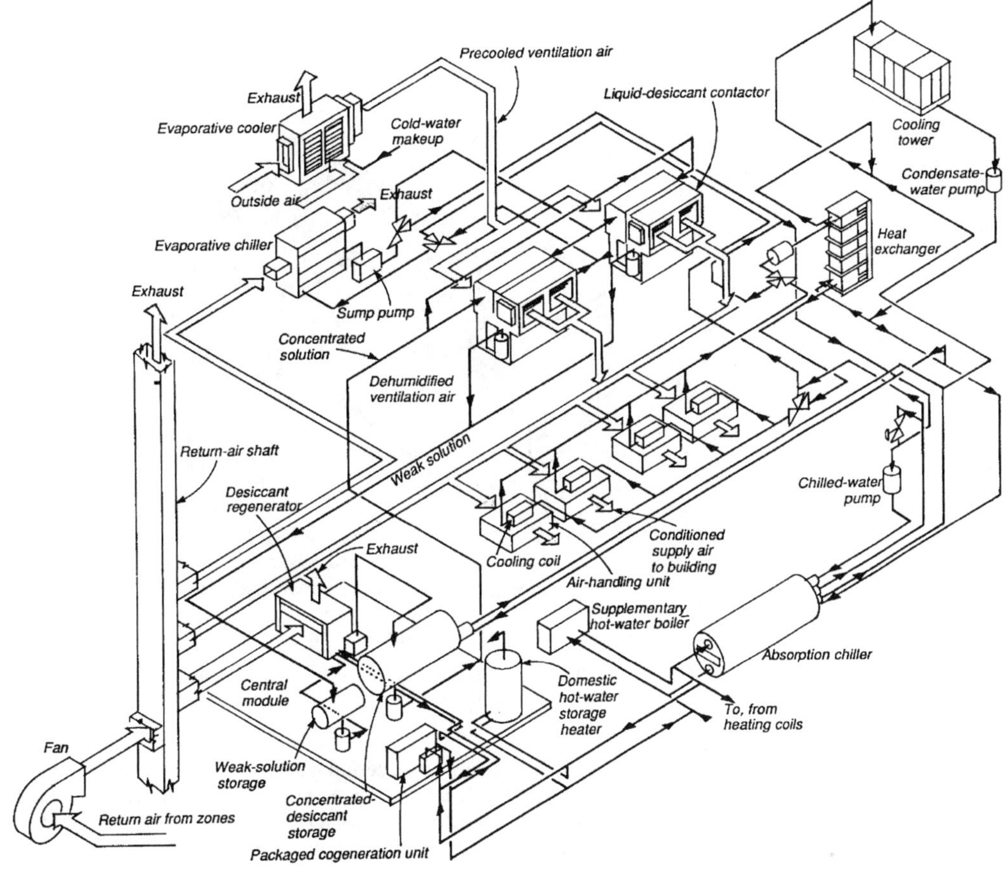

FIGURE 3 Integrated system is a proposed off-peak desiccant/evaporative-cooling configuration with cogeneration capability. (*Power* and *The Meckler Group*.)

here for this system were prepared by The Meckler Group and are based on a thorough engineering and economic evaluation for the Southern California Gas Co. of the desiccant-cooling/thermal-energy-storage/cogeneration system, a proprietary design developed for pre- and post-Title-24 mid-rise office buildings. Title 24 is a section of the State of California Administrative Code that deals with energy-conservation standards for construction applicable to office buildings. A summary of the study was presented in *Power* magazine by Milton Meckler.

In certain climates, office buildings are inviting targets for saving energy via evaporative chilling. When waste heat is plentiful, desiccant cooling and cogeneration become attractive. In coupling the continuously available heat-rejection capacity of packaged cogeneration units, Fig. 4, with continuously operating regenerator demands, the use of integrated components for desiccant cooling, thermal-energy storage, and cogeneration increases. The combination also ensures a reasonable constant, cost-effective supply of essentially free electric power for general building use.

Recoverable internal-combustion engine heat should at least match the heat requirement of the regenerator, Fig. 3. The selected engine size (see a later procedure in this section), however, should not cause the cogeneration system's PURPA (Public Utility Regulatory & Policies Act) efficiency to drop below 42.5 percent. (PURPA efficiency decreases as engine size increases.) An engine size is selected to give the most economical performance and still has a PURPA efficiency of greater than 42.5 percent.

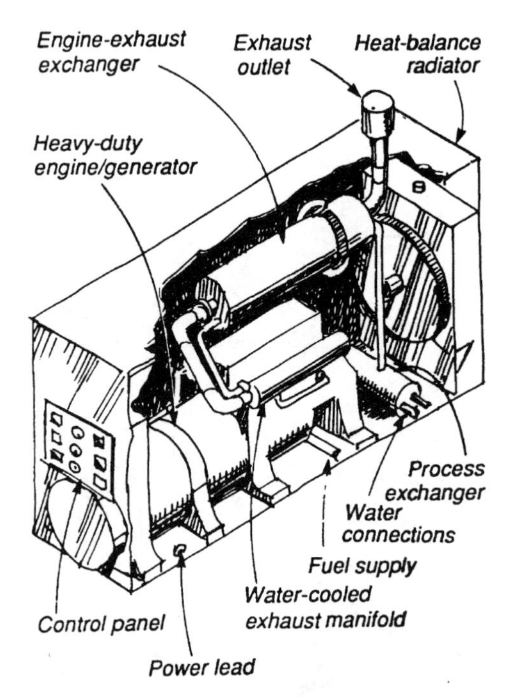

FIGURE 4 Packaged cogeneration I-C engine unit supplies waste heat to desiccant regenerator. (*Power* and *The Meckler Group.*)

The utility study indicated a favorable payout period and internal rate of return both for retrofits of pre-Title-24 office buildings and for new buildings in compliance with current Title-24 requirements (nominal 200 to 500 cooling tons). Although the study was limited to office-building occupancies, it is likely that other building types with high ventilation and electrical requirements would also offer attractive investment opportunities.

Based on study findings, fuel savings ranged from 3300 to 7900 therms per year. Cost savings ranged from $322,000 to $370,000 for the 5-story-building case studies and from $545,000 to $656,000 for 12-story-building case studies where the synchronously powered, packaged cogeneration unit was not used for emergency power.

Where the cogeneration unit was also used for emergency power, the initial cost decreased from $257,000 to $243,000, representing a 31 percent drop in average cost for the 5-story-building cases, and from $513,000 to $432,000, a 22 percent dip in average cost for the 12-story-building cases. The average cost decrease shifts the discounted payback period an average of 5.6 and 5.9 years for the 5- and 12-story-building cases, respectively.

Study findings were conservatively reported, since no credit was taken for potential income resulting from PURPA sales to the serving utility at off-peak hours, when actual building operating requirements fall below rated cogenerator output. This study is another example of the importance of the internal-combustion engine in cogeneration around the world today.

Worldwide there is a movement toward making internal-combustion engines, and particularly diesel engines, cleaner-running. In general, this means reducing particulate emissions from diesel-engine exhaust gases. For cities with large numbers of diesel-powered buses, exhaust emissions can be particularly unpleasant. And some medical personnel say that diesel exhaust gases can be harmful to the health of people breathing them.

The approach to making diesel engines cleaner takes two tacts: (1) improving the design of the engine so that fewer particulates are emitted and (2) using cleaner fuel to reduce the particulate emissions. Manufacturers are using both approaches to comply with the demands of federal and state agencies regulating emissions. Today's engineers will find that "cleaning up" diesel engines is a challenging and expensive procedure. However, cleaner-operating diesels are being introduced every year.

ENERGY EFFICIENCY OF DIESEL GENERATING UNIT

A 3000-kW diesel generating unit performs thus: fuel rate, 1.5 bbl (238.5 L) of 25° API fuel for a 900-kWh output; mechanical efficiency, 82.0 percent; generator efficiency, 92.0 percent. Compute engine fuel rate, engine-generator fuel rate, indicated thermal efficiency, overall thermal efficiency, brake thermal efficiency.

Calculation Procedure:

1. *Compute the engine fuel rate*
The fuel rate of an engine driving a generator is the weight of fuel, lb, used to generate 1 kWh at the generator input shaft. Since this engine burns 1.5 bbl (238.5 L) of fuel for 900 kW at the generator terminals, the total fuel consumption is (1.5 bbl)(42 gal/bbl) = 63 gal (238.5 L), at a generator efficiency of 92.0 percent.

To determine the weight of this oil, compute its specific gravity s from $s = 141.5/(131.5 + °\text{API})$, where °API = API gravity of the fuel. Hence, $s = 141.5(131.5 + 25) = 0.904$. Since 1 gal (3.8 L) of water weighs 8.33 lb (3.8 kg) at 60°F (15.6°C), 1 gal (3.8 L) of this oil weighs (0.904)(8.33) = 7.529 lb (3.39 kg). The total weight of fuel used when burning 63 gal is (63 gal)(7.529 lb/gal) = 474.5 lb (213.5 kg).

The generator is 92 percent efficient. Hence, the engine actually delivers enough power to generate 900/0.92 = 977 kWh at the generator terminals. Thus, the engine fuel rate = 474.5-lb fuel/977 kWh = 0.485 lb/kWh (0.218 kg/kWh).

2. *Compute the engine-generator fuel rate*
The engine-generator fuel rate takes these two units into consideration and is the weight of fuel required to generate 1 kWh at the generator terminals. Using the fuel-consumption data from step 1 and the given output of 900 kW, we see that engine-generator fuel rate = 474.5-lb fuel/900 kWh output = 0.527 lb/kWh (0.237 kg/kWh).

3. *Compute the indicated thermal efficiency*
Indicated thermal efficiency is the thermal efficiency based on the *indicated* horsepower of the engine. This is the horsepower developed in the engine cylinder. The engine fuel rate, computed in step 1, is the fuel consumed to produce the brake or shaft horsepower output, after friction losses are deducted. Since the mechanical efficiency of the engine is 82 percent, the fuel required to produce the indicated horsepower is 82 percent of that required for the brake horsepower, or (0.82)(0.485) = 0.398 lb/kWh (0.179 kg/kWh).

The indicated thermal efficiency of an internal-combustion engine driving a generator is $e_i = 3413/f_i(\text{HHV})$, where e_i = indicated thermal efficiency, expressed as a decimal; f_i = indicated fuel consumption, lb/kWh; HHV = higher heating value of the fuel, Btu/lb.

Compute the HHV for a diesel fuel from HHV = 17,680 + 60 × °API. For this fuel, HHV = 17,680 + 60(25) = 19,180 Btu/lb (44,612.7 kJ/kg).

With the HHV known, compute the indicated thermal efficiency from $e_i = 3{,}413/[(0.398)(19{,}180)] = 0.447$ or 44.7 percent.

4. *Compute the overall thermal efficiency**
The overall thermal efficiency e_a is computed from $e_a = 3413/f_o\,(\text{HHV})$, where f_o = overall fuel consumption, Btu/kWh; other symbols as before. Using the engine-generator fuel rate from step 2, which represents the overall fuel consumption, $e_a = 3413/[(0.527)(19{,}180)] = 0.347$, or 34.7 percent.

---

*Elliott, *Standard Handbook of Power Plant Engineering*, McGraw-Hill, 1989.

5. *Compute the brake thermal efficiency*
The engine fuel rate, step 1, corresponds to the brake fuel rate f_b. Compute the brake thermal efficiency from $e_b = 3413/f_b$ (HHV), where f_b = brake fuel rate, Btu/kWh; other symbols as before. For this engine-generator set, $e_b = 3413/[(0.485)(19{,}180)] = 0.367$, or 36.7 percent.

Related Calculations. Where the fuel consumption is given or computed in terms of lb/(hp · h), substitute the value of 2545 Btu/(hp · h) (1.0 kW/kWh) in place of the value 3413 Btu/kWh (3600.7 kJ/kWh) in the numerator of the e_i, e_o, and e_b equations. Compute the indicated, overall, and brake thermal efficiencies as before. Use the same procedure for gas and gasoline engines, except that the higher heating value of the gas or gasoline should be obtained from the supplier or by test.

DIESEL ENGINE ENERGY EFFICIENCY AND CHARACTERISTICS

A 12×18 in (30.5×44.8 cm) four-cylinder four-stroke single-acting diesel engine is rated at 200 bhp (149.2 kW) at 260 r/min. Fuel consumption at rated load is 0.42 lb/(bhp · h) (0.25 kg/kWh). The higher heating value of the fuel is 18,920 Btu/lb (44,008 kJ/kg). What are the brake mean effective pressure, engine displacement in ft^3/(min · bhp), and brake thermal efficiency?

Calculation Procedure:

1. *Compute the brake mean effective pressure*
Compute the brake mean effective pressure (bmep) for an internal-combustion engine from $bmep = 33{,}000\ bhp_n/LAn$, where $bmep$ = brake mean effective pressure, lb/in^2; bhp_n = brake horsepower output delivered per cylinder, hp; L = piston stroke length, ft; A = piston area, in^2; n = cycles per minute per cylinder = crankshaft rpm for a two-stroke cycle engine, and 0.5 the crankshaft rpm for a four-stroke cycle engine.

For this engine at its rated hbp, the output per cylinder is 200 bhp/4 cylinders = 50 bhp (37.3 kW). Then $bmep = 33{,}000(50)/[(18/12)(12)^2(\pi/4)(260/2)] = 74.8$ lb/in^2 (516.1 kPa). (The factor 12 in the denominator converts the stroke length from inches to feet.)

2. *Compute the engine displacement*
The total engine displacement V_d ft^3 is given by $V_d = LAnN$, where A = piston area, ft^2; N = number of cylinders in the engine; other symbols as before. For this engine, $V_d = (18/12)(12/12)^2(\pi/4)(260/2)(4) = 614$ ft^3/min (17.4 m^3/min). The displacement is in cubic feet per minute because the crankshaft speed is in r/min. The factor of 12 in the denominators converts the stroke and area to ft and ft^2, respectively. The displacement per bhp = (total displacement, ft^3/min)/bhp output of engine = 614/200 = 3.07 ft^3/(min · bhp)(0.12 m^3/kW).

3. *Compute the brake thermal efficiency*
The brake thermal efficiency e_b of an internal-combustion engine is given by $e_b = 2545/(sfc)(\text{HHV})$, where sfc = specific fuel consumption, lb/(bhp · h); HHV = higher heating value of fuel, Btu/lb. For this engine, $e_b = 2545/[(0.42)(18{,}920)] = 0.32$, or 32.0 percent.

Related Calculations. Use the same procedure for gas and gasoline engines. Obtain the higher heating value of the fuel from the supplier, a tabulation of fuel properties, or by test.

I-C ENGINE HORSEPOWER AND MEAN EFFECTIVE PRESSURE

A 500-hp (373-kW) internal-combustion engine has a brake mean effective pressure of 80 lb/in^2 (551.5 kPa) at full load. What are the indicated mean effective pressure and friction mean effective pressure if the mechanical efficiency of the engine is 85 percent? What are the indicated horsepower and friction horsepower of the engine?

Calculation Procedure:

1. ***Determine the indicated mean effective pressure***
Indicated mean effective pressure *imep* lb/in^2 for an internal-combustion engine is found from $imep = bmep/e_m$, where $bmep$ = brake mean effective pressure, lb/in^2; e_m = mechanical efficiency, percent, expressed as a decimal. For this engine, $imep = 80/0.85 = 94.1$ lb/in^2 (659.3 kPa).

2. ***Compute the friction mean effective pressure***
For an internal-combustion engine, the friction mean effective pressure *fmep* lb/in^2 is found from $fmep = imep - bmep$, or $fmep = 94.1 - 80 = 14.1$ lb/in^2 (97.3 kPa).

3. ***Compute the indicated horsepower of the engine***
For an internal-combustion engine, the mechanical efficiency $e_m = bhp/ihp$, where ihp = indicated horsepower. Thus, $ihp = bhp/e_m$, or $ihp = 500/0.85 = 588$ ihp (438.6 kW).

4. ***Compute the friction hp of the engine***
For an internal-combustion engine, the friction horsepower is $fhp = ihp - bhp$. In this engine, $fhp = 588 - 500 = 88$ fhp (65.6 kW).

Related Calculations. Use a similar procedure to determine the *indicated engine efficiency* $e_{ei} = e_i/e$, where e = ideal cycle efficiency; *brake engine efficiency*, $e_{eb} = e_b e$; *combined engine efficiency* or *overall engine thermal efficiency*, $e_{eo} = e_o = e_o e$. Note that each of these three efficiencies is an *engine* efficiency and corresponds to an actual thermal efficiency, e_i, e_b, and e_o.

Engine efficiency $e_e = e_i/e$, where e_t = actual *engine* thermal efficiency. Where desired, the respective *actual* indicated brake, or overall, output can be substituted for e_i, e_b, and e_o in the numerator of the above equations if the ideal output is substituted in the denominator. The result will be the respective engine efficiency. Output can be expressed in Btu per unit time, or horsepower. Also, e_e = actual *mep*/ideal *mep*, and $e_{ei} = imep$/ideal *mep*; $e_{eb} = bmep$/ideal *mep*; e_{eo} = overall *mep*/ideal *mep*. Further, $e_b = e_m e_i$, and $bmep = e_m(imep)$. Where the actual heat supplied by the fuel, HHV Btu/lb, is known, compute $e_i e_b$ and e_o by the method given in the previous calculation procedure. The above relations apply to any reciprocating internal-combustion engine using any fuel.

ENERGY ANALYSIS I-C INDUSTRIAL ENGINE CHOICE

Select an internal-combustion engine to drive a centrifugal pump handling 2000 gal/min (126.2 L/s) of water at a total head of 350 ft (106.7 m). The pump speed will be 1750 r/min, and it will run continuously. The engine and pump are located at sea level.

Calculation Procedure:

1. ***Compute the power input to the pump***
The power required to pump water is hp $8.33GH/33{,}000e$, where G = water flow, gal/min; H = total head on the pump, ft of water; e = pump efficiency, expressed as a decimal. Typical centrifugal pumps have operating efficiencies ranging from 50 to 80 percent, depending on the pump design and condition and liquid handled. Assume that this pump has an efficiency of 70 percent. Then $hp = 8.33(2000)/(350)/[(33{,}000)(0.70)] = 252$ hp (187.9 kW). Thus, the internal-combustion engine must develop at least 252 hp (187.9 kW) to drive this pump.

2. ***Select the internal-combustion engine***
Since the engine will run continuously, extreme care must be used in its selection. Refer to a tabulation of engine ratings, such as Table 1. This table shows that a diesel engine that delivers 275 continuous brake horsepower (205.2 kW) (the nearest tabulated rating equal to or greater than the required input) will be rated at 483 bhp (360.3 kW) at 1750 r/min.

The gasoline-engine rating data in Table 1 show that for continuous full load at a given speed, 80 percent of the tabulated power can be used. Thus, at 1750 r/min, the engine must be rated at

TABLE 1 Internal-Combustion Engine Rating Table

Diesel engines						
Continuous bhp (kW) at given rpm					No. of	
1400	1600	1750	1800	Rated bhp	cylinders	Cooling*
187 (139.5)	214 (159.6)	227 (169.3)	230 (171.6)	300 (223.8)	6	*E*
230 (171.6)	256 (190.0)	275 (205.2)	280 (208.9)	438 (326.7)	12	*R*
240 (179.0)	273 (203.7)	295 (220.0)	305 (227.5)	438 (326.7)	12	*E*
Gasoline engines†						
405 (302.1)	430 (320.8)	450 (335.7)	475 (354.4)	595 (438.9)	12	*R*

**E* = heat-exchanger-cooled, *R* = radiator-cooled.
†Use 80 percent of tabulated power if engine is to run at continuous full load.

252/0.80 = 315 bhp (234.9 kW). A 450-hp (335.7-kW) unit is the only one shown in Table 1 that would meet the needs. This is too large; refer to another builder's rating table to find an engine rated at 315 to 325 bhp (234.9 to 242.5 kW) at 1750 r/min.

The unsuitable capacity range in the gasoline-engine section of Table 1 is a typical situation met in selecting equipment. More time is often spent in finding a suitable unit at an acceptable price than is spent computing the required power output.

Related Calculations. Use this procedure to select any type of reciprocating internal-combustion engine using oil, gasoline, liquefied -petroleum gas, or natural gas for fuel.

HIGH-TEMPERATURE AND HIGH-ALTITUDE I-C ENGINE PERFORMANCE

An 800-hp (596.8-kW) diesel engine is operated 10,000 ft (3048 m) above sea level. What is its output at this elevation if the intake air is at 80°F (26.7°C)? What will the output at 10,000-ft (3048-m) altitude be if the intake air is at 110°F (43.4°C)? What would the output be if this engine were equipped with an exhaust turbine-driven blower?

Calculation Procedure:

1. *Compute the engine output at altitude*
Diesel engines are rated at sea level at atmospheric temperatures of not more than 90°F (32.3°C). The sea-level rating applies at altitudes up to 1500 ft (457.2 m). At higher altitudes, a correction factor for elevation must be applied. If the atmospheric temperature is higher than 90°F (32.2°C), a temperature correction must be applied.

Table 2 lists both altitude and temperature correction factors. For an 800-hp (596.8-kW) engine at 10,000 ft (3048 m) above sea level and 80°F (26.7°C) intake air, hp output = (sea-level hp) (altitude correction factor), or output = (800)(0.68) = 544 hp (405.8 kW).

2. *Compute the engine output at the elevated temperature*
When the intake air is at a temperature greater than 90°F (32.3°C), a temperature correction factor must be applied. Then output = (sea-level hp)(altitude correction factor)(intake-air-temperature correction factor), or output = (800)(0.68)(0.95) = 516 hp (384.9 kW), with 110°F (43.3°C) intake air.

TABLE 2 Correction Factors for Altitude and Temprature

Engine altitude		Engine type		Intake temperature		
ft	m	Nonsuper-charged	Super-charged	°F	°C	Correction factor
7,000	2,134	0.780	0.820	90 or less	32.3 or less	1.000
8,000	2,438	0.745	0.790	95	35	0.986
9,000	2,743	0.712	0.765	100	37.8	0.974
10,000	3,048	0.680	0.740	105	40.6	0.962
12,000	3,658	0.612	0.685	110	43.3	0.950
				115	46.1	0.937
				120	48.9	0.925
				125	51.7	0.913
				130	54.4	0.900

3. ***Compute the output of a supercharged engine***
A different altitude correction is used for a supercharged engine, but the same temperature-correction factor is applied. Table 2 lists the altitude-correction factors for supercharged diesel engines. Thus, for this supercharged engine at 10,000-ft (3048-m) altitude with 80°F (26.7°C) intake air, output = (sea-level hp)(altitude correction factor) = (800)(0.74) = 592 hp (441.6 kW).

At 10,000-ft (3048-m) altitude with 110°F (43.3°C) inlet air, output = (sea-level hp)(altitude correction factor)(temperature correction factor) = (800)(0.74)(0.95) = 563 hp (420.1 kW).

Related Calculations. Use the same procedure for gasoline, gas, oil, and liquefied-petroleum gas engines. Where altitude-correction factors are not available for the type of engine being used, other than a diesel, multiply the engine sea-level brake horsepower by the ratio of the altitude-level atmospheric pressure to the atmospheric pressure at sea level. Table 3 lists the atmospheric pressure at various altitudes.

An engine located below sea level can theoretically develop more power than at sea level because the intake air is denser. However, the greater potential output is generally ignored in engine-selection calculations.

TABLE 3 Atmospheric Pressure at Various Altitudes

Altitude		Pressure	
ft	m	inHg	mm
Sea Level		29.92	759.97
4,000	1,219	25.84	656.3
5,000	1,524	24.89	632.2
6,000	1,829	23.98	609.1
8,000	2,438	22.22	564.4
10,000	3,048	20.58	522.7
12,000	3,658	19.03	483.4

Note: A 500- to 1500-ft altitude is considered equivalent to sea level by the Diesel Engine Manufacturers Association if the atmospheric pressure is not less than 28.25 inHg (717.6 mmHg).

ENERGY ANALYSIS OF I-C ENGINES

An indicator card taken on an internal-combustion engine cylinder has an area of 5.3 in^2 (34.2 cm^2) and a length of 4.95 in (12.7 cm). What is the indicated mean effective pressure in this cylinder? What is the indicated horsepower of this four-cycle engine if it has eight 6-in (15.6-cm) diameter cylinders, an 18-in (45.7-cm) stroke, and operates at 300 r/min? The indicator spring scale is 100 lb/in (1.77 kg/mm).

Calculation Procedure:

1. *Compute the indicated mean effective pressure*

For any indicator card, *imep* = (card area, in^2) (indicator spring scale, lb)/(length of indicator card, in) where *imep* = indicated mean effective pressure, lb/in^2. Thus, for this engine, *imep* = (5.3)(100)/4.95 = 107 lb/in^2 (737.7 kPa).

2. *Compute the indicated horsepower*

For any reciprocating internal-combustion engine, *ihp* = (*imep*)*LAn*/33,000, where *ihp* = indicated horsepower per cylinder; L = piston stroke length, ft; A = piston area, in^2; n = number of cycles/min. Thus, for this four-cycle engine where n = 0.5 r/min, *ihp* = $(107)(18/12)(6)^2(\pi/4)(300/2)/33{,}000$ = 20.6 ihp (15.4 kW) per cylinder. Since the engine has eight cylinders, total *ihp* = (8 cylinders) (20.6 ihp per cylinder) = 164.8 ihp (122.9 kW).

Related Calculations. Use this procedure for any reciprocating internal-combustion engine using diesel oil, gasoline, kerosene, natural gas, liquefied-petroleum gas, or similar fuel.

I-C ENGINE CHARACTERISTICS ANALYSES

What is the piston speed of an 18-in (45.7-cm) stroke 300 = r/min engine? How much torque will this engine deliver when its output is 800 hp (596.8 kW)? What are the displacement per cylinder and the total displacement if the engine has eight 12-in (30.5-cm) diameter cylinders? Determine the engine compression ratio if the volume of the combustion chamber is 9 percent of the piston displacement.

Calculation Procedure:

1. *Compute the engine piston speed*

For any reciprocating internal-combustion engine, piston speed = *fpm* = 2*L*(*rpm*), where L = piston stroke length, ft; rpm = crankshaft rotative speed, r/min. Thus, for this engine, piston speed = 2(18/12)(300) = 9000 ft/min (2743.2 m/min).

2. *Determine the engine torque*

For any reciprocating internal-combustion engine, T = 63,000 (*bhp*)/*rpm*, where T = torque developed, in · lb; *bhp* = engine brake horsepower output; *rpm* = crankshaft rotative speed, r/min. Or T = 63,000(800)/300 = 168,000 in · lb (18.981 N · m).

Where a prony brake is used to measure engine torque, apply this relation: $T = (F_b - F_o)r$, where F_b = brake scale force, lb, with engine operating; F_o = brake scale force with engine stopped and brake loose on flywheel; r = brake arm, in = distance from flywheel center to brake knife edge.

3. *Compute the displacement*

The displacement per cylinder d, in^3 of any reciprocating internal-combustion engine is $d_i = L_iA_i$ where L_i = piston stroke, in; A = piston head area, in^2. For this engine, $d_e = (18)(12)^2(\pi/4)$ = 2035 in^3 (33.348 cm^3) per cylinder.

The total displacement of this eight-cylinder engine is therefore (8 cylinders) (2035 in^3 per cylinder) = 16,280 in^3 (266,781 cm^3).

4. *Compute the compression ratio*

For a reciprocating internal-combustion engine, the compression ratio $r_c = V_b/V_a$, where V_b = cylinder volume at the start of the compression stroke, in^3 or ft^3; V_a = combustion-space volume at the end of the compression stroke, in^3 or ft^3. When this relation is used, both volumes must be expressed in the same units.

In this engine, V_b = 2035 in^3 (33,348 cm^3); V_a = (0.09)(2035) = 183.15 in^3. Then r_c = 2035/183.15 = 11.1:1.

Related Calculations. Use these procedures for any reciprocating internal-combustion engine, regardless of the fuel burned.

COOLING-WATER ENERGY NEEDS OF I-C ENGINES

A 1000-hbp (746-kW) diesel engine has a specific fuel consumption of 0.360 lb/(bhp · h) (0.22 kg/kWh). Determine the cooling-water flow required if the higher heating value of the fuel is 10,350 Btu/lb (24,074 kJ/kg). The net heat rejection rates of various parts of the engine are, in percent: jacket water, 11.5; turbocharger, 2.0; lube oil. 3.8; aftercooling, 4.0; exhaust, 34.7; radiation, 7.5. How much 30-lb/in^2 (abs) (206.8-kPa) steam can be generated by the exhaust gas if this is a four-cycle engine? The engine operates at sea level.

Calculation Procedure:

1. *Compute the engine heat balance*

Determine the amount of heat used to generate 1 bhp · h (0.75 kWh) from: heat rate, Btu/bhp · h) = (*sfc*)(HHV), where *sfc* = specific fuel consumption, lb/(bhp · h); HHV = higher heating value of fuel, Btu/lb. Or, heat rate = (0.36)(19.350) = 6967 Btu/(bhp · h) (2737.3 W/kWh).

Compute the heat balance of the engine by taking the product of the respective heat rejection percentages and the heat rate as follows:

			Btu/(bhp · h)	W/kWh
Jacket water	(0.115)(6967)	=	800	314.3
Turbocharger	(0.020)(6967)	=	139	54.6
Lube oil	(0.038)(6967)	=	264	103.7
Aftercooling	(0.040)(6967)	=	278	109.2
Exhaust	(0.347)(6967)	=	2420	880.1
Radiation	(0.075)(6967)	=	521	204.7
Total heat loss		=	4422	1666.6

Then the power output = 6967 – 4422 = 2545 Btu/(bhp · h) (999.9 W/kWh), or 2545/6967 = 0.365, or 36.5 percent. Note that the sum of the heat losses and power generated, expressed in percent, is 100.0.

2. *Compute the jacket cooling-water flow rate*

The jacket water cools the jackets and the turbocharger. Hence, the heat that must be absorbed by the jacket water is 800 + 139 = 939 Btu/(bhp · h) (369 W/kWh), using the heat rejection quantities computed in step 1. When the engine is developing its full rated output of 1000 bhp (746 kW), the jacket water must absorb [939 Btu/(bhp · h)(1000 bhp) = 939.000 Btu/h (275,221 W).

Apply a safety factor to allow for scaling of the heat-transfer surfaces and other unforeseen difficulties. Most designers use a 10 percent safety factor. Applying this value of the safety factor for this engine, we see the total jacket-water heat load = 939.000 + (0.10)(939.000) = 1.032,900 Btu/h (302.5 kW).

Find the required jacket-water flow from $G = H/500\Delta t$, where G = jacket-water flow, gal/min; H = heat absorbed by jacket water, Btu/h; Δt = temperature rise of the water during passage through the jackets, °F. The usual temperature rise of the jacket water during passage through a diesel engine is 10 to 20°F (5.6 to 11.1°C). Using 10°F for this engine, we find G = 1,032,900/[(500)(10)] = 206.58 gal/min (13.03 L/s), say 207 gal/min (13.06 L/s).

3. *Determine the water quantity for radiator cooling*

In the usual radiator cooling system for large engines, a portion of the cooling water is passed through a horizontal or vertical radiator. The remaining water is recirculated, after being tempered by the cooled water. Thus, the radiator must dissipate the jacket, turbocharger, and lube-oil cooler heat, Fig. 5.

The lube oil gives off 264 Btu/(bhp · h) (103.8 W/kWh). With a 10 percent safety factor, the total heat flow is 264 + (0.10)(264) = 290.4 Btu/(bhp · h) (114.1 W/kWh). At the rated output of 1000 bhp (746 kW), the lube-oil heat load = [290.4 Btu/(bhp · h)](1000 bhp) = 290,400 Btu/h (85.1 kW). Hence, the total heat load on the radiator = jacket + lube-oil heat load = 1,032,900 + 290,400 = 1,323,300 Btu/h (387.8 kW).

Radiators (also called fan coolers) serving large internal-combustion engines are usually rated for a 35°F (19.4°C) temperature reduction of the water. To remove 1,323,300 Btu/h (387.8 kW) with a 35°F (19.4°C) temperature decrease will require a flow of $G = H/(500\Delta t)$ = 1,323,300/[(500)(35)] = 76.1 gal/min (4.8 L/s).

4. *Determine the aftercooler cooling-water quantity*

The aftercooler must dissipate 278 Btu/(bhp · h) (109.2 W/kWh). At an output of 1000 bhp (746 kW), the heat load = [278 Btu/(bhp · h)](1000 bhp) = 278,000 Btu/h (81.5 kW). In general, designers do not use a factor of safety for the aftercooler because there is less chance of fouling or other difficulties.

With a 5°F (2.8°C) temperature rise of the cooling water during passage through the aftercooler, the quantity of water required $G = H/(500\Delta t)$ = 278,000/[(500)(5)] = 111 gal/min (7.0 L/s).

5. *Compute the quantity of steam generated by the exhaust*

Find the heat available in the exhaust by using $H_e = Wc\Delta t_e$, where H_e = heat available in the exhaust, Btu/h; W = exhaust-gas flow, lb/h; c = specific heat of the exhaust gas = 0.252 Btu/(lb · °F) (2.5 kJ/kg); Δt_e = exhaust-gas temperature at the boiler inlet, °F – exhaust-gas temperature at the boiler outlet, °F.

The exhaust-gas flow from a four-cycle turbocharged diesel is about 12.5 lb/(bhp · h) (7.5 kg/kWh). At full load this engine will exhaust [12.5 lb/(bhp · h)](1000 bhp) = 12.500 lb/h (5625 kg/h).

The temperature of the exhaust gas will be about 750°F (399°C) at the boiler inlet, whereas the temperature at the boiler outlet is generally held at 75°F (4I.7°C) higher than the steam temperature to prevent condensation of the exhaust gas. Steam at 30 lb/in^2 (abs) (206.8 kPa) has a temperature of 250.33°F (121.3°C). Thus, the exhaust-gas outlet temperature from the boiler will be 250.33 + 75 = 325.33°F (162.9°C), say 325°F (162.8°C). Then H_e = (12,500)(0.252)(750 – 325) = 1,375,000 Btu/h (403.0 kW).

At 30 lb/in^2 (abs) (206.8 kPa), the enthalpy of vaporization of steam is 945.3 Btu/lb (2198.9 kJ/kg), found in the steam tables. Thus, the exhaust heat can generate 1,375,000/945.3 = 1415 lb/h (636.8 kg/h) if the boiler is 100 percent efficient. With a boiler efficiency of 85 percent, the steam generated = (1415 lb/h)(0.85) = 1220 lb/h (549.0 kg/h), or (1200 lb/h)/1000 bhp = 1.22 lb/(bhp · h) (0.74 kg/kWh).

Related Calculations. Use this procedure for any reciprocating internal-combustion engine burning gasoline, kerosene, natural gas, liquefied-petroleum gas, or similar fuel. Figure 1 shows typical arrangements for a number of internal-combustion engine cooling systems.

When ethylene glycol or another antifreeze solution is used in the cooling system, alter the denominator of the flow equation to reflect the change in specific gravity and specific heat of

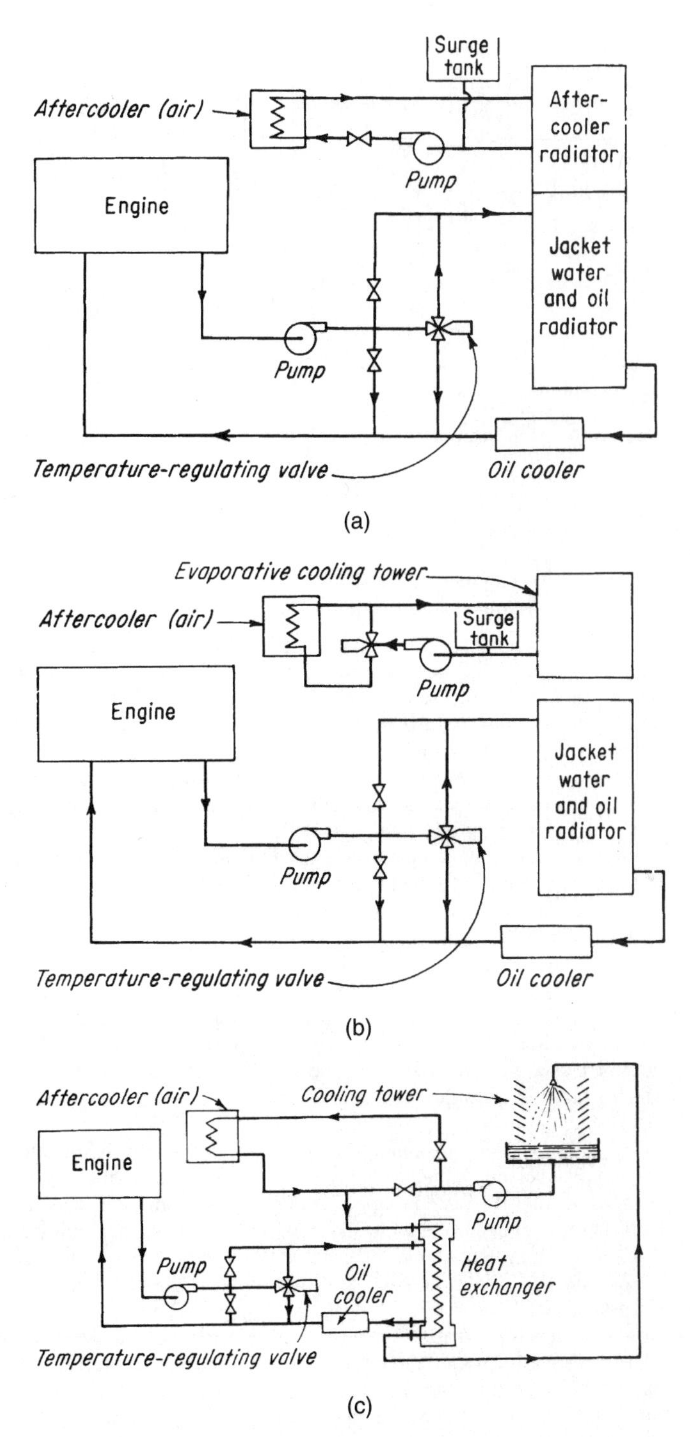

FIGURE 5 Internal-combustion engine cooling system: (*a*) radiator type; (*b*) evaporating cooling tower; (*c*) cooling tower. (*Power*.)

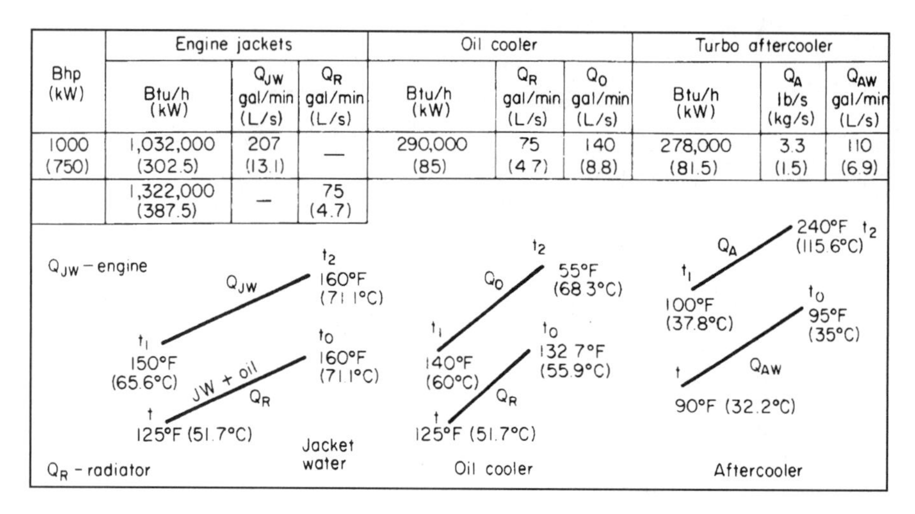

FIGURE 6 Slant diagrams for internal-combustion engine heat exchangers. (*Power.*)

the antifreeze solution, as compared with water. Thus, with a mixture of 50 percent glycol and 50 percent water, the flow equation in step 2 becomes $G = H/(436\Delta t)$. With other solutions, the numerical factor in the denominator will change. This factor = (weight of liquid lb/gal)(60 min/h), and the factor converts a flow rate of lb/h to gal/min when divided into the lb/h flow rate. Slant diagrams, Fig. 6, are often useful for heat-exchanger analysis.

Two-cycle engines may have a larger exhaust-gas flow than four-cycle engines because of the scavenging air. However, the exhaust temperature will usually be 50 to 100°F (27.7 to 55.6°C) lower, reducing the quantity of steam generated.

Where a dry exhaust manifold is used on an engine, the heat rejection to the cooling system is reduced by about 7.5 percent. Heat rejected to the aftercooler cooling water is about 3.5 percent of the total heat input to the engine. About 2.5 percent of the total heat input to the engine is rejected by the turbocharger jacket.

The jacket cooling water absorbs 11 to 14 percent of the total heat supplied. From 3 to 6 percent of the total heat supplied to the engine is rejected in the oil cooler.

The total heat supplied to an engine = (engine output, bhp)[heat rate, Btu/(bhp · h)]. A jacket-water flow rate of 0.25 to 0.60 gal/(min · bhp) (0.02 to 0.05 kg/kW) is usually recommended. The normal jacket-water temperature rise is 10°F (5.6°C); with a jacket-water outlet temperature of 180°F (82.2°C) or higher, the temperature rise of the jacket water is usually held to 7°F (3.9°C) or less.

To keep the cooling-water system pressure loss within reasonable limits, some designers recommend a pipe velocity equal to the nominal pipe size used in the system, or 2 ft/s for 2-in pipe (0.6 m/s for 50.8-mm); 3 ft/s for 3-in pipe (0.9 m/s for 76.2-mm); etc. The maximum recommended velocity is 10 ft/s for 10 in (3.0 m/s for 254 mm) and larger pipes. Compute the actual pipe diameter from $d = (G/2.5v)^{0.5}$, where G = cooling-water flow, gal/min; v = water velocity, ft/s.

Air needed for a four-cycle high-output turbocharged diesel engine is about 3.5 ft^3/(min · bhp) (0.13 m^3/kW); 4.5 ft^3/(min · bhp) (0.17 m^3/kW) for a two-cycle engine. Exhaust-gas flow is about 8.4 ft^3/(min · bhp) (0.32 m^3/kW) for a four-cycle diesel engine; 13 ft^3/(min · bhp) (0.49 m^3/kW) for two-cycle engines. Air velocity in the turbocharger blower piping should not exceed 3300 ft/min (1006 m/min); gas velocity in the exhaust system should not exceed 6000 ft/min (1828 m/min).The exhaust-gas temperature should not be reduced below 275°F (135°C), to prevent condensation.

The method presented here is the work of W. M. Kauffman, reported in *Power*.

I-C ENGINE ROOM VENT SYSTEM DESIGN

A radiator-cooled 60-kW internal-combustion engine generating set operates in an area where the maximum summer ambient temperature of the inlet air is 100°F (37.8°C). How much air does this engine need for combustion and for the radiator? What is the maximum permissible temperature rise of the room air? How much heat is radiated by the engine-alternator set if the exhaust pipe is 25 ft (7.6 m) long? What capacity exhaust fan is needed for this engine room if the engine room has two windows with an area of 30 ft^2 (2.8 m^2) each, and the average height between the air inlet and the outlet is 5 ft (1.5 m)? Determine the rate of heat dissipation by the windows. The engine is located at sea level.

Calculation Procedure:

1. *Determine engine air-volume needs*

Table 4 shows typical air-volume needs for internal-combustion engines installed indoors. Thus, a 60-kW set requires 390 ft^3/min (11.0 m^3/min) for combustion and 6000 ft^3/min (169.9 m^3/min) for the radiator. Note that in the smaller ratings, the combustion air needed is 6.5 ft^3/(min · kW) (0.18 m^3/kW), and the radiator air requirement is 150 ft^3/(min · kW)(4.2 m^3/kW).

TABLE 4 Total Air-Volume Needs*

Set kW	ft^3/min (m^3/min) for combustion	ft^3/min (m^3/min) for radiator	Maximum room temperature rise: Maximum ambient temperature of inlet air, °F (°C)	Maximum room temperature rise: Room air rise °F (°C)
20	130 (3.7)	3000 (84.9)	90 (32.2)	20 (11.1)
30	195 (5.5)	5000 (141.6)	95–105 (35–40.6)	15 (8.3)
40	260 (7.4)	5500 (155.7)	110–120 (43.3–48.9)	10 (5.6)
60	390 (1.0)	6000 (169.9)		

*Power.

2. *Determine maximum permissible air temperature rise*

Table 4 also shows that with an ambient temperature of 95 to 105°F (35 to 40.6°C), the maximum permissible room temperature rise is 15°C (8.3°C). When you determine this value, be certain to use the highest inlet-air temperature expected in the engine locality.

3. *Determine the heat radiated by the engine*

Table 5 shows the heat radiated by typical internal-combustion engine generating sets. Thus, a 60-kW radiator- and fan-cooled set radiates 2625 Btu/min (12.8 W) when the engine is fitted with a 25-ft (7.6-m) long exhaust pipe and a silencer.

4. *Compute the airflow produced by the windows*

The two windows can be used to ventilate the engine room. One window will serve as the air inlet; the other, as the air outlet. The area of the air outlet must at least equal the air-inlet area. Airflow will be produced by the stack effect resulting from the temperature difference between the inlet and outlet air.

The airflow C ft^3/min resulting from the stack effect is $C = 9.4A(h\Delta t_a)^{0.5}$, where A = free air of the air inlet, ft^2; h = height from the middle of the air-inlet opening to the middle of the air-outlet opening, ft; Δt_a = difference between the average indoor air temperature at point H and the temperature of the incoming air, °F. In this plant, the maximum permissible air temperature rise is 15°F (8.3°C), from step 2. With a 100°F (37.8°C) outdoor temperature, the maximum indoor temperature would be 100 + 15 = 115°F (46.1°C). Assume that the difference between the temperature of the incoming and outgoing air is 15°F (8.3°C). Then $C = 9.4(30)(5 \times 15)^{0.5}$ = 2445 ft^3/min (69.2 m^3/min).

TABLE 5 Heat Radiated from Typical Internal-Combustion Units, Btu/min (W)*

	Cooling by radiator and fan		Cooling by radiator, fan, and city water	
Alternator, kW	40	60	40	60
Engine-alternator set, silencer, and 25 ft (7.6 m) of exhaust pipe, Btu/min (W)	1830 (8.94)	2625 (12.8)	1701 (8.3)	2500 (12.2)
Exhaust pipe beyond silencer:				
Length 5 ft (1.5 m)	24 (0.12)	35 (0.17)	20 (0.10)	22 (0.11)
Length 10 ft (3.0 m)	45 (0.22)	65 (0.32)	39 (0.19)	40 (0.20)
Length 15 ft (4.6 m)	65 (0.32)	89 (0.44)	57 (0.38)	55 (0.27)

*Power.

5. *Compute the cooling airflow required*

This 60-kW internal-combustion engine generating set radiates 2625 Btu/min (12.8 W), step 3. Compute the cooling airflow required from $C = HK/\Delta t_a$, where C = cooling airflow required, ft³/min; H = heat radiated by the engine, Btu/min; K = constant from Table 6; other symbols as before, Thus, for this engine with a fan discharge temperature of 111 to 120°F (43.9 to 48.9°C), Table 6, $K = 60$; Δt_a = 15°F (8.3°C) from step 4. Then $C = (2625)(60)/15$ = 10,500 ft³/min (297.3 m³/min).

TABLE 6 Range of Discharge Temperature*

Room fan discharge temperature range			Wind to water gage			
			Wind velocity		Inlet pressure water gage	
°F	°C	K	mph	km/h	in	mm
80–989	26.7–31.7	57	60	96.5	1.75	44.5
90–99	32.6–37.2	58	30	48.3	0.43	10.9
100–110	37.8–43.3	59				
111–120	43.9–48.9	60				
121–130	49.4–54.4	61				

*Power.

The windows provide 2445 ft³/min (69.2 m³/min), step 4, and the engine radiator gives 6000 ft³/min (169.9 m³/min), step 1, or a total of 2445 + 6000 = 8445 ft³/min (239.1 m³/min). Thus, 10,500 – 8445 = 2055 ft³/min (58.2 m³/min) must be removed from the room. The usual method employed to remove the air is an exhaust fan. An exhaust fan with a capacity of 2100 ft³/min (59.5 m³/min) would be suitable for this engine room.

Related Calculations. Use this procedure for engines burning any type of fuel—diesel, gasoline, kerosene, or gas—in any type of enclosed room at sea level or elevations up to 1000 ft (304.8 m). Where windows or the fan outlet are fitted with louvers, screens, or intake filters, be certain to compute the net free area of the opening. When the radiator fan requires more air than is needed for cooling the room, an exhaust fan is unnecessary.

Be certain to select an exhaust fan with a sufficient discharge pressure to overcome the resistance of exhaust ducts and outlet louvers, if used. A propeller fan is usually chosen for exhaust service. In areas having high wind velocity, an axial-flow fan may be needed to overcome the pressure produced by the wind on the fan outlet.

Table 6 shows the pressure developed by various wind velocities. When the engine is located above sea level, use the multiplying factor in Table 7 to correct the computed air quantities for the lower air density.

TABLE 7 Air Density at Various Elevations*

Elevation above sea level			
ft	m	Multiplying factor, *A*	Approximate air density percent compared with sea level for same temperature
4,000	1,219	1,158	86.4
5,000	1,524	1,202	83.2
6,000	1,829	1,247	80.2
7,000	2,134	1,296	77.2
10,000	3,048	1,454	68.8

**Power*.

An engine radiates 2 to 5 percent of its total heat input. The total heat input = (engine output, bhp) [heat rate, Btu/(bhp · h)]. Provide 12 to 20 air changes per hour for the engine room. The most effective ventilators are power-driven exhaust fans or roof ventilators. Where the heat load is high, 100 air changes per hour may be provided. Auxiliary-equipment rooms require 10 air changes per hour. Windows, louvers, or power-driven fans are used. A four-cycle engine requires 3 to 3.5 ft^3/min of air per bhp (0.11 to 0.13 m^3/kW); a two-cycle engine, 4 to 5 ft^3/(min · bhp) (0.15 to 0.19 m^3/kW).

The method presented here is the work of John P. Callaghan, reported in *Power*.

BYPASS COOLING-SYSTEM DESIGN FOR I-C ENGINES

The internal-combustion engine in Fig. 7 is rated at 402 hp (300 kW) at 514 r/min and dissipates 3500 Btu/(bhp · h) (1375 W/kW) at full load to the cooling water from the power cylinders and water-cooled exhaust manifold. Determine the required cooling-water flow rate if there is a 10°F (5.6°C) temperature rise during passage of the water through the engine. Size the piping for the cooling system, using the head-loss data in Fig. 8, and the pump characteristic curve, Fig. 9. Choose a surge tank of suitable capacity. Determine the net positive suction head requirements for this engine. The total length of straight piping in the cooling system is 45 ft (13.7 m). The engine is located 500 ft (152.4 m) above sea level.

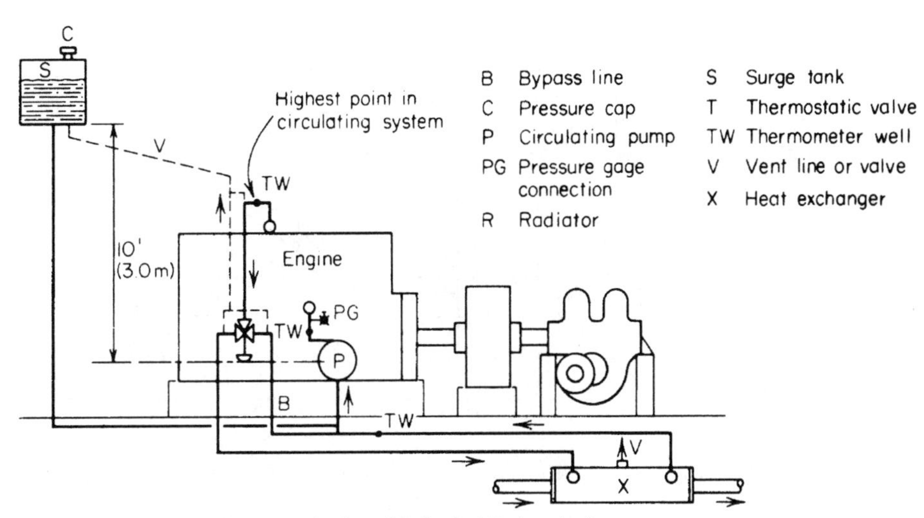

FIGURE 7 Engine cooling-system hookup. (*Mechanical Engineering*.)

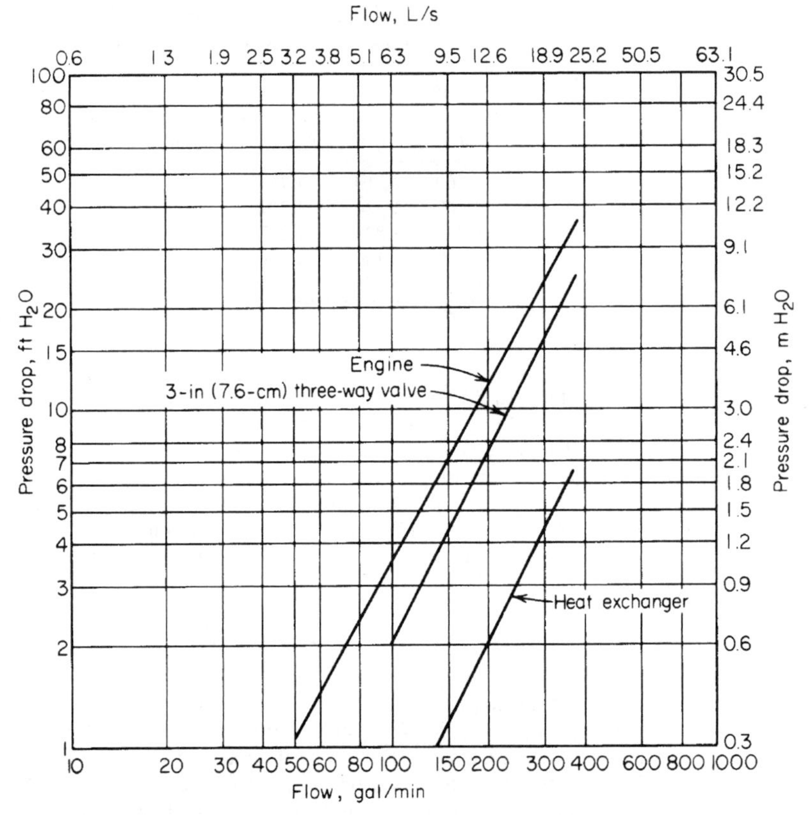

FIGURE 8 Head-loss data for engine cooling-system components. (*Mechanical Engineering*.)

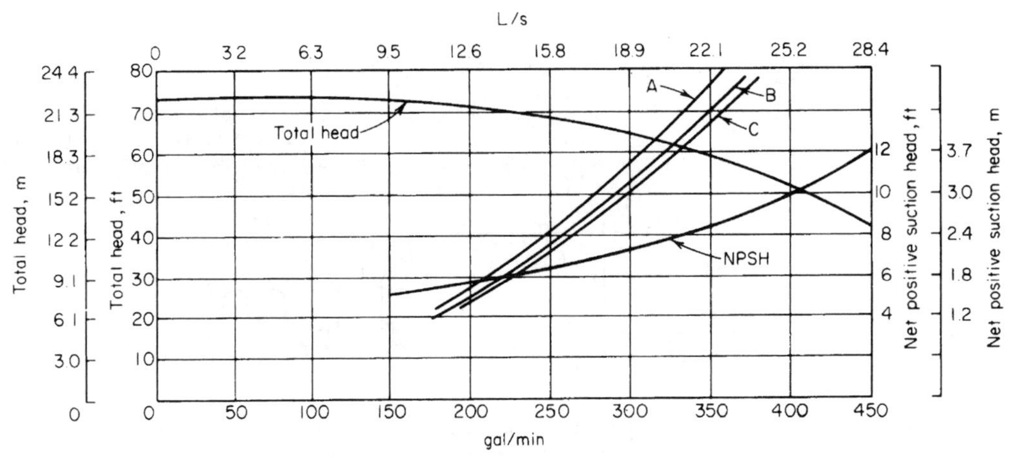

FIGURE 9 Pump and system characteristics for engine cooling system. (*Mechanical Engineering*.)

Calculation Procedure:

1. *Compute the cooling-water quantity required*

The cooling-water quantity required is $G = H/(500\Delta t)$, where G = cooling-water flow, gal/min; H = heat absorbed by the jacket water, Btu/h = (maximum engine hp) [heat dissipated, Btu/(bhp · h)]; Δt = temperature rise of the water during passage through the engine, °F. Thus, for this engine, $G = (402)(3500)/[500(10)] = 281$ gal/min (17.7 L/s).

2. *Choose the cooling-system valve and pipe size*

Obtain the friction head-loss data for the engine, the heat exchanger, and the three-way valve from the manufacturers of the respective items. Most manufacturers have curves or tables available for easy use. Plot the head losses, as shown in Fig. 8, for the engine and heat exchanger.

Before the three-way valve head loss can be plotted, a valve size must be chosen. Refer to a three-way valve capacity tabulation to determine a suitable valve size to handle a flow of 281 gal/min (17.7 L/s). Once such tabulation recommends a 3-in (76.2-mm) valve for a flow of 281 gal/min (17.7 L/s). Obtain the head-loss data for the valve, and plot it as shown in Fig. 8.

Next, assume a size for the cooling-water piping. Experience shows that a water velocity of 300 to 600 ft/min (91.4 to 182.9 m/min) is satisfactory for internal-combustion engine cooling systems. Using the Hydraulic Institute's *Pipe Friction Manual* or Cameron's *Hydraulic Data*, enter at 280 gal/min (17.6 L/s), the approximate flow, and choose a pipe size to give a velocity of 400 to 500 ft/min (121.9 to 152.4 m/min), i.e., midway in the recommended range.

Alternatively, compute the approximate pipe diameter from $d = 4.95\ [gpm/\text{velocity, ft/min}]^{0.5}$. With a velocity of 450 ft/min (137.2 m/min), $d = 4.95(281/450)^{0.5} = 3.92$, say 4 in (101.6 mm). The *Pipe Friction Manual* shows that the water velocity will be 7.06 ft/s (2.2 m/s), or 423.6 ft/min (129.1 m/min), in a 4-in (101.6-mm) schedule 40 pipe. This is acceptable. Using a 3½-in (88.9-mm) pipe would increase the cost because the size is not readily available from pipe suppliers. A 3-in (76.2-mm) pipe would give a velocity of 720 ft/min (219.5 m/min), which is too high.

3. *Compute the piping-system head loss*

Examine Fig. 7, which shows the cooling-system piping layout. Three flow conditions are possible: (*a*) all the jacket water passes through the heat exchanger, (*b*) a portion of the jacket water passes through the heat exchanger, and (*c*) none of the jacket water passes through the heat exchanger—instead, all the water passes through the bypass circuit. The greatest head loss usually occurs when the largest amount of water passes through the longest circuit (or flow condition *a*). Compute the head loss for this situation first.

Using the method given in the piping section of this handbook, compute the equivalent length of the cooling-system fitting and piping, as shown in Table 8. Once the equivalent length of the pipe and fittings is known, compute the head loss in the piping system, using the method given in the piping section of this handbook with a Hazen-Williams constant of $C = 130$ and a rounded-off flow rate of 300 gal/min (18.9 L/s). Summarize the results as shown in Table 8.

The total head loss is produced by the water flow through the piping, fittings, engine, three-way valve, and heat exchanger. Find the head loss for the last components in Fig. 8 for a flow of 300 gal/min (18.9 L/s). List the losses in Table 8, and find the sum of all the losses. Thus, the total circuit head loss is 57.61 ft (17.6 m) of water.

Compute the head loss for 0, 0.2, 0.4, 0.6, and 0.8 load on the engine, using the same procedure as in steps 1, 2, and 3 above. Plot on the pump characteristic curve, Fig. 9, the system head loss for each load. Draw a curve *A* through the points obtained, Fig. 9.

Compute the system head loss for condition *b* with half the jacket water [150 gal/min (9.5 L/s)] passing through the heat exchanger and half [150 gal/min (9.5 L/s)] through the bypass circuit. Make the same calculation for 0, 0.2, 0.4, 0.6, and 0.8 load on the engine. Plot the result as curve *B*, Fig. 9.

Perform a similar calculation for condition *c*—full flow through the bypass circuit. Plot the results as curve *C*, Fig. 9.

TABLE 8 Sample Calculation for Full Flow through Cooling Circuit* (*Fittings and Piping in Circuit*)

Fitting or pipe	Number in circuit	Equivalent length of straight pipe, ft	Equivalent length of straight pipe, m
3-in (76.2-mm) elbow	1	5.5	1.7
3 × 4-in (76.2 × 101.6-mm) reducer	4	7.2	2.2
4-in (101.6-mm) elbow	7	50.4	15.4
4-in (101.6-mm) tee	1	23.0	7.0
3-in (76.2-mm) pipe	—	0.67	0.2
4-in (101.6-mm) pipe	—	45.0	13.7
Total equivalent length of pipe:			
3-in (76.2-mm) pipe, standard weight	—	13.37	4.1
4-in (101.6-mm) pipe, standard weight	—	118.4	36.1

Head-loss calculation: Calculation for a flow rate of 300 gal/min (18.9 L/s) through circuit:

Using the Hazen-Williams friction-loss equation with a C factor of 130 (surface roughness constant), with 300 gal/min (18.9 L/s) flowing through the pipe, the head loss per 100 ft (30.5 m) of pipe is 21.1 ft (6.4 m) and 5.64 ft (1.1 m) for the 3-in (76.2-mm) and 4-in (101.6-mm) pipes, respectively. Thus head loss in piping is†

$$3 \text{ in } \frac{21.1}{100} \times 13.37 = 2.83 \text{ ft } (0.86 \text{ m})$$

$$4 \text{ in } \frac{5.64}{100} \times 118.4 = 6.68 \text{ ft } (2.0 \text{ m})$$

From Fig. 5 the head loss is:	ft	m
Through engine	26.00	7.9
Through 3-in (76.2-mm) three-way valve	17.50	5.3
Through heat exchanger	4.6	1.4
Total circuit head loss	57.61	14.6

**Mechanical Engineering.*

†Shaw and Loomis, *Cameron Hydraulic Data Book*, 12th ed., Ingersoll-Rand Company, 1951, p. 27.

4. *Compute the actual cooling-water flow rate*

Find the points of intersection of the pump total-head curve and the three system head-loss curves A, B, and C, Fig. 9. These intersections occur at 314, 325, and 330 gal/min (19.8, 20.5, and 20.8 L/s), respectively.

The initial design assumed a 10°F (5.6°C) temperature rise through the engine with a water flow rate of 281 gal/min (17.7 L/s). Rearranging the equation in step 1 gives $\Delta t = H/(400G)$. Substituting the flow rate for condition a gives an actual temperature rise of $\Delta t = (402)(3500)/[(500)(314)] =$ 8.97°F (4.98°C). If a 180°F (82.2°C) rated thermostatic element is used in the three-way valve, holding the outlet temperature t_o to 180°F (82.2°C), the inlet temperature t_i will be $\Delta t = t_o - t_i =$ 8.97; $180 - t_i = 8.97$; $t_i = 171.03$°F (77.2°C).

5. *Determine the required surge-tank capacity*

The surge tank in a cooling system provides storage space for the increase in volume of the coolant caused by thermal expansion. Compute this expansion from $E = 62.4g\Delta V$, where E = expansion, gal (L); g = number of gallons required to fill the cooling system: ΔV = specific volume, ft³/lb (m³/kg) of the coolant at the operating temperature – specific volume of the coolant, ft³/lb (m³/kg) at the filling temperature.

The cooling system for this engine must have a total capacity of 281 gal (1064 L), step 1. Round this to 300 gal (1136 L) for design purposes. The system operating temperature is 180°F (82.2°C), and the filling temperature is usually 60°F (15.6°C). Using the steam tables to find the specific volume of the water at these temperatures, we get $E = 62.4(300)(0.01651 - 0.01604) =$ 8.8 gal (33.3 L).

Usual design practice is to provide two to three times the required expansion volume. Thus, a 25-gal (94.6-L) tank (nearly three times the required capacity) would be chosen. The extra volume provides for excess cooling water that might be needed to make up water lost through minor leaks in the system.

Locate the surge tank so that it is the highest point in the cooling system. Some engineers recommend that the bottom of the surge tank be at least 10 ft (3 m) above the pump centerline and connected as close as possible to the pump intake. A 1½- or 2-in (38.1- or 50.8-mm) pipe is usually large enough for connecting the surge tank to the system. The line should be sized so that the head loss of the vented fluid flowing back to the pump suction will be negligible.

6. *Determine the pump net positive suction head*
The pump characteristic curve, Fig. 9, shows the net positive suction head (NSPH) required by this pump. As the pump discharge rate increases, so does the NPSH. This is typical of a centrifugal pump.

The greatest flow, 330 gal/min (20.8 L/s), occurs in this system when all the coolant is diverted through the bypass circuit, Figs. 4 and 5. At a 330-gal/min (20.8-L/s) flow rate through the system, the required NPSH for this pump is 8 ft (2.4 m), Fig. 9. This value is found at the intersection of the 330-gal/min (20.8-L/s) ordinate and the NPSH curve.

Compute the existing NPSH, ft (m), from $\text{NPSH} = H_s - H_f + 2.31(P_s - P_v)/s$, where H_s, = height of minimum surge-tank liquid level above the pump centerline, ft (m); H_f = friction loss in the suction line from the surge-tank connection to the pump inlet flange, ft (m) of liquid; P_s = pressure in surge tank, or atmospheric pressure at the elevation of the installation, lb/in^2 (abs) (kPa); P_v = vapor pressure of the coolant at the pumping temperature, lb/in^2 (abs) (kPa); s = specific gravity of the coolant at the pumping temperature.

7. *Determine the operating temperature with a closed surge tank*
A pressure cap on the surge tank, or a radiator, will permit operation at temperatures above the atmospheric boiling point of the coolant. At a 500-ft (152.4-m) elevation, water boils at 210°F (98.9°C). Thus, without a closed surge tank fitted with a pressure cap, the maximum operating temperature of a water-cooled system would be about 200°F (93.3°C).

If a 7-lb/in^2 (gage) (48.3-kPa) pressure cap were used at the 500-ft (152.4-m) elevation, then the pressure in the vapor space of the surge tank could rise to $P_s = 14.4 + 7.0 = 21.4$ lb/in^2 (abs) (147.5 kPa). The steam tables show that water at this pressure boils at 232°F (111.1°C). Checking the NPSH at this pressure shows that $\text{NPSH} = (10 - 1.02) + 2.31(21.4 - 21.4)/0.0954 = 8.98$ ft (2.7 m). This is close to the required 8-ft (2.4-m) head. However, the engine could be safely operated at a slightly lower temperature, say 225°F (107.2°C).

8. *Compute the pressure at the pump suction flange*
The pressure at the pump suction flange P lb/in^2 (gage) $= 0.433s(H_s - H_f) = (0.433)(0.974)$ $(10{,}00 - 1.02) = 3.79$ lb/in^2 (gage) (26.1 kPa).

A positive pressure at the pump suction is needed to prevent the entry of air along the shaft. To further ensure against air entry, a mechanical seal can be used on the pump shaft in place of packing.

Related Calculations. Use this general procedure in designing the cooling system for any type of reciprocating internal-combustion engine—gasoline, diesel, gas, etc. Where a coolant other than water is used, follow the same procedure but change the value of the constant in the denominator of the equation of step 1. Thus, for a mixture of 50 percent glycol and 50 percent water, the constant = 436, instead of 500.

The method presented here is the work of Duane E. Marquis, reported in *Mechanical Engineering.*

ENERGY RECOVERY VIA HOT WATER FROM I-C ENGINES

An internal-combustion engine fitted with a heat-recovery silencer and a jacket-water cooler is rated at 1000 bhp (746 kW). It exhausts 13.0 lb/(bhp · h) [5.9 kg/(bhp · h)] of exhaust gas at 700°F (371.1°C). To what temperature can hot water be heated when 500 gal/min (31.5 L/s) of jacket water is circulated through the hookup in Fig. 10 and 100 gal/min (6.3 L/s) of 60°F (15.6°C) water is heated? The jacket water enters the engine at 170°F (76.7°C) and leaves at 180°F (82.2°C).

Calculation Procedure:

1. *Compute the exhaust heat recovered*

Find the exhaust-heat recovered from $H_e = Wc\Delta t_e$, where the symbols are the same as in the previous calculation procedures. Since the final temperature of the exhaust gas is not given, a value must be assumed. Temperatures below 275°F (135°C) are undesirable because condensation of corrosive vapors in the silencer may occur. Assume that the exhaust-gas outlet temperature from the heat-recovery silencer is 300°F (148.9°C). The $H_e = (1000)(13)(0.252)(700 - 300) = 1{,}310{,}000$ Btu/h (383.9 kW).

2. *Compute the heated-water outlet temperature from the cooler*

Using the temperature notation in Fig. 10, we see that the heated-water outlet temperature from the jacket-water cooler is $t_z = (w_z/w_1)(t_4 - t_5) + t_1)$, where w_1 = heated-water flow, lb/h; w_z = jacket-water flow, lb/h; the other symbols are indicated in Fig. 10. To convert gal/min of water flow to lb/h, multiply by 500. Thus, w_1 = (100 gal/min)(500) = 50,000 lb/h (22,500 kg/h), and w_z = (500 gal/min)(500) = 250,000 lb/h (112,500 kg/h). Then $t_z = (250{,}000/50{,}000)(180 - 170) + 60 = 110$°F(43.4°C).

3. *Compute the heated-water outlet temperature from the silencer*

The silencer outlet temperature $t_3 = H_e/w_1 + t_z$, or $t_3 = 1{,}310{,}000/50{,}000 + 110 = 136.2$°F (57.9°C).

Related Calculations. Use this method for any type of engine—diesel, gasoline, or gas—burning any type of fuel. Where desired, a simple heat balance can be set up between the heat-releasing and heat-absorbing sides of the system instead of using the equations given here. However, the equations are faster and more direct.

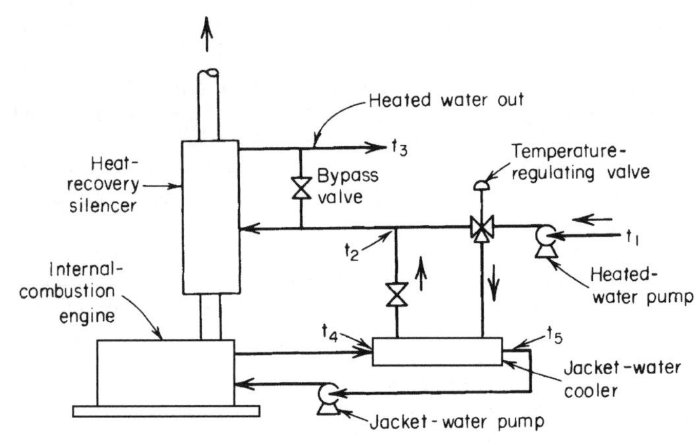

FIGURE 10 Internal-combustion engine cooling system.

FUEL STORAGE CAPACITY AND COST FOR I-C ENGINES

A diesel power plant will have six 1000-hp (746-kW) engines and three 600-hp (448-kW) engines. The annual load factor is 85 percent and is nearly uniform throughout the year. What capacity day tanks should be used for these engines? If fuel is delivered every 7 days, what storage capacity is required? Two fuel supplies are available: a 24° API fuel at $0.0825 per gallon ($0.022 per liter) and a 28° API fuel at $0.0910 per gallon ($0.024 per liter). Which is the better buy?

Calculation Procedure:

1. *Compute the engine fuel consumption*
Assume, or obtain from the engine manufacturer, the specific fuel consumption of the engine. Typical modern diesel engines have a full-load heat rate of 6900 to 7500 Btu/(bhp · h) (2711 to 3375 W/kWh), or about 0.35 lb/(bhp · h) of fuel (0.21 kg/kWh). Using this value of fuel consumption for the nine engines in this plant, we see the hourly fuel consumption at 85 percent load factor will be (6 engines)(1000 hp)(0.35)(0.85) + (3 engines)(600 hp)(0.35)(0.85) = 2320 lb/h (1044 kg/h).

Convert this consumption rate to gal/h by finding the specific gravity of the diesel oil. The specific gravity s = 141.5/(131.5 + °API). For the 24° API oil, s = 141.5/(131.5 + 24) = 0.910. Since water at 60°F (15.6°C) weighs 8.33 lb/gal (3.75 kg/L), the weight of this oil is (0.910)(8.33) = 7.578 lb/gal (3.41 kg/L). For the 28° API oil, s = 141.5/(131.5 + 28) = 0.887, and the weight of this oil is (0.887)(8.33) = 7.387 lb/gal (3.32 kg/L). Using the lighter oil, since this will give a larger gal/h consumption, we get the fuel rate = (2320 lb/h)/(7.387 lb/gal) = 315 gal/h (1192 L/h).

The daily fuel consumption is then (24 h/day)(315 gal/h) = 7550 gal/day (28,577 L/day). In 7 days the engines will use (7 days)(7550 gal/day) = 52,900, say 53,000 gal (200,605 L).

2. *Select the tank capacity*
The actual fuel consumption is 53,000 gal (200,605 L) in 7 days. If fuel is delivered exactly on time every 7 days, a fuel-tank capacity of 53,000 gal (200,605 L) would be adequate. However, bad weather, transit failures, strikes, or other unpredictable incidents may delay delivery. Therefore, added capacity must be provided to prevent engine stoppage because of an inadequate fuel supply.

Where sufficient space is available, and local regulations do not restrict the storage capacity installed, use double the required capacity. The reason is that the additional storage capacity is relatively cheap compared with the advantages gained. Where space or storage capacity is restricted, use 1½ times the required capacity.

Assuming double capacity is used in this plant, the total storage capacity will be (2)(53,000) = 106,000 gal (401,210 L). At least two tanks should be used, to permit cleaning of one without interrupting engine operation.

Consult the National Board of Fire Underwriters bulletin *Storage Tanks for Flammable Liquids* for rules governing tank materials, location, spacing, and fire-protection devices. Refer to a tank capacity table to determine the required tank diameter and length or height depending on whether the tank is horizontal or vertical. Thus, the Buffalo Tank Corporation *Handbook* shows that a 16.5-ft (5.0-m) diameter 33.5-ft (10.2-m) long horizontal tank will hold 53,600 gal (202,876 L) when full. Two tanks of this size would provide the desired capacity. Alternatively, a 35-ft (10.7-m) diameter 7.5-ft (2.3-m) high vertical tank will hold 54,000 gal (204,390 L) when full. Two tanks of this size would provide the desired capacity.

Where a tank capacity table is not available, compute the capacity of a cylindrical tank from capacity = $5.87D^2L$, where D = tank diameter, ft; L = tank length or height, ft. Consult the NBFU or the tank manufacturer for the required tank wall thickness and vent size.

3. *Select the day-tank capacity*
Day tanks supply filtered fuel to an engine. The day tank is usually located in the engine room and holds enough fuel for a 4- to 8-hour operation of an engine at full load. Local laws, insurance

requirements, or the NBFU may limit the quantity of oil that can be stored in the engine room or a day tank. One day tank is usually used for each engine.

Assume that a 4-hour supply will be suitable for each engine. Then the day-tank capacity for a 1000-hp (746-kW) engine = (1000 hp) [0.35 lb/(bhp · h) fuel] (4 hours) = 1400 lb (630 kg), or 1400/7.387 = 189.6 gal (717.6 L), given the lighter-weight fuel, step 1. Thus, one 200-gal (757-L) day tank would be suitable for each of the 1000-hp (746-kW) engines.

For the 600-hp (448-kW) engines, the day-tank capacity should be (600 hp)[0.35 lb/(bhp · h) fuel] (4 hours) = 840 lb (378 kg), or 840/7.387 = 113.8 gal (430.7 L). Thus, one 125-gal (473-L) day tank would be suitable for each of the 600-hp (448-kW) engines.

4. *Determine which is the better fuel buy*

Compute the higher heating value HHV of each fuel from HHV = 17,645 + 54(°API), or for 24° fuel, HHV = 17,645 + 54(24) = 18,941 Btu/lb (44,057 kJ/kg). For the 28° fuel, HHV = 17,645 + 54(28) = 19,157 Btu/lb (44,559 kJ/kg).

Compare the two oils on the basis of cost per 10,000 Btu (10,550 kJ), because this is the usual way of stating the cost of a fuel. The weight of each oil was computed in step 1. Thus the 24° API oil weighs 7.578 lb/gal (0.90 kg/L), while the 28° API oil weighs 7.387 lb/gal (0.878 kg/L).

Then the cost per 10,000 Btu (10,550 kJ) = (cost, $/gal)/[(HHV, Btu/lb)/10,000](oil weight, lb/gal). For the 24° API oil, cost per 10,000 Btu (10,550 kJ) = (cost, $/gal)/[(HHV, Btu/lb)/10,000] (oil weight, lb/gal). For the 24° API oil, cost per 10,000 Btu (10,550 kJ) = $0.0825/[(18.941/10,000)(7.578)] = $0.00574, or 0.574 cent per 10,000 Btu (10,550 kJ). For the 28° API oil, cost per 10,000 Btu = $0.0910/[(19,157/10,000)(7387)] = $0.00634, or 0.634 cent per 10,000 Btu (10,550 kJ). Thus, the 24° API is the better buy because it costs less per 10,000 Btu (10,550 kJ).

Related Calculations. Use this method for engines burning any liquid fuel. Be certain to check local laws and the latest NBFU recommendations before ordering fuel storage or day tanks.

Low-sulfur diesel amendments were added to the federal Clean Air Act in 1991. These amendments required diesel engines to use low-sulfur fuel to reduce atmospheric pollution. Reduction in fuel sulfur content will not require any change in engine operating procedures. If anything, the lower sulfur content will reduce engine maintenance requirements and costs.

The usual distillate fuel specification recommends a sulfur content of not more than 1.5 percent by weight, with 2 percent by weight considered satisfactory. Refineries are currently producing diesel fuel that meets federal low-sulfur requirements. While there is a slight additional cost for such fuel at the time of this writing, when the regulations went into effect, predictions are that the price of low-sulfur fuel will decline as more is manufactured.

Automobiles produce 50 percent of the air pollution throughout the developed world. The Ozone Transport Commission, set up by Congress as part of the 1990 Clear Air Act, is enforcing emission standards for new automobiles and trucks. To date, the cost of meeting such standards has been lower than anticipated. By the year 2003, all new automobiles will be pollution-free—if they comply with the requirements of the act. Stationary diesel plants using low-sulfur fuel will emit extremely little pollution.

ENERGY REQUIREMENTS FOR I-C ENGINE COOLING-WATER AND LUBE-OIL PUMPS

What is the required power input to a 200-gal/min (12.6-L/s) jacket-water pump if the total head on the pump is 75 ft (22.9 m) of water and the pump has an efficiency of 70 percent when it handles freshwater and saltwater? What capacity lube-oil pump is needed for a four-cycle 500-hp (373-kW) turbocharged diesel engine having oil-cooled pistons? What is the required power input to this pump if the discharge pressure is 80 lb/in^2 (551.5 kPa) and the efficiency of the pump is 68 percent?

Calculation Procedure:

1. *Determine the power input to the jacket-water pump*

The power input to jacket-water and raw-water pumps serving internal-combustion engines is often computed from the relation $hp = Gh/Ce$, where hp = hp input; G = water discharged by pump, gal/min; h = total head on pump, ft of water; C = constant = 3960 for freshwater having a density of 62.4 lb/ft^3 (999.0 kg/m^3); 3855 for saltwater having a density of 64 lb/ft^3 (1024.6 kg/m^3).

For this pump handling freshwater, hp = (200)(75)/[(3960)(0.70)] = 5.42 hp (4.0 kW). A 7.5-hp (5.6-kW) motor would probably be selected to handle the rated capacity plus any overloads.

For this pump handling saltwater, hp = (200)(75)/[(3855)(0.70)] = 5.56 hp (4.1 kW). A 7.5-hp (5.6-kW) motor would probably be selected to handle the rated capacity plus any overloads. Thus, the same motor could drive this pump whether it handles freshwater or saltwater.

2. *Compute the lube-oil pump capacity*

The lube-oil pump capacity required for a diesel engine is found from $G = H/200\Delta t$, where G = pump capacity, gal/min; H = heat rejected to the lube oil, Btu/(bhp · h); Δt = lube-oil temperature rise during passage through the engine, °F. Usual practice is to limit the temperature rise of the oil to a range of 20 to 25°F (11.1 to 13.9°C), with a maximum operating temperature of 160°F (71.1°C). The heat rejection to the lube oil can be obtained from the engine heat balance, the engine manufacturer, or *Standard Practices for Stationary Diesel Engines*, published by the Diesel Engine Manufacturers Association. With a maximum heat rejection rate of 500 Btu/(bhp · h) (196.4 W/kWh) from *Standard Practices* and an oil-temperature rise of 20°F (11.1°C), G = [500 Btu/(bhp · h)](1000 hp)/[(200)(20)] = 125 gal/min (7.9 L/s).

By using the *lowest* temperature rise and the *highest* heat rejection rate, a safe pump capacity is obtained. Where the pump cost is a critical factor, use a higher temperature rise and a lower heat rejection rate. Thus, with a heat rejection, the above pump would have a capacity of G = (300)(1000)/[(200)(25)] = 60 gal/min (3.8 L/s).

3. *Compute the lube-oil pump power input*

The power input to a separate oil pump serving a diesel engine is given by $hp = Gp/1720e$, where G = pump discharge rate, gal/min; p = pump discharge pressure, lb/in^2; e = pump efficiency. For this pump, hp = (125)(80)/[(1720)(0.68)] = 8.56 hp (6.4 kW). A 10-hp (7.5-kW) motor would be chosen to drive this pump.

With a capacity of 60 gal/min (3.8 L/s), the input is hp = (60)(80)/[(1720)(0.68)] = 4.1 hp (3.1 kW). A 5-hp (3.7-kW) motor would be chosen to drive this pump.

Related Calculations. Use this method for any reciprocating diesel engine, two- or four-cycle. Lube-oil pump capacity is generally selected 10 to 15 percent oversize to allow for bearing wear in the engine and wear of the pump moving parts. Always check the selected capacity with the engine builder. Where a bypass-type lube-oil system is used, be sure to have a pump of sufficient capacity to handle *both* the engine and cooler oil flow.

Raw-water pumps are generally duplicates of the jacket-water pump, having the same capacity and head ratings. Then the raw-water pump can serve as a standby jacket-water pump, if necessary.

CHOICE OF LUBE-OIL COOLER FOR I-C ENGINES

A 500-hp (373-kW) internal-combustion engine rejects 300 to 600 Btu/(bhp · h) (118 to 236 W/kWh) to the lubricating oil. What capacity and type of lube-oil cooler should be used for this engine if 10 percent of the oil is bypassed? If this engine consumes 2 gal (7.6 L) of lube oil per 24 hours at full load, determine its lube-oil consumption rate.

Calculation Procedure:

1. *Determine the required lube-oil cooler capacity*
Base the cooler capacity on the maximum heat rejection rate plus an allowance for overloads. The usual overload allowance is 10 percent of the full-load rating for periods of not more than 2 hours in any 24-hour period.

For this engine, the maximum output with a 10 percent overload is 500 + (0.10)(500) = 550 hp (410 kW). Thus, the maximum heat rejection to the lube oil would be (500 hp)[600 Btu/(bhp · h)] = 330,000 Btu/h (96.7 kW).

2. *Choose the type and capacity of lube-oil cooler*
Choose a shell-and-tube type heat exchanger to serve this engine. Long experience with many types of internal-combustion engines shows that the shell-and-tube heat exchanger is well suited for lube-oil cooling.

Select a lube-oil cooler suitable for a heat-transfer load of 330,000 Btu/h (96.7 kW) at the prevailing cooling-water temperature difference, which is usually assumed to be 10°F (5.6°C). See previous calculation procedures for the steps in selecting a liquid cooler.

3. *Determine the lube-oil consumption rate*
The lube-oil consumption rate is normally expressed in terms of bhp · h/gal. Thus, if this engine operates for 24 hours and consumes 2 gal (7.6 L) of oil, its lube-oil consumption rate = (24 h)(500 bhp)/2 gal = 6000 bhp · h/gal (1183 kWh/L).

Related Calculations. Use this procedure for any type of internal-combustion engine using any fuel.

DETERMINING SOLIDS INTAKE OF I-C ENGINES

What weight of solids annually enters the cylinders of a 1000-hp (746-kW) internal-combustion engine if the engine operates 24 h/day, 300 days/year in an area having an average dust concentration of 1.6 gr per 1000 ft^3 of air (28.3 m^3)? The engine air rate (displacement) is 3.5 ft^3/(min · bhp) (0.13 m^3/kW). What would the dust load be reduced to if an air filter fitted to the engine removed 80 percent of the dust from the air?

Calculation Procedure:

1. *Compute the quantity of air entering the engine*
Since the engine is rated at 1000 hp (746 kW) and uses 3.5 ft^3/(min · bhp)[0.133 m^3/(min · kW)], the quantity of air used by the engine each minute is (1000 hp)[3.5 ft^3/(min · hp)] = 3500 ft^3/min (99.1 m^3/min).

2. *Compute the quantity of dust entering the engine*
Each 1000 ft^3 (28.3 m^3) of air entering the engine contains 1.6 gr (103.7 mg) of dust. Thus, during every minute of engine operation, the quantity of dust entering the engine is (3500/1000)(0.6) = 5.6 gr (362.8 mg). The hourly dust intake = (60 min/h)(5.6 gr/min) = 336 gr/h (21,772 mg/h).

During the year the engine operates 24 h/day for 300 days. Hence, the annual intake of dust is (24 h/day)(300 days/year)(336 gr/h) = 2,419,200 gr (156.8 kg). Since there is 7000 gr/lb, the weight of dust entering the engine per year = 2,419,200 gr/(7000 gr/lb) = 345.6 lb/year (155.5 kg/year).

3. *Compute the filtered dust load*
With the air filter removing 80 percent of the dust, the quantity of dust reaching the engine is (1.00 – 0.80)(345.6 lb/year) = 69.12 lb/year (31.1 kg/year). This shows the effectiveness of an air filter in reducing the dust and dirt load on an engine.

Related Calculations. Use this general procedure to compute the dirt load on an engine from any external source.

ENERGY PERFORMANCE FACTORS FOR I-C ENGINES

Discuss and illustrate the important factors in internal-combustion engine selection and performance. In this discussion, consider both large and small engines for a full range of usual applications.

Calculation Procedure:

1. *Plot typical engine load characteristics*
Figure 11 shows four typical load patterns for internal-combustion engines. A continuous load, Fig. 11*a,* is generally considered to be heavy-duty and is often met in engines driving pumps or electric generators.

Intermittent heavy-duty loads, Fig. 11*b,* are often met in engines driving concrete mixers, batch machines, and similar loads. Variable heavy-duty loads, Fig. 11*c,* are encountered in large vehicles, process machinery, and similar applications. Variable light-duty loads, Fig. 11*d,* are met in small vehicles like golf carts, lawn mowers, chain saws, etc.

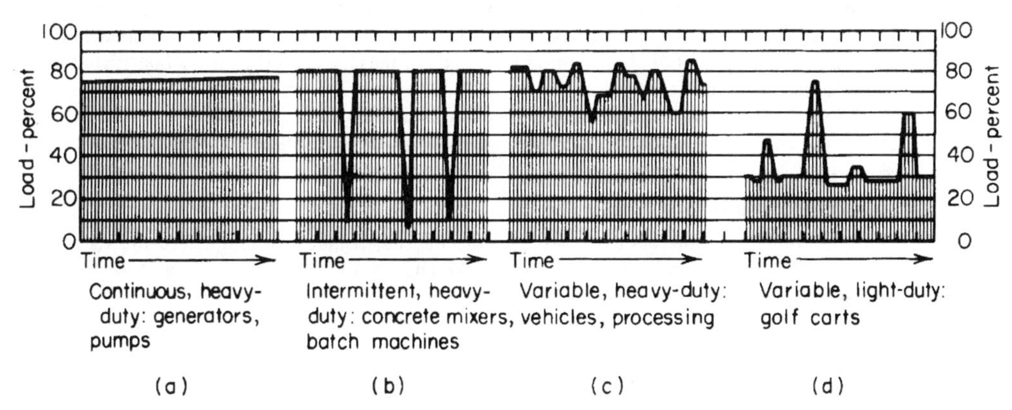

FIGURE 11 Typical internal-combustion engine load cycles; (*a*) continuous, heavy-duty; (*b*) intermittent, heavy-duty; (*c*) variable, heavy-duty; (*d*) variable, light-duty. (*Product Engineering.*)

2. *Compute the engine output torque*
Use the relation $T = 5250$ bhp/(r/min) to compute the output torque of an internal-combustion engine. In this relation, bhp = engine bhp being developed at a crankshaft speed having rotating speed of *rpm*.

3. *Compute the hp output required*
Knowing the type of load on the engine (generator, pump, mixer, saw blade, etc.), compute the power output required to drive the load at a constant speed. Where a speed variation is expected, as in variable-speed drives, compute the average power needed to accelerate the load between two desired speeds in a given time.

4. *Choose the engine output speed*
Internal-combustion engines are classified in three speed categories: high (1500 r/min or more), medium (750 to 1500 r/min), and low (less than 750 r/min).

Base the speed chosen on the application of the engine. A high-speed engine can be lighter and smaller for the same hp rating, and may cost less than a medium-speed or slow-speed engine serving the same load. But medium-speed and slow-speed engines, although larger, offer a higher torque output for the equivalent hp rating. Other advantages of these two speed ranges include longer service life and, in some instances, lower maintenance costs.

Usually an application will have its own requirements, such as allowable engine weight, available space, output torque, load speed, and type of service. These requirements will often indicate that a particular speed classification must be used. Where an application has no special speed requirements, the speed selection can be made on the basis of cost (initial, installation, maintenance, and operating costs), type of parts service available, and other local conditions.

5. *Analyze the engine output torque required*

In some installations, an engine with good lugging power is necessary, especially in tractors, harvesters, and hoists, where the load frequently increases above normal. For good lugging power, the engine should have the inherent characteristic of increasing torque with drooping speed. The engine can then resist the tendency for increased load to reduce the output speed, giving the engine good lugging qualities.

One way to increase the torque delivered to the load is to use a variable-ratio hydraulic transmission. The transmission will amplify the torque so that the engine will not be forced into the lugging range.

Other types of loads, such as generators, centrifugal pumps, air conditioners, and marine drives, may not require this lugging ability. So be certain to consult the engine power curves and torque characteristic curve to determine the speed at which the maximum torque is available.

6. *Evaluate the environmental conditions*

Internal-combustion engines are required to operate under a variety of environmental conditions. The usual environmental conditions critical in engine selection are altitude, ambient temperature, dust or dirt, and special or abnormal service. Each of these, except the last, is considered in previous calculation procedures.

Special or abnormal service includes such applications as fire fighting, emergency flood pumps and generators, and hospital standby service. In these applications, an engine must start and pick up a full load without warmup.

7. *Compare engine fuels*

Table 9 compares four types of fuels and the internal-combustion engines using them. Note that where the cost of the fuel is high, the cost of the engine is low; where the cost of the fuel is low, the cost of the engine is high. This condition prevails for both large and small engines in any service.

8. *Compare the performance of small engines*

Table 10 compares the principal characteristics of small gasoline and diesel engines rated at 7 hp (5 kW) or less. Note that engine life expectancy can vary from 500 to 25,000 hours. With modern, mass-produced small engines it is often just as cheap to use short-life replaceable two-stroke gasoline engines instead of a single long-life diesel engine. Thus, the choice of a small engine is often based on other considerations, such as ease and convenience of replacement, instead of just hours of life. Chances are, however, that most long-life applications of small engines will still require a long-life engine. But the alternative must be considered in each case.

Related Calculations. Use the general data presented here for selecting internal-combustion engines having ratings up to 200 hp (150 kW). For larger engines, other factors such as weight, specific fuel consumption, lube-oil consumption, etc., become important considerations. The method given here is the work of Paul F. Jacobi, as reported in *Product Engineering.*

DIESEL-ENGINE VOLUMETRIC EFFICIENCY

A four-cycle six-cylinder diesel engine of 4.25-in (11.4-cm) bore and 60-in (15.2-cm) stroke running at 1200 rpm has 9 percent CO_2 present in the exhaust gas. The fuel consumption is 28 lb (12.7 kg) per hour. Assuming that 13.7 percent CO_2 indicates an air-fuel ratio of 15 lb of air to 1 lb

TABLE 9 Comparison of Fuels for Internal-Combustion Engines*

	Storage life (quantities)								Weight		Heat content			
	Small	Large	Consistency, Btu/ft^3	Initial cost of engine, relative	Cost of fuel	Residue	Antiknock rating	Filtering necessary	lb/gal	kg/L	Btu/vol	mJ/vol	Btu/lb	mJ/kg
Gasoline	Good	Poor (6 months)	Good	Low	High	High	Best is costly	Medium	6.000	0.714	123,039 Btu/gal	34,291 kJ/L	20,627	47.9
Diesel														
No. 1	Good	Fair (1 year)	Good	High	Low	Low if properly filtered	—	High	6.850	0.815	135,800 But/gal	37,847 kJ/L	19,750	45.9
No. 2	Good	Fair (1 year)	Good	High	Low	Low if properly filtered	—	High	7.020	0.835	139,000 Btu/gal	38,739 kJ/L	19,786	46.0
Natural gas	Not necessary	Not necessary	Poor	Medium	Medium	Low	High	Very little	—	—	1,000 Btu/ft^3	37,250 kJ/m^3		
LPG														
Propane	Good	Good	Poor	Medium	Medium	Low	Good	Very little	4.235	0.504	91,740 Btu/gal	25,568 kJ/L	21,308	49.6
Butane	Good	Good	Poor	Medium	Medium	Low	Good	Very little	4.873	0.580	103,830 Btu/gal	28,937 kJ/L	20,627	47.9

Product Engineering.

TABLE 10 Performance Table for Small Internal-Combustion Engines [Less than 7 hp (5 kW)]*

	Variety of models available	Typical weight lb/hp (kg/kW)	Operating speeds: Typical maximum	Operating speeds: Typical efficient minimum	Lugging ability	Torque output	Relative life expectancy, h	Relative cost	Fuel required	Shaft direction	Noise level	Starters	Integral optional Pto's	Ignition	Cost of operation	Variety options accessor
Lightweight																
2-stork	Narrow	2.1 (1.2:1)	3,600 (governed) to 7,500	2,000 to 3,000	Poor to fair	Fair	500	Lowest	Gasoline oil mixed	Vertical, horizontal, or universal	High	Rope, recoil, impulse	No	Magneto	High	Standard extre low custom wide
4-strok	Wide	6:1: 10:1 (36.1: 6.1:1)	4,000	2,000 to 2,400	Fair to good	Good	500	1 to 2	Gasoline (LPG)	Vertical or horizontal	Moderate	Rope, recoil, impulse, electric	Several	Magneto	Moderate	Standard wide
Heavyweight																
4-stork	Wide	11.1: 20.1 (6.6:1: 12.1:1)	4,000	1,600 to 1,800	Good to excellent	Good	7,500	2 to 4	Gasoline (LPG)	Vertical or horizontal	Moderate	Rope, recoil, impulse, crank, electric	No	Magneto distributor	Moderate	Standard moderate wide
Diesel	Narrow	35.1 (21:1:1)	2,400	1,500	Excellent	Good	25,000	4	Diesel	Horizontal	Moderate to high	Electric	No	Battery, distributor, glow plugs	Low	Narrow

Product Engineering

(6.6 kg to 0.45 kg), calculate the volumetric efficiency of the engine. Intake air temperature is 60°F (15.6°C) and the barometric pressure is 29.8 in (79.7 cm).

Calculation Procedure:

1. *Find the percentage of N_2 in the exhaust gas*
Atmospheric air contains 76.9 percent nitrogen by weight. If an analysis of the fuel oil shows zero nitrogen before combustion, all the nitrogen in the exhaust gas must come from the air. Therefore, with 13.7 percent CO_2 by volume in the dry exhaust the nitrogen content is: N_2 = (76.9/100)(15) = 11.53 lb (5.2 kg) N_2 per lb (0.454 kg) of fuel oil. Converting to moles, 11.53 lb (5.2 kg) N_2/28 lb (12.7 kg) fuel per hour = 0.412 mole N_2 per lb (0.454 kg) of fuel oil.

2. *Compute the weight of N_2 in the exhaust*
Use the relation, percentage of CO_2 in the exhaust gases = (CO_2)/(N_2 + CO_2) in moles. Substituting, (13.7)/(100) = (CO_2)/(N_2 + 0.412). Solving for CO_2, we find CO_2 = 0.0654 mole.

Now, since mole percent is equal to volume percent, for 9 percent CO_2 in the exhaust gases, 0.09 = (CO_2)/(CO_2 + N_2) = 0.0654/(0.0654 + N_2). Solving for N_2, we find N_2 = 0.661 mole. The weight of N_2 therefore = 0.661 × 28 = 18.5 lb (8.399 kg).

3. *Calculate the amount of air required for combustion*
The air required for combustion is found from (N_2) = 18.5/0.769, where 0.769 = percent N_2 in air, expressed as a decimal. Solving, N_2 = 24.057 lb (10.92 kg) per lb (0.454 kg) of fuel oil.

4. *Find the weight of the actual air charge drawn into the cylinder*
Specific volume of the air at 60°F (15.6°C) and 29.8 in (75.7 cm) Hg is 13.03 ft^3 (0.368 m^3) per lb. Thus, the actual charge drawn into the cylinder = (lb of air per lb of fuel)(specific volume of the air, ft^3/lb)(fuel consumption, lb/h)/3600 s/h. Or 24.1(13.02)(28)/3600 = 2.44 ft^3 (0.69 m^3) per second.

5. *Compute the volumetric efficiency of this engine*
Volumetric efficiency is defined as the ratio of the actual air charge drawn into the cylinder divided by the piston displacement. The piston displacement for one cylinder of this engine is (bore area)(stroke length)(1 cylinder)/1728 in^3/ft^3. Solving, piston displacement = $0.785(4.25)^2(6)(1)/1728$ = 0.0492 ft^3 (0.00139 m^3).

The number of suction strokes per minute = rpm/2. The volume displaced per second by the engine = (piston displacement per cylinder)(number of cylinders)(rpm/2)/60 s/min. Substituting, engine displacement = 0.0492(6)(1200/2)/60 = 2.952 ft^3/s (0.0084 m^3/s).

Then, the volumetric efficiency of this engine = actual charge drawn into the cylinder/engine displacement = 2.45/2.952 = 0.8299, or 82.99 percent.

Related Calculations. Use this general procedure to determine the volumetric efficiency of reciprocating internal-combustion engines—both gasoline and diesel. The procedure is also used for determining the fuel consumption of such engines, using test data from actual engine runs.

AIR-COOLED I-C ENGINE CHOICE FOR INDUSTRIAL USES

Choose a suitable air-cooled gasoline engine to replace a 10-hp (7.46-kW) electric motor driving a municipal service sanitary pump at an elevation of 8000 ft (2438 m) where the ambient temperature is 90°F (32.2°C). Find the expected load duty for this engine; construct a typical load curve for it.

Calculation Procedure:

1. *Determine the horsepower (kW) rating required of the engine*
Electric motors are rated on an entirely different basis than are internal-combustion engines. Most electric motors will deliver 25 percent more power than their rating during a period of 1 or 2 hours. For short periods many electric motors may carry 50 percent overload.

Gasoline engines, by comparison, are rated at the maximum power that a new engine will develop on a dynamometer test conducted at an ambient temperature of 60°F (15.6°C) and a sea-level barometric pressure of 29.92 in (759.97 mm) of mercury. For every 10°F (5.56°C) rise in the intake ambient air temperature there will be a 1 percent reduction in the power output. And for every 1-in (2.5-cm) drop in barometric pressure there will be a 3.5 percent power output loss. For every 1000 ft (304.8 m) of altitude above sea level a 3.5 percent loss in power output also occurs.

Thus, for average atmospheric conditions, the actual power of a gasoline engine is about 5 to 7 percent less than the standard rating. And if altitude is a factor, the loss can be appreciable, reaching 35 percent at 10,000-ft (3048-m) altitude.

Also, in keeping with good industrial practice, a gasoline engine is not generally operated continuously at maximum output. This practice provides a factor of safety in the form of reserve power. Most engine manufacturers recommend that this factor of safety be 20 to 25 percent below rated power. This means that the engine will be normally operated at 75 to 80 percent of its standard rated output. The duty cycle, however, can vary with different applications, as Table 11 shows.

For the 10-hp (7.46-kW) electric motor we are replacing with a gasoline engine, the motor can deliver—as discussed—25 percent more than its rating, or in this instance, 12.5 hp (9.3 kW) for short periods. On the basis that the gasoline engine is to operate at not over 75 percent of its rating, the replacement engine should have a rating of 12.5/0.75 = 16.7 hp (12.4 kW).

In summary, the gasoline engine should have a rating at least 67 percent greater than the electric motor it replaces. This applies to both air- and liquid-cooled engines for sea-level operation under standard atmospheric conditions. If the engine is to operate at altitude, a further allowance must be made, resulting—in some instances—in an engine having twice the power rating of the electric motor.

2. *Find the power required at the installed altitude and inlet-air temperature*

As noted above, altitude and inlet-air temperature both influence the required rating of a gasoline engine for a given application. Since this engine will be installed at an altitude of 8000 ft (2438 m), the power loss will be (8000/1000)(3.5) = 28 percent. Further, the increased inlet-air temperature of 90°F (32.2°C) versus the standard of 60°F (15.6°C), or a 30° difference will reduce the power output by (30/10)(1.0) = 3.0 percent. Thus, the total power output reduction will be 28 + 3 = 31 percent. Therefore, the required rating of this gasoline engine will be at least (1.31)(16.7) = 21.87 hp (16.3 kW).

Once the power requirements of a design are known, the next consideration is engine rotative speed, which is closely related to the horsepower and service life. Larger engines, with their increased bearing surfaces and lower speeds, naturally require less frequent servicing. Such engines give longer, more trouble-free life than the smaller, high-speed engines of the same horsepower (kW) rating.

The initial cost of a larger engine is greater but more frequent servicing can easily bring the cost of a smaller engine up to that of the larger one. Conversely, the smaller, higher-speed engine has advantages where lighter weight and smaller installation dimensions are important, along with a relatively low first cost.

Torque is closely associated with engine rotative speed. For most installations an engine with good lugging power is desirable, and in some installations, essential. This is especially true in tractors, harvesters, and hoists, where the load frequently increases considerably above normal.

If the characteristics of the engine output curve are such that the torque will increase with reducing engine speed, the tendency for the increasing load to reduce engine speed is resisted and the engine will "hang on." In short, it will have good lugging qualities, as shown in Fig. 12*a*. If the normal operating speed of the engine is 2000 to 2200 r/min, the maximum lugging qualities will result. Sanitary-pump drives do not—in general—require heavy lugging.

If, however, with the same curve, Fig. 12*a*, the normal operating speed of the engine is held at 1400 r/min or below, stalling of the engine may occur easily when the load is increased. Such an increase will cause engine speed to reduce, resulting in a decrease in torque and causing further reduction in speed until the engine finally stalls abruptly, unless the load can be quickly released.

Figure 12*b* shows performance curves for a typical high-speed engine with maximum power output at top speed. The torque curve for this engine is flat and the engine is not desirable for industrial or agricultural type installations.

TABLE 11 Duty Ratings for Combustion Engine Application

Key: 1—Continuous duty
2—Intermittant heavy duty
3—Variable load duty, heavy
4—Variable load duty, light

INDUSTRIAL SERVICE
- 1—Standby units
- 3—Air compressors
- 3—Floor sanders
- 4—Shop trucks and welders

MUNICIPAL SERVICE
- 3—Street sweepers and flushers
- 3—Sanitary pumps
- 3—Pipe thawing rigs
- 4—Diesel starting units

MINING
- 3—Horizontal & diamond drills
- 3—Rocker shovels

RAILWAY MAINTENANCE
- 3—Tampers
- 3—Tie adzing machines
- 3—Railway maintenance cars
- 3—Rail grinders
- 3—Weed cutters
- 4—Rail leveling machines

HIGHWAY MAINTENANCE
- 1—Road rollers
- 1—Bituminous sprayers
- 2—Concrete cutters

OIL FIELD EQUIPMENT
- 1—Well drills and pumps
- 1—High pressure pumps
- 2—Pipe wrapping machines
- 3—Pipe straightening machines

AGRICULTURAL EQUIPMENT
- 1—Irrigation pumps
- 1—Combine harvesters
- 3—Hay balers, tractors
- 3—Insecticide sprayers
- 3—Rotary tillers
- 3—Potato harvesters
- 3—Mowers
- 3—Spreaders, dusters

MARINE
- 1—Lighthouse units
- 1—Water oxygenation (units)
- 3—Inboard marine engines
- 3—Underwater weed cutters

CONTRUCTION MACHINERY
- 1—Centrifugal pumps
- 2—Concrete mixers
- 3—Concrete vibrators
- 3—Concrete surfacing machines

CONTRUCTION MACHINERY (Cont.)
- 3—Diaphragm pumps
- 4—Hoists and powers saws

SPECIALIZED SERVICE
- 1—Airport service units such as air compressors, hydraulic pumps, and generators
- 1—Weed burners
- 2—Refrigerated trucks
- 2—Paint sprayers
- 2—Portable fire-fighting equipment
- 3—Miniature railways
- 3—Water purification units for armed forces
- 3—Cable reelers
- 3—Lawn mowers and rollers
- 3—Post peelers
- 3—Portable showers for armed forces

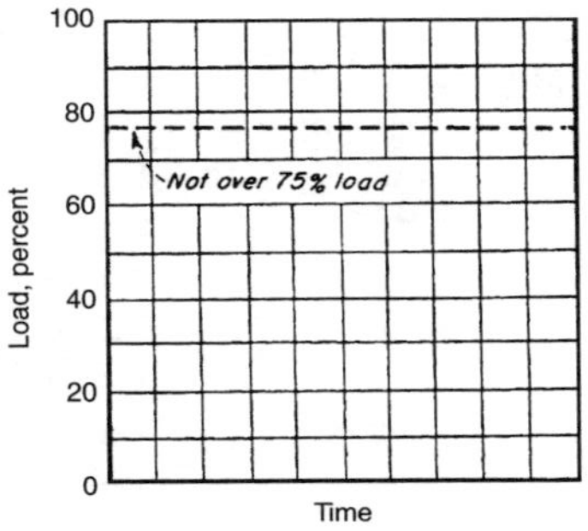

(A) — Continuous Heavy Duty

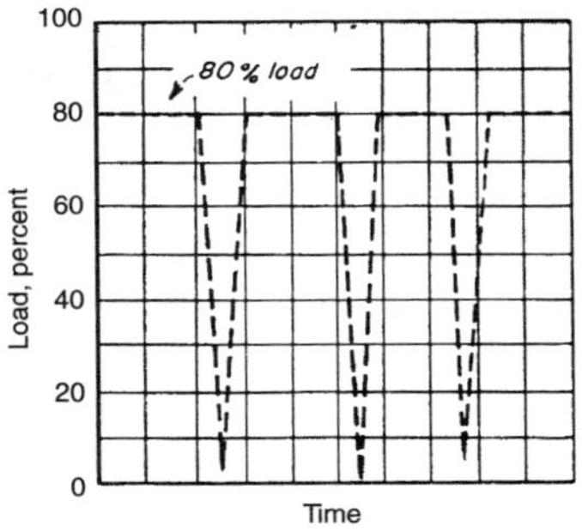

(B) — Typical Intermittent Heavy Duty

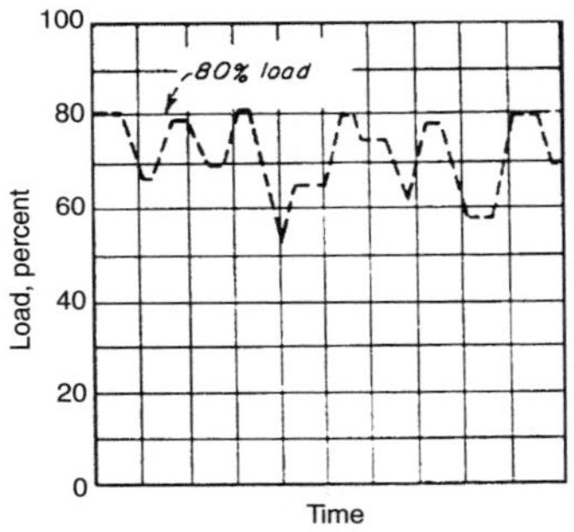

(C) — Typical Variable Load—Heavy

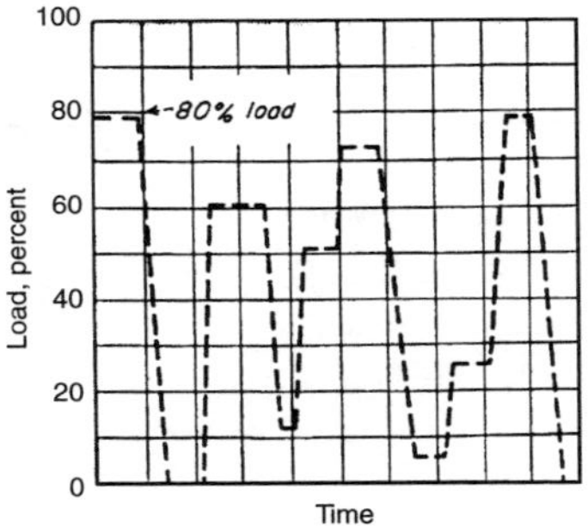

(D) — Typical Variable Load—Light

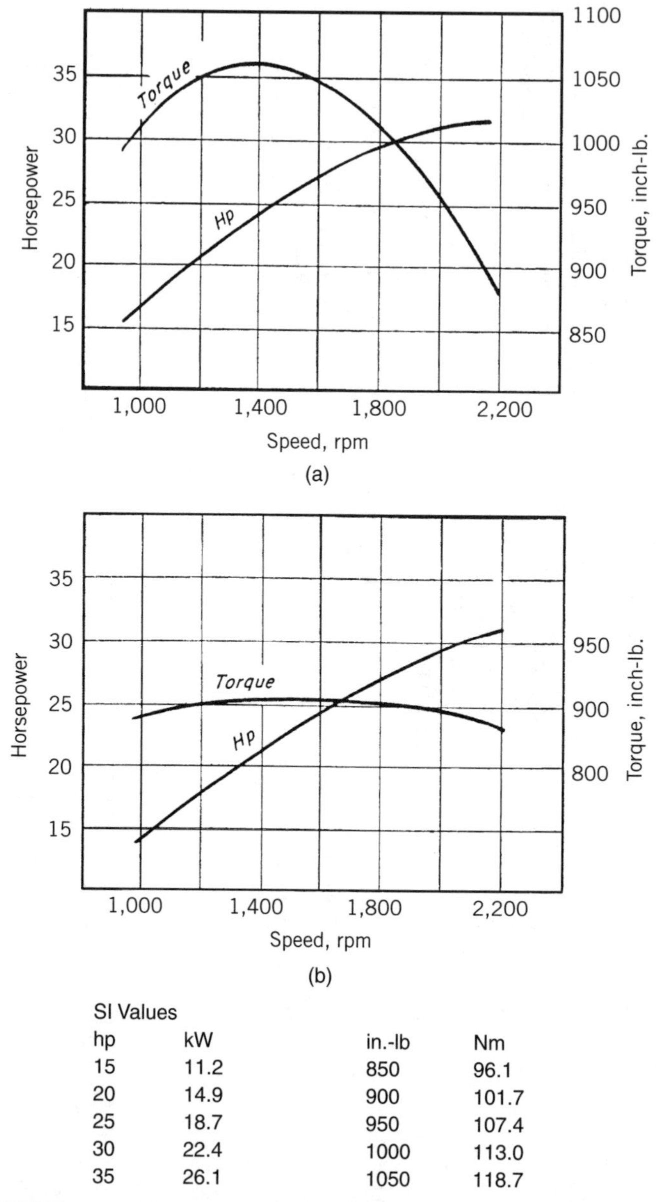

SI Values

hp	kW	in.-lb	Nm
15	11.2	850	96.1
20	14.9	900	101.7
25	18.7	950	107.4
30	22.4	1000	113.0
35	26.1	1050	118.7

FIGURE 12 Torque curves for typical air-cooled internal-combustion engines. (*a*) Engine with good lugging quality will "hang on" as load increases. (*b*) Performance curve for a high-speed engine with maximum power output at top speed.

3. *Determine the duty rating; draw a load curve for the engine*
Refer to Table 11 for municipal service. There you will see that sanitary pumps have a variable-load, heavy-duty rating. Figure 12 shows a plot of the typical load variation in such an engine when driving a sanitary pump in municipal service.

4. *Select the type of drive for the engine*

A variety of power take-offs are used for air-cooled gasoline engines, Fig. 13. For a centrifugal pump driven by a gasoline engine, a flange coupling is ideal. The same is true for engines driving electric generators. Both the pump and generator run at engine speed. When a plain-flange coupling is used, the correct alignment of the gasoline engine and driven machine is extremely important. Flexible couplings and belt drives eliminate alignment problems.

In many instances a clutch is required between the engine and equipment so that the power may be engaged or disengaged at will. A manually engaged clutch is the most common type in use on agricultural and industrial equipment.

Where automatic engagement and disengagement are desired, a centrifugal clutch may be used. These clutches can be furnished to engage at any speed between 500 and 1200 r/min and the load pickup is smooth and gradual. Typical applications for such clutches are refrigerating machines with thermostatic control for starting and stopping the engine.

Clutches also make starting of the engine easier. It is often impossible to start an internal-combustion engine rigidly connected to the load.

There are many applications where a speed reduction between the engine and machine is necessary. If the reduction is not too great, it may be accomplished by belt drive. But often a gear reduction is preferable. Gear reductions can be furnished in ratios up to 4 for larger engines, and up to 6 for smaller sizes. Many of these reductions can be furnished in either enginewise or counter-enginewise rotation, and either with or without clutches.

Related Calculations. Table 11 shows 54 different applications and duty ratings for small air-cooled gasoline engines. With this information the engineer has a powerful way to make a sensible choice of engine, drive, speed, torque, and duty cycle.

Important factors to keep in mind when choosing small internal-combustion engines for any of the 54 applications shown are: (1) Engines should have sufficient capacity to ensure a factor of safety of 20 to 25 percent for the power output. (2) Between high- and low-speed engines, the latter have longer life, but first cost is higher. (3) In take-off couplings, the flexible types are preferred. (4) A clutch is desirable, especially in heavier equipment, to disconnect the load and to make engine starting easier. (5) In operations where the intake air is dusty or contains chaff, intake screens should be used. (6) An oil-bath type air cleaner should always be used ahead of the carburetor. (7) Design engine mountings carefully and locate them to avoid vibration. (8) Provide free flow of cooling air to the flywheel fan inlet and also to the hot-air outlet from the engine. Carefully avoid recirculation of the hot air by the flywheel. (9) If the engine operation is continuous and heavy, Stellite exhaust valves and valve-seat inserts should be used to ensure long life.

Valve rotators are also of considerable value in prolonging valve life, and with Stellite valves, constitute an excellent combination for heavy service.

Exclusive of aircraft, air-cooled engines are usually applied in size ranges from 1 to 30 hp (0.75 to 22.4 kW). Larger engines are being built and, depending on the inherent cooling characteristics of the system, performing satisfactorily. However, the bulk of applications are on equipment requiring about 30 hp (22.4 kW), or less. The smaller engines up to about 8 or 9 hp (5.9 to 6.7 kW) are usually single-cylinder types; from 8 to 15 hp (5.9 to 11.2 kW) two-cylinder engines are prevalent, while above 15 hp (11.2 kW), four-cylinder models are commonly used.

Within these ranges, air-cooled engines have several inherent advantages: they are lightweight, with weight varying from about 14 to 20 lb/hp (8.5 to 12.2 kg/kW) for a typical single-cylinder engine operating at 2600 r/min to about 12 to 15 lb/h (7.3 to 9.1 kg/kW) for a typical four-cylinder unit running at 1800 r/min. Auxiliary power requirements for these engines are low since there is no radiator fan or water pump; there is no danger of the engine boiling or freezing, and no maintenance of fan bearings, or water pumps; and first cost is low.

In selecting an engine of this type, the initial step is to determine the horsepower requirements of the driven load.

On equipment of entirely new design, it is often difficult to ascertain the amount of power necessary. In such instances, a rough estimate of the horsepower range (kW range) is made and one or more sample engines bracketing the range obtained for use on experimental models of the equipment.

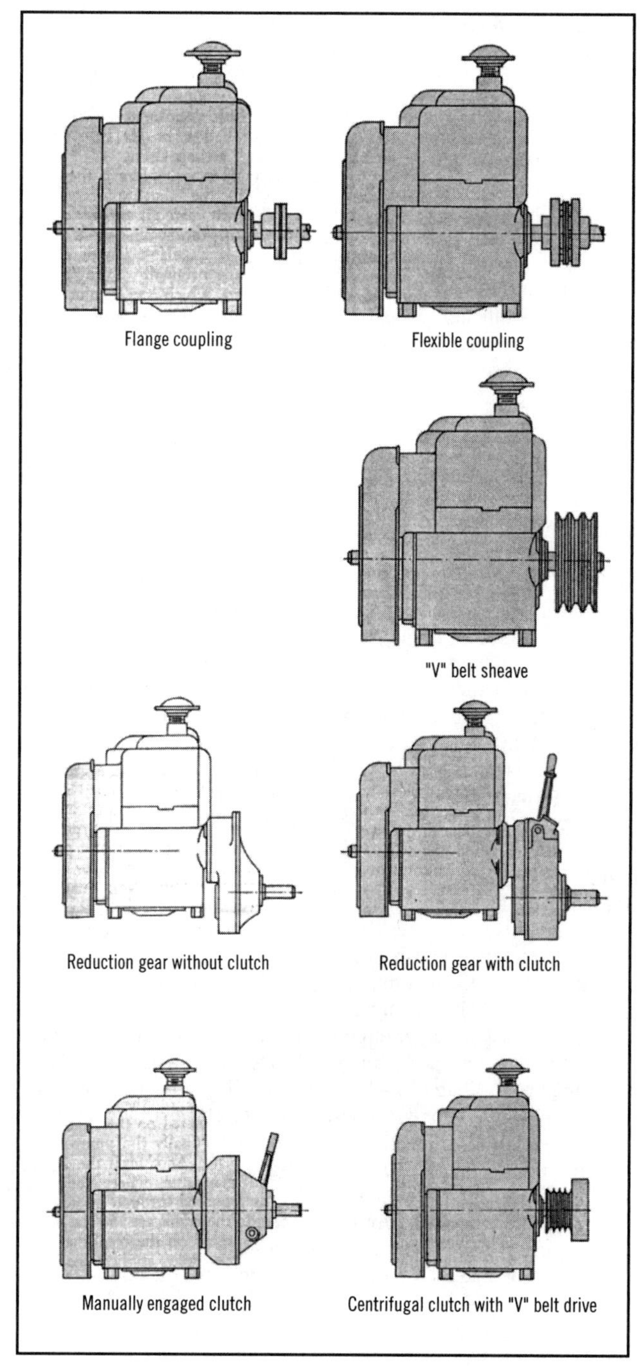

FIGURE 13 Power take-offs for air-cooled engines. Fluid couplings are also used to cushion shock loads in certain specialized applications.

In other applications, it is possible to calculate the torque required, from which the horsepower (kW) can be determined. Or, as is not uncommon, the new piece of equipment may be another size in a line of machines. In this case, the power determination can be made on a proportional basis.

This procedure is the work of A. F. Milbrath, Vice President and Chief Engineer, Wisconsin Motor Corporation, as reported in *Product Engineering* magazine. SI values were added by the handbook editor.

SECTION 5
NUCLEAR ENERGY ENGINEERING

Nuclear Power-Plant Calculation Parameters 5.1
Selecting a Nuclear Power Reactor 5.2
Analysis of Nuclear Power-Plant Cycles 5.4
Fuel Consumption of Nuclear Reactors 5.7
Comparison of Coal and Fissionable Materials as Heat-Generation Sources 5.8
Nuclear Radiation Effects on Human Beings 5.9
Nuclear Power and Its Use in Desalinization 5.11

NUCLEAR POWER-PLANT CALCULATION PARAMETERS

Nuclear power plants offer advantages over conventional fossil-fuel-fired steam plants. The major advantage of a nuclear plant over a fossil-fuel plant is the reduction of greenhouse gases (GHG) offered by nuclear power. Some larger utility companies today are shutting down older fossil-fuel steam plants and replacing them with a larger nuclear power plant.

Balanced against the GHG reduction provided by nuclear plants is the supposed "danger" posed by such facilities. Yet, in the United States, there has not been one significant nuclear accident since the three mile island (TMI) incident in 1979. This record speaks well for the operation of the more than 100 (104 at the time of this writing—69 pressurized-water reactors; 35 boiling-water reactors) nuclear energy plants in the United States. The record is even more spectacular when it is recalled that the first nuclear power plant started delivering power to a utility system grid on July 17, 1955.

Nuclear plant-capacity factors in recent years averaged about 90 percent, a high figure for any generating system. By contrast, coal-fueled steam plants have capacity factors in the 70 percent range. A number of nuclear stations are up-rating their output, adding to the overall output of nuclear plants in the United States. The Nuclear Regulatory Commission (NRC) has numerous up-rate applications it is currently studying for approval. Electricity generated in nuclear plants costs less than that from any other type of generating plant currently operating.

Around the world, at this writing, there are 439 nuclear generating plants in some 30 countries. These plants are rated at 372 GW output. The 104 nuclear plants in the United States have some 100 GW rated output.

Construction of new nuclear generating plants in the United States hinges on government approval of reactor designs by the Nuclear Regulatory Commission. Meanwhile, federal government guarantees for loans needed to build new nuclear plants are being sought by utilities planning to build such facilities. Other countries around the world seem to have fewer constraints on building new nuclear plants. Result? New plants are going on-line at a rapid pace in overseas countries.

The consensus among design engineers and environmentalists is that zero GHG emissions from utility stations can be obtained by switching to nuclear power generation. But the rising costs of new nuclear stations is slowing their approval by both private and government officials. One way to combat the rising cost of large nuclear stations is to use small, mini-stations. Such plants owe their origin to the nuclear submarines built in Russia during the cold war. These mini-stations are liquid-metal-cooled reactors currently rated at 100 MW each. It would take more than 10 of these mini-stations to equal the output of one large nuclear station. There is talk of mounting these mini-stations on barges or ships for easy transport to areas of the world needing emergency power. However, it will take time to see if nuclear mini-stations will catch on with the engineering and consumer population.

Approval of new nuclear reactor designs by the NRC is a slow process. At times an "approved" design is later changed and a new approval process must be started. Again, construction cannot start until the revised design is approved. For this reason, no new plants are under construction in the United States at the time of this writing. For excellent information and the latest updates on nuclear reactor approvals, see the NRC website: www.nrc.gov.

The March, 2011 Japanese nuclear crises caused by the earthquake and tsunami have spread concern around the world about the safety of nuclear reactors used for power generation. While the disaster was triggered by natural causes, engineers are concerned that greater containment protection must be provided for both new and existing reactors.

Further, the design of the regular cooling system for reactors, and the backup and emergency cooling facilities, will get increased scrutiny by engineers worldwide. Protection of cooling systems against a number of emergencies—such as loss of power, scarcity of cooling water supply, failure of the piping system, and so on—is now seen as more important than ever.

Without adequate cooling water for a reactor, the pressure and temperature inside the reactor can rise to dangerous levels. Fuel-rod damage from overheating can lead to fires and explosions, spreading radioactive materials for miles or kilometers.

Meltdown of a reactor caused by insufficient cooling might lead to spillage of molten fuel to the containment building floor. Radioactive materials might then escape to the outside atmosphere through damage to the containment structure and residents in the vicinity of the reactor might be exposed to harmful radiation.

Hence, engineers worldwide are reviewing the entire design of new and existing nuclear plants. These reviews are being done for a variety of both natural and human-caused incidents beyond earthquakes and tsunamis—hurricanes, tornados, floods, terrorism, and so on. Also, every potential failure aspect of the plant design is being reviewed for potential causes and remedies.

SELECTING A NUCLEAR POWER REACTOR

Select a nuclear power reactor to generate 60,000 kW at a thermal efficiency of 35 percent or more. If the selected unit is a 10-ft (3.0-m) diameter reactor that uses a fluidized bed containing 20×10^6 fuel pellets each 0.375 in (9.5 mm) in diameter with a density of 700 lbm/ft^3 (11,213 kg/m^3) and the reactor fluid is pressurized water at 600°F (315.6°C), determine the bed pressure drop when fluidized. Also, compute the reactor fuel volume, the collapsed fuel-bed height, and the density of the pressurized water.

Calculation Procedure:

1. *Select the type of reactor to use*

Table 1 summarizes the operating characteristics of six types of power reactors. Study shows that a pressurized-water reactor will provide the desired thermal efficiency. Further, this type of reactor is successfully used for large-scale power generation. Hence, a pressurized-water reactor will be the first tentative choice for this plant.

2. *Compute the reactor fuel volume*

Use the relation $v_f = nv_p$, where v_f = fuel volume, ft^3; n = number of fuel pellets in the reactor; v_p = volume of each pellet, ft^3. Substituting yields $v_f = 20 \times 10^6\pi(0.375)^3/[6(1728)] = 320$ ft^3 (9.1 m^3).

TABLE 1 Nuclear-Power Reactor Characteristics

Reactor type	Typical thermal efficiency, %	Typical power density, thermal, kW/ft^3(MW/m^3)	Typical reactor pressure, lb/in^2 (gage) (kPa)		Average heat flux, Btu/(h · ft^2) (MW/m^2)	Typical fuel enrichment, %	Reactor coolant
Pressurized-water	36	1,600 (56.5)	1,500	(10,341)	300,000 (945.6)	1.5–30	Light water
Boiling-water	22–30	800 (28.3)	1,000	(6,894)	100,000 (315.2)	1.5	Light water
Gas-cooled	30	200 (7.1)	600–1,000	(4,136–6,894)	—	0.70–2.5	Carbon dioxide
Liquid-metal	33	300 (10.6)	100	(689.4)	—	—	Sodium, bismuth, lead, etc.
Fast-breeder	32	20,000 (706.5)	100	(689.4)	650,000 (2,049)	—	Sodium
Fluid-fueled	30	400 (14.1)	1,000–2,000	(6,894–13,788)	Varies (varies)	Varies	Reactor fuel solution

3. ***Compute the fuel volume in the collapsed form***

With the fuel bed not fluidized, the porosity P with packed spheres is about 0.40. Then collapsed volume $v_c = v_f/(1 - P) = 320/0.60 = 534$ ft³ (15.1 m³).

4. ***Compute the collapsed fuel-bed height***

Use the relation $h = v_c/A_r$, where h = collapsed height of fuel bed, ft; A_r = reactor fuel-bed area, ft². So $h = 534/(\pi 10^2/4) = 6.78$ ft (2.1 m).

5. ***Determine the density of the pressurized water***

Using the steam tables shows $d_w = 42.45$ lb/ft³ (680.0 kg/m³) at 600°F (315.6°C) for saturated liquid.

6. ***Compute the pressure loss through the fluidized bed***

Use the relation $p = 2.9h[(1 - P)d_f + Pd_w]$, where p = pressure loss through fluidized fuel bed, lb/ft²; d_f = fuel density, lbm/ft³; other symbols as before. Substituting, we find $p = 2.9\,[(1 - 0.4)700 + 0.4 \times 42.45] = 1268$ lb/ft² or 8.79 lb/in² (60.6 kPa).

Related Calculations. This general procedure is valid for preliminary selection of the type of nuclear reactor to use for a given power application. Since reactors are expensive, a complete economic analysis must be made of the alternatives available before the final choice is made.

ANALYSIS OF NUCLEAR POWER-PLANT CYCLES

A nuclear power plant using two coolants, Na and NaK, is arranged as shown in Fig. 1. Sodium, the first coolant, enters the reactor at 600°F (315.6°C) and leaves at 1000°F (537.8°C); NaK, the second coolant, enters the intermediate heat exchanger at 550°F (287.8°C) and leaves at 950°F (510.0°C). Neglecting heat and pressure losses in the piping, plot the enthalpy-temperature diagram for the plant if steam leaves the boiler at 1200 lb/in² (8273 kPa). What are the Na and NaK flow rates with the cycle arrangement shown in Fig. 1, a reactor capacity of 400,000 kW of heat energy, and a 155,000-kW turbine output? Determine the plant thermal efficiency if the auxiliary-power needs = 12,000 kW.

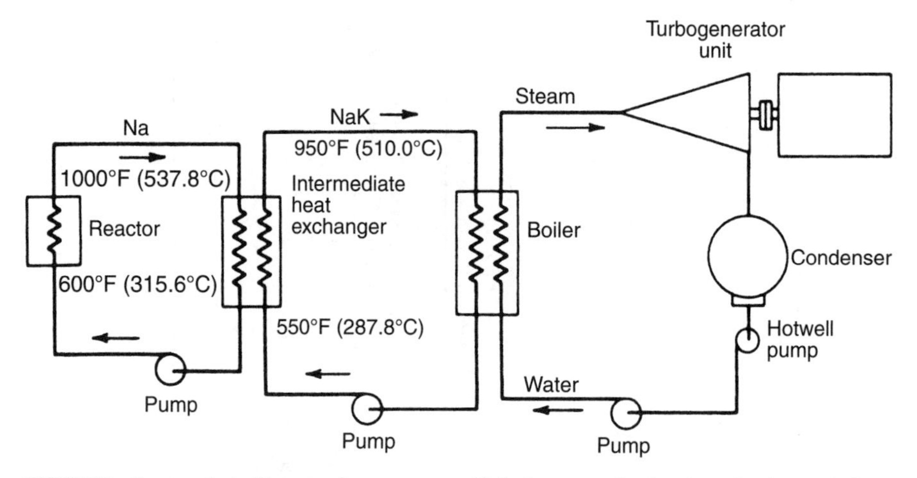

FIGURE 1 Reactor plant with two-coolant system uses Na in the reactor circuit and transfers heat to the intermediate NaK circuit, which acts as a buffer against making the steam circuit radioactive.

Calculation Procedure:

1. ***Determine the steam outlet and saturation temperature***

Figure 1 shows that NaK enters the boiler at 950°F (510.0°C). Draw a horizontal line on the enthalpy-temperature *(h-t)* diagram (Fig. 2), indicating the 950°F (510.0°C) NaK temperature entering the boiler. Also draw a horizontal line on the *h-t* diagram, Fig. 2, at 1000°F (537.8°C), indicating the Na temperature leaving the reactor.

The steam outlet temperature from the boiler will be less than 950°F (510.0°C) because transfer of heat between the NaK and the water and steam in the boiler provides the energy required to convert the water to steam. A temperature difference between the NaK and the steam is needed to produce the desired heat transfer.

Assume a 50°F (27.8°C) temperature difference between the boiler outlet steam and the NaK, which is a typical temperature difference for this type of cycle. With such a temperature difference the outlet steam temperature = 950 − 50 = 900°F (482.2°C). From the steam tables find the saturation temperature of steam at 1200 lb/in^2 (abs) (8273 kPa) as 567.2°F (297.3°C). Hence the steam will be superheated when it leaves the boiler.

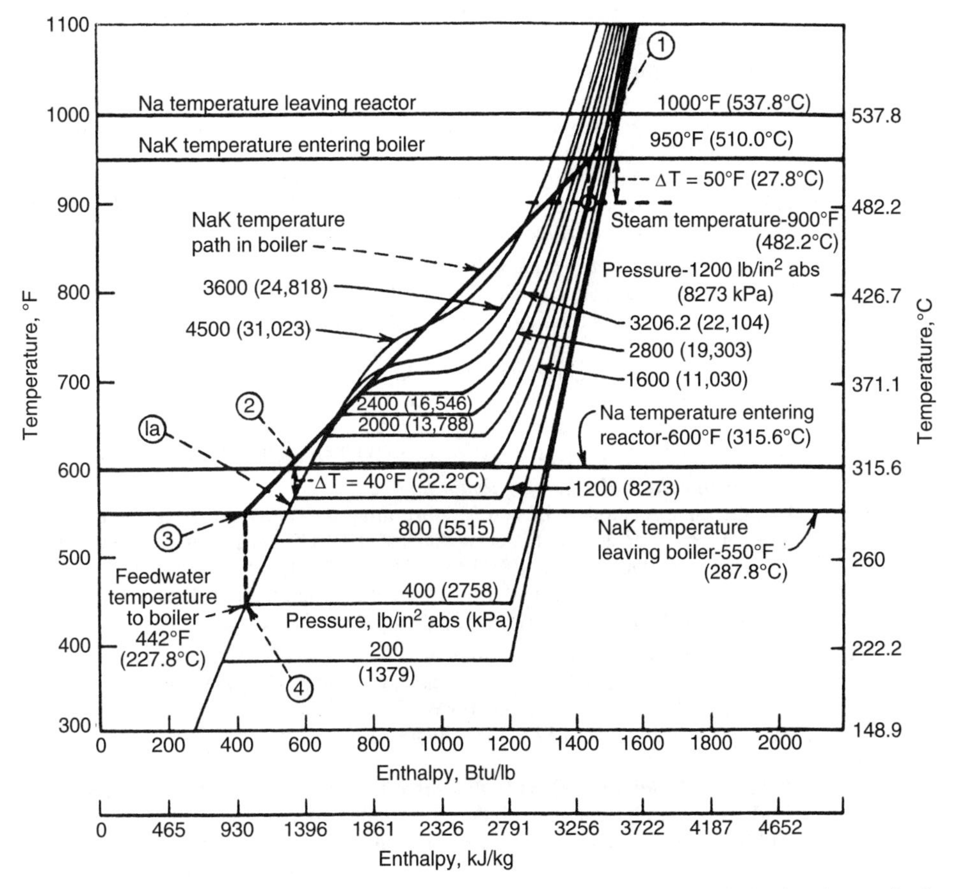

FIGURE 2 Steam-water enthalpy-temperature diagram shows the relation between NaK circuit and steam circuit. Keeping the steam temperature high raises the thermal efficiency of the plant.

2. *Compute the boiler evaporator coolant outlet temperature*

Incoming feedwater enters the boiler evaporator section where it is heated by the NaK before entering the boiler steam section. To provide heat transfer between the NaK leaving the evaporator section of the boiler and the incoming boiler feedwater, a temperature difference between the two fluids is necessary. Assume that the NaK coolant leaves the boiler evaporator section at a temperature 40°F (22.2°C) higher than the incoming feedwater. With the incoming feedwater at the saturation temperature, or 567.2°F (297.3°C), the NaK coolant outlet temperature from the boiler evaporator = 567.2 + 40 = 607.2, say 607°F (319.4°C).

3. *Plot the boiler coolant temperature path*

Locate the boiler outlet steam state on the *h-t* diagram, Fig. 2, on the 1200-lb/in^2 (abs) (8273-kPa) pressure curve and the 900°F (482.2°C) temperature horizontal. From this point, project vertically upward to the 950°F (510°C) NaK temperature horizontal to locate point 1, the temperature of the NaK entering the boiler, Fig. 2.

Next, locate the point l*a* where the liquid enthalpy line of the *h-t* diagram, Fig. 2, intersects the 1200-lb/in^2 (abs) (8273-kPa) evaporation enthalpy line. From point 1*a*, project vertically upward to 607°F (319.4°C), point 2, the temperature of the NaK coolant leaving the boiler evaporator section.

Points 1 and 2 are the NaK *temperature path* in the boiler evaporator and steam-generating sections. Assuming that the NaK has a constant specific heat while flowing through the boiler evaporator and steam-generating sections (a completely valid assumption), draw a straight line between points 1 and 2 and extend it to intersect the 550°F (287.8°C) temperature line at point 3. Note that point 3 represents the temperature of the NaK entering the intermediate heat exchanger.

4. *Determine the boiler feedwater inlet temperature*

Feedwater enters the boiler at a yet unknown temperature. During passage between the boiler inlet and the evaporator section inlet, the feedwater absorbs heat from the NaK coolant, leaving the evaporator at 607°F (319.4°C).

Draw a line vertically downward from point 3 until the liquid enthalpy curve is intersected, point 4. Point 4 represents the boiler feedwater inlet temperature, or 442°F (227.8°C), based on the valid assumption that the feedwater leaving the condenser hot well is in the saturated state.

5. *Compute the reactor coolant flow rate*

Sodium enters the reactor at 600°F (315.6°C) and leaves at 1000°F (537.8°C), Fig. 1. Thus, the temperature rise of the Na during passage through the reactor is 1000 – 600 = 400°F (222.2°C). Also, the average specific heat of Na is 0.306 Btu/(lb · °F) [1.28 kJ/(kg · °C)], found from a tabulation of Na properties in an engineering handbook.

Compute the Na flow from $f = 3413\ kw/\Delta tc$, where f = Na flow rate, lb/h; kw = reactor heat rating, kW; Δt = Na temperature rise during passage through the reactor, °F; c = specific heat of the Na coolant, Btu/(lb · °F). Substituting gives us $f = 3413(400{,}000)/[400(0.306)] = 11{,}130{,}000$ lb/h (1402.4 kg/s).

6. *Compute the boiler heating liquid flow rate*

Use the same relation as in step 5, substituting the temperature change and specific heat of NaK. Since the NaK enters the boiler at 950°F (510.0°C) and leaves at 550°F (287.8°C), its temperature change is 950 – 550 = 400°F (222.2°C). Also, the specific heat of NaK is 0.251 Btu/(lb · °F) [1.05 kJ/(kg · °C)], as found from NaK properties tabulated in an engineering handbook. So $f = 3413(400{,}000)/[400(0.251)] = 13{,}600{,}000$ lb/h (1713.6 kg/s).

7. *Compute the plant thermal efficiency*

The net station output kw = gross output of turbine, kW, minus the total plant auxiliary demand, kW = 155,000 – 12,000 = 143,000 kW. Then overall plant thermal efficiency = net station output, kW/reactor heat output, kW = 143,000/400,000 = 0.357, or 35.7 percent.

Related Calculations. This analysis is valid for a cycle in which the reactor coolant does not do work in the turbine. In general, designers prefer to avoid using the reactor coolant in the turbine. Although the thermodynamic aspects of a nuclear cycle are important, the cost of the plant must also be considered before a final choice of a cycle is made. The method presented is the work of Henry C. Schwenk and Robert H. Shannon, as reported in *Power* magazine.

FUEL CONSUMPTION OF NUCLEAR REACTORS

Determine the amount of fissionable material used in a 500-mW reactor having 3×10^{10} fissions per watt-second. The reactor core has a volume of 1360 ft^3 (38.5 m^3) and the fuel (99.3 percent U 238 plus 0.7 percent U 235) occupies 6 percent of the reactor volume. How much fissionable material is consumed if the plant operates 8760 h/year at an 80 percent load factor and the capture cross-section/fission cross-section ratio = 1.2? What are the maximum allowable atom burnup, the average fuel-cycle time, and the reactor neutron flux?

Calculation Procedure:

1. *Compute the reactor fission rate*
Use the relation $F_r = P_T C$, where F_r = reactor fission rate, fissions/(W · s); P_T = total reactor power, W; C = fissions (W · s). So $F_r = 500 \times 10^6(3 \times 10^{10}) = 1.5 \times 10^{19}$ fissions/s.

2. *Compute the total volume of the fuel*
Since the fuel occupies 6 percent of the reactor volume, the fuel volume $V_f = 0.06 \times 1360 = 81.6$ ft^3 (2.3 m^3). Since reactor fuel quantities are often expressed in cubic centimeters, convert the fuel volume in cubic feet by multiplying by the conversion factor 2.832×10^4, or $V_{fc} = 2.832 \times 10^4(81.6) = 2.31 \times 10^6$ cm^3.

3. *Compute the U 235 nuclei in the reactor*
First determine the uranium nuclei per cm^3 N_U, using the relation N_U = [(uranium density, g/cm^3)/uranium atomic weight] (Avogadro's constant) = $(18.68/238.07)(6.023 \times 10^{23}) = 0.0472 \times 10^{24}$ nuclei/cm^3. In this relation the following constants are used: uranium density = 18.68 g/cm^3, uranium atomic weight = 238.07; Avogadro's constant = $N_m = 6.023 \times 10^{23}$ atoms/(g · atom).

With the uranium nuclei per cm^3 known, compute the U 235 nuclei in the reactor from $N_{U\,235} = 0.007 N_U V_{fc} = 0.007(0.0472 \times 10^{24})(2.31 \times 10^6) = 7.64 \times 10^{26}$ U 235 nuclei in the reactor.

4. *Compute the U 235 fissionable material consumed*
Use the relation $F_{U\,235} = F_r G_m / N_m$, where $F_{U\,235}$ = fissionable U 235 material consumed or burned up for power only, g/s; G_m g/mol of the fissionable material; other symbols as before. Substituting gives $F_{U\,235} = (1.5 \times 10^{19})(235)/6.023 = 5.85 \times 10^{-3}$ g/s.

5. *Compute the annual consumption of fissionable material*
Use the relation $A_c = F_{U\,235} YL/1000$, where A_c = annual consumption of fissionable material, kg; Y = s/year; L = load factor; other symbols as before. Substituting reveals $A_c = 5.85 \times 10^{-3}(3600 \times 8760)(0.8)/1000 = 147.4$ kg/year.

6. *Compute the U 235 annual consumption*
The U 235 is consumed by fissioning for power and is also lost by absorption. The proportion of these two forms of consumption is expressed by α = U 235 total capture cross section/U 235 fission cross section. With $\alpha = 1.2$ for a typical reactor, the total annual U 235 consumption = 1.2(147.4) = 177 kg/year.

7. *Compute the maximum allowable atom burnup*
Both U 235 and U 238 are regarded as reactor fuel. The allowable percentage of burnup depends on the total integrated radiation dosage and radiation energy level, and the effect on fuel material dimensional stability, thermal conductivity, and reduction in effective multiplication factor. Assuming a maximum allowable burnup of 20 percent, which is a typical value, compute B_{ma} = (percentage of burnup)(fuel atoms per cm^3)(total cm^3 of fuel), where B_{ma} = maximum allowable atom burnup, atoms. Substituting gives us $B_{ma} = (0.002)(0.0472 \times 10^{24})(2.31 \times 10^6) = 2.18 \times 10^{26}$ atoms.

8. *Compute the average fuel-cycle time*
Use the relation $A_f = B_{ma}/F_r$, where A_f = average fuel-cycle time, s. Thus $A_f = 2.18 \times 10^{26}/(1.5 \times 10^{19}) = 1.45 \times 10^7$s = 4040 h = 30 weeks, approximately.

9. Compute the reactor neutron flux
Use the reaction $N_f = P_T C/\Sigma f V_f$, where N_f = reactor neutron flux; $\Sigma f = N_{U\,235} \times \sigma_{f\,235}$, where $\sigma_{f\,235}$ = total microscopic absorption cross section for U 235; other symbols as before. So $N_f = 500 \times 10^6 (3 \times 10^{10})/(0.00033 \times 10^{24})(549 \times 10^{-24})(2.31 \times 10^6) = 3.57 \times 10^{13}$. Note that values of $\sigma_{f\,235}$ are obtained from nuclear data sources.

Related Calculations. Use this general method for any reactor designed to generate power. The method presented is the work of Henry C. Schwenk and Robert H. Shannon, as reported in *Power* magazine.

At the time of this writing (1994), the United States is generating more than 22 percent of its power requirements in nuclear plants. Thus, nuclear stations are number two in generating electricity for the United States.

Nuclear power does not pollute the air. Recent studies show that the nuclear plants currently operating annually reduce the amount of carbon dioxide that would be emitted to the atmosphere by some 500 million tons. Likewise, these plants reduce atmospheric pollution by 3.6 million tons of methane and some 2 million tons of nitrous oxides annually. The NO reduction closely approximates the requirements of amendments to the 1990 Clean Air Act.*

COMPARISON OF COAL AND FISSIONABLE MATERIALS AS HEAT-GENERATION SOURCES

How many tons of coal are required to produce the heat equivalent of 1 lb (0.45 kg) of fissionable U 235? If heat is worth 40 cents per million Btu (37.9 cents per 10^6 kJ), what is 1 g of fissionable U 235 worth? One ton of coal contains 24×10^6 Btu (25.3×10^6 kJ).

Calculation Procedure:

1. Compute the heat produced by 1 lb (0.45 kg) of fissionable material
When all the nuclei in the atoms of 1 lb (0.45 kg) of fissionable U 235 fission, about 0.001 lb (0.45 g) of material converts to heat energy. Since by Einstein's mass-energy equation, 1 lbm = 11.3×10^9 kWh of energy, 1 lb (0.45 kg) of fissioning U 235 produces $0.001\,(3413)(11.3 \times 10^9) = 39.5 \times 10^9$ Btu/lb (91.9×10^9 kJ/kg). In this relation, the constant 3413 (3600.9) converts kW to Btu (kJ).

2. Compute the heat equivalent of the fissionable material
Use the relation equivalent tons of coal per pound of U 235 = heat released per pound of U 235, Btu/heat released by 1 ton of coal, Btu = $39.5 \times 10^9/(24 \times 10^6) = 1645$ tons of coal per pound of U 235 (3290.0 t of coal per 1 kg U 235). Thus, it takes 1645 tons of coal to equal the potential heat produced by 1 lb (3290 t/kg) of U 235 in a nuclear reactor.

3. Compute the monetary worth of the nuclear material
Since heat is worth 40 cents per million Btu in this plant, the value of 1 lb (0.45 kg) of U 235 is $(39.5 \times 10^9)(0.4/10^6) = \$15{,}800$, or about $34.80 per gram of U 235.

Related Calculations. Use this general procedure for other fissionable materials used for fuels in nuclear plants. The method presented is the work of Henry C. Schwenk and Robert H. Shannon, as reported in *Power* magazine.

With nuclear-power generation there is always the consideration of what to do with spent fuel. Spent nuclear fuel is still radioactive and hazardous to humans.

*Orval Hansen, President, Columbia Institute.

Spent nuclear fuel is a waste material that requires much more care than ash from coal or SO_2 emitted by a power-plant stack for a coal-fired generating plant. Environmental regulations are equally strong in their control of nuclear waste, stack and boiler-grate effluent, and internal-combustion engine exhausts.

But fossil-fuel-fired generating plants have an option nuclear plants do not have. A fossil-fuel plant can purchase allowances to emit SO_2, as sanctioned by Title IV, the acid-rain provisions, of the 1990 Clean Air Act Amendments (CAAA). No such allowances are permitted for nuclear-fuel waste because spent fuel is much more lethal than SO_2.

At a recent auction one low-sulfur-coal-burning utility bought 85,103 allowances for over $11 million. Each allowance permits a plant to emit 1 ton of SO_2 per calendar year. The utility justified its purchases of the allowances at $135 per ton by comparing it to the cost of installing scrubbers to provide similar reductions, namely $500 per ton. By buying the allowances now the utility believes it can postpone large capital outlays until less costly controls become available in the marketplace.

Under a variety of laws, the Environmental Protection Agency (EPA) seeks to establish a nationwide limit for various stack pollutants. The limits are given in weight of pollutants emitted per year with a targeted reduced annual rate for a future year. Limits established for pollutants vary by the size of a plant in MW terms, and type of plant (larger utilities and cogenerators). Since the regulations are under constant study, and frequent revision, engineers should check with the EPA before finalizing any potential polluting type plant design.

With no allowances available to nuclear plants, the designer must give thought to the eventual disposal of spent fuel. Two approaches can be used in the handling of spent nuclear fuel: (1) storage, (2) reprocessing.

In the first approach, storage, both the heat and radiation of the spent fuel must be contended with during the long-term storage period required. Underground storage of spent fuel is the most common way of handling the waste. Today most spent fuel is buried intact, with no processing before storage. Handling spent nuclear fuel is an ongoing problem for which no final solutions appear available at this time.

In the second approach, reprocessing, a number of usable by-products—plutonium, uranium, and radioisotopes—are obtained. These can be used in agriculture, industry, and medicine to perform beneficial tasks. But even after reprocessing there is a residue of high-level nuclear waste. This residue must be stored in stainless-steel tanks or in solid form. Many different storage options are being studied.

With increasing attention on environmental aspects of nuclear-power generation, the designer has much to contend with. Between federal and state regulators, the environmental demands are enormous. The environment must be "factored into" every engineering cost estimate today. If the environmental aspects are overlooked, the cost estimate will be completely unrealistic, and the time schedule may be off by years.

NUCLEAR RADIATION EFFECTS ON HUMAN BEINGS

What is the total radiation dose in rems for a worker exposed to 0.3 rad of 1.0 MeV beta particles and 0.05 rad of 1.0 MeV neutrons every day? Is the total dose dangerous to this worker? Use National Bureau of Standards data (Tables 2 to 4) in the analysis.

Calculation Procedure:

1. ***Compute the total radiation dose***

Use the relation, total dose, rem = Σ(dose, rad)(RBE), where rem = roentgen equivalent per man; rad = radiation absorbed dose; RBE = relative biological effectiveness. Table 2 lists the RBE values for various types of radiation. By substituting the appropriate values from Table 2, total dose = $(0.3)(1.0) + 0.05(10.5) = 0.825$ rem.

TABLE 2 Conversion: Rad to Rem*†

Radiation effects on humans: Definitions

One r (*r*oentgen) is the quantity of gamma or x-radiation that produces an energy absorption of 83 ergs/g of dry air.
One rep (*r*oentgen *e*quivalent *p*hysical) is the quantity of radiation that produces an energy absorption of 93 ergs/g of aqueous tissue.
One rad (*r*adiation *a*bsorbed *d*ose) is required to deposit 100 ergs/g in any material by any kind of radiation.
One rem (*r*oentgen *e*quivalent *m*an) is the unit of particulate radiation that produces tissue damage in humans.
The conversion factor from rad to rem is the RBE (relative biological effectiveness), i.e., dose in rem = dose in rad × RBE.

Type of radiation	RBE*	Type of radiation	RBE*
X-rays	1	Neutrons, 0.5 MeV	10.2
Gamma rays	1	Neutrons, 1.0 MeV	10.5
Beta particles, 1.0 MeV	1	Neutrons, 10 MeV	6.4
Beta particles, 0.1 MeV	1.08	Protons, 100 MeV	1–2
Neutrons, thermal	2.8	Protons, 1 MeV	8.5
Neutrons, 0.0001 MeV	2.2	Protons, 0.1 MeV	10
Neutrons, 0.005 MeV	2.4	Alpha particles, 5 MeV	15
Neutrons, 0.02, MeV	5	Alpha particles, 1 MeV	20

*Example for total dose: For a given exposure time, a dose of 0.2 rad of γ radiation plus 0.04 rad of thermal neutrons gives a total dose of (0.2 × *I*RBE) + (0.04 × 2.8 RBE) = 0.312 rem.

†Based on most detrimental chronic biological effects for continuous low-dose exposures.

2. *Determine whether the dose is dangerous*
Table 3 lists the exposure tolerance of the human body. This listing shows that a dose of 1 rem/day is believed to cause debilitation within 3 to 6 months and death within 3 to 6 years. Since the daily dose to which this worker is exposed—0.825 rem—is close to the 1.0-rem danger level, the dose is excessive and dangerous.

Table 4 lists the recommended weekly maximum dosage for various types of radiation on different parts of the body. Study of this list also indicates that the radiation to which this worker is exposed is dangerous.

Related Calculations. The effects of radiation can be fatal to all living organisms. Hence, extreme care must be used in computing the dose received by anyone exposed to radiation. Since the allowable

TABLE 3 Exposure Tolerance Values for Humans*

0.001 rem/day	Natural background radiation
0.01 rem/day	Permissible dose range, 1957
0.1 rem/day	Permissible dose range, 1930 to 1950
1 rem/day	Debilitation 3 to 6 months; death 3 to 6 years (projected from animal data)
10 rem/day	Debilitation 3 to 6 weeks; death 3 to 6 years (projected from animal data
100 rem—1 day 150 rem—1 week 300 rem—1 month	Survivable emergency exposure dose but permitting no further exposure for life
25 rem	Single emergency exposure
100 rem	Twenty-year-career allowance
500 rem	Maximum permissible 20-year-career allowance

*Whole-body radiation doses.

TABLE 4 Maximum Weekly Dosage*

Radiation	Skin		Lens of eye	Gonads	Blood-forming organs	Intermediate tissue (0.07–5.0 cm depth)
	Total body	Appendages				
X-rays or γ-rays <3 MeV	0.45	1.5	0.45	0.3	0.4	0.4–0.45
Electrons or β	0.6	1.5	0.3	0.3	0.3	0.3–0.6
Protons	0.6	1.5	0.3	0.3	0.3	0.3–0.6
Fast neutrons	0.3–0.6	0.75–1.5	0.3	0.3	0.3	0.3–0.6
Thermal neutrons	0.5	1.2	0.3	0.1	0.17	0.17–0.5
Alpha particles	1.5	1.5	0.3	0.3	0.3	0.3–1.5
Heavy nuclei (O, N, C, locally generated)	1.5	1.5	0.3	0.3	—	0.3–1.6

*Rems per week.

dose and the effects of various doses are under constant study, be certain to refer to the latest available data from the Nuclear Regulatory Commission before permitting exposure of any worker to radiation of any kind.

Environmental cleanup after a nuclear power-plant accident can be expensive and time consuming. Thus, it took some 14 years to clean up the contamination at Three Mile Island 2 after the accident in March, 1979. An electric evaporator operated for 2 years to boil off some 2.23 10^6 gal (8440 m^3) of contaminated water from Reactor No. 2. Although some contamination still remains in the reactor building, it is confined to the walls and is not thought to pose any danger to the environment.

Since the reactor wall of TMI No. 2 was not punctured, it was possible to leave unreachable radioactive materials inside the reactor. Over time, the radiation level of these materials will decline. Some 150 tons of radioactive wreckage was removed from TMI No. 2 reactor and deposited at the National Engineering Laboratory of the Department of Energy in Idaho. A cooling system malfunction damaged TMI No. 2 reactor core, leading to leakage of radioactive gases. The cleanup has been completed.

NUCLEAR POWER AND ITS USE IN DESALINIZATION

Analyze the feasibility of building and operating nuclear-powered combined electric-generating and water-desalting plants. Sketch the different types of cycles that might be used. Determine the cycle to use for a water production of 100×10^6 gal/day (4.4×10^3 L/s), electric power net output of 500 mW, and a desalting heat performance of 100.

Calculation Procedure:

1. *Draw the cycle diagrams*

Three cycles will be considered: the backpressure, extraction, and multishaft cycles.

Figure 3 shows the backpressure cycle in which the entire exhaust steam flow from the turbine is used to heat brine in the water-desalting system. For a given amount of water produced, this cycle generates large quantities of electric power.

In the extraction cycle (Fig. 4), the steam for brine heating is removed from the turbines at some midpoint during expansion. The exhaust steam goes to a standard condenser. This cycle can have a high product ratio (PR), that is, the ratio of the electric power to desalted water. If desired, large amounts of water can be produced when needed.

The multishaft cycle (Fig. 5) is fundamentally the same as Fig. 3, but it uses parallel condensing and noncondensing turbines. The electric output can vary over a wide range without changing the water-desalting production. Although many other cycles are possible, all are variations of the three basic arrangements described above.

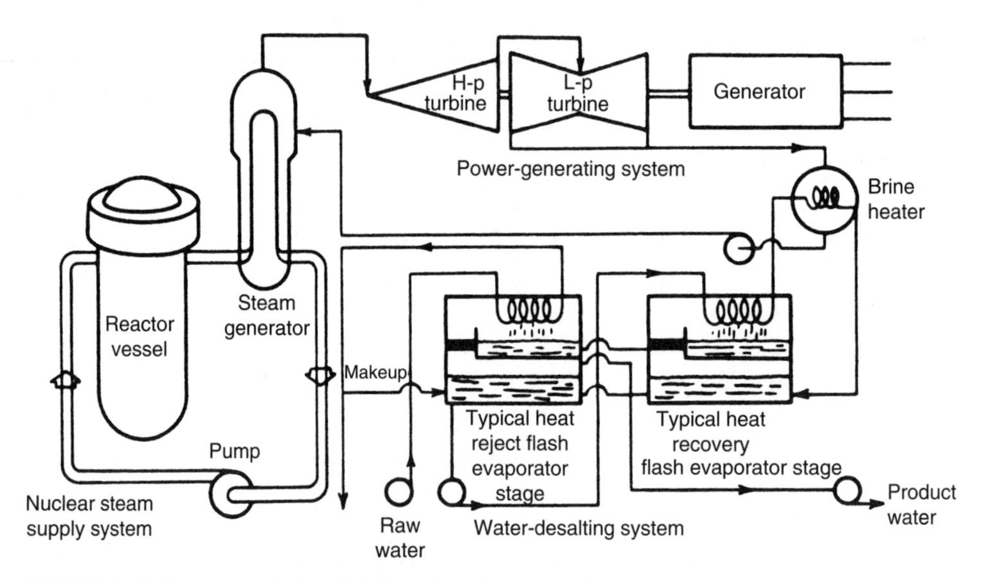

FIGURE 3 Backpressure cycle in which the entire exhaust steam from the turbines is used to heat brine in the water-desalting system.

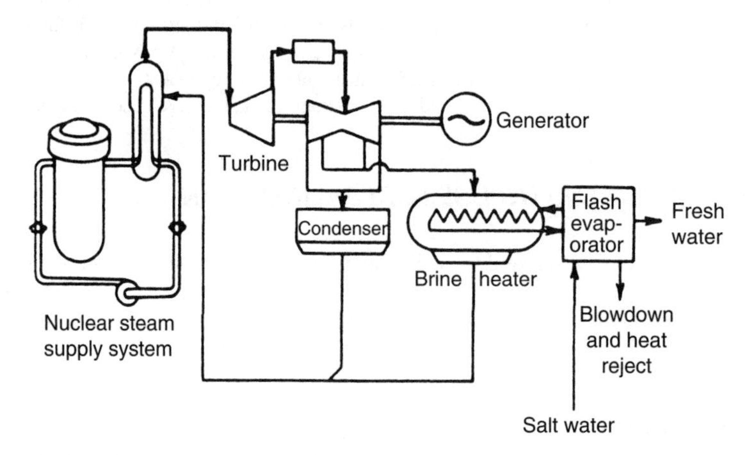

FIGURE 4 Extraction cycle in which the steam for brine heating is removed from the turbines at some midpoint during expansion.

2. *Choose the type of cycle and reactor size to use*

Figure 6 allows quick *estimates* of the type of cycle and reactor size. Any of the four plotted quantities can be determined from Fig. 6 when the other three are known.

Enter Fig. 6 at the bottom at the water production rate of 100×10^6 gal/day (4.4×10^3 L/s), and project vertically upward (1) to the desalting heat performance of 100. From the intersection with the appropriate curve, project horizontally to the left-hand scale of Fig. 6. Next project upward (2) parallel to the index scale. Then project vertically downward (3) from the net electric power output, 500 mW, on the top scale. From the intersection between lines 2 and 3, draw line 4 horizontally to the left-hand scale. At the intersection, read the reactor power as 2250 thermal mW.

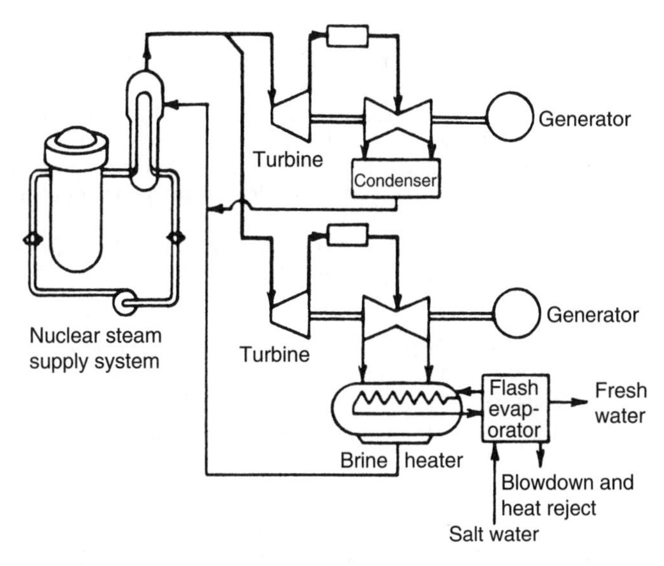

FIGURE 5 Multishaft cycle is the same as backpressure cycle, but it uses parallel condensing and noncondensing turbines.

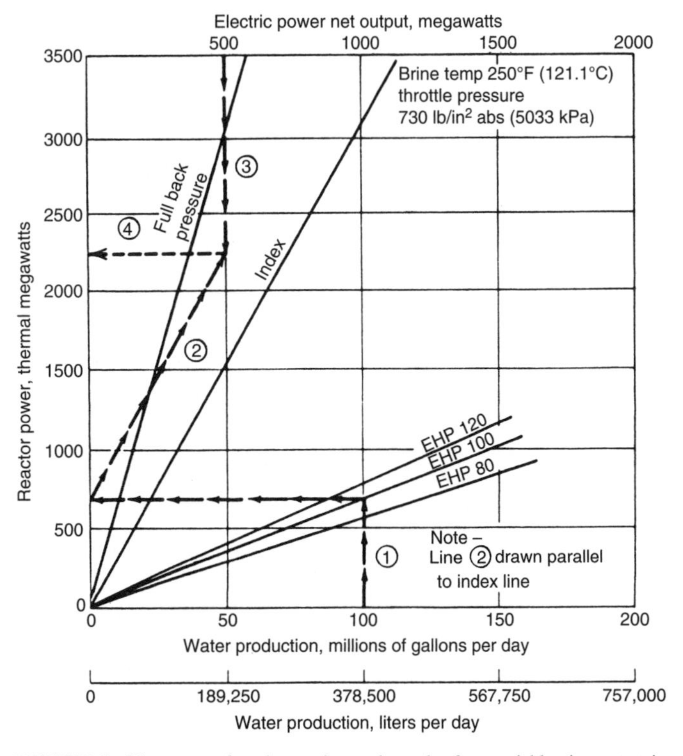

FIGURE 6 Nomogram for plant ratings relates the four variables important in desalting when combined with power generation.

The *type of cycle* is determined by the location of the point of intersection between lines 2 and 3. If lines 2 and 3 intersect to the right of the full back-pressure (FBP) line, the cycle used is the extraction or multishaft type. When the intersection falls directly on the FBP line, a backpressure cycle is indicated. An intersection to the left of the FBP line indicates that some of the steam to the brine heater is bypassed around the turbine regardless of the cycle used. Since the intersection in Fig. 6 occurs to the right of the FBP line, either an extraction or multishaft type of cycle could be used. The final choice of a cycle would depend on the water output required.

Related Calculations. The data presented here were developed by W. H. Comtois, Westinghouse Electric Corp., and were reported in *Mechanical Engineering.* Studies made at Westinghouse show that:

1. The fixed-annual-charge rate exerts the greatest single influence on water cost, increasing the cost by about two-thirds for a factor of 2 increase in the rate. This effect is moderated somewhat for large plant sizes.
2. The plant load factor gives the expected result of decreasing product costs with increasing load factor. The effect is a 1 to 2 percent decrease (increase) for every percentage increase (decrease) in load factor in the range from 75 to 95 percent.
3. Plant design life is of little consequence in the range normally considered (30 to 40 years).
4. The range of maximum brine temperatures studied was 200 to 250°F (93.3 to 121.1°C). Without exception, the computed optimum brine temperature was 250°F (121.1°C).
5. The single-shaft cycles (backpressure or extraction) enjoy a small (5 to 10 percent) water cost advantage over the multishaft cycle.

SECTION 6
HYDROELECTRIC ENERGY POWER PLANTS

Hydroelectric Power-Plant Parameters 6.1
Small-Scale "Clean Energy" Hydro Site Analysis 6.2
Economic Evaluation of Small-Scale Hydro Sites 6.4
Analysis of a Large-Scale Hydroelectric Energy Plant 6.7

HYDROELECTRIC POWER-PLANT PARAMETERS

Hydroelectric plants emit zero, or very small, amounts of greenhouse gases. Thus, hydro plants are an attractive source of power in today's global-warming conscious world. But, according to some experts, every major potential hydro site has been developed; i.e., a hydro plant has been built on the site. The result? Small hydro sites have become very popular, worldwide.

Hydro power is inherently efficient. Efficiencies of 80 percent plus are commonly reported by many hydro plants—both large and small. Thus, a cubic foot per second of water at 62.5 lb/ft^3 falling 8.8 ft is equivalent to 12 hp, and falling 11.8 ft is equivalent to 1 kW.

In selling power from hydro plants, the horsepower-year and kilowatt-year are sometimes used. At 100 percent load factor, 1 hp-year = 0.746 kW-year = 8760 hp-h = 6540 kWh. The power output of any hydro plant is limited by the installed equipment, the available water supply, the head available, and the storage quantity if the plant has pumped-storage facilities.*

Pumped storage is the most widely used energy storage system in the world today. The key to pumped storage is a reversible pump-turbine that can act as a pump during low-energy-demand hours and as a turbine during high-energy-demand hours. Some older hydro stations use separate pumps and turbines to deliver, and use, stored water during off-peak and on-peak hours. Further, the terrain at the installation must be such that there is a significant elevation difference, H, between upper and lower reservoirs, and a minimum horizontal distance, L, to reduce piping energy losses.

Values of L/H less than 2 are considered very favorable. Most existing pumped-storage plants have L/H ratios of between 4 and 6; some are nearly as high as 10. To overcome site problems where there is not enough difference in the reservoir elevations, underground storage may be an alternative option. The upper reservoir is placed at, or near, ground level. To provide sufficient hydraulic head, the lower reservoir is in underground natural caverns, abandoned mines, or other underground cavities. The existing terrain must have such features for such an underground system to be developed.† Use Table 1 as a guide to hydroturbine types and the typical head ranges they are used for.‡

*Davis, *Handbook of Applied Hydraulics,* McGraw-Hill, 1952.
†El-Wakil, *Powerplant Technology,* McGraw-Hill, 1984 and 2002.
‡Avallone and Baumeister III, *Marks' Standard Handbook for Mechanical Engineers,* McGraw-Hill, 2006.

TABLE 1 General Arrangements of Turbine Installations and Usual Head Limits Employed

Type	Setting	Construction	No. of runners	Usual head limits for direct-connected units	
				ft	m
Reaction turbines, 5 to 1000 ft (1.5 to 300 m)	Axial flow	Vertical or horizontal or slanted	1	5–60	1.5–20
head	Encased	Concrete vertical	1	15–130	5–40
		Cast or welded plate steel. Stainless steel. Vertical or horizontal	1 or 2	30–1600	10–500
Impulse turbines, 500 to 6000 ft (150 to 1800 m)	Encased	Horizontal (1–2 nozzles)	2	500–6000	150–1800
head		Vertical (1–6 nozzles)	1	500–6000	150–1800

SMALL-SCALE "CLEAN ENERGY" HYDRO SITE ANALYSIS

A newly discovered hydro site provides a potential head of 65 ft (20 m). An output of 10,000 kW (10 MW) is required to justify use of the site. Select suitable equipment for this installation based on the available head and the required power output.

Calculation Procedure:

1. *Determine the type of hydraulic turbine suitable for this site*
Enter Fig. 1 on the left at the available head, 65 ft (20 m), and project to the right to intersect the vertical projection from the required turbine output of 10,000 kW (10 MW). These two lines intersect in the *standardized tubular unit* region. Hence, such a hydroturbine will be tentatively chosen for this site.

2. *Check the suitability of the chosen unit*
Enter Table 2 at the top at the operating head range of 65 ft (20 m) and project across to the left to find that a tubular-type hydraulic turbine with fixed blades and adjustable gates will produce 0.25 to 15 MW of power at 55 to 150 percent of rated head. These ranges are within the requirements of this installation. Hence, the type of unit indicated by Fig. 1 is suitable for this hydro site.

Related Calculations. Passage of legislation requiring utilities to buy electric power from qualified site developers is leading to strong growth of both site development and equipment suitable for small-scale hydro plants. Environmental concerns over fossil-fuel-fired and nuclear-generating plants make hydro power more attractive. Hydro plants, in general, do not pollute the air, do not take part in the acid-rain cycle, are usually remote from populated areas, and run for up to 50 years with low maintenance and repair costs. Environmentalists rate hydro power as "clean" energy available with little, or no, pollution of the environment.

To reduce capital cost, most site developers choose standard-design hydroturbines. With essentially every high-head site developed, low-head sites become more attractive to developers. Table 2 shows the typical performance characteristics of hydroturbines being used today. Where there is a region of overlap in Table 2 or Fig. 1, site-specific parameters dictate choice and whether to install large units or a greater number of small units.

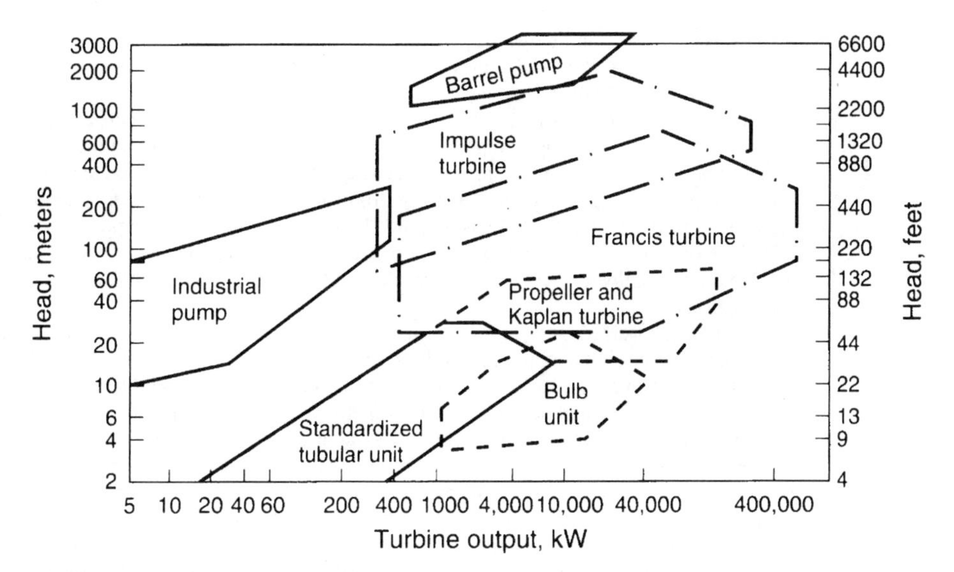

FIGURE 1 Traditional operating regimes of hydraulic turbines. New designs allow some turbines to cross traditional boundaries. (*Power.*)

TABLE 2 Performance Characteristics of Common Hydroturbines

Type	Operating head range: Rated head, ft (m)	Operating head range: % of rated head	Capacity range: MW	Capacity range: % of design capacity
Vertical fixed-blade propeller	7–120 (3–54) and over	55–125	0.25–15	30–115
Vertical Kaplan (adjustable blades and guide vanes)	7–66 (3–30) and over	45–150	1–15	10–115
Vertical Francis	25–300 (11–136) and over	50–150 and over	0.25–15	35–115
Horizontal Francis	25–500 (11–227) and over	50–125	0.25–10	35–115
Tubular (adjustable blades, fixed gates)	7–59 (3–27)	65–140	0.25–15	45–115
Tubular (fixed blades, adjustable gates)	7–120 (3–54)	55–150	0.25–15	35–115
Bulb	7–66 (3–30)	45–140	1–15	10–115
Rim-generator	7–30 (3–14)	45–140	1–8	10–115
Right-angle-driver propeller	7–59 (3–27)	55–140	0.25–2	45–115
Cross flow	20–300 (9–136) and over	80–120	0.25–2	10–115

Source: *Power.*

Delivery time and ease of maintenance are other factors important in unit choice. Further, the combination of power-generation and irrigation services in some installations make hydroturbines more attractive from an environmental view because two objectives are obtained: (1) "clean" power, and (2) crop watering.

Maintenance considerations are paramount with any selection; each day of downtime is lost revenue for the plant owner. For example, bulb-type units for heads between 10 and 60 ft (3 and 18 m) have performance characteristics similar to those of Francis and tubular units, and are often 1 to 2 percent more efficient. Also, their compact and, in some cases, standard design makes for smaller installations and reduced structural costs, but they suffer from poor accessibility. Sometimes the savings arising from the unit's compactness are offset by increased costs for the watertight requirements. Any leakage can cause severe damage to the machine.

To reduce the costs of hydroturbines, suppliers are using off-the-shelf equipment. One way this is done is to use centrifugal pumps operated in reverse and coupled to an induction motor. Although this is not a novel concept, pump manufacturers have documented the capability of many readily available commercial pumps to run as hydroturbines. The peak efficiency as a turbine is at least equivalent to the peak efficiency as a pump. These units can generate up to 1 MW of power. Pumps also benefit from a longer history of cost reductions in manufacturing, a wider range of commercial designs, faster delivery, and easier servicing—all of which add up to more rapid and inexpensive installations.

Though a reversed pump may begin generating power ahead of a turbine installation, it will not generate electricity more efficiently. Pumps operated in reverse are nominally 5 to 10 percent less efficient than a standard turbine for the same head and flow conditions. This is because pumps operate at fixed flow and head conditions; otherwise efficiency falls off rapidly. Thus, pumps do not follow the available water load as well unless multiple units are used.

With multiple units, the objective is to provide more than one operating point at sites with significant flow variations. Then the units can be sequenced to provide the maximum power output for any given flow rate. However, as the number of reverse pump units increases, equipment costs approach those for a standard turbine. Further, the complexity of the site increases with the number of reverse pumps units, requiring more instrumentation and automation, especially if the site is isolated.

Energy-conversion-efficiency improvements are constantly being sought. In low-head applications, pumps may require specially designed draft tubes to minimize remaining energy after the water exists from the runner blades. Other improvements being sought for pumps are: (1) modifying the runner-blade profiles or using a turbine runner in a pump casing, (2) adding flow-control devices such as wicket gates to a standard pump design or stay vanes to adjust turbine output.

Many components of hydroturbines are being improved to reduce space requirements and civil costs, and to simplify design, operation, and maintenance. Cast parts used in older turbines have largely been replaced by fabricated components. Stainless steel is commonly recommended for guide vanes, runners, and draft-tube inlets because of better resistance to cavitation, erosion, and corrosion. In special cases, there are economic tradeoffs between using carbon steel with a suitable coating material and using stainless steel.

Some engineers are experimenting with plastics, but much more long-term experience is needed before most designers will feel comfortable with plastics. Further, stainless steel material costs are relatively low compared to labor costs. And stainless steel has proven most cost-effective for hydroturbine applications.

While hydro power does provide pollution-free energy, it can be subject to the vagaries of the weather and climatic conditions. Thus, at the time of this writing, some 30 hydroelectric stations in the northwestern part of the United States had to cut their electrical output because the combination of a severe drought and prolonged cold weather forced a reduction in water flow to the stations. Purchase of replacement power—usually from fossil-fuel-fired plants—may be necessary when such cutbacks occur. Thus, the choice of hydro power must be carefully considered before a final decision is made.

This procedure is based on the work of Jason Makansi, associate editor, *Power* magazine, and reported in that publication.

ECONOMIC EVALUATION OF SMALL-SCALE HYDRO SITES

A city is considering a small hydro-power installation to save fossil fuel. To obtain the savings, the following steps will be taken: refurbish an existing dam, install new turbines, operate the generating plant. Outline the considerations a designer must weigh before undertaking the actual construction of such a plant.

Calculation Procedure:

1. ***Analyze the available head***
Most small hydro-power sites today will have a head of less than 50 ft (15.2 m) between the high-water level and tail-water level, Fig. 2. The power-generating capacity will usually be 25 MW or less.

2. ***Relate absolute head to water flow rate***
Because heads across the turbine in small hydro installations are often low in magnitude, the tail-water level is important in assessing the possibilities of a given site. At high-water flows, tail-water levels are often high enough to reduce turbine output, Fig. 3*a*. At some sites, the available head at high flow is extremely low, Fig. 3*b*.

The actual power output from a hydro station is $P = HQwe/550$, where P = horsepower output; H = head across turbine, ft; Q = water flow rate, ft^3/s; w = weight of water, lb/ft^3; e = turbine efficiency. Substituting in this equation for the plant shown in Fig. 3*b*, for flow rates of 500 and 1500 m^3/s, we see that a tripling of the water flow rate increases the power output by only 38.7 percent, while the absolute head drops 53.8 percent (from 3.9 to 1.8 m). This is why the tail-water level is so important in small hydro installations.

Figure 3*c* shows how station costs can rise as head decreases. These costs were estimated by the department of energy (DOE) for a number of small hydro-power installations. Figure 3*d* shows that station cost is more sensitive to head than to power capacity, according to DOE estimates. And the prohibitive costs for developing a completely new small hydro site mean that nearly all work will be at existing dams. Hence, any water exploitation for power must not encroach seriously on present customs, rights, and usages of the water. This holds for both upstream and downstream conditions.

3. ***Outline machinery choice considerations***
Small-turbine manufacturers, heeding the new needs, are producing a good range of semistandard designs that will match any site needs in regard to head, capacity, and excavation restrictions.

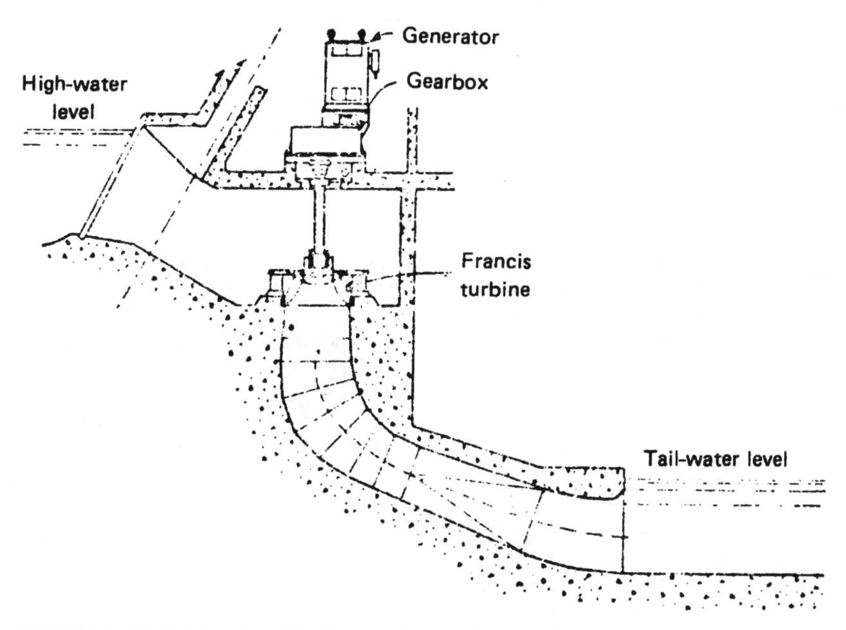

FIGURE 2 Vertical Francis turbine in open pit was adapted to 8-m head in an existing Norwegian dam. (*Power*.)

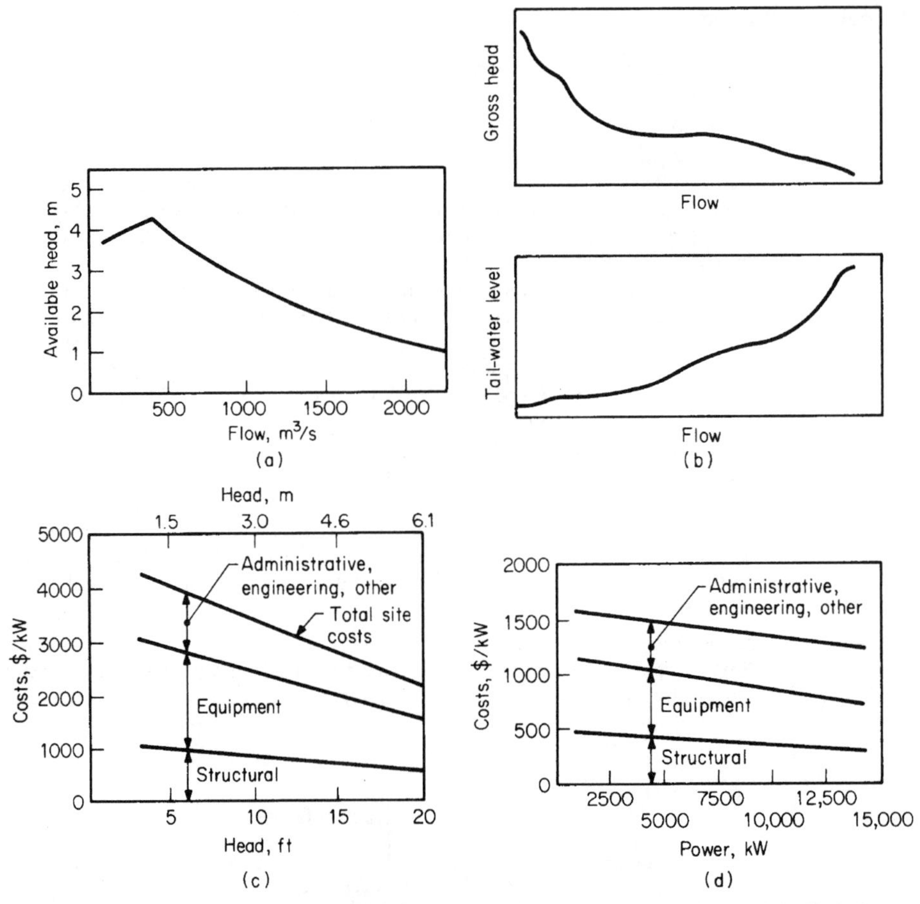

FIGURE 3 (*a*) Rising tail-water level in small hydro projects can seriously curtail potential. (*b*) Anderson-Cottonwood dam head dwindles after a peak at low flow. (*c*) Low heads drive DOE estimates up. (*d*) Linear regression curves represent DOE estimates of costs of small sites. (*Power.*)

The Francis turbine, Fig. 2, is a good example of such designs. A horizontal-shaft Francis turbine may be a better choice for some small projects because of lower civil-engineering costs and compatibility with standard generators.

Efficiency of small turbines is a big factor in station design. The problem of full-load versus part-load efficiency, Fig. 4, must be considered. If several turbines can fit the site needs, then good part-load efficiency is possible by load sharing.

Fitting new machinery to an existing site requires ingenuity. If enough of the old powerhouse is left, the same setup for number and type of turbines might be used. In other installations the powerhouse may be absent, badly deteriorated, or totally unsuitable. Then river-flow studies should be made to determine which of the new semistandard machines will best fit the conditions.

Personnel costs are extremely important in small hydro projects. Probably very few small hydro projects centered on redevelopment of old sites can carry the burden of workers in constant attendance. Hence, personnel costs should be given close attention.

Tube and bulb turbines, with horizontal or nearly horizontal shafts, are one way to solve the problem of fitting turbines into a site without heavy excavation or civil engineering works. Several standard and semistandard models are available.

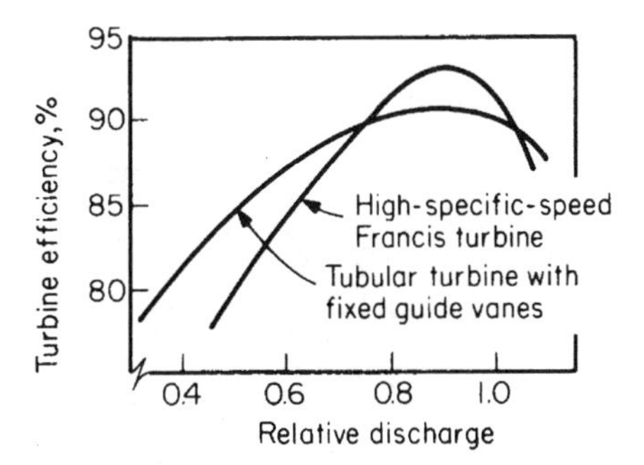

FIGURE 4 Steep Francis-turbine efficiency fall-off frequently makes multiple units advisable.

In low head work, the turbine is usually low-speed, far below the speed of small generators. A speed-increasing gear box is therefore required. A simple helical-gear unit is satisfactory for vertical-shaft and horizontal-shaft turbines. Where a vertical turbine drives a horizontal generator, a right-angle box makes the turn in the power flow.

Governing and control equipment is not a serious problem for small hydro plants.

Related Calculations. Most small hydro projects are justified on the basis of continuing inflation which will make the savings they produce more valuable as time passes. Although this practice is questioned by some people, the recent history of inflation seems to justify the approach.

As fossil-fuel prices increase, small hydro installations will become more feasible. However, the considerations mentioned in this procedure should be given full weight before proceeding with the final design of any plant. The data in this procedure were drawn from an ASME meeting on the subject with information from papers, panels, and discussion summarized by William O'Keefe, senior editor, *Power* magazine, in an article in that publication.

ANALYSIS OF A LARGE-SCALE HYDROELECTRIC ENERGY PLANT

A potential hydroelectric site has an estimated maximum static head for the hydroturbines of 300 ft (91.4 m), between the water source and the downstream pool. Average conduit flow of the water to the turbines is 575 cfs (16.3 m^3/s); maximum flow rate is 675 cfs (19.1 m^3/s). Determine the practical power output of this potential hydro site. Evaluate the various factors that must be considered to make the site economically feasible. Figures 5 through 8 show possible turbine choices.

Calculation Procedure:

1. *Determine the potential power output of the site*

The basic relation for a hydroelectric turbine output, P, in kilowatts is P = QH/11.8, where Q = water flow in cfs (m^3/s), H = static head on the turbine, ft (m). In SI units the denominator = 0.102. Substituting, P = (575 cfs) (300 ft)/11.8 = 14,619 kW at average conduit flow. At maximum flow, using the same relation, P = 17,161 kW.

2. *Evaluate the choice of this site for a hydroelectric plant*

The competitive position of a hydro project must be judged by the cost and reliability of the output at the point of use or market. In most hydro developments, the bulk of the investment is in structures for the collection, control, regulation, and disposal of the water. Electrical transmission frequently adds a substantial financial burden because of remoteness of the hydro site from the market.

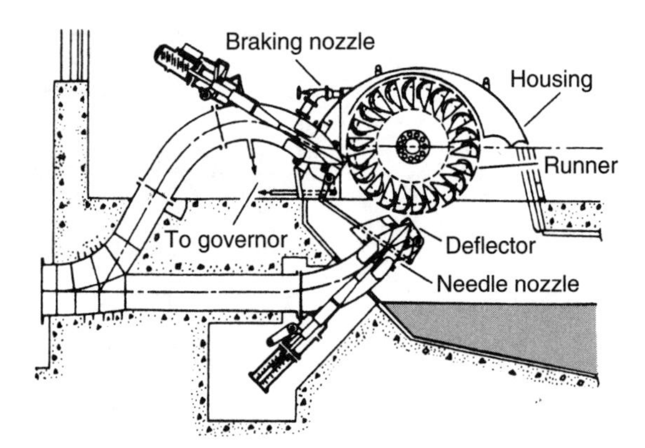

FIGURE 5 Cross section of an impulse (Pelton) type of hydraulic-turbine installation.

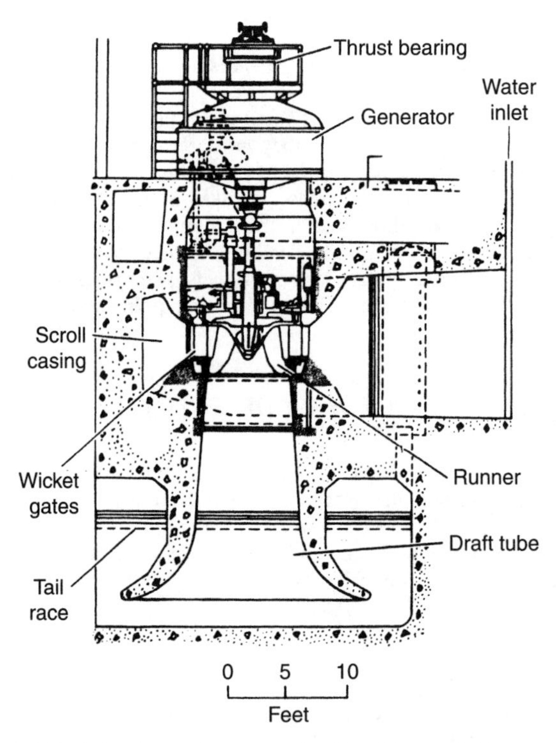

FIGURE 6 Cross section of a reaction (Francis) type of hydraulic-turbine installation. 1 ft = 0.3 m.

The incremental cost for waterwheels, generators, switches, yard, transformers, and water conduit is often a smaller fraction of the total investment than is the cost for the basic structures, real estate, and transmission facilities. Long life is characteristic of hydroelectric installations, and the annual carrying charges of 6 to 12 percent on the investment are a minimum for the power field. Operating and maintenance costs are lower than for other types of generating stations.

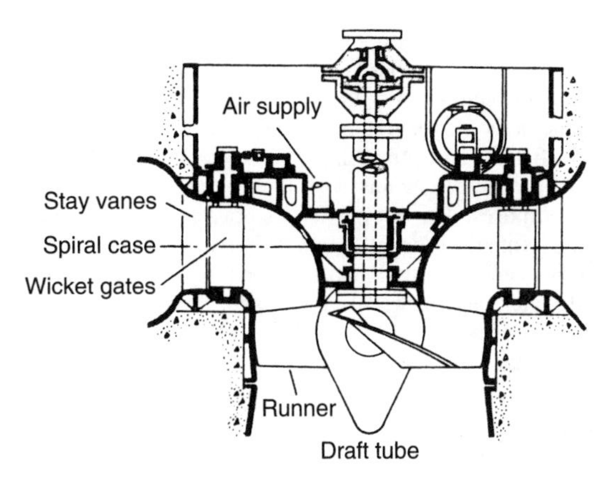

FIGURE 7 Cross section of a propeller (Kaplan) type of hydraulic-turbine installation.

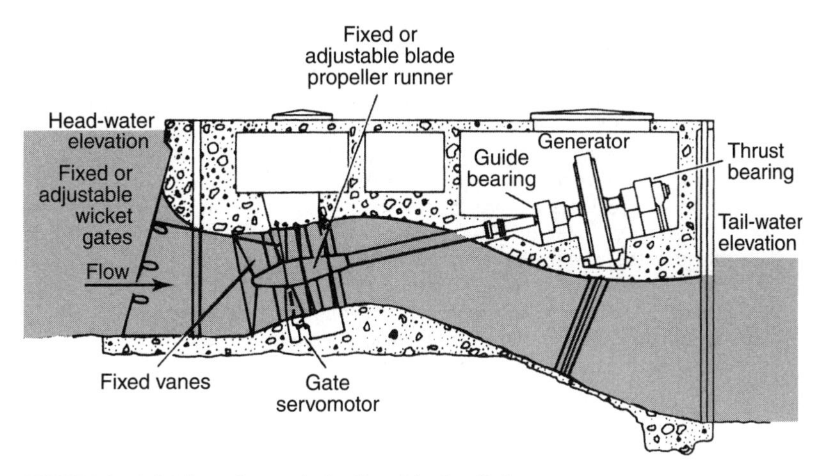

FIGURE 8 Axial-flow tube-type hydraulic-turbine installation.

3. *Assess the site and its runoff and static head*

The fundamental elements of potential power, as given in the equation above, are runoff Q and head H. Despite the apparent basic simplicities of the relation, the technical and economic development of a hydro site is a complex problem. No two sites are alike, so the opportunity for standardization of structures and equipment is nearly nonexistent. The head would appear to be a simple surveying problem based largely on topography. However, geologic conditions, as revealed by core drillings, can eliminate an otherwise economically desirable site. Runoff is complicated, especially when records of flow are inadequate. Hydrology is basic to an understanding of water flow and its variations. Runoff must be related to precipitation and to the disposal of precipitation. It is vitally influenced by climatic conditions, seasonal changes, temperature and humidity of the atmosphere, meteorological phenomena, character of the watershed, infiltration, seepage, evaporation, percolation, and transpiration. Hydrographic data are essential to show the variations of runoff over a period of many years. Reservoirs, by providing storage, reduce the extremes of flow variation, which are often as high as 100 to 1 or occasionally 1000 to 1.

4. *Evaluate the economic factors affecting the generating capacity installed*
The economic factors affecting the capacity to be installed, which must be evaluated on any project, include load requirements, runoff, head, development cost, operating cost, value of output, alternative methods of generation, flood control, navigation, rights of other industries on the stream (such as fishing and lumbering), and national defense. Some of these factors are components of multipurpose developments with their attendant problems in the proper allocation of costs to the several purposes. The prevalence of government construction, ownership, and operation, with its subsidized financial formulas which are so different from those for investor-owned projects, further complicates economic evaluation. Many people and groups are parties of interest in the harnessing of hydro sites, and stringent government regulations prevail, including those of the U.S. Corps of Engineers, Federal Power Commission, Bureau of Reclamation, Geological Survey, and Securities and Exchange Commission.

5. *Assess ways that the selected capacity can be used to serve customers*
Prime capacity is that which is continuously available. Firm capacity is much larger and is dependent upon interconnection with other power plants and the extent to which load curves permit variable-capacity operation. The incremental cost for additional turbine-generator capacity is small, so that many alternatives for economic development of a site must be considered. The alternatives include a wide variety of base load, peak load, run-of-river, and pumped-storage plants. All are concerned with fitting installed capacity, runoff, and storage to the load curve of the power system and to give minimum cost over the life of the installation. In this evaluation it is essential clearly to distinguish capacity (kW) from energy (kWh) as they are not interchangeable. In any practical evaluation of water power in this electrical era, it should be recognized that the most favorable economics will be found with an interconnected electric system where the different methods of generating power are complementary as well as competitive.

As noted above, there is an increasing tendency in many areas to allocate hydro capacity to peaking service and to foster pumped-water storage for the same objective. Pumped storage, to be practical, requires the use of two reservoirs for the storage of water—one at considerably higher elevation, say, 500 to 1000 ft (150 to 300 m). A reversible pump-turbine operates alternatively (1) to raise water from the lower to the upper reservoir during off-peak periods, and (2) to generate power during peak-load periods by letting the water flow in the opposite direction through the turbine. Proximity of favorable sites on an interconnected electrical transmission system reduces the investment burden. Under such circumstances the return of 2 kWh on-peak for 3 kWh pumping off-peak has been demonstrated to be an attractive method of economically utilizing interconnected fossil-fuel, nuclear-fuel, and hydro-power plants.

Related Calculations. The procedures and considerations given here apply to hydroelectric installations anywhere in the world. Since each hydro site is unique, full analysis of the site and its many variables must be made prior to acceptance of the final design.

This procedure is the work of Theodore Baumesister, as reported in the *McGraw-Hill Encyclopedia of Engineering,* 2005, with the numerical procedure provided by the handbook editor.

SECTION 7
WIND-POWER ENERGY DESIGN AND APPLICATION

Wind-Power Calculation Parameters 7.1
Analysis of a Wind Turbine's Power-Generating Capacity 7.2
Choice of Wind-Energy Conversion System 7.4

WIND-POWER CALCULATION PARAMETERS

Wind turbines are appearing on land and sea around the world as more countries and industrial firms embrace this source of alternative renewable energy. In small countries, where electrical distribution distances are short, wind power can have a major impact on reducing the need for GHG-producing coal, oil, or gas plants.

But where transmission distances between the source of wind power and the users of that power are great, problems can emerge. Thus, in the United States, it is estimated that some 20,000 miles (32,000 km) of transmission lines must be installed to bring wind-generated power to major using areas. While wind turbines are relatively easy to install, transmission lines offer a much larger challenge.

A recent estimate of the potentials of wind-power generation, issued by the U.S. Department of Energy, is most promising. The estimate projects that wind power in the lower 48 states could produce more than nine times the current power production in those states. At this writing the annual power production in the United States is some 4 million GWh. It is estimated that wind power could produce 37 GWh. Today's wind turbines are typically 80 m (262 ft) high. At this height, wind velocities are usually higher, enabling each wind turbine to produce more electricity.

Designers today seek to have each wind turbine produce more power, so fewer turbines are needed for a given installation. To achieve greater power output, towers are built higher, blades are made longer, and turbine-generating capacity is increased. With higher towers and longer blades, logistical problems develop in delivering these parts by truck or rail. Some parts are just too large to be transported conveniently.

One of the major problems with wind turbines is getting them to produce power during periods of high demand. Often the time of highest power demand coincides with time of weakest wind velocity. Conversely, wind turbines may produce their highest output at times of weakest power demand.

To overcome these problems and store wind-generated power, a 268-mW system in the midwestern area of the United States feeds excess power to an air compressor that pumps air 3000 ft (914 m) down into porous sandstone. This pressurized air displaces groundwater. When power is needed during peak demand times, air is fed from the ground reservoir into a surface-mounted gas turbine, increasing its efficiency by some 60 percent. Designers hope that such subterranean storage can increase the power supplied by wind turbines in the United States from its present 2 percent to 10 percent, or more. Some subterranean projects can store as much as a 5-month supply of air.

The U.S. Department of Energy sees the possibility of wind turbines generating some 20 percent of the electricity needed in the United States by the year 2030. Wind power could even rival nuclear-power-generating capacity if new nuclear stations are not approved, and built, by that time.

Reports say that wind-energy capacity is growing by almost one-third each year. The only real competitor to wind power in the growth sector is natural gas.

Wind farms containing 100, and more, wind turbines per farm are becoming more popular in the United States. The reason for this is that large-scale wind-turbine installations have a lower cost per unit of capacity than isolated small-number sites. At this writing the usual cost in large sites of wind turbines is $2-million per MW of nameplate capacity.

Again, at this writing, the first offshore wind-farm installation in the United States has been approved. How long it will take to get final approval of this first offshore installation is not known. The United States is far behind Europe in offshore wind farms. The primary reason for this is the large land areas available in the United States for wind farms. Europe, by comparison, has much less land area for wind farms. Hence, designers have installed large quantities of wind capacity offshore.

Wind power has a promising future throughout the world. As wind-turbine capacity rises, these machines will go a long way toward reducing greenhouse gases worldwide.

Two minor problems facing wind farms are birds and noise. In some areas birds have flown into wind turbines and been killed. But with more wind turbines being installed, bird killings appear to have declined. Perhaps the birds have learned to stay away from wind turbines, or their migration routes have changed. It is enough to say that birders have been complaining less vocally in recent years.

Noise complaints about wind turbines have been voiced when wind farms are set up close to populated areas. Some wind farms in the United States reduce the rotational speed of their turbines after 7 p.m. to avoid complaints from nearby residents. And various states set a nighttime noise limit to 45 decibels. In Europe, some wind farms stop their turbines at 10 p.m., rotation resuming at 6 or 7 a.m. Perhaps the easiest "fix" for noise complaints is to locate wind farms at least 2 or more miles (3.2 km) from populated areas. Then, there will not be any noise complaints. While noise and bird injury complaints occur, proponents of wind power do not believe such claims will deter the future installation of more wind turbines throughout the world.

At this writing (2011), Denmark is testing a combined wind-power/wave-power platform that generates power from wind turbines and wave action using a piston pump that delivers pressurized water to a turbine-generator. The growing need for clean power worldwide is increasing the number of wind- and wave-power schemes being tested.

ANALYSIS OF A WIND TURBINE'S POWER-GENERATING CAPACITY

A 10-m/s wind is at 1 standard atm pressure and 15°C temperature. Calculate (1) the total power density in the wind stream, (2) the maximum obtainable power density, (3) a reasonably obtainable power density, all in W/m^2, (4) the total power (in kW) produced if the turbine diameter is 120 m, and (5) the torque and axial thrust N if the turbine were operating at 40 r/min and maximum efficiency.

Calculation Procedure:

1. ***Compute the air density for this wind turbine***

For air, the gas constant $R = 287$ J/(kg · K). 1 atm = 1.01325 × 10^5

Air density
$$\rho = \frac{P}{RT} = \frac{1.01325 \times 10^5}{287(15 + 273.15)} = 1.226 \text{ kg/m}^3$$

2. ***Find the total power density in the wind stream***

$$\frac{P_{\text{tot}}}{A} = \frac{1}{2g_c}\rho V_i^3 = \frac{1}{2 \times 1} 1.226 \times 10^3 = 613 \text{ W/m}^2$$

3. *Calculate the maximum obtainable power density*

$$\frac{P_{\max}}{A} = \frac{8}{27g_c}\rho V_i^3 = \frac{8}{27\times 1}1.226\times 10^3 = 363 \text{ W/m}^2$$

4. *Compute a reasonably obtainable power density*
Assuming $\eta = 40\%$.

$$\frac{P}{A} = 0.4\left(\frac{P_{\text{tot}}}{A}\right) = 0.4\times 613 = 245 \text{ W/m}^2$$

(In English units this corresponds to 22.76 W/ft² at 22.37 mi/h.)

5. *Find the total power produced if the turbine diameter is 120 m*

$$P = 0.245\times\frac{\pi D^2}{4} = 0.245\times\frac{\pi 120^2}{4} = 2770 \text{ kW}$$

6. *What is the torque and axial thrust if the turbine operates at 40 rpm and maximum efficiency*

$$T_{\max} = \frac{2}{27g_c}\frac{\rho D V_i^3}{N} = \frac{2}{27\times 1}\times\frac{1.20\times 1.226\times 10^3}{40/60}$$
$$= 16{,}347 \text{ N } (= 3675 \text{ lb}_f)$$

$$F_{x,\max} = \frac{\pi}{9g_c}\rho D^2 V_i^2 = \frac{\pi}{9\times 1}(1.226\times 120^2\times 10^2)$$
$$= 616{,}255 \text{ N } (= 138{,}540 \text{ lb}_f)$$

There are two types of forces operating on the blades of a propeller-type wind turbine. They are the *circumferential forces* in the direction of wheel rotation that provide the torque and the *axial forces* in the direction of the wind stream that provide an *axial thrust* that must be counteracted by proper mechanical design.

The *circumferential force*, or *torque*, T is obtained from

$$T = \frac{P}{\omega} = \frac{P}{\pi DN}$$

where

T = torque, N or lb_f
ω = angular velocity of turbine wheel, m/s or ft/s
D = diameter of turbine wheel = $\sqrt{4A/\pi}$, m or ft
N = wheel revolutions per unit time, s^{-1}

For a turbine operating at power P, the torque is given by

$$T = \eta\frac{1}{8g_c}\frac{\rho D V_i^3}{N}$$

For a turbine operating at maximum efficiency $\eta_{max} = 16/27$, the torque is given by T_{max}

$$T_{max} = \frac{2}{27g_c}\frac{PDV_i^3}{N}$$

The *axial force,* or axial thrust, is

$$F_x = \frac{1}{2g_c}\rho A(V_i^2 - V_e^2) = \frac{\pi}{8g_c}\rho D^2(V_i^2 - V_e^2)$$

where A = cross-sectional area of stream, m^2 or ft^2

The axial force on a turbine wheel operating at maximum efficiency where $V_e = 1/3\ V_i$ is given by

$$F_{x,max} = \frac{4}{9g_c}\rho AV_i^2 = \frac{\pi}{9g_c}\rho D^2V_i^2$$

The axial forces are proportional to the square of the diameter of the turbine wheel, which makes them difficult to cope with in extremely large-diameter machines. There is thus an upper limit of diameter that must be determined by design and economical considerations.

CHOICE OF WIND-ENERGY CONVERSION SYSTEM

Select a wind-energy conversion system to generate electric power at constant speed and constant frequency in a sea-level area where winds average 18 mi/h (29 km/h), a cut-in speed of 8 mi/h (13 km/h) is sought, blades will be fully feathered (cut out) at wind speeds greater than 60 mi/h (100 km/h), and the system must withstand maximum wind velocities of 150 mi/h (240 km/h). Determine typical costs which might be expected. The maximum rotor diameter allowable for the site is 125 ft (38 m).

Calculation Procedure:

1. *Determine the total available wind power*
Figure 1 shows the total available power in a freely flowing windstream at sea level for various wind speeds and cross-sectional areas of windstream. Since the maximum blade diameter, given that a blade-type conversion device will be used, is 125 ft (38 m), the area of the windstream will be $A = \pi d^2/4 = \pi(125)^2/4 = 12{,}271.9$ ft^2 (1140.1 m^2). Entering Fig. 1 at this area and projecting vertically to a wind speed of 18 mi/h (29 km/h), we see that the total available power is 200 kW.

2. *Select a suitable wind machine*
Typical modern wind machines are shown in Fig. 2. In any wind-energy conversion system there are three basic subsystems: the aerodynamic system, the mechanical transmission system (gears, shafts, bearings, etc.), and the electrical generating system. Figure 2 gives the taxonomy of the more practical versions of wind machines (the aerodynamic system) available today. “Almost any physical configuration which produces an asymmetric force in the wind can be made to rotate, translate, or oscillate—thereby generating power. The governing consideration is economic—how much power for how much size and cost,” according to Fritz Hirschfeld, Member, ASME.

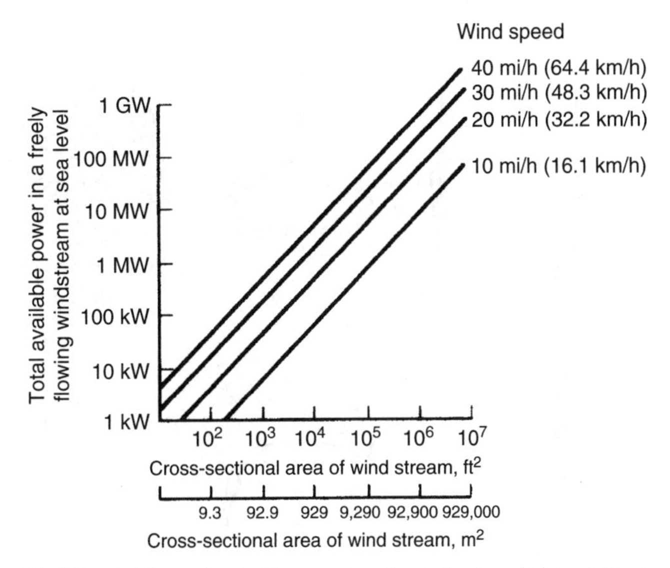

FIGURE 1 The total available power in a freely flowing windstream at sea level versus the cross-sectional area of the windstream and the wind speed. (*Mechanical Engineering.*)

Continuing, Hirschfeld notes, "The power coefficient of an ideal wind machine rotor varies with the ratio of blade tip speed to free-flow windstream speed, and approaches the maximum of 0.59 when this ratio reaches a value of 5 or 6. Experimental evidence indicates that two-bladed rotors of good aerodynamic design—running at high rotational speeds where the ratio of the blade-tip-speed-to-free-flow-speed of the windstream is 5 or 6—will have power coefficients as high as 0.47. Figure 3 outlines the maximum power coefficients obtainable for several rotor designs. Figure 4 plots the typical performance curves of a number of different wind machines."

Choose a horizontal-axis double-bladed rotor wind machine for this application with a power coefficient C_p of 0.375. This type of wind machine is being chosen because (1) the power coefficient is relatively high (0.375), providing efficient conversion of the energy of the wind; (2) the allowable blade diameter, 125 ft (38 m), is suitable for the double-bladed design; (3) a double-bladed rotor will operate well in the average wind speed, 18 mi/h (29 km/h), prevailing in the installation area; and (4) the double-bladed rotor is well-suited for the constant-speed constant-frequency (CSCF) system desired for this installation.

3. *Compute the maximum electric power output of the wind machine*

The power of the wind P_w is converted to mechanical power P_m by the wind machine. In any wind machine, $P_m = C_p P_w$, where C_p = power coefficient. The mechanical power is then converted to electric power by the generator. Since there is an applicable efficiency for each of the systems, that is, C_p for the aerodynamic system, η_m for the mechanical system (gears, usually), or η_g for the generator, the electric power generated is $P_e = P_w \eta_m \eta_g$.

In actual practice, the maximum electric power output in kilowatts of horizontal-axis bladed wind machines geared to a 70 percent efficiency electric generator can be quickly computed from $P_e = 0.38d^2V^3/10^6$, where d = blade diameter, ft (m); V = maximum wind velocity, ft/s (m/s). For this wind machine, $P_e = (0.38)\,(125)^2\,(26.4)^3/10^6 = 109.2$ kW. This result agrees closely with the actual machine on which the calculation procedure is based, which has a rated output of 100 kW.

FIGURE 2 Wind machines come in all shapes and sizes. Some of the more practical design categories are illustrated in this taxonomy. (*Mechanical Engineering.*)

FIGURE 2 (*Continued*)

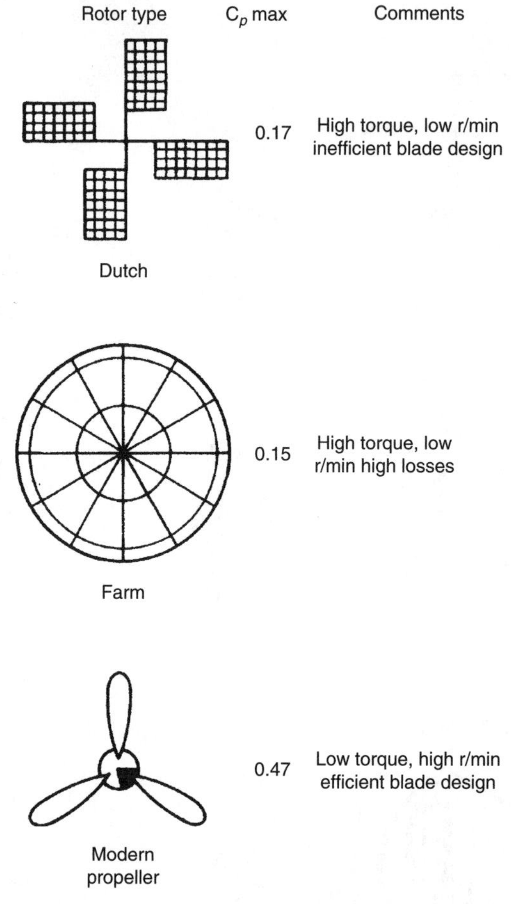

FIGURE 3 The maximum power coefficients for several types of rotor designs. (*Mechanical Engineering.*)

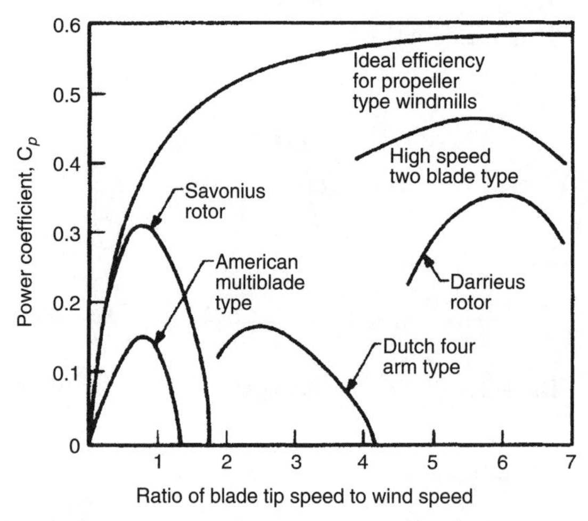

FIGURE 4 Typical performance curves for different types of wind machines. (*Mechanical Engineering.*)

4. *Determine the typical capital cost of this machine*
Figures 5 and 6 show typical capital costs for small conventional wind machines. Larger wind machines, such as the one being considered here, are estimated to have a cost of $150,000 for a 100-kW unit, or $1500 per kilowatt. Such costs may be safely used in first approximations with the base-year cost given in the illustration being suitably adjusted by a factor for inflation.

Related Calculations. Use this general procedure to choose wind machines for other duties—pumping, battery charging, supplying power to utility lines, etc. Be certain to check with manufacturers to determine whether the calculated results agree with actual practice in the field. In general, good agreement will be found to exist.

Wind power is a renewable, nonpolluting energy source in plentiful supply in certain parts of the world. For example, a recent report by the Union of Concerned Scientists points out that four states—Kansas, Nebraska, and North and South Dakota—have enough wind to generate—in theory—all the electricity needed in the United States today.

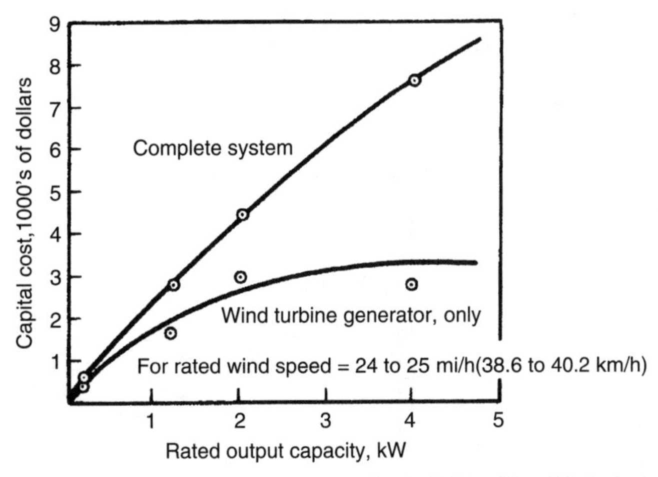

FIGURE 5 Capital cost of small conventional wind machine. (*Mechanical Engineering.*)

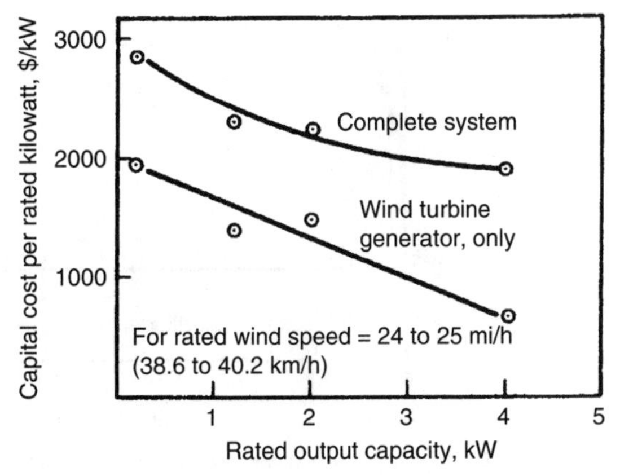

FIGURE 6 Capital cost, per rated kilowatt, for small conventional wind machines. (*Mechanical Engineering.*)

In addition, the agricultural resources of the midwestern part of the United States could provide crops and crop residues to be used as fuel in power plants. Likewise, logging and wood residues could be used to fire boilers to generate electricity. The resulting air pollution would be much less than that produced by coal-fired plants. Crop- and crop-residue-fired plants could eventually supply some 10 percent of the electrical energy needed by the midwestern states.

Current cost to produce a kilowatt of electricity using wind power is less than 6 cents. This compares favorably with the 4–6 cents cost per kilowatt for the typical coal-fired generating station. When the relatively low cost of wind power is combined with its nonpolluting features, this form of electrical generation is extremely attractive to environmentally conscious engineers and scientists.

The advantages cited above apply equally well to many other nations throughout the world. In some areas wind power offers a simple, low-cost solution to energy needs without resorting to complex technical methods. Wind power does have a worldwide future.

The illustrations and much of the data in this procedure are the work of Fritz Hirschfeld, as reported in *Mechanical Engineering* magazine. Also reported in the magazine is a proposal by J. S. Goela of Physical Sciences, Inc., to use kites to extract energy from the wind.

Kites avoid the use of high-capital-cost components such as windmill towers and large rotors. Further, a kite can utilize the full available potential of the wind. As Fig. 7 shows, the earth's boundary layer extends up to 5000 ft (1500 m) above sea level. In this boundary layer, the average wind velocity increases while the air density decreases with altitude. Consequently, the total available

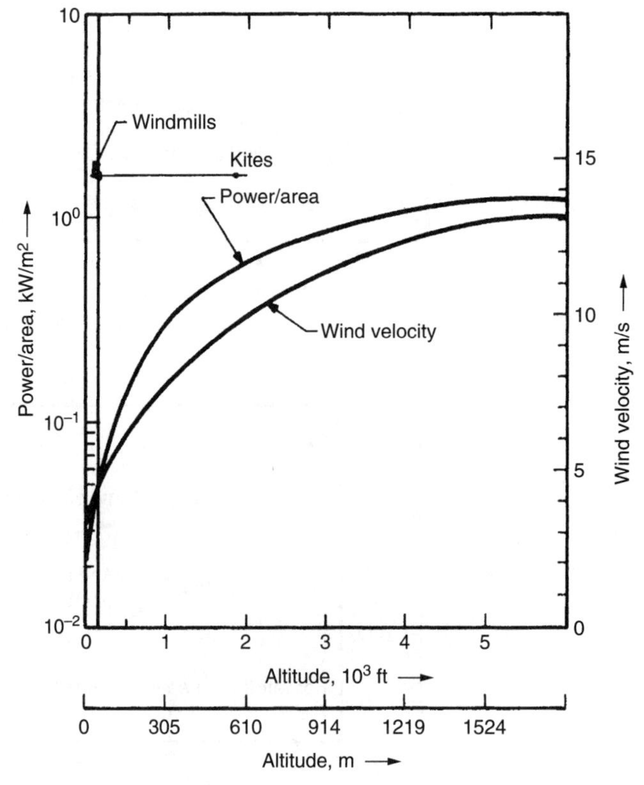

FIGURE 7 Variation of mean annual free-air wind velocity and total available wind power per unit area (= ½ ρV^3) with altitude in New England. (*Mechanical Engineering.*)

wind power per unit area (= ½ ρV^3) increases with altitude until at an altitude of 5000 ft (1500 m) a maximum is reached. The ratio of available wind power in New England at 5000 ft (1500 m) and 150 ft [many wind systems operate at an altitude of 150 ft (50 m) or less] is 25. This is a large factor which makes it very attractive to employ systems that use an energy extraction device located at an altitude of 5000 ft (1500 m). Even at an altitude of 1000 ft (300 m), this ratio is large, approximately equal to 10.

To understand how the proposed scheme will extract energy from the wind, consider the following: The motion of air generates a pull in the rope that holds the kite. This pull is a function of both the angle of attack of the kite and the kite area normal to the wind direction, and by varying any of these we can vary the pull on the rope. On the surface of the earth, this rope will be suitably connected to an energy system which will convert the variation in developed force on the rope to the rotational energy of a rotor.

Whenever a period of calm occurs, the kite will tend to lose its altitude. One solution to this difficulty is to fill the empty spaces in the kite with helium gas such that the upward pull from the helium gas will balance the downward gravity force due to the weight of the kite and its string. Another possibility is to tie a balloon to the kite.

A detailed theoretical analysis of the proposed scheme has been carried out. This analysis indicates that the proposed scheme is scalable, that the drag on the kite string is small in comparison to the pull in the string for large devices, and that approximately 0.38 kW of power theoretically can be obtained from a kite 1.2 yd^2 (1 m^2) in area. In addition, there are no material or system constraints which will prevent the kite from achieving an elevation of 5000 ft (1500 m).

Even though it is difficult to estimate the cost of wind power from the proposed scheme, a rough estimate indicates that for a 100-kW system, the capital cost per unit of energy from the proposed scheme will be less—approximately by a factor of 3—in comparison with the capital cost of one unit of energy produced from other 100-kW wind-energy systems.

The most important application of kite-based energy systems is in developing countries where these systems can be used to pump water from wells, grind grain, and generate electricity. A majority of developing countries do not have an adequate supply of indigenous oil and gas, nor can they afford to buy substantial quantities of fossil fuel at international prices. What these countries prefer is a system that could generate useful energy by using as inputs resources that are available within the country. The simple scheme proposed here is ideally suited for the needs of developing countries.

The kite-based system may also be economically attractive in comparison to a small windmill of less than 1 hp (750 W) which has been used in rural and farm areas in the western United States to pump water, generate electricity, and irrigate land. Another application of the proposed scheme is to generate auxiliary power in large sailboats, motorboats, and ships where conventional wind-energy schemes cannot be employed. The wind-energy system employing kites can also be used as a fuel saver in conjunction with already existing transmission lines.

A kite flying at an altitude of 5000 ft (1500 m) may present a hazard to low-flying airplanes. One way to avoid a collision with the kite or an entanglement with the kite lines is to enhance their visibility by providing flashing lights around the kite structure and along its retaining line. Another approach is to fly the kites at lower altitudes. For instance, even at an altitude of 1000 ft (300 m), the total available wind power is larger by a factor of 10 in comparison with that at 150 ft (50 m), Fig. 7.

Wind energy was welcomed by environmentalists when first introduced on a large-scale basis in the late 70s and early 80s. Today all environmental groups support wind energy—but with less enthusiasm than earlier. The reason for the loss of enthusiasm? Windmills are killing birds that fly into them.

Some of the birds killed by windmills are endangered species, so environmentalists seek a solution to prevent the killing of hundreds of birds a year that crash into the 17,000 100-ft (30.5-m) windmills erected in California where 80 percent of the world's wind power is produced.

A number of states are considering construction of windmill "farms" where thousands of power-generating windmills will be clustered together. Environmentalists are concerned that such farms will raise the death toll among birds. Wind turbine manufacturers estimate that from two to six birds are killed per year per 100 wind turbines.

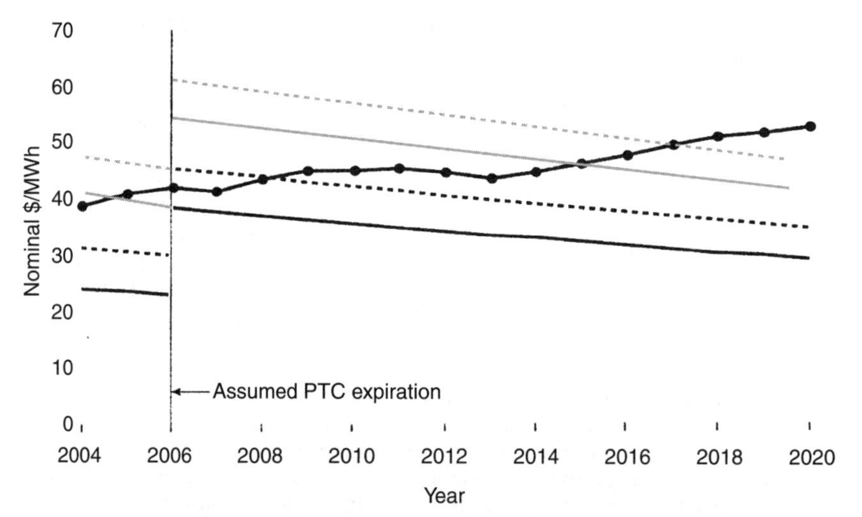

FIGURE 8 The comparative economics of wind power. This graph shows that if the U.S. Renewable Energy Production Tax Credit is allowed to expire again, only wind farms at the breeziest (Classes 5 and 6) sites will be able to compete with conventional generation for many years. Unlike "normal" economic graphs, which show market factors gradually changing trends, this one reflects a policy-driven disconnect, which is why we have a data "gap": On December 31, 2006, Class 5 wind will cost $30; on January 1, 2007, it will cost $45, assuming the PTC expires at the end of 2006. (*Platts Analytics.*)

Further impetus for renewable wind-turbine generation of electricity in Europe comes from the decline in the availability of oil and natural gas supplies. More than 50 percent of the current oil and natural gas used in Europe is imported. The Kyoto Pact seeks to have some 22 percent of the electricity used in Europe generated by renewable energy sources by the year 2010.

In the United States, a number of states that have not yet installed large numbers of wind turbines are considering doing so. Thus, New York and Massachusetts are—at this writing—considering wind-turbine installations in the Atlantic Ocean along their coastlines. There has been some resistance from citizens' groups, but this is expected to subside and the installations to proceed. Such installations will lead to a reduction in the use of fossil fuels. Figure 8 gives the comparative economics of wind power.

Today, and in the future, fossil fuels face enormous regulatory requirements for reducing greenhouse gas emissions. Such gases are believed to lead to global warming. The Kyoto Pact seeks to have power plants reduce their greenhouse gas emissions to less than their pre-1990 count during the period from 2008 through 2012.

SECTION 8
SOLAR-POWER ENERGY APPLICATION AND USAGE

Solar-Energy Power Parameters 8.1
Solar-Powered Electric-Generating System Load and Cost Analysis 8.2
Industrial Solar-Energy System Investment Economics 8.5
Flat-Plate Solar-Energy Heating- and Cooling-System Design 8.6
Solar-Collector Solar-Insolation Computation under Differing Weather Conditions 8.12
Collector Sizing for Solar-Energy Heating Systems for Buildings 8.15
Computing Useful Energy Delivery in Solar Heating Using the F-Chart Method 8.16
Solar-Collector Selection for Domestic Hot-Water Heating Systems 8.22
Design of Passive Solar-Energy Heating Systems for Buildings 8.27
Solar Water Heater Energy Savings Computation 8.32
Choosing Photovoltaic Modules for Electrical Loads in Buildings 8.34
Designing Solar-Powered Pumping Systems 8.37

SOLAR-ENERGY POWER PARAMETERS

The aim of solar-power manufacturers is to produce power at a competitive price to conventionally fired stations using coal, oil, or gas. Traditional power companies worry about a greenhouse gas cap on carbon dioxide and other pollutants. They also worry about a future requirement to produce "clean" power.

Solar energy is about three times more expensive than wind power. And solar energy is also about three times more costly than natural gas. When compared with power generated from coal, solar energy is about six times more expensive.

Current statistics for electric-generation costs on a kWh basis state that most coal-generated electricity costs about 4 cents per kWh. Natural-gas-generated electricity costs about 6 cents per kWh to produce. Solar-energy-produced electricity costs about 24 cents per kWh to produce. While these numbers were valid at the time of this writing, they can change markedly, depending on local conditions, fuel costs, and advances in the manufacture of solar-energy devices. At the time of this writing, solar energy is producing about 1 percent of the electric power in the United States. By comparison, fossil-fuel-fired generation produces about 70 percent of electric power at this time.

Solar manufacturers are aiming at producing panels that cost about $1 per watt generated. Eventually, solar manufacturers hope to compete with gas- or oil-fired plants that produce electricity at 4 to 6 cents per kWh. To achieve this goal, a number of different solar designs are being tested and evaluated.

Solar-energy cost savings come from scale—that is, the larger the solar installation, the lower (usually) the cost of generating a kWh of electricity. Thus, large-scale rooftop installations of photovoltaic devices in California are becoming popular. One utility expects to install 50 MW of capacity on rooftops over a 5-year time span.

Other land-based solar systems include: (*a*) concave dishes that follow the sun and deliver heat to a Stirling engine producing some 25 kW of electricity; (*b*) tower systems, which receive the sun's energy from mirrors, using the heat to generate steam for a turbo-generator, or to heat liquid sodium or a similar material that stores the heat for use at another time; (*c*) trough systems using parabolic mirrors that deliver heat to a pipe containing a heat-transfer medium that produces steam to power a turbo-generator. Much work is being done to develop more efficient steam turbines that can operate at lower pressures and lower steam temperatures. Combined-cycle solar and fossil-fuel plants are also under development. Future developments project 1-GW solar-power towers being used. Tax credits for solar power will help accelerate its wide-scale adoption.

A promising application of solar energy is combining solar power with conventional combined-cycle (steam- and gas-turbine) power plants. The solar power is integrated with the combined-cycle steam plant in several different ways. Some experts are predicting that combining solar energy with fossil-fuel plants will be the eventual answer to reducing greenhouse gas emissions. Working fluids (steam or oil) in solar systems (sometimes called *farms*) are delivering temperatures as high as 1000°F (538°C), and higher. Steam generated by solar energy in solar farms can be used to (*a*) heat feedwater for the steam boiler; (*b*) be fed to a gas-turbine heat-recovery steam generator (HRSG) to increase the steam output to the steam turbine, thus increasing its electrical output; (*c*) feed solar-generated superheated steam to the main steam turbine in the plant to reduce the fossil-fuel consumption. These, and other designs, promise many fuel savings for combined-cycle plants, while reducing the carbon footprint of such plants. Steam produced by various solar-energy plant designs can range from low-pressure saturated steam to superheated steam, giving designers a wide range of design options.

Solar energy, while more expensive than most other sources of alternative energy, will eventually become an important source of power worldwide. Numerous engineers and scientists are working on ways to reduce the cost of converting solar energy to electricity. A major breakthrough may not be far away.

Thus, at this writing (2011), two 280-MW concentrating solar-power stations are planned for the United States—one in Arizona and one in California. These plants will use parabolic troughs that concentrate the sun's rays onto a heat-absorbing pipe that uses a molten-salt fluid to heat water to generate steam for a turbine-generator. Heat can then be stored in the molten salt that is held in storage tanks for use during the night and on cloudy days when sunshine is not available. Some 6 hours of heat storage will be available from the storage tanks. Each plant will have about 2700 parabolic troughs to collect heat from the sun, and each plant will occupy some 1700 acres (688 hectares) of land. New transmission lines are expected to serve each plant.

SOLAR-POWERED ELECTRIC-GENERATING SYSTEM LOAD AND COST ANALYSIS

Analyze the feasibility of a solar electric generating system (SEGS) for a power system located in a subtropical climate. Compare generating loads and costs with conventional fossil-fuel and nuclear generating plants.

Calculation Procedure:

1. ***Determine when a solar electric generating system can compete with conventional power***

Solar electric generation, by definition, requires abundant sunshine. Without such sunshine, any proposed solar electric generating plant could not meet load demands. Hence, such a plant could not compete with conventional fossil-fuel or nuclear plants. Therefore, solar electric generation is, at this time, restricted to areas having high concentrations of sunshine. Such areas are in both the subtropical and tropical regions of the world.

One successful solar electric generating system is located in the Mojave Desert in southern California. At this writing, it has operated successfully for some 12 years with a turbine-cycle efficiency of 37.5 percent for a solar field of more than 2-million ft^2 (1805.802 m^2). A natural-gas backup system has a 39.5 percent efficiency. Both these levels of efficiency are amongst the highest attainable today with any type of energy source.

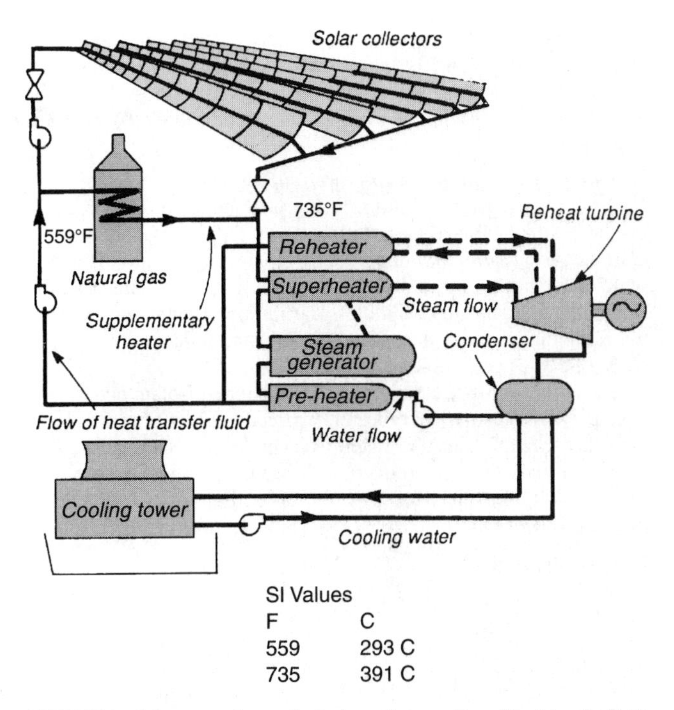

SI Values

F	C
559	293 C
735	391 C

FIGURE 1 Solar-generating-method schematic traces flow of heat-transfer fluid. (*Luz International Ltd.* and *Power*.)

2. *Sketch a typical cycle arrangement*

Technology developed by Luz International Ltd. uses a moderate-pressure state-of-the-art Rankine-cycle steam-generating system using solar radiation as its primary energy source, Fig. 1. In the Mojave Desert plant mentioned above, a solar field comprised of parabolic-trough solar collectors, which individually track the sun using sun sensors and microprocessors, provides heat for the steam cycle.

Collection troughs in the Mojave Desert plant are rear surface mirrors bent into the correct parabolic shape. These specially designed mirrors focus sunlight onto heat-collection elements (HCE). Each mirror is washed every 2 weeks with demineralized water to remove normal dust blown off the desert. The mirrors must be clean to focus the optimal amount of the sun's heat on the HCE.

3. *Detail the sun collector arrangement and orientation*

With the parabolic mirrors described above, sun sensors begin tracking the sun before dawn. Microprocessors prompt the troughs to follow the sun, rotating 180° each day. A central computer facility at the Mojave Desert plant monitors and controls each of the hundreds of individual solar collectors in the field and all of the power plant equipment and systems.

During summer months when solar radiation is strongest, some mirrors must be turned away from the sun because there is too much heat for the turbine capacity. When this occurs, almost every other row of mirrors must be turned away. However, in the winter, when solar radiation is the weakest, every mirror must be employed to produce the required power.

In the Mojave Desert plant, the mirrors focus the collected heat on the HCEs—coated steel pipes mounted inside vacuum-insulated glass tubes. The HCEs contain a synthetic-oil heat-transfer fluid, which is heated by the focused energy to approximately 735°F (390.6°C) and pumped through a series of conventional heat exchangers to generate superheated steam for the turbine-generator.

In the Mojave Desert plant, several collectors are assembled into units called solar collector assemblies (SCA); generally, each 330-ft (100.6-m) row of collectors comprises one SCA. The

SCAs are mounted on pylons and interconnected with flexible hoses. An 80-MW field consists of 852 SCAs arranged in 142 loops. Each SCA has its own sun sensor, drive motor, and local controller, and is comprised of 224 collector segments, or almost 5867 ft^2 (545 m^2) of mirrored surface and 24 HCEs. From this can be inferred that some (5867/80) = 73.3 ft^2 (6.8 m^2) per MW is required at this installation.

4. ***Plan for an uninterrupted power supply***

To ensure uninterrupted power during peak demand periods, an auxiliary natural-gas fired boiler is available at the Mojave Desert plant as a supplemental source of steam. However, use of this boiler is limited to 25 percent of the time by federal regulations. This boiler serves as a backup in the event of rain, for night production when called for, or if "clean sun" is unavailable. According to Luz International, clean sun refers to solar radiation untainted by smog, clouds, or rain. Figure 2 shows the firing modes for typical summer (left) and winter (right) days. Correlation of solar generation to peaking-power requirements is evident.

As shown in the cycle diagram, the balance-of-plant equipment consists of the turbine-generator, steam generator, solar superheater, two-cell cooling tower, and an intertie with the local utility company, Southern California Edison Co. The Mojave Desert installation represents some 90 percent of the world's solar power production. Since installing its first solar electric generating system in 1984, a 13.8-MW facility, Luz has built six more SEGS of 30 MW each. Units 6 and 7 use third-generation mirror technology.

5. ***Determine the costs of solar power***

SEGS are suited to utility peaking service because they provide up to 80 percent of their output during those hours of a utility's greatest demand, with minimal production during low-demand hours.

Cost of Luz's solar-generated power is less than that of many nuclear plants—\$0.08/kWh, down from \$0.24/kWh for the first SEGS, according to company officials. Should the price of oil go up beyond \$90/barrel, solar will become even more competitive with conventional power.

But the advantages over conventional power sources include more than cost-competitiveness. Emissions levels are much lower—10 ppm—because the sun is essentially nonpolluting. SEGS are equipped with the best available technology for emissions cleanup during the hours they burn natural gas, the only time they produce emission.

Related Calculations. Luz International Ltd. has installed more capacity at the Mojave Desert plant mentioned here, proving the acceptance and success of its approach to this important technical challenge. That data in this procedure can be useful to engineers studying the feasibility of solar electric generation for other sites around the world. Luz received an Energy Conservation Award from *Power* magazine, from which the data and illustrations in this procedure were obtained. There are estimates showing that the sunshine impinging the southwestern United States is more than

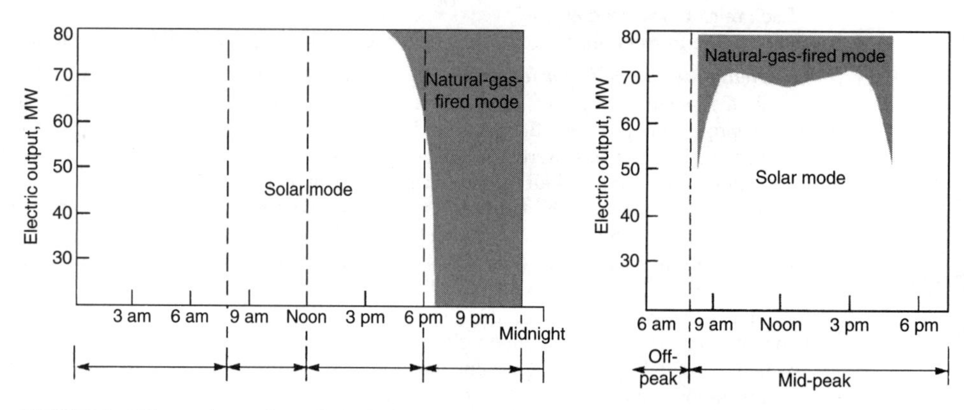

FIGURE 2 Firing modes are shown for typical summer day, left, typical winter day, right. Correlation of solar generation to peaking power requirements is evident. (*Luz International Ltd.* and *Power*.)

enough to generate the entire electrical needs of the country—when efficient conversion apparatus is developed. It may be that the equipment described here will provide the efficiency needed for large-scale pollution-free power generation. Results to date have been outstanding and promise greater efficiency in the future.

INDUSTRIAL SOLAR-ENERGY SYSTEM INVESTMENT ECONOMICS

Determine the rate of return and after-tax present value of a new industrial solar energy system. The solar installation replaces all fuel utilized by an existing fossil-fueled boiler when optimum weather conditions exist. The existing boiler will be retained as an auxiliary unit. Assume a system energy output (E_s) of 3×10^9 Btu/yr (3.17 kJ $\times$ 10^9/yr) an initial cost for the total system of \$503,000 based on a collector area (A_c) of 10,060 ft^2 (934.6 m^2), a depreciation life (DP) of 12 years, a tax rate (τ) of 0.4840, a tax credit (TC) factor of 0.25, a system life of 20 years, an operating cost fraction (OMPI) of 0.0250, an initial fuel cost (P_{f0}) of \$3.11/MBtu (\$3.11/947.9 MJ), and a fuel price escalation rate (e) of 0.1450.

Calculation Procedure:

1. ***Compute unit capacity cost (K_s) in \$/million Btu per year***

$$K_s = \frac{\text{initial cost of system}}{E_s} = \frac{\$503{,}000}{3 \times 10^9 \text{ Btu/yr}}$$

$$= \frac{\$167.67}{1 \text{ million Btu/yr}} (\$167.67/947.9 \text{ MJ/yr}).$$

2. ***Compute levelized coefficient of initial costs (M) over the life of the system***

$$M = \text{OMPI} + \frac{\text{CRF}_{R,N}}{1-\tau}\left[1 - \left(\frac{\text{TC}}{1+R}\right) - (\tau \times \text{DEP})\right]$$

$\text{CRF}_{R,N}$ is the capital recovery factor which is a function of the market discount rate (R)* over the expected lifetime of the system (20 years) and is determined as follows:

$$\text{CRF}_{R,20} = \frac{R}{1-(1+R)^{-20}}$$

DEP is the depreciation which will be calculated by an accelerated method, the sum of the years digits (SOYD), in accordance with the following formula:

$$\text{DEP} = \frac{2}{\text{DP(DP}+1)R}\left(\text{DP} - \frac{1}{\text{CRF}_{R,\text{DP}}}\right)$$

where DP is an allowed depreciation period, or tax life, of 12 years.

Prepare a tabulation (see below) of M values for various market discount rates (R).

3. ***Compute the levelized cost of solar energy (S), for the life cycle of the system in \$/million Btu (\$/MJ)***

Use the relation, $S = (K_s)(M)$. Since M varies with R, refer to the tabulation* of S for various market discount rates.

---

*See tabulation on page 8.6.

4. ***Compute the levelized cost of fuel (F) in \$/million Btu (\$/MJ) and compare to S***

$$F = \frac{P_{f0}}{\eta}\left[\mathrm{CRF}_{R,N}\left(\frac{1+e}{R-e}\right)\left(1-\left[\frac{1+e}{1+R}\right]^N\right)\right]$$

where η is the boiler efficiency for a fossil fuel system which supplies equivalent heat. Referring to the tabulation, the value of F is tabulated at various market discount rates for η values of 70, 80, and 100 percent. The rate of return for the solar installation is that value of R at which $F = S$. For η = 70 percent, R is between 7.5 and 8.0 percent. For η = 80 percent, R is between 6.5 and 7.0 percent. These rates of return should exceed current interest (discount) rates to attain economic feasibility.

5. ***Compute the after-tax present value (PV) of the solar investment if the existing boiler installation has an efficiency of 70 percent***

$$\mathrm{PV} = E_s\left(\frac{1-\tau}{\mathrm{CRF}_{R,N}}\right)(F - S)$$

In order to have a positive value of PV, F must exceed S. Therefore, select a market discount rate (R) from the tabulation which satisfies this criterion. For example, at a 5 percent discount rate,

$$\mathrm{PV} = \frac{3\times 10^9}{10^6}\left(\frac{1-0.484}{0.08024}\right)(20.00 - 13.91) = \$117{,}489.02$$

Note that at 8 percent or higher PV will be negative and the investment proves uneconomical against other investment options.

						F		
R*	$\mathrm{CRF}_{R,N}$	$\mathrm{CRF}_{R,DP}$	$\mathrm{DEP}_{(SOYD)}$	M	S	η = 100%	η = 80%	η = 70%
4.5	0.07688	0.10967	0.8210	0.07915	13.27	14.29	17.86	20.41
5.0	0.08024	0.11283	0.8044	0.08295	13.91	14.00	17.50	20.00
5.5	0.08368	0.11603	0.7882	0.08686	14.56	13.71	17.14	19.59
6.0	0.08718	0.11928	0.7727	0.09093	15.25	13.43	16.79	19.19
6.5	0.09076	0.12257	0.7577	0.09511	15.95	13.16	16.45	18.80
7.0	0.09439	0.12590	0.7431	0.09940	16.67	12.89	16.11	18.41
7.5	0.09809	0.12928	0.7290	0.10382	17.41	12.63	15.79	18.04
8.0	0.10185	0.13270	0.7154	0.10833	18.16	12.38	15.47	17.69

*As used in engineering economics, R, discount rate and interest rate refer to the same percentage. The only difference is that interest refers to a progression in time, and discount to a regression in time. See "Engineering Economics for P.E. Examinations," Max Kurtz, McGraw-Hill, 1966.

Reference

Brown, Kenneth C., "How to Determine the Cost-Effectiveness of Solar Energy Projects," *Power* magazine, March, 1981.

FLAT-PLATE SOLAR-ENERGY HEATING- AND COOLING-SYSTEM DESIGN

Give general design guidelines for the planning of a solar-energy heating and cooling system for an industrial building in the Jacksonville, FL, area to use solar energy for space heating and cooling and water heating. Outline the key factors considered in the design so they may be applied to solar-energy heating and cooling systems in other situations. Give sources of pertinent design data, where applicable.

Calculation Procedure:

1. ***Determine the average annual amount of solar energy available at the site***
Figure 3 shows the average amount of solar energy available, in Btu/(day · ft^2) (W/m^2) of panel area, in various parts of the United States. How much energy is collected depends on the solar panel efficiency and the characteristics of the storage and end-use systems.

Tables available from the National Weather Service and the American Society of Heating, Refrigerating and Air Conditioning Engineers (ASHRAE) chart the monthly solar-radiation impact for different locations and solar insolation [total radiation form the sun received by a surface, measured in Btu/(h · ft^2) (W/m^2); insolation is the sum of the direct, diffuse, and reflected radiation] for key hours of a day each month.

Estimate from these data the amount of solar radiation likely to reach the surface of a solar collector over 1 year. Thus, for this industrial building in Jacksonville, FL, Fig. 3 shows that the average amount of solar energy available is 1500 Btu/(day · ft^2) (4.732 W/m^2).

When you make this estimate, keep in mind that on a clear, sunny day direct radiation accounts for 90 percent of the insolation. On a hazy day only diffuse radiation may be available for collection, and it may not be enough to power the solar heating and cooling system. As a guide, the water temperatures required for solar heating and cooling systems are:

Space heating	Up to 170°F (76.7°C)
Space cooling with absorption air conditioning	From 200 to 240°F (93.3 to 114.6°C)
Domestic hot water	140°F (60°C)

2. ***Choose collector type for the system***
There are two basic types of solar collectors: flat-plate and concentrating types. At present the concentrating type of collector is not generally cost-competitive with the flat-plate collector for normal

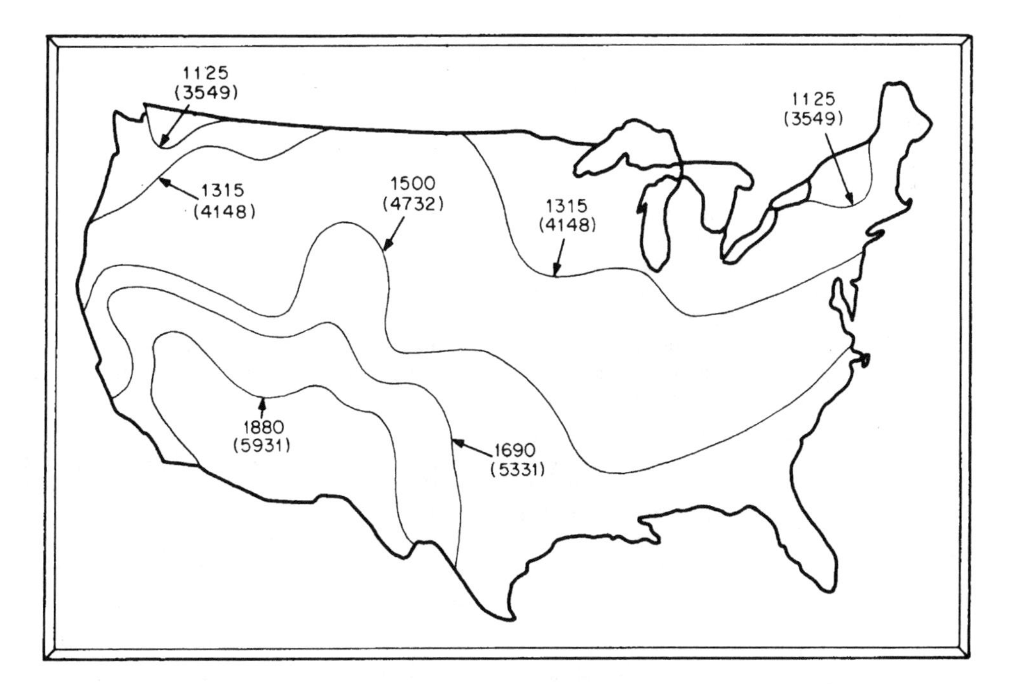

FIGURE 3 Average amount of solar energy available, in Btu/(day · ft^2) (W/m^2), for different parts of the United States. (*Power.*)

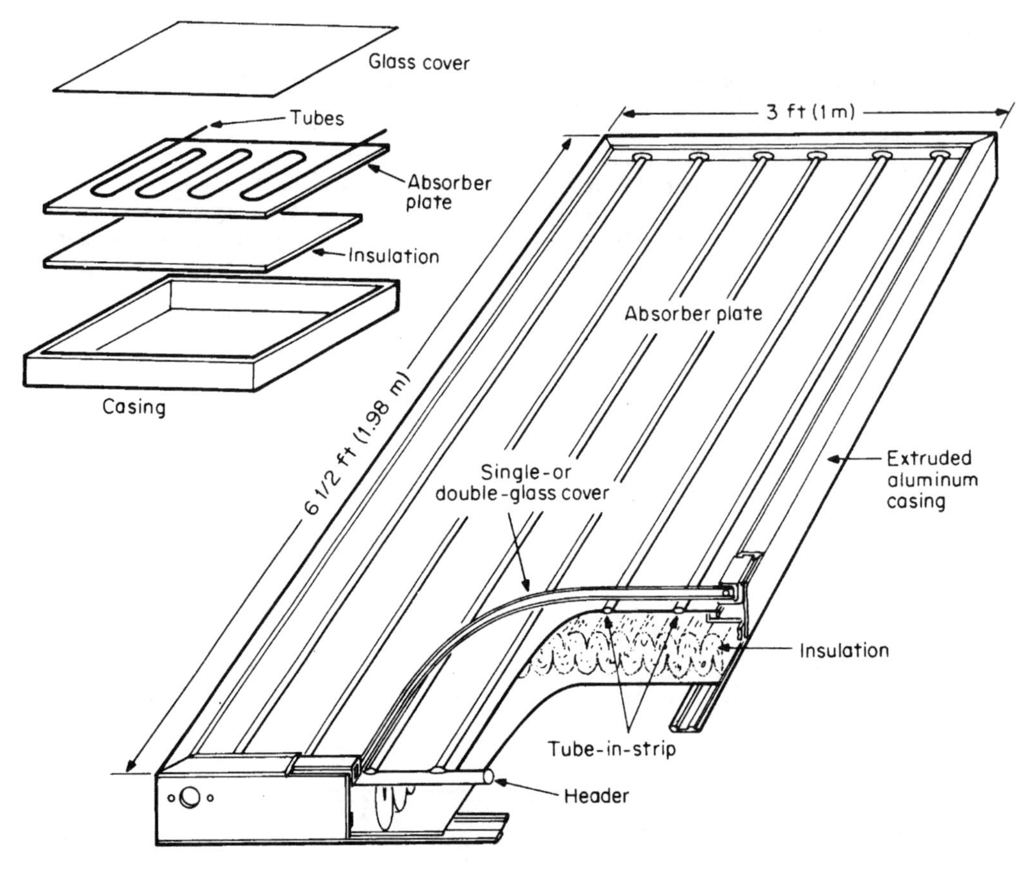

FIGURE 4 Construction details of flat-plate solar collectors. (*Power.*)

space heating and cooling applications. It will probably find its greatest use for high-temperature heating of process liquids, space cooling, and generation of electricity. Since process heating applications are not the subject of this calculation procedure, concentrating collectors are discussed separately in another calculation procedure.

Flat-plate collectors find their widest use for building heating, domestic water heating, and similar applications. Since space heating and cooling are the objective of the system being considered here, a flat-plate collector system will be a tentative choice until it is proved suitable or unsuitable for the system. Figure 4 shows the construction details of typical flat-plate collectors.

3. *Determine the collector orientation*

Flat-plate collectors should face south for maximum exposure and should be tilted so the sun's rays are normal to the plane of the plate cover. Figure 5 shows the optimum tilt angle for the plate for various insolation requirements at different latitudes.

Since Jacksonville, FL, is approximately at latitude 30°, the tilt of the plate for maximum year-round insolation should be 25° from Fig. 5. As a general rule for heating with maximum winter insolation, the tilt angle should be 15° plus the angle of latitude at the site; for cooling, the tilt angle equals the latitude (in the south, this should be the latitude minus 10° for cooling); for hot water, the angle of tilt equals the latitude plus 5°. For combined systems, such as heating, cooling, and hot water, the tilt for the dominant service should prevail. Alternatively, the tilt for maximum year-round insolation can be sued, as was done above.

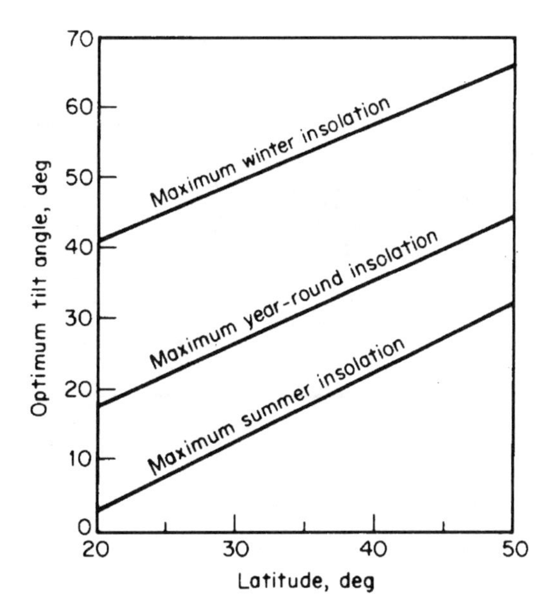

FIGURE 5 Spacing of solar flat-plate collectors to avoid shadowing. (*Power.*)

When collector banks are set in back of one another in a sawtooth arrangement, low winter sun can cause shading of one collector by another. This can cause a loss in capacity unless the units are carefully spaced. Table 1 shows the minimum spacing to use between collector rows, based on the latitude of the installation and collector tilt.

4. *Sketch the system layout*

Figure 6 shows the key components of a solar system using fiat-plate collectors to capture solar radiation. The arrangement provides for heating, cooling, and hot-water production in this industrial building with sunlight supplying about 60 percent of the energy needed to meet these loads—a typical percentage for solar systems.

For this layout, water circulating in the rooftop collector modules is heated to 160°F (71.1°C) to 215°F (101.7°C). The total collector area is 10.000 ft^2 (920 m^2). Excess heated hot water not needed for space heating or cooling or for domestic water is directed to four 6000-gal (22.740-L) tanks for short-term energy storage. Conventional heating equipment provides the hot water needed for heating and cooling during excessive periods of cloudy weather. During a period of 3 hours around noon on a clear day, the heat output of the collectors is about 2 million Btu/h (586 kW), with an efficiency of about 50 percent at these conditions.

For this industrial building solar-energy system, a lithium-bromide absorption air-conditioning unit (a frequent choice for solar-heated systems) develops 100 tons (351.7 kW) of refrigeration for

TABLE 1 Spacing to Avoid Shadowing, ft (m)*

Collector angle, deg	Latitude of installation, deg			
	30	35	40	45
30	9.9 (3.0)	11.1 (3.4)	12.6 (3.8)	13.7 (4.2)
45	10.6 (3.2)	12.2 (3.7)	14.4 (4.4)	16.0 (4.9)
60	10.7 (3.3)	12.6 (3.8)	15.4 (4.7)	17.2 (5.2)

**Power* magazine.

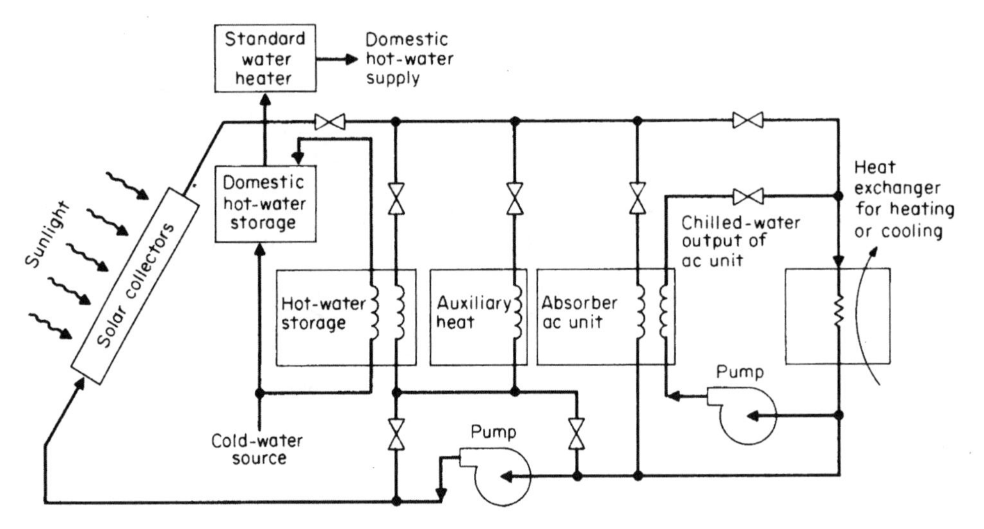

FIGURE 6 Key components of a solar-energy system using flat-plate collectors. (*Power.*)

cooling with a coefficient of performance of 0.71 by using heated water from the solar collectors. Maximum heat input required by this absorption unit is 1.7 million Btu/h (491.8 kW) with a hot-water flow of 240 gal/min (909.6 L/min). Variable-speed pumps and servo-actuated valves control the water flow rates and route the hot-water flow from the solar collectors along several paths—to the best exchanger for heating or cooling of the building, to the absorption unit for cooling of the building, to the storage tanks for use as domestic hot water, or to short-term storage before other usage. The storage tanks hold enough hot water to power the absorption unit for several hours or to provide heating for up to 2 days.

Another—and more usual—type of solar-energy system is shown in Fig. 7. In it a flat-plate collector absorbs heat in a water/antifreeze solution that is pumped to a pair of heat exchangers.

From unit no. 1 hot water is pumped to a space-heating coil located in the duct work of the hot-air heating system. Solar-heated antifreeze solution pumped to unit no. 2 heats the hot water for domestic service. Excess heated water is diverted to fill an 8000-gal (30,320-L) storage tank. This heated water is used during periods of heavy cloud cover when the solar heating system cannot operate as effectively.

5. *Give details of other techniques for solar heating*

Wet collectors having water running down the surface of a tilted absorber plate and collected in a gutter at the bottom are possible. While these "trickle-down" collectors are cheap, their efficiency is impaired by heat losses from evaporation and condensation.

Air systems using rocks or gravel to store heat instead of a liquid find use in residential and commercial applications. The air to be heated is circulated via ducts to the solar collector consisting of rocks, gravel, or a flat-plate collector. From here other ducts deliver the heated air to the area to be heated.

In an air system using rocks or gravel, more space is needed for storage of the solid media, compared to a liquid. Further, the ductwork is more cumbersome and occupies more space than the piping for liquid heat-transfer media. And air systems are generally not suitable for comfort cooling or liquid heating, such as domestic hot water.

Eutectic salts can be used to increase the storage capacity of air systems while reducing the volume required for storage space. But these salts are expensive, corrosive, and toxic, and they become less effective with repeated use. Where it is desired to store thermal energy at temperatures above 200°F (93.3°C), pressurized storage tanks are attractive.

Solar "heat wheels" can be used in the basic solar heating and cooling system in the intake and return passages of the solar system. The wheels permit the transfer of thermal energy from the return to the intake side of the system and offer a means of controlling humidity.

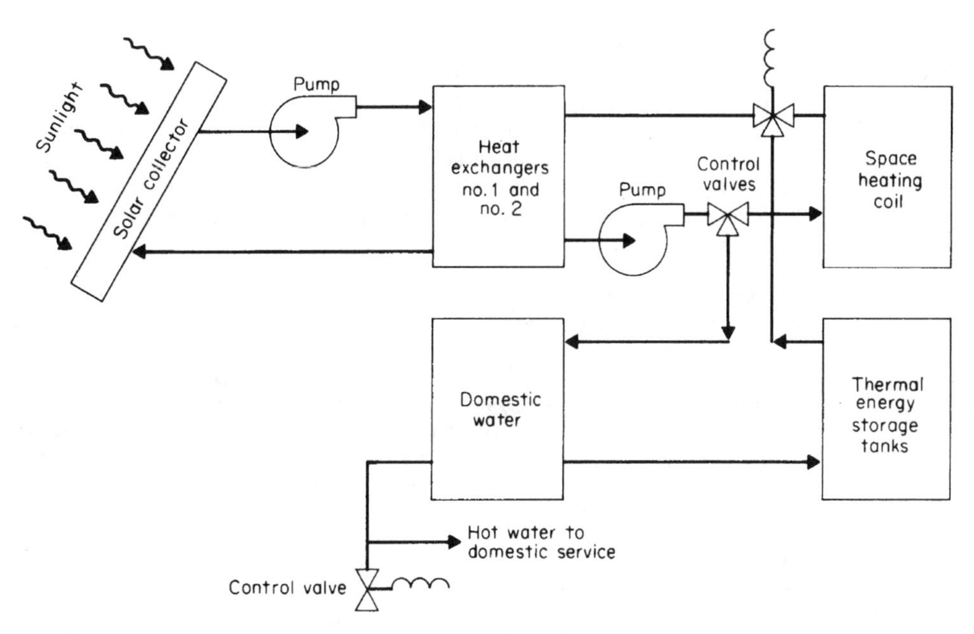

FIGURE 7 Solar-energy system using flat-plate collectors and an antifreeze solution in a pair of heat exchangers. (*Power.*)

For solar cooling, high-performance flat-plate collectors or concentrators are needed to generate the 200 to 240°F (93.3 to 115.6°C) temperatures necessary for an absorption-chiller input. These chillers use either lithium bromide or ammonia with hot water for an absorbent/refrigerant solution. Chiller operation is conventional.

Solar collectors can be used as a heat source for heat-pump systems in which the pump transfers heat to a storage tank. The hot water in the tank can then be used for heating, while the heat pump supplies cooling.

In summary, solar energy is a particularly valuable source of heat to augment conventional space-heating and cooling systems and for heating liquids. The practical aspects of system operation can be troublesome—corrosion, deterioration, freezing, condensation, leaks—but these problems can be surmounted. Solar energy is not "free" because a relatively high initial investment for equipment must be paid off over a long period. And the equipment requires some fossil-fuel energy to fabricate.

But even with these slight disadvantages, the more solar energy that can be put to work, the longer the supply of fossil fuels will last. And recent studies show that solar energy will become more cost-competitive as the price of fossil fuels continues to rise.

6. *Give design guides for typical solar systems*

To ensure the best performance from any solar system, keep these pointers in mind:

a. For space heating, size the solar collector to have an area or, 25 to 50 percent of the building's floor area, depending on geographic location, amount of insulation, and ratio of wall-to-glass area in the building design.

b. For space cooling, allow 250 to 330 ft^2 (23.3 to 30.7 m^2) of collector surface for every ton of absorption air conditioning, depending on unit efficiency and solar intensity in the area. Insulate piping and vessels adequately to provide fluid temperatures of 200 to 240°F (93.3 to 115.6°C).

c. Size water storage tanks to hold between 1 and 2 gal/ft^2 (3.8 to 7.6 L/m^2) of collector surface area.

d. In larger collector installations, gang collectors in series rather than parallel. Use the lowest fluid temperature suitable for the heating or cooling requirements.

e. Insulate piping and collector surfaces to reduce heat losses. Use an overall heat-transfer coefficient of less than 0.04 Btu/(h · ft^2 · °F) [0.23 W/(m^2 · K)] for piping and collectors.

f. Avoid water velocities of greater than 4 ft/s (1.2 m/s) in the collector tubes, or else efficiency may suffer.

g. Size pumps handling antifreeze solutions to carry the additional load caused by the higher viscosity of the solution.

Related Calculations. The general guidelines given here are valid for solar heating and cooling systems for a variety of applications (domestic, commercial, and industrial), for space heating and cooling, and for process heating and cooling, as either the primary or supplemental heat source. Further, note that solar energy is not limited to semitropical areas. There are numerous successful applications of solar heating in northern areas which are often considered to be "cold." And with the growing energy consciousness in all field, there will be greater utilization of solar energy to conserve fossil-fuel use.

Energy experts in many different fields believe that solar-energy use is here to stay. Since there seems to be little chance of fossil-fuel price reductions (only increases), more and more energy users will be looking to solar heat sources to provide some of or all their energy needs. For example, Wagner College in Staten Island, New York, installed, at this writing, 11,100 ft^2 (1032.3 m^2) of evacuated-tube solar panels on the roof of their single-level parking structure. These panels provide heating, cooling, and domestic hot water for two of the buildings on the campus. Energy output of these evacuated-tube collectors is some 3 billion Btu (3.2×10^9 kJ), producing a fuel-cost savings of $25,000 during the first year of installation. The use of evacuated-tube collectors is planned in much the same way as detailed above. Other applications of such collectors include soft-drink bottling plants, nursing homes, schools, etc. More applications will be found as fossil-fuel price increases make solar energy more competitive in the years to come. Table 2 gives a summary of solar-energy collector choices for quick preliminary use.

Data in this procedure are drawn from an article in *Power* magazine prepared by members of the magazine's editorial staff and from Owens-Illinois, Inc.

TABLE 2 Solar-Energy Design Selection Summary

Energy collection device	Typical heat-transfer applications	Typical uses of collected heat
Flat-plate tubed collector	Liquid heating	Space heating or cooling; water heating
Wet collectors	Liquid heating	Water or liquid heating
Rock or gravel collectors	Air heating	Space heating
Flat-plate air heaters	Air heating	Space heating or cooling
Eutectic salts	Air heating	Space heating
Evacuated-tube collectors	Liquid heating to higher temperature levels	Space cooling and heating; process heating
Concentrating collectors	Steam generation	Electric-power generation
Pressurized storage tank	Storage of thermal energy above 200°F (93.3°C)	Industrial processes
Heat wheels	Heat transfer	Humidity control

SOLAR-COLLECTOR SOLAR-INSOLATION COMPUTATION UNDER DIFFERING WEATHER CONDITIONS

A south-facing solar collector will be installed on a building in Glasgow, MT, at latitude 48°13′N. What is the clear-day solar insolation on this panel at 10 a.m. on January 21 if the collector tilt angle is 48°? What is the daily surface total insolation for January 21, at this angle of collector tilt?

Compute the solar insolation at 10:30 a.m. on January 21. What is the actual daily solar insolation for this collector? Calculate the effect on the clear-day daily solar insolation if the collector tilt angle is changed to 74°.

Calculation Procedure:

1. *Determine the insolation for the collector at the specified location*
The latitude of Glasgow, MT, is 48°13′N. Since the minutes are less than 30, or one-half of a degree, the ASHRAE clear-day insolation table for 48° north latitude can be used. Entering Table 3 (which is an excerpt of the ASHRAE table) for 10 a.m. on January 21, we find the clear-day solar insolation on a south-facing collector with a 48° tilt is 206 Btu/(h · ft^2) (649.7 W/m^2). The daily clear-day surface total for January 21 is, from the same table, 1478 Btu/(day · ft^2) (4661.6 W/m^2) for a 48° collector tilt angle.

2. *Find the insolation for the time between tabulated values*
The ASHRAE tables plot the clear-day insolation at hourly intervals between 8 a.m. and 4 p.m. For other times, use a linear interpolation. Thus, for 10:30 a.m., interpolate in Table 3 between 10:00 and 11:00 a.m. values. Or, (249 – 206)/2 + 206 = 227.5 Btu/(h · ft^2) (717.5 W/m^2), where the 249 and 206 are the insolation values at 11 and 10 a.m., respectively. Note that the difference can be either added to or subtracted from the lower, or higher, clear-day insolation value, respectively.

3. *Find the actual solar insolation for the collector*
ASHRAE tables plot the clear-day solar insolation for particular latitudes. Dust, clouds, and water vapor will usually reduce the clear-day solar insolation to a value less than that listed.

To find the actual solar insolation at any location, use the relation $i_A = pi_T$, where i_A = actual solar insolation, Btu/(h · ft^2) (W/m^2): p = percentage of clear-day insolation at the location, expressed as a decimal; i_T = ASHRAE-tabulated clear-day solar insolation, Btu/(h · ft^2) (W/m^2). The value of $p = 0.3 + 0.65(S/100)$, where S = average sunshine for the locality, percent, from an ASHRAE or government map of the sunshine for each month of the year. For January, in Glasgow, MT, the average sunshine is 50 percent. Hence, $p = 0.30 + 0.65(50/100) = 0.625$. Then $i_A = 0.625(1478) = 923.75$, say 923.5 Btu/(day · ft^2) (2913.7 W/m^2), by using the value found in step 1 of this procedure for the daily clear-day solar insolation for January 21.

4. *Determine the effect of a changed tilt angle for the collector*
Most south-facing solar collectors are tilted at an angle approximately that of the latitude of the location plus 15°. But if construction or other characteristics of the site prevent this tilt angle, the effect can be computed by using ASHRAE tables and a linear interpolation.

Thus, for this 48°N location, with an actual tilt angle of 48°, a collector tilt angle of 74° will produce a clear-day solar insolation of i_T = 1578[(74 – 68)/(90 – 68)](1578 – 1478) = 1551.0 Btu/(day · ft^2) (4894.4 W/m^2), by the ASHRAE tables. In the above relation, the insolation values are for solar collector tilt angles of 68° and 90°, respectively, with the higher insolation value for the smaller angle. Note that the insolation (heat absorbed) is greater at 74° than at 48° tilt angle.

Related Calculations. This procedure demonstrates the flexibility and utility of the ASHRAE clear-day solar insolation tables. Using straight-line interpolation, the designer can obtain a number of intermediate clear-day values, including solar insolation at times other than those listed, insolation at collector tilt angles different from those listed, insolation on both normal (vertical) and horizontal planes, and surface daily total insolation. The calculations are simple, provided the designer carefully observes the direction of change in the tabulated values and uses the latitude table for the collector location. Where an exact-latitude table is not available, the designer can interpolate in a linear fashion between latitude values less than and greater than the location latitude.

Remember that the ASHRAE tables give clear-day insolation values. To determine the actual solar insolation, the clear-day values must be corrected for dust, water vapor, and clouds, as shown above. This correction usually reduces the amount of insolation, requiring a larger collector area to produce the required heating or cooling. ASHRAE also publishes tables of the average percentage of sunshine for use in the relation for determining the actual solar insolation for a given location.

TABLE 3 Solar Position and Insolation Values for 48°N Latitude*

Date	Solar time		Solar position		Btu · h/ft² (W/m²) total insolation on surfaces						
							South-facing surface angle with horizontal				
	a.m.	p.m.	Alt.	Azm.	Normal	Horiz.	38	48	58	68	90
Jan 21	8	4	3.5	54.6	37 (116.6)	4 (12.6)	17 (53.6)	19 (59.9)	21 (66.2)	22 (69.4)	22 (69.4)
	9	3	11.0	42.6	185 (583.2)	46 (145.0)	120 (378.3)	132 (416.1)	140 (441.4)	145 (457.1)	139 (438.2)
	10	2	16.9	29.4	239 (753.4)	83 (261.7)	190 (598.9)	206 (649.4)	216 (680.9)	220 (693.6)	206 (649.4)
	11	1	20.7	15.1	261 (822.8)	107 (337.2)	231 (728.2)	249 (784.9)	260 (819.7)	263 (829.1)	243 (766.1)
	12		22.0	0.0	267 (841.7)	115 (362.5)	245 (772.4)	264 (832.3)	275 (866.9)	278 (876.4)	255 (803.9)
	Surface daily totals				1710 (5390.7)	596 (1878.9)	1360 (4287.4)	1478 (4659.4)	1550 (4886.4)	1578 (4974.6)	1478 (4659.4)

*From ASHRAE, 2010, excerpted; used with permission from ASHRAE. Metrication supplied by handbook editor.

COLLECTOR SIZING FOR SOLAR-ENERGY HEATING SYSTEMS FOR BUILDINGS

Select the required collector area for a solar-energy heating system which is to supply 70 percent of the heat for a commercial building situated in Grand Forks, MN, if the computed heat loss is 100,000 Btu/h (29.3 kW), the design indoor temperature is 70°F (21.1°C), the collector efficiency is given as 38 percent by the manufacturer, and collector tilt and orientation are adjustable for maximum solar-energy receipt.

Calculation Procedure:

1. ***Determine the heating load for the structure***

The first step in sizing a solar collector is to compute the heating load for the structure. This is done by using the methods given for other procedures in this handbook in Sec. 13 under Heating, Ventilating and Air Conditioning, and in Sec. 13 under Electric Comfort Heating. Use of these procedures would give the hourly heating load—in this instance, it is 100,000 Btu/h (29.3 kW).

2. ***Compute the energy insolation for the solar collector***

To determine the insolation received by the collector, the orientation and tilt angle of the collector must be known. Since the collector can be oriented and tilted for maximum results, the collector will be oriented directly south for maximum insolation. Further, the tilt will be that of the latitude of Grand Forks, MN, or 48°, since this produces the maximum performance for any solar collector.

Next, use tabulations of mean percentage of possible sunshine and solar position and insolation for the latitude of the installation. Such tabulations are available in ASHRAE publications and in similar reference works. List, for each month of the year, the mean percentage of possible sunshine and the insolation in Btu/(day · ft^2) (W/m^2), as in Table 4.

Using a heating season of September through May, we find total solar energy available from the collector for these months is 103,627.6 Btu/ft^2 (326.9 kW/m^2), found by taking the heat energy per month (= mean sunshine, percent)[total insolation, Btu/(ft^2 · day)](collector efficiency, percent) and summing each month's total. Heat available during the off season can be used for heating water for use in the building hot-water system.

TABLE 4 Solar Energy Available for Heating

Month	Mean sunshine, percent	Total insolation Btu/(ft^2 · day)	Total insolation W/(m^2 · day)	Efficiency of collector, percent	Energy available from collector* Btu/(ft^2 · day)	Energy available from collector* W/(m^2 · day)	Heating-season energy available Btu/(month · ft^2)	Heating-season energy available W/(month · m^2)
Jan.	49	1,478	4,663.1	38	275.2	868.3	8,531.2	26,915.9
Feb.	54	1,972	6,221.7	38	404.7	1,276.8	11,331.6	35,751.2
Mar.	55	2,228	7,029.3	38	465.7	1,469.3	14,436.7	45,547.8
Apr.	57	2,266	7,149.2	38	490.8	1,548.5	14,724.0	46,454.2
May	60	2,234	7,048.3	38	509.4	1,607.2	15,791.4	49,821.9
June	64	2,204	6,953.6	38	NA†	NA	NA	NA
July	72	2,200	6,941.0	38	NA†	NA	NA	NA
Aug.	69	2,200	6,941.0	38	NA†	NA	NA	NA
Sept.	60	2,118	6,682.3	38	482.9	1,523.6	14,487.0	45,706.5
Oct.	54	1,860	5,868.3	38	381.7	1,204.3	11,832.7	37,332.2
Nov.	40	1,448	4,568.4	38	220.1	694.4	6,603.0	20,832.5
Dec.	40	1,250	3,943.8	38	190.0	599.5	5,890.0	18,582.9
						Total = 103,627.6 [326.9 KW/(month · m^2)]		

*Values in this column = (mean sunshine, %)[total insolation, Btu/(ft^2 · day)](efficiency of collector, %).

†Not applicable because not part of the heating season; June, July, and August are ignored in the calculation.

3. ***Find the annual heating season heat load***

Since the heat loss is 100,000 Btu/h (29.3 kW), the total heat load during the 9-month heating season from September through May, or 273 days, is H_a = (24 hours)(273 days)(100,000) = 655,200,000 Btu (687.9 MJ).

4. ***Determine the collector area required***

The calculation in step 2 shows that the total solar energy available during the heating season is 103,627.6 Btu/ft² (326.9 kW/m²). Then the collector area required is A ft² (m²) = H_a/S_a, where S_a = total solar energy available during the heating season, Btu/ft². Or A = 655,200,000/103,627.6 = 6322.64 ft² (587.4 m²) if the solar panel is to supply all the heat for the building. However, only 70 percent of the heat required by the building is to be supplied by solar energy. Hence, the required solar panel area = 0.7(6322.6) = 4425.8 ft² (411.2 m²).

With the above data, a collector of 4500 ft² (418 m²) would be chosen for this installation. This choice agrees well with the precomputed collector sizes published by the U.S. Department of Energy for various parts of the United States.

Related Calculations. The procedure shown here is valid for any type of solar collector—flat-plate, concentrating, or nonconcentrating. The two variables which must be determined for any installation are the annual heat loss for the structure and the annual heat flow available from the solar collector. Once these are known, the collector area is easily determined.

The major difficulty in sizing solar collectors for either comfort heating or water heating lies in determining the heat output of the collector. Factors such as collector tilt angle, orientation, and efficiency must be carefully evaluated before the collector final choice is made. And of these three factors, collector efficiency is probably the most important in the final choice of a collector.

COMPUTING USEFUL ENERGY DELIVERY IN SOLAR HEATING USING THE F-CHART METHOD

Determine the annual heating energy delivery of a solar space-heating system using a double-glazed flat-plate collector if the building is located in Bismarck, ND, and the following specifications apply:

Building

Location: 47°N latitude

Space-heating load: 15,000 Btu/(°F · day) [8.5 kW/(m² · K · day)]

Solar System

Collector loss coefficient: $F_R U_C$ = 0.80 Btu/(h · ft² · °F) [4.5 W/(m² · K)]

Collector optical efficiency (average): $F_R(\overline{\tau\alpha}) = 0.70$

Collector tilt: $\beta = L + 15° = 62°$

Collector area: A_c = 600 ft² (55.7 m²)

Collector fluid flow rate: $\dot{m}_c/A_c$ = 11.4 lb/(h · ft_c^2) (water) [0.0155 kg/(s · m²)]

Collector fluid heat capacity—specific gravity product: c_{pc} = 0.9 Btu/(lb · °F) [3.8 kJ/(kg · K)] (antifreeze)

Storage capacity: 1.85 gal/ft_c^2(water) (75.4 L/m²)

Storage fluid flow rate: $\dot{m}_s/A_c$ = 20 lb/(h · ft²)(water) [0.027 kg/(s · m²)]

Storage fluid heat capacity: c_{ps} = 1 Btu/(lb · °F)(water) [4.2 kJ/(kg · K)]

Heat-exchanger effectiveness: 0.75

Climatic Data

Climatic data from the NWS are tabulated in Table 5.

TABLE 5 Climatic and Solar Data for Bismarck, North Dakota

Month	Average ambient temperature		Heating, degree-days		Horizontal solar radiation, langleys/day
	°F	°C	°F	°C	
Jan.	8.2	−13.2	1761	978	157
Feb.	13.5	−10.3	1442	801	250
Mar.	25.1	−3.8	1237	687	356
Apr.	43.0	6.1	660	367	447
May	54.4	12.4	339	188	550
June	63.8	17.7	122	68	590
July	70.8	21.5	18	10	617
Aug.	69.2	20.7	35	19	516
Sept.	57.5	14.2	252	140	390
Oct.	46.8	8.2	564	313	272
Nov.	28.9	−1.7	1083	602	161
Dec.	15.6	−9.1	1531	850	124

Calculation Procedure:

1. *Determine the solar parameter P_s*

The *F* chart is a common calculation procedure used in the United States to ascertain the useful energy delivery of active solar heating systems. The *F* chart applies only to the specific system designs of the type shown in Fig. 8 for liquid systems and Fig. 9 for air systems. Both systems find wide use today.

The *F*-chart method consists of several empirical equations expressing the monthly solar-heating fraction f_s as a function of dimensionless groups which relate system properties and weather data for a month to the monthly heating requirement. The several dimensionless parameters are grouped into two dimensionless groups called the *solar parameter* P_s and the *loss parameter* P_L.

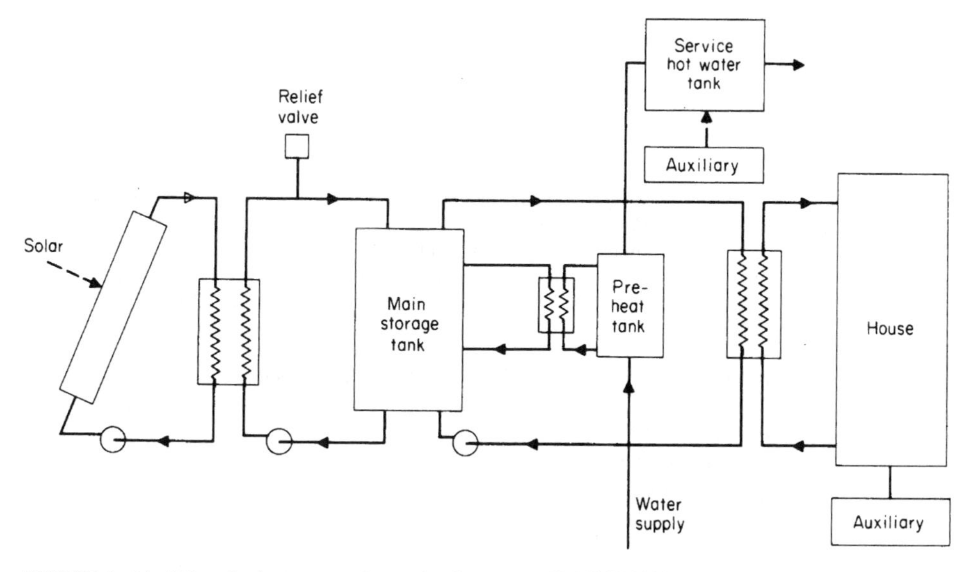

FIGURE 8 Liquid-based solar space- and water-heating system. (*DOE/CS-0011*.)

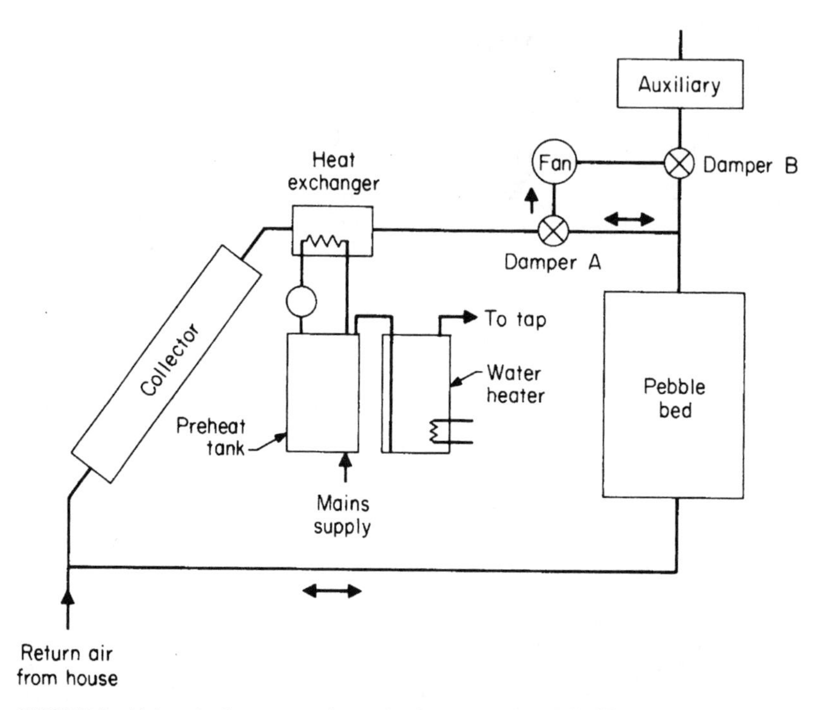

FIGURE 9 Air-based solar space- and water-heating system. (*DOE/CS-0011.*)

The solar parameter P_s is the ratio of monthly solar energy absorbed by the collector divided by the monthly heating load, or

$$P_s = \frac{K_{ldhx} F_{hx} (F_R \overline{\tau\alpha}) \overline{I}_c N}{L}$$

where F_{hx} = heat-exchanger penalty factor (see Fig. 10)
$F_R\tau\alpha$ = average collector optical efficiency = 0.95 × collector efficiency curve intercept $F_R(\tau\alpha)_n$
$\overline{I}_c$ = monthly average insolation on collector surface from a listing of monthly solar and climatic data
N = number of days in a month
L = monthly heating load, *net of any passive system delivery* as calculated by the *P* chart, solar load ratio (SLR), or any other suitable method, Btu/month
K_{ldhx} = load-heat-exchanger correction factor for liquid systems, Table 6

(The *P*-chart method and the SLR method are both explained in Related Calculations below.)

The value of P_s is found for each month of the year by substituting appropriate unit values in the above equation and tabulating the results (Table 7).

2. *Determine the loss parameter P_L*

The loss parameter P_L is related to the long-term energy losses from the collector divided by the monthly heating load:

$$P_L = (K_{\text{stor}} K_{\text{flow}} K_{\text{DHW}}) \frac{F_{hx}(F_R U_c)(T_r - \overline{T}_a)\Delta t}{L}$$

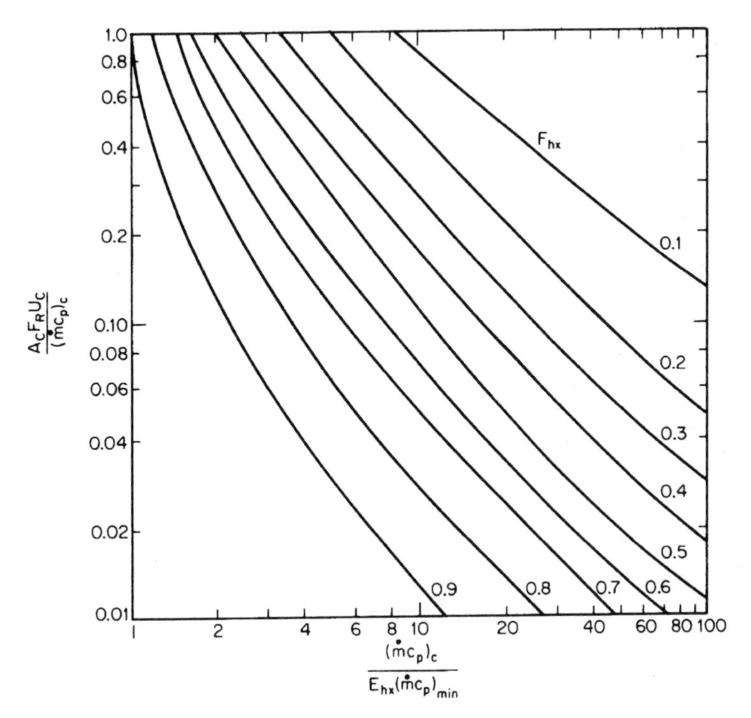

FIGURE 10 Heat-exchanger penalty factor F_{hx}. When no exchanger is present, $F = 1$. (*Kreider*—The Solar Heating Design Process, *McGraw-Hill, 1982.*)

TABLE 6 *F*-Chart *K* Factors

Correction factor	Air or liquid system	Correction factor	Validity range for factor
K_{flow}	A*	$\{[2\ ft^3/(min \cdot ft_c^2)]/\text{actual flow}\}^{-0.28}$ $\{[0.61\ m^3/(min \cdot m^2)]/\text{actual flow}\}^{-0.28}$	$1–4\ ft^3/(min \cdot ft_c^2)$ $[0.03\ to\ 0.11\ m^3/(min \cdot m^2)]$
	L	Small effect included in F_R and F_{hx} only	
K_{stor}	A†	$[(0.82\ ft^3/ft_c^2)/\text{actual volume}]^{0.30}$ $[(0.25\ m^3/m^2)/\text{actual volume}]^{0.30}$	$0.4–3.3\ ft^3/ft_c^2$ $(0.012\ to\ 0.10\ m^3/m^2)$
	L	$[(1.85\ gal/ft_c^2)/\text{actual volume}]^{0.25}$ $[(0.0754\ L/m^2)/\text{actual volume}]^{0.25}$	$(0.9–7.4\ gal/ft_c^2)$ $(36.7\ to\ 301.5\ L/m^2)$
$K_{DHW}^{‡}$	L	$\dfrac{1.18T_{w,o} + 3.86T_{w,i} - 2.32\,\bar{T}_a - 66.2}{212 - \bar{T}_a}$	
$K_{idhx}^{§}$	L	$0.39 + 0.65\exp[-0.139UA/E_L\,(\dot{m}c_p)_{air}]$	$0.5 < \dfrac{E_L(\dot{m}c_p)_{air}}{UA} < 50$
	A	NA	

*User must also include the effect of flow rate in F_R, i.e., in $F_R(\bar{\tau}\bar{\alpha})$ and $F_R U_c$. Refer to manufacturer's data for this.

†For air systems using latent-heat storage, see J. J. Jurinak and S. I. Abdel-Khalik, *Energy,* vol 4, p. 503 (1979) for the expression for K_{stor}.

‡Only applies for liquid storage DHW systems (air collectors can be used, however); $T_{w,o}$ = hot-water supply temperature, $T_{w,i}$ = cold-water supply temperature to water heater, both °F. Applies only to a specific water use schedule; predictions of performance for other schedules will have reduced accuracy.

§UA = unit building heat load, Btu/(h · °F); E_L = load-heat-exchanger effectiveness; $(\dot{m}c_p)_{air}$ = load-heat-exchanger air capacitance rate. Btu/(h · °F) = density × 60 × cubic feet per minute × 0.24.

TABLE 7 *F*-Chart Summary

Month	Collector-plane radiation		Monthly energy demand L		P_L^*	P_s^*	f_s	Monthly delivery	
	Btu/(day · ft²)	W/m²	million Btu	MJ/month				million Btu	MJ/month
Jan.	1506	4749.9	26.41	27.86	2.68	0.72	0.46	12.15	12.82
Feb.	1784	5626.7	21.63	22.82	2.88	0.95	0.60	12.98	13.69
Mar.	1795	5661.4	18.55	19.57	3.50	1.22	0.73	13.54	14.28
Apr.	1616	5096.9	9.90	10.44	5.75	2.00	0.94	9.31	9.82
May	1606	5065.3	5.08	5.36	10.78	>3.00	1.00	5.08	5.36
June	1571	4954.9	1.83	1.93	>20.00	>3.00	1.00	1.83	1.93
July	1710	5393.3	0.27	0.28	>20.00	>3.00	1.00	0.27	0.28
Aug.	1712	5399.6	0.52	0.55	>20.00	>3.00	1.00	0.52	0.55
Sept.	1721	5428.0	3.78	3.99	13.76	>3.00	1.00	3.78	3.99
Oct.	1722	5431.2	8.46	8.93	6.79	2.58	1.00	8.46	8.93
Nov.	1379	4349.4	16.24	17.13	3.79	1.04	0.61	9.91	10.46
Dec.	1270	4005.6	22.96	24.22	2.98	0.70	0.43	9.87	10.41
Annual			135.66	143.1			0.65	87.70	92.52

*$P_s > 3.0$ or $P_L > 20.0$ implies $P_s = 3.0$ and $P_L = 20$, i.e., $f_s = 1.0$. The annual solar fraction f_s is 87.70/135.66, or 6 percent.

where $F_R U_c$ = magnitude of collector efficiency curve slope (can be modified to include piping and duct losses)
$(\bar{T}_a)$ = monthly average ambient temperature, °F (°C)*
Δt = number of hours per month = $24N$
T_r = reference temperature = 212°F (100°C)
K_{stor} = storage volume correction factor, Table 6
K_{flow} = collector flow rate correction factor, Table 6
K_{DHW} = conversion factor for parameter P_L when *only* a water heating system is to be studied, Table 6

3. *Determine the monthly solar fraction*
The monthly solar fraction f_s depends only on these two parameters, P_s and P_L. For liquid heating systems using solar energy as their heat source, the monthly solar fraction is given by

$$f_s = 1.029P_s - 0.065P_L - 0.245P_s^2 + 0.0018P_L^2 + 0.0215P_s^3$$

if $P_s > P_L/12$ (if not, $f_s = 0$).

For air-based solar heating systems, the monthly solar heating fraction is given by

$$f_s = 1.040P_s - 0.065P_L - 0159P_s^2 + 0.00187P_L^2 + 0.0095P_s^3$$

if $P_s > 0.07P_L$ (if not, $f_s = 0$).

Flow rate, storage, load heat-exchanger, and domestic hot-water correction factors for use in the equations for P_s and P_L, in steps 1 and 2 above, are given in Table 11 and Fig. 11.

*Note that $T_a \neq 65°$ – (degree heating days/N), contrary to statements of many U.S. government contractors and in many government reports. The equality is only valid for those months when $T_a < 65$°F every day, i.e., only 3 or 4 months of the year at the most. The errors propagated through the solar industry by assuming the equality to be true are too many to count.

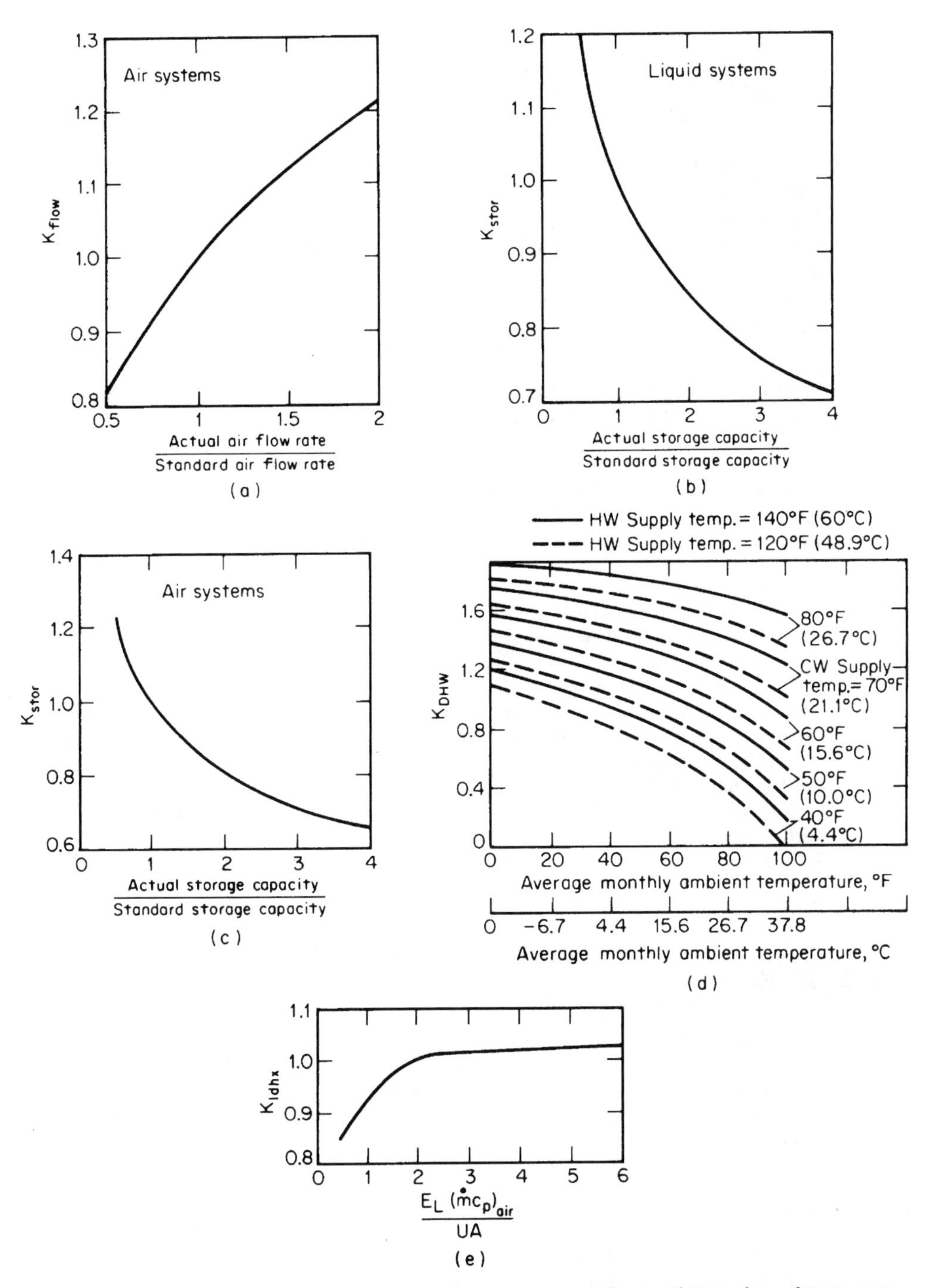

FIGURE 11 *F*-chart correction factors, *(a)* K_{flow} {standard value is 2 standard ft^2/(min · ft^2)[0.01 m^3/(s · m^2)]} *(b)*, *(c)* K_{stor} [standard liquid value is 1.85 gal/ft^2 (75.4 L/m^2); standard rock value is 0.82 ft^3/ft^2 (0.025 m^3/m^2)]. *(d)* K_{DHW}. *(e)* K_{ldhx} (standard value of abscissa is 2.0). (*U.S. Dept. of Housing and Urban Development and Kreider*—The Solar Heating Design Process, *McGraw-Hill, 1982.*)

When you use the F-chart method of calculation for any system, follow this order: collector insolation, collector properties, monthly heat loads, monthly ambient temperatures, and monthly values of P_s and P_L. Once the parameter values are known, the monthly solar fraction and monthly energy delivery are readily calculated, as shown in Table 7.

The total of all monthly energy deliveries is the total annual useful energy produced by the solar system. And the total annual useful energy delivered divided by the total annual load is the annual solar load fraction.

4. *Compute the monthly energy delivery*
Set up a tabulation such as that in Table 7. Using weather data for Bismarck, ND, list the collector-plane radiation, Btu/(day · ft^2), monthly energy demand [= space-heating load, Btu/(°F · day)] (degree days for the month, from weather data), P_L computed from the relation given, P_s computed from the relation given, f_s computed from the appropriate relation (water or air) given earlier, and the monthly delivery found from f_s (monthly energy demand).

Related Calculations. In applying the F-chart method, it is important to use a consistent area basis for calculating the efficiency curve information and the solar and loss parameters, P_s and P_L. The early National Bureau of Standards (NBS) test procedures based the collector efficiency on net glazing area. A more recent and more widely used test procedure developed by ASHRAE (93.77) uses the gross-area basis. The gross area is the area of the glazing plus the area of opaque weather-stripping, seals, and supports. Hence, when ASHRAE test data are used, the solar and loss parameters must be based on gross area. The efficiency curve basis and F-chart basis must be consistent for proper results.

The F-chart method can be used for a number of other solar-heating calculations, including performance of an associated heat-pump backup system, collectors connected in series, etc. For specific steps in these specialized calculation procedures, see Kreider—*The Solar Heating Design Process,* McGraw-Hill. The calculation procedure given here is based on the Kreider book, with numbers and SI units being added to the steps in the calculation by the editor of this handbook.

The P chart mentioned as part of the P_s calculation is a trademark of the Solar Energy Design Corporation. Developed by Arney, Seward, and Kreider for passive predictions of solar performance, the P chart uses only the building heat load in Btu/(°F · day). The P chart will specify the solar fraction and optimum size of three passive systems.

The monthly solar load ratio is an empirical method of estimating monthly solar and auxiliary energy requirements for passive solar systems. For more data on both the P chart and SLR methods, see the Kreider work mentioned above.

SOLAR-COLLECTOR SELECTION FOR DOMESTIC HOT-WATER HEATING SYSTEMS

Select the area for a solar collector to provide hot water for a family of six people in a residential building in Northport, NY, when the desired water outlet temperature is 140°F (60°C) and the water inlet temperature is 50°F (10°C). A pumped-liquid type of domestic hot-water (DHW) system (Fig. 12) is used. Compare the collector area required for 60, 80, and 90 percent of the DHW heading load.

Calculation Procedure:

1. *Find the daily DHW heating load*
A typical family in the United States uses about 20 gal (9.1 L) of hot water per person per day. Hence, a family of six will use a total of 6(20) = 120 gal (54.6 L) per day. Since water has a specific heat of unity (1.0) and weighs 8.34 lb/gal (1.0 kg/dm^3), the daily DHW heating load is L = (120 gal/day)(8.34 lb/gal) [1.0 Btu/(lb · °F)](140 – 50) = 90.072 Btu/day (95,026 kJ/day). This is the 100 percent heating load.

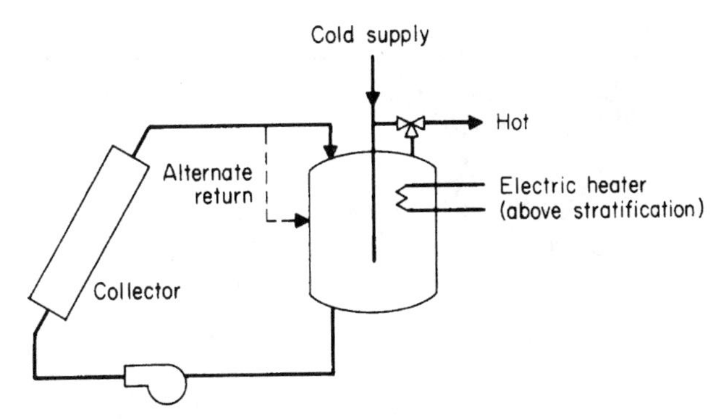

FIGURE 12 Direct-heading pump circulation solar water heater. (*DOE/CS-0011.*)

2. *Determine the average solar insolation for the collector*
Use the month of January because this usually gives the minimum solar insolation during the year, providing the maximum collector area. Using Fig. 13 for eastern Long Island, where Northport is located, we find the solar insolation H = 580 Btu/(ft^2 · day) [1829.3 W/(m^2 · day)] on a horizontal surface. (The horizontal surface insolation is often used in DHW design because it provides conservative results.)

3. *Find the HA/L ratio for this installation*
Use Fig. 14 to find the HA/L ratio, where A = collector area, ft^2 (m^2). Enter Fig. 14 on the left at the fraction F of the annual load supplied by solar energy; project to the right to the tinted area and then vertically downward to the HA/L ratio. Thus, for F = 60 percent, HA/L = 1.0; F = 80 percent, HA/L = 1.5; for F = 90 percent, HA/L = 2.0.

4. *Compute the solar collector area required*
Use the relation HA/L = 1 for the 60 percent fraction. Or, HA/L = 1, A = L/H = 0.6(90,072)/580 = 93.18 ft^2 (8.7 m^2). For HA/L = 1.5 for the 80 percent fraction, A = 0.8(90,072)/580 = 124.24 ft^2 (11.5 m^2). And for the 90 percent fraction, A = 0.9(90,072)/580 = 139.77 ft^2 (12.98 m^2).

Comparing these areas shows that the 80 percent factor area is 33 percent larger than the 60 percent factor area, while the 90 percent factor area is 50 percent larger. To evaluate the impact of the increased area, the added cost of the larger collector must be compared with the fuel that will be saved by reducing the heat input needed for DHW heating.

Related Calculations. Figures 13 and 14 are based on computer calculations for 11 different locations for the month of January in the United States ranging from Boulder, CO, to Boston; New York; Manhattan, KS; Gainesville, FL; Santa Maria, CA; St. Cloud, MN; Washington; Albuquerque, NM; Madison, WI; Oak Ridge, TN. The separate curve above the shaded band in Fig. 14 is the result for Seattle, which is distinctly different from other parts of the country. Hot-water loads used in the computer computations range from 50 gal/day (189.2 L/day) to 2000 gal/day (7500 L/day). The sizing curves in Fig. 14 are approximate and should not be expected to yield results closer than 10 percent of the actual value.

Remember that the service hot-water load is nearly constant throughout the year while the solar energy collected varies from season to season. A hot-water system sized for January, such as that in Fig. 14 with collectors tilted at the latitude angle, will deliver high-temperature water and may even cause boiling in the summer. But a system sized to meet the load in July will not provide all the heat needed in winter. Orientation of the collector can partially overcome the month-to-month fluctuations in radiation and temperature.

Solar-energy water heaters cost from $300 for a roof-mounted collector to over $2000 for a collector mounted on a stand adjacent to the house. The latter are nonfreeze-type collectors fitted with a draindown valve, 50 ft^2 (4.7 m^2) of collector surface area, an 80-gal (302.8-L) water tank, and the

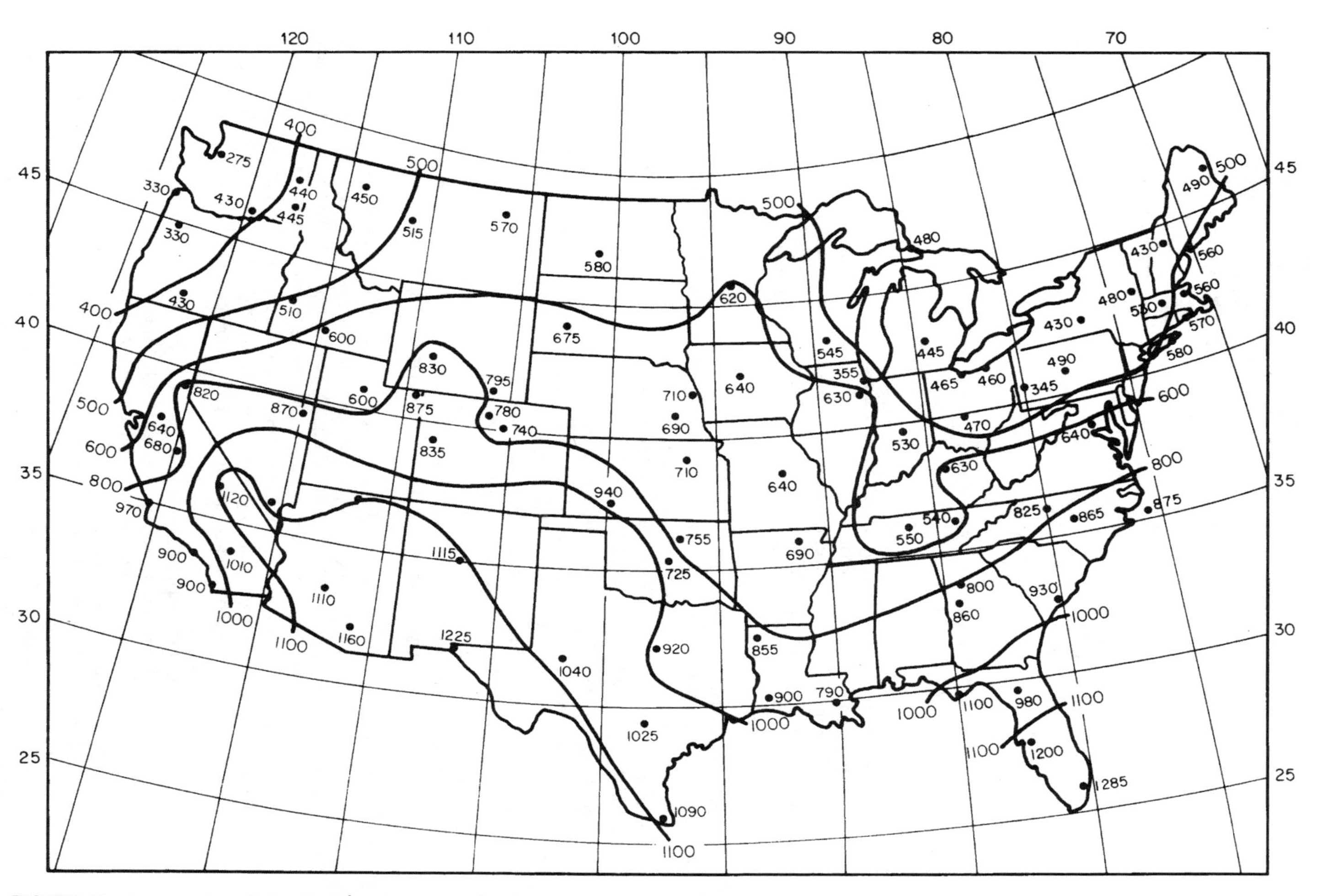

FIGURE 13 Average solar radiation, Btu/ft^2 (× 3.155 = W/m^2), horizontal surface in the month of January. (*DOE/CS-0011.*)

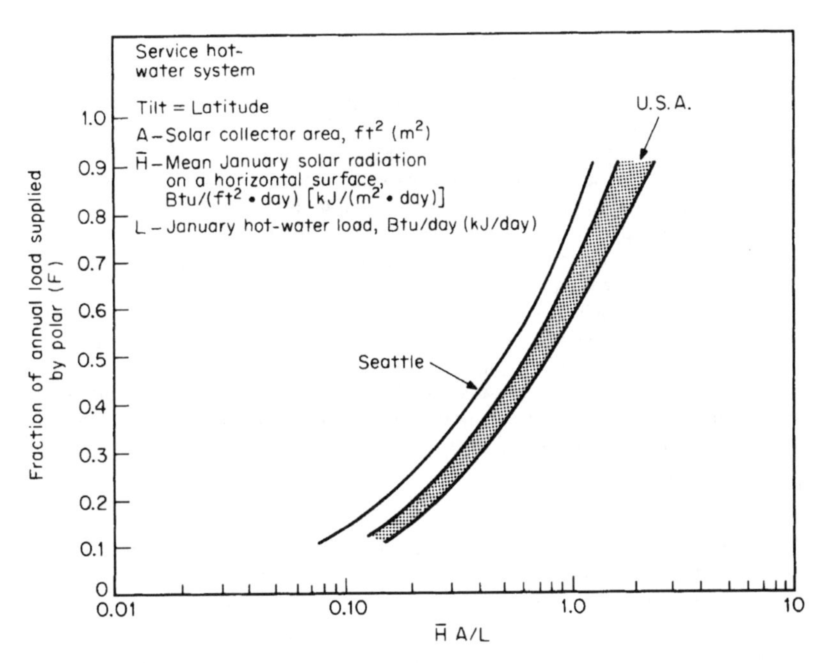

FIGURE 14 Fraction of annual load supplied by a solar hot-water heating system as a function of January conditions. (*DOE/CS-0011.*)

needed pumps and controls. Estimates of the time to recover the investment in such a system can range from as little as 3 to as long as 8 years, depending on the cost of the fuel saved. The charts, used in this procedure, were originally published in DOE/CS-0011, *Introduction to Solar Heating and Cooling—Design and Sizing,* available from the National Technical Information Service, Alexandria, VA, 22312.

DOE/CS-0011 notes that a typical family of four persons requires, in the United States, about 80 gal (302.8 L) of hot water per day. At a customary supply temperature of 140°F (60°C), the amount of heat required if the cold inlet water is at 60°F (15.5°C) is about 50,000 Btu/day (52,750 kJ/day). Further, there is a wide variation in the solar availability from region to region and from season to season in a particular location. There are also the short-term radiation fluctuations owing to cloudiness and the day-night cycle.

Seasonal variations in solar availability result in a 200 to 400 percent difference in the solar heat supply to a hot-water system. In the winter, for example, an average recovery of 40 percent of 1200 Btu/ft^2 (3785 W/m^2) of solar energy of sloping surface would require approximately 100 ft^2 (9.3 m^2) of collector for the 50,000-Btu (52,750-kJ) average daily requirement. Such a design would provide essentially all the hot-water needs on an average winter day, but would fall short on days of less than average sunshine. By contrast, a 50 percent recovery of an average summer radiant supply of 2000 Btu/ft^2 (6308 W/m^2) would involve the need for only 50 ft^2 (4.6 m^2) of collector to satisfy the average hot-water requirements.

If a 50-ft^2 (4.6-m^2) solar collector were installed, it could supply the major part of, or perhaps nearly all, the summer hot-water requirements, but it could supply less than half the winter needs. And if a 100-ft^2 (9.3-m^2) solar collector were used so that winter needs could be more nearly met, the system would be oversized for summer operation and excess solar heat would be wasted. In such circumstances, if an aqueous collection medium were used, boiling in the system would occur and collector or storage venting of steam would have to be provided.

The more important disadvantage of the oversized solar collector (fox summer operation) is the economic penalty associated with investment in a collector that is not fully utilized. Although the cost of the 100-ft^2 (9.3-m^2) solar collector system would not be double that of the 50-ft^2 (4.6-m^2) unit, its annual useful heat delivery would be considerably less than double. It would, of course, deliver about twice as much heat in the winter season, when nearly all the heat could be used. But, in the other seasons, particularly in summer, heat overflow would occur. The net effect of these factors is a lower economic return, per unit of investment, by the larger system. Stated another way, more Btu (kJ) per dollar of investment (hence cheaper solar heat) can be delivered by the smaller system.

If it is sized on average daily radiation in the sunniest months, the solar collector will be slightly oversized and a small amount of heat will be wasted on days of maximum solar input. On partly cloudy days during the warm season, some auxiliary heat must be provided. In the month of lowest average solar energy delivery, typically one-half to one-third as much solar-heated water can be supplied as during the warm season. Thus, fuel requirements for increasing the temperature of solar-heated water to the desired (thermostated) level could involve one-half to two-thirds of the total energy needed for hot-water heating in a midwinter month.

One disadvantage of solar DHW heating systems is the possibility of the water in the collector and associated pipe freezing during unexpectedly cold weather. Since they were introduced on a wide scale, thousands of solar DHW systems have suffered freeze damage, even in relatively warm areas of the world. Such damage is both costly and wasteful of energy.

Three ways are used to prevent freeze damage in solar DHW systems:

1. Pump circulation of warm water through the collector and piping during the night hours reduces the savings produced by the solar DHW heating system because the energy required to run the pump must be deducted from the fuel savings resulting from use of the solar panels.
2. Use an automatic draindown valve or mechanism to empty the system of water during freezing weather. Since the onset of a freeze can be sudden, such systems must be automatic if they are to protect the collector and piping while the occupants of the building are away. Unfortunately, there is no 100 percent reliable draindown valve or mechanism. A number of "fail-safe" systems have frozen during unusually sharp or sudden cold spells. Research is still being conducted to find the completely reliable draindown device.
3. Indirect solar DHW systems use a nonfreeze fluid in the collector and piping to prevent freeze damage. The nonfreeze fluid passes through a heat exchanger wherein it gives up most of its heat to the potable water for the DHW system. To date, the indirect system gives the greatest protection against freezing. Although there is a higher initial cost for an indirect system, the positive freeze protection is felt to justify this additional investment.

There are various sizing rules for solar DHW heating systems. Summarized below are those given by Kreider and Kreith—*Solar Heating and Cooling,* Hemisphere and McGraw-Hill:

Collector area: 1 ft^2/(gal · day) [0.025 m^2/(L · day)]; DHW storage tank capacity: 1.5 to 2 gal/ft^2 (61.1 to 81.5 L/m^2) of collector area; collector water flow rate: 0.025 gal/(min · ft^2) [0.000017 m^3/(s · m^2)]; indirect system storage flow rate: 0.03 to 0.04 gal/(min · ft^2) [0.0002 to 0.00027 m^3/(s · m^2)] of collector area; indirect system heat-exchanger area of 0.05 to 0.1 ft^2/collector ft^2 (0.005 to 0.009 m^2/collector m^2); collector tilt: latitude ±5°; indirect system expansion-tank volume: 12 percent of collector fluid loop; controller turnon ΔT: 15 to 20°F (27 to 36°C); controller turnoff: 3 to 5°F (5.4 to 9°C); system operating pressure: provide 3 lb/in^2 (20.7 kPa) at topmost collector manifold; storage-tank insulation: R-25 to R-30; mixing-valve set point: 120 to 140°F (48.8 to 59.9°C); pipe diameter: to maintain fluid velocity below 6 ft/s (1.83 m/s) and above 2 ft/s (0.61 m/s).

Most domestic solar hot-water heaters are installed to reduce fuel cost. Typically, domestic hot water is heated in an oil-burning boiler or heater. A solar collector reduces the amount of oil needed to heat water, thereby reducing fuel cost. Simple economic studies will show how long it will take to recover the cost of the collector, given the estimated fuel saving.

A welcome added benefit obtained when using a solar collector to heat domestic water is the reduced atmospheric pollution because less fuel is burned to heat the water. All combustion produces carbon dioxide, which is believed to contribute to atmospheric pollution and the possibility of global warming. Reducing the amount of fuel burned to heat domestic water cuts the amount of carbon dioxide emitted to the atmosphere.

Although reduced carbon dioxide emission is difficult to evaluate on an economic basis, it is a positive factor to be considered in choosing a hot-water heating system. With greater emphasis on environmentally desirable design, solar heating of domestic hot water will receive more attention in the future.

DESIGN OF PASSIVE SOLAR-ENERGY HEATING SYSTEMS FOR BUILDINGS

A south-facing passive solar collector will be designed for a one-story residence in Denver, CO. Determine the area of collector required to maintain an average inside temperature of 70°F (21°C) on a normal clear winter day for a corner room 15 ft (4.6 m) wide, 14 ft (4.3 m) deep, and 8 ft (2.4 m) high. The collector is located on the 15-ft (4.6-m) wide wall facing south, and the 14-ft (4.3-m) sidewall contains a 12-ft^2 (1.11 m^2) window. The remaining two walls adjoin heated space and so do not transfer heat. Find the volume and surface area of thermal storage material needed to prevent an unsuitable daytime temperature increase and to store the solar gain for nighttime heating. Estimate the passive solar-heating contribution for an average heating season.

Calculation Procedure:

1. *Compute the heat loss*

The surface areas and the coefficients of heat transmission of collector, windows, doors, walls, and roofs must be known to calculate the conductive heat losses of a space. The collector area can be estimated for purposes of heat-loss calculations from Table 8.

Table 8 lists ranges of the estimated ratio of collector area to floor area, g, of a space for latitudes 36°N or 48°N based on 4°F (2.2°C) intervals of average January temperature and on various types of passive solar collectors. Average January temperatures can be selected from government weather data. Denver has an average January temperature of 32°F (0°C). Choosing a direct-gain system for this installation, read down to the horizontal line for t_o = 32°F (0°C), and then read right to the column for a direct-gain system. To find the estimated ratio of collector area to floor area, use a linear interpolation. Thus for Denver, which is located at approximately 40°N, interpolate between 48°N and 36°N values. Or, $(0.24 - 0.20)/12 \times (40 - 36) + 0.20 = 0.21$, where 0.24 and 0.20 are the ratios at 48°N and 36°N, respectively; 12 is a constant derived from 48 – 36; and 40 is the latitude for which a ratio is sought.

TABLE 8 Estimated Ratio of Collector Area to Floor Area, $g = h_L(65 - t_o)/i_\tau$ for 36 to 48° North Latitude*

Average January temperature T_o °F(°C)	Direct gain g	Water wall g	Masonry wall g
20 (−6.7)	0.27–0.32	0.54–0.64	0.69–0.81
24 (−4.4)	0.25–0.29	0.49–0.58	0.63–0.74
28 (−2.2)	0.22–0.27	0.44–0.52	0.56–0.67
32 (0)	0.20–0.24	0.39–0.47	0.50–0.60
36 (+2.2)	0.17–0.21	0.35–0.41	0.44–0.53
40 (+4.4)	0.15–0.18	0.30–0.35	0.38–0.45
44 (+6.7)	0.13–0.15	0.25–0.30	0.32–0.38

For SI temperature, use the relation $g = h_L(18.33 - t_o)i_\tau$.
*Based on a heat loss of 8 Btu/(day · ft^2 · °F)[0.58 W/(m^2 · K)].

Next, find the collector area by using the relation $A_C = (g)(A_F)$, where A_C = collector area, ft^2 (m^2); g = ratio of collector area to floor area, expressed as a decimal; and A_F = floor area, ft^2 (m^2). Therefore, $A_C = (0.21)(2.10) = 44$ ft^2 (4.1 m^2).

To compute the conductive heat loss through a surface, use the general relation $H_C = UA\ \Delta t$, where H_C = conductive heat loss, Btu/h (W); U = overall coefficient of heat transmission of the surface, Btu/(h · ft^2 · °F)[W/(m^2 · K)]; A = area of heat transmission surface, ft^2 (m^2); and Δt = temperature difference, °F = 65 – t_o (°C = 18.33 – t_o), where t_o = average monthly temperature, °F (°C). The U values of materials can be found in ASHRAE and architectural handbooks.

Since a direct-gain system was selected, the total area of glazing is the sum of the collector and noncollector glazing 44 ft^2 (4.1 m^2) + 12 ft^2 (1.1 m^2) = 56 ft^2 (5.2 m^2). Double glazing is recommended in all passive solar designs and is found to have a U value of 0.42 Btu/(h · ft^2 · °F) [2.38 W/(m^2 · K)] in winter. Thus, the conductive heat loss through the glazing is $H_C = UA\ \Delta t = (0.42)(56)(65 - 32) =$ 776 Btu/h (227.4 W).

The area of opaque wall surface subject to heat loss can be estimated by multiplying the wall height by the total wall length and then subtracting the estimated glazed areas from the total exterior wall area. Thus, the opaque wall area of this space is (8)(15 + 14) – 56 = 176 ft^2 (16.3 m^2). Use the same general relation as above, substituting the U value and area of the wall. Thus, U = 0.045 Btu/(h · ft^2 · °F) [0.26 W/(m^2 · K)], and A = 176 ft^2 (16.3 m^2). Then $H_C = UA\ \Delta t = (0.045)(176)(65 - 32) =$ 261 Btu/h (76.5 W).

To determine the conductive heat loss of the roof, use the same general relation as above, substituting the U value and area of the roof. Thus, U = 0.029 Btu/(h · ft^2 · °F) [0.16 W/(m^2 · K)] and A = 210 ft^2 (19.5 m^2). Then $H_c = UA\ \Delta t = (0.029)(210)(65 - 32) = 201$ Btu/h (58.9 W).

To calculate infiltration heat loss, use the relation $H_i = Vn\ \Delta t/55$, where V = volume of heated space, ft^3 (m^3); n = number of air changes per hour, selected from Table 9. The volume for this space is $V = (15)(14)(8) = 1680$ ft^3 (47.6 m^3). Entering Table 9 at the left for the physical description of the space, read to the right for n, the number of air changes per hour. This space has windows on two walls, so n = 1. Thus, $H_i = (1680)(1)(65 - 32)/55 = 1008$ Btu/h (295.4 W).

TABLE 9 Air Changes per Hour for Well-Insulated Spaces*

Description of space	Number of air changes per hour n
No windows or exterior doors	0.33
Windows or exterior doors on one side	0.67
Windows or exterior doors on two sides	1.0
Windows or exterior doors on three sides	1.33

*These figures are based on spaces with weatherstripped doors and windows or spaces with storm windows or doors. If the space does not have these features, increase the value listed for n by 50%.

The total heat loss of the space is the sum of the individual heat losses of glass, wall, roof, and infiltration. Therefore, the total heat loss for this space is $H_T = 776 + 261 + 201 + 1008 = 2246$ Btu/h (658.3 W). Convert the total hourly heat loss to daily heat loss, using the relation $H_D = 24H_T$, where H_D = total heat loss per day, Btu/day (W). Thus, $H_D = 24(2246) = 53{,}904$ Btu/day (658.3 W).

2. *Determine the daily insolation transmitted through the collector*

Use government data or ASHRAE clear-day insolation tables. The latitude of Denver is 39°50′N. Since the minutes are greater than 30, or one-half of a degree, the ASHRAE table for 40°N is used. The collector is oriented due south. Hence, the average daily insolation transmitted through vertical south-facing single glazing for a clear day in January is i_T = 1626 Btu/ft^2 (5132 W/m^2), or double the half-day total given in the ASHRAE table. Since double glazing is used, correct the insolation transmitted through single glazing by a factor of 0.875. Thus, $i_T = (1626)(0.875) = 1423$ Btu/(day · ft^2) (4490 W/m^2) of collector.

3. *Compute the area of unshaded collector required*
Determine the area of unshaded collector needed to heat this space on an average clear day in January. An average clear day is chosen because sizing the collector for extreme or cloudy conditions would cause space overheating on clear days. January is used because it generally has the highest heating load of all the months.

To compute the collector area, use the relation $A_c = H_D/(E)(i_T)$, where E = a rule of thumb for energy absorptance efficiency of the passive solar-heating system used, expressed as a decimal. Enter Table 10 for a direct gain system to find $E = 0.91$. Therefore, $A_c = 53{,}904/(0.91)(1423) = 42$ ft^2 (3.9 m^2).

TABLE 10 Energy Absorptance Efficiency of Passive Solar-Heating Systems

System	Efficiency E
Direct gain	0.91
Water thermal storage wall or roof pond	0.46
Masonry thermal storage wall	0.36
Attached greenhouse	0.18

If the area of unshaded collector computed in this step varies by more than 10 percent from the area of the collector estimated for heat-loss calculations in step 1, the heat loss should be recomputed with the new areas of collector and opaque wall. In this example, the computed and estimated collector areas are within 10 percent of each other, making a second computation of the collector area unnecessary.

4. *Compute the insolation stored for nighttime heating*
To compute the insolation to be stored for nighttime heating, the total daily insolation must be determined. Use the relation $i_D = (A_C)(i_T)(E)$, where i_D = total daily insolation collected, Btu (J). Therefore, $i_D = (42)(1423)(0.91) = 54{,}387$ Btu (57.4 kJ).

Typically 35 percent of the total space heat gain is used to offset daytime heat losses, requiring 65 percent to be stored for nighttime heating. Therefore, $i_S = (0.65)i_D$, where i_S = insolation stored, Btu (J). Thus, $i_S = (0.65)(54.387) = 35{,}352$ Btu (37.3 kJ). This step is not required for the design of thermal-storage wall systems since the storage system is integrated within the collector.

5. *Compute the volume of thermal storage material required*
For a direct-gain system, use the formula $V_M = i_S/(d)(c_p)(\Delta t_S)(C_S)$, where V_M = volume of thermal storage material, ft^3 (m^3); d = density of storage material, lb/ft^3 (kg/m^3); c_p = specific heat of the material, Btu/(lb · °F) [kJ/(kg · K)]; Δt_S = temperature increase of the material, °F (°C); and, C_S = fraction of insolation absorbed by the material due to color, expressed as a decimal.

Select concrete as the thermal storage material. Entering Table 11, we find the density and specific heat of concrete to be 144 lb/ft^3 (2306.7 kg/m^3) and 0.22 Btu/(lb · °F) [0.921 kJ/(kg · K)], respectively.

TABLE 11 Properties of Thermal Storage Materials

	Density d		Specific heat C_p		Heat capacity	
Material	lb/ft^3	kg/m^3	Btu/(lb · °F)	kJ/(kg · K)	Btu/(ft^3 · °F)	kJ/(m^3 · K)
Water	62.4	999.0	1.00	4184.0	62.40	4180.8
Rock	153	2449.5	0.22	920.5	33.66	2255.2
Concrete	144	2305.4	0.22	920.5	31.68	2122.6
Brick	123	1969.2	0.22	920.5	27.06	1813.0
Adobe	108	1729.1	0.24	1004.2	25.92	1736.6
Oak	48	768.5	0.57	2384.9	27.36	1833.1
Pine	31	496.3	0.67	2803.3	20.77	1391.6

A suitable temperature increase of the storage material in a direct-gain systems is $\Delta t_S = +15°F$ (+8.3°C). A range of +10 to +20°F (+5.6 to 11.1°C) can be used with smaller increases being more suitable. Select from Table 12. In this space, thermal energy will be stored in floors and walls, resulting in a weighted average of $C_S = 0.60$. Thus, $V_M = 35{,}352/(144)(0.22)(15)(0.60) = 124$ ft³ (3.5 m³).

As a rule of thumb for thermal-storage wall systems, provide a minimum of 1 ft³ (0.30 m³) of dark-colored thermal storage material per square foot (meter) of collector for masonry walls or 0.5 ft³ (0.15 m³) of water per square foot (meter) of collector for a water wall. This will provide enough thermal storage material to maintain the inside space temperature fluctuation within 15°F (8.33°C).

TABLE 12 Insolation Absorption Factors for Thermal Storage Material Based on Color

Color/Material	Factor C_S
Black, matte	0.95
Dark blue	0.91
Slate, dark gray	0.89
Dark green	0.88
Brown	0.79
Gray	0.75
Quarry tile	0.69
Red brick	0.68
Red clay tile	0.64
Concrete	0.60
Wood	0.60
Dark red	0.57
Limestone, dark	0.50
Limestone, light	0.35
Yellow	0.33
White	0.18

6. *Determine the surface area of storage material for a direct-gain space*

In a direct-gain system, the insolation must be spread over the surface area of the storage material to prevent overheating. Generally, the larger the surface area of material, the lower the inside temperature fluctuation, and thus the space is more comfortable. To determine this area, enter Fig. 15 at the lower axis to select an acceptable space temperature fluctuation. Project vertically to the curve, and read left to the A_S/A_C ratio. This is the ratio of thermal storage material surface to collector area, where A_S = surface of storage material receiving direct, diffused or reflected insolation, ft² (m²). In this example, 15°F (8.33°C) is selected, requiring $A_S/A_C = 6.8$. Thus, $A_S = (6.8)(42) = 286$ ft² (26.5 m²).

This step is not required for the design of thermal-storage wall systems in which $A_S = A_C$.

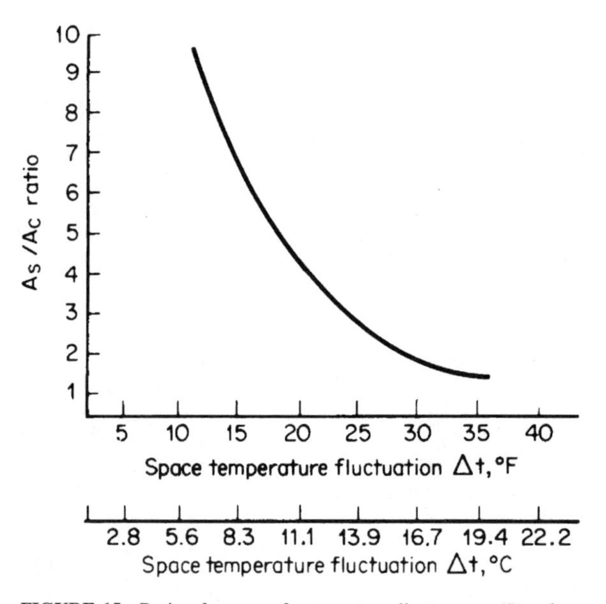

FIGURE 15 Ratio of mass surface area to collector area. (*Based on data in DOE/CS-0127-2* Passive Solar Design Handbook, *vol. 2.*)

7. *Determine the average daily inside temperatures*

To verify that the collector and thermal storage material are correctly sized, the average inside temperature must be determined. Use $t_l = t_a + 5 + (i_D)(65 - t_a)/H_D$, where t_l = average daily inside temperature, °F (°C), and 5°F (2.8°C) is an assumed inside temperature increase owing to internal heat generation such as lights, equipment, and people. Thus, t_l = 32 + 5 + (54,387)(65 – 32)/53,904 = 70.3°F (21.27°C).

To determine the average daily low and high temperatures, use $t_L = t_l - \Delta t/2.5$, and $t_H = t_l, + \Delta t/1.67$, where t_L = minimum average space temperature, °F (°C); t_H = maximum average space temperature, °F (°C); and Δt = inside space temperature fluctuation used in step 6. Thus, t_L = 70.3 – 15/2.5 = 64.3°F (17.9°C), and t_H = 70.3 + 15/1.67 = 79.3°F (26.27°C).

8. *Estimate the passive solar-heating contribution*

To estimate the passive solar-heating contribution (SHC) for an average month, use $SHC_M = 100(i_D)(p)/H_D$, where SHC_M = solar-heating contribution of the total monthly space-heating needs, percent, and p = an insolation factor based on the percentage of clear days, expressed as a decimal. The value of $p = 0.30 + 0.65(S/100)$, where S = average sunshine for the month, percent, from an ASHRAE or government map of sunshine for each month. The average January sunshine for Denver is 67 percent. Hence, p = 0.30 + 0.65(67/100) = 0.74. Thus for this room in January, SHC_M = 100(54,387)(0.74)/53,904 = 74.7 percent of the total average space-heating needs are provided by the passive solar-heating system.

To estimate the average annual solar-heating contribution for a building, repeat steps 1, 2, and 7 for each space for each month of the heating season. Use the collector area computed in step 3 for an average clear day in January to determine i_D for each month unless part of the collector is shaded (in which case, determine the unshaded area and use that figure). Use $SHC_A = 100\Sigma\,(i_D)(p)(D)/\Sigma(H_D)(D)$, where SHC_A = annual passive solar-heating contribution, percent, and D = number of days of the month. The summation of the heat gains for each space for each month of the heating season is divided by the summation of the heat losses for each space for each month.

Related Calculations. These design procedures are suitable for buildings with skin-dominated heat loads such as heat losses through walls, roofs, perimeters, and infiltration. They are not applicable to buildings which have internal heat loads or buildings which are so deep that it is difficult to collect solar heat. Therefore, these procedures generally should be limited to small- and medium-size buildings with good solar access.

These procedures use an average clear-day method as a basis for sizing a passive solar-heating system. Average monthly and yearly data also are used. If the actual weather conditions vary substantially from the average, the performance of the system will vary. For instance, if a winter day is unseasonably warm, the passive solar-heating system will collect more heat than is required to offset the heat loss on that day, possibly causing space overheating. Since passive solar-heating systems rely on natural phenomena, temperature fluctuation and variability in performance are inherent in the system. Adjustable shading, reflectors, movable insulation, venting mechanisms, and backup heating systems are often used to stabilize system performance.

Since passive systems collect, store, and distribute heat through natural physical means, the system must be integrated with the architectural design. The actual efficiency of the system is highly variable and dependent on this integration within the architectural design. Efficiency ratings given in this procedure are rules of thumb. Detailed analyses of many variables and how they affect system performance can be found in DOE/CS-0127-2 and 3, *Passive Solar Design Handbook,* volumes 2 and 3, available from the National Technical Information Service, Alexandria, VA, 22312. *The Passive Solar Energy Book,* by Edward Mazria, available from Rodale Press, Emmaus, PA, examines various architectural concepts and how they can be utilized to maximize system performance.

If thermal collection and storage to provide heating on cloudy days is desired, the collector area can be oversized by 10 percent. This necessitates the oversizing of the thermal storage material to store 75 percent of the total daily heat gain rather than 65 percent, as used in step 4. Oversizing the system will increase the average inside temperature. Step 7 should be used to verify that this higher average temperature is acceptable. Oversizing the system for cloudy-day storage is not recommended for excessively hazy or cloudy climates. Cloudy climates do not have enough clear days in a row to accumulate reserve heat for cloudy-day heating. This increased collector area may increase heat load in these climates. Cloudy-day storage should be considered only for climates with a ratio of several clear days to each cloudy day.

Passive solar-heating systems may overheat buildings if insolation reaches the collector during seasons when heating loads are low or nonexistent. Shading devices are recommended in passive solar-heated buildings to control unwanted heat. Shading devices should allow low-angle winter insolation to penetrate the collector but block higher-angle summer insolation. The shading device should allow enough insolation to penetrate the collector to heat the building during the lower-heating-load seasons of autumn and spring without overheating spaces. If shading devices are used, the area of unshaded collector must be calculated for each month to determine i_D. Methods to calculate the area of unshaded collector can be found in *The Passive Solar Energy Book* and in *Solar Control and Shading Devices,* by V. and A. Olgyay, available from Princeton University Press.

Passive solar-heating systems should be considered only for tightly constructed, well-insulated buildings. The cost of a passive system is generally higher than that of insulating and weatherstripping a building. A building that has a relatively small heat load will require a smaller collection and storage system and so will have a lower construction cost. The cost-effectiveness of a passive solar-heating system is inversely related to the heat losses of the building. Systems which have a smaller ratio of collector area to floor area are generally more efficient.

Significant decreases in the size of the collector can be achieved by placing movable insulation over the collector at night. This is especially recommended for extremely cold climates in more northern latitudes. If night insulation is used, calculate heat loss for the uninsulated collector for 8 hours with the daytime average temperature and for the insulated collector for 16 hours with the nighttime average temperature.

Table 8 is based on a heat loss of 8 Btu/(day · ft^2) of floor area per °F [W/(m^2 · K)]. Total building heat loss will increase with the increase in the ratio of collector to floor area because of the larger areas of glazing. However, it is assumed that this increase in heat loss will be offset by providing higher insulation values in noncollector surfaces. The tabulated values correspond to a residence with a compact plan, 8-ft-high ceilings, R-30 roof insulation, R-19 wall insulation, R-10 perimeter insulation, double glazing, and one air change per hour. It is provided for estimating purposes only. If the structure under consideration differs, the ratio of collector area to floor area, g, can be estimated for heat-loss calculations by using $g = h(65 - t_o)/i_T$, where h_L = estimated heat loss, Btu/(day · ft^2 · °F) [W/(m^2 · K)].

Passive solar heating is nonpolluting and is environmentally attractive. Other than the pollution (air, stream, and soil) possibly created in manufacturing the components of a passive solar heating system, this method of space heating is highly desirable from an environmental standpoint.

Solar heating does not provide carbon dioxide, as does the combustion of coal, gas, oil, and wood. Thus, there is no accumulation of carbon dioxide in the atmosphere from solar heating. It is the accumulated carbon dioxide in the earth's atmosphere that traps heat from the sun's rays and earth reradiation that leads to global warming.

Computer models of the earth's atmosphere and the warming that might be caused by excessive accumulation of carbon dioxide show that steps must be taken to control pollution. Although there is some disagreement about the true effect of carbon dioxide on global warming, most scientists believe that efforts to reduce carbon dioxide emissions are worthwhile. Both a United Nations scientific panel and research groups associated with the National Academy of Sciences recommend careful study and tracking of the possibility of global warming.

For these reasons solar heating will receive more attention from designers. With more attention being paid to the environment, solar heating offers a nonpolluting alternative that can easily be incorporated in the design of most buildings.

SOLAR WATER HEATER ENERGY SAVINGS COMPUTATION

An engineer for a small office building of 2000 ft^2 (186 m^2) area housing 12 people 250 days a year believes the building can more economically heat its required domestic hot water with a 6-ft^2 (0.56-m^2) solar collector water heater. Is this true?

Calculation Procedure:

1. ***Assemble the data needed to analyze the water heating needs***
Data needed for this procedure: (*a*) daily hot-water consumption in gallons (L) per person; (*b*) total solar insolation incident on a vertical surface in the area of the building in Btu/(yr · ft^2) [kJ/(yr · m^2)]; (*c*) percent annual sunshine for the building location; (*d*) difference between incoming city water temperature and the hot-water service temperature *e* (= 100°F [180°C]) for this building. For the typical office building, you can use data from the ASHRAE *Guide.* Assembling the data gives: (*a*) = 1 gal (3.79 L) per person per day; (*b*) 732,000 Btu/(yr · ft^2) [8345 MJ/(yr · m^2)]; (*c*) 59 percent; (*d*) 100°F (180°C).

2. ***Determine if a solar water heater will save energy***
The annual water consumption for this building is 1 gal (3.79 L) per person per day (12 people) (250 days per year) = 3000 gal/yr (11,370 L/yr). The energy required to heat this water, *H* Btu/yr = (gal consumed/yr)(required temperature rise, °F) 8.3, where the constant 9.3 converts gallons to pounds. Or, $H = 3000(100)(8.3) = 2{,}490{,}000$ Btu/yr (2627 MJ/yr).

The usable energy from the solar water heater, *S* = (percent annual sunshine)(solar insolation on collector surface, Btu/yr · ft^2)(solar collect heating absorbing area, ft^2); or $S = 0.59\ (732{,}000)(6) = 2{,}591{,}280$ Btu/yr (2734 MJ/yr).

Comparing the heat required to warm the domestic water for this building, 2490.000 Btu/yr, versus the heat available from the solar water heater, 2,591,280 Btu/yr, shows that the solar water heater could be substituted for the existing water heater in this building. The energy costs saved in heating the water can be used to pay for the solar collector. Figure 16 shows a possible arrangement for a solar water heater used as a booster heater to reduce water heating costs in an occupied building.

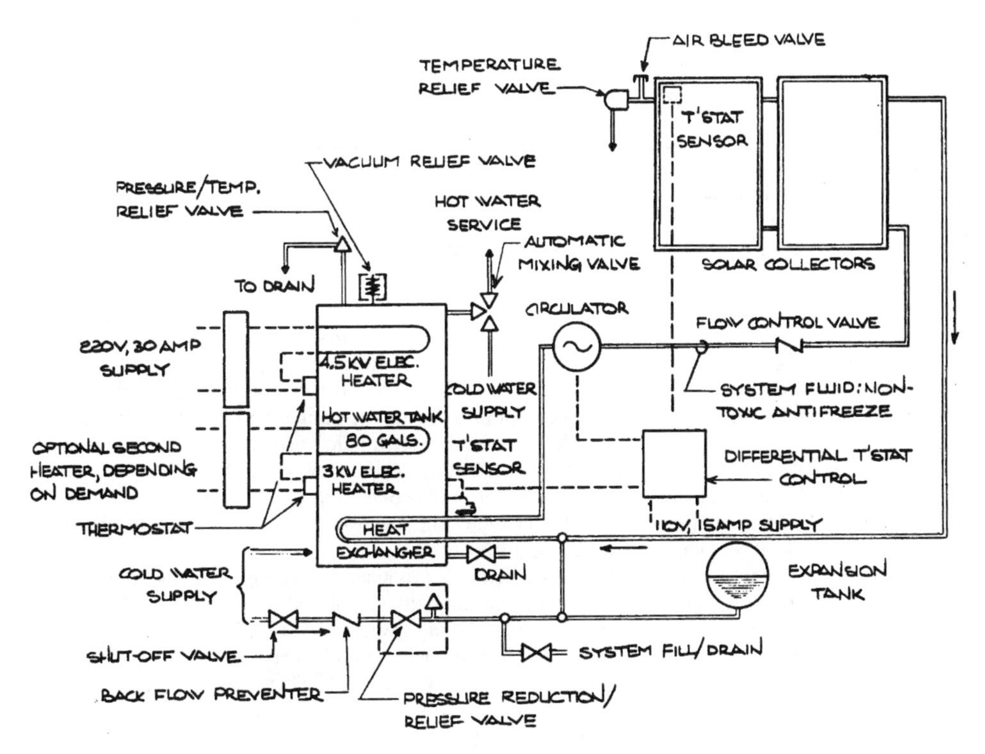

FIGURE 16 Booster solar heater uses nonfreeze glycol solution that is pumped through the heat exchanger in the hot-water tank. (*Jerome F. Mueller.*)

Related Calculations. Use this procedure to analyze the suitability of a substitute water-heating method in any type of building: office, commercial, residential, medical, health-care, hospital, etc. When evaluating a solar water heater, be certain to use accurate data on the availability of sunshine in the area of the structure, and the solar insolation for the area.

This procedure is the work of Jerome F. Mueller, P.E., Mueller Engineering Corp.

CHOOSING PHOTOVOLTAIC MODULES FOR ELECTRICAL LOADS IN BUILDINGS

A small home has the electrical load shown in the tabulation below. Select suitable photovoltaic modules to serve the electrical load for two locations: Albuquerque, NM and Pittsburgh, PA. Show how this load would be serviced by the modules.

Calculation Procedures:

1. *Detail the electrical load of the building*

Construct a tabulation such as that below for the building being considered.

Load	Daily usage, h		Wattage		Energy consumption, Watt-hour (W-h)
Radio	2	×	25	=	50
Television	6	×	60	=	360
Lamps, fluorescent	3	×	27	=	81
VCR	0.5	×	30	=	15
Total daily energy consumption = 506 W-h					

Thus, for this building, a photovoltaic system producing an average daily energy output of 506 W-h is required. Since sunlight (solar energy) is the source of power for photovoltaic systems, we must determine the daily amount of sunlight in the location of the structure. Because photovoltaic systems are rated by peak Watt output, the electricity produced at noon on a clear day is of critical importance in system design.

2. *Show how the system will be hooked up for each location*

Figure 17 shows the typical hookup for a photovoltaic system serving a building. The module contains solar cells which convert the sun's rays to electricity. An inverter and storage batteries are used to change the direct current from the cells to alternating current and to store the direct current.

When the sunlight strikes a typical photovoltaic or solar cell, Fig. 18, photons are absorbed and electrons are freed to flow as shown. Each cell consists of a thin layer of phosphorus-doped silicon in close contact with a layer of boron-doped silicon. By chemically treating silicon in this manner, a permanent electric field is created. The electrons freed when sunlight hits the cell flow through metal contacts to generate electricity.

Photovoltaics, defined as electricity produced solely from the energy of the sun, is modular in design. It can grow with the electrical demands placed on it. An individual cell will produce only a small amount of electricity, but placing several cells together increases the amount of electricity produced. Groups of cells are mounted on a rigid plate and are electrically interconnected to form a photovoltaic *module* which produces a nominal 12 V. Then, groups of modules are mounted together on a permanently attached frame to form a *panel.* Panels are interconnected to form a photovoltaic *array* for differing power levels, Fig. 19.

Photovoltaic systems are a useful form of renewable energy. Thus, the modules have no moving parts, are east to install, require little maintenance, contain no fluids, consume no fuel, produce no pollution, and have a long life span—more than 20 years—and they are equally or more reliable than

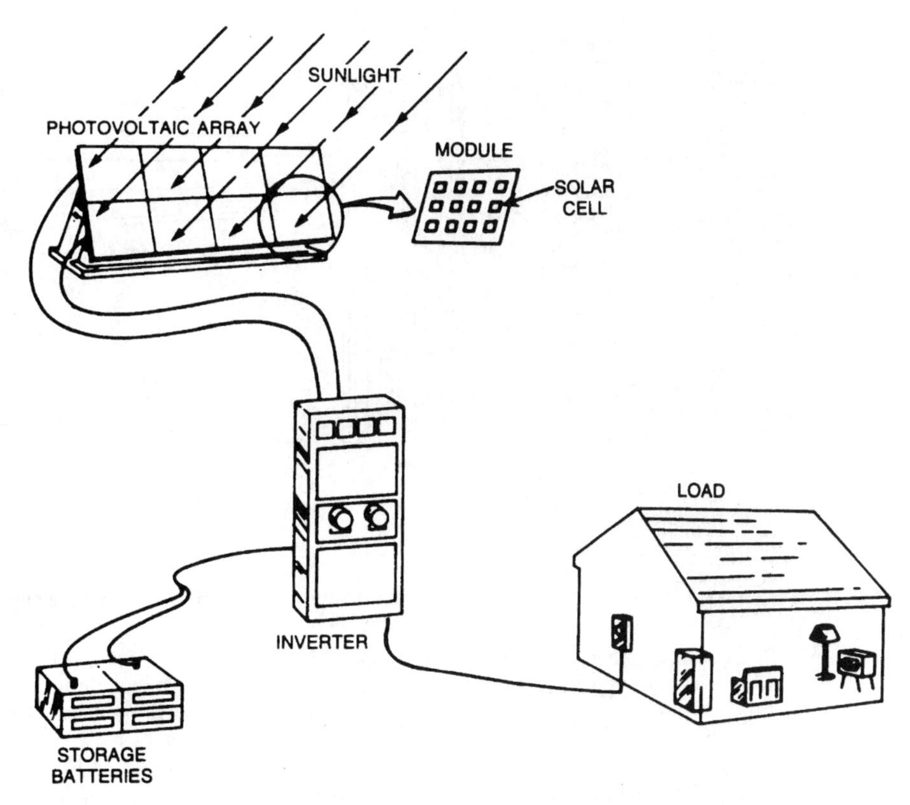

FIGURE 17 Photovoltaic array for providing electricity to a residence. (*U.S. Department of Energy.*)

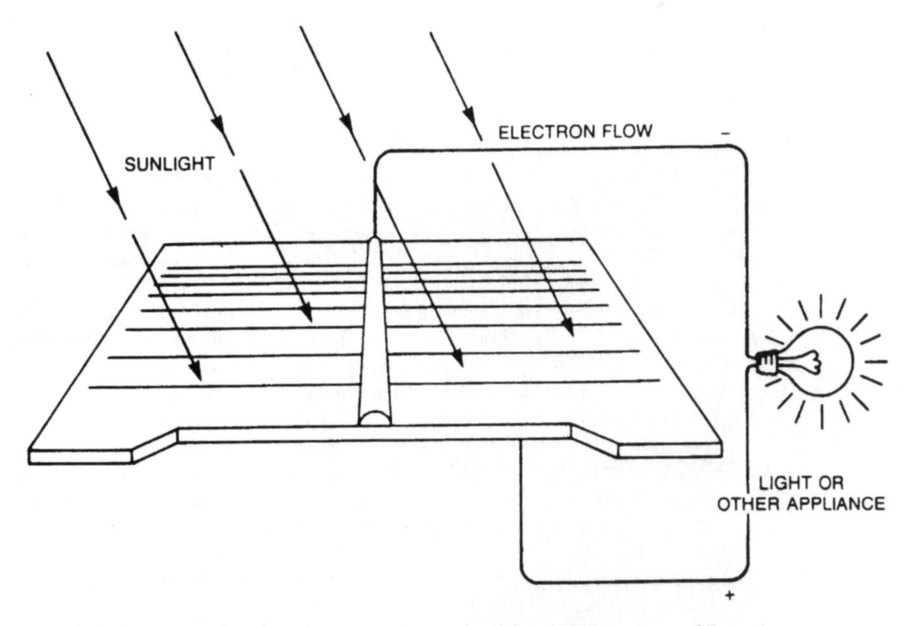

FIGURE 18 Electron flow from the array generates electricity. (*U.S. Department of Energy.*)

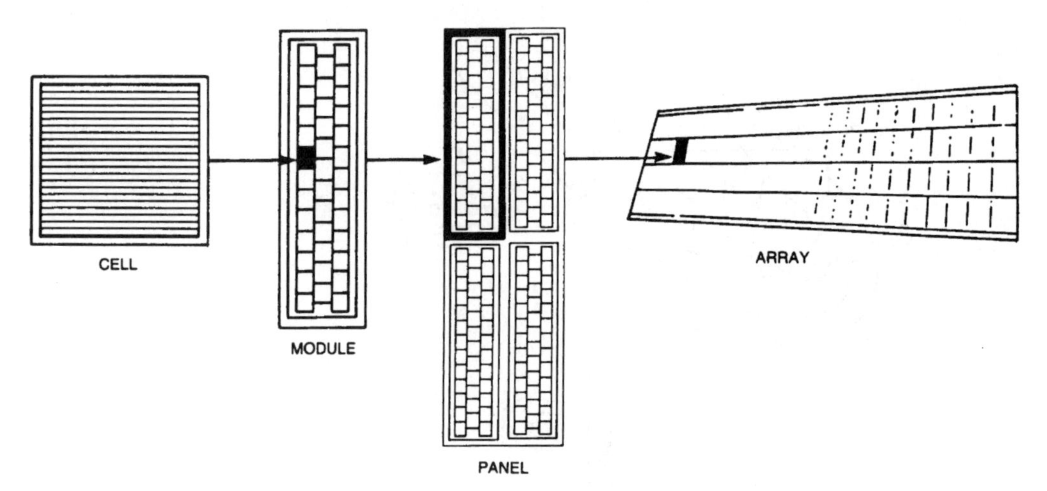

FIGURE 19 Steps in the interconnection of a cell to a module to a panel to an array. (*U.S. Department of Energy.*)

competing power sources, such as batteries and generators. However, components, such as the load and controls, also affect the reliability of the entire system.

3. *Determine the number of modules required*

Reference data show that the winter average peak wattage for a module in Albuquerque, NM, is 6.1 W-h; for Pittsburgh, PA, a cloudier area, the same module will produce 2.4 W-h. To determine the peak Watts of the system, divide the total energy consumption by the W-h for the location. Then divide the result by 0.8 to account for inefficiency; multiply the result by 1.2 (a 20 percent factor of safety) to account for any items you might have overlooked.

Thus, for Albuquerque, NM, we have (506 W-h/6.1/0.8)(1.2) = 124.4 peak Watts. If we're using 50-W modules, a standard and convenient size, we would need 124.4/50 = 2.488 modules. Since modules are made only in whole sizes, three modules would be used for this application.

For Pittsburgh, PA, following the same procedure, we have: (506 W-h/2.4/0.8)(1.2) = 316.3 peak Watts. Using 50-W modules, we have 316.3/50 = 6.32 modules; seven would be chosen.

For precise accuracy in module selection, designers suggest using month-by-month averages for average W-h for a location. Then the lowest monthly average is actually used in the calculations.

Related Calculations. Today, the most common applications for photovoltaics are remote or stand-alone systems. A stand-alone system is not connected to the local utility and is used in remote areas of the world where there are no nearby power lines. Thus, some 50,000 homes, worldwide, are powered by photovoltaics. Connecting such remote homes to a utility costs much more than generating electricity from the sun.

Developing nations use photovoltaics to refrigerate medicines, pump water, or light villages. Developed countries use photovoltaics to light roads, billboards, and road signs. Recreational vehicles and boats also use solar energy to operate various devices.

Other uses of solar power include railway cabooses, airport emergency lighting systems, marine buoys and coastal markers, and electric carts for transporting sightseers.

Hybrid photovoltaic systems use two or more power-generating sources connected together. The secondary source for the solar modules can range from gas or diesel generators to wind generators or hydroelectric generators. However, the most common secondary source is the gas or diesel generator. In the hybrid system, the secondary generator automatically starts when the battery voltage of the photovoltaic array drops below a safe level. By using a hybrid system, the solar energy array can be reduced in size. This allows the secondary source to supply the remaining needed energy during times of minimal sunshine caused by seasonal or weather factors.

Currently, the price of a peak Watt is in the \$3 to \$5 range. As greater use is made of photovoltaics, the price per peak Watt is expected to decline. Since this source of power is more expensive than conventional sources, photovoltaics is still confined to specialized uses.

Data in this procedure are from the U.S. Department of Energy Conservation and Renewable Energy Inquiry and Referral Service.

As the prices of various fuels—oil, coal, gas, etc.—continue to rise, solar energy increases in importance as an alternative energy source. Coupled with the rise in fuel prices, solar energy apparatus continues to fall in price while its efficiency increases. Hence, the future of solar energy as a renewable alternative energy source continues to improve.

More architects and building designers are situating and orienting buildings to take advantage of solar energy. And the green building movement enhances the attractiveness of solar energy.

Studies by private groups and government agencies continue to show that the green building initiative is both economically and environmentally attractive. A dollar spent in making a building green is rewarded by lifetime savings of 2 or 3 times this amount.

Solar energy apparatus can be located on building roofs in otherwise unutilized spaces. Energy delivered to the building cuts utility costs while saving fuel. And with ice or other storage systems for air conditioning, solar energy can become a major green contributor in many buildings, large and small. Hence, mechanical engineers will find many more solar designs crossing their desks in the years ahead.

DESIGNING SOLAR-POWERED PUMPING SYSTEMS

Devise a solar-powered alternative energy source for driving pumps for use in irrigation to handle 10,000 gal/min (37.9 m^3/min) at peak output with an input of 50 hp (37.3 kW). Show the elements of such a system and how they might be interconnected to provide useful output.

Calculation Procedure:

1. *Develop a suitable cycle for this application*
Figure 20 shows a typical design of a closed-cycle solar-energy powered system suitable for driving turbine-powered pumps. In this system a suitable refrigerant is chosen to provide the maximum heat absorption possible from the sun's rays. Water is pumped under pressure to the solar collector, where it is heated by the sun. The water then flows to a boiler where the heat in the water turns the liquid refrigerant into a gas. This gas is used to drive a Rankine-cycle turbine connected to an irrigation pump, Fig. 20.

The rate of gas release in such a closed system is a function of *(a)* the unit enthalpy of vaporization of the refrigerant chosen, (*b*) the temperature of the water leaving the solar collector, and (*c*) the efficiency of the boiler used to transfer heat from the water to the refrigerant. While there will be some heat loss in the piping and equipment in the system, this loss is generally considered negligible in a well-designed layout.

2. *Select, and size, the solar collector to use*
The usual solar collector chosen for systems such as this is a parabolic tracking-type unit. The preliminary required area for the collector is found by using the rule of thumb which states: For parabolic tracking-type solar collectors the required sun-exposure area is 0.55 ft^2 per gal/min pumped (0.093 m^2 per 0.00379 m^3/min) at peak output of the pump and collector. Another way of stating this rule of thumb is: Required tracking parabolic solar collector area = 110 ft^2 per hp delivered (13.7 m^2/kW delivered).

Thus, for a solar collector designed to deliver 10,000 gal/min (37.9 m^3/min) at peak output, the preliminary area chosen for this parabolic tracking solar collector will be, A_p = (10,000 gal/min) (0.55 ft^2/gal/min) = 550 ft^2 (511 m^2). Or, using the second rule of thumb, A_p = (110)(50) = 5500 ft^2 (511 m^2).

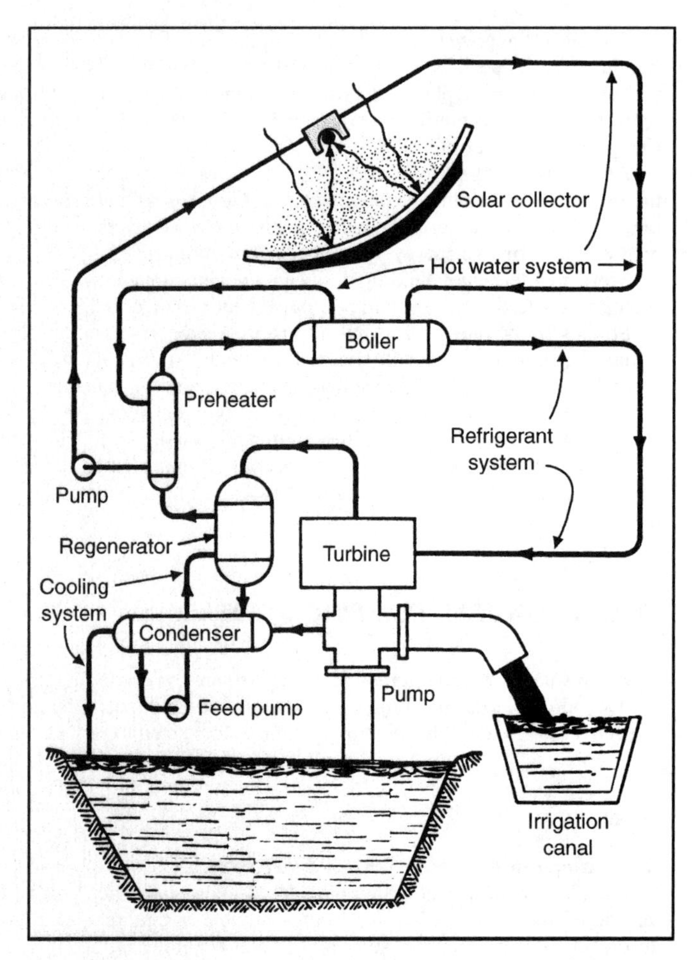

FIGURE 20 Closed-cycle system gassifies refrigerant in boiler to drive Rankine-cycle turbine for pumping water. (*Product Engineering*. Battelle Memorial Institute, and Northwestern Mutual Life Insurance Co.)

Final choice of the collector area will be based on data supplied by the collector manufacturer, refrigerant choice, refrigerant properties, and the actual operating efficiency of the boiler chosen.

In this solar-powered pumping system, water is drawn from a sump basin and pumped to an irrigation canal where it is channeled to the fields. The 50-hp (37.3-kW) motor was chosen because it is large enough to provide a meaningful demonstration of commercial size and it can be scaled up to 200 to 250 hp (149.2 to 186.5 kW) quickly and easily.

Sensors associated with the solar collector aim it at the sun in the morning, and, as the sun moves across the sky, track it throughout the day. These same sensing devices also rotate the collectors to a storage position at night and during storms. This is done to lessen the chance of damage to the reflective surfaces of the collectors. A backup control system is available for emergencies.

3. Predict the probable operating mode of this system

In June, during the longest day of the year, the system will deliver up to 5.6 million gal (21,196 m^3) over a 9.5-hour period. Future provisions for energy storage can be made, if needed.

Related Calculations. Solar-powered pumps can have numerous applications beyond irrigation. Such applications could include domestic water pumping and storage, ornamental fountain water pumping and recirculation, laundry wash water, etc. The whole key to successful solar power for pumps is selecting a suitable application. With the information presented in this procedure the designer can check the applicability and economic justification of proposed future designs.

In today's environmentally conscious design world, the refrigerant must be carefully chosen so it is acceptable from both an ozone-depletion and from a thermodynamic standpoint. Banned refrigerants should not, of course, be used, even if attractive from a thermodynamic standpoint.

This procedure is the work of the editorial staff of *Product Engineering* magazine reporting on the work of Battelle Memorial Institute and the Northwestern Mutual Life Insurance Co. The installation described is located at MMLI's Gila River Ranch, southwest of Phoenix, AZ. SI values were added by the handbook editor.

Designers of solar-powered pumping systems recently found a new way to collect energy from the sun to convert to electricity for the pump's electric motor. The new way of collecting energy from the sun uses closed and capped municipal landfills as sites for solar farms.

New York City, and other large cities and municipalities, are studying the possibility of installing such solar farms. Doing so makes economic sense since the city or municipality already owns the land; hence, it is free. Further, landfills are usually far from residential areas and so there are few visual pollution complaints.

A private company would lease the landfill, install the solar panels, and operate the electrical system. It is estimated that a 250-acre landfill solar farm could produce 50-mW of electricity. New York City is reported to have 3000 acres of landfills. Using closed and capped landfills for solar farms is a win-win for any city or municipality because the cost is small and vacant land is put to beneficial use for the city residents. Further, the solar farm will not interfere with any other power generated by the landfill using gas engines running on methane and other gases from the landfill.

SECTION 9
GEOTHERMAL ENERGY ENGINEERING

Geothermal Calculation Parameters 9.1
Cycle Analysis of a Vapor-Dominated Geothermal Steam Power Plant 9.2
Geothermal and Biomass Power-Generation Analyses 9.5
Flashed-Steam Geothermal Power Plant Analysis 9.9

GEOTHERMAL CALCULATION PARAMETERS

Geothermal energy is considered to be green, semi-renewable,* free of greenhouse gases (GHGs), and an excellent alternative to coal, oil, gas, and similar fossil-fuel energy sources. The major disadvantage of geothermal generation of power is that the energy sources are often great distances from the power users. But as transmission lines are extended this disadvantage decreases, or disappears.

The steam Rankine cycle used in fossil-fueled power plants is also used in geothermal power plants, as reported by Kenneth A. Phair in the Marks reference cited later. Four types of cycles can be used in geothermal plants, Fig. 1. These cycles are: (*a*) Direct steam cycle, Fig. l*a*, in which steam is delivered directly from in-ground production wells to the steam turbine; steam leaving the turbine passes to a condenser where it is converted to water for return to the well; (*b*) In a flash-steam cycle, Fig. l*b*, brine from the well, or a mixture of brine and steam, is delivered to a flash tank in which flash steam is produced that is delivered to the turbine for electric power generation. (*c*) Many geothermal generating plants today are binary-cycle, Fig. l*c*. The plant uses hot water, usually at 275°F to 300°F (135°C to 149°C) to heat a binary fluid (isobutene or isopentane). This hot water comes from drilled wells that range in depth from 4000 to 6000 ft (1219 to 1829 m). The binary fluid is vaporized in a heat exchanger and led to a turbine coupled to an electric generator. (*d*) In a combined-cycle geothermal plant, Fig. 1*d*, the overall thermal efficiency is higher, just as in fossil-fueled combined-cycle plants. Brine and steam enter a separator with steam leaving the separator and being piped to a turbine for production of electricity. Brine leaving the separator enters a heat exchanger where it generates steam in a secondary working fluid for a turbine that produces electricity.

When a geothermal well produces brine or hot water at less than 250°F (121°C), steam power generation of electricity is not normally considered economically possible. But the heated fluid can be used for services such as district heating, food processing, industrial process heating, and so on.

Since geothermal power generation depends on naturally occurring sources, there is a limitation on the number suitable sites. And engineers cannot create new geothermal sites the way they can design, and build, new fossil-fueled power plants. While there are many predictions on wider use of geothermal energy, its ultimate importance is limited by the number of new, acceptable sites that can be found.

*Marks' *Standard Handbook for Mechanical Engineers,* McGraw-Hill, 2006.

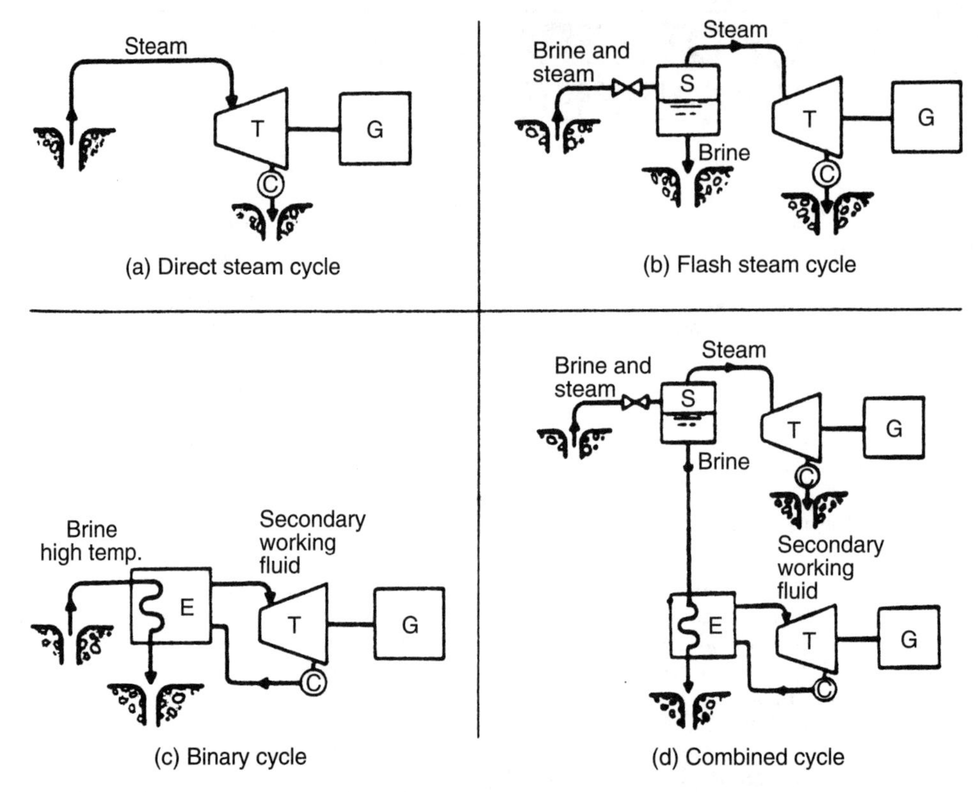

FIGURE 1 Geothermal power cycles. T = turbine. G = generator. C = condenser. S = separator. E = heal exchanger.

Thus, vapor-dominated systems are those* in which water is vaporized into steam that reaches the surface in a relatively dry condition at about 400°F (205°C) and rarely above 100 psig (8 bar). This steam is most suitable for use in turboelectric power plants, with the least cost. It does, however, suffer problems similar to those encountered by all geothermal systems, namely the presence of corrosive gases and erosive material and environmental problems. Vapor-dominated systems, however, are a rarity; there are only five known sites in the world to date.

At this writing (2011), New Zealand recently put a 140-MW geothermal plant online on North Island. As with other geothermal plants, this new plant required huge equipment that had to be transported through nearby forests. Other geothermal projects in New Zealand promise to supply 15 percent, and more, of the nation's power needs in coming years. One of these future plants will have a 250-MW output. These, and other installations around the world, show that engineers will continue to seek viable geothermal sites for many years to come.

CYCLE ANALYSIS OF A VAPOR-DOMINATED GEOTHERMAL STEAM POWER PLANT

A 100-MW vapor-dominated steam power plant as shown in Figs. 2 and 3 uses saturated steam from a geothermal well with a shut-off pressure of 400 psia (2,756 kPa). Steam enters the turbine at 80 psia (551 kPa) and condenses at 2 psia (13.8 kPa). The turbine polytropic efficiency is 0.82, and the

*El-Wakil, *Powerplant Technology,* McGraw-Hill, 1984 and 2002.

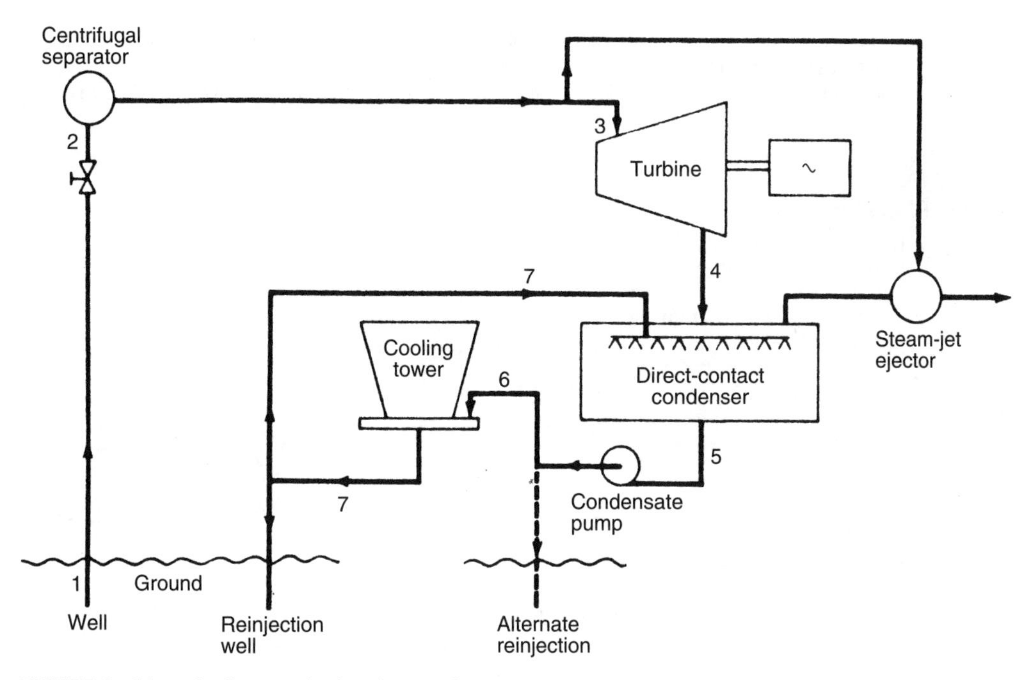

FIGURE 2 Schematic of a vapor-dominated power plant.

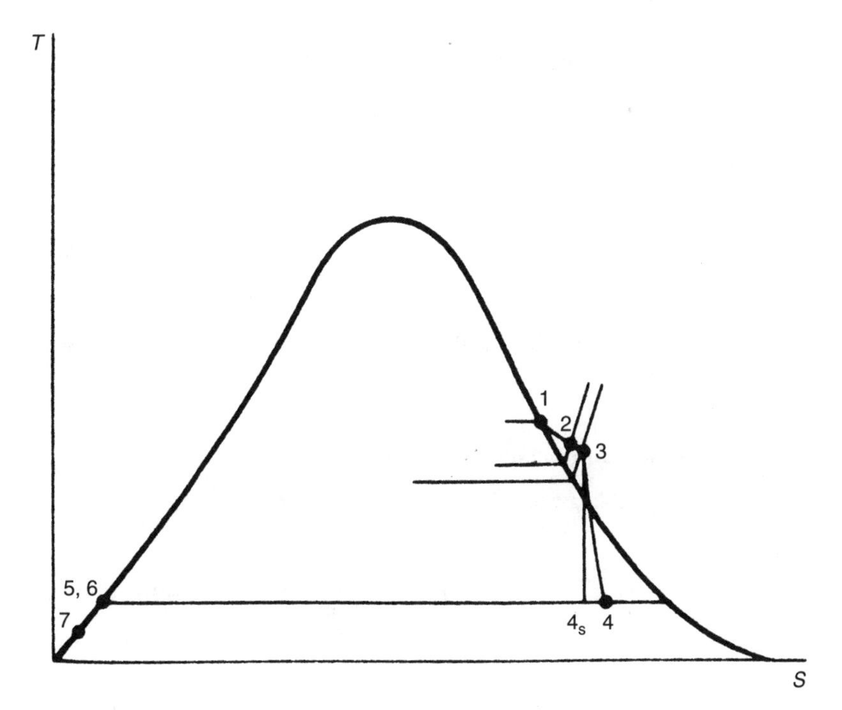

FIGURE 3 *T-S* diagram of the cycle shown in Fig. 2.

turbine-generator combined mechanical and electrical efficiency is 0.90. The cooling-tower cooling-water exit temperature is 70°F (21°C). Calculate the necessary steam flow, lb/h and ft^3/min, the cooling-water flow, lb/h, and the plant efficiency and heat rate, Btu/kWh, if reinjection occurs prior to the cooling tower.

Calculation Procedure:

1. ***Determine the turbine inlet and exhaust steam conditions***

From the steam tables:

$$\text{Inlet:}\quad T_3 = 350°\text{F (38°F superheat) (177°C; 3.3°C)}$$
$$s_3 = 1.6473 \text{ Btu/(lb}_m \cdot °\text{R) (3.83 kJ/kg)}$$
$$v_3 = 5.801 \text{ ft}^3/\text{lb}_m \text{ (3.62 m}^3/\text{kg)}$$

$$\text{Exhaust:}\quad s_{4,s} \text{ at 2 psia} = s_3 = 1.6473 = 0.1750 + x_{4,s}\ (1.7450)$$
$$x_{4,s} = 0.8437$$
$$h_{4,s} = 94.03 + 0.8437\ (1022.1) = 956.4 \text{ Btu/lb}_m \text{ (2,225 kJ/kg)}$$

2. ***Compute the turbine work***

$$\text{Isentropic turbine work} = h_3 - h_{4,s} = 1204.6 - 956.4$$
$$= 248.2 \text{ Btu/lb}_m \text{ (557.3 kJ/kg)}$$

$$\text{Actual turbine work} = 0.82 \times 248.2 = 203.5 \text{ Btu/lb}_m \text{ (473.3 kJ/kg)}$$
$$h_4 = 1204.6 - 203.5 = 1001.1 \text{ Btu/lb}_m \text{ (2,329 kJ/kg)}$$

$$h_{5.6} \text{ (ignoring pump work)} = 94.03 \text{ Btu/lb}_m \text{ (218.7 kJ/kg)}$$
$$h_7 = h_f \text{ at } 70°\text{F} = 38.05 \text{ Btu/lb}_m \text{ (88.5 kJ/kg)}$$

3. ***Find the turbine steam flow***

$$\text{Turbine steam flow} = \frac{100 \times 3.412 \times 10^6}{203.5 \times 0.9} = 1.863 \times 10^6 \text{ lb}_m\text{/h } (0.84 \times 10^6 \text{ kg/h})$$

$$\text{Turbine volume flow} = \frac{1.863 \times 10^6 \times v_3}{60} = 1.8 \times 10^5 \text{ ft}^3\text{/min } (0.051 \times 10^5 \text{m}^3\text{/min})$$

4. ***Compute the cooling-water flow rate***

$$\text{Cooling-water flow } \dot{m}_7\text{: } \dot{m}_7(h_5 - h_7) = \dot{m}_4(h_4 - h_5)$$

$$\dot{m}_7 = \frac{1001.1 - 94.03}{94.03 - 38.05}\dot{m}_4 = 16.2\dot{m}_4 = 16.2 \times 1.863 \times 10^6$$
$$= 30.187 \times 10^6 \text{ lb}_m\text{/h } (13.6 \times 10^6 \text{ kg/h})$$

$$\text{Heat added} = h_1 - h_6 = 1204.6 - 94.03 = 1110.57 \text{ Btu/lb}_m \text{ (2,583 kJ/kg)}$$

5. *Find the geothermal plant efficiency and heat rate*

$$\text{Plant efficiency} = \frac{203.5 \times 0.9}{1110.57} = 0.1649 = 16.49\%$$

$$\text{Plant heat rate} = \frac{3412}{0.1649} = 20{,}690 \text{ But/kWh} \quad (21{,}828 \text{ kJ/jWh})$$

Related Calculations. As indicated previously, vapor-dominated systems are the rarest form of geothermal energy but the most suitable for electricity generation and the most developed of all geothermal systems. They have the lowest cost and the least number of serious problems.

Figures 2 and 3 show a schematic and *T-S* diagram of a vapor-dominated power system. Dry steam from the well (1) at perhaps 400°F (200°C) is used. It is nearly saturated at the bottom of the well and may have a shut-off pressure up to 500 psia (~35 bar). Pressure drops through the well causing it to slightly superheat at the well head (2). The pressure there rarely exceeds 100 psia (~7 bar). It then goes through a centrifugal separation to remove particulate matter and enters the turbine after an additional pressure drop (3). Processes 1-2 and 2-3 are essentially throttling processes with constant enthalpy. The steam expands through the turbine and enters the condenser at (4):

Because turbine flow is not returned to the cycle but reinjected back into the earth (Mother Nature is our boiler), a direct-contact condenser of the barometric or low-level type may be used. Direct-contact condensers are more effective and less expensive than surface-type condensers. (The latter, however, are used in some new units with H_2S removal systems, below.) The turbine exhaust steam at (4) mixes with the cooling water (7) that comes from a cooling tower. The mixture of 7 and 4 is saturated water (5) that is pumped to the cooling tower (6). The greater part of the cooled water at 7 is recirculated to the condenser. The balance, which would normally be returned to the cycle in a conventional plant, is reinjected into the ground either before or after the cooling tower. The mass-flow rate of the reinjected water is less than that originating from the well because of losses in the centrifugal separator, steam-jet ejector (SJE), evaporation, drift and blow-down in the cooling tower, and other losses. No makeup water is necessary.

A relatively large SJE is used to rid the condenser of the relatively large content of noncondensable gases and to minimize their corrosive effect on the condensate system.

This procedure and the data and illustrations presented in Related Calculations are the work of M. M. El-Wakil, as presented in his book *Powerplant Technology*, McGraw-Hill, 1984. At the time of publication he was Professor of Mechanical and Nuclear Engineering at the University of Wisconsin.

GEOTHERMAL AND BIOMASS POWER-GENERATION ANALYSES

Compare the costs—installation and operating—of a 50-MW geothermal plant with that of a conventional fossil-fuel-fired installation of the same rating. Likewise, compare plant availability for each type. Brine available to the geothermal plant free-flows at 4.3 million lb/h (1.95 million kg/h) at 450 lb/in^2 (gage) at 450°F (3100 kPa at 232°C).

Calculation Procedure:

1. *Estimate the cost of each type of plant*

Assuming that the cost of constructing a geothermal plant (i.e., an electric-generating station that uses steam or brine from the ground produced by nature) is in the \$1500 to \$2000 per installed kW range. This cost includes all associated equipment and the development of the well field from which the steam or brine is obtained.

Using this cost range, the cost of a 50-MW geothermal station would be in the range of: 50 MW × (\$1500/kW) × 1000 = \$75 million to 50 MW × (\$2000/kW) × 1000 = \$100 million. Fossil-fuel-fired installations cost about the same—i.e., $1500 to $2000 per installed kW. Therefore, the two types of plants will have approximately the same installed cost.

Department of Energy (DOE) estimates give the average cost of geothermal power at 5.7¢/kWh. This compares with the average cost of 2.4¢/kWh for fossil-fuel-based plants. Advances in geothermal technology are expected to reduce the 5.7¢ cost significantly over the next 40 years.

Because of the simplicity of geothermal plant design, maintenance requirements are relatively low. Some modular plants even run unattended; and because maintenance is limited, plant availability is high. In recent years geothermal-plant availability averaged 97 percent. Thus, the maintenance cost of the usual geothermal plant is lower than a conventional fossil-fuel plant. Further, geothermal plants can meet new emission regulations with little or no pollution-abatement equipment.

2. *Choose the type of cycle to use*

Tapping geothermal energy from liquid resources poses a number of technical challenges—from drilling wells in a high-temperature environment to excessive scaling and corrosion in plant equipment. But DOE-sponsored and private-sector R&D programs have effectively overcome most of these problems. Currently, there are more than 35 commercial plants exploiting liquid-dominated resources. Of the 800 MW of power generated by these plants, 620 MW is produced by flash-type plants and 180 MW by binary-cycle units (Fig. 4).

The flashed-steam plant is best suited for liquid-dominated resources above 350°F (177°C). For lower-temperature sources, binary systems are usually more economical.

In flash-type plants, steam is produced by dropping the pressure of hot brine, causing it to "flash." The flashed steam is then expanded through a conventional steam turbine to produce power. In binary-cycle plants, the hot brine is directed through a heat exchanger to vaporize a secondary fluid which has a relatively low boiling point. This working fluid is then used to generate power in a closed-loop Rankine-cycle system. Because they use lower-temperature brines than flash-type plants, binary units (Fig. 4) are inherently more complex, less efficient, and have higher capital equipment costs.

In both types of plants the spent brine is pumped down a well and reinjected into the resource field. This is done for two reasons: (1) to dispose of the brine—which can be mineral-laden and deemed hazardous by environmental regulatory authorities, and (2) to recharge the geothermal resource.

One recent trend in the industry is to collect noncondensable gases (NCGs) purged from the condenser and reinject them along with the brine. Older plants use pollution-abatement devices to treat NCGs, then release them to the atmosphere. Reinjection of NCGs with brine lowers operating costs and reduces gaseous emissions to near zero.

Major improvements in flashed-steam plants over the past decade centered around are: (1) improving efficiency through a dual-flash process and (2) developing improved water treatment processes to control scaling caused by brines. The pressure of the liquid brine stream remaining after the first flash is further reduced in a secondary chamber to generate more steam. This two-stage process can generate 20 to 30 percent more power than single-flash systems.

Most of the recent improvements in binary-cycle plants have been made by applying new working fluids. The thermodynamic and transport properties of these fluids can improve cycle efficiency and reduce the size and cost of heat-transfer equipment.

To illustrate: By using ammonia rather than the more common isobutane or isopentane, capital cost can be reduced by 20 to 30 percent. It is also possible to improve the conversion efficiency by using mixtures of working fluids, which in turn reduces the required brine flow rate for a given power output.

A flashed-steam cycle will be tentatively chosen for this installation because the brine free-flows at 450°F (232°C), which is higher than the cutoff temperature of 350°F (177°C) for binary systems. An actual plant (Fig. 5), operating with these parameters uses two flashes. The first flash produces 623,000 lb/h (283,182 kg/h) of steam at 100 lb/in^2 (gage) (689 kPa). In the second flash an additional 262,000 lb/h (117,900 kg/h) of steam at 10 lb/in^2 (gage) (68.9 kPa) is produced.

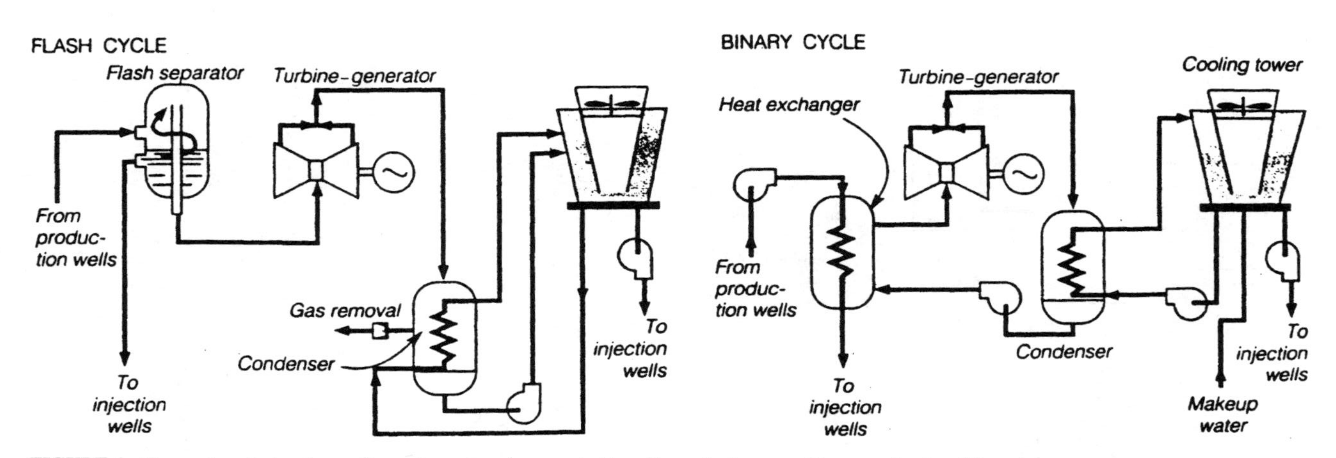

FIGURE 4 Energy from hot-water geothermal resources is converted by either a flash-type or binary-cycle plant (*Power.*)

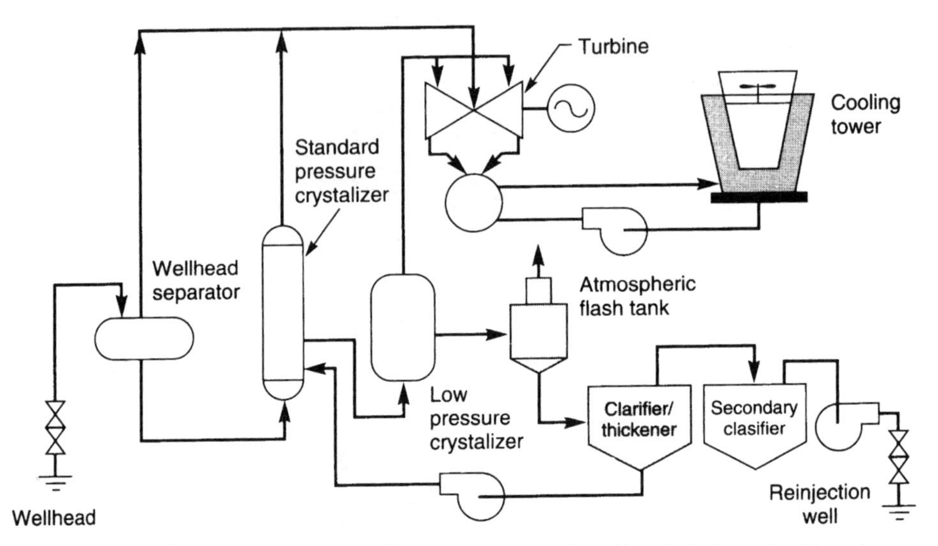

FIGURE 5 Dual-flash process extracts up to 30 percent more power than older, single-flash units. (*Power*.)

Steam is cleaned in two trains of scrubbers, then expanded through a 54-MW, 3600-rpm, dual-flow, dual-pressure, five-stage turbine-generator to produce 48.9 MW. Of this total, 47.5 MW is sold to Southern California Edison Co. because of transmission losses.

The turbine exhausts into a surface condenser, coupled to a seven-cell cooling tower. About 40,000 lb/h (18,000 kg/h) of the high-pressure steam is required by the plant's air ejectors to remove NCGs from the main condenser at a rate of 6500 lb/h (2925 kg/h).

Because the liquid brine from the flash process is supersaturated, various solid compounds precipitate out of solution and must be removed to avoid scaling and fouling of the pumps, pipelines, and injection wells. This is accomplished as the brine flows to the crystallizer and clarifier tanks where, respectively, solid crystals grow and then are separated. The solids are dewatered and used in construction-grade soil cement. The clarified brine is disposed of by pumping it into three injection wells.

Related Calculations. Geothermal generating plants are environmentally friendly because there are no stack emissions from a boiler. Further, such plants do not consume fossil fuel, so they are not depleting the world's supply of such fuels. And by using the seemingly unlimited supply of heat from the earth, such plants are contributing to an environmentally cleaner and safer world while using a renewable fuel.

Another renewable fuel available naturally that is receiving—like geothermal power—greater attention today is *biomass*. The most common biomass fuels used today are waste products and residue left over from various industries, including farming, logging, pulp, paper, and lumber production, and wood-products manufacturing. Wooden and fibrous materials separated from the municipal waste stream also represent a major source of biomass.

Although biomass-fueled power plants currently account only for about 1 percent of the installed generating capacity in the United States, or 8000 MW, they play an important role in solving energy and environmental problems. Since the fuels burned in these facilities are considered waste in many cases, combustion yields the double benefits of reducing or eliminating disposal costs for the seller and providing a low-emission fuel source for the buyer. On a global scale, biomass firing could present even more advantages, such as: (1) there is no net buildup of atmospheric CO_2 and air emissions are lower compared to many coal- or oil-fired plants. (2) Vast areas of deforested or degraded lands in tropical and subtropical regions can be converted to practical use. Because much of the available land is in the developing regions of Latin America and Africa, the fuels produced on these plantations could help improve a country's balance of payments by reducing dependence on imported oil. (3) Industrialized nations could potentially phase out agricultural subsidies by encouraging farmers to grow energy crops on idle land.

The current cost of growing, harvesting, transporting, and processing high-grade biomass fuels is prohibitive in most areas. However, proponents are counting on the successful development of advanced biomass-gasification technologies. They contend that biomass may be a more desirable feedstock for gasification than coal because it is easier to gasify and has a very low sulfur content, eliminating the need for expensive O_2 production and sulfur-removal processes.

One report indicates that integrated biomass-gasification–gas-turbine-based power systems with efficiencies topping 40 percent should be commercially available by year 2000. By 2025, efficiencies may reach 57 percent if advanced biomass-gasification–fuel-cell combinations become viable. Proponents are optimistic because this technology is currently being developed for coal gasification and can be readily transformed to biomass.

Data in this procedure are the work of M. D. Forsha and K. E. Nichols, Barber-Nichols Inc., for the geothermal portion, and Steven Collins, assistant editor, *Power,* for the biomass portion. Data on both these topics was published in *Power* magazine.

FLASHED-STEAM GEOTHERMAL POWER PLANT ANALYSIS

A flashed-steam power plant such as that shown in Figs. 6 and 7 uses a hot-water reservoir that contains water at 460°F (238°C) and 160 psia (1,102 kPa). The separator pressure is 100 psia (689 kPa). Find (1) the mass flow rate of water from the well, and of the reinjected brine per unit of mass flow rate of steam into the turbine, and (2) the ratio of enthalpies of spent brine to steam.

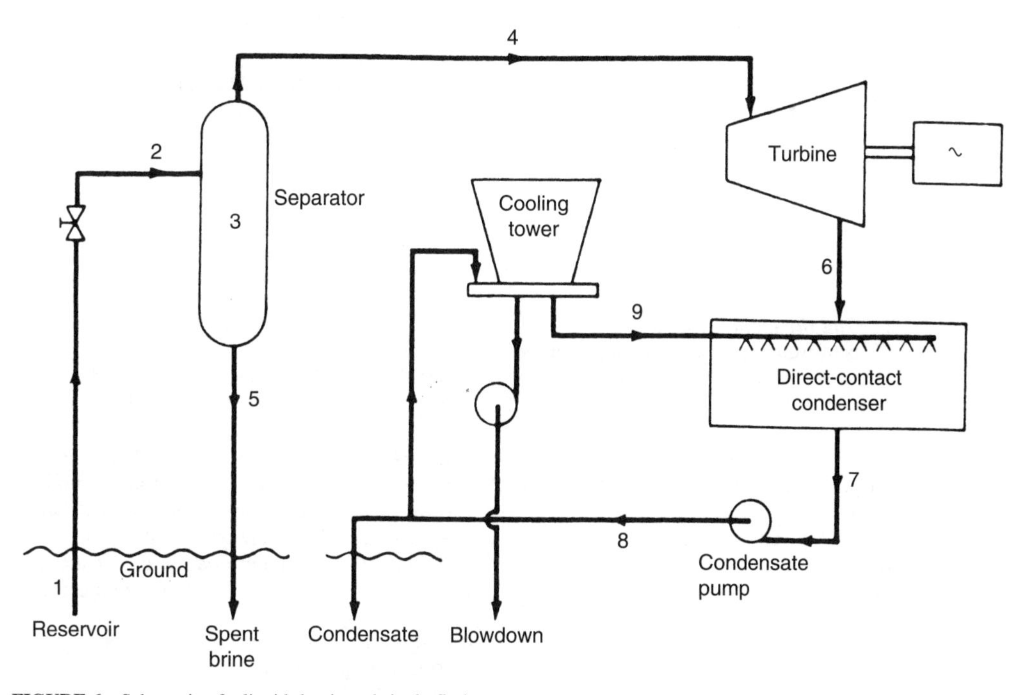

FIGURE 6 Schematic of a liquid-dominated single-flash steam system.

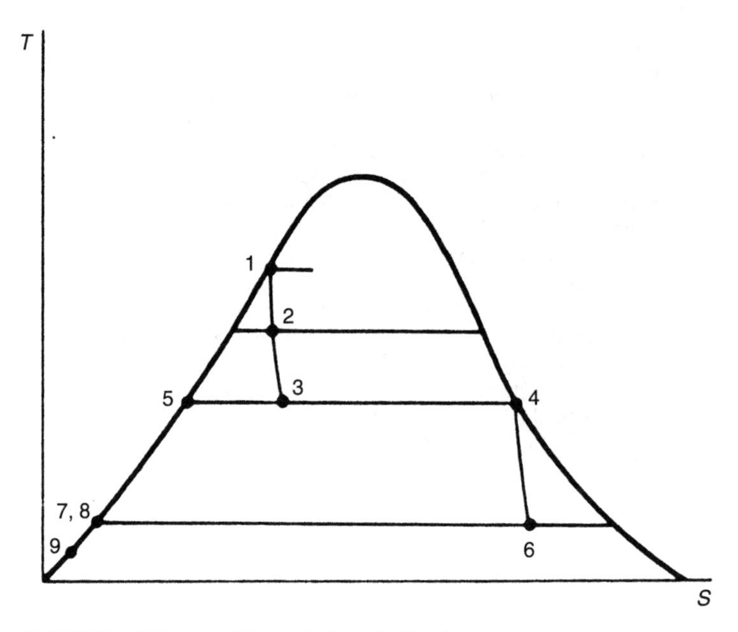

FIGURE 7 *T-S* diagram of the cycle shown in Fig. 6.

Calculation Procedure:

1. *Determine the enthalpy of steam in the reservoir at 460°F (238°C)*

$$h_1 \approx h_f \text{ at } 460°\text{F} = 441.5 \text{ Btu/lb}_m \text{ (1026.9 kJ/kg)}$$
$$h_3 = h_1 = (h_f + x_3 h_{fg})$$
$$441.5 = 298.5 + x_3(888.6)$$

Therefore,

$$x_3 = 0.161$$

2. *Find the enthalpy ratio*
Mass of water from well per unit mass of steam = $1/x_3$ = 6.21.
Mass of reinjected brine per unit mass of steam = 6.21 – 1 = 5.21.
Ratio of enthalpy at 5 to enthalpy at 4 = $5.21(h_5/h_4) = 5.21 \times (298.\ 54/1187.2) = 1.31$.

Related Calculations. The flashed-steam system is reserved for water in the higher-temperature range, and is illustrated by the flow and *T-S* diagrams of Figs. 6 and 7. Water from the underground reservoir at 1 reaches the well head at 2 at a lower pressure. Process 1-2 is essentially a constant enthalpy throttling process that results in a two-phase mixture of low quality at 2. This is throttled further in a flash separator resulting in a still low but slightly higher quality at 3. This mixture is now separated into dry saturated steam at 4 and saturated brine at 5. The latter is reinjected into the ground.

The dry steam, a small fraction of the total well discharge (because of the low quality at 3), and usually at pressures below 100 psig (8 bar), is expanded in a turbine to 6 and mixed with cooling water in a direct-contact condenser with the mixture at 7 going to a cooling tower in the same fashion as the vapor-dominated system. The balance of the condensate after the cooling water is recirculated to the condenser is reinjected into the ground.

The flashed-steam system is a more difficult proposition than the vapor-dominated system for several reasons: (1) much larger total mass-flow rates through the well, as shown by the preceding example; (2) a greater degree of ground surface subsidence as a result of such large flows; (3) a greater degree of precipitation of minerals from the brine, resulting in the necessity for design of valves, pumps, separator internals, and other equipment for operation under scaling conditions; and (4) greater corrosion of piping, well casing, and other conduits.

Flashed-steam systems have been widely used in Japan, New Zealand, Italy, Mexico, and elsewhere.

This procedure and the data and illustrations presented in Related Calculations are the work of M. M. El-Wakil, as presented in his book *Powerplant Technology*, McGraw-Hill, 1984. At the time of publication he was Professor of Mechanical and Nuclear Engineering at the University of Wisconsin.

SECTION 10
OCEAN ENERGY ENGINEERING

Ocean Energy Parameters 10.1
Analysis of an OTEC Claude Cycle Efficiency and Flow Rates 10.2
Wave Calculations for Characteristics and Energy and Power Densities 10.5
Computation of Modulated Single-Pool Tidal System Energy and Power 10.8

*OCEAN ENERGY PARAMETERS**

Energy from the world's oceans has intrigued engineers for more than a century. Three energy conversion methods have been considered for the oceans which make up some 70 percent of the Earth's surface. These methods are:

1. Ocean temperature energy conversion (OTEC).
2. Converting wave energy to useful power.
3. Using the rise and fall of ocean tides to produce power.

In the oceans there is a moderate temperature difference, or gradient, between the surface water and the water at lower depths of the ocean. This temperature gradient, called *ocean temperature energy conversion* (OTEC), can be used in a heat engine to generate power. Because the water temperature difference is small, even in the tropics, OTEC systems have very low efficiencies (usually less than 3 percent). As a result, OTEC systems have very high capital costs.

In the tropics, the ocean surface temperature often exceeds 25°C (77°F). One kilometer below the surface the water temperature is usually no higher than 10°C (50°F). The concept of OTEC is based on the utilization of this temperature difference in a heat engine to generate power, a concept first recognized by the Frenchman d'Arsonval in 1881.

The maximum possible efficiency of a heat engine operating between two temperature limits cannot exceed that of a Carnot cycle operating between the same temperature limits. Because of the temperature drops in steam or other vapor generators, and the condenser in an actual system, inefficiencies in the turbine and pumps, and other inefficiencies, the efficiency of a real OTEC power plant seldom exceeds 2 percent.

The extremely low efficiency of an OTEC system implies extremely large power plant heat exchangers and components. At 2 percent efficiency the heat exchangers must handle 50 times the net output of the plant. Although there are no fuel costs, the capital costs are extremely high, as are the unit capital costs, $/kW. In addition to the large size per unit of power generation, the developmental problems and the uncertainties of market penetration make the financial risks associated with the development of large OTEC technologies so high as to effectively preclude most utilities. The first Calculation Procedure is this section demonstrates OTEC efficiency computation.

*Adapted from M. M. El-Wakil, *Powerplant Technology*, McGraw-Hill, 1984 and 2002.

Ocean waves have long been looked on as a source of power. Like wind power, and OTEC power, ocean and sea waves are caused indirectly by solar energy. Waves are caused by the wind, which in turn is caused by the uneven solar heating and subsequent cooling of the Earth's crust and the rotation of the Earth. Wave energy, at its most active, however, can (like wind energy) be much more concentrated than incident solar energy, even at the latter's peak. Devices that convert energy from waves can therefore produce much higher power densities than solar devices.

The total energy of a wave is the sum of its potential and kinetic energies. Wave energy conversion is done by floats, accumulator machines, dolphin-type generators, dam-atoll conversion devices, multi-pontoon rafts, and others. An accompanying procedure shows a typical computation for the energy and power densities of waves.

The tides are yet another source of energy from the oceans. This energy can be tapped from coastal waters by building dams that entrap the water at high tide and release it back to the sea at low tide. Power can then be obtained by hydraulic turbines from both the in- and the outflows of the water. The amount of energy available is very large but only in a few parts of the world. An accompanying Calculation Procedure shows how to compute the total energy and average power of a modulated single-pool tidal system.

ANALYSIS OF AN OTEC CLAUDE CYCLE EFFICIENCY AND FLOW RATES

A Claude cycle plant, Fig. 1*a*, producing 100 kW (gross) operates on the conditions of Fig. 1*b*. The turbine has a polytropic efficiency of 0.80 and the turbine-generator has a combined mechanical-electrical efficiency of 0.90. Calculate the surface and deep-water flow rates in kg/s and m^3/s, and the gross cycle and plant efficiencies.

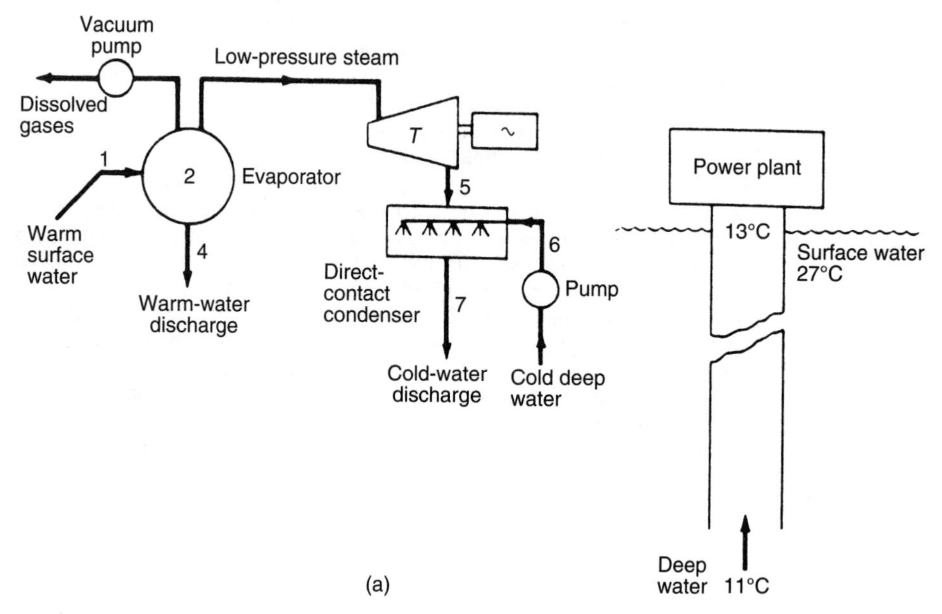

FIGURE 1*a* Flow diagram and schematic of a Claude (open-cycle) OTEC power plant.

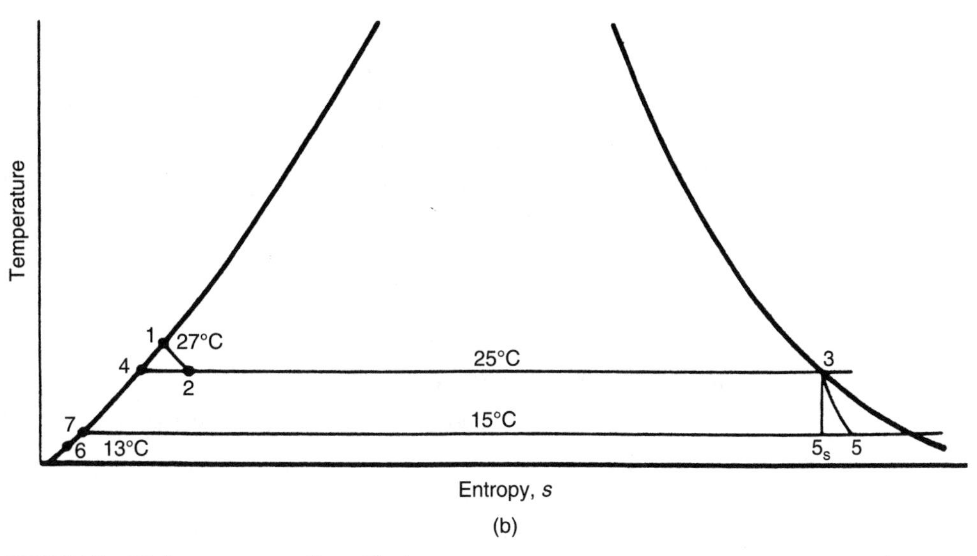

FIGURE 1*b* *T-S* diagram corresponding to Fig. 1*a*.

Calculation Procedure:

1. *Analyze the conditions in the evaporator*

Use the low-temperature steam data (in SI units) from "Thermodynamic and Transport Properties of Steam," The American Society of Mechanical Engineers, 1967, or later editions, for steam temperature, pressure, specific volume, enthalpy, and entropy.

The evaporator

$$h_1 = h_f \text{ at } 27°\text{C} = 113.2 \text{ kJ/kg} \qquad v_1 = 0.0010036 \text{ m}^3\text{/kg}$$

$$h_2 = h_1 = 113.2 = (h_f + x_2 h_{fg}) \text{ at } 25°\text{C}$$

$$= 104.8 + x_2 \times 2442$$

Therefore

$$x_2 = 0.00344, \quad \text{or} \quad 0.344\%$$

2. *Find the actual turbine work output*

Warm-water mass-flow rate per unit turbine mass-flow rate

$$\dot{m}_w = \frac{m_1}{m_3} = \frac{m_2}{m_3} = \frac{1}{x_2} = \frac{1}{0.00344} = 290.7$$

The turbine

$$h_3 = h_g \text{ at } 25°\text{C} = 2550 \text{ kJ/kg}, \qquad s_3 = 8.5570 \text{ kJ/(kg} \cdot \text{K)}$$

For an adiabatic reversible turbine, the expansion is to 5_s

$$S_{5,s} = S_3 = 8.5570 = (S_f + x_{5,s} S_{fg}) \text{ at } 15°\text{C}$$

$$= 0.2244 + x_{5,s} \times 8.5562$$

Therefore

$$x_{5,s} = 0.9739, \qquad \text{or} \qquad 97.39\%$$

$$h_{5,s} = (h_f + x_{5,s}h_{fg}) \text{ at } 15°\text{C} = 62.97 + 0.9739 \times 2465 = 2463.6 \text{ kJ/kg}$$

$$\text{Adiabatic reversible turbine work} = h_3 - h_{5,s} = 2550 - 2463.6$$
$$= 86.4 \text{ kJ/kg}$$
$$\text{Actual turbine work } w_T = 86.4 \times \text{polytropic efficiency}$$
$$= 86.4 \times 0.8$$
$$= 69.1 \text{ kJ/kg}$$

$$h_5 = h_3 - \text{actual work} = 2550 - 69.1 = 2480.9 \text{ kJ/kg}$$

at which $x_5 = 0.9809$ or 98.09% and $v_5 = 76.48 \text{ m}^3/\text{kg}$

$$\text{Turbogenerator output} = 69.1 \times 0.88 = 60.8 \text{ kJ/kg}$$

3. *Find the gross cycle efficiency*
The condenser

$$h_6 \approx h_f \text{ at } 13°\text{C} = 54.60 \text{ kJ/kg} \qquad v_6 = 0.0010007 \text{ m}^3\text{/kg}$$
$$h_7 = h_f \text{ at } 15°\text{C} = 62.97 \text{ kJ/kg}$$

Cold-water mass-flow rate per unit turbine mass-flow rate

$$\dot{m}_c = \frac{h_5 - h_7}{h_7 - h_6} = \frac{2480.9 - 62.97}{62.97 - 54.60} = 288.9$$

The cycle

$$\text{Turbine mass-flow rate } \dot{M}_T = \frac{\text{turbine work}}{w_T} = \frac{100}{69.1}$$
$$= 1.447 \text{ kg/s}$$
$$\text{Turbine volume flow rate at throttle} = \dot{M}_T v_3 = 1.447 \times 43.40$$
$$= 62.8 \text{ m}^3/\text{s}$$
$$\text{Turbine volume flow rate at exhaust} = \dot{M}_T v_5 = 1.447 \times 76.48$$
$$= 110.7 \text{ m}^3/\text{s}$$
$$\text{Warm-water mass-flow rate } \dot{M}_w = \dot{M}_T \dot{m}_w = 1.447 \times 290.7$$
$$= 420.6 \text{ kg/s}$$
$$\text{Warm-water volume flow rate } \dot{V}_w = \dot{M}_w v_1 = 420.6 \times 0.0010036$$
$$= 0.422 \text{ m}^3/\text{s}$$
$$\text{Cold-water mass-flow rate } \dot{M}_c = \dot{M}_T \dot{m}_c = 1.447 \times 288.9$$
$$= 418.0 \text{ kg/s}$$
$$\text{Cold-water volume flow rate } \dot{V}_c = \dot{M}_c v_6 = 418.0 \times 0.0010007$$
$$= 0.418 \text{ m}^3/\text{s}$$
$$\text{Gross cycle efficiency} = \frac{w_T}{q_A} = \frac{h_3 - h_5}{h_3 - h_7} = \frac{69.1}{2487}$$
$$= 0.0278 = 2.78\%$$
$$\text{Gross plant efficiency} = 0.0278 \times 0.9 = 0.0250$$
$$= 2.5\%$$

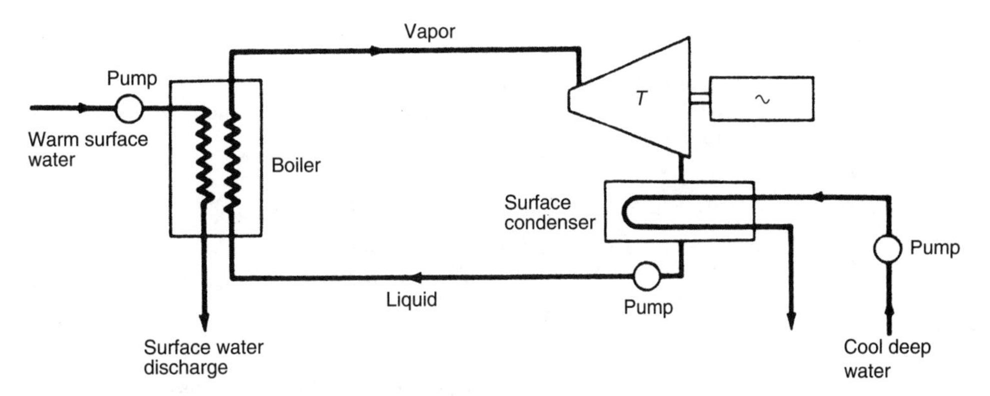

FIGURE 2 Schematic of a closed-cycle OTEC power plant.

Note: The gross plant power, 100 kW, and the gross plant efficiency, 2.5 percent, do not take into account pumping and other auxiliary power inputs to the plant.

It can be seen that very large ocean-water mass and volume flow rates are used in open OTEC systems and that the turbine is a very low-pressure unit that receives steam with specific volumes more than 2000 times that in a modern fossil power plant. Thus the turbine resembles the few last exhaust stages of a conventional turbine and is thus physically large.

Related Calculations. Although the first attempt at producing power from ocean temperature differences was the open cycle of Georges Claude in 1929, d'Arsonval's original concept in 1881 was that of a closed cycle that also utilizes the ocean's warm surface and cool deep waters as heat source and sink, respectively, but requires a separate working fluid that receives and rejects heat to the source and sink via heat exchangers (boiler and surface condenser) (Fig. 2).

The working fluid may be ammonia, propane, or a Freon. The operating (saturation) pressures of such fluids at the boiler and condenser temperatures are much higher than those of water, being roughly 10 bar at the boiler, and their specific volumes are much lower, being comparable to those of steam in conventional power plants.

Such pressures and specific volumes result in turbines that are much smaller and hence less costly than those that use the low-pressure steam of the open cycle. The closed cycle also avoids the problems of the evaporator. It, however, requires the use of very large heat exchangers (boiler and condenser) because, for an efficiency of about 2 percent, the amounts of heat added and rejected are 50 times the output of the plant. In addition, the temperature differences in the boiler and condenser must be kept as low as possible to allow for the maximum possible temperature difference across the turbine, which also contributes to the large surfaces of these units.

This calculation procedure is the work of M. M. El-Wakil given in his excellent text, *Powerplant Technology*, McGraw-Hill, 1984 and 2002. The reader should refer to that text for further insight into energy calculations, along with the derivation of the equations given here. The handbook editor added minor transitional wording to adapt the procedure to the format used in the handbook.

WAVE CALCULATIONS FOR CHARACTERISTICS AND ENERGY AND POWER DENSITIES

A 2-m wave has a 6-s period and occurs at the surface of water 100 m deep. Find the wavelength, the wave velocity, the horizontal and vertical semiaxes for water motion at the surface, and the energy and power densities of the wave. Water density = 1025 kg/m^3.

Calculation Procedure:

1. ***Find the wavelength, velocity, and height***

A two-dimensional progressive wave that has a free surface and is acted upon by gravity (Fig. 3) is characterized by the following parameters:

$$\begin{aligned} \lambda &= \text{wavelength} = c\tau, \text{ m or ft} \\ a &= \text{amplitude, m or ft} \\ 2a &= \text{height (from crest to trough), m or ft} \\ \tau &= \text{period, s} \\ f &= \text{frequency} = 1/\tau, \text{ s}^{-1} \\ c &= \text{wave propagation velocity } \lambda/\tau, \text{ m/s or ft/s} \\ n &= \text{phase rate} = 2\pi/\tau, \text{ s}^{-1} \end{aligned}$$

The period τ and wave velocity c depend upon the wavelength and the depth of water.

The relationship between wavelength and period can therefore be well approximated by

$$\lambda = 1.56\tau^2 \quad (\lambda \text{ in m}, \tau \text{ in s})$$

or

$$\lambda = 5.12\tau^2 \quad (\lambda \text{ in ft}, \tau \text{ in s})$$

Figure 3 shows an isometric of a two-dimensional progressive wave, represented by the sinusoidal simple harmonic wave shown at time 0. Cross sections of the wave are also shown at time 0 and at time θ. That wave is expressed by

$$y = a \sin\left(\frac{2\pi}{\lambda}x - \frac{2\pi}{\tau}\theta\right)$$

or

$$y = a \sin(mx - n\theta)$$

where

$$\begin{aligned} y &= \text{height above its mean level, m or ft} \\ \theta &= \text{time, s} \\ m &= 2\pi/\lambda, \text{ m}^{-1} \text{ or ft}^{-1} \\ (mx - n\theta) &= 2\pi(x/\lambda - \theta/\tau) = \text{phase angle, dimensionless} \end{aligned}$$

Note that the wave profile at time θ has the same shape as that at time 0, except that it is displaced from it by a distance $x = \theta/\tau = \theta(n/m)$. When $\theta = \tau$, $x = \lambda$ and the wave profile assumes its original position.

Then,

$$\begin{aligned} \text{Wavelength } \lambda &= 1.56 \times 6^2 = 56.16 \text{ m} = 184.25 \text{ ft} \\ \text{Wave velocity } c &= \lambda/\tau = 9.36 \text{ m/s} = 30.71 \text{ ft/s} \\ \text{Wave height } 2a &= 2 \text{ m} \qquad \text{Amplitude } a = 1 \text{ m} = 3.28 \text{ ft} \\ m &= 2\pi/\lambda = 2\pi/56.16 = 0.1119 \text{ m}^{-1} \\ \text{At the surface } \eta &= h = 100 \text{ m} \end{aligned}$$

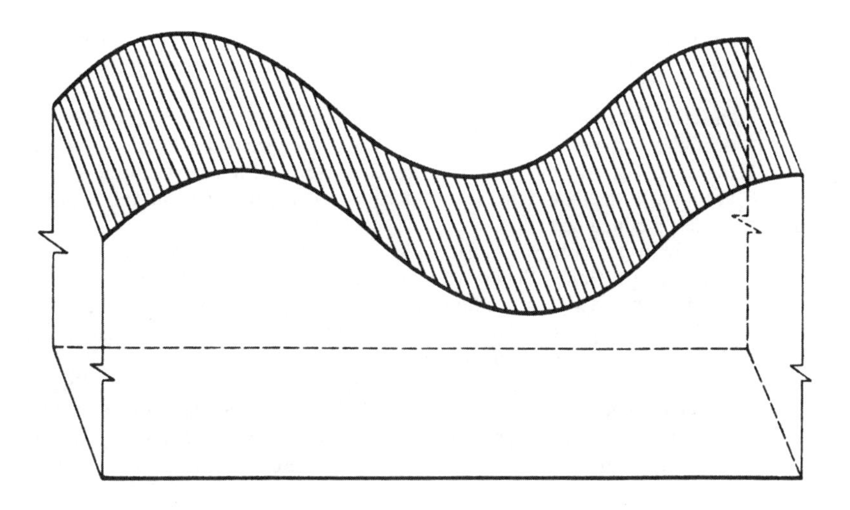

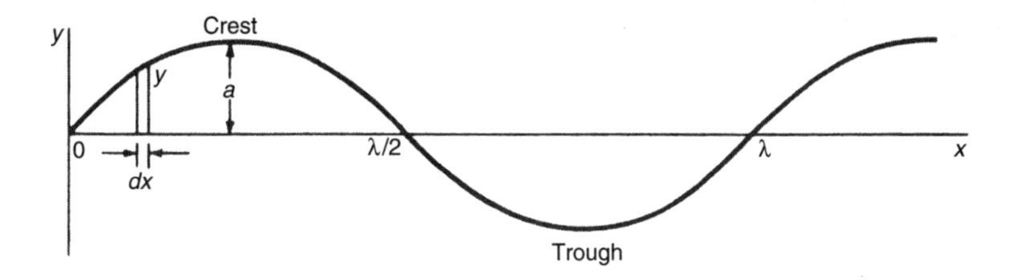

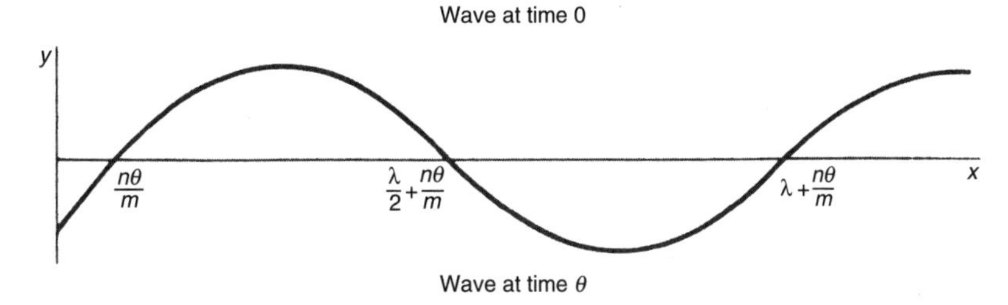

FIGURE 3 A typical progressive wave, a = amplitude, λ = wavelength, showing two-dimensional wave and amplitudes at time 0 and at time θ.

2. *Find the energy and power densities*

$$\text{Horizontal semiaxis } \alpha = 1 \times \frac{\cosh 11.19}{\sinh 11.19} = 1 \text{ m}$$

$$\text{Vertical semiaxis } \beta = 1 \times \frac{\sinh 11.19}{\sinh 11.19} = 1 \text{ m}$$

$$\text{Wave frequency } f = 1/\tau = 1/6 \text{ s}$$

$$\text{Energy density } \frac{E}{A} = \frac{1}{2} \times 1025 \times 1^2 \times \frac{9.81}{1} = 5027.6 \text{ J/m}^2$$

$$= 344.5 \text{ ft} \cdot \text{lb}_f/\text{ft}^2$$

$$\text{Power density } \frac{P}{A} = \frac{E}{A} f = 5027.6 \times \frac{1}{6} = 837.9 \text{ W/m}^2$$

$$= 0.0778 \text{ kW/ft}^2$$

Note: Because of the large depth, the semiaxes are equal, so the motion is circular. Note also that they are small compared with the wavelength, so the water motion is primarily vertical.

Related Calculations. Two-meter waves, of course, do not occur all the time. However, in regions of high wave activity, 2 m is a median with heavier and calmer seas occurring about 50 percent of the time. The total energy and power densities over a period of time should take this spectrum into account. With these densities proportional to a^2, the average densities would be greater than the values obtained here.

It is instructive to compare these values with the *average daily solar incidence* where, in the southwestern United States, a value of 240 W/m² (0.0223 kW/ft²) is often used. Thus, wave-power density is much higher. A complete comparison should take into account the efficiency of conversion to electric energy as well as other factors, like capital costs for land and equipment, operational costs, costs of energy storage, and other factors.

This calculation procedure is the work of M. M. El-Wakil given in his excellent text, *Powerplant Technology,* McGraw-Hill, 1984 and 2002. The reader should refer to that text for further insight into energy calculations, along with the derivation of the equations given here. The handbook editor added minor transitional wording to adapt the procedure to the format used in the handbook.

COMPUTATION OF MODULATED SINGLE-POOL TIDAL SYSTEM ENERGY AND POWER

Calculate the total energy and average power of a modulated single-pool tidal system using the equations below for H and y and the values $R = 12$ m, $a = 0.0625\ \text{h}^{-1}$, $\theta_1 = 1$ h, $\theta_2 = 4$ h, $A = 10{,}000\ \text{km}^2$, and $\rho = 1025\ \text{kg/m}^3$. Compare the results with those for a simple single-pool system. Figure 4 shows a single-pool tidal system.

Calculation Procedure:

1. ***Calculate the total energy***

Use the following equation with the symbols as given below:

$$W = \frac{g}{g_c}\rho A R^2 \left[0.988a\left(-\cos\frac{\pi\theta_2}{6.2083} + \cos\frac{\pi\theta_1}{6.2083} \right) - \frac{a^2}{2}(\theta_2^2 - \theta_1^2) \right]_{\theta_1}^{\theta_2}$$

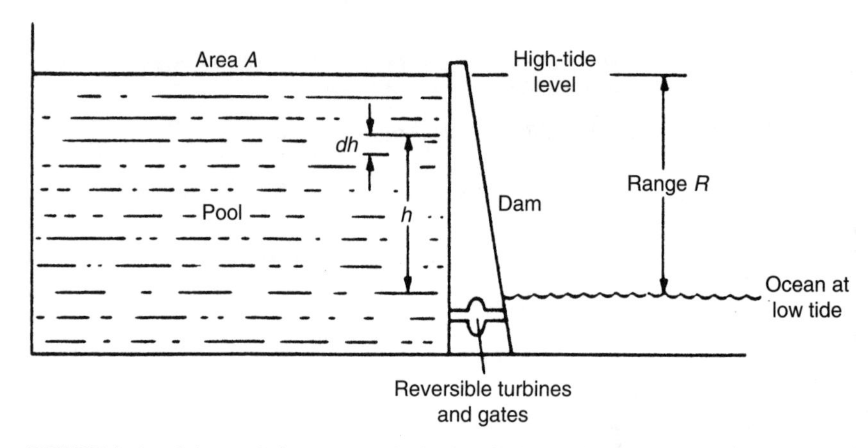

FIGURE 4 Level changes during power production in a single-pool tidal system.

where H = ocean level above mean or other appropriate datum
y = pool level above mean or datum
θ = time

and the other symbols have already been defined. H may be closely approximated by a sinusoidal function of θ such as

$$H = f_1(\theta) = \frac{R}{2}\sin\frac{\pi\theta}{6.2083}$$

where θ is in hours and 6.2083 in hours is one-half of a tidal period, y may be approximated by a linear function of θ, starting at 0 at θ_1 for a constant mass-flow rate such as

$$y = f_2(\theta) = aR(\theta - \theta_1)$$

where a is a constant having the dimension time^{-1}, e.g., h^{-1}, or y could be a function of $h = H - y$ for a constant flow resistance or some other function determined from operational data.

W = work done by the water, ft · lb_f or J
g = gravitational acceleration, 32.2 ft/s^2 or 9.81 m/s^2
g_c = conversion factor, 32.2 lb_m · ft/(lb_f · s^2) or 1.0 kg/(N · s^2)
m = mass flowing through turbine, lb_m or kg
h = head, ft or m
ρ = water density, lb_m/ft^3 or kg/m^3
A = surface area of pool, considered constant, ft^2 or m^2

$$W = \frac{1}{2}\frac{g}{g_c}\rho A R^2$$

Then,

$$W = \frac{9.81}{1} \times 1025 \times 10^{10} \times 12^2\left[0.988 \times 0.0625(0.43795 + 0.87468) - \frac{(0.0625)^2}{2}(16 - 1)\right]$$
$$= 1.448 \times 10^{16}(0.08105 - 0.02930)$$
$$= 7.493 \times 10^{14}\ \text{J}$$

2. *Find the average power*
The average power during the generation period of 4 h is

$$P_{\text{av,gen}} = \frac{7.493 \times 10^{14}}{4 \times 3600} = 5.2 \times 10^{10}\ \text{W} = 5200\ \text{MW}$$

The average power during the total period of 6.2083 h is

$$P_{\text{av}} = \frac{7.493 \times 10^{14}}{6.2083 \times 3600} = 3.35 \times 10^{10}\ \text{W} = 33{,}500\ \text{MW}$$

In the simple single-pool system, the corresponding values are

$$W = \frac{1}{2}\frac{g}{g_c}\rho A R^2 = \frac{1}{2} \times \frac{9.81}{1} \times 1025 \times 10^{10} \times 12^2 = 7.24 \times 10^{15}\ \text{J}$$

and $$P_{\text{av}} = 3.24 \times 10^{11}\ \text{W} \quad \text{or} \quad 324{,}000\ \text{MW}$$

Thus the simple single-pool system produces some 10 times the work and average power of the modulated single-pool system. However, the former does so almost in a "spike," which is very hard on the power grid and requires very large turbines that remain idle most of the time. The latter produces its work over several hours and hence avoids these problems.

The actual work and power above must be multiplied by the efficiency of the system, which is probably in the 25 to 30 percent range.

Related Calculations. In the simple single-pool system (above), two high-peak, short-duration power outputs occur every tidal period. Such peaks necessitate large turbine-generators that remain idle much of the time. The power peaks also occur at different times every day (50 min later each successive day), at times of high and low tides that almost surely will not always correspond to times of peak power demand, and pose a burden on the electric-power grid they are connected to.

The *modulated single-pool tidal system* partially corrects for these deficiencies by generating power more uniformly at a lower average head, though still with some periods of no generation. Because the average head h is lower and work and power are proportional to h^2, the turbine-generators are much smaller and operate over much longer periods. The resulting total work is reduced, however.

In the system, shown by the ocean and pool level and power diagrams of Fig. 5, the reversible turbines are allowed to operate *during* periods of pool filling and emptying instead of at high and low levels only. They cease to operate when the head is too low for efficient operation. Period C_1 begins with both pool and ocean at low-tide level 1, the ocean at the beginning of tide rise, and all

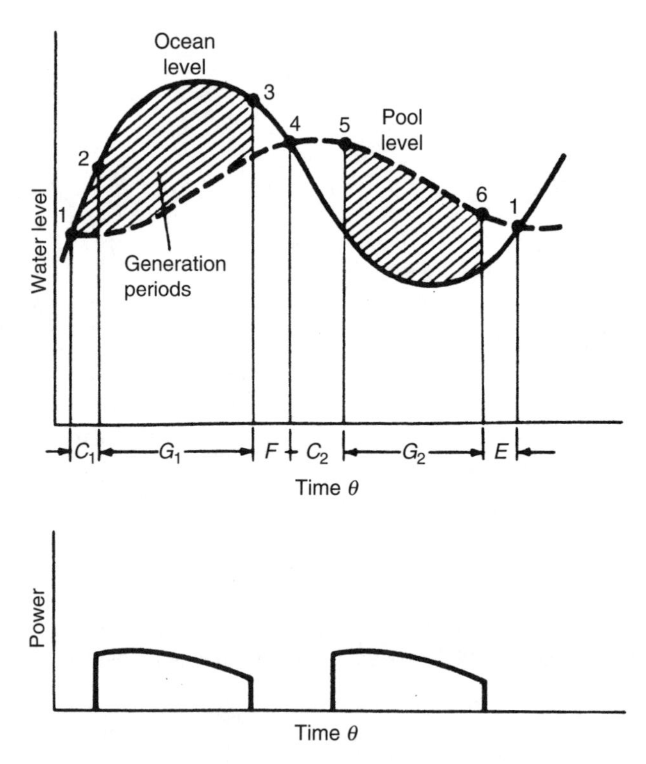

FIGURE 5 Ocean and pool levels in a modulated single-pool tidal system. C = gates closed, G = generation. F = pool filling, E = pool emptying.

gates closed. When the head is sufficient 2, gates to the turbines are opened and water from the ocean is allowed through. Power is generated during period G_1 as both ocean and pool levels rise. The ocean level reaches its peak and begins to decrease but the pool level is still increasing until, at 3, the head is too low for efficient generation. The gates to the turbines are closed and bypass gates are opened so that the pool is allowed to fill up during period F to 4. At 4 all gates are once again closed and the pool level remains constant while the ocean level decreases during period C_2. At 5 the head is once again sufficient to allow for turbine water flow in the opposite direction and a second generation period G_2 begins. At 6 generation ceases but the pool is allowed to empty during period E and the system goes back to point 1, repeating the cycle. The power generation shown is certainly not uniform but much more so than in the case of the simple system.

This calculation procedure is the work of M. M. El-Wakil given in his excellent text, *Powerplant Technology*, McGraw-Hill, 1984 and 2002. The reader should refer to that text for further insight into energy calculations, along with the derivation of the equations given here. The handbook editor added minor transitional wording to adapt the procedure to the format used in the handbook.

SECTION 11
HEAT TRANSFER AND ENERGY CONSERVATION

Heat-Transfer Calculation Parameters 11.1
Heat-Exchanger Choice for Specific Energy Applications 11.2
Sizing Shell-and-Tube Heat Exchangers 11.5
Temperature Determination in Heat-Exchanger Operation 11.7
Selecting and Sizing Heat Exchangers Based on Fouling Factors 11.8
Barometric and Jet Condenser Heat Transfer 11.11
Finned-Tube Heat-Exchanger Selection 11.12
Selection of Spiral-Type Heating Coils 11.14
Industrial-Use Electric Heater Selection and Sizing 11.15
Heat-Transfer Coefficient Determination for Boiler Economizer 11.17
Steam Generating Capacity of Boiler Tubes 11.18
Energy Design Analysis of Shell-and-Tube Heat Exchangers 11.20
Spiral-Plate Heat-Exchanger Design 11.34
Designing Spiral-Tube Heat Exchangers 11.43
Internal Steam Tracing Heat-Transfer Design for Pipelines 11.50
External Steam Tracing Heat-Transfer Design for Pipelines 11.58
Selecting Air-Cooled Heat Exchangers 11.63
Quick Design and Evaluation of Heat Exchangers 11.66

HEAT-TRANSFER CALCULATION PARAMETERS

In the world of energy, heat transfer is of critical importance because the greater the efficiency of heat transfer, the more heat can be recovered from the process being considered. And, in general, the more heat recovered, the more attractive the project is, provided the cost of recovering the heat is not excessive. Greater efficiency in heat recovery is producing more energy savings while reducing carbon-dioxide emissions worldwide. As a result, the overall energy field is benefiting, as is the atmosphere.

To save energy in the heat-transfer process, several requirements must be met: (*a*) a sufficient temperature difference must exist to provide a "heat head" to make heat transfer economical; (*b*) there must be a need and use for the recovered heat; (*c*) the heat transfer must result in a fuel saving that is enough to pay for the needed equipment in an acceptable time frame—usually 3 years, or less.

Factors important in providing effective heat transfer at an economical cost include: (*a*) type of heat exchanger selected for the application, with the type ranging from shell-and-tube, direct-contact with mixing, regenerative-plate type, to double-tube exchanger; (*b*) type of heat transfer occurring—heating, cooling, boiling, or condensing; (*c*) overall heat-transfer coefficient, U, which can vary from 1 to 1000 Btu/(h · °F · ft^2)[W/m^2 · °C], depending on the fluid, type of heat exchanger, relative flow direction, and other factors; (*d*) size of exchanger chosen relative to the amount of heat being transferred—whether the exchanger is over-, under-, or right-sized; (*e*) type of fluid flow selected for the heat exchanger—parallel or counterflow; and (*f*) obtaining, and using, the actual logarithmic mean temperature difference (LMTD) in the heat-transfer calculations. This section of the handbook focuses on energy conservation using heat exchangers of various types to reduce heat losses while recovering the maximum amount of usable heat possible.

HEAT-EXCHANGER CHOICE FOR SPECIFIC ENERGY APPLICATIONS

Determine the type of heat exchanger to use for each of the following applications: (1) heating oil with steam; (2) cooling internal combustion engine liquid coolant; (3) evaporating a hot liquid. For each heater chosen, specify the typical pressure range for which the heater is usually built and the typical range of the overall coefficient of heat transfer U.

Calculation Procedure:

1. *Determine the heat-transfer process involved*
In a heat exchanger, one or more of four processes may occur: heating, cooling, boiling, or condensing. Table 1 lists each of these four processes and shows the usual heat-transfer fluids involved. Thus, the heat exchangers being considered here involve (*a*) oil heater—heating—vapor-liquid; (*b*) internal-combustion engine coolant—cooling—gas-liquid; (*c*) hot-liquid evaporation—boiling—liquid-liquid.

2. *Specify the heater action and the usual type selected*
Using the same identifying letters for the heaters being selected, Table 1 shows the action and usual type of heater chosen. Thus,

Action	Type
a. Steam condensed; oil heated	Shell-and-tube
b. Air heated; water cooled	Tubes in open air
c. Waste liquid cooled; water boiled	Shell-and-tube

3. *Specify the usual pressure range and typical U*
Using the same identifying letters for the heaters being selected. Table 1 shows the action and usual type of heater chosen. Thus,

	Typical U range	
Usual pressure range	Btu/(h · °F · ft^2)	W/(m^2 · °C)
a. 0–500 lb/in^2 (abs) (0 to 3447 kPa)	20–60	113.6–340.7
b. 0–100 lb/in^2 (abs) (0 to 689.4 kPa)	2–10	11.4–56.8
c. 0–500 lb/in^2 (abs) (0 to 3447 kPa)	40–150	227.1–851.7

4. *Select the heater for each service*
Where the heat-transfer conditions are normal for the type of service met, the type of heater listed in step 2 can be safely used. When the heat-transfer conditions are unusual, a special type of heater may be needed. To select such a heater, study the data in Table 1 and make a tentative selection. Check the selection by using the methods given in the following calculation procedures in this section.

Related Calculations. Use Table 1 as a general guide to heat-exchanger selection in any industry—petroleum, chemical, power, marine, textile, lumber, etc. Once the general type of heater and its typical U value are known, compute the required size, using the procedure given later in this section.

TABLE 1 Heat-Exchanger Selection Guide*

	Heat-transfer fluids	Equipment	Action	Type†	Pressure range‡	Typical range of U§	
	Liquid-liquid	Boiler-water blowdown exchanger	Blowdown cooled, feedwater heated	*S*	*M, H*	50–300	(0.28–1.7)
		Laundry-water heat reclaimer	Waste water cooled, feed heated	*S*	*L*	30–200	(0.17–1.1)
		Service-water heater	Waste liquid cooled, water heated	*S*	*L, H*	50–300	(0.28–1.7)
	Vapor-liquid	Bleeder heater	Steam condensed, feedwater heated	*S*	*L, H*	200–800	(1.1–4/5)
		Deaerating feed heater	Steam condensed, feedwater heated	*M*	*L, M*	*DC*	
		Jet heater	Steam condensed, water heated	*M*	*L*	*DC*	
		Process kettle	Steam condensed, liquid heated	*S*	*L, M*	100–500	(0.57–2.8)
		Oil heater	Steam condensed, oil heated	*S*	*L, M*	20–60	(0.11–0.34)
Heating		Service-water heater	Steam condensed, water heated	*S*	*L, M*	200–800	(1.1–4.5)
		Open flow-through heater	Steam condensed, water heated	*M*	*L*	*DC*	
		Liquid-sodium steam superheater	Sodium cooled, steam superheated	*S*	*M, H*	50–200	(0.28–1.1)
	Gas-liquid	Waste-heat water heater	Waste gas cooled, water heated	*T*	*L*	2–10	(0.011–0.05)
		Boiler economizer	Flue gas cooled, feedwater heated	*T*	*M, H*	2–10	(0.011–0.05)
		Hot-water radiator	Water cooled, air heated	*T*	*L*	1–10	(0.0057–0.05)
	Gas-gas	Boiler air heater	Flue gas cooled, combustion air heated	*T, R*	*L*	2–10	(0.011–0.05)
		Gas-turbine regenerator	Flue gas cooled, combustion air heated	*T*	*L*	2–10	(0.011–0.05)
	Vapor-gas	Boiler superheater	Combustion gas cooled, steam superheated	*T*	*M, H*	2–20	(0.011–0.11)
		Steam pipe coils	Steam condensed, air heated	*T*	*L, M*	2–10	(0.011–0.05)
		Steam radiator	Steam condensed, air heated	*T*	*L*	2–10	(0.011–0.05)
Cooling	Liquid-liquid	Oil cooler	Water heated, oil cooled	*S, D*	*L, M*	20–200	(0.11–1.1)
		Water chiller	Refrigerant boiled, water cooled	*S*	*L, M*	30–151	(0.17–0.86)
		Brine cooler	Refrigerant boiled, brine cooled	*S*	*L, M*	30–150	(0.17–0.86)
		Transformer-oil cooler	Water heated, oil cooled	*S*	*L, M*	20–50	(0.11–0.88)
	Vapor-liquid	Boiler desuperheater	Boiler water heated, steam desuperheated	*S, M*	*M, H*	150–800	(0.85–4.5)
Cooling	Gas-liquid	Compressor intercoolers and aftercoolers	Water heated, compressed air cooled	*S*	*L, H*	10–20	(0.057–0.11)
		Internal-combustion-engine radiator	Air heated, water cooled	*T*	*L*	2–10	(0.011–0.05)
		Generator hydrogen, air coolers	Water heated, hydrogen or air cooled	*S*	*L*	2–10	(0.011–0.05)
		Air-conditioning cooler	Water heated, air cooled	*T*	*L*	2–10	(0.011–0.05)
		Refrigeration heat exchanger	Brine heated, air cooled	*T*	*L, M*	2–10	(0.011–0.05)
		Refrigeration evaporator	Refrigerant boiled, air cooled	*T*	*L, M*	2–10	(0.011–0.05)
	Vapor-gas	Boiler desuperheater	Flue gas heated, steam desuperheated	*T*	*L, H*	2–8	(0.011–0.04)
	Liquid-liquid	Hot-liquid evaporator	Waste liquid cooled, water boiled	*S*	*L, H*	40–150	(0.23–0.85)
		Liquid-sodium steam generator	Sodium cooled, water boiled	*S*	*M, H*	500–1000	(2.8–5.7)
	Vapor-liquid	Evaporator (vacuum)	Steam condensed, water boiled	*S*	*L*	400–600	(2.3–3.4)

(*Continued*)

TABLE 1 Heat-Exchanger Selection Guide* (*Continued*)

	Heat-transfer fluids	Equipment	Action	Type†	Pressure range‡	Typical range of U§	
Boiling		Evaporator (high pressure)	Steam condensed, water boiled	*S*	*L, M*	400–600	(2.3–3.4)
		Mercury condenser-boiler	Mercury condensed, water boiled	*S*	*M, H*	500–700	(2.8–4.0)
	Gas-liquid	Waste-heat steam boiler	Flue gas cooled, water boiled	*T*	*L, H*	2–10	(0.011–0.05)
		Direct-fired steam boiler	Combustion gas cooled, water boiled	*T*	*L, H*	2–10	(0.011–0.05)
	Vapor-liquid	Refrigeration condenser	Water heated, refrigerant condensed	*S, D*	*L, M*	80–250	(0.45–1.4)
		Steam surface condenser	Water heated, steam condensed	*S*	*L*	300–800	(1.7–4.5)
		Steam mixing condenser	Water heated, steam condensed	*M*	*L*	*DC*	
Condensing		Intercondenser and aftercondenser	Condensate heated, steam condensed	*S*	*L*	15–300	(0.085–1.7)
	Vapor-gas	Air-cooled surface condenser	Air heated, steam condensed	*T*	*L*	2–16	(0.011–0.09)

**Power.*

†*S*—shell-and-tube exchanger; *M*—direct-contact mixing exchanger; *T*—tubes in path of moving fluid, or exchanger open to surrounding air; *R*—regenerative plate-type or simple plate-type exchanger; *D*—double-tube exchanger.

‡*L*—highest pressure ranges from 0 to 100 lb/in^2 (abs) (0 to 689.4 kPa); *M*—highest pressure from 100 to 500 lb/in^2 (abs) (689.4 to 3447 kPa); *H*—500 lb/in^2 (abs) (3447 kPa) up.

§Values of U represent range of overall heat-transfer coefficients that might be expected in various exchangers. Coefficients are stated in Btu/(h · °F · ft^2) [W/(m^2 · °C)] of heating surface. Total heat transferred in exchanger, in Btu/h, is obtained by multiplying a specific value of U for that type of exchanger by the surface and the log mean temperature difference. *DC* indicates direct exchange of heat.

SIZING SHELL-AND-TUBE HEAT EXCHANGERS

What is the required heat-transfer area for a parallel-flow shell-and-tube heat exchanger used to heat oil if the entering oil temperature is 60°F (15.6°C), the leaving oil temperature is 120°F (48.9°C), and the heating medium is steam at 200 lb/in^2 (abs) (1378.8 kPa)? There is no subcooling of condensate in the heat exchanger. The overall coefficient of heat transfer $U = 25$ Btu/(h · °F · ft^2) [141.9 W/(m^2 · °C)]. How much heating steam is required if the oil flow rate through the heater is 100 gal/min (6.3 L/s), the specific gravity of the oil is 0.9, and the specific heat of the oil is 0.5 Btu/(lb · °F) [2.84 W/(m^2 · °C)]?

Calculation Procedure:

1. ***Compute the heat-transfer rate of the heater***
With a flow rate of 100 gal/min (6.3 L/s) or (100 gal/min)(60 min/h) = 6000 gal/h (22,710 L/h), the weight flow rate of the oil, using the weight of water of specific gravity 1.0 as 8.33 lb/gal, is (6000 gal/h)(0.9 specific gravity)(8.33 lb/gal) = 45,000 lb/h (20,250 kg/h), closely.

Since the temperature of the oil rises 120 – 60 = 60°F (33.3°C) during passage through the heat exchanger and the oil has a specific heat of 0.50, find the heat-transfer rate of the heater from the general relation $Q = wc\ \Delta t$, where Q = heat-transfer rate, Btu/h; w = oil flow rate, lb/h; c = specific heat of the oil, Btu/(lb · °F); Δt = temperature rise of the oil during passage through the heater. Thus, $Q = (45{,}000)(0.5)(60) = 1{,}350{,}000$ Btu/h (0.4 MW).

2. ***Compute the heater logarithmic mean temperature difference***
The LMTD is found from LMTD = $(G - L)/\ln (G/L)$, where G = greater terminal temperature difference of the heater, °F; L = lower terminal temperature difference of the heater, °F; ln = logarithm to the base e. This relation is valid for heat exchangers in which the number of shell passes equals the number of tube passes.

In general, for parallel flow of the fluid streams, $G = T_1 - t_1$ and $L = T_2 - t_2$, where T_1 = heating fluid inlet temperature, °F; T_2 = heating fluid outlet temperature, °F; t_1 = heated fluid inlet temperature, °F; t_2 = heated fluid outlet temperature, °F. Figure 1 shows the maximum and minimum terminal temperature differences for various fluid flow paths.

For this parallel-flow exchanger, $G = T_1 - t_1 = 382 - 60 = 322$°F (179°C), where 382°F (194°C) = the temperature of 200-lb/in^2 (abs) (1379-kPa) saturated steam, from a table of steam properties. Also, $L = T_2 - t_2 = 382 - 120 = 262$°F (145.6°C), where the condensate temperature = the saturated steam temperature because there is no subcooling of the condensate. Then LMTD = $G - L/\ln (G/L)$ = $(322 - 262)/\ln (322/262) = 290$°F (16°C).

3. ***Compute the required heat-transfer area***
Use the relation $A = Q/U \times$ LMTD, where A = required heat-transfer area, ft^2; U = overall coefficient of heat transfer, Btu/(ft^2 · h · °F). Thus, $A = 1{,}350{,}000/[(25)(290)] = 186.4$ ft^2 (17.3 m^2), say 200 ft^2 (18.6 m^2).

4. ***Compute the required quantity of heating steam***
The heat added to the oil = Q = 1,350,000 Btu/h, from step 1. The enthalpy of vaporization of 200-lb/in^2 (abs) (1379-kPa) saturated steam is, from the steam tables, 843.0 Btu/lb (1960.8 kJ/kg). Use the relation $W = Q/h_{fg}$, where W = flow rate of heating steam, lb/h; h_{fg} = enthalpy of vaporization of the heating steam, Btu/lb. Hence, $W = 1{,}350{,}000/843.0 = 1600$ lb/h (720 kg/h).

Related Calculations. Use this general procedure to find the heat-transfer area, fluid outlet temperature, and required heating-fluid flow rate when true parallel flow or counterflow of the fluids occurs in the heat exchanger. When such a true flow does *not* exist, use a suitable correction factor, as shown in the next calculation procedure.

The procedure described here can be used for heat exchangers in power plants, heating systems, marine propulsion, air-conditioning systems, etc. Any heating or cooling fluid—steam, gas, chilled water, etc.—can be used.

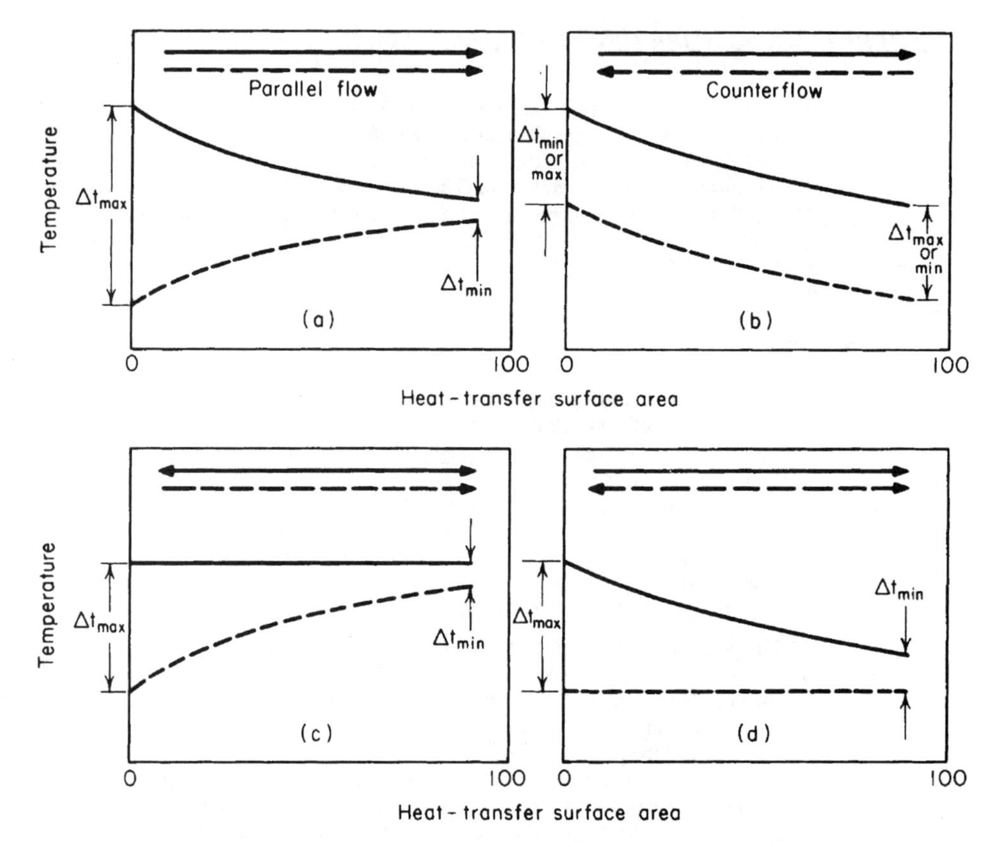

FIGURE 1 Temperature relations in typical parallel-flow and counterflow heat exchangers.

To select a heat exchanger by using the results of this calculation procedure, enter the engineering data tables available from manufacturers at the computed heat-transfer area. Read the heater dimensions directly from the table. Be sure to use the next *larger* heat-transfer area when the exact required area is not available.

When there is little movement of the fluid on either side of the heat-transfer area, such as occurs during heat transmission through a building wall, the arithmetic mean (average) temperature difference can be used instead of the LMTD. Use the LMTD when there is rapid movement of the fluids on either side of the heat-transfer area and a rapid change in temperature in one, or both, fluids. When one of the two fluids is partially, but not totally, evaporated or condensed, the true mean temperature difference is different from the arithmetic mean and the LMTD. Special methods, such as those presented in Perry—*Chemical Engineers' Handbook*, McGraw-Hill, 2007, must be used to compute the actual temperature difference under these conditions.

When two liquids or gases with constant specific heats are exchanging heat in a heat exchanger, the area between their temperature curves, Fig. 2, is a measure of the total heat being transferred. Figure 2 shows how the temperature curves vary with the amount of heat-transfer area for counterflow and parallel-flow exchangers when the fluid inlet temperatures are kept constant. As Fig. 2 shows, the counterflow arrangement is superior.

If enough heating surface is provided, in a counterflow exchanger, the leaving cold-fluid temperature can be raised above the leaving hot-fluid temperature. This cannot be done in a parallel-flow exchanger, where the temperatures can only approach each other regardless of how much surface

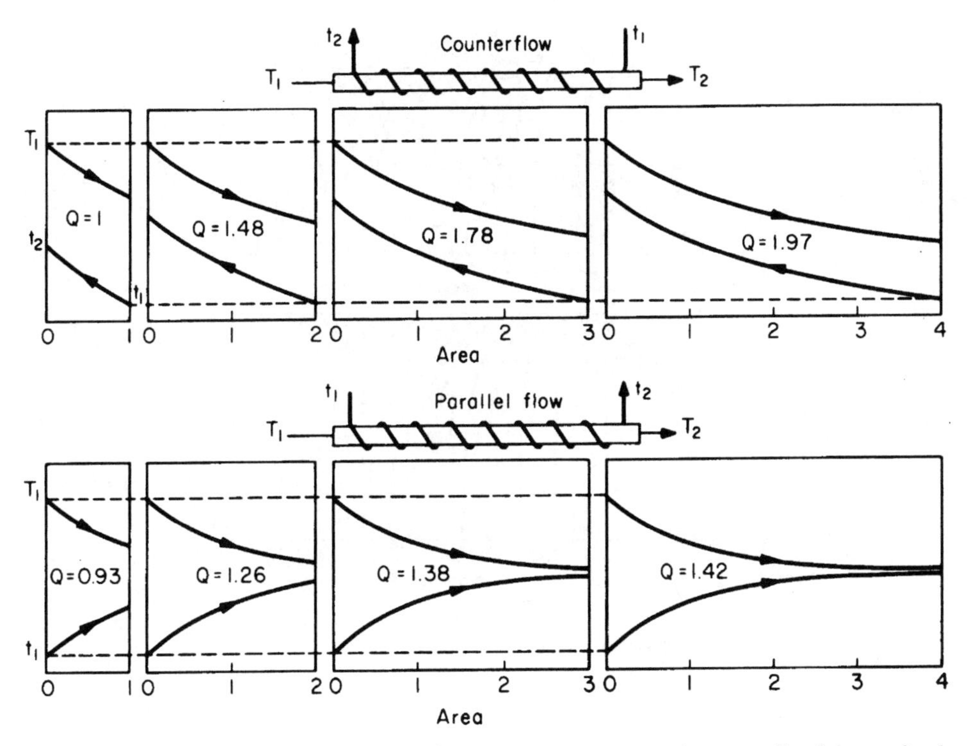

FIGURE 2 For certain conditions, the area between the temperature curves measures the amount of heat being transferred.

is used. The counterflow arrangement transfers more heat for given conditions and usually proves more economical to use.

TEMPERATURE DETERMINATION IN HEAT-EXCHANGER OPERATION

A counterflow shell-and-tube heat exchanger has one shell pass for the heating fluid and two shell passes for the fluid being heated. What is the actual LMTD for this exchanger if T_1 = 300°F (148.9°C), T_2 = 250°F (121°C), t_1 = 100°F (37.8°C), and t_2 = 230°F (110°C)?

Calculation Procedure:

1. *Determine how the LMTD should be computed*

When the numbers of shelf and tube passes are unequal, true counterflow does not exist in the heat exchanger. To allow for this deviation from true counterflow, a correction factor must be applied to the logarithmic mean temperature difference (LMTD). Figure 3 gives the correction factor to use.

2. *Compute the variables for the correction factor*

The two variables that determine the correction factor are shown in Fig. 3 as $P = (t_2 - t_1)/(t_1 - t_1)$ and $R = (T_1 - T_2)/(t_2 - t_1)$. Thus, $P = (230 - 100)/(300 - 100) = 0.65$, and $R = (300 - 250)/(230 - 100) = 0.385$. From Fig. 3, the correction factor is $F = 0.90$ for these values of P and R.

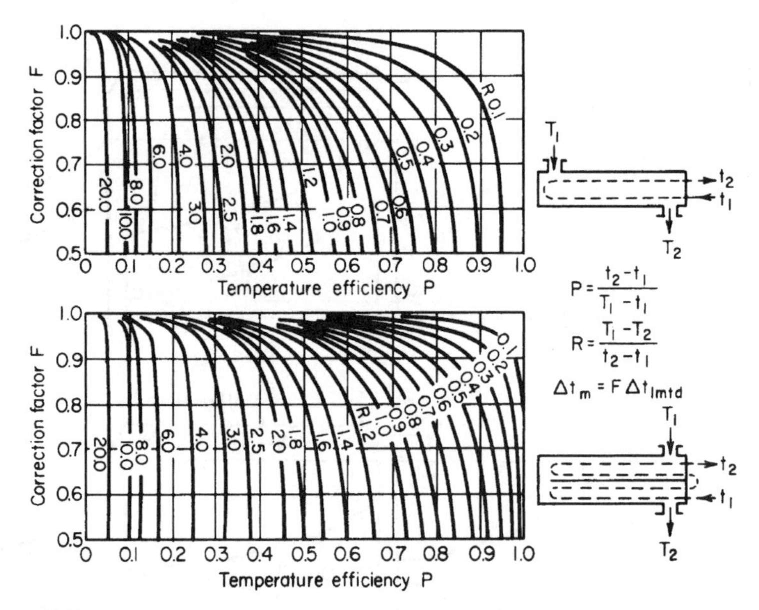

FIGURE 3 Correction factors for LMTD when the heater flow path differs from the counterflow. (*Power.*)

3. *Compute the theoretical LMTD*

Use the relation LMTD = $(G - L)/\ln (G/L)$, where the symbols for counterflow heat exchange are $G = T_2 - t_1$; $L = T_1 - t_2$; ln = logarithm to the base *e*. All temperatures in this equation are expressed in °F. Thus, $G = 250 - 100 = 150$°F (83.3°C); $L = 300 - 230 = 70$°F (38.9°C). Then LMTD = $(150 - 70)/\ln (150/70) = 105$°F (58.3°C).

4. *Compute the actual LMTD for this exchanger*

The actual LMTD for this or any other heat exchanger is $\text{LMTD}_{\text{actual}} = F(\text{LMTD}_{\text{computed}}) = 0.9(105) =$ 94.5°F (52.5°C). Use the actual LMTD to compute the required exchanger heat-transfer area.

Related Calculations. Once the corrected LMTD is known, compute the required heat-exchanger size in the manner shown in the previous calculation procedure. The method given here is valid for both two- and four-pass shell-and-tube heat exchangers. Figure 4 simplifies the computation of the uncorrected LMTD for temperature differences ranging from 1 to 1000°F (–17 to 537.8°C). It gives LMTD with sufficient accuracy for all normal industrial and commercial heat-exchanger applications. Correction-factor charts for three shell passes, six or more tube passes, four shell passes, and eight or more tube passes are published in the *Standards of the Tubular Exchanger Manufacturers Association.*

SELECTING AND SIZING HEAT EXCHANGERS BASED ON FOULING FACTORS

A heat exchanger having an overall coefficient of heat transfer of $U = 100$ Btu/(ft^2 · h · °F) [567.8 W/(m^2 · °C)] is used to cool lean oil. What effect will the tube fouling have on the value of U for this exchanger?

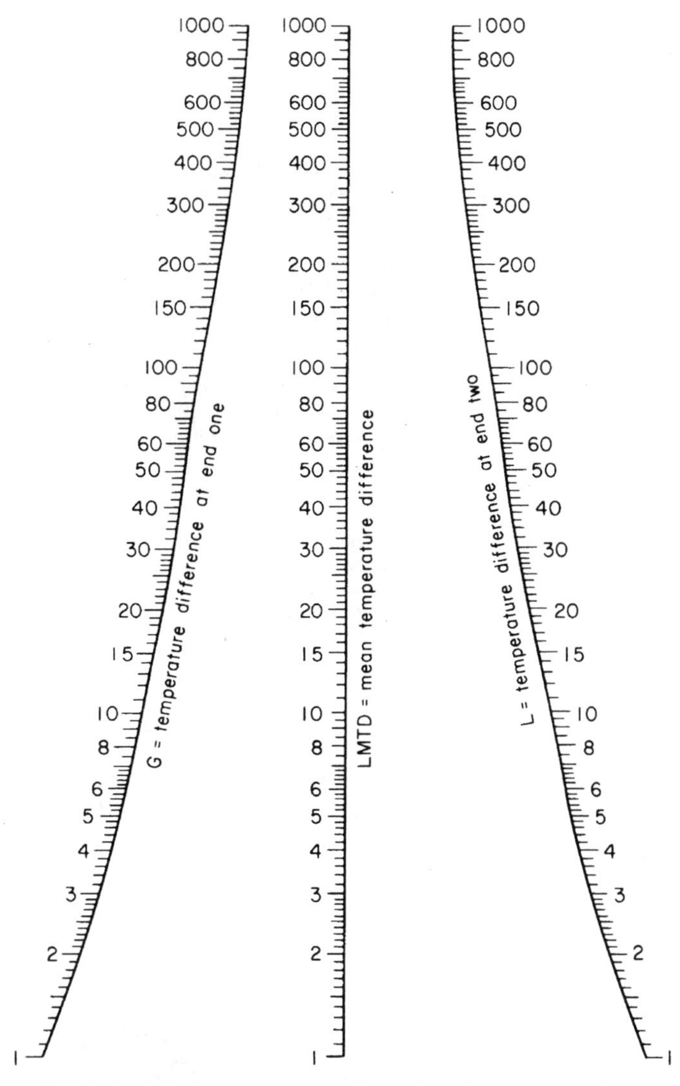

FIGURE 4 Logarithmic mean temperature for a variety of heat-transfer applications.

Calculation Procedure:

1. *Determine the heat exchange fouling factor*

Use Table 2 to determine the fouling factor for this exchanger. Thus, the fouling factor for lean oil = 0.0020.

2. *Determine the actual U for the heat exchanger*

Enter Fig. 5 at the bottom with the clean heat-transfer coefficient of U = 100 Btu/(h · ft^2 · °F) [567.8 W/(m^2 · °C)] and project vertically upward to the 0.002 fouling-factor curve. From the intersection with this curve, project horizontally to the left to read the design or actual heat-transfer coefficient as U_a = 78 Btu/(h · ft^2 · °F) [442.9 W/(m^2 · °C)]. Thus, the fouling of the tubes causes a

TABLE 2 Heat-Exchanger Fouling Factors*

Fluid heated or cooled	Fouling factor
Fuel oil	0.0055
Lean oil	0.0020
Clean recirculated oil	0.0010
Quench oils	0.0042
Refrigerants (liquid)	0.0011
Gasoline	0.0006
Steam-clean and oil-free	0.0001
Refrigerant vapors	0.0023
Diesel exhaust	0.013
Compressed air	0.0022
Clean air	0.0011
Seawater under 130°F (54°C)	0.0006
Seawater over 130°F (54°C)	0.0011
City or well water under 130°F (54°C)	0.0011
City or well water over 130°F (54°C)	0.0021
Treated boiler feedwater under 130°F, 3 ft/s (54°C, 0.9 m/s)	0.0008
Treated boiler feedwater over 130°F, 3 ft/s (54°C, 0.9 m/s)	0.0009
Boiler blowdown	0.0022

*Condenser Service and Engineering Company, Inc.

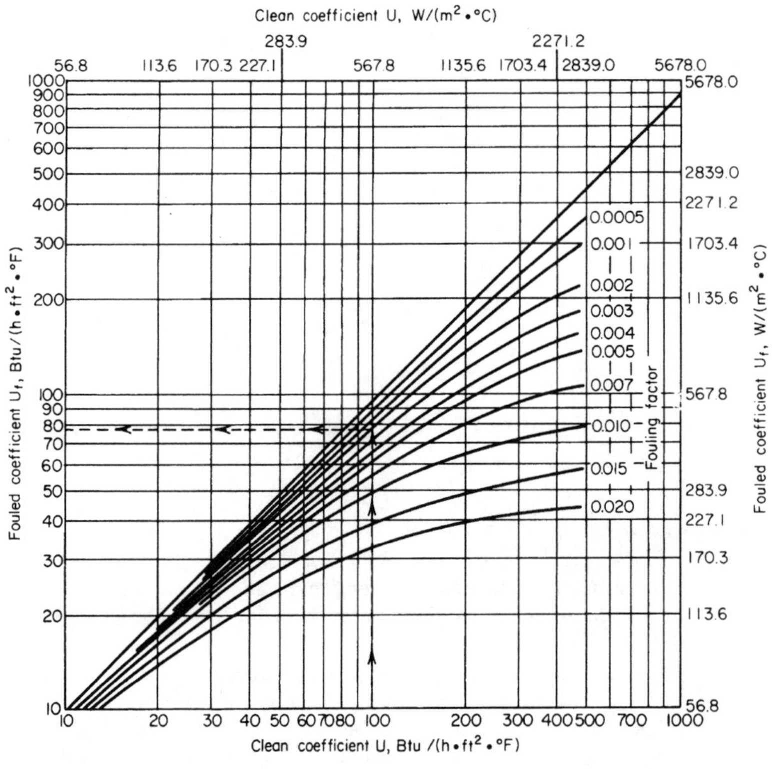

FIGURE 5 Effect of heat-exchanger fouling on the overall coefficient of heat transfer. *(Condenser Service and Engineering Co., Inc.)*

reduction of the U value of 100 – 78 = 22 Btu/(h · ft^2 · °F) [124.9 W/(m^2 · °C)]. This means that the required heat-transfer area must be increased by nearly 25 percent to compensate for the reduction in heat transfer caused by fouling.

Related Calculations. Table 2 gives fouling factors for a wide variety of service conditions in applications of many types. Use these factors as described previously, or add the fouling factor to the film resistance for the heat exchanger to obtain the total resistance to heat transfer. Then U = the reciprocal of the total resistance. Use the actual value U_a of the heat-transfer coefficient when sizing a heat exchanger. The method given here is that used by Condenser Service and Engineering Company, Inc.

BAROMETRIC AND JET CONDENSER HEAT TRANSFER

A counterflow barometric condenser must maintain an exhaust pressure of 2 lb/in^2 (abs) (13.8 kPa) for an industrial process. What condensing-water flow rate is required with a cooling-water inlet temperature of 60°F (15.6°C); of 80°F (26.7°C)? How much air must be removed from this barometric condenser if the steam flow rate is 25,000 lb/h (11,250 kg/h); 250,000 lb/h (112,500 kg/h)?

Calculation Procedure:

1. *Compute the required unit cooling-water flow rate*

Use Fig. 6 as a quick guide to the required cooling-water flow rate for counterflow barometric condensers. Thus, entering the bottom of Fig. 6 at 2-lb/in^2 (abs) (13.8-kPa) exhaust pressure and projecting vertically upward to the 60°F (15.6°C) and 80°F (26.7°C) cooling-water inlet temperature curves

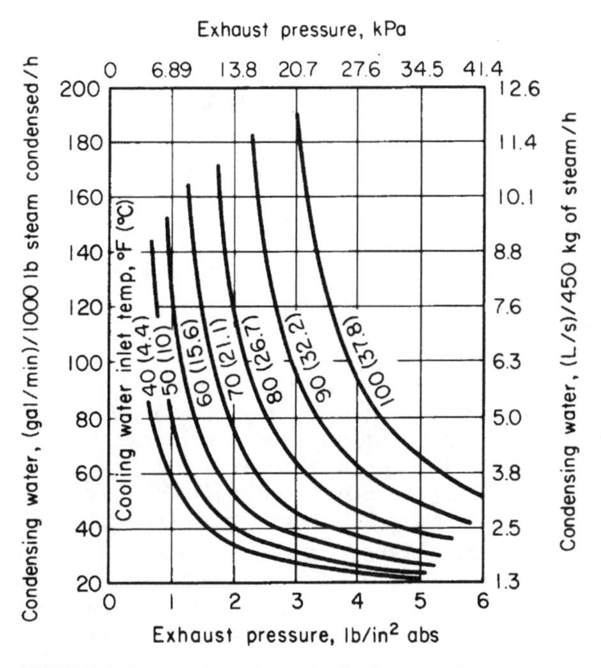

FIGURE 6 Barometric condenser condensing-water flow rate.

show that the required flow rate is 52 gal/min (3.2 L/s) and 120 gal/min (7.6 L/s), respectively, per 1000 lb/h (450 kg/h) of steam condensed.

2. ***Compute the total cooling-water flow rate required***
Use this relation: total cooling water required, gal/min = (unit cooling-water flow rate, gal/min per 1000 lb/h of steam condensed) (steam flow, lb/h)/1000. Or, total gal/min = (52)(250,000/1000) = 13,000 gal/min (820.2 L/s) of 60°F (15.6°C) cooling water. For 80°F (26.7°C) cooling water, total gal/min = (120)(250,000/1000) = 30,000 gal/min (1892.7 L/s). Thus, a 20°F (11.1°C) rise in the cooling-water temperature raises the flow rate required by 30,000 – 13,000 = 17,000 gal/min (1072.5 L/s).

3. ***Compute the quantity of air that must be handled***
With a steam flow of 25,000 lb/h (11,250 kg/h) to a barometric condenser, manufacturers' engineering data show that the quantity of air entering with the steam is 3 ft^3/min (0.08 m^3/min); with a steam flow of 250,000 lb/h (112,500 kg/h), air enters at the rate of 10 ft^3/min (0.28 m^3/min). Hence, the quantity of air in the steam that must be handled by this condenser is 10 ft^3/min (0.28 m^3/min).

Air entering with the cooling water varies from about 2 ft^3/min per 1000 gal/min of 100°F (0.06 m^3/min per 3785 L/min of 37.8°C) water to 4 ft^3/min per 1000 gal/min at 35°F (0.11 m^3/min per 3785 L/min at 1.7°C). Using a value of 3 ft^3/min (0.08 m^3/min) for this condenser, we see the quantity of air that must be handled is (ft^3/min per 1000 gal/min)(cooling-water flow rate, gal/min)(1000, or ft^3/min of air = (3)(13,000/1000) = 39 ft^3/min at 60°F (1.1 m^3/min at 15.6°C). At 80°F (26.7°C) ft^3/min = (3)(30,000/1000) = 90 ft^3/min (2.6 m^3/min).

Hence, the total air quantity that must be handled is 39 + 10 = 49 ft^3/min (1.4 m^3/min) with 60°F (15.6°C) cooling water, and 90 + 10 = 100 ft^3/min (2.8 m^3/min) with 80°F (26.7°C) cooling water. The air is usually removed from the barometric condenser by a two-stage air ejector.

Related Calculations. For help in specifying conditions for parallel-flow and counterflow barometric condensers, refer to *Standards of Heat Exchange Institute—Barometric and Low-Level Jet Condensers.* Whereas Fig. 6 can be used for a first approximation of the cooling water required for parallel-flow barometric condensers, the results obtained will not be as accurate as for counterflow condensers.

FINNED-TUBE HEAT-EXCHANGER SELECTION

Choose a finned-tube heat exchanger for a 1000-hp (746-kW) four-cycle turbocharged diesel engine having oil-cooled pistons and a cooled exhaust manifold. The heal exchanger will be used only for jacket-water cooling.

Calculation Procedure:

1. ***Determine the heat-exchanger cooling load***
The Diesel Engine Manufacturers Association (DEMA) tabulation, Table 3, lists the heat rejection to the cooling system by various types of diesel engines. Table 3 shows that the heat rejection from the jacket water of a four-cycle turbocharged engine having oil-cooled pistons and a cooled manifold is 1800 to 2200 Btu/(bhp · h) (0.71 to 0.86 kW/kW). Using the higher value, we see the jacket-water heat rejection by this engine is (1000 bhp)[2200 Btu/(bhp · h)] = 2,200,000 Btu/h (644.8 kW).

2. ***Determine the jacket-water temperature rise***
DEMA reports that a water temperature rise of 15 to 20°F (8.3 to 11.1°C) is common during passage of the cooling water through the engine. The maximum water discharge temperature reported by DEMA ranges from 140 to 180°F (60 to 82.2°C). Assume a 20°F (11. 1°C) water temperature rise and a 160°F (71.1°C) water discharge temperature for this engine.

TABLE 3 Approximate Rates of Heat Rejection to Cooling Systems*

	Four-cycle engines			
Engine type	Normally aspirated, dry pistons, water-jacketed exhaust manifold, Btu/(bhp · h) (kJ/kWh)	Normally aspirated, oil-cooled pistons, water-jacketed manifold, Btu/(bhp · h) (kJ/kWh)	Turbocharged, oil-cooled pistons, dry manifold, Btu/(bhp · h) (kJ/kWh)	Turbocharged, oil-cooled pistons, cooled manifold, Btu/(bhp · h) (kJ/kWh)
Jacket water	2200–2600 (12.5–14.8)	2000–2500 (11.3–14.2)	1450–1750 (8.2–9.9)	1800–2200 (10.2–12.5)
Lubricating oil	175–350 (1.0–2.0)	300–600 (1.7–3.4)	300–500 (1.7–2.8)	300–500 (1.7–2.8)
Raw water	2375–2950 (13.5–16.7)	2300–3100 (13.1–17.6)	1750–2250 (9.9–12.8)	2100–2700 (11.9–15.3)

	Two-cycle engines		
		Uniflow scavenging oil-cooled pistons	
Engine type	Loop scavenging oil-cooled pistons, Btu/(bhp · h) (kJ/kWh)	Opposed piston, Btu/(bhp · h) (kJ/kWh)	Valve in head, Btu/(bhp · h) (kJ/kWh)
Jacket water	1300–1900 (7.4–10.8)	1200–1600 (6.8–9.1)	1700–2100 (9.6–11.9)
Lubricating oil	500–700 (2.8–4.0)	900–1100 (5.1–6.2)	400–750 (2.3–4.3)
Raw water	1800–2600 (10.2–14.8)	2100–2700 (11.9–15.3)	2100–2850 (11.9–16.2)

*Diesel Engine Manufacturers Association; SI values added by handbook editor.

3. *Determine the air inlet and outlet temperatures*
Refer to weather data for the locality of the engine installation. Assume that the weather data for the locality of this engine show that the maximum dry-bulb temperature met in summer is 90°F (32.2°C). Use this as the air inlet temperature.

Before the required surface area can be determined, the air outlet temperature from the radiator must be known. This outlet temperature cannot be computed directly. Hence, it must be assumed and a trial calculation made. If the area obtained is too large, a higher outlet air temperature must be assumed and the calculation redone. Assume an outlet air temperature of 150°F (65.6°C).

4. *Compute the LMTD for the radiator*
The largest temperature difference for this exchanger is 160 – 90 = 70°F (38.9°C), and the smallest temperature difference is 150 – 140 = 10°F (5.6°C). In the smallest temperature difference expression, 140°F (77.8°C) = water discharge temperature from the engine – cooling-water temperature rise during passage through the engine, or 160 – 20 = 140°F (77.8°C). Then LMTD = (70 – 10)/[ln (70/10)] = 30°F (16.7°C). (Figure 4 could also be used to compute the LMTD.)

5. *Compute the required exchanger surface area*
Use the relation $A = Q/U \times \text{LMTD}$, where A = surface area required, ft^2; Q = rate of heat transfer, Btu/h; U = overall coefficient of heat transfer, Btu/(h · ft^2 · °F). To solve this equation, U must be known.

Table 1 in the first calculation procedure in this section shows that U ranges from 2 to 10 Btu/(h · ft^2 · °F) [56.8 W/(m^2 · °C)] in the usual internal-combustion-engine finned-tube radiator. Using a value of 5 for U, we get A = 2,200,000[(5)(30)] = 14,650 ft^2 (1361.0 m^2).

6. *Determine the length of finned tubing required*
The total area of a finned tube is the sum of the tube and fin area per unit length. The tube area is a function of the tube diameter, whereas the finned area is a function of the number of fins per inch of tube length and the tube diameter.

Assume that 1-in (2.5-cm) tubes having 4 fins per inch (6.35 mm per fin) are used in this radiator. A tube manufacturer's engineering data show that a finned tube of these dimensions has 5.8 ft^2 of area per linear foot (1.8 m^2/lin m) of tube.

To compute the linear feet L of finned tubing required, use the relation $L = A/(ft^2/ft)$, or $L = 14{,}650/5.8 = 2530$ lin ft (771.1 m) of tubing.

7. *Compute the number of individual tubes required*
Assume a length for the radiator tubes. Typical lengths range between 4 and 20 ft (1.2 and 6.1 m), depending on the size of the radiator. With a length of 16 ft (4.9 m) per tube, the total number of tubes required = 2530/16 = 158 tubes. This number is typical for finned-tube heat exchangers having large heat-transfer rates [more than 10^6 Btu/h (100 kW)].

8. *Determine the fan hp required*
The fan hp required can be computed by determining the quantity of air that must be moved through the heat exchanger, after assuming a resistance—say 1.0 in of water (0.025 Pa)—for the exchanger. However, the more common way of determining the fan hp is by referring to the manufacturer's engineering data.

Thus, one manufacturer recommends three 5-hp (3.7-kW) fans for this cooling load, and another recommends two 8-hp (5.9-kW) fans. Hence, about 16 hp (11.9 kW) is required for the radiator.

Related Calculations. The steps given here are suitable for the initial sizing of finned-tube heat exchangers for a variety of applications. For exact sizing, it may be necessary to apply a correction factor to the LMTD. These correction factors are published in Kern—*Process Heat Transfer,* McGraw-Hill, 1997, and McAdams—*Heat Transfer,* McGraw-Hill, 1997.

The method presented here can be used for finned-tube heat exchangers used for air heating or cooling, gas heating or cooling, and similar industrial and commercial applications.

SELECTION OF SPIRAL-TYPE HEATING COILS

How many feet of heating coil are required to heat 1000 gal/h (1.1 L/s) of 0.85-specific-gravity oil if the specific heat of the oil is 0.50 Btu/(lb · °F) [2.1 kJ/(kg · °C)], the heating medium is 65-lb/in^2 (gage) (448.2-kPa) steam, and the oil enters at 60°F (15.6°C) and leaves at 125°F (51.7°C)? There is no subcooling of the condensate.

Calculation Procedure:

1. *Compute the LMTD for the heater*
Steam at 65 + 14.7 = 79.7 lb/in^2 (abs) (549.5 kPa) has a temperature of approximately 312°F (155.6°C), as given by the steam tables. Condensate at this pressure has the same approximate temperature. Hence, the entering and leaving temperatures of the heating fluid are approximately the same.

Oil enters the heater at 60°F (15.6°C) and leaves at 125°F (51.7°C). Therefore, the greater temperature G across the heater is $G = 312 - 60 = 252$°F (140.0°C), and the lesser temperature difference L is $L = 312 - 125 = 187$°F (103.9°C). Hence, the LMTD = $(G - L)/[\ln(G/L)]$, or (252 − 187)/[ln (252/187)] = 222°F (123.3°C). In this relation, ln = logarithm to the base $e = 2.7183$. (Figure 4 could also be used to determine the LMTD.)

2. *Compute the heat required to raise the oil temperature*
Water weighs 8.33 lb/gal (1.0 kg/L). Since this oil has a specific gravity of 0.85, it weighs (8.33)(0.85) = 7.08 lb/gal (0.85 kg/L). With 1000 gal/h (1.1 L/s) of oil to be heated, the weight of oil

heated is (1000 gal/h)(7.08 lb/gal) = 7080 lb/h (0.89 kg/s). Since the oil has a specific heat of 0.5 Btu/(lb · °F) [2.1 kJ/(kg · °C)] and this oil is heated through a temperature range of 125 – 60 = 65°F (36.1°C), the quantity of heat Q required to raise the temperature of the oil is Q = (7080 lb/h) [0.5 Btu/(lb · °F) (65°F)] = 230,000 Btu/h (67.4 kW).

3. ***Compute the heat-transfer area required***
Use the relation $A = Q/(U \times \text{LMTD})$, where Q = heat-transfer rate, Btu/h; U = overall coefficient of heat transfer, Btu/(h · ft^2 · °F). For heating oil to 125°F (51.7°C), the U value given in Table 1 is 20 to 60 Btu/(h · ft^2 · °F) [0.11 to 0.34 kW/(m^2 · °C)]. Using a value of U = 30 Btu/(h · ft^2 · °F) [0.17 kW/(m^2 · °C)] to produce a conservatively sized heater, we find A = 230,000/[(30)(222)] = 33.4 ft^2 (3.1 m^2) of heating surface.

4. ***Choose the coil material for the heater***
Spiral-type tank heating coils are usually made of steel because this material has a good corrosion resistance in oil. Hence, this coil will be assumed to be made of steel.

5. ***Compute the heating steam flow required***
To determine the steam flow rate required, use the relation $S = Q/h_{fg}$, where S = steam flow, lb/h; h_{fg} = latent heat of vaporization of the heating steam, Btu/lb, from the steam tables; other symbols as before. Hence, S = 230,000/901.1 = 256 lb/h (0.03 kg/s), closely.

6. ***Compute the heating coil pipe diameter***
Steam-heating coils submerged in the liquid being heated are usually chosen for a steam velocity of 4000 to 5000 ft/min (20.3 to 25.4 m/s). Compute the heating pipe cross-sectional area a in^2 from $a = 2.4Sv_g/V$, where v_g = specific volume of the steam at the coil operating pressure, ft^3/lb, from the steam tables; V = steam velocity in the heating coil, ft/min; other symbols as before. With a steam velocity of 4000 ft/min (20.3 m/s), a = 2.4(256) (5.47)/4000 = 0.838 in^2 (5.4 cm^2).

Refer to a tabulation of pipe properties. Such a tabulation shows that the internal transverse area of a schedule 40 1-in (2.5-cm) diameter nominal steel pipe is 0.863 in^2 (5.6 cm^2). Hence, a 1-in (2.5-cm) pipe will be suitable for this heating coil.

7. ***Determine the length of coil required***
A pipe property tabulation shows that 2.9 lin ft (0.9 m) of 1-in (2.5-cm) schedule 40 pipe has 1.0 ft^2 (0.09 m^2) of external area. Hence, the total length of pipe required in this heating coil = (33.1 ft^2) (2.9 ft/ft^2) = 96 ft (29.3 m).

Related Calculations. Use this general procedure to find the area and length of spiral heating coil required to heat water, industrial solutions, oils, etc. This procedure also can be used to find the area and length of cooling coils used to cool brine, oils, alcohol, wine, etc. In every case, be certain to substitute the correct specific heat for the liquid being heated or cooled. For typical values of U, consult Perry—*Chemical Engineers' Handbook,* McGraw-Hill, 2007; McAdams—*Heat Transmission,* McGraw-Hill, 1985; or Kern—*Process Heat Transfer,* McGraw-Hill, 1997.

INDUSTRIAL-USE ELECTRIC HEATER SELECTION AND SIZING

Choose the heating capacity of an electric heater to heat a pot containing 600 lb (272.2 kg) of lead from the charging temperature of 70°F (21.1°C) to a temperature of 750°F (398.9°C) if 600 lb (272.2 kg) of the lead is to be melted and heated per hour. The pot is 30 in (76.2 cm) in diameter and 18 in (45.7 cm) deep.

Calculation Procedure:

1. ***Compute the heat needed to reach the melting point***
When a solid is melted, first it must be raised from its ambient or room temperature to the melting temperature. The quantity of heat required is H = (weight of solid, lb) [specific heat of solid,

Btu/(lb · °F)] $(t_m - t_t)$, where H = Btu required to raise the temperature of the solid, °F; t_1 = room, charging, or initial temperature of the solid, °F; t_m = melting temperature of the solid, °F.

For this pot with lead having a melting temperature of 620°F (326.7°C) and an average specific heat of 0.031 Btu/(lb · °F) [0.13 kJ/(kg · °C)], $H = (600)(0.031)(620 - 70) = 10{,}240$ Btu/h (3.0 kW), or (10,240 Btu/h)/(3412 Btu/kWh) = 2.98 kWh.

2. Compute the heat required to melt the solid

The heat H_m Btu required to melt a solid is H_m = (weight of solid melted, lb)(heat of fusion of the solid, Btu/lb). Since the heat of fusion of lead is 10 Btu/lb (23.2 kJ/kg), $H_m = (600)(10) = 6000$ Btu/h, or 6000/3412 = 1.752 kWh.

3. Compute the heat required to reach the working temperature

Use the same relation as in step 1, except that the temperature range is expressed as $t_w - t_m$, where t_w = working temperature of the melted solid. Thus, for this pot, $H = (600)(0.031)(750 - 620) = 2420$ Btu/h (709.3 W), or 2420/3412 = 0.709 kWh.

4. Determine the heat loss from the pot

Use Fig. 7 to determine the heat loss from the pot. Enter at the bottom of Fig. 7 at 750°F (398.9°C), and project vertically upward to the 10-in (25.4-cm) diameter pot curve. At the left, read the heat loss at 7.3 kWh/h.

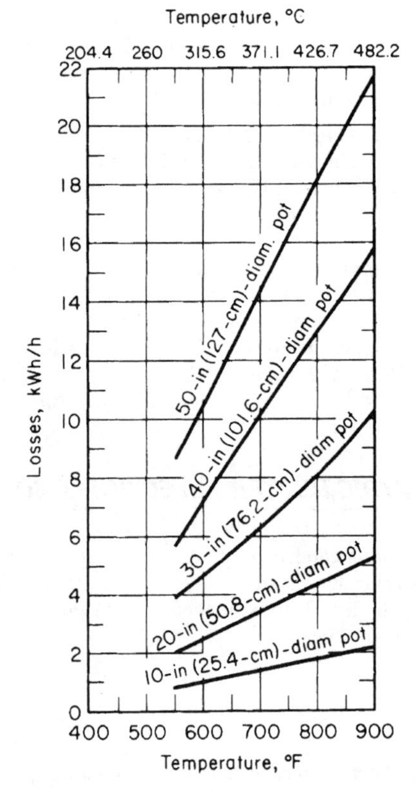

FIGURE 7 Heat losses from melting pots. (*General Electric Co.*)

TABLE 4 Two Methods for Determining Wattage for Heating Buildings Electrically*

	W/ft^3 method	W/m^3 method
1. Interior rooms with no or little outside exposure	0.75 to 1.25	25.6 to 44.1
2. Average rooms with moderate windows and doors	1.25 to 1.75	44.1 to 61.8
3. Rooms with severe exposure and great window and door space	1.0 to 4.0	35.3 to 141.3
4. Isolated rooms, cabins, watchhouses, and similar buildings	3.0 to 6.0	105.9 to 211.9
	The "35" method	
1. Volume in ft^3 for one air change × 0.35 =	0.01 W	
2. Exposed net wall, roof, or ceiling and floor in ft^2 × 3.5 =	0.1 W	
3. Area of exposed glass and doors in ft^2 × 35.0 =	1 W	

*General Electric Company.

5. *Compute the total heating capacity required*

The total heating capacity required is the sum of the individual capacities, or 2.98 + 1.752 + 0.708 + 7.30 = 12.74 kWh. A 15-kW electric heater would be chosen because this is a standard size and it provides a moderate extra capacity for overloads.

Related Calculations. Use this general procedure to compute the capacity required for an electric heater used to melt a solid of any kind—lead, tin, type metal, solder, etc. When the substance being heated is a liquid—water, dye, paint, varnish, oil, etc.—use the relation H = (weight of liquid heated, lb) [specific heat of liquid, Btu/(lb · °F)] (temperature rise desired, °F), when the liquid is heated to approximately its boiling temperature, or a lower temperature.

For space heating of commercial and residential buildings, two methods used for computing the approximate wattage required are the W/ft^3 and the "35" method. These are summarized in Table 4. In many cases, the results given by these methods agree closely with more involved calculations. When the desired room temperature is different from 70°F (21.1°C), increase or decrease the required kilowatt capacity proportionately, depending on whether the desired temperature is higher than or lower than 70°F (21.1°C).

For heating pipes with electric heaters, use a heater capacity of 0.8 W/ft^2 (8.6 W/m^2) of uninsulated exterior pipe surface per °F temperature difference between the pipe and the surrounding air. If the pipe is insulated with 1 in (2.5 cm) of insulation, use 30 percent of this value, or 0.24 ($W/(ft^2 \cdot °F)$ [4.7 $W/(m^2 \cdot °C)$].

The types of electric heaters used today include immersion (for water, oil, plating, liquids, etc.), strip, cartridge, tubular, vane, fin, unit, and edgewound resistor heaters. These heaters are used in a wide variety of applications including liquid heating, gas and air heating, oven warming, deicing, humidifying, plastics heating, pipe heating, etc.

For pipe heating, a tubular heating element can be fastened to the bottom of the pipe and run parallel with it. For large-wattage applications, the heater can be spiraled around the pipe. For temperatures below 165°F (73.9°C), heating cable can be used. Electric heating is often used in place of steam tracing of outdoor pipes.

The procedure presented above is the work of General Electric Company.

HEAT-TRANSFER COEFFICIENT DETERMINATION FOR BOILER ECONOMIZER

A 4530-ft^2 (421-m^2) heating surface counterflow economizer is used in conjunction with a 150,000-lb/h (68,040-kg/h) boiler. The inlet and outlet water temperatures are 210°F (99°C) and 310°F (154°C). The inlet and outlet gas temperatures are 640°F (338°C) and 375°F (191°C). Find the overall heat-transfer coefficient in $Btu/(h \cdot ft^2 \cdot °F)$ [$W/(m^2 \cdot °C)$] [$kJ/(h \cdot m^2 \cdot °C)$].

Calculation Procedure:

1. ***Determine the enthalpy of water at the inlet and outlet temperatures***
From Table 1, Saturation: Temperatures, of the Steam Tables mentioned under Related Calculations of this procedure, for water at inlet temperature, t_1 = 210°F (99°C), the enthalpy, h_1 = 178.14 Btu/lb (414 kJ/kg), and at the outlet temperature, t_2 = 310°F (154°C), the enthalpy, h_2 = 279.81 Btu/lb$_m$ (651 kJ/kg).

2. ***Compute the logarithmic mean temperature difference between the gas and water***
As shown in Fig. 8, the temperature difference of the gas entering and the water leaving, $\Delta t_a = t_3 - t_2$ = 640 – 310 = 330°F (166°C) and for the gas leaving and the water entering, $\Delta t_b = t_4 - t_1$ = 375 – 210 = 165°F (74°C). Then, the logarithmic mean temperature difference, $\Delta t_m = (\Delta t_a - \Delta t_b)/[2.3 \times \log_{10} (\Delta t_a - \Delta t_b)]$ = (330 – 165)/[2.3 × $\log_{10}$ (330/165)] = 238°F (115°C).

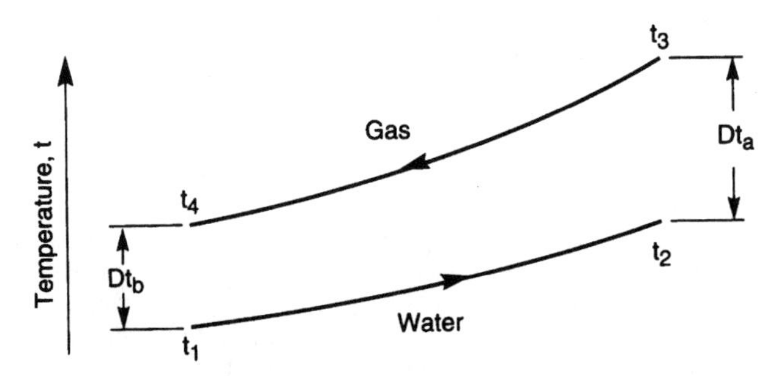

FIGURE 8 Temperature vs surface area of economizer.

3. ***Compute the economizer heat-transfer coefficient***
All the heat lost by the gas is considered to be transferred to the water, hence the heat lost by the gas, $Q = w(h_2 - h_1) = UA\ \Delta t_m$, where the water rate of flow, w = 150,000 lb/h (68,000 kg/h); U is the overall heat-transfer coefficient; heating surface area, A = 4530 ft^2 (421 m^2); other values as before. Then, 150,000 × (279.81 – 178.41) = U(4530)(238). Solving U = [150,000 × (279.81 – 178.14)]/(4530 × 238) = 14.1 Btu/(h · ft^2 · °F) [80 W/(m^2 · °C)] [288 kJ/(h · m^2 · °C)].

Related Calculations. The Steam Tables appear in *Thermodynamic Properties of Water Including Vapor, Liquid, and Solid Phases,* 1969, Keenan, et al., John Wiley & Sons, Inc. Use later versions of such tables whenever available, as necessary.

STEAM GENERATING CAPACITY OF BOILER TUBES

A counterflow bank of boiler tubes has a total area of 900 ft^2 (83.6 m^2) and its overall coefficient of heat transfer is 13 Btu/(h · ft^2 · °F) [73.8 W/(m^2 · K)]. The boiler tubes generate steam at a pressure of 1000 lb/in^2 absolute (6900 kPa). The tube bank is heated by flue gas which enters at a temperature of 2000°F (1367 K) and at a rate of 450,000 lb/h (56.7 kg/s). Assume an average specific heat of 0.25 Btu/(lb · °F) [1.05 kJ/(kg · K)] for the gas and calculate the temperature of the gas that leaves the bank of boiler tubes. Also, calculate the rate at which the steam is being generated in the tube bank.

Calculation Procedure:

1. ***Find the temperature of steam at 1000 lf_f/in^2 (6900 kPa)***
From Table 2, Saturation: Pressures, of the Steam Tables mentioned under Related Calculations of this procedure, the saturation temperature of steam at 1000 lb/in^2 (6900 kPa), t_s = 544.6°F (558 K), a constant value as indicated in Fig. 9.

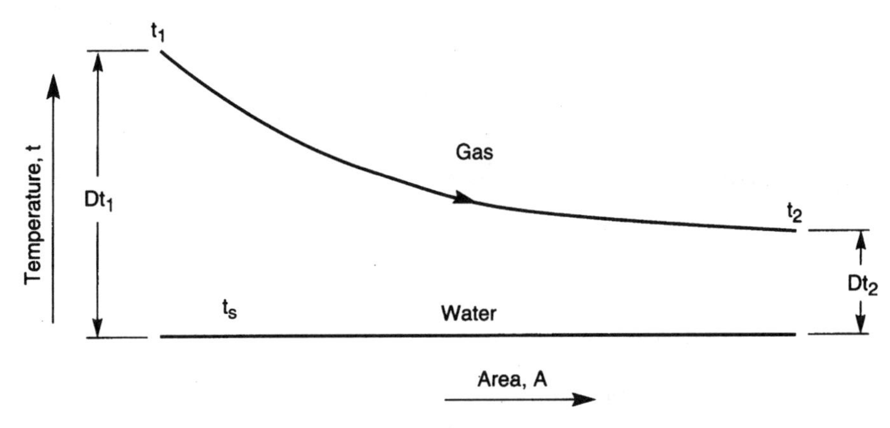

FIGURE 9 Temperature vs surface area of boiler tubes.

2. ***Determine the logarithmic mean temperature difference in terms of the flue-gas leaving temperature***
The logarithmic mean temperature difference, $\Delta t_m = (\Delta t_1 - \Delta t_2)/\{2.3 \times \log_{10} [(t_1 - t_s)/(t_2 - t_s)]\}$, where Δt_1 = flue-gas entering temperature = steam temperature = $(t_1 - t_s)$ = (2000 − 544.6); Δt_2 = flue-gas leaving temperature − steam temperature = $(t_2 - t_s) = (t_2 - 544.6)$; $(\Delta t_1 - \Delta t_2) = [(2000 - 544.6) - (t_2 - 544.6)] = (2000 - t_2)$; $[(t_1 - t_s)/(t_2 - t_s)] = [(2000 - 544.6)/(t_2 - 544.6)]$. Hence, $\Delta t_m = [(2000 - t_2)/\{2.3 \times \log_{10} [(1455.4)/(t_2 - 544.6)]\}$.

3. ***Compute the flue-gas leaving temperature***
Heat transferred to the boiler water, $Q = w_g \times c_p \times (t_1 - t_2) = UA\Delta t_m$, where the flow rate of flue gas, w_g = 450,000 lb/h (56.7 kg/s); flue-gas average specific heat, c_p = 0.25 Btu/(lb/°F) [1.05 kg/(kg · K)]; overall coefficient of heat transfer of the boiler tubes, U = 13 Btu/(h · ft^2 · °F) [73.8 W/(m^2 · K)]; area of the boiler tubes exposed to heat, A = 900 ft^2 (83.6 m^2); other values as before.

Then, $Q = 450{,}000 \times 0.25 \times (2000 - t_2) = 13 \times 900 \times (2000 - t_2)/\{2.3 \times \log_{10} [(1455.4)/(t_2 - 544.6)]\}$. Or, $\log_{10} [(1455.4)/(t_2 - 544.6)] = 13 \times 900/(2.3 \times 450.000 \times 0.25) = 0.0452$. The antilog of 0.0452 = 1.11, hence, $[(1455.4/(t_2 - 544.7)] = 1.11$, and t_2 = (1455.4/1.11) + 544.6 = 1850°F (1280 K).

4. ***Find the heat of vaporization of the water***
From the Steam Tables, the heat of vaporization of the water at 1000 lb/in^2 (6900 kPa), h_{fg} = 649.5 Btu/lb (1511 kJ/kg).

5. ***Compute the steam-generating rate of the boiler tube bank***
Heat absorbed by the water = heat transferred by the flue gas, or $Q = w_s \times h_{fg} = w_g \times c_p \times (t_1 - t_2)$, where the mass of steam generated is w_s in lb/h (kg/s); other values as before. Then, $w_s \times 649.5$ = 450,000 × 0.25 × (2000 − 1850) = 16.9 × 10^6 Btu/h (4950 kJ/s) (4953 kW). Thus, w_s = 16.9 × 10^6/649.5 = 26,000 lb/h (200 kg/s).

Related Calculations. The Steam Tables appear in *Thermodynamic Properties of Water Including Vapor, Liquid, and Solid Phases,* 1969, Keenan, et al., John Wiley & Sons, Inc. Use later versions of such tables whenever available, as required.

ENERGY DESIGN ANALYSIS OF SHELL-AND-TUBE HEAT EXCHANGERS

Determine the heat transferred, shellside outlet temperature, surface area, maximum number of tubes, and tubeside pressure drop for a liquid-to-liquid shell-and-tube heat exchanger such as that in Fig. 10, when the conditions below prevail. This exchanger will be of the single tube-pass and single shell-pass design Table 5, with countercurrent flows of the tubeside and shellside fluids.

Conditions	Tubeside	Shellside			
Flowrate, lb/h	307,500	32,800	kg/h	139,605	14,891
Inlet temperature, °C	105	45	F	221	45
Outlet temperature, °C	unknown	90			
Viscosity, cp	1.7	0.3			
Specific heat, Btu/h/°F	0.72	0.9	kJ/h°C	3.0	3.7
Molecular weight	118	62			
Specific gravity with reference to water at 20°C (68°F)	0.85	0.95			
Allowable pressure drop, psi	10	10	kPa	68.9	68.9
Maximum tube length, ft	12	12	m	5.45	5.45
Minimum, tube dia., in	5/8	5/8	mm	15.9	15.9
Material of construction	steel	($k = 26$)			

Calculation Procedure:

1. *Determine the heat transferred in the heat exchanger*
Use the relation, heat transferred, Btu/h = (flow rate, lb/h)(outlet temperature – inlet temperature) (liquid specific heat)(1.8 to convert from °C to °F). Substituting for this heat exchanger, we have, heat transferred = (32,800)(90 – 45)(0.9)(1.8) = 2,391,120 Btu/h (2522.6 kJ/h). This is the rate of heat transfer from the hot fluid to the cool fluid.

2. *Find the shellside outlet temperature*
The temperature decrease of the hot fluid = (rate of heat transfer)/(flow rate, lb/h)(specific heat) (1.8 conversion factor). Or, temperature decrease = (2,391,120)/(307,500)(0.72)(1.8) = 6°C (10.8°F). Then, the shellside outlet temperature = 105 – 6 = 99°C (210.2°F). Then, the LMTD = (54 – 15)/ln (54/15) = 30.4°C (86.7°F) = ΔT_m.

3. *Make a first-trial calculation of the surface area of this exchanger*
For a first-trial calculation, the approximate surface can be calculated using an assumed overall heat-transfer coefficient, U, of 250 Btu/(h) (ft^2) (°F) (44.1 W/m^2°C). The assumed value of U can be obtained from tabulations in texts and handbooks and is used only to estimate the approximate size for a first trial:

$$A = 2{,}391{,}000/(250 \times 30.4 \times 1.8) = 175\ \text{ft}^2 (16.3\ \text{m}^2)$$

Since the given conditions specify a maximum tube length of 12 ft and a minimum tube diameter of 5/8 in, the number of tubes required is:

$$n = 175(12 \times 0.1636) = 89\ \text{tubes}$$

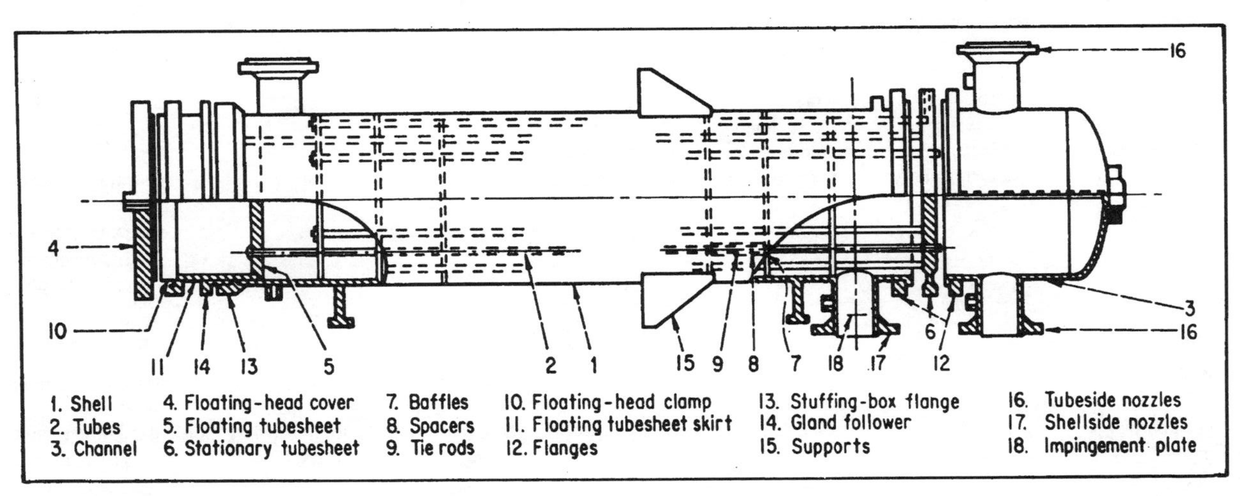

FIGURE 10 Components of shell-and-tube heat exchanger. This unit an outside-packed stuffing box. (*Chemical Engineering.*)

TABLE 5 Design Features of Shell-and-Tube Heat Exchangers

Design features	Fixed tubesheet	Return bend (U-tube)	Outside-packed stuffing box	Outside-packed lantern ring	Pull-through bundle	Inside split-backing ring
Is tube bundle removable?	No	Yes	Yes	Yes	Yes	Yes
Can spare bundles be used?	No	Yes	Yes	Yes	Yes	Yes
How is differential thermal expansion relieved?	Expansion joint in shell	Individual tubes free to expand	Floating head	Floating head	Floating head	Floating head
Can individual tubes be replaced?	Yes	Only those in outside rows without special designs	Yes	Yes	Yes	Yes
Can tubes be chemically cleaned, both inside and outside?	Yes	Yes	Yes	Yes	Yes	Yes
Can tubes be physically cleaned on inside?	Yes	With special tools	Yes	Yes	Yes	Yes
Can tubes be physically cleaned on outside?	No	With square or wide triangular pitch	With square or wide triangular pitch	With square or wide triangular pitch	With square or wide triangular pitch	With square or wide triangular pitch
Are internal gaskets and bolting required?	No	No	No	No	Yes	Yes
Are double tubesheets practical?	Yes	Yes	Yes	No	No	No
What number of tubeside passes are available?	Number limited by number of tubes	Number limited by number of U-tubes	Number limited by number of tubes	One or two	Number limited by number of tubes Odd number of passes requires packed joint or expansion joint	Number limited by number of tubes Odd number of passes requires packed joint or expansion joint
Relative cost in ascending order, least expensive = 1	2	1	4	3	5	6

and the approximate shell diameter will be:

$$D_a = 1.75 \times 0.625 \times 89^{0.47} = 9 \text{ in } (228.6 \text{ mm})$$

With the exception of baffle spacing, all preliminary calculations have been made for the quantities to be substituted into the dimensional equations. For the first trial, we may start with a baffle spacing equal to about half the shell diameter. After calculating the shellside pressure drop, we may adjust the baffle spacing. Also, it is advisable to check the Reynolds number on the tubeside to confirm that the proper equations are being used.

4. *Find the maximum number of tubes for this heat exchanger*

To find the maximum number of tubes (n_{max}) in parallel that still permits flow in the turbulent region (N_{Re} = 12,600), a convenient relationship is $n_{max} = W_i/(2d_iZ_i)$. In this example, $n_{max} = 307.5/(2 \times 0.495 \times 1.7) = 183$. For any number of tubes less than 183 tubes in parallel, we are in the turbulent range and can use Eq. (1)

From Table 6, the appropriate expressions for rating are: Eq. (1) for tubeside, Eq. (11) for shellside, Eq. (18) for tube wall, and Eq. (19) for fouling. Eq. (21) and (25), respectively, are used for tubeside and shellside pressure drops.

5. *Compute the tubeside and shellside heat transfer*

Using the equations from Table 6, Tubeside, Eq. (1):

$$\frac{\Delta T_i}{\Delta T_M} = 10.43\left[\frac{1.7^{0.467} \times 118^{0.222}}{0.85^{0.89}}\right] \times \left[\frac{307.5^{0.2} \times 6}{30.4}\right]\left[\frac{0.495^{0.8}}{89^{0.2} \times 12}\right]$$

$$= 10.43 \times 4.27 \times 0.621 \times 0.0193 = 0.535$$

Shellside, Eq. (11):

$$\frac{\Delta T_o}{\Delta T_M} = 4.28\left[\frac{0.3^{0.267} \times 62^{0.222}}{0.95^{0.89}}\right] \times \left[\frac{32.8^{0.4} \times 45}{30.4}\right]\left[\frac{1^{0.282} \times 5^{0.6}}{89^{0.718} \times 12}\right]$$

$$= 4.28 \times 1.89 \times 5.98 \times 0.00872 = 0.424$$

Tube wall, Eq. (18):

$$\frac{\Delta T_w}{\Delta T_M} = 159\left[\frac{0.72}{26}\right] \times \left[\frac{307.5 \times 6}{30.4}\right]\left[\frac{0.625 - 0.495}{89 \times 0.625 \times 12}\right]$$

$$= 159 \times 0.0277 \times 60.7 \times 0.000195 = 0.052$$

Fouling, Eq. (19):

$$\frac{\Delta T_s}{\Delta T_M} = 3{,}820\left[\frac{0.72}{1{,}000}\right]\left[\frac{307.5 \times 6}{30.4}\right]\left[\frac{1}{89 \times 0.625 \times 12}\right]$$

$$= 3{,}820 \times 0.00072 \times 60.7 \times 0.00150 = 0.250$$

$$(\text{SOP})^* = 0.535 + 0.424 + 0.052 + 0.250 = 1.261$$

Because SOP is greater than 1, the assumed exchanger is inadequate. The surface area must be increased by adding tubes or increasing the tube length, or the performance must be improved by decreasing the baffle spacing. Since the maximum tube length is fixed by the conditions given, the alternatives are increasing the number of tubes and/or adjusting the baffle spacing. To estimate assumptions for the next trial, pressure drops are calculated. See Table 7.

*Sum of the Products—see *Related Calculations* for data.

TABLE 6 Empirical Heat-Transfer Relationships for Rating Heat Exchanger

Eq. no.	Mechanism of restriction	Empirical equation	Numerical factor	Physical-property factor	Work factor	Mechanical-design factor
	Inside the tubes					
(1)	No phase change (liquid), $N_{Re} > 10{,}000$	$\frac{h}{cG} = 0.023\,(N_{Re})^{-0.2}(N_{Pr})^{-2/3}$	$\Delta T_i/\Delta T_M = 10.43$	$\times \frac{(Z_i^{0.467}M_i^{0.22})}{s_i^{0.89}}$	$\times \frac{W_i^{0.2}(t_H - t_L)}{\Delta T_M}$	$\times \frac{d_i^{0.8}}{n^{0.2}L}$
(2)	No phase change (gas), $N_{Re} > 10{,}000$	$h = 0.0144\,G^{0.8}(D_i)^{-0.2}c_p$	$\Delta T_i/\Delta T_M = 9.87$		$\times \frac{W_i^{0.2}(t_H - t_L)}{\Delta T_M}$	$\times \frac{d_i^{0.8}}{n^{0.2}L}$
(3)	No phase change (gas), $2100 < N_{Re} < 10{,}000$	$h = 0.0059[(N_{Re})^{2/3} - 125]$ $[1 + (D/L)^{2/3}](c_p/D_i)(\mu_f/\mu_b)^{-0.14}$	$\Delta T_i/\Delta T_M = 44{,}700$	$\times (Z_f/Z_b)^{0.14}$	$\times \frac{W_i(t_H - t_L)}{\Delta T_M}$	$\times \frac{1}{[(N_{Re})^{2/3} - 125][1 + (d_i N_{PT}/12L)^{2/3}]nL}$
(4)	No phase change (liquid), $2100 < N_{Re} < 10{,}000$	$\frac{h}{cG} = 0.166\left[\frac{(N_{Re})^{2/3} - 125}{N_{Re}}\right]$ $[1 + (D/L)^{2/3}](N_{Pr})^{-2/3}(\mu_f/\mu_b)^{-0.14}$	$\Delta T_i/\Delta T_M = 2260$	$\times \left(\frac{M_i^{0.22}}{s_i^{0.89}Z_b^{1/3}}\right)\left(\frac{Z_f}{Z_b}\right)^{0.14}$	$\times \frac{W_i(t_H - t_L)}{\Delta T_M}$	$\times \frac{1}{[(N_{Re})^{2/3} - 125][1 + (d_i N_{PT}/12L)^{2/3}]nL}$
(5)	No phase change (liquid), $N_{Re} < 2100$	$\frac{h}{cG} = 1.86\,(N_{Re})^{-2/3}(N_{Pr})^{-2/3}$ $(L/D_i)^{-1/3}(\mu_f/\mu_b)^{-0.14}$	$\Delta T_i/\Delta T_M = 17.5$	$\times \left(\frac{M_i^{0.22}}{s_i^{0.89}}\right)\left(\frac{Z_f}{Z_b}\right)^{0.14}$	$\times \frac{W_i^{2/3}(t_H - t_L)}{\Delta T_M}$	$\times \frac{1}{n^{2/3}L^{2/3}(N_{PT})^{1/3}}$
(6)	Condensing vapor, vertical, $N_{Re} < 2100$	$h = 0.925\,k\,(g\rho_i^2/\mu\Gamma)^{1/3}$	$\Delta T_i/\Delta T_M = 4.75$	$\times \frac{(Z_i M_i)^{0.333}}{s_i^2 c_i}$	$\times \frac{W_i^{4/3}\lambda_i}{\Delta T_M}$	$\times \frac{1}{n^{4/3}d_i^{4/3}L}$
(7)	Condensing vapor, horizontal, $N_{Re} < 2100$	$h = 0.76\,k\,(g\rho_i^2/\mu\Gamma)^{1/3}$	$\Delta T_i/\Delta T_M = 2.92$	$\times \frac{(Z_i M_i)^{0.333}}{s_i^2 c_i}$	$\times \frac{W_i^{4/3}\lambda_i}{\Delta T_M}$	$\times \frac{1}{n^{4/3}d_i L^{4/3}}$ (See Note 1)
(8)	Condensate subcooling, vertical	$h = 1.225\,(k/B)\,(cB\Gamma/kL_B)^{5/6}$	$\Delta T_i/\Delta T_M = 1.22$	$\times \left[\frac{(Z_i M_i)^{0.333}}{s_i^2}\right]^{1/6}$	$\times \frac{W_i^{0.222}(t_H - t_L)}{\Delta T_M}$	$\times \frac{1}{(n^{4/3}d_i^{4/3}L)^{1/6}}$
(9)	Nucleate boiling, vertical	$\frac{h}{cG} = 4.02\,(N_{Re})^{-0.3}$ $(N_{Pr})^{-0.6}(\rho_L\sigma/P^2)^{-0.425}\Sigma$	$\Delta T_i/\Delta T_M = 0.352$	$\times \left(\frac{Z_i^{0.3}M_i^{0.2}\sigma_i^{0.425}}{s_i^{1.075}c_i}\right)\left(\frac{\rho_v^{0.7}}{P_i^{0.85}}\right)$	$\times \frac{W_i^{0.3}\gamma_i}{\Delta T_M}$	$\times \left(\frac{1}{n^{0.3}L^{0.3}}\right)\Sigma'$ (See Notes 2 and 5)

Outside the tubes					
(10) Nucleate boiling, horizontal	$\frac{h}{cG} = 4.02(N_{Re})^{-0.3}(N_{Pr})^{0.6}$ $(p_l\sigma/P^2\rho)^{-0.425}\Sigma$	$\Delta T_o/\Delta T_M = 0.352$	$\times\left(\frac{Z_o^{0.3}M_o^{0.2}\sigma_o^{0.425}}{s_o^{1.075}c_o}\right)\left(\frac{\rho_v^{0.7}}{P_o^{0.85}}\right)$	$\times\frac{W_o^{0.3}\lambda_o}{\Delta T_M}$	$\times\left(\frac{1}{n^{0.3}L^{0.3}}\right)\Sigma$
(11) No phase change (liquid), crossflow	$\frac{h}{cG} = 0.33\,(N_{Re})^{-0.4}(N_{Pr})^{-2/3}(0.6)$	$\Delta T_o/\Delta T_M = 4.28$	$\times\frac{Z_o^{0.267}M_o^{0.222}}{s_o^{0.89}}$	$\times\frac{W_o^{0.4}(T_H - T_L)}{\Delta T_M}$	$\times\frac{N_{PT}^{0.282}\ P_B^{0.6}}{n^{0.718}L}$ (See Note 3)
(12) No phase change (gas), crossflow	$h = 0.11\,G^{0.6}D^{-0.4}c_P^{(0.6)}$	$\Delta T_o/\Delta T_M = 7.53$		$\times\frac{W_o^{0.4}(T_H - T_L)}{\Delta T_M}$	$\times\frac{N_{PT}^{0.282}\ P_B^{0.6}}{n^{0.718}L}$ (See Note 4)
(13) No phase change (gas), parallel flow	$h = 0.0144\,G^{0.8}D^{-0.2}c_p(1.3)$	$\Delta T_o/\Delta T_M = 21.7$		$\times\frac{W_o^{0.2}(T_H - T_L)}{\Delta T_M}$	$\times\frac{d_o^{0.8}\ N_{PT}^{0.685}}{n^{0.315}L}$
(14) No phase change (liquid), parallel flow	$\frac{h}{cG} = 0.023\,(N_{Re})^{-0.2}$ $(N_{Pr})^{-2/3}(1.3)$	$\Delta T_o/\Delta T_M = 22.9$	$\times\frac{Z_o^{0.467}M_o^{0.22}}{s_o^{0.89}}$	$\times\frac{W_o^{0.2}(T_H - T_L)}{\Delta T_M}$	$\times\frac{d_o^{0.8}\ N_{PT}^{0.685}}{n^{0.315}L}$
(15) Condensing vapor, vertical, $N_{Re} < 2100$	$h = 0.925\,k(g\rho_L^2/\mu\Gamma)^{1/3}$	$\Delta T_o/\Delta T_M = 4.75$	$\times\frac{(Z_oM_o)^{0.333}}{s_o^2c_o}$	$\times\frac{W_o^{4/3}\lambda_o}{\Delta T_M}$	$\times\frac{1}{n^{4/3}\ d_o^{4/3}\ N_{PT}^{1/3}\ L}$
(16) Condensing vapor, horizontal, $N_{Re} < 2100$	$h = 0.76\,k(g\rho_L^2/\mu\Gamma)^{1/3}$	$\Delta T_o/\Delta T_M = 2.64$	$\times\frac{(Z_o\ M_o)^{0.333}}{s_o^2c_o}$	$\times\frac{W_o^{4/3}\lambda_o}{\Delta T_M}$	$\times\frac{N_{PT}^{0.177}}{n^{1.156}L^{4/3}d_o}$
Tube wall					
(17) Tube wall (sensible-heat transfer)	$h = (24\ k_w)/(d_o - d_i)$	$\Delta T_w/\Delta T_M = 159$	$\times\ c/k_w$	$\times\frac{W(t_H - t_L)}{\Delta T_M}$	$\times\frac{d_o - d_i}{nd_oL}$
(18) Tube wall (latent-heat transfer)	$h = (24k_w)/(d_o - d_i)$	$\Delta T_M/\Delta T_M = 88$	$\times\ 1/k_w$	$\times\frac{W\lambda}{\Delta T_M}$	$\times\frac{d_o - d_i}{nd_oL}$
Fouling					
(19) Fouling (sensible-heat transfer)	h = *assumed*	$\Delta T_s/\Delta T_M = 3820$	$\times\ c/h$	$\times\frac{W(t_H - t_L)}{\Delta T_M}$	$\times\frac{1}{nd_oL}$
(20) Fouling (latent-heat transfer)	h = *assumed*	$\Delta T_s/\Delta T_M = 2120$	$\times\ 1/h$	$\times\frac{W\lambda}{\Delta T_M}$	$\times\frac{1}{nd_oL}$

Notes:

1. If $W_i/(ns_id_i^{2.56}) > 0.3$, multiply $\Delta T_i/\Delta T_M$ by 1.3.
2. Surface-condition factor (Σ') for copper and steel = 1.0; for stainless steel = 1.7; for polished surfaces = 2.5.
3. For square pitch, numerical factor = 5.42.
4. For square pitch, numerical factor = 9.53.
5. $G = W_o\rho_L/(A\rho_v)$.

TABLE 7 Empirical Pressure-Drop Relationship for Rating Heat Exchangers

Eq. No.	Mechanism of restriction	Empirical equation
	Inside the tubes	
(21)	No phase change, $N_{Re} > 10{,}000$	$\Delta P = \dfrac{(Z_i)^{0.2}}{s_i}\left(\dfrac{W_i}{n}\right)^{1.8}\dfrac{N_{PT}[(L_o/d_i)+25]}{(5.4d_i)^{3.8}}$ (See Note 1)
(22)	No phase change, $2100 < N_{Re} < 10{,}000$	$\Delta P = \left(\dfrac{Z_i}{s_i}\right)\left(\dfrac{W_i}{n}\right)\dfrac{N_{PT}[(L_o/d_i)+25][(N_{Re})^{2/3}-25]}{(50.2d_i)^3}$ (See Note 1)
(23)	No phase change, $N_{Re} < 2100$	$\Delta P = \dfrac{(Z_b)^{0.326}(Z_f)^{0.14}}{s_i}\left(\dfrac{W_i}{n}\right)^{4/3}\dfrac{N_{PT}(L_o)^{2/3}}{(5.62d_i)^4}$
(24)	Condensing	$\Delta P = \dfrac{(Z_i)^{0.2}}{s_i}\left(\dfrac{W_i}{n}\right)^{1.8}\dfrac{N_{PT}[(L_o/d_i)+25]}{(5.4d_i)^{2.8}} \times 0.5$ (See Note 1)
	Shellside	
(25)	No phase change, crossflow	$\Delta P = \dfrac{0.326}{s_o}(W_o)^2\dfrac{L_o}{P_B^3 D_o}$
(26)	No phase change, parallel flow	$\Delta P = \dfrac{(Z_o)^{0.2}}{s_o}\left(\dfrac{W_o}{n}\right)^{1.8}\left[\dfrac{n^{0.366}\,L_o}{(N_{PT})^{1.434}(4.912\,d_o)^{4.8}} + \dfrac{0.31n^{0.0414}(W_o)^{0.2}L_o}{d_o(N_{PT})^{1.76}(4.912d_s)^4 Z^{0.2}B_o^2}\right]$ (See Notes 2 and 3)
(27)	Condensing	$\Delta P = \left(\dfrac{0.081}{S_o}\right)(W_o)^2\left(\dfrac{L_o}{P_B^3 D_o}\right)$

Notes:

1. For U-bends, use $[(L_o/d_i) + 16]$ instead of $[(L_o/d_i) + 25]$.
2. B_o is equal to fraction of flow area through baffle.
3. Number of baffles $(N_B) = 0.48\,(L_o/d_o)$.

6. *Make the pressure-drop calculation for the heat exchanger*
Tubeside, Eq. (21):

$$\Delta P = (17^{0.2}/0.85)(307.5/89)^{1.8}[(12/0.495) + 25]/(5.4 \times 0.495)^{3.8}$$
$$= 14.3 \text{ lb/in}^2 (98.5 \text{ kPa})$$

Shellside, Eq. (25):

$$\Delta P = (0.326/0.95)(32.8^2)[12/(5^3 \times 9)] = 3.9 \text{ lb/in}^2 (26.9 \text{ kPa})$$

To decrease the pressure drop on the tubeside to the acceptable limit of 10 lb/in^2 (68.9 kPa), the number of tubes must be increased. This will also decrease the SOP. In addition, shellside performance can be improved by decreasing the baffle spacing, since the pressure drop of 3.9 on the shellside is lower than the allowable 10 psi (68.9 kPa). Before proceeding with successive trials to balance the heat-transfer and pressure-drop restrictions. Table 8 is now set up for clarity.

7. *Perform the second-trial computation for heat-transfer surface and pressure drop*
As a first step in adjusting the heat-transfer surface and pressure drop, calculate the number of tubes to give a pressure drop of 10 lb/in^2 (68.9 kPa) on the tubeside. The pressure drop varies inversely as $n^{1.8}$. Therefore, $14.3/10 = (n/89)^{1.8}$, and $n = 109$.

Each individual product of the factors is then adjusted in accordance with the applicable exponential function of the number of tubes. Since the tubeside product is inversely proportional to the 0.2 power of the number of tubes, the product from the preceding trial is multiplied by $(n_1/n_2)^{0.2}$, where n_1 is the number of tubes used in the preceding trial, and n_2 is the number to be used in the new one. The shellside product of the preceding trial is multiplied by $(n_1/n_2)^{0.718}$, and the tube-wall and fouling products by n_1/n_2. New adjusted products are then calculated as follows:

$$\text{Tubeside product} = (89/109)^{0.2} \times 0.535 = 0.514$$
$$\text{Shellside product} = (89/109)^{0.718} \times 0.424 = 0.367$$
$$\text{Tube-wall product} = (89/109) \times 0.052 = 0.042$$
$$\text{Fouling product} = (89/109) \times 0.250 = 0.204$$
$$\text{SOP} = 1.127$$

8. *Make the last trial calculation*
For the third trial, baffle spacing is decreased to 3.5 in (88.9 mm) from 5 in (127.0 mm). Only the shellside product must be adjusted since only it is affected by the baffle spacing. Therefore, the shellside factor of the previous trial is multiplied by the ratio of the baffle spacing to the 0.6 power:

$$(3.5/5.0)^{0.6} \times 0.367 = 0.296$$

The sum of the products (SOP) for this trial is 1.056. The shellside pressure drop $\Delta P_a = 3.9 \times (5.0/3.5)^3 = 11.4$ lb/in^2 (78.5 kPa).

Because we have now reached the point where the assumed design nearly satisfies our conditions, tube-layout tables can be used to find a standard shell-size containing the next increment above 109 tubes. A 10-in-dia (254 mm) shell in a fixed-tubesheet design contains 110 tubes.

Again, correcting the products of the heat-transfer factors from the previous trial:

$$\text{Tubeside product} = (109/110)^{0.2} \times 0.514 = 0.513$$
$$\text{Shellside product} = (109/110)^{0.718} \times 0.296 = 0.293$$
$$\text{Tube-wall product} = (109/110) \times 0.042 = 0.042$$
$$\text{Fouling product} = (109/110) \times 0.204 = 0.202$$
$$\text{SOP} = 1.050$$

TABLE 8 Results of Trial Calculations

	1st trial	2nd trial	3rd trial	4th trial
Number of tubes	89	109	109	110
Shell diameter, in	9	9	9	10
Baffle spacing, in	5	5	3½	3½
Product of factors:				
Tubeside	0.535	0.514	0.514	0.513
Shellside	0.424	0.367	0.296	0.293
Tube-wall	0.052	0.042	0.042	0.042
Fouling	0.250	0.204	0.204	0.202
Total sum of products	1.261	1.127	1.056	1.050
Tubeside ΔP, psi	14.3	10	10	9.8
Shellside ΔP, psi	3.9	3.9	11.4	10.1

Tubeside pressure drop:

$$\Delta P_i = 10 \times (109/110)^{1.8} = 9.8 \text{ lb/in}^2 (67.5 \text{ kPa})$$

The shellside pressure drop is now corrected for the actual shell diameter of 10 in (25.4 cm) instead of 9 in (22.86 cm).

$$\Delta P_o = 11.4 \times (9/10) = 10.1 \text{ psi } (69.6 \text{ kPa})$$

Any value of SOP between 0.95 and 1.05 is satisfactory as this gives a result within the accuracy range of the basic equations; unknowns in selecting the fouling factor do not justify further refinement. Therefore, the above is a satisfactory design for heat transfer and is within the pressure-drop restrictions specified. The surface area of the heat exchanger is $A = 110 \times 12 \times 0.1636 = 216$ ft^2 (20.1 m^2). The design overall coefficient is $U = 2{,}391{,}000/(30.4 \times 1.8 \times 216) =$ Btu/(h) (ft^2) (°F) (35.6 W/m$^2 \cdot$ °C).

The foregoing example shows that the essence of the design procedure is selecting tube configurations and baffle spacings that will satisfy heat-transfer requirements within the pressure-drop limitations of the system.

Related Calculations. The preceding procedure was for rating a heat exchanger of single tube-pass and single shell-pass design, with countercurrent flows of tubeside and shellside fluids. Often, it will be necessary to use two or more passes for the tubeside fluid. In this case, the LMTD is corrected with the Bowman, Mueller, and Nagle charts given in heat-transfer texts and the TEMA guide. If the correction factor for LMTD is less than 0.8, multiple shells should be used.

Bear in mind that n in all equations is the number of tubes in parallel through which the tubeside fluid flows; N_{PT} is the number of tubeside passes per shell (total number of tubes per shell = nN_{PT}); and L, the total-series length of path, equals shell length (L_o) (N_{PT}) × (number of shells).

The above procedure can be used for any shell-and-tube heat exchanger with sensible-heat transfer—or with no phase change of fluids—on both sides of the tubes. Also, N_{Re} on the tubeside must be greater than 10,000, and the viscosity of the fluid on the shellside must be moderate (500 cp. maximum).

As pointed out, the designer should assume as part of his job the specification of tube arrangement that will prevent the flow in the shell from taking bypass paths either around the space between the outermost tubes and the shell, or in vacant lanes of the bundle formed by channel partitions in multipass exchangers. He should insist that exchangers be fabricated in accordance with TEMA tolerances.

By using the appropriate equations from Tables 6 and 7, the technique described for rating heat exchangers with sensible-heat transfer can be used also for rating exchangers that involve boiling or condensing. The method can also be used in the design of partial condensers, or condensers handling mixtures of condensable vapors and noncondensable gases, and in the design of condensers handling vapors that form two liquid phases. However, for partial condensers and for two-phase liquid-condensate systems, a special treatment is required.

In addition to designing exchangers for specified performances, the method is also useful for evaluating the performance of existing exchangers. Here, the mechanical-design parameters are fixed, and the flow rates and temperature conditions (work factor) are the variables that are adjusted.

The two process variables that have the greatest effect on the size (cost) of a shell-and-tube heat exchanger are the allowable pressure drops of streams, and the mean temperature difference between the two streams. Other important variables include the physical properties of the streams, the location of fluids in an exchanger, and the piping arrangement of the fluids as they enter and leave the exchanger. (See design features in Table 5.)

Selection of optimum pressure drops involves consideration of the overall process. While it is true that higher-pressure drops result in smaller exchangers, investment savings are realized only at the expense of operating costs. Only by considering the relationship between operating costs and investment can the most economical pressure drop be determined.

Available pressure drops vary from a few millimeters of mercury in vacuum service to hundreds of pounds per square inch in high-pressure processes. In some cases, it is not practical to use all the available pressure drop because resultant high velocities may create erosion problems.

Reasonable pressure drops for various levels are listed below. Designs for smaller pressure drops are often uneconomical because of the large surface area (investment) required.

Pressure level	Reasonable ΔP
Subatmospheric	1/10 absolute pressure
1 (6.89 kPa) to 10 lb/in^2 (gage) (68.9 kPa)	½ operating gage pressure
10 lb/in^2 (gage) (68.9 kPa) and higher	5 lb/in^2 (34.5 kPa) or higher

In some instances, velocities of 10 to 15 ft/s (3 to 4.6 m/s) help to reduce fouling, but at such velocities the pressure drop may have to be from 10 (68.9 kPa) to 30 lb/in^2 (206.7 kPa).

Although there are no specific rules for determining the best temperature approach, the following recommendations are made regarding terminal temperature differences for various types of heat exchangers; any departure from these general limitations should be economically justified by a study of alternate system designs:

- The greater temperature difference should be at least 20°C (36°F).
- The lesser temperature difference should be at least 5°C (9°F). When heat is being exchanged between two process streams, the lesser temperature difference should be at least 20°C (36°F).
- In cooling a process stream with water, the outlet-water temperature should not exceed the outlet process-stream temperature if a single body having one shell pass—but more than one tube pass—is used.
- When cooling or condensing a fluid, the inlet coolant temperature should not be less than 5°C (9°F) above the freezing point of the highest freezing component of the fluid.
- For cooling reactors, a 10°C (18°F) to 15°C (27°F) difference should be maintained between reaction and coolant temperatures to permit better control of the reaction.
- A 20°C (36°F) approach to the design air temperature is the minimum for air-cooled exchangers. Economic justification of units with smaller approaches requires careful study. Trim coolers or evaporative coolers should also be considered.
- When condensing in the presence of inerts, the outlet coolant temperature should be at least 5°C (9°F) below the dew point of the process stream.

In an exchanger having one shell pass and one tube pass, where two fluids may transfer heat in either cocurrent or countercurrent flow, the relative direction of the fluids affects the value of the mean temperature difference. This is the log mean in either case, but there is a distinct thermal advantage to counterflow, except when one fluid is isothermal.

In concurrent flow, the hot fluid cannot be cooled below the cold-fluid outlet temperature; thus, the ability of cocurrent flow to recover heat is limited. Nevertheless, there are instances when cocurrent flow works better, as when cooling viscous fluids, because a higher heat-transfer coefficient may be obtained. Cocurrent flow may also be preferred when there is a possibility that the temperature of the warmer fluid may reach its freezing point.

These factors are important in determining the performance of a shell-and-tube exchanger:

Tube Diameter, Length. Designs with small-diameter tubes [5/8 (15.8 mm) to 1 in (25.4 mm)] are more compact and more economical than those with larger-diameter tubes, although the latter may be necessary when the allowable tubeside pressure drop is small. The smallest tube size normally considered for a process heat exchanger is 5/8 in (15.8 mm) although there are applications where ½ (12.7 mm), 3/8 (9.5 mm), or even ¼-in (6.4 mm) tubes are the best selection. Tubes of 1 in (25.4 mm) dia are normally used when fouling is expected because smaller ones are impractical to clean mechanically. Falling-film exchangers and vaporizers generally are supplied with 1½. (38.1 mm) and 2-in (50.8 mm) tubes.

Since the investment per unit area of heat-transfer service is less for long exchangers with relatively small shell diameters, minimum restrictions on length should be observed.

Arrangement. Tubes are arranged in triangular, square, or rotated-square pitch (Fig. 11). Triangular tube-layouts result in better shellside coefficients and provide more surface area in a given shell diameter, whereas square pitch or rotated-square pitch layouts are used when mechanical cleaning of the outside of the tubes is required. Sometimes, widely spaced triangular patterns facilitate cleaning. Both types of square pitches offer lower-pressure drops—but lower coefficients—than triangular pitch.

Primarily, the method given in this calculation procedure combines into one relationship the classical empirical equations for film heat-transfer coefficients with heat-balance equations and with relationships describing tube geometry, baffles, and shell. The resulting overall equation is recast into three separate groups that contain factors relating to physical properties of the fluid, performance or duty of the exchanger, and mechanical design or arrangement of the heat-transfer surface. These groups are then multiplied together with a numerical factor to obtain a product that is equal to the fraction of the total driving force—or log mean temperature-difference (LMTD or ΔT_M)—that is dissipated across each element of resistance in the heat-flow path.

When the sum of the products for the individual resistance equals one, the trial design may be assumed to be satisfactory for heat transfer. The physical significance is that the sum of the temperature

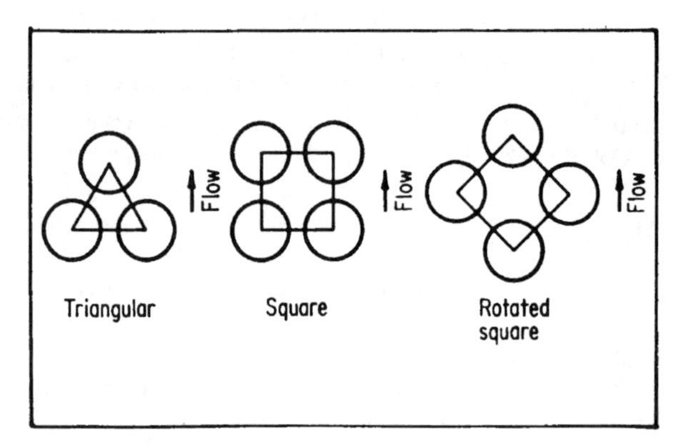

FIGURE 11 Tube arrangements used for shell-and-tube heat exchangers. (*Chemical Engineering.*)

drops across each resistance is equal to the total available LMTD. The pressure drop on both tubeside and shellside must be checked to ensure that both are within acceptable limits. As shown in the sample calculation above, usually several trials are necessary to obtain a satisfactory balance between heat transfer and pressure drop.

Tables 6 and 7, respectively, summarize the equations used with the method for heat transfer and for pressure drop. The column on the left lists the conditions to which each equation applies. The second column lists the standard form of the correlation for film coefficients that is found in texts. The remaining columns then tabulate the numerical, physical-property, work, and mechanical-design factors, all of which together form the recast dimensional equation. The product of these factors gives the fraction of total temperature drop or driving force ($\Delta T_f/\Delta T_M$) across the resistance.

As described above, the addition of $\Delta T_i/\Delta T_M$, tubeside factor; plus $\Delta T_o/\Delta T_M$, shellside factor; plus $\Delta T_s/T_M$, fouling factor; plus $\Delta T_w/\Delta T_M$, tube-wall factor, determine the heat-transfer adequacy. Any combination of $\Delta T_i/\Delta T_M$ and $\Delta T_o/\Delta T_M$ may be used, as long as a horizontal orientation on the tubeside is used with a horizontal orientation on the shellside, and a vertical tubeside orientation has a corresponding shellside orientation.

The units in the pressure-drop equations (Table 7) are consistent with those used for heat transfer. The pressure drop in psi is calculated directly. Because the method is a shortcut approach to design, certain assumptions pertaining to thermal conductivity, tube pitch, and shell diameter are made.

For many organic liquids, thermal conductivity data are either not available or difficult to obtain. Since molecular weights (M) are known, for most design purposes the Weber equation, which follows, yields thermal conductivities with quite satisfactory accuracies:

$$k = 0.86\ (cs^{4/3}/M^{1/3})$$

An important compound for which the Weber equation does not work well is water (the calculated thermal conductivity is less than the actual value). Figure 12 gives the physical-property factor for water (as a function of fluid temperature) that is to be substituted in the equations for sensible-heat transfer with water, or for condensing with steam.

If the thermal conductivity is known, it is best to obtain a pseudomolecular weight by:

$$M = 0.636\ (c/k)^3 s^4$$

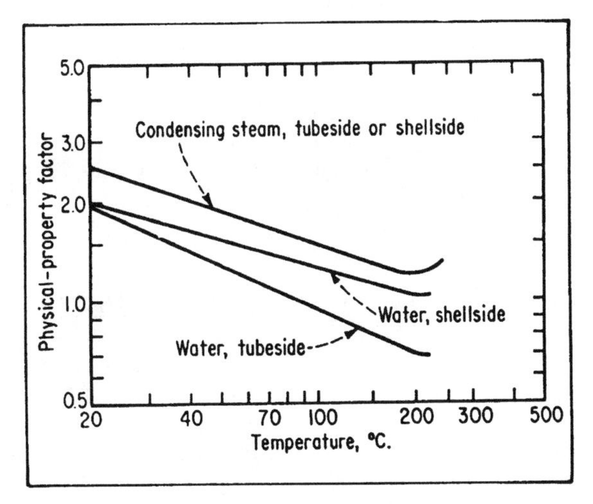

FIGURE 12 Physical-property factors for water and steam vs temperature. (*Chemical Engineering*.)

This value is substituted in the applicable equation to solve for the physical-property factor.

Tube pitch for both triangular and square-pitch arrangements is assumed to be 1.25 times the tube diameter. This is a standard pitch used in the majority of shell-and-tube heat exchanges. Slight deviations do not appreciably affect results.

Shell diameter is related to the number of tubes (nN_{PT}) by the empirical equation:

$$D_o = 1.75\, d_o (nN_{PT})^{0.47}$$

This gives the approximate shell diameter for a packed floating-head exchanger. The diameter will differ slightly for a fixed tubesheet, U-bend, or a multipass shell. For greater accuracy, tube-layout tables can be used to find shell diameters.

The following shows how the design equations are developed for a heat exchanger with sensible-heat transfer and Reynolds number 10,000 on the tubeside, and with sensible-heat transfer and cross-flow (flow perpendicular to the axis of tubes) on the shellside. Equations with other heat-transfer mechanisms are derived similarly.

For the film coefficient or conductance, h, and the heat balance, these equations apply:

Value of h	Heat-balance Eq.
$h_i = \dfrac{0.023\, c_i G_i}{(c_i \mu_i / k_i)^{2/3} (D_i G_i / \mu_i)^{0.2}}$	$W_i c_i (t_H - t_L) = h_i A \Delta T_i$
$h_w = 24\, k_w / (d_o - d_i)$	$W_i c_i (t_H - t_L) = h_w A \Delta T_w$
$h_o = \dfrac{0.33 c_o G_o (0.6)}{(c_o \mu_o / k_o)^{2/3} (D_o G_o / \mu_o)^{0.4}}$	$W_o c_o (T_H - T_L) = h_o A \Delta T_o$
h_s = assumed value	$W_i c_i (t_H - t_L) = h_s A \Delta T_s$

Since the resistances involved in a tube-and-shell exchanger are the tubeside film, the tube wall, the scale caused by fouling, and the shellside film, then:

$$\Delta T_i + \Delta T_w + \Delta T_s + \Delta T_o = \Delta T_M$$

therefore:

$$\Delta T_i / \Delta T_M + \Delta T_w / \Delta T_M + \Delta T_s / \Delta T_M + \Delta T_o / \Delta T_M = 1$$

or:

$$\underset{\text{Tubeside product}}{\frac{W_i c_i (t_H - t_L)}{h_i A \Delta T_M}} + \underset{\text{Tube-wall product}}{\frac{W_i c_i (T_H - t_L)}{h_w A \Delta T_M}} + \underset{\text{Fouling product}}{\frac{W_i c_i (T_H - t_L)}{h_o A \Delta T_M}} + \underset{\text{Shellside product}}{\frac{W_s c_o (T_H - T_L)}{h_o A \Delta T_M}} = 1$$

This last equation is obtained by dividing each heat-balance equation by ΔT_M and solving for $\Delta T_f / \Delta T_M$. The design equations are derived by substituting for h the appropriate correlation for the coefficient; for k, the value obtained from the Weber equation; for A, the equivalent of the surface area in terms of the number of tubes, outside diameter and length, according to the relation $A = \pi n (d_o/12) L$;

for mass velocity on the tubeside, $G_i = 183\ W_i/(d_i^2 n)$; and for mass velocity on the shellside, $G_o = 411.4\ W_o/(d_o n N_{PT}^{0.47} P)$.

The resulting equation is rearranged to separate the physical-property, work, and mechanical-design parameters into groups. To obtain consistent units, the numerical factor in the equation combines the constants and coefficients. The form of the equations shown in Table 6 as Eqs. (1), (11), (18), and (19) omits dimensionless groups such as Reynolds or Prandtl numbers, but includes single functions of the common design parameters such as number of tubes, tube diameter, tube length, baffle pitch, etc.

The individual products calculated from the four equations are added to give the sum of the products (SOP). A valid design for heat transfer should give SOP = 1. If SOP comes out to be less or more than one, the products for each resistance are adjusted by the appropriate exponential function of the ratio of the new design parameter to that used previously.

More sophisticated rating methods are available that make use of complex computer programs; the described method is intended only as a general, shortcut approach to shell-and-tube heat-exchanger selection. Accuracy of the technique is limited by the accuracy with which fouling factors, fluid properties, and fabrication tolerances can be predicted. Nevertheless, test data obtained on hundreds of heat exchangers attest to the method's applicability.

This procedure is the work of Robert C. Lord, Project Engineer, Paul E. Minton, Project Engineer, and Robert P. Slusser, Project Engineer, Engineering Department, Union Carbide Corporation, as reported in *Chemical Engineering* magazine. SI values were added by the handbook editor.

Nomenclature

A	Outside surface area, ft^2	t	Temperature on tubeside, °C
B	Film thickness, $[0.00187Z\Gamma/g_c s^2]^{1/3}$, ft	ΔT_M	Logarithmic mean temperature difference (LMTD), °C
c	Specific heat, Btu/(lb) (°F)	U	Overall coefficient of heat transfer, Btu/[(h)(ft^2) (°F)]
D_i	Inside tube diameter, ft	W	Flowrate, (lb/h)/1000
D_o	Inside shell diameter, in	Z	Viscosity, cp
d	Tube diameter, in	Γ	Tube loading, lb/(h)(ft)
f	Fanning friction factor, dimensionless	λ	Heat of vaporization, Btu/lb
G	Mass velocity, lb/(h)(ft^2 cross-sectional area)	θ	Time, h
g_c	Gravitational constant, (4.18×10^8) ft (h)^2	μ	Viscosity, lb/(h) (ft)
h	Film coefficient of heat transfer, Btu/[(h) (ft^2) (°F)]	ρ_u	Vapor density, lb/ft^3
k	Thermal conductivity, Btu/[(h) (ft^2)]	ρ_L	Liquid density, lb/ft^3
L	Total series length of tubes, ($L_o N_{PT} \times$ number of shells), ft	σ	Surface tension, dynes/cm
L_A	Length of condensing zone, ft	Σ, Σ'	Surface condition factor, dimensionless
	Subscripts		
L_B	Length of subcooled zone, ft	o	Conditions on shellside or outside tubes
L_o	Length of shell, ft	i	Conditions on tubeside or inside tubes
M	Molecular weight, lb/(lb-mol)	b	Bulk fluid properties
N_{PT}	Number of tube passes per shell, dimensionless	f	Film fluid properties
n	Number of tubes per pass (or in parallel) dimensionless	H	High temperature
P	Pressure, lb/in^2 (abs)	L	Low temperature
P_B	Baffle spacing, in	s	Scale or fouling material

(*Continued*)

Nomenclature (*Continued*)

ΔP	Pressure drop, lb/in^2	w	Wall or tube material
Q	Heat transferred, Btu	*Dimensionless groups*	
s	Specific gravity (referred to water at 20°C), dimensionless	N_{Re}	Reynolds Number, DG/μ
T	Temperature on shellside, °C	N_{pr}	Prandtl Number, $c\mu/k$
		N_{st}	Stanton Number, h/cG
		N_{Nu}	Nusselt Number, hD/k

See procedure for SI values for the variables above.

SPIRAL-PLATE HEAT-EXCHANGER DESIGN

Design a liquid-to-liquid spiral-plate heat exchanger for liquids in laminar flow under the conditions given below.

Conditions	Hot side	Cold side			
Flow rate, lb/h	6,225	5,925	kg/h	5925	2690
Inlet temperature, °C	200	60		392°F	140°F
Outlet temperature, °C	120	150.4		248°F	270.7°F
Viscosity, cp	3.35	8			
Specific heat, Btu/lb/°F	0.71	0.66	kJ/h · °C	2.95	3.75
Molecular weight	200.4	200.4			
Specific gravity	0.843	0.843			
Allowable pressure drop, psi	1	1	kPa	6.89	6.89
Material of construction	stainless steel	(k = 10)			
$(Z_f/Z_b)^{0.14}$	1	1			

Determine the heat transferred, the required heat-exchanger surface area, pressure-drop through the exchanger, and the final dimensions of the heat exchanger.

Calculation Procedure:

1. *Find the heat-transfer rate and log mean temperature difference*
Figure 13 shows several possible arrangements of spiral-flow heat exchangers. Using the same relation as in the previous procedure, we have heat-transfer rate, Btu/h = (flow rate, lb/h)(inlet temperature – outlet temperature)(specific heat, Btu/lb · °F)(1.8 conversion factor for temperature). Or, heat-transfer rate = (6225)(200 – 120)(0.71)(1.8) = 636, 444 Btu/h (671.4 kJ/h). Then, the log mean temperature difference, LMTD, T_M = (60 – 49.4)/ln (60/49.4) = 54.5°C (129.9°F).

2. *Find the surface area required for this heat exchanger*
For the first trial, the approximate surface area for this exchanger can be computed using an assumed overall heat-transfer coefficient, U, of 50 Btu/h ft^2 · °F (8.8 W/m^2 · °C). Then, A = 636,444/(50)(54.5)(1.8) = 129.75 ft^2, say 130 ft^2 (12 m^2).

Because at 130 ft^2 (12 m^2), this is a small heat exchanger, we will assume a plate width of 24 in (60.9 cm). Then, the plate length, L = 130/(2)(2) = 32.5 ft (9.9 m). Assume a channel spacing of 3/8 in (0.95 cm) for both fluids.

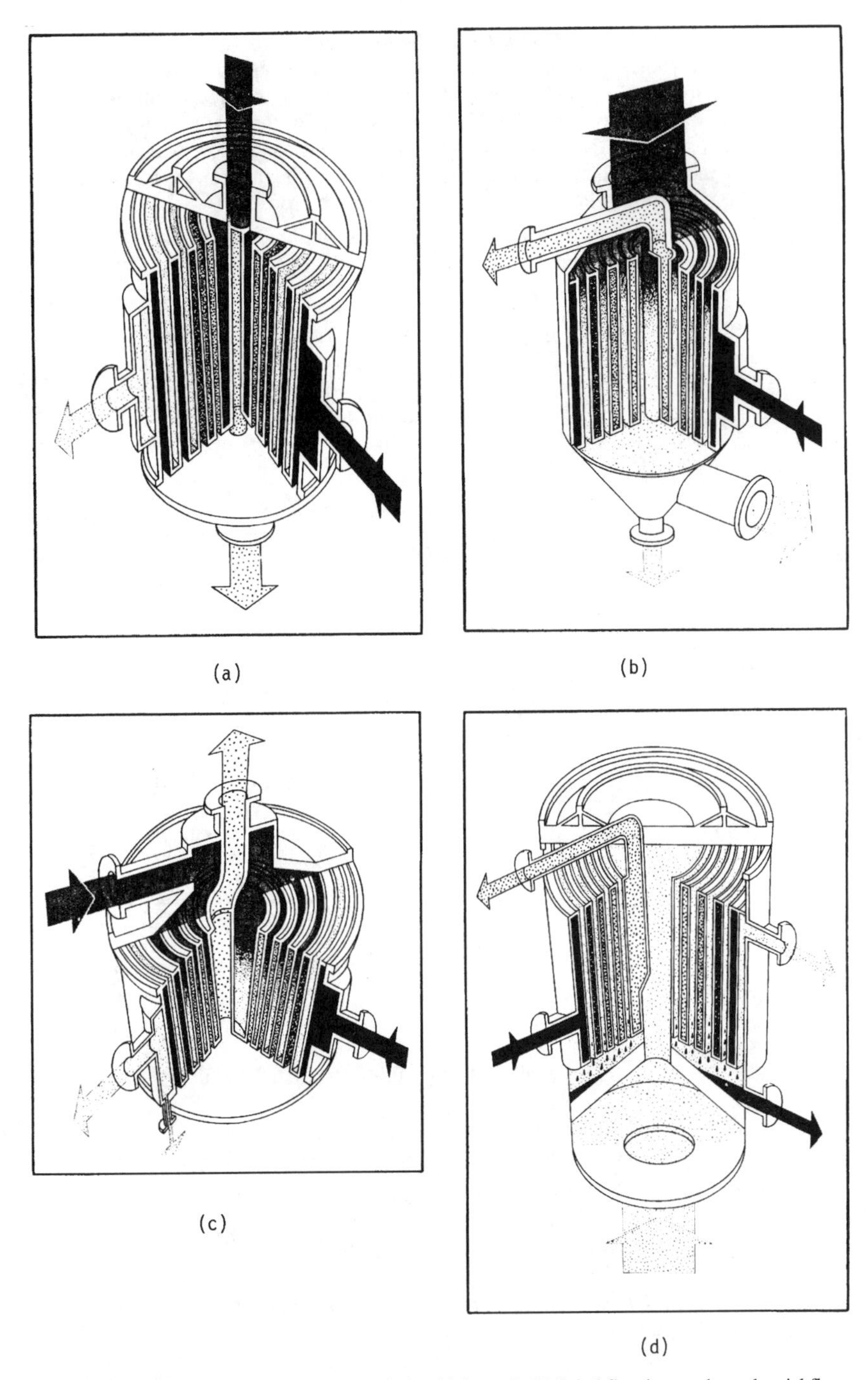

FIGURE 13 (*a*) Spiral flow in both channels is widely used; (*b*) Spiral flow in one channel, axial flow in the other; (*c*) combination flow is used to condense vapors; (*d*) modified combination flow serves on column. (*Chemical Engineering*.)

3. *Find the Reynolds number of the flow conditions*

The Reynolds number for spiral flow can be computed from $N_{Re} = 10{,}000\ (W/HZ)$, where W = flow rate, lb/h/1000; H = channel width, in; Z = fluid viscosity, cp. Substituting, we have, for the hot side, $N_{Re} = (10{,}000)(6225/1000)/(24)(3.35) = 774.4$. For the cold side, $N_{Re} = (10{,}000)(5925/1000)/(24)(8) = 308.6$. Because both fluids are in laminar flow, spiral flow will be chosen for this heat exchanger. From Table 9, the appropriate expressions for rating are Eq. (3) for both fluids, Eq. (10) for the plate, Eq. (12) for fouling, and Eq. (15) for pressure drop.

4. *Make the heat-transfer calculations for the heat exchanger*

Now, substitute values:

Hot side, Eq. (3):

$$\frac{\Delta T_h}{\Delta T_M} = 32.6\left[\frac{200.4^{0.222}}{0.843^{0.889}}\right] \times \left[\frac{6.225^{2/3} \times 80}{54.5}\right]\left[\frac{0.375}{24^{2/3} \times 32.5}\right]$$

$$= 32.6 \times 3.775 \times 4.967 \times 0.001387 = 0.848$$

Cold side, Eq. (3):

$$\frac{\Delta T_c}{\Delta T_M} = 32.6\left[\frac{200.4^{0.222}}{0.843^{0.889}}\right]\left[\frac{5.925^{2/3} \times 90.4}{54.5}\right] \times \left[\frac{0.375}{24^{2/3} \times 32.5}\right]$$

$$= 32.6 \times 3.775 \times 5.431 \times 0.001387 = 0.927$$

Fouling, Eq. (12):

$$\frac{\Delta T_s}{\Delta T_M} = 6{,}000\left[\frac{0.66}{1{,}000}\right]\left[\frac{5.925 \times 90.4}{54.5}\right]\left[\frac{1}{32.5 \times 24}\right]$$

$$= 6{,}000 \times 0.00066 \times 9.828 \times 0.001282 = 0.050$$

Plate, Eq. (10):

$$\frac{\Delta T_w}{\Delta T_M} = 500\left[\frac{0.66}{10}\right]\left[\frac{5.925 \times 90.4}{54.5}\right]\left[\frac{0.125}{32.5 \times 24}\right]$$

$$= 500 \times 0.066 \times 9.828 \times 0.0001603 = 0.052$$

Sum of Products (SOP):

$$\text{SOP} = 0.848 + 0.927 + 0.050 + 0.052 = 1.877\ (18.6\ \text{m})$$

Because SOP is greater than 1, the assumed heat exchanger is inadequate. The surface area must be enlarged by increasing the plate width or the plate length. Because, in all equations, L applies directly, the following new length is adopted:

$$1.877 \times 32.5 = 61\ \text{ft}$$

5. *Compute the pressure drops in the hot and cold sides*

Hot side, Eq. (15):

$$\Delta P = \left[\frac{0.001 \times 61}{0.843}\right]\left[\frac{6.225}{0.375 \times 24}\right] \times \left[\frac{1.035 \times 3.35^{1/2} \times 1 \times 24^{1/2}}{(0.375 + 0.125)\ 6.225^{1/2}} + 1.5 + \frac{16}{61}\right]$$

$$\Delta P = 0.07236 \times 0.6917 \times 9.202 = 0.464\ \text{psi}$$

TABLE 9 Empirical Heat-Transfer and Pressure-Drop Relationships for Rating Spiral-Plate Heat Exchangers

Eq. No.	Mechanism of restriction	Empirical equation—heat transfer	Numerical factor	Physical-property factor	Work factor	Mechanical-design factor	
	Spiral flow						
(1)	No phase change (liquid), $N_{Re} > N_{Rec}$	$h = (1 + 3.54\, D_e/D_H)\, 0.023\, cG\, (N_{Re})^{-0.2} (Pr)^{-2/3}$	$\frac{\Delta T_f}{\Delta T_M} = 20.6$	$\times \frac{Z^{0.467} M^{0.222}}{s^{0.889}}$	$\times \frac{W^{0.2}(T_H - T_L)}{\Delta T_M}$	$\times \frac{d_s}{LH^{0.2}}$	(See Note 1)
(2)	No phase change (gas), $N_{Re} > N_{Rec}$	$h = (1 + 3.54\, D_e/D_H)\, 0.0144\, cG^{0.8} (D_e)^{-0.2}$	$\frac{\Delta T_f}{\Delta T_M} = 19.6$		$\times \frac{W^{0.2}(T_H - T_L)}{\Delta T_M}$	$\times \frac{d_s}{LH^{0.2}}$	(See Note 1)
(3)	No phase change (liquid), $N_{Re} < N_{Rec}$	$h = 1.86\, cG\, (N_{Re})^{-2/3} (Pr)^{-2/3} (L/D_e)^{-1/3} (\mu_f/\mu_b)^{-0.14}$	$\frac{\Delta T_f}{\Delta T_M} = 32.6$	$\times \frac{M^{2/9} (Z_f)^{0.14}}{s^{8/9} (Z_b)^{0.14}}$	$\times \frac{W^{2/3}(T_H - T_L)}{\Delta T_M}$	$\times \frac{d_s}{LH^{2/3}}$	(See Note 1)
	Spiral or Axial Flow						
(4)	Condensing vapor, vertical, $N_{Re} < 2100$	$h = 0.925\, k\, [g_c \rho_L^2 / \mu\Gamma]^{1/3}$	$\frac{\Delta T_f}{\Delta T_M} = 3.8$	$\times \frac{M^{1/3} Z^{1/3}}{cs^2}$	$\times \frac{W^{4/3}\lambda}{\Delta T_M}$	$\times \frac{1}{L^{4/3} H}$	
(5)	Condensing subcooling, vertical, $N_{Re} < 2100$	$h = 1.225\, k/B\, [cB/kL_B]^{5/6}$	$\frac{\Delta T_f}{\Delta T_M} = 1.18$	$\times \frac{M^{1/18} Z^{1/18}}{s^{1/3}}$	$\times \frac{W^{2/9}(T_H - T_L)}{\Delta T_M}$	$\times \frac{1}{H^{1/6} L^{2/9}}$	
	Axial flow						
(6)	No phase change (liquid), $N_{Re} > 10{,}000$	$h = 0.023\, cG\, (N_{Re})^{-0.2} (Pr)^{-2/3}$	$\frac{\Delta T_f}{\Delta T_M} = 167$	$\times \frac{Z^{0.467} M^{0.222}}{s^{0.889}}$	$\times \frac{W^{0.2}(T_H - T_L)}{\Delta T_M}$	$\times \frac{d_s}{HL^{0.2}}$	
(7)	No phase change (gas), $N_{Re} > 10{,}000$	$h = 0.0144\, cG^{0.8}\, (D_e)^{-0.2}$	$\frac{\Delta T_f}{\Delta T_M} = 158$		$\times \frac{W^{0.2}(T_H - T_L)}{\Delta T_M}$	$\times \frac{d_s}{HL^{0.2}}$	
(8)	Condensing vapor, horizontal, $N_{Re} < 2100$	$h = 0.76\, k\, [g_c \rho_L^2 / \mu\Gamma]^{1/3}$	$\frac{\Delta T_f}{\Delta T_M} = 16.1$	$\times \frac{Z^{1/3} M^{1/3}}{cs^2}$	$\times \frac{W^{4/3}\, \lambda}{\Delta T_M}$	$\times \frac{1}{L^{4/3} H^{4/3}}$	
(9)	Nucleate boiling, vertical	$h = 4.02\, cG\, (N_{Re})^{-0.3} (Pr)^{-0.6} (\rho_L \sigma / P^2)^{-0.425}\, \Sigma$	$\frac{\Delta T_f}{\Delta T_M} = 0.619$	$\times \frac{M^{0.2} Z^{0.3} \sigma^{0.425}}{cs^{1.075}} \frac{\rho_r^{\ 0.7}}{P^{0.85}}$	$\times \frac{W^{0.3}\, \lambda}{\Delta T_M}$	$\times \frac{d_s^{0.3} \Sigma'}{L^{0.3} H^{0.3}}$	(See Notes 2 and 3)
	Plate						
(10)	Plate, sensible heat transfer	$h = 12\, k_w/p$	$\frac{\Delta T_w}{\Delta T_M} = 500$	$\times \frac{c}{k_w}$	$\times \frac{W(T_H - T_L)}{\Delta T_M}$	$\times \frac{p}{LH}$	
(11)	Plate, latent heat transfer	$h = 12\, k_w/p$	$\frac{\Delta T_w}{\Delta T_M} = 278$	$\times \frac{1}{k_w}$	$\times \frac{W\lambda}{\Delta T_M}$	$\times \frac{p}{LH}$	

(Continued)

TABLE 9 Empirical Heat-Transfer and Pressure-Drop Relationships for Rating Spiral-Plate Heat Exchangers (*Continued*)

Eq. No.	Mechanism of restriction	Empirical equation—pressure drop	
	Fouling		
(12)	Fouling, sensible heat transfer	$h = assumed$	$\frac{\Delta T_s}{\Delta T_M} = 6{,}000 \times \frac{c}{h} \times \frac{W(T_H - T_L)}{\Delta T_M} \times \frac{1}{LH}$
(13)	Fouling, latent heat transfer	$h = assumed$	$\frac{\Delta T_s}{\Delta T_M} = 3{,}333 \times \frac{1}{h} \times \frac{W\lambda}{\Delta T_M} \times \frac{1}{LH}$
	Spiral flow		
(14)	No phase change, $N_{Re} > N_{Rec}$	$\Delta P = 0.001\frac{L}{s}\left[\frac{W}{d_s H}\right]^2\left[\frac{1.3\,Z^{1/3}}{(d_s + 0.125)}\left(\frac{H}{W}\right)^{1/3} + 1.5 + \frac{16}{L}\right]$	(See Note 1)
(15)	No phase change, $100 < N_{Re} < N_{Rec}$	$\Delta P = 0.001\frac{L}{s}\left[\frac{W}{d_s H}\right]\left[\frac{1.035\,Z^{1/2}}{(d_s + 0.125)}\left(\frac{Z_f}{Z_b}\right)^{0.17}\left(\frac{H}{W}\right)^{1/2} + 1.5 + \frac{16}{L}\right]$	(See Note 1)
(16)	No phase change, $N_{Re} < 100$	$\Delta P = \frac{LsZ}{3{,}385(d_s)^{2.75}}\left(\frac{Z_f}{Z_b}\right)^{0.17}\cdot\left(\frac{W}{H}\right)$	
(17)	Condensing	$\Delta P = 0.005\frac{L}{s}\left[\frac{W}{d_s H}\right]^2\left[\frac{1.3\,Z^{1/3}}{(d_s + 0.125)}\left(\frac{H}{W}\right)^{1/3} + 1.5 + \frac{16}{L}\right]$	
	Axial flow		
(18)	No phase change, $N_{Re} > 10{,}000$	$\Delta P = \frac{4 \times 10^{-5}}{sd_s^2}\left(\frac{W}{L}\right)^{1.8} 0.0115\,Z^{0.2}\frac{H}{d_s} + 1 + 0.03\,H$	
(19)	Condensing	$\Delta P = \frac{2 \times 10^{-5}}{sd_s^2}\left(\frac{W}{L}\right)^{1.8}\left[0.0115\,Z^{0.2}\frac{H}{d_s} + 1 + 0.03H\right]$	

Notes:

1. $N_{Rec} = 20{,}000\,(D_e/D_H)^{0.32}$.
2. $G = W_o\,\rho_L/(AP_u)$.
3. Surface-condition factor (Σ') for copper and steel = 1.0; for stainless steel = 1.7; for polished surfaces = 2.5.

Cold side, Eq. (15):

$$\Delta P = \left[\frac{0.01 \times 61}{0.843}\right]\left[\frac{5.925}{0.375 \times 24}\right] \times \left[\frac{1.035 \times 8^{1/2} \times 1 \times 24^{1/2}}{(0.375 + 0.125)\ 5.925^{1/2}} + 1.5 + \frac{16}{61}\right]$$

$$\Delta P = 0.7236 \times 0.6583 \times 13.55 = 0.645\ \text{lb/in}^2\ (4.4\ \text{kPa})$$

Because the pressure drop is less than the allowable, the spacing can be decreased. For the second trial, ¼ in (0.64 cm) spacing for both channels is adopted.

Because the heat-transfer equation for every factor except the plate varies directly with d_s, a new SOP can be calculated.

$$\Delta T_h/\Delta T_M = 0.848\ (0.25/0.375) = 0.565$$

$$\Delta T_c/\Delta T_M = 0.927\ (0.25/0.375) = 0.618$$

$$\Delta T_s/\Delta T_M = 0.052\ (0.25/0.375) = 0.035$$

$$\Delta T_w/\Delta T_M = 0.050$$

$$\text{SOP} = 0.565 + 0.618 + 0.050 + 0.052 = 1.285$$

$$L = 1.285 \times 32.5 = 41.8\ \text{ft}\ (12.7\ \text{m})$$

$$A = 41.8 \times 2 \times 2 = 167\ \text{ft}^2\ (15.5\ \text{m}^2)$$

The new pressure drop becomes:
Hot side:

$$\Delta P = \left[\frac{0.001 \times 41.8}{0.843}\right] \times \left[\frac{6.225}{0.25 \times 24}\right] \times \left[\frac{1.035 \times 3.35^{1/2} \times 1 \times 24^{1/2}}{0.375 \times 6.225^{1/2}} + 1.5 + \frac{16}{41.8}\right]$$

$$\Delta P = 0.04958 \times 1.037 \times 11.80 = 0.607\ \text{lb/in}^2\ (4.2\ \text{kPa})$$

Cold side:

$$\Delta P = \left[\frac{0.001 \times 41.8}{0.843}\right] \times \left[\frac{5.925}{0.25 \times 24}\right] \times \left[\frac{1.035 \times 8^{1/2} \times 1 \times 24^{1/2}}{0.375 \times 5.925^{1/2}} + 1.5 + \frac{16}{41.8}\right]$$

$$\Delta P = 0.04958 \times 0.9875 \times 17.59 = 0.861\ (5.9\ \text{kPa})$$

The pressure drops are less than the maximum allowable. The plate spacing cannot be less than ¼ in (0.64 cm) for a 23.4-in (59.4-cm) plate width; decreasing the width would result in a higher than allowable pressure drop. Therefore, the design is acceptable.

6. *Establish the final design of the spiral heat exchanger*

The diameter of the outside spiral can now be calculated with Table 10 and the following equation:

$$D_s = [15.36 \times L\ (d_{sc} + d_{sh} + 2p) + C^2]^{1/2}$$

$$D_s = \{15.36\ (41.8)\ [0.25 + 0.25 + 2\ (0.125)] + 8^2\}^{1/2}$$

$$D_s = 23.4\ \text{in}\ (59.4\ \text{cm})$$

TABLE 10 Some Spiral-Plate/Spiral-Tube Exchanger Standards

Plate widths, in	Outside dia., maximum, in	Core dia., in
4	32	8
6	32	8
12	32	8
12	58	12
18	32	8
18	58	12
24	32	8
24	58	12
30	58	12
36	58	12
48	58	12
60	58	12
72	58	12

Channel spacings, in: 3/16 (12 in maximum width), ¼ (48 in maximum width), 5/16, 3/8, ½, 5/8, ¾, and 1. Plate thicknesses: stainless steel, 14-3 U.S. gage; carbon steel, 1/8, 3/16, ¼, and 5/16 in. See procedure for SI values.

Spiral-tube exchanger design		Standards shellside flow area, in^2	Standard lengths, ft		Heat-transfer area, ft^2
No. tubes	Tube spacing, in		Tubeside	Shellside	
Tube O.D., ½ in					
30	5/16	6.30	19.14	22.2	75.0
30	5/16	6.30	27.5	30.8	108.0
30	5/16	6.30	33.41	37.2	132.15

For a spiral-plate exchanger, the best design is often that in which the outside diameter approximately equals the plate width.

Design summary:	
Plate width	24 in (60.9 cm)
Plate length	41.8 ft (12.7 m)
Channel spacing	¼ in (both sides) (0.64 cm)
Spiral diameter	23.4 in (59.4 cm)
Heat-transfer area	167 ft^2 (15.5 m^2)
Hot-side pressure drop	0.607 lb/in^2 (4.2 kPa)
Cold-side pressure drop	0.861 lb/in^2 (5.9 kPa)
U	38.8 Btu/(h) (ft^2) (°F) (220.4 W/m^2 · °C)

Related Calculations. Spiral heat exchangers have a number of advantages over conventional shell-and-tube exchangers: centrifugal forces increase heat transfer; the compact configuration results in a shorter undisturbed flow length; relatively easy cleaning; and resistance to fouling. These curved-flow units (spiral plate and spiral tube) are particularly useful for handling viscous or solids-containing fluids.

A spiral-plate exchanger is fabricated from two relatively long strips of plate, which are spaced apart and wound around an open, split center to form a pair of concentric spiral passages. Spacing is maintained uniformly along the length of the spiral by spacer studs welded to the plates.

For most services, both fluid-flow channels are closed by alternate channels welded at both sides of the spiral plate (Fig. 13). In some applications, one of the channels is left completely open (Fig. 13*d*), the other closed at both sides of the plate. These two types of construction prevent the fluids from mixing.

Spiral-plate exchangers are fabricated from any material that can be cold worked and welded, such as carbon steel, stainless steels, Hastelloy B and C, nickel and nickel alloys, aluminum alloys, titanium, and copper alloys. Baked phenolic-resin coatings, among others, protect against corrosion from cooling water. Electrodes may also be wound into the assembly to anodically protect surfaces against corrosion.

Spiral-plate exchangers are normally designed for the full pressure of each passage. Because the turns of the spiral are of relatively large diameter, each turn must contain its design pressure, and plate thickness is somewhat restricted—for these three reasons, the maximum design pressure is 150 lb/in^2 (1033.5 kPa), although for smaller diameters the pressure may sometimes be higher. Limitations of materials of construction govern design temperature.

The shortcut rating method for spiral-plate exchangers depends on the same technique as that for shell-and-tube heat exchangers (which were discussed by Lord, Minton, and Slusser in the previous procedure).

Primarily, the method combines into one relationship the classical empirical equations for film heat-transfer coefficients with heat-balance equations and with correlations that describe the geometry of the heat exchanger. The resulting overall equation is recast into three separate groups that contain factors relating to the physical properties of the fluid, the performance or duty of the exchanger, and the mechanical design or arrangement of the heat-transfer surface. These groups are then multiplied together with a numerical factor to obtain a product that is equal to the fraction of the total driving force—or log mean temperature difference (ΔT_M or LMTD)—that is dissipated across each element of resistance in the heat-flow path.

When the sum of the products for the individual resistance equals 1, the trial design may be assumed to be satisfactory for heat transfer. The physical significance is that the sum of the temperature drops across each resistance is equal to the total available ΔT_M. The pressure drops for both fluid-flow paths must be checked to ensure that both are within acceptable limits. Usually, several trials are necessary to get a satisfactory balance between heat transfer and pressure drop.

Table 9 summarizes the equations used with the method for heat transfer and pressure drop. The columns on the left list the conditions to which each equation applies, and the second columns gives the standard forms of the correlations for film coefficients that are found in texts. The remaining columns in Table 9 tabulate the numerical, physical-property, work, and mechanical-design factors—all of which together form the recast dimensional equation. The product of these factors gives the fraction of total temperature drop or driving force ($\Delta T_f/\Delta T_M$) across the resistance.

As stated, the sum of $\Delta T_h/\Delta T_M$ (the hot-fluid factor), $\Delta T_c/\Delta T_M$ (the cold-fluid factor), $\Delta T_s/\Delta T_M$ (the fouling factor), and $\Delta T_w/\Delta T_M$ (the plate factor) determines the adequacy of heat transfer. Any combinations of $\Delta T_f/\Delta T_M$ may be used, as long as the orientation specified by the equation matches that of the exchanger's flow path.

The units in the pressure-drop equations are consistent with those used for heat transfer. Pressure drop is calculated directly in psi.

This procedure is the work of Paul E. Minton, Project Engineer, Engineering Department, Union Carbide Corporation, as reported in *Chemical Engineering* magazine. SI values were added by the handbook editor.

Nomenclature

A	Heat-transfer area, ft^2
B	Film thickness $(0.00187\ Z\Gamma/g_c s^2)^{1/3}$, ft
C	Core dia., in
c	Specific heat, Btu/(lb) (°F)
D_e	Equivalent dia., ft
D_H	Helix or spiral dia., ft
D_S	Exchanger outside dia., in
d_s	Channel spacing, in
f	Fanning friction factor, dimensionless
G	Mass velocity, lb/(h) (ft^2)
g_c	Gravitational constant, $ft/(h)^2$ (4.18×10^8)
H	Channel plate width, in
h	Film coefficient of heat transfer, Btu/(h) (ft^2) (°F)
k	Thermal conductivity, Btu/(h) (ft^2) (°F/ft)
L	Plate length, ft
M	Molecular weight, dimensionless
P	Pressure, lb/in^2 (abs)
p	Plate thickness, in
ΔP	Pressure drop, psi
Q	Heat transferred, Btu
s	Specific gravity (referred to water at 20°C)
ΔT_M	Logarithmic mean temperature difference (LMTD), °C
U	Overall heat-transfer coefficient, Btu/(h) (ft^2) (°F)
W	Flow rate, (lb/h)/1000
Γ	Condensate loading, lb/(h) (ft)
Z	Viscosity, cp
θ	Time, h
λ	Heat of vaporization, Btu/lb
μ	Viscosity, lb/(h) (ft)
ρ_L	Liquid density, lb/ft^3
ρ_u	Vapor density, lb/ft^3
Σ, Σ'	Surface condition factor, dimensionless
σ	Surface tension, dynes/cm

Subscripts

b	Bulk fluid properties
c	Cold stream
f	Film fluid properties
H	High temperature
h	Hot stream
L	Low temperature
m	Median temperature (see Fig. 5)
s	Scale or fouling material
w	Wall, plate material

Dimensionless Groups

N_{Re}	Reynolds number
N_{Rec}	Critical Reynolds number
N_{Pr}	Prandtl number

DESIGNING SPIRAL-TUBE HEAT EXCHANGERS

Design a liquid-condensing spiral-tube heat exchanger for the following conditions:

Conditions	Tubeside	Shellside		Shellside	Tubeside
Flow rate, lb/h	30,000	3,422	kg/h	13,620	1554
Inlet temperature, °C	20	121		−6.7°C	49.4°C
Outlet temperature, °C	80	121		26.7°C	49.4°C
Viscosity, liquid, cp	0.55	0.23			
Viscosity, vapor, cp	—	0.013			
Specific heat, (Btu/lb) (°F)	0.998	1.015	kJ/kg°C	4.2	4.25
Thermal conductivity, Btu/(ft^2) (h) (°F/ft)	0.368	0.398			
Specific gravity	0.999	0.95			
Heat of vaporization, Btu/lb	—	945		kJ/kg	2196
Vapor density, lb/ft^3	—	0.0727		kg/m^3	1.17
Material of construction	steel	(k = 26)			
Allowable pressure drop, psi	15	5	kPa	103.3	34.5

Calculation Procedure:

1. *Determine the rate of heat transfer and molecular weight tubeside and shellside for the heat exchanger*

The rate of heat transfer, Btu/h = (flow rate, lb/h) (heat of vaporization of the shellside fluid) = (3422) (945) = 3,233,790 Btu/h (3412 MJ/h).

Use the relation for pseudomolecular weight when the thermal conductivity of one of the fluids is known. This relation is $M = 0.636\ (c/k)^3 s^4$, where the symbols are as given in the Nomenclature list in this procedure. Substituting, we have:

Tubeside:

$$M = 0.636 \times 0.998^3 \times 0.999^4/0.368^3 = 12.63$$

Shellside:

$$M = 0.636 \times 1.015^3 \times 0.95^4/0.398^3 = 8.59 \Delta T_M \text{(or LMTD)}$$
$$= (101 - 41)/\ln(101/41) = 66.5°\text{C}$$

2. *Make a first trial to determine the heat-transfer surface area*

Assume an overall heat-transfer coefficient, U, of 300 Btu/h ft^2 · °F (1.7 W/m^2 · h · °C). Then, A = (3,233,790)/(300)(1.8)(66.5) = 90.05 ft^2 (8.4 m^2).

From a table of heat-exchanger design standards, such as Table 10, choose 300.5-in-outside-diameter tubes, 27.5-ft long, with a net free-flow area of 6.3 in^2; A = heat-transfer area = 108 ft^2; d_s = 0.3125 in. Then:

The maximum number of tubes for turbulent flow can be approximated by the term $W_i/2d_i z_i$:

$$n_{max} = 30/2 \times 0.402 \times 0.55 = 68$$

Because flow will be turbulent for the exchanger selected, the proper heat-transfer equations from Table 11 are: Eq. (1), tubeside; Eq. (8), shellside; Eq. (11), tube wall; and Eq. (13) fouling.

3. *Perform the heat-transfer calculation*

Tubeside, Eq. (1):

$$\frac{\Delta T_i}{\Delta T_M} = 9.07\left[\frac{0.55^{0.467} \times 12.63^{0.222}}{0.999^{0.889}}\right] \times \left[\frac{30^{0.2} \times 60}{66.5}\right]\left[\frac{0.402^{0.8}}{30^{0.2} \times 27.5}\right]$$

$$= 9.07 \times 1.329 \times 1.781 \times 0.008884 = 0.191$$

TABLE 11 Empirical Heat-Transfer and Pressure-Drop Relationships for Rating Spiral-Tube Heat Exchangers

Eq. No.	Mechanism of restriction	Empirical equation—heat transfer	Numerical factor	Physical-property factor	Work factor	Mechanical-design factor	
	Tube side						
(1)	No phase change (liquid), $N_{Re} > N_{Rec}$	$h = (1 + 3.54D_e/D_H)$ $0.023\,cG(N_{Re})^{-0.2}(Pr)^{-2/3}$	$\frac{\Delta T_i}{\Delta T_M} = 9.07$	$\times \frac{Z_i^{0.467} M_i^{0.222}}{s_i^{0.889}}$	$\times \frac{W_i^{0.2}(t_H - t_L)}{\Delta T_M}$	$\times \frac{d_i^{0.8}}{n^{0.2}L}$	(See Note 1)
(2)	No phase change (gas), $N_{Re} > N_{Rec}$	$h = (1 + 3.54D_e/D_H)$ $0.0144\,cG^{0.8}(D_i)^{-0.2}$	$\frac{\Delta Ti}{\Delta T_M} = 8.58$		$\times \frac{W_i^{0.2}(t_H - t_L)}{\Delta T_M}$	$\times \frac{d_i^{0.8}}{n^{0.2}L}$	(See Note 1)
(3)	No phase change (liquid), $N_{Rec} < N_{Rec}$	$h = 1.86cG(N_{Re})^{-2/3}(Pr)^{-2/3}$ $(L/D_e)^{-1/3}(\mu_f/\mu_b)^{-0.14}$	$\frac{\Delta T_i}{\Delta T_M} = 13.0$	$\times \frac{M_i^{0.222} Z_f^{0.14}}{s_i^{0.889} Z_b^{0.14}}$	$\times \frac{W_i^{2/3}(t_H - t_L)}{\Delta T_M}$	$\times \frac{d_i^{1/3}}{n^{2/2}L}$	(See Note 1)
(4)	Condensing vapor, horizontal, $N_{Re} < 2100$	$h = 0.76\,k\,[g_c \rho_L^2/\mu\Gamma]^{1/3}$	$\frac{\Delta T_i}{\Delta T_M} = 2.92$	$\times \frac{Z_i^{1/3} M_i^{1/3}}{c_i s_t^2}$	$\times \frac{W_i^{4/3}\lambda}{\Delta T_M}$	$\times \frac{1}{(nL)^{4/3} d_i}$	
	Shell side						
(5)	No phase change (liquid), $N_{Re} > N_{Rec}$	$h = (1 + 3.54D_e/D_H)$ $0.023\,cG(N_{Re})^{-0.2}(Pr)^{-2/3}$	$\frac{\Delta T_o}{\Delta T_M} = 12.3$	$\times \frac{Z_o^{0.467} M_o^{0.222}}{s_o^{0.889}}$	$\times \frac{W_i^{0.2}(T_H - T_L)}{\Delta T_M}$	$\times \frac{a}{(nd_o)^{1.2} L}$	(See Note 1)
(6)	No phase change (gas), $N_{Re} > N_{Rec}$	$h = (1 + 3.54De/D_H)$ $0.0144\,cG^{0.8}(D_e)^{-0.2}$	$\frac{\Delta T_o}{\Delta T_M} = 11.5$		$\times \frac{W_i^{0.2}(T_H - T_L)}{\Delta T_M}$	$\times \frac{a}{(nd_o)^{1.2} L}$	(See Note 1)
(7)	No phase change (liquid), $N_{Re} < N_{Rec}$	$h = 1.86\,cG\,(N_{Re})^{-2/3}(pr)^{-2/3}$ $(L/D_e)^{-1/3}(\mu_f/\mu_b)^{-0.14}$	$\frac{\Delta T_o}{\Delta T_M} = 15.4$	$\times \frac{M_o^{0.222} Z_f^{0.14}}{s_o^{0.889} Z_b^{0.14}}$	$\times \frac{W_o^{2/3}(T_H - T_L)}{\Delta T_M}$	$\times \frac{a}{(nd_o)^{5/3} L}$	(See Note 1)
(8)	Condensing vapor, horizontal, $N_{Re} < 2100$	$h = 0.76\,k\,[g_c \rho_L^2/\mu\Gamma]^{1/3}$	$\frac{\Delta T_o}{\Delta T_M} = 2.92$	$\times \frac{Z_o^{1/3} M_o^{1/3}}{c_o s_a^2}$	$\times \frac{W_o^{4/3}\lambda}{\Delta T_M}$	$\times \frac{1}{nL^{4/3} d_o}$	
(9)	Nucleate boiling, horizontal	$h = 4.02\,cG\,(N_{Re})^{-0.3}(Pr)^{-0.6}$ $(\rho_L \sigma/P^2)^{-0.425}\,\Sigma$	$\frac{\Delta T_o}{\Delta T_M} = 0.352$	$\times \left[\frac{Z_o^{0.3} M_o^{0.2} \sigma_o^{0.425}}{c_o s_o^{1.075}}\right]\left[\frac{\rho_u^{0.7}}{P_o^{0.85}}\right]$	$\times \left[\frac{W_o^{4/3}\lambda}{\Delta T_M}\right]$	$\times \left[\frac{1}{n^{0.3} d_o^{0.3}}\right]\Sigma'$	(See Notes 2 and 3)
	Tube wall						
(10)	Tube wall, sensible heat	$h = (24k_w)/(d_o - d_t)$	$\frac{\Delta T_w}{\Delta T_M} = 3{,}820$	$\times c/k_w$	$\times W(t_H - t_H)/\Delta T_M$	$\times (d_o - d_i)/nd_o L$	
(11)	Tube wall, latent heat	$h = (24k_w)/(d_o - d_t)$	$\frac{\Delta T_w}{\Delta T_M} = 2{,}120$	$\times c/k_w$	$\times W\lambda/\Delta T_M$	$\times (d_o - d_i)/nd_o L$	
	Fouling						
(12)	Fouling, sensible heat	h = assumed	$\frac{\Delta T_s}{\Delta T_M} = 159$	$\times c/k_w$	$\times W(t_H - t_L)/\Delta T_M$	$\times (d_o - d_i)/nd_o L$	
(13)	Fouling, latent heat	h = assumed	$\frac{\Delta T_s}{\Delta T_M} = 88$	$\times 1k_w$	$\times W\lambda/\Delta T_M$	$\times (d_o - d_i)/nd_o L$	

Eq. No.	Mechanism of restriction	Empirical equation—pressure drop	
	Tube side		
(14)	No phase change, $N_{Re} > N_{Rec}$	$\Delta P = 0.00268\left[\frac{Z_i^{0.15}}{s_i}\right]\left[\frac{W_i}{n}\right]^{1.85}\left[\frac{L/d_i + 16}{d_i^{3.75}}\right]N_{PT}$	(See Note 1)
(15)	No phase change, $100 < N_{Re} < N_{Rec}$	$\Delta P = 0.00195\left[\frac{W_i^{4/3}Z_i^{2/3}Z_f^{0.14}L}{s_i Z_b^{0.14} n^{4/3} d_i^{13/3}}\right]$	(See Note 1)
(16)	No phase change, $N_{Re} < 100$	$\Delta P = 0.000544\left[\frac{LW_iZ_i}{ns_id_i^4}\right]$	
(17)	Condensing	$\Delta P = 0.00134\left[\frac{Z_i^{0.15}}{s_i}\right]\left[\frac{W_i}{n}\right]^{1.85}\left[\frac{L/d_i + 16}{d_i^{3.75}}\right]N_{PT}$	
	Shell side		
(18)	No phase change, $N_{Re} > N_{Rec}$	$\Delta P = 0.00152\left[\frac{Z_o^{0.15}}{s_o}\right][L(nd_o)^{1.15}]\left[\frac{W_o^{1.85}}{a^3}\right]N_{PT}^{1.15}$	(See Note 1)
(19)	No phase change, $100 < N_{Re} < N_{Rec}$	$\Delta P = 0.00011\left[\frac{W_o^{4/3}Z_o^{2/3}(nd_o)^{5/3}LZ_f^{0.14}}{s_oa^3Z_b^{0.14}}\right]N_{PT}^{513}$	(See Note 1)
(20)	No phase change, $N_{Re} < 100$	$\Delta P = 0.000308\left[\frac{LW_oZ_o}{s_o}\right]\left[\frac{(nd_o)^2}{a^3}\right]N_{PT}^2$	
(21)	Condensing (spiral flow)	$\Delta P = 0.000757\left[\frac{Z_o^{0.15}}{s_o}\right]\left[\frac{W_o^{1.85}}{a^8}\right][L(nd_o)^{1.15}]N_{PT}^{1.15}$	
(22)	Condensing (axial flow)	$\Delta P = 4 \times 10^{-5} \times \frac{1}{s_o}[(W_o/Ld_s)^2 n\, N_{PT}]$	

Notes:
1. $N_{Rec} = 20{,}000\,(D_e/D_H)^{0.32}$.
2. $G = W_{rL}/(A\rho_v)$.
3. Surface-condition factor (Σ') for copper and steel = 1.0; for stainless steel = 1.7; for polished surfaces = 2.5.

Shellside, Eq. (8):

$$\frac{\Delta T_O}{\Delta T_M} = 2.92\left[\frac{0.23^{1/3} \times 8.59^{1/3}}{0.95^2 \times 1.015}\right]\left[\frac{3.422^{4/3} \times 945}{66.5}\right] \times \left[\frac{1}{30 \times 27.5^{4/3} \times 0.5}\right]$$

$$= 2.92 \times 1.370 \times 73.28 \times 0.0008032 = 0.235$$

Tube wall, Eq. (11):

$$\frac{\Delta T_s}{\Delta T_M} = 88\left[\frac{1}{26}\right]\left[\frac{3.422 \times 945}{66.5}\right]\left[\frac{0.098}{30 \times 27.5 \times 0.5}\right]$$

$$= 3.385 \times 48.63 \times 0.0002376 = 0.039$$

Fouling, Eq. (13):

$$\frac{\Delta T_w}{\Delta T_M} = 2{,}120\left[\frac{1}{1{,}000}\right]\left[\frac{3{,}422 \times 945}{66.5}\right]\left[\frac{1}{30 \times 27.5 \times 0.5}\right]$$

$$= 2.120 \times 48.63 \times 0.002424 = 0.250$$

Sum of the Products (SOP):

$$\text{SOP} = 0.191 + 0.235 + 0.250 + 0.039 = 0.715$$

The assumed exchanger is adequate with regard to heat transfer. The next step is to check both sides for pressure drop.

4. *Make the pressure-drop computation*

Tubeside, Eq. (14):

$$\Delta P = 0.00268\left[\frac{0.5^{0.15}}{0.999}\right]\left[\frac{30^{1.85}}{30^{1.85}}\right]\left[\frac{(27.5/0.402) + 16}{0.402^{3.75}}\right]$$

$$= 0.00268 \times 0.9151 \times 1.0 \times 2{,}574 = 6\ (43.4\ \text{kPa})$$

Shellside, Eq. (22):

$$\Delta P = 4 \times 10^{-4}\left[\frac{62.4}{0.00727}\right]\left[\frac{3.422}{27.5 \times 0.3125}\right]^2 \times 30$$

$$= 0.00004 \times 858.3 \times 0.1586 \times 30 = 0.163\ \text{lb/in}^2\ (1.1\ \text{kPa})$$

Both the shellside and tubeside pressure drops are much lower than the allowable. Therefore, the next step is to try the next-smaller exchanger, which has a tube length of 19.14 ft.

5. *Correct the heat-transfer factors, if necessary*

The four preceding heat-transfer factors must now be corrected as follows:

$$\text{Tubeside: } 0.191\,(27.5/19.14)^{0.2} = 0.205$$

$$\text{Shellside: } 0.235\,(27.5/19.14)^{4/3} = 0.381$$

$$\text{Fouling: } 0.250\,(27.5/19.14) = 0.359$$

$$\text{Tube wall: } 0.039\,(27.5/19.4) = 0.056$$

The new SOP becomes:

$$\text{SOP} = 0.205 + 0.381 + 0.359 + 0.056 = 1.001$$

This exchanger, which is adequate for heat transfer, will have the following pressure drops:
Tubeside:

$$\Delta P = 6.3\left[\frac{(19.14/0.402) + 16}{(27.5/0.402) + 16}\right] = 4.8 \text{ lb/in}^2 \text{ (abs) (33.1 kPa)}$$

Shellside:

$$\Delta P = 0.163(27.5/19.14)^2 = 0.336 \text{ lb/in}^2 \text{(2.3 kPa)}$$

The second exchanger selected is adequate with respect to heat transfer and pressure drop for both fluids. An exchanger with the same transfer area but with larger tubes would be more expensive. One with fewer or smaller tubes would have an excessive tubeside pressure drop.

6. ***Summarize the final design chosen***

Design summary:

No of tubeside passes	1	
Tube length, ft	19.14	5.8 m
Casing length, ft	22.2	6.8 m
Tube, outside dia., in	0.50	12.7 mm
Tube, inside dia., in	0.402	10.2 mm
Heat-transfer area, ft^2	75.0	6.96 m^2
Tubeside pressure drop, psi	4.8	33.1 kPa
Shellside pressure drop, psi	0.336	2.3 kPa
U, Btu/(h) (ft^2) (°F)	359	2.0 W/m$^2 \cdot$ h $\cdot$ °C

Related Calculations. Spiral-tube heat exchangers (like spiral-plate heat exchangers) offers several advantages over conventional shell-and-tube heat exchangers: secondary flow caused by centrifugal force that increases heat transfer; compact, short, undisturbed flow lengths that make spiral-tube exchangers ideal for heating and cooling viscous fluids, sludges, and slurries; less fouling; relatively easy cleaning.

Spiral-tube exchangers are generally more expensive than shell-and-tube exchangers having the same heat-transfer surface. However, the spiral-tube exchanger's better heat transfer and lower maintenance cost often make it the more profitable choice.

Spiral-tube exchangers consist of one or more concentric, spirally wound coils clamped between a cover plate and a casing. Both ends of each coil are attached to a manifold fabricated from pipe or bar stock (Fig. 14).

The coils, which are stacked on top of each other, are held together by the cover plate and casing. Spacing is maintained evenly between each turn of the coil to create a uniform, spiral-flow path for the shellside fluid.

Coils can be formed from almost any material of construction, with some of the more common ones being carbon steel, copper and copper alloys, stainless steels, and nickel and nickel alloys. Tubes may have extended surfaces. Casings are made of cast iron, cast bronzes, and carbon and stainless steel.

Tubes may be attached to the manifolds by soldering, brazing, welding, or, in some cases, rolling. Draining or venting can be facilitated by various manifold arrangements and casing connections. Flow through both the coil and casing may be single- or multipass (the latter by means of baffling).

FIGURE 14 Spiral-tube heat exchanger. (*Chemical Engineering.*)

Spiral-tube exchangers are available in sizes up to 325 ft^2, and pressures up to 600 psi. Tubeside pressures may be even higher.

The spiral-tube exchanger offers the following advantages over the shell-and-tube exchanger: (1) it is especially suited for low flows or small heat loads; (2) it is particularly effective for heating or cooling viscous fluids; since L/D ratios are much lower than those of straight-tube exchangers, laminar-flow heat transfer is much higher with spiral tubes; (3) its flows can be countercurrent (as with the spiral-plate exchanger, flows are not truly countercurrent, but again, the correction for this can be ignored); (4) it does not present the problems usually associated with differential thermal expansion; and (5) it is compact and easily installed.

The following are the chief limitations of the spiral-tube exchanger: (1) Its manifolds are usually small, making the repair of leaks at tube-to-manifold joints difficult (leaks, however, do not occur frequently); (2) it is limited to services that do not require mechanical cleaning of the inside of tubes (it can be cleaned mechanically on the shellside, and both sides can be cleaned chemically); (3) for some of its sizes, stainless steel coils must be provided with spacers to maintain a uniform shellside flow area—and these spacers increase pressure drop (this increase is not accounted for in the equations presented later).

The shortcut rating method given above for spiral-tube exchangers depends on the same technique as used for shell-and-tube exchangers (which is discussed by Lord, Minton, and Slusser earlier in this section).

Primarily, the method combines into one relationship the classical empirical equations for film heat-transfer coefficients with heat-balance equations and with correlations that describe the geometry of the heat exchanger. The resulting overall equation is recast into three separate groups that contain factors relating to the physical properties of the fluid, the performance or duty of the exchanger, and the mechanical design or arrangement of the heat-transfer surface. These groups are then multiplied together with a numerical factor to obtain a product that is equal to the fraction of the total driving force—or log mean temperature difference (δT_M or LMTD)—that is dissipated across each element of resistance in the heat-flow path.

When the sum of the products for the individual resistances equals 1, the trial design may be assumed satisfactory for heat transfer. The physical significance is that the sum of the temperature drops across each resistance is equal to the total available ΔT_M. The pressure drop for both fluid-flow paths must be checked to ensure that they are within acceptable limits. Usually, several trials are necessary to get a satisfactory balance between heat transfer and pressure drop.

Table 11 summarizes the equations used with spiral-tube exchangers. The column on the left presents the conditions to which each equation applies, and the second column gives the standard form of the film-coefficient correlation found in texts. The remaining columns tabulate the numerical, physical-property, work, and mechanical-design factors—all of which together form the recast dimensional equation. The product of these factors gives the fraction of the total temperature drop or driving force $(\Delta T_f/\Delta T_M)$ across the resistance.

As stated, the sum of $\Delta T_i/\Delta T_M$ (the tubeside factor), $\Delta T_o/\Delta T_M$ (the shellside factor), $\Delta T_s/\Delta T_M$ (the fouling factor), and $\Delta T_w/\Delta T_M$ (the tube-wall factor) determine the adequacy of heat transfer. Any combinations of $\Delta T_f/\Delta T_M$ may be used as long as the orientation specified by the equation matches that of the exchanger's flow path.

The units in the pressure-drop equations are consistent with those used for heat transfer. Pressure drop is calculated directly in psi.

For many organic liquids, thermal conductivity data are either not available or difficult to obtain. Because molecular weights (M) are known, the Weber equation (which follows) yields thermal conductivities whose accuracies are satisfactory for most design purposes:

$$k = 0.86\,(cs^{4/3}/M^{1/3})$$

If, on the other hand, the thermal conductivity is known, a pseudomolecular weight may be used:

$$M = 0.636\,(c/k)^3 s^4$$

This procedure is the work of Paul E. Minton, Project Engineer, Engineering Department, Union Carbide Corporation, as reported in *Chemical Engineering* magazine. In his credits in his article, Mr. Minton thanks Graham Manufacturing Company "for permission to use certain design standards." Also, "he is grateful to the Union Carbide Corp. for permission to publish this article." Further, he notes that "The design method presented in this article is that used by Union Carbide Corp. for the thermal and hydraulic design of spiral-tube exchangers, and is somewhat different from that used by the fabricator." SI values were added by the handbook editor.

Nomenclature	
A	Heat-transfer area, ft^2
a	Net free-flow area, in^2
B	Film thickness $(0.00187\,Z\Gamma/g_c s^2)^{1/3}$, ft
c	Specific heat, Btu/(lb) (°F)
D_e	Equivalent dia., ft
D_H	Helix or spiral dia., ft
D_i	Inside tube dia., ft
d	Tube dia., in
f	Fanning friction factor, dimensionless
G	Mass velocity, lb/(h) (ft^2)
g_c	Gravitational constant, ft/(h)^3 (4.18×10^3)
h	Film coefficient of heat transfer, Btu/(h) (ft^2) (°F)
k	Thermal conductivity, Btu/(h) (ft^2) (°F/ft)
L	Tube length, ft
M	Molecular weight, dimensionless
N_{PT}	No. tube passes/shell, dimensionless
n	No. tube/pass, dimensionless
P	Pressure, lb/in^2 (abs)

(*Continued*)

Nomenclature (*Continued*)

ΔP	Pressure drop, psi
Q	Heat transferred, Btu
s	Specific gravity (referred to water at 20°C)
T	Temperature shellside, °C
t	Temperature tubeside, °C
ΔT_M	Logarithmic mean temperature difference (LMTD), °C
U	Overall heat-transfer coefficient, Btu/(h) (ft^2) (°F)
W	Flow rate, (lb/h)/1000
Γ	Condensate loading, lb/(h) (ft)
Z	Viscosity, cp.
θ	Time, h
λ	Heat of vaporization, Btu/lb
μ	Viscosity, lb/(h) (ft)
ρ_L	Liquid density, lb/ft^3
ρ_u	Vapor density, lb/ft^3
Σ, Σ'	Surface condition factor, dimensionless
σ	Surface tension, dynes/cm

Subscripts

b	Bulk fluid properties
f	Film fluid properties
H	High temperature
i	Tubeside conditions
L	Low temperature
o	Shellside conditions
s	Scale or fouling material
w	Wall, tube material

Dimensionless Groups

N_{Re}	Reynolds number
N_{Rec}	Critical Reynolds number
N_{Pr}	Prandtl number

See procedure for SI values.

INTERNAL STEAM TRACING HEAT-TRANSFER DESIGN FOR PIPELINES

Size a steam-tracing system to maintain an intermittently used fuel-oil unloading line at 140°F (60°C) when the following conditions apply:

Average temperature	20°F (–6.7°C)
Wind speed	20 mi/h (32.2 km/h)
Line length	500 ft (152.4 m)
Size	12-in (304.8-mm) nominal bore
Max. unloading rate	500 tons/h (450 t/h)
Steam available at 65 lb/in^2 gage (447.9 kPa)	
Specific gravity of oil	0.985
Viscosity (see step 5)	
Note: Viscosity in cP × 2.42 = viscosity in lb/(ft)(h)	
Thermal conductivity	0.0788 Btu/(h) (ft^2) (°F/ft) (0.134 W/m°C)
Specific heat	0.47 Btu/(lb) (°F) (1.97 kJ/kg°C)

The maximum internal tracer length normally used is 150 ft (45.7 m).

1. *Choose the type of steam tracing to use for the pipeline*
As this fuel-oil unloading pipeline will be used only intermittently (as when a tanker or tank truck delivers a load of fuel oil), use internal steam tracing and an unlagged (no insulation) pipeline.

When internal-trace pipes are installed, all the available heat-transfer surface is utilized. The disadvantages are: (*a*) reduction in the equivalent internal diameter of the pipeline, (*b*) loss of ability to clean the pipeline by pigging or by using rotary brushes, (*c*) difficulty in cleaning fouled heat-transfer surfaces.

The trace pipe can only be installed in straight lengths of pipeline that are free of valves. Trace-pipe lengths have to be short, to prevent problems in supporting the trace pipe. It must enter and leave the pipeline frequently, increasing the possibility of leaks. Stresses arising due to differential expansion of the trace and the pipeline should be considered.

One possible application of the internal type of trace is for a fuel-oil unloading line, where the operating temperature is such that it is uneconomical to insulate the line, and where the line is used only intermittently. (Insulation is essential for externally traced lines.) Internal tracing is acceptable only if leakage of steam into the product conveyed by the pipeline can be tolerated.

2. *Determine the Reynolds number for the pipeline*
Using the nomenclature given, apply the Reynolds number equation. Or:

Heat loss—Neglecting the effect of the internal tracer at this point:

$$\text{Reynolds No., } N_{Re} = \frac{6.31W}{\mu D_{ip}}$$

where
$$W_i = \text{mass flow rate}$$
$$= 500 \times 2240 \text{ lb/h (1017 kg/h)}$$
$$\mu = \text{viscosity}$$
$$= 200 \text{ cP at } 140°\text{F}$$
$$D_{ip} = \text{pipe dia.}$$
$$= 12 \text{ in}$$
$$\therefore N_{Re} = \frac{6.31 \times 2240 \times 500}{200 \times 12}$$
$$= 2940$$
$$L/D = 500$$

3. *Find the tubeside heat transfer*
Using, Fig. 15, the chart for determining the tubeside heat transfer based on the Reynolds number and Colburn factor,

$$\left(\frac{h_i D_{ip}}{k_a \times 12}\right)\left(\frac{C_p \cdot \mu}{k_a}\right)^{-1/3}\left(\frac{\mu}{\mu_w}\right)^{-0.14} = 8.0$$

whence,

$$h_i = \left(\frac{12k_a}{D_{ip}}\right)\left(\frac{C_p\mu}{k_a}\right)^{1/3}\left(\frac{\mu}{\mu_w}\right)^{0.14} \times 8.0$$

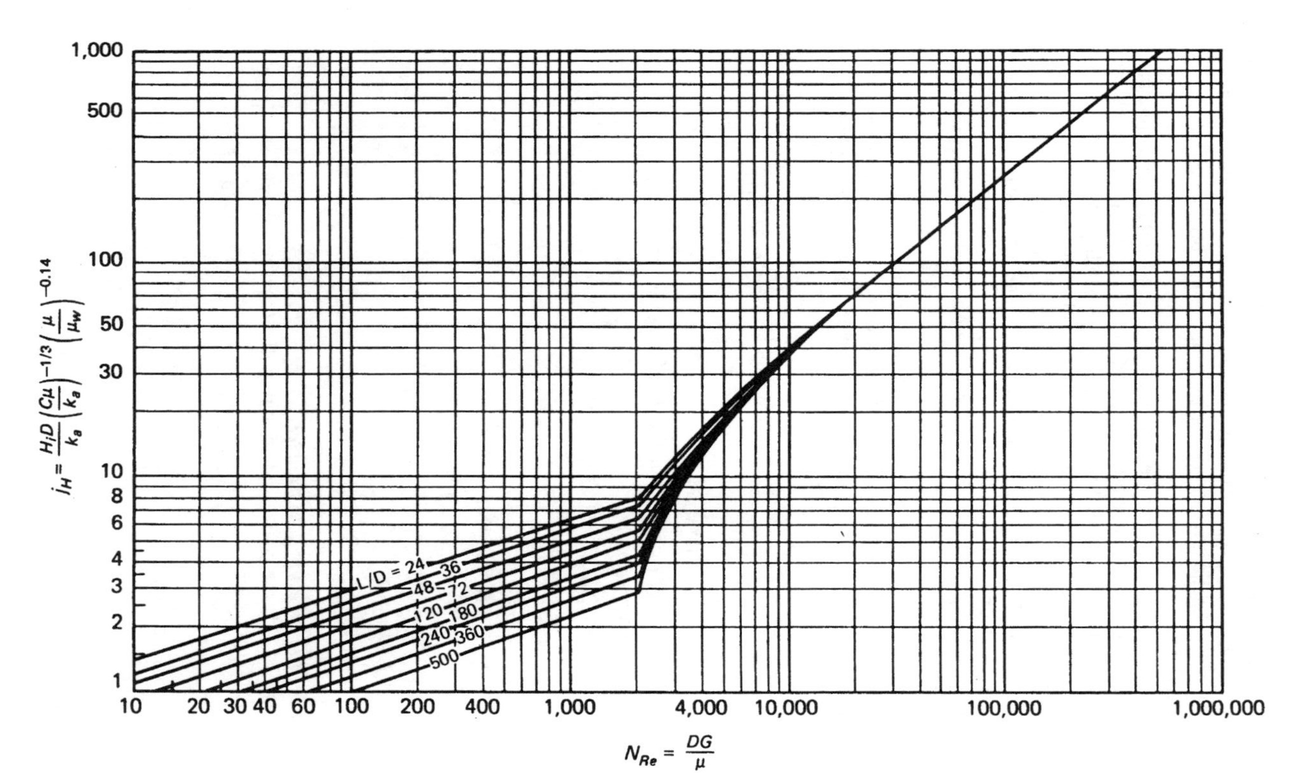

FIGURE 15 Chart for determining tubeside heat transfer, based on Reynolds number and Colburn factor. (*Chemical Engineering*.)

Assuming for the first trial that $t_s = 90°F$:

$$
\begin{aligned}
D_{ip} &= 12 \text{ in} \\
\mu &= 200 \times 2.42 \text{ lb/(ft)(h) [at 140°F] (60°C)} \\
\mu_w &= 520 \times 2.42 \text{ lb/(ft)(h) [at 90°F] (32.2°C)} \\
C_p &= 0.47 \text{ Btu/(lb)(°F)(1.97 kJ/kg} \cdot \text{°C)} \\
\rho &= 0.985 \times 62.4 \text{ lb/ft}^3 \text{ (1000 kg/m}^3\text{)} \\
k_a &= 0.0778 \text{ Btu/(h)(ft}^3\text{)(°F/ft)} \\
\therefore h_i &= \left(\frac{8 \times 12 \times 0.0778}{12}\right)\left(\frac{0.47 \times 200 \times 2.42}{0.0778}\right)^{1/3}\left(\frac{200 \times 2.42}{520 \times 2.42}\right)^{0.14} \\
&= 0.621\,(2.920)^{1/3}\,(0.385)^{0.14} \\
&= 0.621\,(14.3)\,(0.875) \\
&= 7.78 \text{ Btu/(h)(ft}^2\text{)(°F)} \\
\text{take } h &= 8.0 \text{ Btu/(h)(ft}^2\text{)(°F) (45.5 W/m}^2 \cdot \text{°C)}
\end{aligned}
$$

Correcting to O.D. of pipe:

$$h_i = 8.0 \times (12.7/12) = 8.5 \text{ Btu/(h)(ft}^2\text{)(°F)}$$

Hence,

$$
\begin{aligned}
Q_{conv} &= h_i A(t_p - t_s) \\
&= 8.5A\ (140 - t_s) \text{ Btu/(h)(ft}^2\text{)}
\end{aligned}
$$

and

$$Q_a = h_a A(t_s - t_a)$$

From Tables 12 and 13,

$$
\begin{aligned}
h_a &= 1.83 \times 2.76 \\
&= 5.04 \text{ Btu/(h)(ft}^2\text{)(°F)} \\
&\qquad \text{[corrected for wind velocity]} \\
\therefore Q_a &= 5.04A\ (t_s - 20) \text{ Btu/(h)(ft}^2\text{)}
\end{aligned}
$$

4. *Compute the heat loss from the pipeline*
Since $Q_{conv} = Q_a$

$$8.5A\ (140 - t_s) = 5.04A\ (t_s - 20)$$

i.e.,

$$1190 - 8.5\,t_s = 5.04\,t_s - 100.8$$

$$\therefore t_s = 1290/13.54 = 95°F\ (35°C)$$

TABLE 12 Values of h_s, for Pipes in Still Air*

Nominal pipe dia., in	$(t_s - t_a)$, °F [For an unlagged pipe $t_s - t_w$]							
	50	100	150	200	250	300	400	500
½	2.12	2.48	2.76	3.10	3.41	3.75	4.47	5.30
1	2.03	2.38	2.65	2.98	3.29	3.62	4.33	5.16
2	1.93	2.27	2.52	2.85	3.14	3.47	4.18	4.99
4	1.84	2.16	2.41	2.75	3.01	3.33	4.02	4.83
8	1.76	2.06	2.29	2.60	2.89	3.20	3.83	4.68
12	1.71	2.01	2.24	2.54	2.82	3.12	3.83	4.61
24	1.64	1.93	2.15	2.45	2.72	3.03	3.70	4.48

h_a = Btu/(h)(ft^2)(°F) based on still air.
*From McAdams, W. H., *Heat Transmission*, 3rd ed., McGraw-Hill, New York, 1954, p. 179.

TABLE 13 Correction Factor for h_a at Different Wind Velocities*

Wind velocity mi/h	$(t_s - t_a)$, °F [For an unlagged pipe, $t_s - t_w$]				
	100	200	300	400	500
2.5	1.46	1.43	1.40	1.36	1.32
5.0	1.74	1.69	1.64	1.59	1.53
10.0	2.16	2.10	2.02	1.93	1.84
15.0	2.50	2.42	2.33	2.27	2.08
20.0	2.76	2.69	2.58	2.45	2.30
25.0	2.98	2.89	2.78	2.64	2.49
30.0	3.15	3.06	2.94	2.81	2.66
35.0	3.30	3.21	3.10	2.97	2.81

*From Thermon Data Book, Premaberg (GB) Ltd.

There is an insignificant difference in μ_w between 90° (32.2°C) and 95°F; similarly, h_a does not change.

$$\therefore \text{ heat loss} = (8.5)\,\pi(12.7/12)(140 - 95)$$
$$= 1{,}290 \text{ Btu/(h)(ft of pipe) (4465 kJ/h m)}$$

Due to the low temperature of the pipe, radiation heat loss has been neglected.

5. *Calculate the tracer pipe size*

For first trial, assume tracer is 1.5 in I.D. (1.9 in O.D.)

$$\text{Equivalent diameter, } D_e = \frac{144 - (1.9)^2}{12} = 11.7 \text{ in (297.2 mm)}$$

$$N_{Re} = D_e \frac{G}{\mu}$$

where:

G = mass velocity, lb/(h)(ft^2)

$$= \frac{500 \times 2240}{\frac{\pi}{4}\left(1 - \frac{(1.9)^2}{144}\right)}$$

$$= 1.45 \times 10^6 \text{ lb/(h) (ft}^2\text{) } (7.1 \times 10^6 \text{ kg/h m}^2)$$

$$D_e = 0.975 \text{ ft}$$

$$\mu = 200 \text{ cP at } 140°\text{F}$$

$$\therefore N_{Re} = \frac{11.7 \times 1.45 \times 10^6}{12 \times 200 \times 2.42} = 2.8 \times 10^3$$

From Fig. 15,

$$j_H = 9.0; N_{Re} = 2800; L/D_e$$

$$= 150/0.975 = 154$$

Heat-transfer coefficient for heat transferred to the fuel oil based on I.D. of the 12-in (304.8-mm) nominal-bore pipe at the steam tracer surface where the fuel-oil viscosity is:

$$\mu_w = \frac{h_i D_e}{k_a}\left(\frac{C_p \mu}{k_a}\right)^{-1/3}\left(\frac{\mu}{\mu_w}\right)^{-0.34} = p$$

Assuming a 20-lb/in^2 (137.8-kPa) pressure drop along the steam tracer, the average steam pressure = 55 lb/in^2 gage (378.9 kPa) (steam available at 65 lb/in^2 gage) (447.9 kPa).

$$\text{Average steam temperature} = 303°\text{F } (150.6°\text{C})$$

$$\therefore \mu_w = 1.0 \times 2.42 \text{ lb/(ft)(h)[for fuel oil]}$$

and heat-transfer coefficient at steam-tracer outside surface.

$$h_o = 9 \times \left(\frac{0.0778}{0.975}\right)\left(\frac{0.47 \times 200 \times 2.42}{0.0778}\right)^{1/3}\left(\frac{200}{1.0}\right)^{0.14}$$

$$= 21.6 \text{ Btu/(h)(ft}^2\text{)(°F) } (122.6 \text{ W/m}^2 \cdot °\text{C})$$

$$\frac{1}{U_o} = \frac{1}{h_o} + 0.005 = \frac{1}{21.6} + 0.005$$

$$= 0.051 \text{ Btu/(h)(ft}^2\text{)(°F), approx.}$$

$$\therefore U_o = 19.6 \text{ Btu/(h)(ft}^2\text{)(°F)}$$

Heat transfer from steam tracer = heat lost from pipe without steam tracer:

$$Q_p = U_o A_p (t_{st} - t_p)$$

$$\therefore 1290 = 19.6 A_p (303 - 140)$$

Hence,

$$A_p = \frac{1290}{19.6 \times 163}$$

$$= 0.04 \text{ ft}^2\text{/ft run of pipe } (0.12 \text{ m}^2\text{/m})$$

Surface area of 1-in nominal-bore pipe = 0.344 ft^2/ft (0.10 m^2/m). Surface area of $1^1/_2$-in nominal-bore pipe = 0.498 ft^2/ft (0.15 m^2/m). Therefore, $1^1/_2$-in (38.1-mm) nominal-bore tracing will be acceptable.

6. *Check the effect of fluid velocity on the outside film coefficient at the pipe wall*
Since the internal tracer pipe increases the fluid velocity in the pipeline.

$$N_{Re} = \frac{D_e G}{\mu} = 1 \times \frac{1.45 \times 10^6}{2.42 \times 200} = 3010$$

where G is based on the equivalent diameter of the pipe, and μ is at 140°F (60°C), the desired fuel-oil temperature.

$$j_H = 8.05 \text{ [from Fig. 15]}$$

There will be no significant increase in the calculated heat loss.

7. *Find the pressure drop in the pipeline*
Correction factor

$$= \frac{(D_i)^{2.4}(D_i + 0.5D_o')^{1.2}}{(D_i^2 - D_o'^2)^{1.8}}\left(\frac{\mu_w}{\mu}\right)^{0.14}$$

$$= \frac{(12)^{2.4}(12 + 0.95)^{1.2}}{(144 - 3.6)^{1.8}}\left(\frac{520}{200}\right)^{0.14} = 1.33$$

Check steam tracer pressure-drop

Maximum tracer length = 150 ft (45.7 m)
Heat load = 150 × 1290 Btu/h
Latent heat of steam = 901 Btu/lb

$$\therefore \text{Steam load} = \frac{150 \times 1290}{901}$$

= 215 lb/h per tracer (97.6 kg/h)

From Fig. 16, steam pressure drop at inlet conditions:

$$0.55 \text{ lb/(in}^2\text{)(100-ft run)}$$

$$\therefore \text{pressure drop/tracer} = 1.5 \times 0.275 = 0.412 \text{ lb/in}^2 \text{ (2.8 kPa)}$$

∴ the average steam temperature = 312°F (65 lb/in^2 gage) (155.6°C, 447.9 kPa)

$$U_o = 19.6 \text{ Btu/(h)(ft}^2\text{)(°F)}$$

$$\Delta T = 312 - 140 = 172\text{°F (77.8°C)}$$

$$\therefore \text{area of tracer required} = \frac{1290}{(19.6 \times 172)} = 0.383 \text{ ft}^2\text{/ft (0.117 m}^2\text{/m)}$$

The surface area of 1½-in (38.1-mm) nominal-bore pipe is 0.498 ft^2 /ft (0.15 m^2/m).

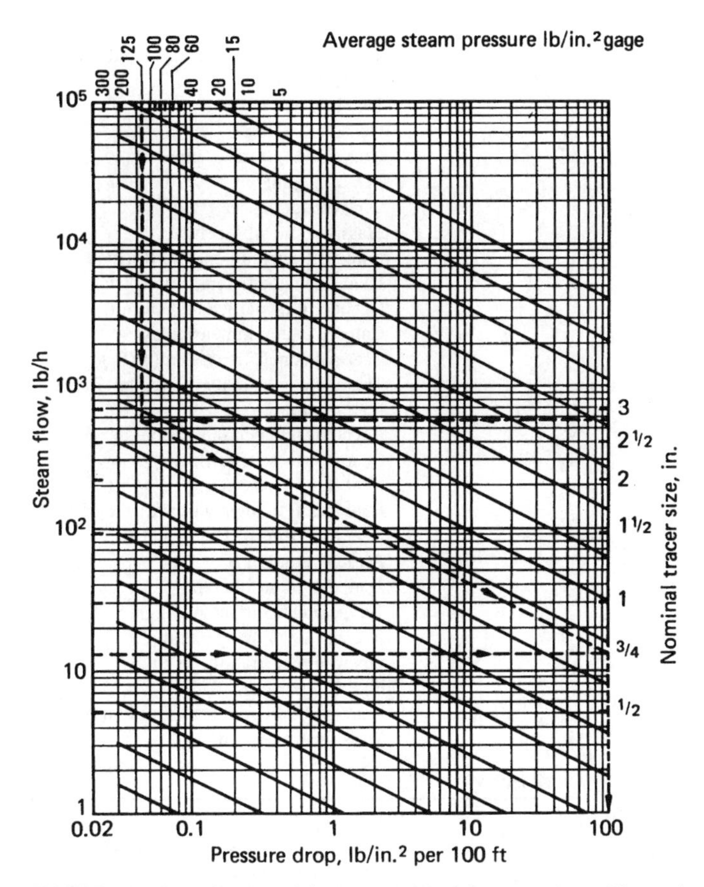

FIGURE 16 Chart for determining pressure drop in steam pipes. (*Chemical Engineering.*)

Related Calculations. The internal heat tracer is less used than the external heat tracer. Reasons given in step 1 above are sufficient enough to discourage designers from choosing internal tracing, except for specialized applications, such as the one considered here. Where the internal heat tracer can be used, it provides maximum heat transfer because the heat source is immersed in the medium that is to be heated. Further, the internal heat tracer does not require out-of-round pipe insulation because the heating line is inside the pipe that is being heated.

This procedure is the work of I. P. Kohli, Consultant, as reported in *Chemical Engineering* magazine. SI values and numbered procedure steps were added by the handbook editor.

Nomenclature

A	Area of pipe, ft^2/ft of length
A_p	Outside area of tracer pipe, ft^2/ft length
A_1	Outside area of pipe lagging, ft^2/ft length
C_p	Specific heat, Btu/(lb)(°F)
D	Dia., in
D_e	Equivalent pipe dia., ft

(*Continued*)

Nomenclature (*Continued*)

D_i	Inside dia., in
D_{ip}	Pipe I.D., in
D_o	Outside dia., in
D'_o	Outside dia. of tracer pipe, in
d_i	I.D. of lagging, in
d_o	O.D. of lagging, in
G	Mass velocity, lb/(h)(ft^2)
h_o	Film coefficient to air, Btu/(h)(ft^2)
h_i	Inside film coefficient, Btu/(h)(ft^2)
h_o	Outside film coefficient, Btu/(h)(ft^2)
j_H	Colburn factor, dimensionless
k_a	Thermal conductivity, Btu/(h)(ft^2)(°F/ft)
k_1	Thermal conductivity of lagging, Btu/(h)(ft^2)(°F/ft)
L	Pipe length, ft
N_{Re}	Reynolds number, dimensionless
Q_a	Heat flux to air, Btu/(h)(ft^2)
Q_{conv}	Heat flux by convection, Btu/(h)(ft^2)
Q_p	Heat loss from pipe without tracer, Btu/(h)(ft^2)
q_2	Heat loss, Btu/(h)(ft of length)
t_a	Air temperature, °F
t_p	Temperature inside pipe, °F
t_s	Temperature, surface of lagging, °F
t_{st}	Average steam temperature, °F
t_w	Pipewall temperature, °F
T_p	Pipe temperature, °F
U_o	Heat-transfer coefficient based on outside surface, Btu/(h)(ft^2)(°F)
W_1	Mass flow rate, lb/h

Greek letters

μ	Viscosity, lb/(ft)(h)
μ_w	Viscosity at wall temperature, lb/(ft)(h)
ρ	Fluid density, lb/ft^3

See procedure for SI values.

EXTERNAL STEAM TRACING HEAT-TRANSFER DESIGN FOR PIPELINES

Size an external steam tracing system, Fig. 17, to maintain a 3-in (76.2-mm) pipeline at 320°F (160°C). The line will be lagged with 1.5 in (38.1 mm) of insulation. Conditions at the pipeline are:

$$k_1 = 0.033 \text{ Btu/h ft}^2 \cdot \text{°F ft } (0.057 \text{ W/m} \cdot \text{°C})$$

$$\text{Wind speed} = 20 \text{ mph } (23.2 \text{ km/h})$$

$$\text{Minimum air temperature, } T_a = 20\text{°F } (-6.7\text{°C})$$

$$\text{Line length} = 100 \text{ ft } (30.5 \text{ m})$$

Steam for tracing is available at 150 lb/in^2 (gage) (1033.5 kPa)

$$\text{Pipe temperature, } t_p = 320\text{°F } (160\text{°C})$$

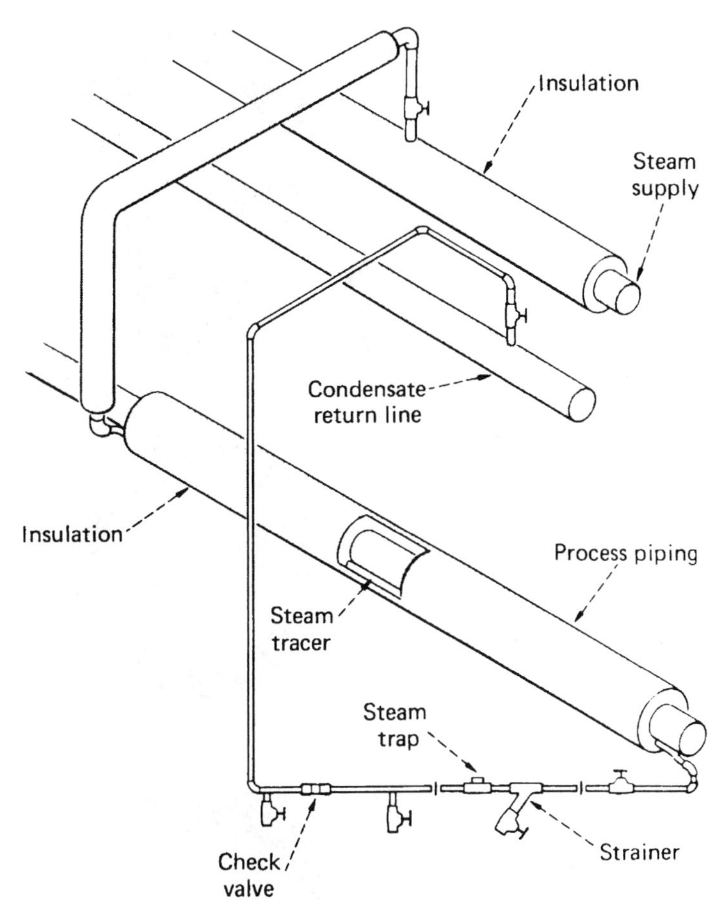

FIGURE 17 Typical steam-tracing system using external tracer pipe. (*Chemical Engineering.*)

Calculation Procedure:

1. *Determine the heat load for the tracer pipe*

Assume that one of the insulation schemes shown in Fig. 18 will be used for this pipeline. Further, assume a 5-lb/in^2 (34.5-kPa) pressure drop along the external steam tracer pipe. Then, the average steam temperature, using the nomenclature in the previous procedure, is:

$$t_{st} = \frac{366 + 363°\text{F}}{2} = 364.5°\text{F, say, } 364°\text{F } (184°\text{C})$$

$$\begin{aligned}\therefore \text{ average temperature} &= 0.5(t_{st} + t_p)\ [°\text{F}] \\ &= 0.5(364 + 320)\ [°\text{F}] \\ &= 342°\text{F } (172°\text{C})\end{aligned}$$

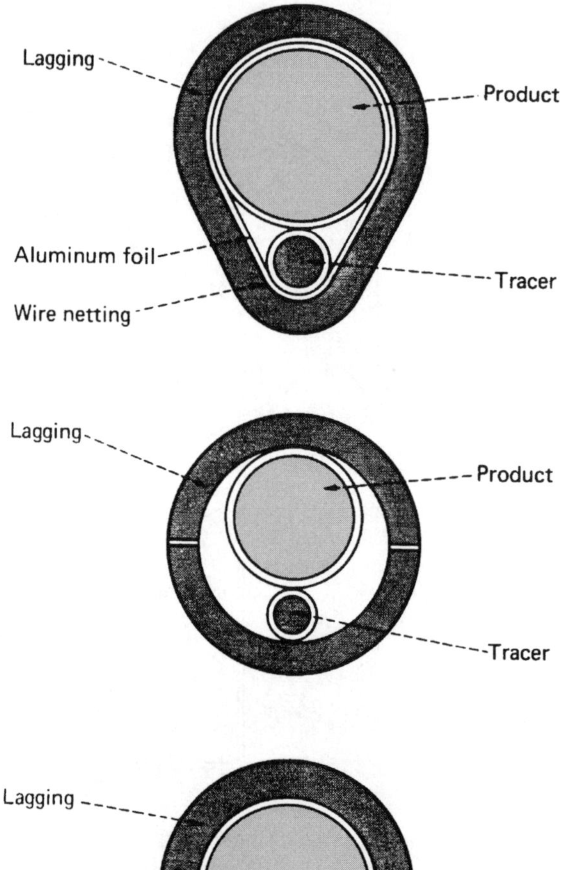

FIGURE 18 Various configurations of tracer piping and insulation lagging. (*Chemical Engineering.*)

Allowing for tracer, I.D. of lagging = 4 in (101.6 mm)

$$\therefore q_2 = \frac{2\pi k_1 (t_p - t_s)}{\log_e \dfrac{d_o}{d_i}} = h_a A_1 (t_s - t_a)$$

$$d_o = 7 \text{ in}, d_i = 4 \text{ in (101.6 mm)}$$

Assuming for the first trial that $h_a = 4.0$ Btu/(h)(ft²)(°F) (22.7 W/m² · °C):

$$q_2 = \frac{2 \times \pi \times 0.033(342 - t_s)}{2.303 \log_{10}(7.0/4.0)}$$
$$= 4.0\pi \times (7.0/12) \times 1(t_s - 20)$$

or

$$\frac{0.0895(342 - t_s)}{0.245} = 7.35(t_s - 20)$$

or

$$0.368(342 - t_s) = 7.35t_s - 147$$

or

$$126 - 0.368t_s = 7.35t_s - 147$$
$$\therefore t_s = \frac{273}{7.718} = 35°\text{F } (1.6°\text{C})$$

Tables 12 through 14 are not sufficiently accurate in this region, but the value of 4.0 taken for h_a is on the safe side, allowing for wind speed.

$$\therefore q_2 = 7.35(t_s - 20) \text{ Btu/(h)(ft run of pipe)}$$
$$= 7.35(35 - 20)$$
$$= 113 \text{ Btu/(h)(ft run of pipe) (391 kJ/h} \cdot \text{m)}$$

2. *Compute the required size of the steam-tracer pipe*
Do this by using the thermal-conductance data on tracer to pipe: Btu/(h)(°F)(ft of pipe). This takes into account the heat-transfer coefficient and pipe surface area. These have been found by experiment.

If one ½-in (12.7-mm) tracer without cement is used, heat transfer

$$= 0.393(t_{st} - t_p)$$
$$= 0.393(364 - 320) = 17.3 \text{ Btu/(h)(ft run) (59.9 kJ/h} \cdot \text{m)}$$

TABLE 14 C_t, Thermal Conductance, Tracer to Pipe*

Tube size	C_t with no heat-transfer cement, Btu/(h)(°F)(ft of pipe)	C_i with heat-transfer cement Btu/(h)(°F)(ft of pipe)
3/8	0.295	3.44
½	0.393	4.58
5/8	0.490	5.73

*See procedure for SI values.

This does not meet the heat load requirement. Two ½-in (12.7-mm) tracers without cement do not overcome the problem. If one ½-in tracer with heat-transfer cement is installed, the heat-transfer rate is:

$$4.58(364 - 320) = 201 \text{ Btu/(h)(ft of run)} \ (695.7 \text{ kJ/h} \cdot \text{m})$$

A 3/8-in (9.5-mm) tracer with cement will yield a heat-transfer rate of:

$$3.44(364 - 320) = 151 \text{ Btu/(h)(ft of run)} \ (522.7 \text{ kJ/h} \cdot \text{m})$$

Standard practice requires the installation of ½-in (12.7-mm) tracer pipes. There is always some overdesign with steam tracers.

Related Calculations. External tracing involves the placing of steam pipes or tubes outside the pipeline. The traced line is then insulated by using preformed sectional insulation.

Heat transfer from trace to pipeline is by conduction, convection in the air space, and radiation. The contact area between the trace and the pipeline is quite small. However, when heat-sensitive liquids are being heated, or when the pipe is plastic-lined, this contact may give rise to undesirable hot spots. In such cases, asbestos packing rings are often used to eliminate any direct contact between the trace and the pipeline.

The simplest method of external tracing is to wrap copper tube around the pipeline, and then to cover the traced pipe with insulation. This is the only way to trace around valves, pipe fittings, and instruments (Fig. 19). This procedure is unsuitable for horizontal runs because steam condensate collects at low points and may freeze during a shutdown.

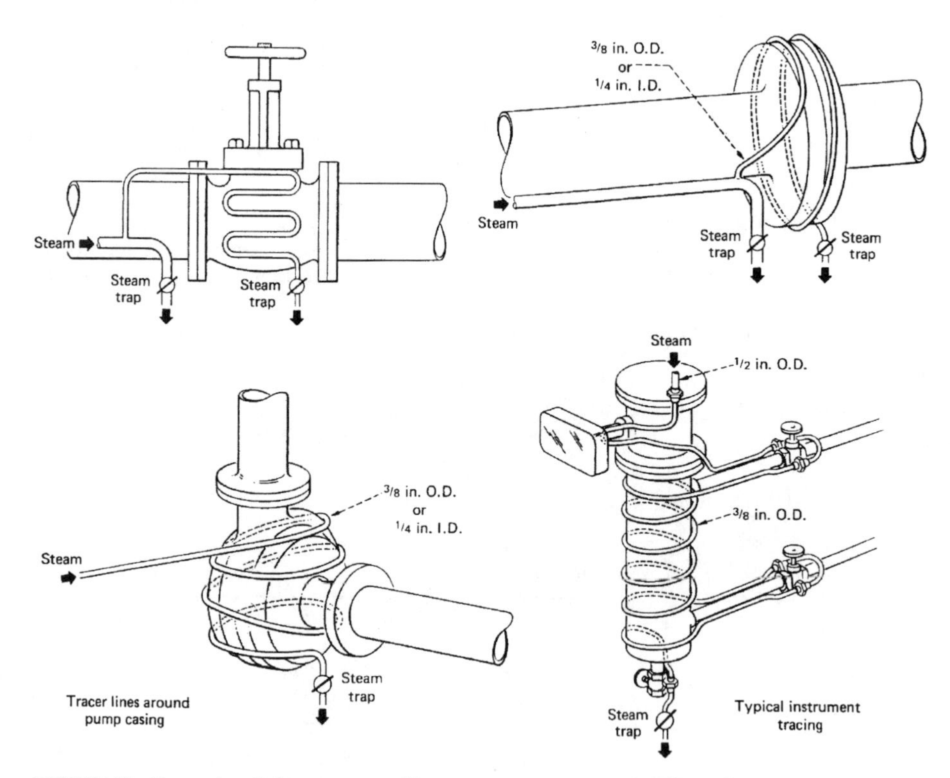

FIGURE 19 How various fittings, pumps, and instruments are steam-traced. (*Chemical Engineering.*)

It is essential to ensure that the trace lines are self-draining. Copper tubing of 0.5-in (12.7-mm) O.D. and 0.035-in (0.89-mm) wall thickness is usually used with straight piping. If the tubing has to be bent into a small radius (as when tracing valves), 0.375-in-O.D. (9.5-mm) tubing may be used.

The length of a single trace tube (from steam supply valve to steam trap) is limited by pressure drop in the trace. The trap should have a condensate drainage capacity to match the heating load. At a steam pressure of 100 lb/in^2 (gage) (689 kPa) or higher, the length of a single trace should not exceed 200 ft (60.9 m). If the steam pressure is lower, a tracer length of 100 ft (30.5 m) is recommended.

To improve contact, the trace tube may be wired to the pipeline. Even then, the conductive heat-transfer rate is quite low. It may be increased by putting a layer of heat-conducting cement (graphite mixed with sodium silicate or other binders) between the trace and the pipeline. This provides much more surface for conductive heat transfer.

When higher heat loads are desirable, straight lengths of $^1/_2$-in (12.7-mm) carbon-steel pipe are clipped along the pipeline. The number of tracer pipes depends upon the heat load—large-diameter pipes carrying liquids that have a high melting point may require up to 10 tracers. Steel bands, fitted by using a packing-case banding machine, help to minimize air gaps between the pipeline and tracers. At pipeline bends, the tracers are also bent.

At valves and flanges, the tracers are formed into loops that also function as expansion joints. The loops are formed in a nearly horizontal plane, to ensure self-drainage. The traced pipeline is covered with larger-sized preformed lagging.

In some cases, the traced line is wrapped with aluminum foil and then covered with shaped lagging (Fig. 18). The foil increases the radiation heat transfer. It is essential that the space between the pipeline and the tracer be kept free of particles of lagging material.

Conductive heat transfer may be enhanced by welding the trace on the pipeline. However, welding causes problems due to differential expansion of the pipeline and trace. Because horizontal pipelines are traced on the lower half, welding is a difficult operation. It is more convenient to use heat-transfer cement, as noted earlier.

This procedure is the work of I. P. Kohli, Consultant, as reported in *Chemical Engineering* magazine. SI values were added by the handbook editor.

SELECTING AIR-COOLED HEAT EXCHANGERS

Kerosene flowing at a rate of 250,000 lb/h (31.5 kg/s) is to be cooled from 160°F (71°C) to 125°F (51.6°C), for a total heat duty of 4.55 million Btu/h (1.33 MW). How large an air cooler (sometimes called a *dry* heat exchanger) is needed for this service if the design dry-bulb temperature of the air is 95°F (35°C)?

Calculation Procedure:

1. *Determine the temperature rise of the air during passage through the cooler*

From Table 15 estimate the overall heat-transfer coefficient for an air cooler handling kerosene at 55 Btu/(h · ft^2 · °F) [312.3 W/(m^2 · K)]. Then the air-temperature rise is $t_2 - t_1 = 0.005U\{[(T_1 + T_2)/2] - t_1\}$, where t = inlet air temperature, °F or °C; t_2 = outlet air temperature, °F or °C; U = overall heat-transfer coefficient, Btu/(h · ft^2 · °F) [W/(m^2 · K)]. T_1 = cooled fluid inlet temperature, °F or °C; T_2 = cooled fluid outlet temperature, °F or °C. Substituting yields $t_2 - t_1 = 0.005(55)\{[(160 + 125)/2] - 95\} = 13.06$°F (7.2°C).

Next, from Fig. 20, the correction factor for a process-fluid temperature rise of 160 – 125 = 35°F (19.4°C) is 0.94. So the corrected temperature rise = $f(t_2 - t_1) = 0.94(13.06) = 12.28$°F (6.8°C). Therefore, $t_2 = 95 + 12.28 = 107.28$°F (41.8°C).

2. *Find the log mean temperature difference (LMTD) for the heat exchanger*

Use the relation LMTD = $(\Delta t_2 - \Delta t_1)/\ln(\Delta t_2 - \Delta t_1)$. Or, LMTD = [(160 – 107.28) – (125 – 95)]/ln [(160 – 107.28)/(125 – 95)] = 40.30. This value of the LMTD must be corrected by using Fig. 21

TABLE 15 Heat-Transfer Coefficients for Air-Cooled Heat Exchangers*

Liquid cooled	Heat-transfer coefficient Btu/(h · ft² · °F)	W/(m² · K)
Diesel oil	45–55	255.5–312.3
Kerosene	55–60	312.3–340.7
Heavy naphtha	60–65	340.7–369.1

Chemical Engineering.

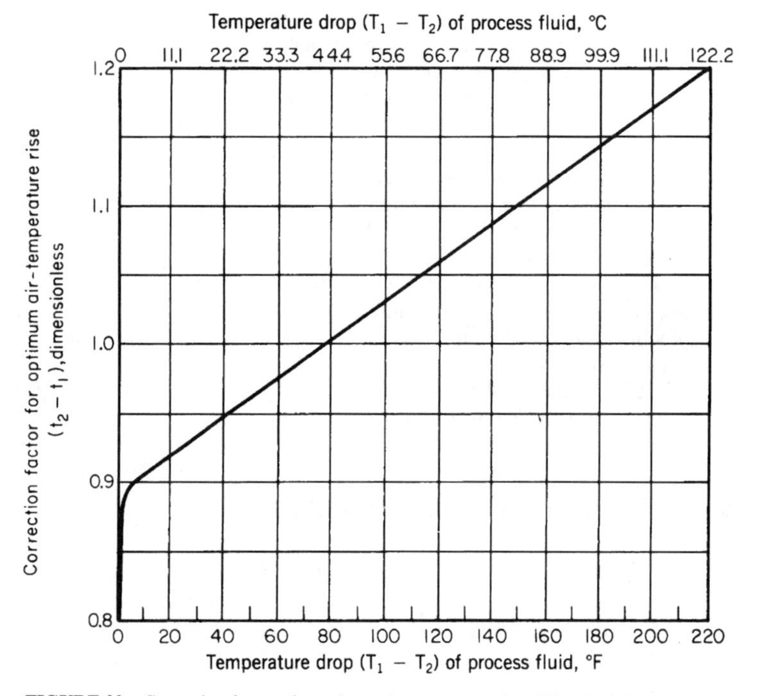

FIGURE 20 Correction factors for estimated temperature rise. (*Chemical Engineering.*)

for temperature efficiency P and a correlating factor R. Thus, $P = (t_2 - t_1)/(T_1 - t_1) = (107.28 - 95)/(160 - 95) = 0.189$. Also, $R = (T_1 - T_2)/(t_2 - t_1) = (160 - 125)/(107.28 - 95) = 2.85$. Then, from Fig. 21, LMTD correction factor = 0.95, and the corrected LMTD = f(LMTD) = 0.95(40.30) = 38.29°F (21.2°C).

3. *Determine the hypothetical bare-tube area needed for the exchanger*

Use the relation $A = Q/U\Delta T$, where A = hypothetical bare-tube area required, ft² (m²); Q = heat transferred, Btu/h (W); ΔT = effective temperature difference across the exchanger = corrected LMTD. Substituting gives A = 4,550,000/55(38.29) = 2160 ft² (200.7 m²).

4. *Choose the cooler size and number of fans*

Enter Table 16 with the required bare-tube area, and choose a 12-ft (3.6-m) wide cooler with either four rows of 40 ft (12 m) long tubes with two fans, for a total bare surface of 2284 ft² (205.6 m²), or five rows of 32 ft (9.6 m) long with two fans for 2288 ft² (205.5 m²) of surface. From Fig. 22, the fan hp for the cooler would be 1.56(2284/100) = 35.63 hp (25.6 kW).

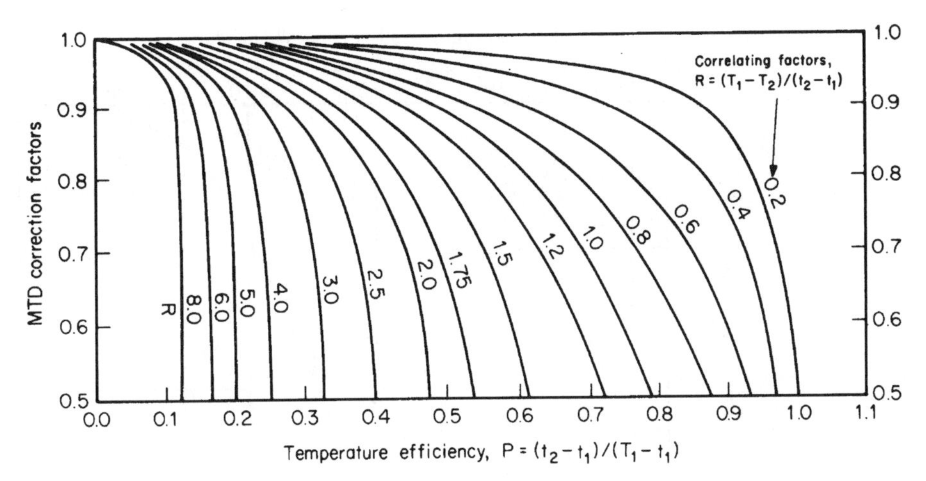

FIGURE 21 MTD correction factors for one-pass crossflow with both shellside and tubeside unmixed. T represents hot-fluid characteristics, and t represents cold-fluid characteristics. Subscripts 1 and 2 represent inlet and outlet, respectively. (*Chemical Engineering*.)

TABLE 16 Typical Air-Cooled Heat-Exchanger Cooling Area, ft^2 (m^2)*

Approximate cooler width		Tube length		Fans per unit	No. of 1-in (2.5-cm) tube rows in depth on 2-3/8-in (6-cm) pitch	
ft	m	ft	m		4	5
12	3.66	32	9.8	2	1827 (169.7)	2288 (212.6)
		36	10.9	2	2056 (191.0)	2574 (239.1)
		40	12.2	2	2284 (212.2)	2861 (265.8)
14	4.27	14	4.3	1	931 (86.5)	1166 (108.3)
		16	4.9	1	1064 (98.8)	1333 (123.8)

*Chemical Engineering.

Related Calculations. Air coolers are widely used in industrial, commercial, and some residential applications because the fluid cooled is not exposed to the atmosphere, air is almost always available for cooling, and energy is saved because there is no evaporation loss of the fluid being cooled.

Typical uses in these applications include process-fluid cooling, engine jacket-water cooling, air-conditioning condenser-water cooling, vapor cooling, etc. Today there are about seven leading design manufacturers of air coolers in the United States.

The procedure given here depends on three key assumptions: (1) an overall heat-transfer coefficient is assumed, depending on the fluid cooled and its temperature range; (2) the air-temperature rise $t_2 - t_1$ is calculated by an empirical formula; (3) bare tubes are assumed and fan hp (kW) is estimated on this basis to avoid the peculiarities of one fin type. By using the empirical formula given in step 1 of this procedure, the size air cooler obtained will be within 25 percent of optimum. This is adjusted for greater accuracy through use of the correction factor shown in Fig. 20.

Since no existing computer program is capable of considering all variables in optimizing air coolers, the procedure given here is useful as a first trial in calculating an optimum design. The flow pattern and correction factors used for this estimating procedure are those for one-pass crossflow with both tube fluid and air unmixed as they flow through the exchanger.

Where additional correction factors are needed for different flow patterns across the exchanger, the designer should consult the standards of the Tubular Exchanger Manufacturers Association (TEMA). Similar data will be found in reference books on heat exchange.

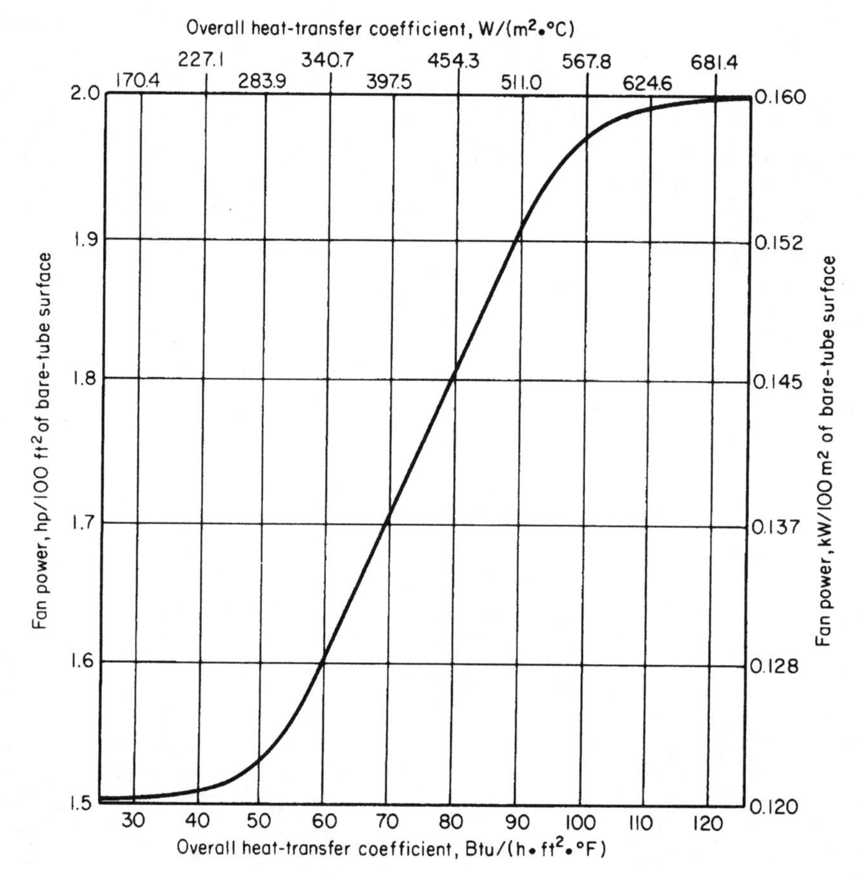

FIGURE 22 Approximate fan power requirements for air coolers. (*Chemical Engineering.*)

The procedure given here is the work of Robert Brown, General Manager, Happy Division, Therma Technology, Inc., as reported in *Chemical Engineering* magazine. Note that the procedure given is for a preliminary selection. The final selection will usually be made in conjunction with advice and guidance from the manufacturer of the air cooler.

QUICK DESIGN AND EVALUATION OF HEAT EXCHANGERS

Find the required surface area and shellside flow rate of a crossflow heat exchanger with four single-pass tube rows being designed to meet the following conditions: tube mass flow rate, $\dot{m}$ = 22,200 lb/h (10,070 kg/h); tube specific heat capacity, c = 0.20 Btu/lb · °F (0.84 kJ/kg · °C); shell specific heat capacity, C = 0.24 Btu/lb · °F (1.00 kJ/kg · °C): tubeside inlet fluid temperature, t_1 = 500°F (260°C); tubeside outlet fluid temperature, t_2 = 320°F (160°C); shellside inlet liquid temperature, T_1 = 86°F (30°C); shellside outlet liquid temperature. T_2 = 131°F (55°C); overall heat-transfer coefficient, U = 9.0 Btu/h · ft² · °F (51.1 W/m² · °C) (183.9 kJ/h · m² · °C).

Another unit, a 1-shell-pass and 2-tube-pass heat exchanger with a vertical shellside baffle for divided flow, performs as follows: $\dot{m}$ = 33,500 lb/h (15,200 kg/h): c = 0.98 Btu/lb · °F

(4.10 kJ/kg · °C); shell mass flow rate, $\dot{M}$ = 50,000 lb/h (22,680 kg/h); C = 0.60 Btu/lb · °F (2.51 kJ/kg · °C); t_1 = 270°F (132.2°C); T_1 = 520°F (271.1°C): U = 12.6 Btu/h · ft^2 · °F (71.5 W/m^2 · °C) (257.4 kJ/h · m^2 · °C): heat-exchanger surface area, A = 2200 ft^2 (204.4 m^2). Evaluate the performance of this unit by finding its thermal effectiveness, its efficiency, the outlet temperature of the tubeside fluid, and the outlet temperature of the shellside vapor.

Calculation Procedure:

1. ***Compute the ratio of shellside liquid to tubeside fluid temperature differences and the thermal effectiveness, or temperature efficiency, of the crossflow unit***

The shell-to-tube ratio of temperature differences is found by $R = (T_1 - T_2)/(t_2 - t_1)$ = (86 – 131)/(320 – 500) = 0.25. Thermal effectiveness, $P = (t_2 - t_1)/(T_1 - t_1)$ = (320 – 500)/(86 – 500) = 0.43.

2. ***Compute the shellside vapor flow rate***

Shellside vapor flow rate is $\dot{M} = \dot{m}c/RC$ = (22,200)(0.20)/[(0.25)(0.24)] = 74,000 lb/h (33,570 kg/h).

3. ***Determine the number of transfer units and heat-exchanger efficiency***

The point where R = 0.25 and P = 0.43 on Fig. 31, shown among several figures appearing after step 7, corresponds to values for the number of transfer units, NTU = 0.62 and heat-exchanger efficiency, F = 0.99. This value for F shows that the design has an efficiency close to that for a pure countercurrent configuration where F = 1.00.

4. ***Compute the heat-exchanger surface area***

To find the area use the formula $A = (\text{NTU})(\dot{m}c)/U$ = (0.62)(22,200)(0.20)/9.0 = 306 ft^2 (28.4 m^2).

5. ***Compute R and NTU for the shell-and-tube unit***

Substitute appropriate values into the following equations, thus $R = \dot{m}c/\dot{M}C$ = (33,500)(0.98)/(50,000)(0.60) = 1.09 and NTU = $UA/\dot{m}c$ = (12.6)(2200)/(33,500)(0.98) = 0.84.

6. ***Determine the thermal effectiveness and heat-exchanger efficiency***

The point where curves for R and NTU intersect on Fig. 26 corresponds to a thermal effectiveness, P = 0.42 and an efficiency, F = 0.89. For the configuration of this unit, the value of F can be considered acceptable.

7. ***Find the exit temperatures of the tubeside fluid and the shellside vapor***

The tubeside fluid exit temperature, $t_2 = P(T_1 - t_1) + t_1$ = 0.42(520 – 270) + 270 = 375°F (190.6°C), and the shellside vapor exit temperature, $T_2 = T_1 - R(t_2 - t_1)$ = 520 – 1.09(375 – 270) = 405°F (297.2°C).

Related Calculations. On the design and performance charts, Figs. 23 through 35, heat-exchanger efficiency, F = true mean temperature difference/logarithmic temperature difference for countercurrent flow and relates the actual rate of heat transfer, Q, to the theoretical rate, $U \times A \times$ logarithmic temperature difference. True mean temperature difference has been solved analytically for each configuration. This, in conjunction with the new NTU curves, eliminates the need for trial-and-error calculations to determine the design and evaluate the performance of specific heat exchangers. By establishing desired conditions which in effect specify any two of the four parameters, the other two parameters may then be read directly from a chart and used to design a unit and/or to evaluate its performance.

The 1-shell-pass–3-tube-pass and 1-shell-pass–2-tube-pass heat exchangers represent conventional shell-and-tube units such as those for steam heating, heating of one process stream by cooling another, and condensation and cooling by a cooling-water utility. Shellside pressure drops through divided-flow units are typically one-eighth of those through conventional shell-and-tube heat exchangers; hence divided flow units are recommended where low shellside pressure drops are required.

Crossflow heat exchangers differ from the conventional and divided-flow shell-and-tube units in that they have tube-bank arrangements over which another stream flows perpendicular to the tubes.

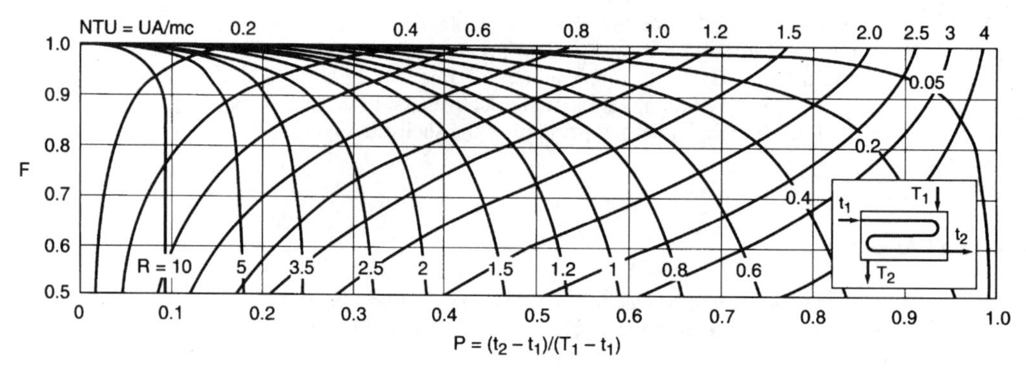

FIGURE 23 Design and performance chart for a 1-shell-pass and 3-tube-pass exchanger. (*Chemical Engineering*.)

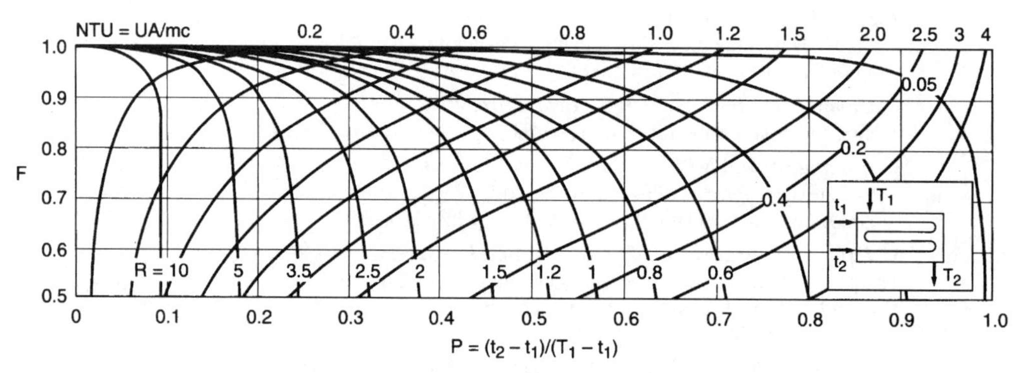

FIGURE 24 Design and performance chart for a 1-shell-pass and 4-tube-pass exchanger. (*Chemical Engineering*.)

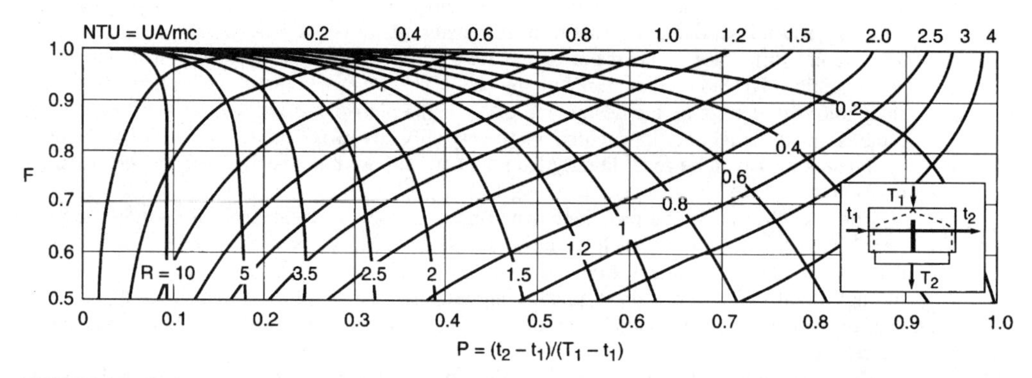

FIGURE 25 Design and performance chart for a 1-shell-pass and 1-tube-pass exchanger with a vertical shellside baffle. (*Chemical Engineering*.)

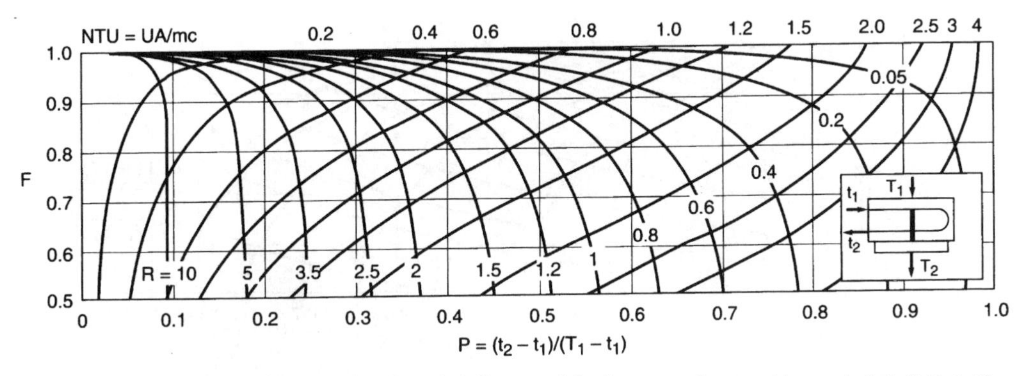

FIGURE 26 Design and performance chart for a 1-shell-pass and 2-tube-pass exchanger with a vertical shellside baffle. (*Chemical Engineering.*)

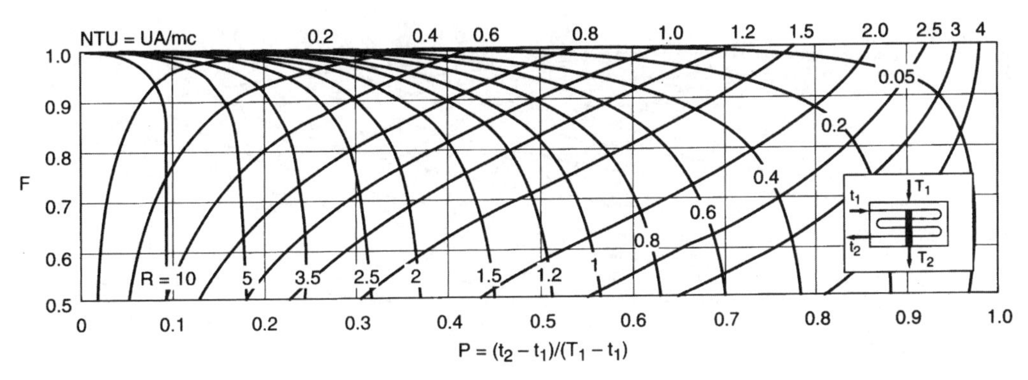

FIGURE 27 Design and performance chart for a 1-shell-pass and 4-tube-pass exchanger with a vertical shellside baffle. (*Chemical Engineering.*)

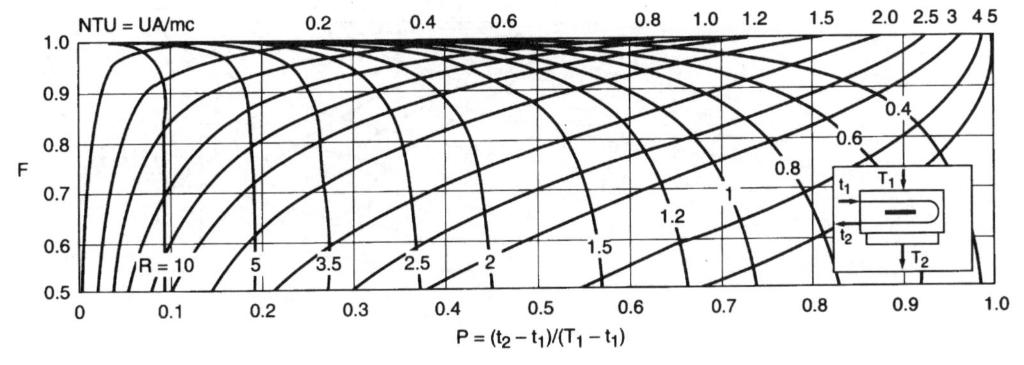

FIGURE 28 Design and performance chart for a 1-shell-pass and 2-tube pass exchanger with a horizontal shellside baffle. (*Chemical Engineering.*)

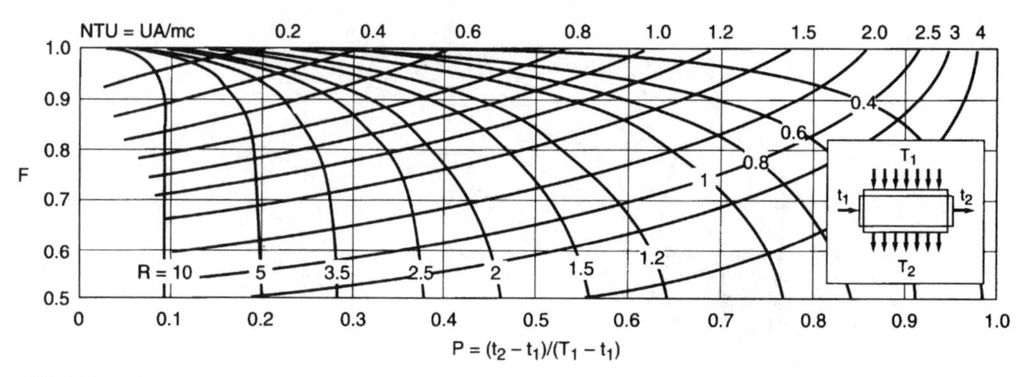

FIGURE 29 Design and performance chart for crossflow exchanger with two single-pass row tubes. (*Chemical Engineering*.)

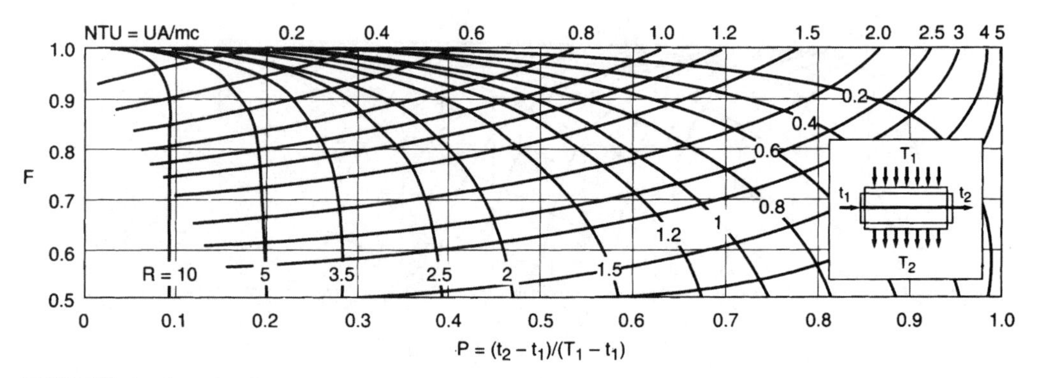

FIGURE 30 Design and performance chart for a crossflow exchanger with three single-pass tube rows. (*Chemical Engineering*.)

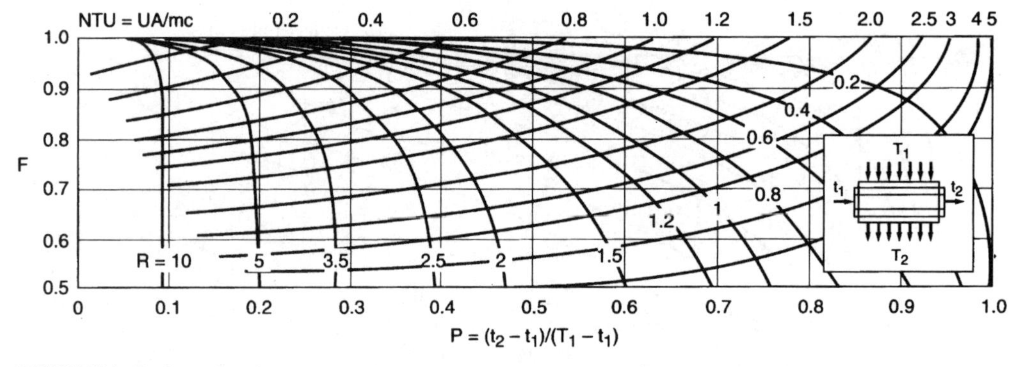

FIGURE 31 Design and performance chart for crossflow exchanger with four single-pass tube rows. (*Chemical Engineering*.)

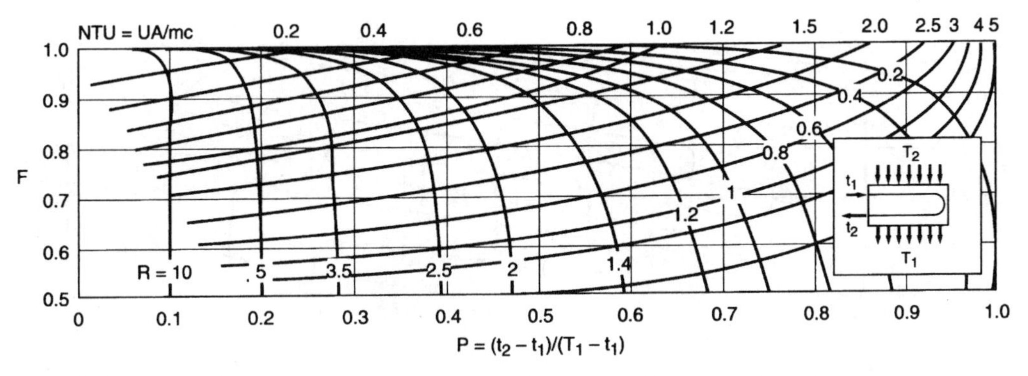

FIGURE 32 Design and performance chart for a crossflow exchanger with a 2-tube pass. (*Chemical Engineering.*)

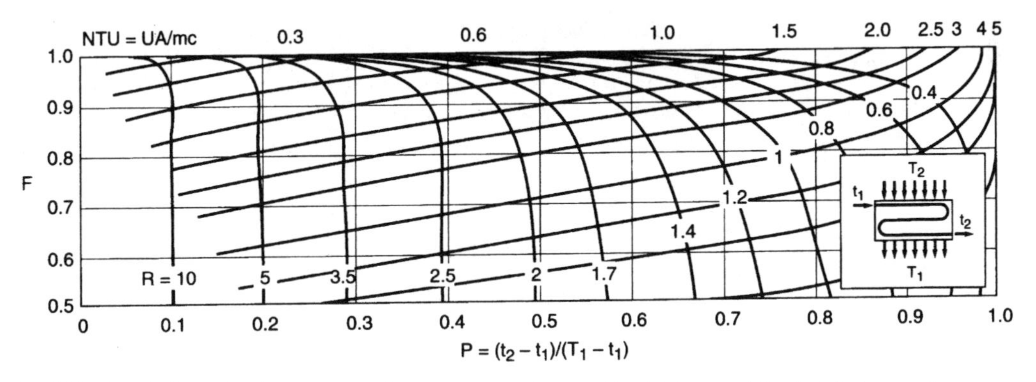

FIGURE 33 Design and performance chart for a crossflow exchanger with a 3-tube pass. (*Chemical Engineering.*)

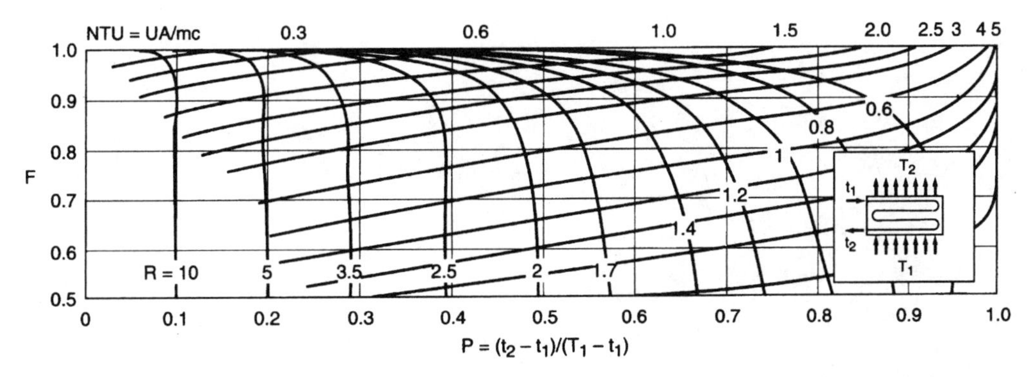

FIGURE 34 Design and performance chart for a crossflow exchanger with a 4-tube pass. (*Chemical Engineering.*)

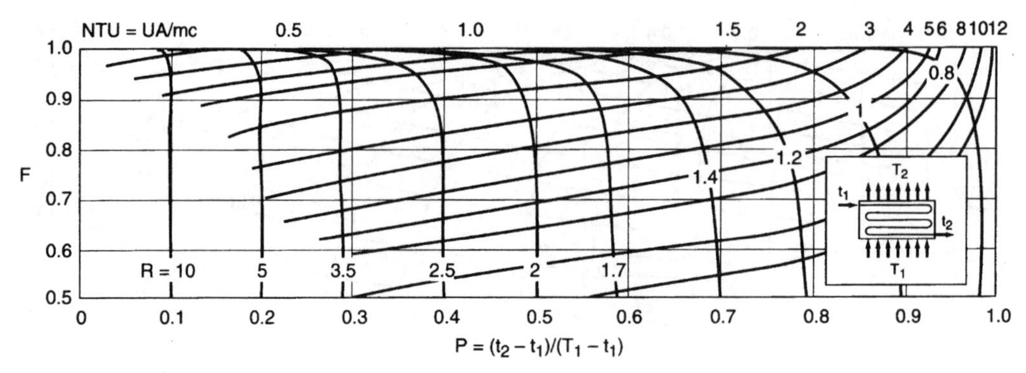

FIGURE 35 Design and performance chart for a crossflow exchanger with a 5-tube pass. (*Chemical Engineering*.)

Typical applications of crossflow units include air cooling of overhead condensate streams and trim-product coolers.

This calculation procedure is based upon the work of Jeff Bowman, E.I. du Pont de Nemours Co., and Richard Turton, assistant professor of chemical engineering at West Virginia University, as reported in *Chemical Engineering* magazine. Note that final selection of a unit will usually be made in conjunction with advice and guidance from the manufacturer of the unit.

SECTION 12
FLUID TRANSFER ENGINEERING

PART 1: PUMP AND PUMPING SYSTEM ENERGY CALCULATIONS 12.2
Pump and Pumping System Energy Calculation Parameters 12.2
Pump Choice to Conserve Energy 12.3
Energy Savings Using Pump Characteristic and System-Head Curves 12.5
Choosing Centrifugal Pump Rotating Speed to Meet Capacity and Head Needs 12.9
Energy Safety and Environmental Analysis for Equipment Criticality 12.9
Affinity Laws for Energy Analysis of Centrifugal Pumps 12.14
Centrifugal-Pump Choice Using Affinity Laws 12.14
Selecting Centrifugal Pumps Based on Specific Speed Analysis 12.15
Choosing Centrifugal Pump Best Rotary Speed 12.17
Vapor-Free Liquid Pump Total Head Computation 12.18
Energy Conservation in Choosing Equipment for a Pumping Installation 12.22
Pump and System Characteristic-Curve Energy Analysis 12.30
Hot-Liquid Pump Net Positive Suction Head 12.36
Steam Power Plant Condensate Pump Selection 12.37
Centrifugal Pump Minimum Safe Fluid Flow Determination 12.40
Viscous Liquid Centrifugal Pump Selection 12.41
Shaft Deflection and Pump Critical Speed 12.43
Liquid Viscosity Effect on Regenerative-Pump Performance 12.45
Liquid Viscosity Effect on Reciprocating Pump Performance 12.46
Viscosity and Dissolved Gas Effects on Rotary Pump Performance 12.47
Materials Selection for Pump Parts 12.48
Hydropneumatic Storage Tank Sizing 12.50
Centrifugal Pump Energy Usage as a Hydraulic Turbine 12.50
Impeller Sizing for Centrifugal Pump Safety Service 12.55
Reducing Energy Consumption and Loss with Proper Pump Choice 12.58
Energy and Economic Competitive Analysis of Power-Plant Condensate Pumps 12.60

PART 2: FLUID FLOW IN PIPING SYSTEMS 12.75
Pressure Surge in a Piping System from Rapid Valve Closure 12.75
Piping Pressure Surge with Different Material and Fluid 12.77
Pressure Surge in Piping System with Compound Pipeline 12.79
Quick Calculation of Flow Rate and Pressure Drop in Piping Systems 12.80
Fluid Head Loss Approximations for All Types of Piping 12.81
Pipe-Wall Thickness and Schedule Number 12.82
Pipe-Wall Thickness Determination by Piping Code Formula 12.84
Determining the Pressure Loss in Steam Piping 12.85
Piping Warm-Up Condensate Load 12.88
Steam Trap Selection for Industrial Applications 12.90
Selecting Heat Insulation for High-Temperature Piping 12.96
Orifice Meter Selection for a Steam Pipe 12.97
Selection of a Pressure-Regulating Valve for Steam Service 12.98
Hydraulic Radius and Liquid Velocity in Water Pipes 12.101
Friction-Head Loss in Water Piping of Various Material 12.101
Chart and Tabular Determination of Friction Head 12.102
Relative Carrying Capacity of Pipes 12.107
Pressure-Reducing Valve Selection for Water Piping 12.107
Sizing a Water Meter 12.109
Equivalent Length of a Complex Series Pipeline 12.110
Equivalent Length of a Parallel Piping System 12.111
Maximum Allowable Height for a Liquid Siphon 12.112
Water-Hammer Effects in Liquid Pipelines 12.113
Specific Gravity and Viscosity of Liquids 12.113

Pressure Loss in Piping Having Laminar Flow 12.114
Determining the Pressure Loss in Oil Pipes 12.115
Flow Rate and Pressure Loss in Compressed-Air and Gas Piping 12.118
Flow Rate and Pressure Loss in Gas Pipelines 12.122
Slip-Type Expansion Joint Selection and Application 12.123
Corrugated-Expansion Joint Selection and Application 12.126
Design of Steam-Transmission Piping 12.129
Steam Desuperheater Analysis 12.138
Steam Accumulator Selection and Sizing 12.139
Selecting Plastic Piping for Industrial Use 12.141
Analyzing Plastic Piping and Linings for Tanks, Pumps, and Other Components for Specific Applications 12.142
Friction Loss in Pipes Handling Solids in Suspension 12.149
Desuperheater Water Spray Quantity 12.150
Sizing Condensate Return Lines for Optimum Flow Conditions 12.151
Estimating Cost of Steam Leaks from Piping and Pressure Vessels 12.153
Quick Sizing of Restrictive Orifices in Piping 12.154
Steam Tracing a Vessel Bottom to Keep the Contents Fluid 12.155
Designing Steam-Transmission Lines without Steam Traps 12.157
Line Sizing for Flashing Steam Condensate 12.160
Determining the Friction Factor for Flow of Bingham Plastics 12.163
Time Needed to Empty a Storage Vessel with Dished Ends 12.166
Time Needed to Empty a Storage Vessel without Dished Ends 12.170
Time to Drain a Storage Tank Through Attached Piping 12.170

PART 1

PUMP AND PUMPING SYSTEM ENERGY CALCULATIONS

PUMP AND PUMPING SYSTEM ENERGY CALCULATION PARAMETERS

Some of the largest energy users today are pumps and pumping systems used in a variety of applications—power plants, water-supply facilities, air-conditioning systems, sewage-treatment plants, well-water systems, large-scale irrigation installations, etc. Today most pumps are electrically driven, be they centrifugal, turbine, reciprocating, and so on. Small steam-turbine-driven pumps are usually confined to power-plant auxiliary services.

A variety of electric motors are used to drive pumps. Today most pump drivers are alternating-current (ac) motors. Such motors use utility-furnished energy from incoming power lines. Industrial-plant pumps are often powered by onsite-generated alternating current.

This section of the handbook features energy calculations for a variety of pumping systems. Such energy calculations are important to engineers because the fluid system size, layout, physical design, and capacity determine the energy requirements—horsepower (kW)—of the pump-drive electric motors.

With the pump being the second most widely used machine in the universe (the electric motor being the first), there are many opportunities for engineers to save energy in pumping system design. Thus, energy can be saved in most pumping systems by applying six simple engineering design tips: 1. *Reduce* the head that the pump must deliver liquids to. 2. *Increase* the internal diameter of the piping system through which the liquid is pumped. 3. *Shorten* the piping system so the liquid is pumped a shorter distance. 4. *Reduce* the number of fittings and valves in the piping system to decrease the friction the pump must overcome. 5. *Select* lower internal-friction piping materials to reduce the lifetime friction energy losses the pumping system will incur. 6. *Size* the pump so its capacity capability is adequate for the system it serves. Undersizing can lead to excessive energy consumption. Oversizing increases the original pump cost and wastes energy during pump operation. In using any of these energy-saving tips, the cost of the change must be balanced against the potential energy savings.

Since most pumping systems have long lives—25 years being typical (with many 50-year-old systems still operating)—engineers must be careful in their design energy decisions. Undersizing piping can lead to years of excessive energy costs. Likewise, excessive piping lengths, too many fittings, and/or inappropriate control devices, can lead to unnecessary increased power input and the waste of expensive energy. The calculation procedures in this section show the design engineer many ways in which energy can be saved by smart system choices.

Since the world is focused on alternative energy and "green" facilities, engineers will have to keep these interests in mind during their design work. Whether one accepts global warming or not, the future of pumping systems will always include the question: "Have you considered all the energy savings possible for this design?" Plant and facilities owners will be looking for a loud "Yes" answer to this energy question. The second most commonly used machine in the universe today must by an energy "penny pincher" if it, and its system, are to be quickly accepted by a plant owner.

PUMP CHOICE TO CONSERVE ENERGY

A new plant addition using special convectors in the heating system requires a system pumping capability of 45 gal/min (2.84 L/s) at a 26-ft (7.9-m) head. The pump characteristic curves for the tentatively selected floor-mounted units are shown in Fig. 1; one operating pump and one standby pump, each 0.75 hp (0.56 kW) are being considered. Can energy be conserved, and how much, with some other pumping arrangement?

Calculation Procedure:

1. *Plot the characteristic curves for the pumps being considered*

Figure 2 shows the characteristic curves for the proposed pumps. Point 1 in Fig. 1 is the proposed operating head and flow rate. An alternative pump choice is shown at Point 2 in Fig. 1. If two of the smaller pumps requiring only 0.25 hp (0.19 kW) each are placed in series, they can generate the required 26-ft (7.9-m) head.

2. *Analyze the proposed pumps*

To analyze properly the proposal, a new set of curves, Fig. 2, is required. For the proposed series pumping application, it is necessary to establish a *seriesed pump curve.* This is a plot of the head and flow rate (capacity) which exists when both pumps are running in series. To construct this curve, double the single-pump head values at any given flow rate.

Next, to determine accurately the flow a single pump can deliver, plot the system-head curve using the same method fully described in the previous calculation procedure. This curve is also plotted on Fig. 2.

Plot the point of operation for each pump on the seriesed curve, Fig. 2. The point of operation of each pump is on the single-pump curve when both pumps are operating. Each pump supplies half the total required head.

When a single pump is running, the point of operation will be at the intersection of the system-head curve and the single-pump characteristic curve, Fig. 2. At this point both the flow and the hp (kW) input of the single pump decrease. Series pumping, Fig. 2, requires the input motor hp (kW) for both pumps; this is the point of maximum power input.

3. *Compute the possible savings*

If the system requires a constant flow of 45 gal/min (2.84 L/s) at 26-ft (7.9-m) head, the two-pump series installation saves (0.75 hp − 2 × 0.25 hp) = 0.25 hp (0.19 kW) for every hour the pumps run. For every 1000 hours of operation, the system saves 190 kWh. Since 2000 hours are generally equal to one shift of operation per year, the saving is 380 kWh per shift per year.

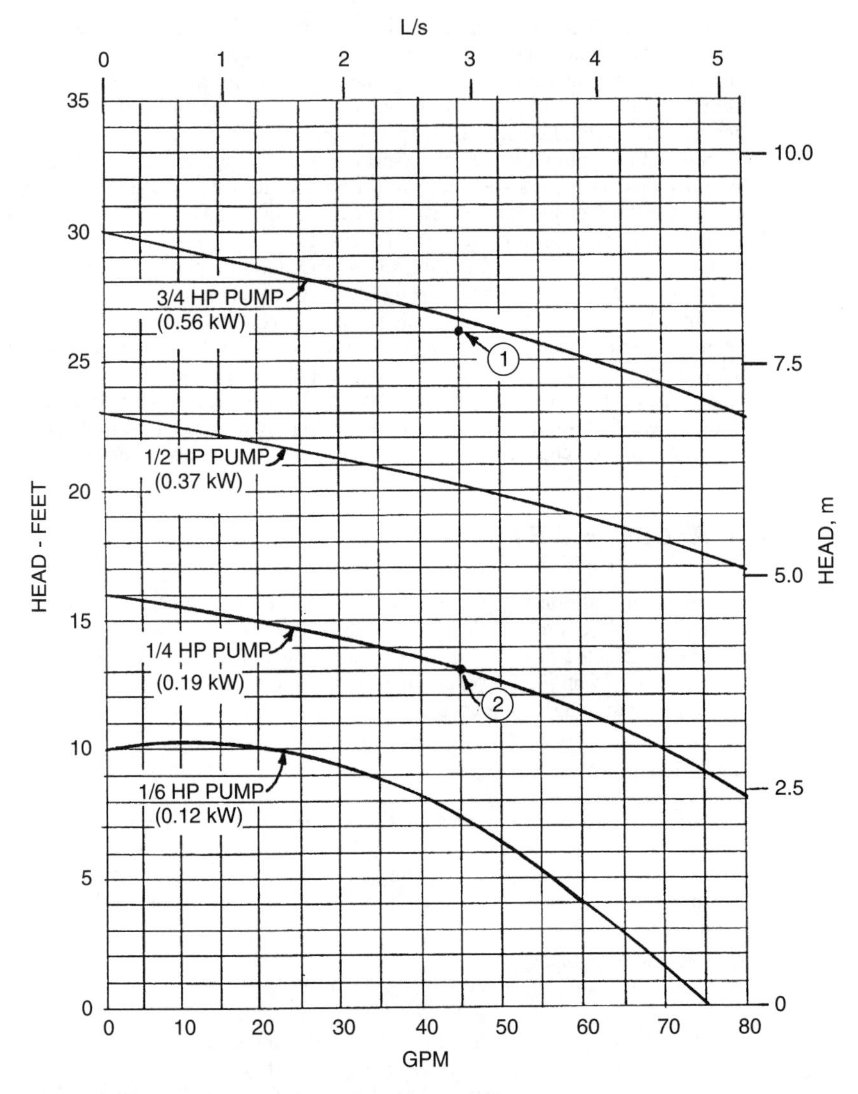

FIGURE 1 Pump characteristic curves for use in series installation.

If the load is frequently less than peak, one-pump operation delivers 32.5 gal/min (2.1 L/s), This value, which is some 72 percent of full load, corresponds to doubling the saving.

Related Calculations. Series operation of pumps can be used in a variety of designs for industrial, commercial, residential, chemical, power, marine, and similar plants. A series connection of pumps is especially suitable when full-load demand is small; i.e., just a few hours a week, month, or year. With such a demand, one pump can serve the plant's needs most of the time, thereby reducing the power bill. When full-load operation is required, the second pump is started. If there is a need for maintenance of the first pump, the second unit is available for service.

This procedure is the work of Jerome F. Mueller, P.E., of Mueller Engineering Corp.

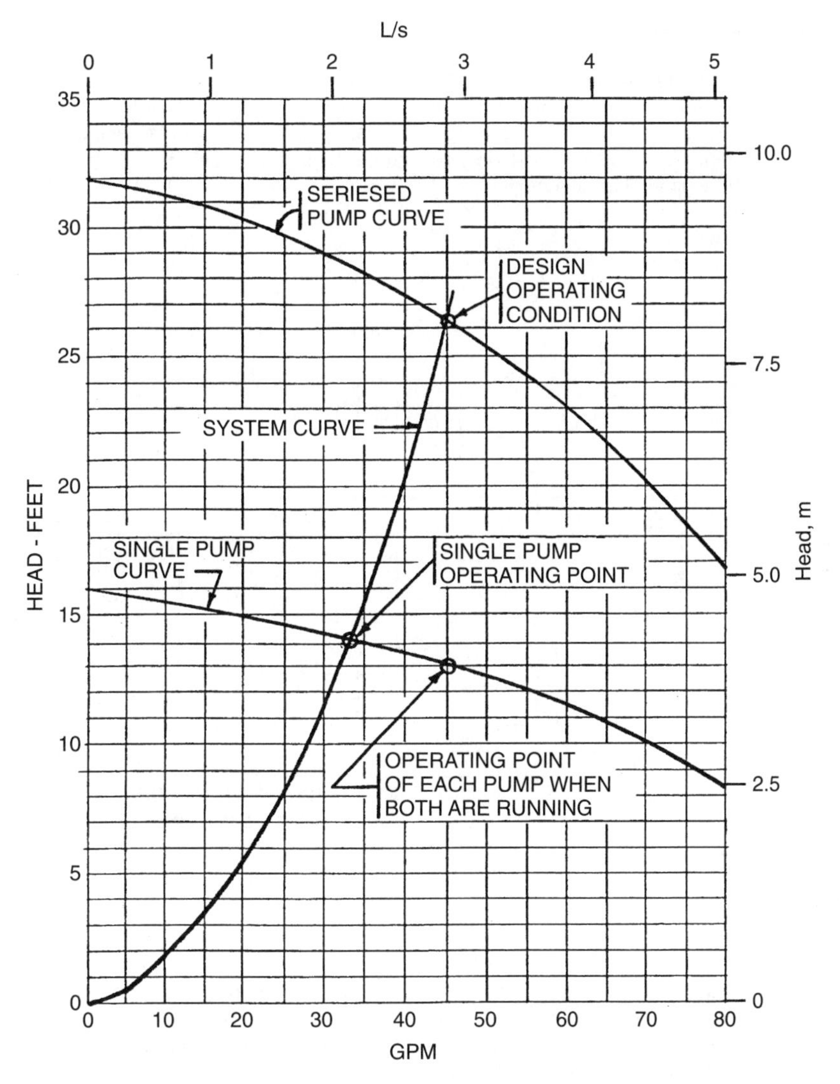

FIGURE 2 Seriesed-pump characteristic and system-head curves.

ENERGY SAVINGS USING PUMP CHARACTERISTIC AND SYSTEM-HEAD CURVES

A system proposed for heating a 20,000-ft^2 (1858-m^2) addition to an industrial plant using hot-water heating requires a flow of 80 gal/min (7.4 L/s) of 260°F (126.5°C) water at a 20°F (11.1°C) temperature drop and a 13-ft (3.96-m) system head. The required system flow can be handled by two pumps, one an operating unit and one a spare unit. Each pump will have a 0.5-hp (0.37-kW) drive motor. Could there be any appreciable energy saving using some other arrangement? The system requires 50 hours of constant pump operation and 40 hours of partial pump operation per week.

Calculation Procedure:

1. *Plot characteristic curves for the proposed system*

Figure 3 shows the proposed hot-water heating-pump selection for this industrial building. Looking at the values of the pump head and capacity in Fig. 3, it can be seen that if the peak load of 80 gal/min (7.4 L/s) were carried by two pumps, then each would have to pump only 40 gal/min (3.7 L/s) in a parallel arrangement.

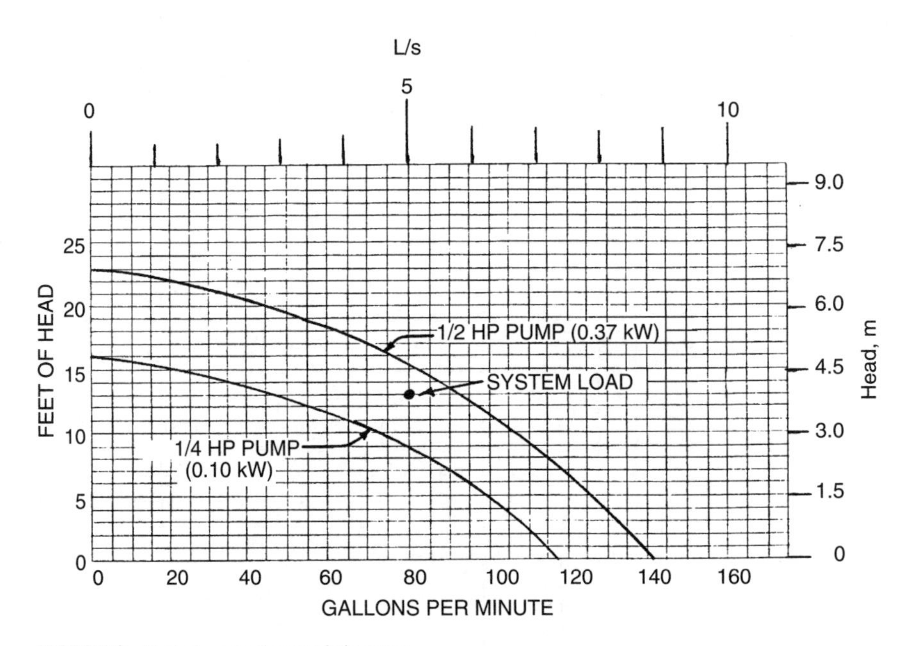

FIGURE 3 Typical pump characteristic curves.

2. *Plot a characteristic curve for the pumps in parallel*

Construct the paralleled-pump curve by doubling the flow of a single pump at any given head, using data from the pump manufacturer. At 13-ft head (3.96-m) one pump produces 40 gal/min (3.7 L/s); two pumps 80 gal/min (7.4 L/s). The resulting curve is shown in Fig. 4.

The load for this system could be divided among three, four, or more pumps, if desired. To achieve the best results, the number of pumps chosen should be based on achieving the proper head and capacity requirements in the system.

3. *Construct a system-head curve*

Based on the known flow rate, 80 gal/min (7.4 L/s) at 13-ft (3.96-m) head, a system-head curve can be constructed using the fact that pumping head varies as the square of the change in flow, or $Q_2/Q_1 = H_2/H_1$, where Q_1 = known design flow, gal/min (L/s); Q_2 = selected flow, gal/min (L/s); H_1 = known design head, ft (m); H_2 = resultant head related to selected flow rate, gal/min (L/s).

Figure 5 shows the plotted system-head curve. Once the system-head curve is plotted, draw the single-pump curve from Fig. 3 on Fig. 5, and the paralleled-pump curve from Fig. 4. Connect the different pertinent points of concern with dashed lines, Fig. 5.

The point of crossing of the two-pump curve and the system-head curve is at the required value of 80 gal/min (7.4 L/s) and 13-ft (3.96-m) head because it was so planned. But the point of crossing of the system-head curve and the single-pump curve is of particular interest.

The single pump, instead of delivering 40 gal/min (7.4 L/s) at 13-ft (3.96-m) head will deliver, as shown by the intersection of the curves in Fig. 5, 72 gal/min (6.67 L/s) at 10-ft (3.05-m) head. Thus,

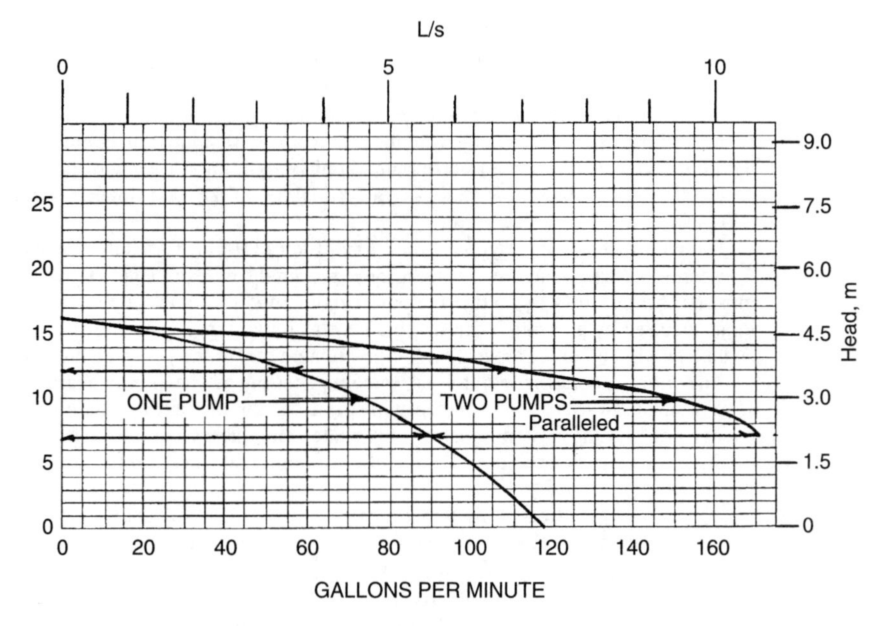

FIGURE 4 Single- and dual-parallel pump characteristic curves.

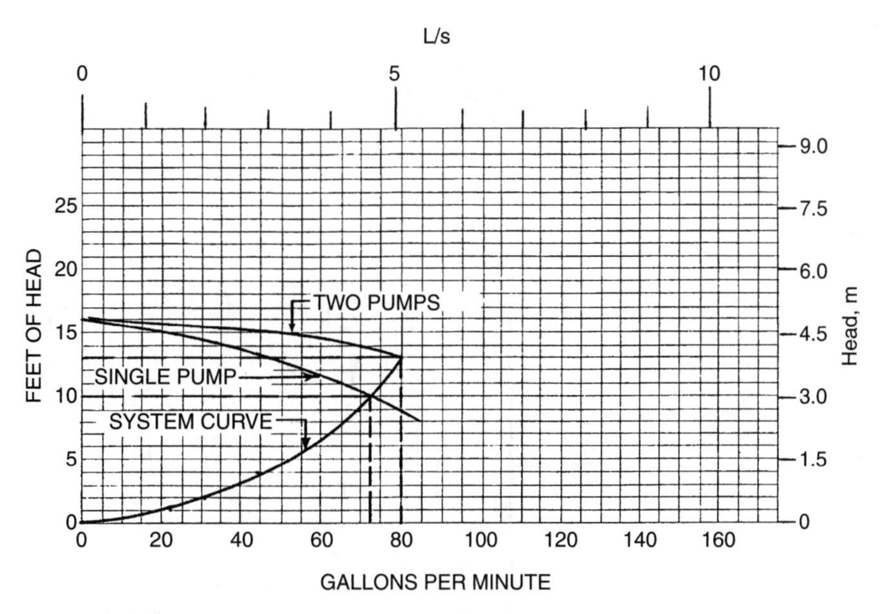

FIGURE 5 System-head curve for parallel pumping.

the single pump can effectively be a standby for 90 percent of the required capacity at a power input of 0.5 hp (0.37 kW). Much of the time in heating and air conditioning, and frequently in industrial processes, the system load is 90 percent, or less.

4. *Determine the single-pump horsepower input*
In the installation here, the pumps are the inline type with nonoverload motors. For larger flow rates, the pumps chosen would be floor-mounted units providing a variety of horsepower (kW) and flow curves. The horsepower (kW) for—say a 200-gal/min (18.6-L/s) flow rate would be about half of a 400-gal/min (37.2 L/s) flow rate.

If a pump were suddenly given a 300-gal/min (27.9-L/s) flow-rate demand at its crossing point on a larger system-head curve, the hp required might be excessive. Hence, the pump drive motor must be chosen carefully so that the power required does not exceed the motor's rating. The power input required by any pump can be obtained from the pump characteristic curve for the unit being considered. Such curves are available free of charge from the pump manufacturer.

The pump operating point is at the intersection of the pump characteristic curve and the system-head curve in conformance with the first law of thermodynamics, which states that the energy put into the system must exactly match the energy used by the system. The intersection of the pump characteristic curve and the system-head curve is the only point that fulfills this basic law.

There is no practical limit for pumps in parallel. Careful analysis of the system-head curve versus the pump characteristic curves provided by the pump manufacturer will frequently reveal cases where the system load point may be beyond the desired pump curve. The first cost of two or three smaller pumps is frequently no greater than for one large pump. Hence, smaller pumps in parallel may be more desirable than a single large pump, from both the economic and reliability standpoints.

One frequently overlooked design consideration in piping for pumps is shown in Fig. 6. This is the location of the check valve to prevent reverse-flow pumping. Figure 6 shows the proper location for this simple valve.

5. *Compute the energy saving possible*
Since one pump can carry the fluid flow load about 90 percent of the time, and this same percentage holds for the design conditions, the saving in energy is $0.9 \times (0.5\ \text{kW} - .25\ \text{kW}) \times 90$ hours per week = 20.25 kWh/week. (In this computation we used the assumption that 1 hp = 1 kW.) The annual savings would be 52 weeks × 20.25 kW/week = 1053 kWh/yr. If electricity costs 5 cents per kWh, the annual saving is \$0.05 × 1053 = \$52.65/yr.

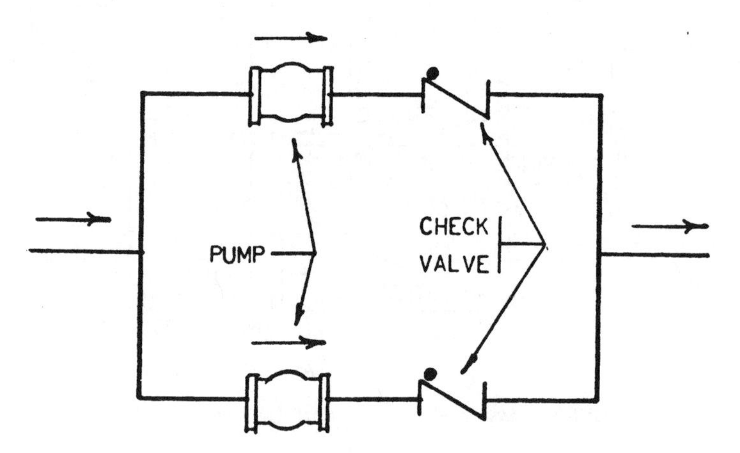

FIGURE 6 Check valve locations to prevent reverse flow.

While a saving of some $51 per year may seem small, such a saving can become much more if: (1) larger pumps using higher horsepower (kW) motors are used; (2) several hundred pumps are used in the system; (3) the operating time is longer—168 hours per week in some systems. If any, or all, these conditions prevail, the savings can be substantial.

Related Calculations. This procedure can be used for pumps in a variety of applications: industrial, commercial, residential, medical, recreational, and similar systems. When analyzing any system, the designer should be careful to consider all the available options so the best one is found.

This procedure is the work of Jerome F. Mueller, P.E., of Mueller Engineering Corp.

CHOOSING CENTRIFUGAL PUMP ROTATING SPEED TO MEET CAPACITY AND HEAD NEEDS

A double-suction condenser circulator handling 20,000 gal/min (75,800 L/min) at a total head of 60 ft (18.3 m) is to have a 15-ft (4.6-m) lift. What should be the rpm of this pump to meet the capacity and head requirements?

Calculation Procedure:

1. *Determine the specific speed of the pump*

Use the Hydraulic Institute specific-speed chart, Fig. 7. Entering at 6-ft (18.3-m) head, project to the 15-ft suction-lift curve. At the intersection, read the specific speed of this double-suction pump as 4300.

2. *Use the specific-speed equation to determine the pump operating rpm*

Solve the specific-speed equation for the pump rpm. Or rpm = $N_s \times H^{0.75}/Q^{0.5}$, where N_s = specific speed of the pump, rpm, from Fig. 7; H = total head on pump, ft (m); Q = pump flow rate, gal/min (L/s). Solving, rpm = $4300 \times 60^{0.75}/20{,}000^{0.5}$ = 655.5 r/min. The next common electric motor rpm is 660; hence, we would choose a motor or turbine driver whose rpm does not exceed 660.

The next lower induction-motor speed is 585 r/min. But we could buy a lower-cost pump and motor if it could be run at the next higher full-load induction motor speed of 700 r/min. The specific speed of such a pump would be: $N_s = [700(20{,}000)^{0.5}]/60^{0.75} = 4592$. Referring to Fig. 7, the maximum suction lift with a specific speed of 4592 is 13 ft (3.96 m) when the total head is 60 ft (18.3). If the pump setting or location could be lowered 2 ft (0.6 m), the less expensive pump and motor could be used, thereby saving on the investment cost.

Related Calculations. Use this general procedure to choose the driver and pump rpm for centrifugal pumps used in boiler feed, industrial, marine, HVAC, and similar applications. Note that the latest Hydraulic Institute curves should be used.

ENERGY SAFETY AND ENVIRONMENTAL ANALYSIS FOR EQUIPMENT CRITICALITY

Rank the criticality of a boiler feed pump operating at 250°F (121°C) and 100 lb/in^2 (68.9 kPa) if its Mean Time Between Failures (MTBF) is 10 months, and vibration is an important element in its safe operation. Use the National Fire Protection Association (NFPA) ratings of process chemicals for health, fire, and reactivity hazards. Show how the criticality of the unit is developed.

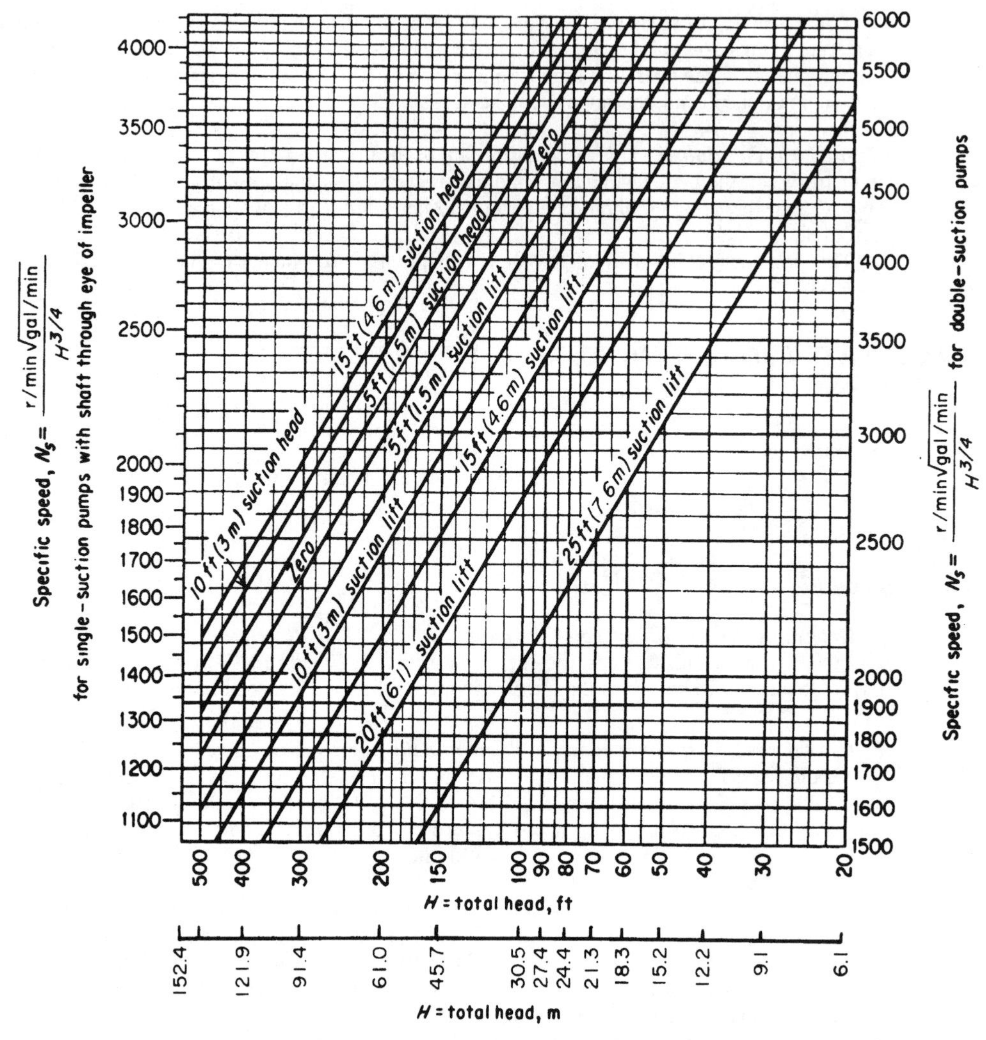

FIGURE 7 Upper limits of specific speeds of single-stage, single- and double-suction centrifugal pumps handling clear water at 85°F (29.4°C) at sea level. (*Hydraulic Institute.*)

Calculation Procedure:

1. *Determine the Hazard Criticality Rating (HCR) of the equipment*

Process industries of various types—chemical, petroleum, food, etc.—are giving much attention to complying with new process safety regulations. These efforts center on reducing hazards to people and the environment by ensuring the mechanical and electrical integrity of equipment.

To start a program, the first step is to evaluate the most critical equipment in a plant or factory. To do so, the equipment is first ranked on some criteria, such as the relative importance of each piece of equipment to the process or plant output.

The Hazard Criticality Rating (HCR) can be determined from a listing such as that in Table 1. This tabulation contains the analysis guidelines for assessing the process chemical hazard (PCH) and the Other Hazards (O). The ratings for such a table of hazards should be based on the findings of an

TABLE 1 The Hazard Criticality Rating (HCR) Is Determined in Three Steps*

Hazard criticality rating
1. Assess the Process Chemical Hazard (PCH) by:
• Determining the NFPA ratings (*N*) of process chemicals for: Health, Fire, Reactivity hazards
• Selecting the highest value of *N*
• Evaluating the potential for an emissions release (0 to 4):
High (RF = 0): Possible serious health, safety, or environmental effects
Low (RF = 1): Minimal effects
None (RF = 4): No effects
• Then, $PCH = N - RF$. (Round off negative values to zero.)
2. Rate Other Hazards (O) with an arbitrary number (0 to 4) if they are:
• Deadly (4), if:
Temperatures > 1000°F
Pressures are extreme
Potential for release of regulated chemicals is high
Release causes possible serious health safety or environmental effects
Plant requires steam turbine trip mechanisms, fired-equipment shutdown systems, or toxic- or combustible-gas detectors†
Failure of pollution control system results in environmental damage†
• Extremely dangerous (3), if:
Equipment rotates at >5000 r/min
Temperatures >500°F
Plant requires process venting devices
Potential for release of regulated chemicals is low
Failure of pollution control system may result in environmental damage†
• Hazardous (2), if:
Temperatures >300°F
Extended failure of pollution control system may cause damage†
• Slightly hazardous (1), if:
Equipment rotates at >3600 r/min
Temperatures >140°F or pressures > 20 lb/in^2 (gage)
• Not hazardous (0), if:
No hazards exist
3. Select the higher value of PCH and O as the Hazard Criticality Rating

**Chemical Engineering*.

†Equipment with spares drop one category rating. A spare is an inline unit that can be immediately serviced or be substituted by an alternative process option during the repair period.

experienced team thoroughly familiar with the process being evaluated. A good choice for such a task is the plant's Process Hazard Analysis (PHA) Group. Since a team's familiarity with a process is highest at the end of a PHA study, the best time for rating the criticality of equipment is toward the end of such safety evaluations.

From Table 1, the NFPA rating, *N*, of process chemicals for Health, Fire, and Reactivity, is $N = 2$, because this is the highest of such ratings for Health. The Fire and Reactivity ratings are 0, 0, respectively, for a boiler feed pump because there are no Fire or Reactivity exposures.

The Risk Reduction Factor (RF), from Table 1, is RF = 0, since there is the potential for serious burns from the hot water handled by the boiler feed pump. Then, the Process Chemical Hazard, $PCH = N - RF = 2 - 0 = 2$.

The rating of Other Hazards, O, Table 1, is O = 1, because of the high temperature of the water. Thus, the Hazard Criticality Rating, HCR = 2, found from the higher numerical value of PCH and O.

2. *Determine the Process Criticality Rating, PCR, of the equipment*

From Table 2, prepared by the PHA Group using the results of its study of the equipment in the plant, PCR = 3. The reason for this is that the boiler feed pump is critical for plant operation because its failure will result in reduced capacity.

TABLE 2 Process Criticality Rating*

	Process criticality rating
Essential (4)	The equipment is essential if failure will result in shutdown of the unit, unacceptable product quality, or severely reduced process yield
Critical (3)	The equipment is critical if failure will result in greatly reduced capacity, poor product quality, or moderately reduced process yield
Helpful (2)	The equipment is helpful if failure will result in slightly reduced capacity, product quality, or reduced process yield
Not critical (1)	The equipment is not critical if failure will have little or no process consequences

**Chemical Engineering.*

3. ***Find the Process and Hazard Criticality Rating, PHCR***
The alphanumeric PHC value is represented first by the alphabetic character for the category. For example, Category A is the most critical, while Category D is the least critical to plant operation. The first numeric portion represents the Hazard Criticality Rating, HCR, while the second numeric part the Process Criticality Rating, PCR. These categories and ratings are a result of the work of the PHA Group.

From Table 3, the Process and Hazard Criticality Rating, PHCR = B23. This is based on the PCR = 3 and HCR = 2, found earlier.

TABLE 3 The Process and Hazard Criticality Rating*

PHC rankings					
	Hazard criticality rating				
Process criticality rating	4	3	2	1	0
4	A44	A34	A24	A14	A04
3	A43	B33	B23	B13	B03
2	A42	A32	C22	C12	C02
1	A41	B31	C21	C11	D01

Note: The alphanumeric PHC value is represented first by the alphabetic character for the category (for example, Category A is the most critical while D is the least critical). The first numeric portion represents the Hazard Criticality Rating, and the second numeric part the Process Criticality Rating.
**Chemical Engineering.*

4. ***Generate a criticality list by rating equipment using its alphanumeric PHCR values***
Each piece of equipment is categorized, in terms of its importance to the process, as: Highest Priority, Category A; High Priority, Category B; Medium Priority, Category C; Low Priority, Category D.

Since the boiler feed pump is critical to the operation of the process, it is a Category B, i.e., High Priority item in the process.

5. ***Determine the Criticality and Repetitive Equipment, CRE, value for this equipment***
This pump has an MTBF of 10 months. Therefore, from Table 4, CRE = bl. Note that the CRE value will vary with the PCHR and MTBF values for the equipment.

6. ***Determine equipment inspection frequency to ensure human and environmental safety***
From Table 5, this boiler feed pump requires vibration monitoring every 90 days. With such monitoring it is unlikely that an excessive number of failures might occur to this equipment.

7. ***Summarize criticality findings in spreadsheet form***
When preparing for a PHCR evaluation, a spreadsheet, Table 6, listing critical equipment, should be prepared. Then, as the various rankings are determined, they can be entered in the spreadsheet where they are available for easy reference.

TABLE 4 The Criticality and Repetitive Equipment Values*

PHCR	CRE values: Mean time between failures, months 0–6	6–12	12–24	>24
A	a1	a2	a3	a4
B	a2	b1	b2	b3
C	a3	b2	c1	c2
D	a4	b3	c2	d1

Chemical Engineering.

TABLE 5 Predictive Maintenance Frequencies for Rotating Equipment Based on Their CRE Values

	Maintenance cycles			
	Frequency, days			
CRE	7	30	90	360
a1, a2	VM*	LT†		
a3, a4		VM	LT	
b1, b3			VM	
c1, d1				VM

*VM: Vibration monitoring.
†LT: Lubrication sampling and testing.
Chemical Engineering.

TABLE 6 Typical Spreadsheet for Ranking Equipment Criticality*

Spreadsheet for calculating equipment PHCRs										
		NFPA rating								
Equipment number	Equipment description	H	F	R	RF	PCH	Other	HCR	PCR	PHCR
TKO	Tank	4	4	0	0	4	0	4	4	A44
TKO	Tank	4	4	0	1	3	3	3	4	A34
PU1BFW	Pump	2	0	0	0	2	1	2	3	B23

Chemical Engineering.

Enter the PCH, Other, HCR, PCR, and PHCR values in the spreadsheet, as shown. These data are now available for reference by anyone needing the information.

Related Calculations. The procedure presented here can be applied to all types of equipment used in a facility—fixed, rotating, and instrumentation. Once all the equipment is ranked by criticality, priority lists can be generated. These lists can then be used to ensure the mechanical integrity of critical equipment by prioritizing predictive and preventive maintenance programs, inventories of critical spare parts, and maintenance work orders in case of plant upsets.

In any plant, the hazards posed by different operating units are first ranked and prioritized based on a PHA. These rankings are then used to determine the order in which the hazards need to be addressed. When the PHAs approach completion, team members evaluate the equipment in each operating unit using the PHCR system.

The procedure presented here can be used in any plant concerned with human and environmental safety. Today, this represents every plant, whether conventional or automated. Industries in which this procedure finds active use include chemical, petroleum, textile, food, power, automobile, aircraft, military, and general manufacturing.

This procedure is the work of V. Anthony Ciliberti, Maintenance Engineer, The Lubrizol Corp., as reported in *Chemical Engineering* magazine.

AFFINITY LAWS FOR ENERGY ANALYSIS OF CENTRIFUGAL PUMPS

A centrifugal pump designed for a 1800-r/min operation and a head of 200 ft (60.9 m) has a capacity of 3000 gal/min (189.3 L/s) with a power input of 175 hp (130.6 kW). What effect will a speed reduction to 1200 r/min have on the head, capacity, and power input of the pump? What will be the change in these variables if the impeller diameter is reduced from 12 to 10 in (304.8 to 254 mm) while the speed is held constant at 1800 r/min?

Calculation Procedure:

1. *Compute the effect of a change in pump speed*

For any centrifugal pump in which the effects of fluid viscosity are negligible, or are neglected, the similarity or affinity laws can be used to determine the effect of a speed, power, or head change. For a *constant impeller diameter,* the laws are $Q_1/Q_2 = N_1/N_2$; $H_1/H_2 = (N_1/N_2)^2$; $P_1/P_2 = (N_1/N_2)^3$. For a *constant speed,* $Q_1/Q_2 = D_1/D_2$; $H_1/H_2 = (D_1/D_2)^2$; $P_1/P_2 = (D_1/D_2)^3$. In both sets of laws, Q = capacity, gal/min; N = impeller rpm; D = impeller diameter, in; H = total head, ft of liquid; P = bhp input. The subscripts 1 and 2 refer to the initial and changed conditions, respectively.

For this pump, with a constant impeller diameter, $Q_1/Q_2 = N_1/N_2$; $3000/Q_2 = 1800/1200$; $Q_2 =$ 2000 gal/min (126.2 L/s). And, $H_1/H_2 = (N_1/N_2)^2 = 200/H_2 = (1800/1200)^2$; $H_2 = 88.9$ ft (27.1 m). Also, $P_1/P_2 = (N_1/N_2)^3 = 175/P_2 = (1800/1200)^3$; $P_2 = 51.8$ bhp (38.6 kW).

2. *Compute the effect of a change in impeller diameter*

With the speed constant, use the second set of laws. Or, for this pump, $Q_1/Q_2 = D_1/D_2$; $3000/Q_2 =$ 12/10; $Q_2 = 2500$ gal/min (157.7 L/s). And $H_1/H_2 = (D_1/D_2)^2$: $200/H_2 = (12/10)^2$; $H_2 = 138.8$ ft (42.3 m). Also, $P_1/P_2 = (D_1/D_2)^3$; $175/P_2 = (12/10)^3$; $P_2 = 101.2$ bhp (75.5 kW).

Related Calculations. Use the similarity laws to extend or change the data obtained from centrifugal pump characteristic curves. These laws are also useful in field calculations when the pump head, capacity, speed, or impeller diameter is changed.

The similarity laws are most accurate when the efficiency of the pump remains nearly constant. Results obtained when the laws are applied to a pump having a constant-impeller diameter are somewhat more accurate than for a pump at constant speed with a changed-impeller diameter. The latter laws are more accurate when applied to pumps having a low specific speed.

If the similarity laws are applied to a pump whose impeller diameter is increased, be certain to consider the effect of the higher velocity in the pump suction line. Use the similarity laws for any liquid whose viscosity remains constant during passage through the pump. However, the accuracy of the similarity laws decreases as the liquid viscosity increases.

CENTRIFUGAL-PUMP CHOICE USING AFFINITY LAWS

A test-model pump delivers, at its best efficiency point, 500 gal/min (31.6 L/s) at a 350-ft (106.7-m) head with a required net positive suction head (NPSH) of 10 ft (3 m) a power input of 55 hp (41 kW) at 3500 r/min, when a 10.5-in (266.7-mm) diameter impeller is used. Determine the performance of the model at 1750 r/min. What is the performance of a full-scale prototype pump with a 20-in (50.4-cm) impeller operating at 1170 r/min? What are the specific speeds and the suction-specific speeds of the test-model and prototype pumps?

Calculation Procedure:

1. ***Compute the pump performance at the new speed***

The similarity or affinity laws can be stated in general terms, with subscripts p and m for prototype and model, respectively, as $Q_p = K_d^3 N_n Q_m$; $H_p = K_d^2 K_n^2 H_m$; $\text{NPSH}_p = K_2^2 K_n^2 \text{NPSH}_m$; $P_p = K_d^5 K_n^5 P_m$, where K_d = size factor = prototype dimension/model dimension. The usual dimension used for the size factor is the impeller diameter. Both dimensions should be in the same units of measure. Also, K_n = (prototype speed, r/min)/(model speed, r/min). Other symbols are the same as in the previous calculation procedure.

When the model speed is reduced from 3500 to 1750 r/min, the pump dimensions remain the same and $K_d = 1.0$; $K_n = 1750/3500 = 0.5$. Then $Q = (1.0)(0.5)(500) = 250$ r/min; $H = (1.0)^2(0.5)^2(350) = 87.5$ ft (26.7 m); NPSH $= (1.0)^2(0.5)^2(10) = 2.5$ ft (0.76 m); $P = (1.0)^5(0.5)^3(55) = 6.9$ hp (5.2 kW). In this computation, the subscripts were omitted from the equations because the same pump, the test model, was being considered.

2. ***Compute performance of the prototype pump***

First, K_d and K_n must be found: $K_d = 20/10.5 = 1.905$; $K_n = 1170/3500 = 0.335$. Then $Q_p = (1.905)^3(0.335)(500) = 1158$ gal/min (73.1 L/s); $H_p = (1.905)^2(0.335)^2(350) = 142.5$ ft (43.4 m); $\text{NPSH}_p = (1.905)^2(0.335)^2(10) = 4.06$ ft (1.24 m); $P_p = (1.905)^5(0.335)^3(55) = 51.8$ hp (38.6 kW).

3. ***Compute the specific speed and suction-specific speed***

The specific speed or, as Horwitz* says, "more correctly, discharge specific speed," is $N_s = N(Q)^{0.5}/(H)^{0.75}$, while the suction-specific speed $S = N(Q)^{0.5}/(\text{NPSH})^{0.75}$, where all values are taken at the best efficiency point of the pump.

For the model, $N_s = 3500(500)^{0.5}/(350)^{0.75} = 965$; $S = 3500(500)^{0.5}/(10)^{0.75} = 13{,}900$. For the prototype, $N_s = 1170(1158)^{0.5}/(142.5)^{0.75} = 965$; $S = 1170(1156)^{0.5}/(4.06)^{0.75} = 13.900$. The specific speed and suction-specific speed of the model and prototype are equal because these units are geometrically similar or homologous pumps and both speeds are mathematically derived from the similarity laws.

Related Calculations. Use the procedure given here for any type of centrifugal pump where the similarity laws apply. When the term *model* is used, it can apply to a production test pump or to a standard unit ready for installation. The procedure presented here is the work of R. P. Horwitz, as reported in *Power* magazine.*

SELECTING CENTRIFUGAL PUMPS BASED ON SPECIFIC SPEED ANALYSIS

What is the upper limit of specific speed and capacity of a 1750-r/min single-stage, double-suction centrifugal pump having a shaft that passes through the impeller eye if it handles clear water at 85°F (29.4°C) at sea level at a total head of 280 ft (85.3 m) with a 10-ft (3-m) suction lift? What is the efficiency of the pump and its approximate impeller shape?

Calculation Procedure:

1. ***Determine the upper limit of specific speed***

Use the Hydraulic Institute upper specific-speed curve, Fig. 7, for centrifugal pumps or a similar curve, Fig. 8, for mixed- and axial-flow pumps. Enter Fig. 7 at the bottom at 280-ft (85.3-m) total head, and project vertically upward until the 10-ft (3-m) suction-lift curve is intersected. From here,

*R. P. Horwitz, "Affinity Laws and Specific Speed Can Simplify Centrifugal Pump Selection." *Power*, November 1964.

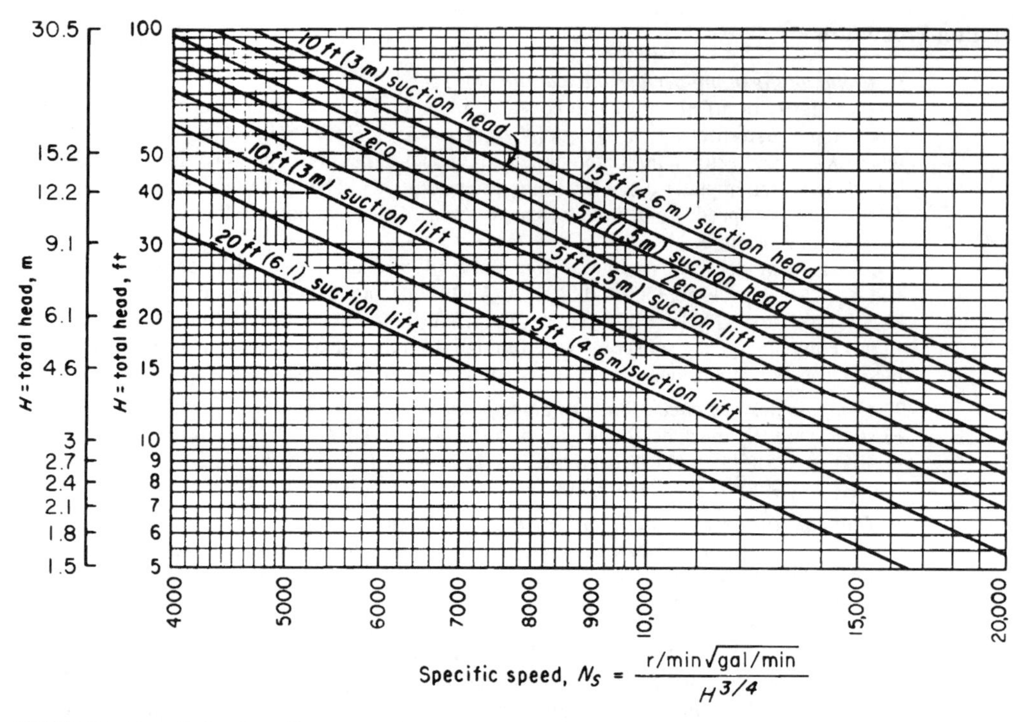

FIGURE 8 Upper limits of specific speeds of single-suction mixed-flow and axial-flow pumps. (*Hydraulic Institute.*)

project horizontally to the right to read the specific speed $N_S = 2000$. Figure 8 is used in a similar manner.

2. *Compute the maximum pump capacity*

For any centrifugal, mixed- or axial-flow pump, $N_S = (gpm)^{0.5}(rpm)/H_t^{0.75}$, where H_t = total head on the pump, ft of liquid. Solving for the maximum capacity, we get $gpm = (N_S H_t^{0.75}/rpm)^2 = (2000 \times 280^{0.75}/1750)^2 = 6040$ gal/min (381.1 L/s).

3. *Determine the pump efficiency and impeller shape*

Figure 9 shows the general relation between impeller shape, specific speed, pump capacity, efficiency, and characteristic curves. At $N_S = 2000$, efficiency = 87 percent. The impeller, as shown in Fig. 9, is moderately short and has a relatively large discharge area. A cross section of the impeller appears directly under the $N_S = 2000$ ordinate.

Related Calculations. Use the method given here for any type of pump whose variables are included in the Hydraulic Institute curves, Figs. 7 and 8, and in similar curves available from the same source. *Operating specific speed,* computed as above, is sometimes plotted on the performance curve of a centrifugal pump so that the characteristics of the unit can be better understood. *Type specific speed* is the operating specific speed giving maximum efficiency for a given pump and is a number used to identify a pump. Specific speed is important in cavitation and suction-lift studies. The Hydraulic Institute curves, Figs. 7 and 8, give upper limits of speed, head, capacity and suction lift for cavitation-free operation. When making actual pump analyses, be certain to use the curves (Figs. 7 and 8) in the latest edition of the *Standards of the Hydraulic Institute.*

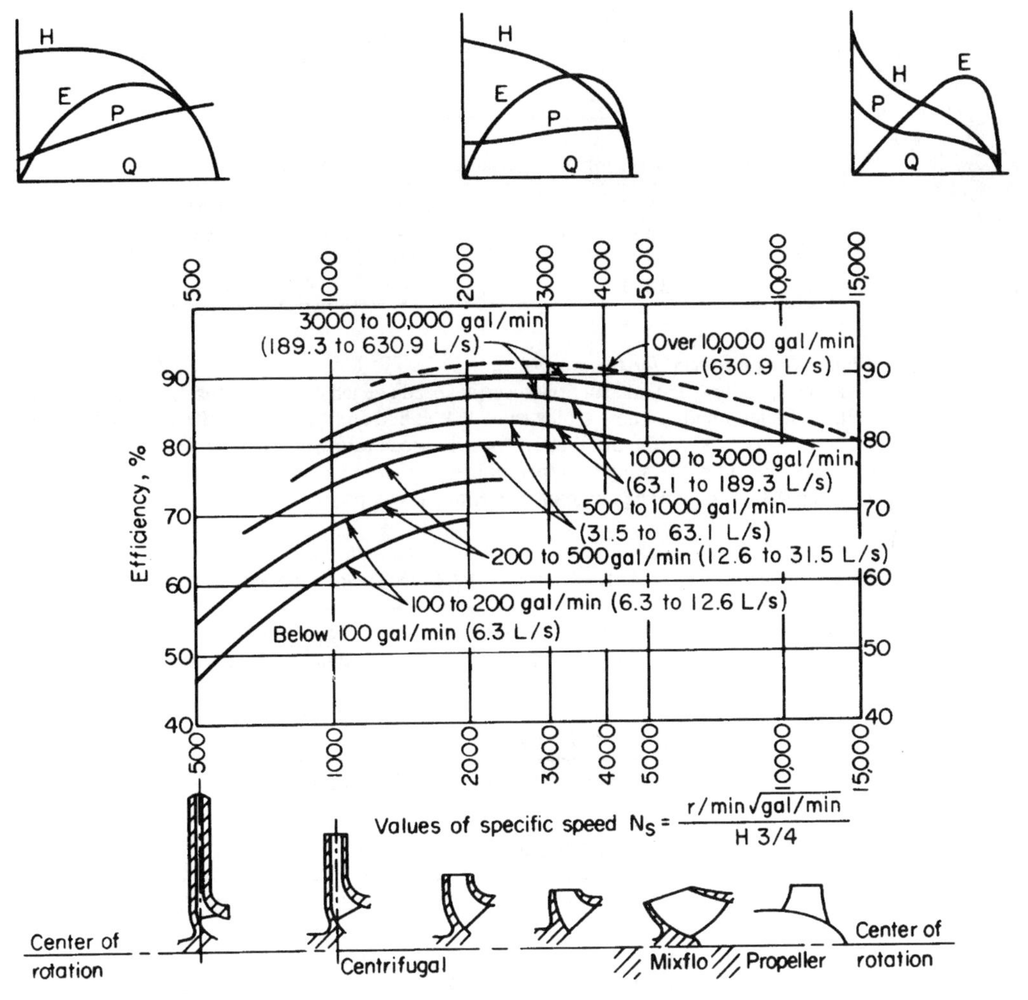

FIGURE 9 Approximate relative impeller shapes and efficiency variations for various specific speeds of centrifugal pumps. (*Worthington Corporation.*)

CHOOSING CENTRIFUGAL PUMP BEST ROTARY SPEED

A single-suction centrifugal pump is driven by a 60-Hz ac motor. The pump delivers 10,000 gal/min (630.9 L/s) of water at a 100-ft (30.5-m) head. The available net positive suction head = 32 ft (9.7 m) of water. What is the best operating speed for this pump if the pump operates at its best efficiency point?

Calculation Procedure:

1. ***Determine the specific speed and suction-specific speed***

Ac motors can operate at a variety of speeds, depending on the number of poles. Assume that the motor driving this pump might operate at 870, 1160, 1750, or 3500 r/min. Compute the specific speed $N_S = N(Q)^{0.5}/(H)^{0.75} = N(10{,}000)^{0.5}/(100)^{0.75} = 3.14N$ and the suction-specific speed $S = N(Q)^{0.5}/(\text{NPSH})^{0.75} = N(10{,}000)^{0.5}/(32)^{0.75} = 7.43N$ for each of the assumed speeds. Tabulate the results as follows:

Operating speed, r/min	Required specific speed	Required suction-specific speed
870	2,740	6,460
1,160	3,640	8,620
1,750	5,500	13,000
3,500	11,000	26,000

2. *Choose the best speed for the pump*

Analyze the specific speed and suction-specific speed at each of the various operating speeds, using the data in Tables 7 and 8. These tables show that at 870 and 1160 r/min, the suction-specific-speed rating is poor. At 1750 r/min, the suction-specific-speed rating is excellent, and a turbine or mixed-flow type pump will be the suitable. Operation at 3500 r/min is unfeasible because a suction-specific speed of 26,000 is beyond the range of conventional pumps.

TABLE 7 Pump Types Listed by Specific Speed*

Specific speed range	Type of pump
Below 2,000	Volute, diffuser
2,000–5,000	Turbine
4,000–10,000	Mixed-flow
9,000–15,000	Axial-flow

*Peerless Pump Division, FMC Corporation.

TABLE 8 Suction-Specific-Speed Ratings*

Single-suction pump	Double-suction pump	Rating
Above 11,000	Above 14,000	Excellent
9,000–11,000	11,000–14,000	Good
7,000–9,000	9,000–11,000	Average
5,000–7,000	7,000–9,000	Poor
Below 5,000	Below 7,000	Very poor

*Peerless Pump Division, FMC Corporation.

Related Calculations. Use this procedure for any type of centrifugal pump handling water for plant services, cooling, process, fire protection, and similar requirements. This procedure is the work of R. P. Horwitz, Hydrodynamics Division, Peerless Pump, FMC Corporation, as reported in *Power* magazine.

VAPOR-FREE LIQUID PUMP TOTAL HEAD COMPUTATION

Sketch three typical pump piping arrangement with static suction lift and submerged, free, and varying discharge head. Prepare similar sketches for the same pump with static suction head. Label the various heads. Compute the total head on each pump if the elevations are as shown in Fig. 10 and

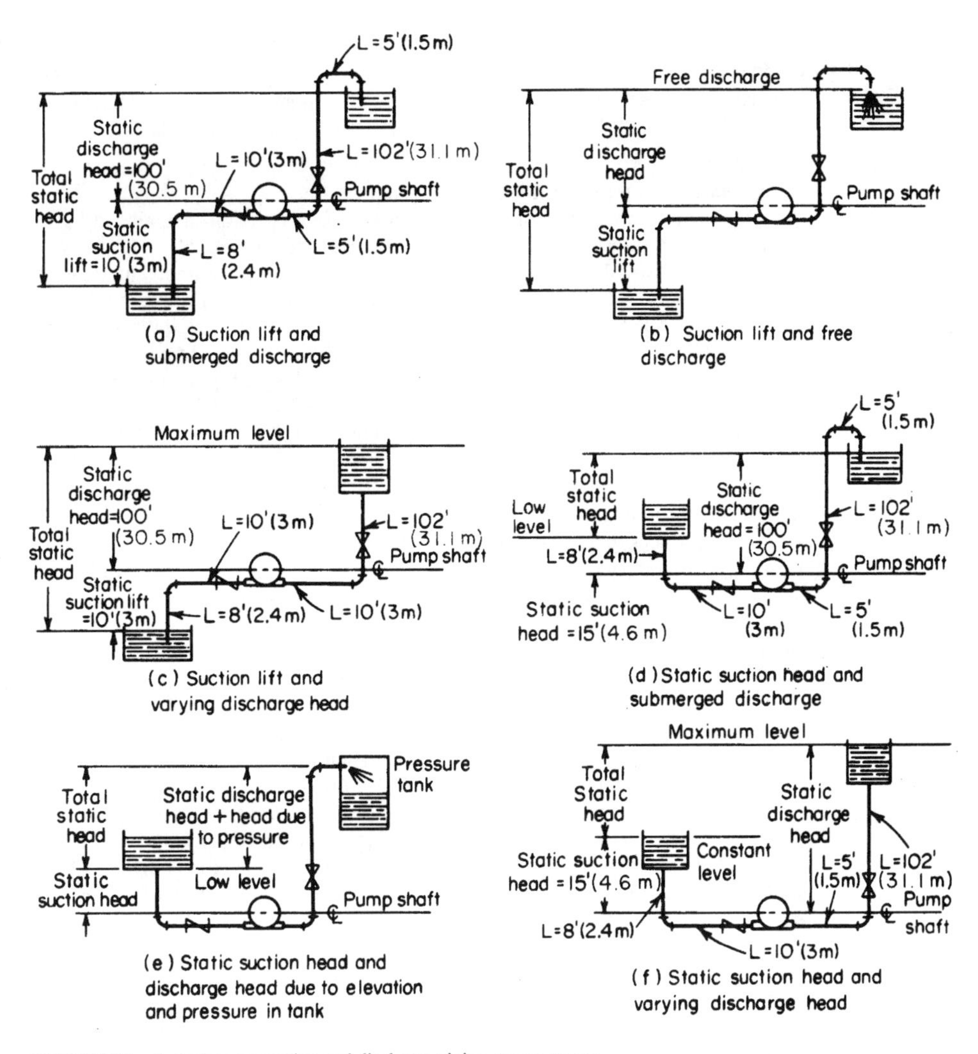

FIGURE 10 Typical pump suction and discharge piping arrangements.

the pump discharges a maximum of 2000 gal/min (126.2 L/s) of water through 8-in (203.2-mm) Schedule 40 pipe. What hp is required to drive the pump? A swing check values is used on the pump suction line and a gate valve on the discharge line.

Calculation Procedure:

1. ***Sketch the possible piping arrangements***

Figure 10 shows the six possible piping arrangements for the stated conditions of the installation. Label the total static head, i.e., the *vertical* distance from the surface of the source of the liquid supply to the free surface of the liquid in the discharge receiver, or to the point of free discharge from the discharge pipe. When both the suction and discharge surfaces are open to the atmosphere, the total static head equals the vertical difference in elevation. Use the free-surface elevations that cause the maximum suction lift and discharge head, i.e., the *lowest* possible level in the supply tank and the *highest* possible level in the discharge tank or pipe. When the supply source is *below* the pump centerline, the vertical distance is called the *static suction lift*; with the supply *above* the pump centerline, the vertical distance is called the *static suction head.* With variable static suction head, use the lowest liquid level in the supply tank when computing total static head. Label the diagrams as shown in Fig. 10.

2. *Compute the total static head on the pump*
The total static head H_{ts} ft = static suction lift, h_{sl} ft + static discharge head h_{sd} ft, where the pump has a suction lift, *s* in Fig. 10*a, b,* and *c.* In these installations, H_{ts} = 10 + 100 = 110 ft (33.5 m). Note that the static discharge head is computed between the pump centerline and the water level with an underwater discharge, Fig. 10*a*; to the pipe outlet with a free discharge, Fig. 10*b*; and to the maximum water level in the discharge tank. Fig. 10*c*. When a pump is discharging into a closed compression tank, the total discharge head equals the static discharge head plus the head equivalent, ft of liquid, of the internal pressure in the tank, or 2.31 × tank pressure, lb/in^2.

Where the pump has a static suction head, as in Fig. 10*d, e,* and *f,* the total static head H_{ts} ft = h_{sd} – static suction head h_{sh} ft. In these installations, H_t = 100 – 15 = 85 ft (25.9 m).

The total static head, as computed above, refers to the head on the pump without liquid flow. To determine the total head on the pump, the friction losses in the piping system during liquid flow must be also determined.

3. *Compute the piping friction losses*
Mark the length of each piece of straight pipe on the piping drawing. Thus, in Fig. 10*a*, the total length of straight pipe L_t ft = 8 + 10 + 5 + 102 + 5 = 130 ft (39.6 m), if we start at the suction tank and add each length until the discharge tank is reached. To the total length of straight pipe must be added the *equivalent* length of the pipe fittings. In Fig. 10*a* there are four long-radius elbows, one swing check valve, and one globe valve. In addition, there is a minor head loss at the pipe inlet and at the pipe outlet.

The equivalent length of one 8-in (203.2-mm) long-radius elbow is 14 ft (4.3 m) of pipe, from Table 9. Since the pipe contains four elbows, the total equivalent length = 4(14) = 56 ft (17.1 m) of straight pipe. The open gate valve has an equivalent resistance of 4.5 ft (1.4 m), and the open swing check valve has an equivalent resistance of 53 ft (16.2 m).

The entrance loss h_e ft, assuming a basket-type strainer is used at the suction-pipe inlet, is h_e ft = $K\upsilon^2/2g$, where K = a constant from Fig. 11; υ = liquid velocity, ft/s; g = 32.2 ft/s^2 (980.67 cm/s^2). The exit loss occurs when the liquid passes through a sudden enlargement, as from a pipe to a tank. Where the area of the tank is large, causing a final velocity that is zero, $h_{ex} = \upsilon^2/2g$.

The velocity υ ft/s in a pipe = $gpm/2.448d^2$. For this pipe, υ = 2000/[(2.448)(7.98)^2] = 12.82 ft/s (3.91 m/s). Then h_e = 0.74(12.82)^2/[2(32.2)] = 1.89 ft (0.58 m), and h_{ex} = (12.82)^2/[(2)(32.2)] = 2.56 ft (0.78 m). Hence, the total length of the piping system in Fig. 10*a* is 130 + 56 + 4.5 + 53 + 1.89 + 2.56 – 247.95 ft (75.6 m), say 248 ft (75.6 m).

Use a suitable head-loss equation, or Table 10, to compute the head loss for the pipe and fittings. Enter Table 10 at an 8-in (203.2-mm) pipe size, and project horizontally across to 2000 gal/min (126.2 L/s) and read the head loss as 5.86 ft of water per 100 ft (1.8 m/30.5 m) of pipe.

The total length of pipe and fittings computed above is 248 ft (75.6 m). Then total friction-head loss with a 2000-gal/min (126.2-L/s) flow is H_f ft = (5.86)(248/100) = 14.53 ft (4.5 m).

TABLE 9 Resistance of Fittings and Valves (length of straight pipe giving equivalent resistance)

Pipe size		Standard ell		Medium-radius ell		Long-radius ell		45° Ell		Tee		Gate valve, open		Globe valve, open		Swing check, open	
in	mm	ft	m	ft	m	ft	m	ft	m	ft	m	ft	m	ft	m	ft	m
6	152.4	16	4.9	14	4.3	11	3.4	7.7	2.3	33	10.1	3.5	1.1	160	48.8	40	12.2
8	203.2	21	6.4	18	5.5	14	4.3	10	3.0	43	13.1	4.5	1.4	220	67.0	53	16.2
10	254.0	26	7.9	22	6.7	17	5.2	13	3.9	56	17.1	5.7	1.7	290	88.4	67	20.4
12	304.8	32	9.8	26	7.9	20	6.1	15	4.6	66	20.1	6.7	2.0	340	103.6	80	24.4

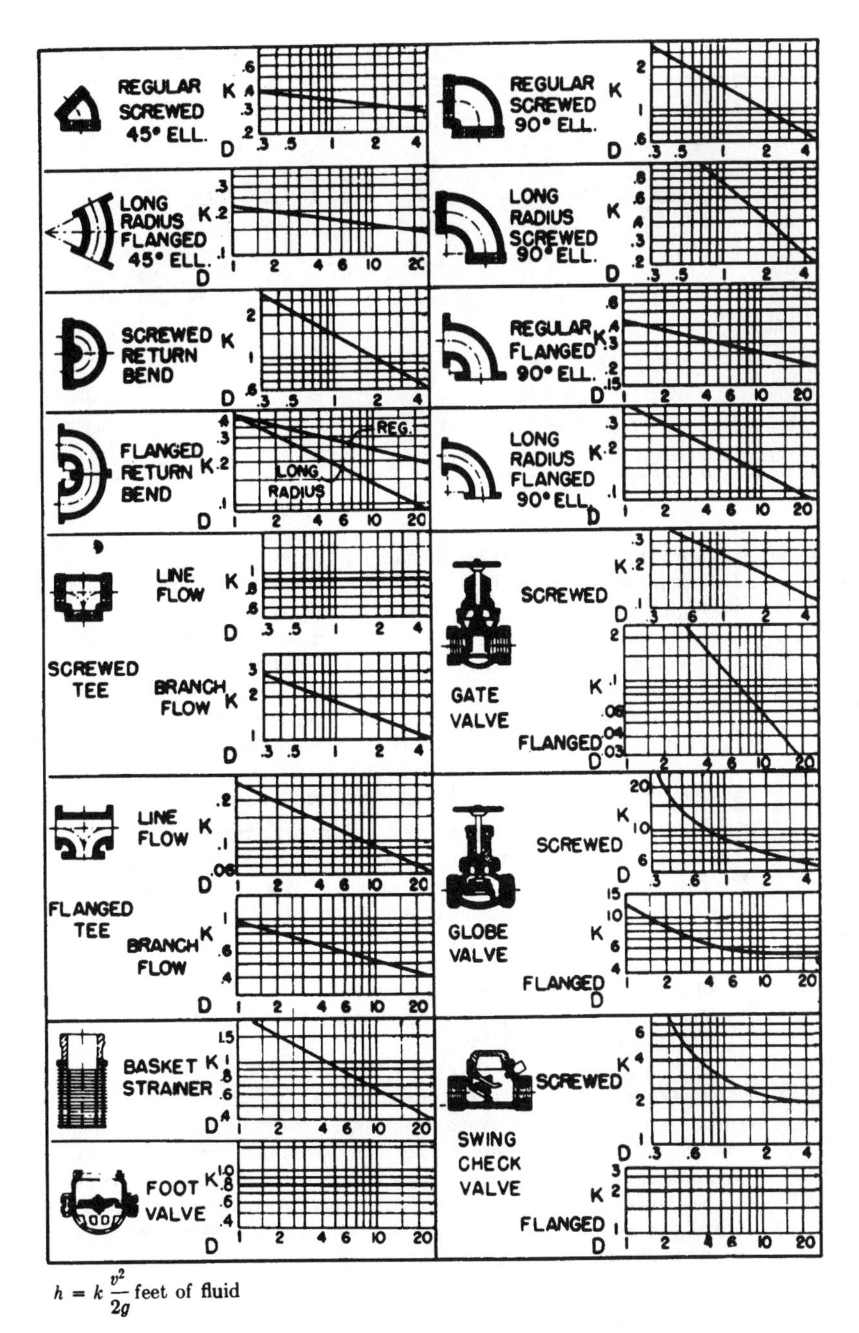

$$h = k \frac{v^2}{2g} \text{ feet of fluid}$$

FIGURE 11 Resistance coefficients of pipe fittings. To convert to SI in the equation for h, v^2 would be measured in m/s and feet would be changed to meters. The following values would also be changed from inches to millimeters: 0.3 to 7.6, 0.5 to 12.7, 1 to 25.4, 2 to 50.8, 4 to 101.6, 6 to 152.4, 10 to 254, and 20 to 508. (*Hydraulic Institute*.)

TABLE 10 Pipe Friction Loss for Water (wrought-iron or steel Schedule 40 pipe in good condition)

Diameter		Flow		Velocity		Velocity head		Friction loss per 100 ft (30.5 m) of pipe	
in	mm	gal/min	L/s	ft/s	m/s	ft water	m water	ft water	m water
6	152.4	1000	63.1	11.1	3.4	1.92	0.59	6.17	1.88
6	152.4	2000	126.2	22.2	6.8	7.67	2.3	23.8	7.25
6	152.4	4000	252.4	44.4	13.5	30.7	9.4	93.1	28.4
8	203.2	1000	63.1	6.41	1.9	0.639	0.195	1.56	0.475
8	203.2	2000	126.2	12.8	3.9	2.56	0.78	5.86	1.786
8	203.2	4000	252.4	25.7	7.8	10.2	3.1	22.6	6.888
10	254.0	1000	63.1	3.93	1.2	0.240	0.07	0.497	0.151
10	254.0	3000	189.3	11.8	3.6	2.16	0.658	4.00	1.219
10	254.0	5000	315.5	19.6	5.9	5.99	1.82	10.8	3.292

4. *Compute the total head on the pump*
The total head on the pump $H_t = H_{ts} + H_f$. For the pump in Fig. 10*a*, $H_t = 110 + 14.53 - 124.53$ ft (37.95 m), say 125 ft (38.1 m). The total head on the pump in Fig. 10*b* and *c* would be the same. Some engineers term the total head on a pump the *total dynamic head* to distinguish between static head (no-flow vertical head) and operating head (rated flow through the pump).

The total head on the pumps in Fig. 10*d*, *c*, and *f* is computed in the same way as described above, except that the total static head is less because the pump has a static suction head. That is, the elevation of the liquid on the suction side reduces the total distance through which the pump must discharge liquid; thus the total static head is less. The static suction head is *subtracted* from the static discharge head to determine the total static head on the pump.

5. *Compute the horsepower required to drive the pump*
The brake hp input to a pump $bhp_i = (gpm)(H_t)(s)/3960e$, where s = specific gravity of the liquid handled; e = hydraulic efficiency of the pump, expressed as a decimal. The usual hydraulic efficiency of a centrifugal pump is 60 to 80 percent; reciprocating pumps, 55 to 90 percent; rotary pumps, 50 to 90 percent. For each class of pump, the hydraulic efficiency decreases as the liquid viscosity increases.

Assume that the hydraulic efficiency of the pump in this system is 70 percent and the specific gravity of the liquid handled is 1.0. Then $bhp_i = (2000)(127)(1.0)/(3960)(0.70) = 91.6$ hp (68.4 kW).

The theoretical or *hydraulic horsepower* $hp_h = (gpm)(H_t)(s)/3960$, or $hp_h = (2000) = (127)(1.0)/3900 = 64.1$ hp (47.8 kW).

Related Calculations. Use this procedure for any liquid—water, oil, chemical, sludge, etc.—whose specific gravity is known. When liquids other than water are being pumped, the specific gravity and viscosity of the liquid, as discussed in later calculation procedures, must be taken into consideration. The procedure given here can be used for any class of pump—centrifugal, rotary, or reciprocating.

Note that Fig. 11 can be used to determine the equivalent length of a variety of pipe fittings. To use Fig. 11, simply substitute the appropriate K value in the relation $h = K\upsilon^2/2g$, where h = equivalent length of straight pipe; other symbols as before.

ENERGY CONSERVATION IN CHOOSING EQUIPMENT FOR A PUMPING INSTALLATION

Give a step-by-step procedure for choosing the class, type, capacity, drive, and materials for a pump that will be used in an industrial pumping system.

Calculation Procedure:

1. ***Sketch the proposed piping layout***

Use a single-line diagram, Fig. 12, of the piping system. Base the sketch on the actual job conditions. Show all the piping, fittings, valves, equipment, and other units in the system. Mark the *actual* and *equivalent* pipe length (see the previous calculation procedure) on the sketch. Be certain to include all vertical lifts, sharp bends, sudden enlargements, storage tanks, and similar equipment in the proposed system.

2. ***Determine the required capacity of the pump***

The required capacity is the flow rate that must be handled in gal/min, million gal/day, ft^3/s, gal/h, bbl/day, lb/h, acre · ft/day, mil/h, or some similar measure. Obtain the required flow rate from the process conditions, for example, boiler-feed rate, cooling-water flow rate, chemical feed rate, etc. The required flow rate for any process unit is usually given by the manufacturer or can be computed by using the calculation procedures given throughout this handbook.

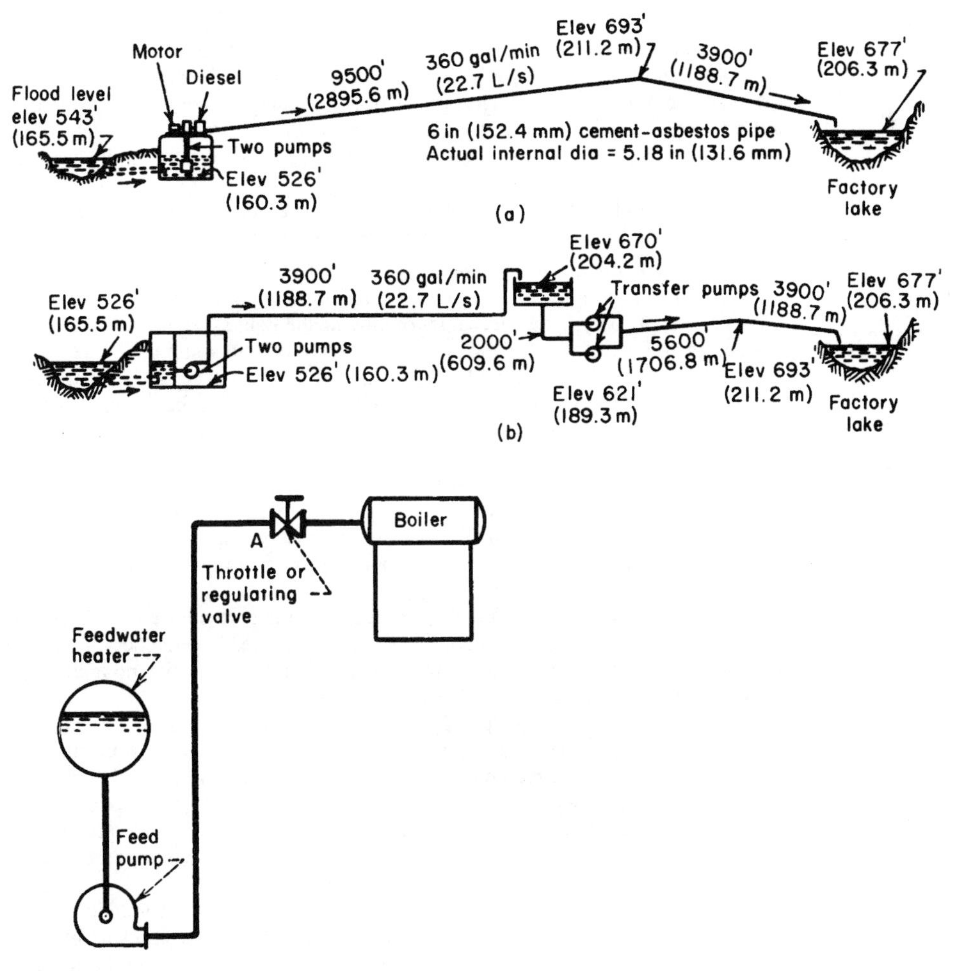

FIGURE 12 (*a*) Single-line diagrams for an industrial pipeline; (*b*) single-line diagram of a boiler-feed system. (*Worthington Corporation.*)

Once the required flow rate is determined, apply a suitable factor of safety. The value of this factor of safety can vary from a low of 5 percent of the required flow to a high of 50 percent or more, depending on the application. Typical safety factors are in the 10 percent range. With flow rates up to 1000 gal/min (63.1 L/s), and in the selection of process pumps, it is common practice to round a computed required flow rate to the next highest round-number capacity. Thus, with a required flow rate of 450 gal/min (28.4 L/s) and a 10 percent safety factor, the flow of 450 + 0.10(450) = 495 gal/min (31.2 L/s) would be rounded to 500 gal/min (31.6 L/s) *before* the pump was selected. A pump of 500 gal/min (31.6 L/s), or larger, capacity would be selected.

3. *Compute the total head on the pump*

Use the steps given in the previous calculation procedure to compute the total head on the pump. Express the result in ft (m) of water—this is the most common way of expressing the head on a pump. Be certain to use the exact specific gravity of the liquid handled when expressing the head in ft (m) of water. A specific gravity less than 1.00 *reduces* the total head when expressed in ft (m) of water, whereas a specific gravity greater than 1.00 *increases* the total head when expressed in ft (m) of water. Note that variations in the suction and discharge conditions can affect the total head on the pump.

4. *Analyze the liquid conditions*

Obtain complete data on the liquid pumped. These data should include the name and chemical formula of the liquid, maximum and minimum pumping temperature, corresponding vapor pressure at these temperature, specific gravity, viscosity at the pumping temperature, pH, flash point, ignition temperature, unusual characteristics (such as tendency to foam, curd, crystallize, become gelatinous or tacky), solids content, type of solids and their size, and variation in the chemical analysis of the liquid.

Enter the liquid conditions on a pump selection form like that in Fig. 13. Such forms are available from many pump manufacturers or can be prepared to meet special job conditions.

5. *Select the class and type of pump*

Three *classes* of pumps are used today—centrifugal, rotary, and reciprocating, Fig. 14. Note that these terms apply only to the mechanics of moving the liquid—not to the service for which the pump was designed. Each class of pump is further subdivided into a number of *types*, Fig. 14.

Use Table 11 as a general guide to the class and type of pump to be used. For example, when a large capacity at moderate pressure is required, Table 11 shows that a centrifugal pump would probably be best. Table 11 also shows the typical characteristics of various classes and types of pumps used in industrial process work.

Consider the liquid properties when choosing the class and type of pump, because exceptionally severe conditions may rule out one or another class of pump at the start. Thus, screw- and gear-type rotary pumps are suitable for handling viscous, nonabrasive liquid, Table 11. When an abrasive liquid must be handled, either another class of pump or another type of rotary pump must be used.

Also consider all the operating factors related to the particular pump. These factors include the type of service (continuous or intermittent), operating-speed preferences, future load expected and its effect on pump head and capacity, maintenance facilities available, possibility of parallel or series hookup, and other conditions peculiar to a given job.

Once the class and type of pump is selected, consult a rating table (Table 12) or rating chart, Fig. 15, to determine whether a suitable pump is available from the manufacturer whose unit will be used. When the hydraulic requirements fall between two standard pump models, it is usual practice to choose the next larger size of pump, unless there is some reason why an exact head and capacity are required for the unit. When one manufacturer does not have the desired unit, refer to the engineering data of other manufacturers. Also keep in mind that some pumps are custom-built for a given job when precise head and capacity requirements must be met.

Other pump data included in manufacturer's engineering information include characteristic curves for various diameter impellers in the same casing, Fig. 16, and variable-speed head-capacity curves for an impeller of given diameter, Fig. 17. Note that the required power input is given in Figs. 15 and 16 and may also be given in Fig. 17. Use of Table 12 is explained in the table.

Summary of Essential Data Required in Selection of Centrifugal Pumps

1. Number of Units Required

2. Nature of the Liquid to Be Pumped
   Is the liquid:
   a. Fresh or salt water, acid or alkali, oil, gasoline, slurry, or paper stock?
   b. Cold or hot and if hot, at what temperature? What is the vapor pressure of the liquid at the pumping temperature?
   c. What is its specific gravity?
   d. Is it viscous or nonviscous?
   e. Clear and free from suspended foreign matter or dirty and gritty? If the latter, what is the size and nature of the solids, and are they abrasive? If the liquid is of a pulpy nature, what is the consistency expressed either in percentage or in lb per cu ft of liquid? What is the suspended material?
   f. What is the chemical analysis, pH value, etc.? What are the expected variations of this analysis? If corrosive, what has been the past experience, both with successful materials and with unsatisfactory materials?

3. Capacity
   What is the required capacity as well as the minimum and maximum amount of liquid the pump will ever be called upon to deliver?

4. Suction Conditions
   Is there:
   a. A suction lift?
   b. Or a suction head?
   c. What are the length and diameter of the suction pipe?

5. Discharge Conditions
   a. What is the static head? Is it constant or variable?
   b. What is the friction head?
   c. What is the maximum discharge pressure against which the pump must deliver the liquid?

6. Total Head
   Variations in items 4 and 5 will cause variations in the total head.

7. Is the service continuous or intermittent?

8. Is the pump to be installed in a horizontal or vertical position? If the latter,
   a. In a wet pit?
   b. In a dry pit?

9. What type of power is available to drive the pump and what are the characteristics of this power?

10. What space, weight, or transportation limitations are involved?

11. Location of Installation
    a. Geographical location
    b. Elevation above sea level
    c. Indoor or outdoor installation
    d. Range of ambient temperatures

12. Are there any special requirements or marked preferences with respect to the design, construction, or performance of the pump?

FIGURE 13 Typical selection chart for centrifugal pumps. (*Worthington Corporation.*)

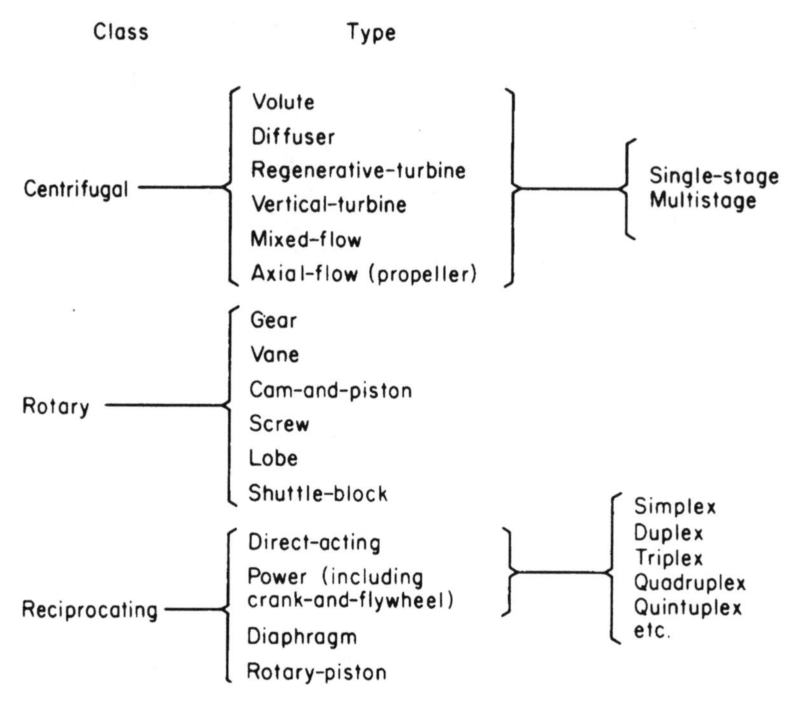

FIGURE 14 Modern pump classes and types.

TABLE 11 Characteristics of Modern Pumps

	Centrifugal		Rotary	Reciprocating		
	Volute and diffuser	Axial flow	Screw and gear	Direct acting steam	Double acting power	Triplex
Discharge flow	Steady	Steady	Steady	Pulsating	Pulsating	Pulsating
Usual maximum suction lift, ft (m)	15 (4.6)	15 (4.6)	22 (6.7)	22 (6.7)	22 (6.7)	22 (6.7)
Liquids handled	Clean, clear, dirty, abrasive; liquids with high solids content		Viscous; nonabrasive	Clean and clear		
Discharge pressure range	Low to high		Medium	Low to highest produced		
Usual capacity range	Small to largest available		Small to medium	Relatively small		
How increased head affects:						
Capacity	Decrease		None	Decrease	None	None
Power input	Depends on specific speed		Increase	Increase	Increase	Increase
How decreased head affects:						
Capacity	Increase		None	Small increase	None	None
Power input	Depends on specific speed		Decrease	Decrease	Decrease	Decrease

TABLE 12 Typical Centrifugal-Pump Rating Table

Size		Total head			
gal/min	L/s	20 ft, r/min—hp	6.1 m, r/min—kW	25 ft, r/min—hp	7.6 m, r/min—kW
3 CL:					
200	12.6	910–1.3	910–0.97	1010–1.6	1010–1.19
300	18.9	1000–1.9	1000–1.41	1100–2.4	1100–1.79
400	25.2	1200–3.1	1200–2.31	1230–3.7	1230–2.76
500	31.5	—	—	—	—
4 C:					
400	25.2	940–2.4	940–1.79	1040–3	1040–2.24
600	37.9	1080–4	1080–2.98	1170–4.6	1170–3.43
800	50.5	—	—	—	—

Example: 1080—4 indicates pump speed is 1080 r/min; actual input required to operate the pump is 4 hp (2.98 kW).
Source: Condensed from data of Goulds Pumps. Inc., SI values added by handbook editor.

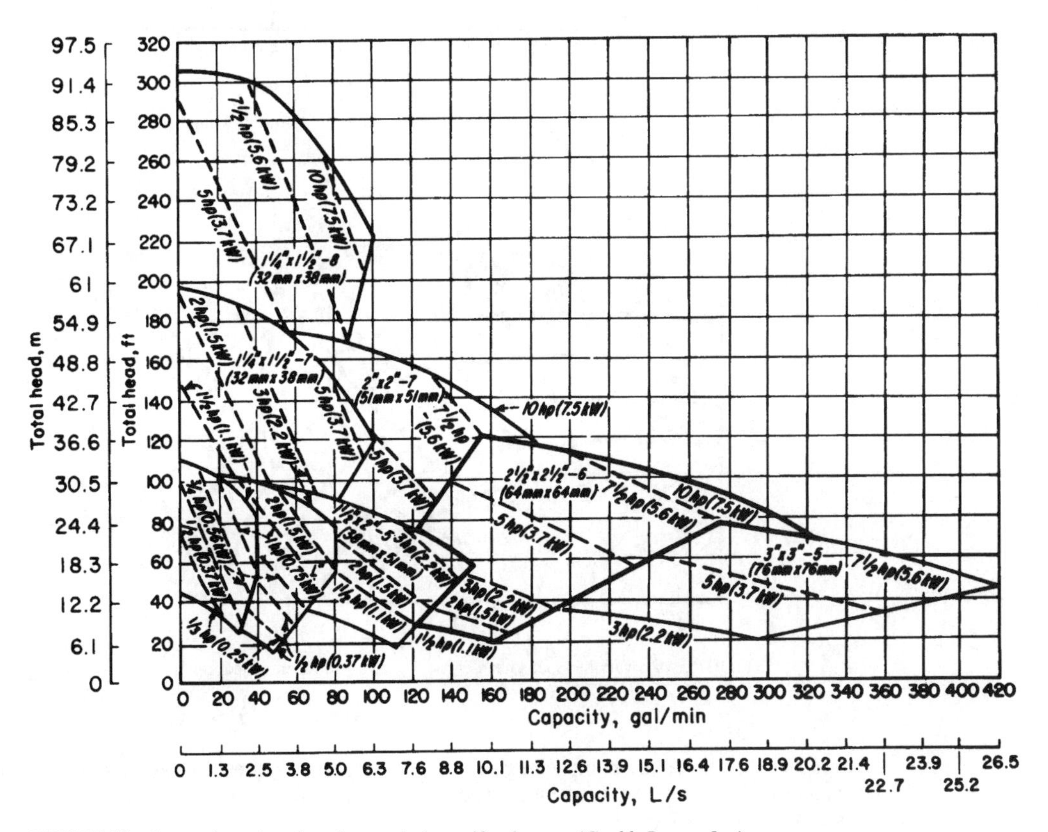

FIGURE 15 Composite rating chart for a typical centrifugal pump. (*Goulds Pumps, Inc.*)

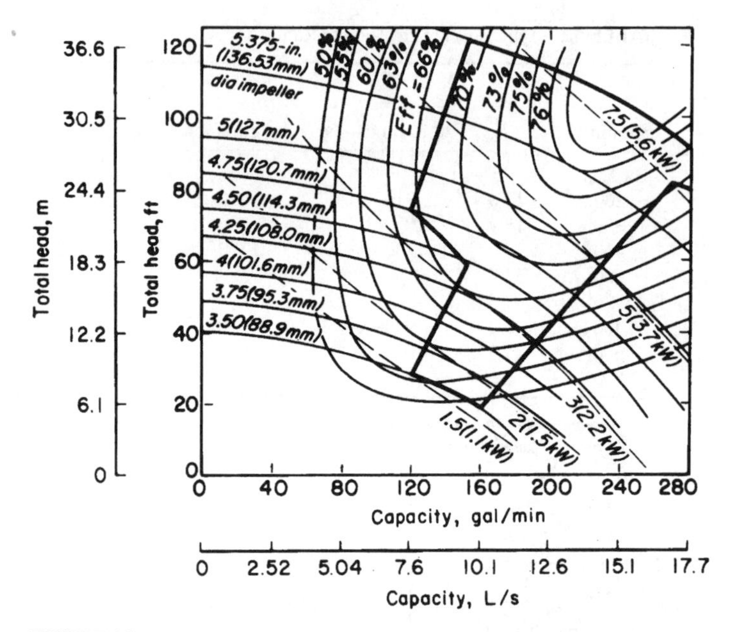

FIGURE 16 Pump characteristics when impeller diameter is varied within the same casing.

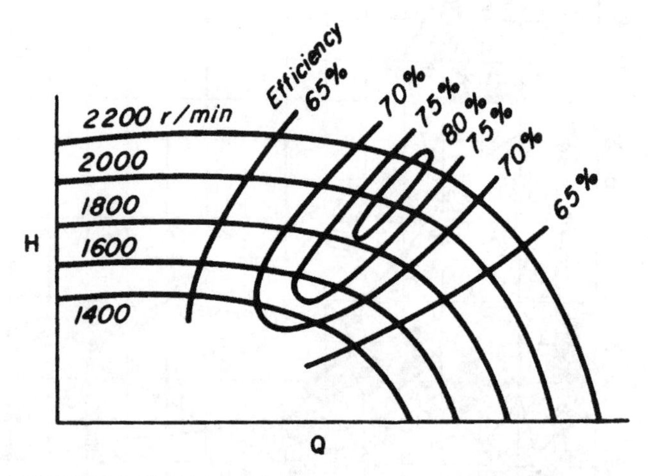

FIGURE 17 Variable-speed head-capacity curves for a centrifugal pump.

Performance data for rotary pumps are given in several forms. Figure 18 shows a typical plot of the head and capacity ranges of different types of rotary pumps. Reciprocating-pump capacity data are often tabulated, as in Table 13.

6. *Evaluate the pump chosen for the installation*

Check the specific speed of a centrifugal pump, using the method given in an earlier calculation procedure. Once the specific speed is known, the impeller type and approximate operating efficiency can be found from Fig. 9.

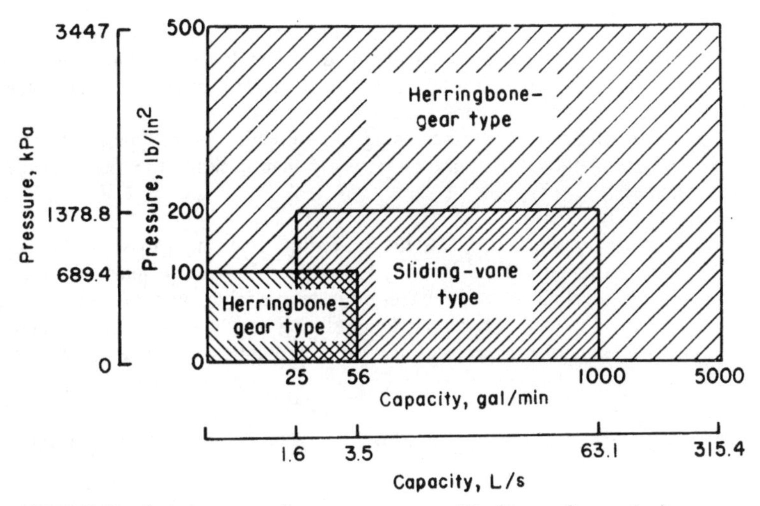

FIGURE 18 Capacity ranges of some rotary pumps. (*Worthington Corporation.*)

TABLE 13 Capacities of Typical Horizontal Duplex Plunger Pumps

Size		Cold-water pressure service			
				Piston speed	
in	cm	gal/min	L/s	ft/min	m/min
6 × 3½ × 6	15.2 × 8.9 × 15.2	60	3.8	60	18.3
7½ × 4½ × 10	19.1 × 11.4 × 25.4	124	7.8	75	22.9
9 × 5 × 10	22.9 × 12.7 × 25.4	153	9.7	75	22.9
10 × 6 × 12	25.4 × 15.2 × 30.5	235	14.8	80	24.4
12 × 7 × 12	30.5 × 17.8 × 30.5	320	20.2	80	24.4

Size		Boiler-feed service					
				Boiler		Piston speed	
in	cm	gal/min	L/s	hp	kW	ft/min	m/min
6 × 3½ × 6	15.2 × 8.9 × 15.2	36	2.3	475	354.4	36	10.9
7½ × 4½ × 10	19.1 × 11.4 × 25.4	74	4.7	975	727.4	45	13.7
9 × 5 × 10	22.9 × 12.7 × 25.4	92	5.8	1210	902.7	45	13.7
10 × 6 × 12	25.4 × 15.2 × 30.5	141	8.9	1860	1387.6	48	14.6
12 × 7 × 12	30.5 × 17.8 × 30.5	192	12.1	2530	1887.4	48	14.6

Source: Courtesy of Worthington Corporation.

Check the piping system, using the method of an earlier calculation procedure, to see whether the available net positive suction head equals, or is greater than, the required net positive suction head of the pump.

Determine whether a vertical or horizontal pump is more desirable. From the standpoint of floor space occupied, required NPSH, priming, and flexibility in changing the pump use, vertical pumps may be preferable to horizontal designs in some installations. But where headroom, corrosion, abrasion, and ease of maintenance are important factors, horizontal pumps may be preferable.

As a general guide, single-suction centrifugal pumps handle up to 50 gal/min (3.2 L/s) at total heads up to 50 ft (15.2 m); either single- or double-suction pumps are used for the flow rates to 1000 gal/min (63.1 L/s) and total heads to 300 ft (91.4 m); beyond these capacities and heads, double-suction or multistage pumps are generally used.

Mechanical seals are becoming more popular for all types of centrifugal pumps in a variety of services. Although they are more costly than packing, the mechanical seal reduces pump maintenance costs.

Related Calculations. Use the procedure given here to select any class of pump—centrifugal, rotary, or reciprocating—for any type of service—power plant, atomic energy, petroleum processing, chemical manufacture, paper mills, textile mills, rubber factories, food processing, water supply, sewage and sump service, air conditioning and heating, irrigation and flood control, mining and construction, marine services, industrial hydraulics, iron and steel manufacture.

PUMP AND SYSTEM CHARACTERISTIC-CURVE ENERGY ANALYSIS

Analyze a set of pump and system characteristic curves for the following conditions: friction losses without static head; friction losses with static head; pump without lift; system with little friction, much static head; system with gravity head; system with different pipe sizes; system with two discharge heads; system with diverted flow; and effect of pump wear on characteristic curve.

Calculation Procedure:

1. ***Plot the system-friction curve***

Without static head, the system-friction curve passes through the origin (0,0), Fig. 19, because when no head is developed by the pump, flow through the piping is zero. For most piping systems, the friction-head loss varies as the square of the liquid flow rate in the system. Hence, a system-friction curve, also called a friction-head curve, is parabolic—the friction head increases as the flow rate or capacity of the system increases. Draw the curve as shown in Fig. 19.

2. ***Plot the piping system and system-head curve***

Figure 20*a* shows a typical piping system with a pump operating against a static discharge head. Indicate the total static head, Fig. 20*b*, by a dashed line—in this installation H_{ts} = 110 ft. Since static head is a physical dimension, it does not vary with flow rate and is a constant for all flow rates. Draw the dashed line parallel to the abscissa, Fig. 20*b*.

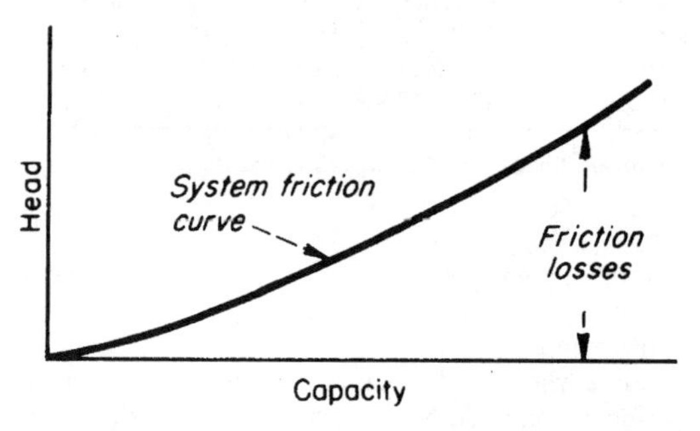

FIGURE 19 Typical system-friction curve.

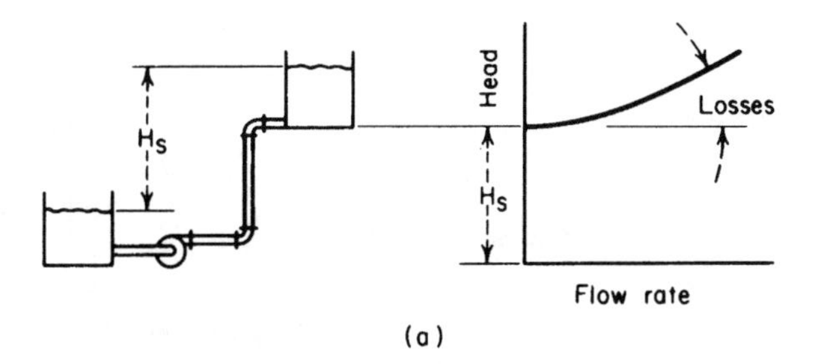

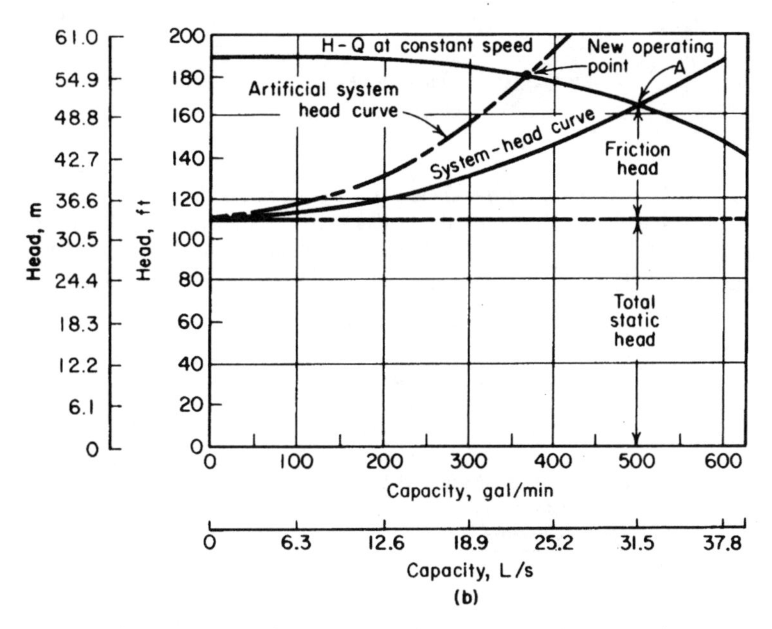

FIGURE 20 (*a*) Significant friction loss and lift; (*b*) system-head curve superimposed on pump head-capacity curve. (*Peerless Pumps*.)

From the point of no flow—zero capacity—plot the friction-head loss at various flow rates—100, 200, 300 gal/min (6.3, 12.6, 18.9 L/s), etc. Determine the friction-head loss by computing it as shown in an earlier calculation procedure. Draw a curve through the points obtained. This is called the *system-head curve*.

Plot the pump head-capacity (*H-Q*) curve of the pump on Fig. 20*b*. The *H-Q* curve can be obtained from the pump manufacturer or from a tabulation of *H* and *Q* values for the pump being considered. The point of intersection *A* between the *H-Q* and system-head curves is the operating point of the pump.

Changing the resistance of a given piping system by partially closing a valve or making some other change in the friction alters the position of the system-head curve and pump operating point. Compute the frictional resistance as before, and plot the artificial system-head curve as shown. Where this curve intersects, the *H-Q* curve is the new operating point of the pump. System-head curves are valuable for analyzing the suitability of a given pump for a particular application.

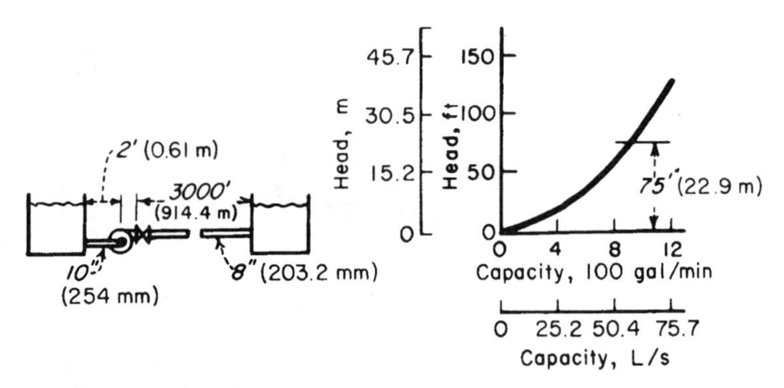

FIGURE 21 No lift; all friction head. (*Peerless Pumps.*)

3. ***Plot the no-lift system-head curve and compute the losses***
With no static head or lift, the system-head curve passes through the origin (0,0), Fig. 21, For a flow of 900 gal/min (56.8 L/s) in this system, compute the friction loss as follows, using the Hydraulic Institute *Pipe Friction Manual* tables or the method of earlier calculation procedures:

	ft	m
Entrance loss from tank into 10-in (254-mm) suction pipe, $0.5\upsilon^2/2g$	0.10	0.03
Friction loss in 2 ft (0.61 m) of suction pipe	0.02	0.01
Loss in 10-in (254-mm) 90° elbow at pump	0.20	0.06
Friction loss in 3000 ft (914.4 m) of 8-in (203.2-mm) discharge pipe	74.50	22.71
Loss in fully open 8-in (203.2-mm) gate valve	0.12	0.04
Exit loss from 8-in (203.2-mm) pipe into tank, $\upsilon^2/2g$	0.52	0.16
Total friction loss	75.46	23.01

Compute the friction loss at other flow rates in a similar manner, and plot the system-head curve, Fig. 21. Note that if all losses in this system except the friction in the discharge pipe were ignored, the total head would not change appreciably. However, for the purposes of accuracy, all losses should always be computed.

4. ***Plot the low-friction, high-head system-head curve***
The system-head curve for the vertical pump installation in Fig. 22 starts at the total static head, 15 ft (4.6 m), and zero flow. Compute the friction head for 15,000 gal/min as follows:

	ft	m
Friction in 20 ft (6.1 m) of 24-in (609.6-mm) pipe	0.40	0.12
Exit loss from 24-in (609.6-mm) pipe into tank, $\upsilon^2/2g$	1.60	0.49
Total friction loss	2.00	0.61

Hence, almost 90 percent of the total head of 15 + 2 = 17 ft (5.2 m) at 15,000-gal/min (946.4-L/s) flow is static head. But neglect of the pipe friction and exit losses could cause appreciable error during selection of a pump for the job.

5. ***Plot the gravity-head system-head curve***
In a system with gravity head (also called negative lift), fluid flow will continue until the system friction loss equals the available gravity head. In Fig. 23 the available gravity head is 50 ft (15.2 m).

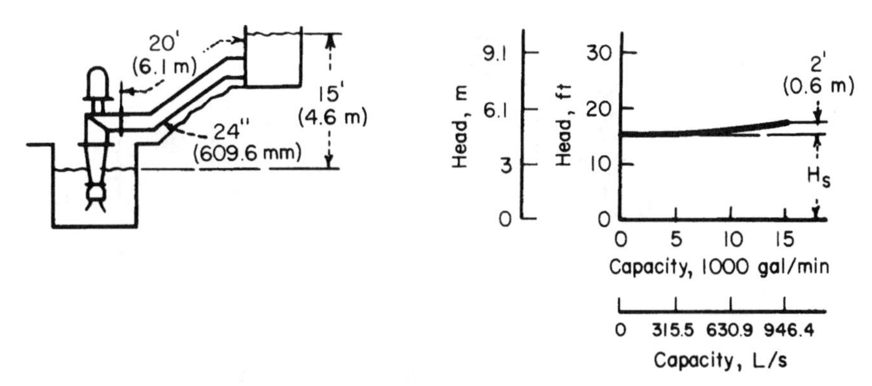

FIGURE 22 Mostly lift; little friction head. (*Peerless Pumps.*)

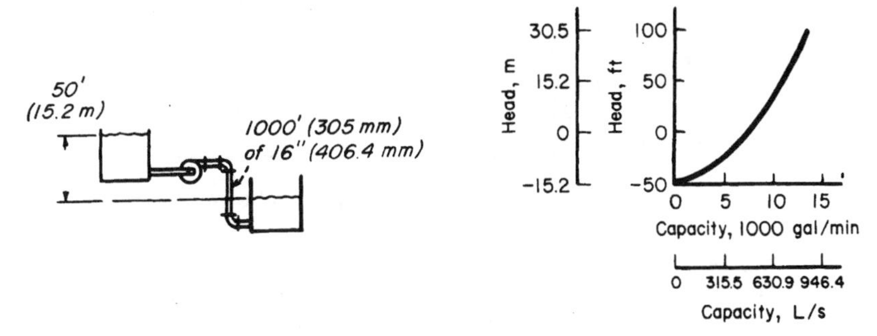

FIGURE 23 Negative lift (gravity head). (*Peerless Pumps.*)

Flows up to 7200 gal/min (454.3 L/s) are obtained by gravity head alone. To obtain larger flow rates, a pump is needed to overcome the friction in the piping between the tanks. Compute the friction loss for several flow rates as follows:

	ft	m
At 5000 gal/min (315.5 L/s), friction loss in 1000 ft (305 m) of 16-in (406.4-mm) pipe	25	7.6
At 7200 gal/min (454.3 L/s), friction loss = available gravity head	50	15.2
At 13,000 gal/min (820.2 L/s), friction loss	150	45.7

Using these three flow rates, plot the system-head curve, Fig. 23.

6. *Plot the system-head curves for different pipe sizes*

When different diameter pipes are used, the friction loss vs. flow rate is plotted independently for the two pipe sizes. At a given flow rate, the total friction loss for the system is the sum of the loss for the two pipes. Thus, the combined system-head curve represents the sum of the static head and the friction losses for all portions of the pipe.

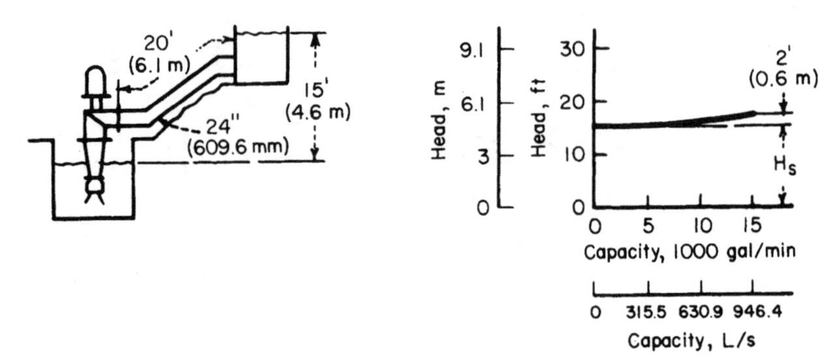

FIGURE 24 System with two different pipe sizes. (*Peerless Pumps.*)

Figure 24 shows a system with two different pipe sizes. Compute the friction losses as follows:

	ft	m
At 150 gal/min (9.5 L/s), friction loss in 200 ft (60.9 m) of 4-in (102-mm) pipe	5	1.52
At 150 gal/min (9.5 L/s), friction loss in 200 ft (60.9 m) of 3-in (76.2-mm) pipe	19	5.79
Total static head for 3- (76.2-) and 4-in (102-mm) pipes	10	3.05
Total head at 150-gal/min (9.5-L/s) flow	34	10.36

Compute the total head at other flow rates, and then plot the system-head curve as shown in Fig. 24.

7. *Plot the system-head curve for two discharge heads*
Figure 25 shows a typical pumping system having two different discharge heads. Plot separate system-head curves when the discharge heads are different. Add the flow rates for the two pipes at the same head to find points on the combined system-head curve, Fig. 25. Thus,

	ft	m
At 550 gal/min (34.7 L/s), friction loss in 1000 ft (305 m) of 8-in (203.2-mm) pipe	10	3.05
At 1150 gal/min (72.6 L/s), friction	38	11.6
At 1150 gal/min (72.6 L/s), friction + lift in pipe 1	88	26.8
At 550 gal/min (34.7 L/s), friction + lift in pipe 2	88	26.8

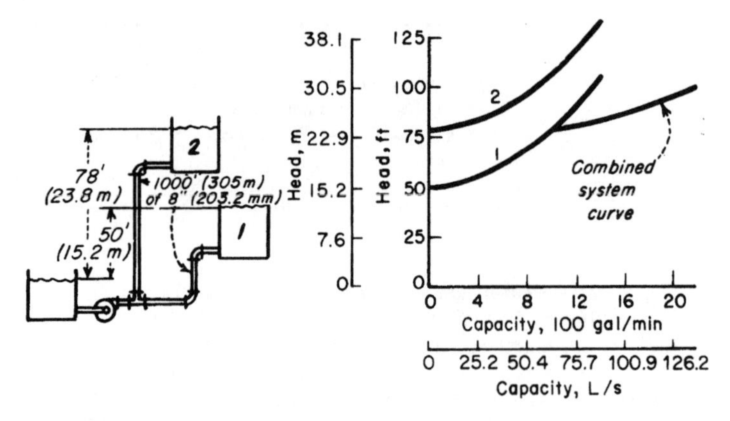

FIGURE 25 System with two different discharge heads. (*Peerless Pumps.*)

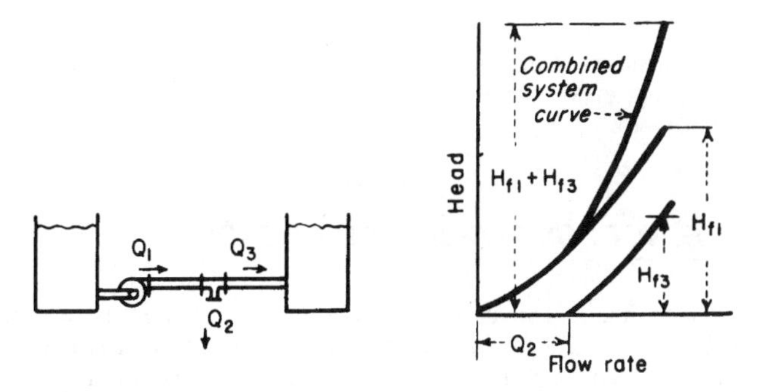

FIGURE 26 Part of the fluid flow is diverted from the main pipe. (*Peerless Pumps.*)

The flow rate for the combined system at a head of 88 ft (26.8 m) is 1150 + 550 = 1700 gal/min (107.3 L/s). To produce a flow of 1700 gal/min (107.3 L/s) through this system, a pump capable of developing an 88-ft (26.8-m) head is required.

8. ***Plot the system-head curve for diverted flow***

To analyze a system with diverted flow, assume that a constant quantity of liquid is tapped off at the intermediate point. Plot the friction loss vs. flow rate in the normal manner for pipe 1, Fig. 26. Move the curve for pipe 3 to the right at zero head by an amount equal to Q_2, since this represents the quantity passing through pipes 1 and 2 but not through pipe 3. Plot the combined system-head curve by adding, at a given flow rate, the head losses for pipes 1 and 3. With Q = 300 gal/min (18.9 L/s), pipe 1 = 500 ft (152.4 m) of 10-in (254-mm) pipe, and pipe 3 = 50 ft (15.2 m) of 6-in (152.4-mm) pipe.

	ft	m
At 1500 gal/min (94.6 L/s) through pipe 1, friction loss	11	3.35
Friction loss for pipe 3 (1500 – 300 = 1200 gal/min) (75.7 L/s)	8	2.44
Total friction loss at 1500-gal/min (94.6-L/s) delivery	19	5.79

9. ***Plot the effect of pump wear***

When a pump wears, there is a loss in capacity and efficiency. The amount of loss depends, however, on the shape of the system-head curve. For a centrifugal pump, Fig. 27, the capacity loss is greater for a given amount of wear if the system-head curve is flat, as compared with a steep system-head curve.

Determine the capacity loss for a worn pump by plotting its *H-Q* curve. Find this curve by testing the pump at different capacities and plotting the corresponding head. On the same chart, plot the *H-Q* curve for a new pump of the same size, Fig. 27. Plot the system-head curve, and determine the capacity loss as shown in Fig. 27.

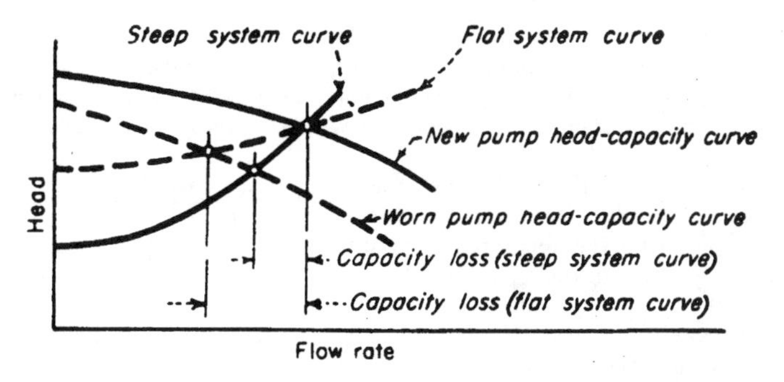

FIGURE 27 Effect of pump wear on pump capacity. (*Peerless Pumps.*)

Related Calculations. Use the techniques given here for any type of pump—centrifugal, reciprocating, or rotary—handling any type of liquid—oil, water, chemicals, etc. The methods given here are the work of Melvin Mann, as reported in *Chemical Engineering*, and Peerless Pump Division of FMC Corp.

HOT-LIQUID PUMP NET POSITIVE SUCTION HEAD

What is the maximum capacity of a double-suction condensate pump operating at 1750 r/min if it handles 100°F (37.8°C) water from a hot well in a condenser having an absolute pressure of 2.0 in (50.8 mm) Hg if the pump centerline is 10 ft (30.5 m) below the hot-well liquid level and the friction-head loss in the suction piping and fitting is 5 ft (1.52 m) of water?

Calculation Procedure:

1. ***Compute the net positive suction head on the pump***

The net positive suction head h_n on a pump when the liquid supply is *above* the pump inlet = pressure on liquid surface + static suction head – friction-head loss in suction piping and pump inlet – vapor pressure of the liquid, all expressed in ft absolute of liquid handled. When the liquid supply is *below* the pump centerline—i.e., there is a static suction lift—the vertical distance of the lift is *subtracted* from the pressure on the liquid surface instead of added as in the above relation.

The density of 100°F (37.8°C) water is 62.0 lb/ft^3 (992.6 kg/m^3), computed as shown in earlier calculation procedures in this handbook. The pressure on the liquid surface, in absolute ft of liquid = (2.0 inHg)(1.133)(62.4/62.0) = 2.24 ft (0.68 m). In this calculation, 1.133 = ft of 39.2°F (4°C) water = 1 inHg; 62.4 = lb/ft^3 (999.0 kg/m^3) of 39.2°F (4°C) water. The temperature of 39.2°F (4°C) is used because at this temperature water has its maximum density. Thus, to convert inHg to ft absolute of water, find the product of (inHg)(1.133)(water density at 39.2°F)/(water density at operating temperature). Express both density values in the same unit, usually lb/ft^3.

The static suction head is a physical dimension that is measured in ft (m) of liquid at the operating temperature. In this installation, h_{sh} = 10 ft (3 m) absolute.

The friction-head loss is 5 ft (1.52 m) of water. When it is computed by using the methods of earlier calculation procedures, this head loss is in ft (m) of water at maximum density. To convert to ft absolute, multiply by the ratio of water densities at 39.2°F (4°C) and the operating temperature, or (5)(62.4/62.0) = 5.03 ft (1.53 m).

The vapor pressure of water at 100°F (37.8°C) is 0.949 lb/in^2 (abs) (6.5 kPa) from the steam tables. Convert any vapor pressure to ft absolute by finding the result of [vapor pressure, lb/in^2 (abs)] (144 in^2/ft^2)/liquid density at operating temperature, or (0.949)(144)/62.0 = 2.204 ft (0.67 m) absolute.

With all the heads known, the net positive suction head is h_n = 2.24 + 10 – 5.03 – 2.204 = 5.01 ft (1.53 m) absolute.

2. ***Determine the capacity of the condensate pump***

Use the Hydraulic Institute curve, Fig. 28, to determine the maximum capacity of the pump. Enter at the left of Fig. 28 at a net positive suction head of 5.01 ft (1.53 m), and project horizontally to the right until the 3500-r/min curve is intersected. At the top, read the capacity as 278 gal/min (17.5 L/s).

Related Calculations. Use this procedure for any condensate or boiler-feed pump handling water at an elevated temperature. Consult the *Standards of the Hydraulic Institute* for capacity curves of pumps having different types of construction. In general, pump manufacturers who are members of the Hydraulic Institute rate their pumps in accordance with the *Standards*, and a pump chosen from a catalog capacity table or curve will deliver the stated capacity. A similar procedure is used for computing the capacity of pumps handling volatile petroleum liquids. When you use this procedure, be certain to refer to the latest edition of the *Standards*.

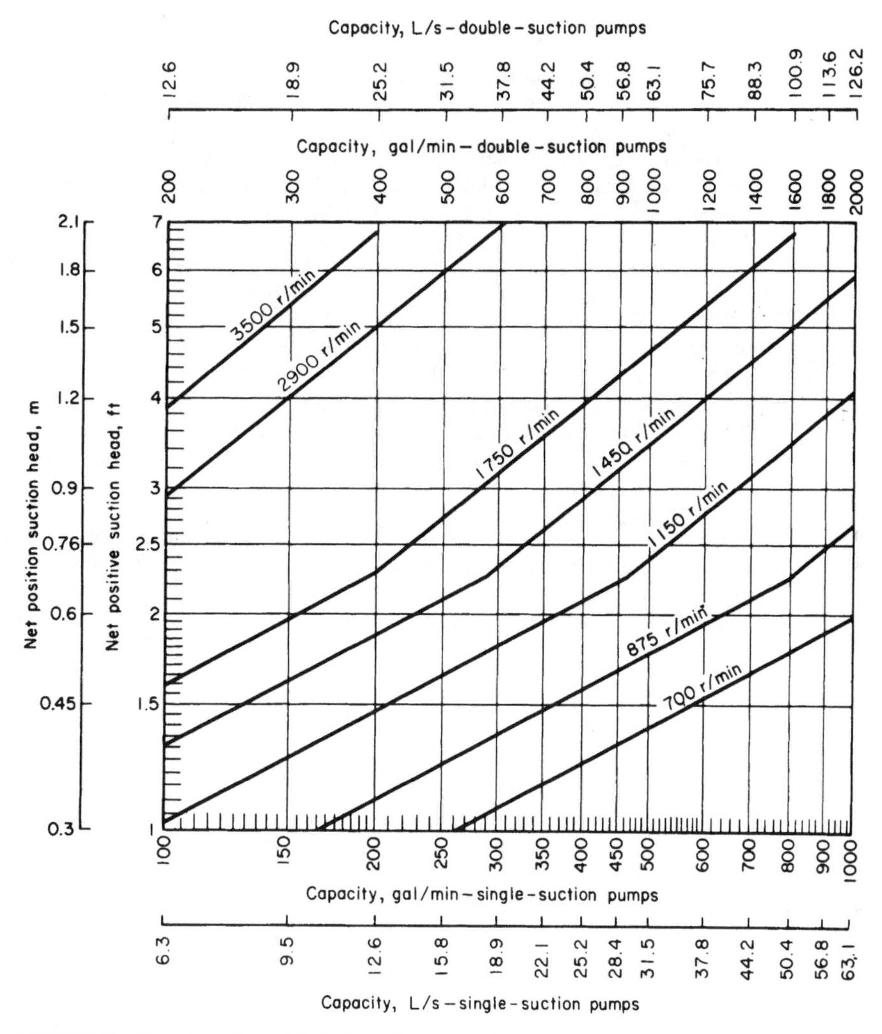

FIGURE 28 Capacity and speed limitations of condensate pumps with the shaft through the impeller eye. (*Hydraulic Institute.*)

STEAM POWER PLANT CONDENSATE PUMP SELECTION

Select the capacity for a condensate pump serving a steam power plant having a 140,000-lb/h (63,000-kg/h) exhaust flow to a condenser that operates at an absolute pressure of 1.0 in (25.4 mm) Hg. The condensate pump discharges through 4-in (101.6-mm) Schedule 40 pipe to an air-ejector condenser that has a frictional resistance of 8 ft (2.4 m) of water. From here, the condensate flows to and through a low-pressure heater that has a frictional resistance of 12 ft (3.7 m) of water and is vented to the atmosphere. The total equivalent length of the discharge piping, including all fittings and bends, is 400 ft (121.9 m), and the suction piping total equivalent length is 50 ft (15.2 m). The inlet of the low-pressure heater is 75 ft (22.9 m) above the pump centerline, and the condenser hotwell water level is 10 ft (3 m) above the pump centerline. How much power is required to drive the pump if its efficiency is 70 percent?

Calculation Procedure:

1. ***Compute the static head on the pump***

Sketch the piping system as shown in Fig. 29. Mark the static elevations and equivalent lengths as indicated.

The total head on the pump $H_t = H_{ts} + H_f$, where the symbols are the same as in earlier calculation procedures. The total static head $H_{ts} = h_{sd} - h_{sh}$. In this installation, h_{sd} = 75 ft (22.9 m). To make the calculation simpler, convert all the heads to absolute values. Since the heater is vented to the atmosphere, the pressure acting on the surface of the water in it = 14.7 lb/in^2 (abs) (101.3 kPa), or 34 ft (10.4 m) of water. The pressure acting on the condensate in the hot well is 1 in (25.4 mm) Hg = 1.133 ft (0.35 m) of water. [An absolute pressure of 1 in (25.4 mm) Hg = 1.133 ft (0.35 m) of water.] Thus, the absolute discharge static head = 75 + 34 = 109 ft (33.2 m), whereas the absolute suction head = 10 + 1.13 = 11.13 ft (3.39 m). Then $H_{ts} = h_{hd} - h_{sh}$ = 109.00 − 11.13 = 97.87 ft (29.8 m), say 98 ft (29.9 m) of water.

2. ***Compute the friction head in the piping system***

The total friction head H_f = pipe friction + heater friction. The pipe-friction loss is found first, as shown below. The heater-friction loss, obtained from the manufacturer or engineering data, is then added to the pipe-friction loss. Both must be expressed in ft (m) of water.

To determine the pipe friction, use Fig. 30 of Part 1 of this section and Table 17 and Fig. 6 of Part 2 of this section in the following manner. Find the product of the liquid velocity, ft/s, and the pipe internal diameter, in, or vd. With an exhaust flow of 140,000 lb/h (63,636 kg/h) to the condenser, the condensate flow is the same, or 140,000 lb/h (63,636 kg/h) at a temperature of 79.03°F (21.6°C), corresponding to an absolute pressure in the condenser of 1 in (25.4 mm) Hg, obtained from the

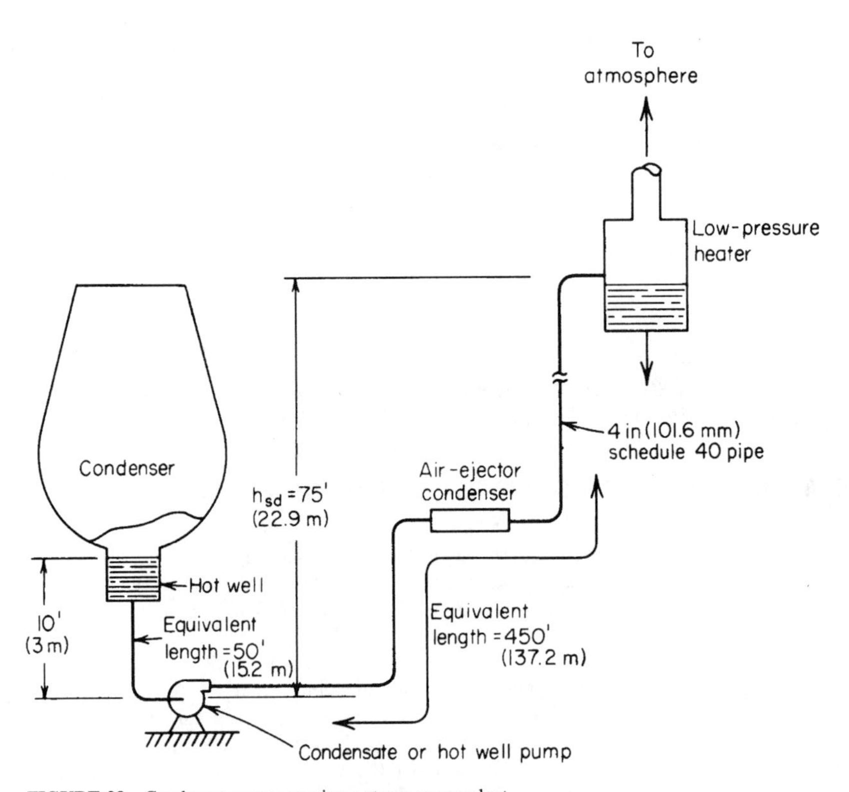

FIGURE 29 Condensate pump serving a steam power plant.

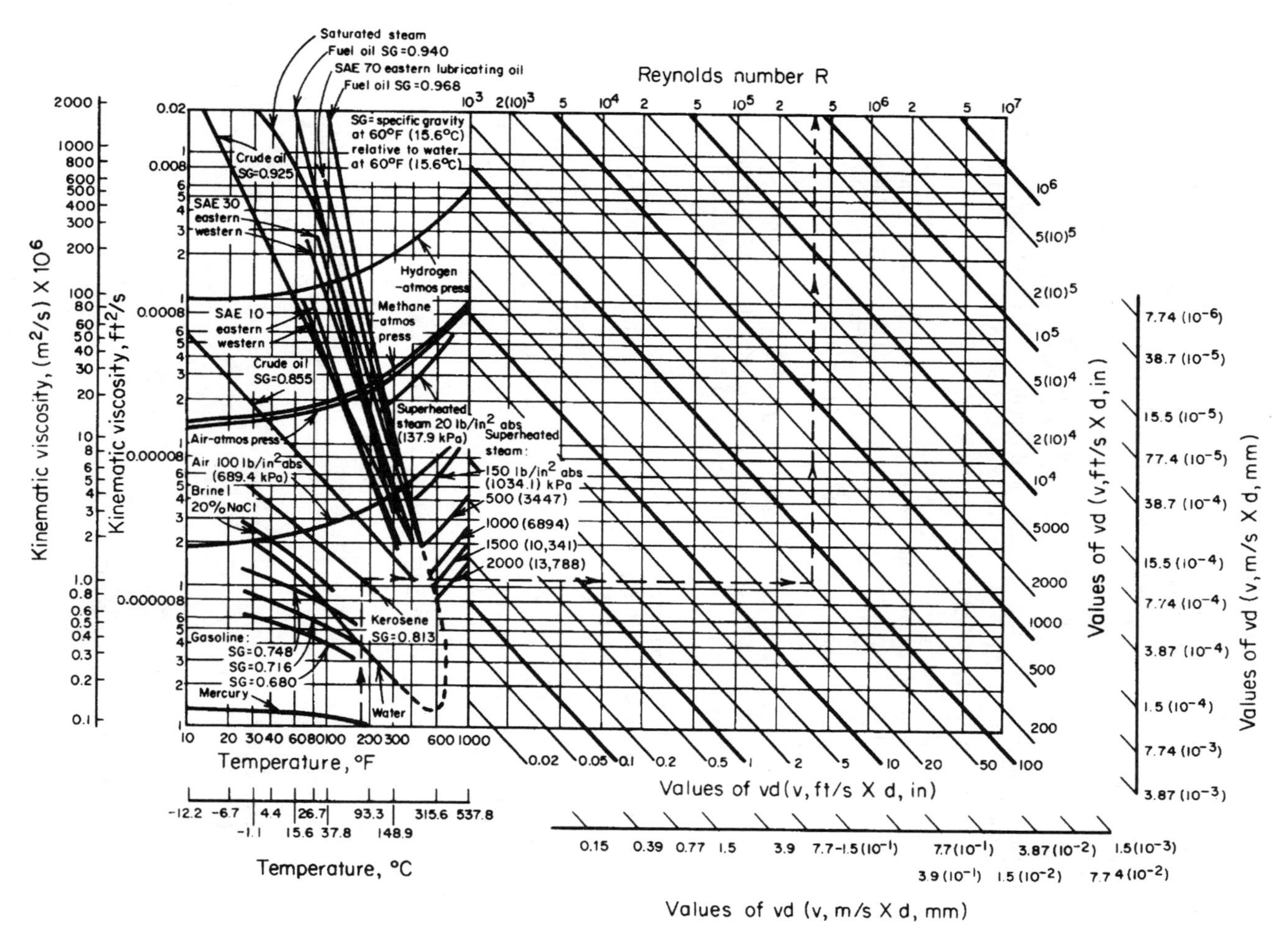

FIGURE 30 Kinematic viscosity and Reynolds number chart. (*Hydraulic Institute.*)

steam tables. The specific volume of the saturated liquid at this temperature and pressure is 0.01608 ft^3/lb (0.001 m^3/kg). Since 1 gal (0.26 L) of liquid occupies 0.13368 ft^3 (0.004 m^3), specific volume, gal/lb, is (0.01608/0.13368) = 0.1202 gal/lb (1.01 L/kg). Therefore, a flow of 140,000 lb/h (63,636 kg/h) = a flow of (140,000)(0.1202) = 16,840 gal/h (63,739.4 L/h), or 16,840/60 = 281 gal/min (17.7 L/s). Then the liquid velocity $\upsilon = gpm/2.448d^2 = 281/2.448(4.026)^2 = 7.1$ ft/s (2.1 m/s), and the product $\upsilon d = (7.1)(4.026) = 28.55$.

Enter Fig. 30 at a temperature of 79°F (26.1°C), and project vertically upward to the water curve. From the intersection, project horizontally to the right to $\upsilon d = 28.55$ and then vertically upward to read $R = 250{,}000$. Using Table 17 and Fig. 6 of Part 2 of this section and $R = 250{,}000$, find the friction factor $f = 0.0185$. Then the head loss due to pipe friction $H_f = (L/D)(\upsilon^2/2g) = 0.0185$ $(450/4.026/12)/[(7.1)^2/2(32.2)] = 19.18$ ft (5.9 m). In this computation, L = total equivalent length of the pipe, pipe fittings, and system valves, or 450 ft (137.2 m).

3. *Compute the other head losses in the system*

There are two other head losses in this piping system: the entrance loss at the square-edged hot-well pipe leading to the pump and the sudden enlargement in the low-pressure heater. The velocity head $\upsilon^2/2g = (7.1)^2/2(32.2) = 0.784$ ft (0.24 m). Using k values from Fig. 11 in this section, $h_e = k\upsilon^2/2g =$ $(0.5)(0.784) = 0.392$ ft (0.12 m); $h_{ex} = k\upsilon^2/2g = 0.784$ ft (0.24 m).

4. *Find the total head on the pump*

The total head on the pump $H_t = H_{ts} + H_f = 97.87 + 19.18 + 8 + 12 + 0.392 + 0.784 = 138.226$ ft (42.1 m), say 140 ft (42.7 m) of water. In this calculation, the 8-ft (2.4-m) and 12-ft (3.7-m) head losses are those occurring in the heaters. With a 25 percent safety factor, total head = (1.25) (140) = 175 ft (53.3 m).

5. *Compute the horsepower required to drive the pump*

The brake horsepower input $bhp_i = (gpm)(H_t)(s)/3960e$, where the symbols are the same as in earlier calculation procedures. At 1 in (25.4 mm) Hg, 1 lb (0.45 kg) of the condensate has a volume of 0.01608 ft^3 (0.000455 m^3). Since density = 1/specific volume, the density of the condensate = 1/0.01608 = 62.25 ft^3/lb (3.89 m^3/kg). Water having a specific gravity of unity weighs 62.4 lb/ft^3 (999 kg/m^3). Hence, the specific gravity of the condensate is 62.25/62.4 = 0.997. Then, assuming that the pump has an operating efficiency of 70 percent, we get $bhp_i = (281)(175) \times (0.997)/[3960(0.70)] =$ 17.7 bhp (13.2 kW).

6. *Select the condensate pump*

Condensate or hot-well pumps are usually centrifugal units having two or more stages, with the stage inlets opposed to give better axial balance and to subject the sealing glands to positive internal pressure, thereby preventing air leakage into the pump. In the head range developed by this pump, 175 ft (53.3 m), two stages are satisfactory. Refer to a pump manufacturer's engineering data for specific stage head ranges. Either a turbine or motor drive can be used.

Related Calculations. Use this procedure to choose condensate pumps for steam plants of any type—utility, industrial, marine, portable, heating, or process—and for combined steam-diesel plants.

CENTRIFUGAL PUMP MINIMUM SAFE FLUID FLOW DETERMINATION

A centrifugal pump handles 220°F (104.4°C) water and has a shut-off head (with closed discharge valve) of 3200 ft (975.4 m). At shutoff, the pump efficiency is 17 percent and the input brake horsepower is 210 bhp (156.7 kW). What is the minimum safe flow through this pump to prevent overheating at shutoff? Determine the minimum safe flow if the NPSH is 18.8 ft (5.7 m) of water and the liquid specific gravity is 0.995. If the pump contains 500 lb (225 kg) of water, determine the rate of the temperature rise at shutoff.

Calculation Procedure:

1. *Compute the temperature rise in the pump*
With the discharge valve closed, the power input to the pump is converted to heat in the casing and causes the liquid temperature to rise. The temperature rise $t = (1 - e) \times H_s/778e$, where t = temperature rise during shutoff. °F; e = pump efficiency, expressed as a decimal; H_s = shut-off head, ft. For this pump, $t = (1 - 0.17)(3200)/[778(0.17)] = 20.4$°F (36.7°C).

2. *Compute the minimum safe liquid flow*
For general-service pumps, the minimum safe flow M gal/min = 6.0(bhp input at shutoff)/t. Or, M = 6.0(210)/20.4 = 62.7 gal/min (3.96 L/s). This equation includes a 20 percent safety factor.

Centrifugal boiler-feed pumps usually have a maximum allowable temperature rise of 15°F (27°C). The minimum allowable flow through the pump to prevent the water temperature from rising more than 15°F (27°C) is 30 gal/min (1.89 L/s) for each 110-bhp (74.6-kW) input at shutoff.

3. *Compute the temperature rise for the operating NPSH*
An NPSH of 18.8 ft (5.73 m) is equivalent to a pressure of $18.8(0.433 \times 0.995) = 7.78$ lb/in^2 (abs) (53.6 kPa) at 220°F (104.4°C), where the factor 0.433 converts ft of water to lb/in^2. At 220°F (104.4°C), the vapor pressure of the water is 17.19 lb/in^2 (abs) (118.5 kPa), from the steam tables. Thus, the total vapor pressure of the water can develop before flashing occurs = NPSH pressure + vapor pressure at operating temperature = 7.78 + 17.19 = 24.97 lb/in^2 (abs) (172.1 kPa). Enter the steam tables at this pressure, and read the corresponding temperature as 240°F (115.6°C). The allowable temperature rise of the water is then 240 – 220 = 20°F (36.0°C). Using the safe-flow relation of step 2, we find the minimum safe flow is 62.9 gal/min (3.97 L/s).

4. *Compute the rate of temperature rise*
In any centrifugal pump, the rate of temperature rise t_r °F/min = 42.4(bhp input at shutoff)/wc, where w = weight of liquid in the pump, lb; c = specific heat of the liquid in the pump, Btu/(lb · °F). For this pump containing 500 lb (225 kg) of water with a specific heat, $c = 1.0$, $t_r = 42.4(210)/[500(1.0)]$ = 17.8°F/min (32°C/min). This is a very rapid temperature rise and could lead to overheating in a few minutes.

Related Calculations. Use this procedure for any centrifugal pump handling any liquid in any service—power, process, marine, industrial, or commercial. Pump manufacturers can supply a temperature-rise curve for a given model pump if it is requested. This curve is superimposed on the pump characteristic curve and shows the temperature rise accompanying a specific flow through the pump.

VISCOUS LIQUID CENTRIFUGAL PUMP SELECTION

Select a centrifugal pump to deliver 750 gal/min (47.3 L/s) of 1000-SSU oil at a total head of 100 ft (30.5 m). The oil has a specific gravity of 0.90 at the pumping temperature. Show how to plot the characteristic curves when the pump is handling the viscous liquid.

Calculation Procedure:

1. *Determine the required correction factors*
A centrifugal pump handling a viscous liquid usually must develop a greater capacity and head, and it requires a larger power input than the same pump handling water. With the water performance of the pump known—from either the pump characteristic curves or a tabulation of pump performance parameters—Fig. 31, prepared by the Hydraulic Institute, can be used to find suitable correction factors. Use this chart only within its scale limits; do not extrapolate. Do not use the chart for mixed-flow or axial-flow pumps or for pumps of special design. Use the chart only for pumps handling

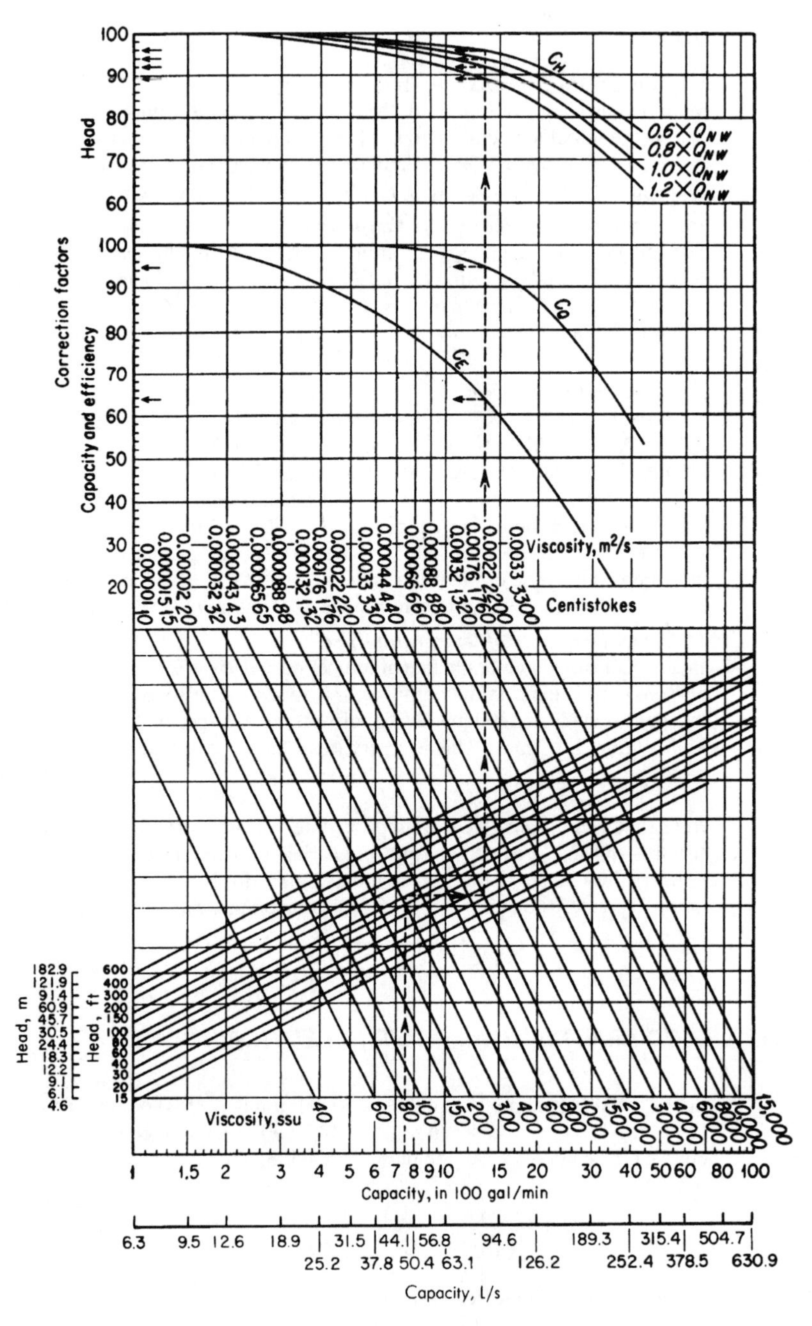

FIGURE 31 Correction factors for viscous liquids handled by centrifugal pumps. (*Hydraulic Institute*.)

uniform liquids; slurries, gels, paper stock, etc., may cause incorrect results. In using the chart, the available net positive suction head is assumed adequate for the pump.

To use Fig. 31, enter at the bottom at the required capacity, 750 gal/min (47.3 L/s), and project vertically to intersect the 100-ft (30.5-m) head curve, the required head. From here project horizontally to the 1000-SSU viscosity curve, and then vertically upward to the correction-factor curves. Read $C_E = 0.635$; $C_Q = 0.95$; $C_H = 0.92$ for $1.0Q_{NW}$. The subscripts E, Q, and H refer to correction factors for efficiency, capacity, and head, respectively; and NW refers to the water capacity at a particular efficiency. At maximum efficiency, the water capacity is given as $1.0Q_{NW}$; other efficiencies, expressed by numbers equal to or less than unity, give different capacities.

2. *Compute the water characteristics required*

The water capacity required for the pump $Q_w = Q_\upsilon/C_Q$ where Q_υ = viscous capacity, gal/min. For this pump, $Q_w = 750/0.95 = 790$ gal/min (49.8 L/s). Likewise, water head $H_w = H_\upsilon/C_H$, where H_υ = viscous head. Or, $H_w = 100/0.92 = 108.8$ ft (33.2 m), say 109 ft (33.2 m) of water.

Choose a pump to deliver 790 gal/min (49.8 L/s) of water at 109-ft (33.2-m) head of water, and the required viscous head and capacity will be obtained. Pick the pump so that it is operating at or near its maximum efficiency on water. If the water efficiency $E_w = 81$ percent at 790 gal/min (49.8 L/s) for this pump, the efficiency when handling the viscous liquid $E_\upsilon = E_w C_E$. Or, $E_\upsilon = 0.81(0.635) = 0.515$, or 51.5 percent.

The power input to the pump when handling viscous liquids is given by $P_\upsilon = Q_\upsilon H_\upsilon s/3960E_\upsilon$, where s = specific gravity of the viscous liquid. For this pump, $P_\upsilon = (750) \times (100)(0.90)/[3960(0.515)] = 33.1$ hp (24.7 kW).

3. *Plot the characteristic curves for viscous-liquid pumping*

Follow these eight steps to plot the complete characteristic curves of a centrifugal pump handling a viscous liquid when the water characteristics are known: (*a*) Secure a complete set of characteristic curves (H, Q, P, E) for the pump to be used. (*b*) Locate the point of maximum efficiency for the pump when handling water. (*c*) Read the pump capacity, Q gal/min, at this point. (*d*) Compute the values of $0.6Q$, $0.8Q$, and $1.2Q$ at the maximum efficiency. (*e*) Using Fig. 31, determine the correction factors at the capacities in steps *c* and *d*. Where a multistage pump is being considered, use the head per stage (= total pump head, ft/number of stages), when entering Fig. 31. (*f*) Correct the head, capacity, and efficiency for each of the flow rates in *c* and *d*, using the correction factors from Fig. 31. (*g*) Plot the corrected head and efficiency against the corrected capacity, as in Fig. 32. (*h*) Compute the power input at each flow rate and plot. Draw smooth curves through the points obtained, Fig. 32.

Related Calculations. Use the method given here for any uniform viscous liquid—oil, gasoline, kerosene, mercury, etc.—handled by a centrifugal pump. Be careful to use Fig. 31 only within its scale limits; *do not extrapolate*. The method presented here is that developed by the Hydraulic Institute. For new developments in the method, be certain to consult the latest edition of the Hydraulic Institute Standards.

SHAFT DEFLECTION AND PUMP CRITICAL SPEED

What are the shaft deflection and approximate first critical speed of a centrifugal pump if the total combined weight of the pump impellers is 23 lb (10.4 kg) and the pump manufacturer supplies the engineering data in Fig. 33?

Calculation Procedure:

1. *Determine the deflection of the pump shaft*

Use Fig. 33 to determine the shaft deflection. Note that this chart is valid for only one pump or series of pumps and must be obtained from the pump builder. Such a chart is difficult to prepare from test data without extensive test facilities.

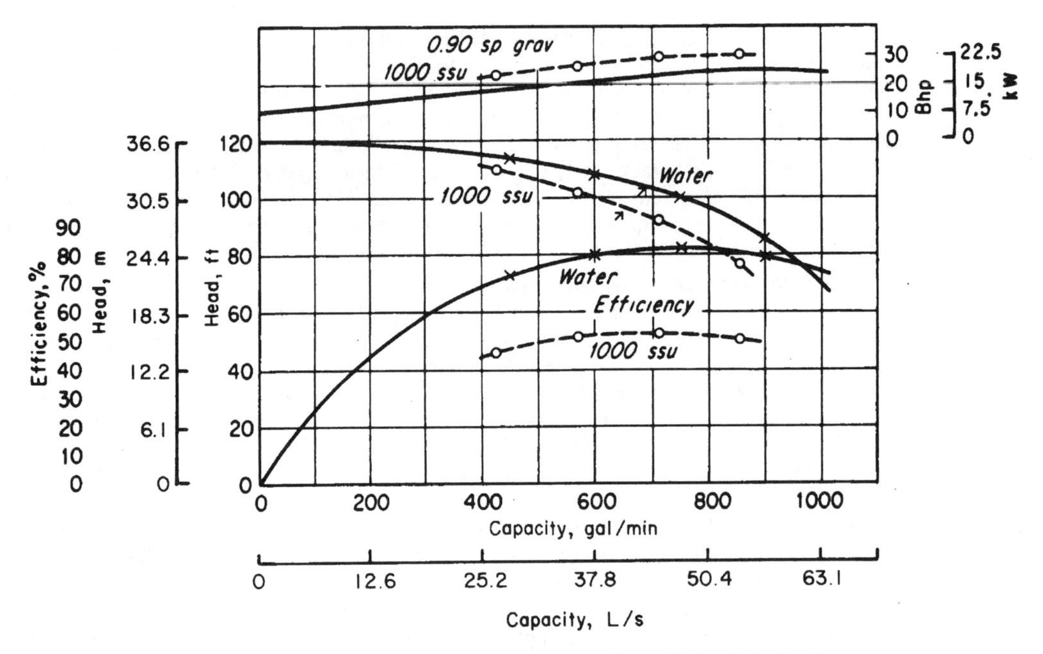

FIGURE 32 Characteristics curves for water (solid line) and oil (dashed line). (*Hydraulic Institute.*)

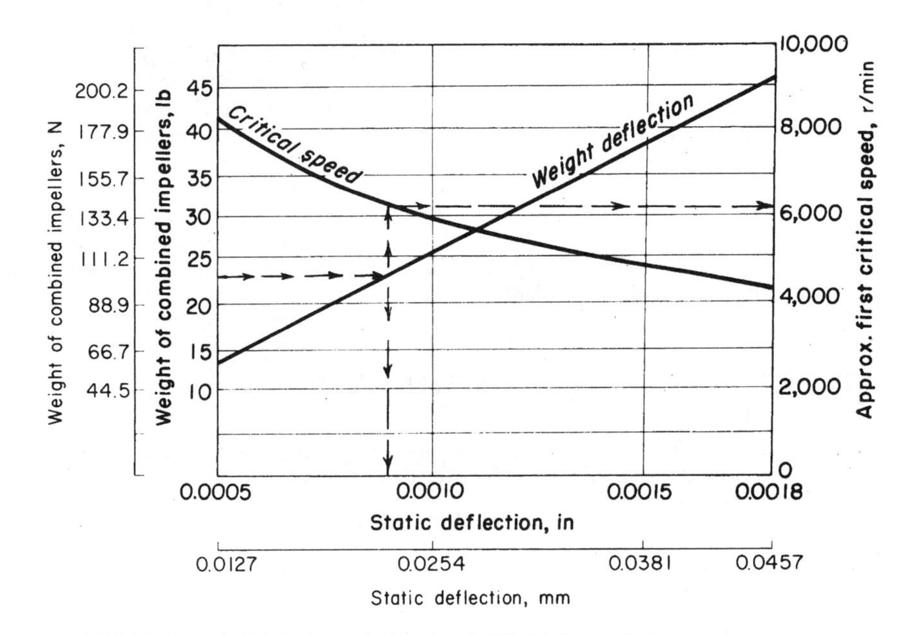

FIGURE 33 Pump shaft deflection and critical speed. (*Goulds Pumps, Inc.*)

Enter Fig. 33 at the left at the total combined weight of the impellers, 23 lb (10.4 kg), and project horizontally to the right until the weight-deflection curve is intersected. From the intersection, project vertically downward to read the shaft deflection as 0.009 in (0.23 mm) at full speed.

2. *Determine the critical speed of the pump*
From the intersection of the weight-deflection curve in Fig. 33 project vertically upward to the critical-speed curve. Project horizontally right from this intersection and read the first critical speed as 6200 r/min.

Related Calculations. Use this procedure for any class of pump—centrifugal, rotary, or reciprocating—for which the shaft-deflection and critical-speed curves are available. These pumps can be used for any purpose—process, power, marine, industrial, or commercial.

LIQUID VISCOSITY EFFECT ON REGENERATIVE-PUMP PERFORMANCE

A regenerative (turbine) pump has the water head-capacity and power-input characteristics shown in Fig. 34. Determine the head-capacity and power-input characteristics for four different viscosity oils to be handled by the pump—400, 600, 900, and 1000 SSU. What effect does increased viscosity have on the performance of the pump?

Calculation Procedure:

1. *Plot the water characteristics of the pump*
Obtain a tabulation or plot of the water characteristics of the pump from the manufacturer or from their engineering data. With a tabulation of the characteristics, enter the various capacity and power points given, and draw a smooth curve through them, Fig. 34.

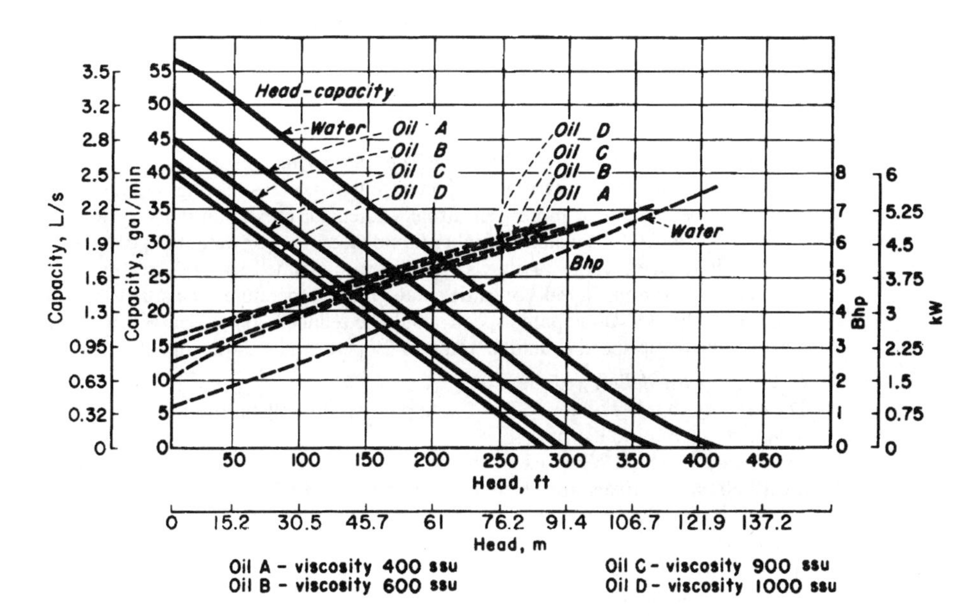

FIGURE 34 Regenerative pump performance when handling water and oil. (*Aurora Pump Division, The New York Air Brake Company.*)

2. *Plot the viscous-liquid characteristics of the pump*
The viscous-liquid characteristics of regenerative-type pumps are obtained by test of the actual unit. Hence, the only source of this information is the pump manufacturer. Obtain these characteristics from the pump manufacturer or their test data, and plot them on Fig. 34, as shown, for each oil or other liquid handled.

3. *Evaluate the effect of viscosity on pump performance*
Study Fig. 34 to determine the effect of increased liquid viscosity on the performance of the pump. Thus at a given head, say 100 ft (30.5 m), the capacity of the pump decreases as the liquid viscosity increases. At 100-ft (30.5-m) head, this pump has a water capacity of 43.5 gal/min (2.74 L/s), Fig. 34. The pump capacity for the various oils at 100-ft (30.5-m) head is 36 gal/min (2.27 L/s) for 400 SSU; 32 gal/min (2.02 L/s) for 600 SSU; 28 gal/min (1.77 L/s) for 900 SSU; and 26 gal/min (1.64 L/s) for 1000 SSU, respectively. There is a similar reduction in capacity of the pump at the other heads plotted in Fig. 34. Thus, as a general rule, the capacity of a regenerative pump decreases with an increase in liquid viscosity at constant head. Or conversely, at constant capacity, the head developed decreases as the liquid viscosity increases.

Plots of the power input to this pump show that the input power increases as the liquid viscosity increases.

Related Calculations. Use this procedure for a regenerative-type pump handling any liquid—water, oil, kerosene, gasoline, etc. A decrease in the viscosity of a liquid, as compared with the viscosity of water, will produce the opposite effect from that of increased viscosity.

LIQUID VISCOSITY EFFECT ON RECIPROCATING PUMP PERFORMANCE

A direct-acting steam-driven reciprocating pump delivers 100 gal/min (6.31 L/s) of 70°F (21.1°C) water when operating at 50 strokes per minute. How much 2000-SSU crude oil will this pump deliver? How much 125°F (51.7°C) water will this pump deliver?

Calculation Procedure:

1. *Determine the recommended change in pump performance*
Reciprocating pumps of any type—direct-acting or power—having any number of liquid-handling cylinders—one to five or more—are usually rated for maximum delivery when handling 250-SSU liquids or 70°F (21.1°C) water. At higher liquid viscosities or water temperatures, the speed—strokes or rpm—is reduced. Table 14 shows typical recommended speed-correction factors for reciprocating pumps for various liquid viscosities and water temperatures. This table shows that with a liquid viscosity of 2000 SSU the pump speed should be reduced 20 percent. When 125°F (51.7°C) water is handled, the pump speed should be reduced 25 percent, as shown in Table 14.

2. *Compute the delivery of the pump*
The delivery capacity of any reciprocating pump is directly proportional to the number of strokes per minute it makes or to its rpm.

When 2000-SSU oil is used, the pump strokes per minute must be reduced 20 percent, or (50)(0.20) = 10 strokes/min. Hence, the pump speed will be 50 – 10 = 40 strokes/min. Since the delivery is directly proportional to speed, the delivery of 2000-SSU oil = (40/50)(100) = 80 gal/min (5.1 L/s).

When handling 125°F (51.7°C) water, the pump strokes/min must be reduced 25 percent, or (50)(0.5) = 12.5 strokes/min. Hence, the pump speed will be 50.0 – 12.5 = 37.5 strokes/min. Since the delivery is directly proportional to speed, the delivery of 125°F (51.7°C) water = (37.5/50)(10) = 75 gal/min (4.7 L/s).

TABLE 14 Speed-Correction Factors

Liquid viscosity, SSU	Speed reduction, %	Water temperature °F	Water temperature °C	Speed reduction, %
250	0	70	21.1	0
500	4	80	26.7	9
1000	11	100	37.8	18
2000	20	125	51.7	25
3000	26	150	65.6	29
4000	30	200	93.3	34
5000	35	250	121.1	38

Related Calculations. Use this procedure for any type of reciprocating pump handling liquids falling within the range of Table 14. Such liquids include oil, kerosene, gasoline, brine, water, etc.

VISCOSITY AND DISSOLVED GAS EFFECTS ON ROTARY PUMP PERFORMANCE

A rotary pump handles 8000-SSU liquid containing 5 percent entrained gas and 10 percent dissolved gas at a 20-in (508-mm) Hg pump inlet vacuum. The pump is rated at 1000 gal/min (63.1 L/s) when handling gas-free liquids at viscosities less than 600 SSU. What is the output of this pump without slip? With 10 percent slip?

Calculation Procedure:

1. ***Compute the required speed reduction of the pump***

When the liquid viscosity exceeds 600 SSU, many pump manufacturers recommend that the speed of a rotary pump be reduced to permit operation without excessive noise or vibration. The speed reduction usually recommended is shown in Table 15.

TABLE 15 Rotary Pump Speed Reduction for Various Liquid Viscosities

Liquid viscosity, SSU	Speed reduction, percent of rated pump speed
600	2
800	6
1,000	10
1,500	12
2,000	14
4,000	20
6,000	30
8,000	40
10,000	50
20,000	55
30,000	57
40,000	60

With this pump handling 8000-SSU liquid, a speed reduction of 40 percent is necessary, as shown in Table 15. Since the capacity of a rotary pump varies directly with its speed, the output of this pump when handling 8000-SSU liquid = (1000 gal/min) × (1.0 – 0.40) = 600 gal/min (37.9 L/s).

2. ***Compute the effect of gas on the pump output***
Entrained or dissolved gas reduces the output or a rotary pump, as shown in Table 16. The gas in the liquid expands when the inlet pressure of the pump is below atmospheric and the gas occupies part of the pump chamber, reducing the liquid capacity.

With a 20-in (508-mm) Hg inlet vacuum, 5 percent entrained gas, and 10 percent dissolved gas, Table 16 shows that the liquid displacement is 74 percent of the rated displacement. Thus, the output of the pump when handling this viscous, gas-containing liquid will be (600 gal/min) (0.74) = 444 gal/min (28.0 L/s) without slip.

3. ***Compute the effect of slip on the pump output***
Slip reduces rotary-pump output in direct proportion to the slip. Thus, with 10 percent slip, the output of this pump = (444 gal/min)(1.0 – 0.10) = 369.6 gal/min (23.3 L/s).

Related Calculations. Use this procedure for any type of rotary pump—gear, lobe, screw, swinging-vane, sliding-vane, or shuttle-block, handling any clear, viscous liquid. Where the liquid is gas-free, apply only the viscosity correction. Where the liquid viscosity is less than 600 SSU but the liquid contains gas or air, apply the entrained or dissolved gas correction, or both corrections.

MATERIALS SELECTION FOR PUMP PARTS

Select suitable materials for the principal parts of a pump handling cold ethylene chloride. Use the Hydraulic Institute recommendation for materials of construction.

Calculation Procedure:

1. ***Determine which materials are suitable for this pump***
Refer to the data section of the Hydraulic Institute *Standards*. This section contains a tabulation of hundreds of liquids and the pump construction materials that have been successfully used to handle each liquid.

The table shows that for cold ethylene chloride having a specific gravity of 1.28, an all-bronze pump is satisfactory. In lieu of an all-bronze pump, the principal parts of the pump—casing, impeller, cylinder, and shaft—can be made of one of the following materials: austenitic steels (low-carbon 18-8; 18-8/Mo; highly alloyed stainless); nickel-base alloys containing chromium, molybdenum, and other elements, and usually less than 20 percent iron; or nickel-copper alloy (Monel metal). The order of listing in the *Standards* does not necessarily indicate relative superiority, since certain factors predominating in one instance may be sufficiently overshadowed in others to reverse the arrangement.

2. ***Choose the most economical pump***
Use the methods of earlier calculation procedures to select the most economical pump for the installation. Where the corrosion resistance of two or more pumps is equal, the standard pump, in this instance an all-bronze unit, will be the most economical.

Related Calculations. Use this procedure to select the materials of construction for any class of pump—centrifugal, rotary, or reciprocating—in any type of service—power, process, marine, or commercial. Be certain to use the latest edition of the Hydraulic Institute *Standards*, because the recommended materials may change from one edition to the next.

TABLE 16 Effect of Entrained or Dissolved Gas on the Liquid Displacement of Rotary Pumps (liquid displacement: percent of displacement)

Vacuum at pump inlet, inHg (mmHg)	Gas entrainment					Gas solubility					Gas entrainment and gas solubility combined				
	1%	2%	3%	4%	5%	2%	4%	6%	8%	10%	1% 2%	2% 4%	3% 6%	4% 8%	5% 10%
5 (127)	99	97½	96½	95	93½	99½	99	98½	97	97½	98½	96½	96	92	91
10 (254)	98½	97¼	95½	94	92	99	97½	97	95	95	97½	95	90	90	88½
15 (381)	98	96½	94½	92½	90½	97	96	94	92	90½	96	93	89½	86½	83½
20 (508)	97½	94½	92	89	86½	96	92	89	86	83	94	88	83	78	74
25 (635)	94	89	84	79	75½	90	83	76½	71	66	85½	75½	68	61	55

For example, with 5 percent gas entrainment at 15 inHg (381 mmHg) vacuum, the liquid displacement will be 90½ percent of the pump displacement, neglecting slip, or with 10 percent dissolved gas liquid displacement will be 90½ percent of the pump displacement; and with 5 percent entrained gas combined with 10 percent dissolved gas, the liquid displacement will be 83½ percent of pump replacement.

Source: Courtesy of Kinney Mfg. Div., The New York Air Brake Co.

HYDROPNEUMATIC STORAGE TANK SIZING

A 200-gal/min (12.6-L/s) water pump serves a pumping system. Determine the capacity required for a hydropneumatic tank to serve this system if the allowable high pressure in the tank and system is 60 lb/in^2 (gage) (413.6 kPa) and the allowable low pressure is 30 lb/in^2 (gage) (206.8 kPa). How many starts per hour will the pump make if the system draws 3000 gal/min (189.3 L/s) from the tank?

Calculation Procedure:

1. *Compute the required tank capacity*

In the usual hydropneumatic system, a storage-tank capacity in gal of 10 times the pump capacity in gal/min is used, if this capacity produces a moderate running time for the pump. Thus, this system would have a tank capacity of (10)(200) = 2000 gal (7570.8 L).

2. *Compute the quantity of liquid withdrawn per cycle*

For any hydropneumatic tank the withdrawal, expressed as the number of gallons (liters) withdrawn per cycle, is given by $W = (\upsilon_L - \upsilon_H)/C$, where υ_L = air volume in tank at the lower pressure, ft^3 (m^3); υ_H = volume of air in tank at higher pressure, ft^3 (m^3); C = conversion factor to convert ft^3 (m^3) to gallons (liters), as given below.

Compute V_L and V_H using the gas law for υ_H and either the gas law or the reserve percentage for υ_L. Thus, for υ_H, the gas law gives $\upsilon_H = p_L\upsilon_L/p_H$, where p_L = lower air pressure in tank, lb/in^2 (abs) (kPa); p_H = higher air pressure in tank lb/in^2 (abs) (kPa); other symbols as before.

In most hydropneumatic tanks a liquid reserve of 10 to 20 percent of the total tank volume is kept in the tank to prevent the tank from running dry and damaging the pump. Assuming a 10 percent reserve for this tank, $\upsilon_L = 0.1\ V$, where V = tank volume in ft^3 (m^3). Since a 2000-gal (7570-L) tank is being used, the volume of the tank is 2000/7.481 ft^3/gal = 267.3 ft^3 (7.6 m^3). With the 10 percent reserve at the 44.7-lb/in^2 (abs) (308.2-kPa) lower pressure, υ_L = 0.9 (267.3) = 240.6 ft^3 (6.3 m^3), where $0.9 = V - 0.1\ V$.

At the higher pressure in the tank, 74.7 lb/in^2 (abs) (514.9 kPa), the volume of the air will be, from the gas law, $\upsilon_H = p_L\upsilon_L/p_H$ = 44.7 (240.6)/74.7 = 143.9 ft^3 (4.1 m^3). Hence, during withdrawal, the volume of liquid removed from the tank will be W_g = (240.6 − 143.9)/0.1337 = 723.3 gal (2738 L). In this relation the constant converts from cubic feet to gallons and is 0.1337. To convert from cubic meters to liters, use the constant 1000 in the denominator.

3. *Compute the pump running time*

The pump has a capacity of 200 gal/min (12.6 L/s). Therefore, it will take 723/200 = 3.6 min to replace the withdrawn liquid. To supply 3000 gal/h (11,355 L/h) to the system, the pump must start 3000/723 = 4.1, or 5 times per hour. This is acceptable because a system in which the pump starts six or fewer times per hour is generally thought satisfactory.

Where the pump capacity is insufficient to supply the system demand for short periods, use a smaller reserve. Compute the running time using the equations in steps 2 and 3. Where a larger reserve is used—say 20 percent—use the value 0.8 in the equations in step 2. For a 30 percent reserve, the value would be 0.70, and so on.

Related Calculations. Use this procedure for any liquid system having a hydropneumatic tank—well drinking water, marine, industrial, or process.

CENTRIFUGAL PUMP ENERGY USAGE AS A HYDRAULIC TURBINE

Select a centrifugal pump to serve as a hydraulic turbine power source for a 1500-gal/min (5677.5-L/min) flow rate with 1290 ft (393.1 m) of head. The power application requires a 3600-r/min speed, the specific gravity of the liquid is 0.52, and the total available exhaust head is 20 ft (6.1 m). Analyze the cavitation potential and operating characteristics at an 80 percent flow rate.

Calculation Procedure:

1. *Choose the number of stages for the pump*

Search of typical centrifugal-pump data shows that a head of 1290 ft (393.1 m) is too large for a single-stage pump of conventional design. Hence, a two-stage pump will be the preliminary choice for this application. The two-stage pump chosen will have a design head of 645 ft (196.6 m) per stage.

2. *Compute the specific speed of the pump chosen*

Use the relation N_s = pump $rpm(Q)^{0.5}/H^{0.75}$ where N_s = specific speed of the pump; rpm = r/min of pump shaft; Q = pump capacity or flow rate, gal/min; H = pump head per stage, ft. Substituting, we get $N_s = 3600(1500)^{0.5}/(645)^{0.75} = 1090$. Note that the specific-speed value is the same regardless of the system of units used—USCS or SI.

3. *Convert turbine design conditions to pump design conditions*

To convert from turbine design conditions to pump design conditions, use the pump manufacturer's conversion factors that relate turbine best efficiency point (bep) performance with pump bep performance. Typically, as specific speed N_s varies from 500 to 2800, these bep factors generally vary as follows: the conversion factor for capacity (gal/min or L/min) C_Q, from 2.2 to 1.1; the conversion factor for head (ft or m) C_H, from 2.2 to 1.1; the conversion factor for efficiency C_E, from 0.92 to 0.99. Applying these conversion factors to the turbine design conditions yields the pump design conditions sought.

At the specific speed for this pump, the values of these conversion factors are determined from the manufacturer to be $C_Q = 1.24$; $C_H = 1.42$; $C_E = 0.967$.

Given these conversion factors, the turbine design conditions can be converted to the pump design conditions thus: $Q_p = Q_t/C_Q$, where Q_p = pump capacity or flow rate, gal/min or L/min; Q_t = turbine capacity or flow rate in the same units; other symbols are as given earlier. Substituting gives Q_p = 1500/1.24 = 1210 gal/min (4580 L/min).

Likewise, the pump discharge head, in feet of liquid handled, is $H_p = H_t/C_H$. So H_p = 645/1.42 = 454 ft (138.4 m).

4. *Select a suitable pump for the operating conditions*

Once the pump capacity, head, and rpm are known, a pump having its best bep at these conditions can be selected. Searching a set of pump characteristic curves and capacity tables shows that a two-stage 4-in (10-cm) unit with an efficiency of 77 percent would be suitable.

5. *Estimate the turbine horsepower developed*

To predict the developed hp, convert the pump efficiency to turbine efficiency. Use the conversion factor developed above. Or, the turbine efficiency $E_t = E_pC_E = (0.77 \times 0.967) = 0.745$, or 74.5 percent.

With the turbine efficiency known, the output brake horsepower can be found from bhp = $Q_tH_tE_ts/3960$, where s = fluid specific gravity; other symbols as before. Substituting, we get bhp = 1500(1290)(0.745)(0.52)/3960 = 198 hp (141 kW).

6. *Determine the cavitation potential of this pump*

Just as pumping requires a minimum net positive suction head, turbine duty requires a net positive exhaust head. The relation between the total required exhaust head (TREH) and turbine head per stage is the cavitation constant σ_r = TREH/H. Figure 35 shows σ_r vs. N_s for hydraulic turbines. Although a pump used as a turbine will not have exactly the same relationship, this curve provides a good estimate of σ_r for turbine duty.

To prevent cavitation, the total available exhaust head (TAEH) must be greater than the TREH. In this installation, N_s = 1090 and TAEH = 20 ft (6.1 m). From Fig. 35, σ_r = 0.028 and TREH = 0.028(645) = 18.1 ft (5.5 m). Because TAEH > TREH, there is enough exhaust head to prevent cavitation.

7. *Determine the turbine performance at 80 percent flow rate*

In many cases, pump manufacturers treat conversion factors as proprietary information. When this occurs, the performance of the turbine under different operating conditions can be predicted from the general curves in Figs. 36 and 37.

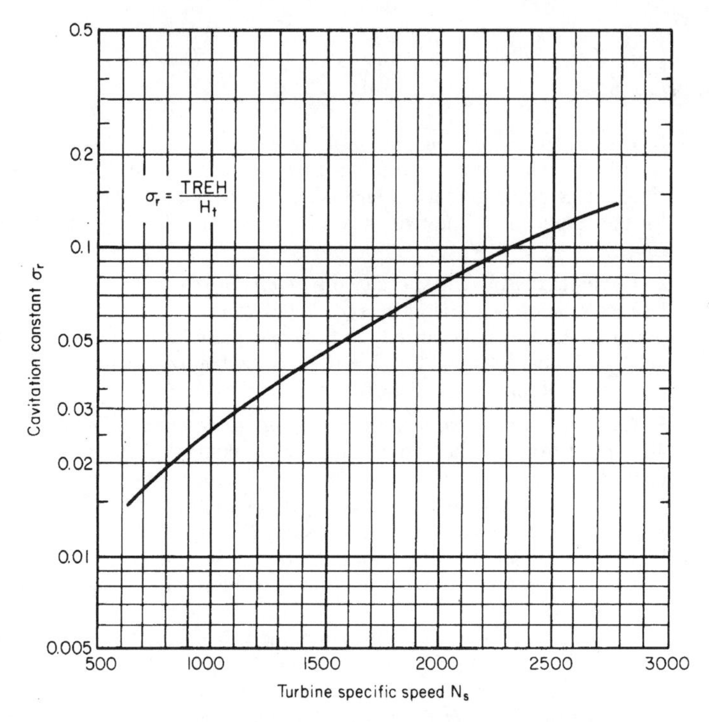

FIGURE 35 Cavitation constant for hydraulic turbines. (*Chemical Engineering*).

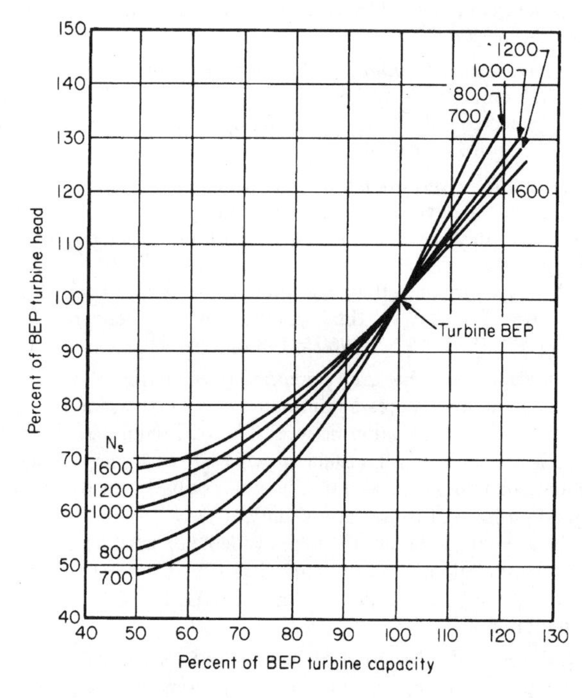

FIGURE 36 Constant-speed curves for turbine duty. (*Chemical Engineering.*)

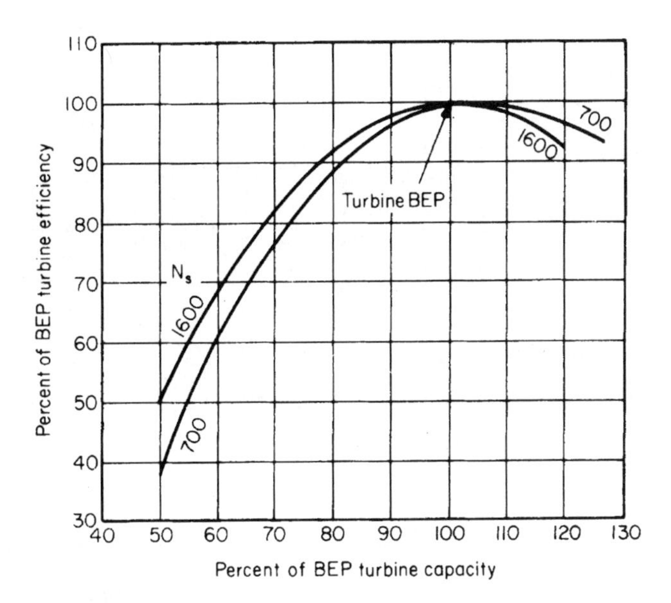

FIGURE 37 Constant-speed curves for turbine duty. (*Chemical Engineering.*)

At the 80 percent flow rate for the turbine, or 1200 gal/min (4542 L/min), the operating point is 80 percent of bep capacity. For a specific speed of 1090, as before, the percentages of bep head and efficiency are shown in Figs. 36 and 37: 79.5 percent of bep head and percent of bep efficiency. To find the actual performance, multiply by the bep values. Or, $H_t = 0.795(1290) = 1025$ ft (393.1 m); $E_t = 0.91(74.5) = 67.8$ percent.

The bhp at the new operating condition is then bhp = 1200(1025)(0.678)(0.52)/3960 = 110 hp (82.1 kW).

In a similar way, the constant-head curves in Figs. 38 and 39 predict turbine performance at different speeds. For example, speed is 80 percent of bep speed at 2880 r/min. For a specific speed of 1090, the percentages of bep capacity, efficiency, and power are 107 percent of the capacity, 94 percent of the efficiency, and 108 percent of the bhp. To get the actual performance, convert

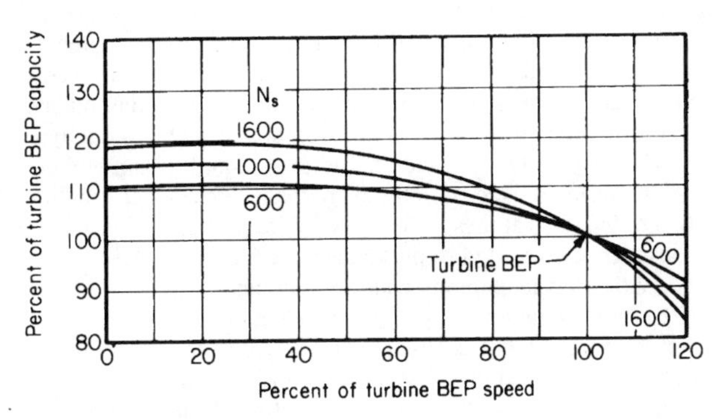

FIGURE 38 Constant-head curves for turbine duty. (*Chemical Engineering.*)

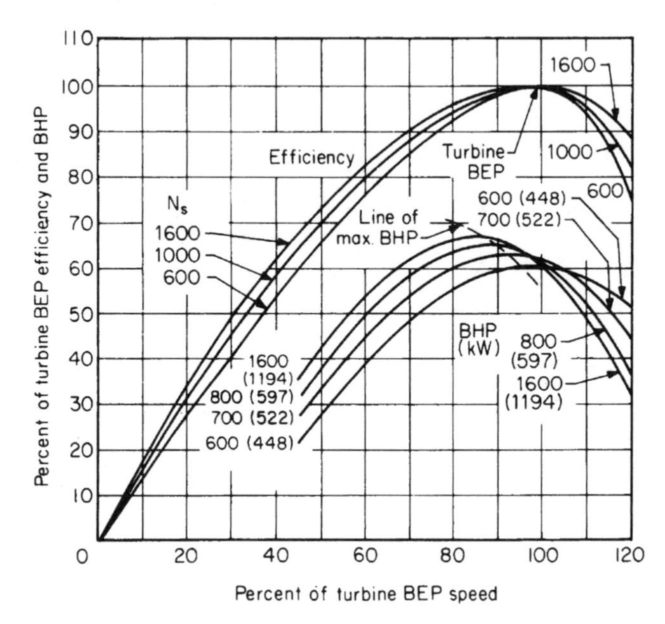

FIGURE 39 Constant-head curves for turbine only. (*Chemical Engineering*.)

as before: Q_t = 107(1500) = 1610 gal/min (6094 L/min); E_t = 0.94(74.5) = 70.0 percent; bhp = 1.08(189) = 206 hp (153.7 kW).

Note that the bhp in this last instance is higher than the bhp at the best efficiency point. Thus more horsepower can be obtained from a given unit by reducing the speed and increasing the flow rate. When the speed is fixed, more bhp cannot be obtained from the unit, but it may be possible to select a smaller pump for the same application.

Related Calculations. Use this general procedure for choosing a centrifugal pump to drive—as a hydraulic turbine—another pump, a fan, a generator, or a compressor, where high-pressure liquid is available as a source of power. Because pumps are designed as fluid movers, they may be less efficient as hydraulic turbines than equipment designed for that purpose. Steam turbines and electric motors are more economical when steam or electricity is available.

But using a pump as a turbine can pay off in remote locations where steam or electric power would require additional wiring or piping, in hazardous locations that require nonsparking equipment, where energy may be recovered from a stream that otherwise would be throttled, and when a radial-flow centrifugal pump is immediately available but a hydraulic turbine is not.

In the most common situation, there is a liquid stream with fixed head and flow rate and an application requiring a fixed rpm: these are the turbine design conditions. The objective is to pick a pump with a turbine bep at these conditions. With performance curves such as Fig. 40, turbine design conditions can be converted to pump design conditions. Then you select from a manufacturer's catalog a model that has its pump bep at those values.

The most common error in pump selection is using the turbine design conditions in choosing a pump from a catalog. Because catalog performance curves describe pump duty, not turbine duty, the result is an oversized unit that fails to work properly.

This procedure is the work of Fred Buse. Chief Engineer, Standard Pump Aldrich Division of Ingersoll-Rand Co., as reported in *Chemical Engineering* magazine.

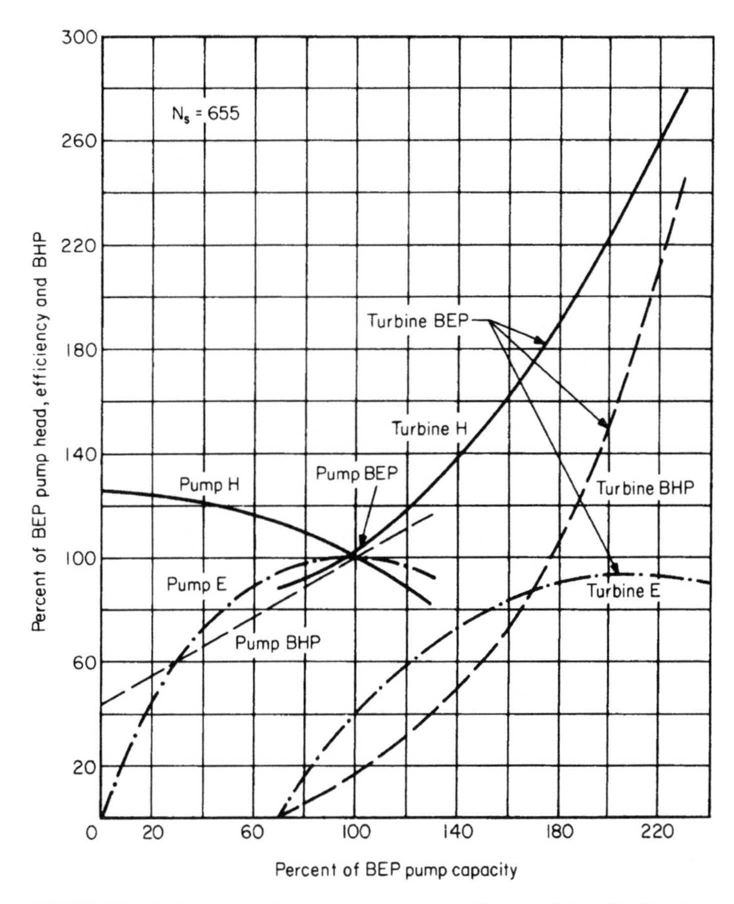

FIGURE 40 Performance of a pump at constant speed in pump duty and turbine duty. (*Chemical Engineering*.)

IMPELLER SIZING FOR CENTRIFUGAL PUMP SAFETY SERVICE

Determine the impeller size of a centrifugal pump that will provide a safe continuous-recirculation flow to prevent the pump from overheating at shutoff. The pump delivers 320 gal/min (20.2 L/s) at an operating head of 450 ft (137.2 m). The inlet water temperature is 220°F (104.4°C), and the system has an NPSH of 5 ft (1.5 m). Pump performance curves and the system-head characteristic curve for the discharge flow (without recirculation) are shown in Fig. 41, and the piping layout is shown in Fig. 42. The brake horsepower (bhp) of an 11-in (27.9-cm) and an 11.5-in (29.2-cm) impeller at shutoff is 53 and 60, respectively. Determine the permissible water temperature rise for this pump.

Calculation Procedure:

1. *Compute the actual temperature rise of the water in the pump*

Use the relation $P_o = P_\upsilon + P_{NPSH}$, where P_o = pressure corresponding to the actual liquid temperature in the pump during operation, lb/in^2 (abs) (kPa); P_υ = vapor pressure in the pump at the inlet

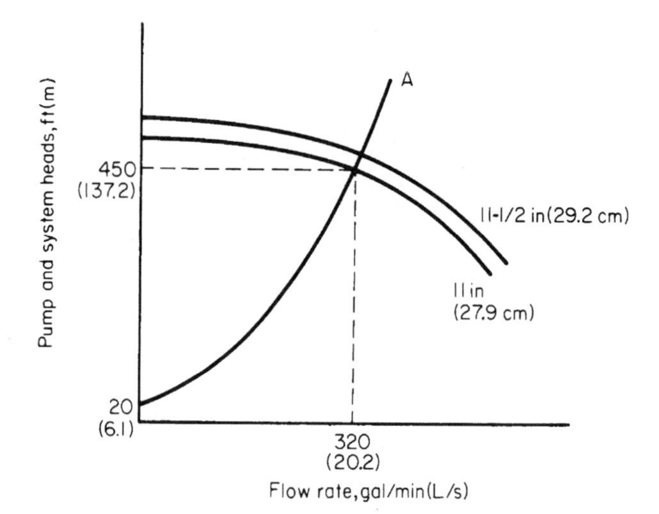

FIGURE 41 System-head curves without recirculation flow. (*Chemical Engineering.*)

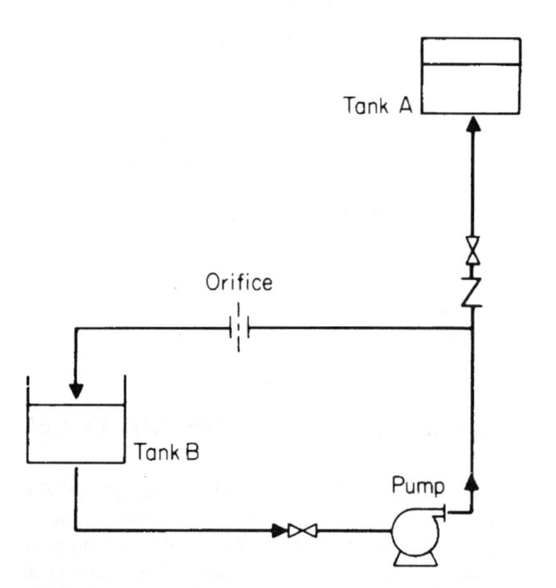

FIGURE 42 Pumping system with a continuous-recirculation line. (*Chemical Engineering.*)

water temperature, lb/in^2 (abs) (kPa); P_{NPSH} = pressure created by the net positive suction head on the pumps, lb/in^2 (abs) (kPa). The head in feet (meters) must be converted to lb/in^2 (abs) (kPa) by the relation lb/in^2 (abs) = (NPSH, ft) (liquid density at the pumping temperature, lb/ft^3)/(144 in^2/ft^2). Substituting yields P_o = 17.2 lb/in^2 (abs) + 5(59.6)/144 = 19.3 lb/in^2 (abs) (133.1 kPa).

Using the steam tables, find the saturation temperature T_s corresponding to this absolute pressure as T_s = 226.1°F (107.8°C). Then the permissible temperature rise is $T_p = T_s - T_{op}$, where T_{op} = water temperature in the pump inlet. Or, T_p = 226.1 − 220 = 6.1°F (3.4°C).

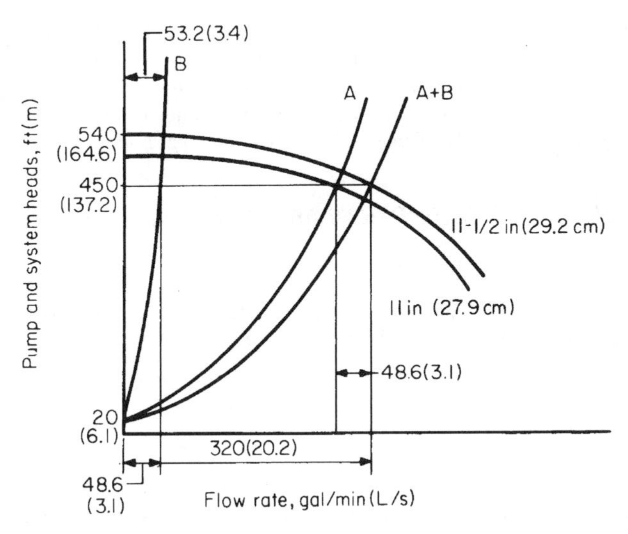

FIGURE 43 System-head curves with recirculation flow. (*Chemical Engineering.*)

2. *Compute the recirculation flow rate at the shut-off head*
From the pump characteristic curve with recirculation, Fig. 43, the continuous-recirculation flow Q_B for an 11.5-in (29.2-cm) impeller at an operating head of 450 ft (137.2 m) is 48.6 gal/min (177.1 L/min). Find the continuous-recirculation flow at shut-off head H_s ft (m) of 540 ft (164.6 m) from $Q_s = Q_B (H_s/H_{op})^{0.5}$, where H_{op} = operating head, ft (m). Or Q_s = 48.6(540/450) = 53.2 gal/min (201.4 L/min).

3. *Find the minimum safe flow for this pump*
The minimum safe flow, lb/h, is given by $w_{min} = 2545bhp/[C_pT_p + (1.285 \times 10^{-3})H_s]$, where C_p = specific head of the water; other symbols as before. Substituting, we find $w_{min} = 2545(60)/[1.0(6.1) + (1.285 \times 10^{-3})(540)]$ = 22,476 lb/h (2.83 kg/s). Converting to gal/min yields $Q_{min} = w_{min}/[(\text{ft}^3/\text{h})(\text{gal/min})(\text{lb/ft}^3)]$ for the water flowing through the pump. Or, Q_{min} = 22,476/[(8.021)(59.6)] = 47.1 gal/min (178.3 L/min).

4. *Compare the shut-off recirculation flow with the safe recirculation flow*
Since the shut-off recirculation flow Q_s = 53.2 gal/min (201.4 L/min) is greater than Q_{min} = 47.1 gal/min (178.3 L/min), the 11.5-in (29.2-cm) impeller is adequate to provide safe continuous recirculation. An 11.25-in (28.6-cm) impeller would not be adequate because Q_{min} = 45 gal/min (170.3 L/min) and Q_s = 25.6 gal/min (96.9 L/min).

Related Calculations. Safety-service pumps are those used for standby service in a variety of industrial plants serving the chemical, petroleum, plastics, aircraft, auto, marine, manufacturing, and similar businesses. Such pumps may be used for fire protection, boiler feed, condenser cooling, and related tasks. In such systems the pump is usually oversized and has a recirculation loop piped in to prevent overheating by maintaining a minimum safe flow. Figure 41 shows a schematic of such a system. Recirculation is controlled by a properly sized orifice rather than by valves because an orifice is less expensive and highly reliable.

The general procedure for sizing centrifugal pumps for safety service, using the symbols given earlier, is this: (1) Select a pump that will deliver the desired flow Q_A, using the head-capacity characteristic curves of the pump and system. (2) Choose the next larger diameter pump impeller to maintain a discharge flow of Q_A to tank *A*, Fig. 41, and a recirculation flow Q_B to tank *B*, Fig. 41. (3) Compute the recirculation flow Q_s at the pump shut-off point from $Q_s = Q_B (H_s/H_{op})^{0.5}$. (4) Calculate the minimum safe flow Q_{min} for the pump with the larger impeller diameter. (5) Compare the recirculation flow Q_s at the pump shut-off point with the minimum safe flow Q_{min}. If $Q_s \geq Q_{min}$, the selection process has been completed. If $Q_s < Q_{min}$, choose the next larger-size impeller and repeat steps 3, 4, and 5 above until the impeller size that will provide the minimum safe recirculation flow is determined.

This procedure is the work of Mileta Mikasinovic and Patrick C. Tung, design engineers, Ontario Hydro, as reported in *Chemical Engineering* magazine.

REDUCING ENERGY CONSUMPTION AND LOSS WITH PROPER PUMP CHOICE

Choose an energy-efficiency pump to handle 1000 gal/min (3800 L/min) of water at 60°F (15.6°C) at a total head of 150 ft (45.5 m). A readily commercially available pump is preferred for this application.

Calculation Procedure:

1. *Compute the pump horsepower required*

For any pump, $bhp_t = (gpm)(H_t)(s)/3960e$, where bhp_i = input brake (motor) horsepower to the pump; H_t = total head on the pump, ft; s = specific gravity of the liquid handled; e = hydraulic efficiency of the pump. For this application where $s = 1.0$ and a hydraulic efficiency of 70 percent can be safely assumed, $bhp_i = (1000)(150)(1)/(3960)(0.70) = 54.1$ bhp (40.3 kW).

2. *Choose the most energy-efficient pump*

Use Fig. 44, entering at the bottom at 1000 gal/min (3800 L/min) and projecting vertically upward to a total head of 150 ft (45.5 m). The resulting intersection is within area 1, showing from Table 17 that a single-stage 3500-r/min electric-motor-driven pump would be the most energy-efficient.

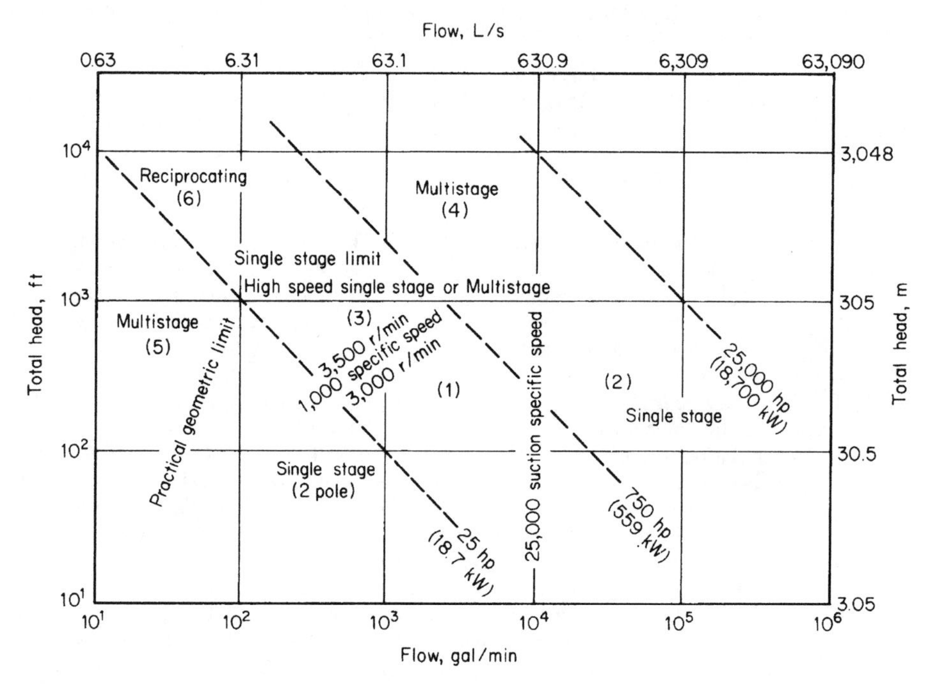

FIGURE 44 Selection guide is based mainly on specific speed, which indicates impeller geometry. (*Chemical Engineering*.)

TABLE 17 Type of Pump for Highest Energy Efficiency*

Area 1:	Single-stage, 3500 r/min
Area 2:	Single-stage, 1750 r/min or lower
Area 3:	Single-stage, above 3500 r/min, or multistage, 3500 r/min
Area 4:	Multistage
Area 5:	Multistage
Area 6:	Reciprocating

*Includes ANSI B73.1 standards, see area number in Fig. 38.

Related Calculations. The procedure given here can be used for pumps in a variety of applications—chemical, petroleum, commercial, industrial, marine, aeronautical, air-conditioning, cooling-water, etc., where the capacity varies from 10 to 1,000,000 gal/min (38 to 3,800,000 L/min) and the head varies from 10 to 10,000 ft (3 to 3300 m). Figure 44 is based primarily on the characteristic of pump specific speed $N_s = NQ^2/H^{3/4}$, where N = pump rotating speed, r/min; Q = capacity, gal/min (L/min); H = total head, ft (m).

When N_s is less than 1000, the operating efficiency of single-stage centrifugal pumps falls off dramatically; then either multistage or higher-speed pumps offer the best efficiency.

Area 1 of Fig. 44 is the densest, crowded both with pumps operating at 1750 and 3500 r/min, because years ago 3500-r/min pumps were not thought to be as durable as 1750-r/min ones. Since the adoption of the AVS standard in 1960 (superseded by ANSI B73.1), pumps with stiffer shafts have been proved reliable.

Also responsible for many 1750-r/min pumps in area 1 has been the impression that the higher (3500-r/min) speed causes pumps to wear out faster. However, because impeller tip speed is the same at both 3500 and 1750 r/min [as, for example, a 6-in (15-cm) impeller at 3500 r/min and a 12-in (30-cm) one at 1750 r/min], so is the fluid velocity, and so should be the erosion of metal surface. Another reason for not limiting operating speed is that improved impeller inlet design allows operation at 3500 r/min to capacities of 5000 gal/min (19,000 L/min) and higher.

Choice of operating speed also may be indirectly limited by specifications pertaining to suction performance, such as that fixing the top suction-specific speed S directly or indirectly by choice of the sigma constant or by reliance on Hydraulic Institute charts.

Values of S below 8000 to 10,000 have long been accepted for avoiding cavitation. However, since the development of the inducer, S values in the range of 20,000 to 25,000 have become commonplace, and values as high as 50,000 have become practical.

The sigma constant, which relates NPSH to total head, is little used today, and Hydraulic Institute charts (which have been revised) are conservative.

In light of today's designs and materials, past restrictions resulting from suction performance limitations should be reevaluated or eliminated entirely.

Even if the most efficient pump has been selected, there are a number of circumstances in which it may not operate at peak efficiency. Today's cost of energy has made these considerations more important.

A centrifugal pump, being a hydrodynamic machine, is designed for a single peak operating-point capacity and total head. Operation at other than this best efficiency point (bep) reduces efficiency. Specifications now should account for such factors as these:

1. A need for a larger number of smaller pumps. When a process operates over a wide range of capacities, as many do, pumps will often work at less than full capacity, hence at lower efficiency. This can be avoided by installing two or three pumps in parallel, in place of a single large one, so that one of the smaller pumps can handle the flow when operations are at a low rate.
2. Allowance for present capacity. Pump systems are frequently designed for full flow at some time in the future. Before this time arrives, the pumps will operate far from their best efficiency points. Even if this interim period lasts only 2 or 3 years, it may be more economical to install a smaller pump initially and to replace it later with a full-capacity one.

3. Inefficient impeller size. Some specifications call for pump impeller diameter to be no larger than 90 or 95 percent of the size that a pump could take, so as to provide reserve head. If this reserve is used only 5 percent of the time, all such pumps will be operating at less than full efficiency most of the time.
4. Advantages of allowing operation to the right of the best efficiency point. Some specifications, the result of such thinking as that which provides reserve head, prohibit the selection of pumps that would operate to the right of the best efficiency point. This eliminates half of the pumps that might be selected and results in oversized pumps operating at lower efficiency.

This procedure is the work of John H. Doolin, Director of Product Development, Worthington Pumps, Inc., as reported in *Chemical Engineering* magazine.

ENERGY AND ECONOMIC COMPETITIVE ANALYSIS OF POWER-PLANT CONDENSATE PUMPS

Evaluate the economic and application feasibility of replacing the vertical condensate pumps in a typical 1100-MW pressurized-water-reactor steam power plant having a feedwater train of two feedwater pumps, two heater drain pumps, and three vertical condensate pumps, with a water-jet pump in combination with a horizontal centrifugal pump. The flow rates, pressure heads, and related characteristics of the plant being considered are shown in Tables 18A and 18B.

Calculation Procedure:

1. ***Develop the performance parameters for the water-jet pump***

During the past two decades, turbine generator sizes increased from about 100 MW in the 1950s to 300 MW in the 1960s, and then up to about 750 MW in the early 1970s. At this writing (1997), generator sizes for both nuclear and fossil-fuel plants are even larger than the 750 MW cited here. This drastic increase in size, plus the introduction of low-pressure nuclear power cycles, brought about an increase in condensate flow to more than 10×10^6 lb/h (4.5×10^6 kg/h). Actually, in a typical 1300-MWe nuclear thermal cycle today, the condensate flow from condenser hotwell may be as high as 12×10^6 lb/h (5.5×10^6 kg/h). Current practice in condensate pumping system design is to either increase the pump capacity or increase the number of pumps operating in parallel to meet the flow requirements. However, these measures may increase:

- Initial fabrication and installation costs
- Probability of pump failures
- Routine maintenance and repair costs (and equally, if not more important, the attendant costs of plant down time).

The purpose of this calculation procedure is to examine the technical feasibility of using a water-jet pump in combination with a horizontal centrifugal pump to replace a vertical centrifugal pump. In comparison with a typical vertical pump installation, the horizontal pump installation appears to offer a relatively higher system availability factor which should stem directly from the greater accessibility offered by the typical horizontal pump installation over the typical vertical pump installation. Marked improvement in both preventive and corrective maintenance times, even where equivalent failures or failure rates are assumed for both types of pump installations, invoke serious economic factors which cannot be overlooked in consideration of the current and inordinately high costs of plant down time. Moreover, in combination with the water-jet pump, the horizontal configuration appears to offer a solution to the related NPSH problems. The combination also appears pertinent in the design of other systems, such as the heater drain systems in the feedwater cycle, where similar conditions may obtain.

TABLE 18A Feedwater System Pressure of a 1100-MW PWR Plant

Percent of max. design load	100	95.9	75	50	25	15
Percent of guarantee load	104.3	100	78.2	52.1	26.1	15.6
Turbine output—kW	1,210,081	1,160,596	907,560	605,040	302,519	181,512
Feedwater flow—lb/h	15,886,500	15,155,582	11,669,947	8,006,481	4,535,192	2,980,230
Feedwater flow—gal/min	36,996	35,181	26,653	17,937	9,889	6,409
Feed pump suction temp.—°F	403	398.8	377	348.1	305.7	279
Steam generator press.—lb/in^2 (abs)	990	990	1,036	1,073	1,091	1,093
Feed pump discharge press.—lb/in^2 (abs)	1,149	1,144	1,156	1,160	1,146	1,136
Feed pump suction press.—lb/in^2 (abs)	459	475	522	559	582	589
Feed pump TDH—lb/in^2	690	669	634	601	564	547
Feed pump TDH—ft	1,856	1,794	1,673	1,555	1,420	1,360
Condensate flow—lb/h	10,500,000	10,150,000	7,820,000	5,475,000	3,125,000	2,200,000
Condensate flow—gal/min	21,212	20,505	15,798	11,061	6,313	4,444
Condensate pump discharge press—lb/in^2	547.2	549.4	565.2	578.1	589.3	593.6
Condensate pump discharge press—ft	1,275	1,280	1,317	1,347	1,373	1,383
Condensate system loss—lb/in^2	88.2	74.4	43.2	19.1	7.3	4.6

Notes: 1. Based on three condensate pumps and two heater drain pumps operating through the full load range.
2. The conversion factors from English Units to SI Units are tabulated below:

English Unit	1 lb/h	1 gal/min	1°F	1 lb/in^2	1 ft
SI unit	1.26×10^{-4} kg/s	6.309×10^{-5} m^3/s	.5556 K	6.895×10^3 Pa	.3048 M

TABLE 18B SI Values for Feedwater System Pressure of an 1100-MW PWR Plant

Percent max. design load	100	95.9	75	50	25	15
Percent guarantee load	104.3	100	78.2	52.1	26.1	15.6
Turbine output, kW	1,210,081	1,160,596	907,560	605,040	302,519	181,512
Feedwater flow, kg/h	7,212,471	6,880,634	5,298,k56	3,634,932	2,058,977	1,353,024
Feedwater flow, L/s	2524	2220	1682	1132	624	404
Feedwater pump suction temp, °C	206	203.8	191.7	175.6	152.1	137.2
Steam generator press, kPa	6821	6821	7138	7393	7517	7531
Feedwater pump disch press, kPa	7917	7882	7965	7992	7896	7827
Feedwater pump suction press, kPa	3163	3273	3997	3852	4010	4058
Feed pump TDH, kPa	4754	4609	4368	4141	3886	3769
Feed pump TDH, m	566	547	510	474	433	415
Condensate flow, kg/h	4,767,000	4,608,100	3,550,280	2,485,650	1,418,750	998,800
Condensate flow, L/s	1338	1294	997	698	398	280
Condensate pump disch press, kPa	3770	3785	3894	3983	4060	4090
Condensate pump disch press, m	389	390	401	411	418	422
Condensate system loss, kPa	608	513	298	132	50	32

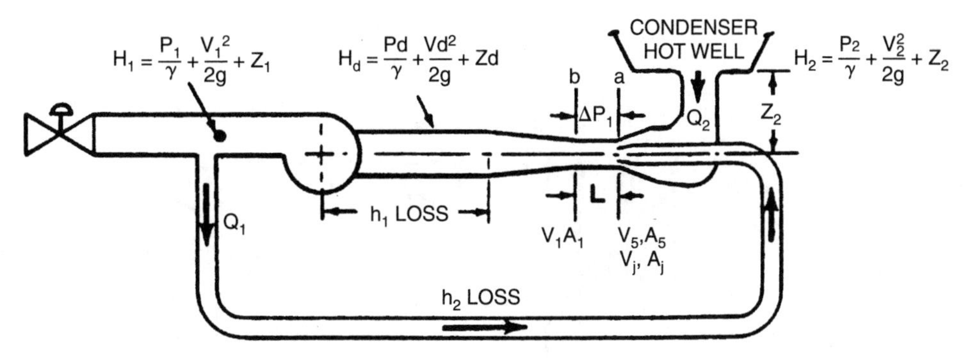

FIGURE 45 Water-jet centrifugal pump.

A water-jet pump, Fig. 45, consists of a centrifugal pump discharging through a nozzle located at the bottom of the condenser hotwell. The operating principle of the water-jet pump is based on the transfer of momentum from one stream of fluid to another.

Water-jet pumps were incorporated into the flow recirculation system of boiling water reactor design in 1965. The pumps were selected in lieu of conventional centrifugal pumps because of their basic simplicity and the economic incentives resulting from the possible reduction in the number of coolant loops and vessel nozzles by placing the water-jet pump inside the pressure vessel. The reduction in the number of coolant loops permits a smaller drywell so that both primary and secondary containment structures can be designed more compactly. Concurrently, the efficiency of the water-jet pump has been markedly improved through extensive development and testing programs pursued by the manufacturer. An efficiency of 41.5 percent has been obtained at a suction flow to driving flow ratio of 2.55 in the manufacturer's second-generation jet pumps (1).

Figure 45 shows the proposed arrangement of a water-jet pump and horizontal centrifugal pump combination to replace the conventional vertical condensate pump. The high momentum jet stream ejected from the recirculation nozzle is mixed with a low momentum stream from the condenser hotwell in the throat. This mixed flow slows down in the diffuser section where part of its momentum (kinetic energy) is converted into pressure. The flow is then led to the centrifugal pump suction through a short piping section. The pressure of the fluid is increased through the pump and a major part of this flow is then directed through the lower pressure heaters and finally to the suction of the feedwater pump, or to the deaerator, to provide the required feedwater flow to the stream generator. The remaining flow is led through the recirculation line back to the jet pump throat to induce the suction flow from the condenser hotwell, and thereby provide for continuous recirculation. A possible turbine cycle arrangement with a water-jet and horizontal centrifugal condensate pump is shown in Fig. 46.

The characteristics of a water-jet pump are defined by the following equations and nomenclature, using data from Fig. 45:

$$Q_1 = A_j V_j, \quad Q_2 = A_s V_s, \quad Q_1 + Q_2 = A_t V_t = Q_t \tag{1}$$

$$\frac{Q_2}{Q_1} = M \tag{2}$$

$$\frac{A_j}{A_t} = R \tag{3}$$

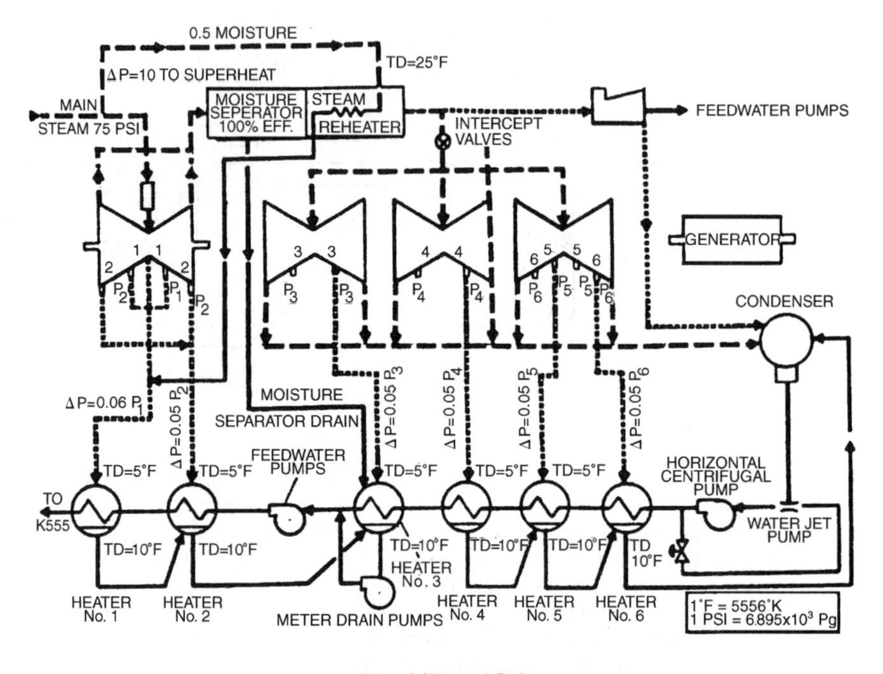

75 psi (516.6 kPa)
25°F (13.8°C)
5°F (2.8°C)
10°F (5.6°C)

FIGURE 46 Thermal cycle arrangement with water-jet centrifugal pump.

$$Q_1 = A_j \sqrt{\frac{2g(H_1 - \text{Pa}/\gamma)}{1 + K_j}} \tag{4}$$

$$Q_2 = A_s \sqrt{\frac{2g(H_2 - \text{Pa}/\gamma)}{1 + K_s}} \tag{5}$$

$$\frac{\text{Pa}}{\gamma} + \frac{V_t^2}{2g} = H_d + K_d \frac{V_t^2}{2g} \tag{6}$$

Following Gosline and O'Brien (3), the head ratio, N, depends upon six parameters; M, R, K_s, K_j, K_t, and K_d, or:

$$N = \frac{H_d - H_2}{H_1 - H_d} = f(M, R, K_s, K_j, K_t, K_d) \tag{7}$$

which vary with the design of the water-jet pump itself and with the length of the connecting pipes. Once the design of the water-jet pump is fixed, these parameters are known functions of flow. Based

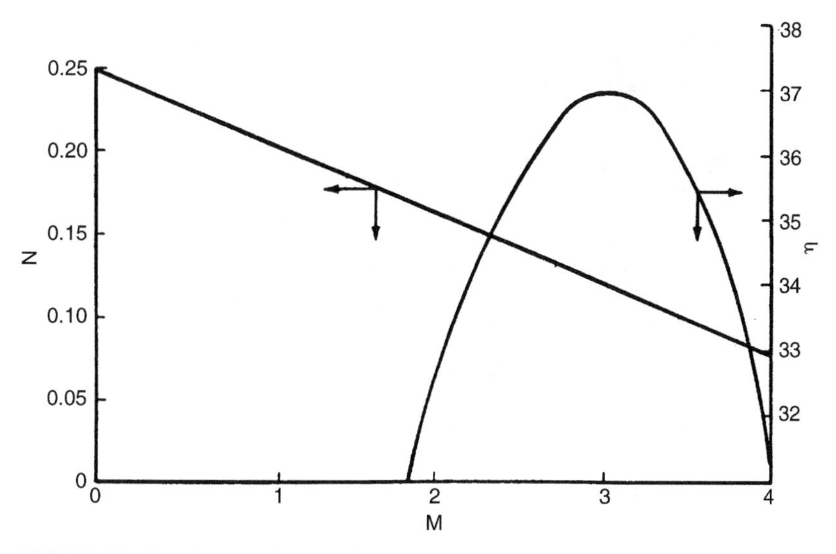

FIGURE 47 Water-jet pump characteristic curve.

upon their extensive testing, the manufacturer suggests the use of the *M-N* curve shown in Fig. 47 for the water-jet pump design evaluation. This *M-N* correlation is essentially a straight line and can be represented by:

$$N = N_0 - \frac{N_0}{M_0} M \tag{8}$$

where N_0 is the value of N at $M = 0$ and M_0 is the value of M when $N = 0$. The *M-N* curve of the water-jet pump shown in Fig. 45 may be represented by:

$$N = .246 - .04125M \tag{9}$$

Nomenclature

A = cross-section area, ft^2 (0.0929 m^2)
Bhp = pump brake horsepower
D = diameter, ft (0.3048 m)
e = centrifugal pump efficiency
f = friction factor
F = brake horsepower ratio
g = acceleration of gravity, 32.2 ft/s^2 (9.815 m/s^2)
H = hydraulic head, ft (0.3048 m)
hp = horsepower
ΔHp = pump total dynamic head, ft (0.3048 m)
K = friction parameter = fL/D
kW = kilowatt
L = length, ft (0.3048 m)
M = induced flow and driving flow ratio = Q_2/Q_1
M_0 = constant

(*Continued*)

Nomenclature (*Continued*)

n = pump speed, rpm
N = head ratio = $H_d - H_2/H_1 - H_d$
N_0 = constant
NHR = net heat rate, Btu/kWh (1054 J/kWh)
NPSH = net positive suction head
P = pressure, lb/ft^2 (47.88 Newton/m^2)
q_{in} = thermal energy input, kW
Q = flow rate, ft^3/s (0.02832 m^3/s)
R = ratio of area of nozzle to area of throat = A_j/A_t
V = velocity, ft/s (0.3048 m/s)
Z = elevation, ft (0.3048 m)
η = jet pump efficiency
γ = specific weight of liquid, lb · ft/ft^3 (16.02 kg/m^3)

Subscripts

a = entrance of throat
b = end of throat
c = centrifugal pump
d = jet pump discharge
j = tip of the nozzle
s = annular area surrounding tip of nozzle
t = throat of mixing chamber
υ = vertical condensate pump
j-c = water-jet–centrifugal pump combination

The water-jet pump efficiency, η, is defined as the ratio of the total energy increase of the suction flow to the total energy decrease of the driving flow, or:

$$\eta = M \cdot N \cdot 100 \tag{10}$$

This definition of efficiency is different from the centrifugal pump efficiency, e_c, which is defined as:

$$e_c = \frac{\text{pump output}}{\text{bhp}} = \frac{Q\gamma H}{550 \times \text{bhp}} \tag{11}$$

2. *Define the performance of the centrifugal pump associated with the water-jet pump* The performance of the horizontal centrifugal pump is defined in the manufacturer supplied pump characteristic curve and pump affinity laws:

$$\frac{Q}{Q_0} = \frac{n}{n_0} \quad \text{and} \quad \frac{H}{H_0} - \left(\frac{n}{n_0}\right)^2 \tag{12}$$

From Fig. 45, the pump discharge head, H_1, is:

$$H_1 = H_d - h_1 + \Delta Hp \tag{13}$$

where h_1 is the head loss from the water-jet pump exit to the centrifugal pump suction.

Hydraulic Horsepower of Water-Jet–Centrifugal Pump. From Fig. 45, the total new, Q_t, through the horizontal centrifugal pump is:

$$Q_t = Q_1 + Q_2 \tag{14}$$

The hydraulic horsepower of the pump is:

$$(\text{Hydraulic hp})_{j\text{-}c} = \frac{Q_t(H_1 - H_d)}{550} \tag{15}$$

From Eqs. (7) and (8):

$$H_d = \frac{\left(N_0 - \dfrac{N_0}{M_0}M\right)H_1 + H_2}{1 + N_0 - \dfrac{N_0}{M_0}M} \tag{16}$$

Substituting Eq. (16) into Eq. (15), we have:

$$(\text{Hydraulic hp})_{j\text{-}c} = \frac{Q_t\gamma}{550}\left(\frac{H_1 - H_2}{1 + N_0 - \dfrac{N_0}{M_0}M}\right) \tag{17}$$

For the conventional vertical condensate pump, the hydraulic horsepower is:

$$(\text{Hydraulic hp})_{\upsilon} = \frac{Q_2\gamma(H_1 - H_2)}{550} \tag{18}$$

From Eqs. (17) and (18):

$$\frac{(\text{Hydraulic hp})_{j\text{-}c}}{(\text{Hydraulic hp})_{\upsilon}} = \frac{1 + \dfrac{1}{M}}{1 + N_0 - \dfrac{N_0}{M_0}M} \tag{19}$$

The hydraulic hp ratio for different values of M is shown in Fig. 48. The manufacturer's suggested M-N curve, Fig. 47, is used in the calculation. From Eq. (11), the brake hp ratio is:

$$\frac{(\text{bhp})_{j\text{-}c}}{(\text{bhp})_{\upsilon}} = \frac{(\text{Hydraulic hp})_{j\text{-}c}}{(\text{Hydraulic hp})_{\upsilon}}\frac{e_c}{e_{j\text{-}c}} \tag{20}$$

Since the centrifugal pump efficiencies within a normal operating range do not change significantly, $e_{j\text{-}c} \approx e_c$, and:

$$\frac{(\text{bhp})_{j\text{-}c}}{(\text{bhp})_{\upsilon}} \approx \frac{1 + \dfrac{1}{M}}{1 + N_0 - \dfrac{N_0}{M_0}M} = F \tag{21}$$

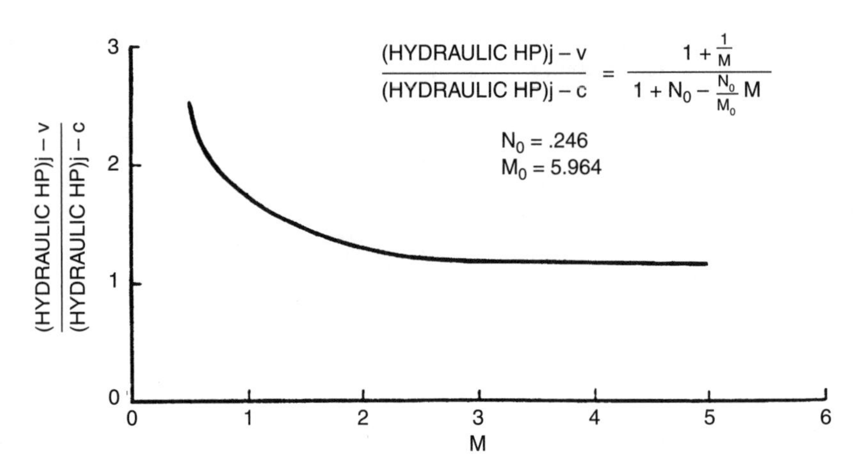

FIGURE 48 Hydraulic hp ratio vs M.

3. ***Compute the effect of the water-jet pump on the net heat rate***
The net heat rate is defined as:

$$\text{NHR} = \text{Net Heat Rate} = \frac{q_{in}}{kW_E - kW_{AUX} - kW_{CON}} \tag{22}$$

where: q_{in} = total thermal energy input
kW_E = generator output
kW_{AUX} = total plant auxiliary power excluding the power to condensate pumps
kW_{CON} = total power required to drive a motor-driven condensate pump

If we further define that $kW = kW_E - kW_{AUX}$,

$$(\text{Net Heat Rate})_\upsilon = \frac{q_{in}}{kW - kW_{CON}} \tag{23}$$

If a water-jet–centrifugal pump combination is used to replace the vertical condensate pump while keeping q_{in} and kW unchanged, the net heat rate can be shown as:

$$(\text{Net Heat Rate})_{j\text{-}c} = \frac{q_{in}}{kW - kW_{CON} \times F} \tag{24}$$

From the foregoing equations:

$$\frac{(\text{NHR})_{j\text{-}c}}{(\text{NHR})_\upsilon} = 1 + \frac{kW_{CON}(F-1)}{kW - kW_{CON} \times F} \tag{25}$$

The effect of M on net heat ratio is calculated according to Eq. (25) and this is shown in Fig. 49. The following data of a 1100-MW PWR plant have been used in the calculation

$$kW = 1{,}160{,}596$$

$$\text{Condensate flow} = 20{,}505 \text{ gal/min } (1.294 \text{ m}^3/\text{s})$$

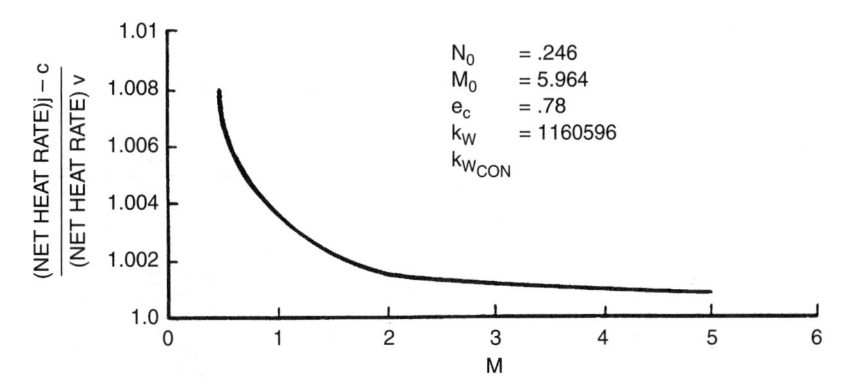

FIGURE 49 Net heat rate ratio vs *M*.

$$\text{Condensate pump TDH} = 1280 \text{ ft } (390.14 \text{ m})$$

$$\text{Pump efficiency} = 0.78$$

$$kW_{CON} = \frac{gpm \times H}{3960 \times e_c} \times 0.746$$
$$= 6339 \text{ kW}$$

As shown in Fig. 49, the increase in heat rate caused by the water-jet pump is approximately 0.1 percent at $M = 3$ and is less than 0.2 percent at $M = 2$. However, the heat rate ratio increases very rapidly with a further decrease in *M*.

4. *Develop the performance calculations for this installation*

The performance calculations of a water-jet–centrifugal pump combination in a power plant must be developed from an overall analysis of the feedwater-condensate system. To initiate the design analysis, the *M-N* relationship developed in Eq. (9) and shown in Fig. 47 is examined and, for obvious reasons, the peak efficiency point ($M = 3$) is selected for the design of the water-jet pump. At this design point, the efficiency of the water-jet pump is about 37 percent.

The head and flow characteristics of the horizontal centrifugal pump are shown in Fig. 50, and the following simplifying assumptions are made in the development of the performance calculations (refer also to Fig. 45).

1. $K_2 = 0$.
2. V_s is small such that $H_2 \approx Pa/\gamma = 5$ lb/in² (34.5 kPa) or 11.65 ft (3.6 m) of water.
3. h_1 loss is 10 lb/in² (68.9 kPa) (or 23.3 ft) (7.1 m) and is a constant under all loading conditions.
4. h_2 loss is 25 lb/in² (172.2 kPa) (or 58.25 ft) (17.8 m) and is a constant under all loading conditions.

From Eqs. (7), (9), and (13), we have:

$$H_1 = H_2 + (1.246 - .04125M) \times (\Delta HP - h_1) \tag{26}$$

To match the vertical condensate pump at maximum design condition, the total dynamic head of the horizontal pump (ΔHp) shall be such that $H_1 = 1275$ ft (388.6 m). With $h_1 = 23.3$ ft (7.1 m), $H_2 = 11.65$ ft (3.6 m), and $M = 3$, the required ΔHp from Eq. (26) is:

$$\Delta Hp = (H_1 - H_2) \div (1.246 - .04125M) + h_1 = 1149 \text{ ft } (350.2 \text{ m})$$

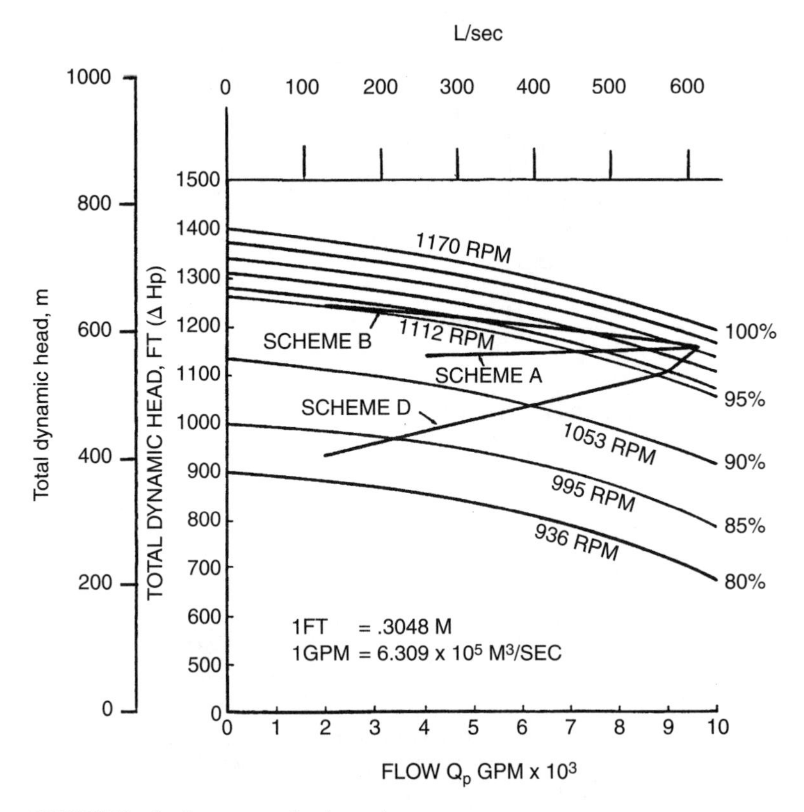

FIGURE 50 Condensate pump head-capacity curve.

and H_d from Eq. (13) is:

$$H_d = H_1 + h_1 - \Delta Hp = 149.3 \text{ ft } (45.5 \text{ m})$$

From Eqs. (2) and (14):

$$Q_t = Q_2\left(1 + \frac{1}{M}\right) \tag{27}$$

with $M = 3$, the required condensate flow of the horizontal centrifugal pump Q_t is:

$$Q_t = 10.5 \times 10^6 \times \left(1 + \frac{1}{3}\right)$$
$$= 14 \times 10^6 \text{ lb} \cdot \text{s/h or } 28.283 \text{ gal/min } (1.784 \text{ m}^3/\text{s})$$

From Fig. 50, the condensate pump characteristic curve, which is generated by pump affinity law, the horizontal centrifugal pump will be running at 1147 r/min.

The jet nozzle area, A_j, can be calculated from:

$$Q_1 = CA_j\sqrt{2g(H_1 - h_2 - H_2)} \tag{28}$$

which is another form of Eq. (4),

Assume that nozzle flow coefficient $C = 0.9$, we have:

$$A_j = .020948 \text{ ft}^2 \ (.00195 \text{ m}^2)$$

or:

$$d_j = 1.96 \text{ in } (.0498 \text{ m})$$

Substitute the value of A_j and C into Eq. (28), we have:

$$Q_1 = .1513\sqrt{H_1 - 69.9} \tag{29}$$

Accordingly, Eqs. (2), (26), (27), and (29) define the water-jet–centrifugal pump performance.

5. *Determine the best drive for the water-jet–centrifugal-condensate pump*

Pump DRIVING Schemes. Four possible schemes of driving the water-jet–centrifugal-condensate pump are examined. For each scheme, the water-jet–centrifugal pump is designed to duplicate the head and capacity performance of the corresponding vertical condensate pump at the maximum design condition and then the schemes are examined for continuous operation at other loading conditions; namely, 100, 75, 50, 25, and 15 percent. Schematic arrangements of each scheme and sample calculations at 75 percent load are shown in Table 19.

Scheme A: Variable Speed Motor Drive and Variable M Ratio. In this scheme, variable speed electric motor is used to drive the water-jet–centrifugal pump so that the condensate flow and the pressure head at the feed pump suction are identical to that of the base case which uses conventional vertical condensate pumps. To satisfy Eq. (4), the water-jet pump flow ratio M changes from 3 at maximum design condition to 0.6021 at 15 percent load as shown in Table 20.

At maximum design condition, the horizontal centrifugal pump is running at 1147 r/min and 9428 gal/min (594.9 L/s). For the partial load operation, the pump will follow the curve labeled Scheme A in Fig. 50. From Eqs. (21) and (25), the increase in net heat rate will be higher at the 25 and 15 percent partial loading operation. If long-term partial load operation is expected, this scheme should be avoided.

Scheme B: Variable Speed Motor Drive and Constant M Ratio. In this scheme, the pump drive is identical to Scheme A except that a flow regulating control valve is installed in the water-jet pump recirculation line to maintain a constant M ratio at all loading conditions. In this case, the speed of the horizontal centrifugal pump will vary according to the curve labeled Scheme B in Fig. 50. The feedwater pump operation is identical to that of Scheme A and the base case.

Scheme C: Constant Speed Motor Drive and Constant M Ratio. In this scheme, the horizontal centrifugal pump is running at a constant speed of 1147 r/min. A control valve is used to keep the flow ratio $M = 3$. In this case, the pressure head at the water-jet–centrifugal pump discharge is higher than that of the vertical condensate pump. Consequently, the feedwater pump will be running at a lower speed and lower total dynamic head to keep the steam generator pressure identical to that of the base case. The required feed pump total dynamic head and corresponding speed are shown as the curve labeled Scheme C in Fig. 51.

TABLE 19 Schematic Arrangements of Each Scheme

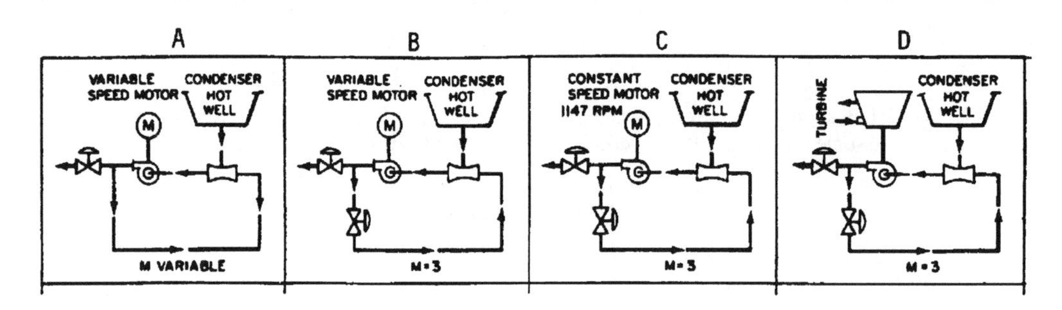

A	B	C	D
From Table 18 @ 75% load Flow = 7.82×10^6 labs/h $Q_2 = 5266$ gal/min $H_1 = 565.2$ lb/in^2 = 1317 ft From Eq. 29 $Q_1 = .1513\sqrt{1317 - 69.9}$ = 5.343 ft^3/s = 2398 gal/min $M = \frac{Q_2}{Q_1} = \frac{5266}{2398} = 2.196$ From Eq.26 $\Delta Hp = \frac{1317 - 11.65}{1.246 - .04125 \times 2.196}$ $+ 23.3 = 1153$ ft $Q_1 = Q_1 + Q_2 = 7664$ gal/min From Eq. 50 Pump speed = 1123 r/min From Eq. 13 $H_d = 1317 + 23.3 - 1153$ = 187.3 ft	From Table 18 @ 75% load $Q_2 = 5266$ gal/min $H_1 = 565.2$ lb/in^2 = 1317 ft $Q_1 = Q_2\left(1 + \frac{1}{M}\right)$ = 7021 gal/min From Eq. 26 $\Delta H = 1186.5$ ft From Eq. 6 Pump speed = 1131 RPM From Eq. 13 $H_d = 1317 + 23.3 - 1186.5$ = 153.8 ft	From Table 18 @ 75% load $Q_2 = 5266$ gal/min $Q_1 = 7021$ gal/min From Fig. 50 @ 1147 RPM $\Delta H_P = 1229$ ft From Eq. 26 $H_1 = 11.65 + 1.12225$ $(1229 - 23.3) = 1364.8$ ft = 858.7 lb/in^2 From Table 18 condensate System head loss = 43.2 lb/in^2 Feed pump suction pressure = 542.3 lb/in^2 Feed pump ΔHp = 1156 − 542.3 = 613.7 lb/in^2 = 1619 ft From Eq. 13 $H_d = 1364.8 + 23.3 - 1229$ = 159.1 ft	Feed pump speed = 4830 RPM Condensate pump speed = 1147 RPM Gear reduction ratio = $\frac{4.2}{1}$ $Q_1 = 7021$ gal/min Iterative procedure is used to find pump running speed Try feed pump speed = 4500 RPM Condensate pump RPM $= \frac{4500}{4.2} = 1071$ rptn From Fig. 50 $\Delta Hp = 1061$ ft From Eq. 26 $H_1 = 1176.2$ ft = 504.8 lb/in^2 Feed pump suction pressure = 505.8 − 43.3 = 461.6 lb/in^2 Feed pump $\Delta H_P = 1156 - 461.6$ = 694.4 lb/in^2 = 1832.0 ft From Fig. 51 Feed pump speed = 4990 rpm Very close to assumed 4500 r/min $Hd = 1176.2 + 23.3$ $- 1061 = 138.5$ ft

(*Continued*)

TABLE 19 Schematic Arrangements of Each Scheme (*Continued*)

SI Values					
A		A		A	
gal/min	L/s	lb/in²	kPa	ft	m
5266	332.3	565.2	3894	1317	401.4
2398	151.3			187.3	57.1
7664	483.6				
B		B		B	
5266	332.3	565.2	3894	1317	401.4
7021	443.0			1186.5	361.6
				153.8	46.9
C		C		C	
5266	332.3	585.7	4035.5	1229	374.6
7021	443.0	43.2	297.6	1364.8	415.6
		542.3	3736.4	1619	493.5
		613.7	4228.4	159.1	48.5
D		D		D	
7021	443.0	504.8	3478.1	1061	323.4
		461.6	3180.4	1832	558.4
		694.4	4784.4	138.5	42.2

TABLE 20 *M* Ratio vs. Load Variation

Load condition	VWO	100%	75%	50%	25%	15%
M Ratio	3	2.894	2.2	1.52	.86	.6021

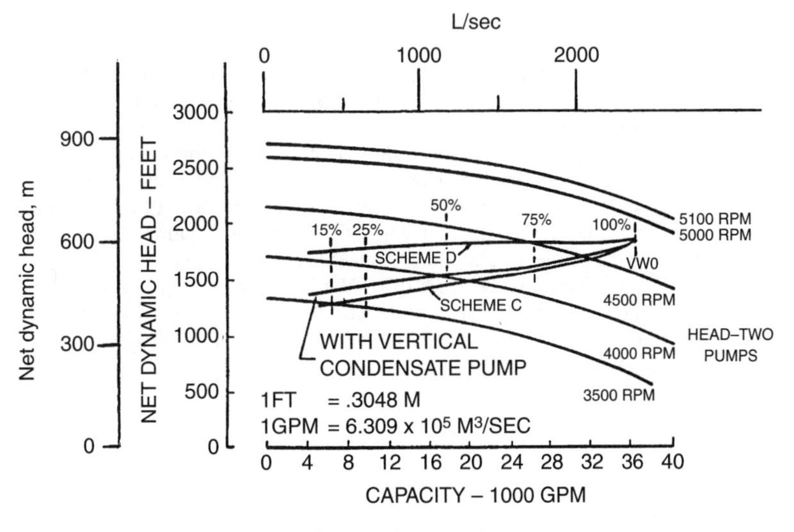

FIGURE 51 Feed pump head-capacity curve (two pumps).

Scheme D: Turbine Drive Jet-Centrifugal Pump and Constant M Ratio. In this scheme, the feedwater pump and water-jet-centrifugal pump are running at a constant speed ratio and both are driven by the auxiliary turbine. A control valve is used to keep the flow ratio $M = 3$. Under these conditions, the water-jet–centrifugal pump and the feedwater pump will follow the curves labeled Scheme D in Figs. 50 and 51, respectively, to produce the identical steam generator conditions in Table 18. It should be noted that the auxiliary turbine driven feedwater pump has been shown to have a better cycle efficiency than a motor driven pump in the same application for large power plants (4); intuitively, the auxiliary turbine driven jet-centrifugal pump arrangement may also provide certain gains in cycle efficiency over other water-jet–centrifugal pump drive schemes.

6. *Summarize the findings for this pump application*

It has been shown that a water-jet–centrifugal pump can be used to replace the conventional vertical condensate pump in a steam power plant feedwater system. All four schemes discussed in the preceding section are feasible means of driving the water-jet–centrifugal pump combination. While the resulting auxiliary power requirements for the jet-centrifugal pump system will be slightly higher, the increase will be insignificant if the flow rate M is kept greater than 2.

The proposed change from conventional vertical pump to a water-jet–centrifugal pump may have advantages:

1. Increased feedwater system reliability and reduced plant downtime
2. Easier maintenance operations and reduced cost of maintenance
3. More flexibility in plant layout which, in turn, may favorably effect condensate system piping costs.

With the present high cost of plant outage, the improvement in system reliability alone may provide sufficient economic incentive for considering the water-jet–centrifugal pump combination.

Related Calculations. While the study here was directed at a PWR steam power plant, the approach used is valid for any steam power plant—utility, industrial, commercial, or marine—using the types of pumps considered. The water-jet pump, developed in the mid-1800s, has many inherent advantages which can be used in today's highly competitive power-generation industry. In every such installation, the condensate pump in the feedwater system of the steam electric generating power plant takes suction from the condenser hotwell and delivers the condensate through the tubeside of the lower pressure feedwater heaters to the deaerator, or to the suction of the feed pump. The continuous operation of the entire plant depends upon the proper functioning of the condensate pumps. It should also be noted that the condensate pumping system consumes a significant portion of the auxiliary power, and represents a measurable portion of the plant first cost.

In power plant applications, multiple parallel pumping arrangements are employed to provide a flexible operational system. Condensate pumps are of constant speed motor-driven, vertical centrifugal type, and are located in a pit near the condenser. The difference in fluid elevations between the condenser hotwell and the first stage of the centrifugal pump is the only NPSH available to the pump because the condensate in the hotwell is always saturated.

This procedure is the work of E. N. Chu, Engineering Specialist, and F. S. Ku, Assistant Chief Mechanical Engineer, Bechtel Power Corporation, as reported in *Combustion* magazine and presented at the IEEE-ASME Joint Power Generation Conference. SI values were added by the handbook editor.

PART 2

FLUID FLOW IN PIPING SYSTEMS

PRESSURE SURGE IN A PIPING SYSTEM FROM RAPID VALVE CLOSURE

Oil, with a specific weight of 52 lb/ft^3 (832 kg/m^3) and a bulk modulus of 250,000 lb/in^2 (1723 MPa), flows at the rate of 40 gal/min (2.5 L/s) through stainless-steel pipe. The pipe is 40 ft (12.2 m) long, 1.5 in (38.1 mm) O.D., 1.402 in (35.6 mm) I.D., 0.049 in (1.24 mm) wall thickness, and has a modulus of elasticity, E, of 29×10^6 lb/in^2 (199.8 kPa $\times 10^6$). Normal static pressure immediately upstream of the valve in the pipe is 500 lb/in^2 (abs) (3445 kPa). When the flow of the oil is reduced to zero in 0.015 s by closing a valve at the end of the pipe, what is: (*a*) the velocity of the pressure wave; (*b*) the period of the pressure wave; (*c*) the amplitude of the pressure wave; and (*d*) the maximum static pressure at the valve?

Calculation Procedure:

1. *Find the velocity of the pressure wave when the valve is closed*

(*a*) Use the equation

$$a = \frac{68.094}{\sqrt{\gamma[(1/K) + (D/Et)]}}$$

Where the symbols are as given in the notation below. Substituting,

$$a = \frac{68.094}{\sqrt{52[(1/25 \times 10^4) + (1.402/29 \times 10^6 \times 0.049)]}}$$

$$= 4228 \text{ ft/s } (1288.7 \text{ m/s})$$

An alternative solution uses Fig. 1. With a D/t ratio = 1.402/0.049 = 28.6 for stainless-steel pipe, the velocity, a, of the pressure wave is 4228 ft/s (1288.7 m/s).

2. *Compute the time for the pressure wave to make one round trip in the pipe*

(*b*) The time for the pressure wave to make one round trip between the pipe extremities, or one interval, is: $2L/a = 2(40)/4228 = 0.0189$ s, and the period of the pressure wave is: $2(2L/a) = 2(0.0189) = 0.0378$ s.

3. *Calculate the pressure surge for rapid valve closure*

(*c*) Since the time of 0.015 s for valve closure is less than the internal time $2L/a$ equal to 0.0189 s, the pressure surge can be computed from:

$$\Delta p = \gamma a V/144g$$

for rapid valve closure.

The velocity of flow, $V = [(40)(231)(4)]/[(60)(\pi)(1.402^2)(12)]$ using the standard pipe flow relation, or $V = 8.3$ ft/s (2.53 m/s).

Then, the amplitude of the pressure wave, using the equation above is:

$$\Delta p = \frac{52 \times 4228 \times 8.3}{144 \times 32.2} = 393.5 \text{ lb/in}^2 \qquad (2711.2 \text{ kPa}).$$

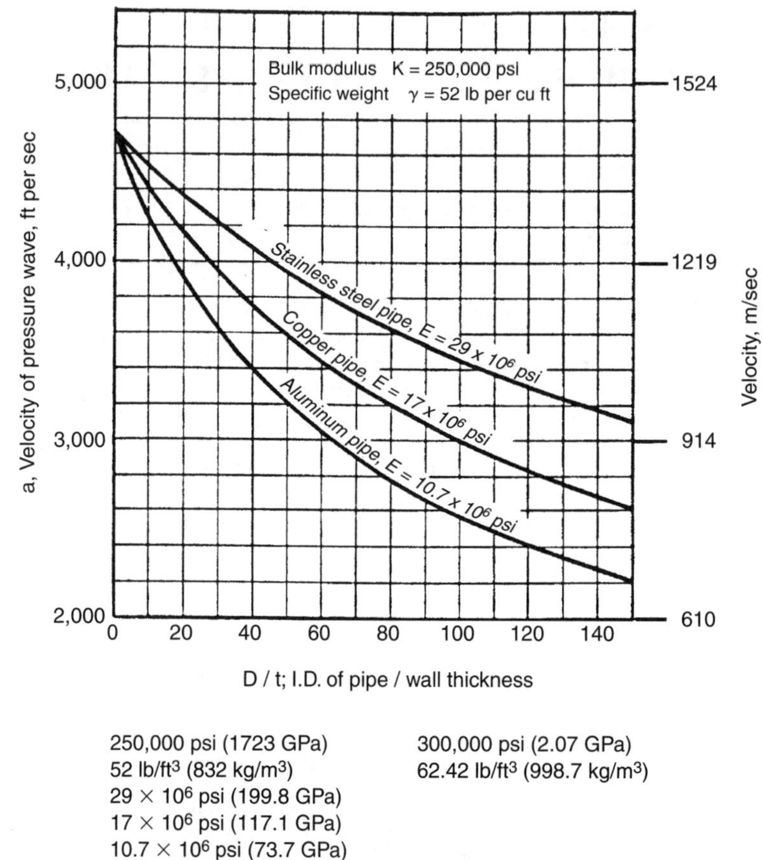

FIGURE 1 Velocity of pressure wave in oil column in pipe of different diameter-to-wall thickness ratios. (*Product Engineering*.)

4. *Determine the resulting maximum static press in the pipe*

(*d*) The resulting maximum static pressure in the line, $p_{max} = p + \Delta p = 500 + 393.5 = 893.5$ lb/in^2 (abs) (6156.2 kPa).

Related Calculations. In an industrial hydraulic system, such as that used in machine tools, hydraulic lifts, steering mechanisms, etc., when the velocity of a flowing fluid is changed by opening or closing a valve, pressure surges result. The amplitude of the pressure surge is a function of the rate of change in the velocity of the mass of fluid. This procedure shows how to compute the amplitude of the pressure surge with rapid valve closure.

The procedure is the work of Nils M. Sverdrup, Hydraulic Engineer, Aerojet-General Corporation, as reported in *Product Engineering* magazine. SI values were added by the handbook editor.

Notation

a = velocity of pressure wave, ft/s (m/s)
a_E = effective velocity of pressure wave, ft/s (m/s)
A = cross-sectional area of pipe, in^2 (mm^2)
A_o = area of throttling orifice before closure, in^2 (mm^2)
c = velocity of sound, ft/s (m/s)

(*Continued*)

Notation (*Continued*)

C_D = coefficient of discharge
D = inside diameter of pipe, in (mm)
E = modulus of elasticity of pipe material, lb/in^2 (kPa)
F = force, lb (kg)
g = gravitational acceleration, 32.2 ft/s^2
K = bulk modulus of fluid medium, lb/in^2 (kPa)
L = length of pipe, ft (m)
m = mass, slugs
$N = T/(2L/a)$ = number of pressure wave intervals during time of valve closure
p = normal static fluid pressure immediately upstream of valve when the fluid velocity is V, lb/in^2 (absolute) (kPa)
Δp = amplitude of pressure wave, lb/in^2 (kPa)
p_{max} = maximum static pressure immediately upstream of valve, lb/in^2 (absolute) (kPa)
p_d = static pressure immediately downstream of the valve, lb/in^2 (absolute) (kPa)
Q = volume rate of flow, ft^3/s (m^3/s)
t = wall thickness of pipe, in (mm)
T = time in which valve is closed, s
υ = fluid volume, in^3 (mm^3)
υ_A = air volume, in^3 (mm^3)
V = normal velocity of fluid flow in pipe with valve wide open, ft/s (m/s)
V_E = equivalent fluid velocity, ft/s (m/s)
V_n = velocity of fluid flow during interval n, ft/s (m/s)
W = work, ft · lb (W)
γ = specific weight, lb/ft^3 (kg/m^3)
ϕ_n = coefficient dependent upon the rate of change in orifice area and discharge coefficient
τ = period of oscillation of air cushion in a sealed chamber, s

PIPING PRESSURE SURGE WITH DIFFERENT MATERIAL AND FLUID

(*a*) What would be the pressure rise in the previous procedure if the pipe were aluminum instead of stainless steel? (*b*) What would be the pressure rise in the system in the previous procedure if the flow medium were water having a bulk modulus, K, of 300,000 lb/in^2 (2067 MPa) and a specific weight of 62.42 lb/ft^3 (998.7 kg/m^3)?

Calculation Procedure:

1. ***Find the velocity of the pressure wave in the pipe***
(*a*) From Fig. 2, for aluminum pipe having a D/t ratio of 28.6, the velocity of the pressure wave is 3655 ft/s (1114.0 m/s). Alternatively, the velocity could be computed as in step 1 in the previous procedure.

2. ***Compute the time for one interval of the pressure wave***
As before, in the previous procedure, $2L/a = 2(40/3655) = 0.02188$ s.

3. ***Calculate the pressure rise in the pipe***
Since the time of 0.015 s for the valve closure is less than the interval time of $2L/a$ equal to 0.02188, the pressure rise can be computed from

$$\Delta p = \gamma a V / 144 g$$

or,

$$\Delta p = \frac{52 \times 3655 \times 8.3}{144 \times 32.2} = 340.2 \text{ lb/in}^2 \qquad (2343.98 \text{ kPa})$$

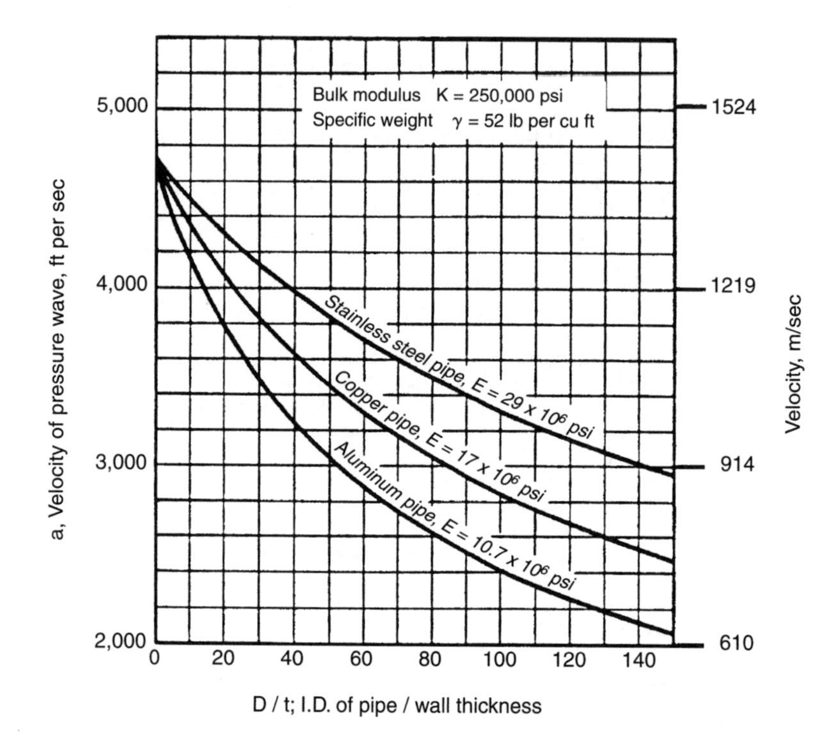

FIGURE 2 Velocity of pressure wave in water column in pipe of different diameter-to-wall thickness ratios. (*Product Engineering*.)

4. *Find the maximum static pressure in the line*
Using the pressure-rise relation, $p_{max} = 500 + 340.2 = 840.2$ lb/in^2 (abs) (5788.97 kPa).

5. *Determine the pressure rise for the different fluid*
(*b*) For water, use Fig. 2 for stainless-steel pipe having a D/t ratio of 28.6 to find $a = 4147$ ft/s (1264 m/s). Alternatively, the velocity could be calculated as in step 1 of the previous procedure.

6. *Compute the time for one internal of the pressure wave*
Using $2L/a = 2(40)/4147 = 0.012929$ s.

7. *Find the pressure rise and maximum static pressure in the line*
Since the time of 0.015 s for valve closure is less than the interval time $2L/a$ equal to 0.01929 s, the pressure rise can be computed from

$$\Delta p = \gamma a V/144g$$

for rapid valve closure. Therefore, the pressure rise when the flow medium is water is

$$\Delta p = \frac{62.42 \times 4147 \times 8.3}{144 \times 32.2} = 463.4 \text{ lb/in}^2 \qquad (3192.8 \text{ kPa})$$

The maximum static pressure, $p_{max} = 500 + 463.4 = 963.4$ lb/in^2 (abs) (6637.8 kPa).

Related Calculations. This procedure is the work of Nils M. Sverdrup, as detailed in the previous procedure.

PRESSURE SURGE IN PIPING SYSTEM WITH COMPOUND PIPELINE

A compound pipeline consisting of several stainless-steel pipes of different diameters, Fig. 3. conveys 40 gal/min (2.5 L/s) of water. The length of each section of pipe is: L_1 = 25 ft (7.6 m); L_2 = 15 ft (4.6 m); L_3 = 10 ft (3.0 m); pipe wall thickness in each section is 0.049 in (1.24 mm); inside diameter of each section of pipe is D_1 = 1.402 in (35.6 mm); D_2 = 1.152 in (29.3 mm); D_3 = 0.902 in (22.9 mm). What is the equivalent fluid velocity and the effective velocity of the pressure wave on sudden valve closure?

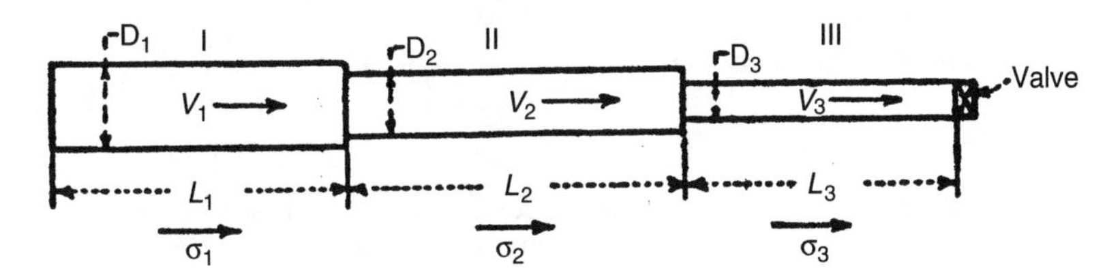

FIGURE 3 Compound pipeline consists of pipe sections having different diameters. (*Product Engineering*.)

Calculation Procedure:

1. *Determine fluid velocity and pressure-wave velocity in the first pipe*

D_1/t_1 ratio of the first pipe = 1.402/0.049 = 28.6. Then, the fluid velocity in the pipe can be found from $V_1 = 0.4085(G_n/(D_n)^2$, where the symbols are as shown below. Substituting, $V_1 = 0.4085(40)/(1.402)^2$ = 8.31 ft/s (2.53 m/s).

Using these two computed values, enter Fig. 2 to find the velocity of the pressure wave in pipe 1 as 4147 ft/s (1264 m/s).

2. *Find the fluid velocity and pressure-wave velocity in the second pipe*

The D_2/t_2 ratio for the second pipe = 1.152/0.049 = 23.51. Using the same velocity equation as in step 1, above $V_2 = 0.4085(40)/(1.152)^2$ = 12.31 ft/s (3.75 m/s).

Again, from Fig. 2, a_2 = 4234 ft/s (1290.5 m/s). Thus, there is an 87-ft/s (26.5-m/s) velocity increase of the pressure wave between pipes 1 and 2.

3. *Compute the fluid velocity and pressure-wave velocity in the third pipe*

Using a similar procedure to that in steps 1 and 2 above, V_3 = 20.1 ft/s (6.13 m/s); s_3 = 4326 ft/s (1318.6 m/s).

4. *Find the equivalent fluid velocity and effective pressure-wave velocity for the compound pipe*

Use the equation

$$V_E = \frac{L_1V_1 + L_2V_2 + \cdots + L_nV_n}{L_1 + L_2 + \cdots + L_n}$$

to find the equivalent fluid velocity in the compound pipe. Substituting,

$$V_E = \frac{25 \times 8.3 + 15 \times 12.3 + 10 \times 20.1}{25 + 15 + 10}$$

$$= 11.9 \text{ ft/s } (3.63 \text{ m/s})$$

To find the effective velocity of the pressure wave, use the equation

$$a_g = \frac{L_1 + L_2 + \cdots L_n}{(L_1/a_1) + (L_2/a_2) + \cdots + (L_n/a_n)}$$

Substituting,

$$a_g = \frac{25 + 15 + 10}{(25/4147) + (15/4234) + (10/4326)}$$

$$= 4209 \text{ ft/s } (1282.9 \text{ m/s})$$

Thus, equivalent fluid velocity and effective velocity of the pressure wave in the compound pipe are both less than either velocity in the individual sections of the pipe.

Related Calculations. Compound pipes find frequent application in industrial hydraulic systems. The procedure given here is useful in determining the velocities produced by sudden closure of a valve in the line.

$L_1, L_2, \ldots, L_n$ = length of each section of pipe of constant diameter, ft (m)
$a_1, a_2, \ldots, a_n$ = velocity of pressure wave in the respective pipe sections, ft/s (m/s)
a_g = velocity of the pressure wave, ft/s
$V_1, V_2, \ldots, V_n$ = velocity of fluid in the respective pipe sections, ft/s (m/s)
V_E = equivalent fluid velocity, ft/s (m/s)
G_n = rate of flow in respective section, U.S. gal/min (L/s)
D_n = inside diameter of respective pipe, in (mm)

The fluid velocity in an individual pipe is

$$V_n = 0.4085 G_n / D_n^2$$

This procedure is the work of Nils M. Sverdrup, as detailed earlier.

QUICK CALCULATION OF FLOW RATE AND PRESSURE DROP IN PIPING SYSTEMS

A 3-in (76-mm) Schedule 40S pipe has a 300-gal/min (18.9-L/s) water flow rate with a pressure loss of 8 lb/in^2 (55.1 kPa)/100 ft (30.5 m). What would be the flow rate in a 4-in (102-mm) Schedule 40S pipe with the same pressure loss? What would be the pressure loss in a 4-in (102-mm) Schedule 40S pipe with the same flow rate, 300 gal/min (18.9 L/s)? Determine the flow rate and pressure loss for a 6-in (152-mm) Schedule 40S pipe with the same pressure and flow conditions.

Calculation Procedure:

1. *Determine the flow rate in the new pipe sizes*

Flow rate in a pipe with a fixed pressure drop is proportional to the ratio of (new pipe inside diameter/known pipe inside diameter)$^{2.4}$. This ratio is defined as the *flow factor*, F. To use this ratio, the exact inside pipe diameters, known and new, must be used. Take the exact inside diameter from a table of pipe properties.

Thus, with a 3-in (76-mm) and a 4-in (102-mm) Schedule 40S pipe conveying water at a pressure drop of 8 lb/in^2 (55.1 kPa)/100 ft (30.5 m), the flow factor $F = (4.026/3.068)^{2.4} = 1.91975$. Then, the flow rate, FR, in the large 4-in (102-mm) pipe with the 8 lb/in^2 (55.1 kPa) pressure drop/100 ft (30.5 m), will be, FR = 1.91975 × 300 = 575.9 gal/min (36.3 L/s).

For the 6-in (152-mm) pipe, the flow rate with the same pressure loss will be $(6.065/3.068)^{2.4} \times 300 = 1539.8$ gal/min (97.2 L/s).

2. *Compute the pressure drops in the new pipe sizes*
The pressure drop in a known pipe size can be extrapolated to a new pipe size by using a *pressure factor*, *P*, when the flow rate is held constant. For this condition, P = (known inside diameter of the pipe/new inside diameter of the pipe)$^{4.8}$.

For the first situation given above, $P = (3.068/4.026)^{4.8} = 0.27134$. Then, the pressure drop, PD_N, in the new 4-in (102-mm) Schedule 40S pipe with a 300-gal/min (18.9-L/s) flow will be $PD_N + P(PD_K)$, where PD_K = pressure drop in the known pipe size. Substituting, $PD_N = 0.27134(8) = 2.17$ lb/in^2/100 ft (14.9 kPa/30.5 m).

For the 6-in (152-mm) pipe, using the same approach, $PD_N = (3.068/6.065)^{4.8}\ (8) = 0.303$ lb/in^2/100 ft (2.1 kPa/30.5 m).

Related Calculations. The flow and pressure factors are valuable timesavers in piping system design because they permit quick determination of new flow rates or pressure drops with minimum time input. When working with a series of pipe-size possibilities of the same schedule number, the designer can compute values for *F* and *P* in advance and apply them quickly. Here is an example of such a calculation for Schedule 40S piping of several sizes:

Nominal pipe size, new/known	Flow factor, *F*	Nominal pipe size, known/new	Pressure factor, *p*
2/1	5.092	½	0.0386
3/2	2.58	2/3	0.150
4/3	1.919	¾	0.271
6/4	2.674	4/6	0.1399
8/6	1.933	6/8	0.267
10/8	1.726	8/10	0.335
12/10	1.542	10/12	0.421

When computing such a listing, the actual inside diameter of the pipe, taken from a table of pipe properties, must be used when calculating *F* or *P*.

The *F* and *P* values are useful when designing a variety of piping systems for chemical, petroleum, power, cogeneration, marine, buildings (office, commercial, residential, industrial), and other plants. Both the *F* and *P* values can be used for pipes conveying oil, water, chemicals, and other liquids. The *F* and *P* values are not applicable to steam or gases.

Note that the ratio of pipe diameters is valid for any units of measurement—inches, cm, mm—provided the same units are used consistently throughout the calculation. The results obtained using the *F* and *P* values usually agree closely with those obtained using exact flow or pressure-drop equations. Such accuracy is generally acceptable in everyday engineering calculations.

While the pressure drop in piping conveying a liquid is inversely proportional to the fifth power of the pipe diameter ratio, turbulent flow alters this to the value of 4.8, according to W. L. Nelson, Technical Editor, *The Oil and Gas Journal*.

FLUID HEAD LOSS APPROXIMATIONS FOR ALL TYPES OF PIPING

Using the four rules for approximating head loss in pipes conveying fluid under turbulent flow conditions with a Reynolds number greater than 2100, find: (*a*) A 4-in (101.6-mm) pipe discharges 100 gal/min (6.3 L/s); how much fluid would a 2-in (50.8-mm) pipe discharge under the same conditions? (*b*) A 4-in (101.6-mm) pipe has 240 gal/min (15.1 L/s) flowing through it. What would be the friction loss in a 3-in (76.2-mm) pipe conveying the same flow? (*c*) A flow of 10 gal/min (6.3 L/s) produces 50 ft (15.2 m) of friction in a pipe. How much friction will a flow of 200 gal/min (12.6 L/s) produce? (*d*) A 12-in (304.8-mm) diameter pipe has a friction loss of 200 ft (60.9 m)/1000 ft (304.8 M). What is the capacity of this pipe?

Calculation Procedure:

1. *Use the rule: At constant head, pipe capacity is proportional to $d^{2.5}$*

(***a***) Applying the constant-head rule for both pipes: $4^{2.5} = 32.0$; $2^{2.5} = 5.66$. Then, the pipe capacity = (flow rate, gal/min or L/s)(new pipe size$^{2.5}$)/(previous pipe size$^{2.5}$) = (100)(5.66)/32 = 17.69 gal/min (1.11 L/s).

Thus, using this rule you can approximate pipe capacity for a variety of conditions where the head is constant. This approximation is valid for metal, plastic, wood, concrete, and other piping materials.

2. *Use the rule: At constant capacity, head is proportional to $1/d^5$*

(***b***) We have a 4-in (101.6-mm) pipe conveying 240 gal/min (15.1 L/s). If we reduce the pipe size to 3 in (76.2 mm) the friction will be greater because the flow area is smaller. The head loss = (flow rate, gal/min or L/s)(larger pipe diameter to the fifth power)/(smaller pipe diameter to the fifth power). Or, head = (240)(4^5)/(3^5) = 1011 ft/1000 ft of pipe (308.3 m/304.8 m of pipe).

Again, using this rule you can quickly and easily find the friction in a different size pipe when the capacity or flow rate remains constant. With the easy availability of handheld calculators in the field and computers in the design office, the fifth power of the diameter is easily found.

3. *Use the rule: At constant diameter, head is proportional to gal/min $(L/s)^2$*

(***c***) We know that a flow of 100 gal/min (6.3 L/s) produces 50-ft (15.2-m) friction, h, in a pipe. The friction, with a new flow will be, h = (friction, ft or m, at known flow rate)(new flow rate, gal/min or L/s)/(previous flow rate, gal/min or L/s^2). Or, h = (50)(200^2)/(100^2) = 200 ft (60.9 m).

Knowing that friction will increase as we pump more fluid through a fixed-diameter pipe, this rule can give us a fast determination of the new friction. You can even do the square mentally and quickly determine the new friction in a matter of moments.

4. *Use the rule: At constant diameter, capacity is proportional to friction, $h^{0.5}$*

(***d***) Here the diameter is 12 in (304.8 mm) and friction is 200 ft (60.9 m)/1000 ft (304.8 m). From a pipe friction chart, the nearest friction head is 84 ft (25.6 m) for a flow rate of 5000 gal/min (315.5 L/s). The new capacity, c = (known capacity, gal/min or L/s)(known friction, ft or m$^{0.5}$)/(actual friction, ft or m$^{0.5}$). Or, c = 5000($200^{0.5}$)/($84^{0.5}$) = 7714 gal/min (486.6 L/s).

As before, a simple calculation, the ratio of the square roots of the friction heads times the capacity will quickly give the new flow rates.

Related Calculations. Similar laws for fans and pumps give quick estimates of changed conditions. These laws are covered elsewhere in this handbook in the sections on fans and pumps. Referring to them now will give a quick comparison of the similarity of these sets of laws.

PIPE-WALL THICKNESS AND SCHEDULE NUMBER

Determine the minimum wall thickness t_m in (mm) and schedule number SN for a branch steam pipe operating at 900°F (482.2°C) if the internal steam pressure is 1000 lb/in^2 (abs) (6894 kPa). Use ANSA B31.1 *Code for Pressure Piping* and the ASME *Boiler and Pressure Vessel Code* valves and equations where they apply. Steam flow rate is 72,000 lb/h (32,400 kg/h).

Calculation Procedure:

1. *Determine the required pipe diameter*

When the length of pipe is not given or is as yet unknown, make a first approximation of the pipe diameter, using a suitable velocity for the fluid. Once the length of the pipe is known, the pressure loss can be determined. If the pressure loss exceeds a desirable value, the pipe diameter can be increased until the loss is within an acceptable range.

Compute the pipe cross-sectional area a in^2 (cm^2) from $a = 2.4W\upsilon/V$, where W = steam flow rate, lb/h (kg/h); υ = specific volume of the steam, ft^3/lb (m^3/kg); V = steam velocity, ft/min (m/min). The only unknown in this equation, other than the pipe area, is the steam velocity V. Use Table 1 to find a suitable steam velocity for this branch line.

Table 1 shows that the recommended steam velocities for branch steam pipes range from 6000 to 15,000 ft/min (1828 to 4572 m/min). Assume that a velocity of 12,000 ft/min (3657.6 m/min) is used in this branch steam line. Then, by using the steam table to find the specific volume of steam at 900°F (482.2°C) and 1000 lb/in^2 (abs) (6894 kPa), $a = 2.4(72{,}000)(0.7604)/12{,}000 = 10.98$ in^2 (70.8 cm^2). The inside diameter of the pipe is then $d = 2(a/\pi)^{0.5} = 2(10.98/\pi)^{0.5} = 3.74$ in (95.0 mm). Since pipe is not ordinarily made in this fractional internal diameter, round it to the next larger size, or 4-in (101.6-mm) inside diameter.

TABLE 1 Recommended Fluid Velocities in Piping

Service	Velocity of fluid	
	ft/min	m/s
Boiler and turbine leads	6,000–12,000	30.5–60.9
Steam headers	6,000–8,000	30.5–40.6
Branch steam lines	6,000–15,000	30.5–76.2
Feedwater lines	250–850	1.3–4 3
Exhaust and low-pressure steam lines	6,000–15,000	30.5–76.2
Pump suction lines	100–300	0.51–1.52
Bleed steam lines	4,000–6,000	20.3–30.5
Service water mains	120–300	0.61–1.52
Vacuum steam lines	20,000–40,000	101.6–203.2
Steam superheater tubes	2,000–5,000	10.2–25.4
Compressed-air lines	1,500–2,000	7.6–10.2
Natural-gas lines (large cross-country)	100–150	0.51–0.76
Economizer tubes (water)	150–300	0.76–1.52
Crude-oil lines [6 to 30 in (152.4 to 762.0 mm)]	50–350	0 25–1.78

2. *Determine the pipe schedule number*

The ANSA *Code for Pressure Piping*, commonly called the *Piping Code*, defines schedule number as SN = 1000 P_i/S, where P_i = internal pipe pressure, lb/in^2 (gage); S = allowable stress in the pipe, lb/in^2, from *Piping Code*. Table 2 shows typical allowable stress values for pipe in power piping systems. For this pipe, assuming that seamless ferritic alloy steel (1% Cr, 0.55% Mo) pipe is used with the steam at 900°F (482°C), SN = (1000)(1014.7)/13,100 = 77.5. Since pipe is not ordinarily made in this schedule number, use the next *highest* readily available schedule number, or SN = 80. [Where large quantities of pipe are required, it is sometimes economically wise to order pipe of the exact SN required. This is not usually done for orders of less than 1000 ft (304.8 m) of pipe.]

3. *Determine the pipe-wall thickness*

Enter a tabulation of pipe properties, such as in Crocker and King—*Piping Handbook*, and find the wall thickness for 4-in (101.6-mm) SN 80 pipe as 0.337 in (8.56 mm).

Related Calculations. Use the method given here for any type of pipe—steam, water, oil, gas, or air—in any service—power, refinery, process, commercial, etc. Refer to the proper section of B31.1 *Code for Pressure Piping* when computing the schedule number, because the allowable stress S varies for different types of service.

TABLE 2 Allowable Stresses (S Values) for Alloy-Steel Pipe in Power Piping Systems* (*Abstracted from ASME Power Boiler Code and Code for Pressure Piping, ASA B31.1*)

Material	ASTM specification	Grade or symbol	Minimum tensile strength		S values for metal temperatures not to exceed†					
			lb/in²	MPa	850°F	454°C	900°F	482°F	950°F	510°F
Seamless ferritic steels:										
Carbon-molybdenum	A335	P1	55,000	379.2	13,150	90.7	12,500	86.2	—	—
0.65 Cr, 0.55 Mo	A335	P2	55,000	379.2	13,150	90.7	12,500	86.2	10,000	68.9
1.00 Cr, 0.55 Mo	A335	P12	60,000	413.6	14,200	97.9	13,100	90.3	11,000	75.8

*Crocker and King—*Piping Handbook*.

†Where welded construction is used, consideration should be given to the possibility of graphite formation in carbon-molybdenum steel above 875°F (468°C) or in chromium—molybdenum steel containing less than 0.60 percent chromium above 975°F (523.9°C).

The *Piping Code* contains an equation for determining the minimum required pipe-wall thickness based on the pipe internal pressure, outside diameter, allowable stress, a temperature coefficient, and an allowance for threading, mechanical strength, and corrosion. This equation is seldom used in routine piping-system design. Instead, the schedule number as given here is preferred by most designers.

PIPE-WALL THICKNESS DETERMINATION BY PIPING CODE FORMULA

Use the ANSA B31.1 *Code for Pressure Piping* wall-thickness equation to determine the required wall thickness for an 8.625-in (219.1-mm) OD ferritic steel plain-end pipe if the pipe is used in 900°F (482°C) 900-lb/in² (gage) (6205-kPa) steam service.

Calculation Procedure:

1. *Determine the constants for the thickness equation*

Pipe-wall thickness to meet ANSA *Code* requirements for power service is computed from $t_m = \{DP/[2(S + YP)]\} + C$, where t_m = minimum wall thickness, in; D = outside diameter of pipe, in; P = internal pressure in pipe, lb/in² (gage); S = allowable stress in pipe material, lb/in²; Y = temperature coefficient; C = end-condition factor, in.

Values of S, Y, and C are given in tables in the *Code for Pressure Piping* in the section on Power Piping. Using values from the latest edition of the *Code*, we get S = 12,500 lb/in² (86.2 MPa) for ferritic-steel pipe operating at 900°F (482°C); Y = 0.40 at the same temperature; C = 0.065 in (1.65 mm) for plain-end steel pipe.

2. *Compute the minimum wall thickness*

Substitute the given and *Code* values in the equation in step 1, or $t_m = [(8.625)(900)/[2(12{,}500 + 0.4 \times 900)] + 0.065 = 0.367$ in (9.32 mm).

Since pipe mills do not fabricate to precise wall thicknesses, a tolerance above or below the computed wall thickness is required. An allowance must be made in specifying the wall thickness found with this equation by *increasing* the thickness by 12½ percent. Thus, for this pipe, wall thickness = 0.367 + 0.125(0.367) = 0.413 in (10.5 mm).

Refer to the *Code* to find the schedule number of the pipe. Schedule 60 8-in (203-mm) pipe has a wall thickness of 0.406 in (10.31 mm), and Schedule 80 has a wall thickness of 0.500 in (12.7 mm). Since the required thickness of 0.413 in (10.5 mm) is greater than Schedule 60 but less than Schedule 80, the higher schedule number, 80, should be used.

3. *Check the selected schedule number*

From the previous calculation procedure, SN = 1000 P_i/S. From this pipe, SN = 1000(900)/12,500 = 72. Since piping is normally fabricated for schedule numbers 10, 20, 30, 40, 60, 80, 100, 120, 140, and 160, the next larger schedule number higher than 72, that is 80, will be used. This agrees with the schedule number found in step 2.

Related Calculations. Use this method in conjunction with the appropriate *Code* equation to determine the wall thickness of pipe conveying air, gas, steam, oil, water, alcohol, or any other similar fluids in any type of service. Be certain to use the correct equation, which in some cases is simpler than that used here. Thus, for lead pipe, $t_n = Pd/2S$, where P = safe working pressure of the pipe, lb/in^2 (gage); d = inside diameter of pipe, in; other symbols as before.

When a pipe will operate at a temperature between two tabulated *Code* values, find the allowable stress by interpolating between the tabulated temperature and stress values. Thus, for a pipe operating at 680°F (360°C), find the allowable stress at 650°F (343°C) [= 9500 lb/in^2 (65.5 MPa)] and 700°F (371°C) [= 9000 lb/in^2 (62.0 MPa)]. Interpolate thus: allowable stress at 680°F (360°C) = [(700°F – 680°F)/(700°F – 650°F)](9500 – 9000) + 9000 = 200 + 9000 = 9200 lb/in^2 (63.4 MPa). The same result can be obtained by interpolating downward from 9500 lb/in^2 (65.5 MPa), or allowable stress at 680°F (360°C) = 9500 – [(680 – 650)/(700 – 650)](9500 – 9000) = 9200 lb/in^2 (63.4 MPa).

DETERMINING THE PRESSURE LOSS IN STEAM PIPING

Use a suitable pressure-loss chart to determine the pressure loss in 510 ft (155.5 m) of 4-in (101.6-mm) flanged steel pipe containing two 90° elbows and four 45° bends. The Schedule 40 piping conveys 13,000 lb/h (5850 kg/h) of 20-lb/in^2 (gage) (275.8-kPa) 350°F (177°C) superheated steam. List other methods of determining the pressure loss in steam piping.

Calculation Procedure:

1. *Determine the equivalent length of the piping*

The equivalent length of a pipe L_e ft = length of straight pipe, ft + equivalent length of fittings, ft. Using data from the Hydraulic Institute, Crocker and King—*Piping Handbook*, earlier sections of this handbook, or Fig. 4, find the equivalent length of a 90° 4-in (101.6-mm) elbow as 10 ft (3 m) of straight pipe. Likewise, the equivalent length of a 45° bend is 5 ft (1.5 m) of straight pipe. Substituting in the above relation and using the straight lengths and the number of fittings of each type, we get L_e = 510 + (2) (10) + 4(5) = 550 ft (167.6 m) of straight pipe.

2. *Compute the pressure loss, using a suitable chart*

Figure 2 presents a typical pressure-loss chart for steam piping. Enter the chart at the top left at the superheated steam temperature of 350°F (177°C), and project vertically downward until the 40-lb/in^2 (gage) (275.8-kPa) superheated steam pressure curve is intersected. From here, project horizontally to the right until the outer border of the chart is intersected. Next, project through the steam flow rate, 13,000 lb/h (5900 kg/h) on scale B, Fig. 5, to the pivot scale C. From this point, project through 4-in (101.6-mm) Schedule 40 pipe on scale D, Fig. 5. Extend this line to intersect the pressure-drop scale, and read the pressure loss as 7.25 lb/in^2 (50 kPa)/100 ft (30.4 m) of pipe.

Since the equivalent length of this pipe is 550 ft (167.6 m), the total pressure loss in the pipe is (550/100)(7.25) = 39.875 lb/in^2 (274.9 kPa), say 40 lb/in^2 (275.8 kPa).

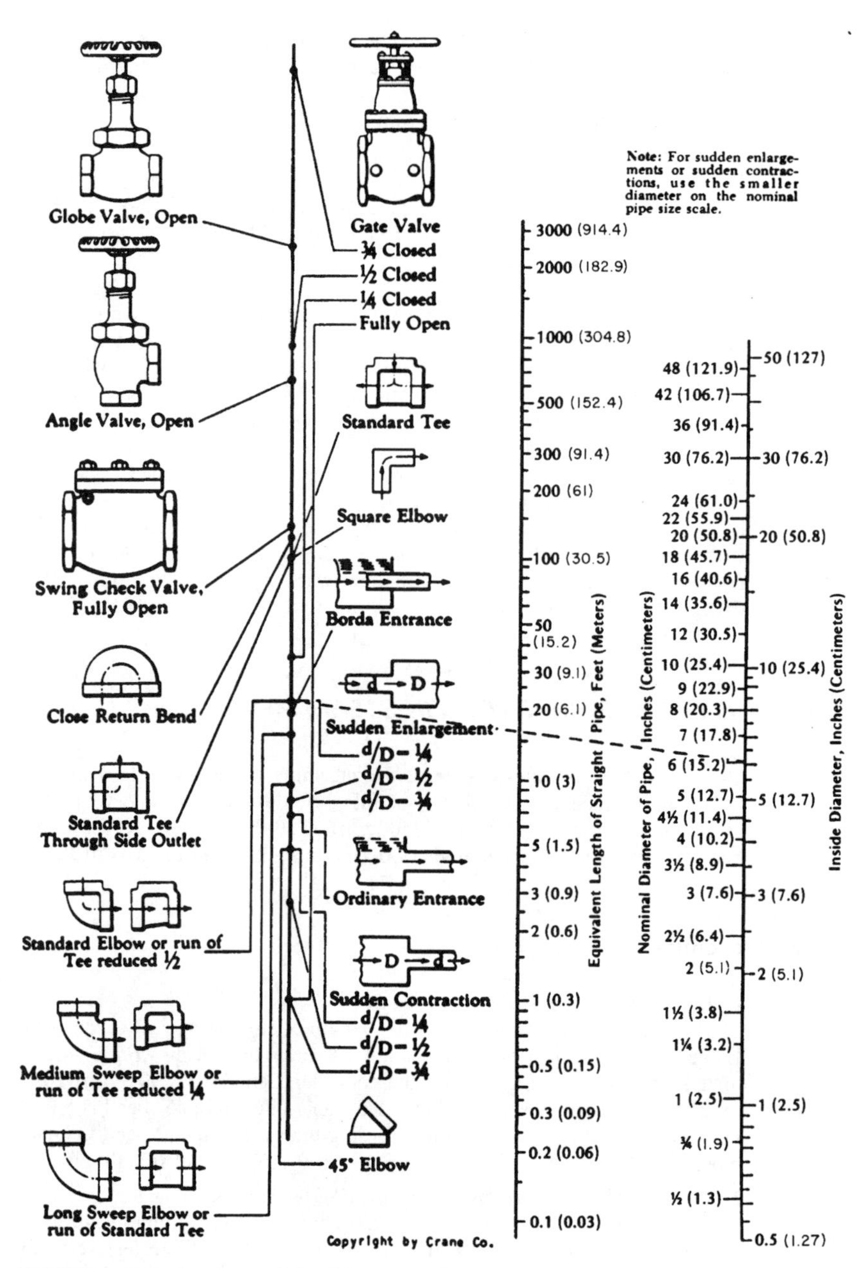

FIGURE 4 Equivalent length of pipe fittings and valves. (*Crane Company*.)

3. *List the other methods of computing pressure loss*

Numerous pressure-loss equations have been developed to compute the pressure drop in steam piping. Among the better known are those of Unwin, Fritzche, Spitz-glass, Babcock, Guttermuth, and others. These equations are discussed in some detail in Crocker and King—*Piping Handbook* and in the engineering data published by valve and piping manufacturers.

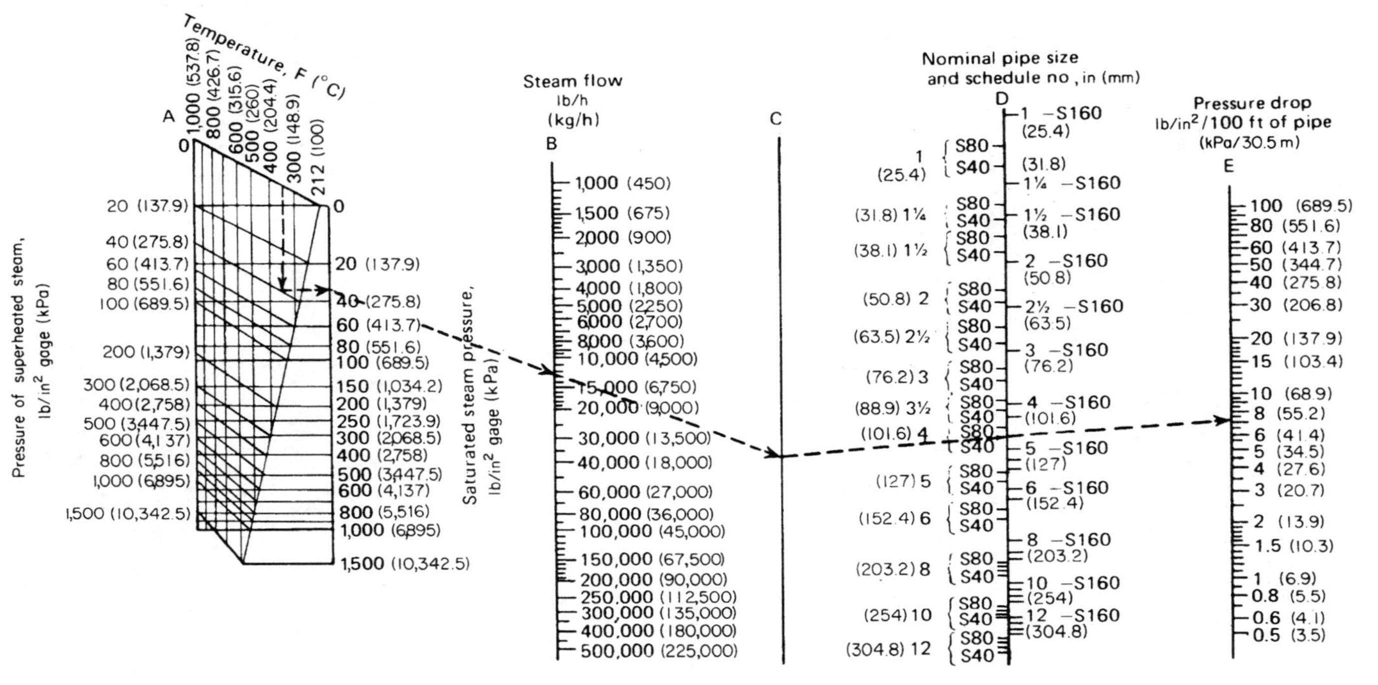

FIGURE 5 Pressure loss in steam pipes based on the Fritzche Formula. (*Power.*)

Most piping designers use a chart to determine the pressure loss in steam piping because a chart saves time and reduces the effort involved. Further, the accuracy obtained insufficient for all usual design practice.

Figure 5 is a popular flowchart for determining steam flow rate, pipe size, steam pressure, or steam velocity in a given pipe. Using this chart, the designer can determine any one of the four variables listed above when the other three are known. In solving a problem on the chart in Fig. 6, use the steam-quantity lines to intersect pipe sizes and the steam-pressure lines to intersect steam velocities. Here are two typical applications of this chart.

Example. What size Schedule 40 pipe is needed to deliver 8000 lb/h (3600 kg/h) of 120-lb/in^2 (gage) (827.3-kPa) steam at a velocity of 5000 ft/min (1524 m/min)?

Solution. Enter Fig. 6 at the upper left at a velocity of 5000 ft/min (1524 m/min), and project along this velocity line until the 120-lb/in^2 (gage) (827.3-kPa) pressure line is intersected. From this intersection, project horizontally until the 8000-lb/h (3600-kg/h) vertical line is intersected. Read the *nearest* pipe size as 4 in (101.6 mm) on the *nearest* pipe-diameter curve.

Example. What is the steam velocity in a 6-in (152.4-mm) pipe delivering 20,000 lb/h (9000 kg/h) of steam at 85 lb/in^2 (gage) (586 kPa)?

Solution. Enter the bottom of the chart, Fig. 6, at the flow rate of 20,000 lb/h (9000 kg/h), and project vertically upward until the 6-in (152.4-mm) pipe curve is intersected. From this point, project horizontally to the 85-lb/in^2 (gage) (586-kPa) curve. At the intersection, read the velocity as 7350 ft/min (2240.3 m/min).

Table 3 shows typical steam velocities for various industrial and commercial applications. Use the given values as guides when sizing steam piping.

PIPING WARM-UP CONDENSATE LOAD

How much condensate is formed in 5 minutes during warm-up of 500 ft (152.4 m) of 6-in (152.4-mm) Schedule 40 steel pipe conveying 215-lb/in^2 (abs) (1482.2-kPa) saturated steam if the pipe is insulated with 2 in (50.8 mm) of 85 percent magnesia and the minimum external temperature is 35°F (1.7°C)?

Calculation Procedure:

1. *Compute the amount of condensate formed during pipe warm-up*

For any pipe, the condensate formed during warm-up C_h lb/h = $60(W_p)(\Delta t)(s)/h_{fg}N$, where W_p = total weight of pipe, lb; Δt = difference between final and initial temperature or the pipe, °F; s = specific heat of pipe material, Btu/(lb · °F); h_{fg} = enthalpy of vaporization of the steam, Btu/lb; N = warm-up time, min.

A table of pipe properties shows that this pipe weighs 18.974 lb/ft (28.1 kg/m). The steam table shows that the temperature of 215-lb/in^2 (abs) (1482.2-kPa) saturated steam is 387.89°F (197.7°C), say 388°F (197.8°C); the enthalpy h_{fg} = 837.4 Btu/lb (1947.8 kJ/kg). The specific heat of steel pipe s = 0.144 Btu/(lb · °F) [0.6 kJ/(kg · °C)]. Then $C_h = 60(500 \times 18.974)(388 - 35)(0.114)/[(837.4)(5)]$ = 5470 lb/h (2461.5 kg/h).

2. *Compute the radiation-loss condensate load*

Condensate is also formed by radiation of heat from the pipe during warm-up and while the pipe is operating. The warm-up condensate load decreases as the radiation load increases, the peak occurring midway (2½ min in this case) through the warm-up period. For this reason, one-half the normal radiation load is added to the warm-up load. Where the radiation load is small, it is often disregarded. However, the load must be computed before its magnitude can be determined.

For any pipe. $C_r = (L)(A)(\Delta t)(H)/h_{fg}$, where L = length of pipe, ft; A = external area of pipe, ft^2/ft of length; H = heat loss through bare pipe or pipe insulation, Btu/(ft^2 · h · °F), from the piping or insulation tables. This 6-in (152.4-mm) Schedule 40 pipe has an external area A = 1.73 ft^2/ft (0.53 m^2/m) of length. The heat loss through 2 in (50.8 mm) of 85 percent magnesia, from insulation tables, is

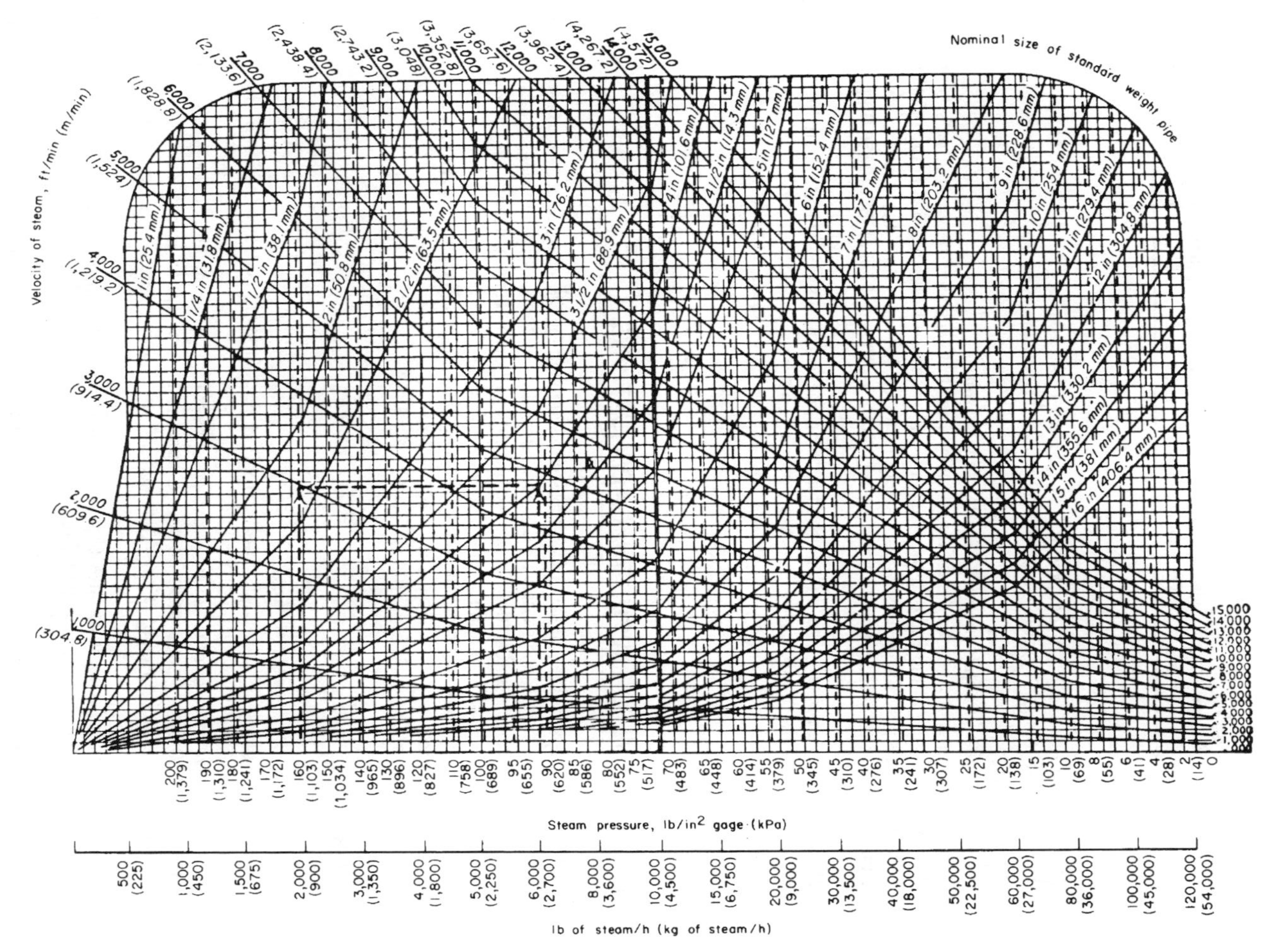

FIGURE 6 Spitzglass chart for saturated steam flowing in Schedule 40 pipe.

TABLE 3 Steam Velocities Used in Pipe Design

Steam condition	Steam pressure		Steam use	Steam velocity	
	lb/in^2	kPa		ft/min	m/min
Saturated	0–15	0–103.4	Heating	4,000–6,000	1,219.2–1,828.8
Saturated	50–150	344.7–1,034.1	Process	6,000–10,000	1,828.8–3,048.0
Superheated	200 and higher	1,378.8 and higher	Boiler leads	10,000–15,000	3,048.0–4,572.0

$H = 0.286$ Btu/(ft^2 · h · °F) [1.62 W/(m^2 · °C)]. Then $C_r = (500) \times (1.73)(388 - 35)\ (0.286)/837.4 =$ 104.2 lb/h (46.9 kg/h). Adding half the radiation load to the warm-up load gives 5470 + 52.1 = 5522.1 lb/h (2484.9 kg/h).

3. *Apply a suitable safety factor to the condensate load*

Trap manufacturers recommend a safety factor of 2 for traps installed between a boiler and the end of a steam main; traps at the end of a long steam main or ahead of pressure-regulating or shut-off valves usually have a safety factor of 3. With a safety factor of 3 for this pipe, the steam trap should have a capacity of at least 3(5522.1) = 16,566.3 lb/h (7454.8 kg/h), say 17,000 lb/h (7650.0 kg/h).

Related Calculations. Use this method to find the warm-up condensate load for any type of steam pipe—main or auxiliary—in power, process, heating, or vacuum service. The same method is applicable to other vapors that form condensate—Dowtherm, refinery vapors, process vapors, and others.

STEAM TRAP SELECTION FOR INDUSTRIAL APPLICATIONS

Select steam traps for the following four types of equipment: (1) the steam directly heats solid materials as in autoclaves, retorts, and sterilizers; (2) the steam indirectly heats a liquid through a metallic surface, as in heat exchangers and kettles, where the quantity of liquid heated is known and unknown; (3) the steam indirectly heats a solid through a metallic surface, as in dryers using cylinders or chambers and platen presses; and (4) the steam indirectly heats air through metallic surfaces, as in unit heaters, pipe coils, and radiators.

Calculation Procedure:

1. *Determine the condensate load*

The first step in selecting a steam trap for any type of equipment is determination of the condensate load. Use the following general procedure.

a. Solid materials in autoclaves, retorts, and sterilizers. How much condensate is formed when 2000 lb (900.0 kg) of solid material with a specific heat of 1.0 is processed in 15 min at 240°F (115.6°C) by 25-lb/in^2 (gage) (172.4-kPa) steam from an initial temperature of 60°F in an insulated steel retort?

For this type of equipment, use $C = WsP$, where C = condensate formed, lb/h; W = weight of material heated, lb; s = specific heat, Btu/(lb · °F); P = factor from Table 4. Thus, for this application, $C = (2000)\ (1.0)\ (0.193) = 386$ lb (173.7 kg) of condensate. Note that P is based on a temperature rise of 240 – 60 = 180°F (100°C) and a steam pressure of 25 lb/in^2 (gage) (172.4 kPa). For the retort, using the specific heat of steel from Table 5, $C = (4000)\ (0.12)\ (0.193) = 92.6$ lb of condensate, say 93 lb (41.9 kg). The total weight of condensate formed in 15 min is 386 + 93 = 479 lb (215.6 kg). In 1 h, 479(60/15) = 1916 lb (862.2 kg) of condensate is formed.

TABLE 4 Factors $P = (T - t)/L$ to Find Condensate Load

Pressure		Temperature		
lb/in² (abs)	kPa	160°F (71.1°C)	180°F (82.2°C)	200°F (93.3°C)
20	137.8	0.170	0.192	0.213
25	172.4	0.172	0.193	0.214
30	206.8	0.172	0.194	0.215

TABLE 5 Use These Specific Heats to Calculate Condensate Load

Solids	Btu/(lb · °F)	kJ/(kg · °C)	Liquids	Btu/(lb · °F)	kJ/(kg · °C)
Aluminum	0.23	0.96	Alcohol	0.65	2.7
Brass	0.10	0.42	Carbon tetrachloride	0.20	0.84
Copper	0.10	0.42	Gasoline	0.53	2.22
Glass	0.20	0.84	Glycerin	0.58	2.43
Iron	0.13	0.54	Kerosene	0.47	1.97
Steel	0.12	0.50	Oils	0.40–0.50	1.67–2.09

A safety factor must be applied to compensate for radiation and other losses. Typical safety factors used in selecting steam traps are as follows:

Steam mains and headers	2–3
Steam heating pipes	2–6
Purifiers and separators	2–3
Retorts for process	2–4
Unit heaters	3
Submerged pipe coils	2–4
Cylinder dryers	4–10

With a safety factor of 4 for this process retort, the trap capacity = (4)(1916) = 7664 lb/h (3449 kg/h), say 7700 lb/h (3465 kg/h).

***b*(1).** *Submerged heating surface and a known quantity of liquid.* How much condensate forms in the jacket of a kettle when 500 gal (1892.5 L) of water is heated in 30 min from 72 to 212°F (22.2 to 100°C) with 50-lb/in² (gage) (344.7-kPa) steam?

For this type of equipment, $C = GwsP$, where G = gal of liquid heated; w = weight of liquid, lb/gal. Substitute the appropriate values as follows: C = (500) (8.33) (1.0) × (0.154) = 641 lb (288.5 kg), or (641)(60/3) = 1282 lb/h (621.9 kg/h). With a safety factor of 3, the trap capacity = (3)(1282) = 3846 lb/h (1731 kg/h), say 3900 lb/h (1755 kg/h).

***b*(2).** *Submerged heating surface and an unknown quantity of liquid.* How much condensate is formed in a coil submerged in oil when the oil is heated as quickly as possible from 50 to 250°F (10 to 121°C) by 25-lb/in² (gage) (172.4-kPa) steam if the coil has an area of 50 ft² (4.66 m²) and the oil is free to circulate around the coal?

For this condition, $C = UAP$, where U = overall coefficient of heat transfer, Btu/(h · ft² · °F), from Table 6; A = area of heating surface, ft². With free convection and a condensing-vapor-to-liquid type of heat exchanger, U = 10 to 30. With an average value of U = 20, C = (20) (50) (0.214) = 214 lb/h (96.3 kg/h) of condensate. Choosing a safety factor 3 gives trap capacity = (3) (214) = 642 lb/h (289 kg/h), say 650 lb/h (292.5 kg/h).

***b*(3).** *Submerged surfaces having more area than needed to heat a specified quantity of liquid in a given time with condensate withdrawn as rapidly as formed.* Use Table 7 instead of step *b*(1) or *b*(2). Find the condensate rate by multiplying the submerged area by the appropriate factor from Table 7.

TABLE 6 Ordinary Ranges of Overall Coefficients of Heat Transfer

Type of heat exchanger	State of controlling resistance				Typical fluid	Typical apparatus
	Free convection, U		Forced convection, U			
Liquid to liquid	25–60	[141.9–340.7]	150–300	[851.7–1703.4]	Water	Liquid-to-liquid heat exchangers
Liquid to liquid	5–10	[28.4–56.8]	20–50	[113.6–283.9]	Oil	
Liquid to gas*	1–3	[5.7–17.0]	2–10	[11.4–56.8]	—	Hot-water radiators
Liquid to boiling liquid	20–60	[113.6–340.7]	50–150	[283.9–851.7]	Water	Brine coolers
Liquid to boiling liquid	5–20	[28.4–113.6]	25–60	[141.9–340.7]	Oil	
Gas* to liquid	1–3	[5.7–17.0]	2–10	[11.4–56.8]	—	Air coolers, economizers
Gas* to gas	0.6–2	[3.4–11.4]	2–6	[11.4–34.1]	—	Steam superheaters
Gas* to boiling liquid	1–3	[5.7–17.0]	2–10	[11.4–56.8]	—	Steam boilers
Condensing vapor to liquid	50–200	[283.9–1136]	150–800	[851.7–4542.4]	Steam to water	Liquid heaters and condensers
Condensing vapor to liquid	10–30	[56.8–170.3]	20–60	[113.6–340.7]	Steam to oil	
Condensing vapor to liquid	40–80	[227.1–454.2]	60–150	[340.7–851.7]	Organic vapor to water	
Condensing vapor to liquid	—	—	15–300	[85.2–1703.4]	Steam-gas mixture	
Condensing vapor to gas*	1–2	[5.7–11.4]	2–10	[11.4–56.8]	—	Steam pipes in air, air heaters
Condensing vapor to boiling liquid	40–100	[227.1–567.8]	—	—	—	Scale-forming evaporators
Condensing vapor to boiling liquid	300–800	[1703.4 –4542.4]	—	—	Steam to water	
Condensing vapor to boiling liquid	50–150	[283.9–851.7]	—	—	Steam to oil	

*At atmospheric pressure.

Note: U–Btu/(h · ft^2 · °F)[W/(m^2 · °C)]. Under many conditions, either higher or lower values may be realized.

TABLE 7 Condensate Formed in Submerged Steel* Heating Elements, lb/(ft^2 · h)[kg/(m^2 · min)]

MTD†		Steam pressure				
°F	°C	75 lb/in^2 (abs) (517.1 kPa)	100 lb/in^2 (abs) (689.4 kPa)	150 lb/in^2 (abs) (1034.1 kPa)	Btu/(ft^2 · h)	kW/m^2
175	97.2	44.3 (3.6)	45.4 (3.7)	46.7 (3.8)	40,000	126.2
200	111.1	54.8 (4.5)	56.8 (4.6)	58.3 (4.7)	50,000	157.7
250	138.9	90.0 (7.3)	93.1 (7.6)	95.7 (7.8)	82,000	258.6

*For copper multiply table data by 2.0; for brass, by 1.6.
†Mean temperature difference, °F or °C equals temperature of steam minus average liquid temperature. Heat transfer data for calculating this table obtained from and used by permission of the American Radiator & Standard Sanitary Corp.

Use this method for heating water, chemical solutions, oils, and other liquids. Thus, with steam at 100 lb/in^2 (gage) (689.4 kPa) and a temperature of 338°F (170°C) and heating oil from 50 to 226°F (10 to 108°C) with a submerged surface having an area of 500 ft^2 (46.5 m^2), the mean temperature difference (*Mtd*) = steam temperature minus the average liquid temperature = 338 – (50 + 226/2) = 200°F (93.3°C). The factor from Table 7 for 100 lb/in^2 (gage) (689.4 kPa) steam and a 200°F (93.3°C) *Mtd* is 56.75. Thus, the condensate rate = (56.75)(500) = 28,375 lb/h (12,769 kg/h). With a safety factor of 2, the trap capacity = (2) (28.375) = 56,750 lb/h (25,538 kg/h).

c. Solids indirectly heated through a metallic surface. How much condensate is formed in a chamber dryer when 1000 lb (454 kg) of cereal is dried to 750 lb (338 kg) by 10-lb/in^2 (gage) (68.9-kPa) steam? The initial temperature of the cereal is 60°F (15.6°C), and the final temperature equals that of the steam.

For this condition, $C = 970(W - D)/h_{fg} + WP$, where D = dry weight of the material, lb; h_{fg} = enthalpy of vaporization of the steam at the trap pressure, Btu/lb. From the steam tables and Table 4, C = 970(1000 – 750)/952 + (1000) (0.189) = 443.5 lb/h (199.6 kg/h) of condensate. With a safety factor of 4, the trap capacity = (4) (443.5) = 1774 lb/h (798.3 kg/h)

d. Indirect heating of air through a metallic surface. How much condensate is formed in a unit heater using 10-lb/in^2 (gage) (68.9-kPa) steam if the entering-air temperature is 30°F (–1.1°C) and the leaving-air temperature is 130°F (54.4°C)? Airflow is 10,000 ft^3/min (281.1 m^3/min).

Use Table 8, entering at a temperature difference of 100°F (37.8°C) and projecting to a steam pressure of 10 lb/in^2 (gage) (68.9 kPa). Read the condensate formed as 122 lb/h (54.9 kg/h) per 1000 ft^3/min (28.3 m^3/min). Since 10,000 ft^3/min (283.1 m^3/min) of air is being heated, the condensate rate = (10,000/1000) (122) = 1220 lb/h (549 kg/h). With a safety factor of 3, the trap capacity = (3)(1220) = 3660 lb/h (1647 kg/h), say 3700 lb/h (1665 kg/h).

Table 9 shows the condensate formed by radiation from bare iron and steel pipes in still air and with forced-air circulation. Thus, with a steam pressure of 100 lb/in^2 (gage) (689.4 kPa) and an initial air temperature of 75°F (23.9°C), 1.05 lb/h (0.47 kg/h) of condensate will be formed per ft^2 (0.09 m^2) of heating surface in still air. With forced-air circulation, the condensate rate is (5) (1.05) = 5.25 lb/(h · ft^2) [25.4 kg/(h · m^2)] of heating surface.

TABLE 8 Steam Condensed by Air, lb/h at 1000 ft^3/min (kg/h at 28.3 m^3/min)*

Temperature difference		Pressure		
°F	°C	5 lb/in^2 (gage) (34.5 kPa)	10 lb/in^2 (gage) (68.9 kPa)	50 lb/in^2 (gage) (344.7 kPa)
50	27.8	61 (27.5)	61 (27.5)	63 (28.4)
100	55.6	120 (54.0)	122 (54.9)	126 (56.7)
150	83.3	180 (81.0)	183 (82.4)	189 (85.1)

*Based on 0.0192 Btu (0.02 kJ) absorbed per ft^3 (0.028 m^3) of saturated air per 33°F (0.556°C) at 32°F (0°C). For 0°F (–17. 8°C), multiply by 1.1.

TABLE 9 Condensate Formed by Radiation from Bare Iron and Steel, lb/(ft^2 · h) [kg/(m^2 · h)]*

Air temperature		Steam pressure			
°F	°C	50 lb/in^2 (gage) (344.7 kPa)	75 lb/in^2 (gage) (517.1 kPa)	100 lb/in^2 (gage) (689.5 kPa)	150 lb/in^2 (gage) (1034 kPa)
65	18.3	0.82 (3.97)	1.00 (5.84)	1.08 (5.23)	1.32 (6.39)
70	21.2	0.80 (3.87)	0.98 (4.74)	1.06 (5.13)	1.21 (5.86)
75	23.9	0.77 (3.73)	0.88 (4.26)	1.05 (5.08)	1.19 (5.76)

*Based on still air; for forced-air circulation, multiply by 5.

TABLE 10 Unit-Heater Correction Factors

Steam pressure		Temperature of entering air		
lb/in^2 (gage)	kPa	20°F (−6.7°C)	40°F (4.4°C)	60°F (15.6°C)
5	34.5	1.370	1.206	1.050
10	68.9	1.460	1.290	1.131
15	103.4	1.525	1.335	1.194

Source: Yarway Corporation; SI values added by handbook editor.

Unit heaters have a *standard rating* based on 2-lb/in^2 (gage) (13.8-kPa) steam with entering air at 60°F (15.6°C). If the steam pressure or air temperature is different from these standard conditions, multiply the heater Btu/h capacity rating by the appropriate correction factor form, Table 10. Thus, a heater rated at 10,000 Btu/h (2931 W) with 2-lb/in^2 (gage) (13.8-kPa) steam and 60°F (15.6°C) air would have an output of (1.290)(10,000) = 12.900 Btu/h (3781 W) with 40°F (4.4°C) inlet air and 10-lb/in^2 (gage) (68.9-kPa) steam. Trap manufacturers usually list heater Btu ratings and recommend trap model numbers and sizes in their trap engineering data. This allows easier selection of the correct trap.

2. *Select the trap size based on the load and steam pressure*

Obtain a chart or tabulation of trap capacities published by the manufacturer whose trap will be used. Figure 7 is a capacity chart for one type of bucket trap manufactured by Armstrong Machine Works. Table 11 shows typical capacities of impulse traps manufactured by the Yarway Company.

To select a trap from Fig. 7, when the condensate rate is uniform and the pressure across the trap is constant, enter at the left at the condensation rate, say 8000 lb/h (3600 kg/h) (as obtained from step 1). Project horizontally to the right to the vertical ordinate representing the pressure across the trap [= Δp = steam-line pressure, lb/in^2 (gage) – return-line pressure with trap valve closed, lb/in^2 (gage)]. Assume Δp = 20 lb/in^2 (gage) (138 kPa) for this trap. The intersection of the horizontal 8000-lb/h (3600-kg/h) projection and the vertical 20-lb/in^2 (gage) (137.9-kPa) projection is on the sawtooth capacity curve for a trap having a 9/16-in (14.3-mm) diameter orifice. If these projections intersected beneath this curve, a 9/16-in (14.3-mm) orifice would still be used if the point were between the verticals for this size orifice.

The dashed lines extending downward from the sawtooth curves show the capacity of a trap at reduced Δp. Thus, the capacity of a trap with a 3/8-in (9.53-mm) orifice at Δp = 30 lb/in^2 (gage) (207 kPa) is 6200 lb/h (2790 kg/h), read at the intersection of the 30-lb/in^2 (gage) (207-kPa) ordinate and the dashed curve extended from the 3/8-in (9.53-mm) solid curve.

To select an impulse trap from Table 11, enter the table at the trap inlet pressure, say 125 lb/in^2 (gage) (862 kPa), and project to the desired capacity, say 8000 lb/h (3600 kg/h), determined from step 1. Table 11 shows that a 2-in (50.8-mm) trap having an 8530-lb/h (3839-kg/h) capacity must be used because the next smallest size has a capacity of 5165 lb/h (2324 kg/h). This capacity is less than that required.

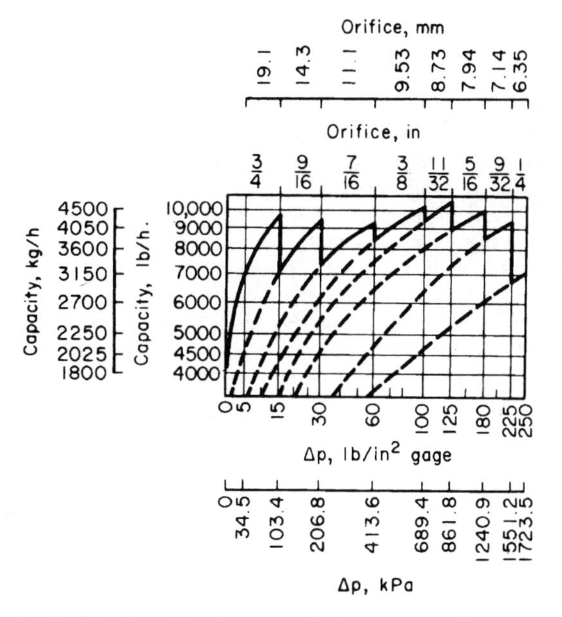

FIGURE 7 Capacities of one type of bucket steam trap. (*Armstrong Machine Works.*)

TABLE 11 Capacities of Impulse Traps, lb/h (kg/h) [*Maximum continuous discharge of condensate, based on condensate at 30°F (16.7°C) below steam temperature.*]

Pressure at trap inlet		Trap nominal size	
lb/in² (gage)	kPa	1.25 in (38.1 mm)	2.0 in (508 mm)
125	861.8	6165 (2774)	8530 (3839)
150	1034.1	6630 (2984)	9075 (4084)
200	1378.8	7410 (3335)	9950 (4478)

Source: Yarway Corporation.

Some trap manufacturers publish capacity tables relating various trap models to specific types of equipment. Such tables simplify trap selection, but the condensate rate must still be computed as given here.

Related Calculations. Use the procedure given here to determine the trap capacity required for any industrial, commercial, or domestic application, including acid vats, air dryers, asphalt tanks, autoclaves, baths (dyeing), belt presses, bleach tanks, blenders, bottle washers, brewing kettles, cabinet dryers, calendars, can washers, candy kettles, chamber dryers, chambers (reaction), cheese kettles, coils (cooking, kettle, pipe, tank, tank-car), confectioners' kettles, continuous dryers, conveyor dyers, cookers (nonpressure and pressure), cooking coils, cooking kettles, cooking tanks, cooking vats, cylinder dryers, cylinders (jacketed), double-drum dryers, drum dryers, drums (dyeing), dry cans, dry kilns, dryers (cabinet, chamber, continuous, conveyor, cylinder, drum, festoon, jacketed, linoleum, milk, paper, pulp, rotary, shelf, stretch, sugar, tray, tunnel), drying rolls, drying rooms, drying tables, dye vats, dyeing baths and drums, dryers (package), embossing-press platens, evaporators, feed waterheaters, festoon dryers, fin-type heaters, fourdriniers, fuel-oil pre-heaters, greenhouse coils, heaters (steam), heat exchangers, heating coils and kettles, hot-break tanks, hot plates, kettle coils, kettles (brewing, candy, cheese, confectioners', cooking, heating, process), kiers, kilns (dry), liquid heaters, mains (steam), milk-bottle washers, milk-can washers, milk dryers, mixers, molding press platens, package dryers, paper dryers, percolators, phonograph-record press platens, pipe coils (still- and circulating-air), platens, plating tanks, plywood press platens, preheaters (fuel-oil), preheating tanks, press platens, pressure cookers, process kettles, pulp dryers, purifiers, reaction chambers, retorts, rotary dryers, steam mains (risers, separators), stocking boarders, storage-tank coils, storage water heaters, stretch dryers, sugar dryers, tank-car coils, tire-mold presses, tray dryers, tunnel dryers, unit heaters, vats, veneer press platens, vulcanizers, and water stills. Hospital equipment—such as autoclaves and sterilizers—can be analyzed in the same way, as can kitchen equipment—bain marie, compartment cooker, egg boiler, kettles, steam table, and urns; and laundry equipment—blanket dryers, curtain dryers, flat-work ironers, presses (dry-cleaning, laundry) sock forms, starch cookers, tumblers, etc.

When using a trap capacity diagram or table, be sure to determine the basis on which it was prepared. Apply any necessary correction factors. Thus, *cold-water capacity ratings* must be corrected for traps operating at higher condensate temperatures. Correction factors are published in trap engineering data. The capacity of a trap is greater at condensate temperatures less than 212°F (100°C) because at or above this temperature condensate forms flash steam when it flows into a pipe or vessel at atmospheric [14.7 lb/in^2 (abs) (101.3 kPa)] pressure. At altitudes above sea level, condensate flashes into steam at a lower temperature, depending on the altitude.

The method presented here is the work of L. C. Campbell, Yarway Corporation, as reported in *Chemical Engineering*.

SELECTING HEAT INSULATION FOR HIGH-TEMPERATURE PIPING

Select the heat insulation for a 300-ft (91.4-m) long 10-in (254-mm) turbine lead operating at 570°F (299°C) for 8000 h/yr in a 70°F (21.1°C) turbine room. How much heat is saved per year by this insulation? The boiler supplying the turbine has an efficiency of 80 percent when burning fuel having a heating value of 14,000 Btu/lb (32.6 MJ/kg). Fuel costs $6 per ton ($5.44 per metric ton). How much money is saved by the insulation each year? What is the efficiency of the insulation?

Calculation Procedure:

1. ***Choose the type of insulation to use***

Refer to an insulation manufacturer's engineering data or Crocker and King—*Piping Handbook*, McGraw-Hill, 1973, for recommendations about a suitable insulation for a pipe operating in the 500 to 600°F (260 to 316°C) range. These references will show that calcium silicate is a popular insulation for this temperature range. Table 12 shows that a thickness of 3 in (76.2 mm) is usually recommended for 10-in (254-mm) pipe operating at 500 to 599°F (260 to 315°C).

2. ***Determine heat loss through the insulation***

Refer to an insulation manufacturer's engineering data to find the heat loss through 3-in (76.2-mm) thick calcium silicate as 0.200 Btu/(h · ft^2 · °F) [1.14 W/(m^2 · °C)]. Since 10-in (254-mm) pipe has an area of 2.817 ft^2/ft (0.86 m^2/m) of length and since the temperature difference across the pipe is 570 – 70 = 500°F (260°C), the heat loss per hour = (0.200) (2.817) (50) = 281.7 Btu/(h · ft) (887.9 W/m^2). The heat loss from bare 10-in (254-mm) pipe with a 500°F (260°C) temperature difference is, from an insulation manufacturer's engineering data, 4.640 Btu/(h · ft^2 · °F) [26.4 W/(m^2 · °C)], or (4.64)(2.817)(500) = 6510 Btu/(h · ft) (6.3 kW/m).

3. ***Determine annual heat saving***

The heat saved = bare-pipe loss, Btu/h – insulated-pipe loss, Btu/h = 6510 – 281.7 – 6228.3 Btu/(h · ft) (5989 W/m) of pipe. Since the pipe is 300 ft (91.4 m) long and operates 8000 h/yr, the annual heat saving = (300) (8000) (6228.3) = 14,940,000,000 Btu/yr (547.4 kW).

TABLE 12 Recommended Insulation Thickness

Normal pipe size		Pipe temperature			
		400–499°F	204–259°F	500–599°F	260–315°C
in	mm	in	mm	in	mm
6	152.4	2½*	63.5	2½	63.5
8	203.2	2½	63.5	3	76.2
10	254.0	2½	63.5	3	76.2
12	304.8	3	76.2	3	76.2
14 and over	355.6 and over	3	76.2	3½	88.9

*Available in single- or double-layer insulation.

4. *Compute the money saved by the heat insulation*
The heat saved in fuel as fired = (annual heat saving, Btu/yr)/(boiler efficiency) = 14,940,000,000/0.80 = 18,680,000,000 Btu/yr (5473 MW). Weight of fuel saved = (annual heat saving, Btu/yr)/(heating value of fuel, Btu/lb) (2000 lb/ton) = 18,680,000,000/[(14,000) (2000)] = 667 tons (605 t). At $6 per ton ($5.44 per metric ton), the monetary saving is ($6) (667) = $4002 per year.

5. *Determine the insulation efficiency*
Insulation efficiency = (bare-pipe loss – insulated-pipe loss)/bare-pipe loss, all expressed in Btu/h, or bare-pipe loss = (6510.0 – 281.7)/6510.0 = 0.957, or 95.7 percent.

Related Calculations. Use this method for any type of insulation—magnesia, fiber-glass, asbestos, felt, diatomaceous, mineral wool, etc.—used for piping at elevated temperatures conveying steam, water, oil, gas, or other fluids or vapors. To coordinate and simplify calculations, become familiar with the insulation tables in a reliable engineering handbook or comprehensive insulation catalog. Such familiarity will simplify routine calculations.

ORIFICE METER SELECTION FOR A STEAM PIPE

Steam is metered with an orifice meter in a 10-in (254-mm) boiler lead having an internal diameter of d_p = 9.760 in (247.9 mm). Determine the maximum rate of steam flow that can be measured with a steel orifice plate having a diameter of d_0 = 5.855 in (148.7 mm) at 70°F (21.1°C). The upstream pressure tap is 1D ahead of the orifice, and the downstream tap is 0.5D past the orifice. Steam pressure at the orifice inlet p_p = 250 lb/in^2 (gage) (1724 kPa), temperature is 640°F (338°C). A differential gage fitted across the orifice has a maximum range of 120 in (304.8 cm) of water. What is the steam flow rate when the observed differential pressure is 40 in (101.6 cm) of water? Use the ASME Research Committee on Fluid Meters method in analyzing the meter. Atmospheric pressure is 14.696 lb/in^2 (abs) (101.3 kPa).

Calculation Procedure:

1. *Determine the diameter ratio and steam density*
For any orifice, meter, diameter ratio = β = meter orifice diameter, in/pipe internal diameter, in = 5.855/9.760 = 0.5999.

Determine the density of the steam by entering the superheated steam table at 250 + 14.696 = 264.696 lb/in^2 (abs) (1824.8 kPa) and 640°F (338°C) and reading the specific volume as 2.387 ft^3/lb (0.15 m^3/kg). For steam, the density = 1/specific volume = d_s = 1/2.387 = 0.4193 lb/ft^3 (6.7 kg/m^3).

2. *Determine the steam viscosity and meter flow coefficient*
From the ASME publication, *Fluid Meters—Their Theory and Application*, the steam viscosity gu_1 for a steam system operating at 640°F (338°C) is gu_1 0.0000141 in · lb/(°F · s · ft^2) [0.000031 N · m/(°C · s · m^2)].

Find the flow coefficient K from the same ASME source by entering the 10-in (254-mm) nominal pipe diameter table at $\beta = 0.5999$ and projecting to the appropriate Reynolds number column. Assume that the Reynolds number $= 10^7$, approximately, for the flow conditions in this pipe. Then $K = 0.6486$. Since the Reynolds number for steam pressures above 100 lb/in^2 (689.4 kPa) ranges from 10^6 to 10^7, this assumption is safe because the value of K does not vary appreciably in this Reynolds number range. Also, the Reynolds number cannot be computed yet because the flow rate is unknown. Therefore, assumption of the Reynolds number is necessary. The assumption will be checked later.

3. *Determine the expansion factor and the meter area factor*
Since steam is a compressible fluid, the expansion factor Y_1 must be determined. For superheated steam, the ratio of the specific heat at constant pressure c_p to the specific heat at constant volume c_υ is $k = c_p/c_\upsilon = 1.3$. Also, the ratio of the differential maximum pressure reading h_w, in of water, to the maximum pressure in the pipe, lb/in^2 (abs) = 120/246.7 = 0.454. From the expansion-factor curve in the ASME *Fluid Meiers*, $Y_1 = 0.994$ for $\beta = 0.5999$ and the pressure ratio = 0.454. And, from the same reference, the meter area factor $F_a = 1.0084$ for a steel meter operating at 640°F (338°C).

4. *Compute the rate of steam flow*
For square-edged orifices, the flow rate, lb/s = $w = 0.0997F_aKd^2Y_1(h_wd_s)^{0.5}$ = (0.0997)(1.0084)(0.6486)(5.855)^2(0.994)(120 × 0.4188)$^{0.5}$ = 15.75 lb/s (7.1 kg/s).

5. *Compute the Reynolds number for the actual flow rate*
For any steam pipe, the Reynolds number $R = 48w/(d_pgu_1)$ = 48(15.75)/[(3.1416)(0.760)(0.0000141)] = 1,750,000.

6. *Adjust the flow coefficient for the actual Reynolds number*
In step 2, $R = 10^7$ was assumed and $K = 0.6486$. For R = 1,750,000, $K = 0.6489$, from ASME *Fluid Meters*, by interpolation. Then the actual flow rate w_h = (computed flow rate)(ratio of flow coefficients based on assumed and actual Reynolds numbers) = (15.75)(0.6489/0.6486)(3.600) = 56,700 lb/h (25,515 kg/h), closely, where the value 3600 is a conversion factor for changing lb/s to lb/h.

7. *Compute the flow rate for a specific differential gage deflection*
For a 40-in (101.6-cm) H_2O deflection, F_a is unchanged and equals 1.0084. The expansion factor changes because h_w/p_p = 40/264.7 = 0.151. From the ASME *Fluid Meters*, $Y_1 = 0.998$. By assuming again that $R = 10^7$, $K = 0.6486$, as before, w = (0.0997) (1.0084)(0.6486)(5.855)^2(0.998)(40 × 0.4188)$^{0.5}$ = 9.132 lb/s (4.1 kg/s). Computing the Reynolds number as before, gives R = (40)(0.132)/[(3.1416)(0.76)(0.0000141)] = 1,014,000. The value of K corresponding to this value, as before, is from ASME—*Fluid Meters:* $K = 0.6497$. Therefore, the flow rate for a 40-in (101.6-cm) H_2O reading, in lb/h = w_h = (0.132)(0.6497/0.6486)(3600) = 32,940 lb/h (14,823 kg/h).

Related Calculations. Use these steps and the ASME *Fluid Meters* or comprehensive meter engineering tables giving similar data to select or check an orifice meter used in any type of steam pipe—main, auxiliary, process, industrial, marine, heating, or commercial, conveying wet, saturated, or superheated steam.

SELECTION OF A PRESSURE-REGULATING VALVE FOR STEAM SERVICE

Select a single-seat spring-loaded diaphragm-actuated pressure-reducing valve to deliver 350 lb/h (158 kg/h) of steam at 50 lb/in^2 (gage) (344.7 kPa) when the initial pressure is 225 lb/in^2 (gage) (1551 kPa). Also select an integral pilot-controlled piston-operated single-seat pressure-regulating valve to deliver 30,000 lb/h (13,500 kg/h) of steam at 40 lb/in^2 (gage) (275.8 kPa) with an initial pressure of 225 lb/in^2 (gage) (1551 kPa) saturated. What size pipe must be used on the downstream side of the valve to produce a velocity of 10,000 ft/min (3048 m/min)? How large should the pressure-regulating valve be if the steam entering the valve is at 225 lb/in^2 (gage) (1551 kPa) and 600°F (316°C)?

Calculation Procedure:

1. *Compute the maximum flow for the diaphragm-actuated valve*
For best results in service, pressure-reducing valves are selected so that they operate 60 to 70 percent open at normal load. To obtain a valve sized for this opening, divide the desired delivery, lb/h, by 0.7 to obtain the maximum flow expected. For this valve then, the maximum flow = 350/0.7 = 500 lb/h (225 kg/h).

2. *Select the diaphragm-actuated valve size*
Using a manufacturer's engineering data for an acceptable valve, enter the appropriate valve capacity table at the valve inlet steam pressure, 225 lb/in^2 (gage) (1551 kPa), and project to a capacity of 500 lb/h (225 kg/h), as in Table 13. Read the valve size as ¾ in (19.1 mm) at the top of the capacity column.

3. *Select the size of the pilot-controlled pressure-regulating valve*
Enter the capacity table in the engineering data of an acceptable pilot-controlled pressure-regulating valve, similar to Table 14, at the required capacity, 30,000 lb/h (13,500 kg/h). Project across until the correct inlet steam pressure column, 225 lb/in^2 (gage) (1551 kPa), is intercepted, and read the required valve size as 4 in (101.6 mm).

Note that it is not necessary to compute the maximum capacity before entering the table, as in step 1, for the pressure-reducing valve. Also note that a capacity table such as Table 14 can be used only for valves conveying saturated steam, unless the table notes state that the values listed are valid for other steam conditions.

TABLE 13 Pressure-Reducing-Valve Capacity, lb/h (kg/h)

Inlet pressure		Valve size		
lb/in^2 (gage)	kPa	½ in (12.7 mm)	¾ in (19.1 mm)	1 in (25.4 mm)
200	1379	420 (189)	460 (207)	560 (252)
225	1551	450 (203)	500 (225)	600 (270)
250	1724	485 (218)	560 (252)	650 (293)

Source: Clark-Reliance Corporation.

TABLE 14 Pressure-Regulating-Valve Capacity

Steam capacity		Initial steam pressure, saturated			
lb/h	kg/h	40 lb/in^2 (gage) (276 kPa)	175 lb/in^2 (gage) (1206 kPa)	225 lb/in^2 (gage) (1551 kPa)	300 lb/in^2 (gage) (2068 kPa)
20,000	9,000	6*(152.4)	4 (101.6)	4 (101.6)	3 (76.2)
30,000	13,500	8 (203.2)	5 (127.0)	4 (101.6)	4 (101.6)
40,000	18,000	—	5 (127.0)	5 (127.0)	4 (101.6)

*Valve diameter measured in inches (millimeters).
Source: Clark-Reliance Corporation.

4. *Determine the size of the downstream pipe*

Enter Table 14 at the required capacity, 30,000 lb/h (13,500 kg/h); project across to the valve *outlet pressure*, 40 lb/in^2 (gage) (275.8 kPa); and read the required pipe size as 8 in (203.2 mm) for a velocity of 10,000 ft/min (3048 m/min). Thus, the pipe immediately downstream from the valve must be enlarged from the valve size, 4 in (101.6 mm), to the required pipe size, 8 in (203.2 mm), to obtain the desired steam velocity.

5. *Determine the size of the valve handling superheated steam*

To determine the correct size of a pilot-controlled pressure-regulating valve handling superheated steam, a correction must be applied. Either a factor or a tabulation of corrected pressures, Table 15, may be used. To use Table 15, enter at the valve inlet pressure, 225 lb/in^2 (gage) (1551.2 kPa), and project across to the total temperature, 600°F (316°C), to read the corrected pressure, 165 lb/in^2 (gage) (1137.5 kPa). Enter Table 14 at the *next highest* saturated steam pressure, 175 lb/in^2 (gage) (1206.6 kPa) project down to the required capacity, 30,000 lb/h (13,500 kg/h), and read the required valve size as 5 in (127 mm).

Related Calculations. To simplify pressure-reducing and pressure-regulating valve selection, become familiar with two or three acceptable valve manufacturers' engineering data. Use the procedures given in the engineering data or those given here to select valves for industrial, marine, utility, heating, process, laundry, kitchen, or hospital service with a saturated or superheated steam supply.

Do not oversize reducing or regulating valves. Oversizing causes chatter and excessive wear.

When an anticipated load on the downstream side will not develop for several months after installation of a valve, fit to the valve a reduced-area disk sized to handle the present load. When the load increases, install a full-size disk. Size the valve for the ultimate load, not the reduced load.

Where there is a wide variation in demand for steam at the reduced pressure, consider installing two regulators piped in parallel. Size the smaller regulator to handle light loads and the larger regulator to handle the difference between 60 percent of the light load and the maximum heavy load. Set the larger regulator to open when the minimum allowable reduced pressure is reached. Then both regulators will be open to handle the heavy load. Be certain to use the actual regulator inlet pressure and not the boiler pressure when sizing the valve if this is different from the inlet pressure. Data in this calculation procedure are based on valves built by the Clark-Reliance Corporation, Cleveland, Ohio.

Some valve manufacturers use the valve flow coefficient C_v for valve sizing. This coefficient is defined as the flow rate, lb/h, through a valve of given size when the pressure loss across the valve is 1 lb/in^2 (6.89 kPa). Tabulations like Tables 13 and 14 incorporate this flow coefficient and are somewhat easier to use. These tables make the necessary allowances for downstream pressure less than the critical pressure (= 0.55 × absolute upstream pressure, lb/in^2, for superheated steam and hydrocarbon vapors; and 0.58 × absolute upstream pressure, lb/in^2, for saturated steam). The accuracy of these tabulations equals that of valve sizes determined by using the flow coefficient.

TABLE 15 Equivalent Saturated Steam Values for Superheated Steam at Various Pressures and Temperatures

Steam pressure		Steam temperature		Total temperature					
				500°F	600°F	700°F	260.0°C	315.6°C	371.1°C
lb/in^2 (gage)	kPa	°F	°C	Steam values, lb/in^2 (gage)			Steam values, kPa		
205	1413.3	389	198	171	149	133	1178.9	1027.2	916.9
225	1551.2	397	203	190	165	147	1309.9	1137.5	1013.4
265	1826.9	411	211	227	200	177	1564.9	1378.8	1220.2

Source: Clark-Reliance Corporation.

HYDRAULIC RADIUS AND LIQUID VELOCITY IN WATER PIPES

What is the velocity of 1000 gal/min (63.1 L/s) of water flowing through a 10-in (254-mm) inside-diameter cast-iron water main? What is the hydraulic radius of this pipe when it is full of water? When is the water depth 8 in (203.2 mm)?

Calculation Procedure:

1. *Compute the water velocity in the pipe*

For any pipe conveying water, the liquid velocity is υ ft/s = gal/min/($2.448d^2$), where d = internal pipe diameter, in. For this pipe, υ = 1000/[2.448(10)] = 4.08 ft/s (1.24 m/s), or (60)(4.08) = 244.8 ft/min (74.6 m/min).

2. *Compute the hydraulic radius for a full pipe*

For any pipe, the hydraulic radius is the ratio of the cross-sectional area of the pipe to the wetted perimeter, or d/4. For this pipe, when full of water, the hydraulic radius = 10/4 = 2.5.

3. *Compute the hydraulic radius for a partially full pipe*

Use the hydraulic radius tables in King and Brater—*Handbook of Hydraulics*, McGraw-Hill, 1996, or compute the wetted perimeter by using the geometric properties of the pipe, as in step 2. From the King and Brater table, the hydraulic radius = Fd, where F = table factor for the ratio of the depth of water, in/diameter of channel, in = 8/10 = 0.8. For this ratio, F = 0.304. Then, hydraulic radius = (0.304)(10) = 3.04 in (77.2 mm).

Related Calculations. Use this method to determine the water velocity and hydraulic radius in any pipe conveying cold water—water supply, plumbing, process, drain, or sewer.

FRICTION-HEAD LOSS IN WATER PIPING OF VARIOUS MATERIAL

Determine the friction-head loss in 2500 ft (762 m) of clean 10-in (254-mm) new tar-dipped cast-iron pipe when 2000 gal/min (126.2 L/s) of cold water is flowing. What is the friction-head loss 20 years later? Use the Hazen-Williams and Manning formulas, and compare the results.

Calculation Procedure:

1. *Compute the friction-head loss by the Hazen-Williams formula*

The Hazen-Williams formula is $h_f = [\upsilon/(1.318CR_h^{0.63})]^{1.85}$, where h_f = friction-head loss per ft of pipe, ft of water; υ = water velocity, ft/s; C = a constant depending on the condition and kind of pipe; R_h = hydraulic radius of pipe, ft.

For a water pipe, υ = gal/min/($2.44d^2$); for this pipe, $\upsilon = 2000/[2.448(10)^2]$ = 8.18 ft/s (2.49 m/s). From Table 16 or Crocker and King—*Piping Handbook*, C for new pipe = 120; for 20-year-old pipe, C = 90; R_h = d/4 for a full-flow pipe = 10/4 = 2.5 in, or 2.5/12 = 0.208 ft (63.4 mm). Then $h_f = [8.18/(1.318 \times 120 \times 0.208^{0.63})]^{1.85}$ = 0.0263 ft (8.0 mm) of water per ft (m) of pipe. For 2500 ft (762 m) of pipe, the total friction-head loss = 2500(0.0263) = 65.9 ft (20.1 m) of water for the new pipe.

For 20-year-old pipe and the same formula, except with C = 90, h_f = 0.0451 ft (13.8 mm) of water per ft (m) of pipe. For 2500 ft (762 m) of pipe, the total friction-head loss = 2500(0.0451) = 112.9 ft (34.4 m) of water. Thus, the friction-head loss nearly doubles [from 65.9 to 112.9 ft (20.1 to 34.4 m)] in 20 years. This shows that it is wise to design for future friction losses; otherwise, pumping equipment may become overloaded.

TABLE 16 Values of C in Hazen-Williams Formula

Type of pipe	C*	Type of pipe	C*
Cement-asbestos	140	Cast iron or wrought iron	100
Asphalt-lined iron or steel	140	Welded or seamless steel	100
Copper or brass	130	Concrete	100
Lead, tin, or glass	130	Corrugated steel	60
Wood stave	110		

*Values of C commonly used for design. The value of C for pipes made of corrosive materials decreases as the age of the pipe increases; the values given are those that apply at an age of 15 to 20 years. For example, the value of C for cast-iron pipes 30 in (762 mm) in diameter or greater at various ages is approximately as follows: new, 130; 5 years old, 120; 10 years old, 115; 20 years old, 100; 30 years old, 90; 40 years old, 80; and 50 years old, 75. The value of C for smaller-size pipes decreases at a more rapid rate.

2. *Compute the friction-head loss from the Manning formula*

The Manning formula is $h_f = n^2 \upsilon^2/2.208/R_h^{4/3}$, where n = a constant depending on the condition and kind of pipe, other symbols as before.

Using $n = 0.011$ for new coated cast-iron pipe from Table 17 or Crocker and King—*Piping Handbook*, we find $h_f = (0.011)^2(8.18)^2/[2.208(0.208)^{4/3}] = 0.0295$ ft (8.9 mm) of water per ft (m) of pipe. For 2500 ft (762 m) of pipe, the total friction-head loss = 2500(0.0295) = 73.8 ft (22.5 m) of water, as compared with 65.9 ft (20.1 m) of water computed with the Hazen-Williams formula.

TABLE 17 Roughness Coefficients (Manning's n) for Closed Conduits

Type of conduit			Manning's n	
			Good construction*	Fair construction*
Concrete pipe			0.013	0.015
Corrugated metal pipe or pipe arch, 2 2/3 × ½ in (67.8 × 12.7 mm) corrugation, riveted:				
Plain			0.024	
Paved invert:				
Percent of circumference paved	25	50		
Depth of flow				
Full	0.021	0.018		
0.8D	0.021	0.016		
0 6D	0.019	0.013		
Vitrified clay pipe			0.012	0.014
Cast-iron pipe, uncoated			0.013	
Steel pipe			0.011	
Brick			0.014	0.017
Monolithic concrete:				
Wood forms, rough			0.015	0.017
Wood forms, smooth			0.012	0.014
Steel forms			0.012	0.013
Cemented-rubble masonry walls:				
Concrete floor and top			0.017	0.022
Natural floor			0.019	0.025
Laminated treated wood			0.015	0.017
Vitrified-clay liner plates			0.015	

*For poor-quality construction, use larger values of n.

For coated cast-iron pipe in fair condition, $n = 0.013$, and $h_f = 0.0411$ ft (12.5 mm) of water. For 2500 ft (762 m) of pipe, the total friction-head loss = 2500(0.0411) = 102.8 ft (31.3 m) of water, as compared with 112.9 ft (34.4 m) of water computed with the Hazen-Williams formula. Thus, the Manning formula gives results higher than the Hazen-Williams in one case and lower in another. However, the differences in each case are not excessive; (73.8 – 65.9)/65.9 = 0.12, or 12 percent higher, and (112.9 – 102.8)/102.8 = 0.0983, or 9.83 percent lower. Both these differences are within the normal range of accuracy expected in pipe friction-head calculations.

Related Calculations. The Hazen-Williams and Manning formulas are popular with many piping designers for computing pressure losses in cold-water piping. To simplify calculations, most designers use the precomputed tabulated solutions available in the previously referenced publications by Crocker and King—*Piping Handbook*, King and Brater—*Handbook of Hydraulics*, and similar publications. In the rush of daily work these precomputed solutions are also preferred over the more complex Darcy-Weisbach equation used in conjunction with the friction factor f, the Reynolds number R, and the roughness-diameter ratio.

Use the method given here for sewer lines, water-supply pipes for commercial, industrial, or process plants, and all similar applications where cold water at temperatures of 33 to 90°F (0.6 to 32.2°C) flows through a pipe made of cast iron, riveted steel, welded steel, galvanized iron, brass, glass, woodstove, concrete, vitrified, common clay, corrugated metal, unlined rock, or enameled steel. Thus, either of these formulas, used in conjunction with a suitable constant, gives the friction-head loss for a variety of piping materials. Suitable constants are given in Tables 16 and 17 and in the above references. For the Hazen-Williams formula, the constant C varies from about 70 to 140, while n in the Manning formula varies from about 0.017 for $C = 70$ to 0.010 for $C = 140$. Values obtained with these formulas have been used for years with satisfactory results. At present, the Manning formula appears the more popular.

CHART AND TABULAR DETERMINATION OF FRICTION HEAD

Figure 8 shows a process piping system supplying 1000 gal/min (63.1 L/s) of 70°F (21.1°C) water. Determine the total friction head, using published charts and pipe-friction tables. All the valves and fittings are flanged, and the piping is 10-in (254-mm) steel, Schedule 40.

Calculation Procedure:

1. *Determine the total length of the piping*

Mark the length of each piping run on the drawing after scaling it or measuring it in the field. Determine the total length by adding the individual lengths, starting at the supply source of the liquid. In Fig. 8, beginning at the storage sump, the total length of piping = 10 + 20 + 40 + 50 + 75 + 105 = 300 ft (91.4 m). Note that the physical length of the fittings is included in the length of each run.

2. *Compute the equivalent length of each fitting*

The frictional resistance of pipe fittings (elbows, tees, etc.) and valves is greater than the actual length of each fitting. Therefore, the equivalent length of straight piping having a resistance equal to that of the fittings must be determined. This is done by finding the equivalent length of each fitting and taking the sum for all the fittings.

Use the equivalent length table in the pump section of this handbook or in Crocker and King—*Piping Handbook*, McGraw-Hill, 1973, Baumeister and Marks—*Standard Handbook for Mechanical Engineers*, McGraw-Hill, 2006, or *Standards of the Hydraulic Institute*, McGraw-Hill, 2011. Equivalent length values will vary slightly from one reference to another.

Starting at the supply source, as in step 1, for 10-in (254-mm) flanged fittings throughout, we see the equivalent fitting lengths are: bell-mouth inlet, 2.9 ft (0.88 m); 90° ell at pump, 14 ft (4.3 m); gate valve, 3.2 ft (0.98 m); swing check valve, 120 ft (36.6 m); 90° ell, 14 ft (4.3 m); tee, 30 ft (9.1 m); 90° ell, 14 ft (4.3 m); 90° ell, 14 ft (4.3 m); globe valve, 310 ft (94.5 m); swing check valve, 120 ft (36.6 m); sudden enlargement = (liquid velocity, ft/s)^2/2g = $(4.07)^2/2(32.2)$ = 0.257 ft (0.08 m),

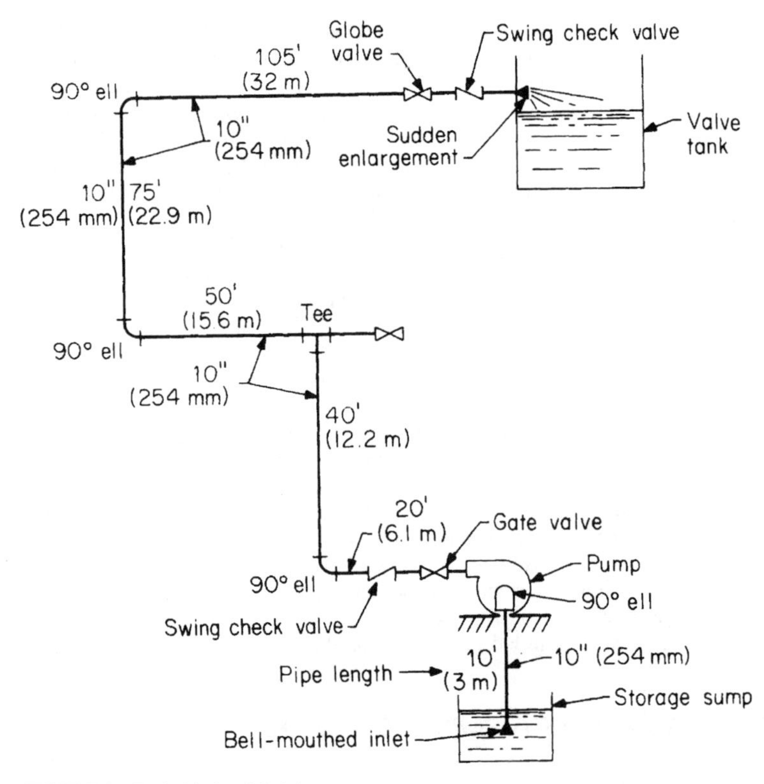

FIGURE 8 Typical industrial piping system.

where the terminal velocity is zero, as in the tank. Find the liquid velocity as shown in a previous calculation procedure in this section. The sum of the fitting equivalent lengths is 2.9 + 14 + 3.2 + 120 + 14 + 30 + 14 + 14 + 310 + 120 + 0.257 = 642.4 ft (159.8 m). Adding this to the straight length gives a total length of 642.4 + 300 = 942.4 ft (287.3 m).

3. ***Compute the friction-head loss by using a chart***

Figure 9 is a popular friction-loss chart for fairly rough pipe, which is any ordinary pipe after a few years' use. Enter at the left at a flow of 1000 gal/min (63.1 L/s), and project to the right until the 10-in (254-mm) diameter curve is intersected. Read the friction-head loss at the top or bottom of the chart as 0.4 lb/in^2 (2.8 kPa), closely, per 100 ft (30.5 m) of pipe. Therefore, total friction-head loss = (0.4)(942.4/100) = 3.77 lb/in^2 (26 kPa). Converting gives (3.77)(2.31) = 8.71 ft (2.7 m) of water.

4. ***Compute the friction-head loss from tabulated data***

Using the *Standards of the Hydraulic Institute* pipe-friction table, we find that the friction head h_f of water per 100 ft (30.5 m) of pipe = 0.500 ft (0.15 m). Hence, the total friction head = (0.500)(942.4/100) = 4.71 ft (1.4 m) of water. The Institute recommends that 15 percent be added to the tabulated friction head, or (1.15)(4.71) = 5.42 ft (1.66 m) of water.

Using the friction-head tables in Crocker and King—*Piping Handbook*, the friction head = 6.27 ft (1.9 m) per 1000 ft (304.8 m) of pipe with C = 130 for new, very smooth pipe. For this piping system, the friction-head loss = (942.4/1000)(6.27) = 5.91 ft (1.8 m) of water.

5. ***Use the Reynolds number method to determine the friction head***

In this method, the friction factor is determined by using the Reynolds number R and the relative roughness of the pipe ε/D, where ε = pipe roughness, ft, and D = pipe diameter, ft.

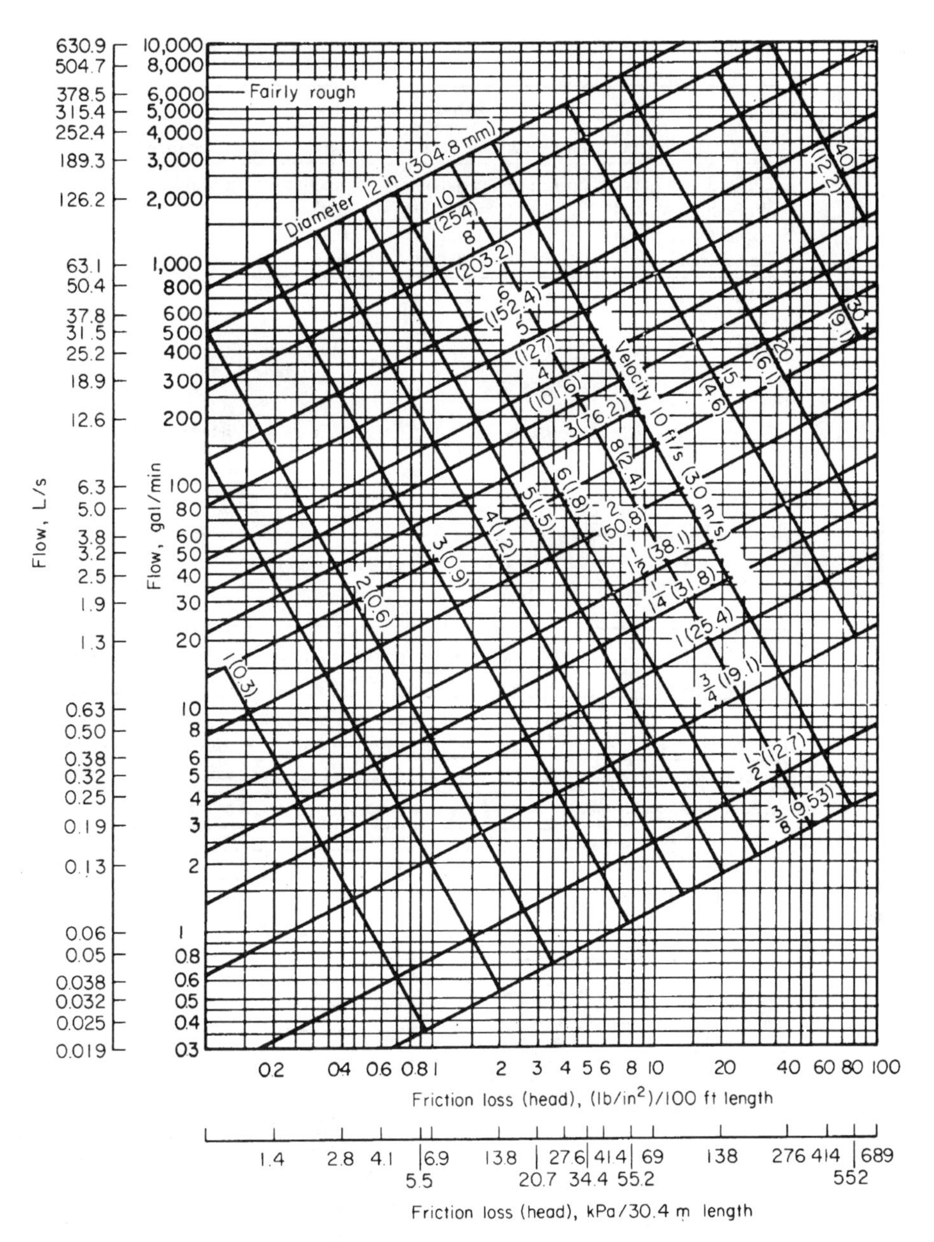

FIGURE 9 Friction loss in water piping.

For any pipe, $R = D\upsilon/\nu$, where υ = liquid velocity, ft/s, and ν = kinematic viscosity, ft²/s. Using King and Brater—*Handbook of Hydraulics*, υ = 4.07 ft/s (1.24 m/s), and ν = 0.00001059 ft²/s (0.00000098 m²/s) for water at 70°F (21.1°C). Then $R = (10/12)(4.07)/0.00001059 = 320.500$.

From Table 18 or the above reference, $\varepsilon = 0.00015$, and $\varepsilon/D = 0.00015/(10/12) = 0.00018$. From the Reynolds-number, relative-roughness, friction-factor curve in Fig. 10 or in Baumeister—*Standard Handbook for Mechanical Engineers*, the friction factor $f = 0.016$.

TABLE 18 Absolute Roughness Classification of Pipe Surfaces for Selection of Friction Factor *f* in Fig. 10

Commercial pipe surface (new)	Absolute roughness ε, ft	Absolute roughness ε, mm	Commercial pipe surface (new)	Absolute roughness ε, ft	Absolute roughness ε, mm
Glass, drawn brass, copper, lead	smooth	Smooth	Cast iron	0.00085	0.26
Wrought iron, steel	0.00015	0.05	Wood stave	0.0006–0.003	0.18–0.91
Asphalted cast iron	0.0004	0.12	Concrete	0.001–0.01	0.30–3.05
Galvanized iron	0.0005	0.15	Riveted steel	0.003–0.03	0.91–9.14

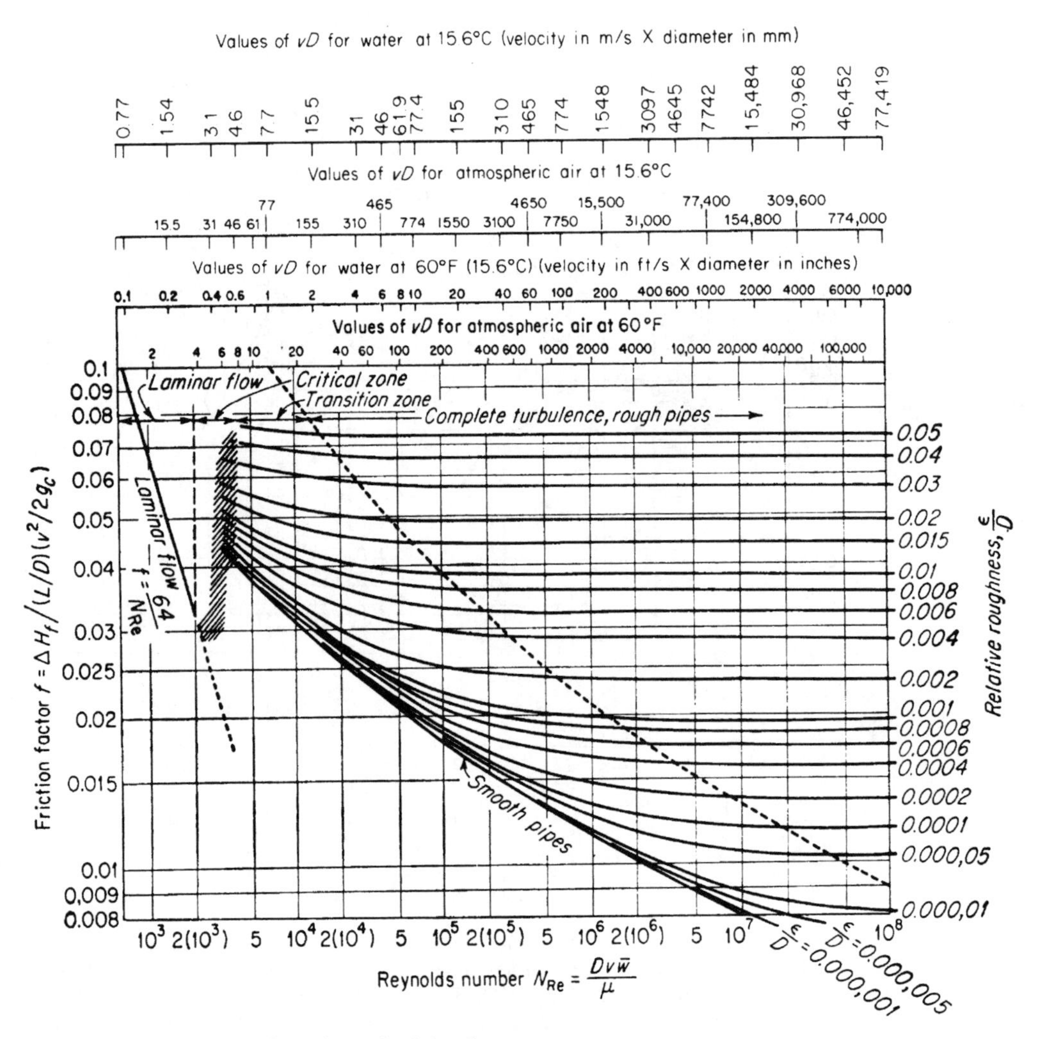

FIGURE 10 Friction factors for laminar and turbulent flow.

Apply the Darcy-Weisbach equation $h_f = f(l/D)(\upsilon^2/2g)$ where l = total pipe length, including the fittings' equivalent length, ft. Then $h_f = (0.016)(942.4/10/12)(4.07)^2/(2 \times 32.2) = 4.651$ ft (1.43 m) of water.

6. *Compare the results obtained*
Three different friction-head values were obtained: 8.71, 5.91, and 4.651 ft (2.7, 1.8, and 1.4 m) of water. The results show the variations that can be expected with the different methods. Actually, the Reynolds number method is probably the most accurate. As can be seen, the other two methods give safe results—i.e., the computed friction head is higher. The *Pipe Friction Manual*, published by the Hydraulic Institute, presents excellent simplified charts for use with the Reynolds number method.

Related Calculations. Use any of these methods to compute the friction-head loss for any type of pipe. The Reynolds number method is useful for a variety of liquids other than water—mercury, gasoline, brine, kerosene, crude oil, fuel oil, and lube oil. It can also be used for saturated and superheated steam, air, methane, and hydrogen.

RELATIVE CARRYING CAPACITY OF PIPES

What is the equivalent steam-carrying capacity of a 24-in (609.6-mm) inside-diameter pipe in terms of a 10-in (254-mm) inside-diameter pipe? What is the equivalent water-carrying capacity of a 23-in (584.2-mm) inside-diameter pipe in terms of a 13.25-in (336.6-mm) inside-diameter pipe?

Calculation Procedure:

1. *Compute the relative carrying capacity of the steam pipes*
For steam, air, or gas pipes, the number N of small pipes of inside diameter d_2 in equal to one pipe of larger inside diameter d_1 in is $N = (d_1^3\sqrt{d_2} + 3.6)/(d_2^3 + \sqrt{d_1} + 3.6)$. For this piping system, $N = (24^3 + \sqrt{10} + 3.6)/(10^3 + \sqrt{24} + 3.6) = 9.69$, say 9.7. Thus, a 24-in (609.6-mm) inside-diameter steam pipe has a carrying capacity equivalent to 9.7 pipes having a 10-in (254-mm) inside diameter.

2. *Compute the relative carrying capacity of the water pipes*
For water, $N = (d_2/d_1)^{2.5} = (23/13.25)^{2.5} = 3.97$. Thus, one 23-in (584-cm) inside-diameter pipe can carry as much water as 3.97 pipes of 13.25-in (336.6-mm) inside diameter.

Related Calculations. Crocker and King—*Piping Handbook* and certain piping catalogs (Crane, Walworth, National Valve and Manufacturing Company) contain tabulations of relative carrying capacities of pipes of various sizes. Most piping designers use these tables. However, the equations given here are useful for ranges not covered by the tables and when the tables are unavailable.

PRESSURE-REDUCING VALVE SELECTION FOR WATER PIPING

What size pressure-reducing valve should be used to deliver 1200 gal/h (1.26 L/s) of water at 140 lb/in^2 (275.8 kPa) if the inlet pressure is 140 lb/in^2 (965.2 kPa)?

Calculation Procedure:

1. *Determine the value capacity required*
Pressure-reducing valves in water systems operate best when the nominal load is 60 to 70 percent of the maximum load. Using 60 percent, we see that the maximum load for this valve = 1200/0.6 = 2000 gal/h (2.1 L/s).

TABLE 19 Maximum Capacities of Water Pressure-Reducing Valves, gal/h (L/s)

Inlet pressure		Valve size		
lb/in² (gage)	kPa	¾ in (19.1 mm)	1 in (25.4 mm)	1¼ in (31.8 mm)
120	827.3	1550 (1.6)	2000 (2.1)	4500 (4.7)
140	965.2	1700 (1.8)	2200 (2.3)	5000 (5.3)
160	1103.0	1850 (1.9)	2400 (2.5)	5500 (5.8)

Source: Clark-Reliance Corporation.

2. *Determine the valve size required*

Enter a valve capacity table in suitable valve engineering data at the valve inlet pressure, and project to the exact, or next higher, valve capacity. Thus, enter Table 19 at 140 lb/in² (965.2 kPa) and project to the next higher capacity, 2200 gal/h (2.3 L/s), since a capacity of 2000 gal/h (2.1 L/s) is not tabulated. Read at the top of the column the required valve size as 1 in (25.4 mm).

Some valve manufacturers present the capacity of their valves in graphical instead of tabular form. One popular chart, Fig. 11, is entered at the difference between the inlet and outlet pressures on the abscissa, or 140 – 40 = 100 lb/in² (689.4 kPa). Project vertically to the flow rate of 2000/60 = 33.3 gal/min (2.1 L/s). Read the valve size on the intersecting valve capacity curve, or on the next curve if there is no intersection with the curve. Figure 11 shows that a 1-in (25.4-mm) valve should be used. This agrees with the tabulated capacity.

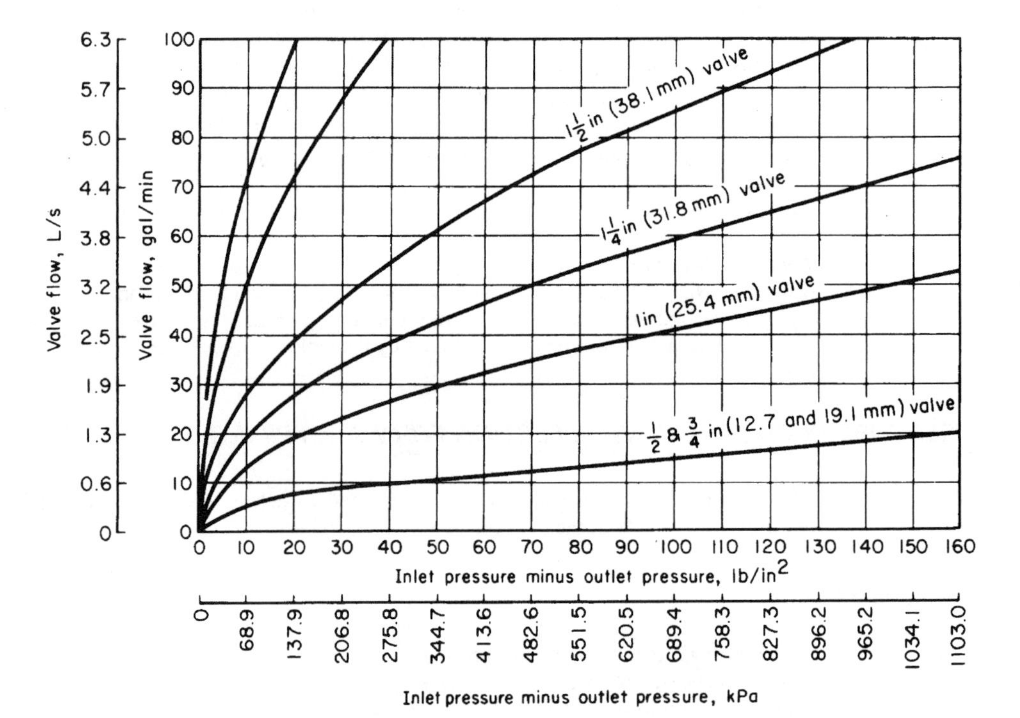

FIGURE 11 Pressure-reducing valve flow capacity. (*Foster Engineering Company.*)

Related Calculations. Use this method for pressure-reducing valves in any type of water piping—process, domestic, commercial—where the water temperature is 100°F (37.8°C) or less. Table 19 is from data prepared by the Clark-Reliance Corporation, Fig. 11 is from Foster Engineering Company data.

Some valve manufacturers use the valve flow coefficient C_v for valve sizing. This coefficient is defined as the flow rate, gal/min, through a valve of given size when the pressure loss across the valve is 1 lb/in^2 (6.9 kPa). Tabulations like Table 19 and flowcharts like Fig. 11 incorporate this flow coefficient and are somewhat easier to use. Their accuracy equals that of the flow coefficient method.

SIZING A WATER METER

A 6 × 4 in (152.4 × 101.6 mm) Venturi tube is used to measure water flow rate in a piping system. The dimensions of the meter are: inside pipe diameter $d_p = 6.094$ in (154.8 mm); throat diameter $d =$ 4.023 in (102.2 mm). The differential pressure is measured with a mercury manometer having water on top of the mercury. The average manometer reading for 1 h is 10.1 in (256.5 mm) of mercury. The temperature of the water in the pipe is 41°F (5.0°C), and that of the room is 77°F (25°C). Determine the water flow rate in lb/h, gal/h, and gal/min. Use the ASME Research Committee on Fluid Meters method in analyzing the meter.

Calculation Procedure:

1. *Convert the pressure reading to standard conditions*

The ASME meter equation constant is based on a manometer liquid temperature of 68°F (20.0°C). Therefore, the water and mercury density at room temperature, 77°F (25°C), and the water density at 68°F (20.0°C), must be used to convert the manometer reading to standard conditions by the equation $h_w = h_m(m_d - w_d)/w_s$, where h_w = equivalent manometer reading, in (mm) H_2O at 68°F (20.0°C); h_m = manometer reading at room temperature, in mercury; m_d = mercury density at room temperature, lb/ft^3; w_d = water density at room temperature, lb/ft^3; w_d = water density at standard conditions, 68°F (20.0°C), lb/ft^3. From density values from the ASME publication *Fluid Meters: Their Theory and Application*, $h_w = 10.1(844.88 - 62.244)/62.316 = 126.8$ in (322.1 cm) of water at 68°F (20.0°C).

2. *Determine the throat-to-pipe diameter ratio*

The throat-to-pipe diameter ratio $\beta = 4.023/6.094 = 0.6602$. Then $1/(1 - \beta^4)^{0.5}$ and $1/(1 - 0.6602^4)^{0.5} =$ 1.1111.

3. *Assume a Reynolds number value, and compute the flow rate*

The flow equation for a Venturi tube is w lb/h = $359.0\,(Cd^2/\sqrt{1-\beta^4})(w_{dp}h_w)^{0.5}$, where C = meter discharge coefficient, expressed as a function of the Reynolds number; w_{dp} = density of the water at the pipe temperature, lb/ft^3. With a Reynolds number greater than 250,000, C is a constant. As a first trial, assume $R > 250{,}000$ and $C = 0.984$ from *Fluid Meters*. Then $w = 359.0(0.984)(4.023)^2(1.1111)$ $(62.426 \times 126.8)^{0.5} = 565{,}020$ lb/h (254,259 kg/h), or 565,020/8.33 lb/gal = 67,800 gal/h (71.3 L/s), or 67,800/60 min/h = 1129 gal/min (71.23 L/s).

4. *Check the discharge coefficient by computing the Reynolds numbers*

For a water pipe, $R = 48w_s/(\pi d_p g u)$, where w_s = flow rate, lb/s = $w/3600$; u = coefficient of absolute viscosity. Using *Fluid Meters* data for water at 41°F (5°C), we find $R = 48(156.95)$ $[(\pi \times 6.094)(0.001004)] = 391{,}900$. Since C is constant for $R > 250{,}000$, use of $C = 0.984$ is correct, and no adjustment in the computations is necessary. Had the value of C been incorrect, another value would be chosen and the Reynolds number recomputed. Continue this procedure until a satisfactory value for C is obtained.

5. *Use an alternative solution to check the results*

Fluid Meters gives another equation for Venturi meter flow rate, that is, w lb/s = 0.525 $(Cd^2/\sqrt{1-\beta^4})[w_{dp}(p_1 - p_2)]^{0.5}$, where $p_1 - p_2$ is the manometer differential pressure in lb/in^2. Using the conversion factor in *Fluid Meters* for converting in of mercury under water at 77°F (25°C) to lb/in^2 (kPa),we get $p_1 - p_2$ = (10.1)(0.4528) = 4.573 lb/in^2 (31.5 kPa). Then w = (0.525) (0.984) $(4.023)^2(1.1111)(62.426 \times 4.573)^{05}$ = 156.9 lb/s (70.6 kg/s), or (156.9)(3600 s/h) = 564,900 lb/h (254,205 kg/h), or 564,900/8.33 lb/gal = 67,800 gal/h (71.3 L/s), or 67,800/60 min/h = 1129 gal/min (71.2 L/s). This result agrees with that computed in step 3 within 1 part in 5600. This is much less than the probable uncertainties in the values of the discharge coefficient and the differential pressure.

Related Calculations. Use this method for any Venturi tube serving cold-water piping in process, industrial, water-supply, domestic, or commercial service.

EQUIVALENT LENGTH OF A COMPLEX SERIES PIPELINE

Figure 12 shows a complex series pipeline made up of four lengths of different size pipe. Determine the equivalent length of this pipe if each size of pipe has the same friction factor.

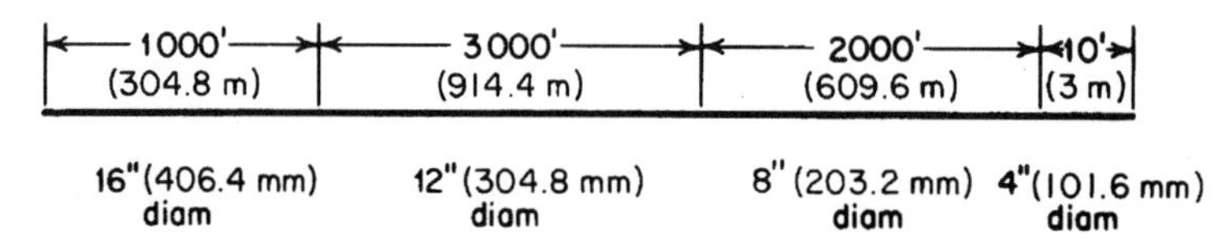

FIGURE 12 Complex series pipeline.

Calculation Procedure:

1. *Select the pipe size for expressing the equivalent length*

The usual procedure in analyzing complex pipelines is to express the equivalent length in terms of the smallest, or next to smallest, diameter pipe. Choose the 8-in (203.2-mm) size as being suitable for expressing the equivalent length.

2. *Find the equivalent length of each pipe*

For any complex series pipeline having equal friction factors in all the pipes, L_e = equivalent length, ft, of a section of constant diameter = (actual length of section, ft) (inside diameter, in, of pipe used to express the equivalent length/inside diameter, in, of section under consideration)^5.

For the 16-in (406.4-mm) pipe, $L_e = (1000)(7.981/15.000)^5$ = 42.6 ft (12.9 m). The 12-in (304.8-mm) pipe is next; for it $L_e = (3000)(7.981/12.00)^5$ = 390 ft (118.9 m). For the 8-in (203.2-mm) pipe, the equivalent length = actual length = 2000 ft (609.6 m). For the 4-in (101.6-mm) pipe, $L_e = (10)(7.981/4.026)^5$ = 306 ft (93.3 m). Then the total equivalent length of 8-in (203.2-mm) pipe = sum of the equivalent lengths = 42.6 + 390 + 2000 + 306 = 2738.6 ft (834.7 m), or, by rounding off, 2740 ft (835.2 m) of 8-in (203.2-mm) pipe will have a frictional resistance equal to the complex series pipeline shown in Fig. 12. To compute the actual frictional resistance, use the methods given in previous calculation procedures.

Related Calculations. Use this general procedure for any complex series pipeline conveying water, oil, gas, steam, etc. See Crocker and King—*Piping Handbook* for derivation of the flow equations. Use the tables in Crocker and King to simplify finding the fifth power of the inside diameter of a pipe. The method of the next calculation procedure can also be used if a given flow rate is assumed.

Choosing a flow rate of 1000 gal/min (63.1 L/s) and using the tables in the Hydraulic Institute *Pipe Friction Manual* give an equivalent length of 2770 ft (844.3 m) for the 8-in (203.2-mm) pipe. This compares favorably with the 2740 ft (835.2 m) computed above. The difference of 30 ft (9.1 m) is negligible and can be accounted for by calculator variations.

The equivalent length is found by summing the friction-head loss for 1000-gal/min (63.1-L/s) flow for each length of the four pipes—16, 12, 8, and 4 in (406, 305, 203, and 102 mm)—and dividing this by the friction-head loss for 1000 gal/min (63.1 L/s) flowing through an 8-in (203.2-mm) pipe. Be careful to observe the units in which the friction-head loss is stated, because errors are easy to make if the units are ignored.

EQUIVALENT LENGTH OF A PARALLEL PIPING SYSTEM

Figure 13 shows a parallel piping system used to supply water for industrial needs. Determine the equivalent length of a single pipe for this system. All pipes in the system are approximately horizontal.

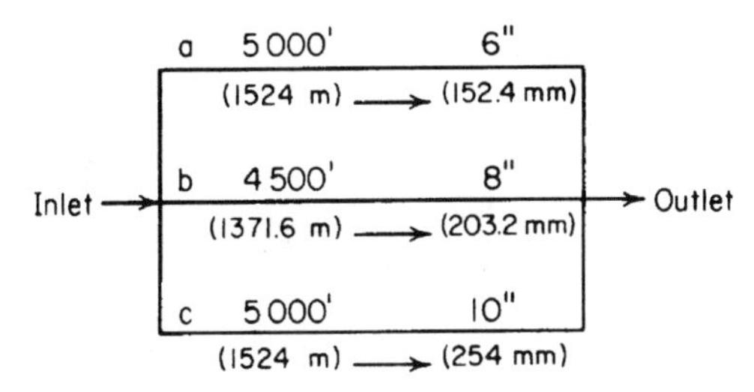

FIGURE 13 Parallel piping system.

Calculation Procedure:

1. *Assume a total head loss for the system:*
To determine the equivalent length of a parallel piping system, assume a total head loss for the system. Since this head loss is assumed for computation purposes only, its value need not be exact or even approximate. Assume a total head loss of 50 ft of water for each pipe in this system.

2. *Compute the flow rate in each pipe in the system*
Assume that the roughness coefficient C in the Hazen-Williams formula is equal for each of the pipes in the system. This is a valid assumption. Using the assumed value of C, compute the flow rate in each pipe. To allow for possible tuberculation of the pipe, assume that $C = 100$.

The Hazen-Williams formula is given in a previous calculation procedure and can be used to solve for the flow rate in each pipe. A more rapid way to make the computation is to use the friction-loss tabulations for the Hazen-Williams formula in Crocker and King—*Piping Handbook*, the Hydraulic Institute—*Pipe Friction Manual,* or a similar set of tables.

Using such a set of tables, enter at the friction-head loss equal to 50 ft (15.2 m) per 5000 ft (1524 m) of pipe for the 6-in (152.4-mm) line. Find the corresponding flow rate Q gal/min. Using the Hydraulic Institute tables, $Q_a = 270$ gal/min (17.0 L/s); $Q_b = 580$ gal/min (36.6 L/s); $Q_c = 1000$ gal/min (63.1 L/s). Hence, the total flow = 270 + 580 + 1000 = 1850 gal/min (116.7 L/s).

3. *Find the equivalent size and length of the pipe*
Using the Hydraulic Institute tables again, look for a pipe having a 50-ft (15.2-m) head loss with a flow of 1850 gal/min (116.7 L/s). Any pipe having a discharge equal to the sum of the discharge rates for all the pipes, at the assumed friction head, is an equivalent pipe.

Interpolating friction-head values in the 14-in (355.6-mm) outside-diameter [13.126-in (333.4-mm) inside-diameter] table shows that 5970 ft (1820 m) of this pipe is equivalent to the system in Fig. 13. This equivalent size can be used in any calculations related to this system—selection of a pump, determination of head loss with longer or shorter mains, etc. If desired, another equivalent-size pipe could be found by entering a different pipe-size table. Thus, 5310 ft (1621.5 m) of 14-in (355.6-mm) pipe [12.814-in (326.5-mm) inside diameter] is also equivalent to this system.

Related Calculation. Use this procedure for any liquid—water, oil, gasoline, brine—flowing through a parallel piping system. The pipes are assumed to be full at all times.

MAXIMUM ALLOWABLE HEIGHT FOR A LIQUID SIPHON

What is the maximum height *h* ft (m), Fig. 14, that can be used for a siphon in a water system if the length of the pipe from the water source to its highest point is 500 ft (152.4 m), the water velocity is 13.0 ft/s (3.96 m/s), the pipe diameter is 10 in (254 mm), and the water temperature is 70°F (21.1°C) if 3200 gal/min (201.9 L/s) is flowing?

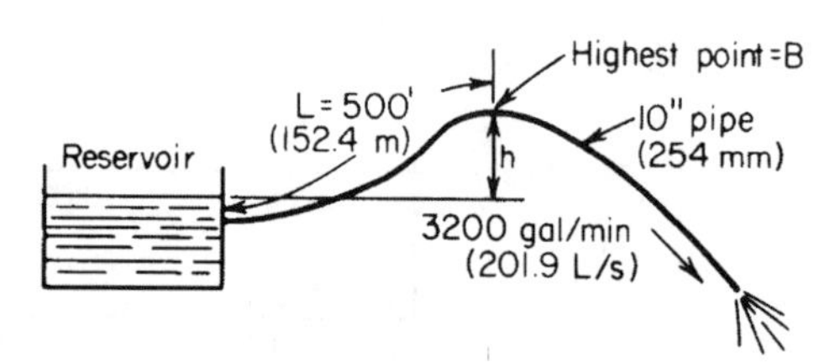

FIGURE 14 Liquid siphon piping system.

Calculation Procedure:

1. ***Compute the velocity of the water in the pipe***
From an earlier calculation procedure, $\upsilon = gpm/(2.448d^2)$. With an internal diameter of 10.020 in (254.5 mm), $\upsilon = 3200/[(2.448)(10.02)^2] = 13.0$ ft/s (3.96 m/s).

2. ***Determine the vapor pressure of the water***
Using a steam table, we see that the vapor pressure of water at 70°F (21.1°C) is $p_\upsilon = 0.3631$ lb/in^2 (abs)(2.5 kPa), or (0.3631)(144 in^2/ft^2) = 52.3 lb/ft^2 (2.5 kPa). The specific volume of water at 70°F (21.1°C) is, from a steam table, 0.01606 ft^3/lb (0.001 m^3/kg). Converting this to density at 70°F (21.1°C), density = 1/0.01606 = 62.2 lb/ft^3 (995.8 kg/m^3). The vapor pressure in ft of 70°F (21.1°C) water is then f_υ = (52.3 lb/ft^2)/(62.2 lb/ft^3) = 0.84 ft (0.26 m) of water.

3. ***Compute or determine the friction-head loss and velocity head***
From the reservoir to the highest point of the siphon, *B*, Fig. 14, the friction head in the pipe must be overcome. Use the Hazen-Williams or a similar formula to determine the friction head, as given in earlier calculation procedures or a pipe-friction table. From the Hydraulic Institute *Pipe Friction Manual*, $h_f = 4.59$ ft per 100 ft (1.4 m per 3.5 m), or (500/100)(4.59) = 22.95 ft (7.0 m). From the same table, velocity head = 2.63 ft/s (0.8 m/s).

4. ***Determine the maximum height for the siphon***
For a siphon handling water, the maximum allowable height *h* at sea level with an atmospheric pressure of 14.7 lb/in^2 (abs)(101.3 kPa) = [14.7 × (144 in^2/ft^2)/(density of water at operating temperature, lb/ft^3) – (vapor pressure of water at operating temperature, ft + 1.5 × velocity head, ft + friction head, ft)].

For this pipe, $h = 14.7 \times 144/62.2 - (0.84 + 1.5 \times 2.63 + 22.95) = 11.32$ ft (3.45 m). In actual practice, the value of h is taken as 0.75 to 0.8 the computed value. Using 0.75 gives $h = (0.75)(11.32) = 8.5$ ft (2.6 m).

Related Calculations. Use this procedure for any type of siphon conveying a liquid—water, oil, gasoline, brine, etc. Where the liquid has a specific gravity different from that of water, i.e., less than or greater than 1.0, proceed as above, expressing all heads in ft of liquid handled. Divide the resulting siphon height by the specific gravity of the liquid. At elevations above atmospheric, use the actual atmospheric pressure instead of 14.7 lb/in^2 (abs) (101.3 kPa).

WATER-HAMMER EFFECTS IN LIQUID PIPELINES

What is the maximum pressure developed in a 200-lb/in^2 (1378.8-kPa) water pipeline if a valve is closed nearly instantly or pumps discharging into the line are all stopped at the same instant? The pipe is 8-in (203.2-mm) Schedule 40 steel, and the water flow rate is 280 gal/min (176.7 L/s). What maximum pressure is developed if the valve closes in 5 s and the line is 5000 ft (1524 m) long?

Calculation Procedure:

1. *Determine the velocity of the pressure wave*

For any pipe, the velocity of the pressure wave during water hammer is found from $\upsilon_w = 4720/(1 + Kd/Et)^{0.5}$, where υ_w = velocity of the pressure wave in the pipeline, ft/s; K = bulk modulus of the liquid in the pipeline = 300,000 for water; d = internal diameter of pipe, in; E = modulus of elasticity of pipe material, lb/in^2 = 30×10^6 lb/in^2 (206.8 Gpa) for steel; t = pipe-wall thickness, in. For 8-in (203.2-mm) Schedule 40 steel pipe and data from a table of pipe properties, $\upsilon_w = 4720/[1 + 300{,}000 \times 7.981/(30 \times 10^6 \times 0.322)]^{0.5} = 4225.6$ ft/s (1287.9 m/s).

2. *Compute the pressure increase caused by water hammer*

The pressure increase p_1 lb/in^2 due to water hammer = $\upsilon_w\upsilon/[32.2(2.31)]$, where υ = liquid velocity in the pipeline, ft/s; 32.2 = acceleration due to gravity, ft/s^2; 2.31 ft of water = 1-lb/in^2 (6.9-kPa) pressure.

For this pipe, $\upsilon = 0.4085\ gpm/d^2 = 0.4085(2800)/(7.981)^2 = 18.0$ ft/s (5.5 m/s). Then $p_i = (4225.6)(18)/[32.2(2.31)] = 1022.56$ lb/in^2 (7049.5 kPa). The maximum pressure developed in the pipe is then p_1 + pipe operating pressure = 1022.56 + 200 = 1222.56 lb/in^2 (8428.3 kPa).

3. *Compute the hammer pressure rise caused by valve closure*

The hammer pressure rise caused by valve closure p_υ lb/in^2 = $2p_i\,L/\upsilon_w\,T$, where L = pipeline length, ft; T = valve closing time, s. For this pipeline, $p_\upsilon = 2(1022.56)(5000)/[(4225.6)(5)] = 484$ lb/in^2 (3336.7 kPa). Thus, the maximum pressure in the pipe will be 484 + 200 = 684 lb/in^2 (4467.3 kPa).

Related Calculations. Use this procedure for any type of liquid—water, oil, etc.—in a pipeline subject to sudden closure of a valve or stoppage of a pump or pumps. The effects of water hammer can be reduced by relief valves, slow-closing check valves on pump discharge pipes, air chambers, air spill valves, and air injection into the pipeline.

SPECIFIC GRAVITY AND VISCOSITY OF LIQUIDS

An oil has a specific gravity of 0.8000 and a viscosity of 200 SSU (Saybolt Seconds Universal) at 60°F (15.6°C). Determine the API gravity and Bé gravity of this oil and its weight in lb/gal (kg/L). What is the kinematic viscosity in cSt? What is the absolute viscosity in cP?

Calculation Procedure:

1. ***Determine the API gravity of the liquid***

For any oil at 60°F (15.6°C), its specific gravity S, in relation to water at 60°F (15.6°C), is $S = 141.5/(131.5 + °\text{API})$, or $°\text{API} = (141.5 - 131.5S)/S$. For this oil, °API = [141.5 − 131.5(0.80)]/0.80 = 45.4 °API.

2. ***Determine the Bé gravity of the liquid***

For any liquid lighter than water, $S = 140/(130 + \text{Bé})$, or $\text{Bé} = (140 - 130S)/S$. For this oil, Bé = [140 − 130(0.80]/0.80 = 45 Bé.

3. ***Compute the weight per gal of liquid***

With a specific gravity of S, the weight of 1 ft^3 of oil = (S)[weight of 1 ft^3 (1 m^3) of fresh water at 60°F (15.6°C)] = (0.80)(62.4) = 49.92 lb/ft^3 (799.2 kg/m^3). Since 1 gal (3.8 L) of liquid occupies 0.13368 ft^3, the weight of this oil is (49.92) (0.13368) = 6.66 lb/gal (0.79 kg/L).

4. ***Compute the kinematic viscosity of the liquid***

For any liquid having an SSU viscosity greater than 100 s, the kinematic viscosity $k = 0.220$ (SSU) = 135/SSU cSt. For this oil, $k = 0.220(200) - 135/200 = 43.325$ cSt.

5. ***Convert the kinematic viscosity to absolute viscosity***

For any liquid, the absolute viscosity, cP = (kinematic viscosity, cSt)(density). Thus, for this oil, the absolute viscosity = (43.325)(49.92) = 2163 cP.

Related Calculations. For liquids *heavier* than water, $S = 145/(145 - \text{Bé})$. When the SSU viscosity is between 32 and 99 SSU, $k = 0.226$ (SSU) − 195/SSU cSt. Modern terminology for absolute viscosity is dynamic viscosity. Use these relations for any liquid—brine, gasoline, crude oil, Kerosene, Bunker C, diesel oil, etc. Consult the *Pipe Friction Manual* and Crocker and King—*Piping Handbook* for tabulations of typical viscosities and specific gravities of various liquids.

PRESSURE LOSS IN PIPING HAVING LAMINAR FLOW

Fuel oil at 300°F (148.9°C) and having a specific gravity of 0.850 is pumped through a 30,000-ft (9144-m) long 24-in (609.6-mm) pipe at the rate of 500 gal/min (31.6 L/s). What is the pressure loss if the viscosity of the oil is 75 cP (0.075 Pa · s)?

Calculation Procedure:

1. ***Determine the type of flow that exists***

Flow is laminar (also termed *viscous*) if the Reynolds number R for the liquid in the pipe is less than 1200. Turbulent flow exists if the Reynolds number is greater than 2500. Between these values is a zone in which either condition may exist, depending on the roughness of the pipe wall, entrance conditions, and other factors. Avoid sizing a pipe for flow in this critical zone because excessive pressure drops result without a corresponding increase in the pipe discharge.

Compute the Reynolds number from $R = 3.162G/kd$, where G = flow rate gal/min (L/s); k = kinematic viscosity of liquid, cSt = viscosity z, cP/specific gravity of the liquid S; d = inside diameter of pipe, in (cm). From a table of pipe properties, $d = 22.626$ in (574.7 mm). Also, $k = z/S = 75/0.85 = 88.2$ cSt. Then $R = 3162(500)/[88.2(22.626)] = 792$. Since $R < 1200$, laminar flow exists in this pipe.

2. ***Compute the pressure loss by using the Poiseuille formula***

The Poiseuille formula gives the pressure drop p_d lb/in^2 (kPa) = $2.73(10^{-4})\,luG/d^4$, where l = total length of pipe, including equivalent length of fittings, ft; u = absolute viscosity of liquid, cP (Pa · s);

TABLE 20 Reynolds Number

		Numerator			Denominator	
Reynolds Number R	Coefficient	First symbol	Second symbol	Third symbol	Fourth symbol	Fifth symbol
$D\upsilon\rho/\mu$	—	ft	ft/s	lb/ft^3	lb mass/(ft · s)	
$124d\upsilon\rho/z$	124	in	ft/s	lb/ft^3	cP	
$50.7G\rho/dz$	50.7	gal/min	lb/ft^3	—	in	cP
$6.32W/dz$	6.32	lb/h	—	—	in	cP
$35.5B\rho/dz$	35.5	bbl/h	lb/ft^3	—	in	cP
$7742d\upsilon/k$	7,742	in	ft/s	—	—	cP
$3162G/dk$	3,162	gal/min	—	—	in	cP
$2214B/dk$	2,214	bbl/h	—	—	in	cP
$22{,}735q\rho/dz$	22,735	ft^3/s	lb/ft^3	—	in	cP
$378.9Q\rho/dz$	378.9	ft^3/min	lb/ft^3	—	in	cP

G = flow rate gal/min (L/s); d = inside diameter of pipe, in (cm). For this pipe, $p_d = 2.73(10^{-4})$ (10,000)(75)(500)/262,078 = 1.17 lb/in^2 (8.1 kPa).

Related Calculations. Use this procedure for any pipe in which there is laminar flow of the liquid. Other liquids for which this method can be used include water, molasses, gasoline, brine, kerosene, and mercury. Table 20 gives a quick summary of various ways in which the Reynolds number can be expressed. The symbols in Table 20, in the order of their appearance, are D = inside diameter of pipe, ft (m); v = liquid velocity, ft/s (m/s); p = liquid density, lb/ft^3 (kg/m^3); μ = absolute viscosity of liquid, lb mass/(ft · s) [kg/(m · s)]; d = inside diameter of pipe, in (cm). From a table of pipe properties, d = 22.626 in (574.7 mm). Also, $k = z/S$ liquid flow rate, lb/h (kg/h); B = liquid flow rate, bbl/h (L/s); k = kinematic viscosity of the liquid, cSt; q liquid flow rate, ft^3 (m^3/s); Q = liquid flow rate, ft^3/min (m^3/min). Use Table 20 to find the Reynolds number for any liquid flowing through a pipe.

DETERMINING THE PRESSURE LOSS IN OIL PIPES

What is the pressure drop in a 5000-ft (1524-m) long 6-in (152.4-mm) oil pipe conveying 500 bbl/h (22.1 L/s) of kerosene having a specific gravity of 0.813 at 65°F (18.3°C), which is the temperature of the liquid in the pipe? The pipe is Schedule 40 steel.

Calculation Procedure:

1. *Determine the kinematic viscosity of the oil*
Use Fig. 15 and Table 21 or the Hydraulic Institute—*Pipe Friction Manual* kinematic viscosity and Reynolds number chart to determine the kinematic viscosity of the liquid. Enter Table 21 at kerosene, and rind the coordinates as X = 10.2, Y = 16.9. Using these coordinates, enter Fig. 15 and find the absolute viscosity of kerosene at 65°F (18.3°C) as 2.4 cP. By the method of a previous calculation procedure, the kinematic viscosity = absolute viscosity, cP/specific gravity of the liquid = 2.4/0.813 = 2.95 cSt. This value agrees closely with that given in the *Pipe Friction Manual.*

2. *Determine the Reynolds number of the liquid*
The Reynolds number can be found from the *Pipe Friction Manual* chart mentioned in step 1 or computed from $R = 2214B/(dk) = 2214(500)/[(6.065)(2.95)] = 61{,}900$.

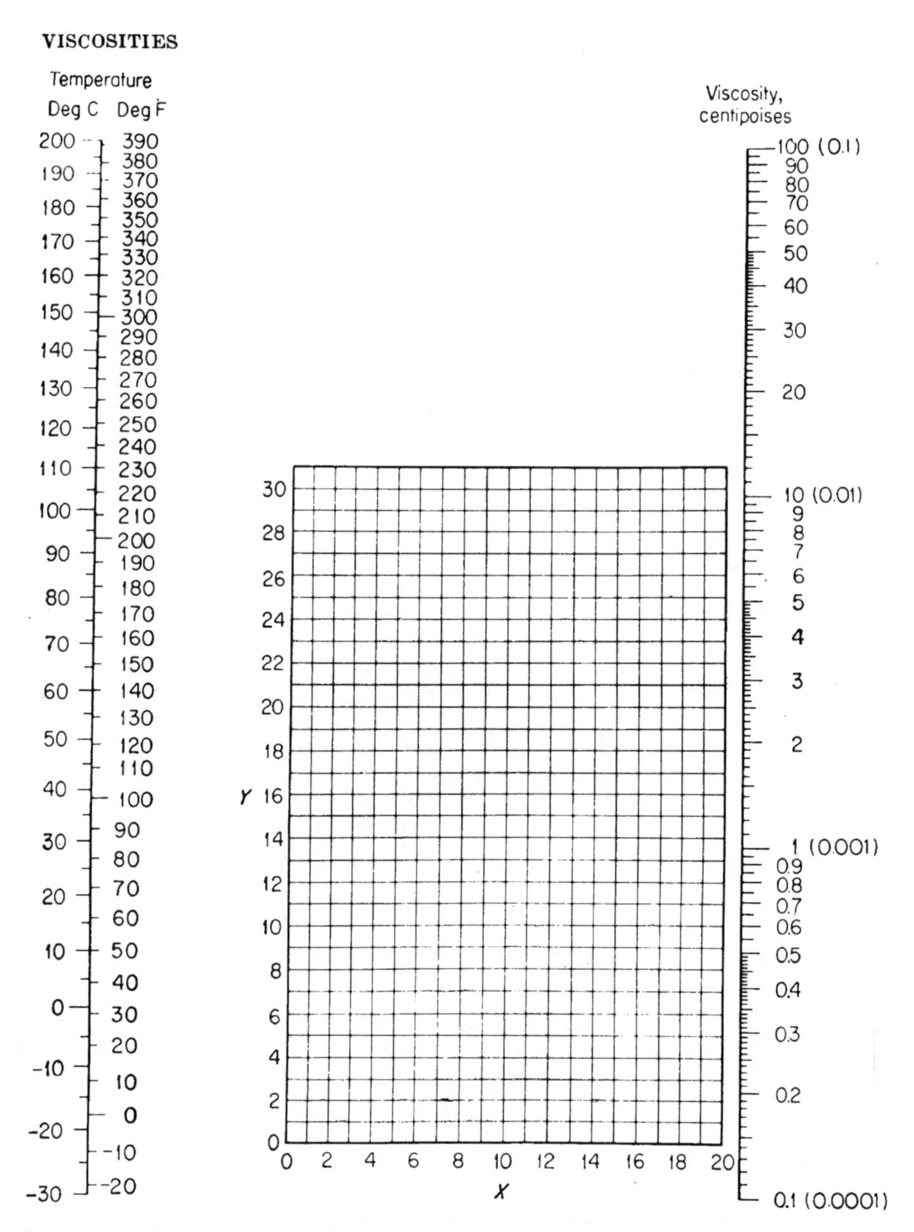

FIGURE 15 Viscosities of liquids at 1 atm. For coordinates, see Table 21.

To use the *Pipe Friction Manual* chart, compute the velocity of the liquid in the pipe by converting the flow rate to ft^3/s. Since there is 42 gal/bbl (0.16 L) and 1 gal (0.00379 L) = 0.13368 ft^3 (0.00378 m^3), 1 bbl = (42)(0.13368) = 5.6 ft^3 (0.16 m^3). With a flow rate of 500 bbl/h (79.5 m^3/h), the equivalent flow = (500)(5.6) = 2800 ft^3/h (79.3 m^3/h), or 2800/3600 s/h = 0.778 ft^3/s (0.02 m^3/s). Since 6-in (152.4-mm) Schedule 40 pipe has a cross-sectional area of 0.2006 ft^2 (0.02 m^2) internally, the liquid velocity = 0.778/0.2006 = 3.88 ft/s (1.2 m/s). Then, the product (velocity, ft/s) (internal

TABLE 21 Coordinates for Use with Figure 15

No.	Liquid	X	Y
1	Acetaldehyde	15.2	4.8
	Acetic acid		
2	100%	12.1	14.2
3	70%	9.5	17.0
4	Acetic anhydride	12.7	12.8
	Acetone		
5	100%	14.5	7.2
6	35%	7.9	15.0
7	Allyl alcohol	10.2	14.3
	Ammonia		
8	100%	12.6	2.0
9	26%	10.1	13.9
10	Amyl acetate	11.8	12.5
11	Amyl alcohol	7.5	18.4
12	Aniline	8.1	18.7
13	Anisole	12.3	13.5
14	Arsenic trichloride	13.9	14.5
15	Benzene	12.5	10.9
	Brine:		
16	$CaCl_2$, 25%	6.6	15.9
17	NaCl, 25%	10.2	16.6
18	Bromine	14.2	13.2
19	Bromotoluene	20.0	15.9
20	Butyl acetate	12.3	11.0
21	Butyl alcohol	8.6	17.2
22	Butyric acid	12.1	15.3
23	Carbon dioxide	11.6	0.3
24	Carbon disulfide	16.1	7.5
25	Carbon tetrachloride	12.7	13.1
26	Chlorobenzene	12.3	12.4
27	Chloroform	14.4	10.2
28	Chlorosulfonic acid	11.2	18.1
	Chlorotoluene:		
29	Ortho	13.0	13.3
30	Meta	13.3	12.5
31	Para	13.3	12.5
32	Cresol, meta	2.5	20.8
33	Cyclohexanol	2.9	24.3
34	Dibromoethane	12.7	15.8
35	Dichloroethane	13.2	12.2
36	Dichloromethane	14.6	8.9
37	Diethyl oxalate	11.0	16.4
38	Dimethyl oxalate	12.3	15.8
39	Diphenyl	12.0	18.3
40	Dipropyl oxalate	10.3	17.7
41	Ethyl acetate	13.7	9.1
	Ethyl alcohol:		
42	100%	10.5	13.8
43	95%	9.8	14.3
44	40%	6.5	16.6
45	Ethyl benzene	13.2	11.5
46	Ethyl bromide	14.5	8.1
47	Ethyl chloride	14.8	6.0
48	Ethyl ether	14.5	5.3
49	Ethyl formate		
50	Ethyl iodide	14.7	10.3
51	Ethylene glycol	6.0	23.6
52	Formic acid	10.7	15.8
53	Freon-11	14.4	9.0
54	Freon-12	16.8	5.6
55	Freon-21	15.7	7.5
56	Freon-22	17.2	4.7
57	Freon-13	12.5	11.4
	Glycerol:		
58	100%	2.0	30.0
59	50%	6.9	19.6
60	Heptene	14.1	8.4
61	Hexane	14.7	7.0
62	Hydrochloric acid, 31.5%	13.0	16.6
63	Isobutyl alcohol	7.1	18.0
64	Isobutyric acid	12.2	14.4
65	Isopropyl alcohol	8.2	16.0
66	Kerosene	10.2	16.9
67	Linseed oil, raw	7.5	27.2
68	Mercury	18.4	16.4
	Methanol:		
69	100%	12.4	10.5
70	90%	12.3	11.8
71	40%	7.8	15.5
72	Methyl acetate	14.2	8.2
73	Methyl chloride	15.0	3.8
74	Methyl ethyl ketone	13.9	8.6
75	Naphthalene	7.9	18.1
	Nitric acid:		
76	95%	12.8	13.8
77	60%	10.8	17.0
78	Nitrobenzene	10.6	16.2
79	Nitrotoluene	11.0	17.0
80	Octane	13.7	10.0
81	Octyl alcohol	6.6	21.1
82	Pentachloroethane	10.9	17.3
83	Pentane	14.9	5.2
84	Phenol	6.9	20.8
85	Phosphorus tribriomide	13.8	16.7
86	Phosphorus trichloride	16.2	10.9
87	Propionic acid	12.8	13.8
88	Propyl alcohol	9.1	16.5
89	Propyl bromide	14.5	9.6
90	Propyl chloride	14.4	7.5
91	Propyl iodide	14.1	11.6
92	Sodium	16.4	13.9
93	Sodium hydroxide, 50%	3.2	25.8
94	Stannic chloride	13.5	12.8
95	Sulfur dioxide	15.2	7.1
	Sulfur acid:		
96	110%	7.2	27.4
97	98%	7.0	24.8
98	60%	10.2	21.3
99	Sulfuryl chloride	15.2	12.4
100	Tetrachloroethane	11.9	15.7
101	Tetrachloroethylene	14.2	12.7
102	Titanium tetrachloride	14.4	12.3
103	Toluene	13.7	10.4
104	Trichloroethylene	14.8	10.5
105	Turpentine	11.5	14.9
106	Vinyl acetate		
107	Water	10.2	13.0
	Xylene:		
108	Ortho	13.5	12.1
109	Meta	13.9	10.6
110	Para	13.9	10.9

diameter, in) = (3.88)(6.065) = 23.75 ft/s. In the *Pipe Friction Manual*, project horizontally from the kerosene specific-gravity curve to the υd product of 23.75, and read the Reynolds number as 61,900, as before. In general, the Reynolds number can be found more quickly by computing it using the appropriate relation given in an earlier calculation procedure, unless the flow velocity is already known.

3. *Determine the friction factor of this pipe*
Enter Fig. 16 at the Reynolds number value of 61,900, and project to the curve 4 as indicated by Table 22. Read the friction factor as 0.0212 at the left. Alternatively, the *Pipe Friction Manual* friction-factor chart could be used, if desired.

4. *Compute the pressure loss in the pipe*
Use the Fanning formula $p_d = 1.06(10^{-4})f\rho/B^2/d^5$. In this formula, ρ = density of the liquid, lb/ft^3. For kerosene, ρ = (density of water, lb/ft^3)(specific gravity of the kerosene) = (62.4)(0.813) = 50.6 lb/ft^3(810.1 kg/m^3). Then $p_d = 1.06(10^{-4}) \times (0.0212)(50.6)(5000)(500)^2/8206 = 17.3$ lb/in^2 (119.3 kPa).

Related Calculations. The Fanning formula is popular with oil-pipe designers and can be stated in various ways: (1) with velocity υ ft/s, $p_d = 1.29(10^{-3})\, f\rho V^2 l/d$; (2) with velocity V ft/min, $p_d = 3.6(10^{-7})f\rho V^2 l/d$; (3) with flow rate in G gal/min, $p_d = 2.15(10^{-4})f\rho lG^2 l/d$; (4) with the flow rate in W lb/h, $p_d = 3.36(10^{-6})flW^2/d^5\rho$.

Use this procedure for any petroleum product—crude oil, kerosene, benzene, gasoline, naphtha, fuel oil, Bunker *C*, diesel oil, toluene, etc. The tables and charts presented here and in the *Pipe Friction Manual* save computation time.

FLOW RATE AND PRESSURE LOSS IN COMPRESSED-AIR AND GAS PIPING

Dry air at 80°F (26.7°C) and 150 lb/in^2 (abs) (1034 kPa) flows at the rate of 500 ft^3/min (14.2 m^3/min) through a 4-in (101.6-mm) Schedule 40 pipe from the discharge of an air compressor. What are the flow rate in lb/h and the air velocity in ft/s? Using the Fanning formula, determine the pressure loss if the total equivalent length of the pipe is 500 ft (152.4 m).

Calculation Procedure:

1. *Determine the density of the air or gas in the pipe*
For air or a gas, $pV = MRT$, where p = absolute pressure of the gas, lb/ft^2 (abs); V = volume of M lb of gas, ft^3; M = weight of gas, lb; R = gas constant, ft · lb/(lb · °F); T = absolute temperature of the gas, °R. For this installation, using 1 ft^3 of air, $M = pV/(RT)$, M = (150)(144)/[(53.33)(80 + 459.7)] = 0.750 lb/ft (12.0 kg/m^3). The value of R in this equation was obtained from Table 23.

2. *Compute the flow rate of the air or gas*
For air or a gas, the flow rate W_h lb/h = (60) (density, lb/ft^3)(flow rate, ft^3/min), or W_h = (60)(0.750)(500) = 22,500 lb/h (10,206 kg/h).

3. *Compute the velocity of the air or gas in the pipe*
For any air or gas pipe, velocity of the moving fluid υ ft/s = 183.4 $W_h/3600\, d^2\rho$, where d = internal diameter of pipe, in; ρ = density of fluid, lb/ft^3. For this system, υ = (183.4)(22,500)/[(3600)(4.026)^2(0.750)] = 94.3 ft/s (28.7 m/s).

4. *Compute the Reynolds number of the air or gas*
The viscosity of air at 80°F (26.7°C) is 0.0186 cP, obtained from Crocker and King—*Piping Handbook*, Perry et al.—*Chemical Engineers' Handbook*, or a similar reference. Then, by using the Reynolds number relation given in Table 20, $R = 6.32W/(dz)$ = (6.32)(22,500)/[(4.026)(0.0186)] = 1.899,000.

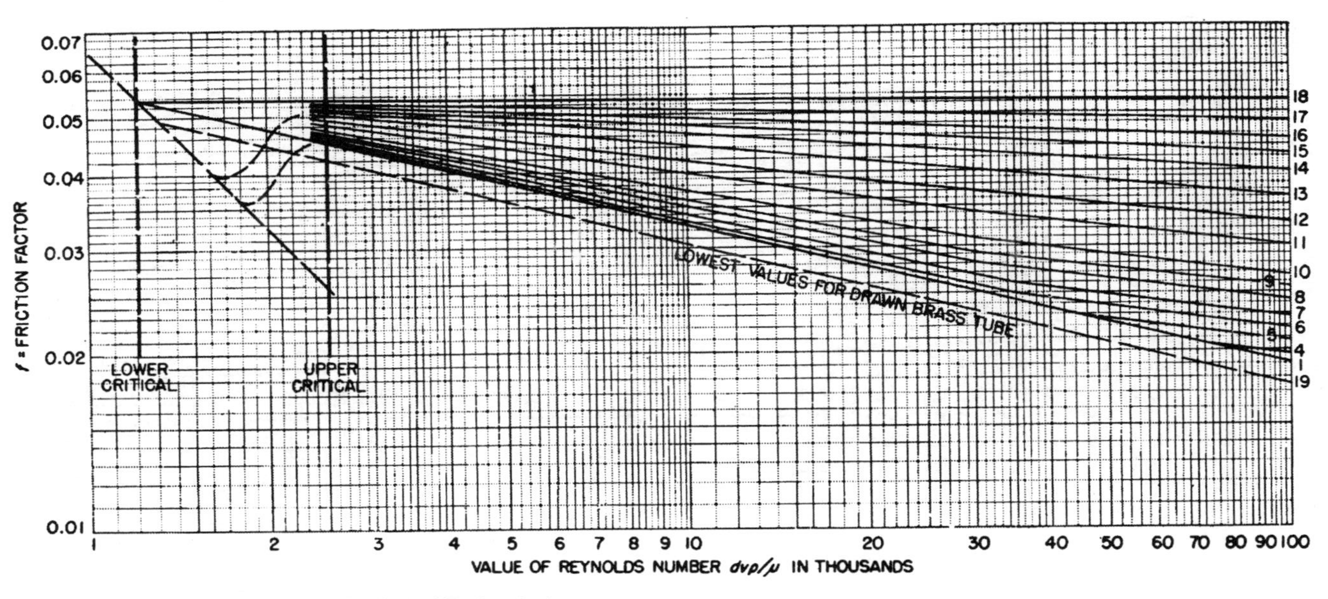

FIGURE 16A Friction-factor curves. (*Mechanical Engineering*.)

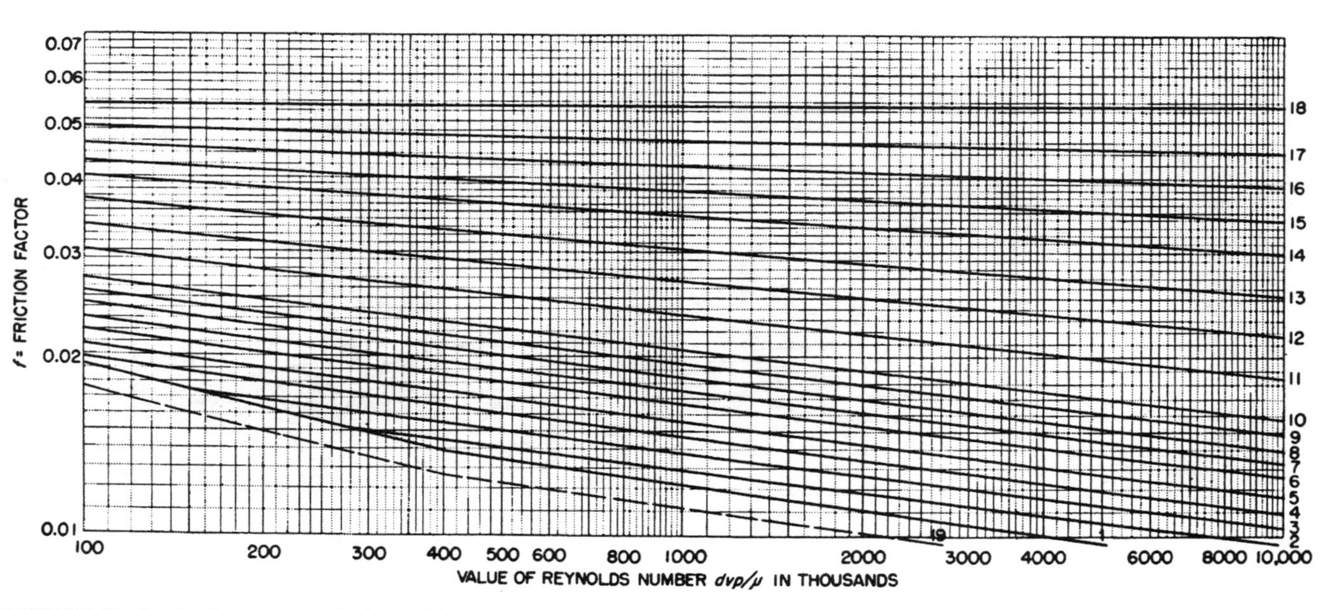

FIGURE 16B Friction-factor curves. (*Mechanical Engineering*.)

TABLE 22 Data for Figure 16

Percentage of roughness	For value of f see curve	Diameter (actual of drawn tubing, nominal of standard-weight pipe)											
		Drawn tubing, brass, tin, lead, glass		Clean steel, wrought iron		Clean, galvanized		Best cast iron		Average cast iron		Heavy riveted, spiral riveted	
		in	mm	in	mm	in	mm	in	mm	in	mm	in	mm
0.2	1	0.35 up	8.89 up	72	1829	—	—	—	—	—	—	—	—
1.35	4	—	—	6–12	152–305	10–24	254–610	20–48	508–1219	42–96	1067–2438	84–204	2134–5182
2.1	5	—	—	4–5	102–127	6–8	152–203	12–16	305–406	24–36	610–914	48–72	1219–1829
3.0	6	—	—	2–3	51–76	3–5	76–127	5–10	127–254	10–20	254–508	20–42	508–1067
3.8	7	—	—	1½	38	2½	64	3–4	76–102	6–8	152–203	16–18	406–457
4.8	8	—	—	1–1¼	25–32	1½–2	38–51	2–2½	51–64	4–5	102–127	10–14	254–356
6.0	9	—	—	¾	19	1¼	32	1½	38	3	76	8	203
7.2	10	—	—	½	13	1	25	1¼	32	—	—	5	127
10.5	11	—	—	¾	9.5	¾	19	1	35	—	—	4	102
14.5	12	—	—	¼	6.4	½	13	—	—	—	—	3	76
24.0	14	0.125	3.18	—	—	¾	9.5	—	—	—	—	—	—
31.5	16	—	—	—	—	¼	6.4	—	—	—	—	—	—
37.5	18	0.0625	1.588	—	—	½	3.2	—	—	—	—	—	—

TABLE 23 Gas Constants

Gas	R		C for critical velocity equation
	Ft · lb/(lb · °F)	J/(kg · K)	
Air	53.33	286.9	2870
Ammonia	89.42	481.1	2080
Carbon dioxide	34.87	187.6	3330
Carbon monoxide	55.14	296.7	2820
Ethane	50.82	273.4	
Ethylene	54.70	294.3	2480
Hydrogen	767.04	4126.9	750
Hydrogen sulfide	44.79	240.9	
Isobutane	25.79	138.8	
Methane	96.18	517.5	2030
Natural gas	—	—	2070–2670
Nitrogen	55.13	296.6	2800
n-butane	25.57	137.6	
Oxygen	48.24	259.5	2990
Propane	34.13	183.6	
Propylene	36.01	193.7	
Sulfur dioxide	23.53	126.6	3870

5. *Compute the pressure loss in the pipe*
Using Fig. 16 or the Hydraulic Institute *Pipe Friction Manual*, we get $f = 0.0142$ to 0.0162 for a 4-in (101.6-mm) Schedule 40 pipe when the Reynolds number = 3,560,000. From the Fanning formula from an earlier calculation procedure and the higher value of f, $p_d = 3.36(10^{-6})flW^2/d^5\rho$, or $p_d = 3.36(10^{-6})\,(0.0162)(500)(22.500)^2/[(4.026)^5(0.750)] = 17.37$ lb/in^2 (119.8 kPa).

Related Calculations. Use this procedure to compute the pressure loss, velocity, and flow rate in compressed-air and gas lines of any length. Gases for which this procedure can be used include ammonia, carbon dioxide, carbon monoxide, ethane, ethylene, hydrogen, hydrogen sulfide, isobutane, methane, nitrogen, *n*-butane, oxygen, propane, propylene, and sulfur dioxide.

Alternate relations for computing the velocity of air or gas in a pipe are $\upsilon = 144\ W_s/a\rho$; $\upsilon = 183.4\ W_s/d^2\rho$; $\upsilon = 0.0509\ W_s\upsilon_g/d^2$, where W_s = flow rate, lb/s; a = cross-sectional area of pipe, in^2; υ_g = specific volume of the air or gas at the operating pressure and temperature, ft^3/lb.

FLOW RATE AND PRESSURE LOSS IN GAS PIPELINES

Using the Weymouth formula, determine the flow rate in a 10-mi (16.1-km) long 4-in (101.6-mm) Schedule 40 gas pipeline when the inlet pressure is 200 lb/in^2 (gage) (1378.8 kPa), the outlet pressure is 20 lb/in^2 (gage) (137.9 kPa), the gas has a specific gravity of 0.80, a temperature of 60°F (15.6°C), and the atmospheric-pressure is 14.7 lb/in^2 (abs) (101.34 kPa).

Calculation Procedure:

1. *Compute the flow rate from the Weymouth formula*
The Weymouth formula for flow rate is $Q = 28.05[(P_i^2 - p_0^2)d^{5.33}/sL]^{0.5}$, where p_i = inlet pressure, lb/in^2 (abs); p_0 = outlet pressure, lb/in^2 (abs); d = inside diameter of pipe, in; s = specific gravity of gas; L = length of pipeline, mi. For this pipe, $Q = 28.05 \times [(214.7^2 - 34.7^2)4.026^{5.33}/0.8 \times 10]^{0.5}$ = 86,500 lb/h (39,925 kg/h).

2. *Determine if the acoustic velocity limits flow*
If the outlet pressure of a pipe is less than the critical pressure p_e lb/in^2 (abs), the flow rate in the pipe cannot exceed that obtained with a velocity equal to the critical or acoustic velocity, i.e., the velocity of sound in the gas. For any gas, $p_e = Q(T_i)^{0.5}/d^2C$, where T_i = inlet temperature, °R; C = a constant for the gas being considered.

Using $C = 2070$ from Table 23, or Crocker and King—*Piping Handbook*, $p_c = (86{,}500)(60 + 460)^{0.5}/[(4.026)^2(2070)] = 58.8$ lb/in^2 (abs) (405.4 kPa). Since the outlet pressure $p_0 = 34.7$ lb/in^2 (abs) (239.2 kPa), the critical or acoustic velocity limits the flow in this pipe because $p_c > p_0$. When $p_c < p_0$, critical velocity does not limit the flow.

Related Calculations. Where a number of gas pipeline calculations must be made, use the tabulations in Crocker and King—*Piping Handbook*, McGraw-Hill, 1973 and Bell—*Petroleum Transportation Handbook*, McGraw-Hill, 1963. These tabulations will save much time. Other useful formulas for gas flow include the Panhandle, Unwin, Fritsche, and rational. Results obtained with these formulas agree within satisfactory limits for normal engineering practice.

Where the outlet pressure is unknown, assume a value for it and compute the flow rate that will be obtained. If the computed flow is less than desired, check to see that the outlet pressure is less than the critical. If it is, increase the diameter of the pipe. Use this procedure for natural gas from any gas field, manufactured gas, or any other similar gas.

To find the volume of gas that can be stored per mile of pipe, solve $V_m = 1.955p_m d^2K$, where p_m = mean pressure in pipe, lb/in^2 (abs) $\approx (p_i + p_0)/2$; $K = (1/Z)^{0.5}$, where Z = super compressibility factor of the gas, as given in Baumeister and Marks—*Standard Handbook for Mechanical Engineers* and Perry et al.—*Chemical Engineer's Handbook*. For exact computation of p_m, use $p_m = (2/3)(p_i + p_0 - p_i p_0/p_i + p_0)$.

SLIP-TYPE EXPANSION JOINT SELECTION AND APPLICATION

Select and size slip-type expansion joints for the 20-in (508-mm) carbon-steel Schedule 40 pipeline in Fig. 17, if the pipe conveys 125-lb/in^2 (gage) (861.6-kPa) steam having a temperature of 380°F (193°C). The minimum temperature expected in the area where the pipe is installed is 0°F (–17.8°C). Determine the anchor loads that can be expected. The steam inlet to the pipe is at A; the outlet is at F.

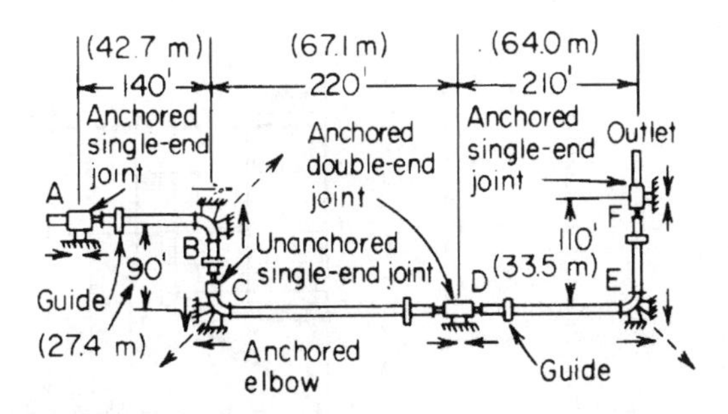

FIGURE 17 Slip-type expansion joints in a piping system. (*Yarway Corporation.*)

Calculation Procedure:

1. *Determine the expansion of each section of pipe*
From Fig. 18, the expansion of steel pipe at 380°F (193°C) with a 0°F (–17.8°C) minimum temperature is 3.4 in (88.9 mm) per 100 ft (30.5 m) of pipe. Expansion of each section of pipe is then

e in = (3.4)(pipe length, ft/100). For AB, e = (3.4)(140/100) = 4.76 in (120.9 mm); for BC, e = (3.4)(90/100) = 3.06 in (77.7 mm); for CD, e = (3.4)(220/100) = 7.48 in (190 mm); for DE, e = (3.4)(210/100) = 71.4 in (1813.6 mm); for EF, e = (3.4)(110/100) = 3.74 in (95 mm).

2. *Select the type and the traverse of each expansion joint*

The slip-type expansion joint at *A* will absorb expansion from only one direction—the right-hand side. This expansion will occur in pipe section *AB* and is 4.76 in (120.9 mm) from step 1. Therefore, a single-end slip-type expansion joint (one that absorbs expansion on only one side) can be used. The traverse—the amount of expansion a slip joint will absorb—is usually given in multiples of 4 in (101.6 mm), that is. 4, 8, and 12 in (101.6, 203.2, and 304.8 mm). Hence, an 8-in (203.2-mm) traverse slip-type single-end joint will be suitable at *A* because the expansion is 4.76 in (120.9 mm). A 4-in (101.6-mm) traverse joint would be unsatisfactory because it could not absorb at 4.76-in (120.9-mm) expansion.

The next joint, at *C*, must absorb the expansion in the vertical pipe *BC*. Since the elbow beneath the joint is anchored, an unanchored joint can be used. With pipe expansion in only one direction—from *B* to *C*—a single-end joint can be used. Since the expansion of section *BC* is 3.06 in (77.7 mm), use a single-end 4-in (101.6-mm) traverse slip-type expansion joint, unanchored at *C*.

The expansion joint at *D* must absorb expansion from two directions—from *C* to *D* and from *E* to *D*. Therefore, a double-end joint (one that can absorb expansion on each end) must be used. The double-end joint must be anchored because the pipe expands *away* from the anchored elbow *C* in section *CD* and *away* from the anchored elbow *E* in section *DE*. In both instances the pipe expands *toward* the expansion joint at *D*.

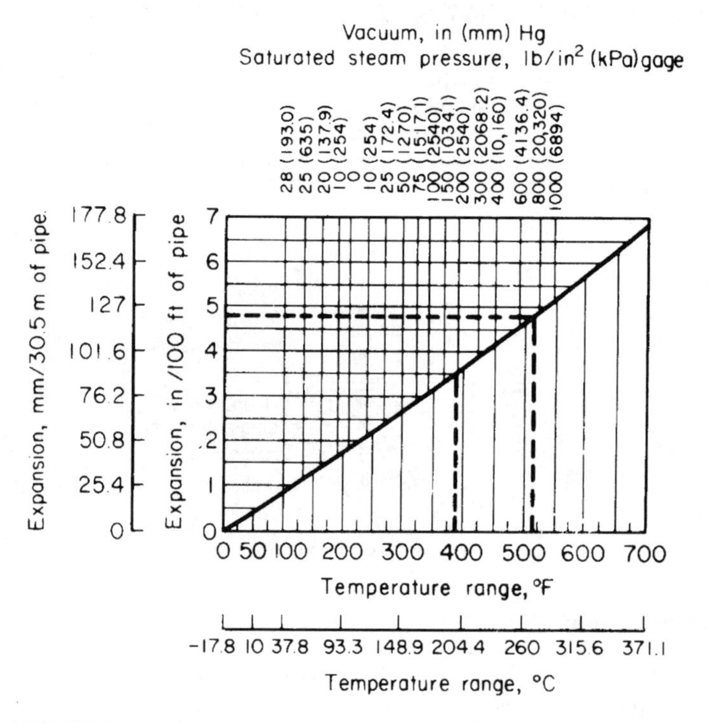

FIGURE 18 Expansion of steel pipe. (*Yarway Corporation.*)

The expansion in section *CD* is, from step 1, 7.48 in (190 mm), whereas the expansion in *DE* is 7.14 in (181.4 mm). Therefore, a double-end anchored joint with an 8-in (203.2-mm) traverse at *each* end will be suitable.

Since the pipe outlet is at *F* and there is no anchor in the pipe at *F*, the expansion joint at this joint must be anchored. The pipe section between *E* and *F* will expand vertically upward into the joint for a distance of 3.74 in (95 mm), as computed in step 1. Therefore, a single-end anchored joint with a 4-in (101.6-mm) traverse will be suitable.

3. *Compute the anchor loads in the pipeline*

Use Fig. 19 to determine the anchor loads on intermediate and end anchors (those where the pipe makes a sharp change in direction). Enter Fig. 19 at the bottom at a pipe size of 20-in (508-mm) diameter, and project vertically upward to the dashed curve labeled *intermediate anchor—all pressures*. At the left read the anchor load at each intermediate anchor, *A*, *D*, and *F*, as 20,000 lb (88.9 kN). Note that the joint expansion load = joint contraction load = 20,000 lb (88.9 kN).

The end anchors, *B*, *C*, and *E*, have, from Fig. 19, a possible maximum load of 58,000 lb (258 kN), found by projecting vertically upward from the 20-in (508-mm) pipe size to 125-lb/in^2 (gage) (862-kPa) steam pressure, which lies midway between the 100- and 150-lb/in^2 (gage) (689.5- and 1034-kPa) curves. Indicate the possible maximum end-anchor loads by the solid arrows at each elbow, as shown in Fig. 17. The resultant *R* of the loads at any end anchor is found by the pythagorean theorem to be $R = (58{,}000^2 + 58{,}000^2)^{0.5} = 82{,}200$ lb (365.6 kN). Indicate the resultant by a dotted arrow, as shown in Fig. 17.

Contraction loads on the end anchors are in the reverse direction and consist only of friction. This friction load equals the joint expansion load, or 20,000 lb (88.9 kN). The resultant of the joint expansion loads is $(20{,}000^2 + 20{,}000^2)^{0.5} = 28{,}350$ lb (126.1 kN).

Locate guides within 25 or 12 ft (7.62 or 3.66 m) of the expansion joint, depending on the type of packing used, Table 24. These guides should allow free axial movement of the pipe into and out of the joint with minimum friction.

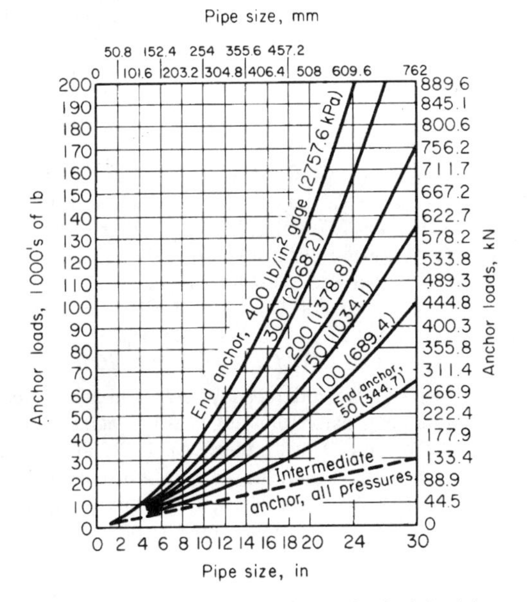

FIGURE 19 End- and intermediate-anchor loads in piping systems. (*Yarway Corporation*.)

TABLE 24 Guide and Support Spacing

Nominal pipe size, in (mm)	Distance between guide and joint, ft (m) — Packing type: Gum	Gland	Distance between guides, ft (m)
18 (457)	24 (7.3)	11 (3.4)	100 (30.5)
20 (508)	25 (7.6)	12 (3.7)	105 (32)
24 (610)	26 (7.9)	12 (3.7)	110 (33.5)

Related Calculations. Use this procedure to choose slip-type expansion joints for pipes conveying steam, water, air, oil, gas, and similar vapors, liquids, and gases. In some instances, the gland friction and pressure thrust are used instead of Fig. 19 to determine anchor loads. With either method, the results are about the same.

CORRUGATED-EXPANSION JOINT SELECTION AND APPLICATION

Select corrugated-expansion joints for the 8-, 6-, and 4-in (203.2-, 152.4-, and 101.6-mm) carbon-steel pipeline in Fig. 20 if the steam pressure in the pipe is 75 lb/in^2 (gage) (517.1 kPa), the steam temperature is 340°F (171°C), and the installation temperature is 60°F (15.6°C).

Calculation Procedure:

1. ***Determine the expansion of each section of pipe***

From a table of thermal expansion of pipe, the expansion of carbon-steel pipe at 340°F (171°C) is 2.717 in/100 ft (2.26 mm/30.5 m) from 0 to 340°F (–17.8 to 171°C). Between 0 and 60°F (–17.8 and 15.6°C) the expansion is 0.448 in/100 ft (50 mm/30.5 m). Hence, the expansion between 60 and 340°F (15.6 and 171°C) is 2.717 – 0.448 = 2.269 in/100 ft (1.89 mm/m). This factor can now be applied to each length of pipe by finding the product of (pipe-section length, ft/100) (expansion, in/100 ft) = expansion of section, in = e.

For section AD, e = (87/100)(2.269) = 1.97 in (50 mm); for DE, e = (78/100)/(2.269) = 1.77 in (45 mm); for EC, e = (83/100)(2.269) = 1.88 in (47.8 mm); for CF, e = (60/100)(2.269) = 1.36 in (34.5 mm); for FG, e = (175/100)(2.269) = 3.97 in (100.8 mm).

In selecting corrugated-expansion joints, the usual practice is to increase the computed expansion by a suitable safety factor to allow for any inaccuracies in temperature measurement. By applying a 25 percent safety* factor: for AD, e = (1.97)(1.25) = 2.46 in (62.5 mm); for DE, e = (1.77)(1.25) = 2.13 in (54.1 mm); for EC, e = (1.88)(1.25) = 2.35 in (59.7 mm); for CF, e = (1.36)(1.25) = 1.70 (43.2 mm); for FG, e = (3.97)(1.25) = 4.96 in (126 mm).

2. ***Select the traverse for, and type of, each expansion joint***

Obtain corrugated-expansion joint engineering data, and select a joint with the next largest traverse for each section of pipe. Thus, traverse $AD \geq 2\frac{1}{2}$ in (63.5 mm); traverse $DE \geq 2\frac{1}{4}$ in (57.2 mm); traverse $EC \geq 2\frac{1}{2}$ in (63.5 mm); traverse $CF \geq 1\frac{3}{4}$ in (44.5 mm); traverse $FG \geq 5.0$ in (127 mm).

Two types of expansion joints are commonly used: free-flexing and controlled-flexing. Free-flexing joints are generally used where the pressures in the pipeline are relatively low and the

*This value is for illustration purposes only. Contact the expansion-joint manufacturer for the exact value of the safety factor to use.

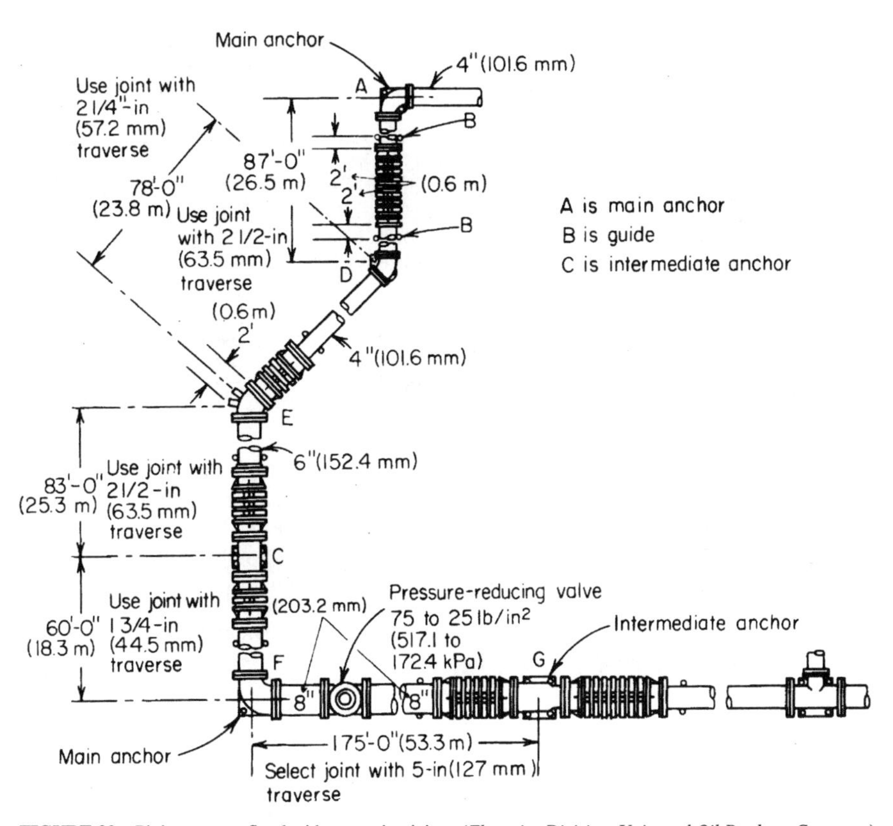

FIGURE 20 Piping system fitted with expansion joints. (*Flexonics Division, Universal Oil Products Company.*)

required motion is relatively small. Controlled-flexing expansion joints are generally used for higher pressures and larger motions. Both types of expansion joints are available in stainless steel in both single and dual units. For precise data on a given joint being considered, consult the expansion-joint manufacturer. Corrugated-expansion joints are characterized by their freedom from any maintenance needs.

3. *Compute the anchor loads in the pipeline*

Main anchors are used between expansion joints, as at F and A, Fig. 20, and at turns such as at F and A. The forced* a main anchor must absorb is given by F_i lb $= F_p + F_e$, where F_p = pressure thrust in the pipe, lb $= pA$, where p = pressure in pipe, lb/in^2 (gage); A = effective internal cross-sectional area of expansion joint, in^2 (see Table 35 for cross-sectional areas of typical corrugated joints); F_e = force required to compress the expansion join, lb = [300 lb/in (52.5 N/mm)] (joint inside diameter, in) for stainless-steel self-equalizing joints, and [200 lb/in (35 N/mm)](joint inside diameter, in) for copper nonequalizing joints. Determining the main anchor force for the 8-in (203.2-mm) pipeline gives $F_i = (75)(85) + (300)(8) = 8775$ lb (39.0 kN). In this equation, the area of 85 in^2 (548.3 cm^2) in the first term is obtained from Table 25.

The total force at a main anchor, as at A and F, Fig. 20, is the vector sum of the forces in each line leading to the anchor. Thus, at F, there is a force of 8775 lb (39.0 kN) in the 8-in (203.2-mm) line

*This is an approximate method for finding the anchor force. For a specific make of expansion joint, consult the joint manufacturer.

TABLE 25 Effective Area of Corrugated-Expansion Joints

Joint inside diameter		Joint effective area	
in	mm	in^2	cm^2
6	152.4	51.0	329.0
8	203.2	85.0	548.4
10	254.0	120.0	774.2
12	304.8	174.0	1122.6
14	355.6	215.0	1387.1
16	406.4	270.0	1741.9
18	457.2	310.0	1999.9
20	508.0	390.0	2516.1
24	609.6	540.0	3483.9

and a force of $F_i = (75)(51) + (300)(6) = 5625$ lb (25 kN) in the 6-in (152.4-mm) line connected to the elbow outlet. Since the elbow at *F* is a right angle, use the pythagorean theorem, or *R* = resultant anchor force, lb = $(8775^2 + 5625^2)^{0.5} = 10.400$ lb (46.3 kN).

Where two lines containing corrugated expansion joints are connected by a bend of other than 90°, as at *D* and *E*, use a force triangle to determine the anchor force after computing F_i for each pipe. Thus, at *E*, F_t for the 6-in (152.4-mm) pipe = 5625 lb (25 kN), and $F_i = (75) \times (23.5) + (300)(4) = 2963$ lb (13.2 kN) for the 4-in (101.6-m) pipe. Draw the force triangle in Fig. 21 with the 6-in (152.4-mm) pipe F_i and the 4-in (101.6-mm) pipe F_i as two sides and the bend angle, 45°, as the

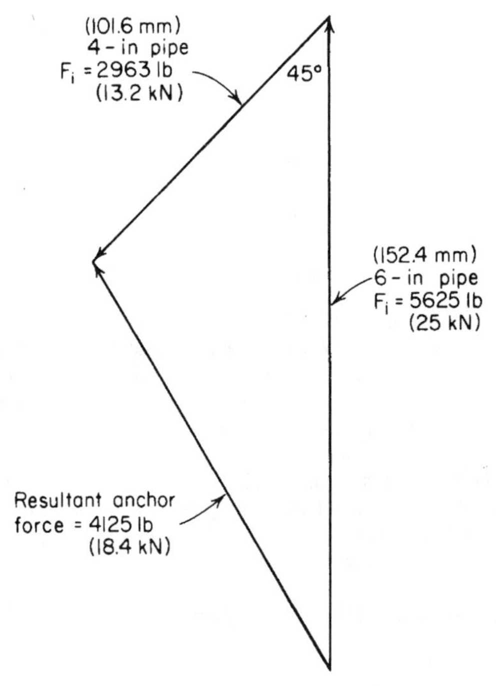

FIGURE 21 Force triangle for determining piping anchor force.

included angle. Connect the third side, or resultant, to the ends of the force vectors, and scale the resultant as 4125 lb (18.4 kN), or compute the resultant from the law of cosines. Find the resultant force at *D* in a similar manner as 2963 lb (13.2 kN).

Intermediate anchors, as at *C* and *G*, must withstand only one force—the unbalanced (differential) spring force. With approximate force calculations*, starting at *C*, for a 6-in (152.4-mm) expansion joint, $F_e = (300)(6) = 1800$ lb (8 kN). At *G*, for an 8-in (203.2-mm) expansion joint, $F_e = (300)(8) = 2400$ lb (10.7 kN). Thus, the loads the intermediate anchors must withstand are considerably less than the main-anchor loads.

Provide the pipe guides at suitable locations in accordance with the joint manufacturer's recommendations and at suitable intervals on the pipeline to prevent any lateral and buckling forces on the joint and adjacent piping. Intermediate anchors between two joints in a straight run of pipe ensure that each joint will absorb its share of the total pipe motion. Slope the pipe in the direction of fluid flow to prevent condensate accumulation. Use enough pipe hangers to prevent sagging of the pipe.

Related Calculations. Use this procedure to choose corrugated-type expansion joints for pipes conveying steam, water, air, oil, gas, and similar vapors, liquids, and gases. When choosing a specific make of corrugated-expansion joint, use the manufacturer's engineering data, where available, to determine the maximum allowable traverse. One popular make has a maximum traverse of 7.5 in (190.5 mm) or a maximum allowable lateral motion of 1.104 in (28.0 mm) in its various joint sizes. The larger the lateral motion, the greater the number of corrugations required in the joint.

In some pipelines there is an appreciable pressure thrust caused by a change in direction of the pipe. This pressure or centrifugal thrust F_c is usually negligible, but the wise designer makes a practice of computing this thrust from $F_c = (2A\rho\upsilon^2/32.2) \times (\sin\theta/2)$ lb, where, A = inside area of pipe, ft^2; ρ = density of fluid or vapor, lb/ft^3; υ = fluid or vapor velocity, ft/s; θ = change in direction of the pipeline.

The number of corrugations required in a joint varies with the expansion and lateral motion to be absorbed. A typical free-flexing joint can absorb 6.25 in (158.8 mm) of expansion and a variable amount of lateral motion, depending on joint size and operating condition. Free-flexing joints are commonly built in diameters up to 48 in (1219 mm), while controlled-flexing joints are commonly built in diameters up to 24 in (609.6 mm). For a more precise calculation procedure, consult the Flexonics Division, Universal Oil Products Company.

DESIGN OF STEAM-TRANSMISSION PIPING

Design a steam-transmission pipe to supply a load that is 1700 ft (518.2 m) from the power plant. The terrain permits a horizontal run between the power plant and the load. Maximum steam flow required by the load is 300,000 lb/h (135,000 kg/h), whereas the average steam flow required is estimated as 150,000 lb/h (67,500 kg/h). The maximum steam pressure at the load must not exceed 150 lb/in^2 (abs) (1034.1 kPa) saturated. Superheated steam at 450 lb/in^2 (abs) (3102.7 kPa) and 600°F (316°C) is available at the power plant. Two schemes are proposed for the line: (1) Reduce the steam pressure to 180 lb/in^2 (abs) (1240.9 kPa) at the line inlet, thus allowing a 180 – 150 = 30-lb/in^2 (206.8-kPa) loss in the 1700-ft (518.2-m) long line. This scheme is called the *nominal pressure-loss line*. (2) Admit high-pressure steam to the line and thereby allow the steam pressure to fall to a level slightly greater than 150 lb/in^2 (abs) (1034.1 kPa). Since 600°F (316°C) steam would probably cause expansion and heat-loss difficulties in the pipe, assume that the inlet temperature of the steam is reduced to 455°F (235°C) in a desuperheater in the power plant. There is a 10-lb/in^2 (68.9-kPa) pressure loss between the power plant and the line, reducing the line inlet pressure to 440 lb/in^2 (abs) (3033.4 kPa). Since the pressure can fall about 440 – 150 = 290 lb/in^2 (1999.3 kPa), this will be called the *maximum pressure-loss line*. During design, determine which line is the most economical.

*Consult the expansion-joint manufacturer or an exact procedure for computing the anchor forces.

Calculation Procedure:

1. ***Determine the required pipe diameter for each condition***

The average steam pressure in the nominal pressure-loss line is (inlet pressure + outlet pressure)/2 = (180 + 150)/2 = 165 lb/in^2 (abs) (1138 kPa). Use this average pressure to determine the pipe size, because the average pressure is more representative of actual conditions in the pipe. Assume that there will be a 5-lb/in^2 (34.5-kPa) pressure drop through any expansion bends and other fittings in the pipe. Then, the allowable friction-pressure drop = 30 – 5 = 25 lb/in^2 (172.4 kPa).

Use the Thomas saturated-steam formula to determine the required pipe diameter, or $d = (80{,}000\ W/P\upsilon)^{0.5}$, where d = inside pipe diameter, in; W = weight of steam flowing, lb/min; P = average steam pressure, lb/in^2 (abs); υ = steam velocity, ft/min. Assuming a steam velocity of 10,000 ft/min (3048 m/min), which is typical for a long steam-transmission line, we get $d = [(80{,}000 \times 300{,}000/60)/(165 \times 10{,}000)]^{0.5} = 15.32$ in (389.1 mm).

The inside diameter of a Schedule 40 16-in (406-mm) outside-diameter pipe is, from a table of pipe properties, 15 in (381 mm). Assume that a 16-in (406-mm) pipe will be used if Schedule 40 wall thickness is satisfactory for the nominal pressure-loss line. Note that the larger flow was used in computing the size of this line because a pipe satisfactory for the larger flow will be acceptable for the smaller flow.

The maximum pressure-loss line will have an average pressure that is a function of the inlet pressure at the pressure-reducing valve at the line outlet. Assume that there is a 10-lb/in^2 (68.9-kPa) drop through this reducing valve. Then steam will enter the valve at 150 + 10 = 160 lb/in^2 (abs) (1103 kPa), and the average line pressure = (440 + 160)/2 = 300 lb/in^2 (abs) (2068 kPa). Using a higher steam velocity [15,000 ft/min (4572 m/min)] for this maximum pressure-loss line than for the nominal pressure-loss line [10,000 ft/min (3048 m/min)], because there is a larger allowable pressure drop, compute the required inside diameter from the Thomas saturated-steam formula because the steam has a superheat of only 456.28 – 455.00 = 1.28°F (2.3°C). Or, $d = [(80{,}000 \times 300{,}000/60)/(300 \times 15{,}000)]^{0.5} = 9.44$ in (239.8 mm). Since a 10-in (254-mm) Schedule 40 pipe has an inside diameter of 10.02 in (254.5 mm), use this size for the maximum pressure-loss line.

2. ***Compute the required pipe-wall thickness***

As shown in an earlier calculation procedure, the schedule number SN = $1000P_i/S$. Assuming that seamless carbon-steel ASTM A53 grade A pipe is used for both lines, the *Piping Code* allows a stress of 12,000 lb/in^2 (82.7 MPa) for this material at 600°F (316°C). Then SN = (1000) × (435)/12,000 = 36.2; use Schedule 40 pipe, the next largest schedule number for both lines. This computation verifies the assumption in step 1 of the suitability of schedule 40 for each line.

3. ***Check the pipeline for critical velocity***

In a steam line, $p_c = W'/Cd^2$, where p_c = critical pressure in pipe, lb/in^2 (abs), W' = steam flow rate. lb/h; C = constant from Crocker and King—*Piping Handbook*; d = inside diameter of pipe, in.

When the pressure loss in a pipe exceeds 50 to 58 percent of the initial pressure, flow may be limited by the fluid velocity. The limiting velocity that occurs under these conditions is called the *critical velocity*, and the coexisting pipeline pressure, the *critical pressure*.

Critical velocity may limit flow in the 10-in (254-mm) maximum pressure-loss line because the terminal pressure of 150 lb/in^2 (abs) (1034 kPa) is less than 58 percent of 440 lb/in^2 (abs) (3033.4 kPa), the inlet pressure. Use the above equation to find the critical pressure. Or, $p_c = (300{,}000)/[(75.15)(10.02)^2] = 39.7$ lb/in^2 (abs) (273.7 kPa), using the constant from the *Piping Handbook* after interpolating for the initial enthalpy of 1205.4 Btu/lb (2804 kJ/kg), which is obtained from steam-table values.

Critical velocity would limit flow if the pipeline terminal pressure were equal to, or less than, 39.7 lb/in^2 (abs) (273.7 kPa). Since the terminal pressure of 150 lb/in^2 (1034.1 kPa) is greater than 39.7 lb/in^2 (abs) (237.7 kPa), critical velocity does not limit the steam flow. With smaller flow rates, the critical pressure will be lower because the denominator in the equation remains constant for a given pipe. Hence, the 10-in (254-mm) line will readily transmit 300,000-lb/h (135,000-kg/h) and smaller flows.

If critical pressure existed in the pipeline, the diameter of the pipe might have to be increased to transmit the desired flow. The 16-in (406.4-mm) line does not have to be checked for critical pressure because its final pressure is more than 58 percent of the initial pressure.

4. *Compute the heat loss for each line*

Assume that 2-in (50.8-mm) thick 85 percent magnesia insulation is used on each line and that the lines will run above the ground in an area having a minimum temperature of 40°F (4.4°C). Set up a computation form as follows:

Pipe size, in (mm)	16 (406.5)	10 (254.0)
Steam temperature, °F (°C)	373 (189)	455 (235)
Air temperature, °F (°C)	40 (4.4)	40 (4.4)
Temperature, difference, °F (°C)	333 (184.6)	415 (184.6)
Insulation heat loss, Btu/(h · ft² · °F)* · [W/(m² · °C)]	1.11 (6.3)	0.704 (3.99)
Heat loss, Btu/(h · lin ft)(W/m)	370 (356)	292 (281)
Heat loss, Btu/h (kW), for 1700 ft (518 m)	629,000 (184)	496,400 (145.6)
Total heat loss, Btu/h (kW), with a 25% safety factor	786,250 (230)	620,500 (182.0)
Heat loss, Btu/lb (W/kg) of steam, for 300,000-lb/h (135,000-kg/h) flow	2.62 (1.7)	2.07 (1.35)
Heat loss, Btu/lb (W/kg), for the average flow of 150,000 lb/h (67,500 kg/h)	5.24 (3.4)	4.14 (2.69)

*From table of pipe insulation, Ehret Magnesia Manufacturing Company.

In this form, the following computations were made for both pipes: heat loss, Btu/(h · lin ft) = [insulation heat loss, Btu/(h · ft · °F)] (temperature difference, °F); heat loss, Btu/h for 1700 ft (518.2 m) = [heat loss, Btu/(h · lin ft)] (1700); total heat loss, Btu/h, 25 percent safety factor = (heat loss, Btu/h) (1700 ft)(1.25); heat loss, Btu/lb steam = (total heat loss, Btu/h, with a 25 percent safety factor)/ (300,000-lb steam).

5. *Compute the leaving enthalpy of the steam in each line*

Acceleration of steam in each line results from an enthalpy decrease of $h_a = (\upsilon_2^2 - \upsilon_1^2)/2g(778)$, where h_a = enthalpy decrease, Btu/lb; υ_2 and υ_1 = final and initial velocity of the steam, respectively, ft/s; $g = 32.2$ ft/s². The velocity at any point x in the pipe is found from the continuity equation $\upsilon_x = (W'\upsilon_g)/3600\,A_x$, where υ_g = steam velocity, ft/s, when the steam volume is υ_g ft³/lb, and A_x is the cross-sectional area of pipe, ft², at the point be considered.

For the 16-in (406.4-mm) nominal pressure-loss line with a flow of 300,000 lb/h (135,000 kg/h) at 180-lb/in² (abs) (1241-kPa) entering and 150-lb/in² (abs) (1034.1-kPa) leaving pressure, using steam and piping table values, $\upsilon_1 = (300{,}000)(2.53)/[(3600)(1.23)] - 171.5$ ft/s (52.3 m/s); $\upsilon_2 = 300{,}000(3.015)/(3600)(1.23)] = 205$ ft/s (62.5 m/s). Then $h_a = [(204.5)^2 - (171.5)^2]/[(64.4)(778)] = 0.2504$ Btu/lb (0.58 kJ/kg), say 0.25 Btu/lb (0.58 kJ/kg).

By an identical calculation, $h_a = 3.7$ Btu/lb (8.6 kJ/kg) for the 10-in (254-mm) maximum pressure-loss line when the leaving steam is assumed to be 150 lb/in² (abs) (1034.1 kPa), saturated.

Enthalpy of the 180-lb/in² (abs) (1241-kPa) saturated steam entering the 16-in (406.4-mm) line is 1196.9 Btu/lb (2784 kJ/kg). Heat loss during 300,000-lb/h (135,000-kg/h) flow is 2.62 Btu/lb (6.1 kJ/kg), as computed in step 4. The enthalpy drop of 0.25 Btu/lb (0.58 kJ/kg) accelerates the steam. Hence, the calculated leaving enthalpy is 1196.9 – (2.62 + 0.25) = 1194.03 Btu/lb (2777.3 kJ/kg). The enthalpy of the leaving steam at 150 lb/in² (abs) (1034.1 kPa) saturated is 1194.1 Btu/lb (2777.5 kJ/kg). To have saturated steam leave the line, 1194.10 –1194.03, or 0.07 Btu/lb (0.16 kJ/kg), must be supplied to the steam. This heat will be obtained from the enthalpy of vaporization given off by condensation of some of the steam in the line.

Make a group of identical calculations for the 10-in (254-mm) maximum pressure-loss line. The enthalpy of 440-lb/in (abs) (3033.4-kPa) 445°F (235°C) entering steam is 1205.4 Btu/lb (2803.8 kJ/kg), found by interpolation in the steam tables. Heat loss during 300,000-lb/h

(135,000-kg/h) flow is 2.07 Btu/lb (4.81 kJ/kg). An enthalpy drop of 3.7 Btu/lb (8.6 kJ/kg) accelerates the steam. Hence, the calculated leaving enthalpy = 1205.4 – (2.07 + 3.7) = 1199.63 Btu/lb (2790.3 kJ/kg).

The enthalpy of the leaving steam at 150-lb/in^2 (abs) (1034-kPa) saturated is 1194.1 Btu/lb (2777.5 kJ/kg). As a result, under maximum-flow conditions, the steam will be superheated from the entering point to the leaving point of the line. The enthalpy difference of 5.53 Btu/lb = 1199.63 – 1194.10) (12.9 kJ/kg) produces this superheat. Because the steam is superheated throughout the line length, condensation of the steam will not occur during maximum-flow conditions.

For most industrial applications, the steam leaving the line may be considered as saturated at the desired pressure. But for precise temperature regulation, some form of pressure-temperature control must be used at the end of long lines.

During average flow conditions of 150,000 lb/h (67,500 kg/h), the line heat loss is 4.14 Btu/lb (9.6 kJ/kg), as computed in step 4. The enthalpy drop to accelerate the steam is 0.925 Btu/lb (2.2 kJ/kg). As in the case of maximum flow, the steam is superheated throughout the length of the 10-in (254-mm) maximum pressure-loss line because the calculated leaving enthalpy is 1205.40 – 5.07 = 1200.33 Btu/lb (2791.9 kJ/kg).

6. *Compute the quantity of condensate formed in each line*

For either line, the quantity of condensate formed, lb/h = $C = W'$ (h_g at leaving pressure – calculated leaving h_g)/outlet pressure h_{fg}.

Using computed values from step 5 and steam-table values, we see the 16-in (406.4-mm) line with 300,000 lb/h (135,000 kg/h) flowing forms C = (300,000)(0.07)/863.6 = 24.35, say 24.4 lb/h (10.9 kg/h) of condensate.

Condensation during an average flow of 150,000 lb/h (67,500 kg/h) is found in the same way. The enthalpy drop to accelerate the steam is neglected for average flow in normal pressure-loss lines because the value is generally small. For the 150,000-lb/h (67,500-kg/h) flow, the calculated leaving enthalpy = 1196.90 – 5.24 = 1191.66 Btu/lb (536.3 kg/h). Hence, C = (150,000)(1194.10 – 1191.66)/ 863.6 = 424 lb/h (190.8 kg/h) say 425 lb/h (191.3 kg/h).

The largest amount of condensate is formed during line warm-up. Condensate-removal equipment—traps and related piping—must be sized up on the basis of the warm-up not the average steam flow. Using a warm-up time of 30 min and the method of an earlier calculation procedure, we see the condensate formed in 16-in (406.4-mm) Schedule 40 pipe weighing 83 lb/ft (122.8 kg/m) is, with a 25 percent safety factor to account for radiation, $C = 1.25 \times (60)(83)(1700)(373 - 40)(0.12)/$ $[(30)(850.8) = 16{,}550$ lb/h (7448 kg/h). Thus, the trap or traps should have a capacity of about 17,000 lb/h (7650 kg/h) to remove the condensate during the 30-min warm-up period.

Condensate does not form in the 10-in (254-mm) maximum pressure-loss line during either maximum or average flow. Warm-up condensate for a 30-min warm-up period and a 25 percent safety factor is $C = 1.25(60)(40.5)(1700)(455 - 40)(0.12)/[(30)(770.0)] = 11{,}120$ lb/h (5004 kg/h). Thus, the trap or traps should have a capacity of about 11,500 lb/h (5175 kg/h) to remove the condensate during the 30-min warm-up period.

In general, traps sized on a warm-up basis have adequate capacity for the condensate formed during the maximum and average flows. However, the condensate formed under all three conditions must be computed to determine the maximum rate of formation for trap and drain-line sizing.

7. *Determine the number of plain U bends needed*

A 1700-ft (518-m) long steel steam line operating at a temperature in the 400°F (204°C) range will expand nearly 50 in (1270 mm) during operation. This expansion must be absorbed in some way without damaging the pipe. There are four popular methods for absorbing expansion in long transmission lines: plain U bends, double-offset expansion U bends, slip or corrugated-expansion joints, and welded-elbow expansion bends. Each of these will be investigated to determine which is the most economical.

Assume that the governing code for piping design in the locality in which the line will be installed requires that the combined stress resulting from bending and pressure S_{bp} not exceed three-fourths the sum of the allowable stress for the piping material at atmospheric temperature S_a and the allowable stress at the operating temperature S_o of the pipe. This is a common requirement. In equation form, $S_{bp} = 0.75(S_a + S_o)$, where each stress is in lb/in^2.

By using allowable stress values from the *Piping Code* or the local code for 16-in (406.4-mm) seamless carbon-steel ASTM A53 grade A pipe operating at 373°F (189°C), $S_{bp} = 0.75(12{,}000 + 12{,}000) = 18{,}000$ lb/in^2 (124.1 MPa).

Determine the longitudinal pressure stress P_L by dividing the end force due to internal pressure F_e lb by the cross-sectional area of the pipe wall a_m in^2, or $P_L = F_e/a_m$. In this equation, $F_e = pa$, where p = pipe operating pressure, lb/in^2 (gage); a = cross-sectional area of the pipe, in^2. Since the 16-in (406.4-mm) line operates at 180 – 14.7 = 165.3 lb/in^2 (gage) (1139.6 kPa) and, from a table of pipe properties, a = 176.7 in^2 (1140 cm^2) and a_m = 24.35 in^2 (157.1 cm^2). P_L = (165.3)(176.7)/24.35 = 1197, say 1200 lb/in^2 (8.3 MPa). The allowable bending stress at 373°F (189°C), the pipe operating temperature, is then $S_{np} - P_L$ = 18,000 – 1200 = 16,800 lb/in^2 (115.8 MPa).

Assume that the expansion U bend will have a radius of seven times the nominal pipe diameter, or (7)(16 in) = 112 in (284.5 cm). The allowable bending stress is 16.800 lb/in^2 (115.8 MPa). Full *Piping Code* allowable credit will be taken for cold spring; i.e., the pipe will be cut short by 50 percent or more of the computed expansion and sprung into position.

Referring to Crocker and King—*Piping Handbook*, or a similar tabulation of allowable U-bend overall lengths for various operating temperatures, and choosing the length for 400°F (204°C), we see that the next higher tabulated temperature greater than the 373°F (189.4°C) operating temperature, an allowable length of 157.0 ft (47.9 m) is obtained for the bend. Plot a curve of the allowable bend length vs. temperature at 200, 300, 400, and 500°F (93.3, 148.9, 204.4, and 260°C). From this curve, the allowable pipe stress of 12,000 lb/in^2 (82.7 MPa) and no cold spring. Since the allowable stress is 16,800 lb/in^2 (115.8 MPa) and maximum cold spring is used, permitting a length 1.5 times the tabulated length, the total allowable length per bend = (175.0) (16,800/12,000)(1.5) = 367.5 ft (112 m). With the total length of pipe between the power plant and a load of 1700 ft (518.2 m), the number of bends required = 1700/367.5 = 4.64 bends. Since only a whole number of bends can be used, the next larger whole number, or five bends, would be satisfactory for this 16-in (406.4-mm) line. Each bend would have an overall length. Fig. 22, of 1700/5 = 340 ft (103.6 m).

Find the actual stress S_a in the pipe when five 340-ft (103.6-m) bends are used by setting up a proportion between the tabulated stress and bend length. Thus, the *Piping Handbook* chart is based on a stress of 12.000 lb/in^2 (82.7 MPa) without cold spring. For this stress, the maximum allowable bend length is 175 ft (53.3 m) found by graphical interpolation of the tabular values, as discussed above. When a 340-ft (103.6-m) bend with maximum cold spring is used, the pipe stress is such that the allowable bend length is 340/1.5 = 226.5 ft (69 m). The actual stress in the pipe is therefore S_a/12,000 = 226.5/175, or S_a = 15.520 lb/in^2 (107 MPa). This compares favorably with the allowable stress of 16,800 lb/in^2 (115.8 MPa). The actual stress is less because the overall bend length was reduced.

Use the *Piping Handbook* calculation procedure to find the anchor reaction forces for these bends. Using the *Piping Handbook* method with graphical interpolation, the anchor reacting force for a 16-in (406.4-mm) schedule 80 bend having a radius of seven times the pipe diameter is 10,550 lb (46.9 kN) at 373°F (189.4°C), based on a 12.000-lb/in^2 (82.7-MPa) stress in the pipe. This tabular reaction must be corrected for the actual pipe stress and for Schedule 40 pipe instead of Schedule 80 pipe. Thus, the actual anchor reaction, lb = (tabular reaction, lb) [(actual stress, lb/in^2) (tabular stress, lb/in^2)] (moment of inertia, Schedule 40 pipe, in^4/moment of inertia of Schedule 80 pipe, in^4) = (10,550)(15,520/12,000)(731.9/1156.6) = 8650 lb (38.5 kN). With a reaction of this magnitude, each anchor would be designed to withstand a force of 10,000 lb (44.5 kN). Good design would locate the bends midway between the anchor points; that is, there would be an anchor at each end of each bend. Adjustment for cold spring is not necessary, because it has negligible effect on anchor forces.

Use the same procedure for the 10-in (254-mm) maximum pressure-loss line. If 100-in (254-cm) radius bends are used, seven are required. The bending stress is 14,700 lb/in^2 (65.4 kN), and the anchor force is 2935 lb (13.1 kN). Anchors designed to withstand 3000 lb (13.3 kN) would be used.

8. *Determine the number of double-offset U bends needed*

By the same procedure and the *Piping Handbook* tabulation similar to that in step 7, the 16-in (406.4-mm) nominal pressure-loss line requires two 850-ft (259.1-m) long 112-in (284.5-cm)

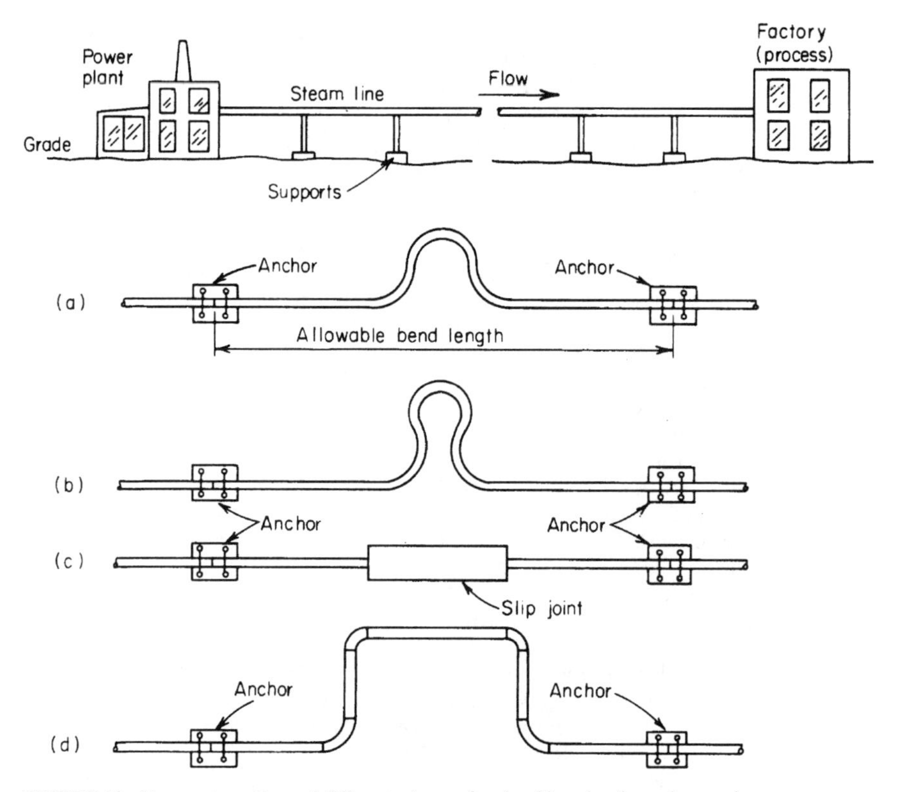

FIGURE 22 Process steam line and different schemes for absorbing pipe thermal expansion.

radius bends. Stress in the pipe is 15,610 lb/in^2 (107.6 MPa), and the anchor reaction is 4780 lb (21.3 kN).

The 10-in (254-mm) maximum pressure-loss line requires five 340-ft (103.6-m) long 70-in (177.8-cm) radius bends. Stress in the pipe is 12,980 lb/in^2 (89.5 MPa), and the anchor reaction is 2090 lb (9.3 kN).

Note that a smaller number of double-offset U bends are required—two rather than five for the 16-in (406.4-mm) pipe and five rather than seven for the 10-in (254-mm) pipe. This shows that double-offset U bends can absorb more expansion than plain U bends.

9. *Determine the number of expansion joints needed*

For any pipe, the total linear expansion e_t, at an elevated temperature above 32°F (0°C) is $e_t = (c_e)(\Delta t)(l)$, where c_e = coefficient of linear expansion, in/(ft · °F); Δt = operating temperature, °F – installation temperature,°F; l = length of straight pipe, ft. Using Crocker and King—*Piping Handbook* as the source for c_e, for both lines, we see the expansion of the 373°F (189°C) 16-in (406.4-mm) line with a 40°F (4.4°C) installation temperature is $e_t = (12)(0.0000069)(373 - 40)(1700) = 46.8$ in (1189 mm). For the 10-in (254-mm) 455°F (235°C) line, $e_t = (12)(0.0000072)(455 - 40)(1700) =$ 61 in (1549 mm). The factor 12 is used in each of these computations because Crocker and King give c_e in in/in; therefore, the pipe total length must be converted to inches by multiplying by 12.

Double-ended slip-type expansion joints that can absorb up to 24 in (609.6 mm) of expansion are available. Hence, the number of joints N needed for each line is: 16-in (406.4-mm) line, $N = 46.8/24$, or 2; 10-in (254-mm) line. $N = 61/24$, or 3.

The joints for each line would be installed midway between anchors, Fig. 22. Joints in both lines would be anchored to the ground or a supporting structure. Between the joints, the pipe must be adequately supported and free to move. Roller supports that guide and permit longitudinal movement are usually best for this service. Whereas roller-support friction varies, it is usually assumed to be about 100 lb (444.8 N) per support. At least six supports per 100 ft (30.5 m) are needed for the 16-in (406.4-mm) line and seven per 100 ft (30.5 m) for the 10-in (254-mm) line. Support friction and the number of rollers required are obtained from Crocker and King—*Piping Handbook* or piping engineering data.

The required anchor size and strength depend on the pipe diameter, steam pressure, slip-joint construction, and type of supports used. During expansion of the pipe, friction at the supports and in the joint packing sets up a force that must be absorbed by the anchor. Also, steam pressure in the joint tends to force it apart. The magnitude of these forces is easily computed. With the total force known, a satisfactory anchor can be designed. Slip-joint packing-gland friction varies with different manufacturers, type of joint, and packing used. Gland friction in one popular type of slip joint is about 2200 lb/in (385.3 N/mm) of pipe diameter. Assuming use of these joints in both lines, compute the anchor force as follows:

	lb	kN
16-in (406.4-mm) nominal pressure-loss line		
Support friction = [(1700 ft)(6 supports per 100 ft) (100 lb per support)]/100	= 10,200	45.3
Gland friction = (2200 lb/in diameter)(16 in)	= 35,200	156.6
Pressure force = [165.3 lb/in^2 (gage)](176.7-in^2 pipe area)	= 29,200	129.9
Total force to be absorbed by anchor	= 74,600	331.8
10-in (254-mm) maximum pressure-loss line		
Support friction = [(1700)(7)(100)]/100	= 11,900	52.8
Gland friction = (2200)(10)	= 22,000	97.9
Pressure force = (425.3)(78.9)	= 33,600	149.5
Total force to be absorbed by anchor	= 67,500	300.2

Comparing these results shows that the 10-in (254-mm) line requires smaller anchors than does the 16-in (406.4-mm) line. However, the 16-in (406.4-mm) line requires only three anchors whereas the 10-in (254-mm) line needs four anchors. The total cost of anchors for both lines will be about equal because of the difference in size of the anchors.

The advantages of slip joints become apparent when the piping layout is studied. Only a minimum of pipe is needed because the pipe runs in a straight line between the point of supply and point of use. The amount of insulation is likewise a minimum.

Corrugated-expansion joints could be used in place of slip-type joints. These would reduce the required anchor size somewhat because there would be no gland friction. The selection procedure resembles that given for slip-type joints.

10. *Select welded-elbow expansion bends*

Use the graphical analysis in Crocker and King—*Piping Handbook* or in any welding fittings engineering data. Using either method shows that three bends of the most economical shape are suitable for the 16-in (406.4-mm) line and four for the 10-in (254-mm) line. The most economical bend is obtained when the bend width, divided by the distance between the anchor points, is 0.50. With these proportions, the longitudinal stress at the top and bottom of the bend is the same. Use of such bends, although desirable, is not always feasible, because existing piping or structures interfere.

When bend dimensions other than the most economical must be used, the maximum longitudinal stress occurs at the top of the bend when the width/anchor distance < 0.5. When this ratio is > 0.5, the maximum stress occurs at the bottom of the bend. Regardless of the bend type—plain U, double-offset U, or welded—the actual stress in the pipe should not exceed 40 percent of the tensile strength of the pipe material.

TABLE 26 Summary of Material Requirements for Various Lines

Means used to absorb expansion	Number of anchors required		Approximate number of supports required		Approximate feet (meters) of pipe and insulation required	
	Pipe size, in (mm)					
	10 (254)	16 (406.4)	10 (254)	16 (406.4)	10 (254)	16 (406.4)
Plain U bends	9	5	127	120	2120 (646.2)	1970 (600.5)
Double-offset U bends	6	3	119	114	1985 (605.0)	1820 (554.7)
Slip joints	4	3	102	102	1700 (518.2)	1700 (518.2)
Welding elbows	5	4	106	106	1760 (536.4)	1760 (536.4)

11. *Determine the materials, quantities, and costs*

Set up tabulations showing the materials needed and their cost. Table 26 shows the materials required. Piping length is computed by using standard bend tables available in the cited references.

Table 27 shows the approximate material costs for each pipeline. The costs used in preparing this table were the most accurate available at the time of writing. However, the actual numerical values given in the table should not be used for similar design work because price changes may cause them to be incorrect. The important findings in such a tabulation are the differences in total cost. These differences will remain substantially constant even though prices change. Hence, if an $8000 difference exists between two sizes of pipe, this difference will not change appreciably with a moderate rise or fall in unit prices of materials.

Study of Table 27 shows that, in general, lines using double-offset U bends or welding elbows have the lowest material first cost. However, higher first costs do not rule out slip joints or plain U bends. Frequently, use of slip joints will eliminate offsets to clear existing buildings or piping because the pipe path is a straight line. Plain U bends have smaller overall heights than double-offset U bends. For this reason, the plain bend is often preferable where the pipe is run through congested areas of factories.

In some cases, past piping practice will govern line selection. For instance, in a factory that has made wide use of slip joints, the slightly higher cost of such a line might be overlooked. Preference might also be shown for plain U bends, double-offset U bends, or welded bends.

The values given in Table 27 do not include installation, annual operating costs, or depreciation. These have been omitted because accurate estimates are difficult to make unless actual conditions are known. Thus, installation costs may vary considerably according to who does the work. Annual costs are a function of the allowable depreciation, nature of process served, and location of the line. For a given transmission line of the type considered here, annual costs will usually be less for the smaller line.

The economic analysis, as made by the pipeline designer, should include all costs relative to the installation and operation of the line. The allowable cost of money and recommended depreciation period can be obtained from the accounting department.

12. *Select the most economical pipe size*

Table 27 shows that from the standpoint of first costs, the smaller line is more economical. This lower first cost is not, however, obtained without losing some large-line advantages.

Thus, steam leaves the 16-in (406.4-mm) line at 150 lb/in^2 (abs) (1034.1 kPa) saturated, the desired outlet condition. Special controls are unnecessary. With the 10-in (254-mm) line, the desired leaving conditions are not obtained. Slightly superheated steam leaves the line unless special controls are used. Where an exact leaving temperature is needed by the process served, a desuperheater at the end of the 10-in (254-mm) line will be needed. Neglecting this disadvantage, the 10-in (254-mm) line is more economical than is the 16-in (406.4-mm) line.

Besides lower first cost, the small line loses less heat to the atmosphere, has smaller anchor forces, and does not cause steam condensation during average flows. Lower heat losses and condensation reduce operating costs. Therefore, if special temperature controls are acceptable, the 10-in (254-mm) maximum pressure-loss line will be a more economical investment.

TABLE 27 Approximate Material Costs for Various Lines

Means used to absorb expansion	Total material cost, $		Condensate removal equipment		Cost of anchors, $		Cost of supports, $		Costs of insulation, $		Cost of pipe and bends or joints, $	
	Pipe size, in (mm)											
	10 (254)	16 (406.4)	10 (254)	16 (406.4)	10 (254)	16 (406.4)	10 (254)	16 (406.4)	10 (254)	16 (406.4)	10 (254)	16 (406.4)
Plain U bends	26,500	51,350	2,000	3,000	1,000	1,800	1,800	2,400	3,700	5,650	18,000	38,500
Double-offset U bends	23,800	44,000	2,000	3,000	600	800	1,700	2,300	3,500	5,400	16,000	32,500
Slip joints	29,650	51,775	2,000	3,000	400	600	1,500	2,000	3,000	4,675	22,750	41,500
Welding elbows	23,800	43,975	2,000	3,000	400	800	1,800	2,300	3,600	5,375	16,000	32,500

Such a conclusion neglects the possibility of future plant expansion. Where expansion is anticipated, installation of a small line now and another line later to handle increased steam requirements is uneconomical. Instead, installation of a large nominal pressure-loss line now that can later be operated as a maximum pressure-loss line will be found more economical. Besides the advantage of a single line in crowded spaces, there is a reduction in installation and maintenance costs.

13. ***Provide for condensate removal***
Fit a condensate drip line for every 100 ft (30.5 m) of pipe, regardless of size. Attach a trap of suitable capacity (see step 6) to each drip line. Pitch the steam-transmission pipe toward the trap, if possible. Where the condensate must flow *against* the steam, the steam-transmission pipe *must* be sloped in the direction of condensate flow. Every vertical rise of the main line must also be dripped. Where water is scarce, return the condensate to the boiler.

Related Calculations. Use this method to design long steam, gas, liquid, or vapor lines for factories, refineries, power plants, ships, process plants, steam heating systems, and similar installation. Follow the applicable piping code when designing the pipeline.

STEAM DESUPERHEATER ANALYSIS

A spray- or direct-contact-type desuperheater is to remove the superheat from 100,000 lb/h (45,000 kg/h) of 300-lb/in^2 (abs) (2068-kPa) 700°F (371°C) steam. Water at 200°F (93.3°C) is available for desuperheating. How much water must be furnished per hour to produce 30-lb/in^2 (abs) (206.8-kPa) saturated steam? How much steam leaves the desuperheater? If a shell-and-tube type of noncontact desuperheater is used, determine the required water flow rate if the overall coefficient of heat transfer $U = 500$ Btu/(h · ft^2· F) [2.8 kW/(m^3 · °C)]. How much tube area A is required? How much steam leaves the desuperheater? Assume that the desuperheating water is not allowed to vaporize in the desuperheater.

Calculation Procedure:

1. ***Compute the heat absorbed by the water***
Water entering the desuperheater must be heated from the entering temperature, 200°F (93.3°C), to the saturation temperature of 300-lb/in^2 (abs) (2068-kPa) steam, or 417.3°F (214°C). Using the steam tables, we see the sensible heat that must be absorbed by the water = h_f at 417.3°F (214°C) – h_f at 200°F (93.3°C) = 393.81 – 167.99 = 255.81 Btu/lb (525.2 kJ/kg) of water used.

Once the desuperheating water is at 417.3°F (214°C), the saturation temperature of 300°F (148.9°C) steam, the water must be vaporized if additional heat is to be absorbed. From the steam tables, the enthalpy of vaporization at 300 lb/in^2 (abs) (2068 kPa) is h_{fg} = 809.0 Btu/lb (1881.7 kJ/kg). This is the amount of heat the water will absorb when vaporized from 417.3°F (214°C).

Superheated steam at 300 lb/in^2 (abs) (2068 kPa) and 700°F (371°C) has an enthalpy of h_g = 1368.3 Btu/lb (3182.7 kJ/kg), and the enthalpy of 300-lb/in^2 (abs) (2068-kPa) saturated steam is h_g = 1202.8 Btu/lb (2797.7 kJ/kg). Thus 1368.3 – 1202.8 = 165.5 Btu/lb (384.9 kJ/kg) must be absorbed by the water to desuperheat the steam from 700°F (371°C) to saturation at 300 lb/in^2 (abs) (2068 kPa).

2. ***Compute the weight of water required for the spray***
The weight of water evaporated by 1 lb (0.45 kg) of steam while it is being desuperheated = heat absorbed by water, Btu/lb of steam/heat required to evaporate 1 lb (0.45 kg) of water entering the desuperheater at 200°F (93.3°C), Btu = 165.5/(225.81 + 809.0) = 0.16 lb (0.07 kg) of water. Since 100,000 lb/h (45,000 kg/h) of steam is being desuperheated, the water flow rate required = (0.16) (100,000) = 16,000 lb/h (7200 kg/h). Water for direct-contact desuperheating can be taken from the feedwater piping or from the boiler.

Note that 16,000 lb/h (7200 kg/h) of additional steam will leave the desuperheater because the superheated steam is not condensed while being desuperheated. Thus, the total flow from the desuperheater = 100,000 + 16,000 = 116,000 lb/h (52,200 kg/h).

3. *Compute the tube area required in the desuperheater*

The total heat transferred in the desuperheater, Btu/h = UAt_m, where t_m = logarithmic mean temperature difference across the heater. Using the method for computing the logarithmic temperature difference given elsewhere in this handbook, or a graphical solution as in Perry et al.—*Chemical Engineers' Handbook*, we find t_m = 134°F (74.4°C) with desuperheating water entering at 200°F (93.3°C) and leaving at 430°F (221.1°C), a temperature about 13°F (7°C) higher than the leaving temperature of the saturated steam, 417.3°F (214°C). Steam enters the desuperheater at 700°F (371°C). Assumption of a leaving water temperature 10 to 15°F (5.6 to 8.3°C) higher than the steam temperature is usually made to ensure an adequate temperature difference so that the desired heat-transfer rate will be obtained. If the graphical solution is used, the greatest temperature difference then becomes 700 – 200 = 500°F (278°C), and the least temperature difference = 430 – 417.3 = 12.7°F (7°C).

Then the heat transferred = (500)(*A*)(134), whereas the heat given up by the steam is, from step 1, (100,000 lb/h)(165.5 Btu/lb)[(45,000 kg/h)(384.9 kJ/kg)]. Since the heat transferred = the heat absorbed, (500)(*A*)(134) = (100,000)(165.5); *A* = 247 ft^2 (22.9 m^2), say 250 ft^2 (23.2 m^2).

4. *Compute the required water flow*

Heat transferred to the water = (500)(247)(134) Btu/h (W). The temperature rise of the water during passage through the desuperheater = outlet temperature, °F – inlet temperature = outlet temperature, °F = 430 – 200 = 230°F (127.8°C). Since the specific heat of water = 1.0, closely, the heat absorbed by the water = (flow rate, lb/h)(230)(1.0). Then the heat transferred = heat absorbed, or (500)(247)(134) = (flow rate, lb/h)(230)(1.0); flow rate = 72,000 lb/h (32,400 kg/h). Since the water and steam do *not* mix, the steam output of the desuperheater = steam input = 100,000 lb/h (45,000 kg/h).

Only about 25 percent as much water, 16,000 lb/h (7200 kg/h), is required by the direct-contact desuperheater as compared with the indirect desuperheater. The indirect type of superheater requires more cooling water because the enthalpy of vaporization, nearly 1000 Btu/lb (2326 kJ/kg) of water, is not used to absorb heat. Some indirect-type desuperheaters are designed to permit the desuperheating water to vaporize. This steam is returned to the boiler. The water-consumption determination and the calculation procedure for this type are similar to the spray-type discussed earlier. Where the water does not vaporize, it must be kept at a high enough pressure to prevent vaporization.

Related Calculations. Use this method to analyze steam desuperheaters for any type of steam system—industrial, utility, heating, process, or commercial.

STEAM ACCUMULATOR SELECTION AND SIZING

Select and size a steam accumulator to deliver 10,000 lb/h (4500 kg/h) of 25-lb/in^2 (abs) (172.4-kPa) steam for peak loads in a steam system. Charging steam is available at 75 lb/in^2 (abs) (517.1 kPa). Room is available for an accumulator not more than 30 ft (9.1 m) long, 20 ft (6.1 m) wide, and 20 ft (6.1 m) high. How much steam is required for startup?

Calculation Procedure:

1. *Determine the required water capacity of the accumulator*

One lb (0.45 kg) of water stored in this accumulator at 75 lb/in^2 (abs) (517.1 kPa) has a saturated liquid enthalpy h_f = 277.43 Btu/lb (645.3 kJ/kg) from the steam tables, whereas for 1 lb (0.45 kg) of water at 25 lb/in^2 (abs) (172.4 kPa), h_f = 208.42 Btu/lb (484.8 kJ/kg). In an accumulator, the stored water flushes to steam when the pressure on the outlet is reduced. For this accumulator, when the pressure on the 75-lb/in^2 (abs) (517.1-kPa) water is reduced to 25 lb/in^2 (abs) (172.4 kPa) by a demand for steam, each pound of stored 75-lb/in^2 (517.1-kPa) water flashes to steam, releasing 277.43 – 208.42 = 69.01 Btu/lb (160.5 kJ/kg).

The enthalpy of vaporization of 25 lb/in^2 (abs) (172.4-kPa) steam is h_{fg} = 952.1 Btu/lb (2215 kJ/kg). Thus, 1 lb (0.45 kg) of 75-lb/in^2 (abs) (517.1-kPa) water will form 69.01/952.1 = 0.0725 lb (0.03 kg) of steam. To supply 10,000 lb/h (4500 kg/h) of steam, the accumulator must store 10,000/0.0725 = 138,000 lb/h (62,100 kg/h) of 75-lb/in^2 (abs) (517.1-kPa) water.

Saturated water at 75 lb/in^2 (abs) (517.1 kPa) has a specific volume of 0.01753 ft^3/lb (0.001 m^3/kg) from the steam tables. Since density = 1/specific volume, the density of 75-lb/in^2 (abs) (517.1-kPa) saturated water = 1/0.01753 = 57 lb/ft^3 (912.6 kg/in^3). The volume required in the accumulator to store 138,000 lb (62,100 kg) of 75-lb/in^2 (abs) (517.1-kPa) water = total weight, lb/density of water = 138,000/57 = 2420 ft^3 (68.5 m^3).

2. ***Select the accumulator dimensions***

Many steam accumulators are cylindrical because this shape permits convenient manufacture. Other shapes—rectangular, cubic, etc.—may also be used. However, a cylindrical shape is assumed here because it is the most common.

The usual accumulator that serves as a reserve steam supply between a boiler and a load (often called a Ruths-type accumulator) can safely release steam at the rate of 0.3 [accumulator storage pressure, lb/in^2 (abs)] lb/ft^2 of water surface per hour [kg/(m^2 · h)]. Thus, this accumulator can release (0.3)(75) = 22.5 lb/(ft^2 · h) [112.5 kg/(m^2 · h)]. Since a release rate of 10,000 lb/h (4500 kg/h) is desired, the surface area required = 10,000/225 = 445 ft^2 (41.3 m^2).

Space is available for a 30-ft (9.1-m) long accumulator. A cylindrical accumulator of this length would require a diameter of 445/30 = 14.82 ft (4.5 m), say 15 ft (4.6 m). When half full of water, the accumulator would have a surface area (30)(15) = 450 ft^2 (41.8 m^2).

Once the accumulator dimensions are known, its storage capacity must be checked. The volume of a horizontal cylinder of d-ft diameter and l-ft length = $(\pi d^2/4)(l) = (\pi \times 15^2/4)(3)$ = 5300 ft^3 (150 m^3). When half full, this accumulator could store 5300/2 = 2650 ft^3 (75 m^3). Since, from step 1, a capacity of 2420 ft^3 (68.5 m^3) is required, a 15 × 30 ft (4.6 × 9.1 m) accumulator is satisfactory. A water-level controller must be fitted to the accumulator to prevent filling beyond about the midpoint. In this accumulator, the water level could rise to about 60 percent, or (0.60)(15) = 9 ft (2.7 m), without seriously reducing the steam capacity. When an accumulator delivers steam from a more-than-half-full condition, its releasing capacity increases as the water level falls to the midpoint, where the release area is a maximum. Since most accumulators function for only short periods, say 5 or 10 min, it is more important that the vessel be capable of delivering the desired rate of flow than that it deliver the last pound of steam in its lb/h rating.

If the size of the accumulator computed as shown above is unsatisfactory from the standpoint of space, alter the dimensions and recompute the size.

3. ***Compute the quantity of charging steam required***

To start an accumulator, it must first be partially filled with water and then charged with steam at the charting pressure. The usual procedure is to fill the accumulator from the plant feedwater system. Assume that the water used for this accumulator is at 14.7 lb/in^2 (abs) and 212°F (101.3 kPa and 100°C) and that the accumulator vessel is half-full at the start.

For any accumulator, the weight of charging steam required is found by solving the following heat-balance equation: (weight of starting water, lb)(h_f of starting water, Btu/lb) + (weight of charging steam, lb)(charging steam h_g, Btu/lb) = (weight of charging steam, lb + weight of starting water, lb) (h_f at charging pressure, Btu/lb). For this accumulator with a 75-lb/in^2 (abs) (517-kPa) charging pressure and 212°F (100°C) starting water, the first step is to compute the weight of water in the half-full accumulator. Since, from step 2, the accumulator must contain 2420 ft^3 (68.5 m^3) of water, this water has a total weight of (volume of water, ft^3)/(specific volume of water, ft^3/lb) = 2420/0.01672 = 144,600 lb (65,070 kg). However, the accumulator can actually store 2650 ft^3 (75 m^3) of water. Hence, the actual weight of water = 2650/0.1672 = 158.300 lb (71.235 kg). Then, with C = weight of charging steam, lb, (158.300)(180.07) + (C)(1181.9) = (C + 158,300)(277.43); C = 17,080 lb (7686 kg) of steam.

Once the accumulator is started up, less steam will be required. The exact amount is computed in the same manner, by using the steam and water conditions existing in the accumulator.

Related Calculations. Use this method to size an accumulator for any type of steam service—heating, industrial, process, utility. The operating pressure of the accumulator may be greater or less than atmospheric.

SELECTING PLASTIC PIPING FOR INDUSTRIAL USE

Select the material, schedule number, and support spacing for a 1-in (25.4-mm) nominal-diameter plastic pipe conveying ethyl alcohol liquid having a temperature of 75°F (23.9°C) and a pressure of 400 lb/in^2 (2758 kPa). What expansion must be anticipated if a 1000-ft (304.8-m) length of the pipe is installed at a temperature of 50°F (10°C)? How does the cost of this plastic pipe compare with galvanized-steel pipe of the same size and length?

Calculation Procedure:

1. *Determine the required schedule number*
Refer to Baumeister and Marks—*Standard Handbook for Mechanical Engineers* or a plastic-pipe manufacturer's engineering data for the required schedule number. Table 28 shows typical pressure ratings for various sizes and schedule number polyvinyl chloride (PVC) (plastic) piping.

Table 28 shows that Schedule 40 normal-impact grade 1-in (25.4-mm) pipe is unsuitable because its maximum operating pressure with fluid at 75°F (24°C) is 310 lb/in^2 (2.13 MPa). Plain-end 1-in (25.4-mm) Schedule 80 pipe is, however, satisfactory because it can withstand pressures up to 435 lb/in^2 (2.99 MPa). Note that threaded Schedule 80 pipe can withstand pressures only to 225 lb/in^2 (1757 kPa). Therefore, plain-end normal-impact grade pipe must be used for this installation. High-impact grade pipe, in general, has lower allowable pressure ratings at 75°F (24°C) because the additive used to increase the impact resistance lowers the tensile strength, temperature, and chemical resistance. Data shown in Table 28 are also presented in graphical form in some engineering data.

2. *Select a suitable piping material*
Refer to piping engineering data to determine the corrosion resistance of PVC to ethyl alcohol. A Grinnell Company data sheet rates PVC normal-impact and high-impact pipe as having excellent corrosion resistance to ethyl alcohol at 72 and 140°F (22.2 and 60°C). Therefore, PVC is a sizable piping material for this liquid at its operating temperature of 75°F (24°C).

3. *Find the required support spacing*
Use a tabulation or chart in the plastic-pipe engineering data to find the required support spacing for the pipe. Be sure to read the spacing under the correct schedule number. Thus, a Grinnell Company plastic-piping tabulation recommends a 5-ft 4-in (162.6-cm) spacing for Schedule 80 1-in (25.4-mm) PVC pipe that weighs 0.382 lb/ft (0.57 kg/m) when empty. The pipe hangers should not clamp the pipe tightly; instead, free axial movement should be allowed.

TABLE 28 Maximum Operating Pressure, PVC Pipe [normal-impact grade, fluid temperature 75°F (23.9°C) or less]

				Schedule 80			
Pipe size		Schedule 40, plain end		Plain end		Threaded	
in	mm	lb/in^2	MPa	lb/in^2	MPa	lb/in^2	MPa
½	12.7	410	2.83	575	3.96	330	2.28
¾	19.1	335	2.31	470	3.24	285	1.97
1	25.4	310	2.14	435	2.99	255	1.76
1½	38.1	230	1.59	325	2.24	205	1.41

TABLE 29 Thermal Expansion of Plastic Pipe

Piping material	Expansion in/(ft · °F)	Expansion cm/(m · °C)
Butyrate	0.00118	0.018
Kralastic	0.00067	0.010
Polyethylene	0.00108	0.016
Polyvinyl chloride	0.00054	0.008
Saran	0.00126	0.019

4. *Compute the expansion of the pipe*
The temperature of the pipe rises from 50 to 75°F (10 to 24°C) when it is put in operation. This is a rise of 75 – 50 = 25°F (14°C). Table 29 shows the thermal expansion of various types of plastic piping.

The thermal expansion of any plastic pipe is found from $E_t = LC\,\Delta t$, where E_t = total expansion, in; L = pipe length, ft; C = coefficient of thermal expansion, in/(ft · °F), from Table 29, Δt = temperature change of the pipe, °F. For this pipe, E_t = (1000)(0.00054)(25) = 13.5 in (342.9 mm) when the temperature rises from 50 to 75°F(10 to 24°C).

5. *Determine the relative cost of the pipe*
Check the prices of galvanized-steel and PVC pipe as quoted by various suppliers. These quotations will permit easy comparison. In this case, the two materials will be approximately equal in per-foot cost.

Related Calculations. Use the method given here for selecting plastic pipe for any service—process, domestic, or commercial—conveying any fluid or gas. Note that the maximum operating pressure of plastic piping is normally taken as about 20 percent of the bursting pressure. The maximum allowable operating pressure decreases with an increase in temperature. The maximum allowable operating temperature is usually 150°F (65.5°C). The pressure loss caused by pipe friction in plastic pipe is usually about one-half the pressure loss in galvanized-steel pipe of the same diameter. Pressure loss for plastic piping is computed in the same way as for steel piping.

ANALYZING PLASTIC PIPING AND LININGS FOR TANKS, PUMPS, AND OTHER COMPONENTS FOR SPECIFIC APPLICATIONS

Choose plastic piping for fire protection, process, and compressed air for an industrial application where corrosive fluids and fumes are likely to be encountered during routine operations. Show how to assemble, and evaluate key data used in choosing suitable materials for such an application. Choose which type of plastic pipe to use, A or B, when the key data you assemble show that the costs associated with each type of piping are as follows:

	Pipe A	Pipe B
First cost, $	80,000	36,000
Salvage value, $	15,000	4000
Life, years	40	15
Annual maintenance cost, $	1000	2300
Annual taxes, $/$100	1.30	1.30
Annual insurance, $/$1000	2	5

If the firm owning this industrial plant earns 6 percent on its invested capital, which type of plastic piping, A or B, is the more economical?

Calculation Procedure:

1. *Assemble the data from which economic choices can be made*
The world of plastic piping and equipment is comprised of a multitude of materials suitable for a variety of industrial, commercial, and power applications. Perhaps the best way to prepare oneself for the important task of materials selection is to obtain several product materials specifications and catalogs from major and specialty plastics manufacturers who make the types of products being considered, in this case, piping, pumps, and tanks.

From the data in these materials specifications and catalogs, prepare a listing, such as that in Table 30 showing the recommended applications for various plastics used for piping, tanks, and pumps. If you do not have access to specifications and catalogs of various manufacturers. Table 30 can serve as a substitute until you do obtain the needed specifications. Studying Table 30 shows that, based on the materials listed, three types of plastic piping are suitable for fire protection (firewater) service, nine types are suitable for process piping, and one for compressed-air piping. This mini-survey, which is current and valid for practical applications in industry today, immediately shows the range of choices open to the designer. Thus, the widest choice is amongst process piping (nine different materials); the narrowest choice (one material) is for compressed-air piping.

Where a difference in cost exists between the choices available, as is almost always the case where more than one material is suitable, an economic study will show which material is the best

TABLE 30 Plastics Are Often Specified for Piping and Equipment That Must Be Corrosion-Resistant and Lightweight*

	Examples of Plastics Applications																		
	Epoxy	Vinyl ester	Polyester	PE	HDPE	PP	PVC	CPVC	DLPVC	PVDF	TFE	PTFE	Polyurethane	FEP	PFA	Furan	Furate	ABS	Bis-A
Firewater piping	•	•			•														
Process piping	•	•	•		•	•	•	•		•					•				
Compressed air piping																		•	
Valves	•	•				•	•	•		•	•				•				
Pumps	•	•	•			•	•	•		•	•				•				
Process area drains manhole & catch basins	•	•	•		•	•	•	•		•									
Tanks	•	•	•		•	•				•					•	•	•		•
Vessels & columns		•	•	•		•		•	•	•				•					
Column trays	•	•	•	•	•	•				•						•	•		•
Strippers & scrubbers	•		•	•		•		•	•	•				•					
Column packing		•													•				
Structural shapes & grating	•	•	•																
Fans & blowers		•	•	•		•	•	•	•	•				•					
Ducts		•	•				•	•											
Stacks & stack liners	•	•	•													•	•		•
Mist eliminators		•	•			•	•												
Fume hoods		•	•				•	•											
Filters & strainers						•	•	•											
Bearing pads											•		•						
Expansion joints												•							
Heat exchanger tubes												•							

**Chemical Engineering.*

selection for the conditions at hand, based on the investment required. Thus, plastic piping is inherently corrosion-resistant. However, other factors, Table 31, must be considered in choosing a piping material. For example, when considering plastic vs. metal, plastic may have a number of advantages, as summarized in Table 31.

TABLE 31 Advantages of Plastic Piping vs. Metal Piping

Plastic piping:
Has high corrosion resistance
Does not require internal or external coating to prevent corrosion
Does not need internal or external cathodic protection to prevent corrosion
Can be welded at lower temperatures where ignition is a problem
Does not usually need welding; if welding it needed, it can be done at lower temperatures
Is nonconducting as manufactured—hence stray electric currents are not a factor in design and installation
Is lighter weight than metal—hence, handling and transportation costs are lower
Has inherently good thermal insulation in itself; in some installations additional thermal insulation may not be required
Has inherent freeze protection and heat retention; outside supplementary protection may not be required

Another consideration where fluids are handled is the permeation of fluids or gases through the plastic material. Table 32 lists permeation rates for selected plastic materials. Some thermoplastics have a high degree of impermeability. However, the level of impermeability may deteriorate with prolonged exposure to ultraviolet radiation, such as from the sun.

TABLE 32 Some Thermoplastics Have a High Degree of Impermeability*

	Permeation of some thermoplastic materials				
Material	Water vapor†	Oxygen‡	Helium‡	Nitrogen‡	CO_2†
PVDF	2	20	600	30	100
PTFE	5	1500	35000	500	15000
FEP	1	2900	18000	1200	4700
PFA	8	—	17000	—	7000
PP	—	25	200	10	100
HDPE	—	30–40	20	18	200
PVC	—	3	16	1	16

*_Chemical Engineering._
†g/m^2 · d at 1 bar and 73°F.
‡Permeability through unreinforced l-mm-thick sheet in cm^3/m^2 · d at 1 bar and 73°F.

2. _Evaluate the relative corrosion resistance of the plastic material selected_
Table 33 lists the relative corrosion resistance of some plastics used for the purposes considered here: piping, tanks, and pumps. Data such as these can be assembled from manufacturer's specifications and catalogs. Or they can be used directly from Table 33 until such time as sufficient contemporary data are compiled.

To use the data in Tables 30 and 33, assume that PVC has been selected for process piping handling corrosive (acidic) liquids and fumes. Table 33 shows that PVC is resistant to caustics but not resistant to acids. Hence, you could choose PVDF (polyvinyl fluoride), which is resistant to acids. You would perform an economic study to see if PVDF was the best choice in this installation, based on the annual cost of each type of piping being considered. These same general principles apply equally well to the corrosion resistance of tanks and pumps.

TABLE 33 Relative Corrosion Resistance of Some Plastics*

	Caustics	Acids	Weak acids	Aliphatic solvent	Aromatic solvents	Alcohols	Halogenated solvents	Ketones	Deionized water
PVC	R	M	R	M	N	M	N	N	R
CPVC	R	M	R	M	N	R	N	N	R
PP	R	N	R	M	N	R	N	R	R
LDPE	R	N	R	M	N	R	N	M	R
HDPE	R	N	R	M	N	R	N	N	R
UHMWPE	R	M	R	M	N	R	N	N	R
PVDF	M	R	R	R	R	R	N	N	R
TFE	R	R	R	R	R	R	R	R	R
FEP	R	R	R	R	R	R	R	R	R
PEEK	M	M	M	—	N	—	—	M	—
Isothalic Polyester	N	N	R	R	M	M	N	N	R
Vinyl Ester	R	M	R	R	R	R	N	N	R
Epoxy Novolac Vinyl Ester	R	R	R	R	R	R	N	M	R
Bisphenol A Fumarate	R	N	R	R	M	M	N	N	R
Furan	R	M	R	R	R	R	M	R	R
	R–resistant			N–not resistant			M–marginal		

Chemical Engineering.

3. *Determine the annual cost of each type of piping being considered*
The first step in determining the annual cost of an asset is computation of the operating and maintenance cost. For Pipe A, the annual operating and maintenance cost, \$ c = maintenance cost per year, \$ + annual taxes, \$ + annual insurance cost, \$. Or, Pipe A, c = \$1000 + (\$1.30/\$100)(\$80,000) + (\$2/\$1000)(\$80,000) = \$2200. For Pipe B, c = \$2300 + (\$1.30/\$100)(\$36,000) + (\$5/\$1000)(\$36,000) = \$2948.

The second, and last, step in computing the annual cost of each type of pipe uses the capital-recovery equation, $A = (P - L)(CR) + Li_t + c$, where A = annual cost, \$; P = initial cost of each type of piping, \$; L = salvage value of each type of piping, \$; CR = capital-recovery factor from compound-interest tables for 6 percent interest rate; i_t = interest rate on the invested capital, 6 percent in this case; c = annual operating and maintenance cost, \$, as computed earlier for each type of pipe.

Substituting for Pipe A, A = (\$80,000 – \$15,000)(0.06646) + \$15,000(0.06) + \$2200 = \$7420. For Pipe B, A = (\$36,000 – \$4000)(0.10296) + \$4000(0.06) + \$2948 = \$6483. Since Pipe B has a lower annual cost, it is the more economical of the two, presuming that the two piping materials have equal, or nearly equal, corrosion resistance properties.

Related Calculations. The same approach given here can be used when choosing the plastic materials for piping, tanks, pumps, ducts, and other components for any industrial, commercial, or residential application. Table 34 gives suggested lining materials for piping, vessels, columns, pumps, and other structures, and gasket materials for joints. The key consideration is obtaining minimum annual cost with the desired level of corrosion or other resistance to deleterious substances. Today, plastic piping is widely used in many applications and is almost universally accepted as superior to metal where corrosion resistance is a primary requirement for the piping. Except for an unfortunate experience with plastic domestic piping installed in single-family homes and some multi-family residences, engineers have had favorable results with plastic piping.

The advantages of plastic piping listed in Table 31 apply to almost every design situation an engineer faces. And as more experience is gained by plastic piping manufacturers, the advantages cited in Table 31 are likely to increase.

For analytical purposes, it is desirable to convert the estimated costs associated with proposed alternatives to an equivalent series of uniform annual payments. The annual payment thus obtained is termed the *annual cost* of each alternative. The interest rate applied in making this conversion is the minimum investment rate considered acceptable by the organization making the investment or incurring the costs. Where alternative schemes are being evaluated on the basis of their annual cost, the usual procedure is to exclude those expenses which are identical for all schemes, since they do not affect the comparison.

Data tables in this procedure were obtained from information prepared by Benjamin S. Fultz, Engineering Supervisor, Nonmetallic Section, Materials and Quality Services Dept., Bechtel Corp.; Robert H. Rogers, Nonmetallic Engineer Specialist, Materials and Quality Services Dept., Bechtel Corp.; J. S. (Steve) Young, Engineering Specialist, Materials and Quality Services Dept., Beclitel Corp.; and Pradip Khaladkar, Materials Consultant, DuPont Engineering (Table 34), as reported in *Chemical Engineering* magazine. The economic study step 3 is the work of Max Kurtz, P.E.

TABLE 34 Which Material to Use for Piping, Vessels, and Equipment*

Linings on carbon steel, concrete or FRP[a]

Thin linings

[<0.025 in (0.64 mm)]

[used when corrosion rate of carbon steel is ≥ 0.010 in (0.25 mm)/yr]

Elastomers—sprayed

Epoxy- and phenolic-based (chemically or heat cured)

Fluoropolymer (sprayed and baked)

Thick linings

[>0.025 in (0.64 mm)]

[used when corrosion rate of carbon steel is >0.010 in (0.25 mm)/yr]

Elastomeric sheet linings[d]

Reinforced vinyl ester, plasticized PVC

Fluoropolymer linings[e]

Self-supporting structures

Vessels and columns

FRP (vinyl ester and furan)

Rotomolded PE

Piping[b]

FRP (vinyl ester, turan, epoxy), PE, PVC, PVDF, FEP

Valves[c]

FRP PVC, CPVC

Dip tubes, agitators, baffles

FRP, PTFE

Others

Gaskets

Elastomers, fluoropolymers

Seals

Elastomers, fluoropolymers

**Chemical Engineering*.

[a]See dual laminate constructions—Table 35.

[b]Loose fluoropolymer linings are typically used for piping. Dual laminate and rotolined piping is also available.

[c]Metal valves and pumps are also lined with plastics for corrosion resistance.

[d]Also used for piping, valves, pumps, agitators, and other applications.

[e]Loose linings are used for piping, molded liners for valves and pumps. Dual laminate and rotolined piping is also available.

TABLE 35 Fluoropolymer Lining Systems*

Lining system and materials	Thickness in (mm)	Maximum size	Design limits	Fabrication†	Repair considerations
Adhesive bonding					
Fabric-backed		No limit	Pressure allowed. Full vacuum only at ambient temperature. Smallest nozzle is 2 in (51 mm). Max. temp. limited by adhesive, typically 275°F (135°C)	Neoprene or epoxy adhesive, sheets welded with cap strips. Heads are thermoformed or welded	Repair is possible but testing is recommended
PVDF	0.06, 0.9 (1.5, 2.3)				
PTFE	0.08, 0.12 (2.0, 3.1)				
FEP	0.06, 0.9 (1.5, 2.3)				
ECTFE	0.06 (1.5)				
ECTFE	0.06, 0.09 (1.5, 2.3)				
PFA	0.09 (2.3)				
Rubber-backed‡					
PVDF	0.05 (1.3)				
Dual laminate					
Except for rubber-backed, same as adhesive bonded		12 ft (3.7 m) dia.,	No pressure allowed. Vacuum rating not determined	Liner fabricated on mandrel by hand and machine welding. FRP built up over liner	Repair is possible but testing is recommended
Sprayed dispersion					
FEP	0.04 (1.0)	8 ft (2.4 m) dia., 40 ft (12.2 m) length	Pressure allowed. Vacuum rating not determined	Primer and multiple coats applied with conventional spray equipment. Each coat is baked	Hot patching is possible but testing is recommended
PFA	0.01–0.04 (0.25–1.0)				
PFA w/mesh and carbon	0.08 (2.0)				
PVDF	0.025–0.03 (0.6–0.76)				
PVDF w/glass or carbon fabric	0.04, 0.09 (1.0, 2.3)				
Electrostatic spray-powder					
ETFE	up to 0.09 (2.3)	8 ft (2.4 m) dia., 40 ft (12.2 m) length	Pressure allowed. Vacuum rating not determined	Primer and multiple coats applied with electrostatic spraying equipment. Each coat is baked	Hot patching is possible but testing is recommended
FEP	0.01 (0.28)				
PFA	0.01 (0.28)				

(Continued)

TABLE 35 Fluoropolymer Lining Systems* (*Continued*)

Lining system and materials	Thickness in (mm)	Maximum size	Design limits	Fabrication†	Repair considerations
ECTFE	0.06, 0.07 (1.5, 1.8)				
PVDF	0.025 (0.64)				
Rotolining					
ETFE PVDF ECTFE	0.1–0.2 (2.5–5.1)	8 ft (2.4 m) dia., 22 ft (6.7 m) length	Pressure allowed. Vacuum rating not determined	Rotationally molded. No seams. No primer used	Hot patching is possible but testing is recommended
Isostatic molding, paste or ram extrusion, or tape wrapping					
PTFE			Pressure allowed. Vacuum rating depends on lining thickness	PTFE is preformed under isostatic pressure, or is paste- or ram-extruded as tubing, and then sintered by heating. Tubing can also be built up by wrapping tape layers on a mandrel and then sintering	Hot patching with PFA possible but testing is recommended
Loose lining					
FEP, PFA	0.06–0.187 (1.5–4.75)	Determined by body flange	Pressure allowed. No vacuum. Gasketing required between liner and flange	Liner with nozzles hand or machine welded, then slipped inside housing	Difficult

**Chemical Engineering*.

†Nondestructive spark testing should be used, along with visual inspection for all systems except loose linings. Adhesive bonding can be done in the shop or field; other systems are shop only.

‡Is rarely used.

FRICTION LOSS IN PIPES HANDLING SOLIDS IN SUSPENSION

What is the friction loss in 800 ft (243.8 m) of 6-in (152.4-mm) Schedule 40 pipe when 400 gal/min (25.2 L/s) of sulfate paper stock is flowing? The consistency of the sulfate stock is 6 percent.

Calculation Procedure:

1. *Determine the friction loss in the pipe*

There are few general equations for friction loss in pipes conveying liquids having solids in suspension. Therefore, most practicing engineers use plots of friction loss available in engineering handbooks. *Cameron Hydraulic Data*, *Standards of the Hydraulic Institute*, 2011, and from pump engineering data. Figure 23 shows one set of typical friction-loss curves based on work done at the University of Maine on the data of Brecht and Heller of the Technical College. Darmstadt, Germany, and published by Goulds Pumps, Inc. There is a similar series of curves for commonly used pipe sizes from 2 through 36 in (50.8 through 914.4 mm).

Enter Fig. 23 at the pipe flow rate, 400 gal/min (25.2 L/s), and project vertically upward to the 6 percent consistency curve. From the intersection, project horizontally to the left to read the friction loss as 60 ft (18.3 m) of liquid per 100 ft (30.5 m) of pipe. Since this pipe is 800 ft (243.8 m) along the total friction-head loss in the pipe = (800/100)(60) = 480 ft (146.3 m) of liquid flowing.

2. *Correct the friction loss for the liquid consistency*

Friction-loss factors are usually plotted for one type of liquid, and correction factors are applied to determine the loss for similar, but different, liquids. Thus, with the Goulds charts, a factor of 0.9 is used for soda, sulfate, bleached sulfate, and reclaimed paper stocks. For ground wood, the factor is 1.40.

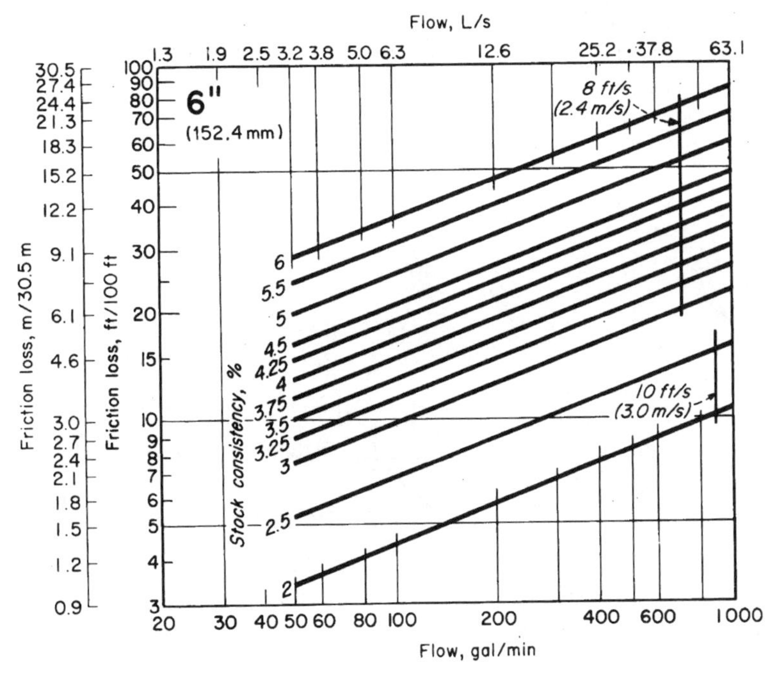

FIGURE 23 Friction loss of paper stock in 4-in (101.6-mm) steel pipe. (*Goulds Pumps, Inc.*)

When the stock consistency is less than 1.5 percent, water-friction values are used. Below a consistency of 3 percent, the velocity of flow should not exceed 10 ft/s (3.05 m/s). For suspensions of 3 percent and above, limit the maximum velocity in the pipe to 8 ft/s (2.4 m/s).

Since the liquid flowing in this pipe is sulfate stock, use the 0.9 correction factor, or the actual total friction head = (0.9)(480) = 432 ft (131.7 m) of sulfate liquid. Note that Fig. 23 shows that the liquid velocity is less than 8 ft/s (2.4 m/s).

Related Calculations. Use this procedure for soda, sulfate, bleached sulfite, and reclaimed and ground-wood paper stock. The values obtained are valid for both suction and discharge piping. The same general procedure can be used tor sand mixtures, sewage, slurries, trash, sludge, and foods in suspension in a liquid.

DESUPERHEATER WATER SPRAY QUANTITY

A pressure- and temperature-reducing station in a steam line is operating under the following conditions: pressure and temperature ahead of the station are 1400 lb/in^2 absolute (5650 kPa), 950°F (510°C); the reduced temperature and pressure after the station are 600°F (315°C), 200 lb/in^2 absolute (1380 kPa). If 450,000 lb/h (3400 kg/s) of steam is required at 200 lb/in^2 (1380 kPa), how much water, which is available at 200 lb/in^2 absolute (1380 kPa) and 635.8°F (335.4°C), must be sprayed in at the superheater? See Fig. 24.

Calculation Procedure:

1. ***Determine the quantity of heat entering the desuperheater via the spray in terms of the amount of water***

The quantity of heat entering the desuperheater via the spray, $Q = w \times h_f$, where the amount of water is w, lb/h (kg/s); from Table 2, Saturation Pressures, of the steam tables mentioned under Related

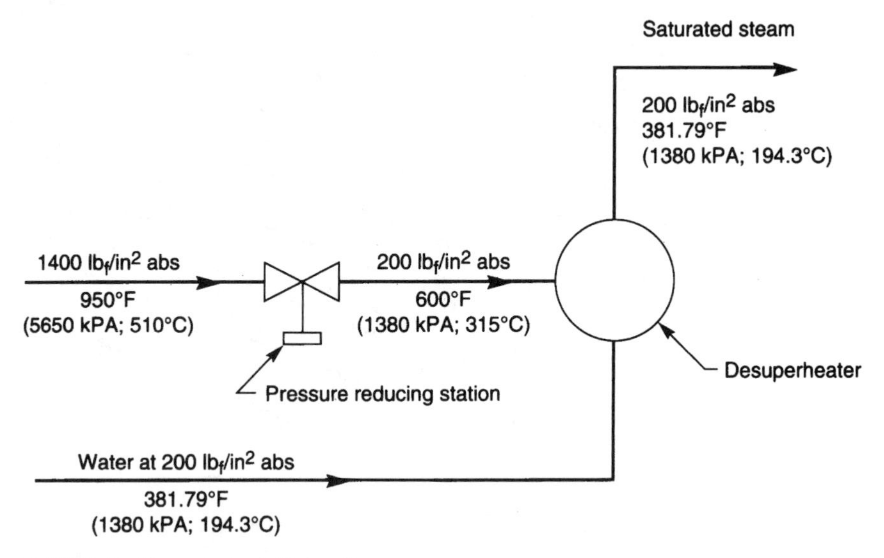

FIGURE 24 Desuperheater fluid flow diagram.

Calculations of this procedure, heat content of water, saturated steam. At 200 lb/in^2 (1380 kPa), h_f = 355.4 Btu/lb$_m$ (826 kJ/kg). Thus $Q = w \times 355.4$ lb/h (kg/s). It should be noted that the steam tables show the saturation temperature to be 381.79°F (194.3°C) at the given pressure, shown on Fig. 24. Obviously, the 635.8°F (335.4°C) given in the problem is not correct, because at that temperature there would either be superheated steam, vapor, or the water would have to be under a pressure of 2000 lb/in^2 (13.8×10^3 kPa).

2. *Find the enthalpy of the superheated steam entering the desuperheater and the enthalpy of the saturated steam leaving*
From Table 3, Vapor, of the steam tables, superheated steam entering the desuperheater at 200 lb/in^2 (1380 kPa) and 600°F (315°C) has an enthalpy, h = 1322.1 Btu/lb (2075 kJ/kg). From Table 2, Saturation Pressures, of the steam tables, saturated steam leaving the superheater at 200 lb/in^2 (1380 kPa) and 381.79°F (194.3°C) at saturated vapor has an enthalpy, h_g = 1198.4 Btu/lb$_m$ (2787 kJ/kg).

3. *Compute the amount of water which must be sprayed into the desuperheater*
The amount of water which must be sprayed into the desuperheater, w_w lb/h (kg/s), can be found by the use of a heat balance equation where, as an adiabatic process, the amount of heat into the superheater equals the amount of heat out. Then, $w \times h_f + (450{,}000 - w)h = 450{,}000 \times h_g$. Or, $w \times 355.4 + (450.000 - w)(1322.1) = 450{,}000 \times 1198.4$. Solving, $w = 450{,}000 \times 123.7/966.7 = 57.580$ lb/h (435.3 kg/s).

Related Calculations. Strictly speaking, the given pressure and temperature conditions before the pressure-reducing station were irrelevant in the Calculation Procedure. Also, the incorrect given saturation temperature of 635.8°F (335.4°C) is an example of possible distractions which should be guarded against while solving such problems. The steam tables appear in *Thermodynamic Properties of Water Including, Liquid, and Solid Phases*, 1969, Keenan, et al., John Wiley & Sons, Inc. Use later versions of such tables whenever available, as necessary.

SIZING CONDENSATE RETURN LINES FOR OPTIMUM FLOW CONDITIONS

An evaporator is condensing 5500 lb/h (2497 kg/h) of steam at 150-lb/in^2 (gage) (1033.5-kPa) supply pressure. During normal operation, a control value maintains a pressure of 85 lb/in^2 (gage) (585.7 kPa) upstream of the steam trap handling the condensate from the evaporator. The condensate discharged by the steam trap is returned to an atmospheric-vented tank. What pipe line size should be used on the steam-trap outlet to provide optimum flow conditions?

Calculation Procedure:

1. *Compute the percentage of flash steam in the steam-trap discharge line*
The percentage of flash steam in the steam-trap discharge line is found from

$$x_{fs} = \frac{(h_{t_1} - h_{t_2})}{\Delta h_{v2}} \times 100$$

where the symbols are as given below.

Nomenclature

A_{req}	= Required cross-sectional area, ft^2 (m^2)
D_u	= Nominal pipe size, based on velocity u, in (mm)
D_{50}	= Nominal pipe size, based on 50 ft/s, in (m/s, mm)

(*Continued*)

Nomenclature (*Continued*)

h_{f1}	= Condensate enthalpy at upstream pressure, P_1, Btu/lb (kJ/kg)
h_{f2}	= Condensate enthalpy at end-pressure, P_2, Btu/lb (kJ/kg)
$\Delta h_{\upsilon 2}$	= Latent heat of vaporization at P_2, Btu/lb (kJ/kg)
Q_υ	= Flash-steam volumetric flow rate, ft³/h
u	= New flash-steam velocity, ft/s (m/s)
$\upsilon_{\upsilon 2}$	= Flash-steam specific volume at P_2, ft³/lb (m³/kg)
W_l	= Condensate formed at P_1, lb/h (kg/h)
W_υ	= Flash steam formed at P_2, lb/h (kg/h)
Z_{fx}	= Flash steam, wt %

Essentially, what this relation does is to convert the difference of condensate enthalpies, out and in, to flash steam using the latent heat of vaporization at the outlet pressure of the steam trap as the flash-heat source. Substituting in the equation above, using steam-table data for the enthalpies, x_{fs} = [(298.4 – 180.1)/1150.4](100) = 10.28 percent flash steam by weight.

2. *Find the weight of flash steam formed at the trap outlet*
Use the relation

$$W_\upsilon = W_l \frac{x_{fs}}{100}$$

Substituting, W_υ = (5500)(10.28/100) = 565.4 lb/h (256.7 kg/h).

3. *Calculate the flash-steam volumetric flow rate*
Use the relation

$$Q_\upsilon = W_\upsilon \upsilon_{\upsilon 2}$$

Substituting, Q_υ = (565.4) (26.8) = 15,152.7 ft³/h (428.8 m³/h).

4. *Determine the required cross-sectional area of the steam trap discharge pipe*
Use the relation

$$A_{req} = \frac{Q_\upsilon}{3,600 \times 50}$$

for a flash-steam velocity of 50 ft/s (15.24 m/s), the usual value used in sizing such pipes.

Substituting, A_{req} = (15,152.7)/(3600)(50) = 0.0842 ft² (0.00782 m²). Converting to in², (0.00842)(144) = 12.12 in² (7820.5 mm²). Note that this is the required internal area of the trap discharge pipe.

5. *Choose the pipe size to use*
Entering a table of pipe properties, we find that a 4-in (101.6-mm) pipe is required to convey the flashing condensate from this steam trap at the chosen velocity.

Related Calculations. Undersized condensate return lines create one of the most common problems met with process (and power-plant) steam traps. Hot condensate passing through a trap orifice loses pressure, which lowers the enthalpy of the condensate. This enthalpy change causes some of the condensate to flash into steam. The volume of the resulting two-phase mixture is usually many times that of the upstream condensate entering the trap.

The trap-outlet or downstream piping must be adequately sized to handle effectively the greater volume of the two-phase mixture. An undersized condensate return line on the trap outlet results in a high flash-steam velocity. This may cause water hammer (due to wave formation), hydrodynamic noise, premature erosion, and high backpressure. An excessively high backpressure reduces the working differential pressure across the trap and, hence, the condensate removal capability of the steam trap. In some traps excessive backpressure causes partial or full failure.

Because of the much greater volume of flash steam compared with unflashed condensate, sizing of the return line is based solely on flash steam. It is assumed that all flashing occurs across the steam trap and that the resulting vapor-liquid mixture can be evaluated at the end-pressure conditions. To ensure the condensate line does not have an appreciable pressure drop, a low flash-steam velocity is assumed, namely, 50 ft/s (15.24 m/s).

Where there is only a small pressure drop in the discharge line, or high subcooling of the condensate, it may be necessary to size the condensate line based on the liquid velocity, generally, a velocity of 3 ft/s (0.91 m/s). For flash-steam velocities other than the 50 ft/s (15.24 m/s) used above, the nominal pipe size can be approximated from

$$D_u = \frac{7.07 D_{50}}{\sqrt{u}}$$

The method presented here yields a single result; thus decision-making is not required on the part of the user, as it is in other methods of steam-trap discharge-line sizing. This method can be used for new construction of all types: industrial commercial, residential, marine, HVAC, etc. It can also be used to check existing line sizes where trap performance is questionable. With minimal training, field maintenance personnel can use this method on the job.

The procedure presented here is the work of Michael V. Calogero, GESTRA, Inc., and Arthur W. Brooks, TECHMAR Engineering, Inc., as presented in *Chemical Engineering* magazine.

ESTIMATING COST OF STEAM LEAKS FROM PIPING AND PRESSURE VESSELS

Steam at 135 lb/in^2 (930 kPa) is leaking from a 0.05-in^2 (0.32-cm^2) opening in a pipe and escaping to the atmosphere where it cannot be recovered. Temperature of the steam in the pipe is 450°F (232°C). Determine the cost of this leak if steam costs \$2.50 per 1000 lb (454 kg) and the pipe operates 8000 h per year.

Calculation Procedure:

1. *Compute the rate of steam loss through the opening*

The rate of steam loss through an opening is given by $L = KA(PD)^{0.5}$, where L = steam loss, lb/h (kg/h); K = factor for steam condition; for saturated steam, K = 1085; for superheated steam, K = 1138; A = opening area, in^2 (cm^2); P = steam pressure on the pressure side of the opening, lb/in^2 (abs) (kPa); D = density of the steam on the pressure side of the opening, lb/ft^3 (kg/m^3).

Substituting for this leak, using the superheated steam factor of 1138 because the steam temperature of 450°F (232°C) is about 100° higher than the saturation temperature of 358°F (181°C) at 150 lb/in^2 (abs) (= 135 lb/in^2 + 14.7 lb/in^2) (1033.5 kPa), gives $L = 1138(0.05)(150[0.3316])^{0.5}$ = 401.3 lb/h (182.2 kg/h).

2. *Determine the cost of the leaking steam*

The cost of a steam leak, C \$ = $LSH/1000$, where C = annual cost of steam leak, \$, for H hours of yearly operation of the pipe; S = cost of steam, \$ per 1000 lb (454 kg). Substituting, C = 401.3 lb/h × \$2.50/1000 lb × 8000 h/yr = \$8026 per year for this leak. This is a significant cost when compared with the low cost of modern materials that can be used to stop such a leak.

Related Calculations. Use the relation given here for steam leaks for openings in pipes, pressure vessels, traps, meters, orifices, nozzles, and other steam apparatus. The relation can also be used for leaks from the open ends of pipes. The one restriction on the use of the relation is that the pressure on the outlet side must be 0.578, or less than the pressure on the inside of the pipe or vessel. This situation prevails in almost every case of a leaking pipe or vessel. The reason for this is that most leaks are from a pressurized source to the atmosphere. Few leaks occur from one pressurized source to another.

The value 0.578 is the critical pressure ratio for steam flow through in orifice. Velocity of the escaping steam is determined by this ratio. The higher this velocity, the larger the amount or steam escaping in a unit time.

There is greater emphasis today than ever before on preventing unnecessary steam losses through neglected leaks in piping and pressure vessels. The reason for this is that engineers and managers now recognize the chain effect of uncontrolled steam leaks. This effect is as follows: Leaking steam that does no work represents wasted fuel that was burned to generate the steam. This wasted fuel produces unnecessary pollution of the atmosphere during its combustion. Further, there is a drain or natural resources because the fuel burned produces no useful work. Thus, the total cycle is environmentally offensive when the steam is wasted by not performing any work before being exhausted to the atmosphere or a condenser. For these reasons, steam leaks are getting more attention than ever before. The same—of course—can be said about leakage of any other valuable liquid or gas—such as oil, compressed air, etc. A different equation must be used to compute such losses because the equation above applies only to steam leaks.

QUICK SIZING OF RESTRICTIVE ORIFICES IN PIPING

Choose the bore size of a restrictive orifice for a 25-gal/min (1.57-L/s) minimum bypass for a pump discharging water at 100 lb/in^2 (gage) (689.4 kPa) and 80°F (26.7°C) into a 50-lb/in^2 (gage) (344.7-kPa) drum when a 0.125-in (0.3175-cm) thick orifice plate is used. What bore size would be required for a 0.25-in (0.635-cm) thick orifice plate? Use the quick-sizing approach.

Calculation Procedure:

1. *Determine the needed parameters—pressure drop, flow rate, and fluid specific gravity*
The pressure drop, $\Delta P = 100 - 50 = 50$ lb/in^2 (gage) (344.7 kPa). Flow rate through the orifice is given as 25 gal/min (1.57 L/s). Specific gravity of water at 80°F (26.7°C) = 1.0.

2. *Compute the restrictive orifice coefficient*
Use the relation $C_{vro} = Q_t/(\Delta P/S_g)^{0.5}$, where C_{vro} = restrictive orifice coefficient, dimensionless; Q_t = liquid flow rate, gal/min (L/s); ΔP = pressure drop through the orifice, lb/in^2 (gage); S_g specific gravity of the water, dimensionless. Substituting, $C_{vro} = 25/(50/1)^{0.5} = 3.536$.

3. *Calculate the required orifice bore diameter*
Use the relation $D = 0.875(C_{vro}/14.0)^{0.5}$, where D = required orifice bore (hole) diameter, in (cm); other symbols as before. Substituting, $D = 0.875(3.536/14.0)^{0.5} = 0.4397$; round to 0.440 in (1.12 cm) for ease of manufacturing.

4. *Determine the orifice bore size for the thicker plate*
When the plate thickness is different from 0.125 in (0.3175 cm), use the relation $D_{corr} = D(L/0.125)^{0.2}$, D_{corr} = bore diameter in (cm) for plate thickness L, in (cm) different from 0.125 in (0.3175 cm); other symbol as before. Substituting, $D_{corr} = 0.40(0.25/0.125)^{0.2} = 0.505$ in (1.28 cm). Thus, as the orifice plate thickness increases beyond 0.125 in (0.3175 cm), the required bore diameter also increases.

Related Calculations. Restrictive orifices can be easily sized, starting with valve coefficients and making some simple assumptions, namely: (1) It is known that for a plate thickness of 0.125 in (0.3175 cm) a straight-bore orifice has a C_υ of 14.0. (2) The Reynolds number is 2300 for turbulent flow. Using these two assumptions, plus the definition of C_υ, one can size any restrictive orifice with a plate thickness of 0.125 in (0.3175 cm).

For plates thicker than 0.125 in (0.3175 cm), the orifice bore can be found by applying a correction factor, since I/D^5 = a constant for turbulent flow, as derived from the pressure-drop formula.

C_υ = the flow in gal/min (L/s), when the medium is water with a specific gravity = 1, and the pressure drop is 1 lb/in^2 (6.89 kPa). When calculating C_υ, use the physical variables of the flow conditions, shown above.

The C_υ = 14.0 for a 0.125-in (0.3175-cm) straight-through sharp-edge orifice is taken from Scientific Apparatus Makers' Association (Washington, DC) data for the discharge of water for different pressures and orifice sizes. The control-valve formulas were published by Cashco Inc. (Ellsworth, KS) and were modified by the author of this procedure (see below) to include the compressibility for gases and the superheat factor. For cases of two-phase flow where there is a change of cavitation, use the methods found in Fisher Controls International, Inc. (Marshalltown, IA) Catalog 10.

The control-valve formulas for gases and steam are: *For gases:* $Q_g = 1360\ (C_\upsilon)(\Delta P[P_1 + P_2]/2\ S_{gIZ})^{0.5}$, where Q_g = gas flow rate, std ft^3/*h* (std m^3/h); ΔP = pressure drop across the orifice, lb/in^2; P_1 = upstream pressure, lb/in^2; P_2 = downstream pressure, lb/in^2; T = gas temperature, R; Z = gas compressibility; other symbols as before. *For steam*: $W = 3\ C_\upsilon/K(\Delta P[P_1 + P_2]/2)^{0.5}$, where $K = 1 + 0.0007\ \Delta T_{sh}$, where W = steam flow, lb/h (kg/h); ΔT_{sh} = degree of superheat, F (C).

The restrictive orifice diameter, D, and correction for plate thickness, D_{corr}, are as given above in steps 3 and 4. If ΔP is less than or equal to 0.5 P_1, then $(\Delta P\ [P_1 + P_2]/2)$ reduces to $P_{1/2}(1.5)^{0.5}$ = 0.6124 P_1, for gases and steam only—this is sonic flow.

The orifice can be union or paddle type. Use stainless steel Type 304 or 316 for the orifice. For corrosive atmospheres, use special materials.

This procedure is the work of Herman E. Waisvisz, as reported in *Chemical Engineering* magazine.

STEAM TRACING A VESSEL BOTTOM TO KEEP THE CONTENTS FLUID

The bottom of a 4-ft (1.22-m) diameter, stainless-steel, solvent-recovery column holds a liquid that freezes at 320°F (160°C) and polymerizes at 400°F (204.4°C). The bottom head must be traced to keep the material fluid after a shutdown. Determine the required pitch of the tracing, using 150-lb/in^2 (gage) (1034-kPa) saturated steam. The ambient temperature is –20°F (–28.9°C), the supply steam temperature T_s = 366°F (185.628°C), thermal conductivity of stainless steel = 9.8 Btu/(h · ft^2 · °F · ft) [16.95 W/(m · K)], insulation thickness = 2 in (5.1 cm), thermal conductivity of insulation = 0.3 Btu/(h · ft^2 · °F · ft) [0.52 W/(m · K)], and wall thickness = 0.375 in (0.95 cm). The heat-transfer coefficient between the insulation and the air is 2.0 Btu/(h · ft^2 · °F) [11.4 W/(m^2 · K)], and the inside convection coefficient is 20.

Calculation Procedure:

1. ***Compute the process-fluid heat-transfer coefficient and the overall heat-transfer coefficient***

Use the relation $1/h_o = 1/h_{air} + x_{ins}/k_{inx}$, where the symbols are as defined in the previous calculation procedure. Substituting, we get $1/h_o$ = ½ + 2/0.3; h_o = 0.14 Btu/(h · ft^2 · °F) [0.79 W/(m^2 · K)].

Then process-fluid heat-transfer coefficient h_t = 20 Btu/(h · ft^2 · °F) [11.36 W/(m^2 · K)], assumed.

2. ***Determine the constants A and B and the ratio B/A***

To determine the value of A, solve $A = (h_o + h_t)/Kt$ = (0.14 + 20)/(9.8)(0.375/12) = 65.6, dimensionless. Also, $B = (h_tT_p + h_oT_{amb})/Kt$ = [(20)(320) + (0.14)(–20)]/(9.8)(0.375/12) = 20,900. Then B/A = 20,900/65.6 = 318.

3. *Compute the tracer-steam outlet temperature*

The tracer steam is supplied at 50 lb/in^2 (gage) (344.7 kPa). Assuming a 15-lb/in^2 (gage) (103.4-kPa) pressure drop in the tracer system, we see the outlet pressure = 150 – 15 = 135 lb/in^2 (gage) (930.8 kPa). The corresponding saturated-steam temperature is, from the steam tables, T_o = 358°F (181°C). This is the tracer outlet steam temperature.

4. *Calculate the adjusted temperature ratio*

Use the relation $(T_{mid} - B/A)/(T_o - B/A) = (320 - 318)/(358 - 318) = 0.05$.

5. *Determine the required tracing pitch*

From Fig. 25, with the adjusted temperature ratio of 0.5, $\sqrt{A}(L) = 3.7$ when $\alpha = 1$. (Here α = a parameter = x/L, where x = distance along the pipe or vessel wall, ft.) Then, by solving for $L = 3.7/65.6^{1/2}$, $L = 0.46$ ft (0.14 m) = 5.5 in (13.97 cm).

The maximum allowable pitch for tracing the bottom of the column, Fig. 26, is 21, or 2(5.5) = 11 in (27.9 cm). A typical tracing layout is shown in Fig. 26.

Related Calculations. This procedure can be used to design steam tracing for a variety of tanks and vessels used in chemical, petroleum, food, textile, utility, and similar industries. The medium heated can be liquid, solid, vapor, etc. As with the previous calculation procedure, this procedure is the work of Carl G. Bertram, Vikram J. Desai, and Edward Interess, the Badger Company, as reported in *Chemical Engineering* magazine.

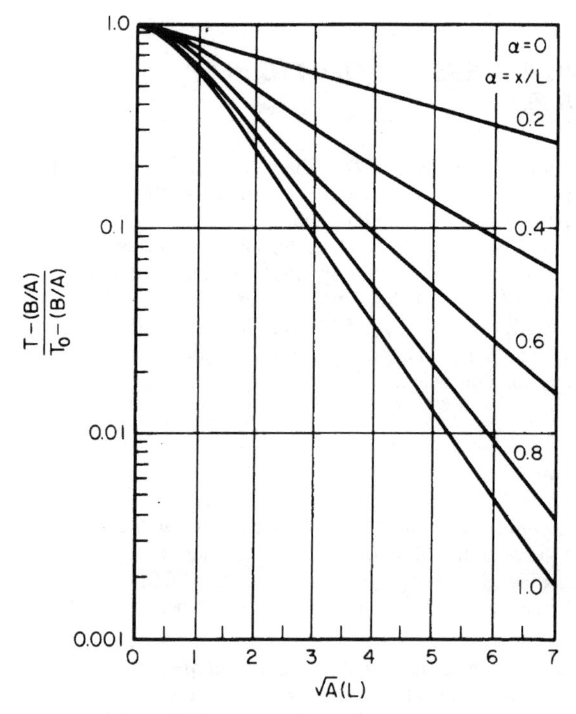

FIGURE 25 Graphical solution for steam-tracing design. (*Chemical Engineering*.)

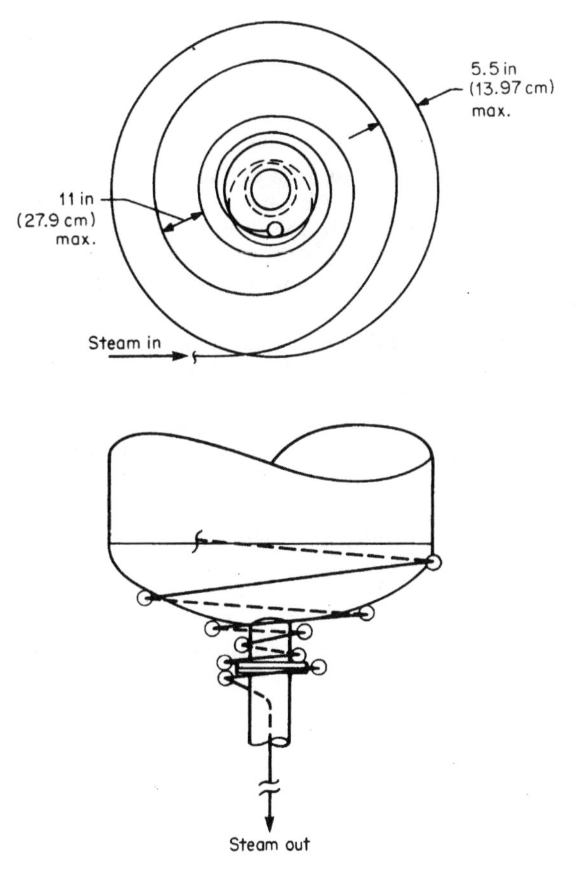

FIGURE 26 Steam-traced vessel bottom. (*Chemical Engineering.*)

DESIGNING STEAM-TRANSMISSION LINES WITHOUT STEAM TRAPS

Design a steam line for transporting a minimum of 6.0×10^5 lb/h (2.7×10^5 kg/h) and a maximum of 8.0×10^5 lb/h (3.6×10^5 kg/h) of saturated steam at 205 lb/in^2 (gage) and 309°F (1413 kPa and 198.9°C). The line is 3000 ft (914.4 m) long, With eight 90° elbows and one gate valve. Ambient temperatures range from – 40 to 90°F (−40 to 32.2°C). The line is to be designed to operate without steam traps. Insulation 3 in (7.6 cm) thick with a thermal conductivity of 0.48 Btu · in/(h · ft^2 · °F) [0.069 W/(m · K)] will be used on the exterior of the line.

Calculation Procedure:

1. ***Size the pipe by using a suitable steam velocity for the maximum flow rate***
The minimum acceptable steam velocity in a transmission line which is not fitted with steam traps is 110 ft/s (33.5 m/s). Assuming, for safety purposes, a steam velocity of 160 ft/s (48.8 m/s) to use in sizing this transmission line, compute the pipe diameter in inches from $d = 0.001295 f\rho\, LV^2/\Delta P$, where f = friction factor for the pipe (= 0.0105, assumed); ρ = density of the steam, lb/ft^3 (kg/m^3)

[= 0.48 (7.7) for this line]; L = length of pipe, ft, including the equivalent length of fittings [= 3500 ft (1067 m) for this pipe]; V = steam velocity, ft/s [= 160 ft/s (48.8 m/s) for this line]; ΔP = pressure drop in the line between inlet and outlet, lb/in^2 [= 25 lb/in^2 (172.4 kPa) assumed for this line]. Substituting yields $d = 0.001295(0.0105)(0.48)(3500)(160)^2/25 = 22.94$ in (58.3 cm); use 24-in (61-cm) Schedule 40 pipe, the nearest standard size.

2. *Check the actual steam velocity in the pipe chosen*

The actual velocity of the steam in the pipe can be found from $V = Q/A$, where V = steam velocity, ft/s (m/s); Q = flow rate of steam, lb/s (kg/s); A = cross-sectional area of pipe, ft^2 (m^2). Substituting gives V = (800,000 lb/h ÷ 3600 s/h)(2.08 ft^3/lb for steam at the entering pressure)/2.94 ft^2 = 157.6 ft/s (48.0 m/s) for maximum-flow conditions; V = (600,000/3600)(2.08)/2.94 = 117.9 ft/s (35.9 m/s) for minimum-flow conditions.

3. *Compute the pressure drop in the pipe for each flow condition*

Use the relation $\Delta P = 0.001295\, f\rho LV^2/D$, where the symbols are the same as in step 1. Substituting, we find $\Delta P = 0.001295(0.0105)(0.48)(3500)(157.6)^2/24 = 23.2$ lb/in^2 (159.9 kPa) for maximum-flow conditions. For minimum-flow conditions by the same relation, $\Delta P = 0.001295(0.0105)(0.48)(3500)(117.9)^2/24 = 13.23$ lb/in^2 (91.2 kPa). The pressure at the line outlet will be 220.0 – 23.2 = 196.5 lb/in^2 (1356.9 kPa) for the maximum-flow condition and 220.0 – 13.2 = 206.8 lb/in^2 (1425.9 kPa) for minimum-flow conditions.

4. *Compute the steam velocity at the pipe outlet*

Use the velocity relation in step 1. Hence, for maximum-flow conditions, V = (800,000/3600)(2.30)/2.94 = 173.8 ft/s (52.9 m/s). Likewise, for minimum-flow conditions, V = (600,000/3600)(2.19)/2.94 = 124.1 ft/s (37.8 m/s).

5. *Determine the enthalpy change in the steam at maximum temperature-difference conditions*

First, the heat loss from the insulated pipe must be determined for the maximum temperature-difference condition from $Q_m = h\Delta tA$, where Q_m = heat loss at maximum flow rate, Btu/h (W); h = overall coefficient of heat transfer for the insulated pipe, Btu · in/(h · ft^2 · °F) [W · cm/(m^2 · °C)]; Δt = temperature difference when the minimum ambient temperature prevails, °F (°C); A = insulated area of pipe exposed to the outdoor air, ft^2 (m^2). Substituting yields Q_m = 0.16 (430)(3362)(6.28) = 1,452,599 Btu/lb (3378.7 MJ/kg). In this relation, 430°F = 390°F steam temperature ± (–40°F) ambient temperature; 3362 = pipe length including elbows and valves, ft; 6.28 = area of pipe per ft of pipe length, ft^2.

The enthalpy change for the maximum temperature difference will be the largest with the minimum steam flow. This change, in Btu/lb (J/kg) of steam, is $\Delta h_{max} = Q_m/F$, where F = flow rate in the line, lb/h, or Δh_{max} = 1,452,599/600,000 = 2.42 Btu/lb (5631 J/kg).

The minimum enthalpy at the pipe line outlet = inlet enthalpy – enthalpy change. For this pipe line, h_{0min} = 1199.60 – 2.42 = 1197.18 Btu/lb (2784.6 kJ/kg).

6. *Determine the enthalpy change in the steam at the minimum temperature-difference conditions*

As in step 5, $Q_{min} = h\Delta tA$, or Q_{min} = 0.16(300)(3362)(6.28) = 1,013,441 Btu/lb (2357.3 mJ/kg). Then Δh_{min} = 1,013,441/800,000 = 1.26 Btu/lb (2946.6 J/kg). Also h_{2max} = 1199.60 – 1.26 = 1198.34 Btu/lb (2787.3 kJ/kg).

7. *Determine the steam conditions at the pipe outlet*

From step 3, the pressure at the transmission line outlet at minimum flow and lowest ambient temperature is 206.8 lb/in^2 (1425.9 kPa), and the enthalpy is 1197.18 Btu/lb (2784.6 kJ/kg). Checking this condition on a Mollier chart for steam, we find that the steam is wet because the condition point is below the saturated-vapor line.

From steam tables, the specific volume of the steam is 2.22 ft^3/lb of total mass, while the specific volume of the condensate is 0.0000342 ft^3/lb of total mass. Thus, the percentage of condensate per volume = 100(0.0000342)/2.22 = 0.00154 percent condensate per volume. The percentage volume of dry steam therefore = 100(1.00000 – 0.00154) = 99.99846 percent dry steam per volume.

Since the velocity under these steam conditions is 124.1 ft/s (37.8 m/s), the steam will exist as a fine mist because such a status prevails when the steam velocity exceeds 110 ft/s (33.5 m/s). In the fine-mist condition, the condensate cannot be collected by a steam trap. Hence, no steam traps are required for the transmission line as long as the pressure and velocity conditions mentioned above prevail.

Related Calculations. Some energy is lost whenever a steam trap is used to drain condensate from a steam-transmission line. This energy loss continues for as long as the steam trap is draining the line. Further, a steam-trap system requires an initial investment and an ongoing cost for routine maintenance. If the energy loss and trap-system costs can be reduced or eliminated, many designers will take the opportunity to do so.

A steam-transmission line carries energy from point 1 to point 2. This energy is a function of temperature, pressure, and flow rate. Along the line, energy is lost through the pipe insulation and through steam traps. A design that would reduce the energy loss and the amount of required equipment would be highly desirable.

The designer's primary concern is to ensure that steam conditions stay as close to the saturated line as possible. The steam state in the line changes according to the change in pressure due to a pressure drop and the change in enthalpy due to a heat loss through insulation. These changes of condition are plotted in Fig. 27, a simplified Mollier chart for steam. Point 1 is defined by P_1 and T_1 steam conditions. Because of the variability of such parameters as flow rate and ambient temperature, the designer should consider extreme conditions. Thus, P_2 would be defined by the minimum pressure drop produced by the minimum flow rate. Similarly, h_2 would be defined by the maximum heat loss produced by the lowest ambient temperature. Point 2 on the *h-s* diagram is defined by the above P_2 and h_2. If point 2 is above or on the saturated-steam line, no condensate is generated and steam traps are not required.

In some cases (small pressure drop, large heat loss), point 2′ is below the saturated-steam line, and some condensate is generated. The usual practice has been to provide trap stations to collect this condensate and steam traps to remove it. However, current research in two-phase flow demonstrates that the turbulent flow, produced by normal steam velocities and reasonable steam qualities, disperses any condensate into a fine mist equally distributed along the flow profile. The trap stations do not collect the condensate, and once again, the steam traps are not required. For velocities greater than 110 ft/s (33.5 m/s) and a steam fraction more than 98 percent by volume, the condensate normally generated in a transmission line exists as a fine mist that cannot be collected by steam-trap stations.

A few basic points should be followed when a steam line is operated without steam traps. All lines must be sloped. If a line is long, several low points may be required. Globe valves are used on

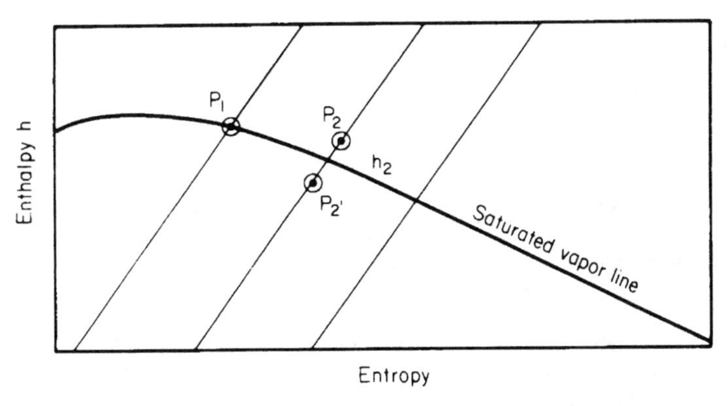

FIGURE 27 Simplified Mollier chart showing changes in steam state in a steam-transmission line.

drains for each low point and for a drain at the end of the line. Since trap stations are not required, drain valves should be located as close to the line as possible to avoid freezing. Vents are placed at all high points. All vents and drains are opened prior to warming the line. Once steam is flowing from all vent valves, they are closed. As each drain valve begins to drain steam only, it is partially closed so that it may still bleed condensate if necessary. When full flow is established, all drain valves are shut. If the flow is shut down, all valves are opened until the pipe cools and are then closed to isolate the line from the environment. The above procedure would be the same if the steam traps were on the line.

In summary, it can be demonstrated that steam traps are not required for steam-transmission lines, provided that one of the following parameters is met:

1. Steam is saturated or superheated.
2. Steam velocity is greater than 110 ft/s (33.5 m/s), and the steam fraction more than 98 percent by volume.

By using the above design, the steam energy normally lost through traps is saved, along with the construction, maintenance, and equipment costs for the traps, drip leg, strainers, etc., associated with each trap station.

This calculation procedure can be used for steam-transmission lines in chemical plants, petroleum refineries, power plants, marine installations, factories, etc. The procedure is the work of Mileta Mikasinovic and David R. Dautovich, Ontario Hydro, and reported in *Chemical Engineering* magazine.

Leaks of hazardous materials from underground piping and tanks can endanger lives and facilities. To reduce leakage dangers, the EPA now requires all underground piping through which hazardous chemicals or petrochemicals flow to be designed for double containment. This means that the inner pipe conveying the hazardous material is contained within an outer pipe, giving the "double-containment" protection.

Likewise, underground tanks are governed by the new Underground storage Tank (UST) laws. The UST laws also cover underground piping. By December 1998, all existing underground piping conveying hazardous materials will have to be retrofitted to double-containment systems to comply with EPA requirements.

Double containment of piping brings a host of new problems for the engineering designer, Expansion of the inner and outer pipes must be accommodated so that there is no interference between the two. While prefabricated double-containment piping can solve some of these problems, engineers are still faced with consideration of soil loading, pipe expansion and contraction, and fluid flow. Careful study of the EPA requirements is needed before any double-containment design is finalized. Likewise, local codes and laws must be reviewed prior to starting and before finalizing any design.

LINE SIZING FOR FLASHING STEAM CONDENSATE

REFERENCES; [1] O. Baker, *Oil & Gas J.*, July 26, 1954; [2] S. G. Bankoff, *Trans. ASME*, Vol. C82, 265 (1960); [3] M. W. Benjamin and J. G. Miller, *Trans. ASME*, Vol. 64, 657 (1942); [4] J. M. Chenoweth and M. W. Martin, *Pet. Ref.*, Vol. 34, 151 (1955); [5] A. E. Dukler, M. Wickes, and R. G. Cleveland, *AIChE J.*, Vol. 10, 44 (1964); [6] E. C. Kordyban, *Trans. ASME*, Vol. D83, 613 (1961); [7] R. W. Lockhart and R. C. Martinelli, *Chem. Eng. Prog.*, Vol. 45, 39 (1949); [8] P. M. Paige, *Chem. Eng.*, p. 159, Aug. 14, 1967.

A reboiler in an industrial plant is condensing 1000 lb/h (0.13 kg/s) of steam of 600 lb/in^2 (gage) (4137 kPa) and returning the condensate to a nearby condensate return header nominally at 200 lb/in^2 (gage) (1379 kPa). What size condensate line will give a pressure drop of (1 lb/in^2)/100 ft (6.9 kPa/30.5 m) or less?

Calculation Procedure:

1. ***Use a graphical method to determine a suitable pipe size***
Flow in condensate-return lines is usually two-phase, i.e., comprised of liquid and vapors. As such, the calculation of line size and pressure drop can be done by using a variety of methods, a number of which are listed below. Most of these methods, however, are rather difficult to apply because they require extensive physical data and lengthy computations. For these reasons, most design engineers prefer a quick graphical solution to two-phase flow computations. Figure 28 provides a rapid estimate of the pressure drop of flashing condensate, along with a determination of fluid velocity. To use Fig. 28, take these steps.

Enter Fig. 28 near the right-hand edge at the steam pressure of 600 lb/in^2 (gage) (4137 kPa) and project downward to the 200-lb/in^2 (gage) (1379-kPa) end-pressure curve.

From the intersection with the end-pressure curve, project horizontally to the left to intersect the 1000-lb/h (0.13-kg) curve. Project vertically from this intersection to one or more trial pipe sizes to find the pressure loss for each size.

Trying the 1-in (2.5-cm) pipe diameter first shows that the pressure loss—[3.0 lb/in^2 (gage)]/100 ft (20.7 kPa/30.5 m) exceeds the desired [1 lb/in^2 (gage)]/100 ft (6.9 kPa/30.5 m). Projecting to the next larger standard pipe size, 1.5 in (3.8 cm), gives a pressure drop of [2 lb/in^2 (gage)]/100 ft (1.9 kPa/30.5 m). This is within the desired range. The velocity in this size pipe will be 16.5 ft/s (5.0 m/s).

2. ***Determine the corrected velocity in the pipe***
At the right-hand edge of Fig. 28, project upward from 600-lb/in^2 (gage) (4137-kPa) to 200-lb/in^2 (gage) (1379-kPa) end pressure to read the velocity correction factor as 0.41. Thus, the actual velocity of the flashing mixture in the pipe = 0.41 (16.5) = 6.8 ft/s (2.1 m/s).

Related Calculations. This rapid graphical method provides pressure-drop values comparable to those computed by more sophisticated techniques for two-phase flow [1–8]. Thus, for the above conditions, the Dukler [5] no-slip method gives [0.22 lb/in^2 (gage)]/100 ft (1.52 kPa/30.5 m), and the Dukler constant-slip method gives [0.25 lb/in^2 (gage)]/100 ft (1.72 kPa/30.5 m).

The chart in Fig. 28 is based on the simplifying assumption of a single homogeneous phase of fine liquid droplets dispersed in the flashed vapor. Pressure drop is computed by Darcy's equation for single-phase flow. Steam-table data were used to calculate the isenthalpic flash of liquid condensate from a saturation pressure to a lower end pressure; the average density of the resulting liquid-vapor mixture is used as the assumed homogeneous fluid density. Flows within the regime of Fig. 28 are characterized as either in complete turbulence or in the transition zones near complete turbulence.

Pressure drops for steam-condensate lines can be determined by assuming that the vapor-liquid mix throughout the lines is represented by the mix for conditions at the end pressure. This assumption conforms to conditions typical of most actual condensate systems, since condensate lines are sized for low-pressure drop, with most flashing occurring across the steam trap or control valve at the entrance.

If the condensate line is to be sized for a considerable pressure drop, so that continuous flashing occurs throughout its length, end conditions will be quite different from those immediately downstream of the trap. In such cases, an iterative calculation should be performed, involving a series of pressure-drop determinations across given incremental lengths.

This iteration is begun at the downstream end pressure and worked back to the trap, taking into account the slightly higher pressure, and thus the changing liquid-vapor mix, in each successive upstream incremental pipe length. The calculation is complete when the total equivalent length for the incremental lights equals the equivalent length between the trap and the end-pressure point. This operation can be performed by using Fig. 28.

Results from Fig. 28 have also been compared to those calculated by a method suggested by a Paige [8] and based on the work of Benjamin and Miller [3]. Paige's method assumed a homogeneous liquid-vapor mixture with no liquid holdup, and thus it is similar in approach to the present method. However, Paige suggests calculation of the liquid-vapor mix based on an isentropic flash,

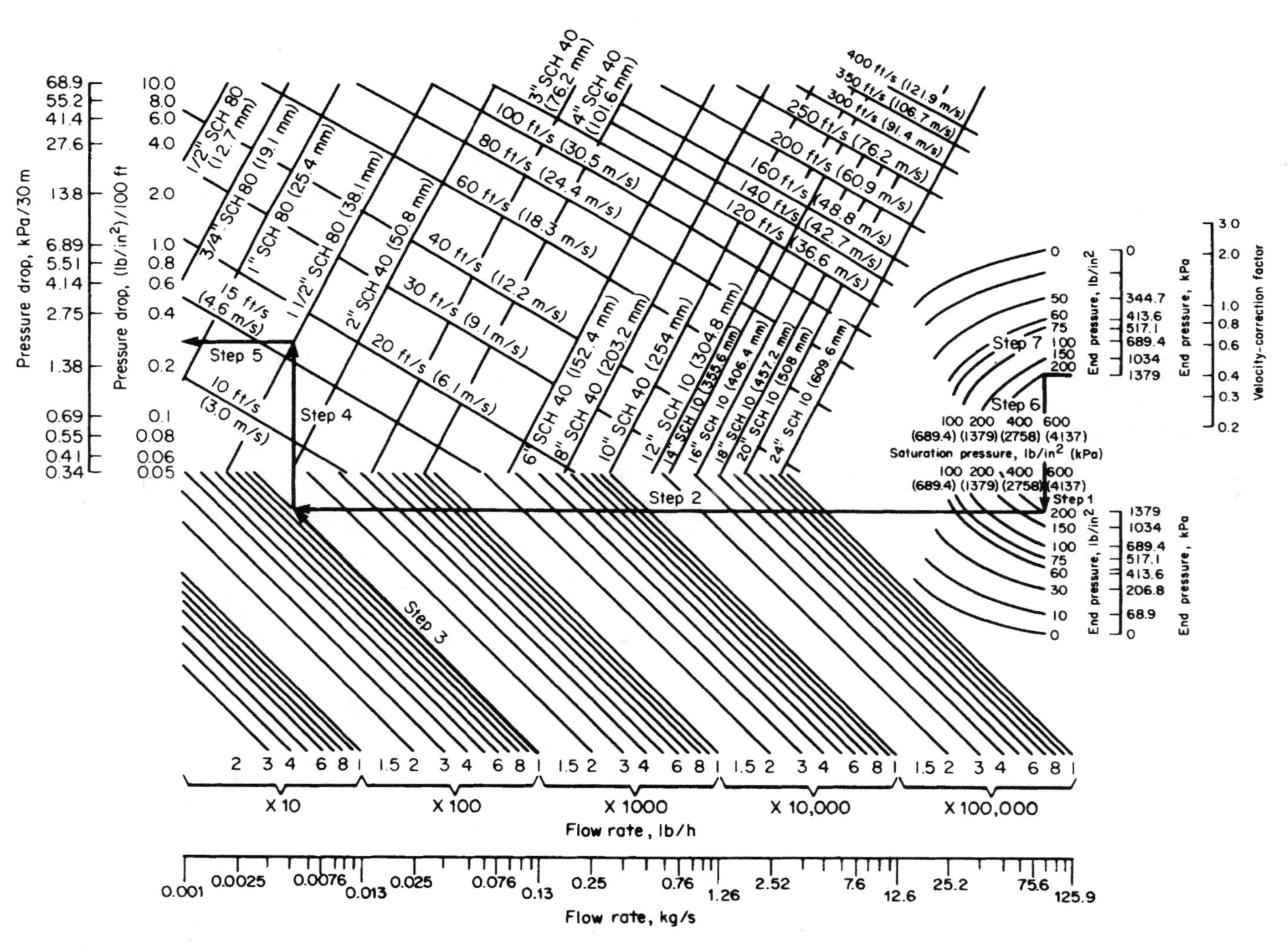

FIGURE 28 Flashing steam condensate line-sizing chart. Divide by 10^4 to obtain numerical values for flow rate measured in kg/s. (*Chemical Engineering*.)

whereas Fig. 28 is based on an isenthalpic flash, and this is believed to be more representative of steam-condensate collecting systems.

For the example, the Paige method gives ($0.26\ lb/in^2$)/100 ft (1.79 kPa/30.5 m) at the terminal pressure and ($0.25\ lb/in^2$)/100 ft (1.72 kPa/30.5 m) at a point 1000 ft (305 m) upstream of the terminal pressure, owing to the slightly higher pressure, which suppresses flashing.

The method given here is valid for sizing lines conveying flashing steam used in power plants, factories, air-conditioning systems, petroleum refineries, ships, heating systems, etc. Further, Fig. 28 is designed so that it covers the majority of steam-condensate conditions met in these applications. This calculation procedure is the work of Richard P. Ruskin, Process Engineer, Arthur G. McKee & Co., as reported in *Chemical Engineering* magazine.

DETERMINING THE FRICTION FACTOR FOR FLOW OF BINGHAM PLASTICS

REFERENCES: [1] E. Buckingham, On Plastic Flow through Capillary Tubes, *ASTM Proc.*, Vol. 21, 1154 (1921); [2] R. W. Hanks and D. R. Pratt, On the Flow of Bingham Plastic Slurries in Pipes and between Parallel Plates, *Soc. Petrol. Eng. J.*, Vol. 1, 342 (1967); [3] R. W. Hanks and B. H. Dadia, Theoretical Analysis of the Turbulent Flow of Non-Newtonian Slurries in Pipes, *AIChE J.*, Vol. 17, 554 (1971); [4] S. W. Churchill, Friction-Factor Equation Spans All Fluid-Flow Regimes, *Chem. Eng.*, Nov. 7, 1977, pp. 91–92; [5] S. W. Churchill and R. A. Usagi, A General Expression for the Correlation of Rates of Transfer and Other Phenomena, *AIChE J.*, Vol. 18, No. 6, 1121–1128 (1972); [6] R. L. Whitmore, *Rheology of the Circulation*, Pergamon Press, Oxford, 1968; [7] N. Casson, A Flow Equation for Pigment-Oil Dispersions of the Printing Ink Type, Ch. 5 in *Rheology of Disperse Systems*, C. C. Mill (ed.), Pergamon Press, Oxford, 1959; [8] R. Darby and B. A. Rogers, Non-Newtonian Viscous Properties of Methacoal Suspensions. *AIChE J.*, Vol. 26, 310 (1980); [9] G. W. Govier and A. K. Azia, *The Flow of Complex Mixtures in Pipes*, Van Nostrand Reinhold, New York, 1972; [10] E. H. Steiner, The Rheology of Molten Chocolate, Ch. 9 in C. C. Mill (ed.), *op. cit.*; [11] R. B. Bird, W. I. Stewart, and E. N. Lightfoot, *Transport Phenomena*, John Wiley & Sons, New York, 1960.

A coal slurry is being pumped through a 0.4413-m (18-in) diameter Schedule 20 pipeline at a flow rate of 400 m^3/h. The slurry behaves as a Bingham plastic, with the following properties (at the relevant temperature): $\tau_0 = 2\ N/m^2$ ($0.0418\ lbf/ft^2$); $\mu_x = 0.03$ Pa · s (30 cP); $\rho = 1500\ kg/m^3$ ($93.6\ lbm/ft^3$). What is the Fanning friction factor for this system?

Calculation Procedure:

1. ***Determine the Bingham Reynolds number and the Hedstrom number***

Engineers today often must size pipe or estimate pressure drops for fluids that are non-newtonian in nature—coal suspensions, latex paint, or printer's ink, for example. This procedure shows how to find the friction factors needed in such calculations for the many fluids that can be described by the Bingham-plastic flow mode. The method is convenient to use and applies to all regimes of pipe flow.

A Bingham plastic is a fluid that exhibits a yield stress; that is, the fluid at rest will not flow unless some minimum stress τ_0 is applied. Newtonian fluids, in contrast, exhibit no yield stress, as Fig. 29 shows.

The Bingham-plastic flow model can be expressed in terms of either shear stress τ versus shear rate γ, as in Fig. 29, or apparent viscosity η versus shear rate:

$$\tau = \tau_0 + \mu_\infty \dot{\gamma} \tag{1}$$

$$\eta = \frac{\tau}{\dot{\gamma}} = \frac{\tau_0}{\dot{\gamma}} + \mu_\infty \tag{2}$$

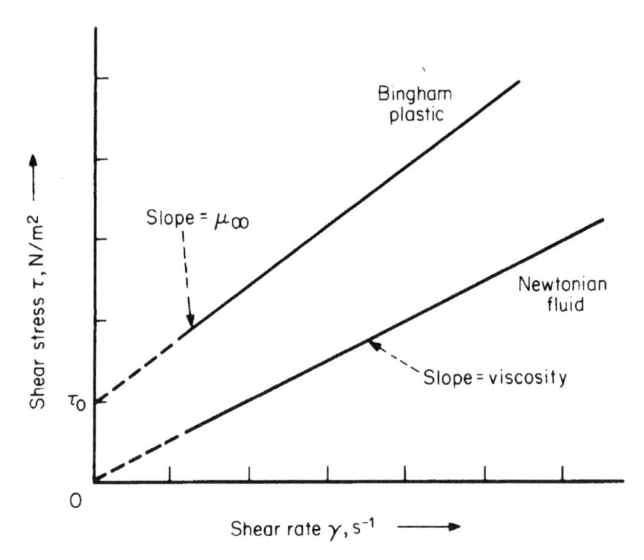

FIGURE 29 Bingham plastics exhibit a yield stress. (*Chemical Engineering.*)

Equation (2) means that the apparent viscosity of a Bingham plastic depends on the shear rate. The parameter μ_∞ is sometimes called the coefficient of rigidity, but it is really viscosity. As Eq. (2) shows, apparent viscosity approaches μ_∞ as shear rate increases indefinitely. Thus the Bingham plastic behaves almost like a newtonian at sufficiently high shear rates, exhibiting a viscosity of μ_∞ at such conditions. Table 36 shows values of τ_0 and μ_∞ for several actual fluids.

For any incompressible fluid flowing through a pipe, the friction loss per unit mass F can be expressed in terms of a Fanning friction factor f:

$$F = \frac{2fL\upsilon^2}{D} \tag{3}$$

Where L is the length of the pipe section, D is its diameter, and υ is the fluid velocity.

An exact description of friction loss for Bingham plastics in fully developed laminar pipe flow was first published by Buckingham [1]. His expression can be rewritten in dimensionless form as follows:

$$f_L = \frac{16}{N_{\text{Re}}}\left(1 + \frac{N_{\text{He}}}{6N_{\text{Re}}} - \frac{N_{\text{He}}^4}{3f_L^3 N_{\text{Re}}^7}\right) \tag{4}$$

TABLE 36 Values of τ_0 and μ_∞

Fluid	τ_0, N/m²	μ_∞, Pa · s	Ref.
Blood (45% hematocrit)	0.005	0.0028	[6]
Printing-ink pigment in varnish (10% by wt.)	0.4	0.25	[7]
Coal suspension in methanol (35% by vol.)	1.6	0.04	[8]
Finely divided galena in water (37% by vol.)	4.0	0.057	[9]
Molten chocolate (100°F)	20	2.0	[10]
Thorium oxide in water (50% by vol.)	300	0.403	[11]

Where N_{Re} is the Bingham Reynolds number (Dup/μ_∞) and N_{He} is the Hedstrom number ($D^2p\tau_0/\mu^2$). Equation (4) is implicit in f_L, the laminar friction factor, but can be readily solved either by Newton's method or by iteration. Since the last term in Eq. (4) is normally small, the value of f obtained by omitting this term is usually a good starting point for iterative solution.

For this pipeline

$$N_{Re} = \frac{4Q\rho}{\pi D\mu_\infty} = \frac{4(400)(1/3)(600)(1500)}{\pi(0.4413)(0.03)} = 16{,}030$$

$$N_{He} = \frac{D^2\rho\tau_0}{\mu_\infty^2} = \frac{(0.4413)^2(1500)(2)}{(0.03)^2} = 649{,}200$$

2. *Find the friction factor f_L for the laminar-flow regime*

Substituting the values for N_{Re} and N_{He} into Eq. (4), we find $f_L = 0.007138$.

3. *Determine the friction factor f_T for the turbulent-flow regime*

Equation (4) describes the laminar-flow sections. An empirical expression that fits the turbulent-flow regimen is

$$f_T = 10^a N_{Re}^{-0.193} \tag{5}$$

where

$$a = -1.378[1 + 0.146 \exp(-2.9 \times 10^{-5} N_{He})] \tag{6}$$

We now have friction-factor expressions for both laminar and turbulent flow. Equation (6) does not apply when N_{He} is less than 1000, but this is not a practical constraint for most Bingham plastics with a measurable yield stress.

When N_{He} is above 300,000, the exponential term in Eq. (6) is essentially zero. Thus $a = -1.378$ here, and Eq. (5) becomes

$$f_T = 10^{-1.378}(16{,}030)^{-0.193}$$
$$= 0.006463$$

4. *Find the friction factor f*

Combine the f_L and f_T expressions to get a single friction factor valid for all flow regimes :

$$f = (f_L^m + f_T^m)\frac{1}{m} \tag{7}$$

Where f_L and f_T are obtained from Eqs. (4) and (5), and the power m depends on the Bingham Reynolds number:

$$m = 1.7 + \frac{40{,}000}{N_{Re}} \tag{8}$$

The values of f predicted by Eq. (7) coincide with Hank's values in most places, and the general agreement is excellent. Relative roughness is not a parameter in any of the equation because the friction factor for non-newtonian fluids, and particularly plastics, is not sensitive to pipe roughness.

Substituting yields $m = 1.7 + 40{,}000/16{,}030 = 4.20$, and $f = [(0.007138)^{4.20} + (0.006463)^{4.20}]^{1/4.20} = 0.00805$.

If m had been very large, the bracketed term above would have approached zero. Generally, when N_{Re} is below 4000, Eq. (8) should be solved by taking f equal to the greater of f_L and f_T.

Related Calculations. This procedure is valid for a variety of fluids met in many different industrial and commercial applications. The procedure is the work of Rob Darby, Professor of Chemical Engineering, Texas A & M University, College of Engineering, and Jeff Melson, Undergraduate Fellow, Texas A & M, as reported in *Chemical Engineering* magazine. In their report they cite works by Hanks and Pratt [2], Hanks and Dadia [3], Churchil [4], and Churchill and Usagi [5] as important in the procedure described and presented here.

For the more general cases of turbulent (nonlaminar) and laminar flow of fluids in pipes of all types, the editor of this handbook has added the following 21 formulas—6 for nonlaminar (turbulent) flow and 15 for laminar flow of fluids in pipes (Table 37). These formulas allow easy conversion when different variables are known for the flow situation. The variables involved are pipe length, fluid velocity, pipe diameter, pressure loss, flow rate, friction factor, and absolute viscosity.

Nomenclature

cSt = cP (lb · s/ft²)
d = pipe inside diameter, in
D = pipe inside diameter, ft
f = friction factor, dimensionless
g = gravity constant, 32.2 ft/s²
H_L = head loss, ft
L = pipe length, ft
L_{in} = pipe length, in
Δp = pressure loss, lb/in²
ΔP = pressure, lb/ft²
q = flow, gal/min
Q = flow, ft³/s
Q_{in^3} = flow, in³/s
R_e = Reynolds number $\rho DV/\mu$. Typical units are:
$3162\, q/dv_{eSt}$ gal/min, in, cSt
$50.6\, q\gamma/d\mu_{cp}$ gal/min, lb/ft³, in, cP
$\gamma DV/g\mu$ lb/ft³, ft, ft/s, lb · s/ft²
v = fluid velocity, in/s
V = fluid velocity, in/s
μ = absolute viscosity, lb · s/ft²
ν = μ/ρ, kinematic viscosity, ft²/s
ρ = mass density, lb · s²/ft⁴ = slugs/ft³
γ = weight density, lb/ft³
ε/D = relative roughness of pipe wall. Values of ε, ft: drawn tubing, 5×10^{-6}; steel or wrought iron, 150×10^{-6}. D = inside diameter, ft

TIME NEEDED TO EMPTY A STORAGE VESSEL WITH DISHED ENDS

A tank with a 6-ft (1.8-m) diameter cylindrical section that is 16 ft (4.9 m) long has elliptical ends, each with a depth of 2 ft (0.7 m), and is half-full with ethanol, a newtonian fluid. How long will it take to empty the tank if it is set horizontally and fitted at the bottom with a drain consisting of a short tube of 2-in (5.1-cm) double extrastrong pipe? How long will it take to empty the tank if it is set vertically and fitted at the bottom with a drainpipe of 2-in (5.1-cm) double extrastrong pipe? The drain system extends 4 ft (1.2 m) below the dished bottom and has an equivalent length of 250 ft (76.2 m).

TABLE 37 Formulas for Flow in Pipes

General empirical relationships: all flows, $H_L = f\frac{L}{D}\frac{V^2}{2g}$; laminar only, $H_L = 32\frac{\mu LV}{pgd^2}$; useful conversions, for oil only (= 55 lb/ft²).

Variables							Formulas		
L	V	D	ΔP	Q	f	μ	Pressure loss	Velocity	Flow
							All flows		
ft	$\frac{ft}{s}$	in	lb/in²	$\frac{ft^3}{s}$	—	—	$\Delta p = f\frac{L}{d}V^2 \times 0.072$	$V = \sqrt{\frac{d\Delta}{fL}} \times 3.73$	$Q = \sqrt{\frac{d^5\Delta p}{fL}} \times 0.0203$
ft	$\frac{ft}{s}$	in	lb/in²	gal/min (q)	—	—	$\Delta p = f\frac{L}{d^5}q^2 \times 0.0123$	$V = \sqrt{\frac{d\Delta p}{fL}} \times 3.73$	$q = \sqrt{\frac{d^5\Delta p}{fL}} \times 9.1$
							Laminar only		
ft	$\frac{ft}{s}$	in	lb/in²	$\frac{ft^3}{s}$	—	cSt	$\Delta p = \nu\frac{L}{d^2}V \times 0.0006$	$V = \frac{d^2\Delta p}{\nu L} \times 1670$	$Q = \frac{d^4\Delta p}{\nu L} \times 9.1$
ft	$\frac{ft}{s}$	in	lb/in²	gal/min	—	cSt	$\Delta p = \nu\frac{L}{d^4}q \times 2.45 \times 10^{-4}$	$V = \frac{d^2\Delta p}{\nu L} \times 1670$	$q = \frac{d^4\Delta p}{\nu L} \times 4080$
in	$\frac{in}{s}$	in	lb/in²	$\frac{in}{s}$	—	$\frac{lb \cdot s}{in^2}$	$\Delta p = \mu\frac{L_{in}}{d^4}Q_{in^3} \times 40.75$	$V = \frac{d^2\Delta p}{\mu L_{in}} \times 0.0312$	$Q_{in^3} = \frac{d^4\Delta p}{\mu L_{in}} \times 0.0245$
ft	$\frac{ft}{s}$	ft	lb/ft²	$\frac{ft^3}{s}$	—	$\frac{lb \cdot s}{ft^2}$	$\Delta p = \mu\frac{L}{D^2}V \times 32$	$V = \frac{D\Delta p}{\mu L} \times 0.0312$	$Q = \frac{d^4\Delta p}{\mu L} \times 0.0248$
ft	$\frac{ft}{s}$	ft	lb/ft²	gal/min	—	$\frac{lb \cdot s}{ft^2}$	$\Delta p = \mu\frac{L}{D^4}q \times 0.091$	$V = \frac{D^2\Delta p}{\nu L} \times 0.0312$	$q = \frac{D^4\Delta P}{\nu L} \times 11$

Calculation procedure:

1. *Determine the discharge coefficient for, and orifice area of, the drain tube*

Figure 30 shows the discharge coefficient is $C_d = 0.80$ for a short, flush-mounted tube. Baumeister, in *Mark's Standard Handbook for Mechanical Engineers*, indicates the internal section area of the tube is $A_n = 1.774$ in^2 (11.4 cm^2).

2. *Compute the discharge time for the tank in a horizontal position*

Substitute the appropriate values in the equation for t_υ shown under the storage tanks in Fig. 31. Thus, $t_p = [(8)^{0.5}]/[3(0.80)(1.774/144)(32.2)^{0.5}]\ \{16[(6)^{1.5} - (6 - 3)^{1.5}] + [2\pi(3)^{1.5}/6][6 - (3/5)(3)]\} =$ 2948 s, or 49.1 min.

3. *Determine the internal diameter and friction factor for the drainpipe*

From Baumeister, *Mark's Standard Handbook for Mechanical Engineers*, the internal diameter of the pipe is $d = 1.503$ in (3.8 cm), or 0.125 ft (0.038 m) and the Moody friction factor is $f = 0.020$ for the equivalent length, $l = 250$ ft (76.2 m), of pipe.

4. *Compute the initial and final height above the drainpipe outlet for the cylindrical section*

Initial height of the liquid is $H_1 = a + b + h_o = [(16/2) + 2 + 4] = 14.0$ ft (4.3 m). Final height is $H_F = b + h_o = 2 + 4 = 6$ ft (1.8 m).

5. *Compute the time required to drain the cylindrical section of the tank*

Substitute the appropriate values in the equation for t_c shown under the storage tanks in Fig. 31. Hence, $t_c = [(6)^2/(0.125)^2]\ \{(2/32.2)[1 + (0.020 \times 250/0.125)]\}^{0.5}[(14)^{0.5} - (6)^{0.5}] = 4751$ s, or 79.2 min.

6. *Compute the initial and final liquid height above the drainpipe outlet for the elliptically dished head*

Initial height of the liquid is $H_1 = b + h_o$ eq $2 + 4 = 6$ ft (1.8 m). Final height is $H_2 = h_o = 4$ ft (1.2 m).

7. *Compute how long it will take to empty the dished bottom of the tank*

In order to solve for t_e it is necessary to determine the following values: $B = h_o + b = 4 + 2 = 6$ ft (1.8 m); $E^2 = h_o^2 + 2bh_o = (4)^2 + 2(2)(4) = 32$ ft^2 (3.0 m^2); $C = [D/(db)]^2\ \{[1/(2g)][1 + (f_1/d)]\}^{0.5} =$ $[6/(0.125 \times 2)]^2\{[1/(2 \times 32.2)][1 + ([0.02 \times 250]/0.125)]\}^{0.5} = 459.6$, s/ft$^{5/2}$ (s/m$^{5/2}$).

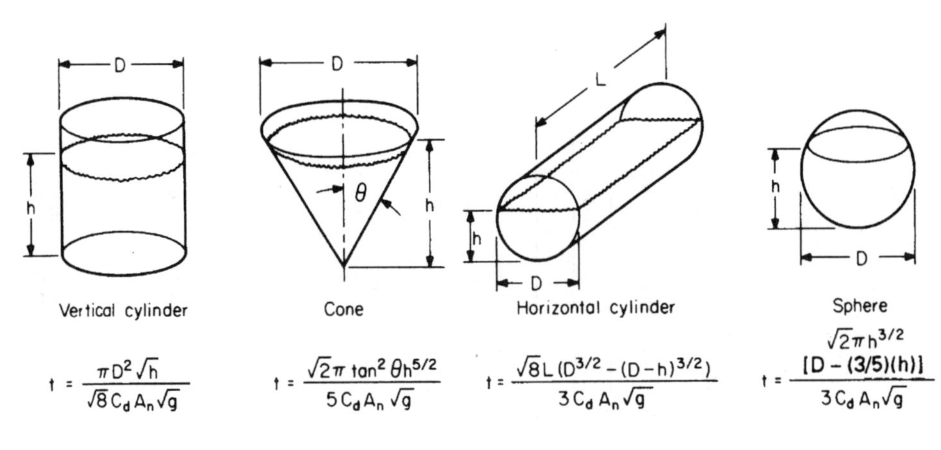

FIGURE 30 Time to empty tanks. (*Chemical Engineering*.)

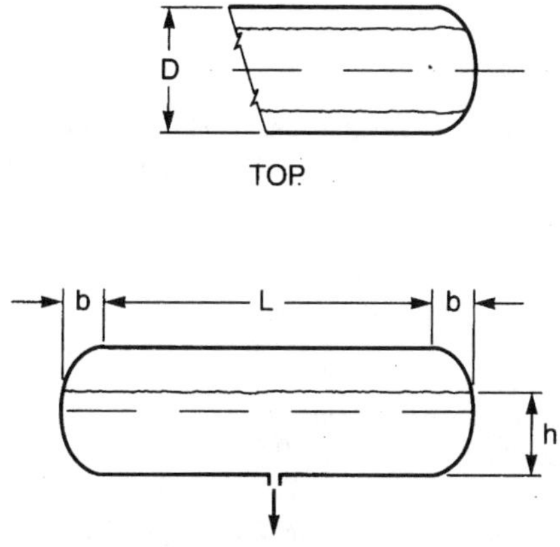

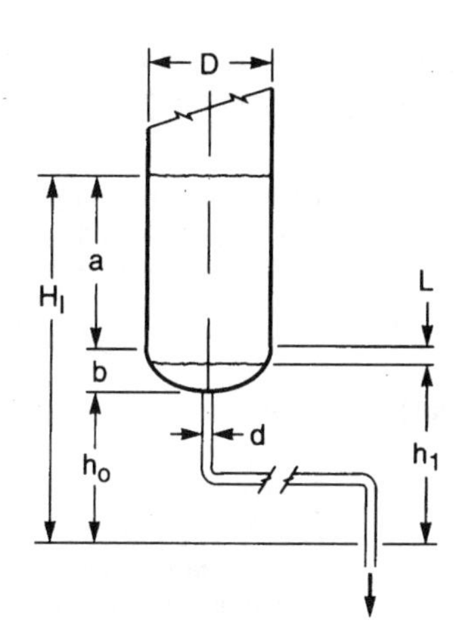

FRONT
Horizontal cylinder with dished ends

Vertical Cylinder with dished end
and drainpipe system

$$t_c = \frac{D^2}{d^2}\{[(2/g)\ (1 + [fl/d])]^{1/2}\ (H_l{}^{1/2} - H_F{}^{1/2})\}$$

$$t_e = C\{[(2 \text{ X } H_2{}^2/5) - (4 \text{ X } B \text{ X } H_2/3) + 2E^2\}(H_2{}^{1/2}) - [(2 \text{ X } H_1{}^2/5) - (4 \text{ X } B \text{ X } H_1/3) + 2E^2](H_1{}^{1/2})\}$$

$$t_p = \frac{\sqrt{8}}{3C_d A_n \sqrt{g}}\ \{L[D^{3/2} - (D - h)^{3/2}] + \frac{bph^{3/2}}{D}[D - (3h/5)]\}$$

FIGURE 31 Time to drain tanks. (*a*) Top and (*b*) front view of horizontal cylinder with dished ends. (*c*) Vertical cylinder with dished-end and drainpipe system. (*Chemical Engineering*.)

Then, use the values for B, E^2, C, and other relevant dimensions to find t_e from the equation shown under the storage tanks in Fig. 31. Thus, $t_e = 459.6[(2 \times 4^2/5) - (4 \times 6 \times 4/3) + 2(32)](4)^{0.5} - [(2 \times 6^2/5) - (4 \times 6 \times 6/3) + 2(32)](6^{0.5}) = 1073$ s, or 17.9 min.

8. *Compute the time it will take to drain the half-full vertical tank*
Total time is $t_t + t_c + t_e = 4751 + 1073 = 5824$ s, or 96.1 min.

Related Calculations. Figure 31 shows the equation for computing the emptying time for horizontal cylindrical tank with elliptically dished ends and equations for calculating the emptying time for a vertical cylindrical tank with an elliptically dished bottom end fitted with a drain system. The symbols A_n, g, t, and C_d are defined as in the previous problem for a storage vessel without dished ends, except that A_n is now the drainpipe internal area.

The term associated with the second pair of brackets in the equation for t_p accounts for the dished ends of the horizontal tank. For hemispherical ends $b = D/2$ and for flat ends, $b = 0$.

When seeking the time required to drain a portion of the cylindrical part of the vertical tank, use the formula for t_c with the appropriate values for H_t and H_F and other applicable variables. To find the time it takes to drain a portion of the dished bottom of the vertical tank, use the formula for t_e with given values of H_1 and H_2 and other applicable variables.

The relations given here are valid for storage tanks used in a variety of applications—chemical and petrochemical plants, power plants, waterworks, ships and boats, aircraft, etc. The procedure for a horizontal cylindrical tank with dished ends is the work of Jude T. Sommerfeld, and the procedure for a vertical cylindrical tank with a dished bottom ends is the work of Mahnoosh Shoael and Jude T. Sommerfeld, and reported in *Chemical Engineering* magazine.

TIME NEEDED TO EMPTY A STORAGE VESSEL WITHOUT DISHED ENDS

How long will it take to empty a 10-ft (3-m) diameter spherical tank filled to a height of 8 ft (2.4 m) with ethanol, a newtonian fluid, if the drain is a short 2-in (5. 1-cm) diameter tube of double extra-strong pipe?

Calculation Procedure:

1. *Determine the discharge coefficient for the drain*

Figure 30 shows that the discharge coefficient is $C_d = 0.80$ for a short, flush-mounted tube.

2. *Compute the discharge time*

Substitute the appropriate values in the equation in Fig. 30 for spherical storage tanks. Or, $t = (2)^{0.5} (\pi)(8)^{1.5} [10 - (0.6 \times 8)]/[3(0.8)(1.774/144)(32.2)^{0.5}] = 3116$ s, or 51.9 min.

Related Calculations. Figure 30 gives the equations for computing the emptying for four common tank geometrics. The discharge coefficient C_d is constant for newtonian fluids in turbulent flow, but the coefficient depends on the shape of the orifice. Water flowing through sharp-edged orifices of 0.25-in (0.64-cm) diameter, or larger, is always turbulent. Thus, the assumption of a constant C_d is valid for most practical applications. Figure 30 lists accepted C_d values.

The relations given here are valid for storage tanks used in a variety of applications—chemical and petrochemical plants, power plants, waterworks, ships and boats, aircraft, etc. This procedure is the work of Thomas C. Foster, as reported in *Chemical Engineering* magazine.

TIME TO DRAIN A STORAGE TANK THROUGH ATTACHED PIPING

Determine the time required to drain a 50-ft (15.2-m) diameter tank to a level of 10 ft (3 m) if the tank is filled with water to a height of 25 ft (7.6 m). The drain pipe system has 474 ft (144.5 m) equivalent length of 4 in (101.6 mm) Schedule 40 pipe with a friction factor, $f = 0.0185$. The pipe outlet elevation is 0 ft (0 m). Compute the drainage time if the outlet pipe elevation is −3 ft (−1 m) instead of 0 ft (0 m).

Calculation Procedure:

1. *Compute the drainage time for the first outlet elevation*

The literature presents numerous equations and nomograms to determine the time to drain a vertical or horizontal cylindrical tank. However, these equations do not normally consider any associated piping through which a tank might be drained.

Using the Bernoulli equation for point 1, the tank liquid surface at any height above the bottom, and point 2, the drainpipe height, gives $t = (D^2/d^2)\ [(2/g)\{fL/d + 1\}]^{0.5}([H_o]^{0.5} - [H_f]^{0.5})$, where t = time required to drain the tank through the attached piping, s; D = tank diameter, ft (m); d = drainpipe diameter, ft (mm); g = gravitational constant, 32.2 ft/s² (9.8 m/s²); f = pipe friction factor; L = equivalent length of piping and any associated fittings, ft (m); H_o = liquid height at any time above the drainpipe outlet, ft (m); H_f = liquid height when tank drainage is completed, ft (m).

For the first situation where the drainpipe outlet is 0 ft (0 m), $t = (50^2/0.336^2)[(2/32.2)\{0.0185(474)/0.336\} + 1]^{0.5}([25])^{0.5} - [10]^{0.5}) = 52{,}496$ s, or 52,496/3600 s/h = 14.58 h.

2. Determine the drainage time when the piping outlet is below the tank bottom
When the tank drainage-pipe outlet is below the tank bottom, H_f is a negative number. For the given situation, $H_f = -3$ ft (−0.9 m). Then, the time to drain this tank, $t = (50^2/0.336^2)[(2/32.2)\{0.0185(474)/0.336\} + 1]^{0.5}([28]^{0.5} - [13]^{0.5}) = 48{,}436$ s, or 48,436/3600 = 13.5 h.

Related Calculations. The equation presented here can be used for any liquid—oil, water, acid, caustic, etc.—and any type of piping—steel, cast iron, plastic, wood, etc., provided the friction factor and equivalent length are adjusted to reflect the fluid and type of piping involved in the calculation.

Where frequent tank drainage is expected, this procedure can be used to determine quickly the difference in drainage time for pipes of various diameters. Then, with the cost of each diameter of piping known, an economic study will show which size piping is most attractive from an investment standpoint when each drainage time is given a relative-importance rating. Where tank drainage will be done infrequently, and drainage time is not an important factor in plant or process operation, the cheapest available drainage piping which provides safe performance is usually the best choice.

This procedure is the work of Nick J. Loiacono, P.E., Wink Engineering, as reported in *Chemical Engineering* magazine. Data on the economic aspects of drainage-pipe sizing and SI values were added by the handbook editor.

SECTION 13

INTERIOR CLIMATE CONTROL ENERGY ECONOMICS

Interior Climate Control—Heating, Ventilating, and Air-Conditioning Energy Parameters 13.2

INTERIOR CLIMATE CONTROL ENERGY ECONOMICS 13.2

Energy Equations for Interior Climate Control 13.2

Fan Horsepower Requirements for Cooling Towers 13.12

Energy Savings with Ice Storage System for Facility Cooling 13.12

Energy Annual Heating and Cooling Costs and Loads 13.21

Energy Heat Recovery Using a Run-Around System of Heat Transfer 13.23

Energy Savings with Rotary Heat Exchanger 13.25

Energy Savings with "Hot Deck" Temperature Reset 13.27

Energy Performance of Air-to-Air Heat Exchanger 13.27

Heating Capacity Requirements for Buildings with Steam and Hot-Water Heating 13.29

Energy and Heating Steam Requirements for Specialized Rooms 13.30

Computation of Carbon Dioxide Buildup in Occupied Spaces 13.31

Analysis of Bypass Air Quantity and Dehumidifier Exit Conditions 13.32

Vibration Potential of Motor-Driven Fans 13.33

Centrifugal-Compressor Power and Energy Input Computation 13.34

Moisture Evaporation from Open Tanks 13.35

Fan and Pump Energy Performance Analyzed from Motor Data 13.36

Air-Bubble Enclosure Choice for Planned Usage 13.37

Expansion-Tank Sizing for Hydronic Systems 13.40

ENERGY ASPECTS OF SYSTEM ANALYSIS AND EQUIPMENT SELECTION 13.48

Heat- and Energy-Loss Determination for Buildings and Structures 13.48

Selection and Analysis of Building Heating Systems 13.49

Unit Heater Capacity Computation 13.52

Heating Apparatus Steam and Energy Consumption 13.57

Air-Heating Coil Selection and Energy Analysis 13.59

Choosing and Sizing Radiant Heating Panels 13.64

Choosing and Sizing Snow-Melting Heating Panels 13.67

Energy Savings in Space Heating with Heat Recovery from Lighting Systems 13.69

Energy and Heat-Load Computation for Air-Conditioning Systems—General Method 13.70

Energy and Heat-Load Computation for Air-Conditioning Systems—Numerical Calculation 13.75

Cooling-Coil Selection for Air-Conditioning Systems 13.79

Energy Analysis of Mixing Two Air Streams 13.86

Energy Aspects of Selecting an Air-Conditioning System for a Known Load 13.87

Equal-Friction Method for Sizing Low-Velocity Air-Conditioning Ducts 13.90

Static-Regain Method for Sizing Low-Velocity Air-Conditioning System Ducts 13.97

Selecting Humidifiers for Chosen Indoor Climate Conditions 13.101

Using the Psychometric Chart in Air-Conditioning Energy Calculations 13.106

Energy Aspects of Designing High-Velocity Air-Conditioning Ducts 13.108

Outlet and Return Air Grille Choice for Air-Conditioning Systems 13.110

Energy Considerations in Choosing Roof Ventilators for Structures 13.115

Air-Conditioner Vibration Isolation Device Selection 13.118

Noise-Reduction Materials Choice for Air-Conditioning Systems 13.121

Designing Door and Window Air Curtains for Various Types of Structures 13.123

INTERIOR CLIMATE CONTROL—HEATING, VENTILATING, AND AIR-CONDITIONING ENERGY PARAMETERS

One of the largest energy consumers throughout the world today is interior climate control—heating, ventilating, and air conditioning (HVAC). Thus, on a recent summer's day in New York City, the local utility—Consolidated Edison—generated 13,141 MW of electricity at 3 p.m. This output exceeds that of many mid-size utility plants for several days.

HVAC energy requirements grow every year as more interior climate control is demanded in almost every structure occupied by humans—office buildings, multifamily apartment houses, hotels, motels, places of religious worship, factories, schools, colleges, universities, stores, supermarkets, and so on. And, unfortunately, the time of peak demand for this energy is often the same time that other large energy demands occur. The severe resultant overloading can cause power outages lasting from a few minutes to more than a week.

Rising power demand by interior climate control devices has led to greater emphasis on system efficiency. And the "green" movement and LEED certification place greater responsibility on HVAC design engineers. Numerous new and improved HVAC design methods and more efficient mechanical and electrical systems are being introduced to save energy while providing needed comfort. Indoor air quality (IAQ) is getting increased attention from engineers and designers, and greater regulation by local and national authorities. Ductless HVAC systems are being used in many older buildings where there is not enough space for conventional ducts. In a ductless installation, two or more roof-mounted HVAC units supply conditioned air to areas needing it. Local indoor air handling units may also be used to condition specific areas. Such installations usually save on costs, are environmentally desirable, and allow a building's structural characteristics to be kept intact.

INTERIOR CLIMATE CONTROL ENERGY ECONOMICS

ENERGY EQUATIONS FOR INTERIOR CLIMATE CONTROL

A variety of calculation procedures are used in designing heating, ventilating, and air-conditioning systems. To help save time for design and application engineers, technicians, and consulting engineers, some 75 design equations are presented at the start of this section of the handbook. These equations are used in both manual and computer-aided design (CAD) applications. And since this handbook is designed for worldwide use, the first group of equations presents USCS and SI versions to allow easy comparisons of the results. Abbreviations used in the equations are presented first.

H_S = sensible heat, Btu/h
H_{SM} = sensible heat, kJ/h
H_L = latent heat, Btu/h
H_{LM} = latent heat, kJ/h
H_T = total heat, Btu/h
H_{TM} = total heat, kJ/h
H = total heat, Btu/h
H_M = total heat, kJ/h
ΔT = temperature difference, °F
ΔT_M = temperature difference, °C
ΔW = humidity ratio difference, gr H_2O/lb DA
ΔW_M = humidity ratio difference, kg H_2O/kg DA

(*Continued*)

Δh = enthalpy difference, Btu/lb DA
Δh = enthalpy difference, kJ/lb DA
CFM = airflow rate, ft^3/min
CMM = airflow rate, m^3/min
GPM = water flow rate, gal/min
LPM = water flow rate, L/min
AC/HR = air change rate per hour, English
AC/HR_M = air change rate per hour, SI
AC/HR = AC/HR_M
VOLUME = space volume, ft^3
$VOLUME_M$ = space volume, m^3
kJ/h = Btu/h × 1.055
CMM = CFM × 0.02832
LPM = GPM × 3.785
kJ/lb = Btu/lb × 2.326
m = ft × 0.3048
$m^2 = ft^2 \times 0.0929$
$m^3 = ft^3 \times 0.02832$
kg = lb × 0.4536
1.0 GPM = 500 lb steam/h
1.0 lb steam/h = 0.002 GPM
1.0 lb H_2/h = 1.0 lb steam/h
$kg/m^3 = lb/ft^3 \times 16.017$ (density)
$m^3/kg = ft^3/lb \times 0.0624$ specific volume
kg H_2O/kg DA = gr H_2O/lb DA/7000 = lb H_2O/lb DA

$$H_S = 1.08 \frac{\text{Btu} \cdot \text{min}}{\text{h} \cdot \text{ft}^3 \cdot {}^\circ\text{F}} \times \text{CFM} \times \Delta T \tag{1}$$

$$H_{SM} = 72.42 \frac{\text{kJ} \cdot \text{min}}{\text{h} \cdot \text{m}^3 \cdot {}^\circ\text{C}} \times \text{CMM} \times \Delta T_\text{M} \tag{2}$$

$$H_L = 0.68 \frac{\text{Btu} \cdot \text{min} \cdot \text{lb DA}}{\text{h} \cdot \text{ft}^3 \cdot \text{gr H}_2\text{O}} \times \text{CFM} \times \Delta W \tag{3}$$

$$H_{LM} = 177{,}734.8 \frac{\text{kJ} \cdot \text{min} \cdot \text{kg DA}}{\text{h} \cdot \text{m}^3 \cdot \text{kg H}_2\text{O}} \times \text{CMM} \times \Delta W_\text{M} \tag{4}$$

$$H_T = 4.5 \frac{\text{lb} \cdot \text{min}}{\text{h} \cdot \text{ft}^3} \times \text{CFM} \times \Delta h \tag{5}$$

$$H_{TM} = 72.09 \frac{\text{kg} \cdot \text{min}}{\text{h} \cdot \text{m}^3} \times \text{CMM} \times \Delta h_\text{M} \tag{6}$$

$$H_T = H_S + H_L \tag{7}$$

$$H_{TM} = H_{SM} + H_{LM} \tag{8}$$

$$H = 500 \frac{\text{Btu} \cdot \text{min}}{\text{h} \cdot \text{gal} \cdot {}^\circ\text{F}} \times \text{GPM} \times \Delta T \tag{9}$$

$$H_\text{M} = 250.8 \frac{\text{kJ} \cdot \text{min}}{\text{h} \cdot \text{L} \cdot {}^\circ\text{C}} \times \text{LPM} \times \Delta T_\text{M} \tag{10}$$

$$\frac{\text{AC}}{\text{HR}} = \frac{\text{CFM} \times 60 \text{ min/h}}{\text{VOLUME}} \tag{11}$$

$$\frac{\text{AC}}{\text{HR}_\text{M}} = \frac{\text{CMM} \times 60 \text{ min/h}}{\text{VOLUME}_\text{M}} \tag{12}$$

$$°\text{C} = \frac{°\text{F} - 32}{1.8} \tag{13}$$

$$°\text{F} = 1.8°\text{C} + 32 \tag{14}$$

Steam Pipe Pressure Drop and Flow Rate Equations

$$\Delta P = \frac{0.01306W^2(1 + 3.6/\text{ID})}{3600 \times D \times \text{ID}^5} \tag{15}$$

$$W = 60\sqrt{\frac{\Delta P \times D \times \text{ID}^5}{0.01306 \times (1 + 3.6/\text{ID})}} \tag{16}$$

$$W = 0.41667VA_{\text{INCHES}}D = 60VA_{\text{FEET}}D \tag{17}$$

$$V = \frac{2.4W}{A_{\text{INCHES}}D} = \frac{W}{60A_{\text{FEET}}D} \tag{18}$$

where ΔP = pressure drop per 100 ft of pipe (psig/100 ft)
W = stream flow rate, lb/h
ID = actual inside diameter of pipe, in
D = average density of stream at system pressure, lb/ft^3
V = velocity of stream in pipe, ft/min
A_{INCHES} = actual cross-sectional area of pipe, in^2
A_{FEET} = actual cross-sectional area of pipe, ft^2

Condensate Piping Equations

$$\text{FS} = \frac{H_{\text{SSS}} - H_{\text{SCR}}}{H_{\text{LCR}}} \times 100 \tag{19}$$

$$W_{\text{CR}} = \frac{\text{FS}}{100} \times W \tag{20}$$

where FS = flash stream, %
H_{SSS} = sensible heat at stream supply pressure, Btu/lb
H_{SCR} = sensible heat at condensate return pressure, Btu/lb
H_{LCR} = latent heat at condensate return pressure, Btu/lb
W = stream flow rate, lb/h
W_{CR} = condensate flow based on percentage of flash steam created during condensing process, lb/h. Use this flow rate in steam equations above to determine condensate return pipe size.

HVAC Efficiency Equations

$$\text{COP} = \frac{\text{BTU OUTPUT}}{\text{BTU INPUT}} = \frac{\text{EER}}{3.413} \tag{21}$$

$$\text{EER} = \frac{\text{BTU OUTPUT}}{\text{WATTS INPUT}} \tag{22}$$

Turndown ratio = maximum firing rate:minimum firing rate (that is, 5:1, 10:1, 25:1)

$$\text{OVERALL THERMAL EFF} = \frac{\text{GROSS BTU OUTPUT}}{\text{GROSS BTU INPUT}} \times 100\% \tag{23}$$

$$\text{COMBUSTION EFF} = \frac{\text{BTU INPUT} - \text{BTU STACK LOSS}}{\text{BTU INPUT}} \times 100\% \tag{24}$$

Overall thermal efficiency range	75%–90%
Combustion efficiency range	85%–95%

Equations for HVAC Equipment Room Ventilation

For completely enclosed equipment rooms:

$$\text{CFM} = 100 \times G^{0.5} \tag{25}$$

where CFM = exhaust airflow rate required, ft^3/min
G = mass of refrigerant of largest system, lb

For partially enclosed equipment rooms:

$$\text{FA} = G^{0.5} \tag{26}$$

where FA = ventilation-free opening area, ft^2
G = mass of refrigerant of largest system, lb

Psychrometric Equations

The following equations are from Carrier Corporation publications.* These equations cover air mixing, cooling loads, sensible heat factor, bypass factor, temperature at the apparatus, supply air temperature, air quantity, and determination of air constants. Abbreviations and symbols for the equations are given below.

Abbreviations	
adp	apparatus dew point
BF	bypass factor
(BF) (OALH)	bypassed outdoor air latent heat
(BF) (OASH)	bypassed outdoor air sensible heat
(BF) (OATH)	bypassed outdoor air total heat

(*Continued*)

Handbook of Air-Conditioning System Design, McGraw-Hill, New York, various dates.

Abbreviations (*Continued*)

Btu/h	British thermal units per hour
cfm, ft^3/min	cubic feet per minute
db	dry-bulb
dp	dew point
ERLH	effective room latent heat
ERSH	effective room sensible heat
ERTH	effective room total heat
ESHF	effective sensible heat factor
°F	degrees Fahrenheit
fpm, ft/min	feet per minute
gpm, gal/min	gallon per minute
gr/lb	grains per pound
GSHF	grand sensible heat factor
GTH	grand total heat
GTHS	grand total heat supplement
OALH	outdoor air latent heat
OASH	outdoor air sensible heat
OATH	outdoor air total heat
rh	relative humidity
RLH	room latent heat
RLHS	room latent heat supplement
RSH	room sensible heat
RSHF	room sensible heat factor
RSHS	room sensible heat supplement
RTH	room total heat
Sat Eff	saturation efficiency of sprays
SHF	sensible heat factor
TLH	total latent heat
TSH	total sensible heat
wb	wet-bulb

Symbol	
cfm_{ba}	bypassed air quantity around apparatus
cfm_{da}	dehumidified air quantity
cfm_{oa}	outdoor air quantity
cfm_{ra}	return air quantity
cfm_{sa}	supply air quantity
h	specific enthalpy
h_{adp}	apparatus dew point enthalpy
h_{cs}	effective surface temperature enthalpy
h_{ea}	entering air enthalpy
h_{la}	leaving air enthalpy
h_m	mixture of outdoor and return air enthalpy
h_{oa}	outdoor air enthalpy
h_{rm}	room air enthalpy
h_{sa}	supply air enthalpy
t	temperature
t_{apd}	apparatus dew point temperature
t_{edb}	entering dry-bulb temperature
t_{es}	effective surface temperature
t_{ew}	entering water temperature

(*Continued*)

Symbol (*Continued*)	
t_{ewb}	entering wet-bulb temperature
t_{ldb}	leaving dry-bulb temperature
t_{lw}	leaving water temperature
t_{lwb}	leaving wet-bulb temperature
t_m	mixture of outdoor and return air dry-bulb temperature
t_{oa}	outdoor air dry-bulb temperature
t_{rm}	room dry-bulb temperature
t_{sa}	supply air dry-bulb temperature
W	moisture content or specific humidity
W_{adp}	apparatus dew point moisture content
W_{ea}	entering air moisture content
W_{es}	effective surface temperature moisture content
W_{la}	leaving air moisture content
W_m	mixture of outdoor and return air moisture content
W_{oa}	outdoor air moisture content
W_{rm}	room moisture content
W_{sa}	supply air moisture content

Air Mixing Equations (Outdoor and Return Air)

$$t_m = \frac{\text{cfm}_{oa} \times t_{oa} + \text{cfm}_{ra} \times t_{rm}}{\text{cfm}_{sa}} \tag{27}$$

$$h_m = \frac{(\text{cfm}_{oa} \times h_{oa}) + (\text{cfm}_{ra} \times h_{rm})}{\text{cfm}_{sa}} \tag{28}$$

$$W_m = \frac{(\text{cfm}_{oa} \times W_{oa}) + (\text{cfm}_{ra} \times W_{rm})}{\text{cfm}_{sa}} \tag{29}$$

Cooling Load Equations

$$\text{ERSH} = \text{RSH} + (\text{BF})(\text{OASH}) + \text{RSHS}^* \tag{30}$$

$$\text{ERLH} = \text{RLH} + (\text{BF})(\text{OALH}) + \text{RLHS}^* \tag{31}$$

$$\text{ERTH} = \text{ERLH} + \text{ERSH} \tag{32}$$

$$\text{TSH} = \text{RSH} + \text{OASH} + \text{RSHS}^* \tag{33}$$

$$\text{TLH} = \text{RLH} + \text{OALH} + \text{RLHS}^* \tag{34}$$

$$\text{GTH} = \text{TSH} + \text{TLH} + \text{GTHS}^* \tag{35}$$

$$\text{RSH} = 1.08^{\dagger} \times \text{cfm}_{sa} \times (t_{rm} - t_{sa}) \tag{36}$$

$$\text{RLH} = 0.68^{\dagger} \times \text{cfm}_{sa} \times (W_{rm} - W_{sa}) \tag{37}$$

$$\text{RTH} = 4.45^{\dagger} \times \text{cfm}_{sa} \times (h_{rm} - h_{sa}) \tag{38}$$

$$\text{RTH} = \text{RSH} + \text{RLH} \tag{39}$$

---

*RSHS, RLHS, and GTHS are supplementary loads due to duct heat gain, duct leakage loss, fan and pump horsepower gains, etc.

†See later for the derivation of these air constants.

$$\text{OASH} = 1.08 \times \text{cfm}_{oa}(t_{oa} - t_{rm}) \quad (40)$$

$$\text{OALH} = 0.68 \times \text{cfm}_{oa}(W_{oa} - W_{rm}) \quad (41)$$

$$\text{OATH} = 4.45 \times \text{cfm}_{oa}(h_{oa} - h_{rm}) \quad (42)$$

$$\text{OATH} = \text{OASH} + \text{OALH} \quad (43)$$

$$(\text{BF})(\text{OATH}) = (\text{BF})(\text{OASH}) + (\text{BF})(\text{OALH}) \quad (44)$$

$$\text{ERSH} = 1.08 \times \text{cfm}_{da}{}^{\ddagger} \times (t_{rm} - t_{adp})(1 - \text{BF}) \quad (45)$$

$$\text{ERLH} = 0.68 \times \text{cfm}_{da}{}^{\ddagger} \times (W_{rm} - W_{adp})(1 - \text{BF}) \quad (46)$$

$$\text{ERTH} = 4.45 \times \text{cfm}_{da}{}^{\ddagger} \times (h_{rm} - h_{adp})(1 - \text{BF}) \quad (47)$$

$$\text{TSH} = 1.08 \times \text{cfm}_{da}{}^{\ddagger} \times (t_{edb} - t_{ldb})^{*} \quad (48)$$

$$\text{TLH} = 0.68 \times \text{cfm}_{da}{}^{\ddagger} \times (W_{ea} - W_{la})^{*} \quad (49)$$

$$\text{GTH} = 4.45 \times \text{cfm}_{da}{}^{\ddagger} \times (h_{ea} - h_{la})^{*} \quad (50)$$

Sensible Heat Factor Equations

$$\text{RSHF} = \frac{\text{RSH}}{\text{RSH} + \text{RLH}} = \frac{\text{RSH}}{\text{RTH}} \quad (51)$$

$$\text{ESHF} = \frac{\text{ERSH}}{\text{ERSH} + \text{ERLH}} = \frac{\text{ERSH}}{\text{ERTH}} \quad (52)$$

$$\text{GSHF} = \frac{\text{TSH}}{\text{TSH} + \text{TLH}} = \frac{\text{TSH}}{\text{GTH}} \quad (53)$$

Bypass Factor Equations

$$\text{BF} = \frac{t_{ldb} - t_{adp}}{t_{edb} - t_{adp}} \qquad 1 - \text{BF} = \frac{t_{edb} - t_{ldb}}{t_{edb} - t_{adp}} \quad (54)$$

$$\text{BF} = \frac{W_{la} - W_{adp}}{W_{ea} - W_{adp}} \qquad 1 - \text{BF} = \frac{W_{ea} - W_{la}}{W_{ea} - W_{adp}} \quad (55)$$

$$\text{BF} = \frac{h_{la} - h_{adp}}{h_{ea} - h_{adp}} \qquad 1 - \text{BF} = \frac{h_{ea} - h_{la}}{h_{ea} - h_{adp}} \quad (56)$$

Temperature Equations at the Apparatus

$$t_{edb}{}^{*} = \frac{(\text{cfm}_{oa} \times t_{oa}) + (\text{cfm}_{ra} \times t_{rm})}{\text{cfm}_{sa}{}^{\dagger}} \quad (57)$$

$$t_{ldb} = t_{adp} + \text{BF}(t_{edb} - t_{adp}) \quad (58)$$

---

*RSHS, RLHS, and GTHS are supplementary loads due to duct heat gain, duct leakage loss, fan and pump horsepower gains, etc.

†See later for the derivation of these air constants.

‡When no air is to be physically bypassed around the conditioning apparatus, $\text{cfm}_{da} = \text{cfm}_{sa}$.

Both t_{ewb} and t_{lwb} correspond to the calculated values of h_{ea} and h_{la} on the psychrometric chart.

$$h_{ea}{}^{*} = \frac{(\text{cfm}_{oa} \times h_{oa}) + (\text{cfm}_{ra} \times h_{rm})}{\text{cfm}_{sa}{}^{\dagger}} \quad (59)$$

$$h_{la} = h_{adp} + \text{BF}(h_{ea} - h_{adp}) \quad (60)$$

Temperature Equations for Supply Air

$$t_{sa} = t_{rm} - \frac{\text{RSH}}{1.08\text{cfm}_{sa}{}^{\dagger}} \quad (61)$$

Air Quantity Equations

$$\text{cfm}_{da} = \frac{\text{ERSH}}{1.08(1 - \text{BF})(t_{rm} - t_{adp})} \quad (62)$$

$$\text{cfm}_{da} = \frac{\text{ERLH}}{0.68(1 - \text{BF})(W_{rm} - W_{adp})} \quad (63)$$

$$\text{cfm}_{da} = \frac{\text{ERTH}}{4.45(1 - \text{BF})(h_{rm} - h_{adp})} \quad (64)$$

$$\text{cfm}_{da}{}^{*} = \frac{\text{TSH}}{1.08(t_{edb} - t_{ldb})} \quad (65)$$

$$\text{cfm}_{da}{}^{*} = \frac{\text{TLH}}{0.68(W_{ea} - W_{la})} \quad (66)$$

$$\text{cfm}_{da}{}^{*} = \frac{\text{GTH}}{4.45(h_{ea} - h_{la})} \quad (67)$$

$$\text{cfm}_{sa} = \frac{\text{RSH}}{1.08(t_{rm} - t_{sa})} \quad (68)$$

$$\text{cfm}_{sa} = \frac{\text{RLH}}{0.68(W_{rm} - W_{sa})} \quad (69)$$

*When t_m, W_m, and h_m are equal to the entering conditions at the cooling apparatus, they may be substituted for t_{adb}, W_{ea}, and h_{ea}, respectively.

†See footnote on page 13-8.

$$\text{cfm}_{\text{sa}} = \frac{\text{RTH}}{4.45(h_{\text{rm}} - h_{\text{sa}})} \tag{70}$$

$$\text{cfm}_{\text{ba}} = \text{cfm}_{\text{sa}} - \text{cfm}_{\text{da}} \tag{71}$$

Note: cfm_{da} will be less than cfm_{sa} only when air is physically bypassed around the conditioning apparatus.

$$\text{cfm}_{\text{sa}} = \text{cfm}_{\text{oa}} + \text{cfm}_{\text{ra}} \tag{72}$$

Derivation of Air Constants

$$1.08 = 0.244 \times \frac{60}{13.5} \tag{73}$$

where 0.244 = specific heat of moist air at 70°F db and 50% rh, Btu/(°F · lb DA)
60 = min/h
13.5 = specific volume of moist air at 70°F db and 50% rh

$$0.68 = \frac{60}{13.5} \times \frac{1076}{7000}$$

where 60 = min/h
13.5 = specific volume of moist air at 70°F db and 50% rh
1076 = average heat removal required to condense 1 db water vapor from the room air
7000 = gr/lb

$$4.45 = \frac{60}{13.5}$$

where 60 = min/h
13.5 = specific volume of moist air at 70°F db and 50% rh

Equations for Steam Trap Selection

The selection of the trap for the steam mains or risers is dependent on the pipe warm-up load and the radiation load from the pipe. Warm-up load is the condensate which is formed by heating the pipe surface when the steam is first turned on. For practical purposes, the final temperature of the pipe is the steam temperature. Warm-up load is determined from

$$C_1 = \frac{W(t_f - t_i)(0.114)}{h_l T} \tag{74}$$

where C_1 = warm-up condensate, lb/h
W = total weight of pipe, lb (from tables in engineering handbooks)

t_f = final pipe temperature, °F (steam temp.)
t_i = initial pipe temperature, °F (usually room temp.)
0.114 = specific heat constant for wrought iron or steel pipe (0.092 for copper tubing)
h_l = latent heat of steam, Btu/lb (from steam tables)
T = time for warm-up, h

The radiation load is the condensate formed by unavoidable radiation loss from a bare pipe. This load is determined from the following equation and is based on still air surrounding the steam main or riser:

$$C_2 = \frac{LK(t_f - t_i)}{h_l} \tag{75}$$

where C_2 = radiation condensate, lb/h
L = linear length of pipe, ft
K = heat transmission coefficient, Btu/(h · lin ft · °F)

The radiation load builds up as the warm-up load drops off under normal operating conditions. The peak occurs at the midpoint of the warm-up cycle. Therefore, one-half of the radiation load is added to the warm-up load to determine the amount of condensate that the trap handles.

Safety Factor

Good design practice dictates the use of safety factors in steam trap selection. Safety factors from 2 to 1 to as high as 8 to 1 may be required, and for the following reasons:

1. The steam pressure at the trap inlet or the backpressure at the trap discharge may vary. This changes the steam trap capacity.
2. If the trap is sized for normal operating load, condensate may back up into the steam lines or apparatus during start-up or warm-up operation.
3. If the steam trap is selected to discharge a full and continuous stream of water, the air could not be vented from the system.

The following guide is used to determine the safety factor:

Design	Safety factor
Draining steam main	3 to 1
Draining steam riser	2 to 1
Between boiler and end of main	2 to 1
Before reducing valve	3 to 1
Before shutoff valve (closed part of time)	3 to 1
Draining coils	3 to 1
Draining apparatus	3 to 1

When the steam trap is to be used in a high-pressure system, determine whether the system is to operate under low-pressure conditions at certain intervals such as nighttime or weekends. If this condition is likely to occur, then an additional safety factor should be considered to account for the lower pressure drop available during nighttime operation.

FAN HORSEPOWER REQUIREMENTS FOR COOLING TOWERS

A cooling tower serving an air-conditioning installation is designed for these conditions: Water flow rate, L = 75,000 gal/min (4733 L/s); inlet water temperature, T_i = 110°F (43.3°C); outlet water temperature, T_o = 90°F (32.2°C); atmospheric wet-bulb temperature, T_w = 82°F (27.8°C); total fan efficiency as given by tower manufacturer, E_T = 75%; recirculation of air in tower, given by tower manufacturer, R_C = 8.5%; total air pressure drop through the tower, as given by manufacturer, ΔP = 0.477 in (1.21 cm) H_2O. What is the required fan horsepower input under these conditions? If the weather changes and the air outlet temperature becomes 102°F (38.9°C) with a wet-bulb temperature of 84°F (28.9°C)?

Calculation Procedure:

1. *Determine the fan horsepower for the given atmospheric and flow condition*

The fan brake horsepower input for a cooling tower is given by the relation: $BHP = L \times \Delta T \times V_{sp} \times R_c \times \Delta P / \Delta H \times 6356 \times E_T$, where $\Delta T = T_o - T_i$; V_{sp} = specific volume of outlet air, ft³/lb (m³/kg); ΔH = difference between enthalpy of outlet air and inlet air, Btu/lb (kJ/kg); other symbols as given earlier. Determine the enthalpy difference between the outlet air and inlet air by referring to a psychrometric chart where you will find that the enthalpy of the outlet air for the first case above, H_o = 72 Btu/lb (167.5 kJ/kg); from the same source the enthalpy of the inlet air, H_i = 46 Btu/lb (107.0 kJ/kg); likewise, V_{sp} = 15.1 ft³/lb (93.7 m³/kg) from the chart. Substituting in the equation above, $BHP = (75{,}000 \times 8.337 \times 20 \times 1.085 \times 15.1 \times 0.477)/(72 - 24) \times 6356 \times 0.75 = 788.51$; say 789 hp (588.3 kW). In the above equation the constants 8.337 and 1.085 are used to convert gal/min to lb/min and air flow to ft³/lb, respectively.

2. *Determine the power input required for the second set of conditions*

For the second set of conditions the air outlet temperature, T_o = 102°F (38.9°C) and the wet-bulb temperature is 84°F (28.9°C). Using the psychometric chart again, $\Delta H = 75 - 48 = 27$ Btu/lb (62.8 kJ/kg), and the specific volume of the air at this temperature—from the chart—15.2 ft³/lb (0.95 m³/kg). Substituting as before, BHP = 764 hp (569.6 kW).

Related Calculations. This procedure can be used with any type of cooling tower employed in air conditioning, steam power plants, internal combustion engines, or gas turbines. The method is based on knowing the tower's air outlet temperature. Use the psychometric chart to determine volumes and temperatures for various air states. As presented here, this method is the work of Ashfaq Noor, Dawood Hercules Chemicals Ltd., as reported in *Chemical Engineering* magazine.

With the greater environmental interest in reducing stream pollution of all types, including thermal, cooling towers are receiving more attention as a viable way to eliminate thermal problems in streams and shore waters. The cooling tower is a nonpolluting device whose only environmental impact is the residue left in its bottom pans. Such residue is minor in amount and easily disposed of in an environmentally acceptable manner.

ENERGY SAVINGS WITH ICE STORAGE SYSTEM FOR FACILITY COOLING

Select an ice storage cooling system for a 100-ton (350-kW) peak cooling load, 10-h cooling day, 75 percent diversity factor, $8.00/month kW demand charge, 12-month ratchet—i.e., the utility term for a monthly electrical bill surcharge based on a previous month's higher peak demand. Analyze the costs for a partial-storage and for a full-storage system.

Calculation Procedure:

1. ***Analyze partial-storage and full-storage alternatives***

Stored cooling systems use the term *ton-h* instead of *tons of refrigeration*, which is the popular usage for air-conditioning loads. Figure 1 shows a theoretical cooling load of 100 tons (350 kW) maintained for 100 h, or a 1000 ton-h (3500 kWh) cooling load. Each of the squares in the diagram represents 10 ton-h (35 kWh).

No building air-conditioning system operates at 100 percent capacity for the entire daily cooling cycle. Air-conditioning loads peak in the afternoon, generally from 2:00 to 4:00 p.m. when ambient temperatures are highest. Figure 2 shows a typical building air-conditioning load profile during a design day.

As Fig. 2 shows, the full 100-ton (350-kW) chiller capacity is needed for only 2 of the 10 h in the cooling cycle. For the other 8 h, less than the total chiller capacity is required. Counting the tinted

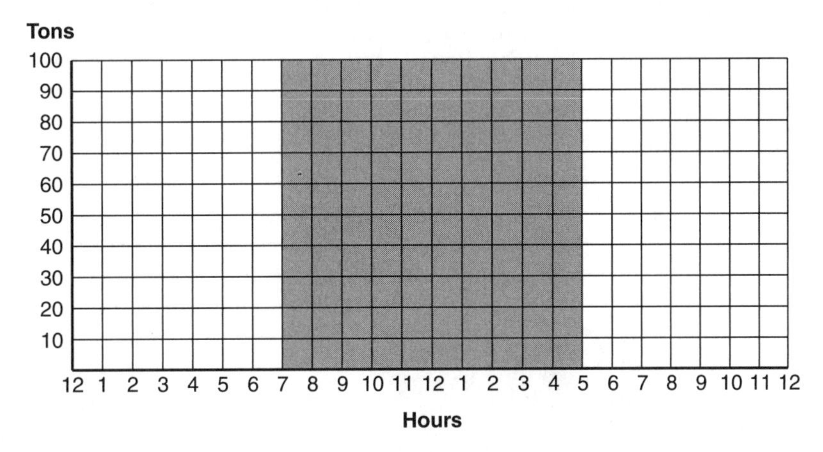

FIGURE 1 Cooling load of 100 tons (351.7 kW) maintained for 10 h, or a 1000 ton-h cooling load. (*Calmac Manufacturing Corporation.*)

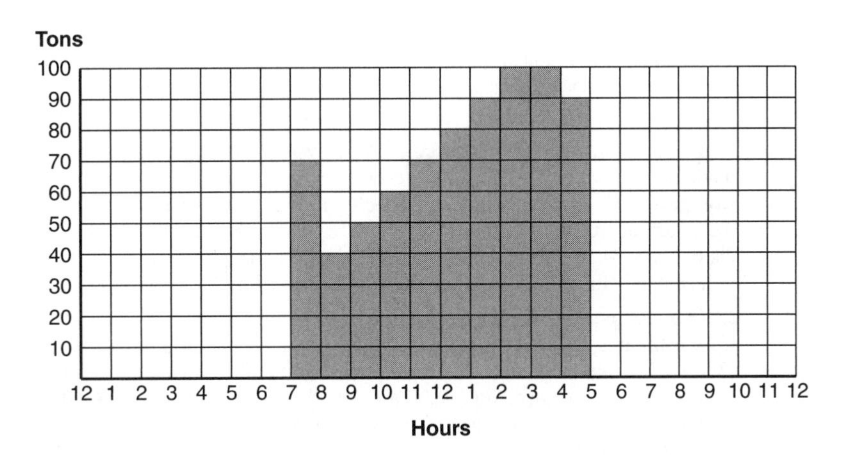

FIGURE 2 Typical building air-conditioning load profile during a design day. (*Calmac Manufacturing Corporation.*)

squares shows only 75, each representing 10 ton-h (35 kWh). The building, therefore, has a true cooling load of 750 ton-h (2625 kWh). A 100-ton (350-kW) chiller must, however, be specified to handle the peak 100-ton (250-kW) cooling load.

The *diversity factor*, defined as the ratio of the actual cooling load to the total potential chiller capacity, or diversity factor, percent = 100 (Actual ton-hours)/total potential ton-hours. For this installation, diversity factor = 100(750)/1000 = 75 percent. If a system's diversity factor is low, its cost efficiency is also low.

Dividing the total ton-hours of the building by the number of hours the chiller is in operation gives the building's average load throughout the cooling period. If the air-conditioning load can be shifted to off-peak hours or leveled to the average load, 100 percent diversity can be achieved, and better cost efficiency obtained.

When electrical rates call for complete load shifting, i.e., are excessively high, a conventionally sized chiller can be used with enough energy storage to shift the entire load into off-peak hours. This is called a *full-storage system* and is used most often in retrofit applications using existing chiller capacity. Figure 3 shows the same building air-conditioning load profile but with the cooling load completely shifted into 14 off-peak hours. The chiller is used to build and store ice during the night. The 32°F (0°C) energy stored in the ice then provides the required 750 ton-h (2625 kWh) of cooling during the day. The average load is lowered to (750 ton-h)/14 h = 53.6 tons (187.6 kW), which results in significantly reduced demand charges.

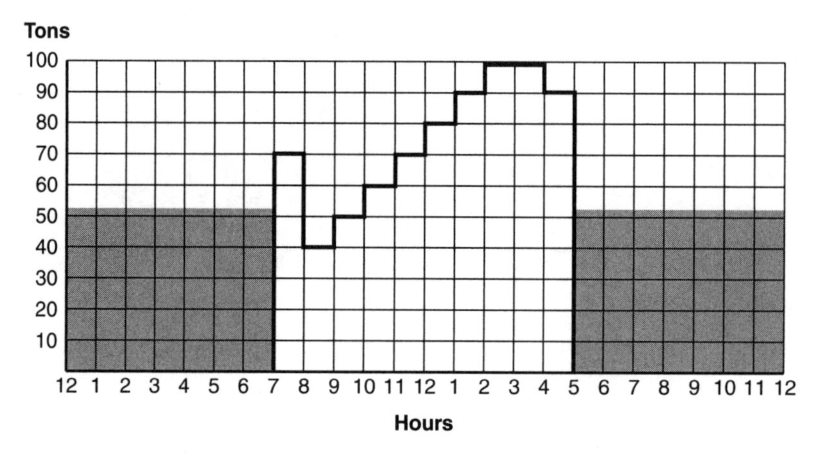

FIGURE 3 Building air-conditioning load profile of Fig. 2 with the cooling load shifted into 14 off-peak hours. (*Calmac Manufacturing Corporation.*)

In new construction, a partial-storage system is the most practical and cost-effective load-management strategy. In this load-leveling method, the chiller runs continuously. It charges the ice storage at night and cools the load directly during the day with help from stored cooling. Extending the hours of operation from 14 to 24 h results in the lowest possible average load, (750 ton-h)/24 hours = 31.25 tons (109.4 kW, as shown by the plot in Fig. 4). Demand charges are greatly reduced and chiller capacity can often be decreased by 50 to 60 percent, or more.

2. *Compute partial-storage demand savings*

Cost estimates for a conventional chilled-water air-conditioning system comprised of a 100-ton (350-kW) chiller with all accessories such as cooling tower, fan coils, pumps, blowers, piping, controls, etc., show a price of \$600/ton, or 100 tons × \$600/ton = \$60,000. The distribution system for this 100-ton (350-kW) plant will cost about the same, or \$60,000. Total cost therefore = \$60,000 + \$60,000 = \$120,000.

With partial-storage using a 40 percent size chiller with ice storage at a 75 percent diversity factor, the true cooling load translates into 750 ton-h (2626 kWh) with the chiller providing 400 ton-h

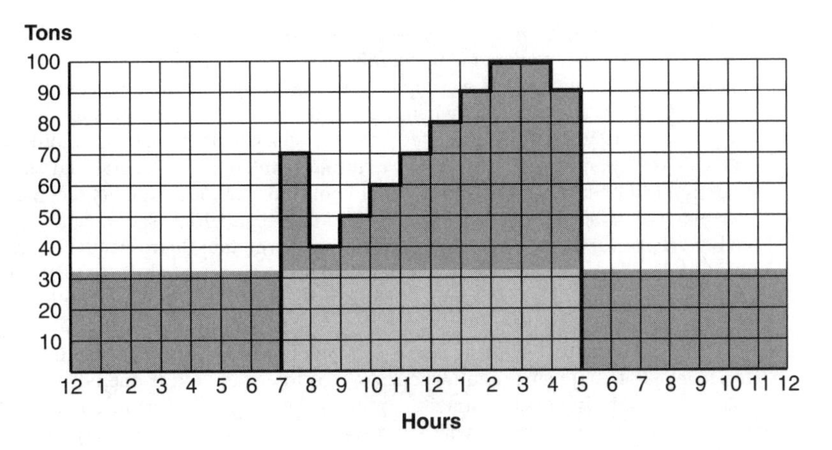

FIGURE 4 Extending the hours of operation for 14 to 24 results in the lowest possible average load = 750 ton-h/24 = 31.25. Demand charges are greatly reduced and chiller capacity can often be decreased 50 to 60 percent, more. (*Calmac Manufacturing Corporation.*)

(1400 kWh) and stored ice the balance, or 750 – 400 = 350 ton-h (1225 kWh). Hence, cost of 40-ton chiller at \$600/ton = \$24,000. From the manufacturer of the stored cooling unit, the installed cost is estimated to be \$60/ton-h, or \$60 × 350 ton-h = \$21,000. The distribution system, as before, costs \$60,000. Hence, the total cost of the partial-storage system will be \$24,000 + \$21,000 + \$60,000 = \$105,000. Therefore, the purchase savings of the partial-storage system over the conventional chilled-water air-conditioning system = \$120,000 – \$105,000 = \$15,000.

The electrical demand savings, which continue for the life of the installation, are: (100 tons – 40 tons chiller capacity)(1.5-kW/ton at peak summer demand conditions, including all accessories)(\$8.00/mo/kW demand charge)(12 mo/yr) = \$8640.

3. *Determine full-storage savings*

With full-storage, 100-ton (350-kW) peak cooling load, 10-h cooling day, 75 percent diversity factor, 1000-h cooling season, \$8.00/mo/kW demand charge, 12-month ratchet, \$0.03/kWh off-peak differential, the chiller cost will be (10 h)(100 tons)(75 percent)(\$60/ton-h, installed) = \$45,000. The demand savings will be, as before, (100 tons)(1.5 kW/ton)(12 mo)(\$8.00/mo/kW demand charge)(\$8.00) = \$14,400/yr. Energy savings are computed using the electric company's off-peak kWh off-peak differential, or (\$0.03/kWh)(1000 h)(100 tons)(1.2 average kW/ton) = \$3,600/yr. The simple payback time for this project = (equipment cost, \$)/(demand savings, \$ + energy savings, \$) = \$45,000/\$18,000 = 2.5 yr. After the end of the payback time there is an annual energy savings of \$18,000/yr. And as rates increase, which they usually do, the annual savings will probably increase above this amount.

Related Calculations. Ice storage systems are becoming more popular for a variety of structures: office buildings, computer data centers, churches, nursing homes, police stations, public libraries, theaters, banks, medical centers, hospitals, hotels, convention centers, schools, colleges, universities, industrial training centers, cathedrals, medical clinics, manufacturing plants, warehouses, museums, country clubs, stock exchanges, government buildings, and courthouses.

There are several reasons for this growing popularity: (1) utility power costs can be reduced by shifting electric power demand to off hours by avoiding peak-demand charges; (2) lower overall electric rates can be obtained for the facility if the kilowatt demand is reduced, thereby eliminating the need for the local utility to build new generating facilities; (3) ice storage can provide uninterrupted cooling in times of loss of outside, or inside, electric generating capability during natural disasters, storms, or line failures—the ice storage system acts like a battery, giving the cooling required until the regular coolant supply can be reactivated; (4) environmental regulations are more readily met because less power input is required, reducing the total energy usage; (5) by making ice at night, the chiller operates when the facilities' electrical demands are lowest and when a utility's generating capacity is underutilized; (6) provision can be made to use more environmentally friendly HCFC-123, thereby complying with current regulations of federal and state agencies; (7) facility design can be planned to include better control of indoor air quality, another environmentally challenging task faced by designers today; (8) new regulations, specifically The Energy Policy Act of 1992 (EPACT), curtails the use of and eliminates certain fluorescent and incandescent lamps (40-W T12, cool white, warm white, daylight white, and warm white deluxe), which will change both electrical demand and replacement bulb costs in facilities, making cooling costs more important in total operating charges.

Designers now talk of "greening" a building or facility, i.e., making it more environmentally acceptable to regulators and owners. An ice storage system is one positive step to greening a facility while reducing the investment required for cooling equipment. The procedure given here clearly shows the savings possible with a typical well-designed ice storage system.

There are three common designs used for ice storage systems today: (1) *direct-expansion ice storage* where ice is frozen directly on metal refrigerant tubes submerged in a water tank; cooled water in the tank is pumped to the cooling load when needed; (2) *ice harvester system* where a thin coat of ice is frozen on refrigerated metal plates and periodically harvested into a bin or water tank by melting the bond of the ice to the metal plates; the chilled water surrounding the ice is pumped to the cooling load when needed; (3) *patented ice bank system* uses a modular, insulated polyethylene tank containing a spiral-wound plastic tube heat exchanger surrounded with water; at night a 26°F (−3.33°C) 75 percent water/25 percent glycol solution from a standard packaged air-conditioning chiller circulates through the heat exchanger, freezing solid all the water in the tank; during the day the ice cools the solution to 44°F (6.66°C) for use in the air-cooling coils where it cools the air from 75°F (23.9°C) to 55°F (12.8°C).

The patented system has several advantages over the first two, namely: (1) ice is the storage medium, rather than water. One pound (0.45 kg) of ice can store 144 Btu (152 J) of energy, while one pound of water in a stratified tank stores only 12 to 15 Btu (12.7 to 15.8 J). Hence, such an ice storage system needs only about one-tenth the space for energy storage. This small space requirement is important in retrofit applications where space is often scarce.

(2) Patented systems are closed; there is no need for water treatment or filtration; pumping power requirements are small; (3) power requirements are minimal; (4) installation of the insulated modular tanks is fast and inexpensive since there are no moving parts; the tanks can be installed indoors or outdoors, stacked or buried to save space. Currently these tanks are available in three sizes: 115, 190, and 570 ton-h (402.5, 665, and 1995 kWh). (5) A low-temperature duct system can be used with 45°F (7.2°C) air instead of the conventional 55°F (12.8°C) air in the air-conditioning system. This can permit further large savings in initial and operating costs. The 45°F (7.2°C) primary air requires much lower air flow [ft^3/min (m^3/m)] than 55°F (12.8°C) air. This reduces the needed size of both the air handler and duct system; both may be halved. Energy savings from the smaller air-handler motors may total 20 percent, even after figuring the additional energy required for the small mixing-box motors.

This procedure is based on data provided by the Calmac Manufacturing Corporation, Englewood, NJ. The economic analysis was provided by Calmac, as were the illustrations in this procedure. Calmac manufactures the *Levload* modular insulated storage tanks mentioned above that are used in their *Ice Bank Stored Cooling System.* Their system, when designed with a low-temperature heat-recovery loop, can also make the chiller into a water-source heat pump for winter heating. Thus, office and similar buildings often require heating warm-up in the morning on winter days, but these same buildings may likewise require cooling in the afternoon because of lights, people, computers, etc. Ice made in the morning to provide heating supplies frees afternoon cooling and melts to be ready for the next day's warm-up. Even on coldest days, low-temperature waste heat (such as cooling water or exhaust air), or off-peak electric heat can be used to melt the ice. Oil or gas connections to the building can thus be eliminated.

Nontoxic eutectic salts are available to lower the freezing point of the water in Calmac Ice Banks to either 28°F (–2.2°C) or 12°F (–11.1°C) and, consequently, the temperature of the resulting ice. Twenty-eight-degree ice, for example, can provide cold, dry primary air for many uses, including extra-low temperature airside applications. Twelve-degree ice can be used for on-ground aircraft cooling, off-peak freezing of ice rinks, and for industrial process applications requiring colder liquids. Other temperatures can be provided for specialized applications, such as refrigerated warehouses.

Figure 5 shows the charge cycle using a partial-storage system for an air-conditioning installation. At night a water-glycol solution is circulated through a standard packaged air-conditioning chiller and the Ice Bank heat exchanger, bypassing the air-handler coil. The cooling fluid is at 26°F (–3.3°C) and freezes the water surrounding the heat exchanger.

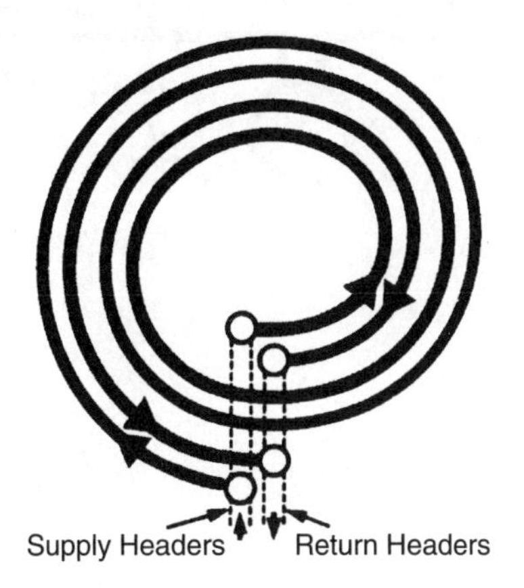

FIGURE 5 Counterflow heat-exchanger tubes used in the Ice Bank. (*Calmac Manufacturing Corporation.*)

During the day, Fig. 6, the water-glycol solution is cooled by the Ice Bank from 52°F (11.1°C) to 34°F (1.1°C). A temperature-modulating valve, set at 44°F (6.7°C) in a bypass loop around the Ice Bank, allows a sufficient quantity of 52°F (11.1°C) fluid to bypass the Ice Bank, mix with 34°F (1.1°C) fluid, and achieve the desired 44°F (6.7°C) temperature. The 44°F (6.7°C) fluid enters the coil, where it cools the air passing over the coil from 75°F (23.9°C) to 55°F (12.8°C). Fluid leaves the coil at 60°F (15.6°C), enters the chiller and is cooled to 52°F (11.1°C).

Note that, while making ice at night, the chiller must cool the water-glycol to 26°F (–3.3°C), rather than produce 44°F (6.7°C) or 45°F (7.2°C) water temperatures required for conventional air-conditioning systems. This has the effect of "derating" the nominal chiller capacity by about 30 percent. Compressor efficiency, however, is only slightly reduced because lower nighttime temperatures result in cooler condenser water from the cooling tower (if used) and help keep the unit operating efficiently. Similarly, air-cooled chillers benefit from cooler condenser entering air temperatures at night.

The temperature-modulating valve in the bypass loop has the added advantage of providing unlimited capacity control. During many mild-temperature days in the spring and fall, the chiller will be capable of providing all the cooling needed for the building without assistance from stored cooling, Fig. 7. When the building's actual cooling load is equal to or lower than the chiller capacity, all of the system coolant flows through the bypass loop, as in Fig. 8.

Using 45°F (7.2°C) rather than 55°F (12.8°C) system air in the air-conditioning system permits further large savings in initial and operating costs. The 45°F (7.2°C) low-temperature air is achieved by piping 38°F (3.3°C) water-glycol solution from the stored cooling Ice Bank to the air handler coil instead of mixing it with bypassed solution, as in Fig. 6. The 45°F (7.2°C) air is used as primary air and is distributed to motorized fan-powered mixing boxes where it is blended with room air to obtain the desired room temperature. Primary 45°F (7.2°C) air requires much lower ft^3/min (m^3/min)

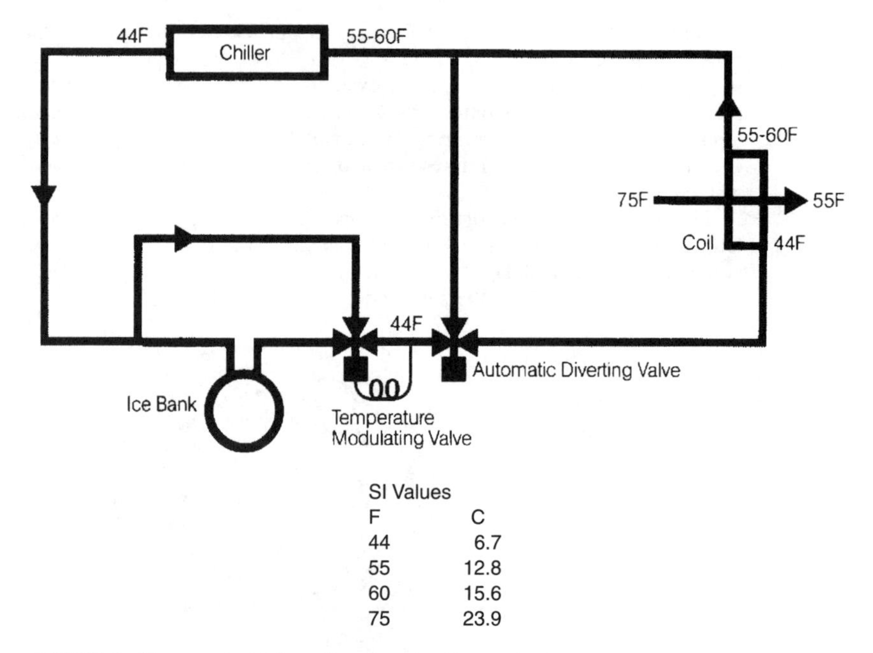

FIGURE 6 Charge cycle. (*Calmac Manufacturing Corporation.*)

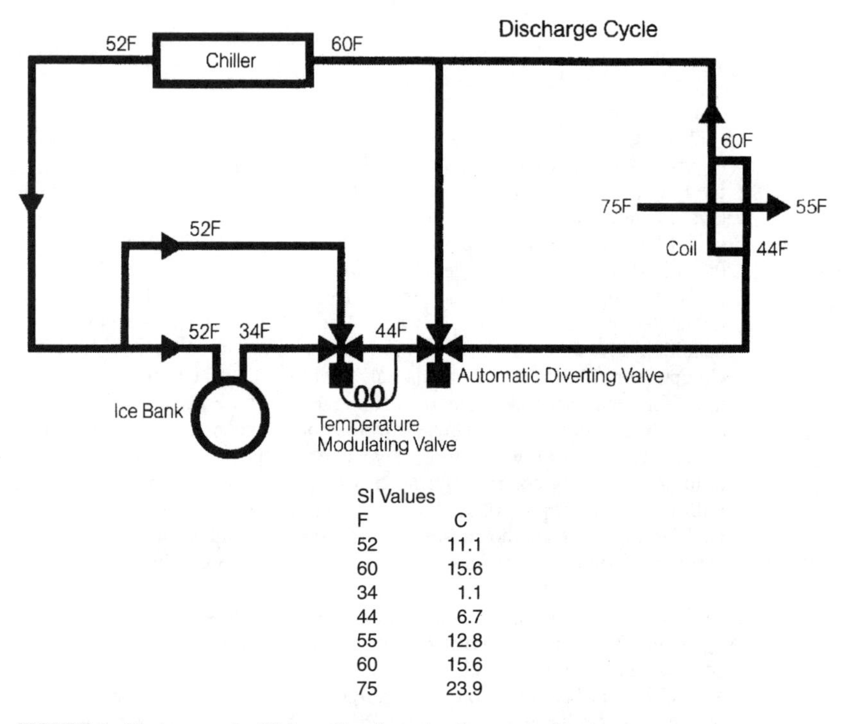

FIGURE 7 Discharge cycle. (*Calmac Manufacturing Corporation.*)

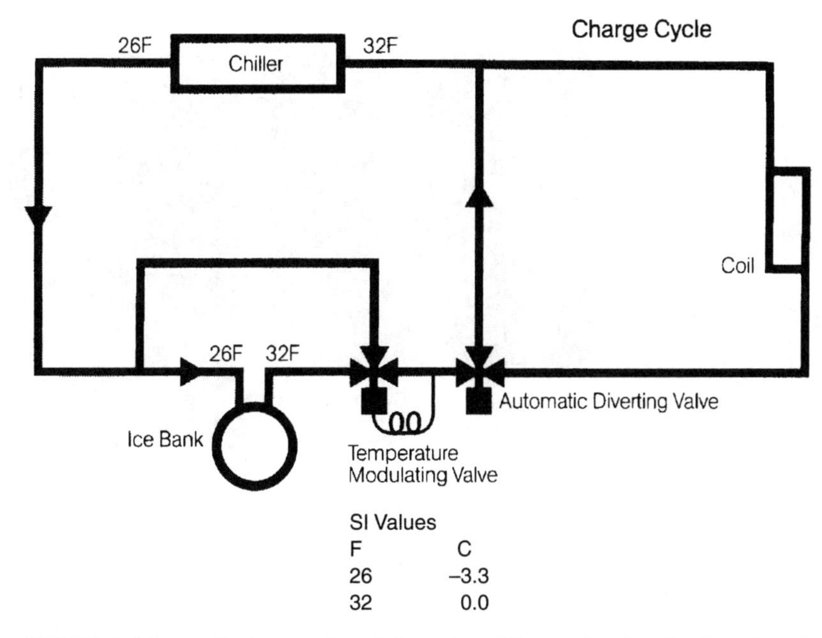

FIGURE 8 When cooling load equals, or is lower than chiller capacity, all coolant flows through bypass loop. (*Calmac Manufacturing Corporation.*)

than 55°F (12.8°C) air. Consequently, the size and cost of the air handler and duct system may be cut in about half. Energy savings of the smaller air handler motors total 20 percent, even counting the additional energy required for the small mixing-box motors.

The recommended coolant solution for these installations is an ethylene glycol-based industrial coolant such as Union Carbide Corporation's UCARTHERM® or Dow Chemical Company's DOWTHERM® SR-1. Both are specially formulated for low viscosity and superior heat-transfer properties, and both contain a multi-component corrosion inhibitor system effective with most materials of construction, including aluminum, copper, solder, and plastics. Standard system pumps, seals, and air-handler coils can be used with these coolants. However, because of the slight difference in the heat-transfer coefficient between water-glycol and plain water, air-handler coil capacity should be increased by about 5 percent. Further, the water and glycol must be thoroughly mixed before the solution enters the system.

Another advantage of ice storage systems for cooling and heating is provision of an uninterrupted power supply (UPS) in the event of the loss of a building's cooling or heating facilities. Such an UPS can be important in data centers, hospitals, research laboratories, and other installations where cooling or heating are critical.

Figure 9*a* shows the conventional "ice builder" and Fig. 9*b* the LEVLOAD Ice Bank. When ice is stored remote from the refrigerating system evaporator, as in Fig. 9*b*, the evaporator is left free to aid the cooling during the occupied hours of a building or other structure. Figure 10 compares the chiller performance of a Partial Storage Ice Bank, Fig. 9*b* (upper curve), with an ice-builder system, Fig. 9*a* (lower curve), on a typical design day. Note that when compressor cooling is done through ice on the evaporator, suction temperatures are low and kW/ton is increased.

With discussions still taking place about chlorofluorocarbon refrigerants suitable for environmental compliance, designers have to seek the best choice for the system chiller. One approach for finding popularity today as an interim solution is to choose a chiller which can use an energy-efficient refrigerant today and the most environmentally friendly refrigerant in the future. Thus, for some chillers, CFC-11

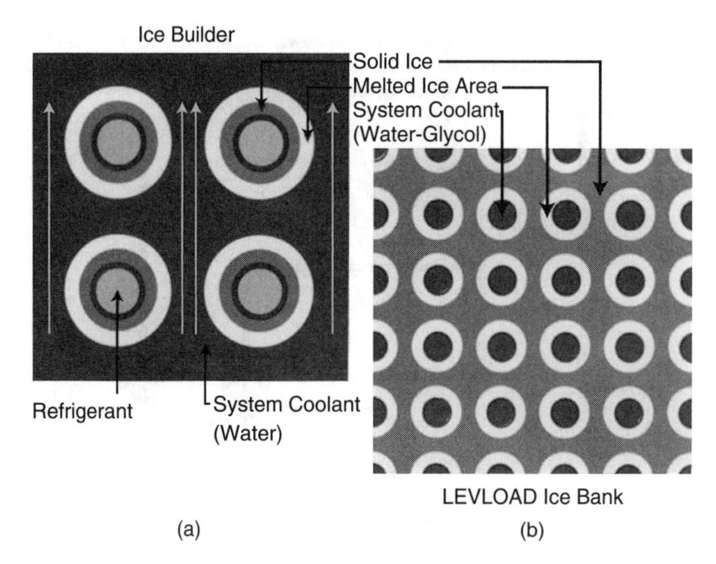

FIGURE 9 (*a*) Typical ice-builder arrangement. (*b*) LEVLOAD Ice Bank method of ice burn-off (ice melting). (*Calmac Manufacturing Corporation.*)

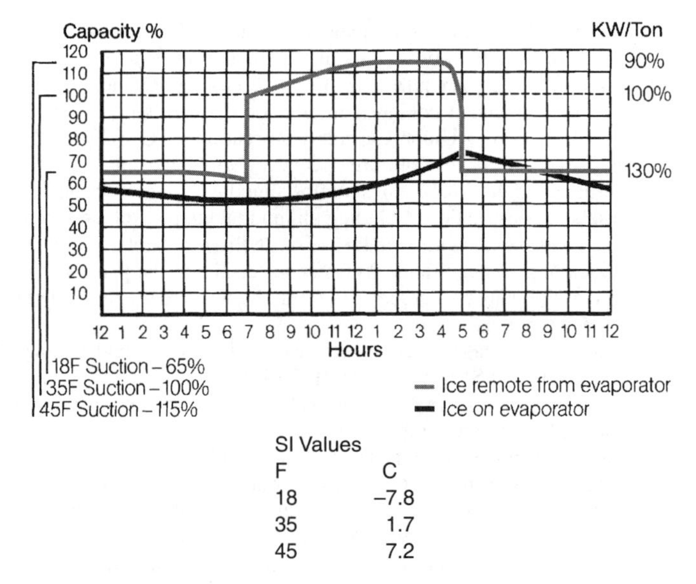

FEGURE 10 Comparison of chiller performance of a Partial Storage Ice Bank (upper curve) and conventional ice-builder system (lower curve). (*Calmac Manufacturing Corporation.*)

is the most energy-efficient refrigerant today. The future most environmentally friendly refrigerant is currently thought to be HCFC-123. By choosing and sizing a chiller that can run on HCFC-123 in the future, energy savings can be obtained today, and, if environmental requirements deem a switch in the future, the same chiller can be used for CFC-11 today and HCFC-123 in the future. It is also possible that the same chiller can be retrofitted to use a non-CFC refrigerant in the future. A number of ice storage systems have adopted this design strategy. Many firms that installed ice storage systems in recent years are so pleased with the cost savings (energy, equipment, ducting, UPS, etc.) that they plan to expand such systems in the future.

Chillers using CFC-11 normally produce ice at 0.64 to 0.75 kW/ton, depending on the amount of ice produced. Power consumption is lower when larger quantities of ice are produced. When HCHC-123 is used, the power input ranges between 0.7 and 0.8 kW/ton, again depending on the number of tons produced. As before, power consumption is lower when larger tonnages are produced. New centrifugal chillers produce cooling at power input ranges close to 0.5 kW/ton.

In all air-conditioning systems, designers must recognize that there are three courses of action open to them when CFC refrigerants are no longer available: (1) continue to use existing CFC-based equipment, taking every precaution possible to stop leaks and conserve available CFC supplies; (2) retrofit existing chilling equipment to use non-CFC refrigerants; this step requires added investment and changed operating procedures; (3) replace existing chillers with new chillers specifically designed for non-CFC refrigerants; again, added investment and changed operating procedures will be necessary.

To avoid CFC problems, new high-efficiency chlorine-free screw chillers are being used. And there are packaged ammonia screw chiller available also. Likewise, a variety of alternative refrigerants are now being produced for new and retrofit refrigeration and air-conditioning uses.

Table 1 shows an economic analysis of typical partial-storage and full-storage installations. A conventional chilled-water air-conditioning system is compared with a partial-storage 40 percent size chiller with Ice Banks in the partial-storage analysis. Full-storage produces the simple payback time of 2.5 years for the investment. Data in this analysis are from Calmac Manufacturing Corporation. When using a similar analysis, be certain to obtain current prices for components, demand charges, and electricity. Values given here are for illustration purposes only.

ENERGY ANNUAL HEATING AND COOLING COSTS AND LOADS

A 2000-ft^2 (185.8-m^2) building has a 100,000-Btu/h (29.3-kW) heat loss in an area where the heating season is 264 days' duration. Average winter outdoor temperature is 42°F (5.6°C); design conditions are 70°F (21.1°C) indoors and 0°F (–17.8°C) outdoors. The building also has a summer cooling load of 7.5 tons (26.4 kW) with an estimated full-load cooling time of 800 operating hours. What are the total winter and summer estimated loads in Btu/h (kW)? If oil is 90 cents/gal and electricity is 7 cents/kWh, what are the winter and summer energy costs? Use a 24-h heating day for winter loads and a boiler efficiency of 75 percent when burning oil with a higher heating value (HHV) of 140,000 Btu/gal (39,018 MJ/L).

Calculation Procedure:

1. *Compute the winter operating costs*

The winter seasonal heating load, WL = N × 24 h/day × Btu/h heat loss × (average indoor temperature, °F – average outdoor temperature, °F)/(average indoor temperature, °F – outside design temperature, °F), where N = number of days in the heating season. For this building, WL = (264 × 24 × 100,000)(70 – 42)/(70 – 0) = 253,440,000 Btu (267.4 MJ).

The cost of the heating oil, CO = Btu seasonal heating load × oil cost $/gal/boiler efficiency × HHV. Or, for this building, CO = (253,440,000)(0.90)/0.75 × 140,000 = $2,172.34 for the winter heating season.

TABLE 1 Economic Analysis of Typical Partial-Storage and Full-Storage Installations*

Partial storage

Assume: 100-ton peak cooling load, 10-h cooling day, 75 percent diversity factor, $8.00/mo/kW demand charge, 12-mo ratchet.*

Conventional Chilled Water Air Conditioning System:

100-ton chiller at $600/ton, installed†	$60,000
Distribution system	60,000
Total	$120,000

Partial storage (40% size chiller with Ice Banks):

At 75 percent diversity factor, the true cooling load translates into 750 ton-h with the chiller providing 400 ton-h and stored cooling the balance, or 350 ton-h. Therefore:

40-ton chiller at $600/ton, installed	$24,000
Stored cooling at $60/ton-h, installed	21,000
Distribution system	60,000‡
Total	$105,000

Purchase savings: $15,000

Demand savings:

60 tons × 1.5 kW/ton§ × 12 mo × $8.00 = $8640/yr

*Utility term for a monthly electrical bill surcharge based on a previous month's higher peak demand.

Specifications	Model 1098	Model 1190	Model 1500
Total ton-hour capacity	115	190	570
Tube surface/ton-h, ft^2	12.0	12.0	12.0
Nominal discharge time, h	6–12	6–12	6–12
Latent storage cap., ton-h	98	162	486
Sensible storage cap., ton-h	17	28	84
Maximum operating temp., °F	100	100	100
Maximum operating press., lb/in^2	90	90	90
Outside diameter, in	89	89	—
Length × width, in	—	—	268 × 96
Height, in	68	101	102
Weight, unfilled, lb	1,060	1,550	4,850
Weight, filled, lb	9,940	16,750	50,450

Full storage

Assume: 100-ton peak cooling load, 10-h cooling day, 75 percent diversity factor, 1000-h cooling season, $8.00/mo/kW demand charge, 12-mo ratchet, $0.03/kWh off-peak differential.

Full storage:

10 h × 100 tons × 75% × $60/ton-h, installed	$45,000

Demand savings:

100 tons × 1.5 kW/ton × 12 mo × $8.00	$14,400/yr

Energy savings:

$0.03/kWh × 1000 h × 100 tons × 1.2 Avg. kW/ton	$3600/yr
Total savings:	$18,000/yr

Simple payback: $45,000 ÷ $18,000 = 2.5 yr.

†The $600/ton includes all accessories, such as cooling tower, fan coils, pumps, blowers, piping, controls, etc.

‡Figure shown is for conventional temperature system. This cost could be reduced by 50 percent by using a low temperature duct system.

§The 1.5 kW/ton is figured at the peak summer demand condition and also includes all accessories.

Specifications	Model 1098	Model 1190	Model 1500
Volume of water/ice, gals.	980	1620	4860
Volume of solution in HX, gals.	90	148	555
Press. drop (25% glycol, 28°F), PSI			
20 gal/min	4.0	—	—
40 gal/min	9.7	5.0	—
60 gal/min	18.8	9.0	—
80 gal/min	—	13.0	2.8
160 gal/min	—	—	7.0
240 gal/min	—	—	13.0

The outlet temperatures from the tanks vary with the rate at which the tanks are discharged. See LEVLOAD Performance Manual for details.

LEVLOAD and CALMAC are registered trademarks of Calmac Manufacturing Corporation. The described product and its application are protected by United States Patents 4,294,078; 4,403,645; 4,565,069; 4,608,836; 4,616,390; 4,671,347; and 4,687,588.

*Calmac Manufacturing Corporation

SI values in procedure text.

2. *Calculate the summer cooling cost*
The summer seasonal electric consumption in kWh, SC = tons of air conditioning × kW/ton of air conditioning × number of operating hours of the system. The kW/design ton factor is based on both judgment and experience. General consensus amongst engineers is that the kW/ton varies from 1.8 for small window-type systems to 1.0 for large central-plant systems. The average value of 1.4 kW/design ton is frequently used and will be used here. Thus, the summer electric consumption, SC = 7.5 × 1.4 × 800 = 8400 kWh. This energy will cost 8400 kWh × $0.07 = $588.00.

The total annual energy cost is the sum of the winter and summer energy costs, $2172.34 + $588.00 = $2760.34.

Related Calculations. Use this procedure to compute the energy costs for any type of structure, industrial, office, residential, medical, educational, etc., having heating or cooling loads, or both. Any type of fuel, oil, gas, coal, etc., can be used for the structure.

This procedure is the work of Jerome F. Mueller, P.E. of Mueller Engineering Corp.

ENERGY HEAT RECOVERY USING A RUN-AROUND SYSTEM OF HEAT TRANSFER

A hospital operating-room suite requires 6000 ft^3/min (169.8 m^3/s) of air in the supply system with 100 percent exhaust and 100 percent compensating makeup air. Winter outdoor design temperature is 0°F (–17.8°C); operating-room temperature is 80°F (26.7°C) with 50 percent relative humidity year-round. How much energy can be saved by installing coils in both the supply and exhaust air ducts with a pump circulating a nonfreeze liquid between the two coils, absorbing heat from the exhaust air and transferring this heat to the makeup air being introduced?

Calculation Procedure:

1. *Choose the coils to use*
In the winter the exhaust air is at 80°F (26.7°C) while the supply air is at 0°F (–17.8°C). Hence, the coil in the exhaust air duct will transfer heat to the nonfreeze liquid. When this liquid is pumped through the coil in the intake-air duct it will release heat to the incoming air. This transfer of otherwise wasted heat will reduce the energy requirements of the system in the winter.

As a first choice, select a coil area of 12 ft^2 (1.1 m^2) with a flow of 6000 ft^3/min (169.8 m^3/s). While a number of coil arrangements are possible, the listing below shows typical coil conditions at face velocities of 500 ft/min (152.4 m/min) and 600 ft/min (182.8 m/min) with a coil having 8-fins/in coil. Entering coolant temperature = 45°F (7.2°C); entering air temperature = 80°F (26.7°C) dry bulb, 67°F (19.4°C) wet bulb. Various manufacturers' values may vary slightly from these values.

Ft/min (m/min) face velocity	Temp rise, °F (°C)	No. of rows	Total MBtu/h (kWh)	Leaving dry bulb, °F (°C)	Leaving wet bulb, °F (°C)
500 (152.4)	12 (21.6)	6	16.6 (4.86)	57.3 (14.1)	56.5 (13.6)
500 (152.4)	12 (21.6)	8	20.1 (5.89)	54.2 (12.3)	53.9 (12.2)
600 (182.9)	10 (18)	4	14.3 (4.19)	62.1 (16.7)	59.7 (15.4)

The middle coil listed above, if placed in the exhaust duct, would produce 20.1 MBtu/h (5.89 kWh) with a 12°F (6.67°C) temperature rise with a leaving air temperature of 53.9°F (12.2°C) when the liquid coolant enters the coil at 45°F (7.2°C) and leaves 57°F (13.9°C).

2. ***Determine the coil heating capacity***

The heating capacity of a coil is the product of (coil face area, ft^2)[heat release, Btu/(h · ft^2) of coil face area]. For this coil, heating capacity = 12 × 20,100 = 241,200 Btu/h (70.7 kWh). The incoming makeup air can be heated to a temperature of: (heating capacity, Btu)/(makeup air flow, ft^3/min) (1.08) = 241,200/6000 × 1.08 = 37.2°F (2.9°C). Hence, the makeup air is heated from 0°F (–17.8°C) to 37.2°F (2.9°C).

The energy saved is—assuming 1000 Btu/lb of steam (2330 kJ/kg)—241,200/1000 = 241.2 lb/h (109.5 kg/h). With a 200-day heating season and 10-h operation/day, the saving will be 200 × 10 × 241.2 = 482,400 lb/yr (219,010 kg/yr). And if steam costs $20/thousand pounds ($20/454 kg), the saving will be (482,400/1000)($20) = $9,648,00/yr. Such a saving could easily pay for the heating coil in one year.

Related Calculations. Use this general approach to choose heating coils for any air-heating application where waste heat can be utilized to increase the temperature of incoming air, thereby reducing the amount of another heating medium that might be required. Most engineers use the 1000 Btu/lb (2300 kJ/kg) latent heat of steam as a safe number to convert quickly from hourly heat savings in Btu (kg) to pounds of steam. This procedure can be used for industrial, commercial, residential, and marine applications.

The procedure is the work of Jerome F. Mueller, P.E., of Mueller Engineering Corp.

Figure 11 is a typical run-around coil detail that is very commonly used in energy recovery systems in which the purpose is to extract heat from air that must be exhausted. Normally about 40 to 60 percent of the heat being wasted can be recovered. This seemingly simple detail has two points that should be carefully noted. The difference in fluid temperatures in this closed system creates small expansion and contraction problems and an expansion tank is required. Most importantly the temperature of the incoming supply air can create a coil temperature so low that the coil in the exhaust air stream begins to ice up. Normally this begins at some 35°F (1.67°C). This is when the

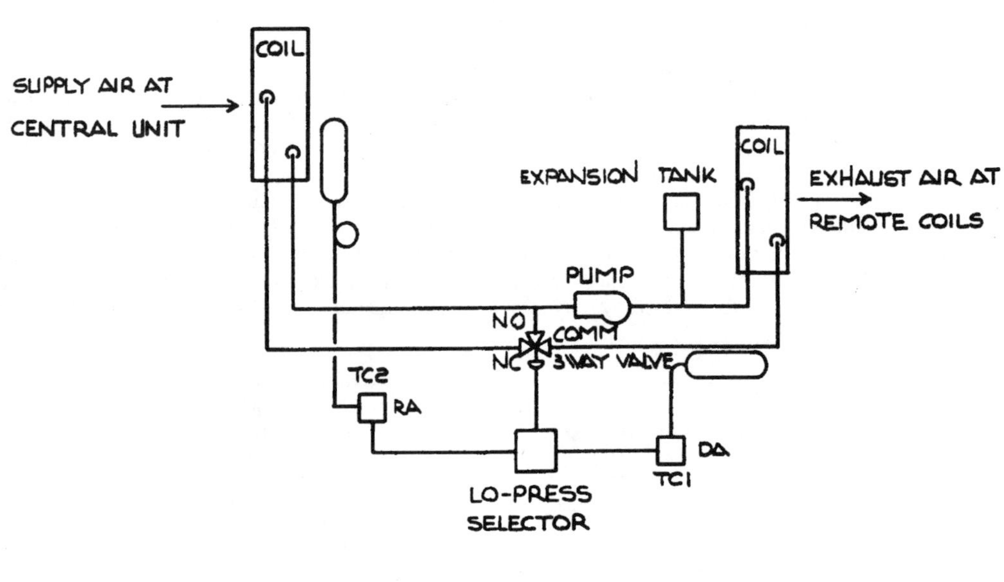

FIGURE 11 Heat-recovery-loop schematic. (*Jerome F. Mueller.*)

three-way bypass valve comes into play and the glycol is not circulated through the outside air coil. Obviously if the outside air temperature is low enough, the system will go into full bypass and the circulating pump should be stopped.

ENERGY SAVINGS WITH ROTARY HEAT EXCHANGER

A hospital heating, ventilating and air-conditioning installation has a one-pass system supplying and exhausting 10,000 ft^3/min (2260 m^3/min) with these operating conditions: Summer outdoor design temperature 95°F dry bulb (35°C), 78°F wet bulb (25.6°C); summer inside exhaust temperature 75°F dry bulb (35°C), 62.5°F wet bulb (16.9°C); winter outdoor design temperature 0°F (–17.8°C); winter outdoor exhaust temperature 75°F (35°C). How much energy can be saved if a rotary heat exchanger (heat or thermal wheel) is used as an energy-saving device?

Calculation Procedure:

1. ***Determine the cooling-load savings***
A rotary heat exchanger generally consists of an all-metallic rotor wheel with radial partitions containing removable heat-transfer media sections made of aluminum, stainless steel, or Monel, Fig. 12. A purge section permits cleaning of the heat-transfer media using water, steam, solvent spray, or compressed air to eliminate bacteria growth, especially in hospitals and laboratories. Data supplied by rotary heat exchanger manufacturers give an average sensible heat transfer efficiency as 80 percent, and an enthalpy efficiency of 65 percent.

From a psychrometric chart the enthalpy of air at 95°F dry bulb (35°C) and 78°F wet bulb (25.6°C) is 41.6 Btu/lb (96.9 kJ/kg); at 75°F dry bulb (23.9°C) and 62.5°F wet bulb (16.9°C) it is 28.3 Btu/lb (65.9 kJ/kg). The specific volume of the air, from the chart, is 13.6 ft^3/lb (0.85 m^3/kg). Then the cooling load heat saving, Btu/h (W) = (heat-wheel efficiency)(air flow, ft^3/min)(60 min/h)(enthalpy

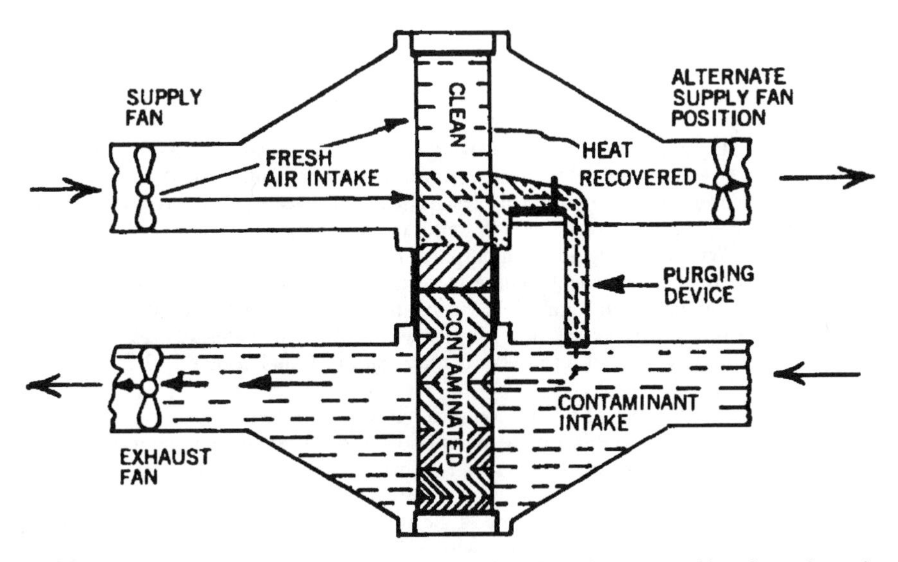

FIGURE 12 Heat or thermal wheel rotates at 1 to 3 r/min and permits recovery of heat from exhaust air. Wheel can also be used to cool incoming air.

difference, Btu/lb)/(specific volume of the air, ft^3/lb). Substituting, heat saving = 0.65(10,000)(60)(41.6 – 28.3)/13.6 = 381,397 Btu/h (111.7 kW), or 31.8 tons of refrigeration.

2. *Find the heat-load saving*

The heat-load saving, Btu/h (W) = (heat-wheel efficiency)(1.08)(air flow, cfm)(indoor temperature – winter outdoor design temperature). Substituting, heat saving = 0.80(1.08)(10,000)(75 – 0) = 648,000 Btu/h (189.9 kWh). Using 1000 Btu/lb (2330 kJ/kg) as the latent heat of steam, the saving will be 648,000/1000 = 648 lb/h (294.2 kg/h).

3. *Find the leaving-air temperature for each condition*

For the summer air cooling load, the temperature of the supply air leaving the rotary heat exchanger = (summer outdoor design temperature) – (sensible heat-transfer efficiency) (summer outdoor design temperature – summer indoor exhaust temperature) = 95 – 0.80 (97 – 75) = 79°F dry bulb (26.1°C).

To determine the summer wet-bulb temperature, use the relation: Summer wet-bulb temperature = temperature at the enthalpy found from (enthalpy at summer outdoor design condition) – (sensible heat-transfer efficiency) (enthalpy at summer outdoor design condition – enthalpy at summer indoor exhaust temperature) = 41.6 – 0.65(41.6 – 28.3) = 32.96 Btu/lb (76.8 kJ/kg). Entering the psychrometric chart, read the wet-bulb temperature as 68.8°F (20.4°C).

For the winter air condition, the supply air leaving the rotary heat exchanger has a temperature of (outdoor design temperature) + (rotary heat exchanger sensible-heat efficiency) (indoor air temperature – outdoor design temperature). Or temperature of supply air leaving the rotary heat exchanger = 0 + 0.80(75 – 0) = 60°F dry bulb (15.6°C).

The winter wet-bulb temperature is found from (indoor air enthalpy) – (efficiency) (indoor air enthalpy – outdoor air enthalpy); once the enthalpy is known, the wet-bulb temperature can be found from the psychometric chart. For this rotary heat exchanger, 28.3 – 0.65(28.3 – 1.0) = 10.56 Btu/lb (kJ/kg). Entering the psychometric chart, find the wet-bulb temperature as 32°F (0°C). In this calculation the enthalpy of the 0°F (–17.8°C) air is taken as 1.0 Btu/lb (2.33 kJ/kg) because that is the value of the enthalpy at the 0°F (–17.8°C) temperature.

Related Calculations. Rotary heat exchangers find use in many applications: industrial, commercial, residential, etc. The key to using any rotary heat exchanger is the tradeoff between heater cost vs. the savings anticipated. Thus, if a rotary heat exchanger can be paid for in either two years, or less, most firms will find the heater acceptable.

Rotary heat exchangers are also called *thermal wheels* and they are popular for energy conservation. Current standard designs can handle clean filtered air from ambient temperature to 500°F (260°C). Wheels designed to handle high-temperature air are rated at air temperatures to 1500°F (816°C). Normal rotative speed is 1 to 3 r/min.

Properly designed heat-recovery thermal wheels can recover 60 to 80 percent of the sensible heat from exhaust air and transmit this heat to incoming outside air. Where both sensible- and latent-heat recovery are desired, specially designed thermal wheels will also recover 60 to 80 percent of the heat in the exhaust air stream and transmit it to the incoming air.

With the increased emphasis on indoor air quality (IAQ), some regulatory groups prohibit use of thermal wheels where there is the possibility of leakage from the exhaust stream contaminating the incoming air. Hence, the designer must carefully check local code requirements before specifying use of a thermal wheel. Both the exhaust stream and the incoming air stream should be filtered to prevent contamination. Usual choice is 2-in-thick (50.4-mm) "roughing" filters.

The thermal wheel can cool incoming air when the exhaust air is at a lower temperature. Thus, with 75°F (24°C) incoming air and 60°F (16°C) exhaust air, the incoming air temperatures can be reduced to, possibly, 65°F (18°C). Hence, the thermal wheel can produce savings in both directions, i.e., heating or cooling incoming supply air.

The procedure given here is the work of Jerome F. Mueller, P.E., Mueller Engineering Corp. Supplementary data on heat wheels is from Grimm and Rosaler—*Handbook of HVAC Design*, McGraw-Hill, 2004.

ENERGY SAVINGS WITH "HOT DECK" TEMPERATURE RESET

An office building has a dual-duct heating and cooling system rated at 30,000 ft^3/min (849 m^3/min). The winter heating season is 37 weeks and the summer cooling season 15 weeks. Following federal guideline suggestions, a decision has been made to reduce (reset) the hot deck by 6°F (–14.4°C) in the summer and by 4°F (–15.6°C) in the winter. How much energy will be saved if the building occupied cycle is 60 h/wk?

Calculation Procedure:

1. ***Compute the summer energy saving***

The energy saved in the summer, *S*, can be found from *S* = ft^3/min(0.5)(1.08)(°F by which hot deck temperature is lowered)(weeks of cooling)(occupied cycle, h/wk). Or, *S* = 30,000(0.5)(1.08)(6)(15)(60) = 87,480,000 (92,291 MW). In this equation the factor 0.5 is used because only one of the dual ducts is the "hot deck."

2. ***Determine the winter energy saving***

Use the same equation, substituting the winter temperature reduction and the duration of the heating season. Or winter saving, *W* = 30,000(0.5)(1.08)(4)(37)(60) = 143,856,000 Btu (151,768 MJ).

3. ***Compute the annual energy saving***

The annual energy saving, *SA*, is the sum of the summer and winter savings, or *SA* = 87,480,000 + 143,856,000 = 231,336,000 Btu (244,059 MJ).

Related Calculations. Use this procedure for any type of building having a dual-duct heating and cooling system: industrial, office, commercial, residential, medical, health-care, etc. Be certain to use the actual reset temperature reduction when analyzing the potential savings.

This procedure is the work of Jerome F. Mueller, P.E., Mueller Engineering Corp.

ENERGY PERFORMANCE OF AIR-TO-AIR HEAT EXCHANGER

A laboratory heating, ventilating, and air-conditioning system requires 100 percent exhaust of its 5000-ft^3/min (141.5-m^3/min) air supply. The operating conditions are: Summer outdoor design temperature 95°F dry bulb (35°C); 78°F wet bulb (25.6°C); Summer laboratory room exhaust temperature 75°F dry bulb (23.9°C); 62.5 wet bulb (16.9°C); Winter outdoor design temperature 0°F (–18°C); Winter laboratory room exhaust temperature 75°F dry bulb (23.9°C); 62.5°F wet bulb (16.9°C). Because the high moisture content of the laboratory room air must be maintained year round, the design requires that an energy-saving device include a moisture-saving feature needing no energy to operate. What is a suitable system? How much savings can be obtained?

Calculation Procedure:

1. ***Evaluate potential systems for this installation***

The design requirements dictate some form of direct heat-transmission interchange which can be achieved by a run-around coil system or a heat wheel. But the further requirement of no energy used in the recovery system dictates a slightly different approach.

There are insulated air-to-air exchangers available using cross-flow cartridges, Fig. 13, which are non-clogging and bacteriostatic. Efficiencies are generally 75 percent for sensible heat and 60 percent enthalpic. The cartridges in such exchangers are constructed of alternating layers of corrugated and flat sheets separating the exhaust and supply air streams. Cross-contamination and leakage are less than 0.3 percent.

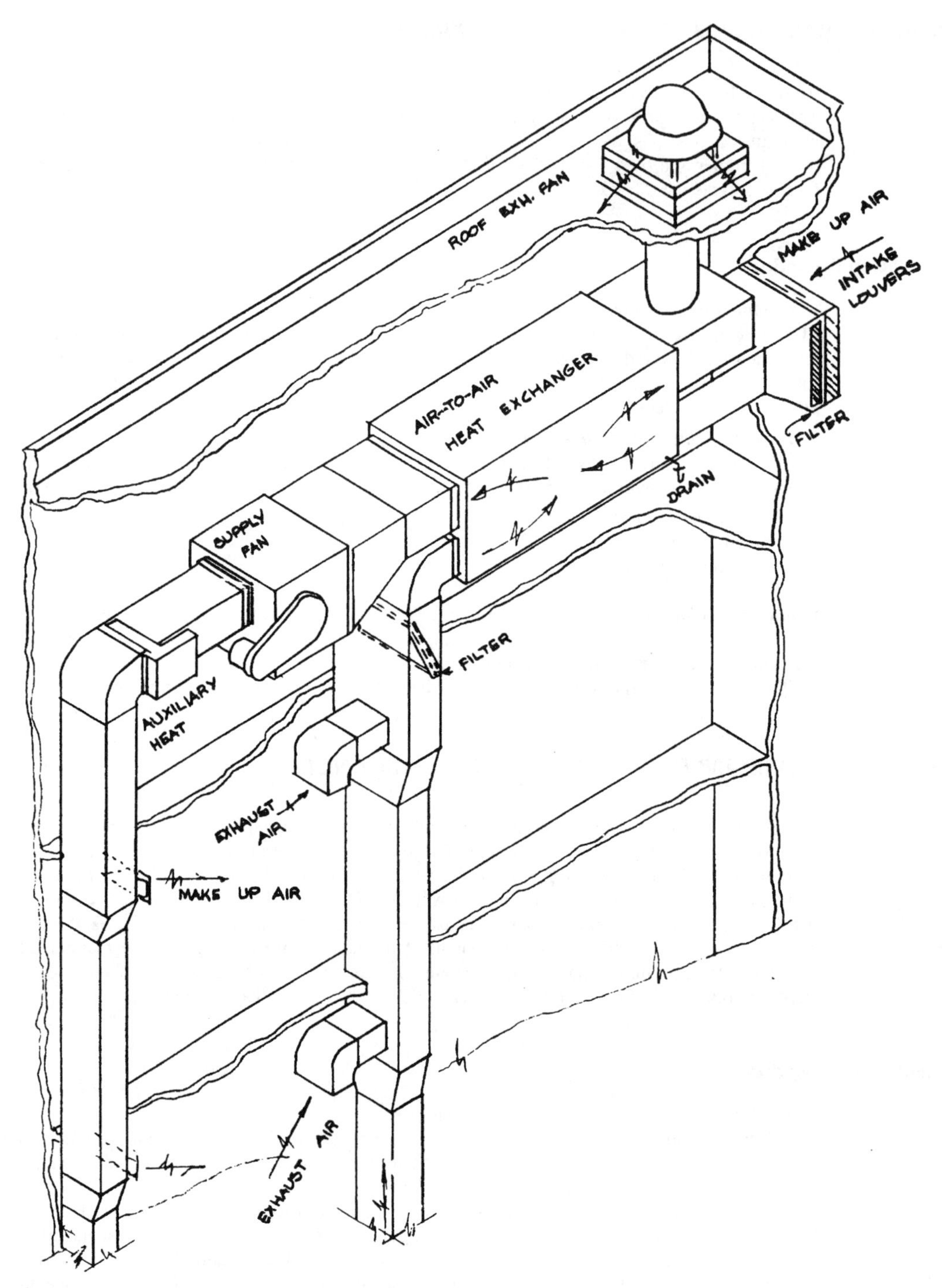

FIGURE 13 Typical air-to-air heat exchanger. System shown is used where there are no contaminants in the exhaust and intake air and a portion of the return air stream is being exhausted. (*Jerome F. Mueller.*)

During summer and winter operation, moisture is entirely in the vapor phase. Permeation of moisture from the humid air stream to the dry air stream is effected without chemical impregnation. In most applications, there is little risk of ice formation even at low winter design temperatures.

To determine if a given application presents a possibility of icing, plot the outside air condition and design air exhaust condition on a psychrometric chart and draw a straight line through these two points. If this straight line between the two points does not intersect the saturation curve on the psychrometric chart, no danger of icing exists.

In the event the saturation curve is intersected, draw a line from the exhaust-air temperature condition tangent to the saturation curve and extend this tangent to the horizontal dry-bulb temperature line. The difference between the temperature value of the horizontal line at this intersection and the actual outside air temperature defines the number of degrees the outside air temperature must be raised by preheating.

2. *Determine the properties of the air*

From a psychrometric chart, the enthalpy at 95°F dry bulb (35°C) and 78°F wet bulb (25.6°C) is 41.6 Btu/lb (96.9 kJ/kg). At 75°F dry bulb (23.9°C) and 62.5°F wet bulb (16.9°C), the enthalpy is 28.3 Btu/lb (65.9 kJ/kg). At 0°F (–17.8°C) the air is very dry and the enthalpy is generally about 1.0 Btu/lb (2.33 kJ/kg).

3. *Compute the cooling-load saving*

The cooling-load saving, C Btu/h = (efficiency)(ft^3/min)(enthalpy difference, Btu/lb)(60 min/h)/specific volume of the air, ft^3/lb. Substituting, using the enthalpic efficiency, $C = 0.60(5000)(41.6 - 28.3)(60)/13.6 = 176{,}029$ Btu/h (51.6 kW), or 14.7 tons of refrigeration.

4. *Find the heating-load savings*

The heating-load saving, H Btu/h (kW) = (sensible-heat efficiency)(1.08)(ft^3/min)(inside room temperature – outside air temperature), or $H = 0.75(1.08)(5000)(75 - 0) = 303{,}750$ Btu/h (88.9 kN), or 303.8 lb (139.8 kg) of steam, using an enthalpy of vaporization of 1000 Btu/lb (2330 kJ/kg) of steam, which is a safe assumption.

5. *Determine summer and winter enthalpies and temperatures*

In the summer, the supply air leaving the heat exchanger has a dry-bulb temperature of: (summer outdoor dry-bulb temperature) – (sensible-heat efficiency)(outdoor dry-bulb – indoor dry-bulb) = $95 - 0.75(95 - 75) = 80$°F (26.7°C).

The summer wet-bulb temperature is found at the enthalpy for: (enthalpy at summer design outdoor wet-bulb) – (enthalpic efficiency)(summer outdoor wet-bulb enthalpy – summer indoor wet-bulb enthalpy); or summer wet-bulb = $41.6 - 0.60(41.6 - 28.3) = 33.62$ Btu/lb (78.3 kJ/kg). On the psychrometric chart, read the wet-bulb temperature at 33.62 Btu/lb (78.3 kJ/kg) as 69.5°F (20.8°C).

In the winter, the supply air dry-bulb temperature leaving the heat exchanger will be at: (outdoor air design temperature) + (sensible-heat efficiency)(winter exhaust temperature – outdoor design temperature); or $0 + 0.75(75 - 0) = 56.25$°F (13.5°C).

The winter wet-bulb temperature is found at the enthalpy for: (indoor wet-bulb enthalpy) – (enthalpic efficiency)(indoor wet-bulb enthalpy – outdoor design temperature enthalpy); or $28.3 - 0.60(28.3 - 1) = 11.92$ Btu/lb (27.8 kJ/kg). On the psychrometric chart, for 11.92 Btu/lb (27.8 kJ/kg), read the wet-bulb temperature as 32°F (0°C).

Related Calculations. Use this general procedure to evaluate the performance of any air-to-air heat exchanger used in an energy-recovery application. Buildings in which such a heat exchanger would be useful include office, factory, commercial, residential, medical, hospitals, etc.

This procedure is the work of Jerome F. Mueller, P.E., Mueller Engineering Corp.

HEATING CAPACITY REQUIREMENTS FOR BUILDINGS WITH STEAM AND HOT-WATER HEATING

A building with a volume of 500,000 ft^3 (14,150 m^3) is to be heated in 0° weather (–17.8°C) to 70°F (21.1°C). The wall and roof surfaces aggregate 28,000 ft^2 (2601.2 m^2) and the glass area aggregates 7000 ft^2 (650.3 m^2). Air in the building is changed three times every hour (a 20-min air change).

Allowing transmission coefficients of 0.25 Btu/(ft^2 · h · °F) of 0.25 for the wall and roof surfaces (1.42 m^2), and 1.13 Btu/(ft^2 · h · °F) (6.42 W/m^2) for the single-paned glass windows, determine the square feet (m^2) of steam and hot-water radiation required if each square foot emits 240 Btu/h [2725.5 kJ/(m^2 · h)] for steam and 150 Btu/h [1703.4 kJ/(m^2 · h)] for hot water.

Calculation Procedure:

1. *Find the wall, roof, and glass heat losses*

The wall and roof losses = (heat-transmission coefficient)(area)(temperature difference). Or, for this building, wall and roof losses = (0.25)(28,000)(70 – 0) = 490,000 Btu/h (143.6 kW). For the glass, using the same relation, heat loss = (1.13)(7000)(70 – 0) = 555,000 Btu/h (162.2 kW).

2. *Compute the ventilation heat load*

The ventilation heat load = (volume of air inflow) (number of air changes/hour) (density of air) (specific heat of air) (temperature rise of entering air). For this building, ventilation heat load = (500,000) (3) (0.075) (0.24) (70 – 0) = 1.89×10^6 Btu/h (553.8 kW).

3. *Calculate the amount of steam and hot-water radiation required*

First sum the various heat loads, namely walls, roof, glass, and ventilation, and divide by the heat emitted by each square foot (m^2) of radiation. Or (490,000 + 555,000 + 1.89×10^6)/240 = 12,229.2 ft^2 of equivalent direct radiation (EDR), or [138,878.4 kJ/(m^2 · h)]. For hot-water heating the required area is 19,567 ft^2 EDR [222,208.7 kJ/(m^2 · h)].

Related Calculations. To find the fuel consumption for a building such as this, divide the total heat load by the efficiency of the heating system times the heating value of the fuel as fired. In such calculations, remember that 4 ft^2 (0.37 m^2) of steam radiation are equivalent to a condensation rate of 1 lb (0.454 kg) of steam/hour for low-pressure heating systems.

ENERGY AND HEATING STEAM REQUIREMENTS FOR SPECIALIZED ROOMS

A control room for an oil refinery unit is to be heated and ventilated by a central duct system. Ventilation is to be at the rate of 3 ft^3/min (0.085 m^3/min) of outside air/square foot (m^2) of floor area. The room to be ventilated is 40 × 60 ft (12.2 × 18.3 m), and is to be pressurized to keep out hazardous gases. Outside design temperature is –10°F (–23.3°C). Determine the steam consumption rate for maximum design conditions with the use of 5-lb/in^2 (gage) (34.5-kPa) saturated steam for heating.

Calculation Procedure:

1. *Compute the amount of outside air needed*

The rate of outside air to be handled by the ventilating system is (floor area) (ft^3/min/ft^2 of floor area) = (40 × 60)(3) = 7200 ft^3/min (203.8 m^3/min).

2. *Find the heating load for this system*

Use the relation, heating load = (ft^3/min) (1.08) (temperature rise of the air). In this relation the constant is 1.08 = (specific heat of air) (minutes/hour)/(specific volume of air). For air at normal atmospheric conditions, (0.24) (60)/(13.3) = 1.0827; this value is normally rounded to 1.08, as given above. Substituting, heating load, Q = (7200) (1.08)(75 – [–10]) = 660,960 Btu/h (193.7 kW).

3. *Calculate the rate of steam consumption*

Saturated steam at 5 lb/in^2 (gage) has a heat of condensation of 960 Btu/lb (2236.8 kJ/kg). The steam rate is then 660,960/960 = 688.6 lb/h (312.6 kg/h).

Related Calculations. Heating systems generally use lower-pressure steam as their heating medium because (1) piping, valve, and fitting costs are lower; (2) the heat of condensation (or latent heat) of lower pressure steam is higher (larger), meaning that more heat is absorbed by the condensation of each pound (kg) of steam. While high-pressure steam may be used under specialized circumstances, the majority of steam-heating systems use low-pressure steam.

COMPUTATION OF CARBON DIOXIDE BUILDUP IN OCCUPIED SPACES

An office space has a total volume of 75,000 ft^3 (2122.5 m^3). Equipment occupies 25,000 ft^3 (707.5 m^3). The space is occupied by 100 employees. If all outside air supply is cut off, how long will it take to render the space uninhabitable?

Calculation Procedure:

1. *Determine the cubage of the space*

For carbon dioxide buildup measurements, the *net volume* or (*cubage*) of the space is used. The net volume of a space = total volume – volume of equipment, files, machinery, etc. For this space, net volume, NV = total volume – machinery and equipment volume = 75,000 – 25,000 = 50,000 ft^3 (1415 m^3).

2. *Compute the time to vitiate the inside air*

Use the relation, $T = 0.04V/P$, where T = time to vitiate the inside air, h; V = net volume, ft^3; P = number of people occupying the space. Substituting, $T = 0.04(50{,}000)/100 = 20$ h. During this time the oxygen content of the air will be reduced from a nominal 21 percent by volume to 17 percent.

It is a general rule to consider that after 5 h, or one-quarter of the calculated time of 20 h, the air would become stale and affect worker efficiency. Atmospheres containing less than 12 percent oxygen or more than 5 percent carbon dioxide are considered dangerous to occupants. The formula used above is popular for determining the time for carbon dioxide to build up to 3 percent with a safety factor.

Related Calculations. In today's environmentally conscious world, smoking indoors is prohibited in office and industrial structures throughout the United States. Much of the Western world is adopting the same prohibition. Part of the reason for prohibiting smoking inside occupied structures is the oxygen depletion of the air caused by smokers.

Today, indoor air quality (IAQ) is one of the most important design considerations faced by engineers. A variety of environmental rules and regulations control the design of occupied spaces. These requirements cannot be overlooked if a building or space is to be acceptable to regulatory agencies.

For years, occupied spaces which were not air conditioned were designed using *general ventilation* rules. In most buildings, exhaust fans located high in the side walls, or on the roof, were used to draw outside air into the building through windows or louvers. The air movement produced an air flow throughout the space to remove smoke, fumes, gases, excess moisture, heat, odors, or dust. A constant inflow of fresh, outside air was relied on for the removal of foul, stale air.

Today, with the increase in external air pollution, combined with the outgassing of construction and furnishing materials, general ventilation is a much more complex design problem. No longer can the engineer rely on clean, unpolluted outside air. Instead, careful choice of the location of outside-air intakes must be made. Other calculation procedures in this handbook deal with this design challenge.

ANALYSIS OF BYPASS AIR QUANTITY AND DEHUMIDIFIER EXIT CONDITIONS

A space to be conditioned has a sensible heat load of 10,000 Btu/min (10,550 kJ/min) and a moisture load of 26,400 gr/min (1,710,667 mg/min); 2300 lb (1044.2 kg) of air are to be introduced each minute to this space for its conditioning to 80°F (26.7°C) dry bulb and 50 percent relative humidity. How much air should be bypassed in this system, Fig. 14, and what is the amount and temperature of the air leaving the dehumidifier?

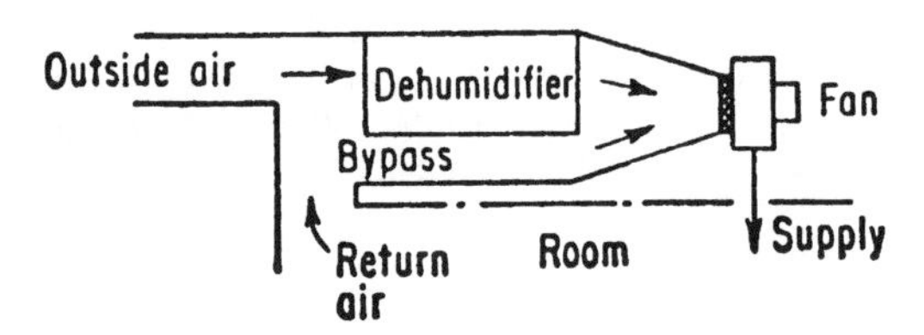

FIGURE 14 Dehumidifier fitted with bypass-air control.

Calculation Procedure:

1. *Set up a listing of the air conditions for this system*
Using a psychrometric chart and a table of air properties, set up the list thus:

	Room conditions	Air leaving dehumidifier
Dry bulb	80°F (26.7°C)	54°F (12.2°C)
Wet bulb	67°F (19.4°C)	54°F (12.2°C)
Dew point	60°F (15.6°C)	54°F (12.2°C)
Relative humidity	50 percent	100 percent
Total heat	31.15 Btu/lb (72.58 kJ/kg)	22.54 Btu/lb (52.52 kJ/kg)
Grains/lb	77.3 (11,033 mg/kg)	62.1 (8863.5 mg/kg)
Specific volume, ft^3/lb	13.84 (0.863 m^3/kg)	13.13 (0.818 m^3/kg)
Lb/min air flow	2300 (1044.2 kg/min)	1610 (730.9 kg/min)

2. *Find the moisture load in the system*
The moisture load = (gr/min)/(7000 gr/lb) (latent heat of air, Btu/lb); or (26,400/7000) (1040) = 3920 Btu/min (4135.6 kJ/min), rounded from 3922.3 Btu/min. The sensible heat load = 10,000 Btu/min (10,550 kJ/min). Total load = 3920 + 10,000 = 13,920 Btu/min (14,685.6 kJ/min).

3. *Using trial and error, find the air quantities in the system*
Solve by trial and error, assuming 53°F (11.7°C) air leaving the dehumidifier. Then, the total heat at 80°F (26.7°C) and 50 percent relative humidity = 31.15 Btu (32.86 kJ); total heat at 53°F (11.7°C) and 100 percent relative humidity = 21.87 Btu (23.07 kJ). The difference in total heat content is the pickup in the dehumidifier. Or, 31.15 – 21.87 = 9.28 Btu (9.79 kJ).

On the basis of our first trial, the weight of air circulated = (total load, Btu/min)/(heat pickup, Btu); or 13,920/9.28 = 1500 lb/min (681 kg/min). Check this result using (lb/min computed) (specific heat of air)(temperature difference, dry bulb – assumed temperature of air leaving the dehumidifier, or 53°F [11.7°C] in this case). Solving, (1500) (0.24) (80 – 53) = 9720 Btu/min (10,254.6 kJ/min). This value is not enough because the sensible heat load is larger, i.e., 10,000 Btu/min (10,550 kJ/min).

Using trial and error again, assume 54°F (12.2°C) air leaving the dehumidifier. Then, as before: Total heat at 80°F (26.7°C) and 50 percent relative humidity = 31.15 Btu (32.86 kJ); total heat at 54°F (12.2°C) and 100 percent relative humidity = 22.54 Btu (23.78 kJ). The difference = 31.15 – 22.54 = 8.61 Btu (9.08 kJ).

The air circulated is now 13,920/8.61 = 1620.2 lb/min (735.6 kg/min). Checking as before, (1620.2) (0.24)(80 – 54) = 10,110 Btu/min (10,666.1 kJ/min). The 10,110 Btu/min is slightly higher than the 10,000 Btu/min required, actually 1.1 percent. This is acceptable for usual design purposes.

4. *Find the amount of air leaving the dehumidifier*
Using the assumed 54°F (12.2°C) leaving temperature, the amount of air leaving the dehumidifier is, from step 3, 1620.2 lb/min (735.6 kg/min).

5. *Compute the quantity of air bypassed*
The quantity of air bypassed = (lb of air introduced/minute – quantity of air leaving the dehumidifier); or air bypassed = 2300 – 1620.2 = 679.8 lb/min (308.6 kg/min). The temperature of the air leaving the dehumidifier is the assumed value of 54°F (12.2°C).

Related Calculations. Strict standards have been introduced governing use of outside air in bypass air-conditioning systems. The reason for this is the increased air pollution in urban areas. In some instances, outside-air intakes have been found close to truck and bus driveways, leading to polluted air being drawn into the air-conditioning unit.

The Environmental Protection Agency publishes guidelines for allowable contaminants in outside air used for air-conditioning units. These guidelines must be used if a design is to be accepted by governing authorities. The guidelines are discussed in other calculation procedures presented in this section of the handbook.

VIBRATION POTENTIAL OF MOTOR-DRIVEN FANS

Determine if the motor-driven fan in Fig. 15 will have excessive vibration. The motor is 110 V, 60 Hz, 2400 r/min; armature weight 40 lb (18.2 kg); radius of gyration = 5 in (12.7 cm). This motor drives a 3-bladed fan weighing 10 lb (4.54 kg) with a radius of gyration of 9 in (22.86 cm); the drive shaft is steel. Is this design acceptable? If not, what changes should be made in the design?

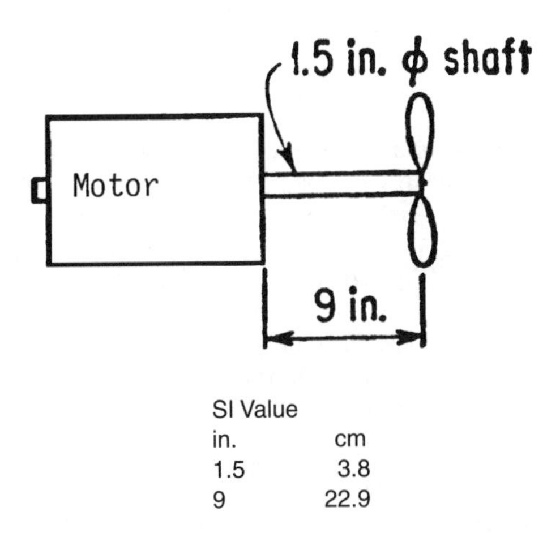

FIGURE 15 Motor-driven fan.

Calculation Procedure:

1. *Find the moment of inertia for each part of the assembly*

The motor and fan are arranged as in Fig. 15. Find the moment of inertia of the fan from I_f = (fan weight/32.2 ft/s · s) (1/12 in/ft) (radius of gyration^2); or I_f = (10 lb/32.2) (1/12)(9 × 9) = 2.094 in^4 (87.15 cm^4). For the motor, I_m = (40 lb/32.2) (1/12)(5 × 5) = 2.587 in^4 (107.68 cm^4).

The torsional constant for the steel shaft, k = (modulus of elasticity of the steel)/(radius of gyration × 0.495). Or, $k = (11.5 \times 10^6)/(9)(0.495) = 6.3 \times 10^5$.

2. *Find the frequency of the assembly*

Use the relation, frequency, $f = (½\pi)[I_f + I_m)(k)/(I_f \times I_m)]^{0.5}$ where the symbols are as defined earlier. Substituting, $f = (½\pi)\ [(2.094 + 2.587)(6.3 \times 10^5)/(2.094 \times 2.587)]^{0.5} = 117.47$ cycles/s.

The motor frequency = 2400 r/min/60 cycles/s = 40 cycles/s. There may be excessive vibration when the system starts or is shut down if the rate of increase or decrease in speed is small—say 40 cycles/s. Recognizing this, the system should be redesigned.

To redesign this system to eliminate the danger of excessive vibration, increase the radius of gyration of both the fan and the motor armature. Also, increase the shaft diameter, or decrease its length until the frequency of the entire system is below 40 cycles/s.

Related Calculations. Use this general approach to design a connected system so there is no danger of excessive vibration. While vibration may be tolerated during the starting and stopping of connected units, it is best to design the assembly so there is no vibration during any of the normal speeds encountered in the design.

CENTRIFUGAL-COMPRESSOR POWER AND ENERGY INPUT COMPUTATION

A centrifugal compressor handling air draws in 12,000 ft^3/min (339.6 m^3/min) of air at a pressure of 14 lb/in^2 (abs) (96.46 kPa) and a temperature of 60°F (15.6°C). The air is delivered from the compressor at a pressure of 70 lb/in^2 (abs) (482.4 kPa) and a temperature of 164°F (73.3°C). Suction-pipe flow area is 2.1 ft^2 (0.195 m^2); area of discharge pipe is 0.4 ft^2 (0.037 m^2) and the discharge pipe is located 20 ft (6.1 m) above the suction pipe. The weight of the jacket water, which enters at 60°F (15.6°C) and leaves at 110°F (43.3°C) is 677 lb/min (307.4 kg/min). What is the horsepower required to drive this compressor, assuming no loss from radiation?

Calculation Procedure:

1. *Determine the variables for the compressor horsepower equation*

The equation for centrifugal-compressor horsepower input is

$$\text{hp} = \frac{w}{0.707}\left[c_p(t_2 - t_1) + \frac{V_2^2 - V_1^2}{50{,}000} + \frac{Z_2 - Z_1}{778}\right] + \left[\frac{w_j(t_o - t_i) + R_c}{0.707}\right]$$

In this equation, we have the following variables: w = weight, lb (kg) of unit flow rate, ft^3/s (m^3/s) through the compressor, lb (kg), where $w = (P_1)(V_1)/R(T_1)$, where P_1 = inlet pressure, lb/in^2 (abs) (kPa); V_1 = inlet volume flow rate, ft^3/s (m^3/s); R = gas constant for air = 53.3; T_1 = inlet air temperature, °R.

The inlet flow rate of 12,000 ft^3/min = 12,000/60 = 200 ft^3/s (5.66 m^3/s); P_1 = 14.0 lb/in^2 (abs) (93.46 kPa); T_1 = 60 + 460 = 520 R. Substituting, w – 14.0(144) (200)/53.3(520) = 14.55 lb (6.6 kg).

The other variables in the equation are: c_p = specific heat of air at inlet temperature = 0.24 Btu/(lb · °F) [1004.2 J/(kg · K)]; t_2 = outlet temperature, °R = 624 R; t_1 = inlet temperature, °R = 520 R;

V_1 = air velocity at compressor entrance, ft/min (m/min); V_2 = velocity at discharge, ft/min (kg/min); Z_1 = elevation of inlet pipe, ft (m); V_2 = elevation of outlet pipe, ft (m); w_j = weight of jacket water flowing through the compressor, lb/min (kg/min); t_j = jacket-water inlet temperature, °F(°C); t_o = jacket-water outlet temperature, °F (°C).

The air velocity at the compressor entrance = (flow rate, ft³/s)/(inlet area, ft²) = 200/2.1 = 95.3 ft/s (29 m/s); outlet velocity at the discharge opening = 200/0.4 = 500 ft/s (152.4 m/s).

2. *Compute the input horsepower for the centrifugal compressor*
Substituting in the above equations, with radiation losses, $R_c = 0$,

$$\text{hp} = \frac{14.55}{0.707}\left[0.24(624-520) + \frac{500^2 - 95.3^2}{50{,}000} + \frac{20}{778}\right] + [677/60 \times (110-60)]/0.707$$

$$= 20.6(24.95 + 4.8 + 0.0256) + 797 = 1409 \text{ hp } (1051 \text{ kW}).$$

Related Calculations. This equation can be used for any centrifugal compressor. Since the variables are numerous, it is a wise procedure to assemble them before attempting to solve the equation, as was done here.

MOISTURE EVAPORATION FROM OPEN TANKS

A paper-mill machine room produces 50 tons (45.5 t) of finished paper/day. Studies show that 1.5 lb (0.68 kg) of water must be evaporated for every pound (kg) of finished paper as the paper goes over the dryer rolls. What capacity exhaust fan is needed if the room conditions are 100°F (37.8°C) and 40 percent relative humidity and tempered air enters the room at 70°F (21.1°C) and 50 percent relative humidity? Determine the air flow and exhaust fan capacity required if the room temperature remains at 100°F (37.8°C), but the exhaust relative humidity of the exhaust air could be raised to 60 percent.

Calculation Procedure:

1. *Determine the amount of water evaporated into the room atmosphere/unit time*
With 50 tons (45.5 t) of paper being produced/24 h, the weight of paper = 50(2000) = 100,000 lb (45,400 kg). Then, the amount of paper produced/minute = (100,000 lb)/(24 h) (60 min/h) = 69.4 lb (31.5 kg)/min. Since water is evaporated at the rate of 1.5 lb/lb (1.5 kg/kg) of paper, the total evaporation rate = 1.5 (69.4) 104.1 lb (47.3 kg)/min.

2. *Find the amount of moisture removed/unit of air flow*
From the psychrometric chart or a table of air properties, find the moisture content of the entering and leaving air for this room. Thus, at an entering air temperature of 70°F (21.1°C) and 50 percent relative humidity, each 100 ft³/min (2.83 m³/min) contains 0.059 lb (0.0268 kg) of moisture. The leaving exhaust air at 100°F (37.8°C) and 40 percent relative humidity contains 0.117 lb (0.053 kg) of moisture/100 ft³/min (2.83 m³/min). The moisture absorbed by the air during passage through the room therefore is 0.117 – 0.059 = 0.058 lb/100 ft³/min (0.026 kg/2.83 m³/min).

3. *Compute the air flow required to remove the moisture generated*
The air flow required = (quantity of moisture to be removed, lb or kg/unit time)/(moisture absorbed/unit time, lb or kg). For this plant, with a total evaporation rate of 104.1 lb (47.3 kg)/min, air flow required = (104.1/0.058)(100) = 179,482.8 ft³/min (5079.3 m³/min). An exhaust fan with a capacity of 180,000 ft³/min (5094 m³/min) would be chosen for this application. If one fan of this capacity was too large for the space available, two fans of 90,000 ft³/min (2547 m³/min) could be

chosen instead. Any other combination of capacities that would give the desired flow could also be chosen.

4. ***Calculate the air flow required with a higher exhaust relative humidity***
With the room temperature remaining at 100°F (37.8°C) but the exhaust air at 60 percent relative humidity, the moisture content would be 0.175 lb (0.079 kg)/100 ft^3/min (2.83 m^3/min). Then, the new absorption, as before, = 0.0175 – 0.058 = 0.117 lb/100 ft^3/min (0.053 kg/2.83 m^3/min). Then, air flow required = (104.1/0.117) (100) = 88,974 ft^3/min (2517.9 m^3/min). Thus, we see that the required air flow is reduced by (179,483 – 88,974) = 90,509 ft^3/min (2561 m^3/min), or 50.4 percent. An exhaust fan capacity of 90,000 ft^3/min (2547 m^3/min) would be chosen for this installation.

Related Calculations. This procedure can be used for any installation where airborne moisture must be removed from a closed space, such as a factory, meeting room, ballroom, restaurant, indoor swimming pool, etc. In every such installation, the required air flow to remove a given quantity of moisture will decrease as the relative humidity of the exhaust air is increased. Reducing the required air flow will save money in several ways: on the initial cost, installation cost, operating cost, and maintenance cost of the exhaust fan(s) chosen. Hence, it is important that the engineer carefully analyze the entering and leaving air relative humidity or moisture content.

Selecting an "environmentally gentle" fan of lower capacity requiring a smaller power input is a design objective of many businesses and institutions today. Hence, careful analysis of conditions in the installation will be rewarded with reduced overall and life-cycle costs while meeting environmental goals.

Where steam or moisture is released to working spaces from open tanks or similar vessels, with resultant high humidity conditions, the engineer is faced with the problem of estimating the rate of evaporation and providing for a reduction in the moisture content of the room air. The procedure above shows one popular way to control the room moisture content.

For steam escaping into a closed space, the escape rate is easily calculated. However, moisture given off by industrial processes is not as easily computed. Further, there isn't much information in the technical literature covering moisture generation by industrial processes. High humidity affects worker comfort and can produce mild to severe condensation problems with moisture dripping from walls and ceilings. Typical industries with this "wet" heat problem and the requirements to reduce it are:

Textile industry: Elimination of fog and condensation in dye houses, bleacheries, and finishing departments. *Paper and pulp mills:* Elimination of high vapor generation in machine, heater, and grinder rooms. *Steel and metal goods:* Elimination of excessive vapor generation in pickling rooms. *Food industries:* Control of large vapor generation in kettle, canning, blanching, bottle washing, and bottle-filling areas. *Process industries:* Control of vapor generation in electroplating, coating, and chemical processing.

FAN AND PUMP ENERGY PERFORMANCE ANALYZED FROM MOTOR DATA

Determine the air flow from a fan driven by a 460-V motor when the current draw is 7 A, the fan delivers air at a pressure of 4-in (10.2-cm) water gage, if the fan and motor efficiencies are 65 and 90 percent, respectively, and the power factor = 0.80. Likewise, determine the efficiency of a pump delivering 90 gal/min (5.7 L/s) at a 1000-lb/in^2 (6890-kPa) pressure differential when the motor current is 100 A, voltage is 460 V, power factor = 0.85, and motor efficiency = 90 percent.

Calculation Procedure:

1. ***Compute the rate of air delivery by the fan***
It can be shown that the relationship between fan horsepower and motor power consumption gives the equation q = (14.757)(motor efficiency)(EI)(cos ϕ)(fan efficiency)/(fan pressure developed, in

water), where q = air delivery rate, ft^3/min (m^3/min); E = motor voltage; I = motor current flow, A; cos ϕ = power factor. Substituting, q = (14.757) (0.90) (460) (7) (0.8) (0.65)/4 = 5559.6 ft^3/min (2624 L/s).

With the computed delivery rate known, we can now easily compare the actual fan efficiency against manufacturer's guarantee data. If the results do not agree with the guarantee, suitable action can be taken.

2. *Calculate the pump efficiency*
It can likewise be shown that pump efficiency, P_e = (flow rate, gal/min)(pressure differential across the pump, lb/in^2)/4(EI)(motor efficiency)(power factor). Or, motor efficiency = (90) (1000)/(4) (460) (100) (0.90) (0.85) = 0.6393; say 64 percent.

Using this computed efficiency, we can refer to pump efficiency curves to see if the plotted efficiency at the flow and head agree. If the computed efficiency is significantly different from the plotted value, then further investigation of the pump performance is warranted.

Related Calculations. For fans used for forced-draft boiler applications, it can be shown, by using the MM Btu (kJ) method of combustion analysis, that for a fixed boiler output the air flow required in lb/h (kg/h) is a constant at a given excess air requirement. Hence, irrespective of density conditions, the same mass flow of air must be delivered to the burner.

Further, a boiler's backpressure in inches (cm) of water is a function of mass flow of air and its density. Hence, as the density of air decreases (as at higher temperatures or altitudes), the pressure head to be delivered to a boiler increases, while the mass flow and head that fan can deliver decreases. Thus, fan performance must be checked at the lowest density conditions. This is an important point that should not be overlooked in practical performance calculations.

A pump delivers the same flow in gal/min (L/s) and head in feet (m) of liquid at any temperature. However, because of changes in fluid density, the flow in lb/h (kg/h), pressure in lb/in^2 (kPa), and brake horsepower (kW) input will change. Thus, a boiler feed pump which must deliver a stated flow in lb/h (kg/h) at a given lb/in^2 (kPa) will require a larger power input as the fluid density decreases. Likewise, if a pump must deliver a given flow in gal/min (L/s) at a given head in feet (m), then as the density of the fluid decreases the brake horsepower input also decreases. However, this is not the situation in boiler applications.

This procedure is the work of V. Ganapathy, Heat Transfer Specialist, ABCO Industries, Inc.

AIR-BUBBLE ENCLOSURE CHOICE FOR PLANNED USAGE

Choose a suitable air-bubble enclosure for a maintenance shop requiring 100,000 ft^2 (9290 m^2) of covered area. The enclosure will house 75 maintenance workers in a southwestern area of the United States where daytime temperatures can reach 110°F (43.3°C) and nighttime temperatures can fall to 0°F (–17.8°C). Determine the size, and number, of blowers required for this enclosure, probable power consumption, enclosure cost, and precautions in construction and use of an air-bubble enclosure.

Calculation Procedure:

1. *Determine the general requirements of the air-bubble enclosure*
Most air-bubble enclosures are rectangular in shape, though there are no restrictions on square and round shapes. Typical plastic-reinforced fabric structures range in size from 80 to 300 ft (24.4 to 91.4 m) wide, 80 to 450 ft (24.4 to 137.2 m) long, and 30 to 75 ft (9.1 to 22.9 m) high. Width constrains the size of the structure, and the limits of material strength determines the structure's width.

The cost of air-bubble structures varies from \$6.00 to \$11.00/ft^2 (\$65 to \$118/m^2), compared to \$36 to \$80/ft^2 (\$388 to \$861/m^2) for a prefabricated light-duty steel building. With 100,000 ft^2 (9290 m^2) enclosed space, and the dimension constraints given above, this enclosure could be 250 ft wide by 400 ft long (76.2 by 122 m), depending on the size of plot available and its access roads. The enclosure height will depend on the clearance required for the machinery needed for the maintenance work performed in the enclosure. For this enclosure, we'll assume that the height of the enclosure is 50 ft (15.2 m).

2. *Find the capacity, and number, of blowers required by the enclosure*
Air leakage from an air-bubble enclosure is typically 6000 to 8000 ft^3/min (2832 to 3776 L/s). This air leakage occurs mainly at doors, air locks, and perimeter anchor points. Electrical operating cost to drive the blower to replace air leakage and keep the enclosure inflated depends on how often the air is changed in the enclosure. The frequency of air changes should be suited to the requirements of the work done in the enclosure; for example, if painting is being done in the enclosure, additional ventilation is necessary.

A positive interior pressure must be maintained to keep the fabric rigid. The enclosure inflation system must automatically respond to wind gusts, equipment failures, power losses, fabric ruptures, and other unpredictable events. For a long-term air-bubble enclosure, such as a maintenance shop, the inflation system should consist of a primary, a secondary, and an emergency blower and a vent system.

When specifying air-change frequency, the design engineer should take into account such considerations as expected air leakage, the inside temperature during the summer, the diluent effect of the inside air volume, and the vehicular traffic, if any, inside the enclosure. Three to four air changes/hour have been shown to be adequate for a 100,000-ft^2 (9290-m^2) air-bubble enclosure.

The primary inflation system should replace normal air losses and high winds. For a typical 100,000-ft^2 (9290-m^2) air-bubble enclosure the internal air pressure is 0.85 to 0.95 in water (21.6 to 24.1 mm). A typical inflation blower for 3 air changes/hour for this enclosure would be a 36,000 ft^3/min (16,992 L/s) unit with a 15-hp (11.2-kW) drive motor. The backup blower would have a capacity equal to the primary blower. An emergency blower of 9000 ft^3/min (4248 L/s) would be chosen to handle somewhat more than the expected air losses.

3. *Determine what types of controls are suitable for the blowers*
Simple, flexible controls are normally specified for air-bubble enclosure blowers. Typically, primary blowers are controlled by an on-off switch, are run continuously, with louvre vents located at a distance from the blowers on the perimeter of the enclosure. Vent cycling open and closed is regulated by a pressure switch.

Design all vents so they are regulated by a pressure switch. Further, all vents should be spring-to-close upon power failure and have grease fittings for easy lubrication. In the event of a sudden large pressure drop within the enclosure, the vents will stay closed to maintain the internal pressure in the enclosure. If the pressure loss is catastrophic, i.e., larger than the primary blowers can handle—the secondary blowers should have switches to turn them on automatically.

Secondary blowers are sometimes activated manually, for example when trucks are operating inside the enclosure, but are usually left on automatic control. Where there is more than one secondary blower, they start automatically in sequence as needed to maintain the needed internal pressure for the enclosure. Each blower has a vent located elsewhere on the perimeter that cycles open and closed as regulated by a pressure switch. Normal controls include a high-pressure shutdown switch and a backup high-pressure shutdown. All blowers should have gravity-operated back-draft dampers to prevent air loss when they are not operating.

4. *Choose a suitable emergency power supply and controls*
The emergency inflation system for an air-bubble enclosure should be completely automatic to ensure the structure's stability in the event of power loss. There has been disagreement among designers and engineers over whether to rely on generators or directly driven internal-combustion-engine blowers, or both. To simplify maintenance, use the least number of units possible in an emergency system. At the same time, the emergency unit must start reliably.

For these reasons, the emergency system often consists of two propane-fueled internal-combustion engines that directly drive the blowers. As noted above, the typical emergency blower has a capacity of 9000 ft^3/min (4248 L/s). The control system includes automatically charged batteries for reliable starting of the engine driving the blower.

To prevent overpressurization, a critical design consideration, a high-pressure switch for shutting down the main and emergency blower motor can be coupled with a low-pressure switch for restarting it. Such a system should be backed up with a completely independent high-pressure switch system having separate air-sensing tubing. Relief vents can present resealing problems and can be accidentally opened during storms or if the structure moves.

5. *Select air locks, fabric, drainage, inside temperature, and lighting*

Size air locks to handle the largest equipment that will enter the enclosure. Thus, some enclosures have air locks permitting the entrance of large trucks and even earth-moving equipment.

For vehicles the air lock usually consists of outer and inner doors, electrically interlocked to prevent them from being opened simultaneously. Each air-lock door should be equipped with a manual chain-operated opener. Specify a heavy roll-up type door. Roll-up flaps allow doors and air locks to be removed. Personnel doors are usually the revolving type. Avoid swinging personnel doors because they add stress to the structure and could leak air.

For a 100,000-ft^2 (9290-m^2) structure, the fabric may have to be three or four sections. The structural design of an air-bubble enclosure should meet local building codes. Such codes may include fabric specifications.

The typical air-bubble enclosure is designed to withstand winds of 80 mi/h (128.8 km/h), with gusts up to 100 mi/h (161 km/h). Snow loads can be up to 10 lb/ft^2 (48.9 kg/m^2) which will be produced by 12 to 14 in (30.5 to 35.5 cm) of snow or 2 in (5.1 cm) of ice. The contour of the structure and its flexibility cause snow and ice to slide off. Also, the internal pressure may be boosted to offset snow or wind loading.

Air-bubble structures can be made from many types of fabrics. If a structure life of more than 2 years is intended, the options become limited because of ultraviolet deterioration. This can be prevented by bonding a polyvinylfluoride film to the exterior of the fabric. While this coating is expensive, it can be justified by the increased service life of the fabric.

Selecting fabrics is not easy because of the variety of reinforcement methods for distributing the structural loading to the anchors. The fabric should be chosen so it resists dry rot, mildew, and weathering. Fabrics for structures over 200-ft (61-m) wide must be more carefully selected than usual because of the higher stresses. Of proven quality is a 28-oz/yd (0.86-kg/m) coated vinyl-polyester fabric.

Sewn seams should be avoided to limit stretching, and air and water leakage. Fusion-welded seams are generally preferred. A double-layered fabric should be considered if the structure will be air conditioned. Manufacturers of air-bubble enclosures and fabrics provide repair kits with complete instructions for plant personnel.

A reinforcement system over the fabric consisting of a metallic net or a web of high-strength plastic material can be used to help distribute the structural loading to the enclosure anchors. The same considerations that apply to fabric selection apply to the reinforcement system. Further, the reinforcement system should not expose the fabric to wear or tear. For example, a cable system, which is preferred for structures over 200-ft (61-m) wide, should be coated with a thermoplastic. The engineer designer should ensure that if one or two cables or webs break, a domino effect will not be created that will lead to a structural failure. Design of the structure should relieve stresses in all directions as much as possible.

Condensate can form on the inside surface of an air-bubble enclosure if it is not double-walled or if the inside is not air conditioned. Specify double walling if moisture drippage cannot be tolerated. For a typical 100,000-ft^2 (9290-m^2) air-bubble enclosure in the southern United States, about 1000 gal/day (3790 L/day) of condensate can form during spring or autumn. Interior gutters can be installed to collect condensate and route it outside.

Specify allowable water and air leakage. Air-bubble enclosures usually allow little leakage. For this 100,000 ft^2 (9290 m^2) enclosure, water leakage has been estimated at less than 10 gal (37.9 L) during a 6-in (15.2-cm) rain storm. Water leaks are usually confined to areas around air locks, doors, and anchors. Such leaks can be minimized by close inspections during construction and start-up.

The inside temperature of an uninsulated and unventilated air-bubble enclosure can be 10°F to 20°F (5.6°C to 11.1°C) higher than outside. Three or four air changes an hour might hold the inside temperature to 10°F (5.6°C) above ambient.

The typical air-bubble fabric is very translucent. On a clear day, the fabric can let in an average of 300–400 footcandles (1028 to 1370 cd/m^2), almost nine times the light required inside a typical warehouse. More light can be let in by placing clear panels in the fabric. However, translucency must be limited as necessary to reduce solar heat gain. Because the enclosure fabric is highly reflective, a few lights are adequate at night.

6. *Choose a suitable anchoring system*
Geological soil data, including soil bearing pressure, at the site are needed for designing the anchoring. Good anchoring helps ensure the structure's reliability and extends its life. Among the anchors used are steel helicoils, concrete piers, and the continuous-grade beam. Helicoils and piers are adequate for short-term structures. The continuous-grade beam is recommended for long-term ones.

Although the continuous-grade beam is the most expensive type of anchor, it minimizes the likelihood of the domino type of simultaneous reinforcement failures. Further, such beams divert rainwater effectively, limiting its entry. This is accomplished by sloping the top of the grade beam at least 1.5 to 2 in (3.8 to 5.1 cm). The beam also permits continuous clamping of the enclosure fabric.

A continuous rope-bead edge on the fabric ensures a structurally sound clamping of the fabric to concrete. The gap between the fabric and concrete is sealed by a gasketed steel angle. Nothing, such as signs, fences, gates, etc., should be closer than 2.5 ft (0.76 m) from the fabric to keep it from being torn during deflations for maintenance work. The surrounding area, outside of the grade beam, should slope away to run off rainwater.

Related Calculations. The general principles presented in this procedure are valid for air-bubble enclosures used for industrial purposes: shops, warehouses, construction shelters, materials-storage areas, conveyor housings, etc. In such applications, they can have important environmental uses because of the Resource Conservation and Recovery Act which regulates pollution resulting from rainfall. Long-term air-bubble enclosures can be insulated and the enclosed space can be heated and air conditioned. Such a structure can last more than 20 years.

Other applications of air-bubble enclosures include sports facilities for tennis, indoor golf, running tracks, swimming pools, etc. The same general principles apply for their design and application. Agricultural uses include greenhouses, farm markets, and farm animal housing.

Air-bubble enclosure manufacturers can be extremely helpful in the planning and layout of these structures. However, since typical enclosures require the usual building services, electric, water, sanitation, heating, and possibly air conditioning, a project engineer should be assigned to supervise the installation. It is the task of the project engineer to coordinate the work of civil, mechanical, electrical, and other engineers assigned to the successful installation of the air-bubble enclosure.

This procedure is the work of Charles W. Hawk, Jr., P.E., Project Manager, Olin Corporation's Southeast Engineering Group, as reported in *Chemical Engineering* magazine. SI values were added by the handbook editor.

EXPANSION-TANK SIZING FOR HYDRONIC SYSTEMS

Determine the needed expansion-tank volume for a hydronic heating system containing 10,000 gal (37,900 L) of water, having a fill pressure of 15 lb/in^2 (gage) (104.4 kPa) at the tank, with the maximum pressure limit set at 25 lb/in^2 (gage) (172.3 kPa), designed to operate over a temperature range of 70°F (21.1°C) at fill to 220°F (104.4°C) during operation. Show the various equations which can be used and the results obtained with each.

Calculation Procedure:

1. *Select the equations which can be used for expansion-tank sizing*
Derivation of equations for sizing expansion tanks in hydronic systems is fundamental if it is assumed that the air cushion in the tank behaves as a perfect gas. For such equations, all the

TABLE 2 Equations for Expansion-Tank Sizing

Equation no.	Equation	Assumptions	Example tank size, gal (L)
1	$V = \dfrac{V_w[(v_2/v_1 - 1) - 3\alpha\Delta t]}{(p_a/p_1 - p_a/p_2)}$	• Air compresses isothermally (t_1) °F (°C) • Water in tank is at temperature t_1 °F (°C) • Initial air charge is atmospheric	3314.6 gal (12,562 L)
2	$V_T = \dfrac{V_w[(v_2/v_1 - 1) - 3\alpha\Delta t]}{(p_a/p_1 - p_a/p_2) - (v_2/v_1 - 1)(1 - p_a/p)}$	• Air compresses isothermally (t_1) °F (°C) • Water in tank is at temperature t_2 °F (°C) • Initial air charge is atmospheric	4033.9 gal (15,288.8 L)
3	$V_T = \dfrac{V_w[(v_2/v_1 - 1) - 3\alpha\Delta t]}{(1 - p_1/p_2)}$	• Air compresses isothermally (t_1) °F (°C) • Initial air charge is at pressure p_1	1639.6 gal (6214.1 L)

where, with volumes in consistent units:

V_w = volume of water in system (piping, heat exchangers, etc.),* gal (L)
V_T = volume of expansion tank, gal (L)
p_a = atmospheric pressure, lb/in^2 (abs) (kPa)
p_1 = pressure at lower temperature, lb/in^2 (abs) (kPa)
p_2 = pressure at higher temperature, lb/in^2 (abs) (kPa)
t_1 = lower temperature, °F (°C)
t_2 = higher temperature, °F (°C)
v_1 = specific volume of water at temperature t_1, °F (°C)
v_2 = specific volume of water at temperature t_2, °F (°C)
α = linear coefficient of thermal expansion, 1/°F (1/1.8 °C)
Δt = higher temperature – lower temperature, °F (°C)

*At t, and not including water in the tank.
HPAC Magazine.

necessary values that are not established as design parameters are readily available from the steam tables.

The only complications that may be met with such equations are those relative to how the tank is used in the system. For example, if it is assumed that the water in the expansion tank always remains at its initial temperature, that compression and expansion of the air in the tank is isothermal, and that the air in the tank was initially compressed from atmospheric pressure in the tank, Eq. 1, from Table 2, is readily derived and applies. Thus, if the designer uses Eq. 1 and anticipates system performance to be in accordance with the design, all possible steps must be taken to ensure that the actual design satisfies the assumptions. This might include leaving the tank uninsulated in such a way that thermal circulation between the tank and piping will be minimal.

To see the results obtainable with Eq. 1, substitute as follows: V_T = 10,000{[(0.01677/0.01606)] – 1 – 3 (0.000006)(220 – 70)}/[(14.7/29/7) – (14.7/39/7)] = 3314.6 gal (12,562 L).

2. *Find the expansion-tank volume with different design assumptions*

If the design assumptions are that (*a*) the initial charge of water in the expansion tank changes temperature with the main volume of water, (*b*) that the air in the tank is at its initial charge temperature and compresses and expands isothermally, and (*c*) that the air in the expansion tank was initially compressed from atmospheric pressure, Eq. 2 results (see Table 2).

Equation 2 becomes a bit more complex if the air in the tank is assumed to increase in temperature with the liquid (note that the total pressure of the gas in the tank is the sum of the partial pressures of the air and water vapor and that a saturated condition always exists). Although not totally accurate, Eq. 2 would be a fair approximation of the condition where a portion of a thermal storage tank is used to provide the expansion cushion.

Substituting in Eq. 2, using data from Eq. 1, we find that $V_T = 4033.9$ gal (15,289 L). Note that the expansion-tank volume given by Eq. 2 is 100 (4033.9 – 3314.6)/4033.9 = 17.8 percent smaller than that computed by Eq. 1. This difference is the result of the change in the assumptions made between the two designs.

3. *Determine expansion-tank volume with a precharged diaphragm*

If the initial charge in the tank is not compressed from atmospheric pressure in the tank itself, but rather is forced into the tank at a design operating pressure (either from a compressed-air system or as a recharged diaphragm type) and the air is assumed to compress and expand isothermally, Eq. 3 results (see Table 2).

Substituting in Eq. 3, we find $V_T = 1639.6$ gal (6214.1 L). Again, this differs from the two earlier computed values because of the changes in the design assumptions made. Further, the first cost of a recharged tank will be higher than that of a simple storage tank with no special attachments or fittings.

Related Calculations. The calculated expansion-tank sizes for the three different conditions show how much the computed sizes can vary. These significant variations indicate that in the selection of an expansion tank for a hydronic system the designer must (1) determine what operating assumptions are to be used in selecting the tank sizing equations, and (2) design the system to achieve the assumed conditions as closely as possible.

While a number of other equations for sizing expansion tanks have been developed by various authorities, the equations, in general, resemble those presented here. The reason for this is that the variables—water volume, pressure, and temperature—enter all the equations developed for the analysis. Further, the results closely approximate those found here.

From the standpoints of thermodynamics and hydraulics, the equations given in this procedure can be used to size expansion tanks with an equal degree of accuracy for hot-water heating systems, chilled-water systems, and dual-temperature water systems. When a tank with a liquid-gas interface is used in a chilled-water system, water logging may occur. Extreme precaution must be taken in design because in such systems there is a continual pumping effect that removes air from the tank by absorption in the water. This causes the small tanks designed by the equations in this procedure to water log frequently.

One option to avoid water logging is to provide an oversize tank to minimize the frequency of needed air charging. Another option is to design such systems to prevent the absorption phenomenon, but this can have numerous other detrimental effects on the system.

The fundamental components of any hydronic system are (*a*) the heat source, (*b*) the load, (*c*) the circulator, (*d*) piping, and (*e*) the expansion tank. Strangely enough, the most complex device of the five, the expansion tank, may appear to be the least complicated.

The expansion tank serves a dual purpose in the hydronic system. It allows for the volumetric changes in the fluid, resulting from fluid temperature changes, to occur between designed pressure limits. Further, the expansion tank establishes the point of constant or known pressure in the hydronic system. And, in many hydronic systems, the tank serves the additional purpose of being an integral part of the air control subsystem.

The correct sizing of expansion tanks is becoming even more critical as larger-volume hydronic systems are being designed and built. Not only are large volumes a result of higher capacity systems, but they are also an integral part of solar systems and other power-conserving installations that utilize thermal storage through liquid-phase temperature changes. The sizing of an expansion tank, as noted by Lockhart and Carlson,* relates not only to the volume of fluid in the hydronic system, the temperature limits, and the pressure limits, but also to how the tank is designed into the system.

This procedure is based on the work of William J. Coad, Vice-President, Charles J. R. McClure & Associates, and Affiliate Professor of Mechanical Engineering, Washington University, as reported in *HPAC* magazine. SI values were added by the handbook editor.

Additional design pointers given by Jerome F. Mueller, P.E., are as follows: In any hydraulic system there is a basic requirement to keep system pressure at a desired normal operational level. The primary

---

*Lockhart, H. A. and Carlson. G. F., "Compression Tank Selection for Hot Water Heating Systems," *Heating/Piping/Air Conditioning*. April, 1953.

objectives of the expansion-tank system are to limit the pressure of all the equipment in the system to the allowable working pressure, to maintain minimum pressure for all normal operating temperatures, to vent air, to prevent cavitation at the pump suction and the boiling of system water, and to accomplish all of this with a minimum addition of water to the system.

In this book the most common basic types of expansion-tank systems are depicted. The first type is a system sized to accommodate the volume created by the water expansion, with sufficient gas space to keep the pressure range within the design limits of the system. Air is normally used and the system must remain as tight as possible because every recharging cycle introduces additional oxygen with the air and thus promotes corrosion. In the initial start-up, the oxygen reacts with the system components, leaving basically a nitrogen atmosphere in the expansion tank. The initial fill pressure set a minimum level of water in the tank. The compression of the gas space because of water expansion determines the maximum system pressure.

An alternative to this arrangement is the diaphragm-type expansion tank, which is precharged to the system fill pressure and sized to accept the normally expected expansion of water. The tank's air charge and the system water are permanently separated by a flexible elastomer diaphragm. For a high temperature hot-water system the pressurization may be either by steam or inert gas, commonly nitrogen. Other details will depict nitrogen pressurization systems for high-temperature hot-water applications.

For all expansion-tank systems the location of the system pump in relation to the expansion-tank connection determines whether the pump pressure is added to or subtracted from the system static pressure. This is due to the fact that the junction of the tank with the system is the point of no pressure change whether the pump is in operation or not.

In our details, which depict common expansion-tank situations, we show the expansion tank on the suction side of the pump. The reason for this location is that when the pump is discharging away from the boiler and the expansion tank, the full pump pressure will appear as an increase at the pump discharge. All points downstream will show a pressure equal to the pump pressure minus the friction loss from the pump at that point. The fill pressure needs to be only slightly higher than the system static pressure. If, on the other hand, the pump discharges into the boiler and the expansion tank, which is common in small residential and small commercial systems, full pump pressure is reduced and system fill pressure is increased. Therefore, until the fill pressure is higher than the pump pressure, a vacuum can be created in the system. Normally, a pump discharging into the boiler and a pressure tank system is used only in low-rise buildings, small systems, or single-family residences in which the pumps need to have only a low total head capability.

Sizing of expansion tanks is generally based on a system determined from the standard ASME formula that is shown in the *ASHRAE Systems Handbook* and other publications. The size of the expansion tank is determined by the volume of water in the system; the range of water temperatures normal to the operation of the system; the pressure of air in the expansion tank when the fill water first enters the tank; the relation of the height of the boiler to the high point of the system, which is usually but not always the item in the system with the lowest working pressure; the characteristics of the expansion tank; the high point of the system; the pressure developed by the circulating pump; and the location of the circulating pump with respect to the expansion tank and the boiler. Finally, it should be noted that any time there is a change in water temperature in a hot- or chilled-water system, an expansion tank should be used. Expansion tanks are therefore needed not only in hot-water systems and run-around coil heat-recovery systems used for energy conservation designs but also in chilled-water systems.

Factory Pressurized Tanks. Figures 16 through 18 illustrate the application of a factory pressurized tank with an expandable diaphragm. Figure 16 shows a typical expansion-tank installation in a system that supplies either hot or chilled water to the system. The boiler pump shown is seemingly inconsistent with the arrangement described in our previous discussion of expansion-tank location, but in this application the pump is performing a specific task. It is used to provide circulation in the boiler under certain special conditions.

A system pump by definition supplies the system. Note that on the chiller side we again have a special circulating pump for the chiller. The object of these two separate pumps is not merely to serve the system but also to balance and maintain the flow from within the chiller and the boiler

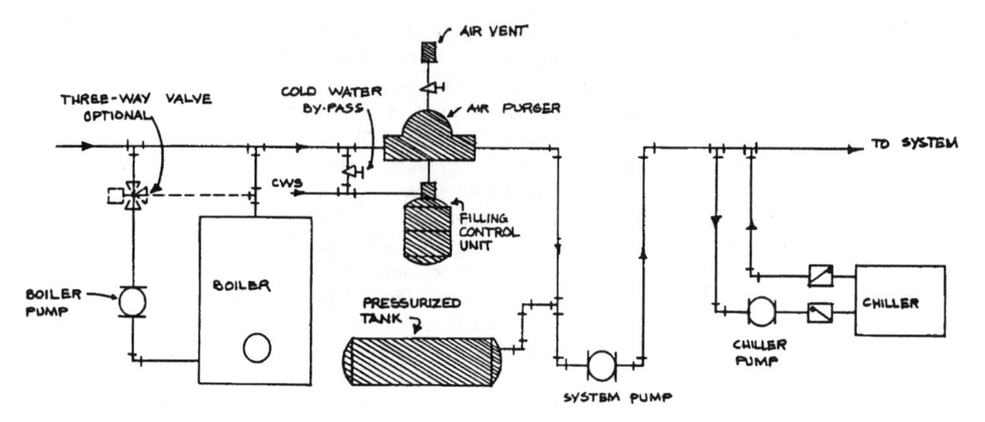

FIGURE 16 Factory-pressurized expansion tank with dual temperatures and separate boiler and chiller pumps. (*Jerome F. Mueller.*)

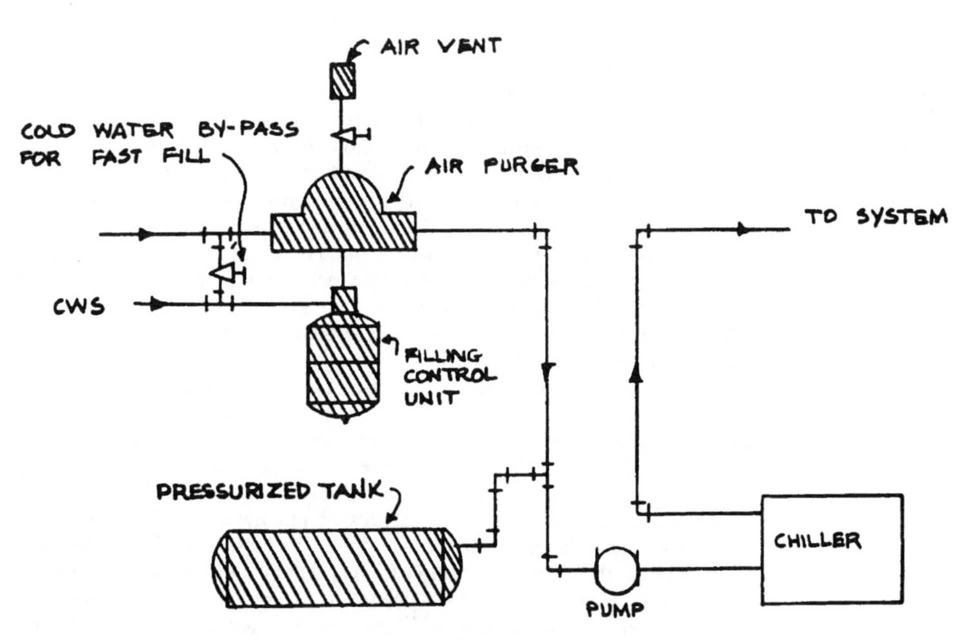

FIGURE 17 Factory-pressurized tank system for chilled-water supply. (*Jerome F. Mueller.*)

under varying overall flow and temperature conditions. Note that in this detail, as in many others, the items related to the expansion tank are described. For example, in the supply of makeup water there is a filling control unit tied into the return side of the system. The system passes through an automatic air purger with a manual or automatic vent. Finally, note that the pressurized tank is tied into the suction side of the system pump.

Chilled-Water Expansion Tank. An expansion tank is needed in a chilled-water system, as well as in a hot-water system. Figure 17 shows the application of an expansion tank to a chilled-water

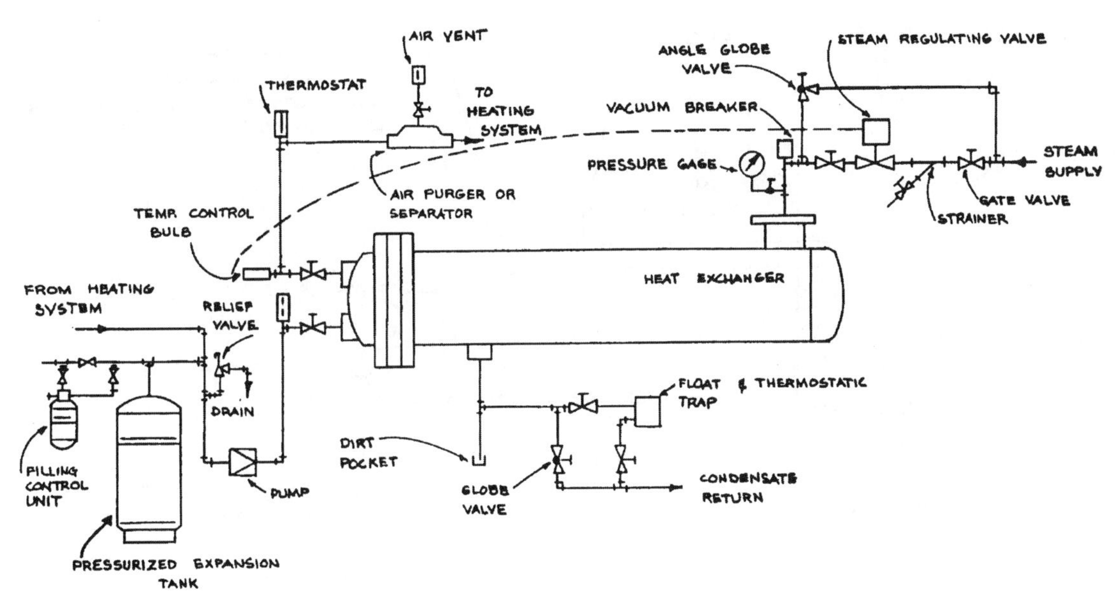

FIGURE 18 Converter/factory-pressurized expansion-tank installation. (*Jerome F. Mueller.*)

system only. This application is common when the chilled water and hot water are in separate piping systems or are otherwise separated. Again, the return fluid is passed through an air purger and vent to remove air. There is an automatic control of makeup water in the filling control unit. The pressurized tank, shown in previous details, is illustrated again in this detail.

In Fig. 18 we show the application of an expansion tank in the usual steam or hot-water converter installation. The converter is a steam converter. We show the basic steam piping connections to the converter. At this point, whether the device is a converter or a boiler, the basic premise of the system is still as shown. System water is pumped into the heat exchanger. Frequently the way to make the heat exchanger perform most efficiently is to create flow under pressure. The vent on the system is at the high point on top of the heat exchanger. It might also be noted that the pumping system for the converter is often on the floor of the equipment room, and the heat exchanger, sometimes separated from the converter by a considerable distance, is at the ceiling.

Air-Source Expansion Tank. In Figs. 19 and 20 we illustrate the air-source expansion tank. Here, there is no diaphragm separating air from water. This arrangement has been common for expansion-tank installation in many applications for a very long time. The boiler pumps and other devices are not clearly shown but are implied with the notation to pump suction and with the obvious notation that the air strainer is on the suction line of the pump. As can be seen, all the connections rise vertically and pitch up to the expansion tank and to the cold-water fill line. The makeup water and expansion-tank lines are tied into the air separator, which is designed for this particular type of system. Both the air separator and the expansion tank have drain valves to drain excess water from the expansion tank. The cold-water fill line shown has a pressure-reducing valve to supply makeup water to the system at an acceptable pressure.

Figure 20 is a nearly identical installation that is used for a special system. When snow-melting or run-around energy conservation coils are used, the system fluid is a combination of glycol and water. The air type of system has the advantage of not having a diaphragm that can be affected by the corrosive nature of this commonly used nonfreeze solution. The detail is similar to the one in Fig. 19 except that there is no cold water makeup. Instead there is a glycol makeup which comes from a special pumped glycol-water solution tank.

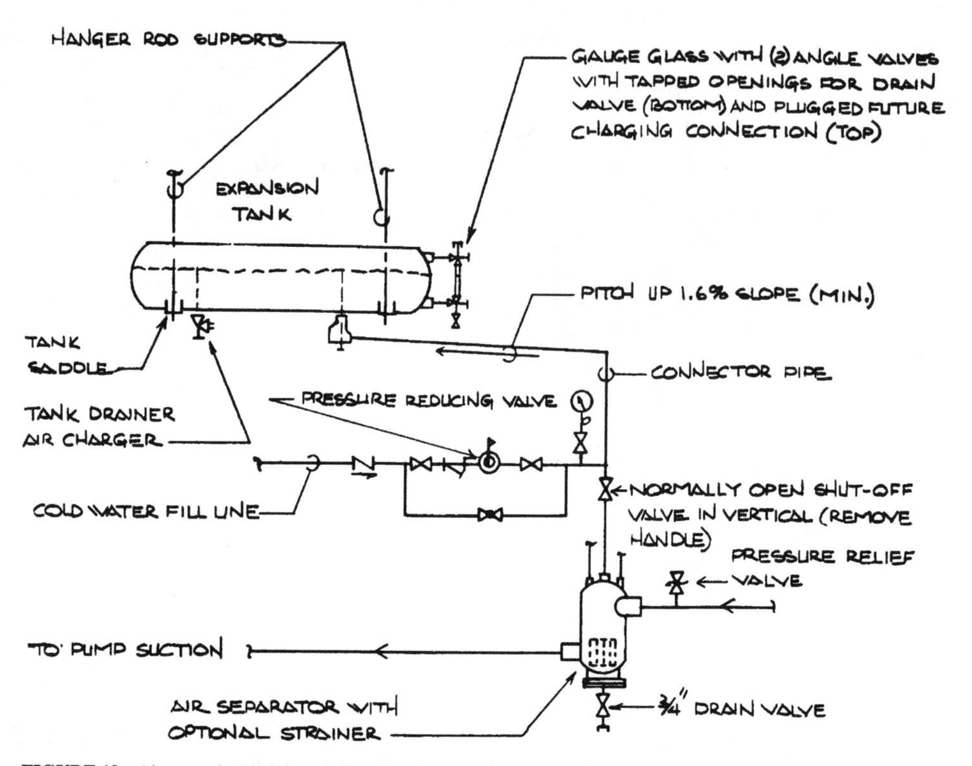

FIGURE 19 Air control and piping connections for water-system expansion tank. (*Jerome F. Mueller.*)

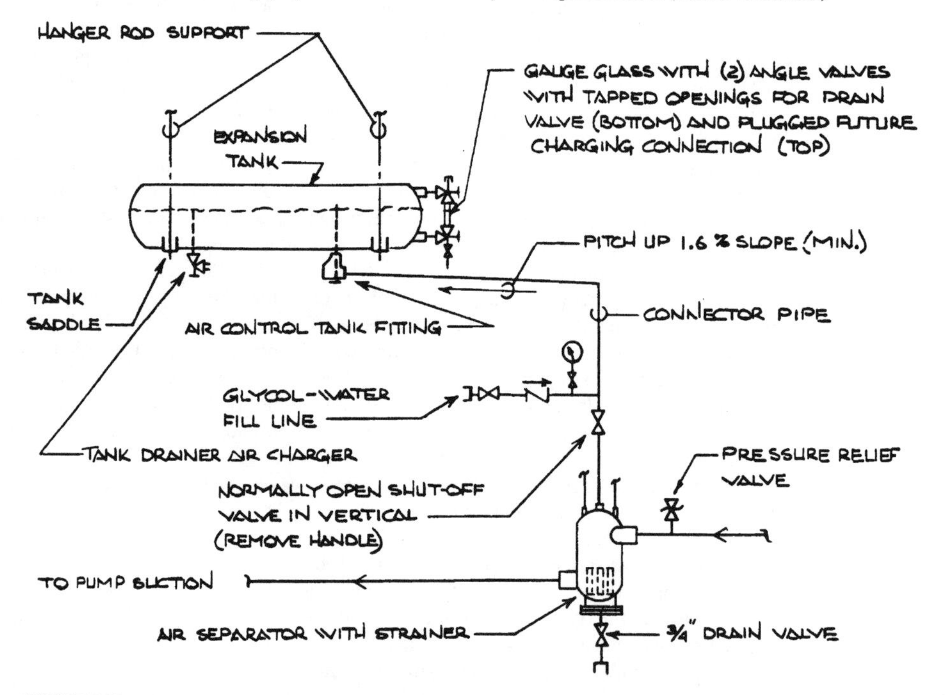

FIGURE 20 Air control and piping connections for glycol-water-system expansion tank. (*Jerome F. Mueller.*)

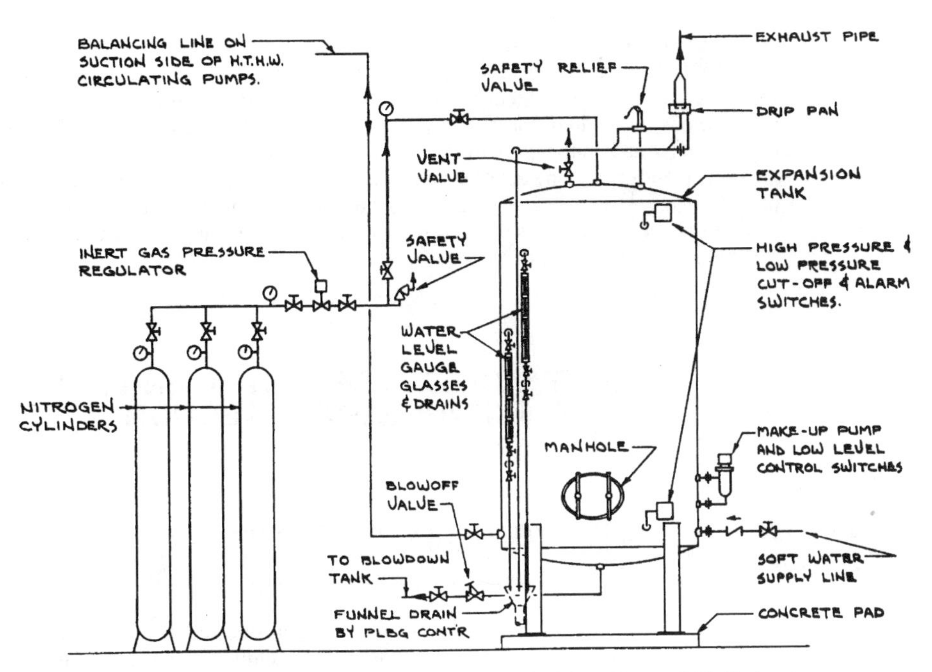

FIGURE 21 Typical high-temperature hot-water expansion-tank piping. (*Jerome F. Mueller.*)

Nitrogen Pressurization. In Fig. 21 we show the complete piping detail around a nitrogen-fed high-temperature hot-water expansion-tank system. This type of system is special to the high-temperature hot-water installation. The piping shown on this system has some unusual points. First, as can be seen, the expansion tank is still on the suction side of the high-temperature hot-water circulating pump. Second, in a high-temperature hot-water system there usually is a soft-water service. We note in our detail where the connection of this treated water goes into our expansion-tank system. Controlling the amount of water in the high-temperature hot-water expansion system is extremely important. In the lower right side of the detail the makeup pump with its low-level switches and controls carefully adds water to the expansion tank. As a further safety precaution there are high and low-pressure cutoffs and alarm switches. Knowing the system's limits is also very important.

Finally, high-pressure water will, if the pressure is released, turn instantly into high-pressure steam. Thus, the pressure-relief valve used is similar to that used in a high-pressure boiler. The connection of the pressure-relief valve drip pan and its exhaust pipe through the roof is nearly identical to that of a 125 lb/in^2 (gage) (861.3 kPa) steam system. If there is a loss of pressure in the expansion tank the result will cause the water to flash into steam. Note the line that goes to a blow-down tank; it provides protection against this occurrence. The pressurization for this system is normally provided by specially piped dry nitrogen cylinders. These cylinders and their output are controlled by a pressure regulator that has a safety valve to ensure that the pressurization to the system is maintained at the proper value at all times. Finally, there are times when excess nitrogen does have to be vented. This is accomplished through a small vent valve in the top of the tank.

The system designer wants to reduce the area of contact between a gas and water, thereby reducing the absorption of gas in the water. This is why the tank is shown installed vertically, which is the generally preferred arrangement.

The ratings of fittings, valves, piping, and equipment generally are based on a minimum pressure, which is about 25 to 50 lb/in^2 (gage) (172.3 to 344.5 kPa) above the maximum saturation pressure. An imposed additional pressure head above the vapor pressure must be sufficient to prevent steaming in the high-temperature hot-water generator at all times, even under unusual flow conditions, such as firing rates at which the created flow of two or more generators is not evenly matched. This is a critical condition since a gas-pressurized system does not have separate safety valves. The pressure varies with changes in water level in the expansion vessel. When the system water volume increases because of a temperature rise, the expansion of the system water into the vessel compresses the inert gas, raising the system pressure. The reverse condition takes place on a drop in system water temperature. The pressure is permitted to vary from a minimum point above saturation to a maximum that is determined by the materials used in the system. The expansion tank itself can be sized for the sum of the volumes required for pressurization plus the volume required for expansion and the volume required for sludge and reserve.

The data and illustrations in this portion of this procedure are the work of Jerome F. Mueller, P.E., Mueller Engineering Corporation.

ENERGY ASPECTS OF SYSTEM ANALYSIS AND EQUIPMENT SELECTION

HEAT- AND ENERGY-LOSS DETERMINATION FOR BUILDINGS AND STRUCTURES

An industrial building has 8-in (203.2-mm) thick uninsulated brick walls, a 2-in (60.8-mm) thick concrete uninsulated roof, and a concrete floor. The building is 150 ft (45.7 m) long, 75 ft (22.9 m) wide, and 15 ft (4.6 m) high. Each long wall contains eight 5 × 10 ft (1.5 × 3 m) double-glass windows, and each short wall contains two 4 × 8 ft (1.2 × 2.4 m) double-glass doors. What is the heat loss of this building per hour if the required indoor temperature is 70°F (21.1°C) and the design outside temperature is 0°F (−17.8°C). How much will the heat loss increase if infiltration causes two air changes per hour?

Calculation Procedure:

1. ***Compute the heat loss through the glass***

The usual heat-loss computation begins with the glass areas of a building. Hence, this procedure is followed here. However, a heat-loss computation can be started with any part of the building, provided each part of the structure is eventually considered.

To compute the heat loss through a building surface, use the general relation $H_L = UA\ \Delta t$, where H_L = heat loss, Btu/h, through the surface; U = overall coefficient of heat transmission for the material, Btu/(h · °F · ft^2); A = area of heat-transmission surface, ft^2; ΔT = temperature difference, $F = t_i - t_o$, where t_i = inside temperature, °F; t_o = outside temperature, °F. Find U from Table 3 for the material in question.

TABLE 3 Typical Overall Coefficients of Heat Transmission, Btu/(ft^2 · h · °F) [W/(m^2 · K)]

Building surface	Type	Type of insulation
Walls	8-in (203.2-mm) brick	0.50 (2.84)
Roof	2-in (50.8-mm) concrete	0.82 (4.66)
Windows	Single glass	1.13 (6.42)
	Double glass	0.45 (2.56)

This building has sixteen 5 × 10 ft (1.5 × 3 m) double-glass windows and four 4 × 8 ft (1.2 × 2.4 m) double-glass doors. Hence, the total glass area = 16 × 5 × 10 + 4 × 4 × 8 = 928 ft^2 (86.2 m^2).

The value of U for double glass is, from Table 3, 0.45. Thus, $H_L = UA\ \Delta t = 0.45(928)(70 - 0) =$ 29,200 Btu/h (8639.8 W).

2. *Compute the heat loss through the building walls*
Use the same relation as in step 1, substituting the wall heat-transfer coefficient and wall area. Thus, $U = 0.50$, and $A = 2 \times 150 \times 15 + 2 \times 75 \times 15 - 928 = 5822$ ft^2 (540.9 m^2). Then $H_L = UA\ \Delta t =$ $0.50(5822)(70 - 0) = 204{,}000$ Btu/h (59,746.5 W).

3. *Compute the heat loss through the building roof*
Use the same relation as in step 1, substituting the roof heat-transfer coefficient and roof area. Thus, $U = 0.82$ and $A = 150 \times 75 = 11{,}250$ ft^2 (1045.1 m^2). Then $H_L = UA\ \Delta t = 0.82 \times (11{,}250)(70 - 0) =$ 646,000 Btu/h (189,197.3 W).

4. *Compute the total heat loss of the building*
The total heat loss of a building is the sum of the individual heat losses of the walls, glass areas, roofs, and floor. In large buildings the heat loss through concrete floors is usually negligible and can be ignored. Hence, the total heat loss of this building caused by transmission through the building surfaces is $H_T = 29{,}200 + 204{,}000 + 646{,}000 = 879{,}200$ Btu/h (257,495.7 W).

5. *Compute the infiltration heat loss*
The building volume is $150 \times 75 \times 15 = 168{,}500$ ft^3 (4773.1 m^3). With two air changes per hour, the volume of infiltration air that must be heated is $2 \times 168{,}500 = 337{,}000$ ft^3/h (9546.1 m^3/h). The heat that must be supplied to raise the temperature of this air is $H_i = (\text{ft}^3/\text{h})(\Delta t)/55 = (337{,}000)(70 - 0)/55 =$ 429,000 Btu/h (125,643.4 W). Thus, the total heat loss of this building, including infiltration, is $H_T =$ $879{,}200 + 429{,}000 = 1{,}308{,}200$ Btu/h (383.1 kW).

Related Calculations. Determine the design outdoor temperature for a given locality from Baumeister—*Standard Handbook for Mechanical Engineers* or the ASHRAE *Guide and Data Book*, published by the American Society of Heating, Refrigerating and Air-Conditioning Engineers. Both these works are also suitable sources of comprehensive listings of U values for various materials and types of building constructions. Since the winter-design outdoor temperature is usually for nighttime conditions, no credit is taken for heat given off by machinery, lights, people, etc., unless the structure will always operate on a 24-h basis. The safest design ignores these heat sources because the machinery in the building can be removed, the operating cycle changed, or the heat sources eliminated in some other way. However, where an internal heat source of any kind will be a permanent part of a building, simply subtract the hourly heat release from the total building heat loss. The result is the net heat loss of the building and is used in choosing the heating equipment for the building.

Most heat-loss calculations for large structures are made on a form available from heat equipment manufacturers. Such a form helps organize the calculations. The steps followed, however are identical to those given above. Another advantage of the calculation form is that it helps the designer remember the various items—walls, roof, glass, infiltration, etc.—to consider.

When some areas in a structure will be kept at a lower temperature than others, compute the heat loss from one area to another in the manner shown above. Substitute the lower indoor temperature for the outdoor-design temperature. For areas exposed to prevailing winds, some manufacturers recommend increasing the computed heat loss by 10 percent. Thus, if the north wall of a building were exposed to the prevailing winds and its heat loss is 50,000 Btu/h (14,655.0 W), this heat loss would be increased to 1.1(50,000) = 55,000 Btu/h (16,120.5 W).

SELECTION AND ANALYSIS OF BUILDING HEATING SYSTEMS

Choose a heating system suitable for an industrial plant consisting of a production area 150 ft (45.7 m) long and 75 ft (22.9 m) wide and an office area 75 ft (22.9 m) wide and 60 ft (18.3 m) long. The heat loss from the production area is 1,396,000 Btu/h (409.2 kW); the heat loss from the office area is 560,00 Btu/h (164.1 kW). Indoor design temperature for both areas is 70°F (21.1°C); outdoor design temperature is 0°F (−17.8°C). What will the fuel consumption of the chosen heating system be if the annual degree-days for the area in which the plant is located is 6000? Compare the annual fuel consumption of gas, oil, and coal.

Calculation Procedure:

1. ***Choose the type of heating system to use***

Table 4 lists the various types of heating systems used today and typical applications. Study of this tabulation shows that steam unit heaters would probably be best for the production area because it is relatively large and open. Either a forced-warm-air or a two-pipe steam heating system could be used for the office area. Since the production area will use steam unit heaters, a two-pipe steam heating system would probably be best for the office area. Since steam unit heaters are almost universally two-pipe, the same method of supply and return is best chosen for the office system.

TABLE 4 Typical Applications of Heating Systems

System type	Fuel*	Typical applications
Gravity warm air	G, O, C	Small residences, wooden or masonry
Forced warm air	G, O, C	Small and large residences, wooden or masonry; small and medium-sized industrial plants, offices
Steam heating:		
One-pipe	G, O, C	Small residences, wooden or masonry
Two-pipe†	G, O, C	Small and large residences, wooden or masonry; small and large industrial plants, offices. High-pressure systems [30 to 150 lb/in^2 (gage) (206.8 to 1034.1 kPa)] may be used in large industrial buildings having unit heaters or fan units. Unit heaters are used for large, open areas
Hot water:		
Gravity	G, O, C	Small residences, wooden or masonry
Forced‡	G, O, C	Small and large residences; small and large industrial plants
Radiant	G, O, C	Small and large residences and plants
Electric	Electricity	Small residences and plants

*G—gas; O—oil; G—coal.
†May be low-pressure, two-pipe vapor; two-pipe vacuum; two-pipe subatmospheric; two-pipe orifice; high-pressure.
‡May be one-pipe; two-pipe.

Note that Table 4 lists six different types of two-pipe steam heating systems. Choice of a particular type of two-pipe system depends on a number of factors, including economics, steam pressure required for nonheating (i.e., process) services in the building, type and pressure rating of boiler used, etc. Where high-pressure steam—30 to 150 lb/in^2 (gage) (206.8 to 1034.1 kPa), or higher—is used for process, the two-pipe system fitted with pressure-reducing valves between the process and heating mains is often an economical choice.

Hot-water heating is unsuitable for this building because unit heaters are required for the production area. An inlet air temperature has less than 30°F (−1.1°C) is generally not recommended for hot-water unit heaters. Since the inlet-air temperature can be as low as 0°F (−17.8°C) in this plant, hot-water unit heaters could not be used.

2. ***Compute the annual fuel consumption of the system***

Use the degree-day method to compute the annual fuel consumption. To apply the degree-day method, substitute the appropriate values in $F_G = DU_gRC$, for gas heating; $F_o = DU_cH_TC/1000$, for oil; $F_c = DU_cH_TC/1000$, for coal, where F = fuel consumption, the type of fuel being identified by the subscript, g for gas, o for oil, c for coal, with the unit of consumption being the therm for gas, gal for oil, and lb for coal; D = degree-days during the heating season; U = unit fuel consumption, the

TABLE 5 Factors for Estimating Fuel Consumption

Fuel	Consumption per degree-day	Heating conditions	
		Steam	Hot water
Gas	Therms [100,000 Btu (105,500 kJ)]	0.00127 [300-ft^2 (28-m^2) EDR] 0.00121 [300–700 ft^2 (28–65 m^2) EDR] 0.0116 [700-ft^2 (65-m^2) EDR]	0.000743 [500-ft^2 (46-m^2) EDR] 0.000709 [500–1000 ft^2 (46–92 m^2) EDR] 0.000675 [1200-ft^2 (111-m^2)EDR]
		Heating-plant efficiency	
Oil [heating value = 141,000 Btu/gal (39,297 MJ/L)]	Gal per 1000-Btu/h heat loss [L/(MJ · h)]	70% 0.00437 (0.01568)	80% 0.00383 (0.01374)
Coal [heating value = 12,000 Btu/lb (27,912 kJ/kg)]	Lb per 1000-Btu/h heat loss [kg/(MJ · h)]	70% 0.0507 (0.02163)	80% 0.0444 (0.01893)

	Outside-design temperature correction factor				
Outside design temperature, °F (°C)	−20 (−28.9)	−10 (−23.3)	0 (−17.8)	+10 (−12.2)	+20 (−6.7)
Correction factor	0.778	0.875	1.000	1.167	1.400

*ASHRAE *Guide and Data Book* with permission.

unit again being identified by the subscript; R = ft^2 EDR (equivalent direct radiation) in the heating system; C = a correction factor for the outdoor design temperatures. Values of U and C are given in Table 5. Note the fuel heating values on which this table is based. For other heating values, see Related Calculations, below.

For gas heating, $F_g = DU_gRC$, therms, where 1 therm = 100,000 Btu (105,500 kJ). To select the correct value for U_g, compute the ft^2 EDR (see the next Calculation Procedure) from $H_T/240$, or ft^2 (m^2) EDR = 1,956,000/240 = 8150(757.1). Hence, U_g = 0.00116 from Table 5. From the same table, C = 1.00 for an outdoor design temperature of 0°F (−17.8°C). Hence, F_g = (6000)(0.00116)(8150)(1.00) = 56,800 therms. Assuming gas having a heating value h_v of 1000 Btu/ft^3 (37,266 kJ/m^3) is burned in this heating system, the annual gas consumption is $F_g \times 10^5/h_v$. Or, 56,800 × 10^5/1000 = 5,680,000 ft^3 (160,801 m^3).

Using Table 5 for oil heating and assuming a heating plant efficiency of 70 percent, U_o = 0.00437 gal per 1000 Btu/h [0.01568 L/(MJ · h)] heat loss, and C = 1.00 for design conditions. Then $F_o = DU_oH_TC/1000$ = (6000)(0.00437)(1,956,000)(1.00)/1000 = 51,400 gal (194,570 L).

Using Table 5 for coal heating and assuming a heating-plant efficiency of 70 percent, we get U_c = 0.0507 lb coal per 1000 Btu/h [0.2163 kg/(MJ · h)] heat loss, and C = 1.00 for the design conditions. Then, $F_c = DU_cH_TC/1000$ = (6000)(0.0507)(1,956,000)(1.00)/1000 = 595,000 lb, or 297 tons of 2000 lb (302 t) each.

Related Calculations. When the outdoor-design temperature is above or below 0°F (−18°C), a correction factor must be applied as shown in the equations given in step 2. The appropriate correction factor is given in Table 5. Fuel-consumption values listed in Table 5 are based on an indoor-design temperature of 70°F (21°C) and an outdoor-design temperature of 0°F (−18°C).

The oil consumption values are based on oil having a heating value of 141,000 Btu/gal (39,296.7 MJ/L). Where oil having a different heating value is burned, multiply the fuel consumption value selected by 141,000/(heating value, Btu/gal, of oil burned).

The coal consumption values are based on coal having a heating value of 12,000 Btu/lb (27,912 kJ/kg). Where coal having a different heating value is burned, multiply the fuel-consumption value selected by 12,000/(heating value, Btu/lb, of coal burned).

For example, if the oil burned in the heating system in step 2 has a heating value of 138,000 Btu/gal (38,460.6 MJ/L), $U_o = 0.00437(141{,}000/138{,}000) = 0.00446$. And if the coal has a heating value of 13.500 Btu/lb (31,401 kJ/kg), $U_c = 0.0507(12{,}000/13{,}500) = 0.0451$.

Steam consumption of a heating system can also be computed by the degree-day method. Thus, the weight of steam W required for the degree-day period—from a day to an entire heating season is $W = 24H_TD/1000$, where all the symbols are as given earlier. Thus, for the industrial building analyzed above, $W = 24(1{,}956{,}000) \times (6000)/1000 = 281{,}500{,}000$ lb (127,954,546 kg) of steam per heating season. The denominator in this equation is based on low-pressure steam having an enthalpy of vaporization of approximately 1000 Btu/lb (2326 kJ/kg). Where high-pressure steam is used and the enthalpy of vaporization is lower, substitute the actual enthalpy in the denominator to obtain more accurate results.

Steam consumption for building heating purposes can also be given in pounds (kilograms) of steam per 1000 ft^3 (28.3 m^3) of building space per degree-day and of steam per 1000 ft^2 (92.9 m^2) of EDR per degree-day. On the building-volume basis, steam consumption in the United States can range from a low of 0.130 (0.06) to a high of 2.07 lb (0.94 kg) of steam per 1000 ft^3/degree-day (28.3 m^3/degree-day), depending on the building type (apartment house, bank, church, department store, garage, hotel or office building) and the building location (southwest or far north). Steam consumption can range from a low of 21 lb (10.9 kg) per 1000 ft^2 (92.9 m^2) EDR per degree-day to a high of 120 lb (54.6 kg) per 1000 ft^2 (92.9 m^2) EDR per degree-day, depending on the building type and location. The ASHRAE *Guide and Data Book* lists typical steam consumption values for buildings of various types in different locations.

UNIT HEATER CAPACITY COMPUTATION

An industrial building is 150 ft (45.7 m) long, 75 ft (22.9 m) wide, and 30 ft (9.1 m) high. The heat loss from the building is 350,000 Btu/h (102.6 kW). Choose suitable unit heaters for this building if 5-lb/in^2 (gage) (30.5-kPa) steam and 200°F (93.3°C) hot water are available to supply heat. Air enters the unit heater at 0°F (–18°C); an indoor temperature of 70°F (21°C) is desired in the building. What capacity unit heaters are needed if 20,000 ft^3/min (566.2 m^3/min) is exhausted from the building?

Calculation Procedure:

1. *Compute the total heat loss of the building*

The heat loss through the building walls and roof is given as 350,000 Btu/h (102.6 kW). However, there is an additional heat loss caused by infiltration of outside air into the building. Compute this loss as follows:

Find the cubic content of the building from volume, ft^3 = LWH, where L = building length, ft; W = building width, ft; H = building height, ft. Thus, volume = (150)(75)(30) = 337,500 ft^3 (9554.6 m^3).

Determine the heat loss caused by infiltration by estimating the number of air changes per hour caused by leakage of air into and out of the building. For the usual industrial building, one to two air changes per hour are produced by infiltration. At one air change per hour, the quantity of infiltration air that must be heated from the outside to the inside temperature = building volume = 337,500 ft^3/h (9554.6 m^3/h). Had two air changes per hour been assumed, the quantity of infiltration air that must be heated = 2 × building volume.

Compute the heat required to raise the temperature of the infiltration air from the outside to the inside temperature from $H_i = (\text{ft}^3/\text{h})(\Delta t)/55$, where H_i = heat required to raise the temperature of the air, Btu/h, through Δt, where $\Delta t = t_i - t_o$, and t_i = inside temperature, °F; t_o = outside temperature, °F.

For this building, H_i = 337,500(70 – 0)/55 = 429,000 Btu/h (125.7 kW). Hence the total heat loss of this building H_t = 350,000 + 429,000 Btu/h (228.3 kW), without exhaust ventilation.

2. *Determine the extra heat load caused by exhausting air*
When air is exhausted from a building, an equivalent amount of air must be supplied by infiltration or ventilation. In either case, an amount of air equal to that exhausted must be heated from the outside temperature to room temperature.

With an exhaust rate of 10,000 ft^3/min = 60(10,000) = 600,000 ft^3/h (19,986 m^3/h), the heat required is $H_e = (ft^3/h)(\Delta t)/55$, where H_e = heat required to raise the temperature of the air that replaces the exhaust air, Btu/h. Or, H_e = 600,000(70 – 0)/55 = 764,000 Btu/h (223.9 kW).

To determine the total heat loss from a building when both infiltration and exhaust occur, add the larger of the two heat requirements—infiltration or exhaust—to the heat loss caused by transmission through the building walls and roof. Since, in this building, $H_e > H_i$, H_t = 350,000 + 764,000 = 1,114,000 Btu/h (326.5 kW).

3. *Choose the location and number of unit heaters*
This building is narrow; i.e., it is half as wide as it is long. In such a building, three vertical-discharge unit heaters (Fig. 22) will provide good distribution of the heated air. With three heaters, the capacity of each should be 764,000/3 = 254,667 Btu/h (74.6 kW) without ventilation and 1,114,000/3 = 371,333 Btu/h (108.8 kW) with ventilation. Once the capacity of each unit heater is chosen, the spread diameter of the heated air discharge by the heater can be checked to determine whether it is sufficient to provide the desired comfort.

4. *Select the capacity of the unit heaters*
Use the engineering data published by a unit-heater manufacturer, such as Table 6, to determine the final air temperature, Btu delivered per hour, cubic feet of air handled, and the quantity of condensate formed. Thus, Table 6 shows that vertical-discharge model D unit heater delivers 277,900 Btu/h (81.5 kW) when the entering air is at 0°F (–17.8°C) and the heating steam is at a pressure of 5 lb/in^2 (gage) (34.5 kPa). This heater discharges 3400 ft^3/min (96.3 m^3/min) of heated air at 76°F (24.4°C) and forms 290 lb/h (131.8 kg/h) of condensate. The capacity table for a horizontal-discharge unit heater is similar to Table 6. When the building is ventilated, a model E unit heater delivering 388,400 Btu/h (113.8 kW) could be used with entering air at 0°F (–17.8°C). This heater, as Table 6 shows, delivers 4920 ft^3/min (139.3 m^3/min) of air at 73°F (22.8°C) and forms 404 lb (183.6 kg) of condensate.

5. *Check the spread diameter produced by the heater*
Table 6 shows the different spread diameters, i.e., diameter of the heated-air blast at the floor level for different mounting heights of the unit heater. Thus, at a 14-ft (4.3-m) mounting level above the floor, model D will produce a spread diameter of 48 ft (14.6 m) or 54 ft (16.5 m), depending on the type of outlet cone used. These spread diameters are based on 2-lb/in^2 (gage) (13.8-kPa) steam and 60°F (15.6°C) room temperature. For 5-lb/in^2 (gage) (34.5-kPa) steam and 70°F (21.1°C) room temperature, multiply the tabulated spread diameter and mounting height by the correction factor shown at the bottom of Table 6. Thus, for model D, spread diameter = 1.07(48) = 51.4 ft (15.7 m) and 1.07(54) = 57.8 ft (17.6 m), whereas the mounting height could be 1.07(14) = 14.98 ft (4.6 m).

Find the spread diameter for model E in the same way, or, 56 ft (17.1 m) and 62 ft (18.9 m) at a 15-ft (4.6-m) height with 2-lb/in^2 (gage) (13.8-kPa) steam. With 5-lb/in^2 (gage) (34.5-kPa) steam, the spread diameters are 1.07(56) = 59.9 ft (18.3 m) and 1.07(62) = 66.4 ft (20.2 m), whereas the mounting height could be 1.07(15) = 16 ft (4.9 m).

6. *Compute the hot-water-heater capacity required*
Study several manufacturers' engineering data to determine the capacity of a suitable hot-water unit heater. This study will show that hot-water unit heaters are generally not available to inlet-air temperatures less than 30°F (–1.1°C). Since the inlet-air temperature in this building is 0°F (–17.8°C), a hot-water unit heater would be unsuitable; hence, it cannot be used. The *minimum* outlet-air temperature often recommended for unit heaters is 95°F (35.0°C).

Related Calculations. Use the same general method to choose horizontal-delivery unit heaters. The heated air delivered by these units travels horizontally or can be deflected down toward the

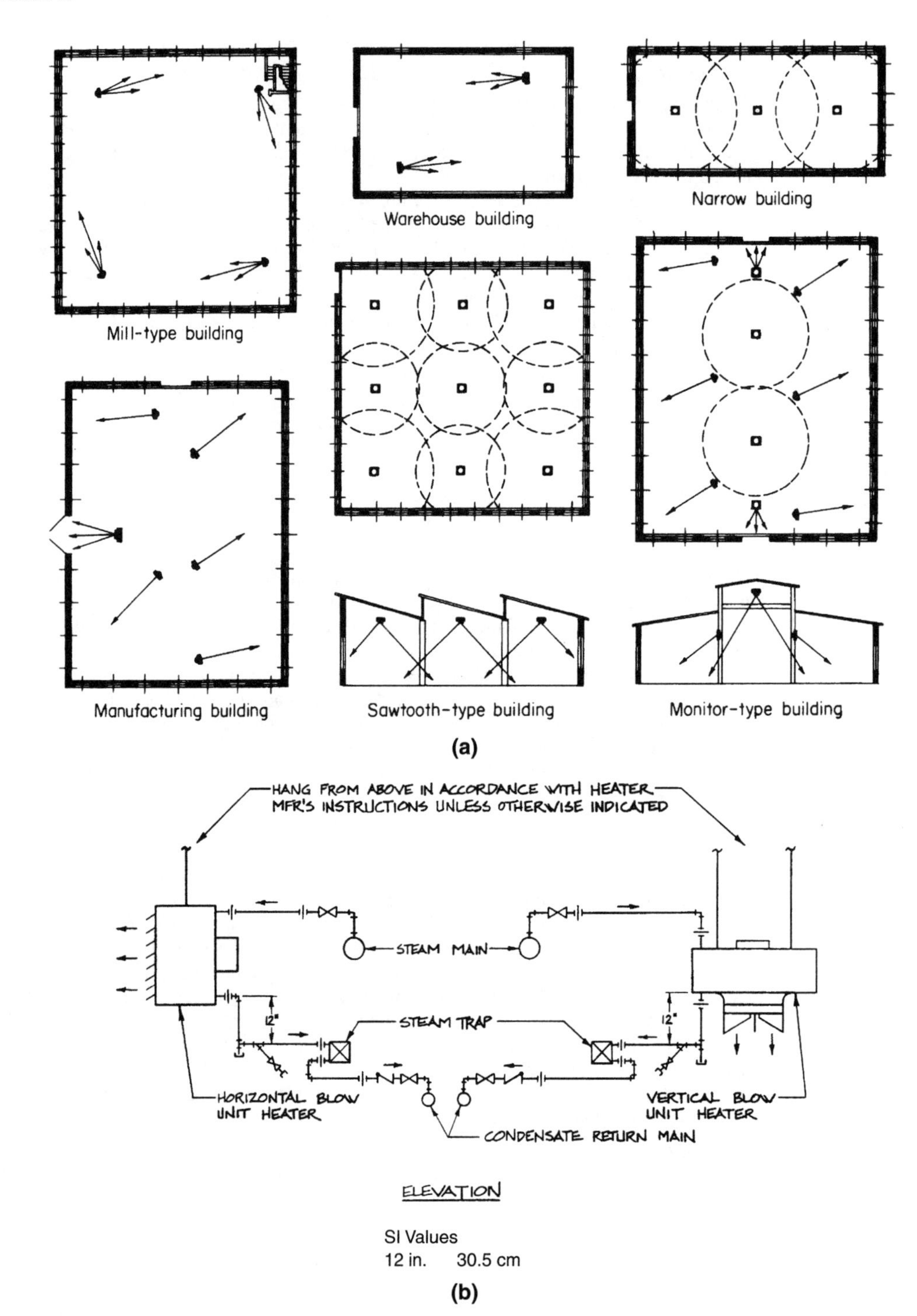

FIGURE 22 (*a*) Recommended arrangements of unit heaters in various types of buildings. (*Modine Manufacturing Company.*) (*b*) Horizontal and vertical blow steam unit-heater piping. (*Jerome F. Mueller.*)

TABLE 6 Typical Vertical-Delivery Unit-Heater Capacities*

					0°F (−17.8°C) entering air			50°F (10°C) entering air		
Mode	Mtg. ht., ft (m)	Spread diam., ft (m)	Motor speed, r/min	ft^3/min† (m^3/min)	Btu/h (kW)	Final temp., °F (°C)	Cond., lb/h (kg/h)	Btu/h (kW)	Final temp., °F (°C)	Condensate, lb/h (kg/h)
D	14 (4.3)	48, 54 (14.8, 16.5)	1,135	3,400 (96)	277,900 (81.5)	76 (24.4)	290 (161)	208,900 (61.2)	111 (43.9)	218 (99)
E	15 (4.6)	56, 62 (17.3, 18.9)	1,135	4,920 (139.3)	388,400 (113.8)	73 (22.8)	404 (183)	292,000 (85.6)	109 (42.8)	304 (151)

Mounting height correction factors

Steam press, lb/in^2 (gage)	(kPa)	Water temp., °F (°C)	Normal room temp., °F (°C) 60 (15.6)	70 (21.1)	80 (26.7)
—	—	210 (98.9)	1.05	1.10	1.20
0–5	(0–34.4)	220 (104.4)	1.00	1.07	1.14
6–15	(41.4–103.4)		0.88	0.94	1.00
16–30	(110.3–206.8)		0.77	0.81	0.86
31–50	(213.7–344.7)		0.70	0.73	0.77

*5-lb/in^2 (gage) (34.5-kPa) steam supply.
†ft^3/min capacity at *final* air temperature. For horizontal-discharge unit heaters, the ft^3/min capacity is usually stated at the *entering* air temperature.

floor. Tables in manufacturers' engineering data list the heat-throw distance for horizontal-delivery unit heaters.

Standard ratings of steam unit heaters are given for 2-lb/in^2 (gage) (13.8-kPa) steam and 60°F (15.6°C) entering air; hot-water unit heaters for 180°F (82.2°C) entering water and 60°F (15.6°C) entering air. For other steam pressures, water temperatures, or entering-air temperatures, *divide* the Btu per hour at stated conditions by the appropriate correction factor from Table 7 to obtain the required rating at *standard* conditions. Thus, a steam unit heater rated at 18,700 Btu/h (5.5 kW) at 20 lb/in^2 (gage) (137.9 kPa) and 70°F (21.1°C) entering air has a standard rating of 18,700/1/178 =

TABLE 7 Unit-Heater Conversion Factors

	Entering air temperature, °F (°C)			
	40 (4.4)	50 (10.0)	60 (15.6)	70 (21.1)
Steam pressure, lb/in^2 (gage) (kPa)	Horizontal-delivery steam heaters			
2 (13.8)	1.153	1.076	1.000	0.927
5 (34.5)	1.209	1.131	1.055	0.981
10 (68.9)	1.288	1.209	1.132	1.057
20 (137.9)	1.413	1.333	1.254	1.178
40 (275.8)	1.593	1.510	1.430	1.351
50 (344.7)	1.664	1.582	1.500	1.421
Entering water temperature, °F (°C)	Horizontal-delivery hot-water heaters			
150 (65.6)	0.911	0.790	0.676	0.568
160 (71.1)	1.027	0.900	0.783	0.670
170 (76.7)	1.142	1.012	0.890	0.773
180 (82.2)	1.262	1.127	1.000	0.880
190 (87.8)	1.384	1.245	1.115	0.990
200 (93.3)	1.503	1.365	1.231	1.102

15,900 Btu/h (4.7 kW), closely, at 2 lb/in^2 (gage) (13.8 kPa) and 60°F (15.6°C) entering-air temperature, using the correction factor from Table 7. Conversely, a steam unit heater rated at 228,000 Btu/h (66.8 kW) at 2 lb/in^2 (gage) (13.8 kPa) and 60°F (15.6°C) will deliver 228,000(1.421) = 324,000 Btu/h (94.9 kW) at 50 lb/in^2 (gage) (344.7 kPa) and 70°F (21.1°C) entering air. Electric and gas-fired unit heaters are also rated on the basis of 60°F (15.6°C) entering air.

When a unit heater is supplied both outside and recalculating air from within the building, use the temperature of the combined airstreams as the inlet temperature. Thus, with 1000 ft^3/min (28.3 m^3/min) of 10°F (–12.2°C) outside air and 4000 ft^3/min (113.2 m^3/min) of 70°F (21.1°C) recirculated air, the temperature of the air entering the heater is $t_e = (t_{oa}cfm_{oa} + t_r cfm_r)/cfm_t$, where t_e = temperature of air entering heater, °F; t_{oa} = temperature of outside air, °F; cfm_{oa} = quantity (cubic *feet* per *minute*) of outside air entering the heater, ft^3/min; t_r = temperature of room air entering heater, °F; cfm_r = quantity of room air entering heater, fr^3/min; cfm_t = total *cfm* entering heater = $cfm_{oa} + cfm_r$. Thus, $t_e = (10 \times 1000 + 70 \times 4000)/5000 = 58$°F (14.4°C). This relation is valid for both steam and hot-water unit heaters.

To find the approximate outlet-air temperature of a unit heater when the capacity cfm, and entering-air temperature are known, use the relation $t_o = t_i + (460 + t_i)/[575 \cdot \text{ft}^3 \cdot \text{h/(min} \cdot \text{Btu)}] - 1$, where t_o = unit-heater outlet-air temperature, °F; t_i = temperature of air entering the heater, °F; *cfm* = quantity of air passing through the heater, ft^3/min; Btu/h = rated capacity of the unit heater. Thus, for a heater rated at 73,500 Btu/h (21.5 kW), 1530 ft^3/min (43.4 m^3/min), with 50°F (10.0°C) entering-air temperature, $t_o = 50 + (460 + 50)/(575 \times 1530/73{,}500) - 1 = 94.1$°F (34.5°C). The unit-heater capacity used in this equation should be the capacity at standard conditions: 2-lb/in^2 (gage) (13.8-kPa) steam or 180°F (82.2°C) water and 60°F (15.6°C) entering-air temperature. Results obtained with this equation are only approximate. In any event, the outlet air temperature should never be less than the room temperature. For the actual outlet temperature of a specific unit heater, refer to the manufacturers' engineering data.

The air discharged by a unit heater should be at a temperature greater than the room temperature because air in motion tends to chill the occupants of a room. Choose the outlet temperature by referring to the heated-air velocity. Thus, with a velocity of 20 ft/min (6.1 m/min), the air temperature should be at least 76°F (24.4°C) at the heater outlet. As the air velocity increases, higher air temperatures are required,

The outlet-air velocity and distance of blow of typical unit heaters are shown in Table 8.

TABLE 8 Unit-Heater Outlet Velocity and Blow Distance

Unit-heater type	Outlet velocity, ft/min (m/min)	Blow distance, ft (m)
Centrifugal fan	1500–2500 (457–762)	20–200 (6–61)
Horizontal propeller fan	400–1000 (122–305)	30–100 (9–30)
Vertical propeller fan	1200–2200 (366–671)	70 (21)

When a unit heater of any type discharges against an external resistance, such as a duct or grille, its heating capacity and air capacity, as compared to standard conditions, are reduced. However, the final air temperature usually increases a few degrees over that at standard conditions.

To convert the rated output of any steam unit heater to ft^2 (m^2) of equivalent direct radiation [abbreviated ft^2 EDR (m^2 EDR)], divide the unit-heater rated capacity in Btu/h (W) by 240 Btu/(h · ft^2 EDR) (70.3 W/m^2 EDR). Thus, a unit heater rated at 240,000 Btu/h (70.34 W) has a heat output of 240,000/240 = 1000 ft^2 EDR (1000 m^2 EDR). For hot-water unit heaters, use the conversion factor of 150 Btu/(h · ft^2 EDR) (43.9 W/m^2 EDR).

To determine the rate of condensate formation in a steam unit heater, divide the rated output in Btu/h (W) by an enthalpy of 930 Btu/lb (2163.2 kJ/kg) of steam. Most unit-heater rating tables list the rate of condensate formation for each heater. Table 9 shows typical pipe sizes recommended for various condensate loads of steam unit heaters, thus, with 1000 lb/h (454.6 kg/h) of condensate and 30-lb/in^2 (gage) (206.8-kPa) steam supply to the unit heater, the supply main should be 2½-in (64-mm) nominal diameter and the return main should be 1¼-in (32-mm) nominal diameter.

TABLE 9 Typical Steam Unit-Heater Pipe Diameters, in (mm)

Condensate, lb/h (kg/h)	Steam-supply pressure, lb/in^2 (gage) (kPa)			
	5 (34.5)		30 (206.8)	
	Supply	Return	Supply	Return
100 (45)	2 (51)	1 (25)	1¼ (32)	¾ (19)
400 (180)	3 (76)	2 (51)	2 (51)	1 (25)
800 (360)	4 (102)	2½ (64)	2½ (64)	1¼ (32)
1000 (450)	5 (127)	2½ (64)	2½ (64)	1¼ (32)

*Gravity return.
Modine Manufacturing Company.

Figure 22*a* shows how unit heaters of any type should be located in buildings of various types. The diagrams are also useful in determining the approximate number of heaters needed once the heat loss is known. Locate unit heaters so that the following general conditions prevail, if possible.

Unit heater type	Desirable conditions
Horizontal delivery	Discharge should wipe exposed walls at an angle of about 30°. With multiple units, the airstreams should support each other.
Vertical delivery	With only vertical units, the airstream should blanket exposed walls with warm air.

The unit-heater arrangements shown in Fig. 22*a* illustrate a number of important principles.* The basic principle of unit-heater location is shown in the *mill-type* building. Here the heated-air flow from each unit heater supports the air flow from the other unit heaters and tends to set up a general circulation of air in the space heated. In the *warehouse* building arrangement, maximum area coverage is obtained with a minimum number of units. The *narrow* building uses vertical-discharge unit heaters that blanket the building walls with warmed air.

In the *manufacturing* building, circular air movement is sacrificed to offset a large roof heat loss and to permit short runouts from a single steam main. Note how a long-throw unit heater blankets a frequently used doorway.

Vertical-discharge unit heaters are used in the medium-height *sawtooth-type* building shown in Fig. 22*a*. The *monitor-type*-building installation combines both horizontal and vertical unit heaters. Horizontal-discharge unit heaters are located in the low-ceiling areas and vertical-discharge units in the high-ceiling areas above the craneway. Much of the data in this procedure were supplied by Modine Manufacturing Company. Figure 22*b* shows typical horizontal- and vertical-blow unit-heater piping.

HEATING APPARATUS STEAM AND ENERGY CONSUMPTION

Determine the probable steam consumption of the non-space-heating equipment in a building equipped with the following: three bain-maries, each 100 ft^2 (9.3 m^2), two 50-gal (189.3-L) coffee urns, one jet-type dishwasher, one plate warmer having a 60-ft^3 (1.7-m^3) volume, two steam tables, each having an area of 50 ft^2 (4.7 m^2), and one water still having a capacity of 75 gal/h (283.9 L/h). The available steam pressure is 40 lb/in^2 (gage) (275.8 kPa); the kitchen equipment will operate at 20 lb/in^2 (gage) (137.9 kPa) and the water still at 40 lb/in^2 (gage) (275.8 kPa).

*Modine Manufacturing Company.

Calculation Procedure:

1. *Determine steam consumption of the equipment*
The general procedure in determining heating-equipment steam consumption is to obtain engineering data from the manufacturer of the unit. When these data are unavailable, Table 10 will provide enough information for a reasonably accurate first approximation. Hence, this tabulation is used here to show how the data are applied.

TABLE 10 Typical Steam Consumption of Heating Equipment

Equipment	Steam, lb/h at 20–40 lb/in^2 (gage) (kg/h at 138–276 kPa)	
Bain-marie, per ft^2 (m^2) of surface	3.0	(14.5)
Coffee urns, per gal (L)	2.5–3.0	(0.30–0.36)
Dishwashers, jet-type	60	(27.0)
Plate warmer, per 20 ft^3 (0.57 m^3)	30	(13.5)
Soup or stock kettle, 60 gal (0.23 L); 40 gal (0.15 L)	60; 45	(27.0; 20.3)
Steam table, per ft^2 (m^2) of surface	1.5	(7.3)
Vegetable steamer, per compartment 5-lb (2.3-kg) press	30	(13.5)
Water still, per gal (L) capacity per h	9	(1.1)

Since the supply steam pressure is 40 lb/in^2 (gage) (275.8 kPa), a pressure-reducing valve will have to be used between the steam main and the kitchen supply main. The capacity of this valve depends on the steam consumption of the equipment. Hence, the valve capacity cannot be determined until the equipment steam consumption is known.

During equipment operation the steam consumption is different from the consumption during start-up. Since the operating consumption must be known before the starting consumption can be computed, the former is determined first.

Using data from Table 10, we see the three 100-ft^2 (9.3-m^2) bain-maries will require (3 lb/h)(3 units) (100 ft^2 each) = 900 lb/h (409.1 kg/h) of steam. The two 50-gal (189.3-L) coffee urns require (2.75 lb/h)(2 units)(50 gal per unit) = 275 lb/h (125 kdg/h), using the average steam consumption. One jet-type dishwasher will require 60 lb/h (27.3 kg/h) of steam. A 60-ft^3 (1.7-m^3) plate warmer will require (60 ft^3)/(1 unit) [(30 lb/h)/20-ft^3 unit] = 90 lb/h (40.9 kg/h). Two 50-ft^2 (4.7-m^2) steam tables require (1.5 lb/h)(2 units) (50 ft^2 per unit) = 150 lb/h. The 75-gal/h (283.9-L/h) water still will consume (9 lb/h)(1 unit)(75 gal/h) = 675 lb/h (306.8 kg/h). Hence, the total operating consumption of 20-lb/in^2 (gage) (137.9-kPa) steam is 900 + 275 + 60 + 90 + 150 = 1475 lb/h (670.5 kg/h). The water still will consume 675 lb/h (306.8 kg/h) of 40-lb/in^2 (gage) (275.8-kPa) steam.

Since the 20-lb/in^2 (gage) (137.9-kPa) steam must pass through the pressure-reducing valve before entering the 20-lb/in^2 (gage) (137.9-kPa) main, the required *operating* capacity of this valve is 1475 lb/h (670.5 kg/h). However, the total steam consumption during operation, without an allowance for condensation in the pipelines, is 1475 + 675 = 2150 lb/h (977.3 kg/h).

2. *Compute the system condensation losses*
Condensation losses can range from 25 to 50 percent of the steam supplied, depending on the type of insulation used on the piping, the ambient temperature in the locality of the pipe, and the degree of superheat, if any, in the steam. Since the majority of the steam used in this building is 20-lb/in^2 (gage) (137.9-kPa) steam reduced in pressure from 40 lb/in^2 (gage) (275.8 kPa) and there will be a small amount of superheating during pressure reduction, a 25 percent allowance for pipe condensation is probably adequate. Hence, the total operating steam consumption = (1.25)(2150) = 2688 lb/h (1221.8 kg/h).

3. *Compute the start-up consumption*
During equipment start-up there is additional condensation caused by the cold metal and, possibly, some cold products in the equipment. Therefore, the start-up steam consumption is different from the operating consumption.

One rule of thumb estimates the start-up steam consumption as two times the operating consumption. Thus, by this rule of thumb, start-up steam consumption = 2(2688) = 5376 lb/h (2443.6 kg/h). Note that this consumption rate is of relatively short duration because the metal parts are warmed rapidly. However, the pressure-reducing valve must be sized for this flow rate unless slower warming is acceptable.

The actual rate of condense formation can be computed if the weight of the equipment, the specific heat of the materials of construction, and initial and final temperatures of the equipment are known. Use the relation steam condensation, lb/h = 60 $Ws(\Delta t)/h_{fg}T$, where W = weight of equipment and piping being heated, lb; s = specific heat of the equipment and piping, Btu/(lb · °F); Δt = temperature rise of the equipment from the cold to the hot state, °F; h_{fg} = latent heat of vaporization of the heating steam, Btu/lb; T = heating period, min. In SI units, condensation is found from the same relationship, except W is in kg; s is in kJ/(kg · °C); t = temperature change, °C; h_{fg} = latent heat of vaporization, kJ/kg; other variables the same.

This relation assumes that the final temperature of the equipment approximately equals the temperature of the heating steam. Where the specific heat of the equipment is different from that of the piping, solve for the steam condensation rate of each unit and sum the results. Where products in the equipment must be heated, use the same relation but substitute the produce weight and specific heat.

Related Calculations. Use the general method given here to compute the steam consumption of any type of industrial equipment for which the unit steam consumption is known or can be determined.

AIR-HEATING COIL SELECTION AND ENERGY ANALYSIS

Select a steam heating coil to heat 80,000 ft^3/m (2264.8 m^3/min) of outside air from 10°F (−12.2°C) to 150° (65.6°C) for steam at 15 lb/in^2 (abs) (103.4 kPa). The heated air will be used for factory space heating. Illustrate how a steam coil is piped. Show the steps for choosing a hot-water heating coil.

Calculation Procedure:

1. *Compute the required face area of the coil*

If the coil-face air velocity is not given, a suitable air velocity must be chosen. In usual air-conditioning and heating practice, the air velocity across the face of the coil can range from 300 to 1000 ft/min (91.4 to 305 m/min) with 500, 800, and 1000 ft/min (152.4, 243.8, and 305 m/min) being common choices. The higher velocities—up to 1000 ft/min (305 m/min)—are used for industrial installations where noise is not a critical factor. Assume a coil face velocity of 800 ft/min (243.8 m/min) for this installation.

Compute the required face area from $A_c = cfm/V_a$, where A_c = required coil face area, ft^2: cfm = quantity of air to be heated by the coil, ft^3/min, at 70°F (21.1°C) and 29.92 in (759.97 mm) Hg; V_a = air velocity through the coil, ft/min. To correct the air quantity to standard conditions, when the air is being delivered at nonstandard conditions, multiply the flow in at the other temperature by the appropriate factor from Table 11. Thus, with the incoming air at 10°F (−12.2°C), ft^3/m at 70°F (21.1°C) and 29.92 in (759.97 mm) Hg = (1.128)(80,000) = 90,400 ft^3/m (2559.2 m^3/min). Hence, A_c = 90,400/800 = 112.8 ft^2 (10.5 m^2).

2. *Compute the coil outlet temperature*

The capacity, final temperature, and condensate formation rate for steam heating coils for air-conditioning and heating systems are usually based on steam supplied at 5 lb/in^2 (abs) (34.5 kPa) and inlet air at 0°F (−17.8°C). At other steam pressures a correction factor must be applied to the tabulated outlet temperature for 5-lb/in^2 (abs) (34.5-kPa) coils with 0°F (−17.8°C) inlet air.

TABLE 11 Air-Volume Conversion Factors*

Air temp., °F (°C)	Factor	Air temp., °F (°C)	Factor
0 (−17.18)	1.152	90 (32.2)	0.964
10 (−12.2)	1.128	100 (37.8)	0.946
20 (−6.7)	1.104	110 (43.3)	0.930
30 (−1.1)	1.082	120 (48.9)	0.914
40 (4.4)	1.060	130 (54.4)	0.898
50 (10.0)	1.039	140 (60.0)	0.883
60 (15.6)	1.019	150 (65.6)	0.869
70 (21.1)	1.000	160 (71.1)	0.855
80 (26.7)	0.981	170 (76.7)	0.841

*ASHRAE

Table 12 shows an excerpt from a typical coil-rating table and excerpts from coil correction-factor tables. To use such a tabulation for a coil supplied steam at 5 lb/in^2 (abs) (34.5 kPa), enter at the air-inlet temperature and coil face velocity. Find the final air temperature equal to, or higher than, the required final air temperature. Opposite this read the number of rows of tubes required. Thus, in a 5-lb/in^2 (abs) (34.5-kPa) coil with 0°F (−17.8°C) inlet air, 800-ft/min (243.8-m/min) face velocity, and a 165°F (73.9°C) final air temperature, Table 12 shows that five rows of tubes would be required. This table shows that the coil forms condensate at the rate of 149.8 lb/(h · ft^2) [732.9 kg/(h · m^2)] of net fin area when the final air temperature is 166°F (74.4°C). The coil thus chosen is the first-trial coil, which must be checked against the actual steam conditions as described below.

TABLE 12 Steam-Heating-Coil Final Temperatures and Condensate-Formation Rate

		Face velocity, 800 ft/min (243.8 m/min)	
Inlet air temp., °F (°C)	Rows of tubes	Final air temp., °F, (°C)	Condensate, lb/h (kg/h)
0 (−17.8)	2	51 (10.6)	83.7 (37.7)
10 (−17.8)	3	124 (51.1)	111.7 (50.3)
0 (−17.8)	4	148 (64.4)	132.9 (59.8)
0 (−17.8)	5	166 (74.4)	149.8 (67.4)
	Temperature-rise correction factor		
	Steam pressure, lb/in^2 (abs) (kPa)		
Actual inlet air temp., °F (°C)	10 (68.9)	15 (103.4)	20 (137.9)
0 (−17.8)	1.054	1.100	1.139
10 (−12.2)	1.010	1.056	1.095
20 (−6.7)	0.966	1.011	1.051
	Condensate correction factors		
0 (−17.8)	1.063	1.117	1.165
10 (−12.2)	1.019	1.072	1.120
20 (−6.7)	0.974	1.027	1.075

When a coil is supplied steam at a pressure different from 5 lb/in^2 (abs) (34.5 kPa), multiply the final air temperature given in the 5-lb/in^2 (abs) (34.5-kPa) table for 0°F (−17.8°C), at the face velocity being used, by the correction factor given in Table 12 for the actual steam pressure and actual inlet-air temperature. Thus, for this coil, which is supplied steam at 15 lb/in^2 (abs) (103.4 kPa) and has an inlet-air temperature of 10°F (−12.2°C), the temperature correction factor from Table 12 is 1.056. Add the product of the correction factor and the tabulated final air temperature to the inlet-air temperature to obtain the actual final air temperature. Several trials may be necessary before the desired outlet temperature is obtained.

The desired final air temperature for this coil is 150°F (65.6°C). Using the 124°F (51.1°C) final air temperature from Table 12 as the first-trial valve, we get the actual final air temperature = (1.056) (124) + 10 = 141°F (60.6°C). This is too low. Trying the next higher final air temperature gives the actual final air temperature = (1.056)(148) + 10 = 166.5°F (74.7°C). This is higher than required, but the steam supply can be reduced to produce the desired final air temperature. Thus, the coil will be four rows of tubes deep, as Table 12 shows. Hence, the five rows of coils originally indicated will not be needed. Instead, four rows will suffice.

3. ***Compute the quantity of condensate produced by the coil***

Use the same general procedure as in step 2, or actual condensate formed, lb/(h · ft^2) = [lb/(h · ft^2)] [condensate from 5-lb/in^2 (abs), 0°F (34.5-kPa, −17.8°C) table] (correction factor from Table 12); or for this coil, (132.9)(1.072) = 142.47 lb/(h · ft^2) [690.1 kg/(h · m^2)]. Since the coil has a net fin face area of 112.8 ft^2 (10.5 m^2), the total actual condensate formed = (112.8)ft^2 (142.47) = 16,100 lb/h (7318.2 kg/h).

4. ***Determine the coil friction loss***

Most manufacturers publish a chart or table of coil friction losses in coils having various face velocities and tube rows. Thus, Fig. 23*a* shows that a coil having a face velocity of 800 ft/min (243.8 m/min) and four rows of tube has a friction loss of 0.45 in (11.4 mm) of water. Figure 23*a* is a typical friction-loss chart and can be safely used for all routine preliminary coil selections. However, when the final choice of heating coil is made, use the friction chart or table prepared by the manufacturer of the coil chosen.

5. ***Determine the coil dimensions***

Refer to the manufacturer's engineering data for the dimensions of the coil chosen. Each manufacturer has certain special construction features. Hence, there will be some variation in dimensions from one manufacturer to another.

6. ***Indicate how the coil will be piped***

The ASHRAE *Guide and Data Book* shows piping arrangements for low- and high-pressure steam heating coils as recommended by various coil manufacturers. Follow the recommendations of the manufacturer whose coil is actually used when the final selection is made. Figure 23*b* shows several typical piping arrangements for steam-heated air-heating coils.

Related Calculations. Typical variables met in heating-coil selection are the *face velocity*, which varies from 300 to 1000 ft/min (91.4 to 304.8 m/min), with the higher velocities being used for industrial applications, the lower velocities for nonindustrial applications; the *final* air temperature, which ranges between 50 and 300°F (10 and 148.9°C), the lower temperatures being used for ventilation, the higher ones for heating; *steam pressures*, which vary from 2 to 150 lb/in^2 (gage) (13.8 to 1034.1 kPa), with the lower pressures—2 to 15 lb/in^2 (gage) (13.8 to 103.4 kPa)—being the most popular for heating.

Hot-water heating coils are also used for air heating. The general selection procedure is: (*a*) Compute the heating capacity, Btu/h, required from (1.08)(temperature rise of air, °F)(cfm heated). (*b*) Compute the coil face area required, ft^2, from cfm/face velocity, ft/min. Assume a suitable face velocity using the guide given above. (*c*) Compute the logarithmic mean-effective-temperature difference across the coil, using the method given elsewhere in this handbook. (*d*) Compute the required hot-water flow rate, gallons per minute (gal/min), from Btu/h heating capacity/(500)(temperature drop of water, °F). The usual temperature drop of the hot water during passage through the

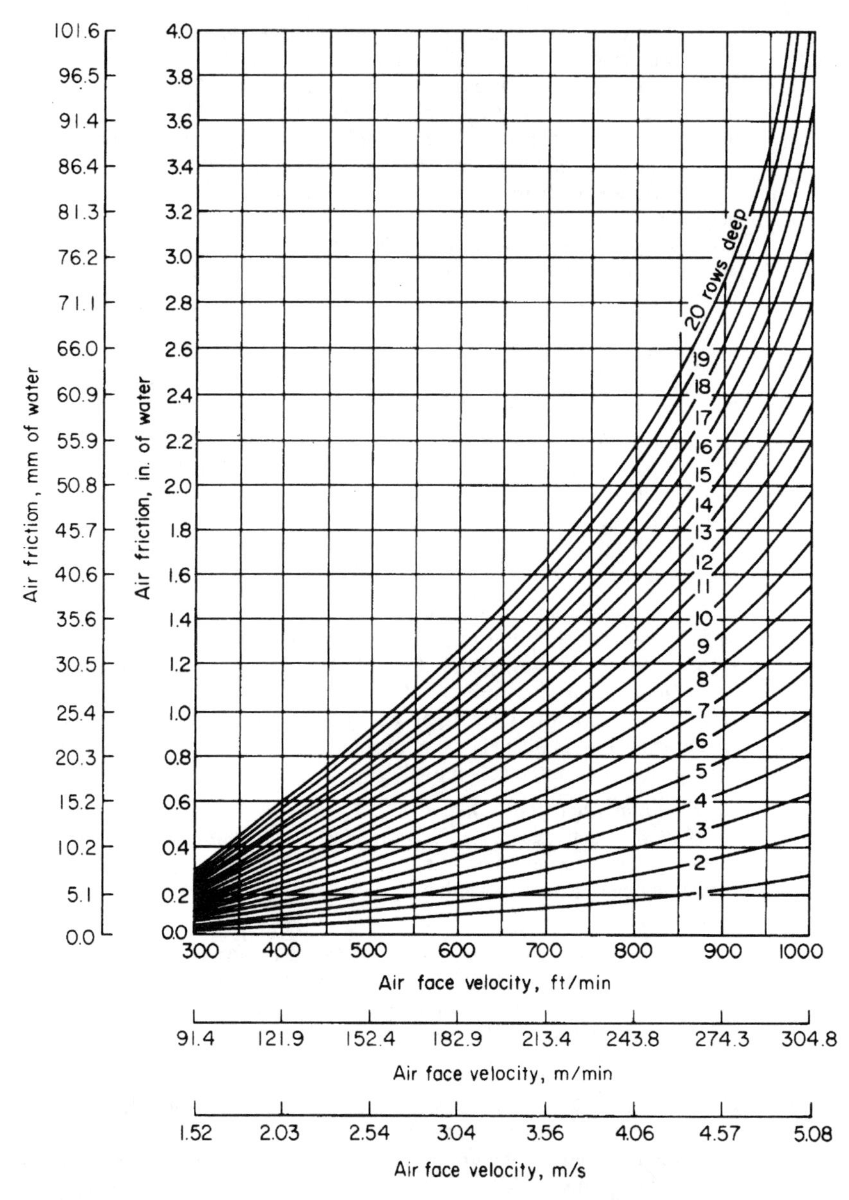

FIGURE 23 (*a*) Heating-coil air-friction chart for air at standard conditions of 70°F (21.1°C) and 29.92 in (76 cm) (*McQuay, Inc.*)

heating coil is 20°F (–6.7°C), with water supplied at 150 to 225°F (65.6 to 107.2°C). (*e*) Determine the tube water velocity, ft/s, from (8.33)(gal/min)/(384)(number of tubes in heating coil). The number of tubes in the coil is obtained by making a preliminary selection of the coil, using heating-capacity tables similar to Table 10. The usual hot-water heating coil has a water velocity between 2 and 6 ft/s (0.6 and 1.8 m/s). (*f*) Compute the number of tube rows required from Btu/h heating

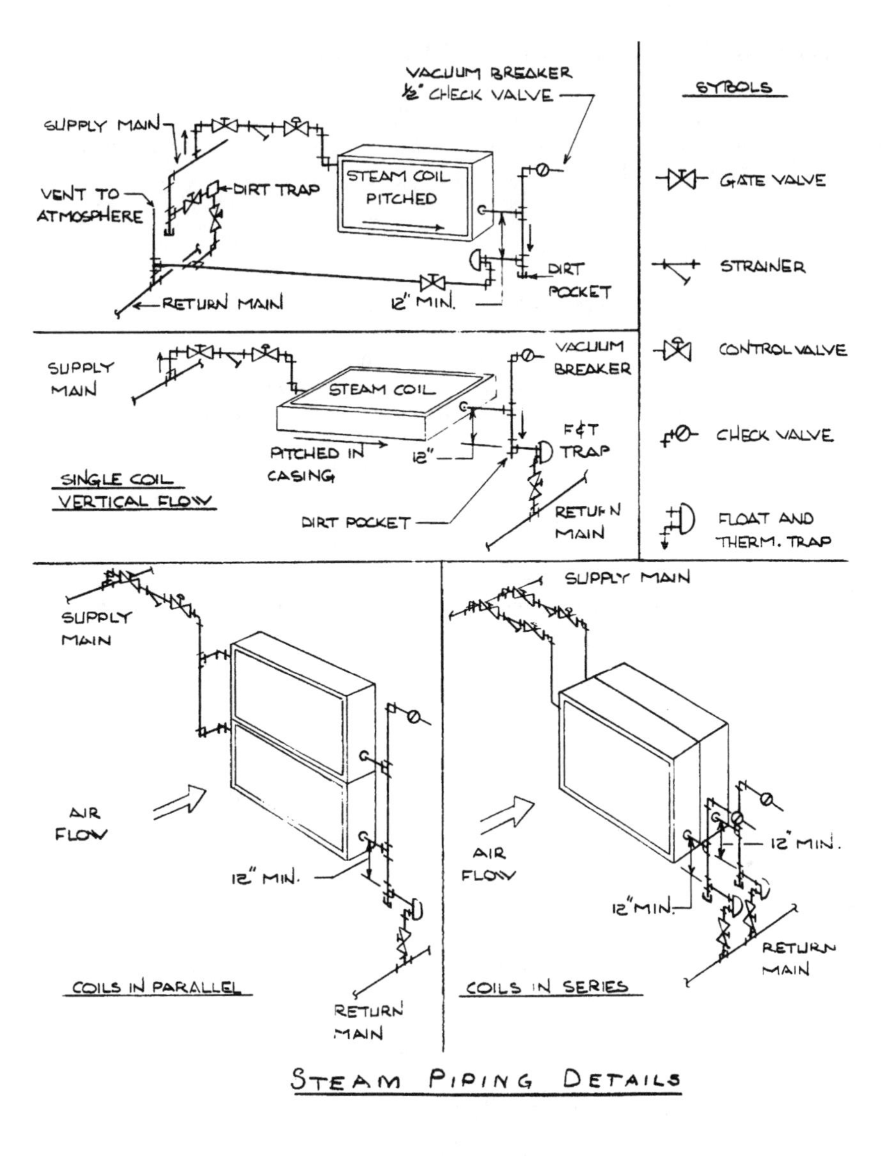

FIGURE 23 (*b*) Several typical piping arrangements for steam-heated air-heating coils. (*Jerome F. Mueller.*)

capacity/(face area ft^2)(logarithmic mean-effective-temperature difference from step 3)(K factor from the manufacturer's engineering data). (g) Compute the coil air resistance or friction loss, using the manufacturer's chart or table. The usual friction loss ranges from 0.375 to 0.675 in (9.5 to 17.2 mm) of water for commercial applications to about 1 in (25 mm) of water for industrial installations.

CHOOSING AND SIZING RADIANT HEATING PANELS

One room of a building has a heat loss of 13,900 Btu/h (4074.1 W). Choose and size a radiant heating panel suitable for this room. Illustrate the trial method of panel choice. The floor of the room is made of wooden blocks.

Calculation Procedure:

1. ***Choose the type and location of the heating coil***

Compute the heat loss for a given room or building, using the method given earlier in this section. Once the heat loss is known, choose the type of heating panel to use—ceiling or floor. In some rooms or buildings, a combination of floor and ceiling panels may prove more effective than either type used alone. Wall panels are also used but not as extensively as floor and ceiling panels.

In general, ceiling panels are embedded in concrete or plaster, as are wall panels. Floor panels are almost always embedded in concrete. Hence, use of another type of floor—block, tile, wood, or metal—may rule out the use of floor panels. Since this room has a wooden-block floor, a ceiling panel will be chosen.

2. ***Size the heating panel***

Table 13 shows the maximum Btu/h (W) heat output of 3/8-in (9.5-mm) copper-tube ceiling panels. The 3/8-in (9.5-mm) size is popular; however; other sizes—½-, 5/8-, and 7/8-in (12.7-, 15.9-, and 22.2-mm) diameter—are also used, depending on the heat load served. Using 3/8-in (9.5-mm) tubing on 6-in (152.4-mm) centers embedded in a plaster ceiling will provide a heat output of 60 Btu/(h · ft^2) (189.1 W/m^2) of tubing, as shown in Table 13. Figure 24 shows a typical piping diagram for radiant ceiling heating and cooling coils.

To obtain the area of the heated panel, A ft^2, use the relation $A = H_L/P$, where H_L = room or building heat loss, Btu/h; P = panel maximum heat output, Btu/(h · ft^2). For this room, A = 13,900/60 = 232 ft^2 (21.6 m^2).

3. ***Determine the total length of tubing required***

Use the appropriate tube-length factor from Table 13, or (2.0)(232) = 464 lin ft (141.4 m) of 3/8-in (9.5-mm) copper tubing.

4. ***Find the maximum panel tube length***

To stay within the commercial limits of smaller hot-water circulating pumps, the maximum tube lengths per panel circuit given in Table 13 are generally used. This tabulation shows that the maximum panel unit tube length for 3/8-in (9.5-mm) tubing on 6-in (152.4-mm) centers is 165 ft (50.3 m). Such a length will not require a pump head of more than 4 ft (1.22 m), excluding the head loss in the mains.

5. ***Determine the number of panels required***

Find the number of panels required by dividing the linear tubing length needed by the maximum unit tube length, or 464/165 = 2.81. Use three panels, the next larger whole number, because partial panels cannot be used. To conserve tubing and reduce the first cost of the installation, three panels, each having 155 lin ft (47.2 m) of piping, would be used. Note that the tubing length chosen for the actual panel must be *less than* the maximum length listed in Table 13.

6. ***Find the required piping main size required***

Determine the number of panels required in the remainder of the building. Use Table 13 to select the proper main size for a pressure loss of about 0.5 ft/100 ft (150 mm/30 m) of the main, including the supply and return lines. Thus, if 12 ceiling panels of maximum length were used in the building, a 1½-in (38.1-mm) main would be used.

7. ***Use the trial method to choose the main size***

If the size of the main for the panels required cannot be found in Table 13, compute the total Btu required for the panels from Table 13. Then, by trial, find the size of the main from

TABLE 13 Heating Panel Characteristics

Maximum Btu/(h · ft²) (W/m²) of 3/8-in (9.5-mm) copper tube embedded in ceiling plaster		Maximum Btu/(h · ft²) (W/m²) of copper tube embedded in concrete floor	
	Approx.		*Approx.*
4½ in (114.3 mm) center to center	75 (236.4)	½ in (12.7 mm) 9 in (228.6 mm) center to center	50 (157.6)
6 in (152.4 mm) center to center	60 (189.1)	¾ in (19.1 mm) 9 in (228.6 mm) center to center	50 (157.6)
9 in (228.6 mm) center to center	45 (141.8)	¾ in (19.1 mm) 12 in (304.8 mm) center to center	50 (157.6)
		1 in (25.4 mm) 12 in (304.8 mm) center to center	50 (157.6)

Total length of tube required, ft (m)*	
2.7 (0.82)	Where 4½-in (114.3-mm) centers are required
2 (0.61)	Where 6-in (152.4-mm) centers are required
1.3 (0.40)	Where 9-in (228.6-mm) centers are required
1 (0.30)	Where 12-in (304.8-mm) centers are required

Maximum panel unit tube length†

Ceilings				Floors			
Nominal size, in (mm)	Centers, in (mm)	Btu/(h · ft) (W/m) of tube	Ft (m)	Nominal size, in (mm)	Centers, in (mm)	Btu/(h · ft) (W/m) of tube	Ft (m)
3/8 (9.5)	4½ (114.3)	27 (25.9)	175 (47.9)	½ (12.7)	9 (228.6)	38 (36.5)	220 (67.1)
3/8 (9.5)	6 (152.4)	30 (28.9)	165 (50.3)	¾ (19.1)	9 (228.6)	38 (36.5)	400 (121.9)
3/8 (9.5)	9 (228.6)	34 (32.7)	150 (45.7)	¾ (19.1)	12 (304.8)	50 (48.1)	350 (106.7)
				1 (25.4)	12 (304.8)	50 (48.1)	550 (167.7)

Number of panel circuits of maximum length‡

Mains	Ceiling	Floor		
Diam, in (mm)	3/8 in — 4½, 6, 9 center-to-center (9.53 mm —114.3, 152.4, 228.6 center-to-center)	½ in—9 center-to-center (12.7 mm—228.6 center-to-center)	¾ in—9, 12 center-to-center (19.1 mm—228.6, 304.8 center-to-center)	1 in—12 center-to-center (25.4 mm—304.8 center-to-center)
2 (50.8)	27	16	8	5
1½ (38.1)	12	7	4	2
1¼ (31.6)	8	5	2	1
1 (25.4)	4	3	1	1
¾ (19.1)	2	1		

*To arrive at the required lin ft of tube per panel, multiply the ft² of heated panel by the factors given.

†To keep within the commercial limits of the smaller pumps, as an example the above maximum tube lengths per panel circuit are suggested. These lengths alone will require not more than about a 4-ft (1.2-m) head. This does not include loss in mains.

‡Use the information given in the section on maximum panel unit tube length, as given above, which can be supplied, allowing about 0.5-ft (150-mm) head required per 100 ft (30 m) of main (supply and return).

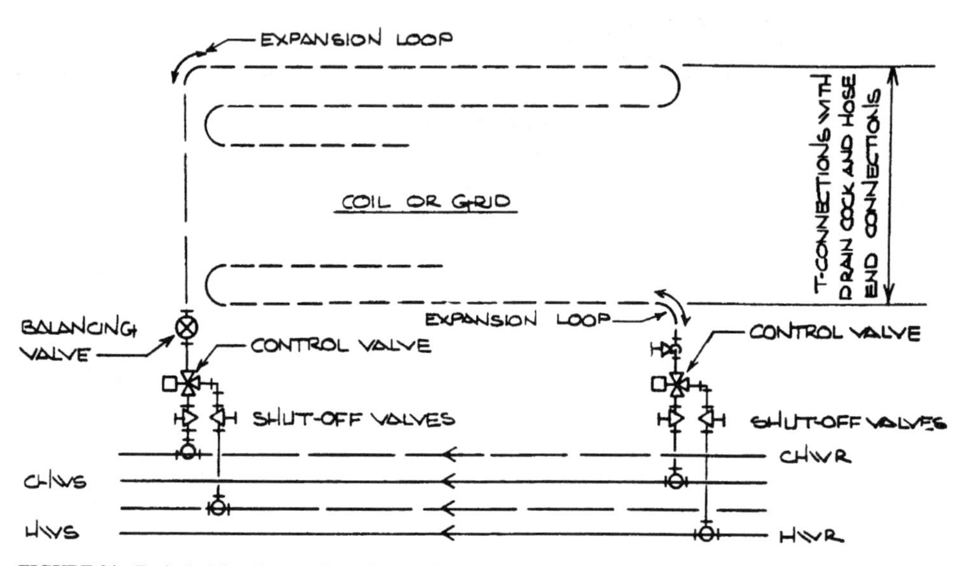

FIGURE 24 Typical piping diagram for radiant ceiling heating and cooling coils. (*Jerome F. Mueller.*)

Table 13 that will deliver approximately, and preferably somewhat more, than this total Btu/h requirement.

For instance, suppose an industrial building requires the following panel circuits:

No. of panel circuits	Tube size and location	Btu (kJ) required from Table 13 (number of circuits × Btu/ft of tube × maximum panel tube length)
4	3/8-in (9.5-mm) tubes on 4½-in (114.3-mm) centers, ceiling	4 × 27 × 175 = 18,900 (19,940)
1	½-in (12.7-mm) tubes on 9-in (228.6-mm) centers, floor	1 × 38 × 220 = 8,360 (8,820)
1	l-in (25.4-mm) tubes on 12-in (304.8-mm) centers, floor	1 × 50 × 550 = 27,500 (29,013)
3	¾-in (19.1-mm) tubes on 9-in (228.6-mm) centers, floor	3 × 38 × 400 = 45,600 (48,108)
9		Total Btu/h (kJ/h) required = 100,360 (105,881)

Trial 1. Assume 100 ft of 1½-in (30 m of 38-mm) main is used for seven floor circuits that are each 220 ft (67.1 m) long and made of ½-in (12.7-mm) tubing on 9-in (228.6-mm) centers. From Table 13 the output of these seven floor circuits is 7 circuits × 38 Btu/(h · ft) of tube (36.5 W/m) × 220 ft (67.1 m) of tubing = 58,520 Btu/h (17.2 kW). Since this output is considerably less than the required output of 100,360 Btu/h (29.4 kW), a 1½-in (38.1-mm) main is not large enough.

Trial 2. Assume 100 ft of 2-in (30 m of 50.8-mm) main for 16 circuits that are each 220 ft (67.1 mm) long and are made of ½-in (12.7-mm) tubing on 9-in (228.6-mm) centers. Then, as in trial 1, 16 × 38 × 220 = 133,760 Btu/h (39.2 kW) delivered. Hence, a 2-in (50.8-mm) main is suitable for the nine panels listed above.

Related Calculations. Use the same general method given here for heating panels embedded in the concrete floor of a building. The liquid used in most panel heating systems is water at about 130°F (54.4°C). This warm water produces a panel temperature of about 85°F (29.5°C) in floors and about 115°F (46.1°C) in ceilings. The maximum water temperature at the boiler is seldom allowed to exceed 150°F (65.6°C).

When the first-floor ceiling of a multistory building is not insulated, the floor above a ceiling panel develops about 17 Btu/(ft^2 · h) (53.6 W/m^2) from the heated panel below. If this type of construction is used, the radiation into the room above can be deducted from the heat loss computed for that room. It is essential, however, to calculate only the heat output in the floor area directly above the heated panel.

Standard references, such as ASHRAE *Guide and Data Book*, present heat-release data for heating panels in both graphical and tabular form. Data obtained from charts are used in the same way as described above for the tabular data. Tubing data for radiant heating are available from Anaconda American Brass Company.

CHOOSING AND SIZING SNOW-MELTING HEATING PANELS

Choose and size a snow-melting panel to melt a maximum snowfall of 3 in/h (76.2 mm/h) in a parking lot that has an area of 1000 ft^2 (92.9 m^2). Heat losses downward, at the edges, and back of the slab are about 25 percent of the heat supplied; also, there is an atmospheric evaporation loss of 15 percent of the heat supplied. The usual temperature during snowfalls in the locality of the parking lot is 32°F (0°C).

Calculation Procedure:

1. *Compute the hourly snowfall weight rate*

The density of the snow varies from about 3 lb/ft^3 at 5°F (48 kg/m^3 at –15°C) to about 7.8 lb/ft^3 at 34°F (124.9 kg/m^3 at 1.1°C). Given a density of 7.3 lb/ft^3 (116.9 kg/m^3) for this installation, the hourly snowfall weight rate per ft^2 is (area, ft)(depth, ft)(density) = (1.0)(3/12)(7.3) = 1.83 lb/(h · ft^2) [8.9 kg/(h · m^2)]. In this computation, the rate of fall of 3 in/h (76.2 mm/h) is converted to a depth in ft by dividing by 12 in/ft.

2. *Compute the heat required for snow melting*

The heat of fusion of melting snow is 144 Btu/lb (334. 9 kJ/kg). Since the snow accumulates at the rate of 1.83 lb/(h · ft^2) [8.9 kg/(h · m^2)], the amount of heat that must be supplied to melt the snow is (1.83)(144) = 264 Btu/(ft^2 · h) (832.1 W/m^2).

Of the heat supplied, the percent lost is 25 + 15 = 40 percent as given. Of this total loss, 25 percent is lost downward and 15 percent is lost to the atmosphere. Hence, the total heat that must be supplied is (1.0 + 0.40(264) = 370 Btu/(h · ft^2) (1166 W/m^2). With an area of 1000 ft^2 (92.9 m^2) to be heated, the panel system must supply (1000)(370) = 370,000 Btu/h (108.6 kW).

3. *Determine the length of pipe or tubing required*

Consult the ASHRAE *Guide and Data Book* or manufacturer's engineering data to find the heat output per ft of tubing or pipe length. Suppose the heat output is 50 Btu/(h · ft) (48.1 W/m) of tube. Then the length of tubing required is 370,000/50 = 7400 ft (2256 m).

Some manufacturers rate their pipe or tubing on the basis of rainfall equivalent of the snowfall and the wind velocity across the heated surface. Where this method is used, compute the heat required to melt the snow as the equivalent amount of heat to vaporize the water. This is $Q_e = 1074(0.002V + 0.055)(0.185 - \upsilon_a)$, where Q_e = heat required to vaporize the water, Btu/(h · ft^2); V = wind velocity over the heated surface, mi/h; υ_a = vapor pressure of the atmospheric air, inHg. Figure 25 shows the piping arrangement for a typical snow-melting system.

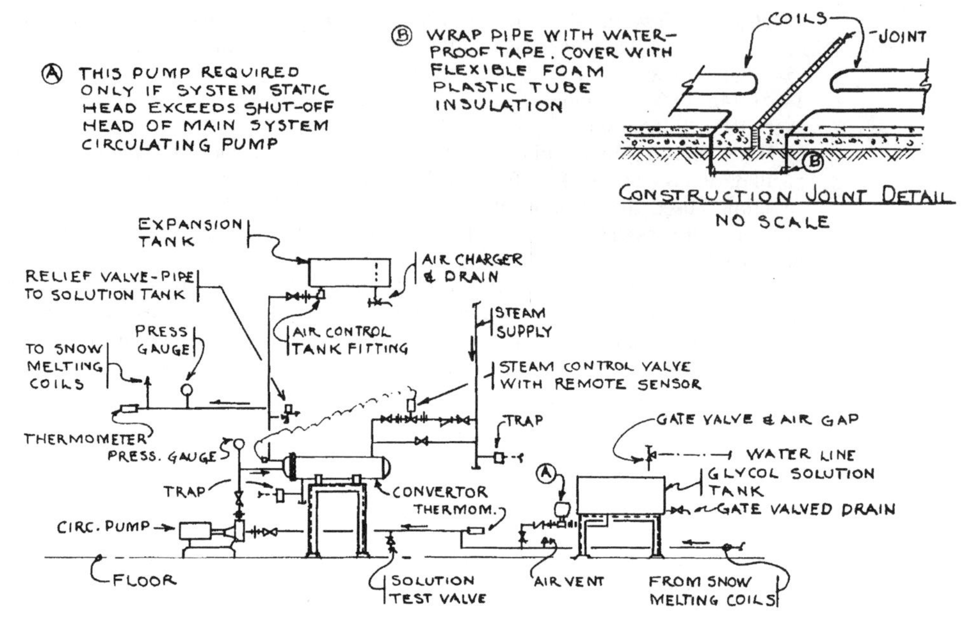

FIGURE 25 Piping arrangement for typical snow-melting system. (*Jerome E Mueller.*)

4. ***Determine the quantity of heating liquid required***

Use the relation gal/min = $0.125\ H_t/dc\ \Delta t$, for ethylene glycol, the most commonly used heating liquid. In this relation, H_t = total heat required for snow melting, Btu/h; d = density of the heating liquid, lb/ft^3; c = specific heat of the heating liquid, Btu/(lb · °F); Δt = temperature loss of the heating liquid during passage through the heating coil, usually taken as 15 to 20°F (−9.4 to −6.7°C).

Assuming that a 60 percent ethylene glycol solution is used for heating, we have d = 68.6 lb/ft^3 (1098.3 kg/m^3); c = 0.75 Btu/(lb · °F) [3140 J/(kg · K)]. Since the piping must supply 370,000 Btu/h (108.5 kW), gal/min = (0.125)(370,000)/(68.6)(0.75)(20) = 45 gal/min (170.3 L/min) when the temperature loss of the heating liquid is 20°F (−6.7°C).

5. ***Size the heater for the system***

The heater must provide at least 370,000 Btu/h (108.5 kW) to the ethylene glycol. If the heater has an overall efficiency of 60 percent, then the required heat input to deliver 370,000 Btu/h (108.5 kW) is 370,000/0.60 = 617,000 Btu/h (180.6 kW).

To avoid a long warm-up time at the start of a snowfall, the usual practice is to operate the system for several hours prior to an expected snowfall. The heating liquid temperature during warm-up is kept at about 100°F (37.8°C) and the pump is operated at half-speed.

Related Calculations. Use this general method to size snow-melting systems for sidewalks, driveways, loading docks, parking lots, storage yards, roads, and similar areas. To prevent an excessive warm-up load on the system, provide for prestorm operation. Without prestorm operation, the load on the heater can be twice the normal hourly load.

Cooper tubing and steel pipe are the most commonly used heating elements. For properties of tubing and piping important in snow-melting calculations, see Baumeiser—*Standard Handbook for Mechanical Engineers.*

ENERGY SAVINGS IN SPACE HEATING WITH HEAT RECOVERY FROM LIGHTING SYSTEMS

Determine the quantity of heat obtainable from 30 water-cooled fluorescent luminaires rated at 200 W each if the entering water temperature is 70°F (21.1°C) and the water flow rate is 1.0 gal/min (3.8 L/min). How much heat can be recovered from 10 air-cooled fluorescent luminaires rated at 100 W each?

Calculation Procedure:

1. *Compute the water-cooled luminaire heat recovery*
Luminaire manufacturers publish heat-recovery data in chart form (Fig. 26). This chart shows that with a flow rate of 0.5 gal/min (1.9 L/min) and a 70°F (21.1°C) entering water temperature, the heat recovery from a luminaire is 74 percent of the total input to the fixture.

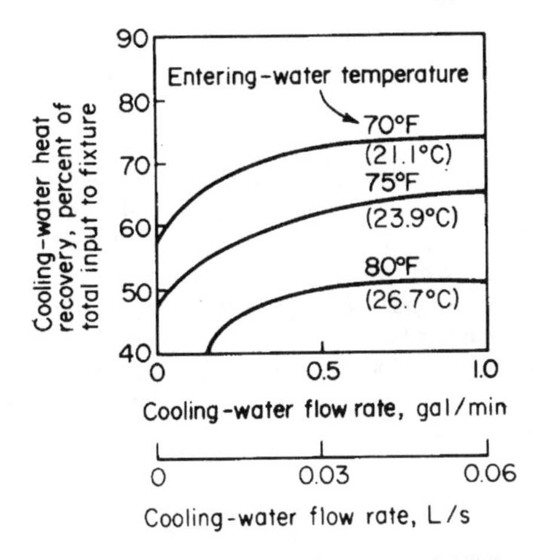

FIGURE 26 Heat recovery in water-cooled lighting fixtures.

For a group of lighting fixtures, total input, W = (number of fixtures) (rating per fixture, W). Or, for this installation, input, W = (30) (200) = 6000 W. Since 74 percent of this input is recoverable by the cooling water, recovered input = 0.74(6000) = 4400 W.

To convert incandescent lighting watts to Btu/h, multiply by 3.4. Where fluorescent lights are used, apply a factor of 1.25 to include the heat gain in the lamp ballast. Thus, the heat available for recovery from these fluorescent lamps is (3.4) (1.25) (4440) = 18,900 Btu/h (5.5 kW).

2. *Compute the temperature of the water*
Find the temperature rise of the water from Δt = (Btu/h)/500 gal/min, where Δt = temperature rise of the water, F; Btu/h = heat available; gal/min = water flow rate through the luminaire. Or, Δt = 18,9000/500 (1.0) = 37.8°F (3.2°C). A temperature rise of this magnitude is seldom used in practice. However, this calculation shows the large amount of heat recoverable with water-cooled lighting fixtures.

3. ***Compute the heat recoverable with air cooling***
In the usual air-cooled luminaire, 50 to 70 percent of the input energy is recoverable. Assuming a 60 percent recovery, with a total input of (10)(100) = 1000 W, we see that the energy recoverable is 0.60(1000) = 600 W. Converting to the heat recoverable gives (3.4)(1.25)(600) = 2545 Btu/h (745.9 W).

Related Calculations. Heat recovery from lighting fixtures is receiving increasing attention for many different structures because substantial fuel savings are possible. The water or air heated by the lighting is used to heat the supply or return air supplied to the conditioned space. Where the heat recovered must be rejected, as in the summer, either a cooling tower (for water) or an air-cooled condenser (for air) may be used.

Other popular sources of heat are refrigeration condensers and electric motors. In many installations the heat is absorbed from the condenser or motors, or both, by air.

ENERGY AND HEAT-LOAD COMPUTATION FOR AIR-CONDITIONING SYSTEMS—GENERAL METHOD

Show how to compute the total heat load for an air-conditioned industrial building fitted with windows having shades, internal heat loads from people and machines, and heat transmission gains through the walls, roof, and floor. Use the ASHRAE *Guide and Data Book* as a data source.

Calculation Procedure:*

1. ***Determine the design outdoor and indoor conditions***
Refer to the ASHRAE *Guide and Data Book* (called *Guide*) for the state and city in which the building is located. Read from the *Guide* table the design outdoor dry-bulb and wet-bulb temperatures for the appropriate city. At the same time, determine from the *Guide* the indoor-design conditions—temperature and relative humidity for the type of application being considered. The *Guide* lists a variety of typical applications such as apartment houses, motels, hotels, industrial plants, etc. It also lists the average summer wind velocity for a variety of locations. Where the exact location of a plant is not tabulated in the *Guide*, consult the nearest local branch of the weather bureau for information on the usual summer outdoor high and low dry- and wet-bulb temperatures, relative humidity, and velocity.

2. ***Compute the sunlight heat gain***
The sunlight heat gain results from the solar radiation through the glass in the building's windows and the materials of construction in certain of the building's walls. If the glass or wall of a building is shaded by an adjacent solid structure, the sunlight heat gain for that glass and wall is usually neglected. The same is true for the glass and wall of the building facing the north.

Compute the glass sunlight heat gain, Btu/h, from (glass area, ft^2)(equivalent temperature difference from the appropriate *Guide* table)(factor for shades, if any are used). The equivalent temperature difference is based on the time of day and orientation of the glass with respect to the points of the compass. A latitude correction factor may have to be applied if the building is located in a tropical area. Use the equivalent temperature difference for the time of day on which the heat-load estimate is based. Several times may be chosen to determine at which time the greatest heat gain occurs. Where shades are used in the building, choose a suitable shade factor from the appropriate *Guide* table and insert it in the equation above.

Compute the sunlight heat gain, Btu/h, for the appropriate walls and the roof from (wall area, ft^2) (equivalent temperature difference from *Guide* for walls) [coefficient of heat transmission for the wall, Btu/(h · ft^2 · °F)]. For the roof, find the heat gain, Btu/h, from (roof area, ft^2) (equivalent temperature difference from the appropriate *Guide* table for roofs) [coefficient of heat transmission for the roof, Btu/(h · ft^2· °F)].

*SI units are given in later numerical procedures.

3. *Compute the transmission heat gain*
All the glass in the building windows is subject to transmission of heat from the outside to the inside as a result of the temperature difference between the outdoor and indoor dry-bulb temperatures. This transmission gain is commonly called the *all-glass gain.* Find the all-glass transmission heat gain, Btu/h, from (total window-glass area, ft^2)(outdoor design dry-bulb temperature, °F—indoor design dry-bulb temperature, °F)[coefficient of heat transmission of the glass Btu/(h · ft^2 · °F), from the appropriate *Guide* table].

Compute the heat transmission, Btu/h, through the shaded walls, if any, from (total shaded wall area, ft^2)(equivalent temperature difference for shaded wall from the appropriate *Guide* table, °F) [coefficient of heat transmission of the wall material, Btu/(h · ft^2 · °F), from the appropriate *Guide* table)].

Where the building has a machinery room or utility room that is not air-conditioned and is next to a conditioned space and the temperature in the utility room is higher than in the conditioned space, find the heat gain, Btu/h, from (area of utility or machine room partition, ft^2)(utility or machine room dry-bulb temperature, °F – conditioned-space dry-bulb temperature, °F)[coefficient of heat transmission of the utility or machine room partition, Btu/(h · ft^2 · °F), from the appropriate *Guide* table].

For buildings having a floor contacting the earth, or over unventilated and unheated basements, there is generally *no* heat gain through the floor because the ground is usually at a lower temperature than the floor. Where the floor is above the ground and in contact with the outside air, find the heat gain, Btu/h, through the floor from (floor area, ft^2)(design outside dry-bulb temperature, °F—design inside dry-bulb temperature. °F) [coefficient of heat transmission of the floor material, Btu/(h · ft^2 · °F), from the appropriate *Guide* table]. When a machine room or utility room is below the floor, use the same relation but substitute the machine room or utility room dry-bulb temperature for the design outside dry-bulb temperature.

For floors above the ground, some designers reduce the difference between the design outdoor and indoor dry-bulb temperatures by 5°F (–15°C); other designers use the shaded-wall equivalent temperature difference from the *Guide.* Either method, or that given above, will provide safe results.

4. *Compute the infiltration heat gain*
Use the relation infiltration heat gain, Btu/h = (window crack length, ft) [window infiltration, ft^3/(ft · min) from the appropriate *Guide* table] (design outside dry-bulb temperature, °F)(1.08). Three aspects of this computation require explanation.

The window crack length used is usually one-half the total crack length in all the windows. Infiltration through cracks is caused by the wind acting on the building. Since the wind cannot act on all sides of the building at once, one-half the total crack length is generally used (but never less than one-half) in computing the infiltration heat gain. Note that the crack length varies with different types of windows. Thus the *Guide* gives, for metal sash, crack length = total perimeter of the movable section. For double-hung windows, the crack length = three times the width plus twice the height.

The window infiltration rate, ft^3/(min · ft) of crack, is given in the *Guide* for various wind velocities. Some designers use the infiltration rate for a wind velocity of 10 mi/h (16.1 km/h); others use 5 mi/h (8.1 km/h). The factor 1.08 converts the computed infiltration to Btu/h (× 0.32 = W).

5. *Compute the outside-air bypass heat load*
Some outside air may be needed in the conditioned space to ventilate fumes, odors, and other undesirables in the conditioned space. This ventilation air imposes a cooling or dehumidifying load on the air conditioner because the heat or moisture, or both, must be removed from the ventilation air. Most air conditioners are arranged to permit some outside air to bypass the cooling coils. The bypassed outdoor air becomes a load within the conditioned space similar to infiltration air.

Determine heat load, Btu/h, of the outside air bypassing the air conditioner from (cfm of ventilation air) (design outdoor dry-bulb-temperature, °F–design indoor dry-bulb temperature, °F)(air-conditioner bypass factor)(1.08).

Find the ventilation-air quantity by multiplying the number of people in the conditioned space by the ft^3/min per person recommended by the *Guide.* The ft^3/min (m^2/min) per person can range from a minimum of 5 (0.14) to a high of 50 (1.42) where heavy smoking is anticipated. If industrial processes within the conditioned space require ventilation, the air may be supplied by increasing the

TABLE 14 Typical Bypass Factors

a. For various applications

Coil bypass factor*	Type of application	Example
0.30–0.50	A *small* total load or a load that is somewhat larger with a low sensible heat factor (high latent load)	Residence
0.20–0.30	Typical comfort application with a *relatively small* total load or a low sensible heat factor with a somewhat larger load	Residence; small retail shop; factory
0.10–0.20	Typical comfort application	Department store; bank; factory
0.05–0.10	Applications with high internal sensible loads or requiring a large amount of outdoor air for ventilation	Department store; restaurant; factory
0–0.10	All outdoor air applications	Hospital; operating room; factory

b. For finned coils

	Without sprays		With sprays*	
	8 fins/in (25.4 mm)	14 fins/in (25.4 mm)	8 fins/in (25.4 mm)	14 fins/in (25.4 mm)
	Velocity, ft/min (m/min)			
Depth of coils, rows	300–700 (91.4–213.4)	300–700 (91.4–213.4)	300–700 (91.4–213.4)	300–700 (91.4–213.4)
2	0.42–0.55	0.22–0.38		
3	0.27–0.40	0.10–0.23		
4	0.19–0.30	0.50–0.14	0.12–0.22	0.03–0.10
5	0.12–0.23	0.02–0.09	0.08–0.14	0.01–0.08
6	0.08–0.18	0.01–0.06	0.06–0.11	0.01–0.05
8	0.03–0.08	—	0.02–0.05	

*The bypass factor with spray coils is decreased because spray provides more surface for contacting the air.
Carrier Air Conditioning Company.

outside air flow, by a local exhaust system at the process, or by a combination of both. Regardless of the method used, outside air must be introduced to make up for the air exhausted from the conditioned space. The sum of the air required for people and processes is the total ventilation-air quantity.

Until the air conditioner is chosen, its bypass factor is unknown. However, to solve for the outside-air bypass heat load, a bypass factor must be applied. Table 14 shows typical bypass factors for various applications.

6. *Compute the heat load from internal heat sources*

Within an air-conditioned space, heat is given off by people, lights, appliances, machines, pipes, etc. Find the sensible heat, Btu/h, given off by people by taking the product (number of people in the air-conditioned space) (sensible-heat release per person, Btu/h, from the appropriate *Guide* table). The sensible-heat release per person varies with the activity of each person (seated, at rest, doing heavy work) and the dry-bulb temperature. Thus, at 80°F (26.7°C), a person doing heavy work in a factory will give off 465 Btu/h (136.3 W) sensible heat; seated at rest in a theater at 80°F (26.7°C) a person will give off 195 Btu/h (57.2 W) sensible heat.

Find the heat, Btu/h, given off by electric lights from (wattage rating of all installed lights) (3.4). Where the installed lighting capacity is expressed in kilowatts, use the factor 3413 instead of 3.4.

For electric motors, find the heat, Btu/h, given off from (total installed motor hp) (2546)/motor efficiency expressed as a decimal. The usual efficiency assumed for electric motors is 85 percent.

Many other sensible-heat-generating devices may be used in an air-conditioned space. These devices include restaurant, beauty shop, hospital, gas-burning, and kitchen appliances. The *Guide* lists the heat given off by a variety of devices, as well as pipes, tanks, pumps, etc.

7. *Compute the room sensible heat*
Find the sum of the sensible heat gains computed in steps 2 (sunlight heat gain), 3 (transmission heat gain), 4 (infiltration heat gain), 5 (outside air heat gain), 6 (internal heat sources). This sum is the room sensible-heat subtotal.

A further sensible-heat gain may result from supply-duct heat gain, supply-duct leakage loss, and air-conditioning-fan horsepower. To the sum of these losses, a safety factor is usually added in the form of a percentage, since all the losses are also generally expressed as a percentage. The *Guide* provides means to estimate each loss and the safety factor. Assuming the sum of the losses and safety factor is x percent, the room sensible heat load, Btu/h, is $(1 + 0.01 \times x)$ (room sensible heat subtotal, Btu/h).

8. *Compute the room latent-heat load*
The room latent-heat load results from the moisture entering the room with the infiltration air and bypass ventilation air, the moisture given off by room occupants, and any other moisture source such as open steam kettles, sterilizers, etc.

Find the infiltration-air latent-heat load, Btu/h from (cfm infiltration) (moisture content of outside air design at design-outdoor conditions, g/lb—moisture content of the conditioned air at the design-indoor conditions, g/lb) (0.68). Use a similar relation for the bypass ventilation air, or Btu/h latent heat = (cfm ventilation)(moisture content of the outside air at design-outdoor conditions, g/lb-moisture content of conditioned air at design-indoor conditions, g/lb)(bypass factor) (0.68).

Find the latent-heat gain from the room occupants, Btu/h, from (number of occupants in the conditioned space) (latent-heat gain, Btu/h per person, from the appropriate *Guide* table). Be sure to choose the latent-heat gain that applies to the activity *and* conditioned-room dry-bulb temperature.

Nonhooded restaurant, hospital, laboratory, and similar equipment produces both a sensible- and latent-heat load in the conditioned space. Consult the *Guide* for the latent heat load for each type of unit in the space. Find the latent-heat load of these units, Btu/h, from (number of units of type) (latent-heat load, Btu/h, per unit).

Take the sum of the latent-heat loads for infiltration, ventilation bypass air, people, and devices. This sum is the room latent-heat subtotal, if water-vapor transmission through the building surfaces is neglected.

Water vapor flows through building structures, resulting in a latent-heat load whenever a vapor-pressure difference exists across a structure. The latent-heat load from this source is usually insignificant in comfort applications and needs to be considered only in low or high dew-point applications. Compute the latent-heat gain from this source, using the appropriate *Guide* table, and add it to the room latent-heat subtotal.

Factors for the supply-duct leakage loss and for a safety margin are usually applied to the above sum. When all the latent-heat subtotals are summed, the result is the room latent-heat total.

9. *Compute the outside-air heat*
Air brought in for space ventilation imposes a sensible- and latent-heat load on the air-conditioning apparatus. Compute the sensible-heat load, Btu/h, from (cfm outside air)(design outdoor dry-bulb temperature – design indoor dry-bulb temperature)(1 – bypass factor)(1.08). Compute the latent-heat load, Btu/h, from (cfm outside air) (design-outdoor moisture content, g/lb – design-indoor moisture content, g/lb)(l – bypass factor)(0.68). Apply percentage factors for return duct heat and leakage gain, pump horsepower, dehumidifier, and piping loss, see the *Guide* for typical values.

10. *Compute the grand-total heat and refrigeration tonnage*
Take the sum of the room total heat and outside-air heat. The result is the grand-total heat load of the space, Btu/h.

Compute the refrigeration load, tons, from (grand-total heat, Btu/h) [12, 000 Btu/(h · ton) (3577 W) of refrigeration]. A refrigeration system having the next higher standard rating is generally chosen.

11. *Compute the sensible-heat factor and the apparatus dew point*
For any air-conditioning system, sensible-heat factor = (room sensible heat, Btu/h)/(room total heat, Btu/h). Using a psychrometric chart and the known room conditions, we find the apparatus dew point. An alternate, and quicker, way to find the apparatus dew point is to use the Carrier *Handbook of Air Conditioning System Design* tables.

12. *Compute the quantity of dehumidified air required*
Determine the dehumidified air temperature rise, °F, from (1 – bypass factor)(design-indoor temperature, °F – apparatus dew point, °F). Compute the dehumidified air quantity, ft^3/min, from (room sensible heat, Btu/h)/1.08 (dehumidified air temperature rise, °F)

Related Calculations. The general procedure given above is valid for all types of air-conditioned spaces—offices, industrial plants, residences, hotels, apartment houses, motels, etc. Use the ASHRAE *Guide and Data Book* or the Carrier *Handbook of Air Conditioning System Design* as a source of data for the various calculations. Application of this method to an actual building is shown in the next calculation procedure.

In actual design work, a calculations form incorporating the calculations shown above is generally used. Such forms are obtainable from equipment manufacturers. Since the usual form does not provide any explanation of the calculations, the present calculation procedure is a useful guide to using the form. Refer to the Carrier *Handbook of Air Conditioning System Design* for one such form.

In using SI units in air-conditioning design calculations, the same general steps are followed as given above. Numerical usage of SI is shown in the next calculation procedure.

With more buildings being built, modernized, and rehabilitated worldwide than ever before in our history, there are a number of important trends influencing the design and layout of these buildings. These trends include the following:

1. Emphasis on "green" designs, LEED,* developed by the U.S. Green Building Council (USGBC), with attention to the conservation of energy—hot water, steam, electricity, heating and cooling, etc. Green designs are proving to be economically wise. Studies show that for every dollar spent on green design, $2 to $3 in savings occur over the life span of the building. Life-cycle studies are now being carried out as part of the standard design procedure for many new, and rehabilitated, buildings.
2. Reduction of moisture within buildings is getting more attention than ever before. School buildings are receiving particular scrutiny for mold and moisture because children may develop health problems when proper indoor air quality is not delivered and controlled. Hence, designers of school mechanical systems must give greater attention to the results that their designs achieve.
3. Under-floor air ducts for air supply and return are becoming more popular because of installation savings and attractive design features—more space for occupants, reduced airflow noise, smaller areas needed for air-handling machinery, etc. The design requirements of under-floor ducts resemble, in general, those of conventional ducts, except possibly for different friction factors, depending on the duct material used.
4. Indoor air quality (IAQ) is high on every HVAC designer's list for improvements that must be made in all air-handling systems. The quality of indoor air is affected by many factors that a designer must consider, such as moisture, carpet outgassing, mold, occupant smoking, etc. With a variety of regulations to be observed, along with tenant requirements, indoor air quality is an important challenge to today's HVAC designer. Schools are particularly demanding in their requirements for better IAQ to benefit the health of their students and faculty.
5. As more buildings and residences are air-conditioned and utilize heat pumps for winter heating, the demand for qualified designers and installation and maintenance technicians is rising. Estimates at the time of this writing show that there's a shortage of some 50,000+ HVAC designers and technicians throughout the United States. Likewise, a similar shortage is believed to exist worldwide. This shortage is projected to increase as time passes, with little hope for a reversal.
6. Heating boiler efficiency is getter greater attention than ever because of green considerations. Benchmarking of boiler efficiency is now contained in a nationwide database showing typical efficiencies of boilers of various manufacturers and similar construction. Factors important in boiler efficiency include control accuracy, excess air regulation, burner cleanliness, etc.

---

*Leadership in Energy and Environmental Design. Many excellent reports on LEED can be found on the Internet under http://www.ofee.gov.

7. Reduction of greenhouse gas emissions from fossil fuels is receiving more attention from national and state agencies. Plants seeking to reduce greenhouse gas emissions from their equipment are being offered loans from the federal government to alter their equipment to produce the desired results.
8. Renewable energy sources for buildings and other structures are drawing greater attention today than in the last decade. Using renewable energy sources will help buildings qualify for LEED certification because their design is more "green." Further, use of renewable energy can produce faster "rent up" for new commercial and residential buildings. This generates revenues for the owner sooner, making the building more justifiable economically.
9. Today's designers must recognize the importance of LEED because it impacts almost every facet of building design and construction. Here are a few actual recent examples of how LEED designs improve a building or a group of them, as reported by the U.S. Department of Health and Human Services:
   - **(a)** A research laboratory reduced energy consumption by 39 percent, energy costs by 24 percent, and greenhouse gas emissions by 27 percent on a square-foot basis.
   - **(b)** A leased campus includes a 10-kW photovoltaic system, cogeneration, variable-frequency drives on chilled and condenser water pumping and cooling-tower fans, reduced lighting loads, variable air-volume systems with variable-frequency drives, demand control ventilation, night setback strategies, and an economizer cycle.
   - **(c)** A tall and narrow building was constructed to take advantage of natural lighting and walls made of architectural concrete that does not require insulation and drywall. Native vegetation was planted to reduce maintenance and irrigation requirements of the landscaping. Reclaimed water will be used for the plants.
   - **(d)** A renovation project for a hospital uses a groundwater-source heat pump for heating and cooling. About 60 heat pumps of various sizes serve the hospital 24 h/day. More than 100 wells beneath the parking lot provide the ground-source temperature exchange. A new direct digital control system provides precise control of each zone. Conservative and preliminary studies estimate that natural gas consumption has been reduced by 30 percent, or $10,000 per year. Maintenance hours have been decreased by 45 percent.
   - **(e)** A cogeneration plant has an approximate efficiency of 85 percent, saving more than 640 million Btu and approximately $3.6 million per year. This plant will reduce greenhouse gas emissions by about 100,000 tons/yr. Other pollutant emissions and particulate matter will be reduced by close to 600 tons/yr.

The five examples just cited give an idea of the range of LEED results. Today's designers must recognize the changes taking place and include them in their design approach.

ENERGY AND HEAT-LOAD COMPUTATION FOR AIR-CONDITIONING SYSTEMS—NUMERICAL CALCULATION

Determine the required capacity of an air-conditioning system to serve the industrial building shown in Fig. 27. The outside walls are 8-in (203.2-mm) brick with an interior finish of 3/8-in (9.50-mm) gypsum lath plastered and furred. A 6-in (152.4-mm) plain poured-concrete partition separates the machinery room from the conditioned space. The roof is 6-in (152.4-mm) concrete covered with ½-in (12.7-mm) thick insulating board, and the floor is 2-in (50.8-mm) concrete. The windows are double-hung, metal-frame locked units with light-colored shades three-quarters drawn. Internal heat loads are: 100 people doing light assembly work; twenty-five 1-hp (746-W) motors running continuously at full load; 20,000 W of light kept on at all times. The building is located in Port Arthur, Tx, at about 30° north latitude. The desired indoor design conditions are 80°F (26.7°C) dry bulb, 67°F (19.4°C) wet bulb, and 51 percent relative humidity. Air-conditioning equipment will be located in the machinery room. Use the general method given in the previous calculation procedure.

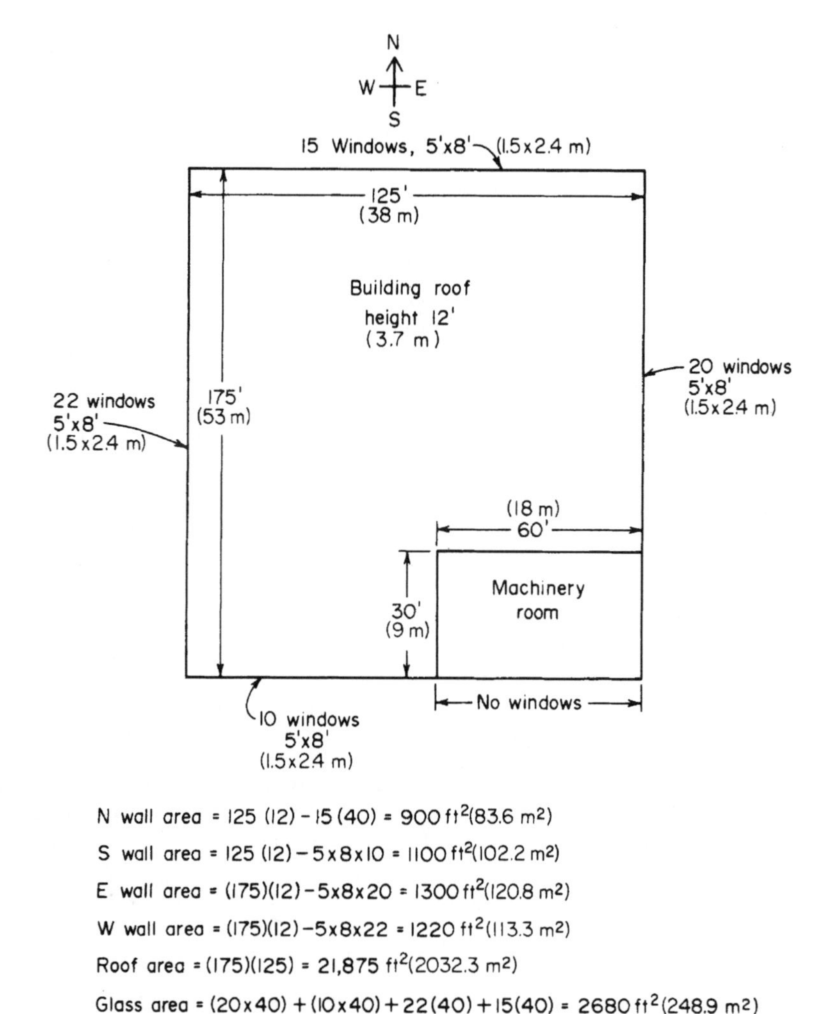

FIGURE 27 Industrial building layout.

Calculation Procedure:

1. ***Determine the design outdoor and indoor conditions***

The ASHRAE *Guide and Data Book* lists the design dry-bulb temperature in common use for Port Arthur, Tx, as 95°F (35°C) and the design wet-bulb temperature as 79°F (26.1°C). Design indoor temperature and humidity conditions are given; if they were not given, the recommended conditions given in the *Guide* for an industrial building housing light assembly work would be used.

2. ***Compute the sunlight heat gain***

The east, south, and west windows and walls of the building are subject to sunlight heat gains. North-facing walls are neglected because the sunlight heat gain is usually less than the transmission heat gain. Reference to the *Guide* table for sunlight radiation through glass shows that the largest amount of heat radiation occurs through the east and west walls. The maximum radiation is 181 Btu/(h · ft^2) (570.5 W/m^2) of glass area at 8 a.m. for the east wall and the same for the west wall at 4 p.m. Radiation through the glass in the south wall never reaches this magnitude. Hence, only the east or west wall need be considered. Since the west wall has 22 windows compared with 20 in the east wall, the west-wall sunlight heat gain will be used because it has a *larger* heat gain. (If both walls had an equal number of windows, either wall could be used.) When the window shades are normally three-quarters drawn, a value 0.6 times the tabulated sunlight radiation can be used. Hence, the west-glass sunlight heat gain = (22 windows)(5 × 8 ft each)(181)(0.6) = 95,600 Btu/h (28,020.4 W).

For the same time of day, 4 p.m., the *Guide* table shows that the east-glass radiation is 0 Btu/(ft^2) and the south glass is 2 Btu/(h · ft^2) (6.3 W/m^2). Hence, the south-glass sunlight heat gain = (10 windows) (5 × 8 ft each)(2)(0.6) = 480 Btu/h (140.7 W).

The same three walls, and the roof, are also subject to sunlight heat gains. Reference to the *Guide* shows that with 8-in (203-mm) walls the temperature difference resulting from sunlight heat gains is 15°F (8.3°C) for south walls and 20°F (–6.7°C) for east and west walls. At 4 p.m. the roof temperature difference is given as 40°F (4.4°C). Hence, sunlight gain, south wall = (wall area, ft^2)(temperature difference, °F) [wall coefficient of heat transfer, Btu/(h · ft^2 · °F)], or (1100)(15) (0.30) = 4950 Btu/h (1450.9 W). Likewise, the east-wall sunlight heat gain = (1300)(20)(0.30) = 7800 Btu/h (2286.2 W); the west-wall sunlight heat gain = (1220)(20)(0.30) = 7,320 Btu/h (2145.5 W); the roof sunlight heat gain = (21,875)(40)(0.33) = 289,000 Btu/h (84.7 kW). Note that the wall and roof coefficients of heat transfer are obtained from the appropriate *Guide* table.

The sum of the sunlight heat gains gives the total sunlight gain, or 405,150 Btu/h (118.7 kW).

3. *Compute the glass transmission heat gain*

All the glass in the building is subject to a transmission heat gain. Find the all-glass transmission heat gain, Btu/h, from (total glass area, ft^2)(outdoor design dry-bulb temperature, °F – indoor design dry-bulb temperature, °F)[coefficient of heat transmission of glass, Btu/(h · ft^2 · °F), from *Guide*], or (2680)(95 – 80)(1.13) = 45,400 Btu/h (13.3 kW).

The transmission heat gain of the south, east, and west walls can be neglected because the sunlight heat gain is greater. Hence, only the north-wall transmission heat gain need be computed. For unshaded walls, the transmission heat gain, Btu/h, is (wall area, ft^2)(design outdoor dry-bulb temperature, °F – design indoor dry-bulb temperature, °F)[coefficient of heat transmission, Btu/ (h · ft^2 · °F)] = (900)(95 – 80)(0.30) = 4050 Btu/h (1187.1 W).

The heat gain from the ground can be neglected because the ground is usually at a lower temperature than the floor. Thus, the total transmission heat gain is the sum of the individual gains, or 66,530 Btu/h (19.5 kW).

4. *Compute the infiltration heat gain*

The total crack length for double-hung windows is 3 × width + 2 × height, or (67 windows)[(3 × 5) + (2 × 8)] = 2077 ft (633.1 m). By using one-half the total length, or 2077/2 = 1039 ft (316.7 m), and a wind velocity of 10 mi/h (16.1 km/h), the leakage ft^3/min is (crack length, ft)(leakage per ft of crack) = (1039)(0.75) = 770 ft^3/min, or 60(779) = 46,740 ft^3/h (1323 m^3/h).

The heat gain due to infiltration through the window cracks is (leakage, ft^3/min) (design outdoor dry-bulb temperature, °F – design indoor dry-bulb temperature, °F)(1.08), or (779)(95 – 80)(1.08) = 12.610 Btu/h (3.7 kW).

5. *Compute the outside-air bypass heat load*

For factories, the *Guide* recommends a ventilation air quantity of 10 ft^3/min (0.28 m^3/min) per person. Local codes may require a larger quantity; hence, the codes should be checked before a final choice is made of the ventilation-air quantity used per person. Since there are 100 people in this factory, the required ventilation quantity is 100(10) = 1000 ft^3/min (28 m^3/min). Next, the bypass factor for the air-conditioning equipment must be chosen.

Table 14 shows that the usual factory air-conditioning equipment has a bypass factor ranging between 0.10 and 0.20. Assume a value of 0.10 for this installation.

The heat load, Btu/h, of the outside air bypassing the air conditioner is (cfm of ventilation air) (design outdoor dry-bulb temperature, °F – design indoor dry-bulb temperature, °F)(air-conditioner bypass factor)(1.08). Hence, (1000)(95 – 80)(0.10)(1.08) = 1620 Btu/h (474.8 W).

6. ***Compute the heat load from internal heat sources***

The internal heat sources in this building are people, lights, and motors. Compute the sensible-heat load of the people, from Btu/h = (number of people in the air-conditioned space)(sensible-heat release per person, Btu/h, from the appropriate *Guide* table). Thus, for this building with an 80°F (26.7°C) indoor dry-bulb temperature and 100 occupants doing light assembly work, the heat load produced by people = (100)(210) = 21,000 Btu/h (6.2 kW).

The motor heat load, Btu/h, is (motor hp)(2546)/motor efficiency. Given an 85 percent motor efficiency, the motor-heat load = (25)(2546)/0.85 = 75,000 Btu/h (21.9 kW). Thus, the total internal-heat load = 21,000 + 68,000 + 75,000 = 164,000 Btu/h (48.1 kW).

7. ***Compute the room sensible heat***

Find the sum of the sensible-heat gains computed in steps 2 through 6. Thus, sensible heat load = 405,150 + 66,530 + 12,610 + 1620 + 164,000 = 649,910 Btu/h (190.5 kW), say 650,000 Btu/h (190.5 kW). Using an assumed safety factor of 5 percent to cover the various losses that may be encountered in the system, we find room sensible heat = (1.05)(650,000) = 682,500 Btu/h (200.0 kW).

8. ***Compute the room latent-heat load***

The room latent load results from the moisture entering the air-conditioned space with the infiltration and bypass air, moisture given off by room occupants, and any other moisture sources.

Find the infiltration-heat load, Btu/h, from (cfm infiltration)(moisture content of outside air at design-outdoor conditions, g/lb – moisture content of the conditioned air at the design-indoor conditions, g/lb)(0.68). Using a psychrometric chart or the *Guide* thermodynamic tables, we get the infiltration latent-heat load = (779)(124 – 78)(0.68) = 24,400 Btu/h (7.2 kW).

Using a similar relation for the ventilation air gives Btu/h latent heat = (cfm ventilation air)(moisture content of outside air at design-outdoor conditions, g/lb – moisture content of the conditioned air at the design-indoor conditions, g/lb)(bypass factor)(0.68). Or, (1000)(124 – 78)(0.10)(0.68) = 3130 Btu/h (917.4 W).

The latent-heat gain from room occupants is Btu/h = (number of occupants in the conditioned space)(latent-heat gain, Btu/h per person, from the appropriate *Guide* table). Or, (100)(450) = 45,000 Btu/h (917.4 W).

Find the latent-heat gain subtotal by taking the sum of the above heat gains, or 24,400 + 3130 + 45,000 = 72,530 Btu/h (21.2 kW). Using an allowance of 5 percent for supply-duct leakage loss and a safety margin gives the latent heat gain = (1.05)(72,530) = 76,157 Btu/h (22.3 kW).

9. ***Compute the outside heat***

Compute the sensible-heat load of the outside ventilation air from Btu/h = (cfm outside ventilation air)(design outdoor dry-bulb temperature – design indoor dry-bulb temperature)(1 – bypass factor)(1.08). For this system, with 1000 ft^3/min (28.3 m^3/min) outside air ventilation, sensible heat = (1000)(95 – 80)(1 – 0.010)(1.08) = 14,600 Btu/h, (4.3 kW).

Compute the latent-heat load of the outside ventilation air from (cfm outside air) × (design-outdoor moisture content, g/lb – design-indoor moisture content, g/lb)(1 – bypass factor)(0.68). Using the moisture content from step 8 gives (1000)(124 – 78)(1 – 0.1) × (0.68) = 28,200 Btu/h (8.3 kW).

10. ***Compute the grand-total heat and refrigeration tonnage***

Take the sum of the room total heat and the outside-air sensible and latent heat. The result is the grand-total heat load of the space, Btu/h (W or kW).

The room total heat = room sensible-heat total + room latent-heat total = 682,500 + 76,157 = 768,657 Btu/h (225.3 kW). Then the grand-total heat = 768,657 + 14,600 + 28,200 = 811,457 Btu/h, say 811,500 Btu/h (237.9 kW).

Compute the refrigeration load, tons, from (grand-total heat, Btu/h)/(12,000 Btu/h per ton of refrigeration), or 811,500/12,000 = 67.6 tons (237.7 kW), say 70 tons (246.2 kW).

The quantity of cooling water required for the refrigeration-system condenser is Q gal/min = 30 × tons of refrigeration/condenser water-temperature rise, °F. Assuming a 75°F (23.9°C) entering water temperature and 95°F (35°C) leaving water temperature, which are typical values for air-conditioning practice Q = 30(70)/95 – 75 = 105 gal/min (397.4 L/min).

11. ***Compute the sensible-heat factor and apparatus dew point***

For any air-conditioning system, sensible-heat factor = (room sensible heat, Btu/h)/(room total heat, Btu/h) = 682,500/768,657 = 0.888.

The *Guide* or the Carrier *Handbook of Air Conditioning Design* gives an apparatus dew point of 58°F (14.4°C), closely.

12. ***Compute the quantity of dehumidified air required***
Determine the dehumidified air temperature rise first from F = (1 – bypass factor) (design indoor temperature, °F – apparatus dew point, °F), or F = (1 – 0.1)(80 – 58) = 19.8°F (–6.8°C).

Next, compute the dehumidified air quantity, ft³/min from (room sensible heat, Btu/h)/1.08 (dehumidified air temperature rise, °F), or (682,500)/1.08(10.8) = 34,400 ft³/min (973.9 m³/min).

Related Calculations. Use this general procedure for any type of air-conditioned building or space—industrial, office, hotel, motel, apartment house, residence, laboratories, school, etc. Use the ASHRAE *Guide and Data Book* or the Carrier *Handbook of Air Conditioning System Design* as a source of data for the various calculations. In comparing the various values from the *Guide* used in this procedure, note that there may be slight changes in certain tabulated values from one edition of the *Guide* to the next. Hence, the values shown may differ slightly from those in the current edition. This should not cause concern, because the procedure is the same regardless of the values used.

COOLING-COIL SELECTION FOR AIR-CONDITIONING SYSTEMS

Select an air-conditioning cooling coil to cool 15,000 ft³/min (424.7 m³/min) of air from 85°F (29.4°C) dry bulb, 67°F (19.4°C) wet bulb, 57°F (13.9°C) dew point, 38 percent relative humidity, to a dry-bulb temperature of 65°F (18.3°C) with cooling water at 50°F (10°C). Suppose the air were cooled below the dew point of the entering air. How would the calculation procedure differ?

Calculation Procedure:

1. ***Compute the weight of air to be cooled***
From a psychrometric chart (Fig. 28) find the specific volume of the entering air as 13.75 ft³/lb (0.86 m³/kg). To convert the air flow in ft³/min to lb/h, use the relation lb/h = 60 *cfm*/u_s, where *cfm* = air flow, ft³/min; u_s = specific volume of the entering air, ft³/lb. Hence for this cooling coil, lb/h = 60(15,000)/13.75 = 65,500 lb/h (29,773 kg/h).

Where a cooling coil is rated by the manufacturer for air at 70°F (21.1°C) dry bulb and 50 percent relative humidity, as is often done, use the relation lb/h = 4.45*cfm*. Thus if this coil were rated for air at 70°F (21.1°C) dry bulb, the weight of air to be cooled would be lb/h = 4.45(15,000) = 66,800 (30,364 kg/h).

Since this air quantity is somewhat greater than when the entering-air specific volume is used, and since cooling coils are often rated on the basis of 70°F (21.1°C) dry-bulb air, this quantity, 66,800 lb/h (30,364 kg/h), will be used. The procedure is the same in either case. [*Note:* 4.45 = 60/135 ft³/lb, the specific volume of air at 70°F (21.1°C) and 50 percent relative humidity.]

2. ***Compute the quantity of heat to be removed***
Use the relation $H_r = ws\Delta t$, where H_r = heat to be removed from the air, Btu/h; w = weight of air cooled, lb/h; s = specific heat of air = 0.24 Btu/(lb · °F); Δt = temperature drop of the air = entering dry-bulb temperature, °F – leaving dry-bulb temperature, °F. For this coil, H_r = (66,800)(0.24) (85 – 65) = 321,000 Btu/h (94.1 kW).

3. ***Compute the quantity of cooling water required***
The quantity of cooling water required in gal/min = $H_r/500\ \Delta t_w$ = leaving-water temperature °F – entering-water temperature, °F. Since the leaving-water temperature is not known, a value must be assumed. The usual temperature rise of water during passage through an air-conditioning cooling coil is 4 to 12°F (2.2 to 6.7°C). Assuming a 10°F (5.6°C) rise, which is a typical value, gal/min required = 321,000/500(10) = 64.2 gal/min (4.1 L/s).

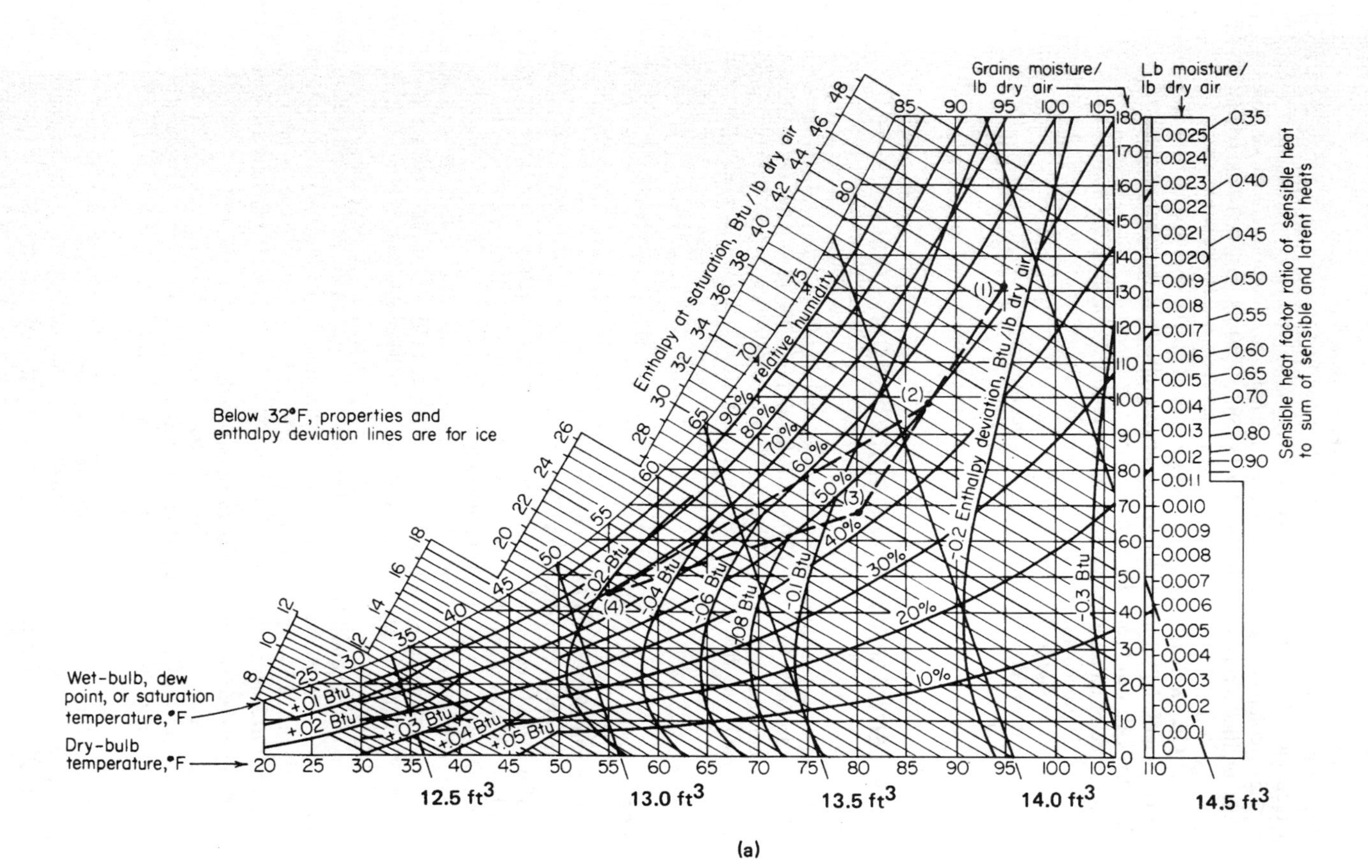

FIGURE 28 (*a*) Psychrometric charts for normal temperatures—USCS version. (*Carrier Air Conditioning Company.*)

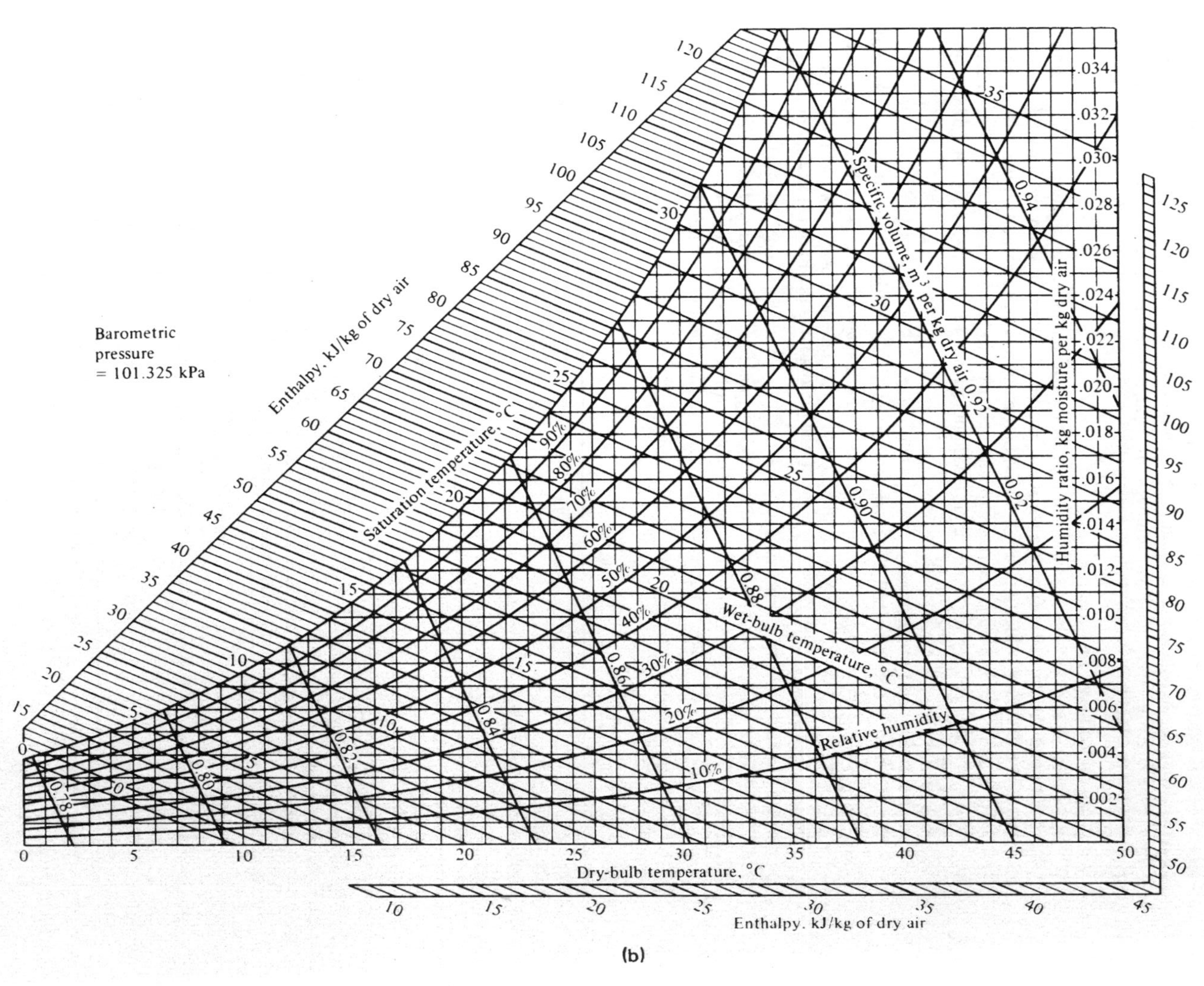

FIGURE 28 (*b*) SI version. (*Stoecker and Jones—Refrigeration and Air Conditioning, McGraw-Hill.*)

4. *Determine the logarithmic mean temperature difference*

Use Fig. 4 in the heat transfer section of this handbook, Section 11, to determine the logarithmic mean temperature difference for the cooling coil. In this chart, greatest terminal difference = entering-air temperature, °F – leaving-water temperature, °F = 85 – 60 = 25°F (13.9°C), and least terminal temperature difference = leaving-air temperature, °F – entering-water temperature, °F = 65 – 50 = 15°F (8.3°C). Entering Fig. 4 at these two temperature values gives a logarithmic mean temperature difference (LMTD) of 19.5°F (10.8°C).

5. *Compute the coil core face area*

The coil core face area is the area exposed to the air flow; it does not include the area of the mounting flanges. Compute the coil core face area from $A_c = cfm/V_a$, where V_a = air velocity through the coil, ft/min. The usual air velocity through the coil, often termed face velocity, ranges from 300 to 800 ft/min (91.4 to 243.8 m/min) although special designs may use velocities down to 200 ft/min (60.9 m/min) or up to 1200 ft/min (365.8 m/min). Assuming a face velocity of 500 ft/min (152.4 m/min) gives A_c = 15,000/500 = 30 ft^2 (2.79 m^2).

6. *Select the cooling coil for the load*

Using the engineering data provided by the manufacturer whose coil is to be used, choose the coil. Table 13 summarizes typical engineering data provided by a coil manufacturer. This table shows that two 15.4-ft^2 (1.43-m^2) coils placed side by side will provide a total face area of 2 × 15.4 = 30.8 ft^2. Hence, the actual air velocity through the coil is $V_a = cfm/A_c$ = 15,000/30.8 = 487 ft/min (148.4 m/min).

7. *Compute the water velocity in the coil*

Table 15 shows that the coil water velocity, ft/min = 3.59*gpm*. Since the required flow is, from step 3, 64.2 gal/min (243 L/min), the water flow for each unit will be half of this, or 64.2/2 = 32.1 gal/min (121.5 L/min). Hence the water velocity = 3.59(32.1) = 115.2 ft/min (35.1 m/min).

TABLE 15 Typical Cooling-Coil Characteristics

Face area, ft^2 (m^2)	14.0 (1.3)	15.4 (1.43)	17.9 (1.66)
Tube length, ft · in (m)	5–6 (1.68)	6–0 (1.83)	7–0 (2.13)
Water velocity, ft/min (m/min)	3.59 gal/min (4.14 L/min)	3.59 gal/min (4.14 L/min)	3.59 gal/min (4.14 L/min)
Coil heat-transfer factors, k = Btu/(h · ft^2 · °F) LMTD row [W/(m^2 · K)]			
	Air velocity, ft/min (m/min)		
Water velocity, ft/min (m/min)	400 (121.9)	500 (152.4)	550 (167.6)
113 (34.4)	154 (46.9)	162 (49.4)	170 (51.8)
115 (35.1)	158 (48.2)	167 (50.9)	175 (53.3)
117 (35.7)	161 (49.1)	172 (52.4)	176 (53.6)
Coil water-pressure drop, ft (m) of water per row			
	Water velocity, ft/min (m/min)		
Tube length, ft · in (m)	90 (27.4)	120 (36.6)	150 (45.7)
5–6 (1.68)	0.16 (0.04)	0.26 (0.08)	0.38 (0.12)
6–0 (1.83)	0.18 (0.05)	0.29 (0.09)	0.42 (0.13)
7–0 (2.13)	0.21 (0.06)	0.32 (0.10)	0.46 (0.14)
Header water-pressure drop, ft (m) of water			
	Water velocity, ft/min (m/min)		
Coil type	90 (27.4)	120 (36.6)	150 (45.7)
A	0.26 (0.08)	0.48 (0.15)	0.72 (0.22)
B	0.34 (0.10)	0.62 (0.19)	0.92 (0.28)

8. *Determine the coil heat-transfer factors*
Table 15 lists typical heat-transfer factors for various water and air velocities. Interpolating between 400- and 500-ft/min (121.9- and 152.4-m/min) air velocities at a water velocity of 115 ft/min (35.1 m/min) gives a heat-transfer factor of $k = 165$ Btu/(h · ft^2 · °F) [936.9 W/(m^2 · K)] LMTD row. The increase in velocity from 15 to 15.2 ft/min (4.63 m/min) is so small that it can be ignored. If the actual velocity is midway, or more, between the two tabulated velocities, interpolate vertically also.

9. *Compute the number of tube rows required*
Use the relation number of tube rows = H_r/(LMTD)(A_c)(k). Thus, number of rows = 321,000/(19.5)(30.8)(165) = 3.24, or four rows, the next larger *even* number.

Water cooling coils for air-conditioning service are usually built in units having two, four, six or eight rows of coils. If the above calculation indicates that an odd number of coils should be used (i.e., the result was 3.24 rows), use the next smaller or larger *even* number of rows after increasing or decreasing the air and water velocity. Thus, to decrease the air velocity, use a coil having a larger face area. Recompute the air velocity and water velocity; find the new heat-transfer factor and the required number of rows. Continue doing this until a suitable number of rows is obtained. Usually only one recalculation is necessary.

10. *Determine the coil water-pressure drop*
Table 15 shows the water-pressure drop, ft (m) of water, for various tube lengths and water velocities. Interpolating between 90 and 120 ft/min (27.4 and 36.6 m/min) for a 6-ft (1.83-m) long tube gives a pressure drop of 0.27 ft (0.08 m) of water per row at a water velocity of 115.2 ft/min (35.1 m/min). Since the coil has four rows, total tube pressure drop = 4(0.27) = 1.08 ft (0.33 m) of water.

There is also a water-pressure drop in the coil headers. Table 15 lists typical values. Interpolate between 90 and 120 ft/min (27.4 and 36.6 m/min) for a B-type coil gives a header pressure loss of 0.57 ft (0.17 m) of water at a water velocity of 115.2 ft/min (35.1 m/min). Hence, the total pressure loss in the coil is 1.08 + 0.57 = 1.65 ft (0.50 m) of water = coil loss + header loss.

11. *Determine the coil resistance to air flow*
Table 16 lists the resistance of coils having two to six rows of tubes and various air velocities. Interpolating for four tube rows gives a resistance of 0.225 in H_2O for an air velocity of 487 ft/min (148.4 m/min). The increase in resistance with a wet tube surface is, from Table 16, 28 percent at a 500-ft/min (152.4-m/min) air velocity. This occurs when the air is cooled below the entering-air dew point and is discussed in step 13.

TABLE 16 Typical Cooling-Coil Resistance Characteristics

	Air-flow resistance, in H_2O for 70°F air (mm H_2O for 21.1°C air)			
	Air-face velocity, ft/min (m/min)			
No. of tube rows	400 (121.9)	500 (152.4)	600 (182.9)	
2	0.081 (2.06)	0.122 (3.10)	0.164 (4.17)	
4	0.162 (4.11)	0.234 (5.94)	0.318 (8.08)	
6	0.234 (5.94)	0.344 (8.74)	0.472 (11.99)	
8	0.312 (7.92)	0.454 (11.53)	0.622 (15.80)	
	Resistance increase due to wet tube surface, percent			
	32	28	24	
Coil cooling capacity, Btu/(h · ft^2) (W/m^2) face area and final air temperature, °F (°C)				
			Entering water temperature, °F (°C)	
Air velocity, ft/min (m/min)	No of tube rows	Entering air temperature, °F (°C)	45 (7.2)	50 (10)
500 (152.4)	4	85 (29.4)	11,900 (36,545) 63 (17.2)	10,300 (32,497) 65 (18.3)

12. ***Check the coil selection in a coil-rating table***
Many manufacturers publish precomputed coil-rating tables as part of their engineering data. Table 16 shows a portion of one such table. This tabulation shows that with an air velocity of 500 ft/min (152.4 m/min), four tube rows, an entering-air temperature of 85°F (29.4°C), and an entering-water temperature of 50°F (10.0°C), the cooling coil has a cooling capacity of 10,300 Btu/(h · ft^2) (32,497 W/m^2), and a final air temperature of 65°F (18.3°C). Since the actual air velocity of 487 ft/min (148.4 m/min) is close to 500 ft/min (152.4 m/min), the tabulated cooling capacity closely approximates the actual cooling capacity. Hence the required heat-transfer area is $A_c = H_r/10{,}300 = 321{,}000/10{,}300 =$ 31.1 ft^2 (2.98 m^2). This agrees closely with the area of 30.8 ft^2 (2.86 m^2) found in step 6.

In actual practice, designers use a coil cooling capacity table whenever it is available. However, the procedure given in steps 1 through 11 is also used when an exact analysis of a coil is desired or when a capacity table is not available.

13. ***Compute the heat removal for cooling below the dew point***
When the temperature of the air leaving the cooling coil is lower than the dew point of hot entering air, H_r = (weight of air cooled, lb)(total heat of entering air at its wet-bulb temperature, Btu/lb – total heat of the leaving air at its wet-bulb temperature, Btu/lb). Once H_r is known, follow all the steps given above except that (*a*) a correction must be applied in step 11 for a wet tube surface. Obtain the appropriate correction factor from the manufacturer's engineering data, and apply it to the air-flow-resistance data, for the coil selected, (*b*) Also, the usual coil-rating table presents only the sensible-heat capacity of the coil. Where the ratio of sensible heat removed to latent heat removed is more than 2:1, the usual coil-rating table can be used. If the ratio is less than 2:1, use the procedure in steps 1 through 13.

Related Calculations. Use the method given here in steps 1 through 11 for any finned-type cooling coil mounted perpendicular to the air flow and having water as the cooling medium where the final air temperature leaving the cooling coil is *higher than* the dew point of the entering air. Follow step 13 for cooling below the dew point of the entering air. Figure 29 shows a typical air-handling

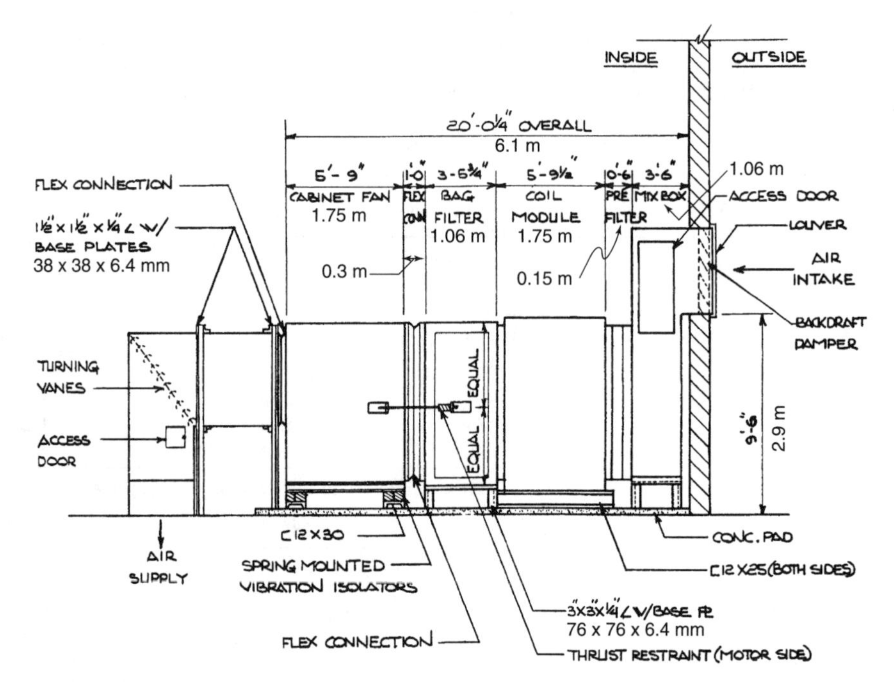

FIGURE 29 Typical air-handling unit for an air-conditioning system and the location of the coil module in it. (*Jerome F. Mueller.*)

unit for an air-conditioning system and the location of the coil module in it. Two piping arrangements for air cooling coils are shown in Fig. 30*a* and *b*.

Cooling and dehumidifying coils used in air-conditioning systems generally serve the following ranges of variables: (1) dry-bulb temperature of the entering air is 60 to 100°F (15.6 to 37.8°C); wet-bulb temperature of entering air is 50 to 80°F (10 to 26.7°C); (2) coil core face velocity can range from 200 to 1200 ft/min (60.9 to 365.8 m/min) with 500 to 800 ft/min (152.4 to 243.8 m/min) being the most common velocity for comfort cooling applications; (3) entering-water-temperature ranges from 40 to 65°F (4.4 to 18.3°C); (4) the water-temperature rise ranges from 4 to 12°F (2.2 to 6.7°C) during passage through the coil; (5) the water velocity ranges from 2 to 6 ft/s (0.61 to 1.83 m/s).

To choose an air-cooling coil using a direct-expansion refrigerant, follow the manufacturer's engineering data. Since most of the procedures are empirical, it is difficult to generalize about which procedure to use. However, the usual range of the volatile refrigerant temperature at the coil suction outlet is 25 to 55°F (−3.9 to 12.8°C). Where chilled water is circulated through the coil, the usual quantity range is 2 to 6 gal/(m · ton) (8.4 to 25 L/t).

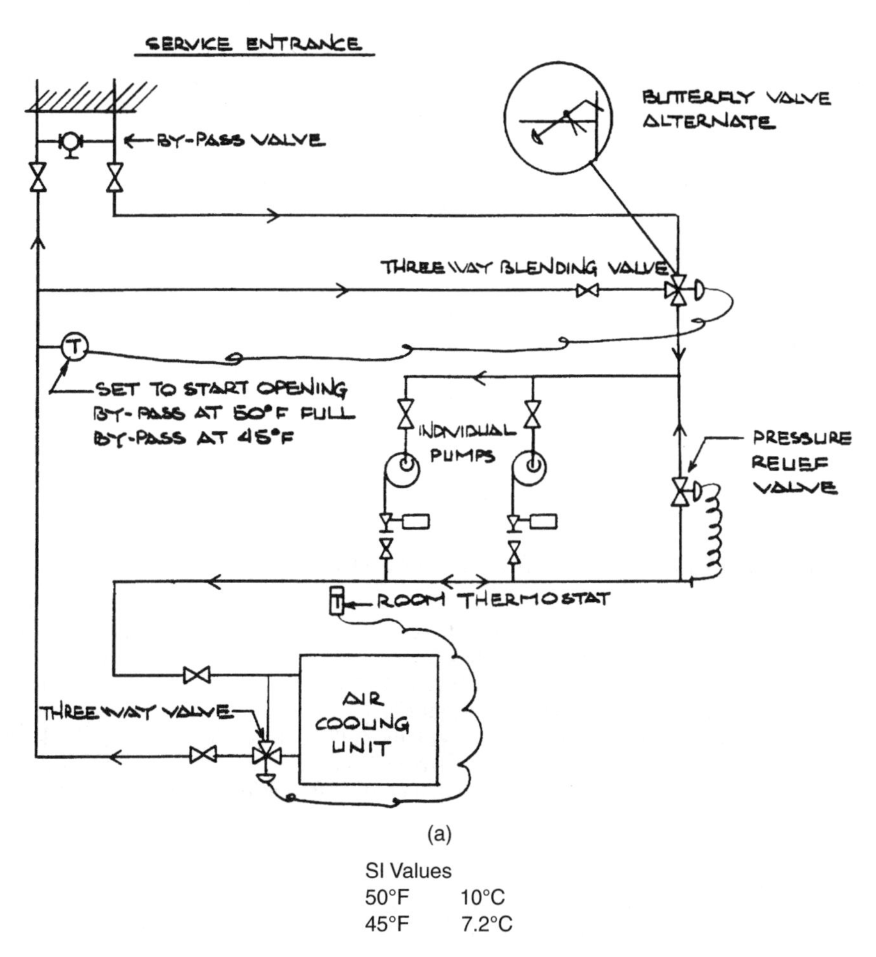

FIGURE 30 (*a*) Chilled-water system designed for constant volume and variable temperature. (*Jerome F. Mueller.*)

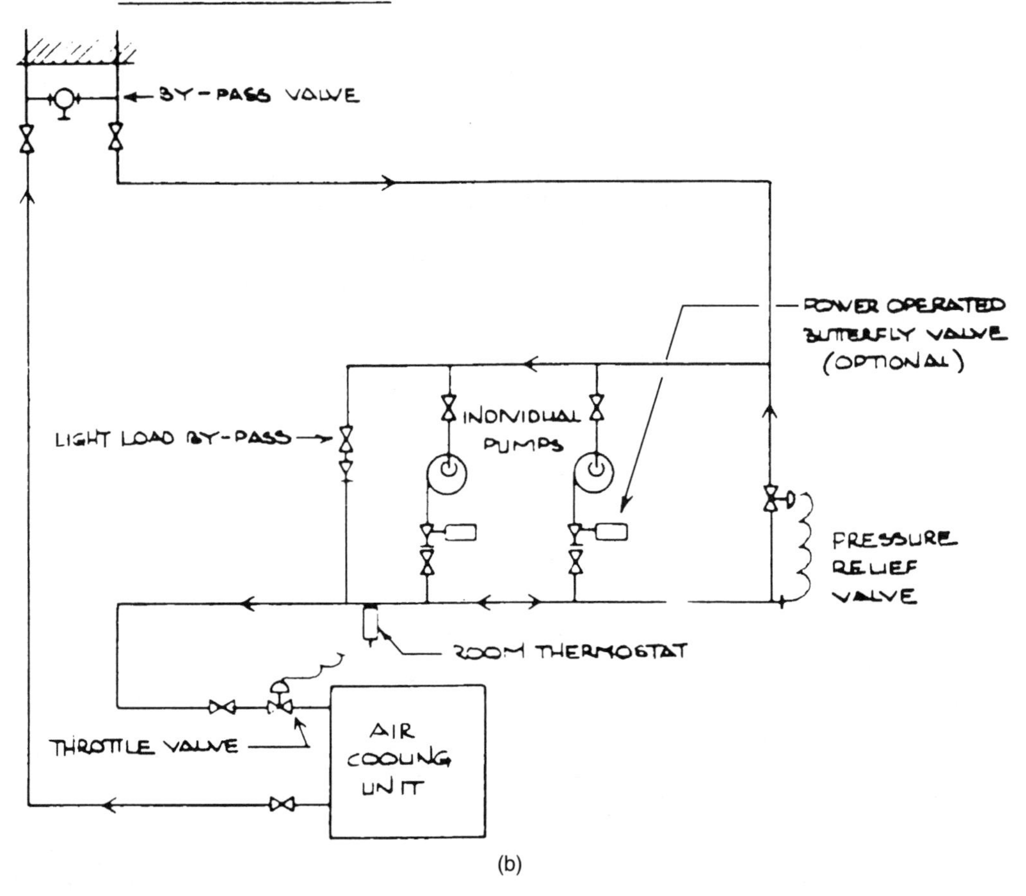

FIGURE 30 (*b*) Chilled-water system designed for constant temperature and variable volume. (*Jerome F. Mueller.*)

ENERGY ANALYSIS OF MIXING TWO AIR STREAMS

An air-conditioning system is designed to deliver 100,000 ft^3/min (2831 m^3/min) of air to a conditioned space. Of this total, 90,000 ft^3/min (2548 m^3/min) is recirculated indoor air at 72°F (22.2°C) and 40 percent relative humidity; 10,000 ft^3/min (283.1 m^3/min) is outdoor air at 0°F (−17.8°C). What are the enthalpy, temperature, moisture content, and relative humidity of the resulting air mixture? If air enters the room from the outlet grille at 60°F (15.6°C) after leaving the apparatus at a 50°F (10°C) dew point and the return air is at 75°F (23.9°C), what proportion of conditioned air and bypassed return air must be used to produce the desired outlet temperature at the grille?

Calculation Procedure:

1. ***Determine the proportions of each air stream***

Use the relations $p_r = r/t$ and $p_0 = o/t$, where p_r = percent recirculated room air, expressed as a decimal; r = recirculated air quantity, ft^3/min; t = total air quantity, ft^3/min; p_0 = percent outside air, expressed as a decimal; o = outside air quantity, ft^3/min. For this system, p_r = 90,000/100,000 = 0.90, or 90 percent; p_0 = 10,000/100,000 = 0.10, or 10 percent. (The computation in SI units is identical.)

2. *Determine the enthalpy of each air stream*
Use a psychrometric chart or table to find the enthalpy of the recirculated indoor air as 24.6 Btu/lb (57.2 kJ/kg).

The enthalpy of the outdoor air is 0.0 Btu/lb (0.0 kJ/kg) because in considering heating or humidifying processes in winter it is always safest to assume that the outdoor air is completely dry. This condition represents the greatest heating and humidifying load because the enthalpy and the water-vapor content of the air are at a minimum when the air is considered dry at the outdoor temperature.

3. *Determine the moisture content of each air stream*
The moisture content of the indoor air at a 72°F (22.2°C) dry-bulb temperature and 40 percent relative humidity is, from the psychrometric chart (Fig. 28), 47.2 gr/lb (6796.8 mg/kg). From a psychrometric table the moisture content of the 0°F (–17.8°C) outdoor air, which is assumed to be completely dry, is 0.0 gr/lb (0.0 mg/kg).

4. *Compute the enthalpy of the air mixture*
Use the relation $h_m = (oh_0 + rh_r)/t$, where h_m = enthalpy of mixture, Btu/lb; h_o = enthalpy of the outside air, Btu/lb; h_r = enthalpy of the recirculated room air, Btu/lb; other symbols as before. Hence, $h_m = (10{,}000 \times 0 + 90{,}000 \times 24.6)/100{,}000 = 22.15$ Btu/lb (51.5 kJ/kg).

5. *Compute the temperature of the air mixture*
Use a similar relation to that in step 4, substituting the air temperature for the enthalpy. Or, $t_m = (ot_o + rt_r)/t$, where t_m = mixture temperature, °F; t_0 = temperature of outdoor air, °F; t_r = temperature of recirculated room air, °F; other symbols as before. Hence, $t_m = (10{,}000 \times 0 + 90{,}000 \times 72)/100{,}000 = 64.9$°F (18.3°C).

6. *Compute the moisture content of the air mixture*
Use a similar relation to that in step 4, substituting the moisture content for the enthalpy. Or, $g_m = (og_0 + rg_r)/t$, where g_m = gr of moisture per lb of mixture; g_0 = gr of moisture per lb of outdoor air; g_r = gr of moisture per lb of recirculated room air; other symbols as before. Thus, $g_m = (10{,}000 \times 0 + 90{,}000 \times 47.2)/100{,}000 = 42.5$ gr/lb (6120 mg/kg).

7. *Determine the relative humidity of the mixture*
Enter the psychrometric chart at the temperature of the mixture, 64.9°F (18.3°C), and the moisture content, 42.5 gr/lb (6120 mg/kg). At the intersection of the two lines, find the relative humidity of the mixture as 47 percent relative humidity.

8. *Determine the required air proportions*
Set up an equation in which x = proportion of conditioned air required to produce the desired outlet temperature at the grille and y = the proportion of bypassed air required. The air quantities will also be proportional to the dry-bulb temperatures of each air stream. Since the dew point of the air leaving an air-conditioning apparatus = dry-bulb temperature of the air, $50x + 75y = 60(x + y)$, or $15y = 10x$. Also, the sum of the two air steams $x + y = 1$. Substituting and solving for x and y, we get x = 60 percent; y = 40 percent. Multiplying the actual air quantity supplied to the room by the percentage representing the proportion of each air stream will give the actual ft^3/min required for supply and bypass air.

Related Calculations. Use this general procedure to determine the properties of any air mixture in which two air streams are mixed without compression, expansion, or other processes involving a marked change in the pressure or volume of either or both air streams.

ENERGY ASPECTS OF SELECTING AN AIR-CONDITIONING SYSTEM FOR A KNOWN LOAD

Choose the type of air-conditioning system to use for comfort conditioning of a factory having a heat gain that varies from 500,000 to 750,000 Btu/h (146.6 to 219.8 kW) depending on the outdoor temperature and the conditions inside the building. Indicate why the chosen system is preferred.

Calculation Procedure:

1. *Review the types of air-conditioning systems available*

Table 17 summarizes the various types of air-conditioning systems *commonly used* for different applications. Economics and special designs objectives dictate the final choice and modifications of the systems listed. Where higher-quality air conditioning is desired (often at a higher cost), certain other systems may be considered. These are dual-duct, dual-conduit, three-pipe induction and fan-coil, four-pipe induction and fan-coil, and panel-air systems.

TABLE 17 Systems and Applications

Applications	Systems*	Applications	Systems*
Single-purpose occupancies		Multipurpose occupancies	
Residential:		Office buildings	2e 2i 2j
Medium	1c	Hotels, dormitories	le 1f 2j 2k
Large	1d le 2i	Motels	le
Restaurants:		Apartment buildings	1f 2j 2k
Medium	1d lh	Hospitals	1f 2h 2j
Large	1d 2f 2g 2h 2i	Schools and colleges	1f 2e 2f 2g 2h
variety and specialty shops	1d	Museums	2h 2i
Bowling alleys	1d 2f	Libraries:	
Radio and TV studios:		Standard	2h 2f 2h 2i
Small	1d 2f 2h 2i	Rare books	2h
Large	1d 2f 2h 2i	Department stores	1d 2f
Country clubs	1d 2f 2h 2i	Shopping centers	1d 2f 2i
Funeral homes	1d 2i	Laboratories:	
Beauty salons	1c 1d	Small	1d 2e 2h 2i
Barber shops	1c 1d	Large building	2e 2g 2j
Churches	1d 2f 2i	Marine	2g 2j
Theaters	2f		
Auditoriums	2f		
Dance and roller skating pavilions	1d 2e 2f		
Factories (comfort)	1d 2f 2h		

*The systems in the table are:

1. Individual room or zone unit systems
   a. DX self-contained
   b. All-water
   c. Room DX self-contained 0.5 to 2 tons (1.76 to 7.0 kW)
   d. Zone DX self-contained 2 tons and over (7.0 kW and over)
   e. All-water room fan-coil recirculating air
   f. All-water room fan-coil with outdoor air
2. Central station apparatus systems
   a. All-air
   b. Air-water
   c. All-air, single airstream
   d. Air-water, primary air systems
   e. All-air, single airstream, variable volume
   f. All-air, single airstream, bypass
   g. All-air, single airstream, reheat at terminal
   h. All-air, single airstream, reheat zone in duct
   i. All-air, single airstream, multizone single duct
   j. Air-water primary air systems, secondary water H-V H-P induction
   k. Air-water primary air systems, room fan-coil with outside air

Systems listed for a particular application are the systems most commonly used. Economics and design objectives dictate the choice and deviations of systems listed above, other systems as listed in note 2, and some entirely new systems.

Several systems are used in many of these applications when higher-quality air conditioning is desired (often at higher expense). They are dual-duct, dual-conduit, three-pipe induction and fan-coil, four-pipe induction and fan-coil, and panel-air systems.

Carrier Air Conditioning—*Handbook of Air Conditioning Systems Design*, McGraw-Hill.

Study of Table 17 shows that four main types of air-conditioning systems are popular: direct-expansion (termed DX), all-water, all-air, and air-water. These classifications indicate the methods used to obtain the final within-the-space cooling and heating. The air surrounding the occupant is the end medium that is conditioned.

2. ***Select the type of air-conditioning system to use***
Table 17 indicates that direct-expansion and all-air air-conditioning systems are *commonly used* for factory comfort conditioning. The load in the factory being considered may range from 500,000 to 750,000 Btu/h (146.6 to 219.8 kW). This is the equivalent of a maximum cooling load of 750,000/12,000 Btu/(h · ton) of refrigeration = 62.5 tons (219.8 kW) of refrigeration.

Where a building has a varying heat load, bypass control wherein neutral air is recirculated from the conditioned space while the amount of cooling air is reduced is often used. With this arrangement, the full quantity of supply air is introduced to the cooled area at all times during system operation.

Self-contained direct-expansion systems can serve large factory spaces. Their choice over an all-air bypass system is largely a matter of economics and design objectives.

Where reheat is required, this may be provided by a reheater in a zone duct. Reheat control maintains the desired dry-bulb temperature within a space by replacing any decrease in sensible loads by an artificial heat load. Bypass control maintains the desired dry-bulb temperature within the space by modulating the amount of air to be cooled. Since the bypass all-air system is probably less costly for this building, it will be the first choice. A complete economic analysis would be necessary before this conclusion could be accepted as fully valid.

Related Calculations. Use this general method to make a preliminary choice of the 11 different types of air-conditioning systems for the 31 applications listed in Table 17. Where additional analytical data for comparison of systems are required, consult Carrier Air Conditioning Company's *Handbook of Air Conditioning System Design* or ASHRAE *Guide and Data Book.*

With the greater emphasis on the environmental aspects of air-conditioning systems, more attention is being paid to natural-gas cooling. There are three types of natural-gas cooling systems used today: (1) engine, (2) desiccant, and (3) absorption. In each type of system, natural gas is used to provide the energy required for the cooling.

In the engine-type natural-gas cooling system, gas fuels an internal-combustion engine which drives a conventional chiller. The advantage of engine-driven chillers is that their fuel consumption of natural gas is low. Further, such engines are clean-burning with minimum atmospheric pollution and easily comply with the Clean Air Act requirements when properly maintained. Such natural-gas fueled engine-driven chiller systems are popular in large office buildings, factories, and similar installations. As a further economy and environmental plus, engine jacket water can be used to heat domestic hot water for the building being cooled. In one installation the engine jacket water is also used to heat water in a swimming pool. This combines natural-gas cooling with cogeneration.

Desiccant systems may use one or more heat wheels to capture heat from the building's exhaust air stream. Where more than one heat wheel is used, the first wheel may be a total-energy recovery unit with a desiccant section, cooling coil, and electric reheat coil. The second wheel may be a polymer-coated sensible-heat-only wheel with a conventional chilled-water or direct-expansion refrigerant coil to cool and dehumidify or heat and humidify the outdoor air, depending on ambient conditions.

With the enormous emphasis on indoor air quality (IAQ) today, desiccant cooling and preconditioning of air is gaining importance. Thus, in a school system originally designed for 15 ft^3/min (0.14 m^3/min) per student, and later redesigned for 15 ft^3/min (0.42 m/min) per student, desiccant preconditioning maintains the indoor relative humidity at 50 to 52 percent in conjunction with packaged HVAC units in a very energy-efficient manner.* Further, the indoor air quality produced removes the risk of mold and mildew resulting from a lack of humidity control in the spaces being served.

Absorption natural-gas cooling can use direct-fired chillers to provide the needed temperature reduction. A variety of direct-fired chillers are available from various manufacturers to serve the

*Smith, James C., "Schools Resolve IAQ/Humidity Problems with Desiccant Preconditioning." *Heating/Piping/Air Conditioning*, April, 1996.

needs of such installations. Such chillers can range in capacity from 30 tons (105.5 kW) to more than 600 tons (2.1 MW). Using natural gas as the heat source provides low-cost cooling and heating for a variety of buildings: office, factory, commercial, etc. Natural gas is a low-pollution fuel and absorption cooling installations using it usually find it easy to comply with the provisions of the Clean Air Act.

EQUAL-FRICTION METHOD FOR SIZING LOW-VELOCITY AIR-CONDITIONING DUCTS

An industrial air-conditioning system requires 36,000 ft³/min (1019.2 m³/min) of air. This low-velocity system will be fitted with enough air outlets to distribute the air uniformly throughout, the conditioned space. The required operating pressure for each duct outlet is 0.20-in (5.1-mm) wg. Determine the duct sizes required for this system by using the equal-friction method of design. What is the required fan static discharge pressure?

Calculation Procedure:

1. ***Sketch the duct system***

The required air quantity, 36,000 ft³/min (1019.2 m³/min), must be distributed in approximately equal quantities to the various areas in the building. Sketch the proposed duct layout as shown in Fig. 31. Locate air outlets as shown to provide air to each area in the building.

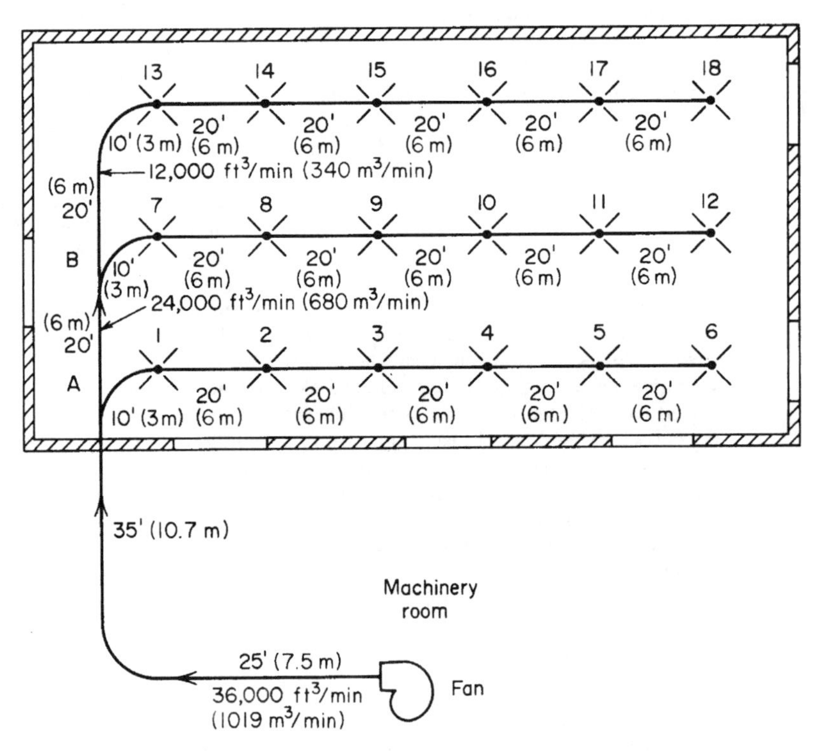

FIGURE 31 Duct-system layout.

Determine the required capacity of each air outlet from air quantity required, ft^3/min per number of outlets, or outlet capacity = 36,000/18 = 2000 ft^3/min (56.6 m^3/min) per outlet. This is within the usual range of many commercially available outlets. Where the required capacity per outlet is extremely large, say 10,000 ft^3/min (283.1 m^3/min), or extremely small, say 5 ft^3/min (0.14 m^3/min), change the number of outlets shown on the duct sketch to obtain an air quantity within the usual capacity range of commercially available outlets. Relocate each outlet so it serves approximately the same amount of floor area as each of the other outlets in the system. Thus, the duct sketch serves as a trial-and-error analysis of the outlet location and capacity.

Where a building area requires a specific amount of air, select one or more outlets to supply this air. Size the remaining outlets by the method described above, after subtracting the quantity of air supplied through the outlets already chosen.

2. *Determine the required outlet operating pressure*
Consult the manufacturer's engineering data for the required operating pressure of each outlet. Where possible, try to use the same type of outlets throughout the system. This will reduce the initial investment. Assume that the required outlet operating pressure is 0.20-in (5.1-mm) wg for each outlet in this system.

3. *Choose the air velocity for the main duct*
Use Table 18 to determine a suitable air velocity for the main duct of this system. Table 18 shows that an air velocity up to 2500 ft/min (762 m/min) can be used for main ducts where noise is the controlling factor; 3000 ft/min (914.4 m/min) where duct friction is the controlling factor. A velocity of 2500 ft/min (762 m/min) will be used for the main duct in this installation.

TABLE 18 Recommended Maximum Duct Velocities for Low-Velocity Systems, ft/min (m/min)

Application	Controlling factor—noise generation, main ducts	Controlling factor—duct friction			
		Main ducts		Branch ducts	
		Supply	Return	Supply	Return
Residences	600 (183)	1000 (300)	800 (244)	600 (183)	600 (183)
Apartments, hotel bedrooms, hospital bedrooms	1000 (300)	1500 (457)	1300 (396)	1200 (366)	1000 (300)
Private offices, directors rooms, libraries	1200 (366)	2000 (610)	1500 (457)	1600 (488)	1200 (366)
Theaters, auditoriums	800 (244)	1300 (396)	1100 (335)	1000 (300)	800 (244)
General offices, high-class restaurants, high-class stores, banks	1500 (457)	2000 (610)	1500 (457)	1600 (488)	1200 (366)
Average stores, cafeterias	1800 (549)	2000 (610)	1500 (459)	1600 (488)	1200 (366)
Industrial	2500 (762)	3000 (914)	1800 (549)	2200 (671)	1500 (457)

Carrier Air Conditioning Company.

4. *Determine the dimensions of the main duct*
The required duct area A ft^2 = (ft^3/min)/(ft/min) = 36,000/2500 = 14.4 ft^2 (1.34 m^2). A nearly square duct, i.e., a duct 46 × 45 in (117 × 114 cm), has an area of 14.38 ft^2 (1.34 m^2) and is a good first choice for this system because it closely approximates the outlet size of a standard centrifugal fan. Where possible, use a square main duct to simplify fan connections. Thus a 46 × 46 in (117 × 117 cm) duct might be the final choice for this system.

5. *Determine the main-duct friction loss*
Convert the duct area to the equivalent diameter in inches d, using $d = 2(144A/\pi)^{0.5} = 2(144 \times 144/\pi)^{0.5} = 51.5$ in (130.8 cm).

Enter Fig. 32 at 36,000 ft^3/min (1019.2 m^3/min) and project horizontally to a round-duct diameter of 51.5 in (130.8 cm). At the top of Fig. 32 read the friction loss as 0.13-in (3.3-mm) wg per 100 ft (30 m) of equivalent duct length.

6. *Size the branch ducts*

For many common air-conditioning systems the equal-friction method is used to size the ducts. In this method the supply, exhaust, and return-air ducts are sized so they have the same friction loss per foot of length for the entire system. The equal-friction method is superior to the velocity-reduction method of duct sizing because the former requires the less balancing for symmetrical layouts.

The usual procedure in the equal-friction method is to select an initial air velocity in the main duct near the fan, using the sound level as the limiting factor. With this initial velocity and the design air flow rate, the required duct diameter is found, as in steps 4 and 5. Once the duct diameter is known, the friction loss is found from Fig. 32, as in step 5. This same friction loss is then maintained throughout the system, and the equivalent round-duct diameter is chosen from Fig. 32.

To expedite equal-friction calculations, Table 19 is often used instead of the friction chart. (It is valid for SI units also.) This provides the same duct sizes. Duct areas are determined from Table 19, and the area found is converted to a round-, rectangular-, or square-duct size suitable for the installation. This procedure of duct sizing automatically reduces the air velocity in the direction of air flow. Hence, the equal-friction method will be used for this system.

Compute the duct areas, using Table 19. Tabulate the results, using the duct run having the highest resistance. The friction loss through all elbows and fittings in the section must be included. The total friction loss in the duct having the highest resistance is the loss the fan must overcome.

Inspection of the duct layout (Fig. 31) shows that the duct run from the fan to outlet 18 probably has the highest resistance because it is the longest run. Tabulate the results as shown.

(1) Duct section	(2) Air quantity, ft^3/min (m^3/min)	(3) ft^3/min (m^3/min) capacity, percent	(4) Duct, percent	(5) Area, ft^2 (m^2)	(6) Duct size, in (cm)
Fan to *A*	36,000 (1,019.2)	100	100	14.4 (1.34)	46 × 45 (117 × 114)
A–B	24,000 (679.4)	67	73.5	10.6 (0.98)	39 × 39 (99 × 99)
B–13	12,000 (339.7)	33	41.0	5.9 (0.55)	30 × 29 (76 × 74)
13–14	10,000 (283.1)	28	35.5	5.1 (0.47)	27 × 27 (69 × 69)
14–15	8,000 (226.5)	22	29.5	4.3 (0.40)	25 × 25 (64 × 64)
15–16	6,000 (169.9)	17	24.0	3.5 (0.33)	23 × 22 (58 × 56)
16–17	4,000 (113.2)	11	17.5	2.5 (0.23)	20 × 18 (51 × 46)
17–18	2,000 (56.6)	6	10.5	1.5 (0.14)	15 × 15 (38 × 38)

The values in this tabulation are found as follows. Column 1 lists the longest duct run in the system. In column 2, the air leaving the outlets in branch *A*, or (6 outlets) (2000 ft^3/min per outlet) = 12,000 ft^3/min (339.7 m^3/min), is subtracted from the quantity of air, 36,000 ft^3/min (1019.2 m^3/min), discharged by the fan to give the air quantity flowing from *A–B*. A similar procedure is followed for each successive duct and air quantity.

Column 3 is found by dividing the air quantity in each branch listed in columns 1 and 2 by 36,000, the total air flow, and multiplying the result by 100. Thus, for run *B*-13, column 3 = 12,000 (100)/36,000 = 33 percent.

Column 4 values are found from Table 19. Enter that table with the ft^3/min capacity from column 3 and read the duct area, percent. Thus, for branch 13-14 with 28 percent ft^3/min capacity, the duct area from Table 19 is 35.5 percent. Determine the duct area, column 6, by taking the product, line

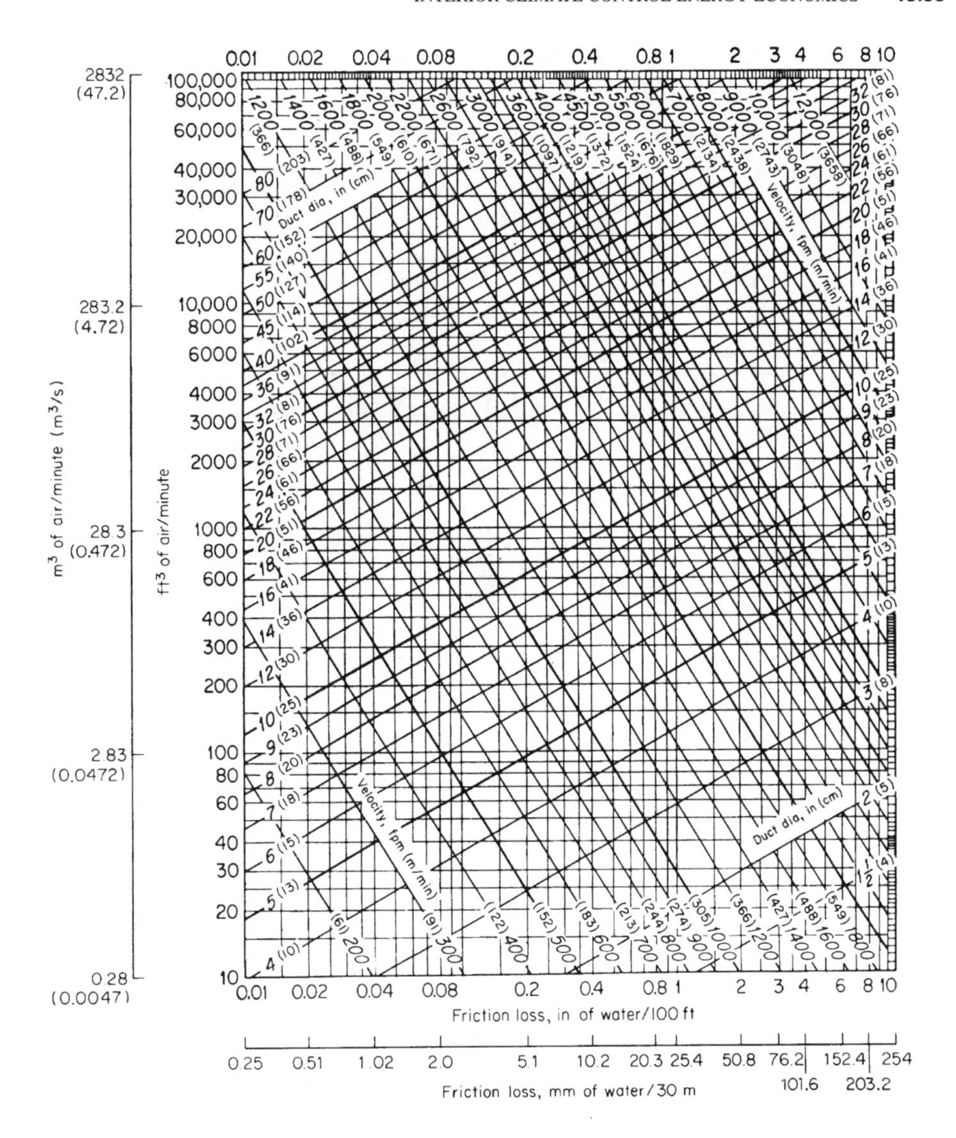

Friction loss for usual air conditions. This chart applies to smooth round galvanized iron ducts. See table below for corrections to apply when using other pipe.

Type of pipe	Degree of roughness	Velocity		Roughness factor (use as multiplier)
		ft/min	m/min	
Concrete	Medium rough	1000–2000	300–610	1.4
Riveted steel	Very rough	1000–2000	300–610	1.9
Tubing	Very smooth	1000–2000	300–610	0.9

FIGURE 32 Friction loss in round ducts.

TABLE 19 Percentage of Section Area in Branches for Maintaining Equal Friction

ft³/min (m³/min) capacity, %	Duct area,* %	ft³/min (m³/min) capacity, %	Duct area, %	ft³/min (m³/min) capacity, %	Duct area, %	ft³/min (m³/min) capacity, %	Duct area, %
1	2.0	26	33.5	51	59.0	76	81.0
2	3.5	27	34.5	52	60.0	77	82.0
3	5.5	28	35.5	53	61.0	78	83.0
4	7.0	29	36.5	54	62.0	79	84.0
5	9.0	30	37.5	55	63.0	80	84.5
6	10.5	31	39.0	56	64.0	81	85.5
7	11.5	32	40.0	57	65.0	82	86.0
8	13.0	33	41.0	58	65.5	83	87.0
9	14.5	34	42.0	59	66.5	84	87.5
10	16.5	35	43.0	60	67.5	85	88.5
11	17.5	36	44.0	61	68.0	86	89.5
12	18.5	37	45.0	62	69.0	87	90.0
13	19.5	38	46.0	63	70.0	88	90.5
14	20.5	39	47.0	64	71.0	89	91.5
15	21.5	40	48.0	65	71.5	90	92.0
16	23.0	41	49.0	66	72.5	91	93.0
17	24.0	42	50.0	67	73.5	92	94.0
18	25.0	43	51.0	68	74.5	93	94.5
19	26.0	44	52.0	69	75.5	94	95.0
20	27.0	45	53.0	70	76.5	95	96.0
21	28.0	46	54.0	71	77.0	96	96.5
22	29.5	47	55.0	72	78.0	97	97.5
23	30.5	48	56.0	73	79.0	98	98.0
24	31.5	59	57.0	74	80.0	99	99.0
25	32.5	50	58.0	75	80.5	100	100.0

*The same duct area percentage applies when flow is measured in m³/min or m³/5.
Carrier Air Conditioning Company.

by line, of column 4 and the main duct area. Thus, for branch 13-14, duct area = (0.355) (14.4) = 5.1 ft^2 (0.47 m^2). Convert the duct area to a nearly square, or a square, duct by finding two dimensions that will produce the desired area.

Duct sections *A* through 6 and *B* through 12 have the same dimensions as the corresponding duct sections *B* through 18.

7. *Find the total duct friction loss*

Examination of the duct sketch (Fig. 31) indicates that the duct run from the fan to outlet 18 has the highest resistance. Compute the total duct run length and the equivalent length of the two elbows in the run thus as shown.

(1) Duct section	(2) System part	(3) Length ft	(3) Length m	(4) Elbow equivalent length ft	(4) Elbow equivalent length m
Fan to *A*	Duct	60	18.3		
	Elbow	. . .		30	9.1
A–B	Duct	20	6.1		
B–13	Duct	30	9.1		
	Elbow	. . .		15	4.6
13–14	Duct	20	6.1		
14–15	Duct	20	6.1		
15–16	Duct	20	6.1		
16–17	Duct	20	6.1		
17–18	Duct	20	6.1		
Total		210	64.0	45	13.7

Note several factors about this calculation. The duct length, column 3, is determined from the system sketch, Fig. 31. The equivalent length of the duct elbows, column 4, is determined from the *Guide* or Carrier *Handbook of Air Conditioning Design.* The total equivalent duct length = column 3 + column 4 = 210 + 45 = 255 ft (77.7 m).

8. *Compute the duct friction loss*

Use the general relation $h_T = Lf$, where h_T = total friction loss in duct, in wg; L = total equivalent duct length, ft; f = friction loss for the system, in wg per 100 ft (30 m). With the friction loss of 0.13 in (3.3 mm) wg per 100 ft (30 m), as determined in step 5, h_r = (229/100)(0.13) = 0.2977-in (7.6-mm) wg; say 0.03-in (7.6-mm) wg.

9. *Determine the required fan static discharge pressure*

The total static pressure required at the fan discharge = outlet operating pressure + duct loss – velocity regain between first and last sections of the duct, all expressed in wg. The first two variables in this relation are already known. Hence, only the velocity regain need be computed.

The velocity υ ft/min of air in any duct is υ = ft³/min/duct area. For duct section *A*, υ = 36,000/14.4 = 2500 ft/min (762 m/min); for the last duct section, 17-18, υ = 2000/1.5 = 1333 ft/min (406.3 m/min).

When the fan discharge velocity is higher than the duct velocity in an air-conditioning system, use this relation to compute the static pressure regain $R = 0.75[(\upsilon_f/4000)^2 - (\upsilon_d/4000)^2]$, where R = regain, in wg; υ_f = fan outlet velocity, ft/min; υ_d = duct velocity, ft/min. Thus, for this system, $R = 0.75[(2500/4000)^2 - (1333/4000)^2 = 0.21$-in (5.3-mm) wg.

With the regain known, compute the total static pressure required as 0.20 + 0.30 – 0.21 = 0.29-in (7.4-mm) wg. A fan having a static discharge pressure of at least 0.30-in (7.6-mm) wg would probably be chosen for this system.

If the fan outlet velocity exceeded the air velocity in duct section *A*, the air velocity in this section would be used instead of the air velocity in the last duct section. Thus, in this circumstance, the last section becomes the duct connected to the fan outlet.

Figure 33 shows details of duct hangers for ducts of various dimensions. Shown in Fig. 34 are details of rectangular duct take-offs for air supply to specific rooms or areas.

Related Calculations. Where the velocity in the fan outlet duct is *higher* than the fan outlet velocity, use the relation $l = 1.1\ [(\upsilon_d/4000)2 - (\upsilon_f/4000)^2]$, where l = loss, in wg. This loss is the additional static pressure required of the fan. Hence, this loss must be *added* to the outlet operating pressure and the duct loss to determine the total static pressure required at the fan discharge.

The equal-friction method does not satisfy the design criteria of uniform static pressure at all branches and air terminals. To obtain the proper air quantity at the beginning of each branch, it is necessary to include a splitter damper to regulate the flow to the branch. It may also be necessary to

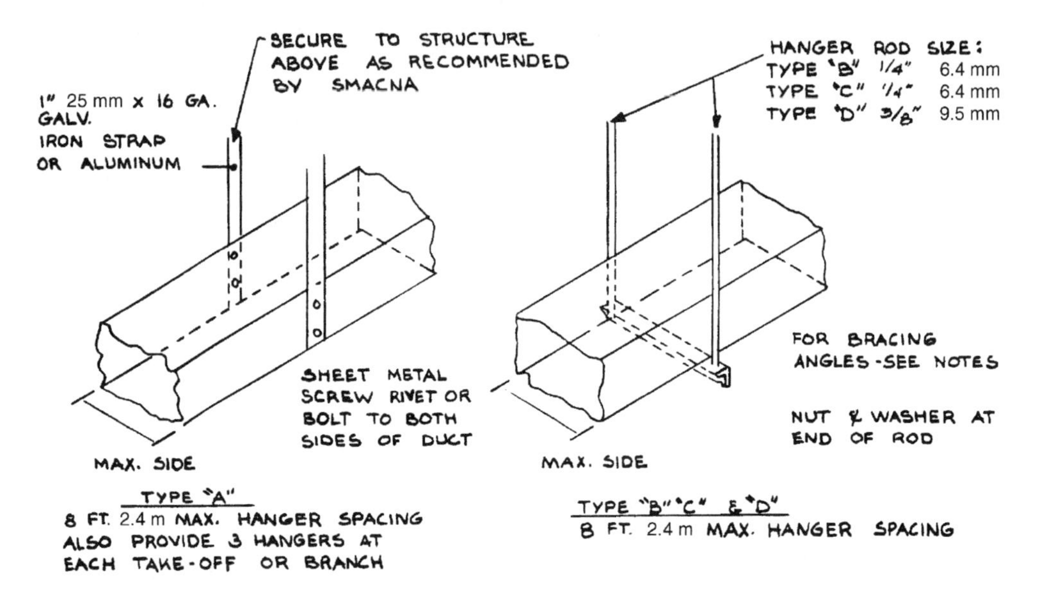

DUCT SCHEDULE				
DUCT DIMENSIONS INCHES		cm		TYPE HANGER
UP THRU 12			31	A
13	18	33	46	A
19	30	48	76	A/B
31	42	79	107	B
43	54	109	137	B
55	60	140	152	B
61	84	155	213	C
85	96	216	244	C
OVER	96	Over	244	D

NOTES:

1. FOR SEVERAL DUCTS ON ONE HANGER TYPE "B"-"C" OR "D" MAY BE USED. SIZE OF HANGER WILL BE SELECTED ON THE SUM OF DUCT WIDTHS EQUAL TO MAX. WIDTH OF DUCT SCHEDULE.

2. SCHEDULE FOR ANGLES FOR BRACING: TYPE "B" 1½" x 1½" x 1/8" 38 x 38 x 3 mm ANGLE. MAX. SPACING 8'-0" CENTERS; TYPE "C" 1½" x 1½" x 3/16" 38 x 38 x 5 mm ANGLE MAX SPACING 8'-0" CENTERS; TYPE "D" 2" x 2" x 1/4 MAX SPACING 51 x 51 x 6 mm 4'-0" 1.2 m CENTERS.

FIGURE 33 Duct hangers for ducts of various dimensions. (*Jerome F. Mueller.*)

have a control device (vanes, volume damper, or adjustable-terminal volume control) to regulate the flow at each terminal for proper air distribution.

The *velocity-reduction method* of duct design is not too popular because it requires a broad background of duct-design experience and knowledge to be within reasonable accuracy. It should be used for only the simplest layouts. Splitters and dampers should be included for balancing purposes.

To apply the velocity-reduction method: (1) Select a starting velocity at the fan discharge. (2) Make arbitrary reductions in velocity down the duct run. The starting velocity should not exceed the values in Table 18. Obtain the equivalent round-duct diameter from Fig. 32. Compute the required duct area from the round-duct diameter, and from this the duct dimensions, as shown in steps 4 and 5. (3) Determine the required fan static discharge pressure for the supply by using the longest run of duct, including all elbows and fittings. Note, however, that the longest run is not necessarily the run with the greatest friction loss, as shorter runs may have more elbows, fittings, and restrictions.

The equal-friction and velocity-reduction methods of air-conditioning system duct design are applicable only to low-velocity systems, i.e., systems in which the maximum air velocity is 3000 ft/min (914.4 m/min), or less.

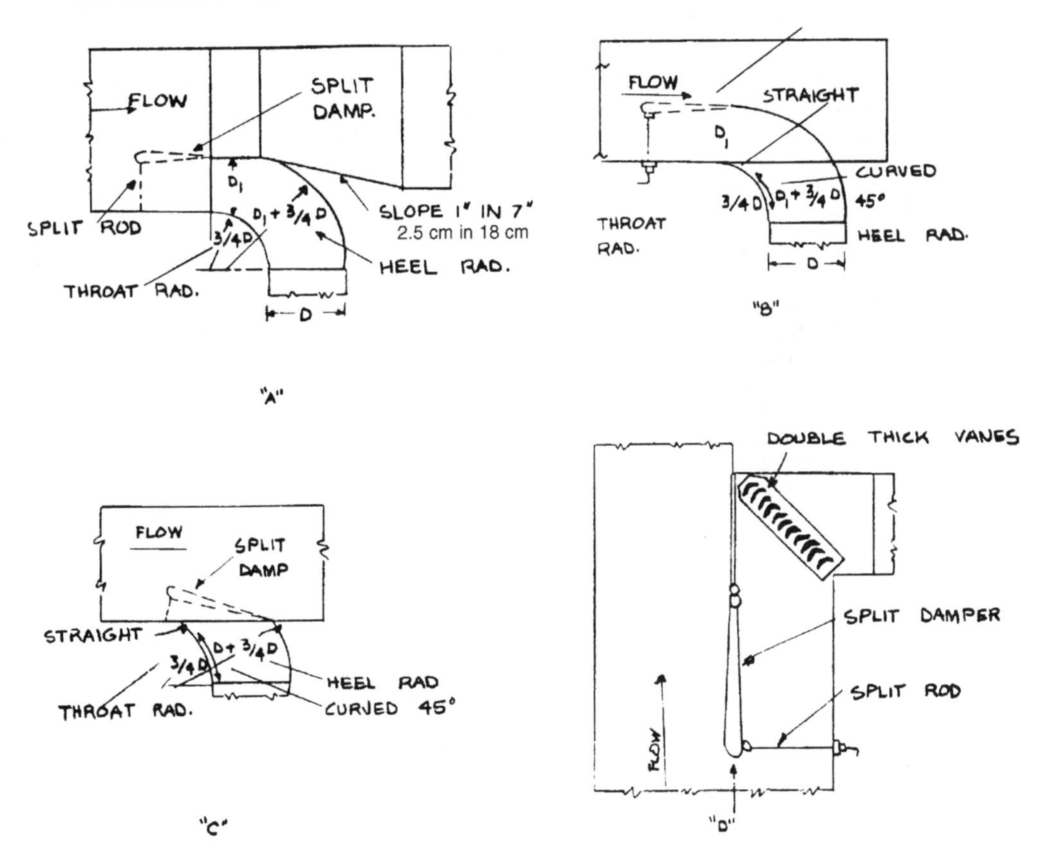

FIGURE 34 Rectangular-duct take-offs for air supply to specific rooms or areas. (*Jerome F. Mueller.*)

STATIC-REGAIN METHOD FOR SIZING LOW-VELOCITY AIR-CONDITIONING SYSTEM DUCTS

Using the same data as in the previous calculation procedure, an air velocity of 2500 ft/min (762 m/min) in the main duct section, an unvaned elbow radius of $R/D = 1.25$, and an operating pressure of 0.20-in (2.5-mm) wg for each outlet, size the system ducts, using the static-regain method of design for low-velocity systems.

Calculation Procedure:

1. ***Compute the fan outlet duct size***

The fan outlet duct, also called the main duct section, will have an air velocity of 2500 ft/min (762 m/min). Hence, the required duct area is $A = 36{,}000/2500 = 14.4$ ft^2 (1.34 m^2). This corresponds to a round-duct diameter of $d = 2(144A/\pi)^{0.5} = 2(144 \times 14.4/\pi)^{0.5} = 51.5$ in (130.8 cm). A nearly square duct, i.e., a duct 46 × 45 in (116.8 × 114.3 cm), has an area of 14.38 ft^2 (1.34 m^2) and is a good first choice for this system because it closely approximates the outlet size of a standard centrifugal fan.

Where possible, use a square main duct to simplify fan connections. Thus, a 46 × 46 in (116.8 × 116.8 cm) duct might be the final choice of this system.

2. *Compute the main-duct friction loss*

Using Fig. 32 find the main-duct friction loss as 0.13-in (3.3-mm) wg per 100 ft (30 m) of equivalent duct length for a flow of 36,000 ft³/min (1019.2 m³/min) and a diameter of 51.5 in (130.8 cm).

3. *Determine the friction loss up to the first branch duct*

The length of the main duct between the fan and the first branch is 25 + 35 = 60 ft (18.3 m). The equivalent length of the elbow is, from the *Guide* or Carrier *Handbook of Air Conditioning Design*, 26 ft (7.9 m). Hence, the total equivalent length = 60 + 30 = 90 ft (27.4 m). The friction loss is then $h_T = L_f = (90/100)(0.13) = 0.117$-in (2.97-mm) wg.

4. *Size the longest duct run*

The longest duct run is from *A* to outlet 18 (Fig. 31). Size the duct using the following tabulation, preparing it as described on the next page.

List in column 1 the various duct sections in the longest duct run, as shown in Fig. 31. In column 2 list the air quantity flowing through each duct section. Tabulate in column 3 the equivalent length of each duct. Where a fitting is in the duct section, as in *B*-13, assume a duct size and compute the equivalent length using the *Guide* or Carrier fitting table. When the duct section does not have a fitting, as with section 13-14, the equivalent length equals the distance between the centerlines of two adjacent outlets.

Next, determine the *L*/*Q* ratio for each duct section, using Fig. 35. Enter Fig. 35 at the air quantity in the duct and project vertically upward to the curve representing the equivalent length of the duct. At the left read the *L*/*Q* ratio for this section of the duct. Thus, for duct section 13-14,

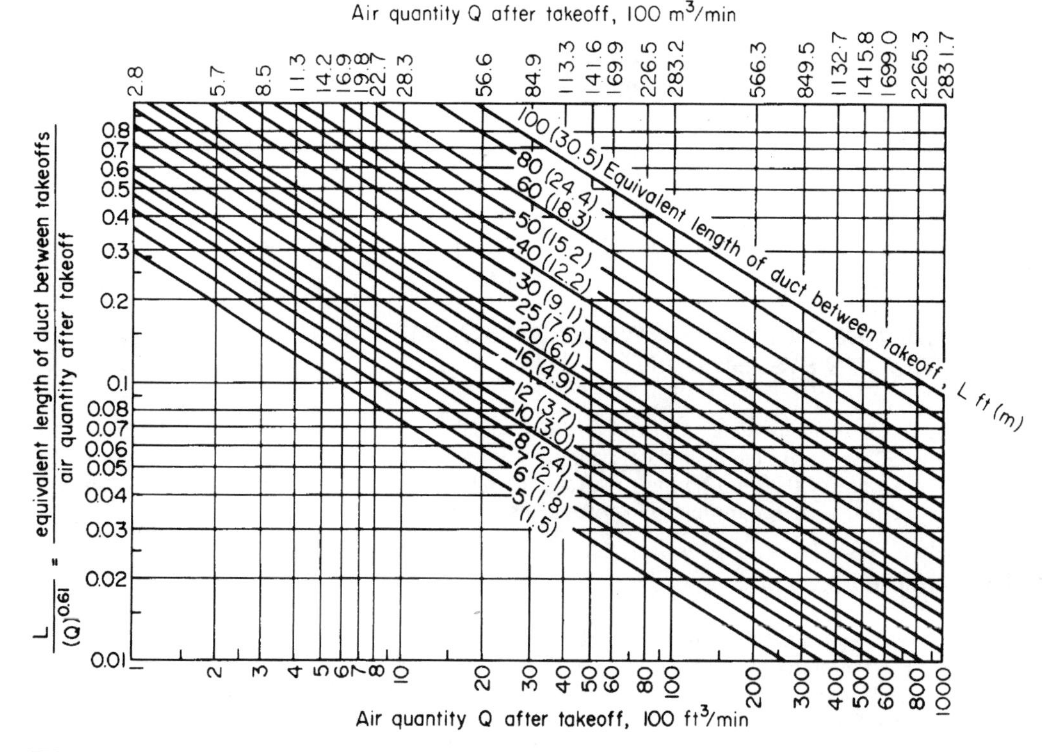

FIGURE 35 *L*/*Q* ratio for air ducts. (*Carrier Air Conditioning Company.*)

Q = 10,000 ft³/min (283.1 m³/min), and L = 20 ft (6.1 m). Entering the chart as detailed previously shows that L/Q = 0.72. Proceed in this manner, determining the L/Q ratio for each section of the duct in the longest duct run.

(1) Section number	(2) Air flow, ft³/min (m³/min)	(3) Equivalent length, ft (m)	(4) L/Q ratio	(5) Velocity, ft/min (m/min)	(6) Duct area, ft² (m²) (2)/(5)	(7) Duct size, in (cm)
Fan to A	36,000 (1,019.2)	86 (26.2)	—	2,500 (762.0)	14.4 (1.34)	46 × 45 (116.8 × 114.3)
A–B	24,000 (679.4)	20 (6.1)	0.034	2,410 (734.6)	9.95 (0.92)	38 × 38 (96.5 × 96.5)
B–13	12,000 (339.7)	26* (7.9)*	0.088	2,200 (670.6)	5.45 (0.51)	28 × 28 (71.1 × 71.1)
13–14	10,000 (283.1)	20 (6.1)	0.072	2,040 (621.8)	4.90 (0.46)	27 × 27 (68.6 × 68.6)
14–15	8,000 (226.5)	20 (6.1)	0.083	1,850 (563.9)	4.33 (0.40)	25 × 25 (63.5 × 63.5)
15–16	6,000 (169.9)	20 (6.1)	0.098	1,700 (518.2)	3.53 (0.33)	24 × 23 (60.9 × 58.4)
16–17	4,000 (113.2)	20 (6.1)	0.130	1,520 (463.3)	2.63 (0.24)	20 × 19 (50.8 × 48.3)
17–18	2,000 (56.6)	20 (6.1)	0.195	1,300 (396.2)	1.54 (0.14)	15 × 15 (38.1 × 38.1)
B–7	12,000 (339.7)	25* (7.6)*	—	—	—	28 × 28 (71.1 × 71.1)
7–8	10,000 (283.1)	20 (6.1)	—	—	—	27 × 27 (68.6 × 68.6)
8–9	8,000 (226.5)	20 (6.1)	—	—	—	25 × 25 (63.5 × 63.5)
9–10	6,000 (169.9)	20 (6.1)	—	—	—	25 × 23 (60.9 × 58.4)
10–11	4,000 (113.2)	20 (6.1)	—	—	—	20 × 19 (50.8 × 48.3)
11–12	2,000 (56.1)	20 (6.1)	—	—	—	15 × 15 (38.1 × 38.1)
A–1	12,000 (339.7)	25* (7.6)	—	—	—	28 × 28 (71.1 × 71.1)
1–2	10,000 (283.1)	20 (6.1)	—	—	—	27 × 27 (68.6 × 68.6)
2–3	8,000 (226.5)	20 (6.1)	—	—	—	25 × 25 (63.5 × 64.5)
3–4	6,000 (169.9)	20 (6.1)	—	—	—	24 × 23 (60.9 × 58.4)
4–5	4,000 (113.2)	20 (6.1)	—	—	—	20 × 19 (50.8 × 48.3)
5–6	2,000 (56.1)	20 (6.1)	—	—	—	15 × 15 (38.1 × 38.1)

*See text.

Determine the velocity of the air in the duct by using Fig. 36. Enter Fig. 36 at the L/Q ratio for the duct section, say 0.072 for section 13-14. Find the intersection of the L/Q curve with the velocity curve for the preceding duct section: 2200 ft/min (670.6 m/min) for section 13-14. At the bottom of Fig. 36 read the velocity in the duct section, i.e., after the previous outlet and in the duct section under consideration. Enter this velocity in column 5. Proceed in this manner, determining the velocity in each section of the duct in the longest duct run.

Determine the required duct area from column 2/column 5, and insert the result in column 6. Find the duct size, column 7, by converting the required duct area to a square- or rectangular-duct dimension. Thus a 27 × 17 in (68.6 × 43.2 cm) square duct has a cross-sectional area slightly greater than 4.90 ft² (0.46 m²).

5. *Determine the sizes of the other ducts in the system*

Since the ducts in runs A and B are symmetric with the duct containing the outlets in the longest run, they can be given the same size when the same quantity of air flows through them. Thus, duct section 7-8 is sized the same as section 13-14 because the same quantity of air, 10,000 ft³/min (4.72 m³/s), is flowing through both sections.

Where the duct section contains a fitting, as B-7 and A-1, assume a duct size and find the equivalent length, using the *Guide* or Carrier fitting table. These sections are marked with an asterisk.

6. *Determine the required fan discharge pressure*

The total pressure required at the fan discharge equals the sum of the friction loss in the main duct plus the terminal operating pressure. Hence the required fan static discharge pressure = 0.117 + 0.20 = 0.317-in (8.1-mm) wg.

Related Calculations. The basic principle of the static-regain method is to size a duct run so that the increase in static pressure (regain due to the reduction in velocity) at each branch or air terminal

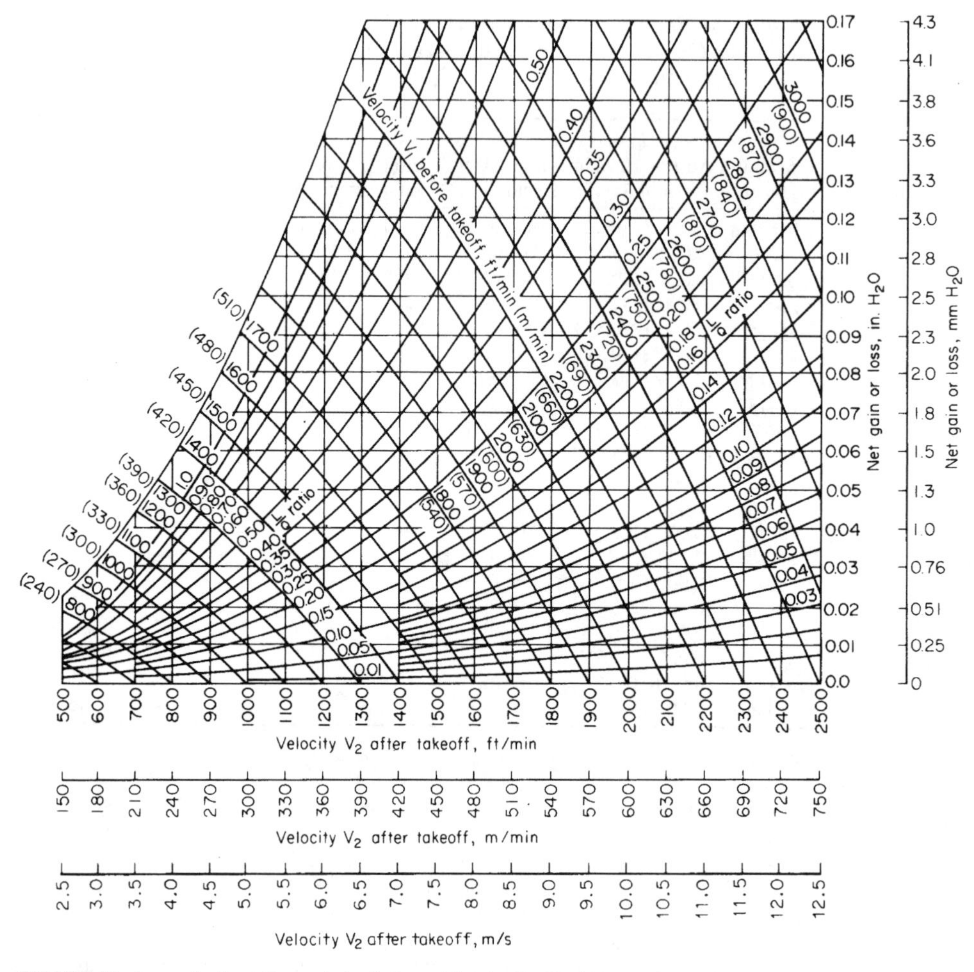

FIGURE 36 Low-velocity static-regain in air ducts. (*Carrier Air Conditioning Company.*)

just offsets the friction loss in the succeeding section of duct. The static pressure is then the same before each terminal and at each branch.

As a *general* guide to the results obtained with the static-regain and equal-friction duct-design methods, the following should be helpful:

	Static regain	Equal friction
Main-duct sizes	Same	Same
Branch-duct sizes	Larger	Smaller
Sheet-metal weight	Greater	Less
Fan horsepower	Less	Greater
Balancing time	Less	Greater
Operating costs	Less	Greater

Note that these tabulated results are *general* and may not apply to every system. The method presented in this calculation procedure is that used by the Carrier Air Conditioning Company at the time of the original writing.

SELECTING HUMIDIFIERS FOR CHOSEN INDOOR CLIMATE CONDITIONS

A paper mill has a storeroom with a volume of 500,000 ft^3 (14,155 m^3). The lowest recorded outdoor temperature in the mill locality is 0°F (–17.8°C). What capacity humidifier is required for this storeroom if a 70°F (21°C) dry-bulb temperature and a 65 percent relative humidity are required in it? Moisture absorption by the paper products in the room is estimated to be 450 lb/h (204.6 kg/h). The storeroom ventilating system produces three air changes per hour. What capacity humidifier is required if the room temperature is maintained at 60°F (15.5°C) and 65 percent relative humidity? The products release 400 lb/h (181.8 kg/h) of moisture. Steam at 25 lb/in^2 (gage) (172.4 kPa) is available for humidification. The outdoor air has a relative humidity of 50 percent and a minimum temperature of 5°F (–15°C).

Calculation Procedure:

1. *Determine the outdoor design temperature*

In choosing a humidifier, the usual procedure is to add 10°F (–12.2°C) to the minimum outdoor recorded temperature because this temperature level seldom lasts more than a few hours. The result is the design-outdoor temperature. Thus, for this mill, design-outdoor temperature = 0 + 10 = 10°F (–12°C).

2. *Compute the weight of moisture required for humidification*

Enter Table 20 at an outdoor temperature of 10°F (–12.2°C), and project across to the desired relative humidity, 65 percent. Read the quantity of steam required as 1.330 lb/h (0.6 kg/h) per 1000 ft^3 (28.3 m^3) of room volume for two air changes per hour. Since this room has three air changes per hour, the quantity of moisture required is (3/2)(1330) = 1.995 lb/h (0.91 kg/h) per 1000 ft^3 (28.3 m^3) of volume.

The amount of moisture in the form of steam required for this storeroom = (room volume, ft^3/1000)(lb/h of steam per 1000 ft^3) = (500,000/1000)(1.995) = 997.5 lb (453.4 kg) for humidification of the air. However, the products in the storeroom absorb 450 lb/h (202.5 kg/h) of moisture.

TABLE 20 Steam Required for Humidification at 70°F (21°C)*

Outdoor temp		Relative humidity desired indoors, percent									
		40		50		60		65		70	
°F	°C	lb	kg	lb	kg	lb	kg	lb	kg	lb	kg
50	10.0	0.045	0.02	0.271	0.12	0.501	0.23	0.616	0.28	0.731	0.33
40	4.4	0.307	0.14	0.537	0.24	0.767	0.35	0.882	0.40	1.000	0.45
30	1.1	0.503	0.23	0.734	0.33	0.964	0.43	1.079	0.49	1.194	0.54
20	–6.7	0.654	0.29	0.883	0.40	1.115	0.50	1.230	0.55	1.345	0.61
10	–12.2	0.754	0.34	0.985	0.44	1.215	0.55	1.330	0.60	1.445	0.65
0	–17.8	0.819	0.37	1.049	0.47	1.279	0.58	1.394	0.63	1.509	0.68
–10	–23.3	0.860	0.39	1.090	0.49	1.320	0.59	1.435	0.65	1.550	0.70
–20	–28.9	0.885	0.30	1.115	0.50	1.345	0.61	1.460	0.66	1.575	0.71

*Pounds (kilograms) of steam per hour required per 1000 ft^3 (28.3 m^3) of space to secure desired indoor relative humidity at 70°F (21°C), with various outdoor temperatures. Assuming two air changes per hour and outdoor relative humidity of 75 percent.

Armstrong Machine Works.

TABLE 21 Humidifier Capacities*

Steam pressure, lb/in² (gage) (kPa)*	Orifice size, in (mm)			
	7/16 (1.11)	3/8 (9.5)	1¼ (31.8)	1-7/64 (28.2)
		—		
5 (34.5)	100†	—	340	42
		—		
		—		
		—		
		—		
10 (68.9)	140	—	610	
		—		
		—		
15 (103.4)	170	138	810	74
20 (137.9)	—	158	980	80
25 (172.4)	—	174	1130	90
30 (206.8)	—	190	1280	100

*Armstrong Machine Works.

†Continuous discharge capacity with steam pressures as indicated. No allowance for pressure drop after solenoid valve opens.

Hence, the total moisture quantity required = moisture for air humidification + moisture absorbed by products = 997.5 + 450.0 = 1447.5 lb/h (657.9 kg/h), say 1450 lb/h (659.1 kg/h) for humidifier sizing purposes.

3. *Select a suitable humidifier*

Table 21 lists typical capacities for humidifiers having orifices of various sizes and different steam pressures. Study of Table 21 shows that one 1¼-in (32-mm) orifice humidifier and two 3/8-in (9.5-cm) orifice humidifiers will discharge 1130 + (2)(174) = 1478 lb/h (671.8 kg/h) of steam when the steam supply pressure is 25 lb/in² (gage) (172.4 kPa). Since the required capacity is 1450 lb/h (659.1 kg/h), these humidifiers may be acceptable.

Large-capacity steam humidifiers usually must depend on existing ducts or large floor-type unit heaters for distribution of the moisture. When such means of distribution are not available, choose a larger number of smaller-capacity humidifiers and arrange them as shown in Fig. 37*c*. Thus, if 3/8-in (9.5-mm) orifice humidifiers were selected, the number required would be (moisture needed, lb/h)/(humidifier capacity, lb/h) = 1450/174 = 8.33, or 9 humidifiers.

4. *Choose a humidifier for the other operating conditions*

Where the desired room temperature is different from 70°F (231°C), use Table 20 instead of Table 21. Enter Table 22 at the desired room temperature, 60°F (15.6°C), and read the moisture content of saturated air at this temperature, as 5.795 gr/ft³ (13.26 mL/dm³). The outdoor air at 5 + 10 = 15°F (9.4°C) contains, as Table 22 shows, 0.984 gr/ft³ (2.25 mL/dm³) of moisture when fully saturated.

Find the moisture content of the air at the room and the outdoor conditions from moisture content, gr/ft³ = (relative humidity of the air, expressed as a decimal) (moisture content of saturate air, gr/ft³). For the 60°F (15.6°C), 65 percent relative humidity room air, moisture content = (0.65)(5.795) = 3.77 gr/ft³ (8.63 mL/dm³). For the 15°F (–9.4°C) 50 percent relative humidity outdoor air, moisture content = (0.50)(0.984) = 0.492 gr/ft³ (1.13 mL/dm³). Thus, the humidifier must add the difference or 3.77 – 0.492 = 3.278 gr/ft³ (7.5 mL/dm³).

This storeroom has a volumc of 500,000 ft³ (14,155 m³) and three air changes per hour. Thus, the weight of moisture that must be added per hour is (number of air changes per hour)(volume, ft³) (gr/ft³ of air)/7000 gr/lb or, for this storeroom, (3)(500,000)(3.278)/7000 = 701 lb/h (0.09 kg/s) excluding the product load. Since the product load is 400 lb/h (0.05 kg/s), the total humdification load is 701 + 400 = 1101 lb/h (0.14 kg/s). Choose the humidifiers for these conditions in the same way as described in step 4.

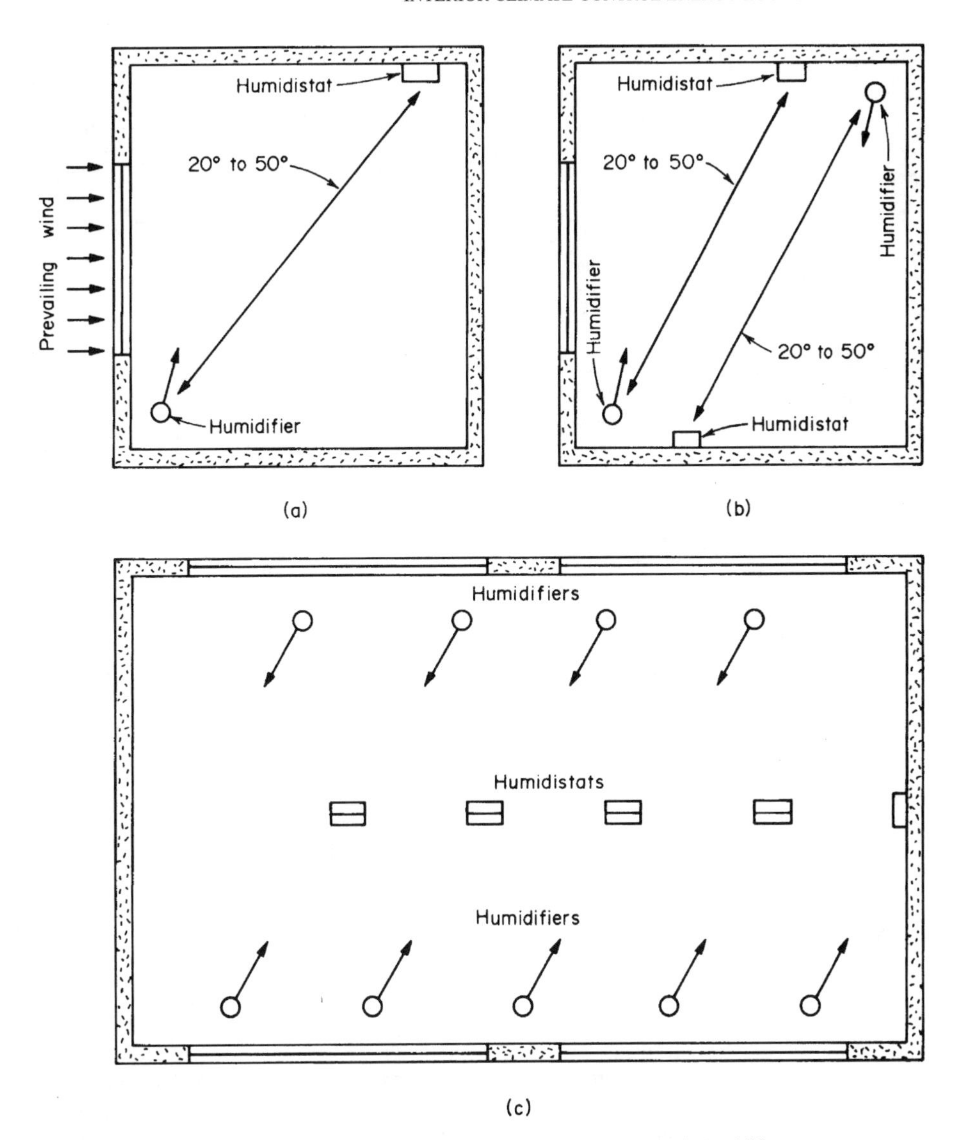

FIGURE 37 Location of (*a*) a single humidifier, (*b*) two humidifiers, (*c*) multiple humidifiers.

TABLE 22 Moisture Content of Saturated Air

		Grains of water	
°F	°C	Per ft^3	Per m^3
15	−9.4	0.984	34.8
20	−6.7	1.242	43.9
40	4.4	2.863	101.1
50	10.0	4.106	145.0
60	15.6	5.795	204.7
70	21.1	8.055	284.5

TABLE 23 Recommended Industrial Humidities and Temperatures

Industry	Degrees °F	Degrees °C	Relative humidity, %
Ceramics:			
Drying refractory shapes	110–150	43–65	50–60
Molding room	80	26	60
Confectionery:			
Chocolate covering	62–65	17–18	50–55
Hard-candy making	70–80	21–27	30–50
Electrical:			
Manufacture of cotton-covered wire, storage, general	60–80	21–27	60–70
Food storage:			
Apple	31–34	–0.5–1.1	75–85
Citrus fruit	32	0	80
Grain	60	16	30–45
Meat ripening	40	4	80
Paper products:			
Binding	70	21	45
Folding	77	25	65
Printing	75	24	60–78
Storage	75–80	24–27	40–60
Textile:			
Cotton carding	75–80	24–27	50–55
Cotton spinning	60–80	16–27	50–70
Rayon spinning, throwing	70	21	85
Silk processing	75–80	24–27	60–70
Wool spinning, weaving	75–80	24–27	55–60
Miscellaneous:			
Laboratory, analytical	60–70	16–21	60–70
Munitions, fuse loading	70	21	55
Cigar and cigarette making	70–75	21–24	55–65

Related Calculations. Use the method given here to choose a humidifier for any normal industrial or comfort application. Table 23 summarizes typical recommended humidities and temperatures for a variety of industrial operations. The relative humidity maintained in industrial plants is extremely important because it can control the moisture content of hygroscopic materials.

Where the number of hourly air changes is not specified, assume two air changes, except in cotton mills where three or four may be necessary. If the plant ventilating system provides more than two air changes per hour, use the actual number of changes in computing the required humidifier capacity.

Many types of manufactured goods and raw materials absorb or release moisture during processing and storage. Since product quality usually depends directly on the moisture content, carefully controlled humidity will often reduce the number of rejects. The room humidifier must supply sufficient moisture for humidification of the air, plus any moisture absorbed by the products or materials in the room. Where these products or materials continuously release moisture to the atmosphere in the room, the quantity released can be subtracted from the moisture required for humidification. However, this condition can seldom be relied on. The usual procedure then is to select the humidifier on the basis of the moisture required for humidification of the air. The humidistat controls the operation of the humidifier, shutting it off when the products release enough moisture to supply the room requirements.

Correct locations for one or more humidifiers are shown in Fig. 37. Proper location of humidifiers is necessary if the design is to take advantage of the prevailing wind in the plant locality. Also, correct location provides a uniform, continuous circulation of air throughout the humidified area.

When only one humidifier is used, it is placed near the prevailing wind wall and arranged to discharge parallel to the wall exposed to the prevailing wind, Fig. 37*a*. Two humidifiers, Fig. 31*b*, are generally located in opposite corners of the manufacturing space and their discharges are used to produce a rotary air motion. Installations using more than two humidifiers generally have a slightly greater number of humidifiers on the windward wall to take advantage of the natural air drift from one side of the room to the other.

Pipe-spray humidifiers are as shown in Fig. 38 unless the manufacturer advises otherwise. Size the return lines as shown in Table 24.

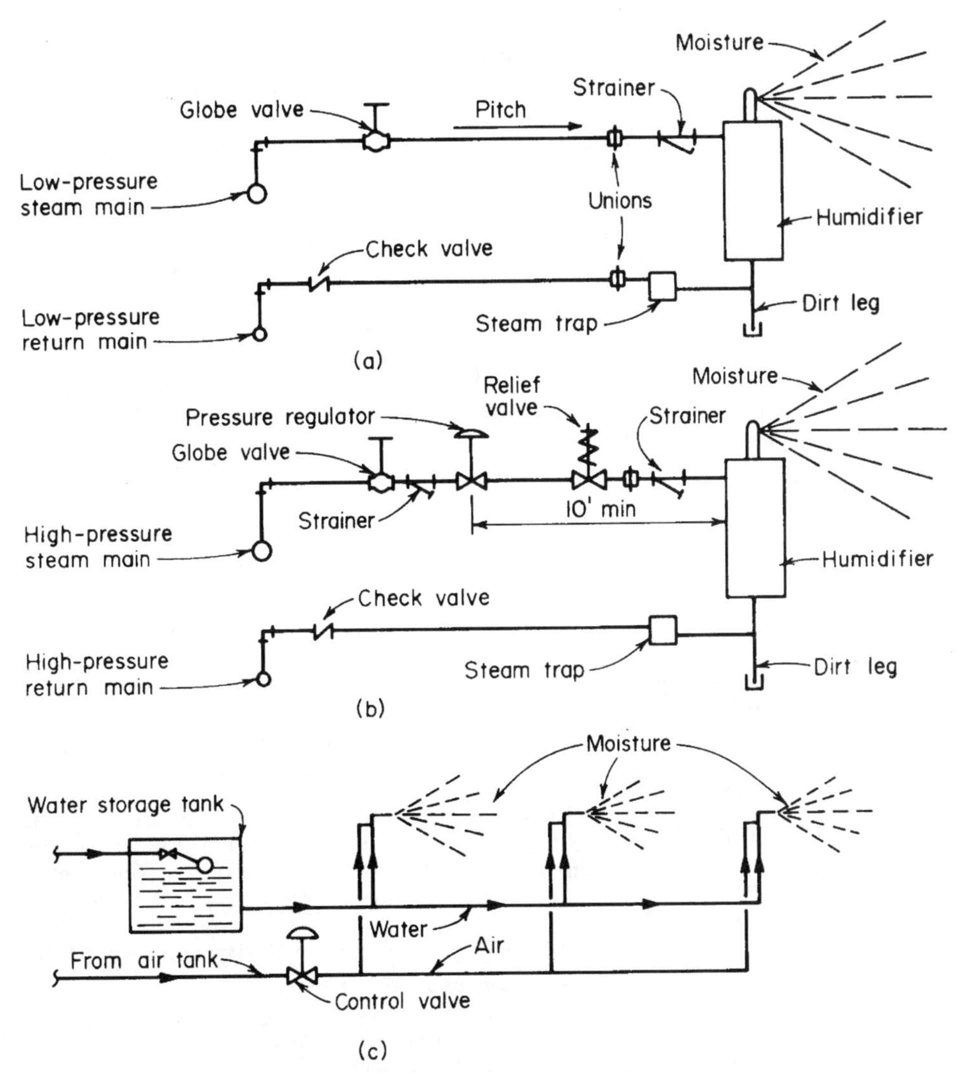

FIGURE 38 Piping for spray-type humidifiers. (*a*) Low-pressure steam; (*b*) high-pressure steam; (*c*) water spray.

TABLE 24 Steam- and Return-Pipe Sizes, in (mm)

Steam or condensate flow, lb/h (kg/h)	Steam pressure, lb/in^2 (gage) (kPa)				Length of return pipe, ft (m)	
	5 (34.5)	10 (68.9)	50 (344.7)	100 (689.4)	100 (30)	200 (60)
100 (45)	1½ (38.1)	1¼ (32)	1 (25)	1 (25)	1 (25)	1 (25)
200 (90)	2 (51)	2 (51)	1¼ (32)	1¼ (32)	1¼ (32)	1¼ (32)
400 (180)	3 (76)	2½ (64)	2 (51)	1½ (38.1)	1½ (38)	2 (51)
500 (225)	3 (76)	2½ (64)	2 (51)	2 (51)	2 (51)	2 (51)
1000 (450)	3½ (88.9)	3 (76)	2½ (64)	2½ (64)	2½ (64)	2½ (64)
2000 (900)	5 (127)	4 (102)	3 (76)	3 (76)	3 (76)	3 (76)
4000 (1800)	6 (152)	5 (127)	4 (102)	4 (102)	4 (102)	4 (102)

Humidistats to start and stop the flow of moisture into the room may be either electrically or air (hygrostat) operated, according to the type of activities in the space. Where electric switches and circuits might cause a fire hazard, use an air-operated hygrostat instead of a humidistat. Locate either type of control to one side of the humidifying moisture stream, 20 to 50 ft (6.1 to 15.2 m) away.

USING THE PSYCHOMETRIC CHART IN AIR-CONDITIONING ENERGY CALCULATIONS

Determine the properties of air at 80°F (26.7°C) dry-bulb (db) temperature and 65°F (18.3°C) wet-bulb (wb) temperature, using the psychrometric chart. Determine the same properties of air if the wet-bulb temperature is 75°F (23.9°C) and the dew-point temperature is 67°F (19.4°C). Show on the psychrometric chart an air-conditioning process in which outside air at 95°F (35°C) db and 80°F (26.7°C) wb is mixed with return air from the room at 80°F (26.7°C) db and 65°F (18.3°C) wb. Air leaves the conditioning apparatus at 55°F (12.8°C) db and 50°F (10°C) wb.

Calculation Procedure:

1. *Determine the relative humidity of the air*

Using Fig. 28, enter the bottom of the chart at the first dry-bulb temperature, 80°F (26.7°C), and project vertically upward until the slanting 65°F (18.3°C) wet-bulb temperature line is intersected. At the intersection, or *state point*, read the relative humidity as 45 percent on the sloping curve. Note that the number representing the wet-bulb temperature appears on the saturation, or 100 percent relative humidity, curve and that the wet-bulb temperature line is a straight line sloping downward from left to right. The relative humidity curves slope upward from left to right and have the percentage of relative humidity marked on them.

When the wet-bulb and dew-point temperatures are given, enter the psychrometric chart at the wet-bulb temperature, 75°F (23.9°C) on the saturated curve. From here project downward along the wet-bulb temperature line until the horizontal line representing the dew-point temperature, 67°F (19.4°C), is intersected. At the intersection, or state point, read the dry-bulb temperature as 94.7°F (34.8°C) on the bottom scale of the chart. Read the relative humidity at the intersection as 40.05 percent because the intersection is very close to the 40 percent relative humidity curve.

2. *Determine the moisture content of the air*

Read the moisture content of the air in grains on the right-hand scale by projecting horizontally from the intersection, or state point. Thus, for the first condition of 80°F (26.7°C) dry bulb and 65°F (18.3°C) wet bulb, projection to the right-hand scale gives a moisture content of 68.5 gr/lb (9.9 gr/kg) of dry air.

For the second condition, 75°F (23.9°C) wet bulb and 67°F (19.4°C) dew point, projection to the right-hand scale gives a moisture content of 99.2 gr/lb (142.9 gr/kg).

3. *Determine the dew point of the air*
This applies to the first condition only because the dew point is known for the second condition. From the intersection of the dry-bulb temperature, 80°F (26.7°C), and the wet-bulb temperature, 65°F (18.3°C), that is, the state point, project horizontally to the left to read the dew point on the horizontal intersection with the saturation curve as 56.8°F (13.8°C). Note that the temperatures plotted along the saturation curve correspond to both the wet-bulb and dew-point temperatures.

4. *Determine the enthalpy of the air*
Find the enthalpy (also called *total heat*) by reading the value on the sloping line on the central scale above the saturation curve at the state point for the air. Thus, for the first condition, 80°F (26.7°C) dry bulb and 65°F (18.3°C) wet bulb, the enthalpy is 30 Btu/lb (69.8 kJ/kg). The enthalpy value on the psychrometric chart includes the heat of 1 lb (0.45 kg) of dry air and the heat of the moisture in the air, in this case, 68.5 gr (98.6 gr) of water vapor.

For the second condition, 75°F (23.9°C) wet bulb and 67°F (19.4°C) dew point, read the enthalpy as 38.5 Btu/lb (89.6 kJ/kg) at the state point.

5. *Determine the specific volume of the air*
The specific volume lines slope downward from left to right from the saturation curve to the horizontal dry-bulb temperature. Values of specific volume increase by 0.5 ft^3/lb (0.03 m^3/kg) between each line.

For the first condition, 80°F (26.7°C) dry-bulb and 65°F (18.3°C) wet-bulb temperature, the stage point lies just to the right of the 13.8 line, giving a specific volume of 13.81 ft^3/lb (0.86 m^3/kg). For the second condition, 75°F (23.9°C) wet-bulb and 67°F (19.4°C) dew-point temperatures, the specific volume, read in the same way, is 14.28 ft^3/lb (0.89 m^3/kg).

The weight of the air-vapor mixture can be found from 1.000 + 68.5 gr/lb of air/(7000 gr/lb) = 1.0098 lb (0.46 kg) for the first condition and 1.000 + 99.2/7000 = 1.0142 lb (0.46 kg). In both these calculations the 1000 lb (454.6 kg) represents the weight of the *dry* air, and 68.5 gr (4.4 gr) and 99.2 gr (6.43 gr) represent the weight of the moisture for each condition.

6. *Determine the vapor pressure of the moisture in the air*
Read the vapor pressure by projecting horizontally from the state point to the extreme left-hand scale. Thus, for the first condition the pressure of the water vapor is 0.228 lb/in^2 (1.57 kPa). For the second condition the pressure of the water is 0.328 lb/in^3 (2.26 kPa).

7. *Plot the air-conditioning process on the psychrometric chart*
Air-conditioning processes are conveniently represented on the psychrometric chart. To represent any process, locate the various state points on the chart and convert the points by means of lines representing the process.

Thus, for the air-conditioning process being considered here, start with the outside air at 95°F (35°C) db and 80°F (26.7°C) wb, and plot point (Fig. 28) at the intersection of the two temperature lines. Next, plot point 3, the return air from the room at 80°F (26.7°C) db and 65°F (18.3°C) wb. Point 2 is obtained by computing the final temperature of two air streams that are mixed, using the method of the calculation procedure given earlier in this section. Plot point 4, using the given leaving temperatures for the apparatus, 55°F (12.8°C) db and 50°F (10°C) wb.

The process in this system is as follows: Air is supplied to the conditioned space along line 4-3. During passage along this line on the chart, the air absorbs heat and moisture from the room. While passing from point 3 to 2, the air absorbs additional heat and moisture while mixing with the warmer outside air. From point 1 to 2, the outside air is cooled while it is mixed with the indoor air. At point 2, the air enters the conditioning apparatus, is cooled, and has its moisture content reduced.

Related Calculations. Use the psychrometric chart for all applied air-conditioning problems where graphic representation of the state of the air or a process will save time. At any given state point of air, the relative humidity in percent can be computed from [partial pressure of the water vapor at the dew-point temperature, lb/in^2 (abs) + partial pressure of the water vapor at saturation corresponding to the dry-bulb temperature of the air, lb/in^2 (abs)](100). Determine the partial pressures from a table of air properties or from the steam tables.

In an *air washer* the temperature of the entering air is reduced. Well-designed air washers produce a leaving-air dry-bulb temperature that equals the wet-bulb and dew-point temperatures of the leaving air. The humidifier portion of an air-conditioning apparatus adds moisture to the air while the dehumidifier removes moisture from the air. In an ideal air washer, adiabatic cooling is assumed to occur.

By using the methods of step 7, any basic air-conditioning process can be plotted on the psychrometric chart. Once a process is plotted, the state points for the air are easily determined from the psychrometric chart.

When you make air-conditioning computations, keep these facts in mind: (1) The total enthalpy, sometimes termed *total heat*, varies with the wet-bulb temperature of the air. (2) The sensible heat of air depends on the wet-bulb temperature of the air; the enthalpy of vaporization, also called the *latent heat*, depends on the dew-point temperature of the air; the dry-bulb, wet-bulb, and dew-point temperatures of air are the same for a saturated mixture. (3) The dew-point temperature of air is fixed by the amount of moisture present in the air.

ENERGY ASPECTS OF DESIGNING HIGH-VELOCITY AIR-CONDITIONING DUCTS

Design a high-velocity air-distribution system for the duct arrangements shown in Fig. 39 if the required total air flow is 5000 ft^3/min (2.36 m^3/s).

Calculation Procedure:

1. ***Determine the main-duct friction loss***

Many high-velocity air-conditioning systems are designed for a main-header velocity of 4000 ft/min (1219.2 m/min) and a friction loss of 1.0 in H_2O per 100 ft (0.08 cm/m) of equivalent duct length. The fan usually discharges into a combined air-diffuser noise-attenuator in which the static pressure

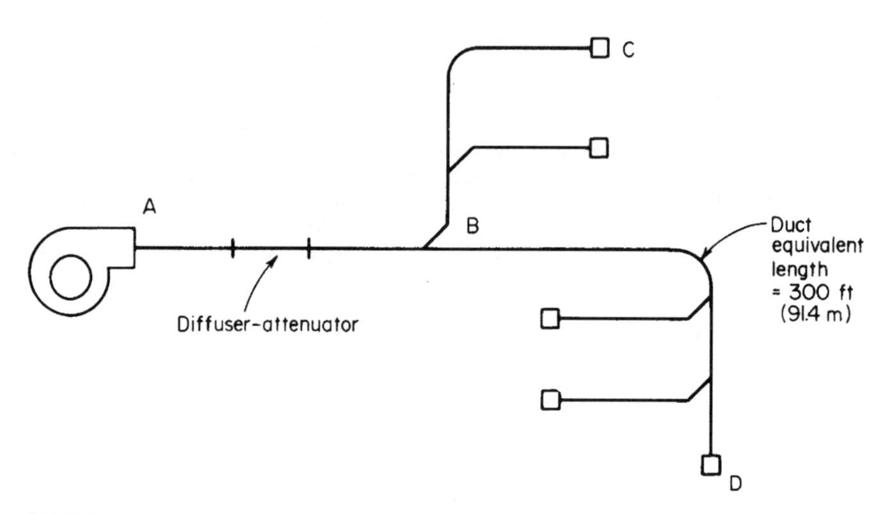

FIGURE 39 High-velocity air-duct-system layout.

of the air increases. This pressure increase must be considered in the choice of the fan-outlet static pressure, but the duct friction loss must be calculated first, as shown below.

Determine the main-duct friction loss by assuming a 1 in/100 ft (8.9 cm/m) static pressure loss for the main duct and a fan-outlet and main-duct velocity of 4000 ft/min (1219.2 m/min). Size the duct by using the equal-friction method. Thus, for the 300-ft (91.4-m) equivalent-length main duct in Fig. 39, the friction pressure loss will be (300 ft)(1.0 in/100 ft) = 3.0 in (76 mm) H_2O.

2. *Compute the required fan-outlet pressure*

The total friction loss in the duct = duct friction, in H_2O + diffuser-attenuator static pressure in H_2O. In typical installations the diffuser-attenuator static pressure varies from 0.3 to 0.5 in (7.6 to 9.7 mm) H_2O. This is the inlet pressure required to force air through the diffuser-attenuator with all outlets open. Using a value of 0.5 in (12.7 mm) give the total friction loss in the duct = 3.0 + 0.5 = 3.5 in (8.9 mm).

At the fan outlet the required static pressure is less than the total friction loss in the main duct because there is static regain at each branch take-off to the outlets. This static regain is produced by the reduction in velocity that occurs at each take-off from the main duct. There is a recovery of static pressure (velocity regain) at the take-off that offsets the friction loss in the succeeding duct section.

Assume that the velocity in branch *C* (Fig. 39) is 2000 ft/min (609.6 m/min). This is the usual maximum velocity in take-offs to terminals. Then, the maximum static regain that could occur $R = (\upsilon_i/4005)^2 - (\upsilon_f/4000)^2$, where R = static regain, in of H_2O; υ_f = initial velocity of the air, ft/min; υ_f = final velocity of the air, ft/min. For this system with an initial velocity of 4000 ft/min (1219.2 m/min) and a final velocity of 2000 ft/min (609.6 m/min), $R = (4000/4005)^2 - (2000/4005)^2 = 0.75$.

The maximum static regain is seldom achieved. Actual static regains range from 0.5 to 0.8 of the maximum. With a value of 0.8, the actual static regain = 0.8(0.75) = 0.60 in (15.2 mm) H_2O. This static regain occurs at point *B*, the take-off, and reduces the required fan discharge pressure to total friction loss in the duct – static region at first take-off = 3.5-0.60 = 2.9 in (73.7 mm) H_2O. Thus, a fan developing a static discharge pressure of 3.0 in (76 mm) H_2O would probably be chosen for this system.

3. *Find the branch-duct pressure loss*

To find the branch-duct pressure loss, find the pressure in the main duct at the take-off point. Use the standard duct-friction chart (Fig. 32) to determine the pressure loss from the fan to the take-off point. Subtract the sum of this loss and the diffuser-attenuator static pressure from the fan static discharge pressure. The result is the pressure available to force air through the branch duct. Size the branch duct by using the equal-friction method.

Related Calculations. Note that the design of a high-velocity duct system (i.e., a system design in which the air velocities and static pressures are higher than in conventional systems) is basically the same as for a low-velocity duct system designed for static regain. The air velocity is reduced at each take-off to the riser and air terminals. Design of any high-velocity duct system involves a compromise between the reduced duct sizes (with a saving in materials, labor, and space costs) and higher fan horsepower.

Class II centrifugal fans (Table 25) are generally required for the higher static pressures used in high-velocity air-conditioning systems. Extra care must be taken in duct layout and construction. The high-velocity ducts are usually sealed to prevent air leakage that may cause objectionable noise. Round ducts are preferred to rectangular ones because of the greater rigidity of the round duct.

TABLE 25 Classes of Construction for Centrifugal Fans

Class	Maximum total pressure, in H_2O (mm H_2O)
I	3¾ (95)—standard
II	6¾ (172)—standard
III	12¾ (324)—standard
IV	More than 12¾ (324)—recommended

TABLE 26 Typical High-Velocity-System Air Velocities

	Velocity, ft/min (m/min)
Header or main duct:	
12-h operation	3000–4000 (914.4–1219.2)
24-h operation	2000–3500 (609.6–1066.8)
Branch ducts:*	
90° conical tee	4000–5000 (1219.1–1524.0)
90° tee	3500–4000 (1066.8–1219.2)

*Branches are defined as a branch header or riser having four to five, or more, take-offs to terminals.
Carrier Air Conditioning Company.

Use as many symmetric duct runs as possible in designing high-velocity duct systems. The greater the system symmetry, the less time required for duct design, layout, balancing, construction, and installation.

The initial starting velocity used in the supply header depends on the number of hours of operation. To achieve an economic balance between first cost and operating cost, lower air velocities in the header are recommended for 24-h operation, where space permits. Table 26 shows typical air velocities used in high-velocity, air-conditioning systems. Use this tabulation to select suitable velocities for the main and branch ducts in high-velocity systems.

Carrier Air Conditioning Company recommends that the following factors be considered in laying out header ductwork for high-velocity air-conditioning systems.

1. The design friction losses from the fan discharge to a point immediately upstream of the first riser take-off from each branch header should be as nearly equal as possible.
2. To satisfy principle 1 above as applied to multiple headers leaving the fan, and to take maximum advantage of the allowable high velocity, adhere to the following basic rule whenever possible: Make as nearly equal as possible the ratio of the total equivalent length of each header run (fan discharge to the first riser take-off) to the initial header diameter (*L/D* ratio). Thus, the longest header run should preferably have the highest air quantity so that the highest velocities can be used throughout.
3. Unless space conditions dictate otherwise, use a 90° tee or 90° conical tee for the take-off from the header rather than a 45° tee. Fittings of 90° provide more uniform pressure drops to the branches throughout the system. Also, the first cost is lower.

OUTLET AND RETURN AIR GRILLE CHOICE FOR AIR-CONDITIONING SYSTEMS

Choose an air grille to deliver 425 ft^3/min (0.20 m^3/s) of air to a broadcast studio having a 12-ft (3.7-m) ceiling height. The room is 10 ft (3.0 m) long and 10 ft (3.0 m) wide. Specify the temperature difference to use, the air velocity, grille static resistance, size and face area.

Calculation Procedure:

1. ***Choose the outlet-grille velocity***

The air velocity specified for an outlet grille is a function of the type of room in which the grille is used. Table 27 lists typical maximum outlet air velocities used in grilles serving various types of

TABLE 27 Typical Air Outlet Velocity

Type of room	Maximum velocity, ft/min (m/min)	
Broadcast studio	300–500	(91.4–152.4)
Apartments, private residences, churches, hotel	500–750	(152.4–228.6)
bedrooms, legitimate theatres, private offices	500–750	(152.4–228.6)
	500–750	(152.4–228.6)
	500–750	(152.4–228.6)
	500–750	(152.4–228.6)
	500–750	(152.4–228.6)
Movie theaters	1000	(304.8)
General offices	1200–1500	(365.8–457.2)
Stores—upper floors	1500	(365.8)
Stores—main floors	2000	(609.6)

rooms. Assuming a velocity of 350 ft/min (106.7 m/min) for the outlet grille in this broadcast studio, compute the grille area required from $A = cfm/\upsilon$, where A = grille area, ft^2; *cfm* = air flow through the grille, ft^3/min; υ = air velocity, ft/min. Hence, $A = 425/350 = 1.214$ ft^2 (0.11 m^2).

2. *Select the outlet-grille size*

Use the selected manufacturer's engineering data, such as that in Table 28. Examination of this table shows that there is no grille rated at 425 ft^3/min (0.20 m^3/s). Hence, the next larger capacity, 459 ft^3/min (0.22 m^3/s) must be used. This grille, as the third column from the right of Table 28 shows, is 24 in wide and 8 in high (60.9 cm wide and 20.3 cm high).

3. *Choose the grille throw distance*

Throw is the horizontal distance the air will travel after leaving the grille. With a *fan spread*, the throw of this grille is 10 ft (3.0 m). Table 28. This throw is sufficient if the duct containing the grille is located at any point in the room, i.e., along one wall, in the center, etc. If desired, the grille can be adjusted to reduce the throw, but the throw cannot be increased beyond the distance tabulated. Hence, a fan-spread grille will be used.

4. *Select the grille-mounting height*

The grille-mounting height is a function of several factors: the difference between the temperature of the entering air and the room air, the room-ceiling height, and the air *drop* (i.e., the distance the air falls from the time it passes through the outlet until it reaches the end of the throw).

By assuming a temperature difference of 20°F (11.1°C) between the entering air and the room air, Table 28 shows that the minimum ceiling height for this grille is 10 ft (3.0 m). Since the room is 12 ft (3.67 m) high, the grille can be mounted at any distance above the floor of 10 ft (3.0 m) or higher.

5. *Determine the actual air velocity in the grille*

Table 28 shows that the actual air velocity in the grille is 375 ft/min (114.3 m/min). Table 27 shows that an air velocity of 300 to 500 ft/min (91.4 to 152.4 m/min) is suitable for broadcast studios. Hence, this grille is acceptable. If the actual velocity at the grille outlet were higher than that recommended in Table 27, a larger grille giving a velocity within the recommended range would have to be chosen.

6. *Determine the grille static resistance*

Table 28 shows that the grille static resistance is 0.01 in (0.25 mm) H_2O. This is within the usual static resistance range of outlet grilles.

7. *Determine the outlet-grille area*

Table 28 shows that the outlet grille has an area of 1.224 ft^2 (0.11 m^2), or 176 in^2 (1135.5 cm^2). This agrees well with the area computed in step 1, or 1.214 ft^2 (0.11 m^2).

TABLE 28 Air-Grille-Selection Table

Air flow, ft^3/min (m^3/s)	Wall area per outlet, ft^2 (m^2)		Throw, ft (m)*	Min. ceiling height, ft (m) for temp. difference of			Air velocity, ft/min (m/min)	Grille static resist., in H_2O (cm H_2O)	Outlet size, in (cm)	Grille face area	
	Max.	Min.		15°F (8.3°C)	20°F (11.1°C)	25°F (13.8°C)				ft^2 (m^2)	in^2 (cm^2)
306 (0.14)	25 (2.32)	8 (0.74)	S:11 (3.35)	13 (3.96)	14 (4.27)	15 (4.57)	250 (76.2)	0.005 (0.01)	24 × 8 (60.9 × 20.3)	1.224 (0.11)	176 (1135.5)
			F:6 (1.83)	10 (3.05)	10 (3.05)	10 (3.05)					
459 (0.22)	57 (5.30)	17 (1.58)	S:20 (6.1)	16 (4.88)	17 (5.18)	18 (5.49)	375 (114.3)	0.01 (0.03)	24 × 8 (60.9 × 20.3)	1.224 (0.11)	176 (1135.5)
			F:10 (3.0)	10 (3.05)	10 (3.05)	11 (3.35)					

*S—straight; F—fan spread.
Waterloo Register.

TABLE 29 Lattice-Type Return-Grille Pressure Drop, in H_2O (mm H_2O)

Free area of grille, percent	Face velocity, ft/min (m/min): 400 (121.9)	Face velocity, ft/min (m/min): 600 (182.9)	Return-intake-air velocities: Intake location	Return-intake-air velocities: Velocity over gross area, ft/min (m/min)
50	0.04 (1.02)	0.09 (2.29)	Above occupied zone	800 and up (243.8 and up)
60	0.03 (0.76)	0.06 (1.52)	In occupied zone:	
70	0.02 (0.51)	0.05 (1.27)	Not near seats	600–800 (182.9–243.8)
80	0.01 (0.25)	0.03 (0.76)	Near seats	400–600 (121.9–182.9)
			Door or wall louvers	500–700 (152.4–213.4)
			Undercut door (through undercut area)	600 (182.9)

ASHRAE *Guide*.

8. ***Select the air-return grille***

Table 29 shows typical air velocities used for return grilles in various locations. By assuming that the air is returned through a wall louvre, a velocity of 500 ft/min (152.4 m/min) might be used. Hence, by using the equation of step 1, grille area A = cfm/υ = 425/500 = 0.85 ft^2 (0.08 m^2).

If a lattice-type return intake having a free area of 60 percent is used, Table 30 shows that the pressure drop during passage of the air through the grille is 0.04 in (1.02 mm) H_2O. Locate the return grille away from the supply grille to prevent short circuiting of the air and excessive noise. The pressure losses in Table 29 are typical for return grilles. Choice of the pressure drop to use is generally left with the system designer. Figure 40 shows details of grille take-offs and installations.

TABLE 30 Approximate Pressure Drop for Lattice Return Intakes, in (mm) H_2O

Percentage of free area	Face velocity, ft/min (m/min): 400 (121.9)	600 (182.9)	800 (243.8)	1000 (304.8)
50	0.06 (1.52)	0.13 (3.30)	0.22 (5.59)	0.35 (8.89)
60	0.04 (1.02)	0.09 (2.29)	0.16 (4.06)	0.24 (6.10)
70	0.03 (0.76)	0.07 (1.78)	0.12 (3.05)	0.18 (4.57)
80	0.02 (0.51)	0.05 (1.27)	0.09 (2.29)	0.14 (3.56)

ASHRAE *Guide and Data Book*.

Related Calculations. Use this general method to choose outlet and return grilles for industrial, commercial, and domestic applications. Be certain not to exceed the tabulated velocities where noise is a factor in an installation. Excessive noise can lead to complaints from the room occupants.

The outlet-table excerpt presented here is typical of the table arrangements used by many manufacturers. Hence, the general procedure given for selecting an outlet is similar to that for any other manufacturer's outlet.

Many modern-design ceiling outlets are built so that the leaving air entrains some of the room air. The air being discharged by the outlet is termed *primary air*, and the room air is termed *secondary air.* The induction ratio R_i = (total air, ft^3/min)/(primary air, ft^3/min). Typical induction ratios run in the range of 30 percent.

For a given room, the total air in circulation, ft^3/min = (outlet ft^3/min)(induction ratio). Also, average room air velocity, ft/min = 1.4 (total ft^3/min in circulation)/area of wall, ft^2, opposite the outlet or outlets. The wall area in the last equation is the *clear* wall area. Any obstructions must be deducted. The multiplier 1.4 allows for blocking caused by the air stream. Where the room circulation factor K must be computed, use the relation K = (average room air velocity,

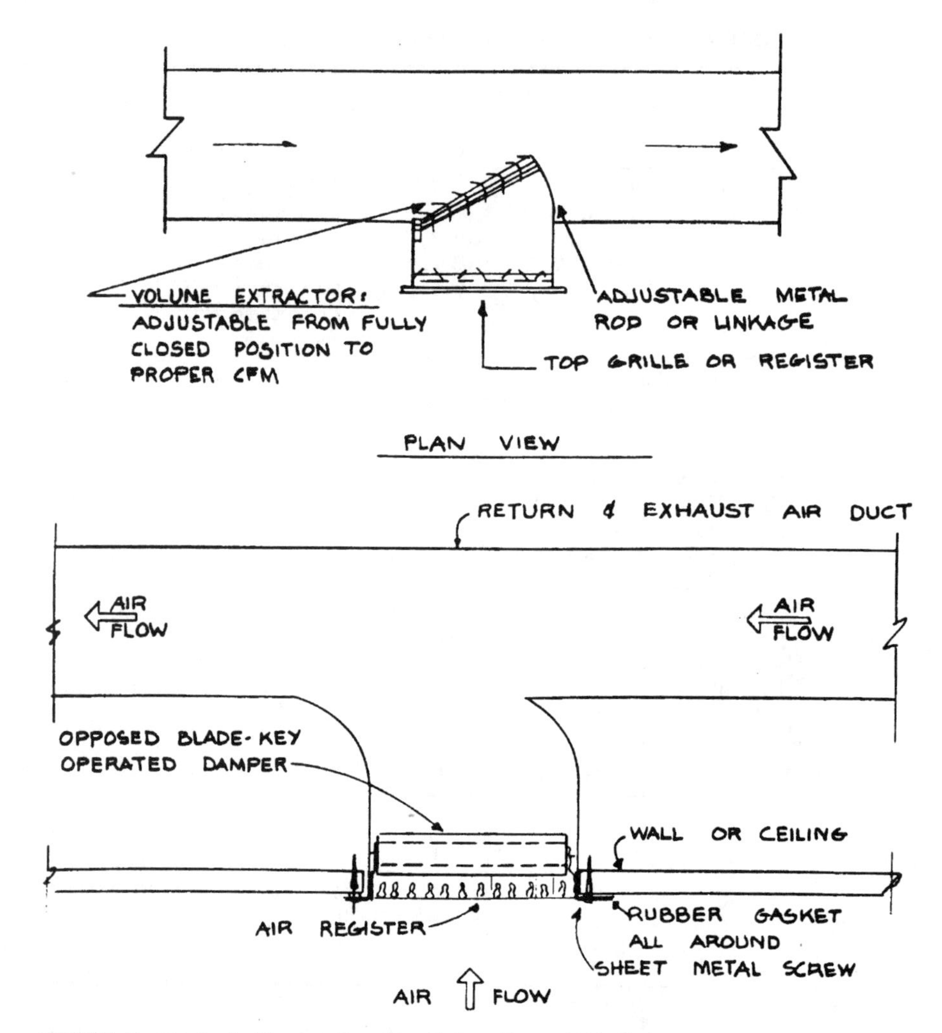

FIGURE 40 Details of grille take-offs and installations. (*Jerome F. Mueller.*)

ft/min)/1.4(induction ratio). The ideal room-air velocity for most applications is 25 ft/min (7.62 m/min). However, velocities up to 300 ft/min (91.4 m/min) are used in some factory air-conditioning applications.

The types of outlets commonly used today are grille (perforated, fixed-bar, adjustable-bar), slotted, ejector, internal induction, pan, diffuser, and perforated ceiling. Choice of a given type depends on the room ceiling height, desired air-temperature difference blow, drop, and spread, as well as other factors that are a function of the room, air quantity, and the activities in the room.

As a general guide to outlet selection, use the following pointers: (1) Choose the number of outlets for each room after considering the quantity of air required, throw or diffusion distance available, ceiling height, obstructions, etc. (2) Try to arrange the outlets symmetrically in the space available as shown by the room floor plan.

ENERGY CONSIDERATIONS IN CHOOSING ROOF VENTILATORS FOR STRUCTURES

A 10-bay building is 200 ft (60.9 m) long, 100 ft (30.5 m) wide, 50 ft (15.2 m) high to the top of the pitched roof, and 35 ft (01.7 m) high to the eaves. The building houses 15 turbine-driven generators and is classed as an engine room. Choose enough roof ventilators to produce a suitable number of air changes in the building. During reduced-load operating periods between 12 midnight and 7 a.m. on weekdays, and on weekends, only half the full-load air changes are required. The prevailing summer-wind velocity against the long side of the building is 10 mi/h (16.1 km/h). The total available open-window area on each long side is 300 ft^2 (27.9 m^2). The minimum difference between the outdoor and indoor temperatures will be 40°F (22.2°C).

Calculation Procedure:

1. ***Determine the cubic volume of the building***
To compute the cubic volume of a pitched-roof building, the usual procedure is to assume an average height from the eaves to the ridge. Since this building has a 15-ft (4.6-m) high ridge from the eaves, the average height = 15/2 = 7.5 ft (4.6 m). Since the height from the ground to the eaves is 35 ft (10.7 m), the building height to be used in the volume computation is 35 + 7.5 = 42.5 ft (13.9 m). Hence, the volume of the building, ft^3 = V = length × width × average height, all measured in ft = 200 × 100 × 42.5 = 850,000 ft^3 (24,064 m^3).

2. ***Determine the number of air changes required***
Table 31 shows that four to six air changes per hour are normally recommended for engine rooms. Using five air changes per hour will probably be satisfactory, and the roof ventilators will be chosen on this basis. During the early morning, and on weekends, 2.5 air changes will be satisfactory, since only half the normal number of air changes is needed during these periods.

3. ***Compute the required hourly air flow***
The required hourly air flow, ft^3/h = Q = (number of air changes per hour) (building volume, ft^3) = (5) (850,000) = 4,250,000 ft^3/h (120,318 m^3/h). During the early morning hours and on weekends when 2.5 air changes are used, Q = 2.5(850,000) = 2,125,000 ft^3/h (60,159 m^3/h).

4. ***Compute the air flow produced by natural ventilation***
The ASHRAE *Guide and Data Book* lists the prevailing winter- and summer-wind velocities for a variety of locations. Usual practice, in designing natural-ventilation systems, is to use one-half of tabulated wind velocity for the season being considered. Since summer ventilation is usually of

TABLE 31 Number of Air Changes Required per Hour

Auditoriums and assembly rooms	10–15	Libraries	3
Boiler rooms	10–15	Machine shops	6
Churches	10–15	Paint shops	10–15
Engine rooms	4–6	Paper mills	15–20
Factory buildings (general)	4	Pump rooms	8–10
Factory buildings (where excessive conditions of fumes, moisture, etc., are present)	15–20	Railroad shops	4
		Schools	10–12
Foundries	12	Textile mills (general)	4
Garages	10–15	Textile mill dye houses	15–20
General offices	3	Theaters	5–8
Hotel dining rooms	4	Waiting rooms	4
Hotel kitchens	10–20	Warehouses	4
Laundries	15–25	Wood-working shops	8

DeBothezat Fans Division, AMETEK Inc.

greater importance than winter ventilation, one-half the prevailing summer-wind velocity is generally used in natural-ventilation calculations. As the prevailing summer wind velocity in this locality is 8 mi/h (12.9 kg/h), a velocity of 8/2 = 4 mi/h (6.4 km/h) will be used to compute the air flow produced by the wind.

Use the relation $Q = VAE$ to find the air flow produced by the wind. In this relation, Q = air flow produced by the wind, ft^3/min; V = design wind velocity, ft/min = 88 × mi/h; A = free area of the air-inlet openings, ft^2; E = effectiveness of the air inlet openings—use 0.50 to 0.60 for openings perpendicular to the wind and 0.25 to 0.35 for diagonal winds.

Assuming $E = 0.50$, we get $Q = VAE = (4 \times 88)(300)\ (0.50) = 52{,}800$ ft^3/min (1494.8 m^3/min) or 60(52,800) = 3,168,000 ft^3/h (89,686.1 m^3/h). Step 3 shows that the required air flow is 4,250,000 ft^3/h (120,318 m^3/h) when all turbines are operating. Hence, the air flow produced by natural ventilation is inadequate for full-load operation. However, since the required flow of 2,125,000 ft^3/h (60,159 m^3/h) for the early morning hours and weekends is less than the natural ventilation flow of 3,168,000 ft^3/h (89,686.1 m^3/h), natural ventilation may be acceptable during these periods.

5. *Determine the number of stationary-type ventilators needed*

A stationary-type roof ventilator (i.e., one that depends on the wind and air-temperature difference to produce the desired air movement) may be suitable for this application. If the stationary-type is not suitable, a power-fan type of roof ventilator will be investigated and must be used. The Breidert-type ventilator is investigated here because the procedure is similar to that used for other stationary-type roof ventilators.

Stationary ventilators produce air flow out of a building by two means: suction caused by wind action across the ventilators and the stack effect caused by the temperature difference between the inside and outside air.

Figure 41 shows the air velocity produced in a stationary Breidert ventilator by winds of various velocities. Thus, with the average 5-mi/h (8.1-km/h) wind assumed earlier for this building, Fig. 41 shows that the air velocity through the ventilator produced by this wind velocity is 220 ft/min (67.1 m/min), closely.

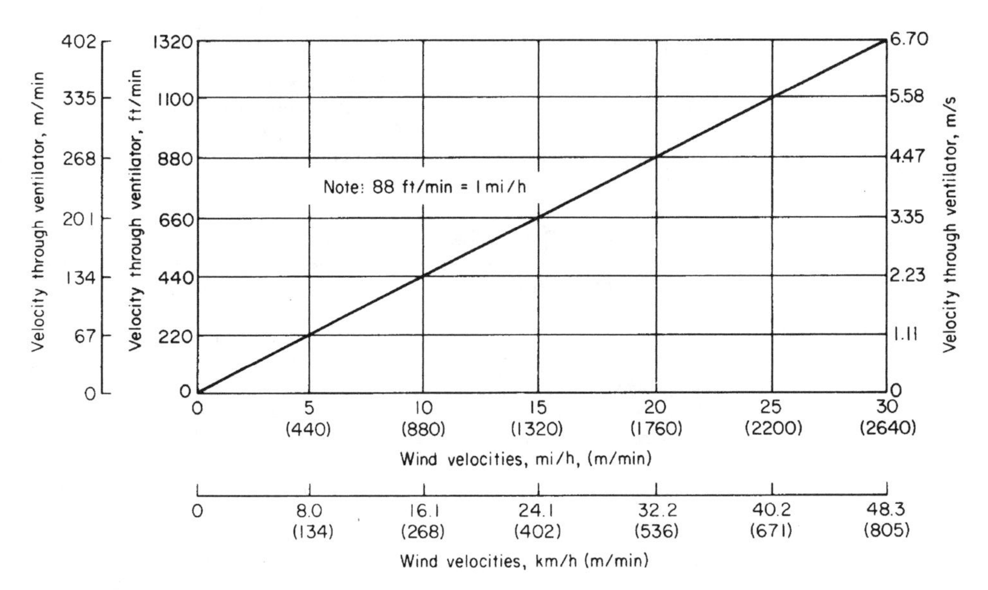

FIGURE 41 Roof-ventilator air-exhaust capacity for various wind velocities. Add the extra velocity for temperature difference given in Table 32. (*G. C. Breidert Co.*)

TABLE 32 Flow of Air in Natural-Draft Flues, $ft^3/(min \cdot ft^2)$ $[m^3/(min \cdot m^2)]$

Difference in temperature, °F (°C)	Height of flue, ft (m), same as height of room or building			
	30 (9.1)	40 (12.2)	50 (15.2)	60 (18.3)
10 (5.6)	188 (57.3)	217 (66.1)	242 (73.7)	264 (80.4)
20 (11.1)	265 (80.6)	306 (93.2)	342 (104.2)	373 (113.7)
30 (16.7)	325 (99.0)	375 (114.3)	419 (127.7)	461 (140.5)
40 (22.2)	374 (113.9)	431 (131.3)	482 (140.2)	529 (161.2)
50 (27.8)	419 (127.7)	484 (147.5)	541 (164.9)	594 (181.0)
60 (33.3)	460 (140.2)	532 (162.1)	595 (181.4)	650 (198.1)

G. C. Breidert Co.

Table 32 shows that a 1.0-ft^2 (0.093-m^2) ventilator installed on a 50-ft (15.2-m) high building having an air-temperature difference of 40°F (22.2°C) will produce, owing to the stack effect, an air-flow velocity of 482 ft/min (146.9 m/min). Hence, the total velocity through this ventilator resulting from the wind and stack action is 220 + 482 = 702 ft/min (213.9 m/min).

Since air flow, ft^3/min = (air velocity, ft/min) (area of ventilator opening, ft^2), an air flow of (702) (1.0) = 702 ft^3/min (19.9 m^3/min) will be produced by each square foot of ventilator-neck or inlet-duct area. Thus, to produce a flow of (4,250,000 ft^3/h)/(60 min/h) = 70,700 ft^3/min (2002 m^3/min) will require a ventilator area of 70,700/702 = 101 ft^2 (9.38 m^2). A 48-in (121.9-cm) Breidert ventilator has a neck area of 12.55 ft^2 (1.17 m^2). Hence, a 101/12.55 = 8.05, or eight ventilators will be required. Alternatively, the Breidert capacity table in the engineering data prepared by the manufacturer shows that a 48-in (121.9-cm) ventilator has a ventilating capacity of 8835 ft^3/min (250.1 m^3/min) when it is used for a 5-mi/h (8.1-km/h), 50-ft (15.2-m) high, 40°F (22.2°C) temperature-difference application. With this capacity, the number of ventilators required = 70,700/8835 = 8.02, say eight ventilators. These ventilators will be suitable for both full- and part-load operation.

6. *Determine the number of powered ventilators needed*

Powered ventilators are equipped with single- or two-speed fans to produce a positive air flow independent of wind velocity and stack effect. For this reason, some engineers prefer powered ventilators where it is essential that air movement out of the building be maintained at all times.

Two-speed powered ventilators are usually designed so that the reduced-speed rpm is approximately one-half the full-speed rpm. The air flow at half-speed is about one-half that at full speed.

Checking the capacity table of a typical powered-ventilator manufacturer shows that ventilator capacities range from about 2100 ft^3/min (59.5 m^3/min) for a 21-in (53.3-cm) diameter unit at a 1/8-in (3.18-mm) static pressure difference to about 24,000 ft/min (679 m^3/min) for a 36-in (91.4-cm) diameter ventilator at the same pressure difference. With a 27-in (68.6-cm) diameter powered ventilator which has a capacity of 14,900 ft^3/min (421.8 m^3/min), the number required is [70,700 ft^3/min (2002 m^3/m)] [14,900 ft^3/min (422 m^3/min) per unit] = 4.76 or 5.

7. *Choose the type of ventilator to use*

Either a stationary or powered ventilator might be chosen for this application. Since a large amount of heat is generated in an engine room, the powered ventilator would probably be a better choice because there would be less chance of overheating during periods of little or no wind.

Related Calculations. Use the general method given here to choose stationary or powered ventilators for any of the 25 applications listed in Table 31. Usual practice is to locate one ventilator in each bay or sawtooth of a building.

With the greater interdiction of smoking in public places (factories, offices, hotels, restaurants, schools, etc.), special exhaust fans—often termed "smoke eaters"—are being installed. These high-velocity fans draw smoke-laden air from a designated smoking area and exhaust it to the atmosphere or to treatment devices.

Local building codes govern smoking in structures of various types, so the engineer must consult the local code before choosing the type of exhaust fan to use for a specific building. Many cities now prohibit all smoking inside a building. In such cities special exhaust fans are not needed to handle cigarette, cigar, or pipe smoke. Restaurants, bowling alleys, billiard rooms, taverns, and similar gathering places prohibit indoor smoking.

Follow the same procedure given above to choose the fan or fans. Be certain to use the required number of air changes specified by any local building code. While an excessive number of changes will increase the winter heating load, many engineers overdesign to be certain they meet clean air requirements. Tobacco smoke must be handled decisively so that all patrons of an establishment are comfortable.

AIR-CONDITIONER VIBRATION ISOLATION DEVICE SELECTION

Choose a vibration isolator for a packaged air conditioner operating at 1800 r/min. What minimum mounting deflection is required if the air conditioner is mounted on a basement floor? On an upper-story floor made of light concrete?

Calculation Procedure:

1. *Determine the suggested isolation efficiency*
Table 33 lists the suggested isolation efficiency for various components used in air-conditioning and refrigeration systems. This tabulation shows that the suggested isolation efficiency for a packaged air conditioner is 90 percent. This means that the vibration isolator or mounting should absorb 90 percent, or more, of the vibration caused by the machine. At this efficiency only 10 percent of the machine vibration would be transmitted to the supporting structure.

2. *Determine the static deflection caused by the vibration*
Use Fig. 42 to find the static deflection caused by the vibration. Enter at the bottom of Fig. 42 at 1800 r/min the disturbing frequency, and project vertically upward to the 90 percent efficiency curve. At the left read the static deflection as 0.11 in (2.79 mm).

TABLE 33 Suggested Isolation Efficiencies

Equipment	Installed efficiency, %
Absorption units	95
Steam generators	95
Centrifugal compressors	98
Reciprocating compressors:	
Up to 15 hp (11.2 kW)	85
20–60 hp (14.9–44.8 kW)	90
75–150 hp (56.0–111.9 kW)	95
Packaged air conditioners	90
Centrifugal fans:	
80 r/min and above; all diameters	90–95
350–800 r/min; all diameters	70–90
200–350 r/min; 48-in (121.9-cm) diameter or smaller	*
200–350 r/min; 54-in (137.2-cm) diameter or larger	70–80
Centrifugal pumps	95
Cooling towers	85
Condensers	80
Fan coil units	80
Piping	95

*Installed for noise isolation only.

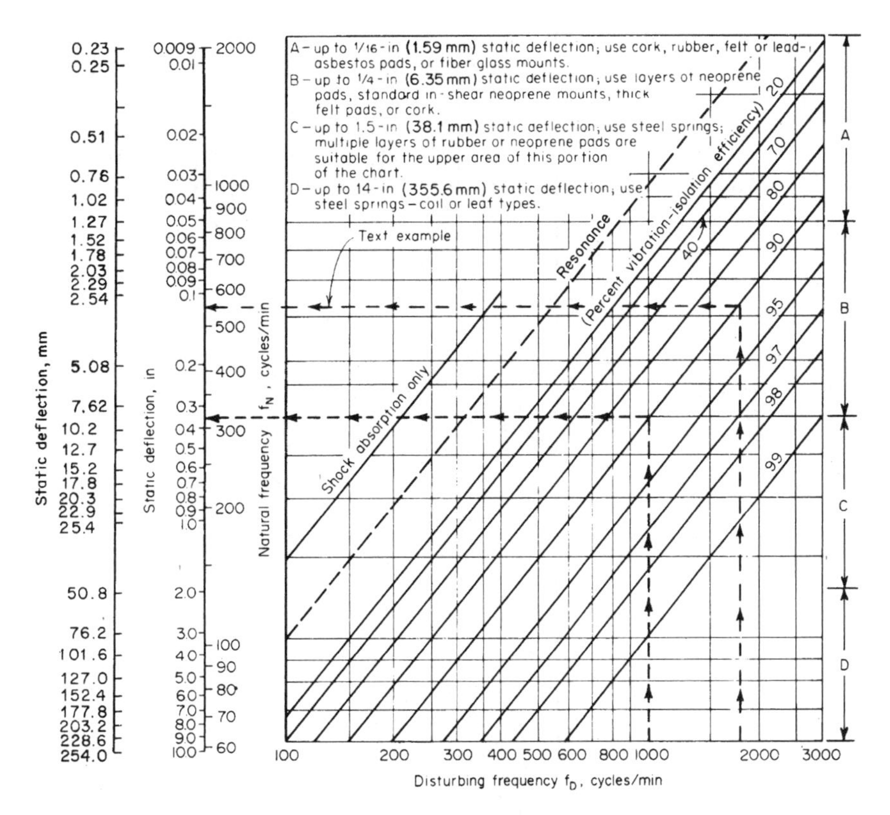

FIGURE 42 Vibration-isolator deflection for various disturbing frequencies. (*Power*.)

3. ***Select the type of vibration isolator to use***

Project to the right from the intersection with the efficiency curve (Fig. 42) to read the type of isolator to use. Thus, neoprene pads or neoprene-in-shear mounts will safely absorb up to 0.25-in (6.35-mm) static deflection. Hence, either type of isolator mounting could be used.

4. ***Check the isolator selection***

Use Table 34 to check the theoretical isolation efficiency of the mounting chosen. Enter at the top at the rpm of the machine, and project vertically downward until an efficiency equal to, or greater than, that desired is intersected.

For this machine operating at 1800 r/min, single-deflection rubber mountings have an efficiency of 94 percent. Since neoprene is also called synthetic rubber, the isolator choice is acceptable because it yields a higher efficiency than required.

5. ***Determine the minimum mounting deflection required***

Table 35 lists the minimum mounting deflection required at various operating speeds for machines installed on various types of floors. Thus, at 1800 r/min, machines mounted on a basement floor must have isolator mountings that will absorb deflections up to 0.10 in (2.54 mm). Since the neoprene mountings chosen in step 3 will absorb up to 0.25 in (6.35 mm) of deflection, they will be acceptable for use on a basement-mounted machine.

For mounting on a light-concrete upper-story floor, Table 35 shows that the mounting must be able to absorb a deflection of 0.80 in (20.3 mm) for machines operating at 1800 r/min. Since the neoprene isolators can absorb only 0.25 in (6.35 mm), another type of mounting is needed if the machine is

TABLE 34 Theoretical Vibration-Isolation Efficiencies

Isolation material	Average static deflection, in (mm)	Average natural frequency	Efficiencies, %									
			350 r/min	500 r/min	600 r/min	800 r/min	1000 r/min	1200 r/min	1500 r/min	1800 r/min	3000 r/min	3600 r/min
2-in (50.8-mm) thick standard-density cork	0.08 (2.03)	By test 1420	—	—	—	—	—	—	—	—	72	82
Type W waffle pad	Curvature corrected, 0.035 (0.89)	1000	—	—	—	—	—	—	20	55	87	92
Two layers of W waffle pad	Curvature corrected, 0.070 (1.78)	710	—	—	—	—	—	46	71	82	93	96
Single-deflection rubber mountings	0.20 (5.08)	420	—	—	—	62	79	86	91	94	98	99
Double-deflection rubber mountings	0.40 (10.16)	300	—	44	67	84	90	93	96	97	99	Almost perfect
Standard spring mountings	1.00 (25.4)	188	70	85	89	94	96	97	98	99	Almost perfect	Almost perfect
Double-deflection rubber and spring mountings	1.40 (35.6)	160	75	89	93	96	97	98	99	Almost perfect	Almost perfect	Almost perfect

Power.

TABLE 35 Minimum Mounting Deflections

Operating speed, r/min	Basement—negligible floor deflection, in (mm)	Rigid concrete floor, in (mm)	Upper story—light-concrete floor, in (mm)	Wood floor, in (mm)
300	1.50 (38.1)	3.00 (76.2)	3.50 (88.9)	4.00 (101.6)
500	0.63 (16.0)	1.25 (31.8)	1.65 (41.9)	1.95 (49.5)
800	0.25 (6.35)	0.60 (15.2)	1.00 (25.4)	1.25 (31.6)
1200	0.20 (5.08)	0.45 (11.4)	0.80 (20.3)	1.00 (25.4)
1800	0.10 (2.54)	0.35 (8.9)	0.80 (20.3)	1.00 (25.4)
3600	0.03 (0.76)	0.20 (5.08)	0.80 (20.3)	1.00 (25.4)
7200	0.03 (0.76)	0.20 (5.08)	0.80 (20.3)	1.00 (25.4)

installed on an upper floor. Figure 42 shows that steel springs will absorb up to 1.5 in (38.1 mm) of static deflection. Hence, this type of mounting would be used for machines installed on upper floors of the building.

Related Calculations. Use this general procedure for engines, compressors, turbines, pumps, fans, and similar rotating and reciprocating equipment. Note that the suggested isolation efficiencies in Table 33 are for air-conditioning equipment located in critical areas of buildings, such as office, hospitals, etc. In noncritical areas, such as basements or warehouses, an isolation efficiency of 70 percent may be acceptable. Note that the efficiencies given in Table 33 are useful as general guides for all types of rotating machinery.

Although much emphasis is placed on atmospheric, soil, and water pollution, greater attention is being placed today on audio pollution than in the past. Audio pollution is the discomfort in human beings produced by excessive or high-pitch noise. One good example is the sound produced by jet aircraft during take-off and landing.

Audio pollution can be injurious when it is part of the regular workplace environment. At home, audio pollution can interfere with one's life-style, making both indoor and outdoor activities unpleasant. For these reasons, regulatory agencies are taking stronger steps to curb audio pollution.

Control of audio pollution almost always reverts to engineering design. For this reason, engineers will be more concerned with the noise their designs produce because it is they who have more control of it than others in the design, manufacture, and use of a product.

NOISE-REDUCTION MATERIALS CHOICE FOR AIR-CONDITIONING SYSTEMS

A concrete-walled test laboratory is 25 ft (7.62 m) long, 20 ft (6.1 m) wide, and 10 ft (3.05 m) high. The laboratory is used for testing chipping hammers. What noise reduction can be achieved in this laboratory by lining it with acoustic materials?

Calculation Procedure:

1. ***Determine the noise level of devices in the room***

Table 36 shows that a chipping hammer produces noise in the 130-dB range. Hence, the noise level of this room can be assumed to be 130 dB. This is rated as deafening by various authorities. Therefore, some kind of sound-absorption material is needed in this room if the uninsulated walls do not absorb enough sound.

TABLE 36 Power and Intensity of Noise Sources

Sound source	Power range, W	Decibel range (10^{-13} W)
Ram jet	100,000.0	180
Turbojet with 7000-lb (31,136-N) thrust	10,000.0	170
Four-propeller airliner	100.0	150
75-piece orchestra, pipe organ; small aircraft engine	10.0	140
Chipping hammer	1.0	130
Piano, blaring radio	0.1	120
Centrifugal ventilating fan at 13,000 ft^3/min (6.14 m^3/s)	0.01	110
Automobile on roadway; vane-axial	0.001	100
ventilating fan	0.0001	90
Subway car, air drill	0.000 01	80
Conversational voice; traffic on street corner	0.000 001	70
Street noise, average radio	0.000 0001	60
Typical office	0.000 000 01	50
Very soft whisper	0.000 000 001	40

Reference values relate decibel scales

dB scale	Definition	Reference quantity
Sound-power level	$\text{PWL} = 10 \log \frac{W}{W_{re}}$	$W_{re} = 10^{-13}$ W
Sound-intensity level	$\text{IL} = 10 \log \frac{I}{I_{re}}$	$I_{re} = 10^{-12}$ W/m^2 $= 10^{-16}$ W/cm^2
Sound-pressure level	$\text{SPL} = 10 \log \frac{P^2}{P_{re}^2} = 20 \log \frac{P}{P_{re}}$	$P_{re} = 0.000{,}02$ N/m^2 $= 0.0002$ μbar $= 0.0002$ dyn/cm^2

Power.

2. *Compute the total sound absorption of the room*

The *sound-absorption coefficient* of bare concrete is 0.1. This means that 10 percent of the sound produced in the room is absorbed by the bare concrete walls.

To find the total sound absorption by the walls and ceiling, find the product of the total area exposed to the sound and the sound-absorption coefficient of the material. Thus, concrete area, excluding the floor but including the ceiling = two walls (25 ft long × 10 ft high) + two walls (20 ft wide × 10 ft high) + one ceiling (25 ft long × 20 ft wide) = 1400 ft^2 (130.1 m^2). Then the total sound absorption = (1400)(0.1) = 140.

3. *Compute the total sound absorption with acoustical materials*

Table 37 lists the sound- or noise-reduction coefficients for various acoustic materials. Assume that the four walls and ceiling are insulated with membrane-faced mineral-fiber tile having a sound absorption coefficient of 0.90, from Table 37.

Then, by the procedure of step 2, total noise reduction = (1400 ft^2) (0.90) = 1260 ft^2 (117.1 m^2).

4. *Compute the noise reduction resulting from insulation use*

Use the relation noise reduction, dB = 10 log (total absorption *after* treatment/total absorption *before* treatment) = 10 log (1260/140) = 9.54 dB. Thus, the sound level in the room would be reduced to 130.00 − 9.54 = 120.46 dB. This is a reduction of 9.54/130 = 0.0733, or 7.33 percent. To obtain

TABLE 37 Noise-Reduction Coefficients

Type	Material	Noise-reduction coefficient range [¾ in (19.1 mm) thick]
1	Regularly perforated cellulose-fiber tile	0.65–0.85
2	Randomly perforated cellulose-fiber tile	0.60–0.75
3	Textured, perforated, fissured, cellulose tile	0.50–0.70
4	Cellulose-fiber lay-in panels	0.50–0.60*
5	Perforated mineral-fiber tile	0.65–0.85
6	Fissured mineral-fiber tile	0.65–0.80
7	Textured, perforated or smooth mineral-fiber tile	0.65–0.85
8	Membrane-faced mineral-fiber tile	0.30–0.90
9	Mineral-fiber lay-in panels	0.20–0.90
10	Perforated-metal pans with mineral-fiber pads	0.60–0.80*
11	Perforated-metal lay-in panels with mineral-fiber pads	0.75–0.85
12	Mineral-fiber tile—fire-resistive assemblies	0.55–0.90
13	Mineral-fiber lay-in panels—fire-resistive units	0.65–0.75†
14	Perforated-asbestos panels with mineral-fiber pads	0.65–0.75‡
15	Sound-absorbent duct lining	0.65–0.75*
16	Special acoustic panels and materials	

*Noise-reduction coefficient 1 in (25.4 mm) thick.
†Noise-reduction coefficient 5/2 in (15.9 mm) thick.
‡Noise-reduction coefficient 15/16 in (23.8 mm) thick.
Acoustical and Insulating Materials Association.

a further reduction of the noise in this room, the floor could be insulated or, preferably, the noise-producing device could be redesigned to give off less noise.

Related Calculations. Use this general procedure to determine the effectiveness of acoustic materials used in any room in a building, on a ship, in an airplane, etc.

DESIGNING DOOR AND WINDOW AIR CURTAINS FOR VARIOUS TYPES OF STRUCTURES

Show how to choose environmental air curtains for shopping malls, supermarkets, hospitals, schools, restaurants, warehouses, service take-out windows, and manufacturing facilities. Detail air-flow requirements to provide (1) thermal barriers to reduce energy consumption, and (2) insect, dust, and odor control.

Calculation Procedure:

1. ***Determine the door and window dimensions***
Commercially available air-curtain units, Fig. 43, are designed for specified door and window widths and heights. Such units consist of an electric-motor-driven fan fitted with suitable deflecting vanes mounted in a metal or plastic casing that is easily installed over the door or window opening. An air heater is usually fitted to units used in colder climates.

Popular standard air-curtain units are available for door widths up to 20 ft (6.1 m), or more, and door heights up to 22 ft (6.7 m). Special high-velocity units can be designed for greater door widths and heights. Both heated and unhealed air-curtain units are available. Heated units can use steam, hot water, gas, or electricity as their heat source.

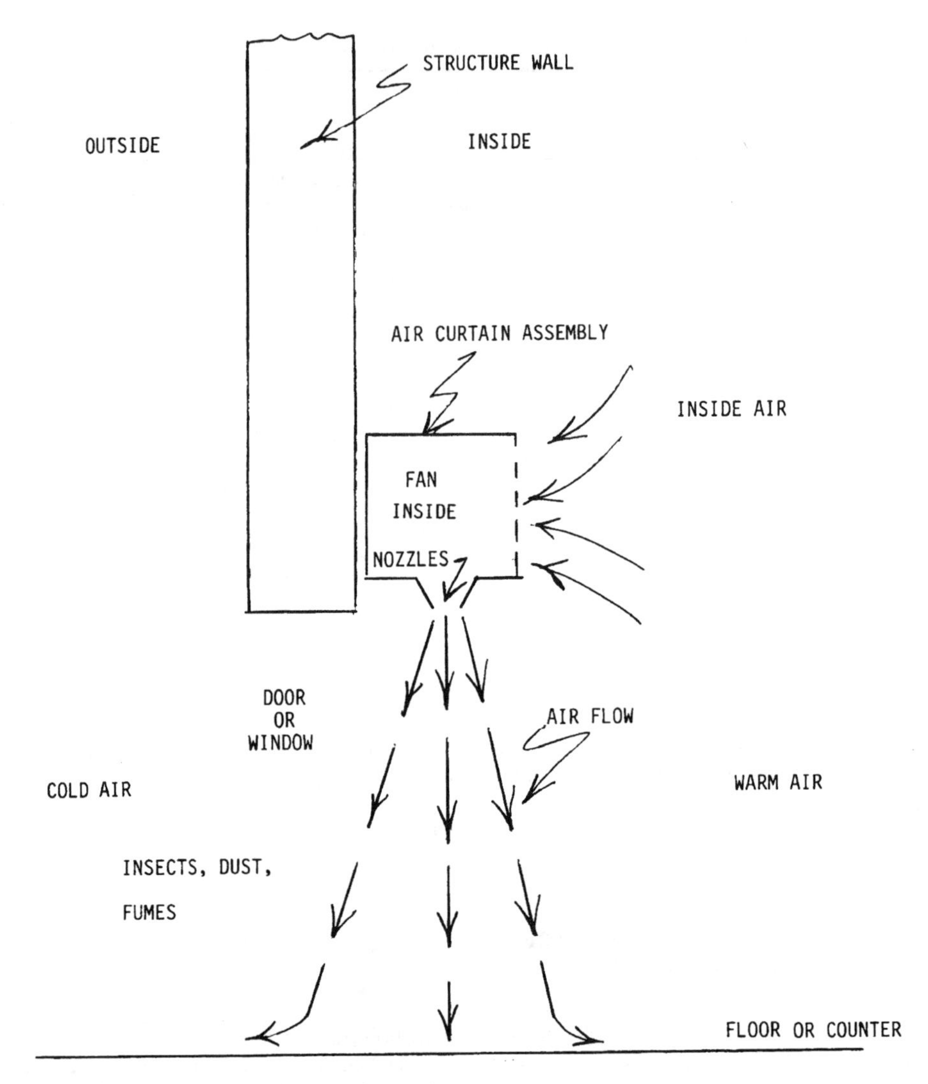

FIGURE 43 Typical air-curtain installation.

Find the door or window width and height from the drawings of the installation. Where a door or window is excessively wide—say more than 14 ft (4.3 m), two or more air-curtain units may be required.

2. ***Choose the type of air-curtain unit to use***

Heated air-curtain units are used where temperature control is required in the installation—termed *climate control* by air-curtain unit manufacturers. Typical climate-control unit applications include customer entrances to shopping areas, receiving/service doors, service take-out windows in fast-food establishments and banks, refrigerated rooms, warehouse doors, hospital entrances, etc.

Where climate control is required—i.e., warmer air is being projected from the ceiling of an opening to the floor or counter of the opening—a heater of some type is required. The type of heater chosen depends on the heating medium available—steam, hot water, gas, or electric. No matter what type of heating medium is chosen, an electric-motor-driven fan fitted with directional vanes at the air outlet nozzles is required. The directional control vanes allow adjustment of the air flow for cold weather, versus the vane setting for control of dust, insects, odors, and fumes. Extra-high-velocity fans are often used for wider doors—those exceeding 12 ft (3.7 m) in height.

To choose the specific air curtain to use, refer to a manufacturer's catalog showing the types of units available. Such catalogs present tabulations of both unheated and heated air-curtain units for various applications. The selection tables also specify which units are best for insect, fume, and dust barriers. Diagrams in the catalogs also suggest installation methods for a variety of door types—track, roll-up, vertical-lift, and sliding. Air-nozzle length and projection angle are also detailed.

An air curtain effectively separates cold (or warm) outside air from the conditioned air inside a structure. Air curtains can also be used inside buildings to separate manufacturing areas from office areas, guard shacks from interior or exterior spaces, etc.

Typical air-flow rates for heavy-duty air curtains separating inside air from outside air can range from about 12,000 cfm (5664 L/s) for a 10- to 14-ft (3- to 4.3-m) high door to 33,000 cfm (15,567 L/s) for a 22-ft (6.7-m) high door. Air velocities for such doors range from about 3500 fpm (1067 m/min) to 6000 fpm (1829 m/min).

Motor horsepower for air curtains ranges from a low of about 0.5 hp (0.37 kW) for the smallest doors to 5 hp (3.7 kW) for the smaller door in the above paragraph to 20 hp (14.9 kW) for the larger door. Weight of air-curtain units varies with the materials of construction of the box enclosure. Typical weights range from 600 lb (272 kg) to 1200 lb (545 kg). Tables in manufacturers' catalogs give detailed information on the door height served, air flow at the nozzle, average and maximum air flow in cfm (L/s), motor horsepower (kW), unit weight, and heating capacity with various heating mediums—hot water, steam, gas, and electric. Using these data, it is an easy task to select a suitable air curtain for specific conditions.

Heating capacity of air-curtain units range from less than 100,000 Btu/h (29.3 kW) to more than 1.5 million Btu/h (439.5 kW). Heated air curtains reduce thermal stratification in buildings by recirculating cold air near the floor to higher levels in the structure where it can mix with warm air in ceiling pockets. Outside winds in the 30 mi/h (48.3 km/h) range can be kept from entering a building when a properly designed air curtain is chosen. In areas of high sustained wind velocity, higher powered air-curtain fans are usually chosen. Many air-curtain package assemblies are installed inside the building door when heat loss and prevention of cold air ingress are desired.

Dual air-curtain assemblies can be mounted side by side for excessively wide doors. Either metal or plastic assembly boxes can be used. Control of flies and other insects is another popular application of air curtains. The air curtain effectively keeps such flying insects out of the conditioned space, where they might cause human infection or other problems.

Related Calculations. While air curtains can be designed from scratch, the best solution to selecting a suitable unit usually is to pick from a premanufactured assembly. Air-curtain manufacturers spend large sums on researching and testing their units so they perform most efficiently. Further, the manufacturers are extremely willing to cooperate on any special design needs you might have for the air-curtain installation you are designing.

By choosing a pre-engineered air curtain, the designer can save time and money for the installation being considered. Off-the-shelf air curtains can be delivered quickly to the site where they are needed, again saving time and energy.

Since the use of air curtains continues to grow, the designer is urged to consult an experienced manufacturer for the latest developments. Data in this procedure were excerpted from information presented in several air-curtain manufacturer's catalogs. Two suppliers of air curtains are Leading Edge, Inc. Miami, FL, and Mars Sales Co., Inc., Gardena, CA, both of whom are long-established and highly experienced air-curtain manufacturers.

SECTION 14

ENERGY CONSERVATION AND ENVIRONMENTAL POLLUTION CONTROL

ENERGY CONSERVATION AND ENVIRONMENTAL POLLUTION 14.2
- Energy Conservation and Environmental Pollution Control Parameters 14.2
- Atmospheric Control System Investment Analysis 14.2
- Environmental Pollution Project Selection 14.3
- Energy-From-Waste Economic Analysis 14.4
- Emission Reduction Using Flue-Gas Heat Recovery 14.8
- Cogeneration Cost Analysis for Alternative Energy Systems 14.13
- Steam Compressor Choice for Diesel-Engine Cogeneration System 14.19
- Cogeneration Heat Savings Analysis Using Need Plots 14.22
- Capital Cost of Cogeneration Heat-Recovery Boilers 14.28
- Chlorofluorocarbon Aspects of Central Chilled-Water Systems Design 14.31
- Sizing an Electrostatic Precipitator for Air-Pollution Control 14.34
- Explosive-Vent Sizing for Industrial Buildings 14.36
- Ventilation Design for Industrial Building Environmental Safety 14.38
- Thermal-Pollution Estimates for Power Plants 14.42
- Flash-Steam Heat Recovery for Cogeneration 14.43
- Energy Conservation and Cost Reduction Design for Flash-Steam Usages 14.45
- Cogeneration Power Plant Cost Separation of Steam and Electricity Using the Energy Equivalence Method 14.50
- Fuel-Cost Allocation for a Cogeneration Plant Using Electricity Cost 14.54
- Environmental and Safety-Regulation Ranking of Equipment Criticality 14.56
- Heat Recovery Energy and Fuel Savings 14.60
- High-Temperature Hot-Water Heating Energy and Fuel Savings 14.62

CONTROLS IN ENVIRONMENTAL AND ENERGY-CONSERVATION DESIGN 14.64
- Energy Process-Control System Selection 14.64
- Process-Energy Temperature Control System Selection 14.67
- Energy Process-Control Valve Selection 14.68
- Control-System Controlled-Volume-Pump Choice 14.71
- Industrial-Steam Boiler Energy Control Choice and Usage 14.72
- Characteristics and Rangeability of Energy Control Valves 14.74
- Steam-Control and Pressure-Reducing Valve Sizing for Maximum Energy Savings 14.75
- Repowering Options for Power-Plant Capacity Additions 14.77
- Energy Aspects of Cooling-Tower Choice 14.85
- Oil-Polluted Beach Cleanup Analysis 14.90

ENERGY CONSERVATION AND ENVIRONMENTAL POLLUTION

ENERGY CONSERVATION AND ENVIRONMENTAL POLLUTION CONTROL PARAMETERS

Today's focus of design and operating engineers in the energy field is on energy conservation and controlling environmental pollution. Rising fuel prices demand that energy needs be reduced so fuel costs can be lowered. Environmental regulations tell us what we can, and cannot, emit to the atmosphere, to rivers, lakes, and oceans.

This section of the handbook covers many forms of practical energy conservation—such as cogeneration, heat energy recovery, fuel-saving methods, energy process-control, repowering options to reduce overall energy needs, energy aspects of cooling-tower choice, plus many others.

Likewise, this section covers ways to reduce environmental and atmospheric pollution, gives methods for emissions reduction, air-pollution control strategies, thermal pollution determination and management, and other ways to reduce the impact of the use of greenhouse-gas-producing fossil fuels. Environmental considerations are sometimes more important in design decisions than energy conservation. Yet plant owners and operators are governed by the bottom line aspects of their facilities. The clash of these two needs—less pollution of all kinds and profits for company shareholders—can make engineering decisions much more difficult.

In the area of energy conservation, cogeneration is a star because it productively uses heat that might otherwise be wasted. This saves fuel, thereby reducing operating costs. A number of cogeneration methods are presented in this section of the handbook.

There are a number of other methods of heat-energy recovery to achieve energy conservation and they are covered in this section. Such methods include flue-gas heat recovery, flash-steam heat utilization, high-temperature hot-water heating, system design and component selection for energy recovery, and energy conservation by wise choice of cooling towers. Engineers using the calculation procedures in this section of the handbook can produce major savings in energy usage and reduce the potential pollution produced. The methods given here can help the engineer thread his or her way through the complexities of modern design while achieving an economic result that is acceptable to regulators and owners.

ATMOSPHERIC CONTROL SYSTEM INVESTMENT ANALYSIS

An engineering atmospheric control to protect the public against environmental pollution will have an incremental operating cost of $100,000. If the pollution were uncontrolled, the damage to the public would have an estimated incremental cost of $125,000. Would this atmospheric control be a beneficial investment?

Calculation Procedure:

1. ***Write the cost-benefit ratio for this investment***

The generalized dimensionless cost-benefit equation is $0 \le C/B \le 1$, where C = incremental operating cost of the proposed atmospheric control, \$, or other consistent monetary units; B = benefit to the public of having the pollution controlled, \$, or other consistent monetary units.

2. ***Compute the cost-benefit ratio for this situation***

Using the values given, $0 \le \$100{,}000/\$125{,}000 \le 1$. Or, $0 \le 0.80 \le 1$. This result means that 80¢ spent on environmental control will yield $1.00 in public benefits. Investing in the control would be a wise decision because a return greater than the cost of the control is obtained.

Related Calculations. In the general cost-benefit equation. $0 \le C/B \le 1$, the upper limit of unity means that $1.00 spent on the incremental operating cost of the atmospheric control will deliver

$1.00 in public benefits. A cost-benefit ratio of more than unity is uneconomic. Thus, $1.25 spent to obtain $1.00 in benefits would not, in general, be acceptable in a rational analysis. The decision would be to accept the environmental pollution until a satisfactory cost-benefit solution could be found.

A negative result in the generalized equation means that money invested to improve the environment actually degrades the condition. Hence, the environmental condition becomes worse. Therefore, the technology being applied cannot be justified on an economic basis.

In applying cost-benefit analyses, a number of assumptions of the benefits to the public may have to be made. Such assumptions, particularly when expressed in numeric form, can be open to change by others. Fortunately, by assigning a number of assumed values to one or more benefits, the cost-benefit ratios can easily be evaluated, especially when the analysis is done on a computer.

ENVIRONMENTAL POLLUTION PROJECT SELECTION

Five alternative projects for control of environmental pollution are under consideration. Each project is of equal time duration. The projects have the cost-benefit data shown in Table 1. Determine which project, if any, should be constructed.

TABLE 1 Project Costs and Benefits

Project	Equivalent uniform net annual benefits, $	Equivalent uniform net annual costs, $	*C/B* ratio
A	200,000	135,000	0.68
B	250,000	190,000	0.76
C	180,000	125,000	0.69
D	150,000	90,000	0.60
E	220,000	150,000	0.68

Calculation Procedure:

1. *Evaluate the cost-benefit (C/B) ratios of the projects*

Setting up the *C/B* ratios for the five projects by the cost by the estimated benefit shows—in Table 1—that all *C/B* ratios are less than unity. Thus, each of the five projects passes the basic screening test of $0 \leq C/B \leq 1$. This being the case, the optimal project must be determined.

2. *Analyze the projects in terms of incremental cost and benefit*

Alternative projects cannot be evaluated in relation to one another merely by comparing their *C/B* ratios, because these ratios apply to unequal bases. The proper approach to analyzing such a situation is: Each project corresponds to a specific *level* of cost. To be justified, *every* sum of money expended must generate at least an equal amount in benefits; the step from one level of benefits to the next should be undertaken only if the incremental benefits are at least equal to the incremental costs.

Rank the projects in ascending order of costs. Thus, Project D costs $90,000; Project C costs $125,000; and so on. Ranking the projects in ascending order of costs gives the sequence D-C-A-E-B.

Next, compute the incremental costs and benefits associated with each step from one level to the next. Thus, the incremental cost going from Project D to Project C is $125,000 – $90,000 = $35,000. And the benefit from going from Project D to Project C is $180,000 – $150,000 = $30,000, using the data from Table 1. Summarize the incremental costs and benefits in a tabulation like that in Table 2. Then compute the *C/B* ratio for each situation and list it in Table 2. This computation shows that Project E is the best of these five projects because it has the lowest cost—75¢ per $1.00 of benefit. Hence, this project would be chosen for control of environmental pollution in this instance.

Related Calculations. There are some situations in which the minimum acceptable *C/B* ratio should be set at some value close to 1.00. For example, with reference to the above projects, assume

TABLE 2 Cost-Benefit Comparison

Step	Incremental benefit, $	Incremental cost, $	*C/B* ratio	Conclusion
D to C	30,000	35,000	1.17	Unsatisfactory
D to A	50,000	45,000	0.90	Satisfactory
A to E	20,000	15,000	0.75	Satisfactory
E to B	30,000	40,000	1.33	Unsatisfactory

that the government has a fixed sum of money that is to be divided between a project listed in Table 1 and some unrelated project. Assume that the latter has a *C/B* ratio of 0.91, irrespective of the sum expended. In this situation, the step from one level to a higher one is warranted only if the *C/B* ratio corresponding to this increment is at least 0.91.

Closely related to cost-benefit analysis and an outgrowth of it is *cost-effectiveness analysis,* which is used mainly in the evaluation of military and space programs. To apply this method of analysis, assume that some required task can be accomplished by alternative projects that differ in both cost and degree of performance. The effectiveness of each project is expressed in some standard unit, and the projects are then compared by a procedure analogous to that for cost-benefit analysis.

Note that cost-benefit analysis can be used in any comparison of environmental alternatives. Thus, cost-benefit analyses can be used for air-pollution controls, industrial thermal discharge studies, transportation alternatives, power-generation choices (windmills vs. fossil-fuel or nuclear plants), cogeneration, recycling waste for power generation, solar power, use of recycled sewer sludge as a fertilizer, and similar studies. The major objective in each comparison is to find the most desirable alternative based on the benefits derived from various options open to the designer.

For example, electric utilities using steam generating stations burning coal or oil may release large amounts of carbon dioxide into the atmosphere. This carbon dioxide, produced when a fuel is burned, is thought to be causing a global greenhouse effect. To counteract this greenhouse effect, some electric utilities have purchased tropical rain forests to preserve the trees in the forest. These trees absorb carbon dioxide from the atmosphere, counteracting that released by the utility.

Other utilities pay lumber companies to fell trees more selectively. For example, in felling the 10 percent of marketable trees in a typical forest, as much as 40 to 50 percent of a forest may be destroyed. By felling trees more selectively, the destruction can be reduced to less than 20 percent of the forest. The remaining trees absorb atmospheric carbon dioxide, turning it into environmentally desirable wood. This conversion would not occur in these trees if they were felled in the usual foresting operation. The payment to the lumber company to do selective felling is considered a cost-benefit arrangement because the unfelled trees remove carbon dioxide from the air. The same is true of the tropical rain forests purchased by utilities and preserved to remove carbon dioxide which the owner-utility emits to the atmosphere.

Recently, a market has developed in the sale of "pollution rights" in which a utility that emits less carbon dioxide because it has installed pollution-control equipment can sell its "rights" to another utility that has less effective control equipment. The objective is to control, and reduce, the undesirable emissions by utilities.

With a potential "carbon tax" in the future, utilities and industrial plants that produce carbon dioxide as a by-product of their operations are seeking cost-benefit solutions. The analyses given here will help in evaluating potential solutions.

ENERGY-FROM-WASTE ECONOMIC ANALYSIS

A municipality requires the handling of 1500 tons/day (1524 mt/day) of typical municipal solid waste. Determine if a waste-to-energy alternative is feasible. If not, analyze the other means by which this solid-waste stream might be handled. Two waste-to-energy alternatives are being

TABLE 3 Estimated Costs of Municipal Solid Waste Disposal Facilities

	Mass burn	Processed fuel
First cost, $	15,000,000	22,500,000
Salvage value, $	1,500,000	2,500,000
Life, years	15	25
Annual maintenance cost, $	750,000	400,000
Annual taxes, $	15,000	15,000
Annual insurance, $	100,000	150,000

considered—mass burn and processed fuel. The expected costs are shown in Table 3. If earnings of 6 percent on invested capital are required, which alternative is more economical?

Calculation Procedure:

1. *Plot the options available for handling typical municipal waste*

Figure 1 shows the options available for handling solid municipal wastes or refuse. The refuse enters the energy-from-waste cycle and undergoes primary shredding. Then the shredded material is separated according to its density. Heavy materials— such as metal and glass—are removed for recovery and recycling. Experience and studies show that recycling will recover no more than 35 percent of the solid wastes entering a waste-to-energy facility. And most such facilities today are able to recycle only about 20 percent of municipal refuse. Assuming this 20 percent applies to the plant facility being considered here, the amount of waste that would be recycled would be 0.20(1500) = 300 tons/day (305 mt/day).

Numerous studies show that complete recycling of municipal waste is uneconomic. Therefore, the usual solution to municipal waste handling today features four primary components: (1) source reduction, (2) recycling, (3) waste to energy, and (4) landfilling. The Environmental Protection Agency (EPA) recently proposed broad policies encouraging recycling and reduction of pollutants at their source.

Using waste-to-energy facilities reduces the volume of wastes requiring disposal while producing a valuable commodity—steam and/or electric power. Combustion control is needed in every waste-to-energy facility to limit the products of incomplete combustion which escape in the flue gas and cause atmospheric pollution. Likewise, limiting the quantities of metal entering the combustor reduces their emission in the ash or flue gas. This, in turn, reduces pollution.

2. *Determine the energy available in the municipal waste*

Usually municipalities generate 1 ton (0.91 Mg) of solid waste per year per capita. About 35 percent of this waste is from residences; 65 percent is from industrial and commercial establishments. The usual heating value of municipal waste is 5500 Btu/lb (12×10^6 J/kg). Table 4 shows typical industrial wastes and their average heating values. Municipalities typically spend $25 or more per ton (0.91 Mg) to dispose of solid wastes.

Because municipal wastes have a variety of ingredients, many plants burn the solid waste as a supplement to coal. The heat in the waste is recovered for useful purposes, such as generating steam or electricity. When burned with high-sulfur coal, the solid waste reduces the sulfur content discharged in the stack gases. The solid waste also increases the retention of sulfur compounds in the ash. The result is reduced corrosion of the boiler tubes by HCl. Further, acid-rain complaints are fewer because of the reduced sulfur content in the stack gases.

Where an existing or future plant burns, or will burn, oil, another approach may be taken to the use of solid municipal waste as fuel. The solid waste is first shredded; then it is partially burned in a rotary kiln in an oxygen-deficient atmosphere at 1652°F (900°C). The gas produced is then burned in a conventional boiler to supplement the normal oil fuel.

Estimates show that about 5 percent of the energy needs of the United States could be produced by the efficient burning of solid municipal wastes in steam plants. Such plants must be located within about 100 mi (160 km) of the waste source to prevent excessive collection and transportation costs.

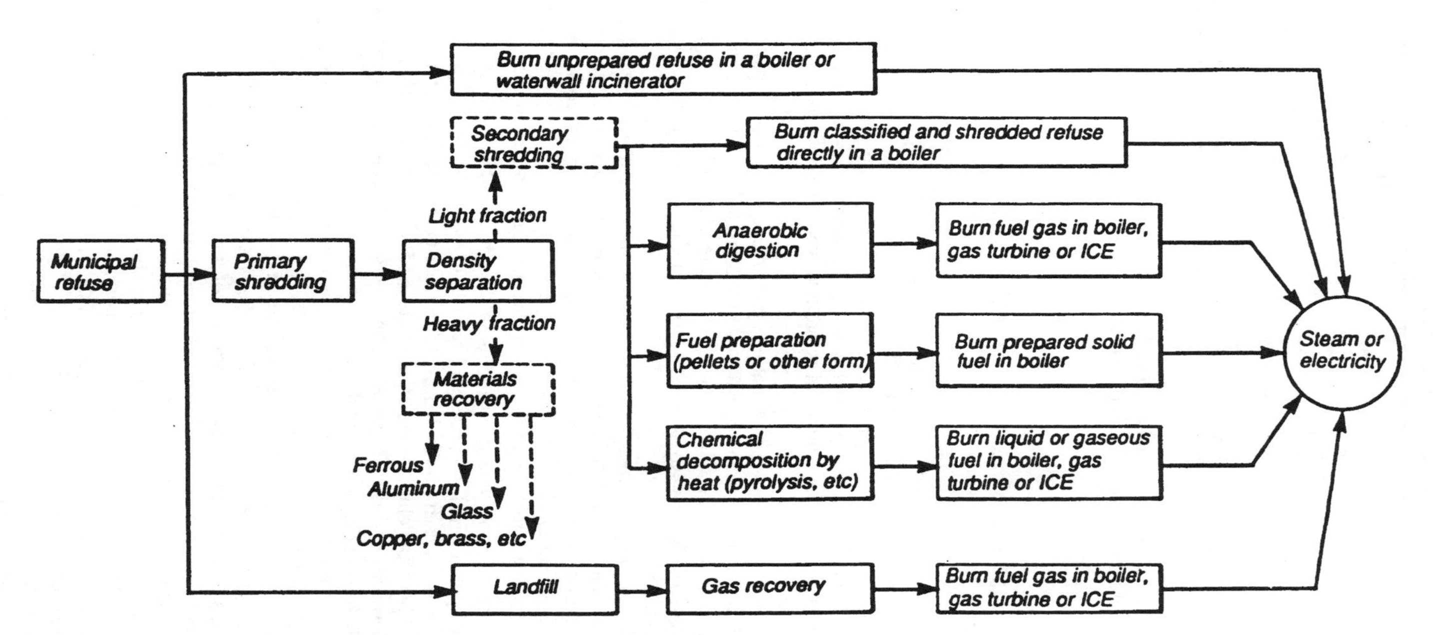

FIGURE 1 Several options for energy extraction from municipal waste. Selection should be based on local variables and economics. (*Power*.)

TABLE 4 Typical Industrial Wastes with Significant Fuel Value

	Average heating value (as fired)	
	Btu/lb	kJ/kg
Waste gases:		
Coke-oven	19,700	45,900
Blast-furnace	1,139	2,654
Carbon monoxide	579	1,349
Liquids:		
Refinery	21,800	50,794
Industrial sludge	3,700–4,200	8,621–9,786
Black liquor	4,400	10,252
Sulfite liquor	4,200	9,786
Dirty solvents	10,000–16,000	23,300–37,280
Spent lubricants	10,000–14,000	23,300–32,620
Paints and resins	6,000–10,000	13,980–23,300
Oily waste and residue	18,000	41,940
Solids:		
Bagasse	3,600–6,500	8,388–15,145
Bark	4,500–5,200	10,485–12,116
General wood wastes	4,500–6,500	10,485–15,145
Sawdust and shavings	4,500–7,500	10,485–17,475
Coffee grounds	4,900–6,500	11,417–15,145
Nut hulls	7,700	17,941
Rice hulls	5,200–6,500	12,116–15,145
Corn cobs	8,000–8,300	18,640–19,339

Power, SI units added by editor.

Combustion of, and heat recovery from, solid municipal wastes reduces waste volume considerably. But there is still ash from the combustion that must be disposed of in some manner. If landfill disposal is used, the high alkali content of the typical municipal ash must be considered. This alkali content often presents leaching and groundwater contamination problems. So, while the solid-waste disposal problem may have been solved, there are still environmental considerations that must be faced. Further, the large noncombustible items often removed from solid municipal waste before combustion—items like refrigerators and auto engine blocks—must still be disposed of in an environmentally acceptable manner. Table 5 shows a number of ash reuse and disposal options available for use today.

3. *Choose between available alternatives*

The two alternatives being considered—*mass burn* and *processed fuel*—have separate and distinct costs. These costs must be compared to determine the most desirable alternative.

TABLE 5 Ash Reuse and Disposal Options

	Treatment required	Use
Bottom ash	Particle-size screening	Coarse highway aggregate, concrete products
Bottom ash		Asphalt paving
Combined ash	Particle-size screening	Artificial reefs
Combined ash	Particle-size screening	Aggregate for paving
Flyash	Particle-size screening	Aggregate for paving
Flyash	Particle-size screening	Fine cement aggregate

Power.

In a mass-burn facility the trash is burned as received, after hand removal of large noncombustible items—sinks, bathtubs, engine blocks, etc. The remaining trash is rough-mixed by a clamshell bucket and delivered to the boiler's moving grate. Some 30 to 50 percent by weight and 5 to 15 percent by volume of the waste burned in a mass-burn facility leaves in the form of bottom ash and flyash.

In a processed-fuel facility [also called a refuse-derived-fuel (RDF) facility], the solid waste is processed in two steps. First, noncombustibles are separated from combustibles. The remaining combustible waste is reduced to uniform-sized pieces in a hammermill-type shredder. The shredded pieces are then delivered to a boiler for combustion.

Using the annual cost of each alternative as the "first cut" in the choice: Operating and maintenance cost = C = maintenance cost per year, $ + annual taxes, $ + annual insurance cost, $. For mass burn, C = $750,000 + $15,000 + 0.002($15,000,000) = $795,000. For processed fuel, C = $400,000 + $15,000 + 0.002($22,500,000) = $460,000.

Next, using a capital-recovery equation for mass burn, the annual cost, A = ($15,000,000 – $1,500,000)(0.06646) + $750,000 + $1,500,000(0.06) + $795,000 = $1,737,210. For processed fuel, A = ($22,500,000 – $2,500,000)(0.06646) + $400,000 + $2,500,000(0.06) = $1,879,200. Therefore, mass burn is the more attractive alternative from an annual-cost basis because it is $1,879,000 – $1,737,210 = $141,990 per year less expensive than the processed-fuel alternative.

Several more analyses would be made before this tentative conclusion was accepted. However, this calculation procedure does reveal an acceptable first-cut approach to choosing between different available alternatives for evaluating an environmental proposal.

Related Calculations. Another source of usable energy from solid municipal waste is landfill methane gas. This methane gas is produced by decomposition of organic materials in the solid waste. The gas has a heating value of about 500 Btu/ft^3 (1.1×10^6 J/kg) and can be burned in a conventional boiler, gas turbine, or internal-combustion engine. Using landfill gas to generate steam or electricity can reduce landfill odors. But such burning does *not* reduce the space and groundwater problems produced by landfills. The cost of landfill gas can range from $0.45 to $5/million Btu ($0.45 to $5/1055 kJ). Much depends on the cost of recovering the gas from the landfill.

Methane gas is recovered from landfills by drilling wells into the field. Plastic pipes are then inserted into the wells and the gas is collected by gas compressors. East coast landfills in the United States have a lifespan of 5 to 7 years. West coast landfills have a lifespan of 15 to 18 years.

Data in this procedure were drawn from *Power* magazine and Hicks, *Power Plant Evaluation and Design Reference Guide,* McGraw-Hill, 1986.

EMISSION REDUCTION USING FLUE-GAS HEAT RECOVERY

A steam boiler rated at 32,000,000 Btu/h (9376 MW) fired with natural gas is to heat incoming feedwater with its flue gas in a heat exchanger from 60°F (15.6°C) to an 80°F (26.7°C) outlet temperature. The flue gas will enter the boiler stack and heat exchanger at 450°F (232°C) and exit at 100°F (37.8°C). Determine the efficiency improvement that might be obtained from the heat recovery. Likewise, determine the efficiency improvement for an oil-fired boiler having a flue-gas inlet temperature of 300°F (148.9°C) and a similar heat exchanger.

Calculation Procedure:

1. *Sketch a typical heat-recovery system hookup*

Figure 2 shows a typical hookup for stack-gas heat recovery. The flue gas from the boiler enters the condensing heat exchanger at an elevated temperature. Water sprayed into the heat exchanger absorbs heat from the flue gas and is passed through a secondary external heat exchanger. Boiler feedwater flowing through the secondary heat exchanger is heated by the hot water from the condensing heat exchanger. Note that the fluid heated can be used for a variety of purposes other than boiler feedwater—process, space heating, unit heaters, domestic hot water, etc.

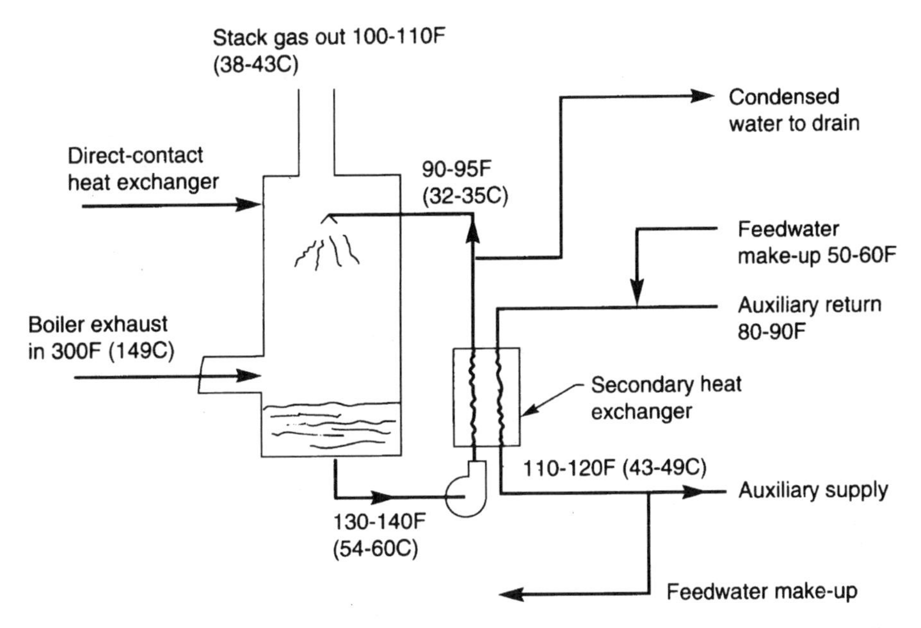

FIGURE 2 Effective condensation heat recovery depends on direct contact between flue gas and cooling medium and low gas-side pressure drop. (*Power.*)

Flue gas from the boiler can enter the condensing heat exchanger at temperatures of 300°F (148.9°C), or higher, and exit at 100 to 120°F (37.8 to 48.9°C). The sensible and latent heat given up is transferred to the spray water. Since this sprayed cooling water may be contaminated by the flue gas, a secondary heat exchanger (Fig. 2) may be used. Where the boiler fuel is clean-burning natural gas, the spray water may be used directly, without a secondary heat exchanger. Since there may be acid contamination from SO_2 in the flue gas, careful analysis is needed to determine if the contamination level is acceptable in the process for which the heated water or other fluid will be used.

2. *Determine the efficiency gain from the condensation heat recovery*

Efficiency gain is a function of fuel hydrogen content, boiler flue-gas exit temperature, spray (process) water temperature, amount of low-level heat needed, fuel moisture content, and combustion-air humidity. The first four items are of maximum significance for gas-, oil-, and coal-fired boilers. Installations firing lignite or high-moisture-content biomass fuels may show additional savings over those computed here. If combustion-air humidity is high, the efficiency improvement from the condensation heat recovery may be 1 percent higher than predicted here.

The inlet-water temperature is normally 20°F (–6.7°C) lower than the flue-gas outlet temperature. And for the usual preliminary evaluation of the efficiency of condensation heat recovery, the flue-gas exit temperature from the heat exchanger is taken as 100°F (37.8°C).

For the natural-gas-burning boiler, flue-gas inlet temperature = 450°F (232°C); cold-water inlet temperature = 60°F (15.6°C); water-outlet temperature = 60 + 20 = 80°F (26.7°C); flue-gas outlet temperature = 80°F (26.7°C). Find the *basic* efficiency improvement, ΔEi, from Fig. 3 as $\Delta Ei = 14.5$ percent by entering at the bottom at the flue-gas temperature of 450°F (232°C), projecting to the gas-fired curve, and reading ΔEi on the left-hand axis.

Next, find the *actual* efficiency improvement, ΔE, from $\Delta E = F(\Delta Ei)$, where F is a factor depending on the flue-gas outlet temperature. Values of F are shown in Fig. 4 for various outlet gas temperatures. With an outlet gas temperature of 80°F (26.7°C), Fig. 4*a* shows $F = 1.19$. Then, $\Delta E = 14.5 \times 1.19 = 17.3$ percent. Table 6 details system temperatures.

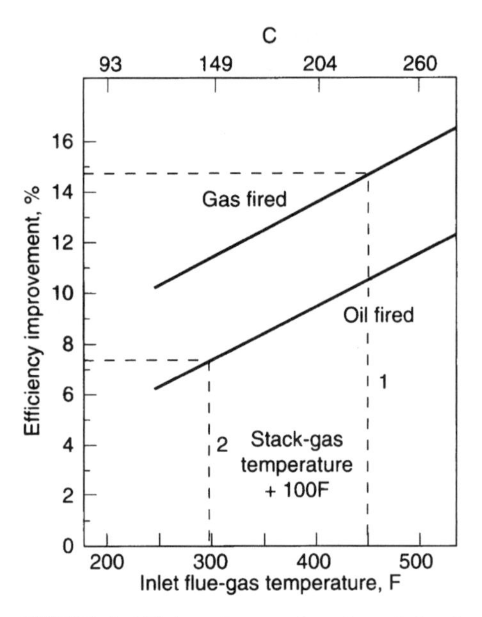

FIGURE 3 Efficiency increase depends on fuel and on temperature of flue gas. (*Power.*)

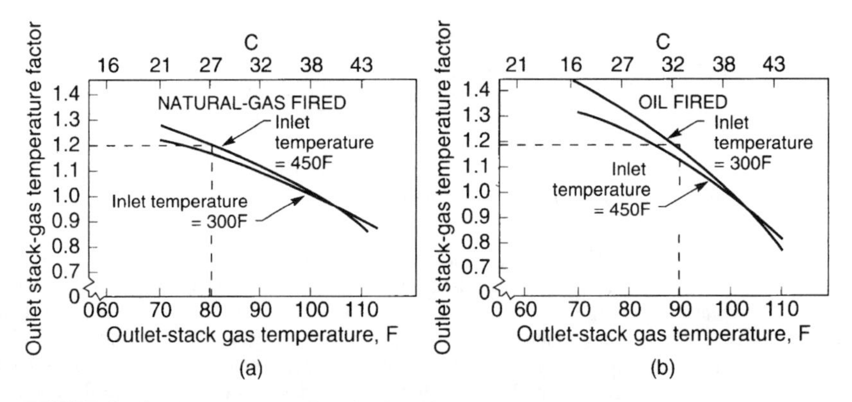

FIGURE 4 Use these curves to allow for the effect of variations in the exit temperature from the recovery unit on efficiency increase possible with heat-recovery unit. (*Power.*)

For the oil-fired boiler, flue-gas inlet temperature = 300°F (148.9°C); cold-water inlet temperature = 70°F (21.1°C); water-outlet temperature = 70 + 20 = 90°F (32.2°C). Find the *basic* efficiency improvement from Fig. 3 as $Ei = 7.2$ percent. Next, find F from Fig. 4*b* as 1.18. Then, the *actual* efficiency = $\Delta E = 1.18(7.2) = 8.5$ percent.

The lower efficiency improvement for oil-fired boilers is generally due to the lower hydrogen content of the fuel. Note, however, that where the cost of oil is higher than natural gas, the dollar saving may be greater.

The efficiency-improvement charts given here assume that all of the low-level heat generated can be used. A plant engineer familiar with a plant's energy balance is in the best position to choose the

TABLE 6 Use of Figs. 3 and 4

1. Natural gas		
Inlet temperature	450°F	232°C
Inlet cold water	60°F	16°C
Outlet temperature = 60 + 20 =	80°F	27°C
ΔEb (from Fig. 3)	14.5%	14.5%
F (from Fig. 4)	1.19	1.19
ΔE (14.5 × 1.19)	17.3%	17.3%
2. Fuel oil		
Inlet temperature	300°F	149°C
Inlet cold water	70°F	21°C
Outlet temperature = 70 + 20 =	90°F	32°C
ΔEb (from Fig. 3)	7.2%	7.2%
F (from Fig. 4)	1.18	1.18
ΔE (7.2 × 1.18)	8.5%	8.5%

optimum level of heat recovery. Typical applications are: makeup-water preheat, low-temperature process load, space heating, and domestic hot water.

Makeup-water-preheat needs depend largely on the amount of condensate that is returned to the boiler. Generally, there is more heat available in the flue gas than can be used to preheat feedwater. If the boiler is operating at 100 percent makeup, only about 60 percent of the available heat can be transferred to the incoming feedwater. One reason for this is the low temperature of the hot water. This limitation can be handled in two ways: (1) Design the heat-recovery unit to take a slip-stream from the flue gas and only recover as much heat as can be used to heat feedwater; or (2) in multiple-boiler plants, install a heat-recovery system on one boiler only and use it to preheat feedwater for all the boilers.

Process hot water, if it is needed, can provide extremely short payback for a condensation heat-recovery unit. In food and textile processes, the hot-water needs account for 15 percent or more of the total boiler load. Any facility with hot-water requirements between 10 and 15 percent of boiler capacity and an operating schedule greater than 4000 h/yr should seriously investigate condensation heat recovery.

Space-heating economics are generally less favorable than makeup or process hot water because of the load variation, limited heating season, and the difficulty of matching demand and supply schedules. The difficulty of retrofitting heat exchangers to an existing heating system also limits the number of useful applications. However, paybacks between 2.5 and 3 years are possible in colder regions and certainly warrant preliminary investigation. Sometimes space heating can be combined with feedwater or process-water heating.

3. ***Estimate the cost of the condensation heat-recovery equipment***

Figure 5 shows an approximate range of costs for equipment and installation. Note that the installed cost may be three times the equipment cost because of retrofit difficulties involved with an existing installation.

Operating costs are primarily fan and pump power consumption. These generally range from 5 to 10 percent of the value of the recovered heat. The lower figure applies to limited distribution of the hot water, while the higher figure applies to systems where hot water is distributed 100 ft (30.5 m) or more from the boiler or where high-pressure-drop heat-recovery units are used.

Figure 6 shows how a heat-recovery unit can be used to heat the feedwater for one or more boilers. The heat recovered, as noted above, may be more than needed to heat the feedwater for just one boiler.

Corrosion in a condensing heat-recovery unit can usually be prevented by using Type 304 or 316 stainless steel or fiberglass-reinforced plastic for the tower pump and secondary exchanger. If the flue gas is unusually corrosive, it may be advisable to do a chemical analysis before planning the recovery unit.

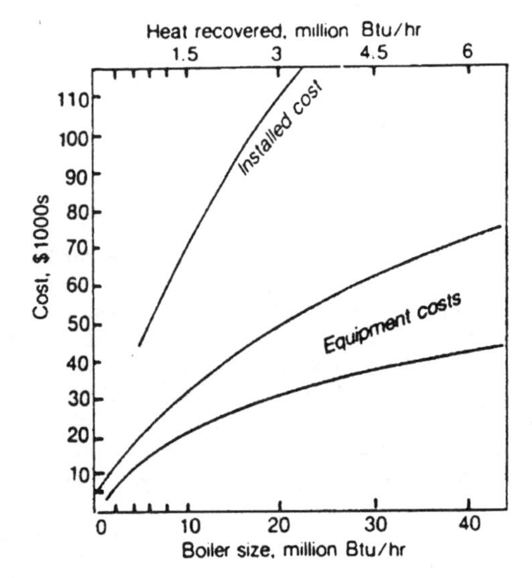

FIGURE 5 Installation cost of condensation heat-recovery unit may run as high as three times equipment cost. (*Power.*)

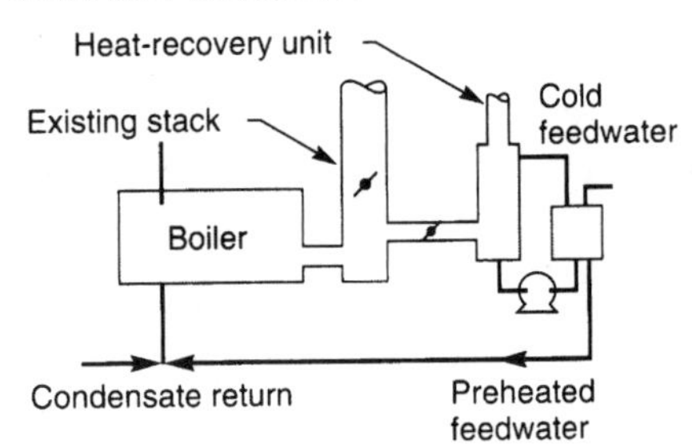

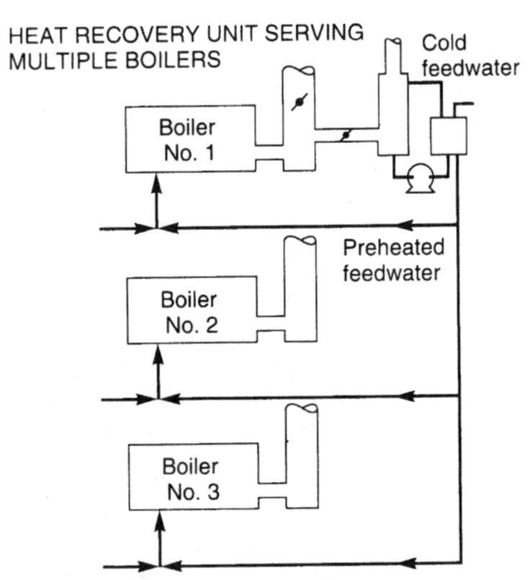

FIGURE 6 Heat recovery from the unit is more than enough to heat feedwater to one boiler. (*Power.*)

A unique feature of condensation heat recovery is that it recovers energy while also reducing emissions. In addition, when natural gas is burned, a small percentage of the NO_x emissions are reduced by condensation of oxides of nitrogen. SO_2 emissions can be reduced significantly by using an alkaline water spray in a pH range of 6 to 8. Natural gas depletes the ozone layer less than other fossil fuels.

The potential emission reduction can have a significant effect in nonattainment areas and could increase allowable plant capacity. But it should be pointed out that the SO_2-emission reduction from the scrubbing and condensation have not been substantiated by independent tests. Such tests should be provided for any installation that depends on emission reduction for its justification.

At the time of the preparation of the revision of this handbook, the Tennessee Valley Authority (TVA) is testing at its Shawnee plant a lime treatment to reduce boiler stack gas SO_2 emissions. Lime, in fine particle form, is suspended in the flue gas before release to the plant smokestack. Sulfur in the flue gas binds to the lime, thereby reducing the potential for acid rain. A cyclone and electrostatic precipitator separate the lime particles from the exiting flue gas. At this time the lime system is believed to have lower equipment and operating costs than other competitive systems.

This procedure is based on the work of R. E. Thompson, KVB, Inc., as reported in *Power* magazine.

COGENERATION COST ANALYSIS FOR ALTERNATIVE ENERGY SYSTEMS

Compare the capital and operating costs of two cogenerational coal-fired industrial steam plants — Option 1: a stoker-fired (SF) plant, Fig. 7; Option 2: a pulverized-coal-fired (PF) plant, Fig. 8. Both plants operate at 600 lb/in^2 (gage) (4134 kPa)/750°F (399°C). The SF plant meets a demand of 200,000 lb/h (90,800 kg/h) of 150-lb/in^2 (gage) (1034-kPa) steam and 10 MW of electric power with the SF boiler and purchased power from a local utility. For the PF boiler the same demands are met using a nonextraction backpressure turbine.

Calculation Procedure:

1. ***Obtain, or develop, the capital costs for the boiler alternatives***

Contact manufacturers of suitable boilers, asking for estimated costs based on the proposed operating capacity, pressure, and temperature. Figure 9 shows a typical plot of the data supplied by manufacturers for the boilers considered here. For coal firing, field-erected boilers are to be used in this plant. (*Note:* The costs given here are for example purposes *only*. Do *not* use the given costs for actual estimating purposes. Instead, obtain current costs from the selected manufacturers.)

From Fig. 9, the SF unit costs $61/lb ($27.70/kg) of steam generated; and the PF unit $72/lb ($32.70/kg) of steam generated per hour. Assuming that SO_2 reduction is not required by environmental considerations for either unit, a dry-scrubber/fabric-filter combination can be used to remove the total suspended particulates (TSP). To cover the cost of this combination to remove TSP, add $7/lb ($3.20/kg) of steam generated. This brings the cost to $68/lb ($30.90/kg) and $79/lb ($35.90/kg) of steam generated.

2. ***Compute the capital cost for the turbine installation***

Before determining the capital cost of the turbine installation—often called a *turbine island*—estimate the potential electric-power generation from the process-steam flow based on the ASME data in Table 7. At 600 lb/in^2 (gage) (4134 kPa)/750°F (399°C) throttle conditions, and 150-lb/in^2 (gage) (1034 kPa) back-pressure, 6127 kW is available, determined as follows: Theoretical steam rate from Table 7 = 23.83 lb/kWh (10.83 kg/kWh); turbine efficiency = 73 percent from Table 7. Then, actual steam rate = theoretical steam rate/efficiency. Or, actual steam rate = 23.83/0.73 = 32.64 lb/kWh (14.84 kg/kWh). Then kW available = steam flow rate, lb/h/steam rate, lb/kWh. Or, kW available = 200,000/32.64 = 6127 kW

Referring to Fig. 10 for a 6-MW nonextraction backpressure turbine shows a capital cost of $380/kW. Summarize the capital costs in tabular form, Table 8. Thus, the SF boiler cost, Option 1 in Table 8, is (200,000 lb/h)($68/lb of steam generated) = $13,600,000. The PC-fired boiler, Option 2, will have a cost of (200,000 lb/h)($79/lb of steam generated) = $15,800,000. Since a turbine is used with the PC-fired unit, its costs must also be included. Or, 6100 kW ($380/kW) = $2,318,000. Computing the total cost for each option shows, in Table 8, that Option 2 costs $18,118,000 – $13,600,000 – $4,518,000 more than Option 1.

3. ***Compare operating costs of each option***

Obtain from the plant owner and equipment suppliers the key data needed to compare operating costs, namely: Operating time, h/yr; boiler efficiency, percent: fuel cost, $/ton ($/tonne); electric-power use, kWh/yr; electric power cost, ¢/kWh; maintenance cost as a percent of the capital investment per year; personnel required; personnel cost; ash removal, tons/yr (tonnes/yr); ash removal cost, $/yr. Using these data, compute the operating cost for each option and tabulate the results as shown in Table 9.

With Option 1, all electric power is purchased; with Option 2, 3900 kW must be purchased. The difference in annual operating cost is $6,288,350 – $4,895,000 = $1,393,350. Since Option 2 costs $4,518,000 more than Option 1, but has a $1,393,350-per-year lower operating cost, the simple payback time for the cogeneration option (2), ignoring the cost of money, is:

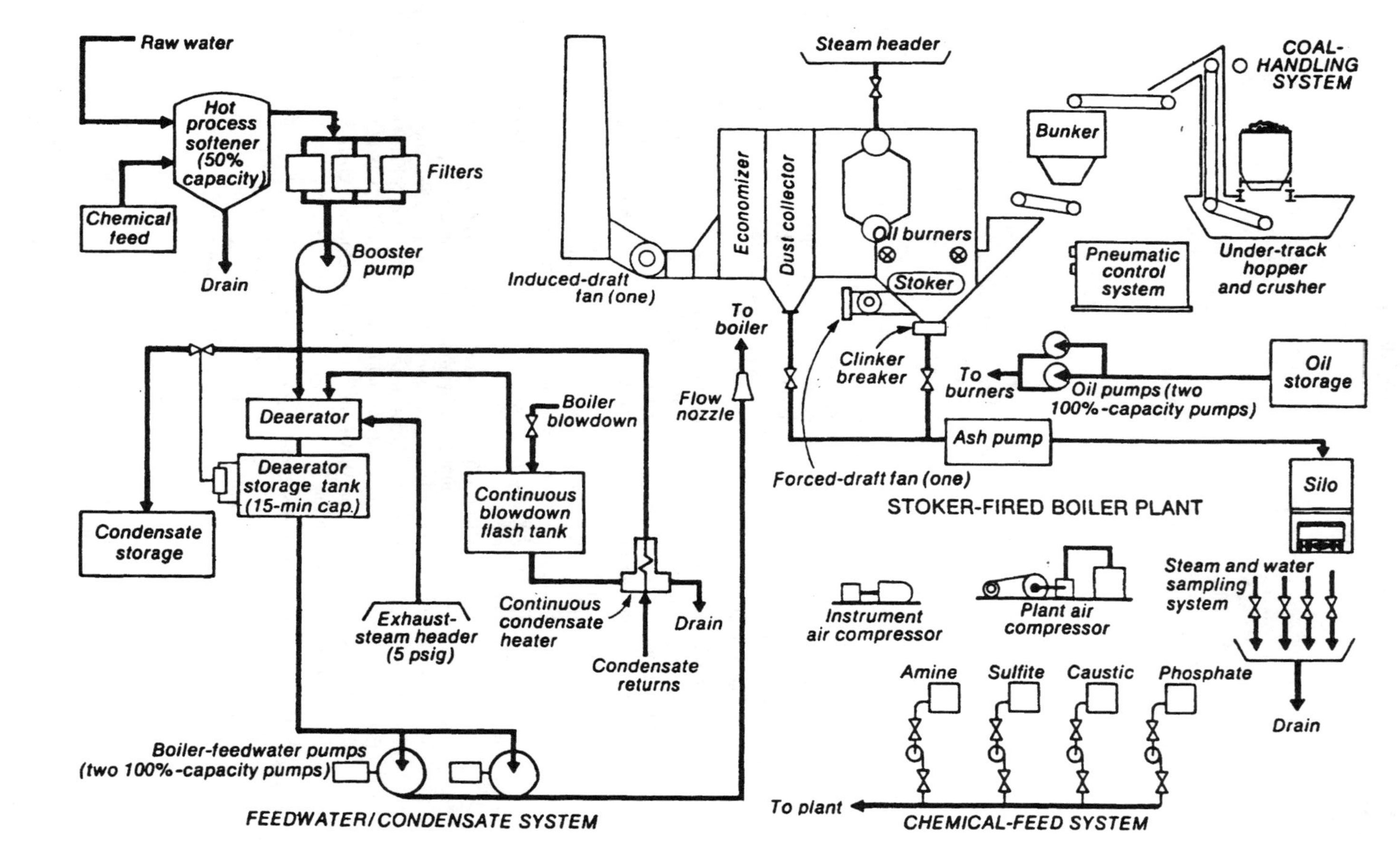

FIGURE 7 Stoker-fired steam plant used in the cost analysis includes all the components shown. (*Power.*)

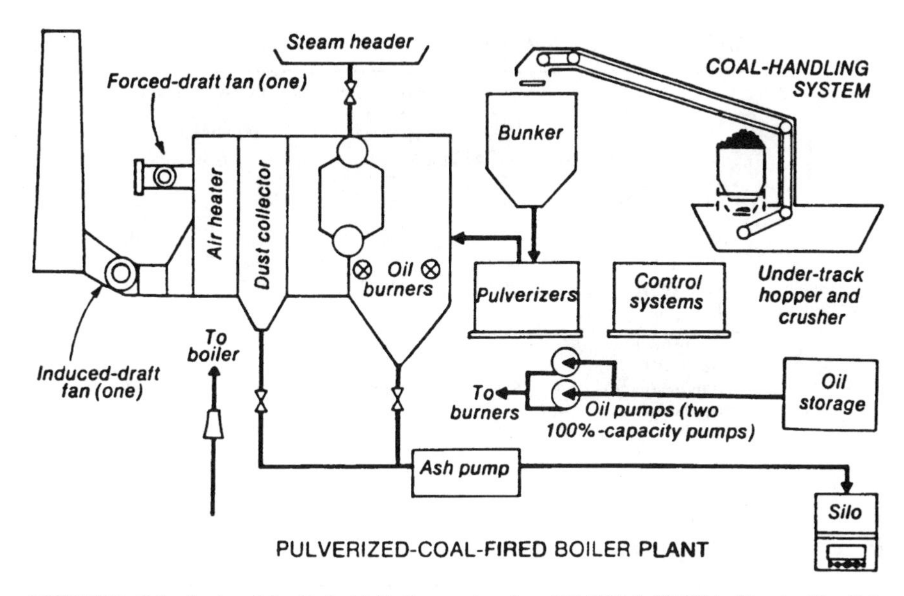

FIGURE 8 Pulverized-coal-fired industrial boilers produce from 200,000 lb/h (90,800 kg/h) to 1 million lb/h (454,000 kg/h) of steam. Unit here does not have an economizer. (*Power.*)

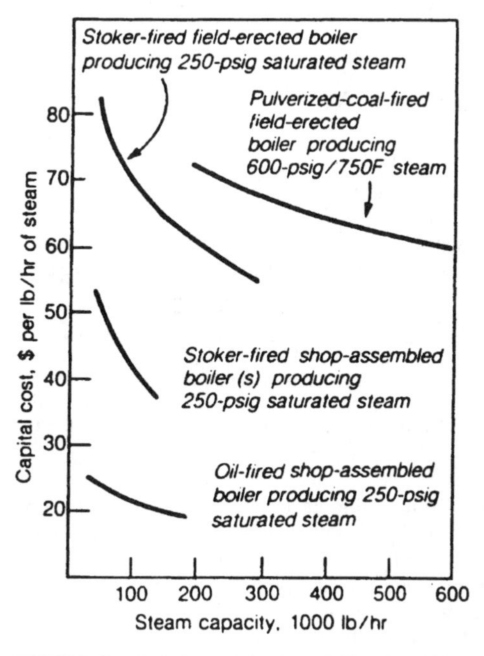

FIGURE 9 Capital costs for three boiler alternatives. (*Power.*)

TABLE 7 ASME Values for Estimating Turbine Steam Rates*

	Theoretical steam rate, lb/kWh (kg/kWh)					
	600 lb/in² (gage) (4134 kPa)		900 lb/in² (gage) (6201 kPa)		1500 lb/in² (gage) (10,335 kPa)	
Exhaust pressure	750°F	(399°C)	900°F	(482°C)	900°F	(482°C)
4.0 inHg (10.2 cmHg)	7.64	3.47	6.69	3.04	6.48	2.94
5 lb/in² (gage) (34.5 kPa)	11.05	5.02	9.21	4.18	8.53	3.87
15 lb/in² (gage) (103.4 kPa)	12.16	5.52	9.98	4.53	9.06	4.11
150 lb/in² (gage) (1034 kPa)	23.83	10.82	16.91	7.68	14.30	6.49
	Turbine efficiency, %					
4.0 inHg		77		78		78
5 lb/in² (gage)		73		76		76
15 lb/in² (gage)		74		76		76
150 lb/in² (gage)		73		74		74

*Actual turbine steam rate equals theoretical steam rate divided by efficiency.

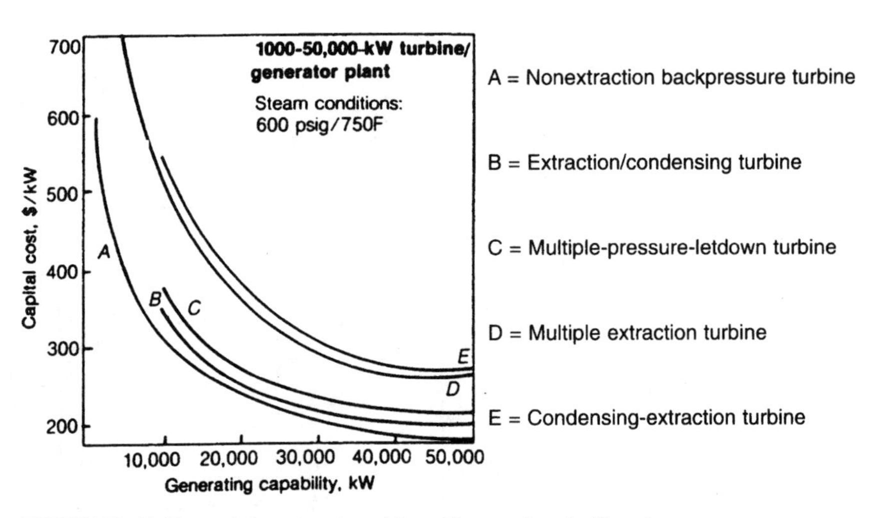

FIGURE 10 Turbine capital cost is estimated from this type of graph. (*Power.*)

Payback time = larger capital cost, $/annual savings, $, of the higher cost option. Or, $4,518,000/$1,393,350 = 3.24 years. Thus, the cogeneration option (2) is attractive because the payback time is relatively short.

Other economic analyses should also be conducted. For example, higher cycle efficiencies can be obtained with higher throttle conditions, but boiler capital and operating costs will be higher. Plants in the 40- to 60-MW range might benefit from an extraction–condensing-turbine arrangement generating all plant electric-power demand. Although not as efficient as a straight extraction machine, the cost, in $/kW, will be less than purchased power.

TABLE 8 Summary of Capital Costs

	Option 1	Option 2
Boiler island		
Stoker-fired unit ($68/lb steam) ($30.90/kg steam)	$13,600,000	—
PC-fired unit ($79/lb steam) ($30.90/kg steam)	—	$15,800,00
Turbine island (6100 kW, $380/kW)	—	2,318,000
Total	$13,600,000	$18,118,000

TABLE 9 Comparison of Operating Costs for Coal-Fired Plants

Operating requirements	Option 1, Stoker-fired	Option 2, PC-fired
Equivalent full-power operating time, h/yr	5,400	5,700
Boiler efficiency, %	82	87
Fuel consumption, tons/yr (tonne/yr)	52,700 (47,798)	62,000 (56,234)
Fuel cost, $/ton ($/tonne)	50 (45.35)	42 (38,09)
Total fuel cost, $/yr	2,640,000	2,600,000
Electric-power use, kWh/yr	3,240,000	8,550,000
Electric-power cost, $/yr @ 5¢/kWh	162,000	427,000
Maintenance cost, $/yr (based on 2.5% of capital investment per year)	340,000	395,000
Personnel per shift	2.5	2.75
Personnel cost, $/yr (based on $30,000/yr for each person)	270,000	330,000
Ash removal, tons/yr (tonne/yr)	5,270 (4780)	6,200 (5,623)
Ash-removal cost, $/yr	26,350	31,000
Total operating cost, $/yr	3,438,350	3,783,500
Total operating costs, steam + purchased power		
Steam	3,438,350	3,783,500
Purchased power @ 5¢/kWh	10 MW 2,850,000	3.9 MW 1,111,500
Total operating cost	$6,288,350	$4,895,000

Related Calculations. The EPA and state environmental bodies favor cogeneration of electricity because it reduces atmospheric pollution while conserving fuel. While the options considered here use steam-powered prime movers, cogeneration installations can use diesel-, gasoline-, or natural-gas-fueled prime movers of many different types—reciprocating, gas-turbine, etc. The principal objective of cogeneration is to wrest more heat from available energy streams by the simultaneous generation of electricity and steam (or some other heated medium), thereby saving fuel while reducing atmospheric pollution.

While the possible choices for boiler fuel are oil, gas, and coal, practical choices are limited to coal for most industrial cogeneration projects. Gas firing in industrial plants is restricted to units 80,000 lb/h (36,364 kg/h) or less by the Industrial Fuel Use Act. Oil firing is allowed in field-erected boilers, but usually gives way to coal on an economic basis. Packaged oil-fired boilers top out at 200,000 lb/h (90,909 kg/h) to permit shipping.

Coal can be burned in either a stoker-fired unit or a pulverized-coal-fired unit, the essential cost differences being in coal preparation and ash handling. PC-fired units produce about 75 percent flyash and 25 percent bottom ash, and stoker-fired units the reverse. Coal supply specifications also vary between the two. Grindability is important to PC-fired units; top size and fines content is critical to stoker-fired units. Coal for stoker-fired units averages \$5/ton (\$5.51/tonne) more for coals of similar heating values.

PC-fired boilers always include an air heater for drying coal upstream of the pulverizer, and usually an economizer for flue-gas heat recovery. The reverse is true of stoker-fired units, but, concerning the air heater, more care is needed to ensure that the grate is not overheated during normal operation.

The greatest overlap in choosing between the two exists in the 200,000 to 300,000 lb/h (90,909 to 136,364 kg/h) size range. Above this range, stoker-fired units are limited by grate size. Below this range, PC-fired units usually do not compete economically. Shop-assembled chain-grate stoker-fired units have been shipped up to a capacity of 45,000 lb/h (20,455 kg/h).

No matter what type of firing is used, industrial power plants must meet EPA emission limits for total suspended particulates (TSP), SO_2, and NO_x. If an on-site coal pile is contemplated, water runoff control must meet National Pollution Discharge Elimination System (NPDES) standards.

NO_x formation is typically limited in the combustion process through careful choice of burners. SO_2 and TSP are usually removed from the flue gas. For TSP reduction, an electrostatic precipitator or a fabric filter is used. These will sometimes be preceded by cyclone collectors for stoker-fired boilers to reduce the total load on the final stage of the ash-collection system.

For boilers under 250-million Btu/h (73.3 MW) heat input, SO_2 formation can be limited by burning low-sulfur coal. Where SO_2 emissions reduction is required to meet the National Ambient Air Quality Standards (NAAQS), a dry-scrubber-fabric-filter combination is a satisfactory strategy. Wet scrubbers, though highly effective, create an additional sludge-disposal problem.

Operating costs for stoker- and pulverized-coal-fired plants designed to produce 200,000 lb/h (90,800 kg/h) of steam from one boiler are shown in Table 9.

To understand more about how operating costs were calculated, look at the entries in Table 9 line by line. First, equivalent full-power operating hours are determined by subtracting 336 h (2 weeks) for maintenance from the total number of hours in a year (8760), and by multiplying the result by both, unit availability (85 percent for stoker, 90 percent for pulverized coal) and the assumed plant load factor—in this case 75 percent.

Fuel consumption is based on typical operating efficiencies for similar plants and a fuel heating value of 12,500 Btu/lb (29,125 kJ/kg) for coal. Note that a premium is paid for stoker coal because a relatively clean fuel of suitable size is needed to maintain efficient operation.

The general procedure given here is applicable to a variety of options because today's emphasis on industrial cogeneration calls for a method of reasonably estimating costs of the many system alternatives. The approach differs from utility cost estimating mainly because steam capacity and power-generation capability are separate design objectives. Either the industrial power plant meets the process-steam demand and then generates whatever power that creates, or else it meets the electric-power demand and generates the required steam.

Choice of approach depends on the steam and electric-power requirements of the facility. Ideally, they balance exactly. In practice, most steam requirements will not generate enough electric power to meet the plant load. Conversely, the steam flow can rarely generate more electric power than the plant needs. Variations in steam conditions and turbine-exhaust pressure lead to many ways of matching the loads. More important, regulated buyback of excess electric power by public utilities now eases the problem of load balancing.

A profile of steam and electric-power consumption is necessary to begin evaluating alternatives. Daily, weekly, monthly, and seasonal variations are all important. During initial evaluation of the balance, use an average of 25 lb (11.4 kg) of steam/kWh as the steam rate of a small steam turbine. It is a conservative number, and the actual value will probably be lower—meaning more electric power for the steam flow—but it will give a rough idea of how close the two demands will match.

In the above procedure, capital costs are separated into costs for the boiler island and costs for the turbine island. Absolute accuracy is to within ±25 percent, not of appropriation quality but good enough to compare different plant designs. In fact, relative accuracy is closer to ±10 percent.

This procedure is the work of B. Dwight Coffin, H. K. Ferguson Co., and was reported in *Power* magazine.

STEAM COMPRESSOR CHOICE FOR DIESEL-ENGINE COGENERATION SYSTEM

Select a suitable steam compressor to deliver an 80-lb/in^2 (gage) (551-kPa) discharge pressure for a cogeneration system using two 1500-kW diesel-engine-generator sets operating at 1200 rpm with 1200°F (649°C) exhaust temperature. Each engine exhaust is vented through a waste-heat (heat-recovery) boiler which generates 3800 lb/h (1725 kg/h) of steam at 110-lb/in^2 (gage) (758-kPa) saturated. The cooling system of each engine generates 5000 lb/h (2270 kg/h) of 15-lb/in^2 (gage) (103-kPa) steam. Choose the compressor to boost the 15-lb/in^2 (gage) (103-kPa) steam pressure to 80-lb/in^2 (gage) (551-kPa) to be used in a distribution system. Steam at 110 lb/in^2 (gage) (758 kPa) is first used in laundry and heat-exchange equipment before being reduced to 80 lb/in^2 (gage) (551 kPa) and combined with the compressor discharge flow. About 16,500 lb/h (7491 kg/h) of 80-lb/in^2 (gage) (551-kPa) steam satisfies the distribution system requirements, except in severe weather when the existing boilers are fired to supplement the steaming requirements.

Calculation Procedure:

1. *Determine the amount of steam that can be generated by each waste-heat boiler*
Exhaust gas from each diesel engine enters the waste-heat boiler at 1200°F (649°C). Using the rule of thumb that a diesel-engine exhaust heat boiler can produce 1.9 lb/h (0.86 kg/h) of 100-lb/in^2 (gage) (689-kPa) saturated steam at full load per rated horsepower, find the amount of steam generated as 1.9 (1500 kW/0.746 hp/kW) = 3820.4 lb/h (1734.5 kg/h). Since the 110-lb/in^2 (gage) (758-kPa) steam required by the cogeneration system needs slightly more heat input, round off the quantity of steam generated to 3800 lb/h (1725 kg/h).

2. *Select the type of steam compressor to use*
The compression ratios used in cogeneration—higher than for many process applications—dictate use of mechanical compressors. Several different thermodynamic paths may be followed during the compression process (Fig. 11). Of the three processes shown, the highest compressor coefficient of performance (COP) is exhibited by direct two-phase compression (Fig. 12). The task becomes one of selecting a suitable unit to follow this path.

Centrifugal and axial compressors, both of the general category of dynamic compressors, work on aerodynamic principles. Though capable of handling large flow rates, they are sensitive to water droplets that may cause blade erosion. Thus, two-phase flow is undesirable. These compressors are also limited to operation within a narrow range because of surging or low efficiency when conditions deviate from design.

Positive-displacement compressors—such as the screw, lobe, or reciprocating variety—are more suited to cogeneration applications. Both the screw compressor and, to a lesser extent, the reciprocating compressor, achieve high-pressure ratios at high COP. Further, the units approach the isothermal condition because work which normally goes into producing sensible heat during compression simply causes additional liquid to evaporate. Intercooling is avoided, and input power requirements are acceptable (Fig. 13).

Cost considerations tend to make reciprocating units the second choice after screw and lobe units, but they are comparable on a technical basis. Note that reciprocating units require large foundations and must be driven at slow speeds. Some manufacturers offer carbon-ring units, requiring no lubrication. Even with the more common units requiring lubrication, use of synthetic lubricants keeps the amount of oil small, making steam contamination a negligible concern.

Based on the above information, two screw compressors will be chosen to boost the 15-lb/in^2 (gage) (103-kPa) steam to 80 lb/in^2 (gage) (551 kPa). Use Fig. 13 to approximate the required horsepower (kW) input to each screw compressor. With 16,500-lb/h (7491-kg/h) process steam required, Fig. 13 shows that the required horsepower input for each screw compressor will be 1250 hp (933 kW).

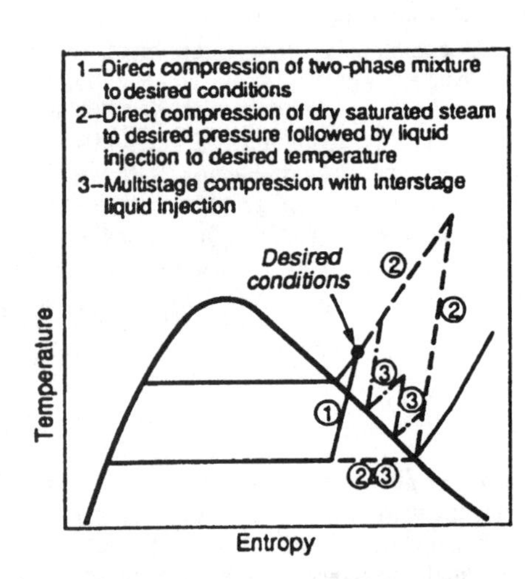

FIGURE 11 Compression can follow several different thermodynamic paths. (*Power.*)

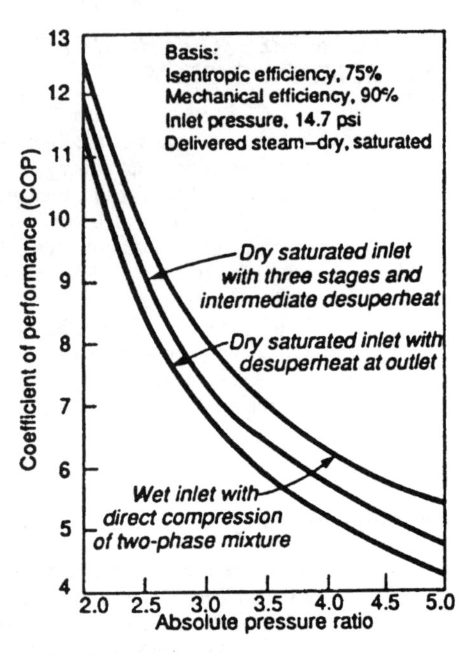

FIGURE 12 Each compression path exhibits a different coefficient of performance. (*Power.*)

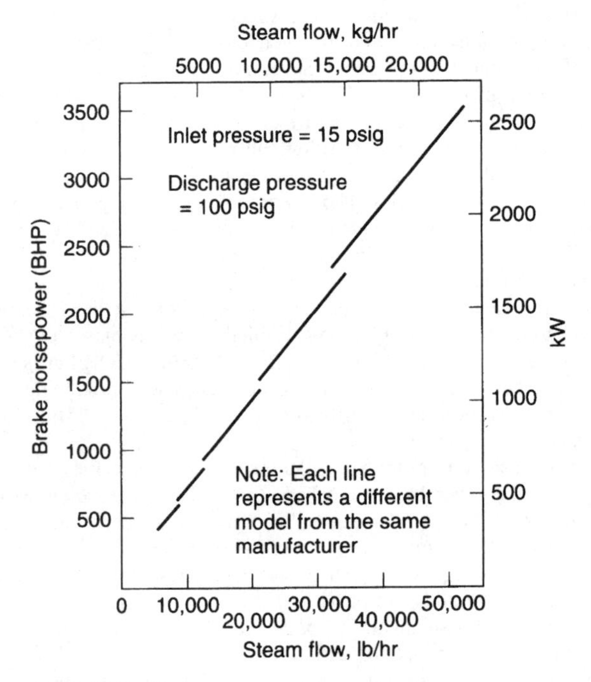

FIGURE 13 Energy input requirements for a screw compressor. (*Power.*)

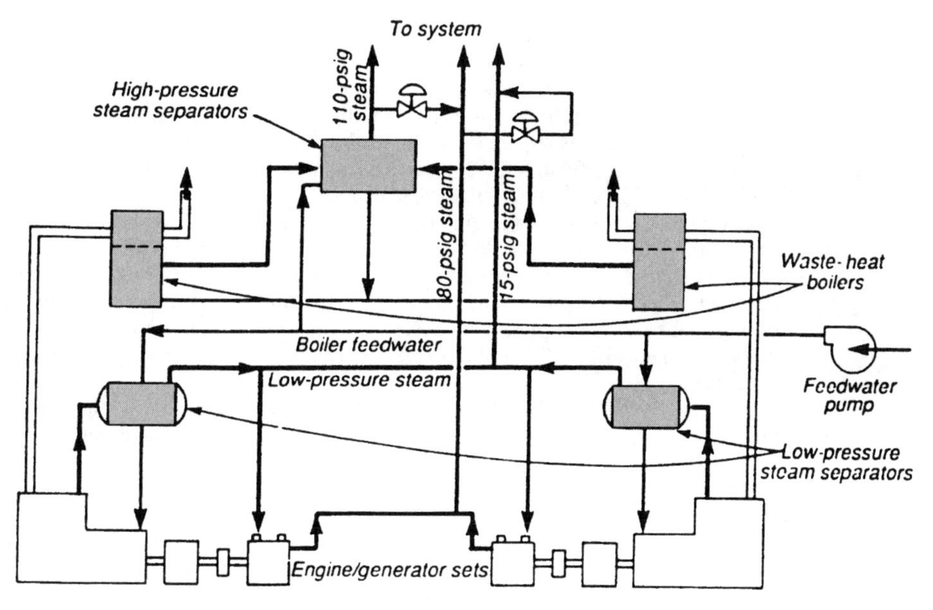

FIGURE 14 System uses common high-pressure steam separator. (*Power.*)

Note: Although Fig. 13 applies to an inlet pressure of 15 lb/in^2 (gage) (103 kPa) and an outlet pressure of 100 lb/in^2 (gage) (689 kPa), the results are accurate enough for an 80-lb/in^2 (gage) (551-kPa) outlet pressure. The required horsepower will be slightly less than that shown.

The two screw compressors will be clutch-connected at the generator end of the two engine-generator sets, as shown in Fig. 14. This diagram also shows the piping layout for the three steam systems—110 (758), 80 (551), and 15-lb/in^2 (gage) (103-kPa). Low-pressure steam separators are used to remove water from the low-pressure cogenerated steam.

In an actual system similar to that shown here, 110-lb/in^2 (gage) (758-kPa) steam is first used in a laundry and in heat-exchange equipment before being reduced to 80 lb/in^2 (gage) (551-kPa) and combined with the compressor discharge flow. In the summer, 15-lb/in^2 (gage) (103-kPa) steam is used to drive a large absorption chiller supplying 600 tons (2112 kW) of refrigeration.

Related Calculations. The recent popularity of reciprocating-engine-based cogeneration has caused a new factor to enter the economic evaluation of these systems: the value of the 15-lb/in^2 (gage) (103-kPa) steam typically recovered from the engine's cooling system. Because uses for 15-lb/in^2 (gage) (103-kPa) steam are limited, boosting the pressure to about 100 lb/in^2 (gage) (689 kPa) multiplies the practical uses of the recovered heat. This concept of pressure boosting is compatible with recent trends in cogeneration to maximize the value of the thermal output through closer coupling of the power-process interface.

Steam recompression has long been an accepted practice in the process industries where large quantities of low-pressure steam can be economically upgraded. A pound (0.45 kg) of steam vented to the atmosphere or condensed represents a loss of about 1000 Btu (1055 kJ) of heat energy. Thus, in many applications, it is less expensive (and more environmentally wise) to boost steam pressure than to produce the equivalent amount in a boiler.

Pressure ratios used to satisfy process requirements are relatively low—about 1.5 to 2. Thermocompressors most economically satisfy these ratios. They use high-pressure steam to boost low-pressure steam to a point in between the two.

Practical limitations on thermocompressors for satisfying higher ratios are two: (1) a large quantity of high-pressure steam is needed, and (2) the heat balance must be such that the steam need not be vented or condensed. For example, if 600-lb/in^2 (gage) (4134-kPa) boiler steam is available to boost 15-lb/in^2 (gage) (103-kPa) steam to 150 lb/in^2 (gage) (1034 kPa), about 12 times the quantity of high-pressure steam is required. So if an engine produces 5000 lb/h (2270 kg/h) of 15-lb/in^2 (gage) (103-kPa) steam, 60,000 lb/h (27,240 kg/h) of high-pressure steam is required to meet a 65,000 lb/h (29,510 kg/h) 150-lb/in^2 (1034-kPa) steam demand.

The typical reciprocating engine rejects 65 to 70 percent of its heat input to exhaust, engine cooling, lube-oil coolers, intercoolers, and radiation. About 20 percent of this heat is recoverable from the exhaust as steam at pressures up to 150 lb/in^2 (gage) (1034 kPa) and beyond, representing about 70 percent of the available heat in the exhaust. Another 28 percent represents engine cooling that can be completely recovered as hot water or as steam at a maximum pressure of 15 lb/in^2 (gage) (103 kPa). Lube-oil heat may also be recoverable, but usually not the intercooler heat.

Although hot water and 15-lb/in^2 (gage) (103-kPa) steam can be used for space heating, as a heat source for absorption refrigeration, or for domestic water heating, there usually is more demand for higher-pressure steam—such as that produced by the engine's exhaust.

As cogeneration systems maximize heat recovery from an internal-combustion engine, it is important to note that the engine exhaust no longer is a simple pipe protruding through the roof of a building. Exhaust explosions, not uncommon to engine operation, thus can be destructive to the often large and complex exhaust systems of cogeneration installations.

For this reason, it is prudent to tighten engine specifications. Partial failure of the ignition system, for example, should not be able to cause a potentially catastrophic exhaust explosion. Further, cross-limiting should be provided when fuel and air are measured at different locations.

Engine manufacturers may require that the exhaust system resist any explosion. A preferred alternative to this requirement is to insist that the engine manufacturer design to minimize exhaust explosions so that the cogeneration plant designer can confidently specify lightweight preformed exhaust ducting.

The data presented in this procedure were drawn from the work of Paul N. Garay, FMC Associates, a division of Parsons Brinckerhoff Quade & Douglas Inc., as reported in *Power* magazine.

COGENERATION HEAT SAVINGS ANALYSIS USING NEED PLOTS

An industrial process plant's heat needs are dominated by distillation. Its heat needs are represented by the fired-heat composite curve (FHCC) shown in Fig. 15. Five distinct heat sources are used in this process plant: two furnaces and steam supplied at three different pressure levels, as shown in Table 10. A gas turbine with the exhaust profile shown in Fig. 15 can supply all the heat needs of the process. Determine the annual fuel savings and payback time.

Calculation Procedure:

1. *Analyze the fired-heat composite curve*

Heat obtained from a cogeneration system generally displaces heat from other sources that can be traced back to direct fuel firing. Even though the fuel may not be fired at the point of use, it is almost always fired somewhere, such as in a boiler.

All these heating needs can be represented by a single FHCC. This curve of heat quantity (H) vs. temperature (T) represents the overall heating duties that, as far as possible, must be satisfied by the cogeneration system. The exhaust-heat profile of the cogeneration system can also be represented by a curve of heat quantity vs. temperature.

To see how such curves are developed, consider the three heat-acceptance profiles in Fig. 16. Profile (a) is a simple constant-heat-capacity profile, typical of heating duties in which no phase

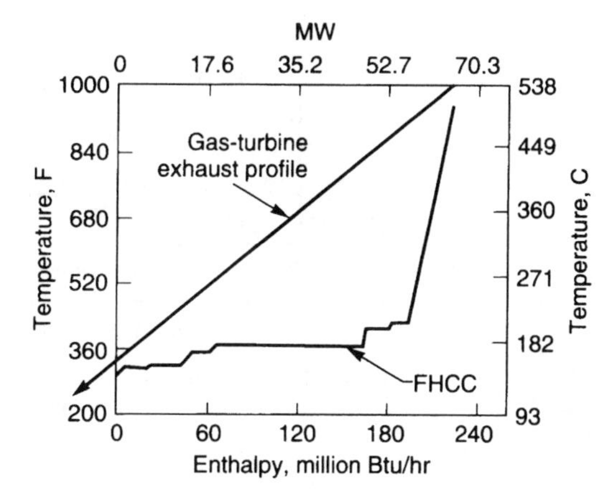

FIGURE 15 Fired-heat composite curve for BTX plant matches well with exhaust profile of gas turbine. (*Power.*)

TABLE 10 Heat Loads

Heat source	Heat load, million Btu/h (MW)
280-lb/in^2 (gage) (1.93-kPa) steam	25.1 (7.35)
140-lb/in^2 (gage) (0.96-kPa) steam	88.1 (25.8)
70-lb/in^2 (gage) (0.48-kPa) steam	23.9 (7.0)
Furnace duty, 242–954°F (117–512°C) nonlinear	51.8 (15.2)
Furnace duty, 356–360°F (180–182°C)	19.4 (5.7)
Total	208.3 (61.05)

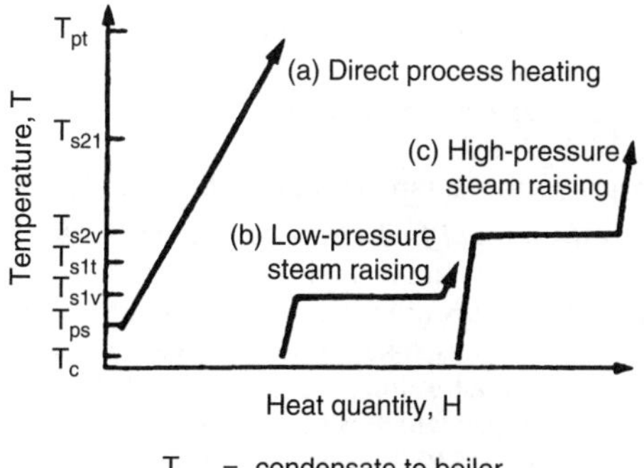

T_c = condensate to boiler
T_{ps} = process supply
T_{s1v} = steam level 1 vaporization
T_{s1t} = steam level 1 target
T_{s2v} = steam level 2 vaporization
T_{s2t} = steam level 2 target
T_{pt} = process target

FIGURE 16 Every heating duty has characteristic heat-acceptance profile. (*Power.*)

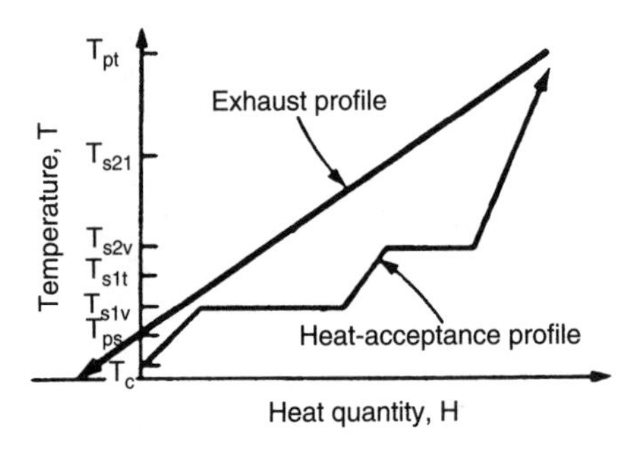

FIGURE 17 Total process heat-acceptance profile is matched with prospective exhaust profile. (*Power.*)

change occurs. The process stream is heated from its supply temperature to its target temperature and the heat load varies linearly between these points.

Profile (b) represents low-pressure steam raising. The first linear part of this curve corresponds to preheat, the horizontal plateau to vaporization, and the final linear section to superheating. Profile (c) represents high-pressure steam raising.

Heat loads, unlike temperature, are additive. Thus it is possible to add the three profiles of Fig. 16 to obtain a combined heat-acceptance profile (Fig. 17). This is the FHCC and it shows total heating needs in terms of the quantity of heat required and the temperature at which it is needed.

The exhaust profile of the proposed cogeneration plant is also shown in Fig. 17. In this case it represents heat in the gas-turbine exhaust and is a straight line, neglecting the effect of condensation. Note that the exhaust profile lies above the heat-acceptance curve, implying that heat can be transferred from the exhaust stream to the process. The vertical separation between the two profiles is a measure of the available thermal driving force for heat transfer. Residual heat in the exhaust system, after the process duties have been satisfied, overhangs the heat-acceptance curve (at the left-hand end) and is lost up the stack.

Composite curves and profile matching provide a convenient way of representing the thermodynamics of heat recovery in cogeneration systems. Implicit within the construction of Fig. 17 are the requirements of the first law of thermodynamics, which demand a heat balance, and those of the second law, which lead to a relationship between the temperatures at which heat is required and the efficiency of the cogeneration system.

Analysis of the FHCC in Fig. 15 shows that all the needed process heat can be supplied by the gas turbine exhaust. Hence, a further evaluation of the proposed cogeneration installation is justified.

2. *Determine the annual fuel saving and payback period*

Assemble the financial data in Table 11 from information available in plant records and estimates. These data show, for *this* proposed cogeneration installation, that the savings that can be obtained are: (*a*) boiler fuel savings, $4.1 million per year; (*b*) credit for cogenerated power, $13.1 million per year; total savings = $4.1 million + $13.1 million = $17.2 million per year. The additional cost is that for the cogeneration gas which is burned in the gas turbine, or $12.2 million. Thus, the net savings will be $17.2 million − $12.2 million = $5.0 million per year.

The payback time = installed cost, $/annual savings, $. Or, payback time = $15.8 million/$5.0 = 3.16, say 3.2 years. This is a relatively short payback time that would be acceptable in most industries.

Related Calculations. Reciprocating internal-combustion engines are also often considered where gas turbines appear to be a possible choice. The reason for this is that about 20 percent of the heat content of fuel fired in a reciprocating engine is rejected in the exhaust gases and the heat-rejection profile is similar to that of a gas turbine. And even more heat, about 30 percent, is removed in cooling water at a temperature of 160°F (71°C) to 250°F (115.6°C). A further 5 percent is available in the

TABLE 11 Parameters Used to Evaluate Cogeneration Process

Displaced furnace fuel cost, $/million Btu	2
Furnace efficiency, %	85
Boiler fuel savings, $ million/yr	4.1
Displaced or exported power, $/kWh	0.045
Gas for cogeneration system, $/million Btu	3.50
Cogeneration gas post, $ million/yr	12.2
Operating hours per year	8000
Power output, MW	36.3
Credit for cogenerated power, $ million/yr	13.1
Cogeneration efficiency, %	78.1
Installed cost, $ million	15.8
Total cash benefit, $ million/yr	5
Estimated payback, years	3

lubricating oil, usually below 180°F (82°C). The heat-rejection profile of a reciprocating engine that closely matches the composite curve of the plant's process is also shown in Fig. 18.

A reciprocating engine has a higher overall efficiency than a gas turbine and therefore generates a greater cash benefit for the plant owner. For the scale of operation we are considering here, it would be necessary to use several engines and the capital cost would be substantially greater than that of a single gas turbine. As a result, payback periods for the two systems are about the same.

Gas turbines are often mated with steam turbines in combined-cycle cogeneration plants. In its basic form the combined-cycle power plant has the gas turbine exhausting into a heat-recovery steam generator (HRSG) that supplies a steam-turbine cycle. This cycle is the most efficient system for generating steam and/or electric power commercially available today. The cycle also has significantly lower capital costs than competing nuclear and conventional fossil-fuel-fired steam/electric stations. Other advantages of the combined-cycle plant are low air emissions, low water consumption, reduced space requirements, and modular units which allow phased-in construction. And from an efficiency standpoint, even in a simple-cycle configuration, gas turbines now exhibit efficiencies of between 30 and 35 percent, comparable to state-of-the-art fossil-fuel-fired power stations.

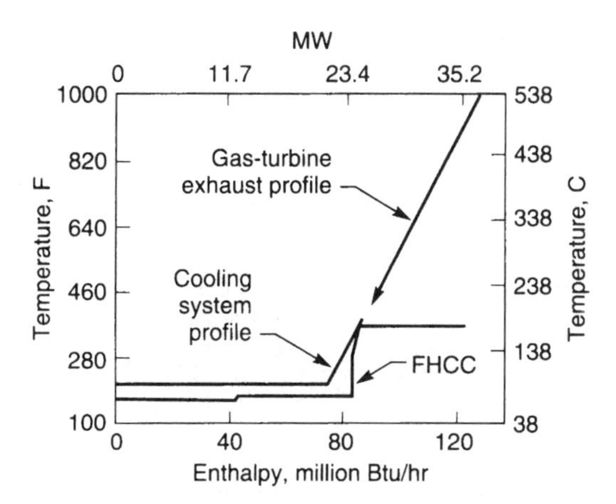

FIGURE 18 Exhaust-heat profile of reciprocating engine is good fit with fired-heat composite curve of textile mill. (*Power.*)

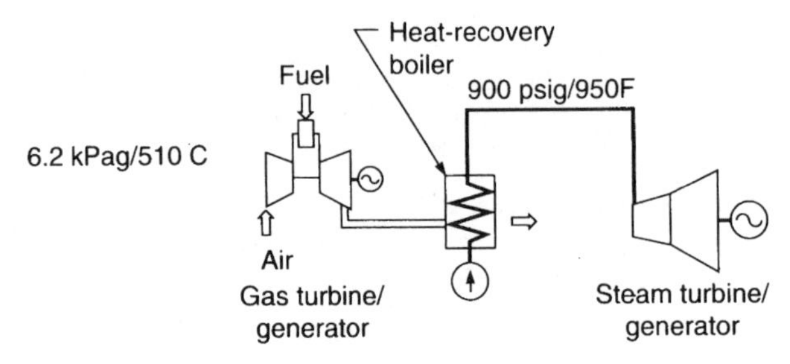

FIGURE 19 Combined-cycle gas-turbine cogeneration with single-pressure non-reheat cycle. (*Power.*)

Cogeneration, which is the simultaneous production of useful thermal energy and electric power from a fuel source, or some variant thereof, is a good match for combined cycles. Experience with cogeneration and combined-cycle power plants has been most favorable. Figures 19 through 21 show a variety of combined-cycle cogeneration plants using reheat in an HRSG to provide steam for a steam-turbine generator. Flexibility is extended as gas turbines, steam turbines, and HRSGs are added to a system. Reheat can improve thermal efficiency and performance by several percentage points, depending on how it is integrated into the combined cycle.

Aeroderivative gas turbines, as part of a combined cycle, increasingly are finding application in cogeneration in the under 100-MW capacity range. Cogeneration has the airline and defense industries to thank for the rapid development of high-efficiency, long-running gas turbines at extremely low research cost.

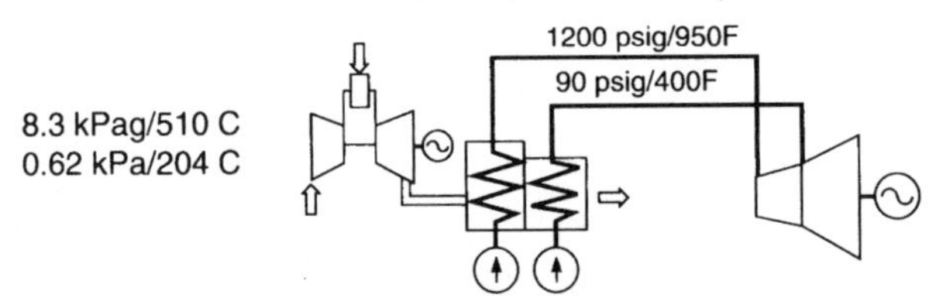

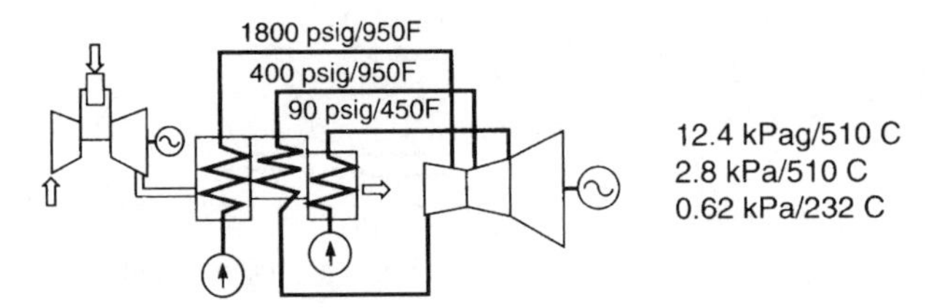

FIGURE 20 Combined-cycle gas-turbine cogeneration with dual-pressure non-reheat cycle. (*Power.*)

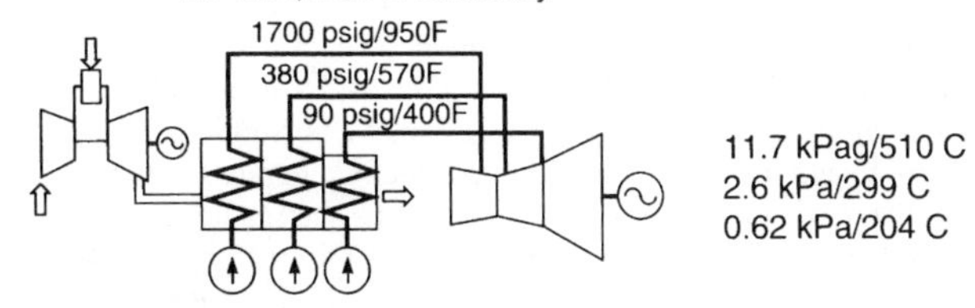

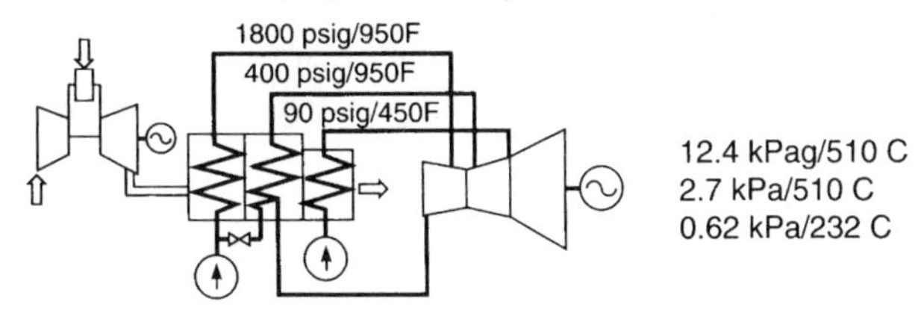

FIGURE 21 Combined-cycle gas-turbine cogeneration with triple-pressure non-reheat cycle. (*Power.*)

The new large gas turbines have exhaust temperatures high enough to justify reheat in the steam cycle without supplementary firing in a boiler. Depending on how the reheat cycle is configured, thermal performance at rated conditions can vary by up to three percentage points.

The Public Utilities Regulatory Policies Act (PURPA) passed by Congress to help manage energy includes incentives for efficient cogeneration systems. Cogeneration plants are allowed to sell power to local electric utilities to increase the return on investment earned from cogeneration.

A whole new energy-saving industry—termed nonutility generation (NUG)—has developed. At this writing NUG plants in the 200- to 300-MW range are common. And the pipeline industry which supplies natural-gas fuel for gas turbines is being restructured under the Federal Energy Regulatory Commission (FERC). Lower fuel costs are almost certain to result.

While lower electricity and energy costs are in the offing, these must be balanced against increased environmental requirements. The Clean Air Act Amendments of 1990 require better cleaning of stack emissions to provide a cleaner atmosphere. Yet this same 1990 act allows utilities to meet the required sulfur standard by installing suitable scrubber cleaning equipment, or by switching to a low-sulfur fuel.

A utility may buy—from another utility which exceeds the required sulfur standard—allowances to exhaust sulfur to the atmosphere. Each allowance permits a utility to emit 1 ton (tonne) of sulfur to the atmosphere. Public auctions of these allowances are now being held periodically by the Chicago Board of Trade.

Active discussions are under way at present over the suitability of selling sulfur allowances. Some opponents to sulfur pollution allowances believe that their use will delay the cleanup that ultimately must take place. Further, these opponents say, the pollution allowances delay the installation of sulfur-removal equipment. Meanwhile, sulfuric acid rain (also called acid rain) continues to plague communities in the path of a utility's sulfur effluent.

Challenging the above view is the Environmental Defense Fund. Its view is that there are too few allowances available to prevent the ultimate cleanup required by law.

The calculation data in this procedure are the work of A. P. Rossiter and S. H. Chang, ICI/Tensa Services as reported in *Power* magazine, along with John Makansi, executive editor, reporting in the same publication. Data on environmental laws are from the cited regulatory agency or act.

CAPITAL COST OF COGENERATION HEAT-RECOVERY BOILERS

Use the Foster-Pegg method to estimate the cost of the gas-turbine heat-recovery boiler system shown in Fig. 22 based on these data: The boiler is sized for a Canadian Westinghouse 251 gas turbine; the boiler is supplementary fired and has a single gas path; natural gas is the fuel for both the gas turbine and the boiler; superheated steam generated in the boiler at 1200 lb/in^2 (gage) (8268 kPa) and 950°F (510°C) is supplied to an adjacent chemical process facility; 230-lb/in^2 (gage) (1585-kPa) saturated steam is generated for reducing NO_x in the gas turbine; steam is also generated at 25 lb/in^2 (gage) (172 kPa) saturated for deaeration of boiler feedwater; a low-temperature economizer preheats underaerated feedwater obtained from the process plant before it enters the deaerator. Estimate boiler costs for two gas-side pressure drops: 14.4 in (36.6 cm) and 10 in (25.4 cm), and without, and with, a gas bypass stack. Table 12 gives other application data. *Note:* Since cogeneration will account for a large portion of future power generation, this procedure is important from an environmental standpoint. Many of the new cogeneration facilities planned today consist of gas turbines with heat-recovery boilers, as does the plant analyzed in this procedure.

Calculation Procedure:

1. *Determine the average LMTD of the boiler*

The average log mean temperature difference (LMTD) of a boiler is indicative of the relative heat-transfer area, as developed by R. W. Foster-Pegg, and reported in *Chemical Engineering* magazine. Thus, $LMTD_{avg} = Q_t/C_t$, where Q_t = total heat exchange rate of the boiler, Btu/s (W); C_t = conductance, Btu/s · °F (W). Substituting, using data from Table 12, $LMTD_{avg}$ = 81,837/1027 = 79.7°F (26.5°C).

2. *Compute the gas pressure drop through the boiler*

The gas pressure drop, ΔP inH_2O (cmH_2O) = $5C_t/G$, where G = gas flow rate, lb/s (kg/s). Substituting, ΔP = 5(1027/355.8) with a gas flow of 355.8 lb/s (161.5 kg/s), as given in Fig. 12; then ΔP = 14.4 inH_2O (36.6 cmH_2O). With a stack and inlet pressure drop of 3 inH_2O (7.6 cmH_2O) and a supplementary-firing pressure drop of 3 inH_2O (7.6 cmH_2O) given by the manufacturer, or determined from previous experience with similar designs, the total pressure drop = 14.4 + 3.0 + 3.0 = 20.4 inH_2O (51.8 cmH_2O).

3. *Compute the system costs*

The conductance cost component, $Cost_{ts}$, is given by $Cost_{ts}$, in thousands of \$ = $5{,}65[(C_{sh}^{0.8} + C_1^{0.8} + \ldots + (C_n^{0.8}) + 2(C_n^{0.8})]$, where C = conductance, Btu/s · F(W), and the subscripts represent the boiler elements listed in Table 12. Substituting, $Cost_{ts}$ = 5.65(404.37) = \$2,285,000 in 1985 dollars. To update to present-day dollars, use the ratio of the 1985 *Chemical Engineering* plant cost index (310) to the current year's cost index thus: Current cost = (today's plant cost index/310)(cost computed above).

The steam-flow cost component, $Cost_w$, in thousands of \$ = $4.97(W_1 + W_2 + \ldots + W_n)$, where $Cost_w$ = cost of feedwater, \$; W = feedwater flow rate, lb/s (kg/s); the subscripts 1, 2, and n denote different steam outputs. Substituting, $Cost_w$ = 4.97(59.14) = \$294,000 in 1985 dollars, with a total feedwater flow of 59.14 lb/s (26.9 kg/s).

The cost for gas flow includes connecting ducts, casing, stack, etc. It is proportional to the sum of the separate gas flows, each raised to the power of 1.2. Or, cost of gas flow, $Cost_g$, in thousands of \$ = $0.236(G_1^{1.2} + G_2^{1.2} + \ldots + G_n^{1.2})$. Substituting, $Cost_g = 0.236(355.8)^{1.2}$ = \$272,000 with a gas flow of 355.8 lb/s (161.5 kg/s) and no bypass stack.

The cost of a supplementary-firing system for the heat-recovery boiler in 1985 dollars is additional to the boiler cost. Typical fuels for supplementary firing are natural gas or No. 2 fuel oil, or both. The supplementary-firing system cost, $Cost_f$, in thousands of \$ = $B/1390 + 30N + 20$, where B = boiler firing capacity in Btu (kJ) high heating value; N = number of fuels burned. For this installation with *one* fuel, $Cost_f$ = 16,980/1390 + 30 + 20 = \$62,000, rounded off. In this equation the 16,980 Btu/s (17,914 kJ/s) is the high heating value of the fuel and N = 1 since only *one* fuel is used.

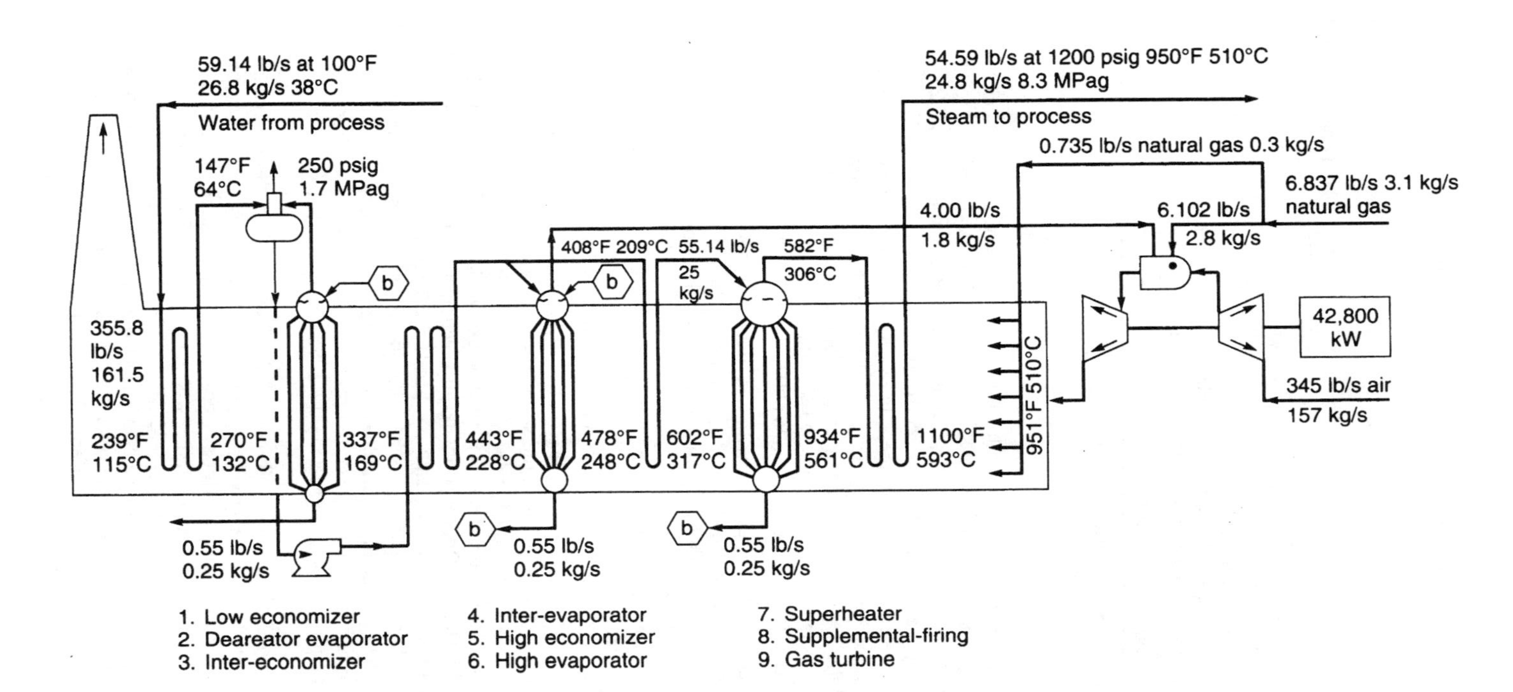

FIGURE 22 Gas-turbine and heat-recovery boiler system. (*Chemical Engineering*.)

TABLE 12 Data for Heat-Recovery Boiler*

	LMTD, °F	Q, Btu/s	C, Btu/s · °F	$C^{0.8}$ Btu/s · °F
Superheater	237	16,098	67.92	29.22
High evaporator	116	32,310	278.53	90.34
High economizer	40	11,583	290.3	93.39
Inter-evaporator	50.5	3,277	64.89	28.17
Inter-economizer	37	9,697	169.82	60.81
Deaerator evaporator	46	6,130	134.43	50.44
Low economizer	131	2,742	20.93	11.39
Additional for superheater material				29.22
Additional for low-economizer material				11.39
Total	81,837		1,027	404.37

*See procedure for SI values in this table.
Chemical Engineering.

The total boiler cost (with base gas ΔP and no gas bypass stack) = total material cost + erection cost, or \$2,285,000 + 294,000 + 272,000 + 62,000 = \$2,913,000 for the materials. A *budget estimate* for the cost of erection = 25 percent of the total material cost, or 0.25 × \$2,913,000 = \$728,250. Thus, the budget estimate for the erected cost = \$2,913,000 + \$728,250 = \$3,641,250.

The estimated cost of the entire system—which includes the peripheral equipment, connections, startup, engineering services, and related erection—can be approximated at 100 percent of the cost of the major equipment delivered to the site, but not erected. Thus, the total cost of the boiler ready for operation is approximately twice the cost of the major equipment material, or 2(boiler material cost) = 2(\$2,913,000) = \$5,826,000.

4. *Determine the costs with the reduced pressure drop*

The second part of this analysis reduces the gas pressure drop through the boiler to 10 inH_2O (25.4 cmH_2O). This reduction will increase the capital cost of the plant because much of the equipment will be larger.

Proceeding as earlier, the total pressure drop, $\Delta P = 10 + 3 + 3 = 16$ inH_2O (40.6 cmH_2O). The pressure drop for normal solidity (i.e., normal tube and fin spacing in the boiler) is $\Delta P_1 = 14.4$ inH_2O (36.6 cmH_2O). For a different pressure drop, ΔP_2, the surface cost, C_s (\$), is at ΔP_2, $C_s = [1.67(\Delta P_1/\Delta P_2)^{0.28} - 0.67)$ (C_s at P_1), Substituting, $C_s = 1.67(14.4/10)^{0.28} - 0.67 = 1.18 \times$ base cost from above. Hence, the surface cost for a pressure drop of 10 inH_2O (25.4 cmH_2O) = 1.18 × (\$2,285,000) = \$2,696,300.

The total material cost will then be \$2,696,300 + \$272,000 + \$62,000, using the data from above, or \$3,324,300. Budget estimate for erection, as before = 1.25(\$3,324,300) = \$4,155,375. And the estimated system cost, ready to operate = 2(\$3,324,300) = \$6,648,600.

Adding for a gas bypass stack, the gas-flow component is the same as before, \$272,000. Then the budget estimate of the installed cost of the gas bypass stack = 1.25(\$272,000) = \$340,000. And the total cost of the boiler ready for operation at a gas-pressure drop of 10 inH_2O (25.4 cmH_2O) with a gas bypass stack = 2(53,324,000 + \$272,000) = \$7,192,600.

Related Calculations. To convert the costs found in this procedure to current-day costs, assume that the *Chemical Engineering* plant cost index today is 435, compared to the 1985 index of 310. Then, today's cost, \$ = (today's cost index/1985 cost index)(1985 plant or equipment cost, \$). Thus, for the first installation, today's cost = (435/310)(\$5,826,000) = \$8,175,194. And for the second installation, today's cost = (435/310)(\$7,192,600) = \$10,092,842.

Boilers for recovering exhaust heat from gas turbines are very different from conventional boilers, and their cost is determined by different parameters. Because engineers are becoming more involved with cogeneration, the differences are important to them when making design and cost estimates and decisions.

In a conventional boiler, combustion air is controlled at about 110 percent of the stoichiometric requirement, and combustion is completed at about 3000°F (1649°C). The maximum temperature of the water (i.e., steam) is 1000°F (538°C), and the temperature difference between the gas and water is about 2000°F (1093°C). The temperature drop of the gas to the stack is about 2500°F (1371°C), and the gas/water ratio is consistent at about 1.1.

By contrast, the exhaust from a gas turbine is at a temperature of about 1000°F (538°C), and the difference between the gas and water temperatures averages 100°F (56°C). The temperature drop of the gas to the stack is a few hundred degrees, and the gas/water ratio ranges between 5 and 10. Because the airflow to a heat-recovery boiler is fixed by the gas turbine, the air varies from 400 percent of the stoichiometric requirement of the fuel to the turbine (unfired boiler) to 200 percent if the boiler is supplementary fired.

In heat-recovery boilers, the tubes are finned on the outside to increase heat capture. Fins in conventional boilers would cause excessive heat flux and overheating of the tubes. Although the lower gas temperatures in heat-recovery boilers allow gas enclosures to be uncooled internally insulated walls, the enclosures in conventional boilers are water-cooled and refractory-lined.

Because the exhaust from a gas turbine is free of particles and contaminants, gas velocities past tubes can be high, and fin and tube spacings can be close, without erosion or deposition. Because the products of combustion in a conventional boiler may contain sticky residues, carbon, and ash particles, tube spacing must be wider and gas velocities lower. Because of its configuration and absence of refractories, the heat-recovery boiler used with gas turbines can be shop-fabricated to a greater extent than conventional boilers.

These differences between conventional and heat-recovery boilers result in different cost relationships. With both operating on similar clean fuels, a heat-recovery boiler will cost more per pound of steam and less per square foot (m^2) of surface area than a conventional boiler. The cost of a heat-recovery boiler can be estimated as the sum of three major parameters, plus other optional parameters. Major parameters are: (1) the capacity to transfer heat ("conductance"), (2) steam flow rate, and (3) gas flow rate. Optional parameters are related to the optional components of supplementary firing and a gas bypass stack. The optional parameters will vary by the size of the installation, its use, and expected life.

This procedure is the work of R. W. Foster-Pegg, Consultant, as reported in *Chemical Engineering* magazine. Note that the costs computed by the given equations are in 1985 dollars. Therefore, they must be updated to current costs using the Chemical Engineering plant cost index.

CHLOROFLUOROCARBON ASPECTS OF CENTRAL CHILLED-WATER SYSTEMS DESIGN

Choose a suitable storage tank size and capacity for a thermally stratified water-storage system for a large-capacity thermal-energy storage system for off-peak air conditioning for these conditions: Thermal storage capacity required = 100,000 ton-h (35,169 kWh); difference between water inlet and outlet temperatures = T = 20°F (11.1°C); allowable nominal soil bearing load in one location is 2500 lb/ft^2 (119.7 kPa); in another location 4000 lb/ft^2 (191.5 kPa). Compare tank size for the two locations.

Calculation Procedure:

1. ***Compute the required tank capacity in gallons (liters) to serve this system***

Use the relation $C = 1800S/\Delta T$, where C = required tank capacity, gal; S = system capacity, ton-h; ΔT = difference between inlet and outlet temperature, °F (°C). For this installation, C = 1800(100,000)/20 = 9,000,000 gal (34,065 m^3).

2. ***Determine the tank height and diameter for the allowable soil bearing loads***

Depending on the proposed location of the storage tank, either the height or diameter may be a restricted dimension. Thus, tank height may be restricted by local zoning laws or possible interference

TABLE 13 Typical Thermal Storage Tank Sizes, Heights, Capacities*

Rated thermal energy storage capacity, ton-hours			Tank shell height (and nominal soil bearing load)					
			64-ft shell height (4000-lb/ft^2 soil)		40-ft shell height ($2500\ \text{lb/ft}^2$)		24-ft shell height (1500-lb/ft^2 soil)	
ΔT, 10°F	ΔT, 15°F	ΔT, 20°F	Gross volume, gal	Tank diameter, ft	Gross volume, gal	Tank diameter, ft	Gross volume, gal	Tank diameter, ft
40,000	60,000	80,000	6,880,000	135	7,200,000	175	7,880,000	236
50,000	75,000	100,000	8,610,000	151	9,000,000	196	9,850,000	264
60,000	90,000	120,000	10,330,000	166	10,800,000	214	11,820,000	290

Chicago Bridge & Iron Company.
*See calculation procedures for SI values.

with aircraft landing or takeoff patterns. Tank diameter may be restricted by the ground area available. And the allowable nominal soil bearing load will determine if the required amount of water can be stored in one tank or if more than one tank will be required.

Starting with 2500-lb/ft^2 (119.7-kPa) bearing-load soil, assume a standard tank height of 40 ft (12.2 m). Then, the required tank volume will be $V = 0.134C$, where $0.134 = \text{ft}^3/\text{gal}$; or $V = 0.134(9{,}000{,}000) = 1{,}206{,}000\ \text{ft}^3$ ($34{,}130\ \text{m}^3$). The tank diameter is $d = (4V/\pi h)^{0.5}$, where d = diameter in feet (m). Or $d = [4(1{,}206{,}000/\pi 40)]^{0.5} = 195.93$ ft; say 196 ft (59.7 m). This result is consistent with the typical sizes, heights, and capacities used in actual practice, as shown in Table 13.

Checking the soil load, the area of the base of this tank is $A = \pi d^2/4 = \pi(196)^2/4 = 30{,}172\ \text{ft}^2$ ($2803\ \text{m}^2$). The weight of the water in the tank is $W = 8.35C = 75{,}150{,}000$ lb (34,159 kg). This will produce a soil bearing pressure of $p = W/A\ \text{lb/ft}^2$ (kPa). Or, $p = 75{,}150{,}000/30{,}172 = 2491\ \text{lb/ft}^3$ (119.3 kPa). This bearing load is within the allowable nominal specified load of $2500\ \text{lb/ft}^2$.

Where a larger soil bearing load is permitted, tank diameter can be reduced as the tank height is increased. Thus, using a standard 64-ft (19.5-m) high tank with the same storage capacity, the required diameter would be $d = [4(1{,}206{,}000)/\pi 64)]^{0.5} = 154.9$ ft (47.2 m). Soil bearing pressure will then be $W/A = 75{,}140{,}000/[\pi(154.9)^2/4] = 3987.8\ \text{lb/ft}^2$ (190.9 kPa). This is within the allowable soil bearing load of $4000\ \text{lb/ft}^2$ (191.5 kPa).

By reducing the storage capacity of the tank 4 percent to 8,610,000 gal ($32{,}589\ \text{m}^3$), the diameter of the tank can be made 151 ft (46 m). This is a standard dimension for 64-ft (19-m) high tanks with a 4000-lb/ft^2 (191.5-kPa) soil bearing load.

Related Calculations. Thermal energy storage (TES) is environmentally desirable because it uses heating, ventilating, and air-conditioning (HVAC) equipment and a storage tank to store heated or cooled water during off-peak hours, allowing more efficient use of electric generating equipment. The stored water is used to serve HVAC or industrial process loads during on-peak hours.

To keep investment, operating, and maintenance costs low, one storage tank can be used to store both cool and warm water. Thermal stratification permits a smaller investment in the tank, piping, insulation, and controls to produce a higher-efficiency system. Lower-density warm water is thermally stratified from higher-density cool water without any mechanical separation in a full storage tank.

While systems using 200,000 gal ($757\ \text{m}^3$) of stored water are feasible, the usual minimum size storage tank is 500,000 gal ($1893\ \text{m}^3$). Tanks as large as 4.4 million gal ($16{,}654\ \text{m}^3$) are currently in use in TES for HVAC and process needs. TES is also used for schools, colleges, factories, and a variety of other applications. Where chlorofluorocarbon (CFC)-based refrigeration systems must be replaced with less, environmentally offensive refrigerants, TES systems can easily be modified because the chiller (Fig. 23) is a simple piece of equipment.

As an added environmental advantage, the stored water in TES tanks can be used for fire protection. The full tank contents are continuously available as an emergency fire water reservoir. With

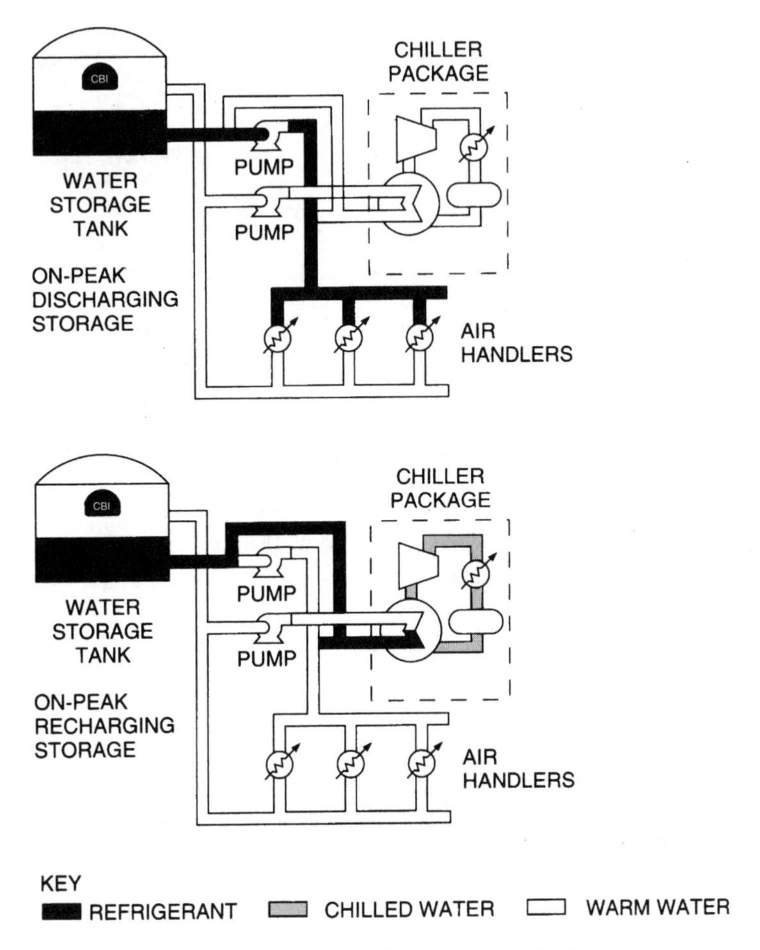

FIGURE 23 On-peak and off-peak storage discharging and recharging of thermally stratified water-storage system. (*Chicago Bridge & Iron Company.*)

such a water reserve, the capital costs for fire-protection equipment can be reduced. Likewise, fire-insurance premiums may also be reduced. Where an existing fire-protection water tank is available, it may be retrofitted for TES use.

Above-ground storage tanks (Fig. 24) are popular in TES systems. Such tanks are usually welded steel, leak-free with a concrete ringwall foundation. Insulated to prevent heat gain or loss, such tanks may have proprietary internal components for proper water distribution and stratification.

Some TES tanks may be installed partially, or fully, below grade. Before choosing partially or fully below-grade storage, the following factors should be considered: (1) system hydraulics may be complicated by a below-ground tank; (2) the tank must be designed for external pressure, particularly when the tank is empty; (3) soil and groundwater conditions may make the tank more costly; (4) local and national regulations for underground tanks may increase costs; (5) the choice of water-treatment methods may be restricted for underground tanks; (6) the total cost of an underground tank may be twice that of an above-ground tank.

The data and illustrations for this procedure were obtained from the Strata-Therm Thermal Systems Group of the Chicago Bridge & Iron Company.

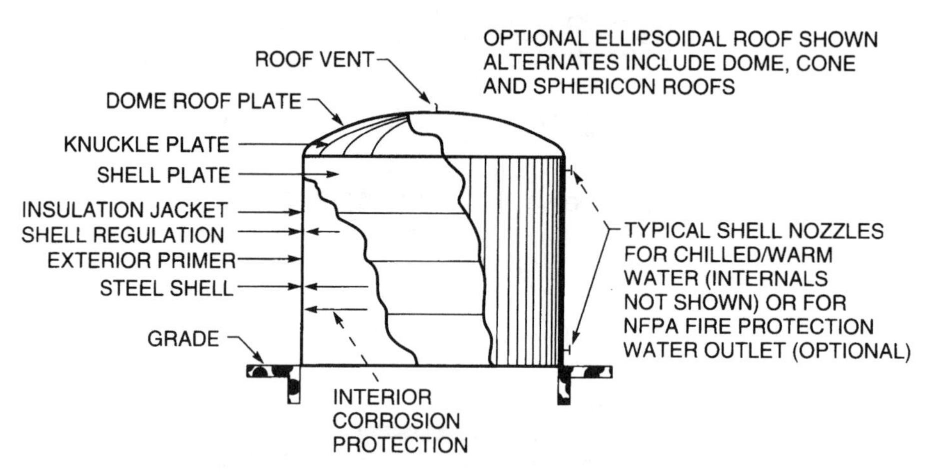

FIGURE 24 Typical chilled-water thermal-energy storage tank installation. (*Chicago Bridge & Iron Company.*)

SIZING AN ELECTROSTATIC PRECIPITATOR FOR AIR-POLLUTION CONTROL

A duct-type electrostatic precipitator is to be used to clean 100,000 actual ft^3/min (47.2 actual m^3/s) of an industrial gas stream containing particulates. The proposed design of the precipitator consists of three bus sections (fields) arranged in series, each having the same amount of collection surface. The inlet loading has been measured as 17.78 gr/ft^3 (628 gr/m^3), and a maximum outlet loading of 0.08 gr/ft^3 (2.8 gr/m^3) (both volumes corrected to dry standard conditions and 50 percent excess air) is allowed by the local air-pollution regulations. The drift velocity for the particulates has been experimentally determined in a similar installation, with the following results:

First section (inlet): 0.37 ft/s (0.11 m/s)

Second section (middle): 0.35 ft/s (0.107 m/s)

Third section (outlet): 0.33 ft/s (0.10 m/s)

Calculate the total collecting surface required. And find the total mass flow rate of particulates captured in each section.

Calculation Procedure:

1. ***Calculate the required total collection efficiency E based on the given inlet and outlet loadings*** The equation is

$$E = 1 - (\text{outlet loading})/(\text{inlet loading})$$

Thus,

$$E = 1 - 0.08/17.78 = 0.9955, \text{ or } 99.55\%$$

2. *Calculate the average drift velocity w*
Thus,

$$w = (0.37 + 0.35 + 0.33)/3 = 0.35 \text{ ft/s } (0.107 \text{ m/s}).$$

3. *Calculate the total surface area required*
Use the Deutsch-Anderson equation:

$$E = 1 - \exp(-wA/q)$$

where E = collection efficiency
w = average drift velocity
A = required surface area
q = gas flow rate

Rearrange the equation as follows:

$$A = \ln(1 - E)/(w/q)$$

For consistency between w and q, convert q from ft^3/min to ft^3/s:

$$100{,}000/60 = 1666.7 \text{ ft}^3\text{/s}$$

Then

$$A = -\ln(1 - 0.9955)/(0.35/1666.7) = 25.732 \text{ ft}^2 \ (2393 \text{ m}^2)$$

4. *Calculate the collection efficiencies of each section*
Use the Deutsch-Anderson equation (from step 3) directly. For the first section, $E_1 = 1 - \exp[-(25{,}732)(0.37)/(3)(1666.7)] = 0.851$. Similarly, E_2 for the second section is found to be 0.835, and E_3 for the third section 0.817.

5. *Calculate the mass flaw rate $\dot{m}$ of particulates captured by each section*
For the first section, the equation is

$$\dot{m} = (E_1)(\text{inlet loading})(q)$$

Thus,

$$\dot{m} = (0.851)(17.78)(100{,}000) = 1.513 \times 10^6 \text{ gr/min} = 216.1 \text{ lb/min}(1.635 \text{ kg/s})$$

For the second section, the equation is

$$\dot{m} = (1 - E_1)(E_2)(\text{inlet loading})(q)$$

Thus,

$$\dot{m} = (1 - 0.851)(0.835)(17.78)(100{,}000) = 2.212 \times 10^5 \text{ gr/min}$$
$$= 31.6 \text{ lb/min } (0.239 \text{ kg/s})$$

And for the third section, the equation is

$$\dot{m} = (1 - E_1)(1 - E_2)(E_3)(\text{inlet loading})(q)$$

which yields 5.10 lb/min (0.039 kg/s).

The total mass captured is the sum of the amounts captured in each section, i.e., 252.8 lb/min (1.91 kg/s). It is not surprising that a full 85 percent of the mass is captured in the first section.

This procedure is the work of Louis Theodore, Eng. Sc. D., Professor, Department of Chemical Engineering, Manhattan College, Bronx, NY, as presented in Chopey, *Handbook of Chemical Engineering Calculation*, 3rd ed., McGraw-Hill, 2004.

EXPLOSIVE-VENT SIZING FOR INDUSTRIAL BUILDINGS

Choose the size of explosion vents to relieve safely the maximum allowable overpressure of 0.75 lb/in^2 (5.2 kPa) in the building shown in Fig. 25 for an ethane/air explosion. Specify how the vents will be distributed in the structure.

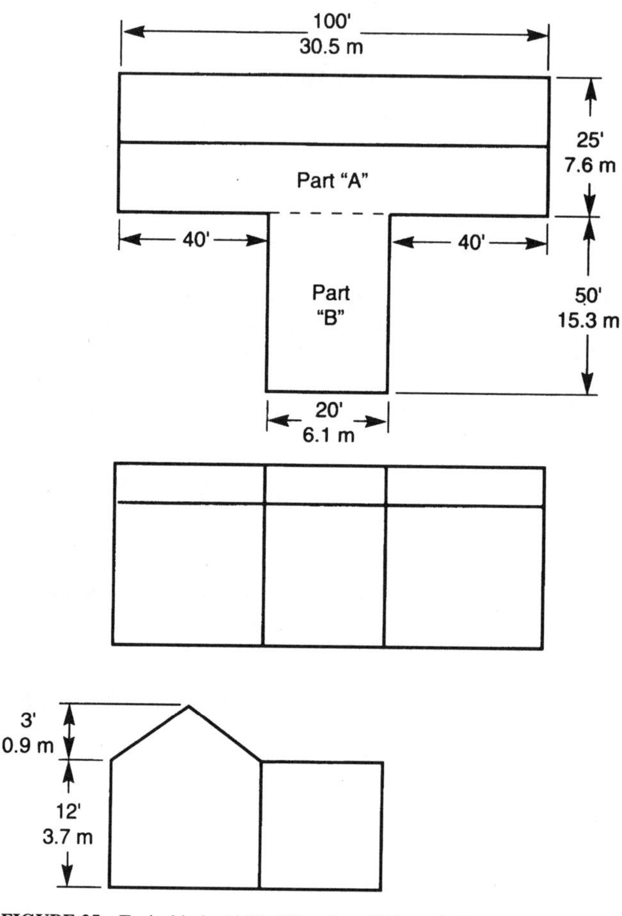

FIGURE 25 Typical industrial building for which explosion vents are sized.

Calculation Procedure:

1. *Determine the total internal surface area of Part A of the building*
Using normal length and width area formulas for Part A, we have: Building floor area = 100 × 25 = 2500 ft^2 (232.3 m^2); front wall area = 12 × 100 = 1200 ft^2 – 12 × 20 = 960 ft^2 (89.2 m^2); rear wall area = 12 × 100 = 1200 ft^2 (111.5 m^2); end wall area = 2 × 25 × 12 + 2 × 25 × 3/2 = 675 ft^2 (62.7 m^2); roof area = 2 × 3 × 100 = 600 ft^2 (55.7 m^2). Thus, the total internal surface area of Part A of the building is 2500 + 960 + 1220 + 600 = 5935 ft^2 (551.4 m^2).

2. *Determine the total internal surface area of Part B of the building*
Using area formulas, as before: Floor area = 50 × 20 = 1000 ft^2 (92.9 m^2); side wall area = 2 × 50 × 12 = 1200 ft^2 (111.5 m^2); front wall area = 20 × 12 = 240 ft^2 (22.3 m^2); roof area = 50 × 20 = 1000 ft^2 (92.9 m^2); total internal surface area of Part B is 1000 + 1200 + 240 + 1000 = 3440 ft^2 (319.6 m^2).

3. *Compute the vent area required*
Using the relation $A_{\upsilon} = CA_s/(P_{red})^{0.5}$, where A_{υ} = required vent area, m^2; C = deflagration characteristic of the material in the building, (kPa)$^{0.5}$, from Table 14: A_s = internal surface area of the structure to be protected, m^2. For this industrial structure, A_{υ} = 0147(551.4 + 319.6)/(5.17)$^{0.5}$ = 180.1-m^2 (1939-ft^2) total vent area.

The required vent area should be divided proportionately between Part A and Part B of the building, or Part A vent area = 180.1(551.4/871.0) = 114 m^2 (11227 ft^2); Part B vent area = 180.1(319.6/871.0) = 66.1 m^2 (712 ft^2).

The required vent area should be distributed equally over the external wall and roof areas in each portion of the building. Before making a final choice of the vent areas to be used, the designer should consult local and national fire codes. Such codes may require different vent areas, depending on a variety of factors such as structure location, allowable overpressure, and gas mixture.

Related Calculations. This procedure is the work of Tom Swift, a consultant reported in *Chemical Engineering*. In his explanation of his procedure he points out that the word *explosion* is an imprecise term. The method outlined above is intended for those explosions known as deflagrations—exothermic reactions that propagate from burning gases to unreacted materials by conduction, convection, and radiation. The great majority of structural explosions at chemical plants are deflagrations.

TABLE 14 Parameters for Vent Area Equation*

Material	$\frac{S_u \rho_u}{G'}$	$\frac{P_{max}}{P_0}$
Methane	1.1×10^{-3}	8.33
Ethane	1.2×10^{-3}	9.36
Propane	1.2×10^{-3}	9.50
Pentane	1.3×10^{-3}	9.42
Ethylene	1.9×10^{-3}	9.39

Material	C, (kPa)$^{1/2}$
Methane	0.41
Ethane	0.47
Propane	0.48
Pentane	0.51
Ethylene	0.75
ST 1 dusts	0.26
ST 2 dusts	0.30

*Article cited in *Related calculation*.

The equation used in this procedure is especially applicable to "low-strength" structures widely used to house chemical processes and other manufacturing operations. This equation is useful for both gas and dust deflagrations. It applies to the entire subsonic venting range. Nomenclature for Table 14 is given as follows:

A_s Internal surface area of structure to be protected, m^2
A_v Vent area, m^2
B Dimensionless constant
C Deflagration characteristic, $(kPa)^{1/2}$
C_D Discharge coefficient
G' Maximum subsonic mass flux through vent, $kg/m^2 \cdot s$
P_f Overpressure, kPa
P_{max} Maximum deflagration pressure in a sealed spherical vessel, kPa
P_0 Initial (ambient) pressure, kPa
P_{red} Maximum reduced explosion pressure that a structure can withstand, kPa
S_u Laminar burning velocity, m/s
γ_b Ratio of specific heats of the combustion gases
ρ_u Density of the unburnt gases, kg/m^3
λ Turbulence enhancement factor

With increased interest in the environment by regulatory authorities, greater attention is being paid to proper control and management of industrial overpressures. Explosion vents that are properly sized will protect both the occupants of the building and surrounding structures. Therefore, careful choice of explosion vents is a prime requirement of sensible environmental protection.

VENTILATION DESIGN FOR INDUSTRIAL BUILDING ENVIRONMENTAL SAFETY

Determine the ventilation requirements to maintain interior environmental safety of a pump and compressor room in an oil refinery in a cool-temperate climate. Floor area of the pump and compressor room is 2000 ft^2 (185.8 m^2) and room height is 15 ft (4.6 m); gross volume = 30,000 ft^3 (849 m^3). The room houses two pumps—one of 150 hp (111.8 kW) with a pumping temperature of 350°F (177°C), and one of 75 hp (55.9 kW) with a pumping temperature of 150°F (66°C). Also housed in the room is a 1000-hp (745.6-kW) compressor and a 50-hp (37.3-kW) compressor.

Calculation Procedure:

1. *Determine the hp-deg for the pumps*

The hp-deg = pump horsepower × pumping temperature. For these pumps, the total hp-deg = (150 × 350) + (75 × 150) = 63,750 hp-deg (19.789 kW-deg). Enter Fig. 26 on the left axis at 63,750 and project to the diagonal line representing the ventilation requirements for pump rooms in cool-temperate climates. Then extend a line vertically downward to the bottom axis to read the air requirement as 7200 ft^3/min (203.8 m^3/min).

The compressors require a total of 1050 hp (782.9 kW). Enter Fig. 26 on the right-hand axis at 1050 and project horizontally to cool-temperate climates for compressor and machinery rooms. From the intersection with this diagonal project vertically to the top axis to read 2200 ft^3/min (62.3 m^3/min) as the ventilation requirement.

Since the ventilation requirements of pumps and compressors are additive, the total ventilation-air requirement for this room is 7200 + 2200 = 9400 ft^3/min (266 m^3/min).

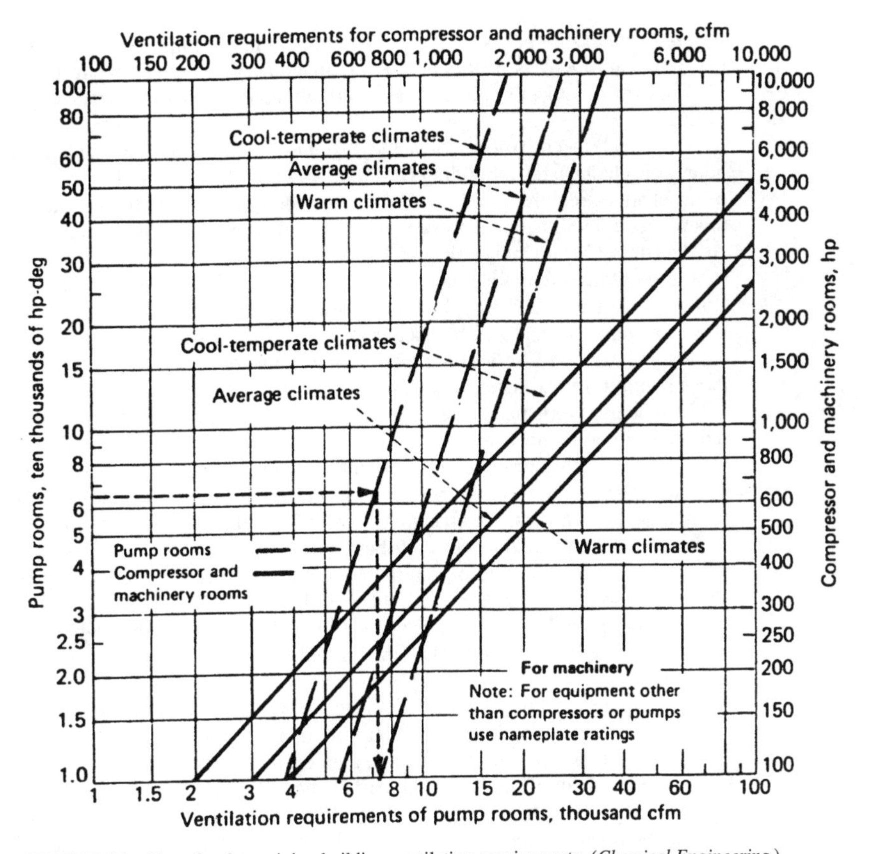

FIGURE 26 Chart for determining building ventilation requirements. (*Chemical Engineering.*)

2. *Check to see if the computed ventilation flow meets the air-change requirements*
Use the relation $N = 60F/V$, where N = number of air changes per hour; F = ventilating-air flow rate, ft^3/min (m^3/min); V = room volume, ft^3 (m^3). Using the data for this room, $N = 60(9400)/30{,}000 = 18.8$ air changes per hour.

Figure 26 is based on a minimum of 10 air changes per hour for summer and 3 air changes per hour for winter. Since the 18.8 air changes per hour computed exceeds the minimum of 10 changes per hour on which the chart is based, the computed air flow is acceptable.

In preparing the chart in Fig. 26 the climate lines are based on ASHRAE degree-day listings, namely: *Cool, temperate climates*, 5000 degree-days and up; *average climates*, 2000 to 5000 degree-days; *warm climates*, 2000 degree-days maximum.

3. *Select the total exhaust-fan capacity*
An exhaust fan or fans must remove the minimum computed ventilation flow, or 9400 ft^3/min (266 m^3/min) for this room. To allow for possible errors in room size, machinery rating, or temperature, choose an exhaust fan 10 percent larger than the computed ventilation flow. For this room the exhaust fan would therefore have a capacity of $1.1 \times 9400 = 10{,}340$ ft^3/min (292.6 m^3/min). A fan rated at 10,500 or 11,000 ft^3/min (297.2 or 311.3 m^3/min), depending on the ratings available from the supplier, would be chosen.

Related Calculations. Ventilation is environmentally important and must accomplish two goals: (1) Removal of excess heat generated by machinery or derived from hot piping and other objects; (2) removal of objectionable, toxic, or flammable gases from process pumps, compressors, and piping.

The usual specifications for achieving these goals commonly call for an arbitrary number of hourly air changes for a building or room. However, these specifications vary widely in the number of air changes required, and use inconsistent design methods for ventilation. The method given in this procedure will achieve proper results, based on actual applications.

Because of health and explosion hazards, workers exposed to toxic or hazardous vapors and gases should be protected against dangerous levels [threshold limit values (TLV)] and explosion hazards [lower explosive limit (LEL)] by diluting workspace air with outside air at adequate ventilation rates. If a workspace is protected by adequate ventilation rates for health (i.e., below TLV) purposes, the explosion hazard (LEL) will not exist. The reason for this is that the health air changes far exceed those required for explosion prevention.

To render a workspace safe in terms of TLV, the number of ft^3/min (m^3/min) of dilution air, A_d required can be found from: $A_d = [1540 \times S \times T/(M \times \text{TLV})]K$, or in SI, A_{dm} = m^3/min = $0.0283A_d$, where S = gas or vapor expelled over an 8-h period, lb (kg); M = molecular weight of vapor or gas; TLV = threshold limit value, ppm; T = room temperature, absolute °R (K); K = air-mixing factor for nonideal conditions, which can vary from 3 to 10, depending on actual space conditions and the efficiency of the ventilation-air distribution system.

If the space temperature is assumed to be 100°F (37.8°C) (good average summer conditions), the above equation becomes $A_d = [862{,}400 \times S/(M \times \text{TLV})]K$.

For every pound (kg) of gas or vapor expelled of an 8-h period, when $S = 1$, the second equation becomes $A_d = [862{,}400/(M \times \text{TLV})]K$. For values of S less or greater than unity, simple multiplication can be used.

It is only for ideal mixing that $K = 1$. Hence, K must be adjusted upward, depending on ventilation efficiency, operation, and the particular system application.

If mixing is perfect and continuous, then each air change reduces the contaminant concentration to about 35 percent of that before the air change. Perfect mixing is seldom attainable, however, so a room mixing factor, K, ranging from 3 to 10 is recommended in actual practice.

The practical mixing factor for a particular workspace is at best an estimate. Therefore, some flexibility should be built into the ventilation system in anticipation of actual operations. For small enclosures, such as ovens and fumigation booths, K-values range from 3 to 5. If you are not familiar with efficient mixing within enclosures, use a K-factor equal to 10. Then your results will be on the safe side. Figure 26 is based on a K-value equal to 8 to 10. Table 15 gives K-factors for ventilation-air distribution systems as indicated.

In some installations, heat generated by rotating equipment (pumps, compressors, blowers) process piping, and other equipment can be calculated, and the outside-air requirements for dilution ventilation determined. In most cases, however, the calculation is either too cumbersome and time consuming or impossible.

Figure 26 was developed from actual practice in the chemical-plant and oil-refinery businesses. The chart is based on a closed processing system. Hence, air quantities found from the chart are not recommended if (1) the system is not closed or (2) if abnormal operating conditions prevail

TABLE 15 *K*-values for Various Ventilation-Air Distribution Systems

K-values	Distribution system
1.2–1.5	Perforated ceiling
1.5–2.0	Air diffusers
2.0–3.0	Duct headers along ceiling with branch headers pointing downward
3.0 and up	Window fans, wall fans, and the like

Chemical Engineering.

that permit the escape of excessive amounts of toxic and explosive materials into the workplace atmosphere.

For these situations, special ventilation measures, such as local exhaust through hoods, are required. Vent the exhaust to pollution-control equipment or, where permitted, directly outdoors.

Figure 26 and the procedure for determining dilution-air ventilation requirements were developed from actual tests of workspace atmospheres within processing buildings. Design and operating show that by supplying outside air into a building near the floor, and exhausting it high (through the roof or upper outside walls), safe and comfortable conditions can be attained. Use of chevron-type storm-proof louvers permits outside air to enter low in the room.

The chevron feature causes the air to sweep the floor, picking up heat and diluting gases and vapors on the way up to the exhaust fan (Fig. 27).

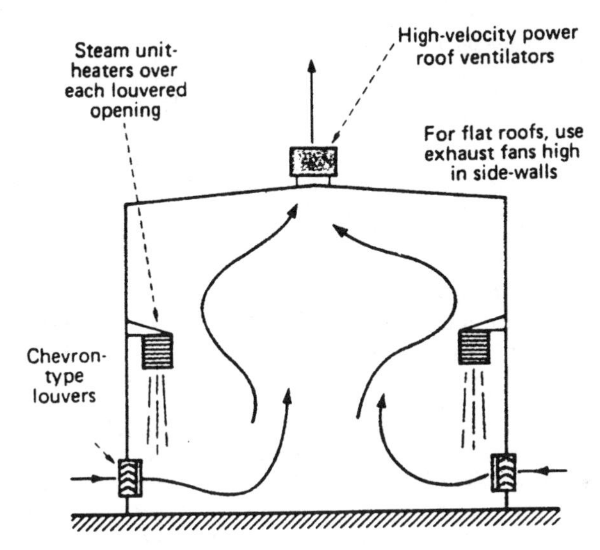

FIGURE 27 Ventilation system for effective removal of plant heat loads. (*Chemical Engineering.*)

In the system shown in Fig. 27 there are a number of features worth noting. With low-level distribution and adequate high exhaust, only the internal plant heat load (piping, equipment) is of importance in maintaining desirable workspace conditions. Wall and transmission heat loads are swept out of the building and do not reach the work areas. Even the temperature rise caused by the plant load occurs above the work level. Hence, low-level distribution of the supply air maintains the work area close to supply-air temperatures.

For any installation, it is good practice to check the ratio of hp-deg/ft^2 (kW-deg/m^2) of floor area. When this ratio exceeds 100, consider installing a totally enclosed ventilation system for cooling. This should be complete with ventilating fans taking outside air, preferably from a high stack, and discharging through ductwork into a sheet metal motor housing.

The result is the greater use of outside air for cooling through a confined system at a lower ventilation rate. Ventilation air flow needs may be obtained from the equipment manufacturer or directly from the chart (Fig. 26). The remainder of the building may be ventilated as usual, based either on the absence or equipment or on any equipment outside the ventilation enclosure. When designing the duct system, take care to prevent moisture entrainment with the incoming airstream.

This procedure is the work of John A. Constance, P.E., consultant, as reported in *Chemical Engineering*.

THERMAL-POLLUTION ESTIMATES FOR POWER PLANTS

A steam power plant has a 1000-MW output rating. Find the cooling-water thermal pollution by this power plant when using a once-through cooling system for the condensers if the plant thermal efficiency is 30 percent.

Calculation Procedure:

1. *Determine the amount of heat added during plant operation*
The general equation for power-plant efficiency is $E = W/Q_A$, where E = plant net thermal efficiency, percent; W = plant net output, MW; Q_A = heat added, MW. For this plant, $Q_A = W/E = 1000/0.30 = 3333$ MW.

2. *Compute the heat rejected by this plant*
The general equation for heat rejected is $Q_R = (W/E - W)$, Q_R = heat rejected, MW. For this power plant, $Q_R = (1000/0.3 - 1000) = 2333$ MW. Thus, this plant will reject 2333 MW to the condenser cooling water.

The heat rejected to the cooling water will be absorbed by the river, lake, or ocean providing the water pumped through the condenser. Depending on the thermal efficiency of the plant, the required cooling-water flow for the condenser will range from 250×10^6 lb/h, or 65,000 ft^3/min (30 m^3/s) to 400×10^6 lb/h or 100,000 ft^3/min (50 m^3/s). The discharged water in a once-through cooling system will be 20 to 25°F (11 to 14°C) higher in temperature than the entering water.

3. *Assess the effects of this thermal pollution*
Warm water discharged in large volume to a restricted water mass may affect the ecosystem in a deleterious way. Fish and plant life, larvae, plankton, and other organisms can be damaged or have a high mortality rate. If chlorine is used to control condenser scaling, the effect on the ecosystem can be more damaging.

If the warm condenser cooling water is discharged into a large body of water, such as a major river or ocean, the effect on the ecosystem can be more beneficial than deleterious. Thus, well-planned cooling-water outlets can be used to increase fish production in hatcheries. In agriculture, the warm-water discharge can be used to markedly increase the output of greenhouses and open fields in cold climates. Thus, the overall effect of thermal pollution can be positive, if the pollution energy is used in an antipolluting manner.

Related Calculations. Since thermal and atmospheric pollution are associated with the generation of electricity, environmental engineers are seeking ways to reduce electricity use. Personal computers (PCs) are big users of electricity today. At the time of this writing, PCs consume some 5 percent of commercial energy used in the United States.

Typical PCs use 150 to 200 W of power when in use, or just on but not in use. Some 30 to 40 percent of PCs are left on overnight and during weekends. The extra electricity which must be generated to carry this PC load leads to more thermal and air pollution.

New PCs have "sleep" circuitry which reduces the electrical load to 30 W when the computer is not being used. Such microprocessors will reduce the electrical load caused by PCs. This, in turn, will reduce thermal and air pollution produced by power-generating plants.

Internal-combustion engines—diesel, gas, and gas turbines—produce both thermal and air pollution. To curb this pollution and wrest more work from the fuel burned, cogeneration is being widely applied. Heat is extracted from the internal-combustion engine's cooling water for use in process or space heating. In addition, exhaust gases are directed through heat exchangers to extract more heat from the internal-combustion engine exhaust. Thus, environmental considerations are met while conserving fuel. This is one reason why cogeneration is so popular today.

Compute the heat recovery from cogeneration using the many concepts given earlier in this section. Try to combine both heat recovery and pollution reduction; then the required investment will be easier to justify from an economic standpoint.

FLASH-STEAM HEAT RECOVERY FOR COGENERATION

Fifty steam traps of various sizes in an industrial plant discharge a total of 95,000 lb/h (11.96 kg/s) of condensate from equipment operating at 150 lb/in^2 (gage) (1034.3 kPa) to a flash tank maintaining a pressure of 5 lb/in^2 (gage) (34.5 kPa) at a temperature close to the steam temperature. The remaining condensate is discharged. Determine the quantity, available heat, and temperature of the flash steam formed. What quantity of water would be heated by this steam in a hot-water heater having an overall efficiency of 85 percent if the temperature is raised from 40°F (4.4°C) to 140°F (55.6°C)? Determine the effect on flash steam and condensate outlet temperature for the flash tank if the terminal temperature difference (flash-down) is 25°F (13.9°C). What would the effect on flash steam be if the condensate in the steam traps is subcooled 13°F (7.2°C)?

Calculation Procedure:

1. ***Sketch the complete condensate and flash-steam recovery system***
Refer to Fig. 28 for a typical installation.

2. ***Determine the percent of flash steam formed***
In Table 16, locate an initial steam pressure of 150 lb/in^2 (1034.3 kPa). Cross to the right to the 5-lb/in^2 (gage) (34.5-kPa) flash tank pressure column and read 14.8 percent of the condensate forms flash steam.

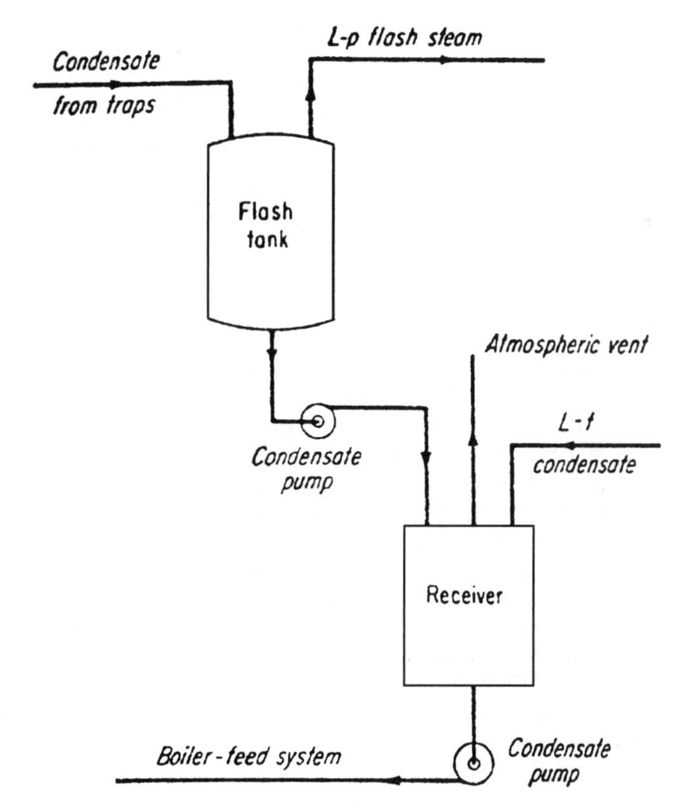

FIGURE 28 Complete condensate and flash-steam recovery system. (*Chemical Engineering.*)

TABLE 16 Percent Flash Steam Formed

Initial steam pressure		Sat. temp		Flash-tank pressure, lb/in^2 (gage) (kPa)						
lb/in^2 (gage)	(kPa)	°F	°C	0 (0)	5 (34.5)	10 (68.9)	50 (344.5)	100 (689)	125 (861.3)	150 (1033)
125	(861.1)	353	577.8	14.8	13.4	12.2	6.3	1.7	0	0
150	(1034.3)	366	601.2	16.8	14.8	13.7	7.8	2.3	1.6	0
175	(1206.5)	377	621.0	17.4	16.0	15.0	9.0	4.6	3.0	1.5
200	(1378.8)	388	640.8	18.7	17.5	16.2	10.4	6.0	4.4	2.8
				Total heat of flash steam, Btu/lb (kPa)						
				1500 (2674.9)	1156 (2693.9)	1160 (2702.8)	1179 (2747.1)	1189 (2770.4)	1193 (2779.7)	1195 (2784.4)
				Latent heat of evaporation, Btu/lb (kJ/kg)						
				970 (2260.1)	960 (2232.9)	952 (2218.2)	912 (2125)	881 (2052.7)	868 (2022.4)	857 (1996.8)
				Heat of liquid, Btu/lb (kJ/kg)						
				180 (419.4)	196 (456.7)	208 (484.6)	267 (622.1)	309 (719.9)	324 (754.9)	338 (787.5)
				Saturated water temperature, °F (°C)						
				212 (100)	228 (108.9)	240 (115.5)	298 (147.8)	338 (170)	353 (178.3)	366 (185.6)
				Volume of flash steam, ft^3/lb (m^3/kg)						
				26.8 (1.67)	20.0 (1.25)	16.3 (1.02)	6.6 (0.41)	3.9 (0.24)	3.2 (0.20)	2.7 (0.17)

3. ***Compute the quantity of flash steam formed***

This equals the percent of flash steam formed multiplied by the condensate discharge from the steam traps, or (0.148)(95,000) = 14,060 lb/h (1.8 kg/s).

4. ***Compute the available heat in the flash steam formed***

This equals the latent heat of evaporation for a flash tank pressure of 5 lb/in^2 (gage) (34.5 kPa) multiplied by the quantity of flash steam formed. From Table 16, the latent head of evaporation is 960 Btu/lb (2232.9 kJ/kg) at 5 lb/in^2 (gage) (34.5 kPa). Hence, the available heat is (960) (14.060) = 13,500,000 Btu/h (3955.5 kW).

5. ***Determine the flash-steam temperature***

This equals the saturated water temperature corresponding to the saturated flash tank pressure of 5 lb/in^2 (gage) (34.5 kPa). From Table 16 this value is shown as 228°F (108.9°C).

6. ***Compute the quantity of water heated in the hot-water heater***

If water were heated with the energy from the flash steam (assuming all of the flash steam could be used), this quantity would be equivalent to the (flash-steam available heat)(efficiency of the hot-water heater)(temperature increase) or (13,500,000)(0.85)(100) = 114,750 lb/h (14.4 kg/s). (*Note:* Additional heat is available in the condensed flash steam.)

7. ***Compute the effect on flash steam and remaining condensate temperature for the flash tank with a terminal temperature difference (flash down) of 25°F (13.9°C)***

The temperature of the remaining condensate at the flash tank outlet equals the saturated water temperature plus flashdown, or 228 + 25 = 253°F (122.8°C). (*Note:* The flashdown represents a loss to the system, and may be necessary due to sizing considerations.) The flashdown process will continue across the flashtank outlet, and until the temperature of the remaining condensate is 228°F (108.9°C),

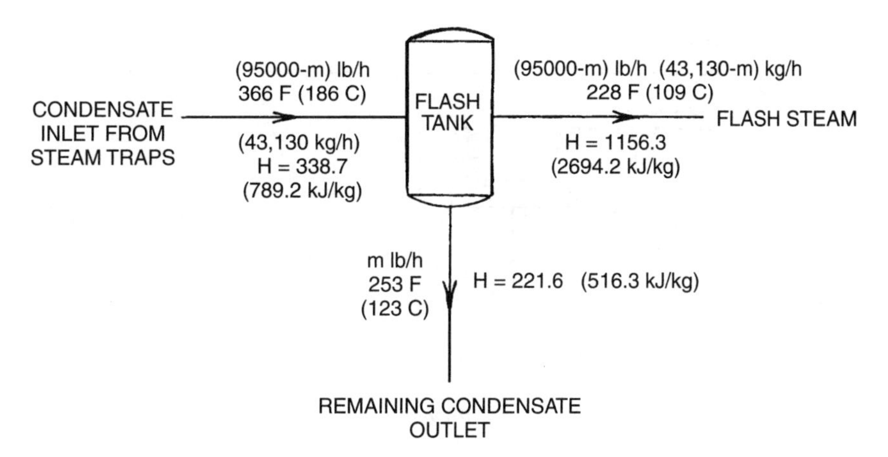

FIGURE 29 Flash-tank heat balance. (*Chemical Engineering.*)

the quantity of flash steam formed as a consequence of flashdown will be reduced. Refer Fig. 29 for a schematic of the flash tank energy balance. Note that the values shown for enthalpy (H) are determined from steam-table data. Hence, since energy input equals energy output, 95,000(338.7) = (m)(221.6) + (95,000 – m)(1156.3). Solving for m, the quantity of remaining condensate is 83,100 lb/h (10.5 kg/s). Therefore, the quantity of flash steam is 95,000 – 83,100 or 11,900 lb/h (1.5 kg/s). And the flashdown reduces the flash steam quantity by the following: (14,060 – 11,900)/14,060 = 0.1536 or about 15.4 percent.

8. ***Compute the effect on flash-steam quantity if the condensate in the steam traps is subcooled by 13°F (7.2°C)***

From Table 16, read a saturated water temperature at 150 lb/in^2 (gage) as 366°F (185.6°C). Therefore, subcooling by 13°F (7.2°C) will reduce the condensate temperature to 366 – 13 = 353°F (178.3°C). Cross to 353°F (178.3°C) in the 5-lb/in^2 (gage) (34.5 kPa) flash-tank-pressure saturation-temperature column and read 13.4 percent of the condensate forms flash steam. The quantity of flash steam may then be computed as (95,000)(0.134) = 12,730 lb/h (1.6 kg/s). Hence, the subcooling reduces the flash-steam quantity by 14,060 – 12,730/14,060 = 0.095 or 9.5 percent. Note that interpolation may be used for intermediate temperature values.

Related Calculations. This general procedure can be used for analyzing steam flow in flash tanks used in commercial, industrial, marine, and similar applications. With the great emphasis on conserving energy to reduce fuel costs, flash tanks are receiving greater attention than every before. Since flash steam contains valuable heat, every effort possible is being made to recover this heat, consistent with the investment required for the recovery.

The method given here is the work of T. R. MacMillan, as reported in *Chemical Engineering* magazine.

ENERGY CONSERVATION AND COST REDUCTION DESIGN FOR FLASH-STEAM USAGES

A plant has the steam layout shown in Fig. 30. Determine the dollar value of the flashed steam and what can be done about reducing the energy loss, if any. In this plant, steam from the boiler is condensed in the heat exchanger at 100 lb/in^2 (gage) (689 kPa) and 338°F (170°C). Process water

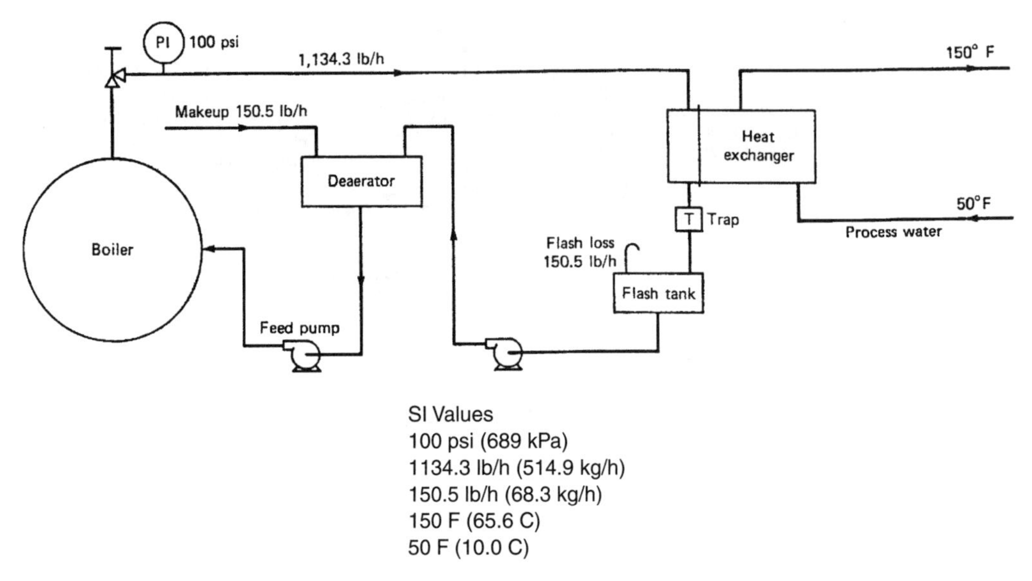

FIGURE 30 With steam valued at 58 per million Btu, Venting flash steam results in an annual loss of almost 512,000. (*Chemical Engineering.*)

is heated from 50°F (10°C) to 150°F (65.6°C), with heat transferred at the rate of 1-million Btu/h (293 kW). Condensate drains through a trap to a flash tank, where the flash steam is vented to the atmosphere. Analyze the benefits of reducing the supply steam pressure, and the financial benefits of recovering the flashed steam.

Calculation Procedure:

1. *Determine the amount of heat lost in the flashed steam*

From the steam tables, each pound of condensate at 100 lb/in^2 (gage) (69 kPa) contains 309.0 Btu (718.7 kJ/kg). At atmospheric pressure, each pound of condensate holds 180.2 Btu (419.2 kJ/kg) as sensible heat. The surplus, 309.0 – 180.2 = 128.8 Btu/lb (299.6 kJ/kg) flashes off 128.8 Btu/(970.6 Btu/lb) = 0.1327 pounds of steam per pound of condensate (0.06 kg/kg), or 13.27 percent. In this relation the value 970.6 is the latent heat of the condensate at 14.7 lb/in^2 (gage) (101.3 kPa). Since the flash steam carries its total heat with it, 13.27 percent of the 1150.8 Btu (1214.1 J) is vented per pound of condensate, or 152.7 Btu/lb (355.2 kJ/kg).

Because 1-million Btu/h (1055 kJ) is transferred in the heat exchanger, and the latent heat of 100-lb/in^2 (gage) steam is 881.6 Btu/lb (2050.6 kJ/kg), the steam flow from the boiler is (1,000,000 Btu/h)/(881.6 Btu/lb) = 1134.3 lb/h (514.9 kg/h). The heat vented from the flash tank is (152.7 Btu/lb) (1134.3 lb/h) = 173.207 Btu/h (50.7 kW). Makeup water at 50°F (10°C) brings in 18 Btu (18.9 J) with each 13.27 percent of 1134.3 lb (514.9 kg), or (18 Btu/lb)(150.5 lb/h) = 2709.0 Btu/h (851.8 W). Thus, the heat loss is 173,207 Btu/h – 2709 Btu/h = 170,498 Btu/h (49.9 kW). This is more than 17 percent of the useful heat transferred to the exchanger, i.e., 170,498/1,000,000 = 0.170498.

2. *Compute the annual dollar cost of the lost heat*

To determine the dollar cost multiply the cost of steam, per million Btu, by the hour loss, Btu, and the number of operating hours per year. Assuming continuous 24-h operation of this plant, the annual dollar cost with steam priced at $8/million Btu is: (170,498 Btu/h)(8760 h/yr)($8/1,000,000 Btu steam cost) = $11,949 per year. This is nearly $1000 per month lost from the flash steam.

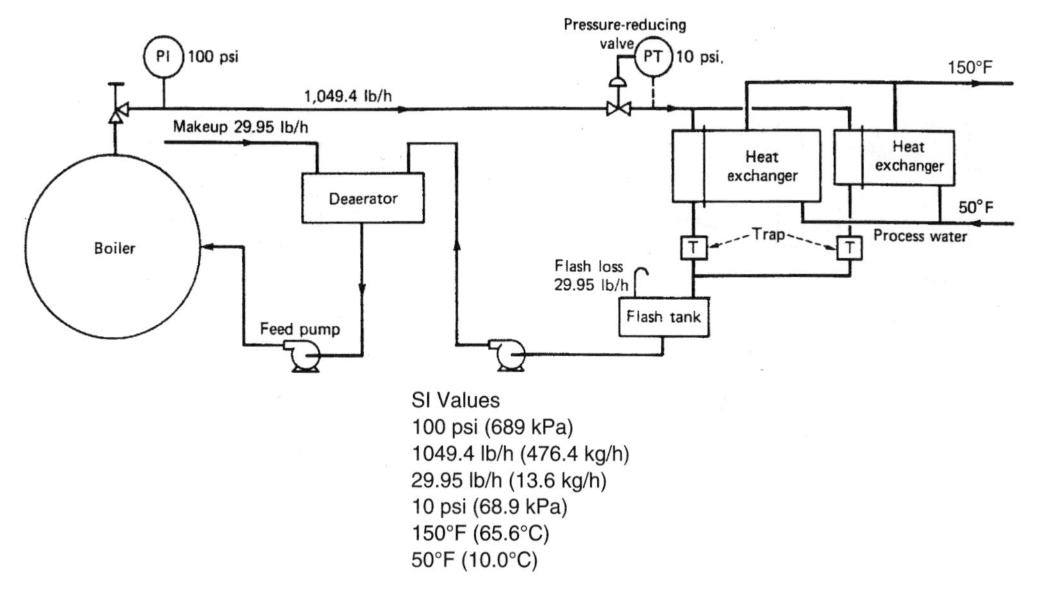

SI Values
100 psi (689 kPa)
1049.4 lb/h (476.4 kg/h)
29.95 lb/h (13.6 kg/h)
10 psi (68.9 kPa)
150°F (65.6°C)
50°F (10.0°C)

FIGURE 31 Lowering the steam pressure by using a pressure-reducing valve cuts operating cost but increases capital cost. (*Chemical Engineering.*)

3. *Analyze an alternative plant layout to reduce the cost of the lost heat*
To reduce the annual loss, an arrangement such as that in Fig. 31 is often proposed. A pressure-reducing valve (PT) in the diagram, in the steam-supply line lowers the boiler steam pressure to, say, 10 lb/in^2 (gage) (68.9 kPa) instead of the 100 lb/in^2 (gage) (689 kPa) used in step 1. The latent heat at this pressure increases to 952.9 Btu/lb (2216.5 kJ/kg), as shown in the steam tables.

With 1,000,000 Btu/h (1055 kJ) transferred in the heat exchanger, as earlier, the steam flow rate will be (1,000,000)/(952.9 Btu/lb) = 1049.4 lb/h (476.4 kg/h). At 10 lb/in^2 (gage) (68.9 kPa) the sensible heat is 207.9 Btu/lb (483.6 kJ/kg) from the steam tables. At 0 lb/in^2, i.e., atmospheric pressure to which the flash tank exhausts, the sensible heat is 180.2 Btu/lb (419.2 kJ/kg). Then, the flash-steam percentage will be (207.9 Btu/lb − 180.2 Btu/lb)/(970.6 Btu/lb) = 0.02854, or 2.854.

The rate of flow of flash steam will be 0.02854 × 1049.4 = 29.95 lb/h (13.6 kg/h). The heat loss of this flash steam will be 29.95 lb/h × 1150.8 Btu/lb = 34,466 Btu/h (10.1 kW). Makeup water containing 18 Btu/lb (41.9 kJ/kg) enters at a rate of 29.95 lb/h (13.6 kg/h) providing 18 × 29.95 = 539.1 Btu/h (158 W). Then, the net heat loss = 34,466 − 539 = 33,927 Btu/h (9.94 kW), or 3.39 percent of the heat transferred in the exchanger.

Such a reduction in heat loss, amounting to about 136,571 Btu/h (40 kW), an annual saving of about $9570 (using the same steam cost as earlier), represents a substantial saving.

But at least one additional factor must be considered before installing the pressure-reducing valve. The heat exchanger that was large enough when supplied with 100-lb/in^2 (gage) (689-kPa) 338°F (170°C) steam would be too small when supplied with 10-lb/in^2 (gage) (68.9-kPa) 240°F (115.6°C) steam.

Temperatures at the exchanger for these two cases (assuming for simplicity that arithmetic mean temperature differences are sufficiently accurate) are shown in Fig. 32. The Δt across the system has dropped from 238°F (114°C) to 140°F (60°C). If the U value remains the same, the surface area of the exchanger would have to be increased by 238°F/140°F = 1.7 times.

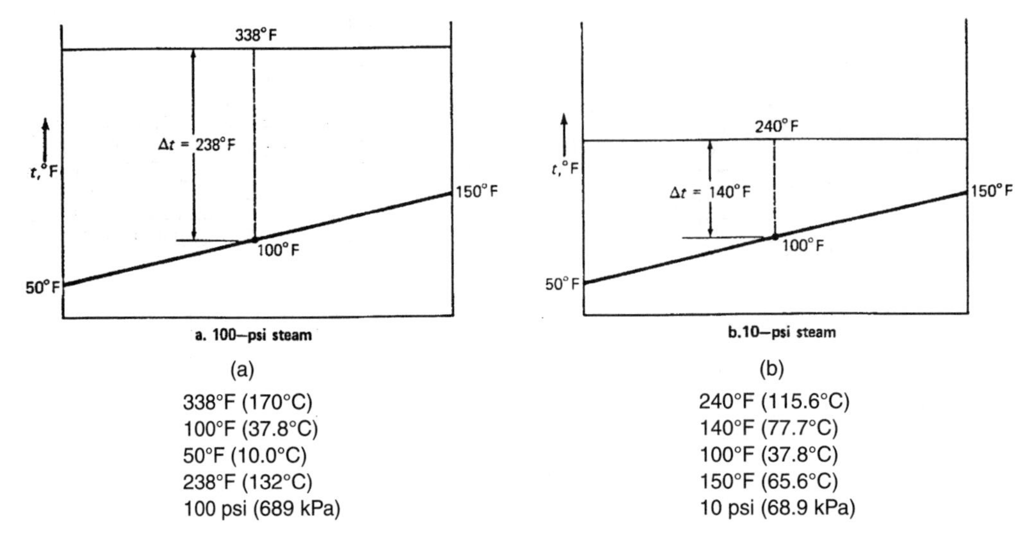

FIGURE 32 Temperatures around the heat exchanger are indicated for before and after the reduction in steam pressure. (*Chemical Engineering.*)

If U = 150 Btu/(ft^2 · °F · h), from $Q = UA\ \Delta t$; A = 1,000,000/(150)(238) = 28 ft^2 (2.6 m^2). Hence, at the lower steam pressure, the exchanger surface area would have to be increased by 0.7 × 28 = 19.6 ft^2 (1.82 m^2). Such an exchanger would cost about $550, $1,200 when installed. This would be in addition to the cost of about $600 for buying and installing the pressure-reducing valve. If such alterations must be made to produce the annual saving of $9570, other alternatives should also be considered.

4. *Determine if recovering the flash steam is economically worthwhile*

The minimum heat-transfer area is needed if as much as possible of the heat flow is to be from the 100-lb/in^2 (gage) (689-kPa) 338°F(170°C) steam. Suppose, however, that the high-pressure condensate were not discharged to a flash venting tank but to a flash-steam recovery vessel. (For maximum economy, let the recovery system operate at atmospheric pressure.) If the flash steam were passed to a supplementary condenser fitted at the inlet side of the main exchanger (thus serving as a preheater), the atmospheric-pressure condensate would flow to the return pump without further heat loss, and the installation would be as shown in Fig. 33, with the temperature diagram shown in Fig. 34.

In Fig. 33, the latent heat of the 100-lb/in^2 (689-kPa) steam is 881.6 Btu/lb (2051 kJ/kg); flash steam at 0 lb/in^2 (gage) (0 kPa) has a latent heat of 970.6 Btu/lb (2258 kJ/kg). Each pound (0.45 kg) of condensate at 100 lb/in^2 (689 kPa) contains 309.0 Btu (718.7 kJ/kg); at atmospheric pressure the sensible heat of the condensate is 180.2 Btu/lb (419.1 kJ/kg). The surplus heat, 309.0 − 180.2 = 128.8 Btu/lb (299.6 kJ/kg) flashes off 128.8/970.6 = 0.1327 lb of steam per lb of condensate (0.06 kg/kg), or 13.2 percent of the condensate. The latent heat transferred = 881.6 + 128.8 = 1010.4 Btu (1065.9 J). The total steam flow with 1,000,000 Btu/h (293 kW) heat transfer is 1,000,000/1010.4 = 989.7 lb (449.3 kg). The proportion of latent heat in the flash steam is 128.8/1010.4 = 0.1275, or 12.75 percent. Temperature rise in the preheater = (0.1275 × 100°F) = 12.75°F (22.95°C). From this, the other temperatures in Fig. 34 can be derived.

Assuming again that the U value of 150 is maintained, the preheater surface area = (127,000 Btu/h)/[150 Btu/(ft^2 · °F · h)](155.6°F) = 5.5 ft^2 (0.51 m^2). Hence, the exchanger is slightly oversized.

In retrofitting, the 28-ft^2 (2.6-m^2) exchanger would allow the steam pressure to be less than 100 lb/in^2 (689 kPa). In a new installation, an exchanger having a surface area of (0.1275 × 1,000,000)/(1,000,000 × 28) = 25 ft^2 (2.32 m^2) could be installed.

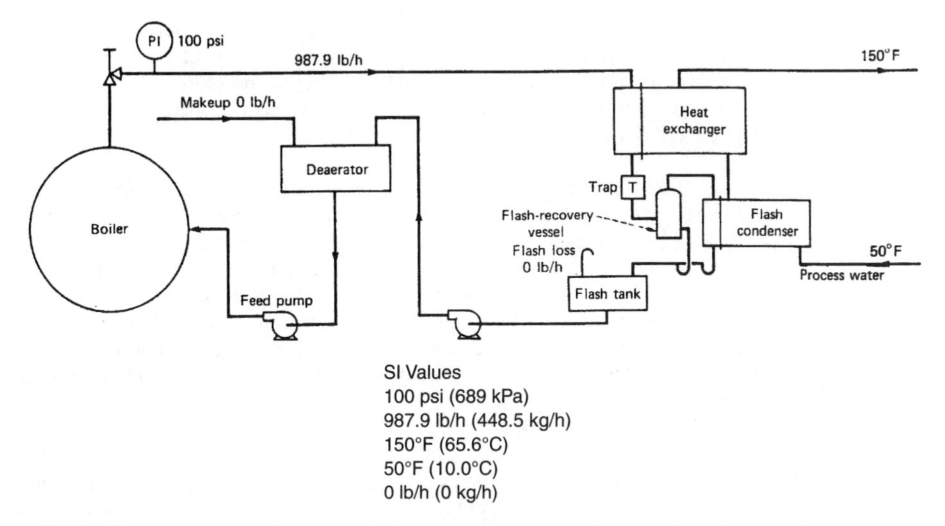

FIGURE 33 Recovering the flash steam eliminates the venting loss and reduces the additional capital investment. (*Chemical Engineering.*)

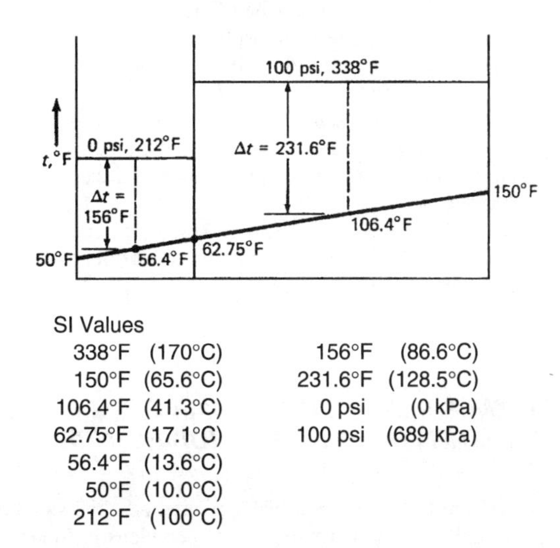

FIGURE 34 Temperatures of exchanger and preheater with flash-steam recovery. (*Chemical Engineering.*)

The cost of the installed preheater would be about \$800, to which must be added the cost of the flash-recovery vessel at about \$600. The following summarizes the choices:

Installation	Extra capital cost, \$	Operating cost, \$/yr
Existing	0	11,964
Add pressure-reducing valve and extra 20 ft² (1.85 m²) of heat exchanger	1,800	2,394
Add flash-recovery vessel and extra 5.5 ft² (0.5 m²) of exchanger	1,400	0

In an actual installation, some of the simplifications made here would be reassessed. However, it will generally remain the case that, with the recovery of flash steam, heat exchangers can operate at the greater efficiency afforded by higher-pressure, higher-temperature steam. Flash steam, recovered at the lowest practical pressure, can be used in a preheater, or in a separate, unconnected load.

Related Calculations. With greater emphasis on lowering environmental air pollution, reducing flash-steam costs takes on more importance in plant analyses. During initial design studies, a choice must often be made between high- and low-pressure operating steam at both the boiler and at any heat exchangers in the system. For new installations, the economics of generating moderately high-pressure, rather than low-pressure, steam are usually fairly obvious.

Low-pressure boilers are physically larger and more costly than boilers producing the same quantity of steam at higher pressure. Also, steam produced at low pressure is often wetter than high-pressure steam, leading to lower heat-transfer rates in exchangers, even if water hammer is avoided.

The choice between high- and low-pressure steam may not always be so clear. Using high-pressure steam to get the benefits of lower capital costs of smaller heat-transfer areas can mean higher operating costs. Steam losses from flash-tank vents are greater with high-pressure steam. The value of the steam lost can quickly exceed the capital-cost savings. Such considerations often lead to choosing lower operating steam pressures.

Since flash-steam losses represent fuel used to generate this steam, every effort possible should be made to control flash steam. When flash losses are reduced, fuel consumption is cut. With lower fuel consumption, there is less atmospheric pollution. Reduced pollution lowers the cost of handling flash, sulfur compounds, and other boiler effluents. So there are many good reasons for limiting flash-steam losses in any plant.

The procedure presented here can be used for any plant using steam for processes, heating, power generation, or other heat-transfer purposes. Such plants include chemical, food, textile, manufacturing, marine, cogeneration, and central-station. All can benefit from reducing flash-steam losses, as described above.

This procedure is the work of Albert Armer, Technical Adviser and Sales Specialist, Spirax Sarco, Inc., as reported in *Chemical Engineering* magazine. SI values were added by the handbook editor to the calculations and illustrations.

COGENERATION POWER PLANT COST SEPARATION OF STEAM AND ELECTRICITY USING THE ENERGY EQUIVALENCE METHOD

Allocate—using the energy equivalence method—the steam and electricity costs in a power plant having a double automatic-extraction, noncondensing steam turbine for process steam and electric generation. Turbine throttle steam flow is 800,000 lb/h (100.7 kg/s) at 865 lb/in^2 (abs) (5964.1 kPa). Process steam is extracted from the turbine in the amounts of 100,000 lb/h (12.6 kg/s) at 335 lb/in^2 (abs) (2309.8 kPa) and 200,000 lb/h (25.2 kg/s) at 150 lb/in^2 (abs) (1034.3 kPa) and is delivered to process plants. A total of 500,000 lb/h (62.9 kg/s) is exhausted at 35 lb/in^2 (abs)(241.3 kPa) with 100,000 lb/h (12.6 kg/s) of this exhaust steam for deaerator heating in the cycle and 400,000 lb/h (50.4 kg/s) sent to process plants. The turbine has a gross electric output of 51,743 kW, and the heat balance for the dual-purpose turbine cycle is shown in Fig. 35. Efficiency of the steam boiler is 85.4 percent, while the fuel is priced at $0.50 per 10^6 Btu ($0.47 per MJ). If a condensing turbine is used, an attainable backpressure is 1.75 inHg (abs) [43.75 mmHg (abs)], while the assumed turbine efficiency is 82 percent and the exhaust enthalpy is h_f = 1032 Btu/lb (2400.4 MJ/kg). Figure 36 shows the expansion-state curve of the turbine on a Mollier diagram. Final feedwater enthalpy is 228 Btu/lb (530.3 mJ/kg). Allocate the fuel cost to each energy use by using the energy equivalence method.

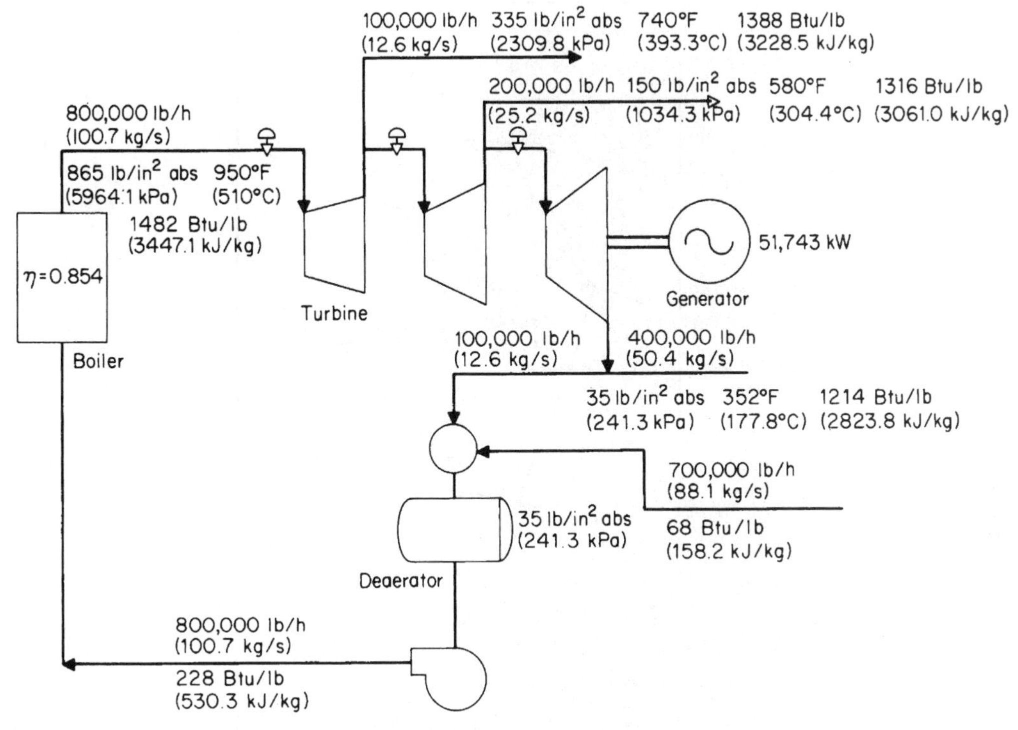

FIGURE 35 Dual-purpose turbine heat balance. (*Combustion.*)

Calculation Procedure:

1. *Compute the hourly total fuel cost*

The total fuel cost for this plant per hour is C_f = (l/boiler efficiency)(0.50/10^6)(throttle steam flow rate m_t, lb/h)($h_t - h_{fw}$), where h_t = throttle enthalpy, Btu/lb, and h_{fw} = feedwater enthalpy, Btu/lb. Substituting gives C_f = (1/0.854)(0.50/10^6)(800,000)(1482 – 228) = $587.35 per hour.

2. *Compute the nonextraction ultimate electric output*

The ultimate electric output is $E_u = m_t(h_t - h_f)/3413$, where m_h = total turbine inlet steam flow, lb/h; h_i = turbine initial enthalpy, Btu/lb; other symbols as before. Substituting, we find E_u = (800,000)(1482 – 1032)/3413 = 105,480 kW.

3. *Determine the actual electric output of the dual-purpose turbine*

Use the relation $E_a = W$(actual)/3413 = the work done by the extraction steam between the throttle inlet and the extraction point, plus the work done by the non-extraction steam between the throttle and the exhaust. Or, from the turbine expansion curve in Fig. 36, E_a = [100,000(1482 – 1388) + 200,000(1482 – 1316) + 500,000(1482 – 1214)]/3413 = 51,743 kW.

4. *Compute the extraction steam kilowatt equivalence*

Again from Fig. 36, $E_{x1} = (h_{x1} - h_f)/3413$ = 100,000(1388 – 1032)/3413 = 10,432 kW; E_{x2} = 200,000(1316 – 1032)/3413 = 16,642 kW; E_{x3} = 500,000(1214 – 1032)/3413 = 26,663 kW. Hence, the nonextraction turbine ultimate electric output = $E_a + E_{x1} + E_{x2} + E_{x3}$ = 51,743 + 10,432 + 16,642 + 26,663 = 105,480 kW.

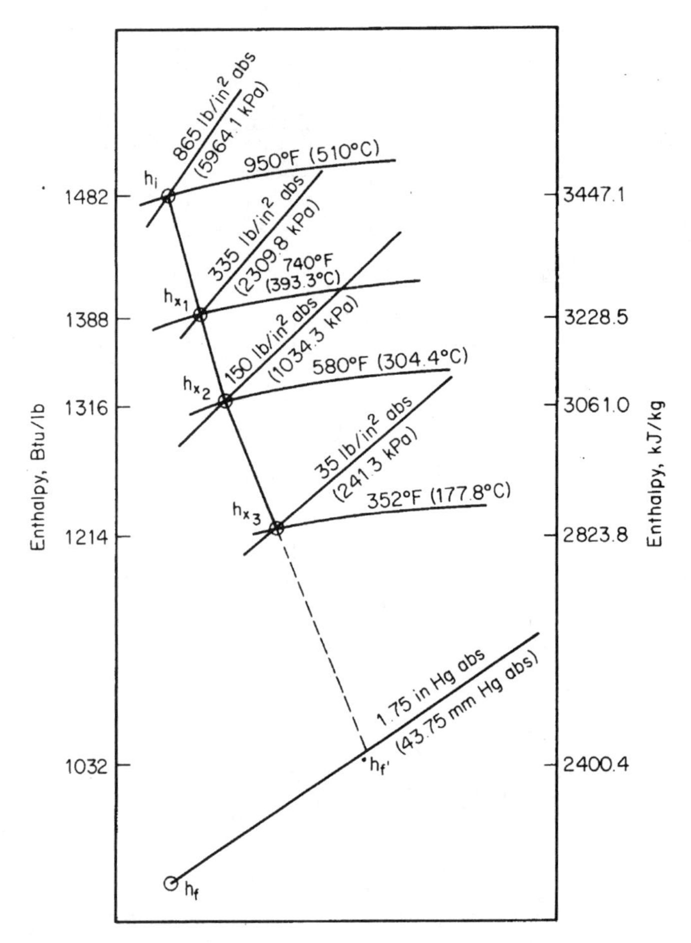

FIGURE 36 Turbine expansion curve. (*Combustion*.)

5. *Determine the base fuel cost of electricity and steam*

The base fuel cost of electricity = C_f/E_u, or \$587.35/105,480 = \$0.005568 per kilowatthour, or 5.568 mil/kWh.

Now the base fuel cost of the steam at the different pressures can be found from (kW equivalence) (base cost of electricity, mil)/(rate of steam use, lb/h). Thus, for the 335-lb/in² (abs) (2309.8-kPa) extraction steam used at the rate of 100,000 lb/h (12.6 kg/s), base fuel cost = 10,432(5,568)/100,000 = \$0.5808 per 1000 lb (\$0.2640 per 1000 kg). For the 150-lb/in² (abs) (1034.3-kPa) steam, base fuel cost = 16.642(5.568)/200,000 = \$0.4633 per 1000 lb (\$0.21059 per 1000 kg). And for the 35-lb/in² (abs) (241.3-kPa) steam, base fuel cost = 26,663(5.568)/500,000 = \$0.2969 per 1000 lb (\$0.13495 per 1000 kg). Since 100,000 lb/h (12.6 kg/s) of the 500,000 lb/h (62.9 kg/s) is used for deaerator heating, the cost of this heating steam = (100,000)/1000)(\$0.2969) = \$29.69 per hour.

Since 100,000 lb (45,000 kg) of steam utilizes its energy for deaerator heating within the cycle, its equivalent electric output of 26,663/5 = 5333 kW should be deducted from the 26,663-kW electric energy equivalency of the 35-lb/in² (abs) (241.3-kPa) steam. The remaining equivalent energy of

21,330 kW (= 26,663 – 5333) represents 35-lb/in^2 (abs) (241.3-kPa) extraction steam to be delivered to process plants.

6. *Determine the added unit fuel cost*
The deaerator-steam fuel cost of $29.69 per hour would be shared by both process steam and electricity in terms of energy equivalency as 105,480 – 5333 = 100,147 kW. Using this output as the denominator, we see that the added unit fuel cost for electricity based on sharing the cost of this heat energy input to the deaerator is $26.69/100,147 = $0.000296 per kilowatthour, or 0.296 mil/kWh.

Likewise, added fuel cost of 335-lb/in^2 (abs) (2309-kPa) steam = 10,432(100.000)/0.296 = $0.031 per 1000 lb (450 kg); added fuel cost of 150-lb/in^2 (abs) (1034-kPa) steam = $0.0246 per 1000 lb (450 kg); added fuel cost of 35-lb/in^2 (abs) (241-kPa) steam = $0.0158 per 1000 lb (450 kg). The fuel-cost allocation of steam and electricity is summarized in Table 17.

Related Calculations. The energy equivalence method is based on the fact that the basic energy source for process steam and electricity is the heat from fuel (combustion or fission). The cost of the fuel must be charged to the process steam and electricity. Since the analysis does not distinguish between types of fuels or methods of heat release, this procedure can be used for coal, oil, gas, wood, peat, bagasse, etc. Also, the procedure can be used for steam generated by nuclear fission.

Cogeneration is suitable for a multitude of industries such as steel, textile, shipbuilding, aircraft, food, chemical, petrochemical, city and town district heating, etc. With the increasing cost of all types of fuel, cogeneration will become more popular than in the past. This calculation procedure is the work of Paul Leung of Bechtel Corporation, as reported at the 34th Annual Meeting of the American Power Conference and published in *Combustion* magazine. Since the procedure is based on thermodynamic and economic principles, it has wide applicability in a variety of industries. For a complete view of the allocation of costs in cogeneration plants, the reader should carefully study the Related Calculations in the next calculation procedure.

TABLE 17 Energy Equivalence Method of Fuel Cost Allocation

Utility	Base	+	Unit cost heating steam	=	Total	Total fuel cost	Percent of total
Electricity, 51.743 kW*	5.568		0.296		$5.864 mil/kWh	$303.40	51.65
Steam @ 335 lb/in^2 (abs) (2309 kPa) 100,000 lb/h (45,000 kg/h)	0.5808		0.031		$0.6118 per 1000 lb (450 kg)	$61.18	10.41
Steam @ 150 lb/in^2 (abs) (1034 kPa) 200,000 lb/h (90,000 kg/h)	0.4633		0.0246		$0.4879 per 1000 lb (450 kg)	$97.57	16.62
Steam @ 35 lb/in^2 (abs) (241 kPa) 400,000 lb/h (180,000 kg/h)	0.2969		0.0158		$0.3127 per 1000 lb (450 kg)	$125.20	21.32
					Total:	$587.35	100.00

*Net kW delivered to process plants should be delivered after deducting fixed mechanical and electrical losses of the alternator. Electricity unit cost charged to production would be slightly higher after this adjustment.

Combustion.

FUEL-COST ALLOCATION FOR A COGENERATION PLANT USING ELECTRICITY COST

A turbine of the single-purpose type, operating at initial steam conditions identical to those in the previous calculation procedure, and a condenser backpressure of 1.75 in (43.75 mm) Hg (abs), would have a turbine heat rate of 9000 Btu/kWh (9495 kJ/kWh). Compute the fuel cost allocation to that of steam by using the established-electricity-cost method.

Calculation Procedure:

1. ***Compute the unit cost of the electricity***
The unit cost of the electricity is F_e = (fuel price, \$)(turbine heat rate, Btu/kWh)/(boiler efficiency). For this plant, $F_e = (0.5/10^6)(9000)/(0.854)$ = \$0.00527 per kilowatthour, or 5.27 mil/kWh.

2. ***Determine where the deaerator heating steam should be charged***
The turbine heat rate of 9000 Btu/kWh is a reasonable and economically justifiable heat rate of a regenerative cycle with a certain degree of feedwater heating. Hence, in this case, the deaerator heating steam should not be charged to the electricity. Instead, this portion of the deaerator-heating-steam cost should be charged to the process steam.

3. ***Allocate the fuel cost to steam***
The total fuel cost from the previous calculation procedure is \$587.35 per hour. The electricity cost allocation = (kW generated)(cost \$/kWh) = (51,743)(0.00527) = \$273. Hence, the fuel cost to the steam is \$587.35 – \$273.00 = \$314.35.

4. ***Compute the power equivalence of the steam***
From the previous calculation procedure, $E_x = E_{x1} + E_{x2} + E_{x3}$, where E_x = equivalent electric output of the extraction steam, kW; $E_{x1}, \ldots$ = equivalent electric output of the various extraction steam flows, kW. Hence, ΣE_x = 10,432 + 16,642 + 26,663 = 53,737 kW.

5. ***Determine the ratio of each extraction steam flow to the total extraction steam flow***
The ratio for any flow is $E_x/\Sigma E_x$. Thus, $E_{x1}/\Sigma E_x$ = 10,432/53,737 = 0.194; $E_{x2}/\Sigma E_x$ = 16,663/53,737 = 0.310; and $E_{x3}/\Sigma E_x$ = 26,663/53,737 = 0.496.

6. ***Compute the base unit fuel cost of steam***
Use the relation $(E_x/\Sigma E_x)$(fuel cost to steam)/m, where m = (steam flow rate, lb/h)/1000. Hence, for 335-lb/in^2 (abs) (2309.8-kPa) steam, base unit fuel cost = (0.194)(\$314.35)/100 = \$0,610 per 1000 lb (\$0.277 per 1000 kg); for 150-lb/in^2 (abs) (1034.3-kPa) steam, base unit fuel cost = (0.310) (\$314.35)/200 = \$0.487 per 1000 lb (\$0.2213 per 1000 kg); for 35-lb/in^2 (abs) (241.3-kPa) steam, base unit fuel cost = (0.496)(\$314.35)/500 = \$0.312 per 1000 lb (\$0.1418 per 1000 kg). Since the deaeration steam is at 35 lb/in^2 (abs) (241 kPa), the cost of this steam = (100,000/1000)(\$0.312) = \$31.20 per hour.

7. ***Determine the unit fuel cost from sharing the cost of the deaerator heating steam***
If the 5333-kW power equivalence of the deaerator heating steam is deducted from the electric power equivalence of the extraction steam, the kilowatt equivalence of all steam to production centers becomes 53,737 – 5333 = 48,404 kW. The unit fuel cost from sharing the cost of the deaerator heating steam is then (\$31.20/h)/(48,404) = \$0.000644 per kilowatthour, or 0.644 mil/kWh.

8. ***Compute the added fuel cost of steam at each pressure***
The added fuel cost at each pressure is (kW output at that pressure/steam flow rate, lb/h)(0.644). Thus, added fuel cost of 335-lb/in^2 (abs) (2309.8-kPa) steam = (10,431/100,000)(0.644) = \$0,067 per 1000 lb (\$0.03045 per 1000 kg); added fuel cost for 150-lb/in^2 (abs) (1034.3-kPa) steam = \$0,053 per 1000 lb (\$0.02409 per 1000 kg); added fuel cost for 35-lb/in^2 (abs) (241.3-kPa) steam = \$0,034 per 1000 lb (\$0.01545 per 1000 kg). Table 18 summarizes the fuel-cost allocation of steam and electricity by using this approach.

TABLE 18 Established-Electricity-Cost Method of Fuel-Cost Allocation

Utility	Base	+	Unit cost heating steam	=	Total	Total fuel cost	Percent of total
Electricity 51,743 kW*	5.27		0		$5.27 mil/kWh	$273.00	46.48
Steam @ 335 lb/in^2 (abs) (2309 kPa) 100,000 lb/h (45,000 kg/h)	0.610		0.067		$0.677 per 1000 lb (450 kg)	$67.70	11.53
Steam @ 150 lb/in^2 (abs) (1034 kPa) 200,000 lb/h (90,000 kg/h)	0.487		0.053		$0.541 per 1000 lb (450 kg)	$108.25	18.43
Steam @ 35 lb/in^2 (abs) (241 kPa) 400,000 lb/h (180,000 kg/h)	0.312		0.034		$0.346 per 1000 lb (450 kg)	$138.40	23.56
					Total:	$587.35	100.00

*Net kW delivered to process plants should be delivered after deducting fixed mechanical loss and electrical loss of the alternator. Electricity unit cost charged to production would be slightly higher after this adjustment.
Combustion.

Related Calculations. The established-electricity-cost method is based on the assumption (or existence) of a reasonable and economically justifiable heat rate of the cycle being considered or used. The cost of the fuel must be charged to the process steam and electricity. Since the analysis does not distinguish between types of fuels or methods of heat release, this procedure can be used for coal, oil, gas, wood, peat, bagasse, etc. Also, the procedure can be used for steam generated by nuclear fission.

Cogeneration is suitable for a multitude of industries such as steel, textile, shipbuilding, aircraft, food, chemical, petrochemical, city and town district heating, etc. With the increasing cost of all types of fuels, cogeneration will become more popular than in the past.

Other approaches to cost allocations for cogeneration include: (1) capital cost segregation, (2) capital cost allocation by cost separation of major functions, (3) cost separation of joint components, (4) capital cost allocation based on single-purpose electric generating plant capital cost, (5) unit cost based on fixed annual capacity factor, and (6) unit cost based on fixed peak demand. Each method has its advantages, depending on the particular design situation.

In the two examples given here (the present and previous calculation procedures), water return to the dual-purpose turbine cycle is assumed to be of condensate quality. Hence, no capital and operating costs of water have been included. In actual cases, a cost account should be set up based on the quantity of the returned condensate. Special charges would be necessary for the unreturned portion of the water. Although the examples presented are for a fossil-fueled cycle, the methods are equally valid for a nuclear steam-turbine cycle. For a contrasting approach and for more data on where this procedure can be used, review the Related Calculations portion of the previous calculation procedure.

This calculation procedure is the work of Paul Leung of Bechtel Corporation, as reported at the 34th Annual Meeting of the American Power Conference and published in *Combustion* magazine. Since the procedure is based on thermodynamic and economic principles, it has wide applicability in a variety of industries.

With utility power plants—some 3500—reaching their 30th birthday within the next few years, designers are evaluating ways of repowering. When a plant is repowered, emissions are reduced, efficiency rises, as do reliability, output, and service life. So repowering has many attractions, including environmental benefits. More than 20 GW of capacity are estimated candidates for repowering.

Repowering replaces older facilities with new or different equipment. Several types of repowering are used today: (1) *Partial repowering*—which combines an existing plant system, infrastructure, and new equipment to provide increased output. *Example:* Combined-cycle repowering using a heat-recovery steam generator (HRSG) that recovers waste heat from the exhaust of a new gas-turbine/generator. *Example:* New gas-turbine/generator exhausts into existing boilers eliminating combustion-air-forced-draft needs while increasing the efficiency of the steam-generation cycle. Capital requirements are smaller than for an HRSG. This form of repowering is popular in Europe.

(2) *Station repowering*—reuses existing buildings, water-treatment systems, and fuel-handling system—but *not* the original steam cycle. New generating capability is installed to replace the existing steam plant—usually in the form of one or more gas turbines.

(3) *Site repowering*—uses an existing site but none of the equipment, such as boilers or turbines. Reusing an existing site eases permitting requirements, compared to developing a new site. To reduce overall project costs, it may be possible to reuse the infrastructure supporting the plant—such as power line and water- and fuel-delivery systems.

Specific methods for repowering include: (1) Combined-cycle repowering uses a new gas turbine and an HRSG to repower an existing facility by replacing or augmenting an existing boiler. (2) Gas turbines serving multiple-pressure HRSG provide power output and steam to existing steam-turbine generators. The gas-turbine power output goes directly to the utility's power lines. Natural gas fuels the gas turbine. (3) Pressurized fluidized-bed boilers are installed in place of existing boilers. Hot gases from the new boiler are used to drive a gas-turbine generator to increase the overall plant output. (4) Hot windbox repowering (also called the turbocharged boiler) adds a gas-turbine/generator to an existing plant. The high-temperature exhaust from the gas turbine is used as combustion air in the existing boiler. This eliminates—in most cases—the need for a forced-draft fan while the gas turbine is operating. Plants using this method of repowering, which is prevalent in Europe, boost efficiency by 10 to 15 percent and output by 20 to 33 percent.

Data presented here on repowering were reported by Steven Collins, assistant editor, in *Power* magazine.

ENVIRONMENTAL AND SAFETY-REGULATION RANKING OF EQUIPMENT CRITICALITY

Rank the criticality of a boiler-feed pump operating at 250°F (121°C) and 100 lb/in^2 (68.9 kPa) if its Mean Time Between Failures (MTBF) is 10 months, and vibration is an important element in its safe operation. Use the National Fire Protection Association (NFPA) ratings of process chemicals for health, fire, and reactivity hazards. Show how the criticality of the unit is developed.

Calculation Procedure:

1. *Determine the Hazard Criticality Rating (HCR) of the equipment*

Process industries of various types—chemical, petroleum, food, etc.—are giving much attention to complying with new process safety regulations. These efforts center on reducing hazards to people and the environment by ensuring the mechanical and electrical integrity of equipment.

To start a program, the first step is to evaluate the most critical equipment in a plant or factory. To do so, the equipment is first ranked on the basis of some criteria, such as the relative importance of each piece of equipment to the process or plant output.

The Hazard Criticality Rating (HCR) can be determined from a listing such as that in Table 19. This tabulation contains the analysis guidelines for assessing the Process Chemical Hazard (PCH)

TABLE 19 The Hazard Critically Rating (HCR) Is Determined in Three Steps*

Hazard criticality rating
1. Assess the process chemical hazard (PCH) by:
• Determining the NFPA ratings (*N*) of process chemicals for:
Health, fire, reactivity hazards
• Selecting the highest value of *N*
• Evaluating the potential for an emissions release (0 to 4):
High (RF = 0): Possible serious health, safety, or environmental effects
Low (RF = 1): Minimal effects
None (RF = 4): No effects
• *Then, PCH = N − RF*. (Round off negative values to zero.)
2. Rate other hazards (O) with an arbitrary number (0 to 4) if they are:
• Deadly (4), if:
Temperatures >1000°F
Pressures are extreme
Potential for release of regulated chemicals is high
Release causes possible serious health safety or environmental effects
Plant requires steam turbine trip mechanisms, fired-equipment shutdown systems, or toxic- or combustible-gas detectors*
Failure of pollution control system results in environmental damage*
• Extremely dangerous (3), if:
Equipment rotates at >5000 rpm
Temperatures >500°F
Plant requires process venting devices
Potential for release of regulated chemicals is low
Failure of pollution control system may result in environmental damage*
• Hazardous (2), if:
Temperatures >300°F;
Extended failure of pollution control system may cause damage*
• Equipment rotates at >3600 rpm
• Temperatures >140°F or pressures 20 psig
• Not hazardous (0), if:
No hazards exist
3. Select the higher value of PCH and O as the hazard criticality rating

Equipment with spares drop one category rating. A spare is an inline unit that can he immediately serviced or be substituted by an alternative process option during the repair period.

**Chemical Engineering*.

and the Other Hazards (O). The rankings for such a table of hazards should be based on the findings of an experienced team thoroughly familiar with the process being evaluated. A good choice for such a task is the plant's Process Hazard Analysis (PHA) Group. Since a team's familiarity with a process is highest at the end of a PHA study, the best time for ranking the criticality of equipment is toward the end of such safety evaluations.

From Table 19, the NFPA rating, N, of process chemicals for Health, Fire, and Reactivity, is $N = 2$, because this is the highest of such ratings for Health. The Fire and Reactivity ratings are 0, 0, respectively, for a boiler-feed pump because there are no Fire or Reactivity exposures.

The Risk Reduction Factor (RF), from Table 19, is RF = 0, since there is the potential for serious burns from the hot water handled by the boiler-feed pump. Then, the Process Chemical Hazard, PCH = N − RF = 2 − 0 = 2.

The rating of Other Hazards, O, Table 19, is O = 1, because of the high temperature of the water. Thus, the Hazard Criticality Rating, HCR = 2, found from the higher numerical value of PCH and O.

TABLE 20 The Process Criticality Rating (PCR)

	Process criticality rating
Essential (4)	The equipment is essential if failure will result in shutdown of the unit, unacceptable product quality, or severely reduced process yield
Critical (3)	The equipment is critical if failure will result in greatly reduced capacity, poor product quality, or moderately reduced process yield
Helpful (2)	The equipment is helpful if failure will result in slightly reduced capacity, product quality, or process yield
Not critical (1)	The equipment is not critical if failure will have little or no process consequences

Chemical Engineering.

2. *Determine the process criticality rating, PCR, of the equipment*
From Table 20, prepared by the PHA Group using the results of its study of the equipment in the plant, PCR = 3. The reason for this is that the boiler-feed pump is critical for plant operation because its failure will result in reduced capacity.

3. *Find the process and hazard criticality ranking, PHCR*
The alphanumeric PHC value is represented first by the alphabetic character for the category. For example, Category A is the most critical, while Category D is the least critical to plant operation. The first numeric portion represents the Hazard Criticality Rating, HCR, while the second numeric part represents the Process Criticality Rating, PCR. These categories and ratings are a result of the work of the PHA Group.

From Table 21, the Process and Hazard Criticality Ranking, PHCR = B23, This is based on the PCR = 3 and HCR = 2, found earlier.

4. *Generate a criticality list by ranking equipment using its alphanumeric PHCR values*
Each piece of equipment is categorized, in terms of its importance to the process, as: Highest Priority, Category A; High Priority, Category B; Medium Priority, Category C; Low Priority, Category D.

Since the boiler-feed pump is critical to the operation of this process, it is a Category B, i.e., High Priority item in the process.

5. *Determine the Criticality and Repetitive Equipment, CRE, value for this equipment*
This pump has an MTBF of 10 months. Therefore, Table 22, CRE = bl. Note that the CRE value will vary with the PCHR and MTBF values for the equipment.

TABLE 21 The Process and Hazard Criticality Rating

	PHC rankings				
Process criticality rating	Hazard criticality rating				
	4	3	2	1	0
4	A44	A34	A24	A14	A04
3	A43	B33	B23	B13	B02
2	A42	A32	C22	C12	C02
1	A41	B31	C21	CD11	D01

Note: The alphanumeric PHC value is represented first by the alphabetic character for the category (for example, Category A is the most critical while D is the least critical). The first numeric portion represents the Hazard Criticality Rating, and the second numeric part the Process Criticality Rating.
Chemical Engineering.

TABLE 22 The Criticality and Repetitive Equipment Values

	CRE values			
	Mean time between failures, months			
PHCR	0–6	6–12	12–24	>24
A	a1	a2	a3	a4
B	a2	b1	b2	b3
C	a3	b2	c1	c2
D	a4	b3	c2	d1

Chemical Engineering.

TABLE 23 Predictive Maintenance Frequencies for Rotating Equipment Based on Their CRE Values

	Maintenance cycles Frequency, days			
CRE	7	30	90	360
a1, a2	VM	LT		
a3, a4		VM	LT	
b1, b3			VM	
c1, d1				VM

LT = Lubrication sampling and testing, VM = vibration monitoring.

Chemical Engineering.

6. ***Determine equipment inspection frequency to ensure human and environmental safety***
From Table 23, this boiler feed pump requires vibration monitoring every 90 days. With such monitoring it is unlikely that an excessive number of failures might occur to this equipment.

7. ***Summarize criticality findings in spreadsheet form***
When preparing for a PHCR evaluation, as spreadsheet, Table 24, listing critical equipment, should be prepared. Then, as the various rankings are determined, they can be entered in the spreadsheet where they are available for easy reference.

Enter the PCH, Other, HCR, PCR, and PHCR values in the spreadsheet, as shown. These data are now available for reference by anyone needing the information.

TABLE 24 Typical Spreadsheet for Ranking Equipment Criticality

Spreadsheet for calculating equipment PHCRS										
		NFPA rating								
Equipment number	Equipment description	H	F	R	RF	PCH	Other	HCR	PCR	PHCR
TKO	Tank	4	4	0	0	4	0	4	4	A44
TKO	Tank	4	4	0	1	3	3	3	4	A34
PUIBFW	Pump	2	0	0	0	2	1	2	3	B23

Chemical Engineering.

Related Calculations. The procedure presented here can be applied to all types of equipment used in a facility: fixed, rotating, and instrumentation. Once all the equipment is ranked by criticality, priority lists can be generated. These lists can then be used to ensure the mechanical integrity of critical equipment by prioritizing predictive and preventive maintenance programs, inventories of critical spare parts, and maintenance work orders in case of plant upsets.

In any plant, the hazards posed by different operating units are first ranked and prioritized based on a PHA. These rankings are then used to determine the order in which the hazards need to be addressed. When the PHAs approach completion, team members evaluate the equipment in each operating unit using the PHCR system.

The procedure presented here can be used in any plant concerned with human and environmental safety. Today, this represents every plant, whether conventional or automated. Industries in which this procedure finds active use include chemical, petroleum, textile, food, power, automobile, aircraft, military, and general manufacturing.

This procedure is the work of V. Anthony Ciliberti, Maintenance Engineer, The Lubrizol Corp., as reported in *Chemical Engineering* magazine.

HEAT RECOVERY ENERGY AND FUEL SAVINGS

Determine the primary-fuel saving which can be produced by heat recovery if 150 M Btu/h (158.3 MJ/h) in the form of 650-lb/in^2 (gage) (4481.1-kPa) steam superheated to 750°F (198.9°C) is recovered. The projected average primary-fuel cost (such as coal, gas, oil, etc.) over a 12-year evaluation period for this proposed heat-recovery scheme is \$0.75 per 10^6 Btu (\$0.71 per million joules) lower heating value (LHV). Expected thermal efficiency of a conventional power boiler to produce steam at the equivalent pressure and temperature is 86 percent, based on the LHV of the fuel.

Calculation Procedure:

1. *Determine the value of the heat recovered during 1 year*

Entering Fig. 37 at the bottom at 1 year and project vertically to the curve marked \$0.75 per 10^6 LHV. From the intersection with the curve, project to the left to read the value of the heat recovered as \$5400 per year per M Btu/h recovered (\$5094 per MJ).

2. *Find the total value of the recovered heat*

The total value of the recovered heat = (hourly value of the heat recovered, \$/10^6 Btu)(heat recovered, 10^6 Btu/h)(life of scheme, years). For this scheme, total value of recovered heat = (\$5400)(150 × 10^6 Btu/h) (12 years) = \$9,720,000.

3. *Compute the total value of the recovered heat, taking the boiler efficiency into consideration*

Since the power boiler has an efficiency of 86 percent, the equivalent cost of the primary fuel would be \$0.75/0.86 – \$0.872 per 10^6 Btu (\$0,823 per million joules). The total value of the recovered heat if bought as primary fuel would be \$9,720,000(\$0.872/\$0.75) = \$11,301,119. This is nearly \$1 million a year for the 12-year evaluation period—a significant amount of money in almost any business. Thus, for a plant producing 1000 tons/day (900 t/day) of a product, the heat recovery noted above will reduce the cost of the product by about \$3.14 per ton, based on 258 working days per year.

Related Calculations. This general procedure can be used for any engineered installation where heat is available for recovery, such as power-generating plants, chemical-process plants, petroleum refineries, marine steam-propulsion plants, nuclear generating facilities, air-conditioning and refrigeration plants, building heating systems, etc. Further, the procedure can be used for these and any other heat-recovery projects where the cost of the primary fuel can be determined. Offsetting the value of any heat saving will be the cost of the equipment needed to effect this saving. Typical equipment used for heat savings include waste-heat boilers, insulation, heat pipes, incinerators, etc.

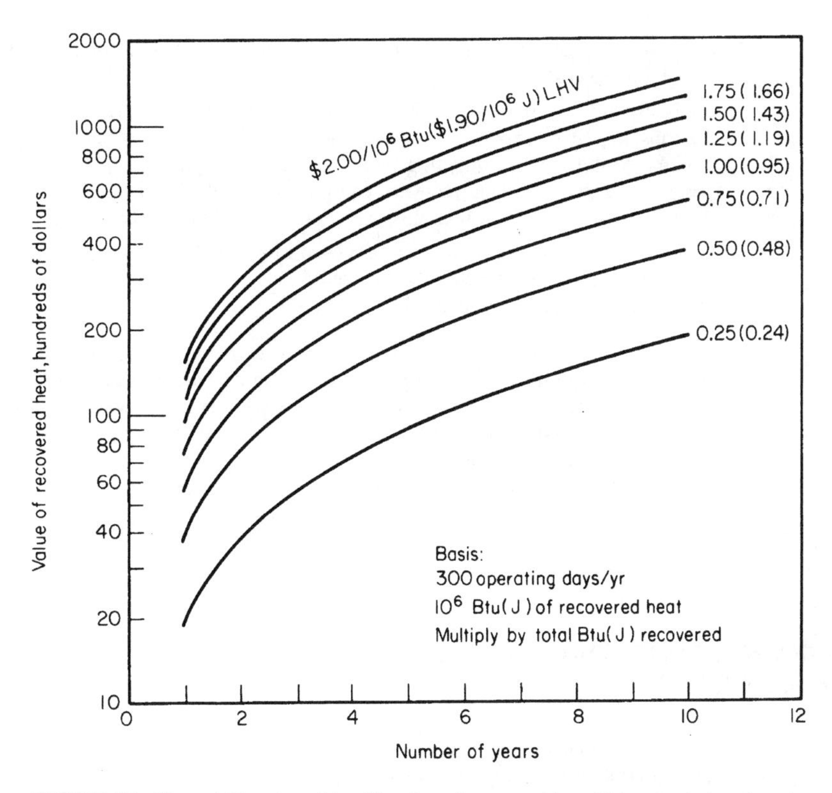

FIGURE 37 Chart yields value of 1 million Btu of recovered heat. This value is based on the projected average costs for primary fuel. (*Chemical Engineering.*)

With the almost certain continuing rise in fuel costs, designers are seeking new and proven ways to recover heat. Ways which are both popular and effective include the following:

1. Converting recovered heat to high-pressure steam in the 600- to 1500-lb/in^2 (gage) (4137- to 10,343-kPa) range where the economic value of the steam is significantly higher than at lower pressures.
2. Superheating steam using elevated-temperature streams to both recover heat and add to the economic value of the steam.
3. Using waste heat to raise the temperature of incoming streams of water, air, raw materials, etc.
4. Recovering heat from circulating streams of liquids which might otherwise be wasted.

In evaluating any heat-recovery system, the following facts should be included in the calculation of the potential savings:

1. The economic value of the recovered heat should exceed the value of the primary energy required to produce the equivalent heat at the same temperature and/or pressure level. An efficiency factor must be applied to the primary fuel in determining its value compared to that obtained from heat recovery. This was done in the above calculation.
2. An economic evaluation of a heat-recovery system must be based on a projection of fuel costs over the average life of the heat-recovery equipment.

3. Environmental pollution restrictions must be kept in mind at all time because they may force the use of a more costly fuel.
4. Many elevated-temperature process streams require cooling over a long temperature range. In such instances, the economic analysis should credit the heat-recovery installation with the savings that result from eliminating non-heat-recovery equipment that normally would have been provided. Also, if the heat-recovery equipment permits faster cooling of a stream and this time saving has an economic value, this value must be included in the study.
5. Where heat-recovery equipment reduces primary-fuel consumption, it is possible that plant operations can be continued with the use of such equipment whereas without the equipment the continued operation of a plant might not be possible.

The above calculations and comments on heat recovery are the work of J. P. Fanaritis and H. J. Streich, both of Struthers Wells Corp., as reported in *Chemical Engineering* magazine.

Where the primary-fuel cost exceeds or is different from the values plotted in Fig. 37, use the value of $1.00 per 10^6 Btu (J) (LHV) and multiply the result by the ratio of (actual cost, dollars per 10^6 Btu/$1)($/J/$1). Thus, if the actual cost is $3 per 10^6 Btu (J), solve for $1 per 10^6 Btu (J) and multiply the result by 3. And if the actual cost were $0.80, the result would be multiplied by 0.8.

HIGH-TEMPERATURE HOT-WATER HEATING ENERGY AND FUEL SAVINGS

Determine the fuel savings possible by using high-temperature-water (HTW) heating instead of steam if 50,000 lb/h (6.3 kg/s) of steam at 150 lb/in^2 (gage) (1034 kPa) is to be produced for delivering heat to equipment 1000 ft (305 m) from the boiler. The saturation temperature of the steam is 360°F (182°C); specific volume = 2.75 ft^3/lb (0.017 m^3/kg); enthalpy of evaporation = 857 Btu/lb (1996.8 kJ/ kg); enthalpy of saturated vapor = 1195.6 Btu/lb (2785.7 kJ/kg); ambient temperature = 70°F (21.1°C); steam velocity = 5000 ft/min (1524 m/s); density of water at 240°F (171.1°C) = 56 lb/ft^3 (896.6 kg/m^3).

Calculation Procedure:

1. ***Compute the required pipe cross-sectional area***

The required pipe cross-sectional area is A ft^2 (m^2) = Wv/V, where W = steam flow rate, lb/h (kg/h); v = specific volume of the steam, ft^3/lb (m^3/kg); V = steam velocity, ft/h (m/h). Or, A = 50,000(2.75)/[5000(60)] = 0.46 ft^2 or 66 in^2 (429 cm^2).

2. ***Choose the size of the steam pipe***

For a 5-lb/in^2 (34.4-kPa) pressure drop in the 1000-ft (305-m) pipeline, standard pressure-loss calculations given elsewhere in this handbook show that a 10-in (25.4-cm) diameter pipe would be suitable when used in conjunction with a 5-in (12.7-cm) condensate-return line. The 10-in (25.4-cm) line would have 2-in (5.1-cm) thick calcium silicate insulation, while the 5-in (12.7-cm) line would have 1-in (2.5-cm) thick insulation of the same material.

3. ***Compute the heat losses in the two lines***

Using the insulation heat-loss calculation methods given elsewhere in this handbook, we find the heat loss in the 10-in (25.4-cm) line is 183,200 Btu/h (53,678 W), while the heat loss in the 5-in (12.7-cm) condensate line is 78,100 Btu/h (22,833 W). Summing these, we see the total heat loss for the steam system is 261,300 Btu/h (76.6 kW).

4. *Determine the amount of condensate formed*
The amount of condensate formed w_c = (steam-line heat loss, Btu/h)/(enthalpy of vaporization, Btu/lb), or w_c = 183,200/857 = 214 lb/h (97.3 kg/h).

5. *Compute the amount of heat delivered to the load*
The amount of heat delivered to the load is H Btu/h = (steam flow rate, lb/h – condensate formation rate for heat loss, lb/h) (enthalpy of vaporization of the steam, Btu/lb). Or, for this steam system, H = (50,000 – 214)857 = 42,666,602 Btu/h (12,501 kW).

6. *Determine the condensate flash-out losses*
If the flash vapor is produced when the condensate is flashed out to atmospheric pressure in the return line and condensate receiver, the losses from flash-out will equal the enthalpy of the saturated water at 365°F (185°C) minus the enthalpy of the saturated water at 212°F (100°C), or 338.5 – 180 = 158.5 Btu/lb (369 kJ/kg).

To produce 857 Btu (904 J) of latent heat per pound of steam, the boiler must supply 1195.6 – 180 = 1015.6 Btu/lb (2366.3 kJ/kg), assuming that the condensate is returned to the boiler at 212°F (100°C). Hence, condensate losses from flash-out = (158.5/1015.6)100 = 15.6 percent. In addition, there is an approximate 5 percent loss due to leakage of steam and condensate, plus blow-down losses, which brings the total losses to 15.6 + 5.0 = 20.6, say 20 percent.

7. *Compute the total boiler heat input required*
With a condensate loss of 20 percent, as computed above, the amount of condensate returned to the boiler = 0.80(50,000 lb/h of steam) = 40,000 lb/h (5.04 kg/s). Hence, the enthalpy of the feedwater to the boiler, including makeup water, is 40,000 lb (18,181.8 kg) of condensate at 212°F (100°C) = 40,000(180 Btu/lb) = 7,200,000 Btu (7596 kJ); 10,000 lb (4545.4 kg) of makeup water at 50°F (10°C), 18 Btu/lb (41.9 kJ/kg), is 10,000(18) = 180,000 Btu (189,900 J); the sum = 7,380,000 Btu (7785 kJ). The boiler must therefore produce 50,000 × 1195.6 – 7,380,000 = 52,400,000 Btu/h (15,353.2 kW).

Assuming 75 percent boiler efficiency for this unit (a valid assumption for the usual steam heating boiler), we find the adjusted total amount of energy needed for steam heat = 52,400,000/0.75 = 69,867,000 Btu (73,710 kJ).

8. *Compute the hourly water flow rate*
To deliver 42,666,600 Btu/h (45,013.2 kJ) to the equipment, assume a 40°F (4.4°C) temperature drop between the supply and the return. Then the hourly flow rate = Btu/h heat required/temperature drop of the water, °F = 42,666,600/40 = 1,066,700 lb/h (134.3 kg/s).

9. *Choose the size pipe to use for the supply and return*
Assume a water flow velocity of 10 ft/s (3.05 m/s). Then the pipe area needed, from the relation in step 1 of this procedure, is 1,066,700/(3600)(10)(56) = 0.529 ft^2 (0.049 m^2). This area requires a 10-in (25.4-cm) pipe.

10. *Compute the heat loss in the piping*
The supply and return lines would require 2000 ft (609.6 m) of 10-in (25.4-cm) pipe with 2-in (5.1-cm) thick calcium silicate insulation. If the supply temperature is 360°F (182.2°C) and the return temperature is 320°F (160°C), the mean temperature would be (360 + 320)/2 = 340°F (171.1°C). Using the insulation heat-loss calculation methods given elsewhere in this handbook, we see that the heat loss in the supply and return lines is 326,800 Btu/h (96.8 kW). Hence, the total amount of heat which must be supplied to the water is 326,800 + 42,666,600 = 42,993,400 Btu/h (12,597 kW) before allowance is made for the efficiency of the boiler.

11. *Compare the steam and hot-water systems*
A typical hot-water heating boiler for a system such as this will have an operating efficiency of 77 percent. Using this value, we find the heat which must be supplied by the fuel = 42,993,400/0.77 = 55,835,600 Btu/h (16,360 kW).

As computed earlier, the heat required by the steam system exceeds that required by the hot-water system by 69,867,000 – 55,835,600 = 14,031,400 Btu/h (4111 kW), or 20 percent. This means that the high-temperature hot-water system will use 20 percent less fuel than the steam system for this installation.

Related Calculations. Use this approach when comparing or designing HTW systems for airports, military installations, hospitals, shopping centers, multifamily dwellings, garden apartments, industrial plants, central heating for large districts, university campuses, chemical-process plants, and similar installations. High-temperature-water systems are those using water in the 250 to 420°F (121.1 to 215.5°C) range, corresponding to a steam pressure of 300 lb/in^2 (gage) (2068.5 kPa). Mechanical problems caused by high water pressures above 420°F (215.5°C) make this temperature the practical upper limit. HTW systems can produce fuel savings 20 percent greater than systems using steam.

Studies show that conversion from steam to HTW is attractive—particularly for systems rated at 20,000,000 Btu/h (5860 kW) or higher. At this rating the conversion cost can usually be paid off in about 2 years. Smaller HTW systems, from 500,000 to 15,000,000 Btu/h (1470 to 4395 kW), are only marginally more economical to operate than steam, but they are still favored because they provide much more accurate and uniform temperature control.

HTW systems can give fuel savings of 20 to 50 percent, compared to an equivalent steam heating system. For new installations, the total capital investment is about the same for both steam and HTW systems. However, the savings in fuel costs and maintenance make the payout period for a new HTW system shorter than for conversion of an existing steam system.

Many plants use their steam boilers for both process and space heating. Cascade (direct-contact) heaters can generate up to 350°F (176.7°C) water from 150-lb/in^2 (1034-kPa) steam [or 400°F (204.4°C) from a 250-lb/in^2 (1724-kPa) boiler]. This water temperature is adequate for the rolls, presses, extruders, evaporators, conveyors, and reactors used in many industrial plants. Steam-pressure reducing valves are not needed to maintain the different temperature levels required by each machine.

Plants having steam boilers can convert to HTW heating simply and quickly by installing direct-contact water heaters in, or adjacent to, the boiler room. Such heaters can also serve as heat reservoirs, absorbing sudden peak loads and allowing the boilers to operate at fairly constant loads. HTW systems can easily supply water at elevated temperatures for process loads and water at lower temperatures for process loads and space-heating loads. Distribution efficiency of such systems approaches 95 percent overall.

For process applications requiring extremely close temperature control, the water circulating rate through a secondary loop can be designed to limit the difference between inlet and outlet temperatures to ±2°F (±1.11°C). The greater heat capacity of hot water over steam and the narrower pipelines required are other advantages. The usual HTW line need be only one or two sizes larger than the condensate line required in a steam system. The ratio of absolute heat-storing capacity is 42 to 1 in favor of HTW over steam. Where steam is needed in an all-HTW system, it can be obtained easily by flashing some of the water to steam. This calculation procedure is the work of William M. Teller, William Diskant, and Louis Malfitani, all of American Hydrotherm Corporation, as reported in *Chemical Engineering* magazine.

CONTROLS IN ENVIRONMENTAL AND ENERGY-CONSERVATION DESIGN

ENERGY PROCESS-CONTROL SYSTEM SELECTION

A continuous industrial process contains four process centers, each of which has two variables that must be controlled. If a fast process-reaction rate is required with only small to moderate dead time, select a suitable mode of control. The system contains more than two resistance-capacity pairs. What type of transmission system would be suitable for this process?

Calculation Procedure:

1. ***Compute the number of process capacities***

The number of process capacities = (number of process centers)(number of variables per center), or, for this system, $4 \times 2 = 8$ process capacities. This is defined as a *multiple* number of process capacities because the number controlled is greater than unity.

2. ***Analyze the process-time lags***

A small to moderate dead time is allowed in this process-control system. With such a dead-time allowance and with two or more resistance-capacity pairs in the system, a mode of control that provides for any number of process-time lags is desirable.

3. ***Select a suitable mode of control***

Table 25 summarizes the forms of control suited to processes having various characteristics. This table is a *general guide*—it provides, at best, an *approximate* aid in selecting control modes. Hence it is suitable for tentative selection of the mode of control. Final selection must be based on actual experience with similar systems.

Inspection of Table 25 shows that for a multiple number of processes with small to moderate dead time and any number of resistance-capacity pairs, a proportional plus reset mode of control is probably suitable. Further, this mode of control provides for any (i.e., fast or slow) reaction rate. Since a fast reaction rate is desired, the proportional plus reset method of control is suitable because it can handle any process-reaction rate.

4. ***Select the type of transmission system to use***

Four types of transmission systems are used for process control today: pneumatic, electric, electronic, and hydraulic. The first three types are by far the most common.

Pneumatic transmission systems use air at 3 to 20 lb/in^2 (gage) (20.7 to 137.9 kPa) to convey the control signal through small-bore metal tubing at distances ranging to several thousand feet. The air used in pneumatic systems must be clean and dry. To prevent a process from getting out of control, a constant supply of air is required. Pneumatic controllers, receivers, and valve positioners usually have small air-space volumes of 5 to 10 in^3 (81.0 to 163.9 cm^3). Air motors of the diaphragm or piston type have relatively large volumes: 100 to 5000 in^2 (1639 to 81,935 cm^3).

Pneumatic control systems are generally considered to be spark-free. Hence, they find wide use in hazardous process areas. Also, control air is readily available, and it can be "dumped" to the atmosphere safely. The response time of pneumatic control systems may be slower than that of electric or hydraulic systems.

Electric and electronic control systems are fast-response with the signal conveyed by a wire from the sensing point to the controller. In hazardous atmospheres the wire must be protected against abrasion and breakage.

Hydraulic control systems are also rapid-response. These systems are capable of high power actuation. Slower-acting hydraulic systems use fluid pressures in the 50 to 100 lb/in^2 (344.8 to 689.5 kPa) range; fast-acting systems use fluid pressures to 5000 lb/in^2 (34,475 kPa).

Dirt and fluid flammability are two factors that may be disadvantages in certain hydraulic-control-system applications. However, new manufacturing techniques and nonflammable fluids are overcoming these disadvantages.

Since a fast response is desired in this process-control system, electric, electronic, or hydraulic transmission of the signals would be considered first. With long distances between the sensing points [say 1000 ft (305 m) or more], an electric or electronic system would probably be best.

Next, determine whether the systems being considered can provide the mode of control (step 2) required. If a system cannot provide the necessary mode of control, eliminate the system from consideration.

Before a final choice of a system is made, other factors must be considered. Thus, the relative cost of each type of system must be determined. Should an electric system prove too costly, the slightly slower response time of the pneumatic system might be accepted to reduce the initial investment.

TABLE 25 Process Characteristics versus Mode of Control

Number of process capacities	Process reaction rate	Process time lags		Load changes		Suitable mode of control
		Resistance capacity (*RC*)	Dead time (transportation)	Size	Speed	
Single	Slow	Moderate to large	Small	Any	Any	Two-position; two-position with differential gap
				Moderate	Slow	Multiposition; proportional input
Single (self-regulating)	Fast	Small	Small	Any	Slow	Floating modes: Single speed, multispeed
					Moderate	Proportional-speed floating
Multiple	Slow to moderate	Moderate	Small	Small	Moderate	Proportional position
Multiple	Moderate	Any	Small	Small	Any	Proportional plus rate
Multiple	Any	Any	Small to moderate	Large	Slow to moderate	Proportional plus reset
Multiple	Any	Any	Small	Large	Fast	Proportional plus reset plus rate
Any	Faster than that of the control system	Small or nearly zero	Small to moderate	Any	Any	Wideband proportional plus fast reset

Considine—*Process Instruments and Controls Handbook*, McGraw-Hill.

Other factors influencing the choice of the type of a control system include type of controls, if any, currently used in the installation, skill and experience of the operating and maintenance personnel, type of atmosphere in which, and type of process for which, the controls will be used. Any of these factors may alter the initial choice.

Related Calculations. Use this general method to make a preliminary choice of controls for continuous processes, intermittent processes, air-conditioning systems, combustion-control systems, etc. Before making a final choice of any control system, be certain to weigh the cost, safety, operating, and maintenance factors listed above. Last, the system chosen *must be* able to provide the mode of control required.

PROCESS-ENERGY TEMPERATURE CONTROL SYSTEM SELECTION

A water-storage tank (Fig. 38) contains 500 lb (226.8 kg) of water at 150°F (65.6°C) when full. Water is supplied to the tank at 50°F (10°C) and is withdrawn at the rate of 25 lb/min (0.19 kg/s). Determine the process-time constant and the zero-frequency process gain if the thermal-sensing pipe contains 15 lb (6.8 kg) of water between the tank and thermal bulb and the maximum steam flow to the tank is 8 lb/min (0.060 kg/s). The steam flow to the tank is controlled by a standard linear regulating valve whose flow range is 0 to 10 lb/min (0 to 0.076 kg/s) when the valve operator pressure changes from 5 to 30 lb/in^2 (34.5 to 206.9 kPa).

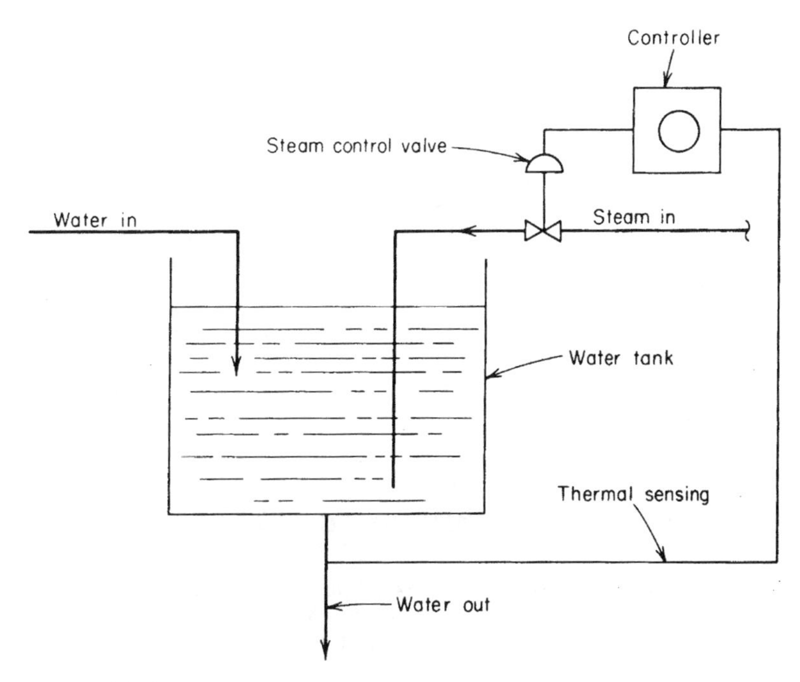

FIGURE 38 Temperature control of a simple process.

Calculation Procedure:

1. ***Compute the distance-velocity lag***

The time in minutes needed for the thermal element to detect a change in temperature in the storage tank is the *distance-velocity lag*, which is also called the *transportation lag*, or *dead time*. For this process, the distance-velocity lag d is the ratio of the quantity of water in the pipe between the tank and the thermal bulb—that is, 15 gal (57.01 L)—and the rate of flow of water out of the tank—that is, 25 lb/min (0.114 kg/s)—or $d = 15/25 - 0.667$ min.

2. ***Compute the energy input to the tank***

This is a *transient-control process;* i.e., the conditions in the process are undergoing constant change instead of remaining fixed, as in *steady-state conditions*. For transient-process conditions the heat balance is $H_{in} = h_{out} + H_{stor}$, where H_{in} = heat input, Btu/min; H_{out} = heat output, Btu/min; H_{stor} = heat stored, Btu/min.

The heat input to this process is the enthalpy of vaporization h_{fg} Btu/(lb · min) of the steam supplied to the process. Since the regulating valve is linear, its sensitivity s is (flow-rate change, lb/min)/(pressure change, lb/in^2). Or, by using the known valve characteristics, $s = (10 - 0)/(30 - 5) = 0.4$ (lb/min)/(lb/in^2) [0.00044 kg/(kPa · s)].

With a change in steam pressure of p lb/in^2 (p' kPa) in the valve operator, the change in the rate of energy supply to the process is $H_{in} = 0.4$ (lb/min)/(lb/in^2) $\times p \times h_{fg}$. Taking h_{fg} as 938 Btu/lb (2181 kJ/kg) gives $h_{in} = 375p$ Btu/min ($6.6p'$ kW).

3. ***Compute the energy output from the system***

The energy output H_{out} = lb/min of liquid outflow × liquid specific heat, Btu/(lb · °F) × (T_a – 150°F), where T_a = tank temperature, °F, at any time. When the system is in a state of equilibrium, the temperature of the liquid in the tank is the same as that leaving the tank or, in this instance, 150°F (65.6°C). But when steam is supplied to the tank under equilibrium conditions, the liquid temperature will rise to 150 + T_r, where T_r = temperature rise, °F (T_r, °C), produced by introducing steam into the water. Thus, the above equation becomes H_{out} = 25 lb/min × 1.0 Btu/(lb · °F) × $T_r = 25T_r$ Btu/min ($0.44T_r$ kW).

4. ***Compute the energy stored in the system***

With the rapid mixing of the steam and water, H_{stor} = liquid storage, lb × liquid specific heat, Btu/(lb · °F) × $T_r q = 500 \times 1.0 \times t_r q$, where q = derivative of the tank outlet temperature with respect to time.

5. ***Determine the time constant and process gain***

Write the process heat balance, substituting the computed values in $H_{in} = H_{out} + H_{stor}$, or $375p = 25T_r + 500T_r q$. Solving gives $T_r/p = 375/(25 + 500q) = 15/(1 + 20q)$.

The denominator of this linear first-order differential equation gives the process-system time constant of 20 min in the expression $1 + 20q$. Likewise, the numerator gives the zero-frequency process gain of 15°F/(lb/in^2) (1.2°C/kPa).

Related Calculations. This general procedure is valid for any liquid using any gaseous heating medium for temperature control with a single linear lag. Likewise, this general procedure is also valid for temperature control with a double linear lag and pressure control with a single linear lag.

ENERGY PROCESS-CONTROL VALVE SELECTION

Select a steam-control valve for a heat exchanger requiring a flow of 1500 lb/h (0.19 kg/s) of saturated steam at 80 lb/in^2 (gage) (551.6 kPa) at full load and 300 lb/h (0.038 kg/s) at 40 lb/in^2 (gage) (275.8 kPa) at minimum load. Steam at 100 lb/in^2 (gage) (689.5 kPa) is available for heating.

Calculation Procedure:

1. ***Compute the valve-flow coefficient***

The valve-flow coefficient C_υ is a function of the maximum steam flow rate through the valve and the pressure drop that occurs at this flow rate. In choosing a control valve for a process-control system, the usual procedure is to assume a maximum flow rate for the valve based on a considered judgment of the overload the system may carry. Usual overloads to not exceed 25 percent of the maximum rated capacity of the system. Using this overload range as a guide, assume that the valve must handle a 20 percent overload, or 0.20 (1500) = 300 lb/h (0.038 kg/s). Hence, the rated capacity of this valve should be 1500 + 300 = 1800 lb/h (0.23 kg/s).

The pressure drop across a steam-control valve is a function of the valve design, size, and flow rate. The most accurate pressure-drop estimate usually available is that given in the valve manufacturer's engineering data for a specific valve size, type, and steam-flow rate. Without such data, assume a pressure drop of 5 to 15 percent across the valve as a first approximation. This means that the pressure loss across this valve, assuming a 10 percent drop at the maximum steam-flow rate, would be $0.10 \times 80 = 8$ lb/in^2 (gage) (55.2 kPa).

With these data available, compute the valve flow coefficient from $C_\upsilon = WK/3(\Delta p\ P_2)^{0.5}$, where W = steam flow rate, lb/h; $K = 1 + (0.007 \times$ °F superheat of the steam); p = pressure drop across the valve at the maximum steam-flow rate, lb/in^2; P_2 = control-valve outlet pressure at maximum steam flow rate, lb/in^2 (abs). Since the steam is saturated, it is not superheated and $K = 1$. Then $C_\upsilon = 1500/3(8 \times 94.7)^{0.5} = 18.1$.

2. ***Compute the low-load steam flow rate***

Use the relation $W = 3(C_\upsilon\ \Delta p\ P_2)^{0.5}/K$, where all the symbols are as before. Thus, with a 40-lb/in^2 (gage) (275.8-kPa) low-load heater inlet pressure, the valve pressure drop is 80 − 40 = 40 lb/in^2 (gage) (275.8 kPa). The flow rate through the valve is then $W = 3(18.1 \times 40 \times 54.7)^{0.5}/1 = 598$ lb/h (0.75 kg/s).

Since the heater requires 300 lb/h (0.038 kg/s) of steam at the minimum load, the valve is suitable. Had the flow rate of the valve been insufficient for the minimum flow rate, a different pressure drop, i.e., a larger valve, would have to be assumed and the calculation repeated until a flow rate of at least 300 lb/h (0.038 kg/s) was obtained.

Related Calculations. The flow coefficient C_υ of the usual 1-in (2.5-cm) diameter double-seated control valve is 10. For any other size valve, the approximate C_υ valve can be found from the product $10 \times d^2$, where d = nominal body diameter of the control valve. Thus, for a 2-in (5.1-cm) diameter valve, $C_\upsilon = 10 \times 2^2 = 40$. By using this relation and solving for d, the nominal diameter of the valve analyzed in steps 1 and 2 is $d = (d_\upsilon/10)^{0.5} = (18.1/10)^{0.5} = 1.35$ in (3.4 cm); use a 1.5-in (3.8-cm) valve because the next smaller standard control valve size, 1.25 in (3.2 cm), is too small. Standard double-seated control-valve sizes are ¾, 1, 1¼, 1½, 2, 2½, 3, 4, 6, 8, 10, and 12 in (1.9, 2.5, 3.2, 3.8, 5.1, 6.4, 7.6, 10.2, 15.2, 20.3, 25.4, 30.5 cm). Figure 39 shows typical flow-lift characteristics of popular types of control valves.

To size control valves for liquids, use a similar procedure and the relation $C_\upsilon = V(G/\Delta p)$, where V = flow rate through the valve, gal/min; Δp = pressure drop across the valve at maximum flow rate, lb/in^2; G = specific gravity of the liquid. When a liquid has a specific gravity of 100 SSU or less, the effect of viscosity on the control action is negligible.

To size control valves for gases, use the relation $C = Q(GT_a)^{0.5}/1360(\Delta p\ P_2)^{0.5}$, where Q = gas flow rate, ft^3/h at 14.7 lb/in^2 (abs) (101.4 kPa) and 60°F (15.6°C); T_a = temperature of the flowing gas, °F abs = 460 + °F; other symbols as before. When the valve outlet pressure P_2 is less than $0.5P_1$, where P_1 = valve inlet pressure, use the value of $P_1/2$ in place of $(\Delta p\ P_2)^{0.5}$ in the denominator of the above relation.

To size control valves for vapors other than steam, use the relation $C_\upsilon = W(\upsilon_2/\Delta p)^{0.5}/63.4$, where W = vapor flow rate, lb/h; υ_2 = specific volume of the vapor at the outlet pressure P_2, ft^3/lb; other symbols as before. When P_2 is less than $0.5P_1$, use the value of $P_1/2$ in place of Δp and use the corresponding value of υ_2 at $P_1/2$.

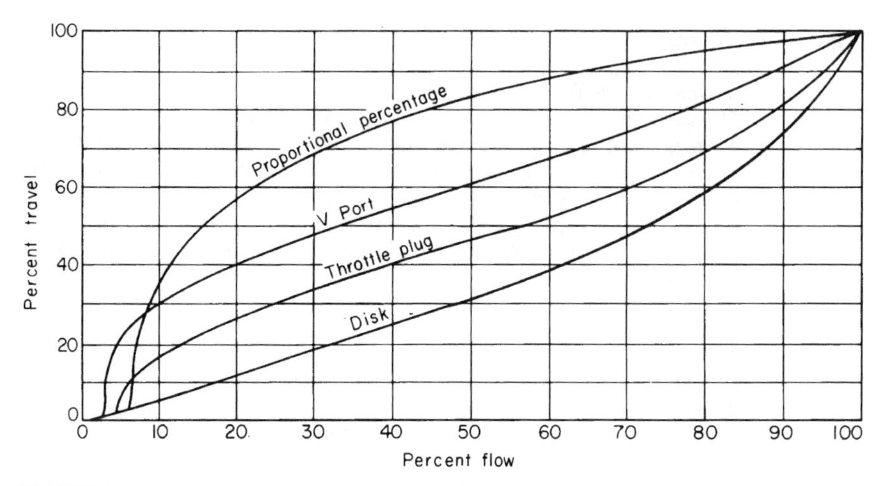

FIGURE 39 Flow-lift characteristics of control valves. (*Taylor Instrument Process Control Division of Sybron Corporation.*)

When the control valve handles a flashing mixture of water and steam, compute C_v by using the relation for liquids given above after determining which pressure drop to use in the equation. Use the *actual* pressure drop or the *allowable* pressure drop, whichever is smaller. Find the allowable pressure drop by taking the product of the supply pressure and the correction factor R, where R is obtained from Fig. 40. For a further discussion of control-valve sizing, see Considine—*Process Instruments and Controls Handbook*, McGraw-Hill, and G. F. Brockett and C. F. King—"Sizing Control Valves Handling Flashing Liquids," Texas A & M Symposium.

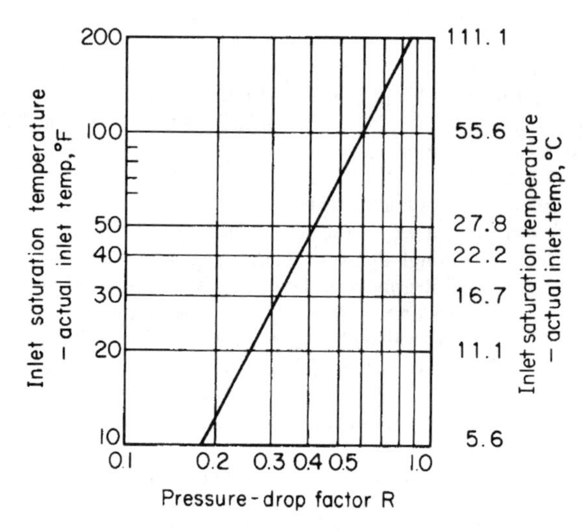

FIGURE 40 Pressure-drop correction factor for water in the liquid state. (*International Engineering Associates.*)

CONTROL-SYSTEM CONTROLLED-VOLUME-PUMP CHOICE

Select a controlled-volume pump to deliver 80 gal/h (0.084 L/s) of 100°F (37.8°C) distilled water to a chemical-feed system operating at 2000 lb/in^2 (abs) (13,790 kPa). What is the net positive suction head (NPSH) at the beginning of the pump suction stroke if the supply tank produces a 2-lb/in^2 (gage) (13.8-kPa) suction head at the pump centerline? Compute the minimum allowable NPSH for this pump if the length of the 1.5-in (3.8-cm) pipe between the pump and suction tank is 30 ft (9.1 m).

Calculation Procedure:

1. *Choose the general type of pump to use*

Controlled-volume pumps serve two functions when used as the final control elements in a control loop: to deliver liquid at the required pressure and to deliver liquid in the required quantities. In its second role, the pump also serves as a meter.

Two types of controlled-volume pumps are popular: plunger and diaphragm. The plunger pump is of somewhat simpler construction and is often used where contact of the plunger and liquid handled is not objectionable. Since distilled water is a relatively bland liquid, a plunger pump will be the tentative first choice for this control application.

2. *Determine the pump dimensions and speed*

The capacity, dimensions, speed, and efficiency of plunger-type controlled-volume pumps are given by $Q = D^2LNE/K$, where Q = pump capacity, gal/h (L/s); D = plunger diameter, in (cm); L = plunger stroke length, in (cm); N = number of strokes per minute, i.e., the pump speed; E = volumetric efficiency of the pump; K = dimensional constant = 4.92 for Q gal/h (0.0052 L/s), 295 for pump capacity in gal/min (18.6 L/s), and 0.0013 for pump capacity in mL/h.

Assume a pump speed of 50 strokes/min. This is a typical speed for plunger-type controlled-volume pumps. The usual efficiency of such a pump is 90 percent. With a 3-in (7.6-cm) stroke, the plunger diameter is $D = (QK/LNE)^{0.5} = [(80 \times 4.92)/(3 \times 50 \times 0.9)]^{0.5} = 1.71$ in, say 1.75 in (4.4 cm). This is a standard pump-plunger diameter.

3. *Compute the pump NPSH*

Use the relation NPSH $= P_a \pm P_h - P_\upsilon$, where NPSH = pump net positive suction head, lb/in^2 (abs) (kPa); P_a = atmospheric pressure at pump location = 14.7 lb/in^2 (abs) (101.4 kPa) at sea level; P_h = pressure head of liquid column above (+) or below (−) the centerline of the pump suction, lb/in^2 (gage) (kPa); P_υ = vapor pressure of the liquid at the pumping temperature, lb/in^2 (abs) (kPa).

From the steam tables, P_υ = 0.949 lb/in^2 (abs) (6.54 kPa) for water at 100°F (37.8°C). With an atmospheric pressure of 14.7 lb/in^2 (abs) (101.4 kPa), NPSH = 14.7 + 2.0 − 0.949 = 15.751 lb/in^2 (abs) (108.6 kPa).

4. *Compute the pump minimum NPSH*

Use the relation $\text{NPSH}_{\min} = sL_p/LN^2D^2/120{,}000D_p^2$, where $\text{NPSH}_{\min}$ = minimum net positive suction head with which the pump can operate, lb/in^2 (abs) (kPa); s = specific gravity of liquid handled; L_p = length of suction pipe, ft (m); D_p = section pipe inside diameter, in (cm); other symbols as given before. Assuming a specific gravity of 1.0 $(\text{NPSH}_{\min}) = (1.0 \times 30 \times 3 \times 50 \times 50 \times 1.5 \times 1.5)/(120{,}000 \times 1.5 \times 1.5) = 1.87$ lb/in^2 (abs) (12.9 kPa).

Since the available NPSH [15.751 lb/in^2 (abs) (108.6 kPa), step 3] is greater than the minimum NPSH required [1.87 lb/in^2 (abs) (12.9 kPa), step 4], the pump will operate satisfactorily and without cavitation.

Related Calculations. Use this general procedure to choose controlled-volume metering pumps for control systems requiring flows ranging from 1 mL/h to 20 gal/min (1 mL/h to 1.3 L/s) or more, at pressures ranging to 50,000 lb/in^2 (344,750 kPa). Typical applications for which this procedure is valid include chemical feed, ratioing, proportioning, and control of process variables.

INDUSTRIAL-STEAM BOILER ENERGY CONTROL CHOICE AND USAGE

Choose a suitable feedwater regulator and combustion control for an industrial boiler serving the following loads: heating, 18,000 lb/h (2.3 kg/s); process, 100,000 lb/h (12.6 kg/s); miscellaneous uses, 12,000 lb/h (1.5 kg/s). The boiler will have a maximum overload of 20 percent, and wide load fluctuations are expected at frequent intervals during operation. Pulverized-coal fuel is used to fire the boiler.

Calculation Procedure.

1. ***Determine the required boiler rating***
Find the sum of the individual loads on the boiler, or 18,000 + 100,000 + 12,000 = 130,000 lb/h (16.4 kg/s). With a 20 percent overload, the boiler rating must be 1.2(130,000) = 156,000 lb/h (19.7 kg/s). With a 10 percent additional reserve capacity to provide for unusual loads, the rated boiler capacity should be 1.1(156,000) × 171,500 lb/h, say 175,000 lb/h (22.0 kg/s) for selection purposes.

2. ***Choose the type of feedwater regulator to use***
Table 26 summarizes typical feedwater regulators used for boilers of various capacities. Study of Table 26 shows that a boiler in the 75,000 to 200,000 lb/h (9.4 to 25.2 kg/s) capacity range can use a relay-operated regulator with one or two elements when the load fluctuations are reasonable. With wide load swings, the relay-operated three-element regulator is a better choice. Since this boiler will encounter wide load swings, a three-element regulator is a wise and safe choice.

3. ***Choose the type of combustion-control system***
Table 27 summarizes the important selection features of four types of combustion-control systems. Study of Table 27 shows that a stream flow–air flow type of combustion-control system would probably be best for the fuel and load conditions in this plant. Hence, this type of control system will be chosen.

TABLE 26 Boiler-Feedwater-Regulator Selector Chart*

	Type of feedwater regulator		
Boiler capacity	Self-operated single-element	Relay-operated single- or two-element	Relay-operated three-element
Below 75,000 lb/h (9.4 kg/s)	For steady loads (building heating or continuous processes)	For irregular loads (batch processes, hoists, rolling mills, etc.)	
75,000–200,000 lb/h (9.4–25.2 kg/s)	Use only in special cases	For all steady and fluctuating loads	For extreme load and water conditions and boilers with steaming economizers
Above 200,000 lb/h (25.2 kg/s)	—	Use only on steady loads	For all types of loads

From Kallen—*Handbook of Instrumentation and Controls*, McGraw-Hill, 1961.

*Excess pressure ahead of feedwater regulator should be at least 50 lb/in^2 (344.8 kPa) and should be controlled by regulation of the feed pump. Use excess-pressure valves only when excess pressure varies more than plus or minus 30 percent. Where drum level is unsteady owing to high solids concentration or boiler feed or other causes, use next-higher-class freed regulator.

TABLE 27 Classification of Combustion-Control Systems

	A, series-fuel	*B*, series-air	*C*, parallel	*D*, calorimeter or steam flow–air flow
Action	Temperature- or pressure-actuated master adjusts fuel rate; fuel meter adjusts air flow	Temperature- or pressure-actuated master adjusts air flow; air-flow meter adjusts fuel flow	Temperature- or pressure-actuated master adjusts fuel flow and air flow simultaneously	Pressure-actuated master adjusts fuel flow; steam flow adjusts air flow
Relative speed of control	Master adjusted for fast response because fuel-rate fluctuations caused by fluctuating pressure or temperature on master do not have correspondingly fast effect on that controlled variable	Master adjusted for slow response, because air-flow fluctuations following fast fluctuating master signal have a rapid effect on controlled variable and may cause hunting action if airflow response is too fast	Master adjusted for slow response for same reason as in series-air	Master adjusted for fast response for same reason as in series-fuel. Steam flow–air flow control can be relatively rapid since steam-flow fluctuations are not so rapid as pressure variations
Used on fuels	Easily metered fuels such as oil and gas	All fuels. Oil, gas, and coal, either solid or burned in suspension	Primarily on solid fuels (grate firing)	Fuels hard to meter or fuels burned simultaneously. Commonly used on pulverized-coal-fired boilers
Advantages	When fuel may be in short supply, eliminates possibility of carrying high excess air for long period	Eliminates possibility of explosive mixture in combustion space when air fails. Eliminates need of fuel cutback for this purpose	Relatively inexpensive control system. No metering necessary	Ensures proper air-fuel ratio, even though fuel cannot be accurately metered or is of varying heat content Ensures this condition even when burning a mixture of different fuels at the same time

From Kallen—*Handbook of Instrumentation and Controls*, McGraw-Hill, 1961.

Related Calculations. Any control system selected for a boiler by using this procedure should be checked out by studying the engineering data available from the control-system manufacturer. The procedure given here is valid for heating, industrial, power, marine, and similar boilers.

CHARACTERISTICS AND RANGEABILITY OF ENERGY CONTROL VALVES

A flow-control valve will be installed in a process system in which the flow may vary from 100 to 20 percent while the pressure drop in the system rises from 5 to 80 percent. What is the required rangeability of the control valve? What type of control-valve characteristic should be used? Show how the effective characteristic is related to the pressure drop that the valve should handle.

Calculation Procedure:

1. *Compute the required valve rangeability*

Use the relation $R = (Q_1/Q_2)(\Delta P_2/\Delta P_1)^{0.5}$, where R = valve rangeability; Q_1 = valve initial flow, percentage of total flow; Q_2 = valve final flow, percentage of total flow; P_1 = initial pressure drop across the valve, percentage of total pressure drop; P_2 = percentage of final pressure drop across the valve.

Substituting gives $R = (100/20)\,(80/5)^{0.5} = 20$.

2. *Select the type of valve characteristic to use*

Table 28 lists the typical characteristics of various control valves. Study of Table 28 shows that as equal-percentage valve must be used if a rangeability of 20 is required. Such a valve has equal stem movements for equal-percentage changes in flow at a constant pressure drop based on the flow occurring just before the change is made.* The equal-percentage valve finds use where large rangeability is desired and where equal-percentage characteristics are necessary to match the process characteristics.

TABLE 28 Control-Valve Characteristics

Valve type	Typical flow rangeability	Stem movement
Linear	12–1	Equal stem movement for equal flow change
Equal-percentage	30–1 to 50–1	Equal stem movement for equal-percentage flow change*
On-off	Linear for first 25% of travel: on-off thereafter	Same as linear up to on-off range

*At constant pressure drop.

3. *Show how the valve effective characteristic is related to pressure drop*

Figure 41 shows the inherent and effective characteristics of typical linear, equal-percentage, and on-off control valves. The inherent characteristic is the theoretical performance of the valves.* If a valve is to operate at a constant load without changes in the flow rate, the characteristic of the valve is important, since only one operating point of the valve is used.

Figure 41*b* and *c* gives definite criteria for the amount of pressure drop the control valve should handle in the system. This pressure drop is not an arbitrary value such as 5 lb/in^2 (34.4 kPa) but rather a percentage of the total dynamic drop. The control valve should take at least 33 percent of the total dynamic system pressure drops* if an equal-percentage valve is used and is to retain its inherent characteristics. A linear valve should not take less than a 50 percent pressure drop if its linear properties are desired.

*E. Ross Forman. "Fundamentals of Process Control," *Chemical Engineering*, June 21, 1965.

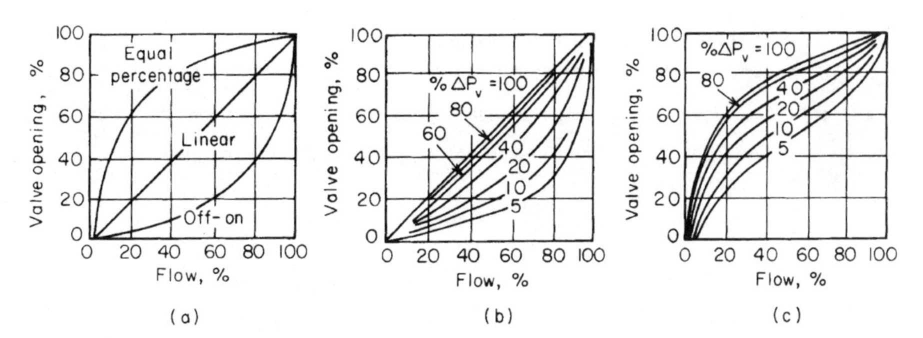

FIGURE 41 (*a*) Inherent flow characteristics of valves at constant pressure drop; (*b*) effective characteristics of a linear valve; (*c*) effective characteristics of a 50:1 equal-percentage valve.

There is an economic compromise in the selection of every control valve. Where possible, the valve pressure drop should be as high as needed to give good control. If experience or an economic study dictates that the requirement of additional horsepower to provide the needed pressure is not worth the investment in additional pumping or compressor capacity, the valve should take less pressure drop with the resulting poorer control.

STEAM-CONTROL AND PRESSURE-REDUCING VALVE SIZING FOR MAXIMUM ENERGY SAVINGS

Dry saturated steam at 30 lb/in^2 (abs) (206.9 kPa) will flow at the rate of 1000 lb/h (0.13 kg/s) through a single-seat pressure-reducing throttling valve. The desired exit pressure is 20 lb/in^2 (abs) (137.9 kPa) at the valve outlet. Select a valve of suitable size.

Calculation Procedure:

1. *Determine the critical pressure for the valve*

Critical pressure exists in a valve and piping system when the pressure at the valve outlet is 58 percent, or less, of the absolute inlet pressure for saturated steam (55 percent for hydrocarbon vapors and superheated steam). Thus, for this system the critical outlet pressure is $P_c = 0.58P_i$, where P_c = critical pressure for the system, lb/in^2 (abs) (kPa); P_i = inlet pressure, lb/in^2 (abs) (kPa). Or, $P_c = 0.58(30) =$ 17.4 lb/in^2 (abs) (119.9 kPa).

Since the outlet pressure, 20 lb/in^2 (abs) (137.9 kPa), is greater than the critical pressure, the flow through the valve is noncritical.

2. *Find the density of the outlet steam*

Assume adiabatic expansion of the steam from 30 lb/in^2 (abs) (206.9 kPa) to 20 lb/in^2 (abs) (137.9 kPa). (This is a valid assumption for a throttling process such as that which takes place in a pressure-reducing valve.) Using the steam tables, we find the density of the steam at the outlet pressure of 20 lb/in^2 (abs) (137.9 kPa) is 0.05 lb/ft^3 (0.8 kg/m^3).

3. *Compute the valve flow coefficient c_υ*

Use the relation $C_\upsilon = W/63.5\sqrt{(P_i - P_2)\rho}$, where C_υ = valve flow coefficient, dimensionless; W = steam (or vapor or gas) flow rate, lb/h (kg/s); P_i = valve inlet pressure, lb/in^2 (abs); p = density of the vapor or gas flowing through the valve, lb/ft^3 (kg/m^3). For this valve, $C_\upsilon = 1000/63.5\sqrt{(30-20)0.05} = 22.3$, say 22.0 because C_υ valves are usually stated in even numbers for larger-size valves.

TABLE 29 Flow Coefficients for Steam-Control Valves

Size		Straight-through Throttling		Straight-through on-off	Straight-through regulators	
in	cm	Single Seat	Double seat	Single seat	Single seat	Double seat
1/8	0.32	0.23				
¼	0.64	0.78				
3/8	0.95	1.7				
½	1.27	3.2				
¾	1.91	5.4	7.2	7	3.6	4.3
1	2.54	9	12	12	6	7.2
1¼	3.18	14	18	18	9	10.8
1½	3. 81	21	28	27	14	16.8
2	5.08	36	48	42	24	28.8
2½	6.35	54	72	65	36	432
3	7.62	75	100	93	50	60
4	10.2	124	165	170	83	99
6	15.2	270	360	380	180	216
8	20.3	480	640	660	320	384
10	25.4	750	1000	1100	500	600
12	30.5	1080	1440	1550	720	864

Chemical Engineering.

4. ***Select the control valve to use***

At normal operating conditions, most engineers recommend that the flow through the valve not exceed 80 percent of the maximum flow possible. Thus the valve selected should have a C_υ equal to or greater than the computed $C_\upsilon/0.80$. Thus, for this valve, choose a unit having a C_υ equal to or greater than $22/0.80 = 27.5$. From Table 29 choose a 2-in (5.08-cm) single-seat valve having a C_υ of 36.

The operating C_υ of any valve is $C_{\upsilon o} = C_{\upsilon f}/C_{\upsilon s}$, where $C_{\upsilon f} = C_\upsilon$ value computed by the formula in step 3 and $C_{\upsilon s} = C_\upsilon$ of actual valve selected. Or, for this valve, $C_{\upsilon o} = 22/36 = 0.61$.

To avoid wire drawing, which occurs when the valve plug operates too close to the valve seat, $C_{\upsilon o}$ values of less than 0.10 should not be used. Since $C_{\upsilon o} = 0.61$ for this valve, wire drawing will not occur.

Related Calculations. To speed up the determination of C_υ, Fig. 42 can be used instead of the formula in step 3. This is a performance-tested chart valid for steam-control valves for blast-heating coils, tank heaters, pressure-reducing stations, and any other installations—stationary, mobile, or marine—where steam flow and pressure are to be regulated. The approach can also be used for valves handling gases other than steam.

The valve coefficient C_υ is conventional; it equals the gallons per minute (liters per second) of clear cold water at 60°F (15.6°C) that will pass through the flow restriction (valve or orifice) while undergoing a pressure drop of 1 lb/in^2 (7.0 kPa). The C_υ value is the same for liquids, gases, and steam. Tables listing C_υ values versus valve size and type are published by the various valve manufacturers. General C_υ values not limited to any manufacturer are given in Table 29 for a variety of valve types and sizes.

Note in Fig. 42 the relations for the density of the steam of various valve outlet pressures. In these relations P_c = critical pressure, lb/in^2 (abs) (kPa), as defined earlier. The solution given in Fig. 42 is for a flow rate of 200 lb/h (0.025 kg/s) of steam having a density of 0.08 lb/ft^3 (1.25 kg/m^3) at the valve outlet with a 10-lb/in^2 (abs) (68.9-kPa) pressure drop through the valve, giving a C_υ of 3.8.

This procedure is the work or John D. Constance, P.E., as reported in *Chemical Engineering* magazine.

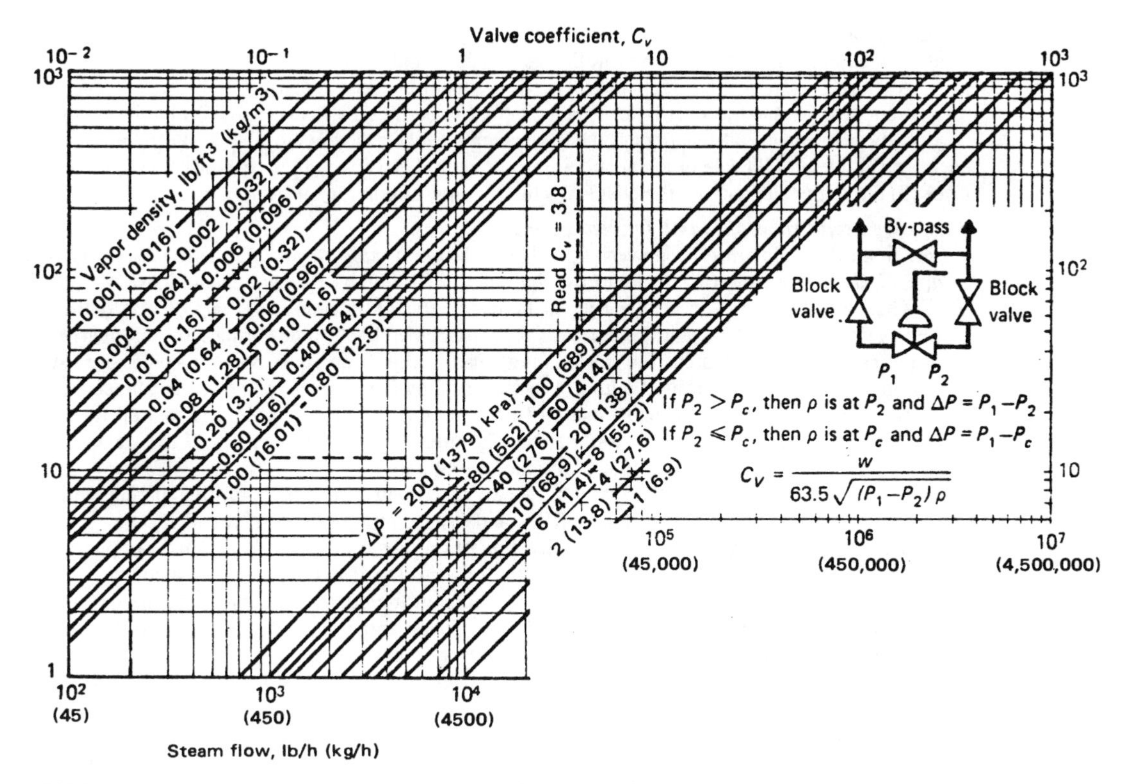

FIGURE 42 C_v valves for steam-control and pressure-reducing valves. (*Chemical Engineering.*)

REPOWERING OPTIONS FOR POWER-PLANT CAPACITY ADDITIONS

Evaluate repowering strategies for an old (20 years +), but still serviceable, steam-cycle electric-generating station to improve significantly its efficiency and/or expand capacity, while maintaining a more favorable environmental profile. Consider the various options available for existing coal-fired and gas-fired stations. Choose a suitable system for preventing damage to a gas turbine from ice ingestion.

Calculation Procedure:

1. *Outline the types of repowering options available*

The types of repowering options considered include (*a*) *partial repowering*—replacing a boiler with a heat-recovery (HRSG) coupled to a gas turbine; (*b*) *station repowering*—use of existing infrastructure but not the original steam cycle; (*c*) *site repowering*—reusing an existing site but none of the original equipment.

While repowering with natural gas and gas turbines gets attention today, other options are worth considering. New schemes use coal gasification, diesel engines, district heating, ultra-high-temperature steam-turbine modules, and adding a steam bottoming cycle to an existing gas turbine.

Integrated resource plans (IRP) and the creation of exempt wholesale generators (EWG) by the National Energy Policy Act of 1992 are often seen as repowering-friendly. Using existing assets and an existing site generally evaluates more favorably, based on raw economics, than a new site with all-new equipment, transmission or utility tie-in, fuel supply, etc.

An EWG operating an efficient, repowered site in an open electric-supply market may have inherent cost advantages over a new-plant EWG. Converting a site into an EWG could involve (1) a utility and nonutility generator (NUG) working together as, say, developer and operator of the asset; (2) the outright sale of the generating asset to an NUG for a quick cash infusion to the utility; or (3) the utility challenging the NUGs for the least-cost option in an IRP.

3. *Give useful rules of thumb for repowering*

The first rule of thumb for repowering is: *The more an existing steam cycle is relied on, the more difficult the repowering will be.* Reasons why this rule of thumb exists are given below.

The basic challenge in repowering is to match the old steam-cycle parameters to the new (repowered) part of the plant. In the case of gas-turbine/HRSG repowerings, the thermal output of the HRSG must be matched to the existing steam-turbine generator. Steam turbines are usually custom-designed for the desired output. Gas turbines, by comparison, come in discrete sizes, each characterized by a specific exhaust energy flow available to generate steam. How well this steam flow, or combination of steam flows, matches the present-day characteristics of the steam turbine determines the efficiency and output of the repowered unit.

Figure 43 shows two different approaches for repowering steam-turbine plants with gas turbines—(*a*) use of a single-pressure HRSG, and (*b*) use of a triple-pressure HRSG.

Ways of getting around the discrete size limitation of gas turbines are: (1) use supplementary firing of the HRSG; (2) inject steam into the gas turbine if excess steam is available; (3) use evaporative coolers and/or inlet-air chillers to augment the power available from the gas turbine, depending on ambient temperatures. All add to the cost and complexity of the retrofit and influence the ultimate unit heat rate. Steam or water injection into the gas turbine, for example, imposes higher maintenance costs on the gas turbine and greater water-treatment needs. So the second rule of thumb is: *Avoiding discrete gas-turbine-size constraints can lead to higher capital and maintenance costs.*

The third rule of thumb for repowering is: *The heat-rejection needs of the new cycle must match the capacity of the existing system.* If the heat-rejection needs do not meet, then extensive upgrades of the condenser and cooling towers may be required.

Two other options for repowering are: (*a*) the hot windbox, Fig. 44*a*, and (*b*) feedwater heating repowering, Fig. 44*a*. In the hot-windbox option, Fig. 44*b*, a high-temperature duct must take gas-turbine exhaust to the boiler. Addition of the high-temperature O_2-rich stream poses substantial changes to the temperature and flow profiles in the boiler. Virtually the entire boiler must be reevaluated for its ability to handle the changes. These changes dovetail into the steam cycle, also. Extensive duct, air heater, and economizer changes may be needed to make this option viable.

The feedwater-heating option, Fig. 44*b*, directs the gas-turbine exhaust energy to the existing feedwater heating circuit—displacing steam available for expansion through the full length of the steam turbine. The more feedwater heating steam displaced, the more efficient the repowered cycle. This cycle is often constrained by technical or environmental limitations in the heat-rejection circuit.

Additional rules of thumb growing out of these findings are: *Matching the old steam-cycle parameters to the new, repowered part of the plant, is the most basic challenge in repowering designs.* And, *Discrete-size gas-turbine matching of steam flows with the existing steam turbine(s) determines the efficiency and output of the repowered unit.* Further, *Old steam-turbine-generators post greater design challenges and costs for repowering.*

3. *Compute the theoretical gains from repowering*

Although highly site-specific, the potential gains from repowering are enormous. The gas-turbine/HRSG repowering options, by one analysis, can provide new capacity at an incremental capital cost of about \$600/kW, a 155 percent increase in net plant output, and a net plant heat-rate improvement of 35 percent. For the feedwater-heating option, the figures are estimated to be \$700/kW, 55 percent greater plant output, and 6 percent heat-rate improvement. With the hot windbox option, the estimated figures are \$800/kW, 42 percent plant output gain, and 8 percent net improvement in heat rate.

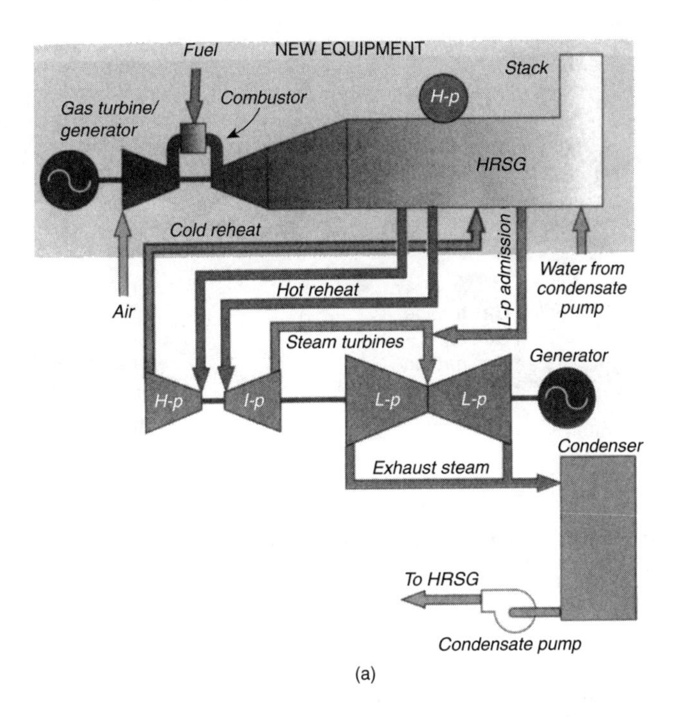

(a)

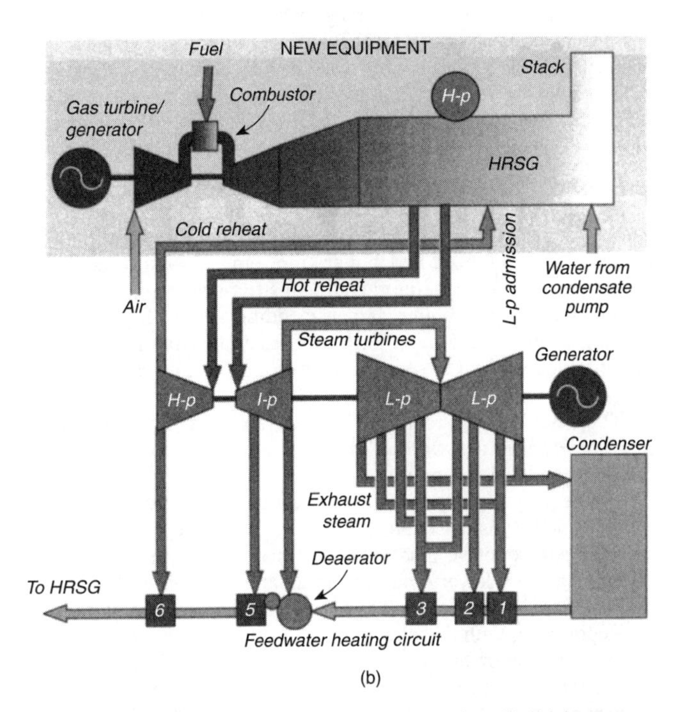

(b)

FIGURE 43 (*a*) Single-pressure HRSG in repowering. (*b*) Triple-pressure HRSG in repowering. (*Power.*)

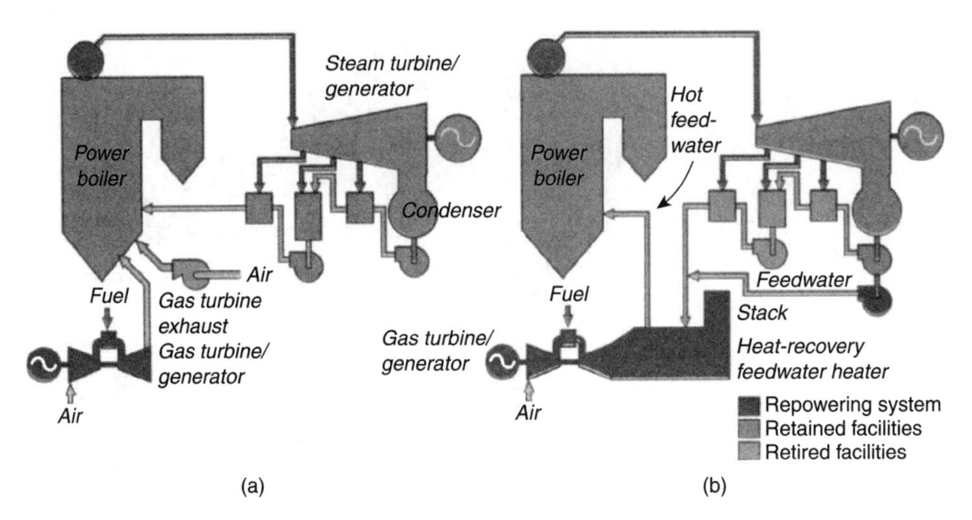

FIGURE 44 (*a*) Hot-windbox repowering injects gas-turbine exhaust gas directly into the boiler, replacing the forced-draft system, (*b*) Feedwater-heating repowering directs gas-turbine exhaust into the feedwater circuit, replacing all, or part of, the feedwater heat exchangers. (*Power.*)

Reductions in emissions are also impressive. When a gas-fired gas-turbine/HRSG replaces the existing boiler, emissions decrease markedly. Most of today's advanced gas turbines achieve NO_x emissions in the 9- to 25-ppm range with no downstream cleanup or steam/water injection into the gas-turbine combustor. And if the original boiler was firing oil with appreciable sulfur content, SO_2 emissions decline substantially, too. Higher efficiency means lower CO_2 emissions for equivalent output. Even in the hot-windbox repowering option, NO_x emissions can be significantly reduced. Aggregate emissions may not be as favorable, depending on output gain and capacity factor for the repowered plant.

4. *Evaluate other options for repowering*

Other forms of repowering include: integrating a coal gasifier or pressurized fluidized-bed (PFB) combustor into the gas-turbine/HRSG-repowered steam turbine; replacing a pulverized-coal-fired boiler with a fluidized-bed boiler; replacing a fossil-fired boiler with a municipal solid-waste (MSW)-fired steam generator; adding district-heating capability; combining diesel engines exhausting into a fossil-fired boiler; and even adding an ultra-high-temperature steam-turbine module to the existing steam-turbine train. Each of these approaches requires careful analysis of the economics of the proposed scheme to see that it earns the required rate of return for the organization sponsoring the project.

Thus, MSW may be possible when an electric utility teams with a waste-to-energy firm to repower. Besides repowering, such a plan might help solve specific solid-waste disposal problems faced by the local community.

Where very high overall thermal efficiency is desired, repowering can include modifying the steam cycle for district heating. Many large electric generating plants installed in Europe in the last decade feed extensive district-heating networks. While including district heating in a repowering scheme does not lower emissions in an absolute sense, it avoids a separate emissions source—and fuel and hardware expenses—to separately generate energy for heating or cooling.

Repowering with diesel engines has attracted the interest of some. Here, the idea is to combine the high efficiency and great fuel flexibility of engines with the economics of existing coal-fired generation, Fig. 45. In one, scheme, oil-gas-fired engines exhaust into a coal-fired boiler modified to burn micronized coal. Heat rate in the range of 9000 Btu/kWh (8550 kJ/kWh) has been projected for the cycle.

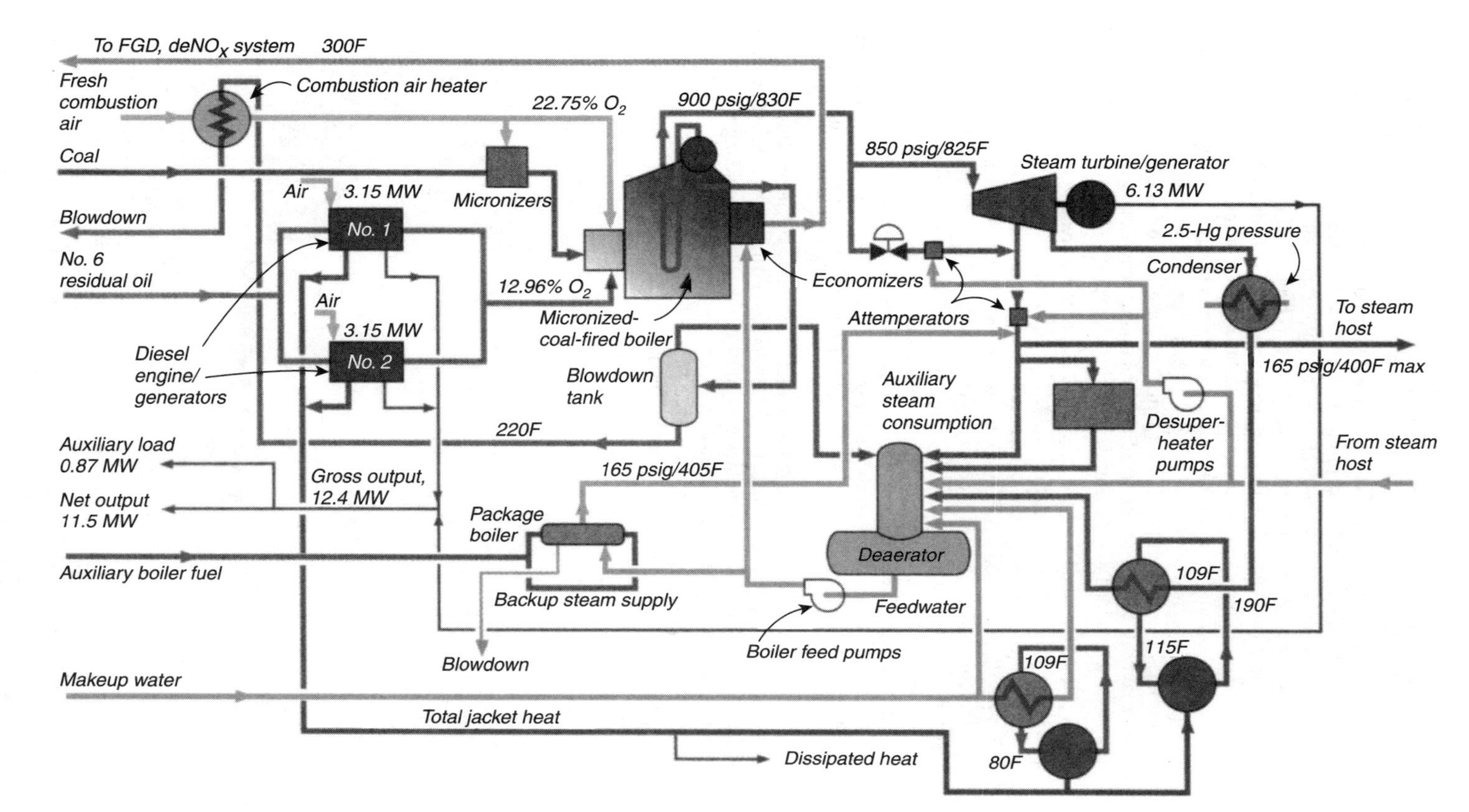

FIGURE 45 Diesel engines are used here to repower a conventional small boiler plant. The same concept can be applied to large steam turbines. (*Power.*)

Such a scheme requires back-end (exhaust) cleanup to achieve respectable emissions levels. Assuming a back-end cleanup system is installed, it can do double-duty—removing pollutants from both the diesel and the Rankine-cycle portions of the plant—and perhaps allow lower-quality, less expensive fuels to be fired in the engine. Other potential advantages noted relative to gas turbines are: greater flexibility matching prime mover to existing steam cycle; less impact on performance from ambient temperature or ambient-air conditions, especially at coastal sites; greater fuel flexibility; less arduous operation and maintenance problems because of the ruggedness of engines compared to gas turbines; and better capability for meeting radically changing thermal and electric loads.

An ultra-high-temperature steam-turbine/generator module, still in the development stage, may be an upcoming option for repowering without changing the fuel basis for the plant. Such a steam turbine, designed for 1300°F (704°C) steam, makes use of alloys and design techniques used in gas turbines. At that temperature, it could increase unit output by 22 percent and lower heat rate by 12 percent. Similar materials of construction are envisioned for making the boiler tubes that would have to be added to the existing steam generator.

5. *Design a suitable system to prevent gas-turbine damage from ice ingestion*

Ice ingestion at a gas-turbine (GT) inlet is not uncommon in northern climates. For many GTs a typical ice-ingestion event can cause primary- and secondary-compressor damage and lead to total repair bills exceeding $250,000, not including plant downtime.

Icing conditions can occur at a GT inlet at ambient temperatures well above freezing. Some GTs, for example, encounter icing at roughly 40°F (4.4°C) and 70 percent relative humidity—primarily caused by the temperature depression that ensues as the filtered air accelerates in the compressor inlet. Thus, potential icing conditions are determined by temperature, humidity, and compressor intake velocity. Figure 46 shows for one particular gas turbine a typical psychrometric chart, which is used to determine the moisture content of the ambient air, and to define warning zones where icing can occur.

Four inlet-heating design solutions are available for preventing icing at GT inlets—(1) water/glycol; (2) compressor bleed air; (3) low-pressure steam; (4) exhaust-heat systems. Each has advantages and disadvantages the designer must evaluate before making a final choice. Here are pertinent facts to help in the design choice.

Water/glycol systems often see service in plants with heat-recovery systems, such as cogeneration applications and where inlet-air chilling may be desired. The system allows a single coil, Fig. 47, in the air filter to serve for both inlet heating and cooling. Mixture ratio for most water/glycol systems is 50/50. Because GT power output is maximized at a particular inlet air temperature, it is important to control the heating system accurately, even while the ambient temperature varies.

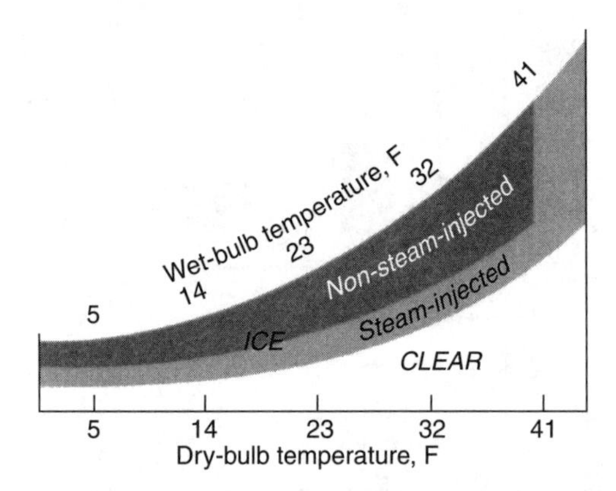

FIGURE 46 Psychrometric chart plots inlet icing conditions. The steam-injected and non-steam-injected areas show where potential icing conditions exist. (*Power.*)

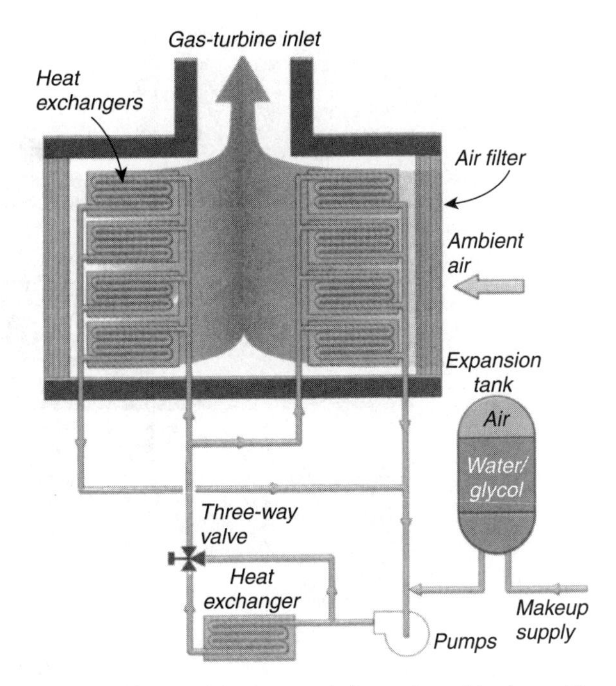

FIGURE 47 Water/glycol system is frequently used in plants with heat-recovery systems. (*Power.*)

Compressor bleed air, Fig. 48, is a popular choice for simple-cycle installations and peaking units. It is a well-proven technology and usually quite reliable and controllable. The biggest drawback with bleed-air systems is the associated performance penalty which, depending on ambient temperature and application, can be 2–5 percent of the total power output. Plants with GTs that operate less than about 200 h/yr under potentially icing conditions should consider this system.

Low-pressure steam is used in many cogeneration plants to provide heat for the anti-icing system. Steam-coil heat exchangers, Fig. 49, are located ahead of the first filter stage and regulated by an upstream control valve. Because the plant is typically recirculating steam otherwise wasted in this application, the heating cost is minimal.

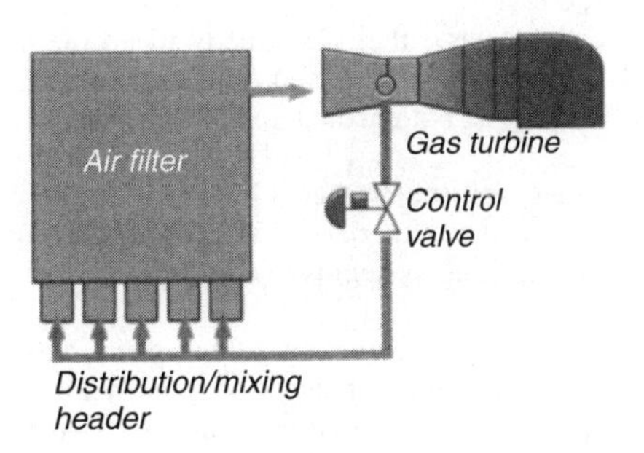

FIGURE 48 Compressor bleed-air system is a popular choice for simple-cycle and peaking units. (*Power.*)

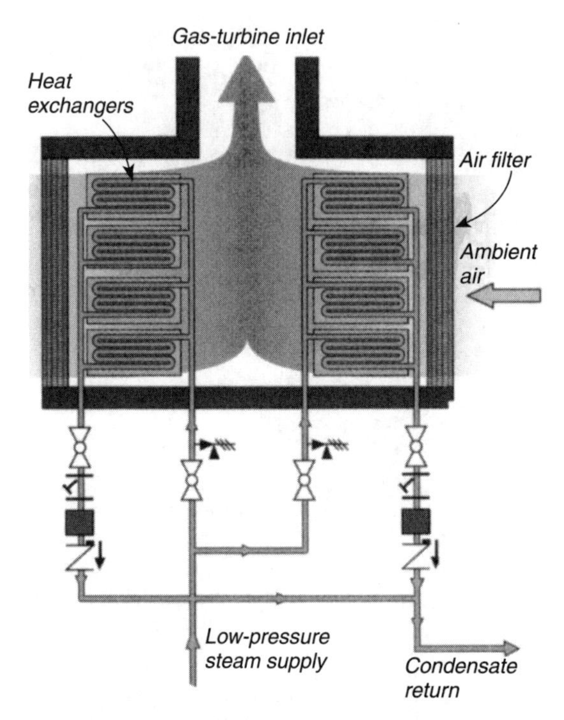

FIGURE 49 In low-pressure system, steam coils located ahead of the first filter stage are regulated by an upstream control valve. (*Power.*)

A key drawback with steam-coil systems is this: Some cogeneration plants encounter difficulties in controlling the modulated temperature for power augmentation. This is attributable, in part, to the demands of accurately modulating the low-pressure steam supply and return pressures. To avoid these difficulties with power augmentation, GT manufacturers are likely to recommend a different inlet-air heating system.

Exhaust-heat systems, Fig. 50, are becoming increasingly attractive because of the desire to lower the performance penalty. Those plants requiring extended anti-icing operation, such as in locations in Canada, will probably find this option more cost-effective than a bleed-air system, provided operation exceeds 200 h/yr.

Exhaust-heat systems either (1) directly bleed the heat from the exhaust stream back into the intake, or (2) use heat exchangers and forced-air ducts. The latter option is desirable because accelerated filter blockage is not encountered and the risk of unburned fuels being ingested into the GT is eliminated.

A performance penalty of less than 0.1 percent—because of added backpressure of heat exchangers in the exhaust duct—will be experienced year-round. The fan-motor parasitic load only adds to operating cost while the heating system is in operation.

Related Calculations. Repowering offers plant designers some of the greatest opportunities for creative solutions to environmentally sound expansion of existing power facilities. When the designer evaluates the many options presented here—using traditional thermodynamic and economic analyses discussed in this part of this handbook—it will be seen that the opportunities for making major savings are enormous. This procedure gives a comprehensive review of the typical options that can be implemented today.

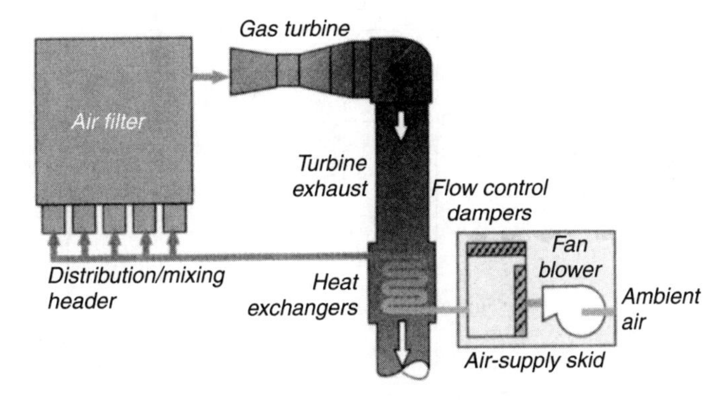

FIGURE 50 Exhaust-heat systems are attractive for reducing the performance penalty. (*Power.*)

Data in this procedure are the work of Jason Makansi, Executive Editor, *Power* magazine, through step 4, and, for step 5, William Calvert, Stewart & Stevenson Services Inc., as reported in *Power* magazine. SI values were added by the handbook editor.

ENERGY ASPECTS OF COOLING-TOWER CHOICE

Select the type of cooling tower to use to cool condenser circulating water for a 450-MW steam central station located where the relative humidity rarely falls below 35 percent and it is desired to keep piping, electrical wiring, and controls to the minimum. The area in which the plant is located has relatively short summers.

Calculation Procedure:

1. *Compare the installed costs of the available cooling towers*

For central stations the two usual choices for cooling towers are (1) induced-draft, and (2) natural-draft. To make the cost comparison, designers base the tower capacity requirements on a constant annual heat rate. This eliminates unnecessary variables in the comparison. Where fuel costs do not change seasonally, average-annual-heat-rate evaluation is valid. We will use this approach in this procedure because it has been found suitable in a variety of installations.

Obtain cost estimates from tower manufacturers for a plant of 450-MW capacity. Here are typical costs for such an installation:

	Induced-draft	Natural-draft
First cost of towers, including fans and drives	$2,000,000	$3,400,000
Electrical controls, wiring	460,000	Nil
Incremental piping	700,000	Base
Capitalized operating costs	1,040,000	Nil
Total cost	$4,200,000	$3,400,000

Plot the results of such a study on a chart as that in Fig. 51, which gives actual data for a 225-MW unit and illustrates a standoff. Varying local conditions for this unit might move the standoff point down to about 200 MW.

Based on the data in the above cost study for the 450-MW unit, a natural-draft hyperbolic cooling tower appears to be the most economic choice. However, other variables must also be considered.

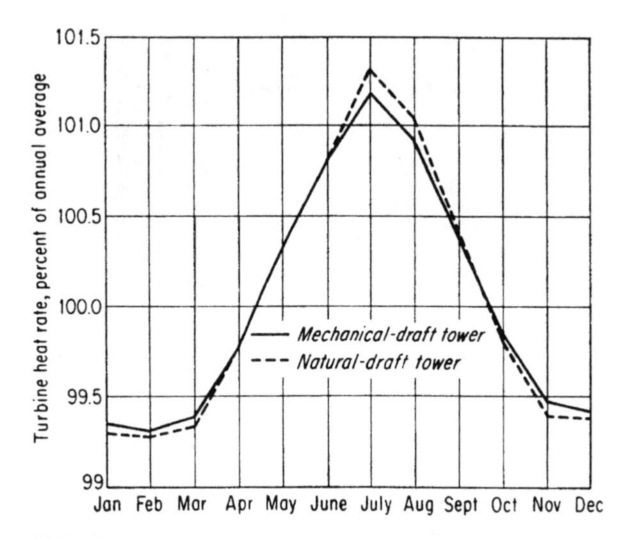

FIGURE 51 Average annual heat rate is the normal evaluation basis for comparing cooling-tower types. Winter gains in heat rate balance summer dryness. (*Power.*)

2. *Evaluate other advantages of the hyperbolic cooling-tower tentative choice*

Hyperbolic cooling towers are natural-draft. As such there is no fan-horsepower (kW) operating cost. In induced-draft mechanical cooling towers fan horsepower can amount to 0.5 percent of installed generating capacity. Thus, a 200-MW unit might need mechanical towers having some 14 cells, each with a 100-hp (75-kW) fan motor. (Normal induced-draft cooling-tower cells handle 15 to 20 MW each.) The fan power becomes available for sale in a natural-draft cooling-tower installation.

As this brief evaluation indicates, several tangible factors can offset the higher first cost of hyperbolic natural-draft towers. There are also several intangibles, difficult to assign exact dollar values but nevertheless important.

Natural-draft hyperbolic towers, Fig. 52, need far less space than comparable induced-draft arrangements. Instead of having to be widely separated to prevent recirculation of heated water vapor, natural-draft towers can be placed on centers 1.5 times their base diameter. They can be located right beside the power station since their operation is not affected by prevailing winds. A recent space-requirement comparison for a 700-MW station made up of several units showed that an induced-draft installation needs about 72 acres (29.2 ha), while a natural-draft installation needs only 15 acres (6 ha)—an 80 percent reduction. Figure 53 shows the relative cost of mechanical-draft and natural-draft cooling towers for generating plants of various capacities.

Where land is expensive, or impossible to obtain, the reduced space requirements of natural-draft towers alone might tip the choice. For example, there are many "downtown" power plants operating a multiplicity of old, small generating units on induced-draft cooling towers. Replacement with large modem units might be impossible if induced-draft towers were again used. But existing land might well be adequate for a natural-draft installation. In a big-city location, ground fogging might also be a distinct problem with induced-draft towers. The height of natural-draft towers lifts vapor discharge, so there is no ground fogging.

3. *Compare the relative humidity constraints for each type of tower*

Hyperbolic natural-draft towers operate satisfactorily under low-humidity (under 50 percent) conditions. Thus, there are many such towers at work in India. But there is an economic lower limit of application—probably about 35 percent relative humidity for design conditions. Below this value of relative humidity, cooling-tower size and cost increase rapidly.

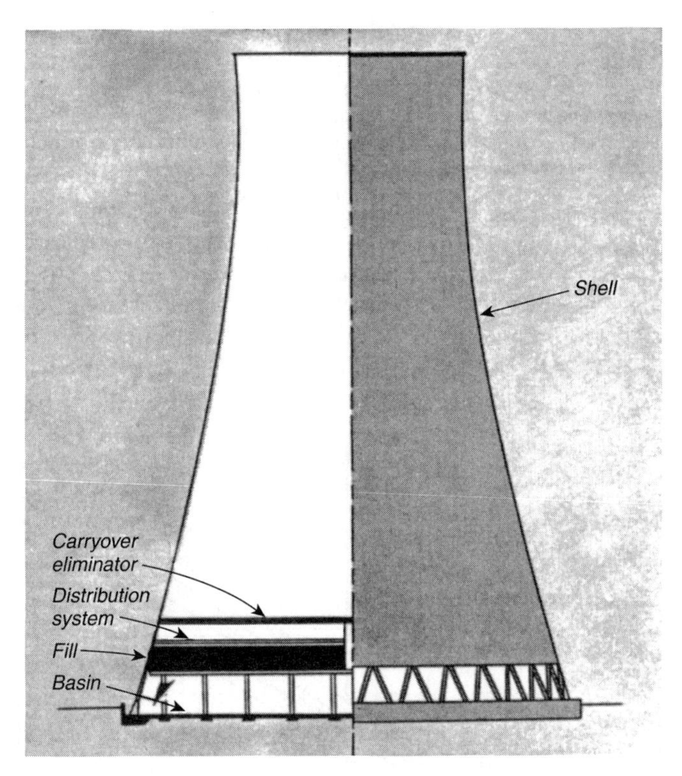

FIGURE 52 Concrete shell of hyperbolic natural-draft cooling lower is curved to agree with the vena contracta formed by air flow through the tower shell. (*Power.*)

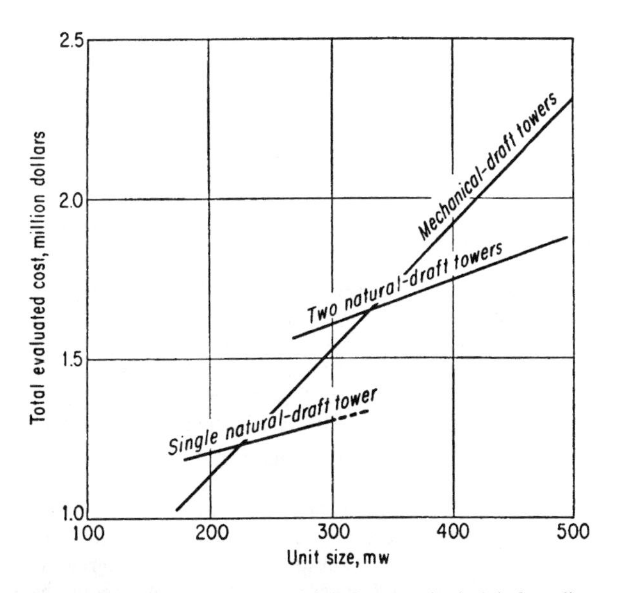

FIGURE 53 Evaluation of natural-draft vs. mechanical-draft cooling towers varies with unit size and number of towers needed. (*Power.*)

Further, if a natural-draft tower is forced to operate at very low relative humidities (5 to 6 percent with the dry-bulb temperature higher than the incoming hot water), air inversion can occur. This will reverse the draft direction in the tower. Such inversion is not a steady-state condition; hunting takes place, reducing the tower's efficiency.

Natural-draft towers perform best when the difference between the cold-water (outlet-water) and air wet-bulb temperatures is equal to, or greater than, the difference between the hot-water (tower inlet water) and cold-water (tower outlet water) temperatures. In cooling-tower language, when the *approach* is equal to, or greater than, the *range*, the larger the tower required.

Operation of a natural-draft tower varies with seasonal changes in dry-bulb temperatures and relative humidity. In winter, air flow increases, producing cooler outlet water, but lower wet-bulb temperatures cut the heat-transfer force and reduce the effect to some extent. Figure 54 shows the higher capacity of natural-draft cooling towers at higher humidities.

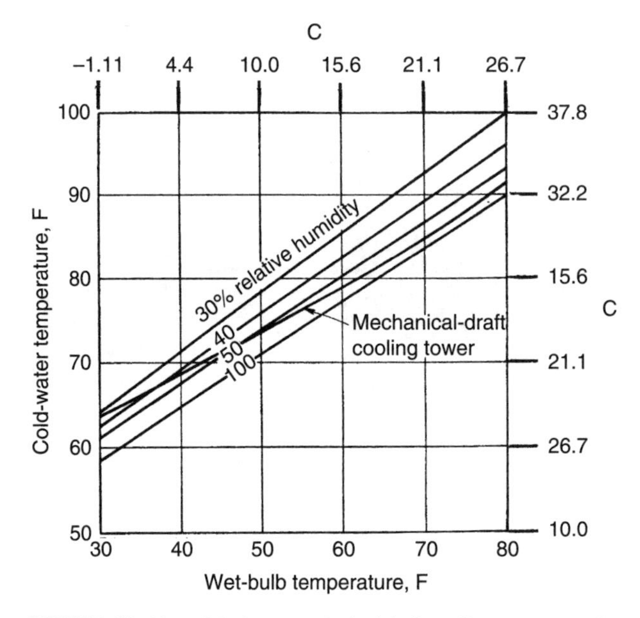

FIGURE 54 Natural-draft vs. mechanical-draft cooling-tower operating curves show higher capacity of natural-draft towers at higher humidities. (*Power.*)

Air-density differences produce the air flow in natural-draft towers. Thus, factors such as dry-bulb temperatures and relative humidity play an important part in tower performance. In operation, heavier outside air displaces the lighter saturated air in the tower, forcing it up and out the top. This is much like chimney operation except that water saturation rather than heat causes the change in air density. Draft losses in typical natural-draft towers range between 0.15 and 0.25 in (3.8 and 6.4 mm) of water.

Unlike the mechanical-draft tower, whose fan produces a fixed air-volume flow regardless of density, a natural-draft tower's air flow varies with changing atmospheric conditions. Studies show that natural-draft towers operate better with high relative humidity at a given wet-bulb temperature.

4. *Evaluate tower construction alternatives*

Natural-draft cooling-tower shells are circular in plan, hyperbolic in profile. Thus, the structure is a double curvature. From a thermal point of view a tower could be cylindrical. But the momentum of the entering air forms a vena contracta whose dimensions vary with the ratio of tower diameter to height of air inlet at the base. Considerable saving in materials and costs can be achieved by tapering the shell to the diameter of the vena contracta. The hyperbolic shape of the tower stiffens the concrete shell against wind forces—an added safety factor.

Today's natural-draft towers have a steel-reinforced concrete shell about 18-in (46-cm) thick at the base. The concrete thickness tapers to 4 to 5 in (10 to 13 cm) at about one-fifth the tower height and continues at this thickness to the top. Vertical ribs are sometimes poured integrally on the outside of the tower shell to break up wind forces by creating turbulence. This permits up to 40 percent theoretical increased stress.

Common tower shell design is based on 72-mph (116-km/h) wind (base reaction), the equivalent of 135 mph (217 km/h) (projected-area basis) for normal chimney design. Towers of 260-ft diameter and 340-ft height (79- to 104-m) are not unusual today.

The entire weight of the concrete shell and the wind-reaction load is supported by reinforced-concrete columns. They are inclined in the same plane as the bottom of the shell and provide a support between the base ring of the shell and the foundation at grade. Open spaces between inclined columns are air-inlet ports to the packing.

A cold-water basin beneath the tower forms the foundation in some designs; in others the basin may be separate. The circular basin can be provided with a conical bottom to collect silt from the cooling water. Blow-down valves remove concentrated silt.

Splash fill for natural-draft cooling towers uses wooden bars spaced at frequent intervals in staggered rows, Fig. 55. Water falls from row to row and breaks into droplets, exposing maximum

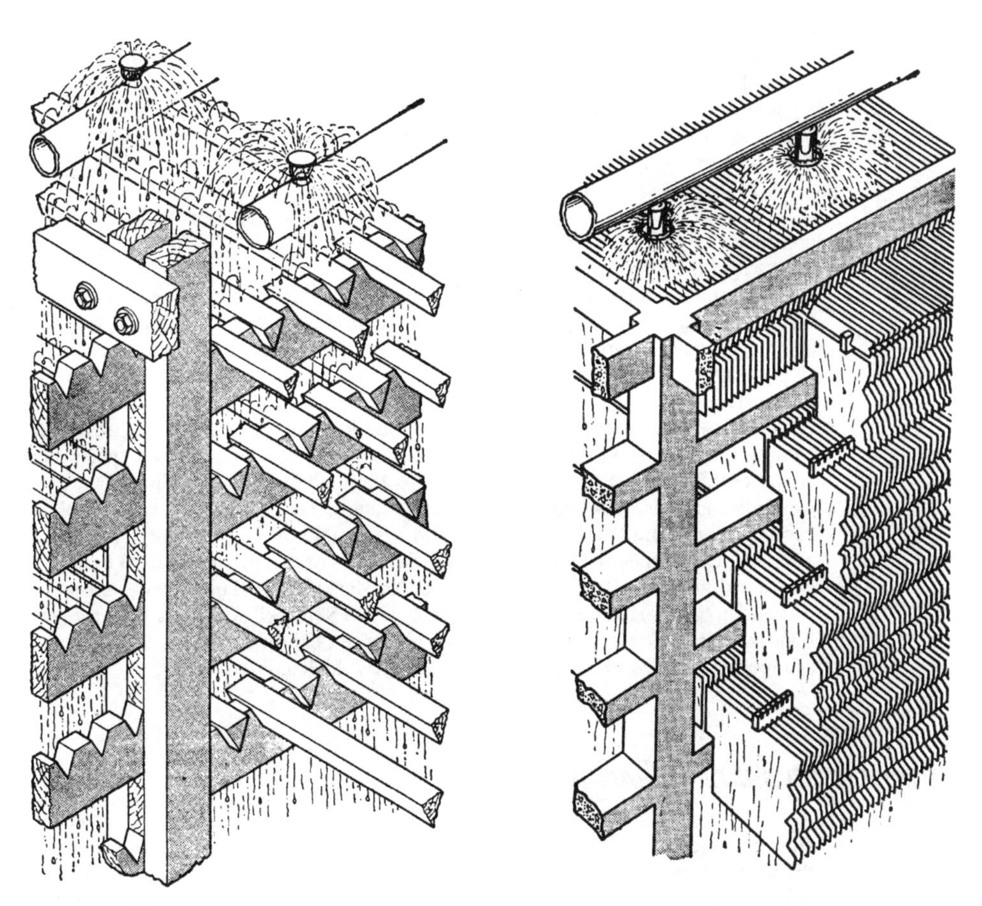

FIGURE 55 Treated-wood fill breaks up falling water droplets continually, *left*. Water film is formed on thin asbestos-cement fill sheets to give surface for cooling, *right*. (*Power.*)

surface to the upflowing air. Typical spacing puts the wooden bars 6 in (15 cm) apart horizontally and 9 in (23 cm) vertically with a resultant depth fill of about 15 ft (4.6 m). Stop logs can be used to cut out half the water-distribution flume (and associated radial pipes) for cleaning with the other half of the tower in operation.

Usual splash fill is California redwood or Scandinavian softwoods. Recent studies show that structurally stronger Douglas fir is the equal of redwood when properly treated. Species chosen depends on lumber-market activity affecting costs at the time of construction.

Water-film fill uses large sheets of 1/8-in (3-mm) thick compressed asbestos cement hung from a series of concrete supports, Fig. 55. Teakwood or plastic spacers separate the sheets evenly. A thin film of water slides down the surface of the asbestos-cement sheets, exposing maximum water surface for evaporative cooling.

Both types of fill (wood and compressed asbestos cement) are specifically designed to cut draft losses and promote high air volume in the tower. Typical fill water loading runs 2 to 4 gal/min per ft^2 (1.4 to 2.7 $L/s/m^2$). This is slightly less than current mechanical-draft tower practice.

Related Calculations. Hyperbolic natural-draft cooling towers have a considerably higher capital cost than mechanical-draft cooling towers. However, the operating cost of natural-draft towers is less than for mechanical-draft towers, as is the cost of piping, electrical wiring, and controls. The inherent advantages of natural-draft towers is leading to their wider adoption throughout the world in both central-station and large industrial power plants.

A recent example of the usefulness of the hyperbolic natural-draft tower is a central station on the Gulf coast of Florida. Two hyperbolic "helper" natural-draft towers are used in conjunction with a bank of mechanical-draft towers, Fig. 56. These helper natural-draft towers reduce the thermal impact that the existing once-through cooling system has on water discharged to the Gulf of Mexico. The complex—when it became operational—was the largest concrete, saltwater hyperbolic helper-tower installation in the United States.

To ensure the highest reliability, unusual cooling-tower features include the use of high-quality precast concrete for the structure, a nonclog splash fill, and a CRT-based digital system to monitor and control the towers and all auxiliary systems.

The primary current use of hyperbolic natural-draft cooling towers is in condenser cooling water temperature reduction. In this application these towers find worldwide use in both warm and temperate climates. European nuclear plants use hyperbolic natural-draft towers almost exclusively. And the use of these towers will increase as more power capacity is installed throughout the world.

In summary, as a guide to tower choice, the following guidelines are offered here: Hyperbolic natural-draft towers are most often selected over the mechanical-draft type when: (1) operating conditions couple low wet-bulb temperature and high relative humidity; (2) a combination of low wet-bulb and high inlet- and exit-water temperatures exists—that is, a broad cooling range and long cooling approach; (3) a heavy winter load is possible, and (4) a long amortization period can be arranged. Economics also tends to favor hyperbolics over mechanical-draft units as power-plant size grows and the possibility of erecting fewer (but larger) hyperbolic towers exists.

Data in this procedure are the work of B. G. A. Skrotzki, Associate Editor, *Power* magazine, Joe Lander, Florida Power Corp., and Gray Christensen, Black & Veatch, (Florida power-plant details) as reported in *Power*. SI values were provided by the handbook editor.

OIL-POLLUTED BEACH CLEANUP ANALYSIS

How much relative work is required to clean a 300-yd (274-m) long beach coated with heavy oil, if the width of the beach is 40 yd (36.6 m), the depth of oil penetration is 20 in (50.8 cm), the beach terrain is gravel and pebbles, the oil coverage is 60 percent of the beach, and the beach contains heavy debris?

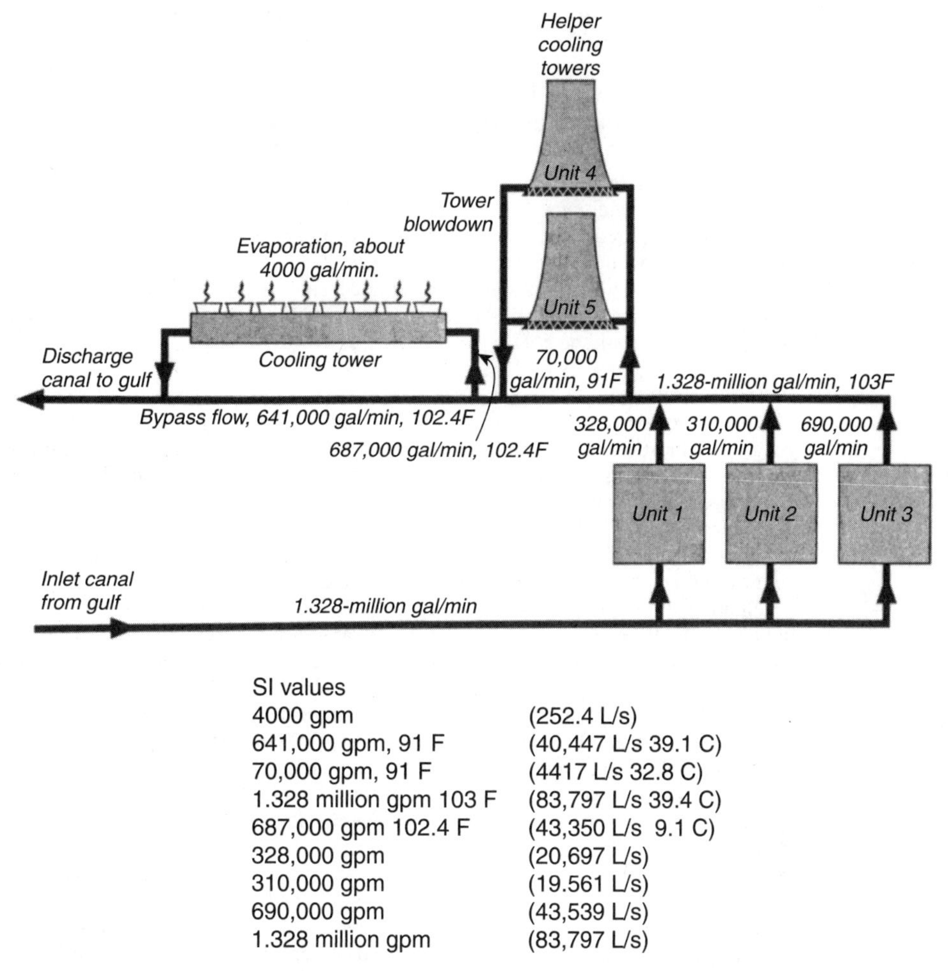

FIGURE 56 Helper system withdraws a portion of discharge-canal water and directs it to cooling towers. Cooled water from towers is returned to canal, where the two streams are mixed. (*Power.*)

Calculation Procedure:

1. *Establish a work-measurement equation from a beach model*

After the *Exxon Valdez* ran aground on Bligh Reef in Prince William Sound, a study was made to develop a model and an equation that would give the relative amount of work needed to rid a beach of spilled oil. The relative amount of work remaining, expressed in clydes, is defined as the amount of work required to clean 100 yd (91.4 m) of lightly polluted beach. As the actual cleanup progressed, the actual work required was found to agree closely with the formula-predicted relative work indicated by the model and equation that were developed.

The work-measurement equation, developed by on-the-scene Commander Peter C. Olsen, U.S. Coast Guard Reserve, and Commander Wayne R. Hamilton, U.S. Coast Guard, is $S = (L/100)(EWPTCD)$, where S = standardized equivalent beach work units, expressed in clydes; L = beach-segment length in yards or meters (considered equivalent because of the rough precision of the model); E = degree of contamination of the beach expressed as: light oil = 1; moderate oil = 1.5; heavy oil = 2;

random tar balls and very light oil = 0.1; W = width of beach expressed as: less than 30 m = 1; 30 to 45 m = 1.5; more than 45 m = 2; P = depth of penetration of the oil expressed as: less than 10 cm = 1; 10 to 20 cm = 2; more than 30 cm = 3; T = terrain of the beach expressed as: boulders, cobbles, sand, mud, solid rock without vertical faces = 1; gravel/pebbles = 2; solid rock faces = 0.1; C = percent of oil coverage of the beach expressed as: more than 67 percent coverage = 1; 50 to 67 percent = 0.8; less than 50 percent = 0.5; D = debris factor expressed as: heavy debris = 1.2: all others = 1.

2. *Determine the relative work required*
Using the given conditions, $S = (300/100)(2 \times 1.5 \times 1 \times 1 \times 0.8 \times 1.2) = 8.64$ clydes. This shows that the work required to clean this beach would be some 8.6 times that of cleaning 100 yd of lightly oiled beach. Knowing the required time input to clean the "standard" beach (100 yd, lightly oiled), the approximate time to clean the beach being considered can be obtained by simple multiplication. Thus, if the cleaning time for the standard lightly oiled beach is 50 h, the cleaning time for the beach considered here would be 50(8.64) = 432 h.

Related Calculations. The model presented here outlines—in general—the procedure to follow to set up an equation for estimating the working time to clean any type of beach of oil pollution. The geographic location of the beach will not in general be a factor in the model unless the beach is in cold polar regions. In cold climates more time will be required to clean a beach because the oil will congeal and be difficult to remove.

A beach cleanup in Prince William Sound was defined as eliminating all gross amounts of oil, all migratory oil, and all oil-contaminated debris. This definition is valid for any other polluted beach be it in Europe, the Far East, the United States, etc.

Floating oil in the marine environment can be skimmed, boomed, absorbed, or otherwise removed. But oil on a beach must either be released by (1) scrubbing or (2) steaming and floated to the nearby water where it can be recovered using surface techniques mentioned above.

Where light oil—gasoline, naphtha, kerosene, etc.—is spilled in an accident on the water, it will usually evaporate with little damage to the environment. But heavy oil—No. 6, Bunker C, unrefined products, etc.—will often congeal and stick to rocks, cobbles, structures, and sand. Washing such oil products off a beach requires the use of steam and hot high-pressure water. Once the oil is freed from the surfaces to which it is adhering, it must be quickly washed away with seawater so that it flows to the nearby water where it can be recovered. Several washings may be required to thoroughly cleanse a badly polluted beach.

The most difficult beaches to clean are those comprised of gravel, pebbles, or small boulders. Two reasons for this are: (1) the surface areas to which the oil can adhere are much greater, and (2) extensive washing of these surface areas is required. This washing action can carry away the sand and the underlying earth, destroying the beach. When setting up an equation for such a beach, this characteristic should be kept in mind.

Beaches with larger boulders having a moderate slope toward the water are easiest to clean. Next in ease of cleaning are sand and mud beaches because thick oil does not penetrate deeply in most instances.

Use this equation as is, and check its results against actual cleanup times. Then alter the equation to suit the actual conditions and personnel met in the cleanup.

The model and equation described here are the work of Commander Peter C. Olsen, U.S. Coast Guard Reserve and Commander Wayne R. Hamilton, U.S. Coast Guard, as reported in government publications.

INDEX

Affinity laws for centrifugal pumps, 12.14
Air-bubble enclosure choice for planned usage, 13.37
Air-conditioner vibration isolation device selection, 13.118
Air conditioning and interior climate control, 13.1–13.123
  energy and load computations for, 13.70, 13.75
Air-cooled heat exchanger selection, 11.63
Air-cooled i-c engine choice for industrial uses, 4.33
Air curtain design for structures, 13.123
Air grille choice for air-conditioning systems, 13.110
Air-heating coil selection and energy analysis, 13.59
Air-to-air heat exchanger energy performance, 13.27
Analysis of nuclear power-plant cycles, 5.4
Analyzing gas-turbine cycle efficiency and output, 3.19
Atmospheric control system energy investment analysis, 14.2
Axial-flow tube-type hydro turbine, 6.9

Barometric and jet condenser heat transfer, 11.11
Binary steam plant energy analysis, 2.15
Biomass-geothermal power-generation analysis, 9.5
Boiling-water nuclear reactor, 5.3
Bubble enclosure, choice of, 13.37
Building heat requirements with steam and hot-water heating, 13.29
Building heating system selection and analysis, 13.48
Bulb hydro turbine, 6.3
Bypass air quantity analysis and dehumidifier exit conditions, 13.32
Bypass cooling system for i-c engines, 4.19

Calculating HRSG temperature profiles, 3.32–3.36
Carbon dioxide buildup computation for occupied spaces, 13.31
Centrifugal compressor power and energy input, 13.34
Centrifugal pump choice using affinity laws, 12.14
Centrifugal pump energy usage as a hydraulic turbine, 12.50
Centrifugal pump minimum safe flow rate, 12.40
Choosing centrifugal pump best rotary speed, 12.17
Choosing centrifugal pump best speed to meet capacity/head needs, 12.9
"Clean energy" hydro site analysis, 6.2
Coal-fuel combustion, 1.1
Cogeneration, 2.7–2.37, 14.28
  capital cost of heat-recovery boilers, 14.28
  cost analysis for alternative energy systems, 14.13
  flash-steam heat recovery for, 14.43
  fuel-cost allocation using electricity cost, 14.54
  heat savings analysis using needs plots, 14.22
  power-plant cost separation of steam and electricity, 14.50
Cogeneration, economics using i-c engines, 4.2
  heat rate of turbogenerator, 2.17
  kW output of extraction turbine, 2.37

Cogeneration, economics using i-c engines (*Cont.*):
mechanical-drive turbine output analysis, 2.29
moisture content of turbine exhaust, 2.7
performance of reheat-steam cycle, 2.13
regenerative bleed-steam cycle analysis, 2.18
reheat-regenerative cycle steam rate, 2.13
reheat-regenerative turbogenerator, 2.36
reheating-regenerative steam turbine cycle analysis, 2.12
steam compressor choice for system, 14.19
steam flow for steam-turbine no-load and partial-load operations, 2.8
system energy efficiency, 2.2
Combined-cycle plant extraction turbine output, 3.10
Combustion calculations, 1.1–1.17
calculation parameters, 1.1
of coal fuel, 1.1
estimating temperature of products of combustion, 1.15
of fuel oil, 1.5
molal method of combustion analysis, 1.13
of natural gas, 1.7
percent excess air while burning coal, 1.4
savings by preheating combustion air, 1.16
using million Btu method, 1.17
of wood fuel, 1.11
Comparison of coal and fissionable materials heat generation, 5.8
Compressed-air piping, pressure loss in, 12.118
Condensate pump economic analysis, 12.60
Controlled-volume pump control system choice, 14.71
Cooling-coil selection for air-conditioning systems, 13.79
Cooling-tower choice, energy aspects of, 14.85
Cooling-water energy needs of i-c engines, 4.13
Corrugated-type expansion joint selection, 12.126
Cross-flow turbine, 6.3
Curtain, design of, 13.123
Cycle analysis of vapor-dominated geothermal power plant, 9.2–9.3

Designing spiral-tube heat exchangers, 11.43
door and window air curtains, 13.123
Desuperheater water spray quantity, 12.150
Diesel engine, efficiency and characteristics of, 4.8
energy analysis of, 4.9
energy performance factors of, 4.29
energy recovery via hot water, 4.24
energy requirements for cooling-water and lube-oil pumps, 4.26
fuel storage capacity and cost, 4.25
volumetric efficiency of, 4.30
Domestic hot-water heating by solar collector, 8.22
Door and window air curtain design, 13.123

Electronic precipitator, sizing for pollution control, 14.34
Emission reduction using flue-gas heat recovery, 14.8
Energy analysis of i-c engine choice, 4.9
Energy and economic analysis of power-plant condensate pumps, 12.60
Energy annual heating and cooling costs and loads, 13.21
Energy conservation and environmental pollution control, 14.1–14.90
atmospheric control system energy analysis, 14.2
capital cost of cogeneration heat-recovery boilers, 14.28
characteristics and rangeability of energy control-valves, 14.74
chlorofluorocarbon aspects of central chilled-water systems, 14.31
in choosing pumping installation equipment, 12.22
cogeneration cost analysis for alternative energy systems, 14.13
cogeneration heat savings analysis using need plots, 14.22
cogeneration power plant cost separation of steam, electricity, energy equivalence method, 14.50
control-system controlled-volume-pump choice, 14.71
emission reduction using flue-gas heat recovery, 14.8
energy aspects of cooling-tower choice, 14.85

Energy conservation and environmental pollution control (*Cont.*):
energy conservation and cost reduction design for flash-steam usages, 14.45
energy-from-waste economic analysis, 14.4
energy process-control system selection, 14.64
energy process-control valve selection, 14.68
environmental and safety-regulation ranking of equipment criticality, 14.56
environmental pollution project selection, 14.3
explosive vent sizing for industrial buildings, 14.36
flash-steam heat energy recovery for cogeneration, 14.43
fuel-cost allocation for cogeneration plant using electricity cost, 14.54
heat recovery energy and fuel savings, 14.60
high-temperature hot-water heating energy and fuel savings, 14.62
industrial-steam boiler energy control choice and usage, 14.72
oil-polluted beach cleanup analysis, 14.90
parameters, 14.2
process-energy temperature control system selection, 14.67
repowering options for power-plant capacity additions, 14.77
sizing an electrostatic precipitator for air-pollution control, 14.34
steam compressor choice for diesel-engine cogeneration system, 14.19
steam-control and pressure-reducing valve sizing for maximum energy savings, 14.75
thermal-pollution estimates for power plants, 14.42
ventilation design for industrial building environmental safety, 14.38
Energy control valves, characteristics of, 14.68
Energy conversion engineering, 1.1–1.18
coal fuel combustion, 1.1
combustion calculations, 1.1–1.18
estimating temperature of final products, 1.15
fuel oil combustion, 1.5
Energy conversion engineering (*Cont.*):
million Btu (1.055 MJ) calculation method, 1.18
molal method of analysis, 1.13
natural gas combustion, 1.7
percent excess air, 1.4
savings produced by preheating air, 1.18
wood fuel combustion, 1.11
Energy design of shell-and-tube heat exchangers, 11.20
Energy equations for interior climate control, 13.2
Energy-from-waste economic analysis, 14.4
Energy heat recovery with run-around system, 13.23
Energy parameters for gas turbines, 3.1
Energy process-control systems selection, 14.64
Energy saving in using pump characteristic and system-head curves, 12.5
Energy savings with ice storage for facility cooling, 13.12
Engine-room vent design for i-c engines, 4.17
Environmental pollution project analysis, 14.3
Equal-friction method for sizing low-velocity air ducts, 13.90
Equipment criticality analysis, 12.9
Estimating final temperature of combustion products, 1.15
Expansion-tank sizing for hydronic systems, 13.40
Explosion-vent sizing for industrial buildings, 14.36
External steam tracing design for pipelines, 11.58
Extraction turbine output in combined-cycle plant, 3.10

F-chart method of useful energy computation, 8.16
Fan energy performance analysis, 13.12
Fan horsepower requirements for cooling towers, 13.12
Finned-tube heat exchanger selection, 11.12
Flash-steam heat energy recovery for cogeneration, 14.43
Flashed-steam geothermal power analysis, 9.1, 9.6, 9.7, 9.9
enthalpy of steam, 9.10
enthalpy ratio, 9.10

Flat-plate solar-energy heating- and cooling-system design, 8.6
Fluid flow in piping systems, 12.1–12.172
Fluid transfer engineering, 12.1–12.172
affinity laws for centrifugal-pump energy analysis, 12.14
analyzing plastic piping and linings for specific applications, 12.142
centrifugal-pump choice using affinity laws, 12.14
centrifugal pump energy usage as a hydraulic turbine, 12.50
centrifugal-pump minimum safe fluid flow determination, 12.40
chart and tabular determination of friction head, 12.102
choosing centrifugal pump best rotary speed, 12.17
choosing centrifugal-pump rotating speed, 12.9
corrugated expansion joint selection and application, 12.126
design of steam transmission piping, 12.129
designing steam-transmission lines without out steam traps, 12.157
desuperheater water spray quantity, 12.150
determining pressure loss in oil piping, 12.115
determining the friction factor for Bingham plastics, 12.163
determining the pressure loss in steam piping, 12.85
energy and economic analysis of power-plant condensate pumps, 12.60
energy and environmental analysis for equipment criticality, 12.9
energy conservation in pumping installation equipment choice, 12.22
energy savings using pump characteristic and system-head curves, 12.5
equivalent length of a complex series pipeline, 12.110
equivalent length of a parallel piping system, 12.111
estimating cost of steam leaks from piping and pressure vessels, 12.153
flow rate and pressure loss in compressed-air and gas piping, 12.118
flow rate and pressure loss in gas pipelines, 12.122
Fluid transfer engineering (*Cont.*):
fluid head approximations for all types of piping systems, 12.81
friction heads loss in water piping, 12.101
friction loss in pipes handling solids in suspension, 12.149
hot-liquid pump net positive suction head, 12.36
hydraulic radius and liquid velocity in water pipes, 12.101
hydropneumatic storage tank sizing, 12.50
impeller sizing for centrifugal pump safety service, 12.55
line sizing for flashing-steam condensate, 12.160
liquid viscosity effect on reciprocating pump performance, 12.46
liquid viscosity effect on regenerative pump performance, 12.45
materials selection for pump parts, 12.48
maximum allowable height for a liquid siphon, 12.112
orifice meter selection for a steam pipe, 12.97
pipe-wall thickness and schedule number, 12.82
pipe-wall thickness determination by piping code formula, 12.84
piping pressure surge with different material and fluid, 12.77
piping warm-up condensate load, 12.88
pressure loss in piping having laminar flow, 12.114
pressure-regulating valve selection for water piping, 12.107
pressure surge in piping system compound pipeline, 12.79
pressure surge in piping system from rapid valve closure, 12.75
pump and system characteristic curve energy analysis, 12.30
pump choice to conserve energy, 12.3
pumps and pumping system calculations, 12.2–12.60
quick calculation of flow rate and pressure drop in piping systems, 12.80
quick sizing of restrictive orifices in piping, 12.154

Fluid transfer engineering (*Cont.*):
reducing energy consumption with proper pump choice, 12.58
relative carrying capacity of pipes, 12.107
selecting centrifugal pumps based on specific speed, 12.15
selecting heat insulation for high-temperature piping, 12.96
selecting plastic piping for industrial use, 1.141
selection of pressure-regulating valve for steam service, 12.98
shaft deflection and pump critical speed, 12.43
sizing a water meter, 12.109
sizing condensate return line for optimum flow conditions, 12.151
slip-type expansion joint selection and application, 12.123
specific gravity and viscosity of liquids, 12.113
steam accumulator selection and sizing, 12.139
steam desuperheater analysis, 12.138
steam-power plant condensate pump selection, 12.37
steam tracing a vessel bottom to keep the contents fluid, 12.155
steam trap selection for industrial applications, 12.90
time needed to empty a storage vessel with dished ends, 12.166
time needed to empty a storage vessel without dished ends, 12.170
time to drain a storage tank through attached piping, 12.170
vapor-free liquid-pump pump and system characteristic curve total head computation, 12.18
viscosity and dissolved gas effects on rotary pump performance, 12.47
viscous-liquid centrifugal pump selection, 12.41
water-hammer effects in a liquid pipeline, 12.113
Francis turbine, 6.3, 6.5, 6.7
Fuel-oil combustion, 1.5

Gas, natural, combustion of, 1.7
Gas piping, pressure loss in, 12.122
Gas-turbine power generation, 3.1–3.36
analyzing gas-turbine cycle efficiency and output, 3.19
calculating HRSG temperature profiles, 3.36
calculation parameters, 3.1
energy parameters, 3.1
gas-turbine-plant oxygen and fuel input, 3.30
gas-turbine steam-turbine energy analysis, 3.2
heat-recovery steam generator (HRSG) choice, 3.17
HRSG tube-bundle bundle vibration, 3.26
industrial gas turbine life-cycle cost model, 3.23
output of extraction turbine in combined-cycle plant, 3.19
regenerative-gas-turbine cycle, 3.7
selecting method for increasing combined-cycle plant output, 3.11
Geothermal energy engineering, 9.1–9.9
binary cycle, 9.1, 9.7
calculation parameters, 9.1
combined cycle, 9.1
cycle analysis of vapor-dominated plant, 9.2
direct steam cycle, 9.2
dual-flash process, 9.8
flashed-steam cycle, 9.1, 9.7, 9.9
flashed-steam geothermal power analysis, 9.9
geothermal and biomass power-generation analysis, 9.5
liquid-dominated single-flash system, 9.9
vapor-dominated power plant, 9.2, 9.3
Grille choice for air-conditioning systems, 13.110

Heat and energy loss determination for buildings, 13.48
Heat exchanger choice for specific applications, 11.2
Heat recovery energy and fuel savings, 14.60
Heat-recovery steam generator (HRSG) choice, 3.32–3.36
steam bundle vibration, 3.26
Heat transfer, 11.1
air-cooled heat exchanger selection, 11.63
barometric and jet condenser heat transfer, 11.11

Heat transfer (*Cont.*):
calculation parameters, 11.1
designing spiral-tube heat exchangers, 11.43
energy design analysis of shell-and-tube exchangers, 11.20
external steam tracing design for pipelines, 11.58
finned-tube heat exchanger selection, 11.12
heat exchanger choice, 11.2
heat-transfer coefficient determination, 11.17
industrial-use electric heater selection and sizing, 11.15
internal steam tracing design for pipelines, 11.50
parameters, 11.1
quick design and evaluation of heat exchangers, 11.66
selecting and sizing heat exchangers on fouling factors, 11.8
sizing shell-and-tube heat exchangers, 11.5
spiral-plate heat exchanger design, 11.34
spiral-type heating coil selection, 11.14
steam generating capacity of boiler tubes, 11.18
temperature determination in heat-exchanger operation, 11.7
Heat transfer coefficient determination, 11.17
Heating apparatus steam and energy consumption, 13.57
Heating-steam requirements for specialized rooms, 13.30
High-temperature high-altitude i-c engine performance, 4.10
High-velocity air-conditioning duct design, 13.108
"Hot deck" temperature reset energy savings, 13.27
Hot-liquid pump net positive suction head, 12.36
HRSG (heat-recovery steam generator), 3.32–3.36
Humidifier selection for indoor climate conditions, 13.101
Hydro-pneumatic storage tank sizing, 12.50
Hydroelectric energy power plants, 6.1–6.10
analysis of large-scale hydroelectric energy plant, 6.7
axial-flow tube-type turbine, 6.9
bulb turbine, 6.3
calculation parameters, 6.1
cross-flow turbine, 6.3
economic evaluation of small-scale hydro sites, 6.4
Francis turbine, 6.3, 6.5, 6.7
impulse turbines, 6.3
Kaplan turbine, 6.3
Pelton-type turbine, 6.8
propeller (Kaplan) turbine, 6.9
reaction turbines, 6.2
small-scale "clean energy" hydro site analysis, 6.2
tubular turbines, 6.9
vertical Francis turbine, 6.5

Impeller sizing for centrifugal pump safety service, 12.55
Impulse turbines, 6.3
Industrial gas turbine life-cycle cost model, 3.23
Industrial solar-energy system economics, 8.5
Industrial steam boiler energy control choice and usage, 14.72
Industrial-use heater selection and application, 11.15
Interior climate control, 13.1–13.123
Interior climate control and air conditioning, 13.1–13.123
air-bubble enclosure choice for planned usage, 13.37
air-conditioner vibration isolation device selection, 13.118
air-heating coil selection and energy analysis, 13.59
air-to-air heat exchanger energy performance, 13.27
annual energy heating and cooling costs and loads, 13.21
bypass air quantity and dehumidifier exit conditions, 13.32
centrifugal compressor power and energy input, 13.34
choosing and sizing radiant heating panels, 13.64

Interior climate control and air conditioning (*Cont.*):
choosing and sizing snow-melting panels, 13.67
computation of CO_2 buildup in occupied spaces, 13.31
cooling-coil selection for air-conditioning systems, 13.79
designing door and window air curtains, 13.123
energy analysis of mixing two air streams, 13.86
energy and heat-load computation for air-conditioning systems, general method, 13.70
energy and heat-load computations for air-conditioning systems, numerical methods, 13.75
energy and heating steam requirements for specialized rooms, 13.30
energy aspects of designing high-velocity air-conditioning ducts, 13.108
energy aspects of selecting an air-conditioning system, 13.87
energy considerations in choosing roof ventilators, 13.115
energy equations for interior climate control, 13.2
energy heat recovery with run-around heat recovery, 13.23
energy savings in space heating with lighting heat recovery, 13.69
energy savings with ice storage system for facility cooling, 13.12
energy savings with rotary heat exchanger, 13.25
equal-volume method of sizing low-velocity air-conditioning ducts, 13.90
expansion-tank sizing for hydronic systems, 13.40
fan and pump energy performance analysis from motor data, 13.36
fan horsepower requirements for cooling towers, 13.12
heat and energy loss determination for buildings, 13.48
heating apparatus steam and energy consumption, 13.57
heating capacity requirements for buildings with steam and hot-water heating, 13.29

Interior climate control and air conditioning (*Cont.*):
hot-deck temperature reset energy savings, 13.27
moisture evaporation from open tanks, 13.35
noise-reduction materials choice for air conditioning, 13.121
outlet and return air grille choice for air conditioning, 13.110
selecting humidifiers for chosen indoor climate conditions, 13.101
selection and analysis of building heating systems, 13.49
static-regain method for sizing low-velocity air-conditioning system ducts, 13.97
unit heater capacity computation, 13.52
using the psychometric chart in air-conditioning energy calculations, 13.106

Internal-combustion engines, 4.1
air-cooled i-c engine choice for industrial uses, 4.33
bypass cooling system design for i-c engines, 4.19
calculation parameters, 4.1
choice of lube-oil cooler for i-c engines, 4.27
cogeneration economics using i-c engines, 4.2
cooling-water energy needs of i-c engines, 4.13
determining solids intake of i-c engines, 4.28
diesel-engine efficiency and characteristics, 4.8
diesel-engine volumetric efficiency, 4.30
energy analysis of i-c engine choice, 4.9
energy analysis of i-c engines, 4.12
energy efficiency of diesel generating unit, 4.7
energy performance factors for i-c engines, 4.29
energy recovery via hot water from i-c engines, 4.24
energy requirements for i-c engine cooling-water and lube-oil pumps, 4.26
fuel storage capacity and cost for i-c engines, 4.25

Internal-combustion engines (*Cont.*):
high-temperature and high-altitude i-c engine performance, 4.10
i-c engine characteristics analyses, 4.12
i-c engine horsepower and mean effective pressure, 4.8
i-c engine room vent system design, 4.17
lube-oil cooler choice for i-c engines, 4.27
Internal steam tracing design for pipelines, 11.50

Kaplan turbine, 6.3

Laminar flow, piping pressure loss, 12.114
Large-scale hydro site analysis, 6.7
Lighting systems, heat recovery from, 13.69
Liquid viscosity effect on reciprocating pump performance, 12.46
on regenerative pump performance, 12.45

Materials selection for pump parts, 12.48
Mixing two air streams, energy analysis for, 13.86
Modulated single-pool tidal system energy and power, 10.8
average power, 10.8
total energy, 10.8
Moisture evaporation from open tanks, 13.35
Motor-driven fan vibration potential, 13.33

Noise-reduction materials selection for air conditioning, 13.121
Nuclear energy engineering, 5.1–5.14
analysis of nuclear power-plant cycles, 5.4
boiling-water reactor, 5.3
calculation parameters, 5.1
comparison of coal and fissionable materials, 5.8
desalinization using nuclear power, 5.11
fast-breeder reactor, 5.3
fluid-fueled reactor, 5.3
fuel consumption of nuclear reactors, 5.7
gas-cooled reactor, 5.3
nuclear power use in desalinization, 5.11
Nuclear energy engineering (*Cont.*):
nuclear radiation effects on human beings, 5.9
power-plant cycle analysis, 5.4
pressurized-water reactor, 5.3
selecting a nuclear power reactor, 5.2
Nuclear radiation effects on human beings, 5.9

Ocean energy engineering, 10.1–10.8
analysis of OTEC Claude cycle efficiency, 10.2
ocean energy parameters, 10.1
evaporator conditions, 10.3
gross cycle efficiency, 10.4
modulated single-pool tidal system energy and power, 10.8
OTEC Claude cycle analysis, 10.2
turbine work output, 10.3
wave characteristics and energy and power, 10.5
Oil, combustion of, 1.5
Oil piping, pressure loss in, 12.118
Oil-polluted beach cleanup analysis, 14.90
OTEC Claude cycle efficiency and flow rates, 10.2
Outlet grille choice for air-conditioning systems, 13.110
Oxygen and fuel input to gas turbines, 3.30

Passive solar energy heating for buildings, 8.27
Pelton-type turbine, 6.8
Percent excess air in combustion products, 1.4
Piping systems for fluid flow, 12.1–12.113
affinity laws for energy analysis of centrifugal pumps, 12.14
analyzing plastic piping and linings, 12.81
chart and tabular determination of friction head, 12.102
corrugated expansion joint selection and application, 12.126
design of steam transmission piping, 12.129
designing steam transmission lines without steam traps, 12.157
desuperheater water spray quantity, 12.150
determining friction factor for Bingham plastics flow, 12.163

Piping systems for fluid flow (*Cont.*):
determining pressure loss in oil pipes, 12.115
determining pressure loss in steam piping, 12.85
equivalent length of a parallel piping system, 12.111
equivalent length of complex series pipeline, 12.110
estimating cost of steam leaks from piping and vessels, 12.153
flow rate and pressure loss in compressed-air and gas piping, 12.118
flow rate and pressure loss in gas pipelines, 12.122
fluid head loss approximations for all types of piping systems, 12.81
friction-head loss in water piping of various materials, 12.101
friction loss in pipes handling solids in suspension, 12.149
hydraulic radius and liquid velocity in water pipes, 12.101
line sizing for flashing steam condensate, 12.160
maximum allowable height for a liquid siphon, 12.112
orifice meter selection for steam pipes, 12.97
pipe-wall thickness and schedule number, 12.82
pipe-wall thickness determination by piping code formula, 12.84
pressure loss in piping having laminar flow, 12.114
pressure-reducing valve selection for water piping, 12.107
pressure-regulating valve selection for steam service, 12.98
pressure surge from rapid valve closure, 12.75
pressure surge in compound pipeline, 12.79
pressure surge with different materials and fluids, 12.77
quick calculation of flow rate and pressure drop, 12.80
quick sizing of restrictive orifices in piping, 12.154
relative carrying capacity of pipes, 12.107
selecting heat insulation for high-temperature piping, 12.96

Piping systems for fluid flow (*Cont.*):
selecting plastic piping for industrial use, 12.141
sizing a water meter, 12.109
sizing condensate return liners for optimum flow, 12.151
slip-type expansion joint selection and application, 12.123
specific gravity and viscosity of pumped fluids, 12.113
steam accumulator selection and sizing, 12.139
steam desuperheater analysis, 12.138
steam tracing a vessel bottom to keep contents fluid, 12.155
steam trap selection for industrial applications, 12.90
time needed to empty a storage vessel with dished ends, 12.166
time needed to empty a storage vessel without dished ends, 12.170
time to drain a storage tank through attached piping, 12.170
warm-up condensate load in piping, 12.88
water hammer effects in liquid pipelines, 12.113
Plastic piping, selection of, 12.141
Process control valve selection, 14.68
Process-energy temperature control system selection, 14.67
Propeller and Kaplan turbine, 6.9
Psychometric chart use in air-conditioning calculations, 13.106
Pump and system characteristic-curve energy analysis, 12.30
Pump choice to conserve energy, 12.3
Pump energy performance analysis, 13.36
Pumping systems, solar-powered, 8.37
Pumps and pumping systems, 12.1–12.170
affinity laws for centrifugal pumps, 12.14
centrifugal pump choice using affinity laws, 12.14
centrifugal pump energy usage as a hydraulic turbine, 12.50
centrifugal pump minimum safe fluid flow determination, 12.40
choosing centrifugal pump best rotary speed, 12.17
choosing centrifugal pump rotating speed to meet capacity/head needs, 12.9

Pumps and pumping systems (*Cont.*):
energy and economic analysis of power-plant condensate pumps, 12.60
energy conservation in choosing pumping installation equipment, 12.22
energy saving using pump characteristic and system-head curves, 12.5
equipment criticality analysis, 12.9
hot-liquid pump net positive suction head, 12.36
hydro-pneumatic storage tank sizing, 12.50
impeller sizing for centrifugal pump safety service, 12.55
liquid viscosity effect on reciprocating pump performance, 12.46
liquid viscosity effect on regenerative pump performance, 12.45
materials selection for pump parts, 12.48
pump and system characteristic-curve energy analysis, 12.30
pump choice to conserve energy, 12.3
reducing energy consumption and loss with pump choice, 12.58
selecting centrifugal pumps on specific speed analysis, 12.17
shaft deflection and pump critical speed, 12.43
steam power plant condensate pump selection, 12.37
vapor-free liquid pump total head computation, 12.18
viscosity and dissolved gas effects on rotary pump performance, 12.47
viscous liquid centrifugal pump selection, 12.41

Quick design and evaluation of heat exchangers, 11.6

Radiant heating panel choice and sizing, 13.64
Reaction turbines, 6.2
Reducing energy consumption and loss with pump choice, 12.58
Regenerative-gas-turbine cycle, 3.7
Repowering options for power-plant capacity additions, 14.77
Return-air grille choice for air-conditioning systems, 13.110
Roof ventilator choice for structures, 13.115
Rotary heat exchanger energy savings, 13.25
Run-around system of heat transfer, 13.23

Savings produced by preheating combustion air, 1.16
Selecting a nuclear power reactor, 5.2
Selecting and sizing heat exchangers on fouling factors, 11.8
Selecting centrifugal pumps on specific speed analysis, 12.17
Shaft deflection and pump critical speed, 12.43
Shell-and-tube heat exchanger selection and sizing, 11.2
Single-pool tidal system energy and power, 10.8
Sizing shell-and-tube heat exchangers, 11.5
Slip-type expansion joint selection, 12.123
Small-scale hydro site analysis, 6.2
Snow-melting heating panel choice and sizing, 13.67
Solar-power energy, 8.1–8.37
calculation parameters, 8.1
collector sizing for solar building heating, 8.15
design of passive solar heating systems for buildings, 8.27
designing solar-powered pumping systems, 8.37
electric-generating system load and cost analysis, 8.2
F-chart solar heating computation method, 8.16
flat-plate heating- and cooling-system design, 8.6
industrial solar energy economics, 8.5
photovoltaic module choice for electric loads in buildings, 8.34
solar collector selection for domestic hot-water heating, 8.22
solar-collector solar-insolation computation, 8.12
solar water heater energy savings comparison, 8.32
Solar-powered pumping system design, 8.37
Solids in suspension, pressure loss in piping, 12.149
Solids intake of i-c engines, 4.28

Space heating energy savings with heat recovery from lighting systems, 13.69
Spiral-type heating coil selection, 11.14
Static-gain method for sizing low-velocity air ducts, 13.97
Steam accumulator selection, 12.138
Steam-control and pressure-reducing valve sizing, 14.75
Steam desuperheater, analysis of, 12.138
Steam generating capacity of boiler tubes, 11.18
Steam power generation, 2.1–2.37
  binary cycle steam plant energy analysis, 2.15
  calculation parameters, 2.1
  cogeneration system energy efficiency, 2.17
  energy efficiency and heat rate of steam turbogenerator, 2.2
  energy output analysis of condensing steam turbine, 2.31
  energy performance of reheat-steam cycle, 2.22
  energy test data for steam power plant performance, 2.10
  kW output of extraction turbine, 2.37
  mechanical-drive turbine energy output analysis, 2.29
  moisture content of turbine exhaust, 2.7
  regenerative bleed-steam cycle analysis, 2.18
  reheat-regenerative cycle steam rate, 2.36
  reheat-regenerative turbogenerator, 2.36
  reheating-regenerative steam turbine cycle analysis, 2.12
  steam flow for steam-turbine no-load and partial load operation, 2.8
  steam rate for turbogenerator at various loads, 2.11
Steam power plant condensate pump selection, 12.37
Steam transmission piping, design of, 12.129
  without steam traps, 12.157

Temperature determination in heat exchanger operation, 11.7
Temperature profiles of HRSG, 3.36
Tubular turbines, 6.9

Unit heater capacity computation, 13.52

Vapor-dominated geothermal steam power plant, 9.1–9.5
  cooling-water flow rate, 9.4
  dual-flash process, 9.8
  plant efficiency, 9.5
  turbine inlet and exhaust steam conditions, 9.4
  turbine steam flow, 9.4
  turbine work output, 9.4
Vapor-free liquid pump head computation, 12.18
Vertical Francis turbine, 6.5
Vibration potential of motor-driven fans, 13.33
Viscosity and dissolved gases effects on rotary pumps, 12.47
Viscous liquid centrifugal pump selection, 12.41

Wave calculations for energy and power, 10.5
Wave energy and power densities, 10.7
Wavelength, velocity, and height, 10.6
Wind power energy, 7.1–7.4
  air density, 7.2
  analysis of wind turbine power generation, 7.2
  calculation parameters, 7.1
  choice of wind-energy conversion system, 7.4
  comparative economics of wind power, 7.12
  maximum obtainable power density, 7.3
  power density in wind stream, 7.2
  selecting suitable wind machine, 7.4
  torque and axial thrust, 7.3
  typical capital cost, 7.9
Wood fuel combustion, 1.11